abel
19.57

20,00

THE HEPATICAE AND
ANTHOCEROTAE
OF NORTH AMERICA

RUDOLF M. SCHUSTER

THE HEPATICAE
AND
ANTHOCEROTAE
OF
NORTH AMERICA

EAST OF THE
HUNDREDTH MERIDIAN

VOLUME II

COLUMBIA UNIVERSITY PRESS

NEW YORK AND LONDON 1969

Rudolf M. Schuster is Professor of Botany and Director of the Herbarium at the University of Massachusetts.

PREFACE

Volume II of this work continues in uninterrupted sequence from Volume I, with sequence of families as outlined on pages 381–386 of that volume. As before, in order to make this book useful to a diversity of workers—nonprofessional taxonomist as well as specialist—technical and supplementary details are given in small type or in footnotes. I have made a great effort to keep the citation of specimens and/or published localities within reasonable bounds (within the limits established on page 6 of Volume I), thus restricting myself even more severely in citation of personally collected materials and of frequent taxa. In many cases a specimen that has been illustrated (thus cited in the figure captions) is not again included in the list of cited specimens. In the technical families Calypogeiaceae and Lophoziaceae, in which previously cited collections have often been incorrectly determined, I have often sharply curtailed references to prior reports, if these reports are almost undoubtedly errone-ous. Numerous routine corrections to published citations have been made. I have not always mentioned earlier (published or unpublished) erroneous determinations, but when the errors are egregious (e.g., see *Nardia geoscyphus*, page 867) or "establish" rare species in peripheral areas where they fail to occur (as in Peary Land; see page 892), complete documentation cannot be avoided.

References are to the bibliography in Volume I; recently published papers (mostly published after 1962) and any additional pertinent references will be found in a supplementary bibliography that will ter-minate the last volume.

The basic approach to Volume I is retained in Volume II; thus introductory material to that volume applies equally to this one. Research on the initial draft and most revisionary work on this volume have been done away from my home institution. I am grateful to Professor Thelma Howell, Director of the Highlands Biological Station at Highlands, North Carolina, for providing research space at the station during several summers. Many of the data on arctic taxa were worked up during sum-mers in Greenland, when I was aided by grants from the Arctic Institute of North America (AINA-58, 1961; ONR 384, 1966). The last grant also made possible research at the Botanical Institute in Copenhagen. I am much indebted to Mr. Kell Damsholt of that institution for his help in locating critical materials; he was also a valued colleague during the 1966 expedition to west Greenland (the jointly collected materials are cited simply as *RMS & KD*). I am further indebted to the governing board of the Danish Arctic Station (Professors Böcher, Salomonsen, and Köie) at Godhavn, Disko I., for making a profitable stay there possible.

Final revision of generic and familial diagnoses and discussions could not have been brought up to their present level of comprehensiveness without the experience of field work and research in the Antipodes—which did much to expunge any holarctic "bias" initially absorbed from the works of my Northern Hemisphere colleagues. A grant from the National Science Foundation (NSF G-7114) supported my work in Latin America; Drs. O. Kuehnemann, I. Gamundi de Amos, and R. Singer facilitated field work there. Subsequently a stay in New Zealand was made possible by a Fulbright award, and in 1967 I was able to return to New Zealand through the generosity of the John Simon Guggenheim Memorial Foundation, which awarded me a second fellowship (much initial research on this work was done with the aid of a fellowship granted in 1955–1956). The friendly and stimulating atmosphere at the University of Otago in Dunedin, N.Z., created by Prof. G. T. S. Baylis and his colleagues, contributed much to the high productivity achieved there.

I also wish to acknowledge my gratitude to the various institutions which provided library and/or herbarium facilities; much synonymy and bibliographical data were checked at the Conservatoire Botanique, Geneva, with the aid of Dr. C. E. B. Bonner. Numerous trips during a ten-year interval were made to the Farlow Herbarium, Harvard University, and to the New York Botanical Garden. I wish to express my debt to Drs. I. MacKenzie Lamb and W. C. Steere of those institutions.

Many illustrations (bearing his initials) were prepared with great skill and intelligence by Mr. Joseph Palazolla.

Finally, I wish to acknowledge again the constant and loyal help of my wife, Olga M. Schuster, in all phases of the development of this volume.

A generous grant from the National Science Foundation has aided in defraying the production costs of this book.

December, 1968   Rudolf M. Schuster

CONTENTS

CONTENTS

CONTENTS

*Specialized Morphological and
Taxonomic Treatment*

Suborders Lepidoziinae
and Jungermanniinae

SUBORDER LEPIDOZIINAE
Schust. (1963)

Suborder Ptilidiinae Schust., Amer. Midl. Nat. 49(2):526, 1953 (*p.p.*).
Suborder Lepidoziinae Schust., Jour. Hattori Bot. Lab. no. 26:228, 1963.

Gametophyte growing by a tetrahedral apical cell, exceedingly variable, erect to procumbent to closely prostrate or creeping, from (primitively) essentially isophyllous to (commonly) strongly bilateral to (*Zoopsis, Pteropsiella*) ± thallose, but *retaining distinct ventral merophytes 2 or more cells broad* (at least 0.2 the perimeter of the stem) *which develop small to large underleaves bearing 2 or more lobes.* Axis often with cortical cells enlarged and hyaline; *branching plastic:* lateral branches often rather regular, 1–2-pinnate, with terminal branching of the *Frullania* (less often also of the *Microlepidozia* or *Acromastigum*) type, and with postical- (rarely, i.a., in *Drucella, Pseudocephalozia, Hyalolepidozia,* also lateral-) intercalary, axillary branching; *branches often running out into flagella.* Main axis normally maintaining dominance, the branching excurrent (but in the Bazzanioideae branching ± pseudodichotomous and deliquescent). Leaves variable in form, from 3–12-lobed to bifid to (*Calypogeia* spp.) entire, *incubous varying to transverse* to (rarely; never in ours) succubous, symmetric to asymmetric (then dorsal lobes usually largest); lobe margins usually entire. *Ventral merophytes always distinct,* almost always *developing underleaves* (highly reduced in the Zoopsidoideae and some *Telaranea* and *Arachniopsis* spp.; lost in *Mytilopsis* spp.), which are *usually 2–8-lobed* and large, exceptionally entire. *Rhizoids confined to underleaf bases.*[1] Cells variable: from strongly collenchymatous to ± equally (and sometimes strongly) thick-walled to large, delicate, and hyaline; cuticle smooth or verruculose; usually with 2–many botryoidal to glistening or papillose oil-bodies (absent in the *Lembidium- Chloranthelia-Megalembidium* complex and in *Psiloclada*). Asexual reproduction infrequently present: *by caducous leaves or* (Calypogeiaceae) 1–2-celled *fasciculate gemmae* from uppermost, reduced leaves.

Dioecious, less often monoecious. *Sexual organs usually on abbreviated, reduced branches, postical- (infrequently lateral-) intercalary in origin;* androecia usually *sharply defined,* the *bracts small, often monandrous,* often hyaline;

[1] Except in *Metacalypogeia schusterana* and *Phycolepidozia.*

bracteoles flat, *without antheridia.*[2] Gynoecia usually on *abbreviated postical branches* (less often on elongated, leafy postical branches;[3] very rarely terminal on leading stems) without normal leaves; *bracts isophyllous, and bracteole identical to bracts* (except Calypogeiaceae) in general contrast to the ± anisophyllous leafy sterile axes. Archegonia numerous. With fertilization either an *erect, fusiform-trigonous perianth* (Lepidoziaceae) *or a pendulous, rhizoidous marsupium* (Calypogeiaceae) *develops.* Sporophyte with seta of numerous cell rows (Calypogeiaceae) or with 4, 8, or 16 (±1-2) epidermal and many (exceptionally only 4) smaller medullary cell rows. Capsule *ellipsoidal to cylindrical, 4-valved to base;* valves 2-5(6)-stratose. *Epidermal cells with a "two-phase" development:* primary longitudinal walls, and all (primary) cross walls remaining colorless and (usually) devoid of secondary, pigmented thickenings, *the secondary 1-3 longitudinal cell walls which subdivide the large, subquadrate primary cells developing strong, pigmented, sheetlike to nodular thickenings.* Inner cells with radial (nodular) to U-shaped (semiannular), weak to distinct transverse bands. Spores usually 1-3 × elaters in diam., never showing intracapsular germination, strongly papillose to covered with vermiculate ridges (formed in part by coalescence of papillae) to areolate; elaters free, *usually rather blunt at the little tapered ends, usually 2-spiral.*

Type Family. Lepidoziaceae Limpr.

The Lepidoziinae include only the families Phycolepidoziaceae, Lepidoziaceae[4] and Calypogeiaceae. Possibly the Cephaloziaceae evolved from succubous-leaved Lepidoziaceae, but the scattered rhizoids and lack of deeply lobed underleaves, the restricted branching modes, and the different seta anatomy suggest a remote affinity. This topic is explored in more detail elsewhere (p. 14); see also Schuster (1965b).

The group was initially placed as a derivative element within the Ptilidiinae. My reasons were as follows. (1) Some Lepidoziinae are essentially isophyllous, and many show polymorphism in leaf-lobe number, with quadrilobed leaves and underleaves commonly produced. (2) The plasticity in branching, within the single family Lepidoziaceae, suggests that the latter is extremely primitive.

---

[2] Except in the antipodal *Neogrollea* and in *Pseudocephalozia paludicola* Schust. (see Schuster, 1965b); in these, postical-intercalary androecia are, at least at times, isophyllous and have inflated bracteoles with 1-2 antheridia.

[3] In the primitive *Pseudocephalozia paludicola* and a few other antipodal taxa, some, many, or rarely all gynoecia occur on lateral-intercalary branches.

[4] As is subsequently developed, recognition of the families Zoopsidaceae Nakai (1943) and Bazzaniaceae Buch (1954), as distinct from the Lepidoziaceae, seems unwarranted. The highly reduced West Indian, monotypic *Phycolepidozia* Schust. (Schuster, 1966a) with polyseriate, essentially leafless axes (the leaves are reduced to 1-2-celled papillae), is of uncertain affinity. The 4 + 4 seriate seta is as in *Cephaloziella*, but the perianth and bracts are Lepidozioid.

(3) The more reduced Lepidoziaceae (such as *Telaranea* and *Microlepidozia* spp.) closely approach such reduced genera of Herbertinae as *Blepharostoma* in aspect and gametophytic organization. (4) In a number of Lepidoziinae androecia may be intercalary on main stems, and gynoecia may occur terminal on main stems (in, e.g., *Telaranea apiahyna*; see J. Taylor, 1961); thus the generally diagnostic restriction and reduction of sexual branches are not exclusively expressed in Lepidoziaceae. (5) The rhizoid distribution is restricted. (6) Similarities in facies and in the erect growth exist between such primitive species as *Lepidozia spinosissima* of New Zealand, with quadrifid leaves and similarly sized quadrifid underleaves, and the isophyllous Herbertinae and Ptilidiinae.

Nevertheless, the group as a whole represents a sharply defined evolutionary sequence, isolated by (*a*) early and very nearly universal reduction in size of the sexual branches, and their general restriction to the postical merophytes; (*b*) the sharply defined manner in which epidermal cells of the capsule undergo a two-phase development (see, e.g., p. 21). These specializations are linked with retention of fantastically diverse modes of vegetative branching, showing reduction to one or two types only in isolated genera. Noteworthy is the fact that within the single, highly polymorphic family Lepidoziaceae there is evolution from isophyllous erect taxa (e.g., *Lepidozia spinosissima*) to thallose types, such as *Zoopsis flagelliforme* Col. and *Pteropsiella frondiformis* Spr. (see, i.a., Vol. I, p. 437, Figs. 30–32). It is also remarkable that in the plastic and ancient *Pseudocephalozia paludicola* Schust. of Tasmania the androecia may still remain isophyllous, with the concave bracteoles bearing antheridia (Schuster, 1965b); similar isophyllous androecia occur in *Neogrollea* Hodgs.

Relationships

The Lepidoziinae show apparent affinities to the Blepharostomaceae in the Herbertinae and to the family Cephaloziaceae. Affinities to Blepharostomaceae become most obvious when such Lepidozioid taxa as *Micrisophylla* Fulf. (particularly the generitype) are compared with *Temnoma* of the Blepharostomaceae. Since, in another place, I have emphasized the obvious similarities between *Temnoma* and the Trichocoleaceae (here placed in the Ptilidiinae!), it is clear that, when the least derivative taxon of each of these groups is studied, obvious convergences appear. Nevertheless, the Lepidoziinae represent a major evolutionary sequence, in which there has been a distinct advance over the Herbertinae in (*a*) the relatively constant restriction of the sex organs to small, weak, abbreviated branches (exceptions admittedly occurring); (*b*) the general development of microphyllous, long stolons or flagella; (*c*) the apparently constant "two-phase" development of the epidermal cells of the capsule, with individual epidermal cells narrowly rectangular, often lying in at least partial tiers. Bridges admittedly occur; thus *Telaranea apiahyna* has gynoecia terminal, at least at times, on main axes (Taylor, 1961). Also, in *Trichotemnoma* Schust.

(Schuster, 1964a), clearly a member of the Blepharostomaceae, abbreviated androecial branches occur. Finally, both of the allied genera *Isophyllaria* Hodgs. and Allis. and *Herzogiaria* Fulf. (Blepharostomaceae), tend to develop geotropic, microphyllous branches (Schuster, 1966b). Nevertheless, distinctions between Lepidoziinae and Herbertinae are, in general, obvious, in contrast to the situation involving the Lepidoziaceae and Cephaloziaceae.

I have reviewed (Schuster, 1965b) the problems attending a separation of the Lepidoziaceae and Cephaloziaceae. In the final analysis, the Lepidozioid complex differs from the Cephaloziaceae chiefly in (*a*) the polymorphism in branching; (*b*) the rhizoids restricted to the bases of the (admittedly sometimes much reduced) underleaves; (*c*) the single, terminal slime papilla per lobe of the underleaves; (*d*) the tendency for *both*, or at least one or the other, of leaves and underleaves to be 3–4-lobed; (*e*) the predominantly incubous leaves of the Lepidoziinae vs. the constantly succubous leaves of the Cephaloziaceae. The separation of the Lepidoziaceae and Cephaloziaceae is treated in detail under the first family.

Key to Families of Lepidoziinae

1. Leaves usually (always in ours) conspicuously 2–4(or more)-lobed (and/or dentate); branching polymorphic, normally in large part lateral-terminal; gynoecial branches isophyllous, bracts and bracteole large and well developed, identical in size and form; a distinct, fusiform-prismatic perianth developed; asexual reproduction absent or by caducous leaves; seta with 8 or 16 ($\pm$1–2) epidermal rows of large cells, surrounding a medulla of 4–many smaller cell rows; capsule ovoid, with straight valves. . . . . . . Lepidoziaceae
1. Leaves entire or minutely bidentate at apex; branches usually all postical-intercalary (rarely some lateral-terminal); gynoecial branches with small to vestigial bracts (and no bracteoles), thus bilateral; no perianth, a fleshy marsupium developing; asexual reproduction by gemmae; seta with many rows of epidermal cells, similar to the internal cell rows in size; capsule el-lipsoidal to cylindric, usually with linear, spirally twisted valves.
   Calypogeiaceae

Family LEPIDOZIACEAE *Limpr.*

Family Lepidozieae Limpr., *in* Cohn, Kryptogamen-Flora von Schlesien, p. 310, 1875; Dědeček, Arch. Naturw. Landesdurchforschn. Böhmen 5(4):11, 1886.
Lepidoziaceae Massal., Atti Reale Ist. Veneto Sci., Lettere Arti:72:103, 1913; H. W. Arnell, *in* Holmberg, Skand. Fl. IIa, 1928.
Zoopsidaceae Nakai, *in* A List of Prof. Nakai's Papers, etc. Tokyo, 1943.
Bazzaniaceae Buch, *in* Rapport Congr. Intern. Bot., 1954.

Plants $\pm$ erect, or prostrate to procumbent in growth, *rarely closely adnate*, usually forming loose mats or patches, whitish or pure green to brown or reddish brown, rarely glaucous, *without bright red or vinaceous pigmentation.* Shoots rather small (250–500 μ wide) to *more commonly robust* (1–6 mm wide), *often freely* 1–2(3)-pinnately or pseudodichotomously *branched*, often ± plumosely branched. *Branching plastic*: branches, at least the gynoecial and sometimes some vegetative and androecial, *in large part postical in origin*, usually intercalary and from axils of underleaves (sometimes terminal, replacing half of underleaf); *vegetative branches largely or exclusively lateral in origin*, usually terminal, *normally of the Frullania type*, very rarely intercalary, normally replacing the ventral half of a lateral leaf (in *Microlepidozia* in part replacing the dorsal half of a lateral leaf). Stems with a unistratose cortical cell layer of relatively short cells (mostly 2–5 × as long as broad), rectangulate, in 6–50 rows, depending on genus; cortical cell *diam. normally slightly to greatly superior to that of medullary cells*, although not necessarily forming a sharply defined hyalodermis; defined endodermal layer absent; medullary cells generally of small caliber, strongly elongated (usually 4–8 × as long as wide), sometimes strongly thick-walled, except at the ends; without mycorrhizal medullary infestation. *Some lateral and/or postical branches running* out into rhizoidous, *microphyllous flagella.* Rhizoids frequent to infrequent, *at bases of underleaves. Leaves and underleaves primitively rather similar in size and form*: the ventral merophytes well preserved, only rarely (*Arachniopsis, Mytilopsis, Zoopsis*) reduced to a width of 2 cells. *Lateral leaves* alternate, usually slightly to strongly *incubous in insertion and orientation*, rarely transverse or ± succubous, *usually fundamentally* 3–4 (6–12)-*lobed or dentate*, less often bilobed or bidentate to entire, the lobes often long and slender and the sinuses frequently descending to the leaf middle or beyond; *lobes normally entire. Underleaves usually large*, the ventral merophytes commonly occupying ca. ¼ the perimeter of the stem, the *underleaf area usually 0.25–0.75 that of leaves; underleaves usually lobed or dentate like lateral leaves*, rarely vestigial or absent (*Zoopsis, Mytilopsis, Arachniopsis*). *Cells usually somewhat to slightly thick-walled, medium-sized*, occasionally with discrete to rather large, rarely strongly bulging trigones, *generally firm*, the cuticle smooth or weakly papillose; oil-bodies usually present (absent in *Lembidium, Chloranthelia*, etc.), homogeneous (or virtually so) and glistening. *Asexual reproduction absent or by caducous leaves.*[5]

Dioecious or autoecious, *never paroecious.* Androecia chiefly on *short,*

---

[5] Early reports (Hooker; Nees) of gemmae in *Lepidozia reptans* and *Microlepidozia setacea* have been repeated many times and apparently were the basis for the statement in Degenkolbe (1938) that gemmae occur in Lepidoziaceae. Until a credible modern report of the actual occurrence of gemmae in the family appears, it seems pointless to maintain that gemmae occur in Lepidoziaceae. The most recent report of gemmae is in Hodgson (1956), for a New Zealand species.

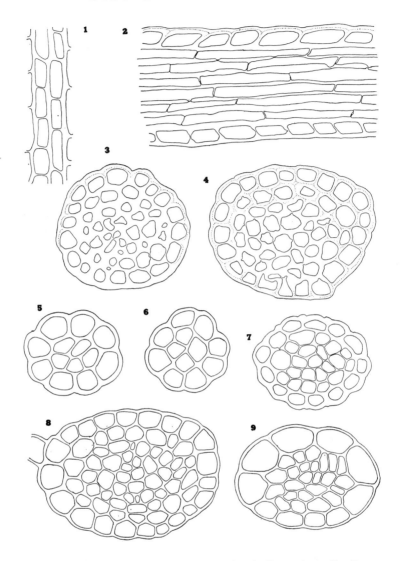

FIG. 85. Lepidoziaceae, stem anatomy. (1–3) *Bazzania nudicaulis:* dorsolateral cortical cells (×247); 2, longisection of mature stem (×247); 3, cross section of stem (×247). (4) *Bazzania denudata*, stem cross section (×247). (5–6) *Microlepidozia sylvatica*, stem cross sections (×420). (7) *Microlepidozia setacea*, stem cross section (×247). (8) *Lepidozia reptans*, stem cross section (×230). (9) *Telaranea nematodes*, stem cross section (×265). [1–3, *RMS 38017*, Mt. Rogers, Va.; 4, *RMS 38016*, Mt. Rogers, Va.; 5–6, *RMS 24109* and *28415*, respectively; 7, *RMS H3402*, Bergen Swamp, N.Y.; 8, *RMS 11990;* 9, *RMS 28622*.]

determinate, postical branches, rarely terminal on elongated lateral branches; androecial branches not innovating apically (exc. in *Chloranthelia*); bracts usually somewhat similar to leaves *but very much smaller;* bracteoles usually ± large, but *without antheridia* (exc. in *Pseudocephalozia* sp.); antheridia usually 1–2 per bract, their stalks 1–2-seriate,[6] quite short. *Gynoecia almost uniformly on abbreviated postical branches* (in *Telaranea* and *Hygrolembidium* rarely and sporadically terminal on short or elongate lateral shoots or leading branches), *with isophyllous bracts,* usually without normal vegetative leaves; bracts commonly in 2–several series, usually closely imbricate, the inner at least usually larger than leaves; *bracteoles similar to bracts.* Perianth cylindrical below, *usually large and quite elongated and gradually tapering distally,* often ovoid-fusiform, *trigonous above; marsupium absent.* Sporophyte with seta strongly elongated, *with 8(9) or (15)16 usually regularly oriented rows of epidermal cells, normally much larger than the numerous* (rarely only 4) *rows of interior cells. Capsule ovoid* to ovoid-cylindrical, regularly 4-valved to base; *valves straight;* wall varying from 2-stratose to 3–4-stratose to at least 5(6)-stratose. Epidermal layer usually with nodular thickenings along adjacent faces of the secondary longitudinal walls, the *juxtaposed faces of primary longitudinal (and all transverse) walls nearly or quite without thickenings.* Inner (or innermost) layer with semiannular, transverse bands (which may be locally or almost completely reduced on the tangential faces, leading to development of weak nodular thickenings) on all longitudinal walls, not on transverse walls. *Spore-elater diam. ratio 1–1.5 : 1* (3 : 1 in *Bazzania tricrenata, teste* Müller); spores papillate or verrucose or vermiculate to delicately areolate. Elaters free, *normally bispiral, tapering only slightly at the ends.* Chromosome no.: $n = 9(7 + H + h)$, in *Bazzania, Lepidozia, Zoopsis* (Tatuno 1941, 1947, 1948). Sporeling development of the *Nardia* type.

Type. Lepidozia Dumort.

The Lepidoziaceae are a large and natural family,[7] including only four regional genera (*Lepidozia, Telaranea, Microlepidozia, Bazzania*),

[6] Attempts to limit the Lepidoziaceae to forms with biseriate antheridial stalks (Müller, 1954, in 1951–58) are untenable, as shown in Schuster & Blomquist (1955), who demonstrated a uniseriate antheridial stalk for *Telaranea* and *Microlepidozia.*

[7] For a discussion of branching see Taylor (1962); Schuster (1963, 1965b) discusses the family in some detail, dealing especially with antipodal elements that deviate from holarctic taxa. Fulford and Taylor (1959a) treat species of *Microlepidozia* with succubous leaves, and *Psiloclada.* Schuster and Blomquist (1955) deal in detail with the generitype of *Telaranea.* Other references to recent literature are given under the specific genera.

The definition and circumscription of the Lepidoziaceae in Fulford (1963, 1963a) are untenable, since she eliminates from the family the almost exclusively antipodal *Zoopsis-Pteropsiella* element, derived from *Bonneria-* or *Paracromastigum-*like antecedents. This element definitely which agree in having lateral, terminal branching, and in developing postical (and sometimes lateral) flagella; they usually have incubously inserted and oriented leaves, although *Microlepidozia*, among others, has transverse leaves. Our taxa retain large underleaves, except for *Telaranea* (subg. *Telaranea*), in which they are reduced. The few regional members give only a limited idea of the tremendous variability of the Lepidoziaceae. Most of our taxa show little reduction and few specializations.

The bulk of genera and species of Lepidoziaceae are developed in the Antipodes, with more species in the colder or temperate portions than in the tropical regions. Most genera and species occur in regions with mild—but not hot—climates which lack strong seasonal restriction of rainfall, such as New Zealand, Tasmania, various other oceanic islands (chiefly Polynesia, the Malayan Archipelago, etc.). There is also a diversity of taxa in wet, montane tropical forests, as in the northern portions of the Cordillera of South America. In the Northern Hemisphere the family is particularly poorly represented in continental regions and scarcely penetrates into the Arctic. The species occur largely on soil or on rocks; some are helophytes, and a limited number become arboreal; with the occasional exception of *Hygrolembidium* Schust. and *Microlepidozia* (Spr.) Joerg., almost none are aquatic. Few taxa tolerate calcareous substrates.

Once antipodal taxa are studied, delimitation of the family becomes difficult on purely gametophytic bases. I have demonstrated (Schuster, 1963) that *Zoopsis* Hook., a genus with vestigial leaves and flattened stems (Figs. 30–32) previously placed in the Cephaloziaceae (Evans, 1939, etc.), must go in the Lepidoziaceae; similarly, *Pteropsiella* Spr. (Fig. 32:7–10) and *Arachniopsis* Spr., genera placed by, i.a., Evans (1939) in the Cephaloziaceae, belong in the Lepidoziaceae, in spite of the fact that they usually possess only postical-intercalary branching. With the discovery of extreme types such as *Pseudocephalozia drucei* Schust. (Schuster, 1965b), in which branching is normally intercalary from all three rows of merophytes, and of *Hygrolembidium* Schust. (Schuster, 1963), in which branching is exclusively intercalary, it is now clear that a wide assemblage of groups bears exclusively intercalary branching—not only the South American genera *Mytilopsis* Spr. and *Micropterygium* Lindenb. and the New Zealand *Drucella* Hodgs. (Hodgson, 1963). A considerable number of divergent antipodal genera have been discovered which do not "fit" into the ordinary concept of the Lepidoziaceae inevitably acquired by the student of the holarctic flora. These genera include *Chloranthelia* Schust. and *Isolembidium* Schust. (Schuster, 1963), as well as *Paracromastigum* (Fulford and

---

belongs to the Lepidoziaceae (Schuster, 1963, 1965b); I demonstrated a "two-phase" development of epidermal capsule cells in *Zoopsis*, exactly as in other Lepidoziaceae; a Lepidozioid seta, with large epidermal and small internal cells; underleaves bifid to base with rhizoids limited to their base. The evolution of this fascinating element, usually heretofore erroneously included in the Cephaloziaceae, is treated on p. 437 (Figs. 30–32) of Vol. I.

Taylor, 1961), *Pseudocephalozia* (Schuster, 1965b), and *Mastigopelma* Mitt. (of Samoa); they lack the ordinary facies of the Lepidoziaceae, which is the result of regular, lateral-terminal branching, linked with the normal development of 3–4-lobed or toothed leaves, incubous in insertion.

With the large number of deviant genera and types with deviant aspects, an attempt has been made to arrange them into five subfamilies as follows:

1. SUBFAMILY LEPIDOZIOIDEAE LIMPR. (*Lepidozia, Microlepidozia, Telaranea* regionally; *Sprucella* Steph., *Psiloclada* Mitt., *Arachniopsis* Spr., *Megalembidium* Schust., and (?) *Drucella*, chiefly in the Antipodes). Axis with leaf insertion *attaining dorsal stem midline* (exc. in *Arachniopsis*; *Telaranea*, in part) leaves incubous to transverse to (rarely) succubous, usually deeply (1–3)4–8(12)-lobed, never canaliculate to complicate; underleaves rarely reduced, remaining at least bifid. Branching variable, but never *Acromastigum* type or pseudodichotomous. Oil-bodies usually distinct (exc. *Megalembidium, Psiloclada*). The succubous-leaved taxa (*Psiloclada*, etc.) still forming a foreign element. The seta remains "Lepidozioid" except in the reduced genus *Arachniopsis*, where only 4 internal cell rows occur.

2. SUBFAMILY BAZZANIOIDEAE RODWAY (*Bazzania;* *Acromastigum* Evans, largely in the Antipodes and tropics; *Mastigopelma* Mitt., with only intercalary branching doubtfully here). The group has been treated as a distinct family, the Bazzaniaceae, by Buch (1954) but is not recognized as such by other workers. Axis with leaf insertion attaining stem midline dorsally; leaves incubous to transverse, never succubous, usually shallowly 2–3-lobed to toothed to unlobed, never canaliculate to complicate; underleaves large, 0–2–3-dentate or shallowly lobed; branching pseudodichotomous, from only one side of main axis (the branch vigorous, laterally displacing main stem, growth deliquescent); sometimes with postical-terminal branching; oil-bodies usually distinct.

3. SUBFAMILY LEMBIDIOIDEAE SCHUST. (*Lembidium* Mitt., *Neogrollea* Hodgs., *Isolembidium* Schust., *Chloranthelia* Schust., *Hygrolembidium* Schust., almost all antipodal). Axis with leaf insertion attaining stem midline dorsally; leaves transverse to feebly succubous to feebly incubous, unlobed to emarginate to 3–4-dentate (rarely 3–4-lobed), never complicate but sometimes strongly concave; underleaves large, usually similar to leaves but often smaller; cells usually equally thick-walled or leptodermous, with oil-bodies absent or locally present in a few cells only; leafy axes with branching usually reduced, normally largely (sometimes exclusively) postical, sometimes with irregular, *Frullania*-type branching. Often fleshy and soft-textured plants with hyaline cells.

Insertio foliorum ad lineam caulis mediam dorsaliter extendens; folia plerumque transversa ad leniter succuba, nunquam complicata, 0-lobata ad 3–4-dentata; cellulae plerumque non-collenchymatosae; rami plerumque intercalares, raro sporadice typi Frullaniae; plantae plerumque molles et saepe succulentae. *Typus: Lembidium* Mitt.

4. SUBFAMILY MICROPTERYGIOIDEAE GROLLE. (*Micropterygium* Lindenb., *Mytilopsis* Spr., both neotropical). Axis with leaf insertion attaining stem midline dorsally; leaves transverse to transverse-incubous, complicate-canaliculate to complicate-bilobed with winged keel; only with intercalary postical branching;

cells firm to rigid, often tuberculate. Firm, often very rigid, usually opaque and often brownish plants.

5. SUBFAMILY ZOOPSIDOIDEAE SCHUST. (*Zoopsis* Hook., *Pteropsiella* Spr., *Hyalolepidozia* S. Arn., *Pseudocephalozia* Schust., *Bonneria* Fulf. & Tayl., ? *Paracromastigum* Fulf. & Tayl., almost exclusively antipodal). Elements were placed as a separate family, the Zoopsidaceae, by Nakai (1943), but certainly on inadequate grounds. Axis with *leaf insertion commonly not attaining stem midline dorsally* (2 or more cell rows "leaf-free"); leaves ± succubous, rarely transverse verging barely to incubous, never complicate-bilobed, usually of few (often 2–20) very large, hyaline, pellucid cells, grading to a thallose organization; stem almost always with a conspicuous hyaloderm; cells large to very large, hyaline, thin- to weakly thick-walled, no trigones, with oil-bodies present. Branching as polymorphous as in Lepidozioideae, primitively often of the *Acromastigum* type, varying to exclusively intercalary—usually monopodial, never plumose, usually irregular and sparing. The neotropical, monotypic *Protocephalozia* Spr., with the sterile gametophyte reduced to a persistent protonema but with Lepidozioid, leafy sexual branches, probably belongs here, or in a separate subfamily.

Insertio foliorum ad lineam caulis mediam plerumque non dorsaliter extendens; folia 2–4(5)-lobata dentave, plerumque ± succube inserta, e cellulis non-collenchymatosis membranas tenues ad crassas habentibus constantia, plerumque cellulas paucas magnasque habentia, in ordinationem thalliformem abeuntia; ramificatio maxime variabilis. *Typus: Zoopsis* Hook.

This last subfamily, as here conceived, may still be heterogeneous. Its more primitive taxa (e.g., *Pseudocephalozia paludicola*) still retain, at times, bracteolar antheridia; the less advanced genera (*Pseudocephalozia, Bonneria, Hyalolepidozia, Zoopsis*) all retain the "Lepidozioid" seta of the preceding subfamilies; in *Pteropsiella* a "Cephalozioid" seta with 8 epidermal and 4 internal cell rows has evolved (Schuster, 1965b), as also in *Zoopsidella* Schust. and *Protocephalozia* Spr.

This classification of the Lepidoziaceae stands in contrast to that in Fulford (1963a; see fig. 203, p. 83). There is not space here to attempt a critique of Fulford's arrangement; it suffices to say that the positions assigned a variety of groups (*Neolepidozia* vs. *Telaranea, Microlepidozia* vs. *Dendrolembidium*, and "*Micrisophylla*," *Psiloclada, Lembidium*, etc.) appear untenable. Fulford nowhere describes the crucial criterion of leaf origin: whether the leaf originates so as to attain the stem midline, or (most Zoopsidoideae) is so inserted that a dorsal strip of the stem remains free of the leaves; compare, i.a., Figs. 30:10 and Fig. 31:2, 10–12. The attempt to place *Marsupidium* Mitt. in the Lepidoziaceae (Fulford, p. 83) appears to be a lapsus.

Of these five subfamilies, only the first two occur regionally.

Several other recently proposed genera are not accepted, such as *Neolepidozia* Fulf. & Tayl. (1959) (a subgenus of *Telaranea*, which see) *Micrisophylla* Fulf. (1962a), *Dendrolembidium* Herz. (see p. 43) and *Lepidoziopsis* Hodgs. (see p. 20). With these omitted from consideration, approximately 25 genera are assignable to the family.

When certain specialized genera are excluded from consideration, the basic ensemble of characters which delimit the Lepidoziaceae includes the following. (1) A great variability in branching modes, involving general retention of lateral-terminal and postical-intercalary branching; deviations from this basic pattern are sufficiently numerous to suggest a plasticity for the Lepidoziaceae which is regarded as primitive. In certain taxa individual plants may show a very extraordinary amplitude in branching patterns; e.g., *Bonneria furcifolia* (Steph.) Schust., has lateral- and postical-terminal branching (*Frullania*, *Microlepidozia*, and *Acromastigum* types), as well as postical-intercalary branching. (2) A general retention of distinct, often large underleaves, with rhizoids restricted to their bases even when, as in *Zoopsis*, the underleaves become vestigial. (3) No matter how anisophyllous and bilateral vegetative portions of the gametophyte become, gynoecial branches remaining isophyllous. (4) A long, usually slender, trigonous-prismatic perianth, bluntly keeled, contracted at the apex; no marsupium or perigynium. (5) Flagella or stolons of some type always developed. (6) Cladogynous perianths, the sexual branches rarely elongating. (7) Sporophyte with seta—with isolated exceptions—with a limited suite (8 or 16 $\pm$ 1) of large epidermal cell rows and a variable number of much smaller, internal cell rows.[8] (8) Capsule wall with epidermal cells undergoing a two-phase cycle of development. As a consequence, primary and secondary cell walls distinct, the former usually lacking, the latter always with pigmented secondary thickenings.

Of this basic assemblage of characters, among the most important are the last two derived from the sporophyte. The form of the seta allows a nearly definite, unambiguous circumscription of the family, separating it from both Calypogeiaceae and (with few exceptions) from Cephaloziaceae, the only others which offer certain obvious points of contact.

The Lepidoziaceae, in spite of showing considerable deviations from "the norm" in exotic taxa, are easily separated regionally on the basis of an ensemble of characters that, by and large, also serves to distinguish exotic genera of the family. Among these are (*a*) only moderate anisophylly, with the ventral merophytes always developing rather large underleaves that most often approach the leaves in form; (*b*) branching never confined to a particular series of merophytes, and, almost without exception, both terminal and intercalary on the same plant; (*c*) incubous leaved (*Microlepidozia* with leaves transverse to succubous); (*d*) usually 3–4-lobed or -dentate form of the leaves (only in the regional Ptilidiaceae and Lophoziaceae do similarly lobed leaves occur); (*e*) development of a sharply defined, restricted rhizoid-initial area at the base of

[8] Müller's diagnosis (1954, *in* 1951–58, p. 556) indicating two concentric rings of small internal cells in the seta of the Lepidoziaceae is incorrect; there is much more variation than Müller admits. However, no considerable deviations occur in the form of the epidermal cells, which are almost always in 8 ($\pm$1) or 16 ($\pm$1) series. Exceptionally, 10 or 12 epidermal cells may be found, by secondary subdivision of 8 primary cells. Schuster (1965b) discusses variations in seta anatomy of the Lepidoziaceae in detail.

the underleaves; rhizoids never arising elsewhere on the stem; (*f*) a trigonous perianth normally occurring on short, ventral branches; (*g*) frequent development of flagella or shoots that run out into flagella; (*h*) absence of gemmae (but occasional occurrence of caducous leaves) as a vegetative means of reproduction.

The family is largely a natural and sharply definable one. Its origin is obscure, although possibly to be found in plastic, unspecialized types similar to *Temnoma* subg. *Eotemnoma* (Schuster, 1963b), in which a quadrifid leaf and underleaf are linked with a variable leaf insertion, ranging from transverse to feebly incubous to feebly succubous. The plasticity in branching of such taxa also makes derivation of the Lepidoziaceae possible.

The Lepidoziaceae are related most closely to the Calypogeiaceae but differ from them in the retention of a perianth, in the usually multifid leaves, in the absence of gemmae, in the presence (generally) of flagella, and in the usually more opaque and chlorophyllose leaves and normally firmer cells, as well as in the ovoid capsule with straight, elliptical valves and, generally, in the more specialized form of the seta.

At one time the Lepidoziaceae (and allied Calypogeiaceae) were united with the succubous-leaved Cephaloziaceae (and such presumably allied groups as the Cephaloziellaceae) into the single family Trigonantheae of Spruce (1885, *in* 1884–85), sometimes emended to Trigonanthaceae (Verdoorn, 1932a), an inclusive portmanteau group embracing leafy Hepaticae with a trigonous perianth, in which the third keel is postical. Another basis for uniting these divergent groups into a single family was the assumption that leaf insertion and orientation did not allow a sharp separation of this complex into natural groups.

If the Cephaloziellaceae are left out of consideration, the three *modern* families Lepidoziaceae, Calypogeiaceae, and Cephaloziaceae unquestionably show definite affinities other than those suggested by the similarities in the abbreviated sexual branches. The most significant of these criteria —and one which has not been previously employed in tracing phylogeny in the Jungermanniales—is the developmental history of the capsule wall. In the three families in question, the epidermal cells of the capsule wall always appear to undergo a distinguishable "two-phase" development (which will be described in detail later) with, basically, restriction of secondary, pigmented thickenings to the secondary cell walls formed, which are usually exclusively longitudinal. Minor deviations from this pattern occur throughout all three families. Notwithstanding such deviations, the basic developmental pattern seems significant in suggesting some affinity among these families.[9]

[9] By contrast, the more common mode of development of the epidermal cells does not involve such a two-step process, and as a consequence the epidermal cells are all more or less uniform in developing thickenings (or lacking them), the thickenings almost always present on the transverse walls of the epidermal cells; see Vol. I, pp. 591, 601.

Although leaf insertion would at first glance suggest that no affinity is likely between the Lepidoziaceae and Cephaloziaceae, study of antipodal groups of Lepidoziaceae demonstrates that the concept of the Lepidoziaceae as a family with exclusively incubous or transverse leaves is erroneous. In *Hygrolembidium* Schust. (Schuster, 1963) the leaves are transverse to weakly succubous; in *Pseudocephalozia* Schust. (1965b), a genus of *Cephalozia*-like taxa previously placed by Stephani in *Cephalozia*, there are decidedly succubously inserted leaves; *Psiloclada* Mitt. and certain species of *Microlepidozia* have succubous leaves (Fulford and Taylor, 1959a). Thus, in critical cases, only two criteria can be used to separate the Lepidoziaceae from the Cephaloziaceae: (*a*) the restriction of rhizoids to underleaf bases; (*b*) the sharp distinction between large epidermal cells of the seta and interior, smaller cells. Although it may prove sound to transfer the Cephaloziaceae to the Lepidoziinae as an advanced family of the suborder, I have left them, with some misgivings, in the Jungermanniinae, in the absence of developmental studies of the capsules of a large enough array of genera.

I would not wish, however, to exclude the possibility that the succubous-leaved "Trigonantheae" (Cephaloziaceae, Odontoschismaceae) may be *derived* from Lepidozioid ancestors. The common possession of the following ensemble of features is suggestive. (1) Seta (*Cephalozia*) with few, large epidermal cells surrounding usually four interior cells; the seta of *Cephaloziella* can be derived by reduction. (2) Capsules with the epidermal cells narrowly oblong, showing apparently an identical two-phase development, with the primary cells longitudinally subdivided, and thickenings confined to the longitudinal secondary radial walls. (3) Stem "tending" to develop large cortical cells in few rows. (4) Remote similarities in branching. The more primitive Cephaloziaceae, and such genera as *Anthelia* and *Pleuroclada*, retain a regular ability to develop lateral, *Frullania*-type branching as well as to develop ventral-intercalary stolons. (5) A tendency for the sexual branches to become reduced, abbreviated, and restricted to the ventral merophytes in origin. (6) A tendency for retention of isophylly in the gynoecial shoot, even though the vegetative shoots are normally more or less strongly anisophyllous. These similarities and the trigonous perianth form which originally gave rise to the association of all these groups in the Trigonantheae are impressive.

Key to Genera of Lepidoziaceae

1. Branching regularly to irregularly pinnate, the lateral branches divergent at a nearly 90° angle; leaves clearly (2)3–4-lobed for at least 0.5 their length; never with asexual reproduction; flagelliform branches (occasionally rare or absent), not exclusively ventral-intercalary. . . . . . . 2.
    2. Lateral branches all of the *Frullania* type (occupying ventral half of segment); cells of leaves with oil-bodies; capsule wall (2)3–5-stratose; plants light to clear green, at least main axes with leaves ± incubously inserted. 3.
        3. Stem without a well-marked hyaloderm, with over 18 rows of cortical cells; leaves asymmetric (antical margin longest), with disc many cells high; leaf lobes, except at tips, several-many cells wide; cells

thick-walled, rigid, noncollapsing, dull; underleaves large, typically
3-4-lobed. Spores papillose. . . . . *Lepidozia* (p. 16)

3. Stem with a well-defined hyaloderm, cortical cells (in ours) in 10–12
rows only; leaves symmetric, with disc less than 1.5 cells high (in ours);
leaf lobes 1-seriate, filiform, except at or near juncture with disc;
cells large, pellucid, often collapsing in drying, nitid; underleaves
small (in ours; of 2 minute segments only). Spores finely areolate.
Telaranea (p.29)

2. Lateral branches on one side of axis of *Frullania* type, on opposed side of
Microlepidozia type (branch occupying dorsal half of segment); all or
most cells of leaves lacking oil-bodies; capsule wall 2-stratose (in ours at
least); plants deep or dull green to brownish, with transverse leaves.
No distinct hyaloderm (in ours), the cortical cells in up to 12 rows;
leaves symmetric or nearly so, with slender leaf lobes 2–4 cells wide toward
base. . . . . *Microlepidozia* (p. 41)

1. Branching pseudodichotomous: leaves shallowly 2–3-dentate at apex, occasionally obsoletely so; sometimes with caducous leaves; flagelliform
branches exclusively ventral-intercalary, whiplike. . . . *Bazzania* (p. 63)

LEPIDOZIA Dumort. emend. *Joerg.*

Pleuroschisma sect. *Lepidozia* Dumort., Syll. Jungerm. Eur., p. 69, 1831.
Mastigophora Nees, Naturg. Eur. Leberm. 1:95, 1833 (in part).
Lepidozia Dumort., Rec. d'Obs., p. 19, 1835; Joergensen, Bergens Mus. Skrifter 16:303, 1934
(emend.; excl. *Microlepidozia*).
Herpetium sect. *Lepidozia* Nees, Naturg. Eur. Leberm. 3:31, 1838.
Lepidoziopsis Hodgs., Rec. Dominion Mus. (N.Z.) 4:105, 1962.

Generally medium-sized, 0.5–2.5(4) mm wide, yellowish or whitish to
olive-green, rarely glaucous, *not brownish-pigmented, freely pinnately or bipinnately branched*, prostrate to procumbent to erect, *never closely attached to
substrate*. Stems oval in cross section, normally with 18–24 (or more) rows
of cortical cells, which are weakly to moderately or strongly thick-walled,
*subequal to somewhat larger than medullary cells in diam., never forming a hyalo-
dermis; medullary cells in numerous rows*, rather thick-walled or somewhat
collenchymatous, varying to thin-walled. Branching of two kinds: (1)
lateral-monopodial, the branches limited in length, uniformly of the
Frullania type, *replacing the ventral half of a leaf* (the dorsal half usually bifid,
rarely trifid), *from both series of lateral merophytes* (those of one side homodromous and of the other, antidromous, to the main axis); (2) *postical-
intercalary* from underleaf axils (*including usually all gynoecial* and all or
some androecial); lateral and postical branches in part running out as
whiplike, microphyllous, rhizoidous flagella (rare or perhaps absent in some
species); branches with the first appendage (underleaf) bifid, the branch
superficially appearing to originate in its axil. Rhizoids frequent or
infrequent, from bases of underleaves. *Leaves and underleaves rather similar,*

the latter usually somewhat smaller and arising from narrower merophytes (which are usually 4–6 or more cells wide, however). *Leaves ± asymmetric (dorsal half larger), incubously inserted and incubously shingled, commonly (3)4–5-lobed for 0.35–0.8 their length,* often with deflexed or suberect *slender lobes* arising from a spreading base, then handlike; margins entire to spinose-dentate; *lobes triangular, usually (2)3–5 cells wide or more at base.* Underleaves transverse, somewhat or *hardly smaller than leaves, ± lobed* like leaves. Cells *medium-sized, usually thick-walled,* with small or no trigones; *oil-bodies present, conspicuous, commonly 5–16 per cell,* small to medium-sized, homogeneous (or nearly) to coarsely segmented. *Asexual reproduction unknown* or occurring rarely by 1–2-celled gemmae.[10]

Monoecious or dioecious. Androecia occurring either as short postical branches or terminal on short to somewhat elongated lateral branches, sometimes running out into flagella; bracts concave, lobed like (but often less deeply than) normal leaves, usually much *smaller than normal leaves;* antheridial stalk 2-seriate. *Gynoecia on abbreviated postical branches;* bracts in several pairs, *inner much larger than leaves,* loosely sheathing perianth, *generally subentire or at least more shallowly lobed than vegetative leaves;* bracteole similar to bracts. Perianth ± pedicellate, elongated and tapering distally, obtusely *trigonous,* at least in distal half. Seta (where known) of *12 or more rows of epidermal cells at least slightly larger* than the numerous interior cell rows. Capsule oval, the *wall* (in examined species) *3–5-stratose;* outer layer with nodular thickenings of all secondary longitudinal walls, the primary longitudinal and all transverse walls free or nearly free of thickenings (or bearing weak and slightly pigmented ones); inner layer with semiannular bands which are often incomplete. Spores brown, *granulate-papillate* (at least in generic type and immediate allies). Chromosome no.: $n = 9$ (Tatuno, 1941, 1947).

Type. Lepidozia reptans (L.) Dumort.

The genus *Lepidozia* is here accepted in the restricted fashion, after exclusion of elements now placed in *Microlepidozia* (following Joergensen, 1934), *Telaranea* (following, i.a., Howe, 1902, and Hodgson, 1962) and *Drucella* (Hodgson, 1963). By 1922 Stephani (1898–1924) had assigned here 297 species including those now referable to *Microlepidozia* and *Telaranea.* Several have been described since, in particular by Herzog with the total number of species names now well over 300. It is doubtful whether more than half of these are valid. The vast majority of species

---

[10] Neither Müller nor I have seen gemmae. Degenkolbe (1938, p. 87) reports them, without detail, for *L. reptans.* I question the correctness of this report.

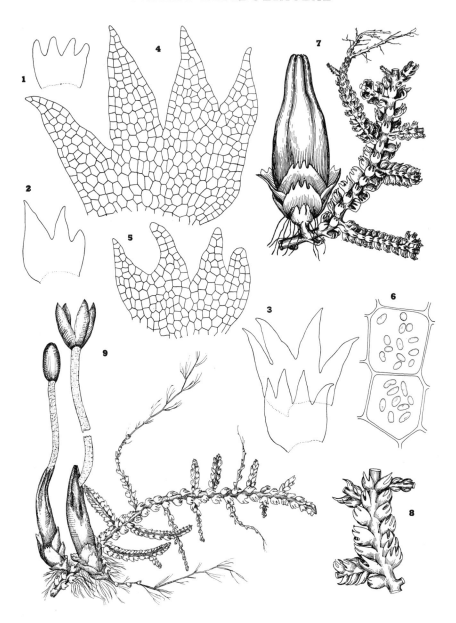

FIG. 86. *Lepidozia reptans*. (1) Underleaf (×40). (2) Abnormal (branch?) leaf (×40). (3) Two leaves (×40). (4–5) Leaf and underleaf from main stem (×85). (6) Two cells of lobe base with oil-bodies (×615). (7) Shoot with perianth, postical aspect (×ca. 12). (8) Shoot sector, showing *Frullania*-type branching with bifid supporting stem leaves (×ca. 15). (9) Fertile plant (×7.5). [1–6, *RMS 37976*, Roan Mt., N.C.; 7–8, near Ithaca, N.Y., *RMS*; 9, after Müller, 1905–16.]

are tropical or subtropical and antipodal, in particular in montane rain forests, etc. By contrast, the genus becomes rare in the Taiga and is evidently lacking in true Tundra.[11]

The genus is in a state of taxonomic chaos. Although Spruce as early as 1884–85 suggested the segregation of *Telaranea* and formally divided *Lepidozia* into two subgenera (*Microlepidozia* and *Eulepidozia*), Stephani (1909, *in* 1898–1924) failed to take into account Spruce's divisions and used a primary division into symmetric- and asymmetric-leaved groups. Evans (1912) observed that in *Microlepidozia* lateral branches from the dorsal half of the lateral merophytes occurred, as well as "normal" branches from the ventral half of the lateral merophytes, suggesting that the group deserved formal recognition as a genus, a step finally taken by Joergensen (1934). *Telaranea* has been generally accepted as a distinct genus, although Müller (1954, 1955, *in* 1951–58) erroneously synonymized *Microlepidozia* under it. As Schuster and Blomquist (1955) have shown, branching in *Telaranea* is the same as in *Lepidozia* and is not of the *Microlepidozia* type, a strong indication that *Telaranea* deserves continued recognition as a genus. Schuster and Blomquist (1955) also showed that *Telaranea* differs from *Lepidozia s. str.* in the strongly developed hyalodermis, the symmetric leaves, the large and often hyaline leaf cells, the areolate spores, and other characters. It is similar to *Lepidozia* in the 3–5-stratose capsule wall (that of *Microlepidozia* being usually 2-stratose).[12]

If the generally accepted segregration of *Microlepidozia* and *Telaranea* is maintained, *Lepidozia s. str.* still remains a large and unwieldy taxon. Fortunately, taxonomic problems hardly exist regionally, since only a single species is definitely known to occur. [Two other species, *L. sandvicensis* Lindenb. and *L. filamentosa* (Lehm. & Lindenb.) G., L., & N., occur in the oceanic parts of western North America.]

The genus has also been redefined by exclusion of taxa with symmetric leaves, rather large and hyaline cells, a distinct hyaloderm of the stem, and narrow-based leaf lobes. These taxa, constituting, in essence, Stephani's Section Symmetricae, were elevated to generic standing by Fulford and

---

[11] An apparent exception is the dubious plant known as *Lepidozia groenlandica* Lehm., Stirp. Pugil. 10:7, 1857. Described from material supposedly from Greenland (*leg.* "Breutel & Curie"), it has never been found since. It is subsequently shown (p. 22) that *L. groenlandica* is an antipodal taxon! According to Frye & Clark (1937–47, p. 656), the authority for this plant's occurrence in Greenland is Gottsche, Lindenberg, & Nees (1844–47). This superficially appears to be an example of scientific clairvoyance, since the plant was not described for another decade! However, perusal of Gottsche, Lindenberg, & Nees fails to reveal any mention of it, leading to the assumption that an error by Frye & Clark is involved.

[12] Müller (1939) accepted the separation of *Microlepidozia* from *Lepidozia* on the erroneous assumption that the former differed from the latter in lacking oil-bodies in the cells. Schuster and Blomquist (1955) have shown that both *M. setacea* and *M. sylvatica* may possess oil-bodies.

Taylor (1959) as *Neolepidozia*. The group is closely allied to *Telaranea*, as is obvious from the fact that one species placed by Fulford and Taylor (1959) in *Neolepidozia*, *Lepidozia tetradactyla*, is placed by Fulford (1963) in *Telaranea*! Because of the close proximity, Hodgson (1962) classified *Neolepidozia* as a synonym of *Telaranea*. Schuster (1963), placing the "Neolepidozia" element as a subgenus of *Telaranea*, also segregated the taxa with 8–12-lobed, symmetric or subsymmetric leaves that Stephani had placed in *Lepidozia* (comprising the antipodal *L. mooreana* element) as a subgenus of *Telaranea*. Thus, in effect, recent studies have broadened the concept of *Telaranea* at the expense of *Lepidozia*. The alteration in concept of *Telaranea* is dealt with on p. 30.

Hodgson (1963) also segregated the antipodal *Lepidozia integristipula* element from *Lepidozia*, as *Drucella*. This segregation appears well founded, *Drucella* differing from *Lepidozia* not only in the axial anatomy but also in the lack of terminal-lateral branching, the lateral branches apparently being consistently intercalary. However, the almost simultaneous attempt by Hodgson (1962) to divorce the *Lepidozia glaucophylla* element from *Lepidozia*, as *Lepidoziopsis* Hodg., seems untenable, and I regard *Lepidoziopsis* as forming merely a section within *Lepidozia*.

Lepidozia admittedly bears relationships to *Microlepidozia* and *Telaranea*. It differs from the former in the 3–5-stratose capsule wall, the strongly obliquely inserted and incubous leaves, the absence of branches from the dorsal end of the acroscopic portion of the lateral segments (all lateral branches arising from the vental half, thus replacing ventral halves of leaves), and the generally less deeply divided leaves and underleaves, usually with less linear lobes (although this distinction hardly holds universally). Differences between *Lepidozia* and *Telaranea* are chiefly derived from the spores (granulate or papillose in both *Lepidozia* and *Microlepidozia*; minutely areolate in *Telaranea*), the stem anatomy (typically no hyaloderm in *Lepidozia*; a generally well-developed hyalodermis in *Telaranea*; compare Fig. 85:8 and 9), the generally less deeply divided and asymmetric leaves with broader-based lobes of *Lepidozia* (those of *Telaranea* deeply divided, symmetric, with uniseriate to narrow-based lobes), and the much less delicate texture of *Lepidozia*, as contrasted to *Telaranea*. Study of the entire range of variation in *Lepidozia* may reduce the number of these differences appreciably.[13]

---

[13] One of the differences cited most often, the incubous leaves of *Lepidozia* vs. the transversely inserted leaves of *Telaranea* (see, i.a., Spruce, 1884–85; Frye & Clark, 1946, *in* 1937–47, p. 663) does not hold. Stephani (1909, *in* 1898–1924, p. 533) already pointed out that the leaf insertion in *Telaranea* is also oblique. This is not always readily evident on branch leaves but is obvious, e.g., from Fig. 90:10.

Sporophyte Anatomy

The basic anatomy of the capsule wall in *Lepidozia reptans* agrees closely
with that in *Telaranea* and *Bazzania*, in so far as the capsule-wall anatomy
of the last two genera is known (Schuster and Blomquist, 1955; Fulford,
1936a; Schuster, 1963). In these three genera the capsule wall is always
more than 2-stratose, varying from 2–5-stratose (*Telaranea*) to 3–5-stratose
(*Lepidozia*) and to 4–5-stratose (*Bazzania*), differing thus from *Microlepidozia*, in which a 2-stratose capsule wall generally prevails. In all
three genera, in contrast to *Microlepidozia*, the epidermal layer is thicker
than any of the inner layers and the thickenings of the epidermal cells are
confined largely or exclusively to longitudinal radial walls. Furthermore,
there is a sharply defined tendency for nodular thickenings of the radial
walls to be more or less united (thus forming brownish, sinuous nodular
walls in profile), rather than to develop as sharply discrete vertical bands
separated from adjacent thickenings by thin, unpigmented wall layers.
[This distinction is most marked, among studied taxa, in "*Neolepidozia*"
(*tetradactyla*), in which longitudinal walls bearing pigmented secondary
thickenings have them laid down as a nearly even sheet, as in some species
of *Calypogeia*.] Perhaps even more characteristic is the fact that thickenings
do not occur equally on all longitudinal walls, but tend to occur along
both faces of some adjoining walls, being absent on others.

This restriction of thickenings to certain longitudinal walls appears to be
related to a two-step development of cells in the epidermal layer. An initial
series of short-oblong to subquadrate cells, which do not normally develop
pigmented secondary thickenings, is cut off. These "primary cells" are secondarily subdivided by 1 or 3 (rarely 2) secondary, longitudinal walls, *both faces
of which develop secondary thickenings*. Thus, depending upon whether 1 or 3
secondary walls develop, two distinct patterns are found in the Lepidoziaceae:
either (type 1) there is an alternation of longitudinal walls without thickenings
or with faint, colorless thickenings (*w*) with walls both of whose faces bear
strongly pigmented, nodular thickenings (*t*); or (type 2) three walls with
thickenings (*t*) alternate with a wall without thickenings (*w*), resulting in the
following two patterns:

Type 1   *w t w t w t w t w t w t w*

Type 2   *w t t t t w t t t t w t t t w*
                 1   2   3

These two patterns intergrade, sometimes even on the same valve, since
even in species with walls 1, 2, 3, etc., free of thickenings (type 1) isolated weak
thickenings tend to occur. Inversely, in species tending to have thickenings on

walls 1, 2, 3 (type 2), these thickenings are often weaker and frequently less pigmented. Furthermore, occasional supplementary longitudinal walls are often produced, tending to destroy the periodicity of ornamentation of the walls. These patterns are most distinct in the middle of the valves, often becoming relatively indistinct near the apices.

The 2–4 strata of thinner and often somewhat narrower, interior cells are also very similar in all genera examined. The strongly elongated cells, which may be slightly sigmoid, have numerous vertical bands (nodular thickenings), usually 7–12 per cell, on each face, and these vertical bands tend to extend slightly across the tangential faces, forming partial or only occasionally complete semiannular bands (*Telaranea*, *Lepidozia*), or commonly extend wholly across the tangential free walls, forming complete or subcomplete semiannular bands (*Bazzania*).

The relatively firm capsule wall varies from 30–32 μ thick (*Lepidozia* when 3-stratose; *Telaranea*) to 35–36 μ (*Lepidozia* when 4-stratose) to 40–42 μ ("*Neolepidozia*" when 4-stratose), to 48–52 μ (*Bazzania* when 5-stratose).[14] In robust taxa of *Lepidozia* (e.g., *L. setigera* of New Zealand) the capsule wall may be 70–75 μ thick, and is then 5-stratose.

The spores are rather varied within species examined. In *Telaranea* they are delicately areolate or labyrinthine (Schuster and Blomquist, 1955), similar to those of "*Neolepidozia*" (Schuster, 1963) and of *Bazzania* (Fulford, 1936a). In *Lepidozia* and *Microlepidozia* the spores are finely to coarsely granulose-papillate to verruculose-papillate, the verrucae often slightly vermiform. In the latter two genera the spore-elater diam. ratio is ca. 1.5:1. In *Bazzania* it varies widely, Müller (1948a, p. 3) reporting it as 1:1 in *B. trilobata*, 3:1 in *B. tricrenata*.

Taxonomy

Only the generitype of *Lepidozia*, *L. reptans*, occurs in eastern North America.

A second species of the genus, *L. groenlandica* Lehm., Stirp. Pugil. 10:7, 1857, is cited from Greenland. Frye and Clark (1937–47, p. 656) erroneously cite the type as being collected by Breutel and Curie. However, I have seen a fragmentary specimen, evidently the type, labeled "Grönland, *Wormskiold*, Hb. Lehmann," in the "Lindenberg Hepat." in Vienna.[15] The broken stems of the *Lepidozia* are mixed with a stem of the purely antipodal *Temnoma quadripartitum*. Hence, it seems impossible that *L. groenlandica* represents a Greenland species; it evidently was based on a plant from the Antipodes (southern South America or possibly New Zealand).

---

[14] Fulford (1936a) gives the capsule wall in *Bazzania trilobata* as 5 μ thick, surely in error.

[15] The confusion with regard to *Lepidozia groenlandica* appears to reside, first, in the confusion of an antipodal plant as coming from Greenland, and, second, in the misunderstanding, by Frye and Clark, of *Jungermannia groenlandica* Nees (in G., L., & N., Syn. Hep., p. 114). The latter taxon, which is cited by Gottsche, Lindenberg, & Nees as having been collected by Breutel and Curie, represents *Lophozia groenlandica*; the explicit diagnosis leaves no room for misinterpretation.

LEPIDOZIA REPTANS (L.) Dumort.

[Figs. 86, 87]

Jungermannia reptans L., Spec. Pl., p. 1133, 1753.
Pleuroschisma reptans Dumort., Syll. Jungerm. Eur., p. 69, 1831.
Lepidozia reptans Dumort., Rec. d'Obs., p. 19, 1835; K. Müller, Rabenh. Krypt.-Fl. 6(2):281, figs. 81, 82a, 85, 1914; Macvicar, Studs. Hdb. Brit. Hep. ed. 2:331, figs. 1–4, 1926; Schuster, Amer. Midl. Nat. 49(2):535, pl. 64, fig. 6, 1953; Hattori & Mizutani, Jour. Hattori Bot. Lab. no. 19:79, fig. 1, 1958.
Lepidozia obliqua Steph., Bull. Herb. Boissier 5:94, 1897.
Lepidozia subalpina Hatt., Jour. Hattori Bot. Lab. no. 2:24, fig. 9, 1947.

In *loose*, flat, infrequently compact, *light green* (rarely yellow-green) to green patches, occasionally scattered among mosses. Shoots 1.5–3 cm long × 450–700 mm wide (rarely only 200–250 μ wide). Stems prostrate to procumbent, usually *freely and rather regularly 1-pinnate* to somewhat 2-pinnate, the branches widely but usually slightly acutely patent, *often flagelliferous at apex*; postical flagelliform branches elongate, usually scarce and only near bases of stems. Stems to ca. 200–225(260–350) μ broad × 150–175(200) μ high, slightly oval in cross section; cortical cells somewhat thick-walled, in up to 20–24 rows, the dorsal and lateral cortical cells larger in diam. [ca. (26)30–40 μ], compared to the ca. 5 rows of cortical cells of the ventral merophytes (hardly larger than the medullary cells) and the small, somewhat thick-walled, numerous medullary cells [in 6–9 tiers and averaging only 16–20(23) μ]. Rhizoids frequent below, rare on younger portions of stem, abundant on flagella, confined to bases of underleaves. Leaves rather laxly imbricate to contiguous, rarely somewhat remote, obliquely inserted and oriented, *subquadrate* to *obtrapezoidal* when flattened, weakly to moderately asymmetrical, ca. 500–625 μ long × 450–800 μ broad, (3)4-*lobed for 0.35–0.55 their length, in situ decurved, the leaf thus handlike;* antical base not strongly dilated, at most somewhat rounded, extending in the extreme to the midline of the stem when *in situ*; lobes triangular, acute (with their tips formed by 1–2 superposed cells), or abruptly blunt, decurved, *generally* (4)5–8 *cells broad at base*. Cells quadrate to hexagonal, variable in size, ca. 20–26(30) μ, equally thick-walled, although usually only slightly so, the trigones obscure; cuticle smooth; oil-bodies chiefly (8)10–16(25) per cell, polymorphous, varying from globular to elliptical to bacilliform to geniculate, pale, with discrete wall and nearly or quite homogeneous contents, or appearing 2–3-segmented or as if with buds, (2.5)3–4 μ to 2.5–4 × 4–7 μ, a few to 9–10 μ long. Underleaves somewhat spreading, ca. ⅔ *the lateral leaves in size*, quadrate to obtrapezoidal, ca. 380–450 μ broad × 325–375 μ long, *normally 4-lobed for 0.3–0.5 their length*, the rather short lobes obtuse to subacute at the tips. Without asexual reproduction [but, according to Degenkolbe (1938, p. 87) with fasciculate few-celled gemmae; this report highly questionable].

Autoecious. Androecia normally forming short, determinate postical branches, 1, sometimes 2, arising *from the axils of underleaves;* bracts very small, pellucid, in 4–8 closely imbricate pairs, strongly concave, one-third 2–3-lobed (antical lobe often merely toothlike); lobes subacute; antheridia usually single per bract. Gynoecia on short postical branches (occasionally from same underleaf

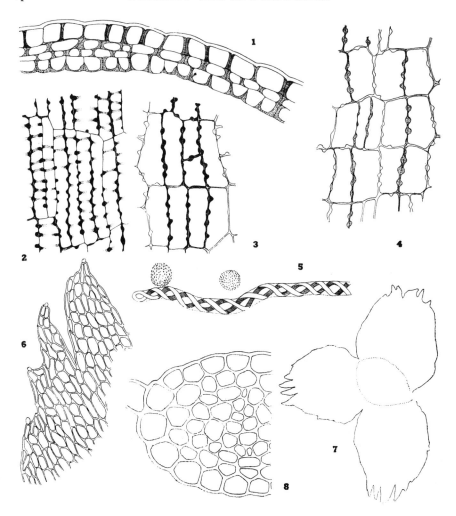

Fig. 87. *Lepidozia reptans.* (1) Capsule wall cross section (×400). (2) Inner cells, capsule wall (×400). (3–4) Epidermal cells, capsule wall (×400; pigmented thickenings drawn dark, others light). (5) Spores and an elater (×400). (6) Part of lobe of bract in fig. 7 (×98). (7) Bracts and bracteole (×24). (8) Stem cross section (×253). [1–7, from plants from N.C., *RMS;* 8, *RMS 11990.*]

axil as an androecium); *bracts much larger than leaves,* almost identical in size and form to bracteole, broadly ovate, from 860–975 μ broad × 900–1100 μ long, margined by irregular, rather strongly elongated cells; in basal half and middle subentire, except for the very tip, which is irregularly (2)3–6-dentate, the teeth small, acute. *Perianth long and prominent, slenderly cylindrical-fusiform, gradually narrowed distally* to the small mouth, obtusely trigonous above, terete below, bistratose near base; mouth trilobed and denticulate. Seta with 12 epidermal cell rows, ca. 16 inner rows. Capsule yellow-brown, cylindrical-oblong, the wall usually 4- (sometimes 3-)stratose. Epidermal cells ca. 12–15 μ high × (10)12–16(20) μ wide × 30–38 μ long, normally with alternation between longitudinal walls without pigmented thickening and three longitudinal walls with strong, ± confluent, nodular thickenings (in some cases with local alternation between walls with and walls without thickenings), the transverse walls usually without local thickenings [occasionally with solitary weak nodular thickenings]; nodular thickenings strong, commonly 2–5 per face, ± coalescent to form a sinuous, sheetlike band of thickening on the radial walls. Inner cells in 2–3 layers, the interior 1–2 each 5–7 μ thick, the innermost layer ca. 10 μ thick (total valve thickness thus 30–36 μ). Innermost layer of narrowly rectangulate cells commonly (9)10–12 μ wide × 65–80 μ long, with rather small but numerous nodular thickenings of all longitudinal walls, which are ± extended as weak, usually incomplete, tangential extensions, only with isolated, complete semiannular bands. Internal 1–2 strata with rather weak, yellowish, nodular thickenings. Elaters (9)10–11(12) μ in diam., only slightly tapering at the free ends, with two red-brown spirals ca. 3 μ wide. Spores (11)13–15 μ, reddish to yellowish brown, closely verruculose-papillate.

Type. Northern Europe.

Distribution

Widespread in the Northern Hemisphere, apparently circumboreal, except in the Arctic, from which it appears to be uniformly absent. In Europe from northern Italy, northwest Spain, southern France, to Belgium, Hungary, Germany, Poland, Czechoslovakia, eastward to Bulgaria and to Asia Minor; northward to Scandinavia (Sweden, Norway, Finland, although rarer northward) and to Ireland and to the Shetlands in the British Isles. Eastward to Russia and Siberia (common, i.a., as far as the Yenisey R., Amur area, Sakhalin) and also reported from the Himalayas, central China, Japan (alpine region and in Hokkaido; *fide* Hattori, 1952, 1955a), Formosa.

In the Atlantic region extending southward to the Azores and Madeira (Allorge, 1955, etc.).

Absent on the Faroes, Iceland, and Greenland.[16]

In North America presumably transcontinental in the Taiga; from Alaska! (central and east coast southward; see Evans, 1914b) to British Columbia!, Yukon!, Washington, Oregon, and California! to Alberta, Montana, Idaho, and New Mexico! (Shields, 1954).[17] In our area as

---

[16] Frye and Clark also cite old and incorrect reports from Hawaii, Java, and Juan Fernandez Island, based partly on Gottsche, Lindenberg, & Nees (1844–47), Montagne (1835), and Evans (1891). They also cite *L. reptans* from South America, supposedly (but erroneously) on the basis of a report by Howe (1917).

[17] Gottsche (1863, p. 126) reports both typical *L. reptans* ("Hovedformen *a*") and a var. "*b, australis*" from Mexico.

follows (with only a sampling of reports and examined specimens):

NEWFOUNDLAND. "Common throughout except for the most exposed and barren areas" (Buch & Tuomikoski, 1955); Witlers Bay (*Waghorne 195, p.p.!*); Channel!; Bay of Islands! MIQUELON I. *Fide* Buch & Tuomikoski (1955). PRINCE EDWARD I. Brackley Pt. (Macoun, 1902). QUEBEC. Manitounuck Sd., Hudson Bay Coast (Wynne & Steere, 1943);[18] Ste-Anne des Monts R. (Macoun, 1902); L. Mistassini; Oka; Pont-Rouge; Tadoussac!; Ste-Anne de la Pocatière; Bic!; R. du Loup; R. Causapscal; Mt. Albert; vic. Carleton; Beauceville; Montmorency R., etc. (Lepage, 1944–45). ANTICOSTI I. Jupiter R. (Macoun, 1902); Ellis Bay (Lepage, 1945). ONTARIO. Hayes I., Moose R., James Bay (Wynne & Steere, 1943); Ottawa; Algonquin Park; Nipissing; L. Nipigon (Macoun, 1902); Oxbow L., W. of Algonquin Park (*Cain 1363!*); Royston Park, Owen Sd.! NORTHWEST TERRITORIES. Belcher Isls., Hudson Bay (Wynne & Steere, 1943). NOVA SCOTIA. Yarmouth!; Digby!; Truro; Cape Breton I. (Macoun, 1902); Baddeck (Brown, 1936a); Halifax (*Brown!*); W. of Ingonish, Cape Breton! NEW BRUNSWICK. Grand Manan; Southern Head (Lorenz, 1923); Pt. Mouton!; St. John's!

MAINE. Mt. Katahdin (*RMS*); Matinicus I.; Mt. Desert (*Lorenz*); Beech Mt., Mt. Desert I.!; Streaked Mt.!; Prospect Harbor! VERMONT. Leicester (*Dutton!*); Newfane!; Stratton!; Haystack Mt., Windham Co.! Mt. Mansfield! NEW HAMPSHIRE. Mt. Washington (*RMS*); Franconia Mts. (Lorenz, 1908c); Glen Ellis, White Mts. (*Farlow!*); Mt. Prospect (*Grout!*); Mt. Adams! MASSACHUSETTS. Hawley Bog, W. Hawley; Worcester; Leicester; Mt. Greylock (Andrews, 1902); Mt. Everett!; Magnolia! RHODE ISLAND. Evans (1903b). CONNECTICUT. Litchfield, Hartford, Windham, New Haven Cos. (Evans & Nichols, 1908); "Wint. Falls," New Haven (*D.C. Eaton, 1878!*).

NEW YORK. Mt. Jo, Adirondack Mts.; Whiteface Mt.!; Little Moose L., Herkimer Co. (*Haynes!*); Washington Co. (*Burnham!*); Forest of Dean L., Rockland Co. (*RMS*); Undercliff, Essex Co. (*Haynes*); Tompkins, Cortland, Madison, Ontario, Onondaga, Cattaraugus, Genesee, Chenango Cos. (Schuster, 1949a); Slide, Wittenberg, and Cornell Mts., Catskill Mts. (*RMS*); Lyon Mt., Ellenburg Mt. near Ledgers Corners, Clinton Co. (*RMS*); Arnold L., Mt. Marcy (*Wilson!*); East Kill, Greene Co.! NEW JERSEY. "On ground in deep shaded ravines"; Austin, Hep. Bor.-Amer. Exsic. no. 75, 1873; Closter (*Austin!*) PENNSYLVANIA. Butler, Clarion, Clinton, Lawrence, McKean, Potter, Venango, Westmoreland Cos. (*Lanfear*); Sayre! DISTRICT OF COLUMBIA. *Fide* Plitt (1908). WEST VIRGINIA. Greenbrier, Ohio, Pocahontas, Preston, Tucker Cos. (Ammons, 1940); Cranberry Glades, Pocahontas Co. (*RMS 61217*).

VIRGINIA. White Top Mt., 5678 ft, Smyth Co. (Evans, 1893a; *RMS*); Mt. Rogers, 5715 ft, Grayson Co. (*RMS*); Mountain L. and Butts Mt., Giles Co. (*Blomquist!*; Patterson, 1949; *RMS*); Surrey Co. (Patterson, 1951). NORTH CAROLINA. Haywood Co.: Mt. Sterling (*Anderson*); Plott Balsam, near summit (*Blomquist!*); Balsam Cone Mt., Mt. Mitchell, Mt. Craig, and Mt. Clingman, Black Mts., Yancey Co. (*RMS*; *Anderson*); Nigger Mt., 4600 ft, near Jefferson, Ashe Co. (*RMS 29600*); Jones Knob, Jackson Co. (*Watson!*); Roan Mt., Mitchell Co. (*RMS*); Grandfather Mt., Avery Co. (*Welch!*) (W. S. Sullivant, Musci Allegh. no. 254!); Indian Gap, Haywood Co. (*Anderson*); Macon Co.: Dry Falls, near Highlands (*RMS*; *Sharp!*); Chattooga R., near Camp Ammons (*Anderson!*); Linville Gorge, Burke Co. (*RMS 29004, 28887a*); Balsam Gap, Blue Ridge Parkway (*RMS 19071, 19148*); Grandfather Mt., Caldwell Co. (*RMS*

---

[18] The specimen on which this, the northernmost supposed station in eastern North America, is based is in the NYBG (*Marr 395*). It contains no *Lepidozia* or *Lophocolea minor* (the other hepatic it is supposed to contain).

30180, p.p.). TENNESSEE. Blount Co. (Sharp, 1939); Roan Mt., Carter Co. (*RMS*); Clingmans Dome and Mt. LeConte, Sevier Co. (*RMS*). KENTUCKY. Powell Co. (Fulford, 1934). OHIO. Athens, Fairfield, Hocking, Lake, Portage Cos. (Miller, 1964). INDIANA. Fern, Putnam Co. (*Underwood, 1892!*). MICHIGAN. Ann Arbor; Isle Royale, Keweenaw Co.; Amygdaloid I., Isle Royale (*RMS*); Copper Harbor, Keweenaw Co. (*RMS*); Marquette, Ontonagon, Cheboygan!, Luce, Chippewa Cos. (Steere, 1937, etc.); [Au]Train Pt., L. Superior (*H. Gillman, 1867!*). WISCONSIN. Sand I., Apostle Isls., Bayfield Co. (*RMS*); Lafayette, Vilas, Oneida, Lincoln, Bayfield, Iron, Ashland, Douglas, Superior, Barron, Grant Cos. (Conklin, 1929). MINNESOTA. Anoka, Isanti, Chisago, Clearwater, Cook, Itasca, Lake, Lake of the Woods, St. Louis Cos. (Schuster, 1953). IOWA. Winneshiek, Allamakee, Clayton!, Hardin, Dubuque, Linn Cos. (Conard, 1945); N. of Homestead, Iowa Co.!

The species is common across northern North America, south of the Tundra margin; it becomes occasional in the Deciduous Forest Panclimax in the eastern United States. Southeastward it is limited, in the Appalachians, to higher elevations (chiefly 3600–6600 feet), with only isolated stations at ca. 2500 feet.

Ecology

Concurrently with wide dispersal, the species is exceptionally varying in its occurrence. In general on humus or on decaying wood or peaty soil, although never in true bogs, less often as a pioneer on rocks (although often a secondary species on rocks, after deposition of a humus or soil layer). Generally in shaded and humid sites, thus distinctly mesophytic, but unable to tolerate flooding. Occasionally on dry stumps or logs where partly insolated; then often as a small, hardly typical phase (mod. *parvifolia-laxifolia*). According to Buch and Tuomikoski (1955), generally with "*Cephalozia media, Blepharostoma trichophyllum, Calypogeia meylanii*, and *Lophozia incisa*, and with mosses such as *Dicranum fuscescens, Tetraphis pellucida, Dolichotheca turfacea, Plagiothecium laetum*, etc." Schuster (1949a) also considered it a "pronounced humicole, like *Bazzania trilobata*, and frequently occurring with it," although occasionally "on wet quartz-conglomerate ledges . . . with . . . *Bazzania denudata, Anastrophyllum michauxii, Anastrophyllum minutum*, etc." As Schuster (1953, p. 535) emphasized, saxicolous occurrences are rarely on calcareous rocks or sandstones. Northward the species is "common on cliffs and ledges, in moist, sheltered situations, and is ubiquitous in moist, deeply shaded woods. The species has a wide tolerance as regards pH northwards, but southward appears more restricted. In the north the species may occur on soil over basic rocks, where the pH is 6.0–6.5, but becomes much more common on humus-covered or peaty soil, where the pH is 3.7–6.0." In addition to the species listed by Buch and Tuomikoski, other common associates, on humus or soil-covered ledges, are *Diplophyllum taxifolium, Anastrophyllum*

minutum, Lophozia ventricosa, Bazzania trilobata, Tritomaria quinquedentata.
On logs *Lepidozia reptans* is found with a wide variety of primary and
secondary species, usually *Blepharostoma, Jamesoniella, Nowellia, Calypogeia
suecica,* occasionally the rarer *Scapania glaucocephala* and *S. apiculata* (at a
pH of 4.6–5.2).

I have summarized its ecological requirements as follows (Schuster,
(1953):

> The distribution appears to be restricted chiefly by the inability of the species
> to tolerate very basic conditions, as well as extremely acid conditions . . . , and
> by the relatively high and constant moisture requirements, as well as by the
> low toleration of light. The species is rarely able to compete successfully in
> bright sunlight and occurs commonly under light conditions of 3.6–12 FC.

In the moist Southern Appalachians, at 4500–6600 feet, the species, like so
many other saxicoles and humicoles, "ascends" trees and may occur as a
secondary species at the bases of *Abies fraseri, Picea rubens,* and rarely other trees,
associated with *Bazzania denudata, Anastrophyllum michauxii, Blepharostoma, Tritomaria exsecta,* and occasionally the more nearly pioneer *Herberta* and *Bazzania
nudicaulis.*

Differentiation

A stenotypic species, at least in our area, hardly to be confused with any
other. The green color (sometimes light or yellowish-tinged in insolated
extremes); the regularly pinnate-bipinnate branching; the incubous,
deflexed, handlike leaves with ordinarily four fingerlike lobes all serve to
give the plant an unmistakable facies. Occasional small, xeromorphic,
simply pinnate extremes are found which are hardly typical in having
largely 3-lobed lateral leaves and 2–3-lobed underleaves. Such forms,
however, are unlikely to be confused with other species.[19]

Lepidozia reptans is allied to two west European species, *L. pinnata* (Hook.)
Dum. and *L. pearsoni* Spr., and is particularly closely affiliated to the latter
taxon. Both of these allies are dioecious, whereas *L. reptans* is clearly autoecious.
The sexual branches of *L. reptans* are usually small and rather inconspicuous
at the time of fertilization; the androecia are short, compact spikes with usually
4–6 pairs of small, usually monandrous bracts. The gynoecia are represented
by equally abbreviated branches, bearing 2–3 pairs of small but gradually
larger bracts at the times of fertilization (when the inner, large bracts are not
mature). The gynoecia and androecia are various in position: often a single
sexual branch occurs in the axil of an underleaf, but occasionally two sexual

---

[19] "On very dry logs or stumps . . . the species often occurs as a reduced, small-leaved,
strongly imbricate modification, then only 0.2–0.25 mm wide. Such xerophytic extremes
(found . . . with *Odontoschisma denudatum* and the moss *Dicranella*) are quite atypical in appearance.
Such depauperate extremes usually have only 3-lobed leaves and largely 2-lobed underleaves"
(Schuster, 1953, p. 536). These plants, of admittedly deviant aspect, are to be regarded as
persistently juvenile manifestations, restricted to "difficult" sites. Similar depauperate forms
occasionally occur in montane sites; such a reduced phase was the basis for *Lepidozia subalpina*
Hatt. (see Hattori & Mizutani, 1958).

branches occur, and sometimes an androecium originates next to a gynoecium. In such cases it is very easy to demonstrate the bisexual nature of the inflorescence of this species.

TELARANEA Spr. *ex Schiffn. emend.*
Schust. (*1963*)

Lepidozia p.p., auct.

Telaranea Spr., Trans. Proc. Bot. Soc. Edinb. 15:358, 360, 365, 1885 [*nomen propositum*; type, *L. chaetophylla* Spr.]; Schiffner, *in* Engler & Prantl, Nat. Pflanzenfam. 1(3):103, 1895; Howe, Bull. Torrey Bot. Club 29:284, 1902; Frye & Clark, Univ. Wash. Publ. Biol. 6(4):662, 1946; Schuster & Blomquist, Amer. Jour. Bot. 42:592, 1955; Fulford, Brittonia 15:85, 1963.

Lepidozia subg. *Telaranea* K. Müll., *in* Rabenh. Krypt.–Fl. 6(2):276, 1914.

Neolepidozia Fulf. & Tayl., Brittonia 11:81, 1959.

Acrolepidozia Schust., Jour. Hattori Bot. Lab. no. 26:254, 1963.

Confervoid to ± *robust*, always *delicate*, prostrate to ascending or creeping among other bryophytes over soil, *pale to whitish green*, often pellucid, *lax*, *soft-textured*. Stems with cortical cells in (6)10–19(28) rows (2–5 or more rows of which form the ventral merophytes), *very large*, forming a ± distinct hyalodermis except sometimes ventrally, the *medullary cells much smaller*, in 3–24 or more rows. Rhizoids restricted to underleaf bases (never from leaf bases or leaf lobes). Branching loosely to ± closely pinnate (to bipinnate), the *vegetative branches terminal-lateral* (*replacing the ventral half of a lateral leaf*); the *sexual branches all or at least in part postical*; ventral flagella present or absent, rarely with lateral branches flagelliferous. Leaves alternate, *deeply and ± symmetrically 2–4–6(8–13)-lobed*, the juvenile leaves and those of branch bases *never monocrural*, varying from *incubously to virtually transversely inserted, dorsally inserted to stem midline, divided to within 0.3–2.5(4–6) cells of the base into 2–6(8–13) subulate and filiform lobes;* thus usually with vestigial or rather low discus; lobes mostly 5–7(9) cells long, *formed of narrow, elongate, leptodermous cells* (2–4.5 × as long as wide), which often appear ± inflated and tend to collapse in drying, *uniseriate throughout*, or 2(3) cells broad for the basal 1 (rarely 2–3) cell; cuticle of leaves *smooth* or faintly papillose; *oil-bodies present*, ± *small*, finely botryoidal, several per cell. Underleaves ca. 0.65 the size of the leaves or less, divided virtually to base into 2–3 (or 3–4) to 6–8 subulate cilia that are uniseriate to base; rhizoids frequent at underleaf bases.

Autoecious or dioecious. Archegonia usually on a short postical branch, rarely terminal on main axis or on elongated lateral branch. ♀ Bracts and bracteole similar, free from each other, from deeply, laciniately, 3–4-parted (the narrow lobes 2–4 cells broad, divided or fringed with cilia) to plurilobate, to shallowly 3–4-lobed at the apex and *Lepidozia*-like.

Perianth terete below, obtusely trigonous above, setose-ciliate at mouth. *Seta with 8 large epidermal + many small, delicate internal cell rows.* Capsule ovoid, ca. 1.6–1.8 × as long as broad, the wall (2) 3–5-stratose. Epidermal cells not or little higher than cells of strata lying within, narrowly rectangular, the longitudinal walls with low and not sharply defined, often confluent vertical bands (in profile the walls undulate-nodular); middle layer or layers with vertical bands that may be stronger; innermost layers with moderate to strong vertical bands, slightly to strongly extended over the free tangential walls (thus partly semiannular), varying to connected and extended across the entire face, becoming semiannular. Spores 1.0–1.5 × the diam. of the elaters, delicately but *distinctly areolate.* Elaters little tortuous, 2(3)-spiral, spirals narrow, brown; elater apices rather blunt. Antheridia solitary, *stalk 1-seriate.*

Type. Telaranea chaetophylla Spr. [= *T. nematodes* (G. ex Aust.) Howe]. Identical, according to S. Arnell, with *T. sejuncta* (Ångstr.) Arn.

Nomenclature

The species currently placed in *Telaranea* were once almost all placed in *Lepidozia. Telaranea*, as a generic name, dates from 1885 (Spruce, Trans. Proc. Bot. Soc. Edinb. 15(2):358, 359–360, 365),[20] based on *Lepidozia chaetophylla* = *Telaranea nematodes* = ? *T. sejuncta.*

Telaranea had only two species assigned to it until recently (Howe, 1902; Frye and Clark, 1937–47; Schuster and Blomquist, 1955). With study of the tropical and antipodal floras, the genus has gradually been broadened by the accretion of new taxa, necessitating the above emended diagnosis.

Hodgson (1962) has stated that *Neolepidozia* Fulf. & Tayl. (1959), a segregate from *Lepidozia*, cannot be kept distinct from *Telaranea*, placing *Neolepidozia* in simple synonymy under *Telaranea*.[21] Independently, I had gradually arrived at the conclusion that the largely tropical and antipodal *Neolepidozia* could well be treated as a subgenus of *Telaranea* (Schuster, 1963), and I then also transferred to *Telaranea* the *Lepidozia mooreana* element, in which the leaves are

---

[20] The history of *Telaranea* as a genus is exceedingly complex. Spruce (*loc. cit.*, p. 358) mentions "*Telaranea* nob." and on pp. 359–360 gives a Latin diagnosis, under subg. *Microlepidozia*, at the end of which he parenthetically asks: "*Telaranea* nobis nov. gen.?" On p. 365 he states: "*Telaranea chaetophylla* Spruce Mst. nov. gen.," followed by a diagnosis of "*Lepidozia chaetophylla* Spruce." The genus thus has dubious antecedents, and one can only arbitrarily solve the problem of whether and when and where it was first validated. The description given under subg. *Microlepidozia* on pp. 359–360 of Spruce can hardly serve to diagnose the *genus Telaranea*. Müller (1951–58) cites p. 365 (in Spruce, *loc. cit.*) as the valid point of publication, but here *Telaranea* is cited under the *genus Lepidozia*. The first unambiguous treatment and diagnosis are those of Schiffner (1893–95).

[21] Grolle (1964c, p. 170) also regards *Neolepidozia* as a group *within* the genus *Telaranea* but does not specify a hierarchial rank. I would now also reduce my genus *Acrolepidozia* (Schuster, 1963) to a subgenus of *Telaranea*; it appears to have evolved from *Neolepidozia*. It was based on *Lepidozia longitudinalis* Herz., Trans. Brit. Bryol. Soc. 1(4):312, fig. 29, 1950 [= *Telaranea longitudinalis* (Herz.) Schust., comb. n.].

8–13-parted. *Telaranea* thus has greatly changed in scope. As redefined, it differs from *Lepidozia* chiefly on the basis of criteria already alluded to: the essentially symmetric leaves, with a low or virtually no disc; the uniseriate antheridial stalks; the uniseriate lobes, at best 2(–3)-seriate at the juncture with disc; the delicately areolate spores;[22] the development of a distinct hyaloderm of the stem; the larger, softer cells, with the consequence that even the largest species of *Telaranea*, such as *T. mooreana*, give a soft and often silky impression.

The treatment of *Telaranea* as identical with *Microlepidozia* in Müller (1956, *in* 1951–58) is based on lack of understanding of branching patterns, axial anatomy, spore exine morphology, and other criteria, as shown by Schuster and Blomquist (1955). Other recent workers concur in recognizing both *Microlepidozia* and *Telaranea*. Müller's *Telaranea* (as defined to include *Microlepidozia*) is stated to differ from *Lepidozia* in the (*a*) "wholly different leaves" that (*b*) are "nearly transversely inserted"; (*c*) lack of oil-bodies; (*d*) unistratose perianths; (*e*) different seta cross section; (*f*) 2–3-stratose capsule wall. Of these criteria only *d* is possibly valid. *Telaranea* has incubously inserted leaves, at least on robust stems; *Telaranea*, as well as all of the species of *Microlepidozia* studied, possesses oil-bodies, thus agreeing with *Lepidozia*; the seta, although stated by Müller to possess 8 epidermal and only 4–16 small, inner cell rows, may have as many as 10 rows of epidermal cells and "16–24 or more" rows of small, interior cells (Schuster and Blomquist, 1955); the capsule wall in *Telaranea nematodes* is 3-stratose, as it often is in *Lepidozia reptans*. Indeed, if only Müller's criteria (p. 1132) were to be used, a separation of *Microlepidozia* and *Telaranea* from *Lepidozia* could not be maintained, and we would have to agree with Hodgson (1956), who separates them very artificially as subgenera of *Lepidozia*.

However, as has been pointed out (Schuster and Blomquist, *loc. cit.*), a whole series of criteria, some of which appear to be fundamental, exists to separate *Telaranea* from *Microlepidozia*. Among these are the type of lateral branching (*Microlepidozia* type in *Microlepidozia*; only *Frullania* type in *Lepidozia* and *Telaranea*); spore exine (papillate in *Microlepidozia*; areolate in *Telaranea*); capsule wall (2–3-stratose in *Microlepidozia*; [2] 3–5-stratose in *Telaranea*); ♀ bracts (oval and entire to 2–4-lobed in *Microlepidozia*; usually 4- or more parted and often laciniate in *Telaranea*); stem anatomy (usually obscure or no hyalodermis in *Microlepidozia*; often conspicuous hyalodermis in *Telaranea*).

Telaranea, through the subg. *Telaranea*, shows a close approach to the exclusively tropical genus *Arachniopsis* Spr. Schuster (1965b) treats the separation of the two, and this is also briefly discussed under subg. *Telaranea* (p. 35).

I have divided *Telaranea* into four subgenera, as follows (Schuster, 1963);

---

[22] The statement (Fulford, 1963a, p. 69) that the spores of *Telaranea* are "finely punctuate" is clearly incorrect, as shown by the figures of Schuster & Blomquist (1955). I have also checked antipodal taxa, such as *T. tetradactyla* and *T. mooreana*, and find that without exception at least the outer spore face is always delicately reticulate, being covered by a close-meshed web of fine, anastomosing ridges.

Similarly, the statement in Fulford (1963a, p. 84) that the capsule-wall of *Telaranea* has "two layers" is incorrect, as is obvious from fig. 1 in Schuster & Blomquist (1955) and as is clear from Figs. 88:9 and 90:1 in this volume. However, the New Zealand *T. exigua* Schust. (Schuster, 1968, p. 458) has a 2-layered capsule wall; it is unique, among described taxa, in this respect.

one of these (*Acrolepidozia*) I had regarded, on insufficient bases, as generically distinct; a fifth has just been described by Grolle (1966i).

Key to Subgenera of Telaranea

a. Leaves with 2–6 lobes, inserted on a symmetric lamina 0.3–3(4) or more cells high; underleaves with 2–6 segments; stem with hyaloderm very distinct, usually formed of 6–18 cell rows. Capsule wall normally 3-stratose . . . . *b*.

    b. Leaves strongly incubously inserted, with a distinct disc, usually 4 or more cells high and (on main axes) usually with 4 leaf lobes; underleaves of main axes usually 4-lobed; leaf lobes usually 2–3 cells wide at base, rarely uniseriate throughout. Supporting stem leaf associated with terminal branches always 2-fid . . . . *c*.

        c. Without sharp dimorphism of branches (both leading axes and primary branches typically with similar 4-lobed leaves and 4-lobed underleaves); leading stem with leaves and underleaves never loosely appressed and scalelike . . . . . subg. *Neolepidozia* (Fulf. & Tayl.) Schust.

        c. Shoot system sharply dimorphic: leading stems with reduced, loosely appressed 4-lobed leaves and underleaves; branches all with longer, horizontal, patent, 3-lobed, oblique leaves and very small to minute 2-lobed underleaves. [Branches of one side of main axis tending to be leafy, ascending; of the other, microphyllous and flagelliform.]. . . . . subg. *Acrolepidozia* (Schust.) Schust., comb. n.

    b. Leaves subtransverse to weakly incubous, with disc vestigial, usually 0.3–1.5(4) cells high; leaf-lobe number variable, from 2–6; underleaves also 2–6-lobed; leaf and underleaf lobes usually 1-seriate throughout (rarely 2-seriate at base). Supporting stem leaf associated with branch 1- or 2-fid. . . . . *d*.

        d. Perianth smooth externally. . . subg. *Telaranea* Spr. ex Schiffn.

        d. Perianth surface armed with ciliiform or setaceous appendages. . . . . .subg. *Chaetozia* Grolle.

a. Leaves with 8–12(13) uniseriate segments, inserted on a lamina that is often slightly asymmetrical and 4–6 cells high; underleaves with (6)8–9 segments or more; leaf and underleaf segments uniseriate; stem with ca. 24–28 rows of cortical cells, forming a weakly differentiated hyaloderm. Capsule wall 5-stratose. . . . . . . subg. *Tricholepidozia* Schust.

The groups *Neolepidozia*, *Acrolepidozia*, and *Tricholepidozia* do not occur regionally; they are discussed in Schuster (1963). Grolle (1966i, p. 280) describes *Chaetozia*, type *T. chaetocarpa* (Pears.) Grolle, of New Caledonia. Perhaps only a section of subg. *Telaranea* is at hand.

Subgenus *TELARANEA* Spr. ex Schiffn.

Delicate, often confervoid, creeping, light to clear green, *often glistening and with a silky texture*, usually irregularly and sparingly 1(2)-pinnate. Stems with *cortical cells in 6–12 (15–18) rows, forming a distinct hyaloderm*; with ventral

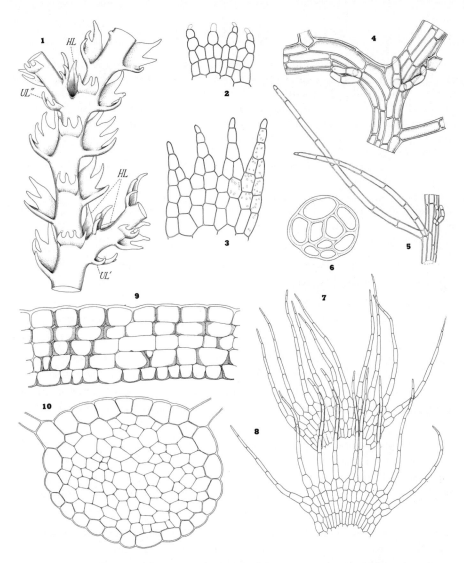

Fig. 88. *Telaranea*, subgeneric characters and extremes. (1–3) Subg. *Neolepidozia* (Fulf. & Tayl.) Schust. (4–6) Subg. *Telaranea*. (7–10) Subg. *Tricholepidozia* Schust. (1–3) *Telaranea remotifolia* (Hodgs.) Hodgs.: 1, shoot sector in ventral aspect; note the *Frullania*-type branches with stem "half leaves" at HL, and the trilobed primary branch underleaves at UL¹ and UL² (×32); 2, underleaf (×70); 3, leaf; note slight asymmetry (×70). (4–6) *Telaranea herzogii* (Hodgs.) Hodgs. *s. lat.*: 4, shoot sector, ventral aspect, to show a *Frullania*-type branch towards the left; note equally bifid underleaves, even at branch base (×137); 5, leaf and underleaf, *in situ* (×68); 6, stem cross section (×295). (7–10) *Telaranea mooreana* (Steph.) Schust.: 7–8, two underleaves from main stem; note that leaves are similar but feebly larger, with 11–13 segments (×36.5); 9, capsule wall cross section (×340); 10, stem cross section (×106). [All figures based on antipodal material; 1–3, *RMS 67–401*, Falls Creek, Fiordland, N.Z.; 4–6, *RMS 60650*, New Zealand; 7–10, *RMS 50364*, Camp Creek, Surprise Valley, Tasmania.]

vegetative branches occasional or frequent, becoming leafy or remaining flagelliform. Leaves with insertion subtransverse to distinctly incubous,[23] but at least always slightly incubously inclined, remote to barely contiguous, symmetric or somewhat asymmetrical (then ventral lobe usually smaller), *divided to within 0.3–2(3–4) cells of base into 2–4 (5–6) uniseriate segments* that may be 2(3) cells wide at the base for at most 0.5–1.5 × cell length; *disc obsolete* or at most 2–3(4) cells high, symmetric. Leaf lobes filiform, mostly 5–7 cells long, formed of narrow, often delicate cells that may collapse in drying; cuticle smooth; cells large, thin-walled, rather hyaline, each with several small, faintly botryoidal-granular, colorless oil-bodies. Underleaves variable: from vestigial and formed of two short segments each 1–3 cells long to 3–4–6-segmented, with the segments approaching those of the leaves in length. Capsule wall (2) 3-stratose.

Type. Telaranea chaetophylla Spr. [= *T. nematodes* (G. ex Aust.) Howe = ?*T. sejuncta* (Ångstr.) S. Arn.]

A small group, except for the widespread *T. nematodes* exclusively antipodal. Seven species from South America are referred here by Fulford (1963a), and three species appear to occur in the New Zealand-Tasmania region [*T. herzogii* (Hodgs.) Hodgs.; *T. nivalis* Schust.; *T. exigua* Schust.]

Telaranea subg. *Telaranea* is approximately equivalent to *Telaranea sensu* Howe (1902), *sensu* Schuster and Blomquist (1955), and *sensu* Fulford (1963a), as a genus. For some time only *T. nematodes* was recognized in the group, although Howe (1902) referred *T. bicruris* (Steph.) Howe, from Brazil, to *Telaranea*; this is regarded by Fulford (1963a) as a synonym of *T. sejuncta* (= *T. nematodes*). However, according to Howe (1902, p. 287), *T. bicruris* differs from *T. nematodes* in the almost constantly bifid leaves, the frequent extension of the stems into leafless flagella, and the simpler ♀ bracts, as well as in probable dioecism.

Howe (1902) and Frye and Clark (1946, p. 651) would separate *Telaranea s. str.* from *Lepidozia* (incl. *Microlepidozia*) because the former supposedly has the leaves divided to within 0.5 cell of the base (Fig. 90:11) or even deeper, and because the lobes are uniseriate, except at the very base (where the lobes may be 2 cells broad). However, main stems of robust plants show (Fig. 90:10) that the sinuses may extend down only to within 1.3 cells of the base, and that the basal 2 cell rows may be 2 cells broad. On this basis no clear distinction can be made between the two genera.

Telaranea varies from a nearly confervoid extreme, in which the plants have bifid leaves and underleaves (*T. bicruris*, *T. herzogii*) associated with a stem formed of only 6 cortical and 3 medullary rows, to a midpoint with the leaves 3–4-lobed and the underleaves 2(3)-lobed (*T. sejuncta*), to robust extremes

---

[23] The statement in Fulford (1963, p. 66) and other authors that *Telaranea* has the "leaf insertion transverse" is incorrect, as already noted in Schuster & Blomquist (1955). Even Fulford's own drawing (fig. 133) shows a clearly incubous insertion in the generitype.

illustrated by *T. apiahyna*. In this last species the plants are regularly bipinnate, with the branches progressively reduced in vigor; main stems possess quadrifid leaves and underleaves and the stem has 18–19 rows of enlarged cortical cells surrounding a large number of small medullary cells. Such species, in stem anatomy and branching, approach the simple species of "*Neolepidozia*," except that they regularly have bifid initial branch underleaves. They also have more delicate cells, with the leaf cells tending to collapse in drying; this is a rare phenomenon in *Lepidozia* and "*Neolepidozia*" (although Hodgson states it may occur in *T. tetradactyla*).

The most reduced species (such as *T. herzogii*), with minute underleaves, bifid underleaves, and a stem with only 6 outer and 3 inner cell rows, closely approach *Arachniopsis* Spr. They are fundamentally distinct, however, in having the usual Lepidozioid seta with many small internal cell rows (*Arachniopsis* has only 4 internal cell rows within the 8 epidermal rows); in never developing rhizoids from the penultimate cells of the leaf lobes; in the regular occurrence of terminal-lateral branching; in the insertion of the leaves to the stem midline dorsally; in the consistent absence of monocrural leaves. These differences are elaborated at length in Schuster (1965b).

Although the filiform leaf lobes of *Telaranea* subg. *Telaranea* suggest *Blepharostoma*, the latter has a more simply organized stem, without a hyalodermis, and shows almost uniform lateral branching. It also has a simpler seta with only 4 inner cell rows and a capsule wall that is only 2 cell layers thick. Capsule-wall anatomy of the two genera *Telaranea* and *Blepharostoma* is also fundamentally distinct (compare Figs. 90:1–3 and 76:1, 2, 4). Hence I maintain my earlier opinion that the two genera have evolved, through parallel reduction, a purely superficial similarity and must be placed in different families.

TELARANEA NEMATODES (G. ex Aust.) Howe
[Figs. 89, 90]

Jungermannia nematodes G., *in* Hep. Cubensis Wrightianae (no date; after 1885).
Cephalozia nematodes Aust., Bull. Torrey Bot. Club 6:302, 1879.
Lepidozia chaetophylla Spr., Trans. Proc. Bot. Soc. Edinb. 15:365, 1885.
Lepidozia nematodes Spr., *ibid.* 15:366, 1885.
Telaranea chaetophylla Spr., "Mst. nov. gen.," Trans. Proc. Bot. Soc. Edinb. 15:365, 1885 [as synonym]; Schiffner, *in* Engler & Prantl, Nat. Pflanzenfam. 1(3):103, 1895.
Lepidozia chaetophylla tenuis Pears., Vidensk. Selsk. Forh. Kristiana 1886(3):7, 1886; Evans, Bull. Torrey Bot. Club 20:308, 1893.
Blepharostoma antillarum Besch. & Spr., Bull. Soc. Bot. France 36, Suppl. clxxxiii, 1889.
Blepharostoma nematodes Underw., Bull. Torrey Bot. Club 23:383, 1896.
Telaranea nematodes Howe, *ibid.* 29:284, 1902.
Telaranea nematodes antillarum Howe, *ibid.* 29:286, 1902.
Telaranea nematodes longifolia Howe, *ibid.* 29:286, 1902; Haynes, The Bryologist 8:97, figs. 1–2, 1905.

Arachnoid, 300–450 μ wide, prostrate, *excessively delicate, filmy, pale to whitish green*, in drying with the cells usually more or less collapsing, becoming *strongly shiny*, quite transparent. Stem 120–150 μ wide × 90–105 μ high; cortical cells of main stems normally *in 10–12 rows*; dorsal cortical cells 40–45 μ wide, much larger than medullary cells [which are 10–18(20) μ], both cortical and medullary cells somewhat evenly thick-walled. Leaves on main stems *3–4-lobed*, 350–500, rarely 600–900, μ long, the linear lobes *uniseriate* (*except*, at most, *for the basal 1–2 cells*), (4)5–8 cells long. Cells thin-walled, *delicate*, in middle of lobes 20–24(28) μ wide × *90–110 μ long* (ratio, 1:2.5–4.5); oil-bodies subglobular, glistening, minute, and appearing homogeneous, several per cell; cuticle smooth. Underleaves *smaller than leaves*, 2–3-lobed to base, their divisions 2–3 cells long, incurved distally. Without asexual reproduction.

Autoecious. ♂ Inflorescences usually on short lateral branches; ♂ bracts like vegetative leaves, monandrous. ♀ Inflorescence terminal, usually on very short postical branches, rarely on elongate lateral branches; the bracts and bracteoles in 2–3 series; bracts deeply 3–4-lobed, to within 1–3 cells of base, the *lobes resolved into long tapering cilia or laciniae* (whose cells are 110–140 × 20–40 μ); lobes of bracts 3–4 cells wide at base, cells at base 40–45 × 120–140 μ; bracteole similar to bracts, of like size. Perianth terete below, somewhat narrowed in the obtusely trigonous distal portion; cells below cilia rectangular, 22–25 × 90–120 μ; mouth very longly ciliate, cilia sharp, occasionally 2–3-furcate, *setose, tapering*; cells 23–36 × 100–130 μ. Capsule with wall 3-stratose. Epidermal cells narrowly rectangular, ca. 10–13 × 43–65 μ, both faces of longitudinal walls with irregular vertical bands (in profile, walls somewhat sinuous-nodular), except every fourth longitudinal wall and all transverse walls lacking thickenings, or with vestigial ones; epidermal cells ca. 10–13 μ thick (subequal to innermost cell layer in thickness); intermediate cell layer with strong nodular (radial) thickenings, 7–8 μ thick; inner cell layer of narrowly rectangulate cells, ca. 10–14 μ wide × 55–80 μ long × 12–14 μ thick, bearing rather strong vertical bands not or weakly extended across the free tangential face (thus with incomplete tangential bands), very occasional tangential extensions extending completely across the face of the cell. Spores 14–16(17) μ, pale brown, *delicately areolate* (the irregular, polyhedral areoles only 1–1.8 μ in diam., defined by slightly elevated narrow ridges), except on the inner faces (where verruculose or vermiculate-marked). Elaters ca. 180–240 μ long × 9–13 μ in diam. (9–10 μ when 2-spiral; 12–13 μ in the exceptional 3-spiral cases); spirals red-brown, 4 μ wide, rather tightly wound.

Type. Cuba (Wright).

Distribution

An austral to tropical species, of wide distribution in oceanic regions. Occurring from Natal, South Africa, northward to Spain (San Sebastian), the Azores (*fide* Persson), and Ireland (Kerry; *leg.* Verdoorn; see Buch, 1938). Also on Tristan da Cunha [Persson, *in* Proc. 7th Intern. Congr. Bot. (1950): 846, 1953]. In the New World more widespread, occurring

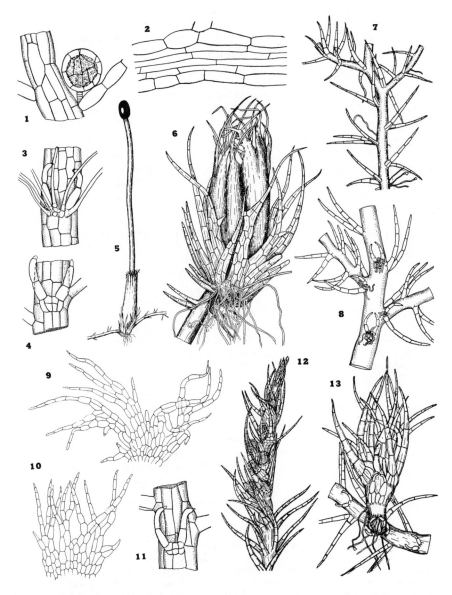

FIG. 89. *Telaranea nematodes.* (1) Antheridium, *in situ* (×112). (2) Longitudinal section of small stem (×112). (3–4, 11) Shoot sectors, showing underleaves (×112). (5) Perianth and sporophyte (×6.5). (6) Perianth and bracts, showing postical origin (×ca. 25). (7) Shoot, antical view, showing branching (×ca. 25). (8) Shoot, postical view, showing branching (×ca. 25). (9) Bract (×ca. 25). (10) Bracteole (×ca. 25). (12) Androecium (×ca. 25). (13) ♀ Inflorescence, showing postical origin (×ca. 50). [1–7, 11–13, drawn by H. L. Blomquist; 1–7, 11–13, from plants from North Carolina; 8–10, *RMS F-101.*]

from northern Brazil and the Peruvian Andes (Spruce, 1884–85; Howe, 1902), to British Guiana, the West Indies (Guadaloupe; Jamaica; Puerto Rico; Cuba, type) to Bermuda (Howe, 1902), the southeastern United States, northward as far as New York.

The distribution—and, indeed, taxonomic limits—of *T. nematodes* remain to be clarified. Arnell (1957, p. 17) stated that this species was identical with *Blepharostoma sejuncta* Ångstr. (Vet.-Akad. Förhandl. 1876–77:78, 1876), and this synonymy is accepted by Fulford (1963a). She also places in synonymy *Lepidozia bicruris* Steph. (Hedwigia 24:166, pl. 3, 1885), a plant Howe (1902) and Schuster & Blomquist (1955) had retained as distinct. If the above two taxa are included as synonyms, *T. nematodes s. lat.* = *T. sejuncta* acquires a very wide range, covering as well tropical and southern Africa, Peru, Bolivia, and Brazil and northward in South America.[24]

FLORIDA. Highlands Hammock State Park, W. of Sebring, Highlands Co. (*RMS 20099, F-101, 102; c. caps; plate*); Midway, Gadsden Co. (*Schornherst 84*); Bristol (*McFarlin*); Mossyhead (*McFarlin*); Sanford, Seminole Co. (*Rapp*); Alexander Springs, Lake Co. (*RMS*); Juniper Springs, Marion Co. (*RMS*); Collier, Manatee, and Polk Cos. (Redfearn, 1952). GEORGIA. Thomasville (*M. S. Brown*); Brunswick (*Austin, Apr. 1878*). MISSISSIPPI. Ocean Springs, Harrison Co. (*Pennebacker*); near Van Cleave, Jackson Co. (*RMS 1902a, 26504c*); 3–5 mi E. of Oxford, Lafayette Co. (*RMS*). SOUTH CAROLINA. N.W. of Sumter, Sumter Co. (*RMS 18709*). NORTH CAROLINA. Near Roper, Washington Co. (*RMS 28350*); E. of St. Paul, Robeson Co. (*Anderson 3802*); Richmond Co. (*Anderson 65353*); Black R. on Rte. 41, Sampson Co. (*Anderson 3862*); White L., Bladen Co. (*Gray 7439*); L. View, Moore Co. (*Blomquist 7244*); Pinehurst (*C. C. Haynes 855*); E. of Varina, Wake Co. (*Blomquist 11019*); E. of Hamlet, Richmond Co. (*Correll 6954*); near Southern Pines, Moore Co. (*Blomquist 7341*); W. of Durham, Durham Co. (*RMS & Blomquist 28622*).

VIRGINIA. Charles City; near Chickahominy R. (Patterson, 1950). NEW JERSEY. Highlands, Monmouth Co. (*Haynes; distr. in Krypt. Exsic. Mus. Vindobonese no. 1571*). NEW YORK. Arlington, Staten I., New York City; Freeport, Long I. (*Howe, 1898; type of T. nematodes longifolia Howe*); Hither Hills State Park, near Montauk, Long I. (*RMS 22074*); Fishers I. (Evans, 1926; northernmost report!).

The species is largely of extreme Coastal Plain distribution but extends inward occasionally into the outer Piedmont (as in North Carolina) or

---

[24] The synonymy accepted by Fulford (1963a) is not accepted here, since there is sufficient confusion in her account to cast doubt on her conclusions. She illustrates a plant (figs. 132–133, 135–141) supposedly "drawn from a portion of the type" (Herbier Boissier) and cites the type locality of *Telaranea sejuncta* as "Caldas, Brazil." No corresponding specimen and none studied by Ångstrom are in the Herbier Boissier. The question arises as to the identity of the so-called "type" she illustrates.

This is relevant because this "type" fails to correspond to *T. nematodes* as illustrated in Schuster & Blomquist (1955). In true *T. nematodes* the mature stem *always* has 10–12 rows of cortical cells and the leaves are inserted on 3–4 rows of these cells; leaf segments are 2 cell rows broad at base. In Fulford's "type" leaf segments are uniseriate to base and the leaves are inserted on only 2 cell rows—thus the stem has only 6 rows of cortical cells. Other discrepancies (compare the form and size of the terminal cells of the leaf segments of Fulford's "type" with Fig. 90:11–12) suggest that the proposed synonymy still needs adequate proof. Until this is at hand, I prefer the less confusing course of retaining *T. nematodes* as a species.

upper Coastal Plain (north Mississippi; there over 300 miles from the coast). It is common only in the coastal swamps.

Ecology

A species largely of hygrophytic sites, particularly frequent in coastal swamps and wet pocosins, associated with *Magnolia virginiana*, *M. grandiflora*, *Persea*, *Gordonia lasianthus*, etc.; most frequent along slow-moving coastal streams or deeply shaded springs cutting through sand, and there abundant on mossy banks, often creeping over *Sphagnum*. Almost invariably associated are *Pallavicinia*, *Odontoschisma prostratum*, *Cephalozia catenulata* (occasionally *C. macrostachya* and *C. connivens*). Less frequently one finds consociated *Microlepidozia sylvatica*, *Calypogeia fissa* and *peruviana*, *Lophozia capitata*, *Scapania nemorosa*, *Lophocolea martiana*, *Riccardia multifida synoica* and *Aneura pinguis*. In the occasional Piedmont stations, *Calypogeia sullivantii*, *C. fissa*, and *C. peruviana* may all be associated.

The occurrence over wet, often peaty or sandy-peaty substrates, under strongly shaded conditions around springs or streams, and in deep swamps, is striking. In swamps the species may inhabit the bases of Cypress knees, and along small ponds the edges may be carpeted, particularly in the shade of *Cephalanthus occidentalis*. Under all these conditions, the species is a pronounced oxylophyte.

Differentiation

One of our most characteristic hepatics, hardly to be confused with any other. Superficially most similar to *Blepharostoma*, but differing in (*a*) the tendency for the cells to shrink in drying; (*b*) the shiny and glistening appearance when dry; (*c*) the much smaller underleaves, only 2–3 cells long, averaging at most 0.3–0.5 as long as the leaves; (*d*) the smooth cuticle; (*e*) the larger leaf cells, averaging 90–110 μ long in the middle of the lobes; (*f*) the leaf lobes, at the very base, 2 cells broad on mature leaves; (*g*) perianths always on highly abbreviated, usually postical branches.

Of these characteristics, *a*, *b*, *d*, and *e* serve equally well to separate the *Telaranea* from *Microlepidozia*. The ♀ bracts and stem anatomy of *Telaranea* are very different from those of *Microlepidozia*.

Our plants were separated by Howe (1902) as the variety *longifolia*, supposedly distinct from the Cuban type in these characteristics: more rigid leaves, of greater length (400–800, rarely 900, μ long); leaf cells 2–4.5 × (rather than 2–3 ×) as long as wide; leaves 5–8 cells long (4–6 cells in the type). However, much of the North American material has leaves only 400–500 μ long and 5–6 cells high. These plants break down Howe's distinctions, which

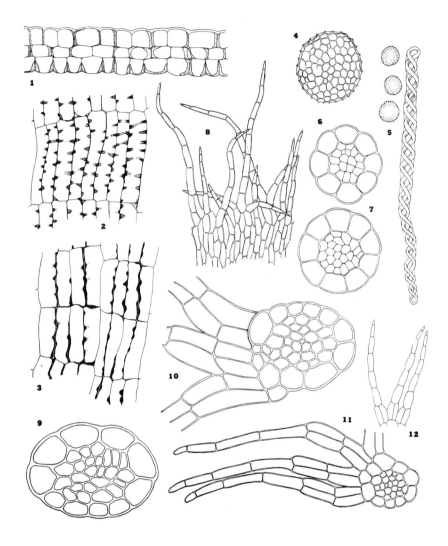

Fɪɢ. 90. *Telaranea nematodes.* (1) Cross section of capsule wall (×375). (2) Inner cells of capsule wall, tangential view (×312). (3) Epidermal cells of capsule wall, tangential view (×312). (4) Spore (×1180). (5) Spores and elater (×312). (6) Seta (×100). (7) Seta (×73). (8) Part of perianth mouth (×65). (9) Stem cross section (×230). (10) Stem cross section and leaf base, inverted (×230). (11) Stem cross section and leaf (×115). (12) Small, trilobed leaf (×65). [1–8, *RMS F-101;* 9–12, *RMS 28622.*]

are probably based on environmentally controlled rather than inherited features. Until more collections are available from the whole range of the species, the attempt to differentiate intraspecific taxa must remain in abeyance.

Telaranea nematodes shows much the same pattern of variation as does *Microlepidozia sylvatica*. Robust plants have largely 4-lobed leaves on main stems; weaker plants, largely 3- or often 2-lobed leaves. Similarly, the seta of robust plants (300 μ in diam.) show 9–10 rows of large epidermal cells (64–76 μ in diam.) and as many as 24 rows of small (23–25 μ) inner cells. In less robust individuals of the same clone, with the seta 200 μ in diam., the epidermal cells are in only 8 rows, the interior in 16 rows (a central group of 4, surrounded by a ring of 12 cells).

Howe (1902) investigated the polymorphism which this species exhibits regarding the position of archegonial branches. Although some or many of these may be lateral in origin, the majority are short and postical in origin.

MICROLEPIDOZIA (*Spr.*) *Joerg. emend. Schust.*[25]

Kurzia v. Martens, Flora 53(27):417, pl. 5, 1870.
Lepidozia subg. *Microlepidozia* Spr., Jour. Bot. London, N.S., 5:165, 1876; Spruce, Trans. Proc. Bot. Soc. Edinb. 15:359, 1885 (exc. *Telaranea*).
Microlepidozia Joerg., Bergens Mus. Skrifter 16:303, 1934.
Telaranea K. Müll., Rabenh. Krypt.–Fl. ed. 3, 6(2):1133, 1956, *p.p.* (not of Spruce).
Dendrolembidium Herz., Ark. f. Bot. 1(13):497, 1951.
Micrisophylla Fulf., Brittonia 14:124, 1962, at least in part (new synonymy).[26]

Isophyllous to anisophyllous, very small or moderate-sized, usually ascending, often caespitose, *rarely dendritic*, sometimes creeping individually over other bryophytes, usually *dull textured, deep green* or olive-green or brownish as a rule, occasionally infuscated.[27] Stems regularly to irregularly 1–2-pinnate, branches often (but not consistently) abbreviated; lateral branches all terminal, of *Frullania* type (branch replacing ventral half of leaf) on one side of axis, of *Microlepidozia* type (branch replacing dorsal half of leaf) on opposite side of axis; ventral branches, vegetative and sexual, always intercalary; lateral and ventral branches and sometimes apices of leading stems may be or become flagelliform and microphyllous, rhizoidous. Stem elliptical in cross section, with (8)12–16(18–24) rows of

[25] The first formal *mention* of *Microlepidozia* as a genus occurs in Pearson (1893, p. 5); however, since there is no allusion to a specific species, or diagnosis, or even an indirect reference to *Microlepidozia* in its former status, this does not suffice to establish Pearson as the author of a genus *Microlepidozia*.
  Since the completion of this manuscript, Grolle (1964c) has adopted *Kurzia* as the name for this genus; he is unique in having done so. Bonner and Schuster (1964) have given reasons why *Microlepidozia* must be conserved. Fulford (1966) also rejects *Kurzia* as the valid epithet.
[26] Grolle (1964c) apparently regards the *Micrisophylla* element as a clear-cut synonym of *Microlepidozia* (*Kurzia*). In any event, species which Fulford (1962a) assigns to *Micrisophylla*, e.g. *M. saddlensis*, are assigned by Grolle to *Microlepidozia* (*Kurzia*). I tentatively agree in regarding the *Microlepidozia* and *Micrisophylla* elements as being basically congeneric, although the type of *Micrisophylla* may represent a distinct genus.
[27] In subg. *Nanolepidozia* (monotypic), whitish, rather glistening.

cortical cells which are usually firm, ± *thick-walled* (*thus hardly forming a hyaloderm*), the *walls thicker than those of medullary cells, which are usually inferior in size*. Leaves with *insertion transverse to succubous*, the lamina (usually low; in subgenera *Kurzia* and *Nanolepidozia* less than 0.5 cell high) generally spreading, the lobes often erect or even incurved, the *leaves thus often handlike*, symmetric to moderately asymmetric, imbricate, *rather small*; *leaves usually* (*2–3*) *4-lobed* (but never bisbifid) for 0.5–0.95 their length,[28] with segments usually subequal to somewhat unequal, acuminate to longly acute, (1)2–5(10) cells wide at base; leaf lamina 3–8–12–16 cells wide at base. Underleaves transverse, varying from 0.2 to 0.95 the size of lateral leaves, also (2)3–4-lobed, symmetrical or asymmetric through abortion of 1–2 segments. Cells relatively small, *usually rigid*, quadrate to oblong, usually faintly *verrucose to striolate*, without defined trigones, *usually* ± *thick-walled*; oil-bodies of mature leaves *often lacking*, isolated cells excepted; cortical cells also often lacking oil-bodies, but the medullary cells usually with them present. No asexual reproduction.

Dioecious. ♂ Branches usually abbreviated, postical (exceptionally on long or leading lateral branches); bracts and bracteoles in 2–6–8 (or more) series; ♂ bracts 1–2-androus, *stalk 1-seriate* (as far as known). ♀ Inflorescences on short postical branches, sometimes on long, leafy branches (which may be lateral-terminal), lacking innovations, isophyllous, the bracts and bracteoles in 3–5 progressively larger series; innermost bracts and bracteoles 2–4-lobed, the lobes (and margins) often dentate or ciliate. Perianth long, usually slenderly fusiform-prismatic, *unistratose*, the contracted mouth crenulate to ciliate. Capsule ellipsoidal, *2(–3)-stratose;* secondary (*alternating*) longitudinal walls of epidermal cells (thus along half the walls) with nodular thickenings that may be subconfluent in sinuous sheets; inner (or innermost) layer with semiannular bands. *Spores verruculose to papillose*, 1–1.5 × the diam. of the 2-spiral elaters.

Type. *Microlepidozia setacea* (Web.) Joerg., by original designation. *Microlepidozia capillaris* (Sw.) Fulf. = *Jungermannia capillaris* Swartz, Prodr. 144, 1788, is proposed as lectotype by Fulford (1962a). (See under subg. *Microlepidozia*.)

Microlepidozia is broadly defined to include, in essence, plants normally with firm leaf cells; transverse to succubous leaves that are usually 3–4-parted; a rigid stem generally without a defined hyaloderm; few or no oil-bodies (when these are present, they are confined to a few leaf cells

---

[28] An exception is *Microlepidozia quadrifida* fo. *pallescens* (Grolle) Schust., comb. n. [= *Kurzia pallescens* Grolle, Rev. Bryol. et Lichén. 32(1963):177, 1964] of New Zealand; it has sexfid leaves.

and some cells of the axis; they rarely occur in all leaf cells); a generally roughened cuticle; a generally prostrate but sometimes ascending to suberect growth, but rarely a truly dendritic organization; a lack of collenchyma, the leaf cells equally thick-walled and—most important— branching of the *Microlepidozia* type. Thus defined, it is unquestionably a later synonym of *Kurzia* v. Martens, with *K. crenacanthoidea* as type (Flora 53:417, 1870).

Kurzia was published as "eine neue Alge" and given its species name in the mistaken belief that it had an "auffallende Ähnlichkeit mit Kützings *Crenacantha orientalis!*" Goebel (1889a, p. 15; 1891, p. 37) pointed out *Kurzia* is unquestionably a *Lepidozia*, identical with *L. gonyotricha* Sande-Lacoste (Syn. Hep. Javan., p. 38), or "nahe verwandt" with this species, and concludes that, therefore, "ist die Gattung *Kurzia* zu streichen."[29]

At the time of Goebel, *Microlepidozia* as a genus had not been validly published and thus his observations were, strictly, correct. However, *Kurzia* agrees in essentials with the element of *Lepidozia s. lat.* now segregated as the genus *Microlepidozia*.

For this reason *Microlepidozia* must be conserved as a genus (Bonner & Schuster, 1964). It has been accepted as one by Evans (1939), Schuster and Blomquist (1955), Schuster (1963), Hattori (1952a), Hattori and Mizutani (1958), Fulford (1962a, 1966), as well as by Hodgson (1962) and others. Its suppression in favor of *Kurzia*—a genus not mentioned since Goebel discarded it—would necessitate numerous new combinations. Furthermore, the name *Microlepidozia* has given us the basis for a well-known and widely discussed branching type; the suppression of *Microlepidozia* would be futile and profitless.

It is uncertain how many species are to be attributed to *Microlepidozia* as defined: a minimum of 12 occur in Latin America; 4 (all in subg. *Microlepidozia*) in the Holarctic; 5–6 are reported from New Zealand (Schuster, 1963, p. 257); 2 from Africa (Arnell, 1963).[30] The genus probably does not include more than 30–40 taxa, of which the bulk are antipodal. None is restricted to the Arctic, and only one (*M. setacea*) extends into the subarctic and arctic regions.

Schuster (1963, p. 257) points out that "in most ways, including branching modes, *Microlepidozia* is most closely allied to [the antipodal] *Dendrolembidium s. str.*" It is emphasized that *Microlepidozia compacta* (Steph.) Schust., when "erect-growing and vigorous, forms a virtual transition between the two genera. Thus *Dendrolembidium* might be regarded as a dendritic and luxuriant subgenus of *Microlepidozia*." The deeply lobed leaves of *Dendrolembidium*; the subisophyllous organization, with transverse leaf insertion; the presence in isolated cells

[29] Schiffner (1893a, p. 254) definitely states that *Kurzia crenacanthoidea* and *Lepidozia gonyotricha* are identical. Comparing authentic material of the latter with Goebel's drawings of *Kurzia*, I come to the same conclusion. I have not been able to locate a type of *Kurzia* to date.

[30] One of the two species referred by Arnell (1963) to *Microlepidozia*, *M. succulenta* (Sim) S. Arn., has a distinct hyaloderm. This plant belongs to *Telaranea*.

of small, vestigial oil-bodies; the firm, thick-walled cells; the verruculose cuticle; and the axial anatomy are all criteria which are identical or comparable in *Microlepidozia* and *Dendrolembidium*. In fact, *Micrisophylla* (Fulf.) Schust. in its vigor, its approach to isophylly, and its broader leaf lobes approaches *Dendrolembidium* and is in some ways intermediate between *Microlepidozia s. str.* and *Dendrolembidium*. Thus, for several reasons, I would now reduce *Dendrolembidium* to a subgenus of *Microlepidozia*.[31] On the other hand, the tendencies toward succubous leaf insertion, the thick-walled, rigid cells of the leaves, usually devoid of oil-bodies, and, to some extent, the facies suggest that *Microlepidozia* is allied to the monotypic antipodal genus *Psioclada* Mitt. *Microlepidozia* thus has, seemingly, no immediate relatives in the Northern Hemisphere flora.

These conclusions as to affinities are in direct contrast to those expressed in fig. 203 in Fulford (1963a, p. 83), where *Psioclada* and *Dendrolembidium* are, respectively, juxtaposed to *Telaranea* and *Lembidium* and far removed from *Microlepidozia*.

I regard *Micrisophylla* Fulf. (1962a) as doubtfully generically distinct from *Microlepidozia*, in its largest part at best subgenerically distinct. The tendency toward erect growth, rather vigorous size, many-celled leaves, and subdendroid facies of *M. compacta* (alluded to above), which suggest that *Microlepidozia* must be united with *Dendrolembidium*, are exactly the criteria which characterize *Micrisophylla*. As Hodgson (1956, p. 616) points out, in this species we find leaves with the lobes 4–5 up to 5–7 cells wide at base, as in *Micrisophylla*, yet the plants are not fully isophyllous.[32] The variation *within* the genus *Telaranea* in this respect is greater; we go from strongly anisophyllous taxa with uniformly bifid leaves and vestigial underleaves to vigorous subisophyllous taxa with 8–13-lobed leaves and 6–12-lobed underleaves, such as *T. mooreana* (Steph.) Schust. of Tasmania. Yet, within this superficial and even striking diversity

---

[31] Since the preceding was written, Grolle (1964c, p. 167) has stated that *Dendrolembidium* is better regarded as a subgenus or section of *Microlepidozia* (*Kurzia sensu* Grolle). Grolle, however, erroneously assigns to *Microlepidozia* (*Kurzia*) the "*Dendrolembidium*" *insulanum* element for which I proposed the genus *Megalembidium* (Schuster, 1963). *Megalembidium* has a much more complex stem anatomy, involving 3 strata: a collapsing hyaloderm; 2–3 internal strata of small, thick-walled, brownish cells; a medulla formed of numerous rows of relatively leptodermous, hyaline cells. It also totally lacks all trace of oil-bodies in the cells and has incubous, shallowly lobed leaves. *Microlepidozia* subg. *Dendrolembidium* (Herz.) Schust., comb. *n.* (= *Dendrolembidium* Herz., Ark. f. Bot. 1(13):497, 1951) is also distinct in the basiscopically directed "comb" of cellular, fingerlike appendages of stem leaves and underleaves. However, the New Zealand *M. appendiculata* Schust. has similar, if fewer, appendages of robust leaves.

[32] Fulford (1962a, p. 126) refers to the "conspicuous and distinctive" radial symmetry of *Micrisophylla* and claims that "this symmetry is a primitive character not previously associated with the family Lepidoziaceae." This is incorrect. For example, *Telaranea mooreana*, described by Stephani as a *Lepidozia*, is almost perfectly radially symmetrical. *Isolembidium* Schust. of New Zealand is also perfectly radially symmetrical. Furthermore, "*Micrisophylla*" *saddlensis* is very definitely not consistently isophyllous, if fig. 83 in Fulford (*loc. cit.*) is accurate. Admittedly, there is a tendency for larger underleaves in *Micrisophylla*, but the separation of a new genus based on a tendency is hardly warranted. I think, however, that perhaps the type *M. setiformis* stands further apart from *Microlepidozia* than preceding accounts make clear. The brownish color and the irregular and almost *Temnoma*-like branching and facies, without the rather regularly pinnate branching typical of *Microlepidozia*, are suggestively different. The status of *Micrisophylla* thus remains an open question.

there is a basic unity that characterizes *Telaranea*. In the same sense, a basic unity characterizes *Microlepidozia s. lat.* Admittedly, the isophyllous to sub-isophyllous subg. *Micrisophylla* retains certain primitive traits that are lost in the more advanced, more reduced "typical" species of subg. *Microlepidozia*, e.g., isophylly; perianths and androecia often on long branches; and leaves with relatively broad lobes, formed of many cells. However, a similar retention occurs in some primitive *Telaranea* species, e.g., *T. apiahyna* (see Taylor, 1961), and no attempt was made by Fulford to divorce this species from *Telaranea*.

Microlepidozia, thus broadly defined, can be tentatively divided into six subgenera: *Micrisophylla* (Fulf.) Schust. of the Antipodes; *Microlepidozia* of world-wide range; *Macrophylla* Fulf. of the Antipodes; *Kurzia* (v. Mart.) Schust. of Indonesia; *Nanolepidozia* Schust. of New Caledonia; and *Dendrolembidium* (Herz.) Schust. A synopsis of these groups is presented in Schuster (1969).

As previously noted, *Dendrolembidium* Herz. may, with some justice, be regarded as a sixth subgenus. Of the groups here taken to form the genus *Microlepidozia*, only subg. *Microlepidozia* occurs regionally. *Kurzia* is Indo-Malayan, as far as known; *Macrophylla* and *Micrisophylla* are both antipodal. The latter, in the isophylly—or approach to it, the occasional brownish coloration, the very plastic and irregular branching (as in certain *Temnoma* species, terminal and specifically *Microlepidozia*-type branches may be only sporadically developed), the often elongated sexual branches, and other features, represents the most primitive element within the genus —and one of the most primitive elements within the family.

Nanolepidozia, based on a single New Caledonian species, has the irregular branching of *Micrisophylla*—irregularly pinnate and diffuse, lacking the regular alternation of *Microlepidozia*- and *Frullania*-type branching. For example, some plants may have 2–3 *Frullania*-type branches of one side of the stem, yet entirely lack branches of the other side; hence *Microlepidozia* branches are not always present. In other respects (trifid mature leaves, with the *dorsal* lobe conspicuously longer; stem with 8–10 cortical cell rows constituting a conspicuous hyaloderm, surrounding 10–12 much smaller medullary cell rows; spores, with scattered high, tumid, hemispherical papillae alternating with remote, much smaller papillae), *Nanolepidozia* is highly specialized.

Subgenus *MICROLEPIDOZIA* (Spr.) Joerg.

Lepidozia subg. *Microlepidozia* Spr., Jour. Bot. London, N.S., 5:165, 1876; Spruce, Trans. Proc. Bot. Soc. Edinb. 15:359, 1885 (exc. *Telaranea*); K. Müller, Rabenh. Krypt.–Fl. 6(2):286, 1914; Macvicar, Studs. Hdb. Brit. Hep. ed. 2:335, 1926.
Microlepidozia Joerg., Bergens Mus. Skrifter no. 16:303, 1934; Schuster & Blomquist, Amer. Jour. Bot. 42:595, 1955.
Telaranea K. Müll., Rabenh. Krypt.–Fl. ed. 3, 6(2):1133, 1956, *p.p.* (incl. *Microlepidozia* as synonym).

Very small or minute, firm, regularly to irregularly *1–2-pinnately branched,* forming interwoven, decumbent to ascending mats or patches, or creeping among other bryophytes, 200–400(450) μ wide; *texture usually dull.* Stem with *cortical cells firm, in ca. 8–12–16 rows (2–4 rows of which form the ventral merophytes;* the lateral merophytes usually 3–4 cell rows broad), hardly or no larger than medullary cells, but usually more *distinctly thick-walled, not forming a distinct hyalodermis;* medulla mostly (2)3–4 cell rows across, leptodermous or slightly thick-walled. *Vegetative branches,* except for occasional ventral flagella, *of the Microlepidozia type:* with branches on one side replacing the dorsal half of the stem leaf, on the other side replacing the ventral half of the stem leaf; sexual branches, or at least the *gynoecial, abbreviated and postical,* from axils of underleaves; leading stems or sometimes lateral branches occasionally running out into *slender, microphyllous flagella* (the flagella often in part, or largely, intercalary, from axils of underleaves). Rhizoids sparingly developed, except on flagella, confined to a small area at bases of underleaves (on flagella, also at bases of the vestigial leaves). *Leaves transversely inserted, very small* (to 200–300 μ long and wide), *rather firm and rigid, subpalmately divided almost to base into (3)4(6) narrow,* lanceolate to subulate segments that are merely *(1)2–3 (rarely 3–4–5) cells wide at base;* bases of lobes occasionally with obscure 1-celled accessory teeth; undivided leaf base only 2–5(6) cells high. *Underleaves very similar to lateral leaves, (3)4-lobed* to within 1–3 cells of base, generally slightly smaller than leaves and often with 1 or 2 divisions aborted. *Cells thick-walled,* generally 1–2(3) × as long as wide and oblong to subquadrate, nearly smooth to *distinctly verruculose,* chlorophyllose and *opaque,* the walls often somewhat brownish, *dull and firm,* not collapsing in drying; *oil-bodies absent from all or most leaf cells,* basal and occasionally lobar cells with a few small, spherical to ovoid, nearly homogeneous oil-bodies; cortical stem cells often free of oil-bodies, but medullary cells apparently always having them.

Dioecious. Androecia budlike to spicate, indiscriminately on abbreviated, leafless postical branches and on elongate lateral stems; bracts in 2–4(6) pairs, subpalmately 2–4-lobed for 0.5–0.65 their length, the lobes (or their bases) with isolated spinous teeth, the undivided bases strongly concave; monandrous; antheridial stalk 1-seriate, short, the antheridial body globose. Gynoecia on short postical branches, without vegetative leaves; bracts and bracteoles much larger than leaves, essentially identical, *oblong-ovate, and considerably longer than broad, 2(3)-lobed to the middle or merely emarginate at apex,* the upper margins denticulate to short-ciliate. Perianth obtusely trigonous, ± cylindrical to fusiform, the mouth lobulate and crenulate to ciliate, 1-stratose. Seta with 8(9–10) rows of *large, hyaline* epidermal cells, surrounding two rings of *much smaller* interior cell rows (the outer ring of ca. 8–18 cell rows; the inner of 1–2 or more; at times, with secondary subdivision, forming numerous interior cell rows, occasionally up to 36). Capsule wall 2- (rarely 3-) stratose; the inner layer (in *M. trichoclados* sometimes 2 layers?) with numerous complete, semiannular bands; epidermal cells equal in diam. and usually in length to cells of inner layer, *with nodular thickenings of both faces* of alternating longitudinal walls (1 : 1 ratio between walls with and without thickenings). Spores verruculose, ca. 10–14.5 μ; elaters ca. 9–12 μ (ratio ca. 1–1.5 : 1).

Type. Jungermannia setacea Web. [= *M. setacea* (Web.) Joerg., by original designation]. *Jungermannia capillaris* Swartz (Prodr. 144, 1788) = *Microlepidozia capillaris* (Sw.) Fulf. is given as the generitype by Fulford (Brittonia 14:122, 1962).

The subg. *Microlepidozia* consists of small, freely and rather regularly 1–2-pinnately branched, subisophyllous plants, bearing leaves and underleaves that are almost perfectly transversely inserted and divided for virtually their entire length into 4 (on weak shoots often only 3) lobes. The subgenus has been variously interpreted—and misinterpreted—in the literature. Schuster and Blomquist (1955) have emphasized that the unique mode of branching, together with a series of other criteria, serves to separate the group from *Telaranea*, as well as from *Lepidozia*.

As is stressed in this paper, Evans has shown *Microlepidozia* has branches "restricted to the anodic segment-halves, but while this brings them in the ventral segment-halves, on one side of a branching axis, it brings them in the dorsal-halves on the other." Consequently, examination of a ramified shoot of *Microlepidozia* shows the branches of one side of the axis originating from the dorsal halves of lateral segments, while the branches from the opposite side of the axis issue from the ventral halves of lateral segments. Branching in *Microlepidozia* is therefore homodromous (i.e., the *direction of the spiral*, resulting from the helical manner of segmentation from the apical cell, is identical on all shoots), as in *Bazzania*, but the homodromy is achieved in an entirely different fashion (in *Bazzania* it arises by suppression of branching along one side of the axis).

By contrast, both *Telaranea s. str.* and *Lepidozia* possess lateral branching of the *Frullania* type; hence on both sides of the axis the branches uniformly replace the *ventral* half of the associated stem leaf, and the branches along one side of the axis are homodromous, whereas on the other they are antidromous, with reference to the direction of the spiral of the main axis.

Since, in spite of this fundamental distinction, Müller (1956, p. 1133, *in* 1951–58) placed *Microlepidozia* in synonymy under *Telaranea*, other distinctions between the two genera are reviewed (see also *Telaranea*). Chief among these are the oblong to ovate-oblong, more shallowly to hardly lobed, ♀ bracts of *Microlepidozia* (those of *Telaranea* are deeply divided, often *Blepharostoma*-like and multifid-laciniate); the transverse leaf insertion (incubous on at least the leading stems in *Telaranea*); the firm, usually roughened cells (delicate, collapsing, and nitid in subg. *Telaranea*); the verrucose spores (areolate on their external faces in *Telaranea*); the slighter anisophylly, with the underleaves 3- or 4-lobed and at least half as large as the lateral leaves in many cases (underleaves usually reduced, bifid in most species of subg. *Telaranea*); the thick-walled cortical stem cells (pellucid and forming a hyalodermis in *Telaranea*). The traditional difference often cited between the two groups (leaf lobes uniseriate in *Telaranea* vs. biseriate in *Microlepidozia*) fails to hold. Of the preceding distinctions, the type of branching and leaf insertion also separate *Microlepidozia* from *Lepidozia s. str.*

The capsule wall anatomy also separates *Microlepidozia* from the other regional genera of Lepidoziaceae. The epidermal cells, as in other Lepidoziaceae, cut off primary cells that appear to be devoid of all thickenings; these cells undergo secondary longitudinal division (see also discussion under family diagnosis (p. 14) and under *Bazzania* and *Lepidozia*). As previously pointed out (p. 21), in other local genera (*Bazzania*, *Telaranea*, and *Lepidozia*) the capsule wall is 3- or more stratose, and the cells of the epidermal layer tend to develop 3 secondary longitudinal walls that divide each primary cell into 4 cells; each of these secondary longitudinal walls usually bears nodular (radial) thickenings, often confluent in partial sheets. Thus a 3:1 ratio arises between walls with and walls without thickenings. Only in *Lepidozia* (which see) is the median of the three longitudinal secondary walls often provided with weak or vestigial thickenings, leading to an approach to a 1:1 ratio. By contrast, in *Microlepidozia* the wall is 2-stratose, and the primary epidermal cells (i.e., those devoid of thickenings in the walls) in studied species undergo a single secondary longitudinal division, giving rise to a regular 1:1 ratio. Minor deviations which occur in this pattern (such deviations are notoriously frequent in the capsule walls of all Jungermanniales) fail to weaken the significance of this distinction.[33]

Microlepidozia is a group restricted to acid habitats, occurring variously in bogs, on peat or humus (often over rocks), less often at the bases of trees, particularly in the cove forests of the Southern Appalachians; it is very similar, ecologically, to *Bazzania* and *Lepidozia*. How many species should be assigned to it is uncertain because of the almost universal confusion with *Lepidozia*, even in recent years. Only four species apparently occur in the Northern Hemisphere, of which *M. trichoclados* appears confined to Europe, while *M. makinoana* is possibly limited to Japan.

Key to Species of Microlepidozia

1. Underleaves normally 3-lobed, with 1 or 2 lobes usually aborted (and ending in slime papillae); ♀ bracts bifid for ca. ⅙–⅓, armed on margins with teeth and short cilia; perianth mouth strongly narrowed, armed with short teeth and cilia 1–3(4) cells long; never in peat bogs.   *M. sylvatica* (p. 49)

---

[33] The exact same 1:1 ratio between thickened and hyaline longitudinal walls occurs also in subg. *Nanolepidozia*. I have not seen enough fertile material of other groups within the genus, *s. lat.*, to feel free to predict that such a pattern of alternation will characterize the genus as a whole.

1. Underleaves of mature stems usually rather regularly and symmetrically (3)4-lobed, with none of the divisions strongly aborted; ♀ bracts divided for ⅓–½ into (2)3–4 narrow, acuminate, ciliate lobes separated by very narrow sinuses, the lobes ± laciniform and lacerate; perianth mouth usually more longly ciliate, wide open; in peat bogs. *M. setacea* (p. 57)

MICROLEPIDOZIA SYLVATICA (Evans) Joerg.

[Figs. 85:5–6, 91–93]

Lepidozia setacea Auct. (in part).
Lepidozia sylvatica Evans, Rhodora 6:186, pl. 57, 1904; Frye & Clark, Univ. Wash. Publ. Biol. 6(4):658, 1946.
Lepidozia exigua Steph., Spec. Hep. 3:626, 1909.
Microlepidozia sylvatica Joerg., Bergens Mus. Skrifter 16:304, 1934.
Lepidozia silvatica (*sic*!) K. Müll., Rabenh. Krypt.–Fl. 6(2):291, fig. 90, 1914.
Telaranea silvatica (*sic*!) K. Müll., *ibid.* ed. 3, 6(2):1136, fig. 431, 1956.
Microlepidozia makinoana Hatt. & Mizutani, Jour. Hattori Bot. Lab. no. 19:88, fig. 3:16–33, 1958 (as regards American plants; not *Lepidozia makinoana* Steph., Bull. Herb. Boissier 5:94, 1897).
Kurzia makinoana Grolle, Rev. Bryol. et Lichén. 32(1963):171, 1964 (p.p.; as regards American and European plants only).

In *dull or deep green to brownish green*, dense, aromatic, interwoven, low tufts or patches (occasionally creeping as scattered stems among other bryophytes). Stems 5–20 mm long, threadlike, ca. 55–80 μ in diam., creeping to ascending, rather irregularly to ± regularly pinnately to bipinnately branched, occasionally terminating in flagella; leafy branches usually lateral, occasionally postical; flagella usually postical. Rhizoids sparsely developed, at the bases of the lower underleaves, more frequent on flagella. Leaves barely *contiguous to imbricate, transversely inserted*, arising from merophytes 4 cells broad, the basal portion spreading but lobes usually ± suberect to erect, or even slightly incurved, the *leaves thus typically handlike, somewhat cupped*. Stem leaves averaging 200–250 μ wide × 180–225 μ long, subquadrate when flattened, *essentially symmetrical*, usually deeply (3)*4-lobed to within* 2(3) *cells of base;* lobes entire, subulate to narrowly lanceolate, usually somewhat incurved (more rarely straight or squarrose on juvenile leaves), 2 *cells broad at base* (35–40 μ broad), more rarely 3 cells broad at base of occasional lobes, terminated by a row of 2–3 (more rarely 1 or 4) superimposed cells; *dorsal lobe not conspicuously reduced;* sinuses descending ca. ⅔–⅞ the length of the leaf, narrow and acute; antical margin above base often with a 1-celled tooth. Branch leaves similar to stem leaves but smaller and more often 2–3-fid. Cells at base of lobes (14)15–20 × (16–20)23–26 μ, at base of leaf 21 × 32–35 μ, the walls equally thickened; cuticle *dull when dry*, almost smooth or very delicately verruculose; oil-bodies absent, except in medullary (and less often in cortical) stem cells, occasionally a few in some leaf cells, when present small, almost homogeneous, subspherical, glistening. Chloroplasts large and numerous, the *cells quite opaque*. Underleaves of stem usually 3-lobed, very rarely 4-lobed, to within 1–2 cells of base, somewhat similar to leaves, 115–225 μ long × 65–120 μ wide, the segments when well developed identical to those of leaves, *but regularly 1 or 2 segments slightly to strongly reduced in size* (*aborted*) to a length of 1–2 (3–4) cells, terminated by an

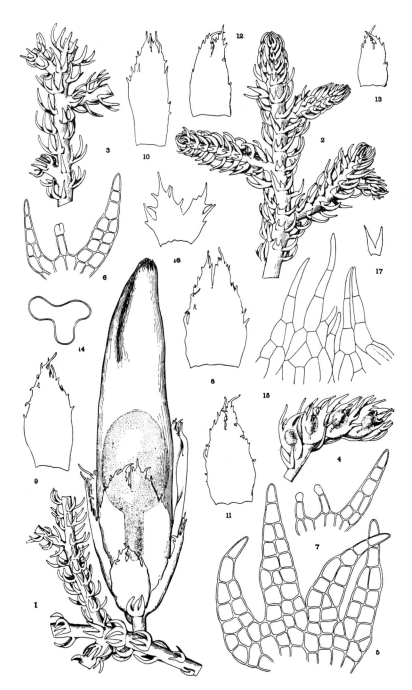

FIG. 91. *Microlepidozia sylvatica*. (1) Shoot sector, postical aspect, with perianth-bearing branch (×35). (2) Sterile shoot apex, postical aspect (×50). (3) Same, dorsal aspect (×50). (4) Androecium, lateral view (×60). (5) Stem leaf (×220). (6–7) Underleaves (×220). (8–10) Innermost ♀ bracts (×35). (11–13) ♀ Bracts of second cycle (×35). (14) Cross section of upper one-third of perianth (×35). (15)

ephemeral slime papilla; underleaves of branches often reduced and only 2–3-lobed.

Dioecious. ♂ Inflorescence on short postical (more rarely lateral) branches, usually proliferating distally and continuing by vegetative growth; bracts in 4–5 pairs, strongly concave, ovate to subquadrate, ca. 300 μ long × 240 μ wide, *divided 0.5–0.65 their length into two ovate to triangular, acuminate lobes, whose margins are sharply spinose-dentate* with 1–2-celled teeth; bracteoles usually bifid, with subulate lobes; monandrous, the antheridia ± oval. ♀ Inflorescence on a very short postical branch, usually with no leaves except for the closely sheathing bracts; bracts and bracteoles in 2–3 series, closely applied to perianth, the innermost bracts and bracteoles almost identical in form and size, ovate to ovate-lingulate, ca. 1000 μ long × 350–500 μ wide, *usually bifid 0.15–0.3 their length*, the *lobes acuminate, denticulate to ciliate*, the dentition extending down to about the middle of the bracts and bracteoles; sinus narrow, acute *(rarely vestigial)*. Cells of bracts longer, with thinner walls, than in leaves; the walls more distinctly striolate or verruculose. Cells of marginal cilia of bracts ca. 30 μ long. Perianth narrowly ovoid-cylindrical to subfusiform, terete and cylindrical below, emergent for half or more its length, gradually narrowed to mouth, obtusely trigonous above, becoming ± hexaplicate near mouth, to 2700 μ long × 600 μ wide; mouth lobulate and ciliate, the cilia 1–4 cells long, with cells *almost always verruculose to striolate;* cells of cilia narrow, the distal varying from 10–14 × 35–60 μ to 8–12 × 115 μ, and from 3 to 5(7) × as long as wide; cells immediately below mouth averaging 14–16 μ wide, rather thin-walled. Capsule oval, ca. 900 μ long × 500 μ broad, yellowish brown; seta to 1–1.5 cm long, 135–210 μ in diam., with 8 (more rarely 10) rows of large epidermal cells, and with 2–3 concentric rings of inner cells, *much smaller in size*, in 9–36 rows; epidermal cells of seta ca. 35 × 60 μ in cross section, the interior ca. 20 μ; the epidermal cells densely chlorophyllose, the interior devoid of chloroplasts. Epidermal cells of capsule mostly 11–13(14) × 20–26 μ, length-width ratio ca. 2:1, regularly rectangular; adjacent faces of *alternate* longitudinal walls each with (1)2–3 ± confluent nodular thickenings (often weakly extended tangentially), both adjoining faces of alternate longitudinal walls, and all transverse walls, usually without thickenings. Inner cell layer of narrowly rectangulate cells, ca. 11–13(14) × 50–54 μ, each with ca. 6–7 semiannular bands which are almost always complete. Spores 10–12 to 12–14.5 μ in diam., reddish brown, delicately papillose; elaters ca. (9)10–12 μ in diam., with 2 reddish brown spirals.

Type. Westville, Conn. (*A. W. Evans, 1903*).

Distribution

Abundant and widely distributed in eastern North America, in the region of the Appalachian Plateau and the Piedmont and Coastal Plain regions lying eastward and southward, ranging from Nova Scotia to subtropical (central) Florida, westward to eastern Texas. Also a scattered

---

Teeth of perianth mouth (×220). (16–17) ♂ Bract and bracteole, respectively (×60). [All after Evans, 1904; drawn from the type, from Westville, Conn., *Evans.*]

distribution in western Europe, where it is evidently less frequent. The species is evidently suboceanic in range, not penetrating the continental land mass of North America to any extent.

In eastern Asia represented by the allied *M. makinoana* (Steph.) Hatt., found throughout most of Japan (Hokkaido, Honshu, Shikoku, Kyushu, Ryuku), questionably southward to Formosa (Hattori and Mizutani, 1958).

The following are representative stations:

ST. PIERRE. *Fide* Gallo (1951). NOVA SCOTIA. 2 mi S. of Lower Argyle, Yarmouth Co. (*RMS 43062a*); Hubbard, Halifax Co. (Brown, 1936a). MAINE. Cape Elizabeth. NEW HAMPSHIRE. White Mts. (Evans, 1904). MASSACHUSETTS. Worcester; Long Pond, Nantucket; Woods Hole, Amesbury, and West Newbury (Evans, 1904); Artichoke R., W. Newbury!; Pelham Hills, 5 mi E. of Amherst (*RMS 43989a*, an atypical form; see p. 62). RHODE ISLAND. Westerley; Hopkinton. CONNECTICUT. Manchester, Hartford Co.; Stafford, Tolland Co.; Canterbury, Windham Co.; East Haven, Hamden, Naugatuck, New Haven, Orange, Oxford, New Haven Co.; Killingworth, Middlesex Co. (Evans & Nichols, 1908); Westville (type of *L. sylvatica*); Wintergreen Falls, New Haven (*D.C. Eaton, 1877!*); West Rock, New Haven (Verdoorn, Hep. Select. et Crit. no. 134); Huggins Gorge, Granby!

NEW YORK. Fishers I. (*Evans*); Long I. (the *Lepidozia setacea* of Jelliffe, 1899, probably belongs here; *fide* Grout, 1906, p. 26); Wildwood State Park, Long I. (*RMS*); Cold Spring Harbor, Long I.; Forest of Dean L., near Ft. Montgomery (*RMS 24234, 24255; plate*). Virtually or quite absent from central and western New York (Schuster, 1949a); Todt Hill and Richmond Valley, Staten I.!; Ft. Montgomery!; Stonykill Falls, Ulster Co.! NEW JERSEY. Quaker Bridge (Evans, 1904); Highlands (*Haynes; in* Krypt. Exsic. Mus. Vindob. no. 1567!); Navesink! near Englewood!. PENNSYLVANIA. Delaware Water Gap, near Stroudsburg (*RMS, Dec. 23, 1943*); Clarion, Fayette, Mifflin, Somerset, Westmoreland Cos. (Lanfear); Stony Creek!; Lehigh Gap!. MARYLAND. Vic. of Baltimore (Plitt, 1908); Suitland!; Little Paint Branch!. DISTRICT OF COLUMBIA. Washington! (Evans, 1904). DELAWARE. Near Claymount!; Wilmington!. WEST VIRGINIA. Tibbs Run (Evans, 1904); Braxton, Greenbrier, Monongalia, Pocahontas, Preston Cos. (Ammons, 1940). OHIO. Athens, Hocking, Vinton Cos. (Miller, 1964).

VIRGINIA. Dickey's Cr.! (Evans, 1904); The Cascades, near Mountain L., Giles Co. (*RMS 40243c*); Jericho Canal, Dismal Swamp; 3–5 mi NE. of Cypress Chapel, W. edge Dismal Swamp, Nansemond Co. (*RMS 34501*); Norfolk and Smyth Cos. (Patterson, 1949); Jackson; Nick's Creek! NORTH CAROLINA. Crabtree Cr., Umstead State Park, Wake Co. (*RMS 37697*); below the Narrows, Chattooga R., Jackson Co. (*RMS 39470b*); Laurel Hill, on New Hope Cr., Orange Co. (*RMS 28480*); N. of Chowan R., 1 mi N. of Winton, Gates Co. (*RMS 34581a, 34578b*); Soco Falls, E. of Cherokee (*RMS 24070a, 24618*); Dry Falls, NW. of Highlands, Macon Co. (*RMS 25225*); above Glenn Falls, Highlands, Macon Co. (*RMS 40608*); Linville Gorge, ca. 0.5 mi below falls, McDowell Co. (*RMS 28982, 28891a*); Bluff Mt., near West Jefferson, Ashe Co., 4600 ft (*RMS 30069*); Bridal Veil Falls, NW. of Highlands, Macon Co. (*RMS 29502*); Cascades, near Hanging Rock State Park, Stokes Co. (*RMS 28255, 28276a, 28269*); Neddie Cr., near jct. with E. fork of Tuckaseegee R., Jackson Co. (*RMS 29450b*); The Caves, New Hope Cr., WSW. of Durham, Orange Co. (*RMS 28534*); Eno R., 4 mi N. of Durham, Durham Co. (*RMS 28415a*); Alleghany, Bladen, Burke, Forsyth, Haywood, Watauga, Wilkes Cos. (Blomquist, 1936); Pinehurst, Moore

Co.!; Negro Mt., Ashe Co. (Sullivant, Musci Allegh. no. 243!; as *Jungermannia setacea*).

KENTUCKY. Elliot, Morgan, Powell, Lewis Cos. (Fulford, 1934, 1936).   TENNESSEE. Alum Cave, Mt. LeConte, Sevier Co. (*RMS 24109, 24176a*); Blount and Morgan Cos. (Sharp, 1939); Van Buren Co. (Clebsch, 1954).   SOUTH CAROLINA. NW. of Sumter, Sumter Co. (*RMS 18710*).   GEORGIA. Tallulah Falls, Rabun Co. (*RMS 34381a, 34364a*); Brasstown Bald, 4500–4600 ft, Towns Co. (*RMS 34308, Aug. 14, 1952; c. caps.*); Thomas Co.; Dade Co. (Carroll, 1945); summit of Rabun Bald, 4700 ft, Rabun Co. (*RMS 40691*).

FLORIDA. Sanford, Seminole Co. (*Rapp, 1916*); Polk and Volusia Cos. (Redfearn, 1952); Enterprise (Evans, 1904); Eustis, Lake Co.!; Port Orange!; Alexander Springs, Ocala Ntl. Forest, Lake Co. (*RMS 33302*); Alum Bluff, Apalachicola R., Liberty Co. (*RMS 33528*).   ALABAMA. Near Mobile (Sullivant, Musci Allegh. no. 244!; as *Jungermannia setacea* var.).   MISSISSIPPI. Tishomingo State Park, Tishomingo Co. (*RMS 19717, 19740, 19741*); 6–8 mi NE. of Vancleave, W. bank of Pascagoula R., Jackson Co., p.p., among *Lophocolea martiana* (*RMS 19205a*); near Handsboro, W. of Biloxi, Harrison Co. (*RMS 20394*).   TEXAS. Ammons (1940).

Ecology

Microlepidozia sylvatica is primarily a mesophytic species of partly to strongly shaded sites; it is absent from calcareous sites and rare in areas with even weakly basic sedimentary rocks. Within these limitations it occurs under a wide variety of conditions. The following discussion illustrates but hardly circumscribes the limits under which it is found.

Northward and on the Coastal Plain southward the species occurs frequently over acid, rather peaty soil in ditches, associated with *Nardia insecta, Cladopodiella francisci, Calypogeia sullivantii,* and *Drosera rotundifolia* (coastal Nova Scotia; *RMS 43062a*). A very similar occurrence much further south is on the sandy-loamy sides of a ditch, in oak-pine woods in southeastern Virginia (*RMS 34501*); here the species occurs with *Nardia lescurii,* a close ally of *N. insecta* and ecologically nearly equivalent to it, and with *Calypogeia sullivantii, Cephalozia macrostachya, Odontoschisma prostratum,* and *Scapania nemorosa.* On Long I., N.Y., on wet sides of a pond, under *Cephalanthus,* again associated with *Cephalozia macrostachya* and *Odontoschisma prostratum,* as well as with *Telaranea.*

Southward the species is also frequently found in densely wooded, partly evergreen swamp and hammock forests, in moist, springy, or swampy areas. For example, it occurs on wet peaty soil in *Chamaecyparis* swamps and at the bases of *Taxodium distichum,* with *Bazzania trilobata, Telaranea nematodes, Cephalozia macrostachya,* and *Odontoschisma prostratum* (Gates Co., N.C.; *RMS 34578b*). The recurrence in coastal Mississippi (*RMS 20394*) of *Microlepidozia* with *Cephalozia macrostachya, Calypogeia peruviana, Lophocolea martiana,* and *Telaranea* (and *Cephalozia connivens*) is remarkable. Quite

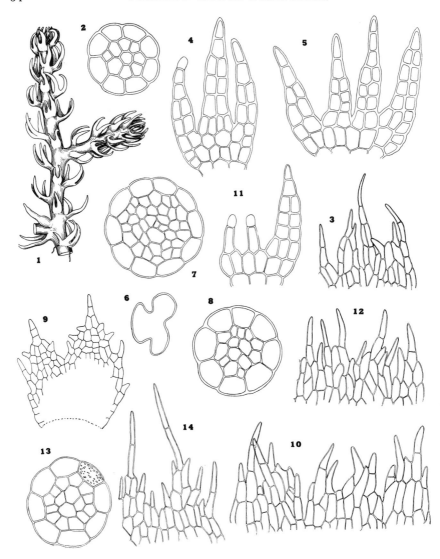

FIG. 92. *Microlepidozia sylvatica*. (1) Shoot sector, postical view (× ca. 45). (2) Seta of simplest type seen (×195). (3) Perianth mouth apex sector (×135). (4–5) Underleaf and leaf, respectively (×180). (6) Perianth cross section (×24). (7) Seta, maximal complexity (×135). (8) Seta, from same clone as that in fig. 7 (×135). (9) ♂ Bract (×110). (10) Cells, perianth mouth (×160). (11) Underleaf (×275). (12) Perianth mouth cells (×175). (13) Seta cross section (×195). (14) Perianth mouth cells, long-ciliate extreme (×155) [1, Delaware Water Gap, Pa., *RMS;* 2–3, Dept, Seine-et-Oise, France, *Douin,* from Schiffner's Hep. Eur. Exsic. No. 681; 4–6, 14, *RMS 24176a,* Mt. LeConte, Tenn.; 7–10, *RMS 24039,* Mt. LeConte, Tenn.; 11, *RMS 24234,* Forest of Dean Lake, New York; 12–13, *Haynes, 1910,* Navesink. Monmouth Co., N.J.]

analogous are occurrences in hammock forests in central Florida (*RMS 33432, 33302*), where the species occurs over peaty humus with *Cephalozia connivens, Pallavicinia lyellii, Odontoschisma prostratum*, and *Calypogeia peruviana*. The same complex of associates is characteristic of sites in warm, humid ravines of the Southern Appalachian Escarpment (i.e., *Bazzania trilobata, Calypogeia peruviana, Cephalozia connivens, Pallavicinia lyellii, Odontoschisma prostratum*); only *Telaranea* and *Lophocolea martiana* are lacking there.

In the interior occurring chiefly over acidic rocks, or over peat or *Sphagnum* near such rock outcrops, and ascending nearly to the summits of the highest peaks in the Southern Appalachians (near summit of Mt. LeConte, Tenn., at 6400 feet alt., there with *Nardia scalaris, Calypogeia muelleriana, Diplophyllum apiculatum*). The species is by far the most abundant at median and lower elevations in the Appalachian Mts., from Pennsylvania to Georgia and Alabama, paralleling the distribution of *Rhododendron maximum* and *catawbiense*. It occurs there most frequently with *Diphyscium sessile, Calypogeia muelleriana* and *fissa*, and *Diplophyllum apiculatum* on damp to dry, shaded banks on acid humus (often under *Tsuga, Quercus*, or *Rhododendron*).

Occasionally occurring on thin soil over rock outcrops along montane cascading brooks (as at Soco Falls, N.C.; with *Diplophyllum apiculatum, Calypogeia neesiana* and *muelleriana, Bazzania trilobata*). The species is only very exceptionally a pioneer on damp rocks, usually undergoing ecesis only after soil deposition.

Although with an extensive distribution from xeric sites on dry banks (there almost invariably with *Diplophyllum apiculatum* and *Diphyscium sessile*, as at Wildwood State Park, Long I., N.Y.) in oak-hickory forests to the dampest, most shaded sites in the Mixed Mesophytic Forest, the species never appears able to undergo ecesis in peat bogs (although it may occur amidst *Sphagnum* when the latter grows on seepage-moistened ledges, as on Mt. LeConte). In its absence from peat bogs it differs diagnostically in behavior from the otherwise similar *M. setacea*. The wide tolerance of the species results in an extremely diversified occurrence along its center of nearctic distribution, in the Appalachian Mts. and the peripheral elevated regions.

Differentiation

The small size, generally dull and brownish to fuscous-green color, and pinnate to bipinnate, rather regular branching are characteristic in the field. Confusion by the beginner is most likely with *Blepharostoma*, a more northern plant (from which the present species differs in the less filamentous leaf lobes, which are 2, rarely 3, cells broad at the base and not as strongly elongated), or with *Telaranea* (from which *Microlepidozia* differs in the usual absence of oil-bodies in the leaf cells, in the less filamentous leaf lobes, and in the duller green, never shining appearance). In the

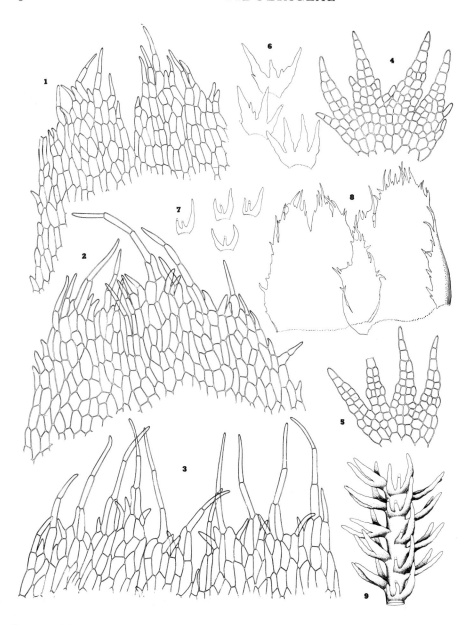

Fɪɢ. 93. *Microlepidozia sylvatica* (extreme approaching *M. setacea*). (1) Apex of ♀ bract in fig. 8 (×105). (2) Apex of another ♀ bract (×140). (3) Cells of perianth mouth (×140). (4–5) Leaves (×148). (6–7) Leaves and underleaves, respectively (×60). (8) Two ♀ bracts (×42.5). (9) Shoot sector, ventral view (×80). [All from *RMS 43989a*, Pelham, Mass; see p. 62.]

field occasionally confused with *Cephaloziella* (which differs at once from our plant in the bilobed leaves and the small or obsolete underleaves).

Under the microscope, however, confusion is possible only with the very closely related *M. setacea* (and with the European *M. trichoclados*, a species which may yet be found at high elevations in our eastern mountains). Supposedly, *M. sylvatica* differs from the European *M. trichoclados* on the basis of several gametophytic characters, among them (*a*) ♀ bracts are ± emarginate or bilobed to ⅕–⅓, with lobes and margins below them short-ciliate (in *M. trichoclados* the entire or barely emarginate bracts are obscurely dentate; only occasional teeth that end in hyaline papillae become 2–3 cells long); (*b*) the perianth mouth in *M. sylvatica* is ciliate with teeth 2–4 cells long (in *trichoclados* the teeth are at most 1–2 cells long); (*c*) in *M. sylvatica* androecial bracts are bilobed, with broad lobes whose margins are ± ciliate-dentate (in *trichoclados* ♂ bracts normally trifid, with the narrow lobes somewhat dentate at base).

Other differences have been cited between *M. sylvatica* and *trichoclados* which I have not been able to verify. (1) *M. sylvatica* is supposed to have a smooth or nearly smooth cuticle; *M. trichoclados*, a delicately verruculose cuticle. In reality, both species have the cuticle faintly to rather distinctly verruculose (on leaves, bracts, and perianth). (2) *M. sylvatica* has spores 10–12 μ in diam., while *M. trichoclados* supposedly has them mostly 12–14.5 μ. Müller (1914, *in* 1905–16) also attempts to separate these two species on the basis of differences in seta anatomy, but his figure of the seta of *M. sylvatica* appears to be based upon confusion with some other plant (as has been pointed out by C. Douin, 1916).

The entire problem of the variation of *M. sylvatica* in the direction of *M. trichoclados*, *M. setacea*, and *M. makinoana* is examined in Schuster (1968), in which a key to the four species is presented. The closest affinities may be to the east Asiatic *M. makinoana*, but I cannot agree with Hattori and Mizutani (1958) that *M. sylvatica* is identical with the last.

MICROLEPIDOZIA SETACEA (*Web*). *Joerg.*
[Fig. 94:1–10]

Jungermannia setacea G. H. Web., Spic. Fl. Goetting., p. 155, 1778.
Jungermannia pauciflora Dicks., Fasc. Pl. Crypt. Brit. 2:15, pls. 5, fig. 9, 1790.
Jungermannia doelaviensis Spreng., Fl. Halens. Tent. Nov. 314, 1806 (*fide* Spreng., 1832).
Jungermannia schultzii Spreng., Plant. Pug. 1:64, 1813.
Jungermannia setacea var. *schultzii* and var. *sertularioides* Hüben., Hep. Germ., p. 51, 1834.
Blepharostoma setacea Dumort., Rec. d'Obs., p. 18, 1835.
Lepidozia quadridigitata Griff., Not. Pl. Asiaticas 2:314, 1849?
Lepidozia setacea Mitt., Jour. Linn. Soc. Bot. 5:103, 1861; Evans, Rhodora 6:185, 1904 (not of Evans, Bull. Torrey Bot. Club 20:308, 1893); Frye & Clark, Univ. Wash. Publ. Biol. 6(4):660, figs. 1–11, 1946.
Lepidozia sphagnicola Evans, Bull. Torrey Bot. Club 20:307, pl. 162, 1893.
Lepidozia (subg. *Microlepidozia*) *setacea* K. Müll., Rabenh. Krypt.–Fl. 6(2):287, figs. 82b, 88, 1914; Macvicar, Studs. Hdb. Brit. Hep. ed. 2:338, figs. 1–5, 1926.
Microlepidozia setacea Joerg., Bergens Mus. Skrifter 16:303, 1934; Schuster, Amer. Midl. Nat. 42(3):529, pl. 2, fig. 1, 1949.
Lepidozia setacea var. *sphagnicola* Hesselbo, Rept. 5th Thule Exped. 2(2):10, 1937.
Kurzia setacea Grolle, Rev. Bryol. et Lichén. 32(1963):171, 1964.
Kurzia pauciflora Grolle, *ibid.*, 32 (1963):171, 1964.

Very small, delicate, scattered among and creeping over peat mosses, then usually light green, or forming thin and intricate to compact patches over peat or peat mosses, then dark green, in sun becoming dull brownish or fuscous. Shoots ca. *200-325* μ wide × 5-30 mm long, prostrate to creeping or occasionally ascending, filiform, *freely but irregularly pinnately* (and partly bipinnately) branched; branches diverging at 45-65° angle, *occasionally terminating as reduced-leaved, rhizoidous flagella;* postical flagella absent or few, usually from basal portions of plants only. Stems to 125-135 μ in diam., often only 80-100 μ, flexuous and filiform, often rather rigid, to 6-7 cells high, *cortical cells thick-walled*, often conspicuously so, in ca. (10)12-13 rows, averaging slightly to somewhat larger than medullary, up to 20-26 μ wide on leading stems; on more slender stems often only (13)14-17 μ wide × 32-56 μ long to 16-17 × 80-90(112) μ long; medullary cells in numerous rows, in 4-5 tiers, averaging only (13)15-18 μ, thin-walled. Rhizoids few to frequent on leafy stems, abundant on flagella, often swollen apically. Leaves essentially transversely inserted and oriented, *remote to barely contiguous*, usually with basal undivided portion erect-spreading to spreading, but the elongate lobes erect to incurved, the *leaves thus cupped*, when flattened quadrate to obtrapezoidal in outline, to ca. 185-300 μ long and wide, normally *deeply quadrifid, to within 2-3(4) cells of base; lobes fingerlike, 2-3(4) cells* broad at base, occasionally with a 1-2-celled tooth near base (fo. *dentigera*), remaining 2 cells broad to within 2-3 cells of the tip, the uniseriate apices acute; sinus descending 0.8-0.9 the leaf length. Cells mostly rectangulate, walls slightly thickened, without trigones, median cells ca. (14)15-19 × (18)20-25 μ, near base 15-19 × 27-34 μ long; distal 1-2 cells of lobes 14-15 × 32-36 μ. Oil-bodies absent in most leaf cells, except near base, but present in cells of most younger leaves, small, spherical to ovoid, homogeneous in appearance; mostly 3-5 per cell, 1.5-2.5(4.5) μ to 3 × 4.5 μ; more frequently well developed in medullary cells of stem, 3-6 or more per cell, but absent in cortical cells. *Underleaves like leaves*, 3-4-parted to within 1-2(3-4) cells of base, ca. 90-120 μ long × 140-160 μ broad, averaging ca. 0.5-0.6 the area of the lateral leaves, the segments somewhat incurved, 1-2 of them often terminated in slime papillae, *but not or only slightly aborted, thus usually almost symmetric.* Asexual reproduction (*teste* Frye and Clark) via gemmae of "upper leaves of young branches, angular, pale." [34]

Dioecious. Androecia forming short postical branches, less often terminal on lateral branches; bracts in 2-5 pairs, broader than long, imbricate, *2-3-lobed to the middle*, strongly concave; *lobes broad and dentate*, dorsal margin with a basal tooth; antheridia single, globose, body ca. 100 μ, on a short uniseriate stalk. Gynoecia on abbreviated postical branches. Bracts larger than leaves, 3-4 times their length, to 2.5 mm long × 1 mm wide, narrowly oblong-ovate, *at apex 3-4-lobed for 0.3-0.5 their length, the lanceolate, lacinia-like lobes acuminate and longly ciliate-fimbriate.* Bracteole similar, usually bidentate or bilobed, the acuminate lobes similarly fimbriate. Perianth ovoid-cylindrical, to 4 mm long × 1 mm wide, often curved or falcate, ca. 0.5-0.6 emergent, obtusely plicate near the

[34] The reports of gemmae are based on Hooker and Nees. No modern student of the species has found gemmae; thus one assumes that for this species, as well as for *Lepidozia reptans*, reports of gemmae are based on some error.

rather weakly contracted mouth, 2-stratose basally but becoming 1-stratose distally; mouth *ciliate-fimbriate, the uniseriate cilia (1)2–3(4) cells long, slender*, formed of cells mostly 4–7× as long as wide. Seta in cross section with 8 large rows of epidermal cells surrounding ca. 16 rows of smaller interior cells. Capsule ovoid, reddish brown, its wall 2-stratose.[35] Epidermal cells with nodular thickenings, inner cell layer with semiannular bands. Elaters 8–14 μ in diam. × 100–170 μ long; spores 8–14 (mostly 10–12) μ in diam., closely papillate or verruculose, yellowish brown.

Type. Central European: Baumannshöle, Harz, Germany.[36]

Distribution

A characteristic species of the Spruce-Fir Panclimax with disjunct stations in the Arctic (the report of Hesselbo, 1937, p. 10, from northern Baffin I. is remarkable). Originally not definitely separated from *M. sylvatica*, so that the earlier reports in part are uncertain. With a range almost wholly complementary to the southern and southeastern *M. sylvatica*, and perhaps with slightly oceanic "tendencies," as noted by Buch and Tuomikoski (1955). In Europe largely Atlantic and subatlantic [widespread in central Europe, where a characteristic species, but ranging from northern Spain and southern France (Allorge, 1955) and northern Italy to Scotland, Ireland, Norway, Sweden, and Finland, east to Czechoslovakia, the Tatra and Carpathian Mts., and the Baltic area]. Also supposedly on the Azores (Allorge, 1948) and Madeira (Allorge, 1955).

Reported from Japan and Formosa (Allorge, 1955). However, Hattori and Mizutani (1958) show that the reports from Japan refer to *M. makinoana*.

In North America generally a rather rare species, confined almost wholly to peat bogs.[37] In the West reported from the eastern and western coasts of Alaska (Evans, 1900) and eastward as follows:

[35] *Teste* Müller (1948, p. 15). Frye and Clark (1946, p. 661) cite "its wall of several layers of cells" and speak of "epidermal" and "innermost" layers, implying a minimal thickness of 3 cell layers.

[36] Grolle (1964c) argues, probably correctly, that on the basis of the *presumed* environment of the type station of *J. setacea* this plant cannot be the oxylophytic, bog-inhabiting *Microlepidozia setacea* (auct.), but must be identical to the plants called *M. sylvatica* or *M. trichoclados*. He adopts for the present species the next valid epithet, *pauciflora* Dicks., and drops *setacea* as a "*nomen dubium*." Since no type of *J. setacea* seems to exist today and since in the last half-century that epithet has been employed quite uniformly for the present taxon, I retain the customary usage. This is especially desirable since Grolle admits that the isotypes of Dickson's *J. pauciflora* he has seen are "völlig steril, so dass ihre Artzugehörigkeit morphologisch nicht absichern konnte." On the basis of the *supposed* habitat—surely a tenuous basis, some 174 years after the fact—he assumes that this plant is identical with *J. setacea* auct. (nec. Weber). In the absence of better data, I prefer the status quo.

[37] The reports of Allorge (1955) from on quartzite and from "rochers granitiques ombragés, avec *Plagiochila*," from the French and Spanish Pyrenees, wholly deviate from the normal pattern of occurrence in northern and central Europe and North America. An error in identification is suggested.

BAFFIN I. Pond Inlet, N. Baffin [Hesselbo, 1937, as *L. setacea* var. *sphagnicola;* Evans (*in* Polunin, 1947) expresses the opinion that *L. sphagnicola* "should be regarded as a simple synonym of *L. setacea*"]. NEWFOUNDLAND. Hogan's Pond, Avalon Distr.; Pushthrough, Rencontre W., Burgeo, Grand Bruit, Port aux Basques, all on S. coast; Stephenville Crossing, W. Nfld.; Port au Choix and Cook Harbour, N. Pen.; Kitty's Brook, C. Nfld. (Buch & Tuomikoski, 1955); S. Spread Eagle!; Broad Cove! MIQUE-LON I. (Lepage, 1944–45). NOVA SCOTIA. Bog near Barrasois R.! and Barrasois Barrens; Ferguson's Cove; Hubbard (Brown, 1936a); Pt. Joli (*Brown, 1923!*).

MAINE. Mt. Katahdin, bog near Roaring Brook (*RMS, 1954*); Cumberland (*Chamberlain*); Matinicus I.; Mt. Desert I. (*Lorenz*); Aunt Betty's Pond, Mt. Desert I.! VERMONT. Brandon (*Dutton, 1921*); Franklin (*Lorenz, 1912*). NEW HAMPSHIRE. Lonesome L., Franconia Mts. (Lorenz, 1908c); Waterville (Lorenz, 1908c). MASSACHUSETTS. Woods Hole (*Evans, 1904*); Nantucket I.; Hawley Bog, Hawley, Franklin Co. (*RMS 43953, 43954, 43959*). CONNECTICUT. Bethany, New Haven Co., in Lebanon Swamp (*Evans, 1893*; issued in Underwood & Cook, Hep. Amer. no. 168 as *Lepidozia sphagnicola!*, type locality of *L. sphagnicola*). NEW YORK. Cold Spring Harbor, Long I. (B. L. Andrews, 1931; doubtful!); Bergen Swamp, Genesee Co. (Schuster, 1949a). NEW JERSEY. Bergen Co. Westward occurring as follows: INDIANA. Lawrence Co. (report needing verification). MICHIGAN. Mud L.! and bog bordering Little L. Sixteen, Cheboygan Co.; Eagle Harbor, Keweenaw Co. (*Steere*). OHIO. Athens, Clark, Fairfield, Highland Cos. (Miller, 1964; reports requiring confirmation).[38]

Ecology

A bog species, giving the impression of being an obligate helophyte, but occurring both in strongly acidic bogs (pH below 3.8) and over *Sphagnum* in bogs underlain by pure marl (as at Bergen Swamp, N.Y.; *fide* Schuster, 1949a). The plants occur usually on the more elevated *Sphagnum* tussocks, associated with *Vaccinium oxycoccus* and/or *macrocarpon*, *Ledum*, *Chamaedaphne*, and other bog ericads. It is found here associated with various members of the *Mylia-Cladopodiella* associule, most often *Cephaloziella spinigera* (more rarely *C. elachista*), *Cephalozia connivens* (and the closely similar *C. compacta*), *C. loitlesbergeri*, *C. pleniceps*, *Lophozia marchica* (or *capitata*), rarely *L. ventricosa*, *Mylia anomala*, *Calypogeia*

[38] There are additional reports from farther south, in part based on old reports, which must be disregarded, including the following: DISTRICT OF COLUMBIA (*Ward, in* Plitt, 1908). VIRGINIA [Smyth Co., see Evans, 1893a, who later (1904) refers the plants to *L. sylvatica*; however, based on these reports are those of Frye & Clark, 1937–47, and Patterson, 1949, still attributing *L. setacea* to Virginia]. WEST VIRGINIA [an "Examination" listed from there by Frye & Clark, 1937–47, based on an Ammons collection made in 1929; yet Ammons (1940) specifically states that the species has not been found in West Virginia]. NORTH CAROLINA [reported by Blomquist, 1936, but the pertinent specimen (in bog, near Glenville, Jackson Co., *Anderson 639!*) is clearly *M. sylvatica!*]. Also recently reported by Redfearn (1952) from Polk Co., Fla.; this report is certainly based on an error.

Frye and Clark (p. 662) also erroneously cite the species from Rhode Island, attributing this report to Evans (1904), who, however, specifically reports it from only Connecticut and Massachusetts. Equally ambiguous are reports in Frye and Clark from Mexico [attributed to Gottsche (1863); there is no mention of *M. setacea* in the work of Gottsche (1863)], Juan Fernandez (attributed to Montagne, 1835), and the East Indies (attributed to Mitten).

sphagnicola (occasionally the systematically problematical "*C. tenuis*"), more rarely *C. integristipula* (in marl underlain bogs), *Scapania paludicola*, occasionally *Cladopodiella fluitans*. In such sites decaying logs may also be invaded by the species, then accompanied (Schuster, 1949a, p. 530) by *Calypogeia neesiana* (or *integristipula*), *Cephalozia connivens*, *Lophozia porphyroleuca*, and *L. incisa*.

The plants either creep among *Sphagnum* and are then lax and green (mod. *viridis-laxifolia*), or form brownish to fuscous, scorched-appearing tufts over dead peat (when in sun, on the crests of exposed *Sphagnum* hummocks), then forming a dense phase (mod. *densifolia-colorata*).

As in such associated helophytes as *Mylia anomala* and *Cephalozia connivens*, the rhizoid tips in this species are usually opaque and in large part capitately swollen at the apices; they are infected by a mycorrhizal fungus. Hill (1912) theorized that the swollen tips of the rhizoids function as "holdfasts," a theory which was soon disproved by the researches of Müller (1914, *in* 1905–16, p. 289), who showed that the swollen rhizoid apices were inhabited by a fungus which was able to fix atmospheric N; it evidently occurs as a symbiont within the *Microlepidozia*, aiding in the N metabolism of the latter. Similar swollen-tipped rhizoids occur commonly also in *Mylia anomala* and *Calypogeia sphagnicola*,

The ecology of this species is quite distinct from that of the allied *M. sylvatica*, a species of humus, loamy banks, rocks, etc.[39]

Differentiation

The only plant, restricted to peat bogs, in which are found (3)4-lobed transverse leaves. It can hardly be mistaken for any other species except *M. sylvatica*. However, careless examination may lead to confusion with *Blepharostoma* or *Telaranea*, which differ in the wholly or largely uniseriate leaf lobes, the deeply plurifid ♀ bracts, the elongated leaf cells, and other characters. The constantly helophytic occurrence of *M. setacea* eliminates potential confusion with *M. sylvatica*, *Blepharostoma trichophyllum*, and *Telaranea nematodes*. The smooth, large, exceedingly delicate leaf cells, which collapse in part on drying, and the shiny cuticle (when dry) at once eliminate *Telaranea*, since *Microlepidozia* has somewhat thickened, rigid and firm cell walls which are dull when dry and covered with delicate papillae, and cells that are also considerably smaller. The more slender and uniformly uniseriate leaf lobes, the much longer divisions of the leaves and particularly of the underleaves, the free development of perianths,

---

[39] The report of *M. setacea* on "loam, on roots of trees" in Frye & Clark (p. 662) further attests to their confusion regarding the two species. Exceptionally, *M. sylvatica*, in a form closely approaching *M. setacea*, may occur in acid, sandy-peaty ground on hillside slopes (as in *RMS 43989a*); associated then are *Cephalozia macrostachya*, *Cephaloziella elachista*, *Lophozia capitata*, and tracheophytes such as *Kalmia latifolia*, *K. angustifolia*, *Vaccinium macrocarpon*, *Epigaea repens*, and *Drosera rotundifolia*.

and the greatly different ♀ bracts at once separate *Blepharostoma* from *Microlepidozia setacea*.

However, *M. setacea* is closely similar to *M. sylvatica* and not separable from it in the field (except by the differences in ecology). Under the microscope, *M. setacea* differs from *M. sylvatica* in a number of features, mostly only quantitative, the chief of which are (*a*) the nearly uniform lobes of the 3–4-parted underleaves; (*b*) the deeply 3–4-lobed ♀ bracts with the margins laciniate to ciliate; (*c*) the more distinctly verruculose cuticle. As Evans (1904, p. 187) has also emphasized:

Under favorable conditions *L. setacea* [*Microlepidozia setacea*] is more robust, and its leaves are more regularly quadrifid; in many cases the antical segment bears an accessory tooth on its free margin, a condition which is exceedingly rare in *L. sylvatica* . . . . The underleaves of *L. sylvatica* are usually trifid but are occasionally quadrifid on very robust axes and are not infrequently bifid on slender branches. One or two of the divisions are tipped with the remains of hyaline papillae and are thereby aborted in their growth and reduced to one or two cells in length . . . . In *L. setacea* quadrifid underleaves are the rule on principal axes . . . . Here again the remains of hyaline papillae may be detected on the tips of the divisions; apparently, however, they do not interfere to any great extent with the development of the segments, which never exhibit the extreme disparity in size found in *L. sylvatica*. Even on slender forms of *L. setacea* this difference in the underleaves seems to be constant.

The most fundamental distinction between the two species lies in the ♀ bracts. In *M. setacea* they are deeply 3–4-lobed, with narrow sinuses, the erect and closely juxtaposed lobes being acuminate and distinctly ciliate (Fig. 94:9). By contrast, in *M. sylvatica* the bracts vary from shallowly bilobed to virtually undivided, with the margins ciliolate or shortly ciliate.[40]

[40] Unfortunately, occasional gatherings from intermediate habitats break down most of the distinctions between *M. setacea* and *M. sylvatica*. Material from a steep, sandy-peaty, springy, insolated slope, characterized by a scrubby tracheophyte flora (*Kalmia latifolia*, *K. angustifolia*, *Comptonia peregrina*, *Quercus ilicifolia*, *Lycopodium obscurum* and *tristachyum*, *Epigaea repens*, and *Drosera rotundifolia*), is illustrative (*RMS 43989a*). Sterile plants had all of the characteristics of *M. sylvatica*: plants rather densely leafy and compact; underleaves regularly with 1 or 2 of the 3 divisions aborted and bearing a slime papilla. However, bracts and perianth are more nearly characteristic of *M. setacea*: bracts are often divided for 0.2–0.3 their length into 2 primary lobes, each of which is armed externally with a smaller lobe; all 4 of the resulting lobes are distinctly ciliate, although clearly less so than in typical material (such as in Fig. 94:9); the perianth mouth is plurilobulate, each narrow lobule ending in a long cilium, which usually terminates in 3–4 superposed cells that are narrow and very elongate (terminal cell ca. 7.5–10 × 65–78 *μ*). The ♀ bracts are distinctly intermediate in the *degree* of development of the marginal cilia; in some cases the bracts are primarily 2-lobed, each lobe, laterally, with 1–2 elongate cilia 2–3 cells wide at base—hence not lobelike. Thus transitions from 2-lobed and ciliate to shallowly 4-lobed bracts occur. Furthermore, in no case does the sinus descend more than 0.3 the bract length.

The fortunately rare occurrence of such clearly annectant plants, from clearly annectant habitats, almost suggests that the two taxa are not fully distinct.

Müller (1939) used the supposed absence of oil-bodies in *Microlepidozia* to separate it from *Lepidozia s. str.* However, Hattori (1951b) showed that the Japanese *M. makinoana* has oil-bodies. Recent study of *M. setacea* (which Müller characterizes as lacking oil-bodies) has demonstrated that this species also possesses them in the basal 2–3 tiers of leaf cells (but apparently rarely or never in mature cells of the distal portions of the lobes). They occur mostly 3–5 per cell, are spherical and 1.5–2.5 (rarely to 4.5) μ or occasionally oval and up to $3 \times 4.5 \mu$; they appear virtually homogeneous and smooth under the oil-immersion system of the microscope. The large, elongate, pellucid cells of the stem appear to uniformly lack oil-bodies in the cortex, but medullary cells are provided with them; 3–6 or more occur in each cell. The medullary oil-bodies are similar in size and form to those of the leaf cells or average barely larger.

Robust plants of *M. setacea* have nearly uniformly 4-parted underleaves on the main stems. They also show lateral leaves with the lobes often 3–4 cells broad at base (2–3 cells broad at base in *M. sylvatica*). Furthermore, the leaves occasionally have a lateral, sharp 1–2-celled tooth on one of the outer lobes, or may show 1–2 sharp but small, 1-celled teeth of one or two of the median lobes. Such a tendency towards production of teeth on the leaf lobes (of main stems) is well developed in plants that may be separated as a distinct form:

Microlepidozia setacea fo. dentigera, fo. n.

Forma a *M. setacea* typica differens ut nonnulla omniave folia caulis matura lobos 1–2 dentibus acutis praeditos habent.

Similar to "typical" *M. setacea*, but mature stem leaves have lobes (or some of them) bearing 1–2 sharp, 1-celled teeth (the outer lobes sometimes with an aciculate, longer subbasal tooth 2–3 cells long).

Type. Bog near Roaring Brook campground, Mt. Katahdin, Me. (*RMS 32951, Aug. 9, 1954*).

Other variations of *M. setacea* appear to be wholly environmental. For example, the var. *sertularioides* (Hüben.)[41] is a strongly elongated, slender phase with very remote, strongly incurved leaves. According to Macvicar (1926, p. 339), it exists as a deep green phase with merely slightly pinnate branching, and as a paler, taller, and more rigid phase, with the stems pinnate to bipinnate. Apparently *M. setacea* var. *sphagnicola* is the same. The "typical" species appears to occupy the midground between this extreme and the small, compact, xeromorphic phase, the "var. *schultzii* (Hüben.)." The last is admittedly a plant of dry ground; it is small, irregularly branched, with short and broad leaf lobes and largely 2–3-lobed leaves. This plant is not unlike many phases of *M. sylvatica*.

BAZZANIA S. F. Gray corr. Carr.

Bazzanius S. F. Gray, Nat. Arr. Brit. Pl. 1:704, 1821.
Pleuroschisma Dumort., Syll. Jungerm. Eur., p. 68, 1831.
Herpetium Nees, Naturg. Eur. Leberm. 1:96, 1833.

[41] *Lepidozia setacea* var. *sertularioides* (Hüben.) Cooke, Hdb. Brit. Hep., p. 91, 1894. (*Jungermannia sertularioides* Linn. f., Suppl. Pl. Syst. Veg., p. 449, 1781, is not identical; it is *Blepharostoma trichophyllum*.)

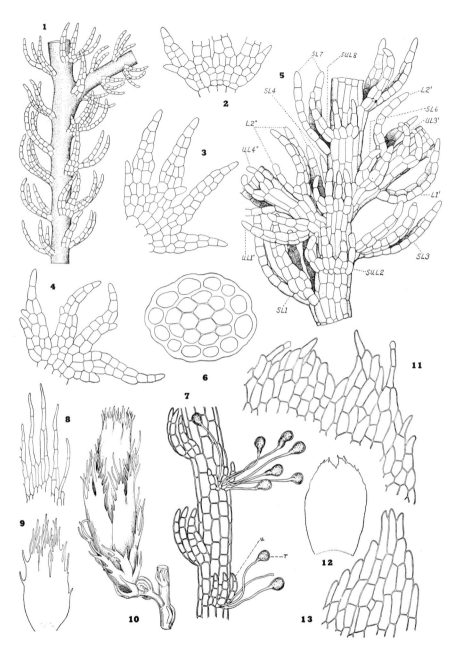

FIG. 94. *Microlepidozia setacea* (1–10) and *M. trichoclados* (11–13). (1) Shoot sector, antical view, with a *Frullania*-type branch (×ca. 60). (2) Underleaf base (×ca. 150). (3–4) Leaves (×ca. 150). (5) Shoot sector, postical view (×120); for details, see p. 451 of Vol. I. (6) Stem cross section (×230). (7) Latero-ventral aspect of shoot, underleaf at *u*, at *r* the mycorrhizal, swollen rhizoids (×ca. 120). (8) Perianth mouth (×ca. 72). (9) Bract (×18). (10) Perianth-bearing ventral branch (×18). (11) ♀ Bract apex (×ca. 110). (12) ♀ Bract (×28). (13) Perianth mouth sector (×ca. 110). [1, 6–7,

Herpetium sect. *Mastigobryum* Nees, *ibid.* 3:43, 1838.
Mastigobryum Nees, *in* G., L., & N., Syn. Hep., p. 214, 1845.
Bazzania S. F. Gray corr. Carr., Trans. Proc. Bot. Soc. Edinb. 10:309, 1870.

Moderate in size or robust, firm, in large tufts or in low, interwoven mats usually, varying from bright green to olive to yellow- or chestnut-brown, *opaque*, usually 1–6 mm wide. Stems *rigid*, with branching of two types: *lateral and usually pseudodichotomous* (less often in part monopodial), from ventral half of a lateral segment (replacing the ventral half of a lateral leaf; the latter is therefore acute and lanceolate and abnormal in form, usually situated in the fork of the "dichotomy"); branches *usually arising from one side of axis*, homodromous, usually subequal to main axis in vigor, the shoot system deliquescent, the branches usually widely spreading (but rarely at as much as 90° to each other); *postical branches frequent, usually flagelliform* and with reduced leaves (but occasionally developing normal leaves and underleaves), *intercalary in origin and from axil of a normal underleaf.* Stem anatomy simple: cells showing little differentiation; *cortical cells often ± larger in diam. but never forming a hyalodermis, short, thick-walled;* medullary cells ± equally thick-walled, strongly elongated, but with pits and thin end walls. *Rhizoids usually sparse* on leafy branches, from bases of underleaves, frequent at bases of reduced leaves and underleaves of the postical flagella. Leaves clearly incubously inserted and oriented, alternate, rarely subopposite, the line of insertion straight to strongly acroscopically curved at its upper end; leaves distant to imbricate, plane or with apices deflexed, *usually asymmetrical*, ovate-rectangular to ovate to lanceolate, the antical base sometimes cordate, the postical base rarely auriculate, the often deflexed *apex usually 2–3-dentate* (rarely 4-dentate; on weak shoots occasionally entire); leaf margins entire or crenulate, rarely (exotic species) serrate or spinose-dentate near base. Cells of leaves usually regularly *quadrate to hexagonal, subisodiametric*, more or less uniform, only moderately larger toward base, rarely (exotic species) vittate with several series of differentiated cells; trigones small to bulging and coalescent; cuticle smooth to verruculose; *oil-bodies usually* (*1*)*2–6* (*7–16*) *per cell*, moderately large for the cell size, but not occluding lumen, *generally homogeneous to few-segmented and highly refractive*, rarely opaque, little refractive, and formed of minute but discrete spherules. *Underleaves large*, inserted on broad merophytes ca. 0.25–0.3 × the stem perimeter, inserted transversely, but sometimes the lateral margins decurrent (line of insertion

---

RMS H3402, Bergen Swamp, N.Y.; 2–5, fo. *dentigera, RMS 32951*, Mt. Katahdin, Me; 8–10, from Macvicar, 1912; 11–13, from Schiffner's Hep. Eur. Exsic. No. 684, Schwarzwald, Germany.]

then approaching an inverted U), distant to imbricate, variable, ranging from transversely rectangulate to quadrate to orbicular, sometimes cordate at base, the apex truncate to rounded-truncate, entire or *2–4-toothed or lobed*, the margins laterally sometimes dentate. *Asexual reproduction, when present, by means of caducous leaves and underleaves.*

Dioecious, often sterile; *rarely with sporophytes. Sexual branches intercalary, short, postical, from the axils of underleaves,*[42] lacking vegetative leaves, not proliferating or with innovations. *Androecia small, compact, spicate*, inconspicuous, hidden from above; bracteoles smaller than bracts; bracts in 4–6 pairs or more, concave, ovate, sometimes subconvolute, their apices truncate or entire to bilobed or bidentate, rarely dentate or denticulate; 1–2-androus. Gynoecia with bracts and bracteoles similar, in several (4–5 or more) imbricate, erect to erect-appressed series, the innermost largest; bracts ovate to ovate-lanceolate, usually at least somewhat lobed at apex, often 2–3–4-lobed or parted, the margins usually variously crenulate to denticulate or ciliate or laciniate-ciliate. Perianth elongate, *subfusiform-trigonous*, terete below and cylindrical, trigonous above, gradually narrowed toward the mouth, unistratose above but usually pluristratose below; mouth 3-lobed, the lobes dentate to ciliate, usually contracted; perigynium lacking. Sporophytes with seta robust, usually consisting of ca. *16 large epidermal cell rows and a medulla of numerous much smaller cell rows.* Capsule ovoid or oblong-ovoid, with elliptical, straight valves; *walls usually 4–6-stratose;* epidermal layer of relatively large, rectangulate cells whose longitudinal walls (or both faces of alternating longitudinal walls) bear coarse nodular, radial thickenings; innermost layer with numerous transverse, semiannular bands extending across the inner tangential walls and both radial walls. Elaters slender, elongate, bispiral; spores vermiculate-marked to verruculose, brown, small, *1–1.25 to 2–3× the diam. of the elaters.* Chromosome no.: $n = 9$ (9 Japanese spp.; Tatuno, 1941, 1947, 1948) or 9/10 (Heitz, 1927; *B. trilobata*).

Type. Bazzania trilobata (L.) Gray.

Bazzania is a large and complex genus, many of whose species exhibit a wide range of environmentally induced variation. As in such genera as *Plagiochila*, reproductive characters are rarely developed and species characters are derived almost exclusively from the vegetative plant. Stephani (1908, *in* 1898–1924) lists 335 species for the world, adding (1924) an additional 109 species. A few more have been described since by Herzog. Judging from the work of Fulford (1946, 1963), which treats

---

[42] Sometimes also from axils of the reduced leaves and underleaves of flagella; see discussion under *B. trilobata* (p. 89).

only a portion of the South and Central American species, many of the Stephani species are clearly synonyms. Probably no more than 250 of the over 450 currently described species are valid.

The genus is largely pantropical and antipodal in distribution, with only a small minority of species occurring in cold or temperate regions. Only one species penetrates north of the Taiga. Six species are known from North America, two of which (*B. pearsoni* and *B. ambigua*) are known in the Western Hemisphere only from the northwest; Stephani, however, credited only three species (*B. pearsoni*, *B. trilobata*, *B. triangularis*) to the North Temperate Zone, under the genus *Mastigobryum*. The first critical revision of our species was that of Evans (1923e), with the review of Fulford (1936a) closely following.

The species of our region center about the Appalachian system, only one (*B. trilobata*) being widespread in the area peripheral to the montane portions of eastern North America. All the species are generally limited to rocky habitats; again only *B. trilobata* and *B. nudicaulis* occur widespread on a series of organic substrates. Of our other species, *B. nudicaulis* occurs frequently, *B. trilobata* and *B. denudata* occasionally, on the trunks of trees. In the tropics many of the species are arboreal. Apparently almost all species of the genus "avoid" calcareous sites.

The genus *Bazzania* is perhaps most closely allied to the pantropical genus *Acromastigum* Evans. The latter differs from *Bazzania* chiefly in that postical branches arise terminally, replacing the right or left half of an underleaf, rather than intercalary and axillary as in *Bazzania*. Among regional genera confusion is possible with sterile plants of *Calypogeia*, a genus which differs fundamentally from *Bazzania* in gynoecial structure, sporophyte, mode of asexual reproduction, seta anatomy, and stem anatomy. The more pellucid, delicate plants of *Calypogeia* differ at once from those of *Bazzania* in the entire or faintly bidentate leaves, the delicate and larger leaf cells with segmented oil-bodies, the free development of rhizoids, the absence (normally) of flagella, the monopodial, postical-intercalary branching, and the lack of secondary pigmentation.

In vegetative plants these distinctions are not always abundantly clear; occasional species of *Calypogeia*, but especially *Metacalypogeia*, may approach *Bazzania* in facies, e.g., *M. schusterana*. In this taxon oil-bodies are botryoidal, underleaves bilobed, leaf apices usually edentate, and flagella lacking, as is typical of the Calypogeiaceae. However, *Bazzania*-like branching is present. In this deviant species one additional noteworthy feature occurs: the unusually frequent rhizoids not only arise from the underleaf bases, but also occur scattered over the ventral merophytes below the insertion of the rhizoids, and sometimes down almost to the next underleaf (in whose axil they appear to arise). This condition is almost unique among the Lepidoziinae (Fig. 102:8).

Classification

Bazzania was divided by Stephani (1885–86, p. 244) into 11 sections. In 1908 (*in* 1898–1924) he revised this classification and divided the 11 sections into 4 subgenera (*Integrifolia*, *Bidentata*, *Inaequilatera*, and *Tridentata*); the last of the subgenera was divided into 7 sections. Fulford (1946) in a treatise on the South American taxa criticized this classification as partly artificial; the subg. *Integrifolia*, as regards American species, is identical with *Tridentata*, or at least no American species can be referred to the subgenus, if there is any basis for its recognition. The subg. *Inaequilatera* was transferred by Evans (1934) to *Acromastigum*. This leaves only the subg. *Bidentata*, whose species are regarded by both Stephani and Fulford as sufficiently homogeneous as not to warrant subdivision into sections, and *Tridentata*. The last "subgenus" is divided by Fulford into 5, rather than 7, sections (Grandistipulae, Connatae, Appendiculatae, Fissistipulae, and Vittatae). If this classification is used, our species would all fall into the subg. *Tridentata*, sectio Grandistipulae.

This classification is adhered to in Fulford (1959),[43] in which her earlier species concepts (1946) are in part revised. The taxonomy of the genus as revised by Fulford remains subjective. I am unimpressed by the separation of the species with predominantly bilobed leaves (subg. *Bidentata*) from those with mostly tridentate leaves (subg. *Tridentata*). Although in 1946 Fulford states (p. 8) that the Bidentatae "form a well defined subgenus," in 1959 she concludes (p. 309) that "most of the species appear to be rather closely related to species in the Section Tridentatae." Although her reference to the Tridentatae as a "section" may have been a lapsus at this point, I am of the opinion that the differences between Tridentatae and Bidentatae are merely of sectional value. This opinion is reinforced by Fulford's (1959, p. 313) admission that in the species she places in the Bidentatae, such as *B. phyllobola*, both trifid and bifid leaves are found "alternating on the same stem" and that "the problem of where to assign plants . . . with both bifid and trifid leaves on a single stem is a difficult one. Such transitional forms are present in most species of the subgenus Bidentatae."

For these reasons, I suggest that, unless and until more basic differences can be found, subgeneric division of *Bazzania* be avoided. The basis for subdivision given in Fulford leads to ambiguity; hence I regard the Bidentatae as a mere section and suggest we drop the subgeneric name *Tridentata* altogether. The student need merely study such American species as *B. ambigua*, *B. tricrenata*, *B. nudicaulis*, and *B. denudata*, all obviously interrelated on the basis of the diagram given in Fulford (1936a), to satisfy himself that leaf lobe (or tooth) number is an arbitrary criterion for subgeneric division.

Whatever classification is adopted, all North American species, except the deviant, western *B. pearsoni*, belong to a single section, the Grandistipulae. *Bazzania trilobata* differs markedly from all our other species of this section and is surely phylogenetically remote from them.

___

[43] The Stephanian ending *a* is, however, altered to *ae*.

Any clear understanding of the species and their classification is likely to come about only as a consequence of study of the sporophyte. The following discussion may therefore prove a starting point for future work.

The anatomy of the capsule wall in *Bazzania* is distinctive; it deserves further study in other taxa known with sporophytes, since it has been virtually ignored. The diagnoses given in some recent works are rote and not very useful with regard to the pattern of thickening of the capsule wall. The capsule wall in *Bazzania trilobata* is fairly representative of the less reduced type found in the family Lepidoziaceae. The 5–6 layers of the wall represent apparently the maximal thickness found. Epidermal cells tend to be somewhat irregular but are mostly rectangulate and show an orientation identical to that in some Calypogeiaceae, such as *C. sullivantii*. The walls of the primary cells generally lack thickenings, but these primary cells undergo 1-3 longitudinal subdivisions, with the 1-3 secondary longitudinal walls bearing marked nodular thickenings on both faces. As a consequence there is, with maximal subdivision, a regular alternation between 1-3 walls with nodular brown thickenings and a hyaline wall (rarely with 1-3 faint thickenings). Very frequently the primary cells also undergo transverse division by a secondary wall which also bears thickenings, although the primary transverse walls lack thickenings. The innermost cell layer, of more or less narrowly rectangulate cells, bears numerous bandlike semiannular thickenings, exactly as in the other Lepidoziaceae and Calypogeiaceae.

The history undergone by the primary cells of the epidermal layer suggests a close affinity between the Calypogeiaceae and Lepidoziaceae. The Cephaloziaceae were once placed with the Lepidozioid-Calypogeioid complex (the Trigonantheae or Trigonanthaceae of, i.a., Spruce). The capsule-wall anatomy of *Microlepidozia* would give support to such an assumed affinity. In many Lepidoziaceae, except as we have seen for *Microlepidozia*, a 1:3:1:3:1 alternation of hyaline and thickening-bearing walls (primary and secondary walls) typically develops. As discussed under the Cephaloziaceae (Vol. III), the epidermal cells in that family often possess a regular 1:1:1:1 alternation of hyaline and thickening-bearing walls. This identical pattern also recurs in *Microlepidozia* and sometimes in *Bazzania*.[44]

The spores of *B. trilobata* show a distinct differentiation in markings between the convex outer face and the inner faces, although at maturity the spores are spherical, without delimitation of these faces. The free faces bear vermiculate,

---

[44] Evans (1923f, p. 62) describes a similar pattern for the western *Bazzania ambigua*, in which the epidermal cells possess "thickenings, usually one to three in each cell, [which] tend to be restricted to alternate longitudinal walls, the intervening longitudinal walls and the transverse walls being free from thickenings or nearly so." In *B. trilobata*, as we have seen, there may be a 1:1:1:1 or a 1:3:1:3:1 alternation.

anastomosing, delicate ridges; the inner faces bear fine asperulations, only rarely confluent to form partial vermiculate lines. As we have seen, in *Microlepidozia* the spores are merely papillose or verruculose, whereas in *Telaranea* they are delicately areolate. Possibly good generic characters can be derived from these differences, although many more observations are needed. The elaters in *B. trilobata*, as in other members of the family, are regularly bispiral, with the rather closely wound, red-brown spirals ca. 3 μ broad. The elaters range from 9 to 10 μ in diam.[45] Since the spores are (12)13–16(18) μ in diam., the spore-elater ratio varies from ca. 1.3: to 2.0:1. In *B. ambigua*, a western species, Evans (1923f) reports the spores as 12–14 μ in diam. and the bispiral elaters as 8–10 μ in diam.; the spore-elater ratio here is thus ca. 1.2–1.75:1. In contrast, *B. tricrenata* has spores 15–20 μ in diam. and the elaters are only about 7 μ in diam., the spore-elater ratio varying from about 2: to 3:1.

The highly reduced sexual branches of *Bazzania* are usually limited in their occurrence to the axils of underleaves of leafy stems; they are then strictly postical in origin. However, as is shown in detail under *B. trilobata* (p. 89), in at least this species androecia and gynoecia are often produced on flagella, and the gynoecia, if fertilized, may mature sporophytes. Such sexual branches originating on flagella do not show restriction to a single series of merophytes; at least the antheridial branch originates from all three merophytes. Fulford (1936a, 1946), merely states that sexual branches issue from the "axils of the underleaves."

Key to Species of Bazzania

1. Pure green to (rarely) olive-green, the leaves without distinct brownish pigmentation; leaves plane or slightly convex, apices not strongly deflexed, usually (1)2–3-dentate; oil-bodies homogeneous or weakly 2–3(4–5)-segmented, (3)4–14(16) per cell, or granular . . . . 2.
   2. Robust, normally 3–6 mm wide; usually with leaves persistent; underleaves strongly plurilobulate-pluridentate; antical leaf bases somewhat ampliate at base, the leaves, *in situ*, usually hiding stem from view (in antical aspect) . . . . *B. trilobata* (p. 83)
   2. Slender, mostly 1–2.2 mm wide, underleaves obscurely dentate or bifid; antical leaf bases not strongly dilated at base, the stem (in dorsal aspect) ± exposed . . . . 3.
      3. Leaves with marginal cells almost all isodiametric; leaves caducous, relatively broad at apex, where at least many leaves are 2–3-dentate; underleaves about twice as wide as stem, at least in part, crenate-lobate, with ca. four shallow teeth or rounded lobes; oil-bodies homogeneous, with age in part 2–3-barred, 5–14(16) per cell. Widespread . . . . *B. denudata* (p. 91)
      3. Leaves with marginal cells ± distinctly, tangentially elongate; leaves persistent, narrow, acuminate at apex, which is 1–2-dentate; underleaves little wider than stem, bidentate or bilobed; oil-bodies granular,

---

[45] According to Müller (1951–58, p. 1153), the spores are 14–16 μ, the elaters 12–14 μ, in diam.; thus the spore-elater ratio varies from 1–1.33:1.

not becoming segmented, 2–5(6) per cell. Rare; on calcareous rocks
or in basic fens . . . . See *Metacalypogeia* (p. 103)

1. Usually deep green to reddish brown in color; leaves slightly to strongly
convex, their apices usually strongly deflexed, largely acute and 1–2-dentate
(on robust shoots strongly asymmetrically 3-dentate); plants small, 0.7–2
mm wide; oil-bodies (1)2–4 per cell . . . . 4.

    4. Leaves and underleaves persistent; leaves commonly 2–3-dentate
at apex; underleaves small, 1–1.5× as wide as stem, wider than
long, mostly distant, 3–4-dentate at apex; often with sex organs;
oil-bodies glistening, homogeneous or 2–3-segmented . . . . . .
. . . . *B. tricrenata* (p. 72)

    4. Leaves and underleaves freely caducous; leaves commonly ovate-
lanceolate and merely acute at apex, more rarely 2–3-dentate;
underleaves large, 2–3× as wide as stem, suborbicular and as long
as wide, often nearly contiguous, entire or subentire; uniformly
sterile; oil-bodies (1)2–3(4) per cell, opaque, delicately granulose.
. . . . *B. nudicaulis* (p. 79)

Supplementary Key to Species of Bazzania

1. Small to moderate-sized, leafy shoots 0.7–2.2 mm wide; marginal rows of
leaf cells nearly equal in size to those within or sometimes larger; leaves
obliquely, narrowly 1–2- or 2–3-dentate at tip, or acute; spore-elater ratio
2–3:1 where known; dorsal leaf base not or slightly auriculate, extending
merely 0.3–0.6 across stem . . . . 2.

    2. Leaves narrow, acute and slender distally, persistent, deflexed when dry;
leaves sharply, asymmetrically 1–2- or 2–3-toothed at tip; underleaves
1–1.5× as wide as stem . . . . 3.

        3. Dark green to brown, with generally 2–3-dentate, strongly deflexed
leaf apices; oil-bodies homogeneous or (with age) weakly 2–3-
segmented; underleaves at the apices generally distinctly 3–4-
lobulate, the lateral margins often crenate-dentate; flagella common
. . . . *B. tricrenata* (p. 72)

        3. Light yellow-green, with leaves acute or merely bidentate at the weakly
to moderately deflexed apices; oil-bodies botryoidal; underleaves
generally bilobed, with lateral margins rarely dentate or crenate;
without flagella . . . . See *Metacalypogeia* (p. 103)

    2. Leaves potentially caducous, often large portions of stems becoming
devoid of leaves; leaves not or scarcely falcate, the ventral margin
straight or convex; lateral margins of underleaves entire (or plant green);
always sterile; leaves often largely entire and acute at tip, or merely
2-dentate, few leaves 3-dentate; underleaves ±2× as wide as stem. . 4.

        4. Reddish brown to blackish at maturity, largely denuded except at
apices of branches; underleaves of main stems about as long as
wide, orbicular, with entire margins; stem cells all ca. 12–20 μ in
diam.; cells of leaf middle and apex ca. 20–25 μ, cells of base
22–25 μ; oil-bodies (1)2–3(4) per cell, ± granular . . . . .
. . . . *B. nudicaulis* (p. 79)

4. Green or yellowish green in life, with many (sometimes all) leaves
persisting on mature shoots; underleaves wider than long, with
margins in part sinuate-crenate to denticulate; stem cells about
20–25 μ in diam.; cells of leaf middle about 25–31 $\times$ 26–36 μ,
cells of apex about 24–26 $\times$ 25–27 μ, cells of base about (26)30–50
μ; oil-bodies mostly 5–14 per cell, homogeneous or few-segmented
. . . . B. denudata (p. 91)

1. Large, normally (2.7)3–6 mm wide; stems and leaves green, persistent
(rarely caducous); leaves at apex at least half as wide as at middle, truncate-
tridentate; marginal row of leaf cells, at least near antical base, smaller
than rows within; spore-elater ratio 1–2:1; dorsal leaf base strongly
auriculate, extending nearly across stem . . . . B. trilobata (p. 83)

BAZZANIA TRICRENATA (Wahl.) Trev.

[Figs. 95, 96]

Jungermannia tricrenata Wahl., Fl. Carpath., p. 364, 1814.
Jungermannia deflexa Mart., Fl. Crypt. Erlang., p. 135, 1817.
?*Jungermannia triangularis* Schleich., Pl. Crypt. Exsic. Helvetiae, Cent. 2, no. 61, 1831 (*nomen nudum*).
Pleurochisma tricrenatum Dumort., Syll. Jungerm. Eur., p. 70, 1831.
P. parvulum, P. deflexum, and *P. ?flaccidum* Dumort., *ibid.,* p. 71, 1831.
Herpetium deflexum Nees, Naturg. Eur. Leberm. 3:57, 1838.
Mastigobryum deflexum Nees, in G., L., & N., Syn. Hep., p. 231, 1845.
Bazzania triangularis Lindb., Acta Soc. F. et Fl. Fennica 10:499, 1874.[46]
Bazzania tricrenata Trev., Mem. R. Ist. Lomb., ser. 3, Cl. Sci., 4:415, 1877; Fulford, Amer.
   Midl. Nat. 17:398, figs. 4–6, 1936; Müller, Rabenh. Krypt.-Fl. ed. 3, 6:1154, 1956.
Bazzania deflexa Underw., Bull. Ill. State Lab. Nat. Hist. 2:83, 1884; Mitten, Jour. Proc. Linn.
   Soc. Lond. 22:322, 1887.
Bazzania triangularis var. *proliferum* Roell, Hedwigia 32:399, 1893.
Pleuroschisma triangulare Loeske, Moosfl. Harzes, p. 96, 1903.
Mastigobryum triangulare Steph., Spec. Hep. 3:475, 1908.
Mastigobryum fissifolium Steph., *ibid.* 3:502, 1908 (*fide* Hattori, 1957, p. 109).
Mastigobryum hamatum Steph., *ibid.* 3:440, 1908 (*fide* Hattori, 1957, p. 109).
Mastigobryum subhamatum Beauv. in Stephani, Spec. Hep. 6:465, 1924 [= *M. hamatum* Steph.,
   msc., nec *in* Spec. Hep. 3:440, 1908 (*fide* Hattori, 1957a, p. 109)].
Bazzania fissifolia var. *subsimplex* Steph. ex Hatt., Bot. Mag. Tokyo 58:66, fig. 24, 1944 [= *M.
   subsimplex* Steph., msc. (*fide* Hattori, 1957a, p. 109)].
Mastigobryum jishibae Steph., Spec. Hep. 6:469, 1924 (*fide* Hattori, 1957a, p. 109).

In loose, $\pm$ flocculent patches or tufts, 2–5(10) cm deep, or scattered, *deep
green to olive-green* (Appalachian phases, in large part) *to brownish or reddish brown.*
Shoots procumbent, (0.75–1.5)1.6–2.0(2.25) mm wide, the stems (120)160–240
μ in diam., (2)3–5(8) cm long. Stems rather rigid, filiform; epidermal cells
similar in diam. to medullary cells, ca. 17–20 $\times$ (25)27–48(50) μ long, thick-
walled (in surface view with rounded angles); medullary cells ca. 125–175 μ

---

[46] The "*Bazzania triangularis*" of Pearson (1902) and Macvicar (1912, 1926) is, according to
Müller (1956, *in* 1951–58), identical with *B. denudata*. However, Jones (1958a, p. 368) emphasizes
that the British plant is "certainly not identical with *B. denudata*," agreeing with other British
workers in assigning it to a "slender form of *B. tricrenata* . . . [which] appears in fact always to
be male." Mizutani (1967, p. 77) has just placed *Pleuroschisma alpinum, Mastigobryum perrottetii,
M. longanum,* all Stephani spp. under *B. tricrenata.*

long; with rather free development of lateral branches and with occasional (but not abundant) postical stolons; *lateral branches forming acute angles with stem.* Rhizoids few, restricted to flagella and occasionally a few at underleaf bases. Leaves ± *convex*, slightly imbricate to remote (particularly in the green, slender Appalachian phases) to weakly imbricate, inserted by only a slightly oblique line, *strongly asymmetrically triangular-ovate to ovate-falcate, persistent*, very variable in size (to 600 μ wide × 1000 μ long; rarely to 1500 μ long), widest at base, widely patent (70–85°, occasionally 90° or more), their apices usually deflexed, often strongly so (particularly in drying), at the antical base only slightly ampliate and extending merely 0.5–0.65 (1.0) across the stem; *leaf apices much narrowed, very obliquely* 2–3-dentate; on weaker phases *usually strongly bidentate*, the sinuses narrow; *leaves asymmetric* and obliquely terminated, the lobelike teeth ± unequal, *the anterior usually larger;* antical margin strongly arched, especially at base; ventral margin slightly convex to somewhat concave. Cells quadrate to rounded quadrate-hexagonal, with slightly thickened walls and distinct (but not bulging) trigones; cuticle smooth to barely verruculose; subapical cells (20)21–25(26) μ, the median up to 22–28 × 28–35 μ (but in the delicate Appalachian phases often as little as 20–22 × 20–25 μ), (16)17–20 μ on the margins; *oil-bodies 2–4(5) per cell, glistening and smooth*, homogeneous or (under oil) barely discernibly formed of minute, obscure spherules, spherical to subovoid and 5–6(7) μ to 4–6 × 7 μ to ellipsoidal and 5 × 8–10–11 to 6 × 13 μ, with age sometimes 2–3-segmented. Underleaves distant to subapproximate, *transversely quadrate-orbicular*, to 300–350 μ long × 450 μ broad, often squarrose, the rounded-truncate apex varying from subentire to more commonly *irregularly 3–4-lobulate or sinuate-dentate* for 0.15–0.25 their length, the lateral margins entire to sinuous or occasionally obscurely sinuous-dentate. *Asexual reproduction lacking.*[47]

Almost constantly sterile. Gynoecial and androecial branches apparently uniformly from axils of underleaves of leafy stems. Androecial branches spicate, often several per axil; bracts quadrate-rotundate, divided for 0.15–0.35 their length into 2–4 blunt to rounded teeth; bracteole smaller but essentially similar; monandrous. Gynoecia with bracts in several series, the lower mostly ovate (ca. 350 μ wide × 400 μ long), lobed or toothed at apex only; inner series of bracts ovate, elongate, to 1500 μ long × 950 μ broad, usually 2–3-lobed to ¼, the lobes broad and distantly ciliate-dentate; margins of bracts crenulate to ciliate-dentate, especially above; bracteole similar to bracts. Perianth slenderly ovoid-cylindrical, fusiform above, gradually narrowed to mouth, 3–4-stratose below, 2-stratose to middle, usually ca. 0.8–1 mm in diam. × 4(6) mm high. *Sporophytes rare:* seta with 14–16 rows of large, hyaline, epidermal cells surrounding ca. 16–25 rows of smaller interior cells; capsule oblong-ovoid, its walls 4–5-stratose. Inner layer with nodular thickenings and, in part, incomplete semiannular bands; epidermal layer with strong nodular thickenings. Elaters (5) 6–8 μ in diam. × 260–500 μ long, hardly

---

[47] American plants, without exception, lack caducous leaves. In my experience, they are usually ♀. Macvicar (1912, 1926) points out that the slender European phase, which he attempted to maintain as "*B. triangularis*," may possess caducous leaves; on this basis Müller (1956, *in* 1951–58) incorrectly attempted to equate these plants with *B. denudata*, while Jones (1958a) claims that they are apparently a slender ♂ phase of *B. tricrenata*.

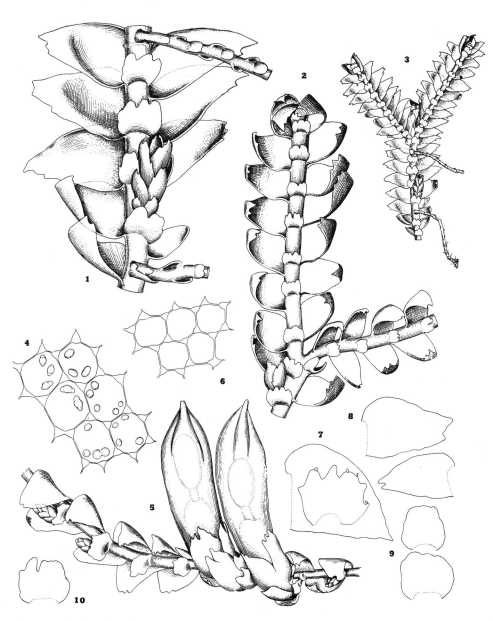

FIG. 95. *Bazzania tricrenata*. (1) Shoot sector, ventral aspect, with two ventral-intercalary flagella and a young gynoecium (×26). (2) Shoot, postical aspect, young gynoecium below (×20). (3) Robust extreme, postical aspect, somewhat artificially flattened (×7). (4) Median cells with oil-bodies (×453). (5) Posterolateral aspect of shoot with two perianths (×ca. 17.5). (6) Median cells (×325). (7) Stem leaf and underleaf (×46). (8–9) Two branch leaves and branch underleaves (×46). (10) Underleaf (×46). [1–3, *RMS 24450*, Mt. Craig, Black Mts., N.C., from plants admixed with *B. nudicaulis*, drawn in Fig. 97:1–4; 4, *RMS 24412*, Cornell Mt., Catskill Mts., N.Y.; 5, from Müller, 1905–16; 6–10, *RMS H2143*, Mt. Washington, N.H.]

tapering towards the ends, bispiral. Spores $(15)17-20$ μ, papillose, vinaceous or purplish.

Type. Carpathians, Europe.

Distribution

Evidently circumboreal; the most "northern" of our species of the genus and the only one whose range extends northward into the Arctic. Widespread in Europe (southward ·in the central European Alps, in Austria, Germany, Switzerland, eastward to the Tatra Mts.); northward from Ireland and Wales to the Faroes, Norway, Sweden and southern Finland; extending southward to the Pyrenees of France and Spain, northern Italy, Istria, near Reggio in Calabria; also in the Trapezunt, on the Black Sea, in Asia Minor, and in the Azores. In Asia extending from China (Shen-si Prov. and Yunnan) to the West-Himalayas, to Japan (Hattori, 1957a, p. 109), Korea and Taiwan (Grolle, 1966h.)

In North America occurring in the Far West, from the Aleutian Isls. (Clark and Frye, 1949) to continental Alaska, especially on the southwest coast; southward to British Columbia, Washington, Oregon and northern California, and in the interior to Idaho,[48] and eastward as follows:

ELLESMERE I. Harbour Fiord (Bryhn, 1906); the only, questionable, report from the Tundra of eastern North America. NEWFOUNDLAND. West Ranton (Evans, 1923e); Pushthrough and Hare Bay, S. coast (Buch & Tuomikoski, 1955); Channel (Fulford, 1936a). NOVA SCOTIA AND CAPE BRETON I. Cape Breton I. (Nichols, 1916). NEW BRUNSWICK. *Fide* Fulford (1936a). QUEBEC. Mt. Albert, Gaspé Co. (Evans, 1923e); L. Mistassini (Lepage, 1945); La Tuque; St.-Simon de Rimouski (Lepage, 1944–45); Percé.

MAINE. Mt. Katahdin (*RMS*; Evans, 1923e). NEW HAMPSHIRE. Mt. Washington (*RMS*); L. of the Clouds, Mt. Monroe, Mt. Adams, Carter Notch and Dome, Crystal Cascade, and elsewhere in the White Mts. (Evans, 1923e; *RMS*; Underwood & Cook, Hep. Amer. no. 53!); Kings Ravine (Evans, 1923e); Bears Cave, Franconia Notch (Evans, 1923e); Crawford Notch (*T. P. James!*); The Flume, Franconia Mts. (Lorenz, 1908c). VERMONT. Mt. Mansfield (Evans, 1923e). MASSACHUSETTS. Reported from Abnecount's I., Nantucket (as the synonym *B. triangularis*), but this almost surely an error; also reported from Greenville (= *B. denudata*). CONNECTICUT. Early reports (*in* Evans & Nichols, 1908) = *B. denudata*; see Evans (1923e). NEW YORK. Little Moose L., Herkimer Co. (*Haynes*); Slide Mt., Wittenberg Mt., Cornell Mt., Catskill Mts. (*RMS 17615, 17591a, 24305, 17620, 17583, 24412, 17571, 24683,* etc.); Martiny Rocks, W. branch of Four Mile Rd., Allegany, Cattaraugus Co. (Boehner, 1943; the report not verified by additional collections in Schuster, 1949a, hence questionable). PENNSYLVANIA. Bull Hill, Sheffield, Warren Co. (*Lanfear*). OHIO. Hocking Co. (Miller, 1964).

Occurring (usually at elevations of 4000–6600 ft) as a disjunct in the following: WEST VIRGINIA. Hunterville (Fulford, 1936a); Monongalia, Pocahontas, Randolph Cos. (Ammons, 1940). VIRGINIA: Summit of White Top Mt. (Evans, 1923f); Mt.

---

[48] As is pointed out in Evans (1923f, p. 59) many of the early reports of *B. tricrenata* from western North America refer, in actuality, to *B. ambigua* (Lindenb.) Trev.

Rogers, Grayson Co., near summit, ca. 5700 ft (*RMS*).   KENTUCKY. McCreary Co. (Fulford, 1936a).   NORTH CAROLINA. Mt. Craig and Mt. Mitchell, Yancey Co. (*RMS 23682a, 23160a, 23180a, 24774,* etc.); Roan Mt., Mitchell Co. (*RMS*); Clingmans Dome and Andrews Bald, Swain Co., in Smoky Mt. Ntl. Park (*RMS*); Mt. Pisgah, Grandfather Mt., and summit of Jones Knob (Evans, 1923f; *RMS 30188a*); Andrews Bald, Swain Co. (*M. S. Taylor*; *RMS*); Avery, Jackson, Stokes, Watauga, Yancey Cos. (Blomquist, 1936).   TENNESSEE. Roan Mt., Carter Co. (*RMS*); Myrtle Pt., Mt. Le Conte (*Sharp 4131*); Mt. Collins (*Sharp*).

Bazzania tricrenata recurs as a disjunct in the Highlands of Mexico-Guatemala, at 9300 feet, in Chimaltenango, Guatemala (Fulford, 1959).[49]

The species occurs farther northward than our other species of the genus: at its southern limits in our region, in North Carolina and Tennessee, it also is restricted to higher elevations (usually over 4500 feet) than the distantly allied *B. denudata*. It is possible the Ellesmere I. report is based on *Metacalypogeia schusterana*, a plant with a slight superficial similarity to *B. tricrenata*.

Ecology

Bazzania tricrenata occurs almost uniformly on soil-covered, damp to moist rocks, particularly on shaded, acidic ledges; it is apparently uniformly absent from calcareous areas in our region (although Müller states that it rarely occurs on calcareous cliffs). Exceptionally on decaying wood. According to Hattori (1957a), it occurs also on serpentine.

On cliffs the species is usually found with *Lepidozia reptans*, *Bazzania denudata* and occasionally *B. nudicaulis*, and *Plagiochila sullivantii*, *P. tridenticulata*, *Anastrophyllum minutum*, and *Diplophyllum taxifolium*, less often *Anastrophyllum michauxii*. Northward it also is often found with *Mylia taylori*, *Lophozia attenuata*, *Lepidozia reptans*, *Bazzania trilobata*, *Herberta adunca tenuis*, *Lophozia silvicola*, and *Cephalozia media* (as, i.a., in the Catskill Mts., N.Y.). The consociation, on boulders and rocks, with *Mylia taylori* and *Anastrophyllum michauxii* is apparently a very frequent one, being also noted for Newfoundland (Buch and Tuomikoski, 1955); it recurs often in the White Mts., N.H. and on Mt. Mansfield, Vt.

Variation and Differentiation

Up to the time of the careful study of Evans (1923e), *B. tricrenata* was almost constantly confused with the more widespread and common *B. denudata*. This confusion was partly due to the fact that in Europe the distinctions between *B. tricrenata* and *B. denudata* do not appear to be as marked—at least plants known as "*B. triangularis*" show features that are intermediate in some respects. In addition, the pattern of variation of the European plants fails to match that of the American plants in certain

[49] A report from Jackson Co., Ill. (Hatcher, 1952), must surely represent an error in determination.

significant aspects. The polymorphism of the European phases led Nees (1838, *in* 1833–38) to recognize five varieties and seven lesser forms.

In North America, however, *B. tricrenata* is easily separated by an ensemble of features, including (*a*) leaves always persistent, always more or less deflexed, especially in drying; (*b*) leaves sharply and asymmetrically 2–3-dentate at the apex, somewhat falcate; (*c*) color usually at least locally brownish, in extreme shade forms deep green, but never a clear, light green as in *B. denudata;* (*d*) underleaves, at least where well developed, distinctly dentate-lobulate apically and tending to be at least sinuate laterally, always broader than long. When well developed, the brownish color and the postically deflexed leaves strongly suggest *B. nudicaulis,* a species which often occurs admixed (in the high mountains of the Southern Appalachians), and at the same time serve to separate the species from the two more nearly plane, more nearly pure green species, *B. denudata* and *B. trilobata.*

Unfortunately, the distinctions emphasized above are not always apparent in the plants from the southern highlands. As has already been emphasized (Evans, 1923f, p. 64), almost all Appalachian plants are slender, with rather remote leaves, often with smaller leaf cells (minimal figures cited in diagnosis), with the less falcate leaves more narrowly lanceolate and usually sharply bidentate at the apex. Isolated leaves, indeed, are merely slenderly acute distally, without any indication of dentition. In these plants the underleaves are hardly or no wider than the stem and very remote. The Appalachian plants are also all deep green, rather conspicuously subnitid, and rarely show the brownish coloration (typical even of shade forms northward) supposedly distinctive of the species. Such plants might be confused with *B. denudata.*[50]

Since slender, deep green Appalachian phases of *B. tricrenata* often occur admixed with *B. denudata* and are distinct at a glance (by their more convex, decurved, deep green or brown leaves, by sharper dentition of the more narrowed leaf tips, and by the uniformly persistent, often more remote leaves), it is clear that the two taxa are distinct. Appalachian phases of *B. tricrenata* approach rather closely in some respects Macvicar's concept (1926) of *B. triangularis,* which is characterized as differing from *B. tricrenata* in being "*smaller* and more slender than *B. tricrenata* . . . olive-green to golden-yellow in colour . . . with few flagella . . . . *Leaves small,* to 1 mm long, *distant or approximate,* sometimes caducous, *convex* with apex decurved . . . the apex narrowed,

---

[50] The distinctions relied upon by Müller (1951–58, p. 1151) to separate *B. tricrenata* from *B. denudata* fail with Appalachian plants. Müller states the leaves in *B. tricrenata* are imbricate, hemispherical, and 1.5× as long as wide when flattened; the median cells are 25–30 μ, the leaf tips are oblique and 3-dentate, the stem 250–330 μ thick. In Appalachian *B. tricrenata* the leaves may be 310 × 620 μ (exactly twice as long as broad; thus exactly the "doppelt so lang wie breit" assigned to *B. denudata* by Müller); the leaves are nearly flat, though deflexed distally; the leaf tips are 2- or rarely 1-dentate at apex; the stem is only 140 μ in diam.; the leaves are remote and the median cells are 20–25 μ—all characters Müller assigns to *B. denudata.* It is obvious that Müller relies, in his key, on the less salient features to separate these species and appears to have confused the two taxa (Jones, 1958a).

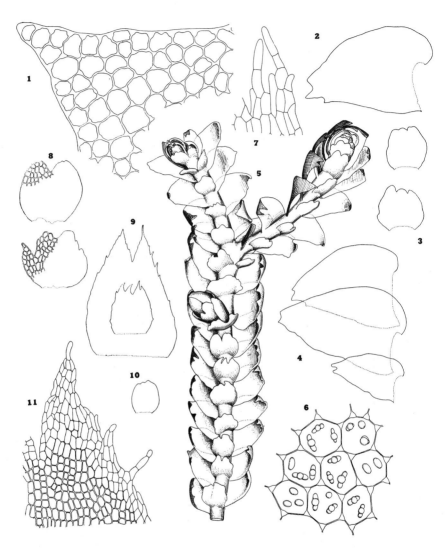

FIG. 96. *Bazzania tricrenata*. (1) Leaf apex (×300). (2) Large leaf (×43). (3) Two underleaves (×43). (4) Leaves, showing variation in size and form (×43). (5) Shoot, postical aspect, with gynoecium (×26). (6) Median cells with oil-bodies (×500). (7) Sector of perianth mouth (×140). (8) Two ♂ bracts (×70). (9) ♀ Bract, innermost series (×24), and (within it) from intermediate series (×18). (10) ♀ Bract, outermost series (×18). (11) Lobe of ♀ bract of innermost series (×70). [1–5, *RMS H2143*, Mt. Washington, N.H.; 6, *RMS 24409*, Wittenberg Mt., Catskill Mts., N.Y.; 7–11, from Fulford, 1936a.]

2-dentate or entire and acute . . . cells 16–25 μ . . . ." Macvicar's figures and the above excerpt from his diagnosis (*italics his*) strongly suggest our Appalachian plant; however, the statement that the leaves are sometimes caducous fails to correspond with our material. It may eventually be necessary to segregate the Appalachian phase as a distinct subspecies; the plants from New York and New England northward correspond much more closely to the typical forms of *B. tricrenata* and should cause no problems in identification. However, even they occasionally produce slender phases of the type discussed above, and such a slender mod. *viridis-parvifolia* is frequent at least as far north as the Catskill Mts., N.Y.

BAZZANIA NUDICAULIS Evans
[Figs. 85:1–3, 97, 100:9.]

Bazzania nudicaulis Evans, The Bryologist 26:62, 1923; Fulford, Amer. Midl. Nat. 17(2):406, figs. 7–8, 1936.

Usually in *loose, rather flocculent tufts or patches*, often on vertical rock or bark faces (then frequently somewhat pendulous), *yellowish brown to reddish brown or piceous, often nearly denuded of leaves and then appearing to consist largely of interwoven, wiry, rigid stems;* mature leafy shoots (750)*975–1550* μ *wide*. Stems *slender, filiform, wiry*, mostly 8–30 mm long × 140–200(225) μ in diam., ca. 7–9 cells in diam.; cortical cells rather short, ca. 2–3(4)× as long as wide, ca. (12)15–20 (24) μ in diam. × 30–45 μ long; medullary cells more elongate, ca. 5–8(10)× as long as wide, ca. 12–15 μ in diam. × 65–125 μ long; both cortical and medullary cells prominently thick-walled, with pits, but the end walls thin. Branches lateral, rather infrequent and distant, widely (80–90°) divergent; *postical flagella infrequent*, occasionally developing normal leaves and underleaves. Rhizoids scarce, on bases of leaves of flagella and occasionally on bases of normal underleaves. Leaves inserted by a nearly straight line, *distant* to (rarely) subimbricate, *slightly to moderately convex, with the apices deflexed*, especially when dry, usually rather widely (65–85°) spreading, *ovate to ovate-lanceolate and virtually symmetrical*, from 500–650 μ long × 300–450 μ wide to (occasional maximum) 760–800 μ long × 525 μ wide, often much smaller on branches, widest just above the somewhat narrowed base; dorsal and ventral margins both somewhat convex, the dorsal margin usually rounded at base, the leaf narrowed gradually from the basal one-fourth to an *often merely acute apex;* on robust shoots the apex narrowly truncate and distinctly 2–3-dentate. Cells of leaf middle from ca. 20 × 20 μ to 21–23 × 25–31 μ, cells at leaf base not appreciably larger; cuticle smooth to almost imperceptibly verruculose; cell walls thin or very slightly thickened, with small but distinct, concave-sided trigones; *oil-bodies (1)2–3(4) per cell, somewhat opaque and of low refractive index, faintly but discernibly formed of minute spherules*, subspherical to ovoid and ellipsoid, 4.5 × 5 to 5 × 6–10 μ up to 6 × 8 μ and 4 × 11 μ, much larger than chloroplasts (which are ca. 3.5–4.5 μ), disintegrating soon after drying. *Underleaves usually distant*, occasionally contiguous, *large, 2.2–3.0× as wide as stem, suborbicular to quadrate-orbicular*, from 400–450 μ wide and long up to a maximum of 580 μ wide and long, rounded but not auriculate at base, the lateral margins convex

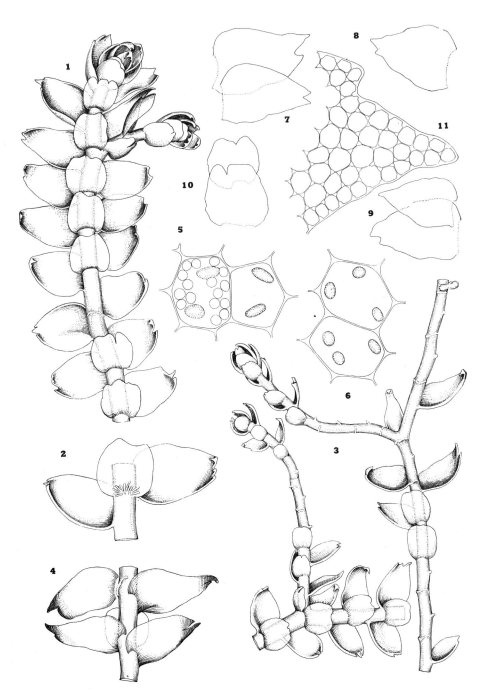

FIG. 97. *Bazzania nudicaulis*. (1) Vigorous shoot with persistent leaves, ventral aspect; growing admixed with *B. tricrenata* in Fig. 95: 1–3 (×26). (2) Same, short shoot sector (×30). (3) Parts of a single, typical plant with freely caducous leaves, postical aspect (×26). (4) Shoot sector, antical aspect, the uppermost leaf deciduous (×32). (5) Two cells with oil-bodies and (at left) chloroplasts (×710). (6) Cells with oil-bodies

but only weakly arched, *entire, the apex usually truncate and subentire to retuse to sinuous*, occasionally obscurely 2–4-lobed, with rounded lobes and shallow sinuses; rarely with isolated underleaves more distinctly 2–3-lobed. *Asexual reproduction ubiquitous*, by caducous leaves and underleaves.

Habitually sterile; sex organs unknown.

Type. High bluff of Roan Mt., N.C., at over 6000 feet (*A.L. Andrews 65*).

Distribution

Endemic to the Spruce-Fir Climax forests of the summits of the highest peaks of the Southern Appalachians, ranging in altitude from ca. 4500 to 6680 feet.

VIRGINIA. White Top Mt., 5675 ft (*J. K. Small 54, p.p.*!; *RMS 38123, 38130*); Mt. Rogers, ca. 5300–5715 ft (*RMS 38003, 38009, 38012, 38017*).   NORTH CAROLINA. Roan Mt., above 6000 ft (*Andrews 65*, type; *RMS 36954, 36949, 36936a, 36937*, etc.); Grandfather Mt., 5964 ft, Caldwell Co. (*RMS*; same locality without county indicated, *Schallert 19, p.p.*); Black Mts., ca. 6275 ft, Yancey Co. (*Lesquereux, 1850*!); Mt. Craig, Black Mts., ca. 6500 ft (*RMS 24450*); Mt. Mitchell, 6400–6640 ft (*RMS 23142a, 23687a*); Mt. Clingman, 6100 ft (*RMS*); Clingmans Dome, Swain Co. (*RMS 28117d*). TENNESSEE. Clingmans Dome, near N.C. state line, 6200 ft, Blount Co. (*RMS 34717*); Mt. LeConte, 4500 and 6590 ft (*Sharp*) and near Myrtle Point, 6200 ft, Sevier Co. (*RMS*); Roan Mt., ca. 6100 ft, Carter Co. (*RMS 36943, 36940a, 36940*, etc.).[51]

Ecology

Occurring only on the fog-shrouded summits of the highest peaks of the Appalachian system; rare below 5500 feet. The species occurs indiscriminately either on the trunks of mature trees of *Abies fraseri* or on shaded, damp to rather dry acidic rocks. In both sites it is frequently associated with various members of the *Herberta-Plagiochila tridenticulata* community, i.e., with *H. adunca tenuis, Plagiochila tridenticulata, Leptoscyphus (Anomylia) cuneifolius*, and *Frullania asagrayana*, more rarely (intertwining) with *Lejeunea ulicina* and *Cephaloziella pearsoni*. On rocks, particularly, the species often occurs admixed with the closely allied *B. tricrenata*.

Occasionally occurring in purple-black patches over relatively exposed, cold, only very intermittently moist cliff faces (as on Roan Mt.), there associated with *Herberta, Cetraria islandica*, etc. Also, occasionally occurring at later stages, successively, over damp peaty soil on rocks, then with *Cetraria islandica, Herberta, Anastrophyllum michauxii, Lophozia ventricosa, L.*

---

[51] *Bazzania nudicaulis* was reported from West Virginia by Frye & Clark (1937–47, p. 673), supposedly on the basis of a collection by Ammons. Ammons (1940), however, fails to include the species as a member of the West Virginia flora.

(×650). (7–9) Leaves (×34). (10) Underleaves (×34). (11) Leaf apex (×258). [1–4, *RMS 24450*, Mt. Craig, Black Mts., N.C.; 5–6, *RMS 34717*, Clingmans Dome, Tenn.; 7–11, *RMS 38074*, White Top Mt., Va.]

incisa, Cephalozia media, Blepharostoma trichophyllum and (only on Roan Mt.!) the disjunct *Anastrophyllum saxicolus*. Such humicolous occurrences are rather unusual for the species and appear restricted to exposed lofty mountain cliffs where there is a high incidence of fog and mist.

The initial, pioneer populations of this species, undergoing ecesis on the bark of Fraser fir, often consist of minute, atypical plants that are hardly recognizable. Such plants can readily be mistaken, in the field, for *Leptoscyphus (Anomylia) cuneifolius*. Since they often occur with this species (as well as with *Frullania asagrayana, Plagiochila tridenticulata, Herberta adunca tenuis,* and *Paraleucobryum longifolium;* see Schuster and Patterson, 1957), potential confusion is obvious.

Differentiation

The small size, usually dull reddish brown coloration, and deflexed leaves all closely suggest *B. tricrenata*, a species which frequently occurs in admixture. However, *B. nudicaulis* clearly differs in a series of characteristics from *B. tricrenata*, among them (*a*) the marked tendency to shed leaves and underleaves; (*b*) the large and subentire underleaves, rotundate in outline; (*c*) the constant absence of all trace of sex organs; (*d*) the high incidence of ovate-lanceolate, entire leaves; (*e*) the ability to develop adventitious shoots. By contrast, *B. tricrenata*, when found intermingled (as in *RMS 24450*, Mt. Craig, N.C.), shows persistent leaves; transverse, clearly dentate underleaves that are but little broader than the stem; free development of gynoecia; a high incidence of 3-dentate leaves; lack of any tendency toward formation of adventitious shoots.

The high incidence of collections in which the two species grew admixed suggests, at first glance, that *B. nudicaulis* is merely an "atypical" sterile phase of *B. tricrenata*. Such an interpretation appears quite unwarranted, since (*a*) no physical connection between "*tricrenata*" and "*nudicaulis*" shoots could be demonstrated—and no transition from one type of shoot to the other; (*b*) many collections from the Southern Appalachians have been seen in which pure mats of either species occurs, without any trace of the other present; (*c*) there is total absence, elsewhere in its North American range, of any tendency to produce caducous-leaved shoots in *B. tricrenata*. The fact that the two species may occur together without any suggestion of transitional shoots appears to demonstrate conclusively that two discrete taxa are at hand.

Like *B. denudata* and *trilobata, B. nudicaulis* differs greatly in the degree to which caducous leaves are produced. Occasional robust shoots with rather short internodes (Fig. 97:1) virtually or quite lack caducous leaves and underleaves; such plants may attain a width of fully 1.5 mm. These plants occur under exceptionally optimal conditions, forming deep, luxuriant, but loose and somewhat flocculent masses on shaded, damp rocks. Under conditions of more intermittent high humidity and moisture, as on exposed bark, the plants are usually smaller and develop a higher incidence of leaves which are acute or merely bidentate at their apices (Fig. 97:4, 7-9); such plants may be only 750–1200 μ wide and have distant leaves and underleaves, which are very freely deciduous. As a consequence of the free dropping of the appendages, such phenotypes are reduced with maturity to a wiry, interwoven mass of almost completely denuded stems, only the apices of the shoots retaining a few leaves and underleaves.

The deciduous leaves and underleaves give rise to new shoots, at times while still attached to the parent plant. Such regeneration also occurs in other species of *Bazzania*. For instance, Kreh (1909) experimentally induced it from the cortical cells of the stem in *B. trilobata* and *tricrenata*. Fulford (1936a) discusses the mode of formation of the regenerants in *B. nudicaulis*.

BAZZANIA TRILOBATA (L.) S. F. Gray
[Figs. 98–99, 100:1]

Jungermannia trilobata L., Spec. Pl., p. 1133, 1753.
Jungermannia tridenticulata Michx., Fl. Bor.-Amer. 2:278, 1803.
Bazzania trilobata S. F. Gray, Nat. Arr. Brit. Pl. 1:704, 1821 (as *Bazzanius*); Fulford, Amer. Midl. Nat. 17:392, figs. 2–3, 1936; Schuster, *ibid.* 49(2):537, pl. 64, figs. 7–9, 1953.
Pleurochisma trilobatum Dumort., Syll. Jungerm. Eur., p. 70, 1831.
Herpetium trilobatum Nees, Naturg. Eur. Leberm. 3:49, 1838.
Mastigobryum trilobatum Nees, *in* G., L., & N., Syn. Hep., p. 230, 1845.
Mastigobryum tridenticulatum Lindenb., *ibid.*, p. 230, 1845.
Bazzania tridenticulata Trev., Mem. R. Ist. Lomb., Ser. 3, Cl. Sci., 4:415, 1877.

Robust, typically forming *dense, deep tufts or polsters* to 6–20 cm deep, *grass-green to olive-green,* occasionally yellowish green. Shoots (1.8–2.5)*3–6 mm wide, erect or ascending* (when corticolous or saxicolous occasionally creeping), stout, the stems usually 400–500 μ in diam., sparingly pseudodichotomously furcate, the branches homodromous, replacing the ventral half of a leaf,[52] widely spreading; with numerous postical flagelliform branches which bear vestigial leaves but numerous rhizoids (these, rarely, reverting to leafy branches). Stem with cortical cells and medullary cells subequal in diam., ca. 20–27 μ; cortical cells thick-walled, elliptical to oblong-elliptical in surface view, *ca.* 60–100 μ long; medullary cells ca. 170 μ long, thick-walled except for pits and thin end walls. Rhizoids few or none on leafy stems, abundant on flagella. Leaves usually clearly imbricate, often rather closely shingled, the *stem not exposed in antical aspect,* very widely spreading (typically 90–100°), not decurrent, rather strongly convex basally but somewhat flatly deflexed distally (as a consequence the entire plant characteristically convex in the median third; the *distal halves of the leaves deflexed,* particularly in drying, when usually postically connivent); *leaves strongly asymmetric, somewhat falcate,* narrowly ovate, to 3 mm long × 1.5 mm wide, the antical margin strongly falcately arched and *the antical base*

[52] Frye and Clark (*loc. cit.*, p. 667) erroneously state "from ventral portion of leaf axil." The lateral branches are never intercalary and axillary in origin in any species of the genus.

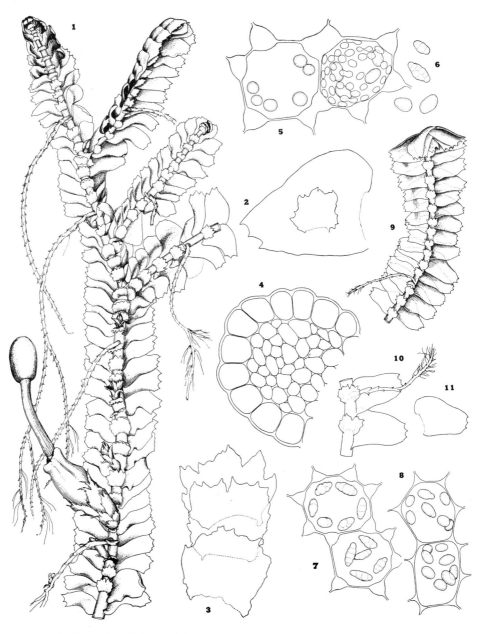

Fɪɢ. 98. *Bazzania trilobata.* (1) Sporophyte-bearing shoot, postical aspect (×6.5).
(2) Leaf and underleaf (×21). (3) Underleaves (×35). (4) Seta cross section, sector
with five epidermal cells omitted (×100). (5) Two cells, leaf middle, with oil-bodies,
and, at right, chloroplasts (×640). (6) Individual oil-bodies (×640). (7) Cells, leaf
middle, smaller-celled phase (×640). (8) Cells, leptodermous extreme (×605).
(9) Shoot in postical aspect, showing the postical curvature of shoot tip (×ca. 5).
(10) Shoot sector with ventral-intercalary flagellum (×ca. 6). (11) Leaf (×ca. 6).
[1–4, *RMS 42398*, Mountain Lake, Va.; 5–6, Big Creek, Ga., *RMS*, large-celled phase

strongly ampliate although not distinctly auriculate, *extending nearly or quite across the stem;* postical margin nearly straight distally, although the *base often somewhat dilated and subauriculate;* apex narrowed and subtruncate, *the truncation square and not oblique, almost constantly tridentate,* occasionally 2- or 4-dentate, the *teeth subrectangular and sharp,* 5–8 cells long, the sinuses obtuse to rectangulate, less often rounded, shallow. Cells rather variable in size, the marginal locally smaller than the intramarginal, at least near antical base (there often 18–20 μ), elsewhere on margins ca. 22–25 μ, tangentially measured; subapical cells ca. (24)25–30(32) × 28–35 μ; median cells from 20–24 × 25–32 μ to ca. (28)32–38 × 32–40(42) μ; basal cells ca. (30)32–36 × (42)45–60 μ; *trigones usually strong, hyaline, often moderately bulging;* cuticle smooth. *Oil-bodies 4–8 per median cell,* glistening, spherical to ellipsoidal, (4.5)5–7.5 μ to 4 × 8–10 μ to 5–6 × 9–12 μ, homogeneous or weakly to distinctly 2–5(8)-segmented. Underleaves remote to subcontiguous, somewhat spreading to (at least on margins) reflexed, somewhat arcuately inserted but hardly decurrent, broadly orbicular-quadrate to transversely oblong-quadrate, from 900–1000 μ wide × 750–800 μ long up to 1400 μ wide × 800–1000 μ long, irregularly 4–5-lobed at apex for 0.2–0.25 their length (in isolated instances one sinus descending 0.3–0.5 the underleaf length), and with accessory sharp to blunt teeth, the nearly straight lateral margins sinuous to dentate, occasionally appendiculate or coarsely 1–2-dentate; *lobes and teeth usually sharp,* the sinuses frequently acute and sharp; 1–2 rows of underleaf cells often decolorate. *Asexual reproduction usually absent,* rarely present and by means of tardily caducous leaves.

Not infrequently with sex organs but rarely with capsules. Androecial branches sessile, very small, budlike, whitish, arising from axils of underleaves (and, on flagella, from axils of both the reduced and undifferentiated leaves and underleaves), often developed in abundance; bracts ca. 1000 μ long × 800 μ wide, strongly saccate, sharply 2–4-dentate at apex, 1-androus; bracteoles ca. 400 μ long × 570 μ broad, bilobed or trilobed at apex. Gynoecia often produced in abundance, frequently a series along a stem (sometimes 2 from axil of a single underleaf); inner bracts hyaline, narrowly ovate to ovate-lanceolate, formed of elongate, thin-walled cells, ca. 2.2 mm long × 750 μ wide, the margins irregularly ciliate-laciniate, the apex divided into 2–4 slender, sometimes tortuous, toothed or ciliate laciniae; bracteole like bracts. Perianth firm, conspicuous, to 6 mm long, ovoid-fusiform, gradually narrowed to mouth; perianth mouth shallowly trilobate, the lobes short and broad, dentate to ciliolate. Seta in section with 16 large rows of epidermal cells and ca. 50 smaller internal cell rows. Sporangium anatomy, spores, and elaters described under generic discussion.

Type. European. "Habitat in Suecia, Anglia, Italia."

Distribution

Imperfectly holarctic, although suboceanic and becoming rare or absent in the interior portions of the continents; essentially of temperate

---

drawn to scale of fig. 7; 7–8, *RMS 40830a*, North Carolina; 9–11, Ithaca, N.Y.; from Schuster, 1949a.]

and warm-temperate range, becoming rare or disappearing well before the Taiga-Tundra ecotone is reached.

The species is widespread in Europe, although, according to Müller (1956, *in* 1951–58) "selten" in northern Europe; ranging from there to northern Italy, northern Spain, Hungary, the Balkans, to southern Finland, Latvia, and Estonia. It is apparently absent from the Mediterranean region, Russia proper, and Siberia. Southward it extends to the Azores and Madeira. It reappears in eastern Asia, ranging from the Amur region to Sakhalin, southward to northern Japan (Hattori, 1952, 1955a, p. 88).

In North America reported for Alaska (Verdure Creek, Evans, 1914b), on the Pacific Coast, and widely distributed in the eastern half of the continent; only selected collections and citations are listed:

GREENLAND? *Fide* Buch & Tuomikoski (1955).[53]   LABRADOR. *Fide* Fulford (1936a) and Buch & Tuomikoski (1955).   NEWFOUNDLAND. Avalon Distr., S. coast, W. Nfld., N. Pen. (to St. Barbe Bay), and C. and NE. Nfld. (Buch & Tuomikoski, 1955). MIQUELON ISL. PRINCE EDWARD I. Brackley Pt. (Macoun, 1902).   S. QUEBEC. Oka et comté d'Argenteuil; Red R., Argenteuil; Nicolet; Montreal; Iberville; Waterloo; La Tuque; Pont-Rouge; Beauceville; Montmorency R.; Tadoussac; Bic; S. of Rimouski; Percé; Jupiter R., Anticosti, etc. (Lepage, 1944–45).   NOVA SCOTIA. Cape Breton (*RMS*); Sandy Cove (Lowe, 1909); Halifax, Truro, Baddeck, Margaree (Brown, 1936a); North Sydney, Cape Breton I. (*Howe & Lang, 1902*).   NEW BRUNSWICK. Grand Manan; Fish Head and Southern Head; Tobique R. (Macoun, etc.).   ONTARIO. Ottawa; Belleville; Owen Sd.; L. Nipissing; L. Nipigon; Algonquin Park (Macoun, 1902; etc.)

MAINE. Mt. Katahdin (*RMS*); Orono!; Bridgton (*Young, 1934*); Mt. Desert I. (*Lorenz*); Matinicus I.   NEW HAMPSHIRE. Mt. Washington (*RMS*); Franconia Mts. (*Lorenz, 1908*).   VERMONT. Mt. Mansfield (*RMS*); Stratton, Windham Co.! MASSACHUSETTS. Roaring Brook, Shutesbury (*RMS*); Mt. Toby, Sunderland (*RMS*); Hawley Bog, W. Hawley, Franklin Co. (*RMS*); Worcester; Oxford, Holden; Mt. Greylock (Andrews, 1902).   RHODE I. *Fide* Evans (1913). CONNECTICUT. Litchfield, Hartford, Tolland, Windham, Fairfield, New Haven, Middlesex, New London Cos. (Evans & Nichols, 1908).

NEW YORK. Tompkins, Onondaga, Cattaraugus, Hamilton, Madison, Tioga, Cortland, Genesee, Ontario, Chenango Cos. (Schuster, 1949a); Wittenberg, Cornell, and Slide Mts., Catskill Mts. (*RMS*); Cold Spring Harbor, Long I.; Indian Reservation, Lawtons; L. George; Little Moose L., Herkimer Co. (*Haynes*); Onondaga Co. (Underwood & Cook, Hep. Amer. no. 12).   NEW JERSEY. Closter, etc. (Austin, Hep. Bor.-Amer. nos. 78, 79).   PENNSYLVANIA. Delaware Water Gap (*RMS*); Center Co.; Cambria, Clarion, Clinton, Crawford, Elk, Forest, Lawrence, McKean, Potter, Somerset, Warren, Westmoreland Cos., etc. (Lanfear, 1939).   DISTRICT OF COLUMBIA. Washington, Rock Creek Park (*Holzinger*).   MARYLAND: *Fide* Plitt (1908) near Baltimore.

VIRGINIA. Cascades, Salt Pond Mt., near Mountain L. (*RMS*); White Top Mt., Mt. Rogers (*RMS*); Dickey Cr., Smyth Co. (Evans, 1893a).   WEST VIRGINIA. Barbour,

---

[53] The citation from Greenland must represent an error; I have combed the Greenland literature and found no other reference to the species, nor have I located a specimen in the rich Greenland collections at Copenhagen.

Clay, Fayette, Grant, Hancock, Kanawha, Marion, Mercer, Mineral, Monongalia, Morgan, Nicholas, Pocahontas, Preston, Randolph, Tucker, Upshur Cos. (Ammons, 1940); Cranberry Glades, Pocahontas Co. (*RMS*). NORTH CAROLINA. Throughout mountain counties into Piedmont (Durham, Stokes, Orange, Wake Cos.), rarely on Coastal Plain (N. of Chowan R., in *Chamaecyparis* swamp, near Winton, *RMS*); see Blomquist (1936); over 90 personal colls. TENNESSEE. Roan Mt., Carter Co. (*RMS*); Mt. LeConte, Mt. Kephart, etc., Sevier Co. (*RMS*); Clingmans Dome, Swain Co. (*RMS*); Blount, Cocke, Grainger, Hamilton, Morgan Cos. (Sharp, 1939); Knox Co. GEORGIA. Big Cr., S. of Highlands, Rabun Co. (*RMS*); Talullah Falls, Rabun Co. (*RMS*); Brasstown Bald, White Co. (*RMS*); Dade Co. (Carroll, 1945). SOUTH CAROLINA. Whitewater R. gorge and Thompson R. gorge, above Jocassee, Oconee Co. (*RMS*); Estatoe R. gorge, Pickens Co. (*RMS*). ALABAMA. *Fide* Mohr (1901); occasional along Rte. 90 in coastal Alabama (*RMS*), general in the northern counties. MISSISSIPPI. Tishomingo State Park, Tishomingo Co. (*RMS*). FLORIDA. Johnson's Juniper Swamp, 8 mi S. of Bristol (Kurz & Little, 1933); rarely in W. Florida, in *Magnolia virginiana-Persea* swamps (*RMS*).

Westward to the following: KENTUCKY. Carter, Clay, Elliot, Lewis, Letcher, McCreary, Powell, Bell, Nicholas Cos. (Fulford, 1934, 1936). OHIO. Ashtabula, Athens, Cuyahoga, Fairfield, Franklin, Hocking, Lake, Vinton to Champaign Cos. (Miller, 1964). MICHIGAN. Amygdaloid and Captain Kidd Isls., Isle Royale (*RMS*); Copper Harbor, Keweenaw Co. (*RMS*); Ann Arbor (*Kauffman*); Tobin Harbor, Ryan Isls., etc., Isle Royale; Leelanau, Marquette, Chippewa, Cheboygan Cos. (*Steere*; *Darlington*, etc.). WISCONSIN. Sand I., Apostle Isls. (*RMS*); Ashland, Vilas, Oneida, Lincoln, Marathon, Bayfield, Douglas, Sawyer, Superior, Barron Cos. (Conklin, 1929); Pikeville, Anderson Co. (*Drexler; in* Frye & Clark, 1937–47). MINNESOTA. Anoka, Isanti, Carlton, Cook, Lake, St. Louis Cos. (Schuster, 1953). INDIANA. Crawford and Putnam Cos. (*Underwood, Parker*, etc.). ILLINOIS. Jackson Co. (Hatcher, 1952). IOWA. Clayton Co. (*Conard*). ARKANSAS. Madison Co. (Redfearn, 1966).

Ecology

Concurrently with its wide distribution, *B. trilobata* occurs under a very wide range of conditions and, associated with this, varies rather markedly in growth, form, and appearance. Like our other species, *B. trilobata* is not found over calcareous substrates, although often occurring on humus or decaying wet logs in calcareous regions.[54] The species typically occurs on shaded banks, either on humus or peaty soil (often over ledges or boulders); here it frequently forms large, extended, luxuriant, deep tufts often yards across—particularly when occurring in the shade of evergreens and conifers (spruce and fir, northward; hemlock, *Kalmia*, *Rhododendron maximum*, and *Leucothöe*, southward). Often in such sites it also carpets much-decayed logs. Northward the species may occur as a

---

[54] I have pointed out (Schuster, 1949a, 1953) that this species "is a decided calciphobe," not occurring on rocks in calcareous regions, while "it is rare under conditions where the soil has not been thoroughly leached of lime." When growing "on basic rocks, a dense mat of acid humus must be formed by more tolerant species before the *Bazzania* is able to invade." Hattori (1955a, p. 88) reports the species also from "serpentine and on humus among serpentine rocks, often associated with *Macrodiplophyllum plicatum*."

humicole, on banks, often with *Sphagnum*, under a pH of 3.8–4.8 (Schuster, 1953, 1957), associated with *Calypogeia sphagnicola, Lepidozia reptans, Mylia anomala, Lophozia incisa*, etc. Here the species may "grow under conditions of physiological drought . . . but occurs most luxuriantly and forms deep tufts only when there is an abundant moisture supply."

Southward, *B. trilobata* is often common in gorges and ravines, under similar but warmer and even more humid conditions; it may then be associated, on soil-covered banks, with *Odontoschisma prostratum, Microlepidozia sylvatica, Calypogeia integristipula* (and sometimes *C. peruviana*), and such mosses as *Plagiothecium, Hypnum*, and *Thuidium* (as in *RMS 40721*, Big Creek, Ga.).

Bazzania trilobata frequently occurs as a pioneer, on acid rocks and even on the bark of trees. It is then much less robust, grows prostrate, and does not form tufts. For example, it is found straggling amidst *Bazzania denudata* and *Bryoxiphium norvegicum*, on thinly soil-covered ledges and rock walls (along the Chattooga R., Macon Co., N.C.). Northward it occurs often on moist to dry ledges in spruce-fir forest, associated (under drier conditions) with lichens, *Ptilidium ciliare*, and *Anastrophyllum michauxii;* on less exposed ledges it is associated with *Ptilidium ciliare, Mylia taylori, Bazzania denudata, Diplophyllum taxifolium, Lophozia attenuata, L. ventricosa* var. *silvicola, Sphagnum*, and *Anastrophyllum michauxii.*

The species is almost constantly restricted to terrestrial situations northward (an exception being noted in Schuster, 1949a, for central New York; here it is found "as a depauperate modification, with somewhat caducous leaves, close to forma *depauperata* of Müller" on bark); northward it is corticolous only "in swamps and bogs where the humidity is sufficiently high" (Schuster, 1953). Southward the species often becomes arboreal, particularly on the trunks of old and large trees of *Tsuga canadensis* in the Escarpment gorges of the Southern Appalachians. Then associated with *Bazzania denudata, Nowellia curvifolia, Anastrophyllum michauxii*, rarely *Anastrophyllum hellerianum* (under which see), *Frullania asagrayana, Cephalozia media, Lejeunea ulicina, Harpalejeuna ovata*, etc.

The species is also rather sporadically distributed in the outer Coastal Plain, from North Carolina to northern Florida, southern Alabama, and Mississippi. It is then almost constantly restricted to bark, less often to logs. On bark it occurs, e.g., up to 10–12 feet high on old trees of *Chamaecyparis thyoides*, associated with *Frullania asagrayana* and *Jamesoniella autumnalis* (as on the Chowan R., N.C.; *RMS*).

Variation and Differentiation

Bazzania trilobata is readily recognizable in all of its multifarious variations. It is the largest local species of the genus and, well developed, is 2–4 × as broad as our other species. Under maximal conditions the plants may form, on damp humus, extensive, pure, deep green tufts that attain a height of 10–15, rarely 20, cm. By contrast, on the bark of trees, in the southeastern Coastal Plain and in the Appalachian gorges, extremely reduced, xerophytic forms, only 1-1.5 mm wide, occur; these may be confused with *B. denudata*, particularly since, in some cases, they develop caducous leaves. However, all but depauperate extremes of *B. trilobata* possess imbricate leaves with dilated antical bases, which are so shingled that the stem is completely hidden from view, antically.

As Fulford (1936a) admits, *B. trilobata* is occasionally caducous-leaved, in spite of the fact that she "keys" the species as lacking such leaves. Caducous-leaved phases occur chiefly at the crests of ledges and high cliffs, where they form rather deep tufts, often associated with *Cladonia* and other lichens. Under such conditions a rather large, albeit xeromorphic, phase is developed which quite regularly drops the older leaves. This phase is not an artifact; it can be readily observed in the field. For some strange reason, such caducous-leaved phases appear confined to the Appalachian Mts., from the Catskill Mts. of New York southward to Virginia, North Carolina, and Tennessee, where they may be very common at high elevations.[55]

Fulford (1936a) and, following her, Frye and Clark (1937-47, p. 668) describe the sexual branches as "on special branches arising in axils of the underleaves." Although this is often—and in many plants uniformly the case— it is by no means exclusively so. For example, I have seen numerous fertile plants from the upper reaches of Little Stony Creek, Mountain Lake, Va. (*RMS 42398*), in which the gynoecial branches originated, in part, in the "normal" position in the axils of the underleaves. On the same or neighboring plants, some or all of the gynoecia arose *from the flagella*. Such gynoecia produced normal bracts, perianths, and matured sporophytes (although these seemed to be slightly smaller than typical). The position of the gynoecia on the stolons did not appear to be restricted to any definite pattern, although this could not be ascertained.

By contrast, the androecial branches were very often present in abundance on both the leading, leafy axes and the stolons or flagella. In several cases, indeed, the stolons were abundantly androecial, with 4-5 ivory-white, budlike androecia produced from each of several stolons, yet the leafy axis was quite devoid of sexual branches. The opposite conditions could also be noted: leafy shoots freely producing androecia, although the many, branched stolons were uniformly devoid of them. As a consequence, little significance accrues to the position of these organs. It is, however, of interest that androecial branches may originate on a single stolon from all three rows of merophytes! In other words, restriction of the sexual branches on the stolons to one of the three merophyte rows does not apply; hence, strictly speaking, sex organs are not confined to the postical segments of the axis, but (when dorsiventrality does not

---

[55] Further study may show that the caducous-leaved phases of *B. trilobata*—which may even be dominant at high elevations, as on Mt. Rogers, Va.—should be segregated taxonomically. They are conspicuous when seen in the field.

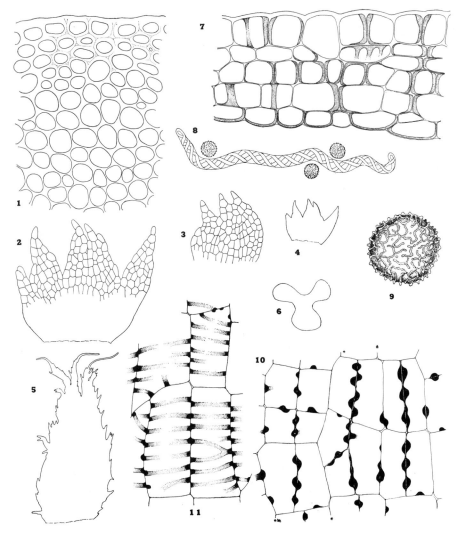

FIG. 99. *Bazzania trilobata.* (1) Dorsal half of median portion of stem cross section (×185). (2) ♂ Bract (×ca. 65). (3) ♂ Bracteole (×ca. 65). (4) ♂ Bract (×ca. 24). (5) ♀ Bract, innermost series (×20). (6) Perianth cross section (×ca. 15). (7) Capsule wall cross section (×ca. 600). (8) Elater and spores (×275). (9) Spore (×1430). (10) Epidermal cells, capsule wall, at asterisks the primary longitudinal walls which are free, or nearly free, of secondary thickenings (×ca. 475). (11) Innermost cells of capsule wall (×ca. 475). [1, 7–11, *RMS 42398*, Mountain Lake, Va.; 2–6, from Fulford, 1936a.]

appear, as on the flagella), the sexual branches originate from all segment rows. Both androecial and gynoecial branches, whatever their origin, are very small and compact and nearly or quite decolorate. The immature sporophytes, however, even if produced deep down in a luxuriant mat, are strongly chlorophyllose.

The development of sporophytes in *B. trilobata* occurs very sporadically, yet sex organs are found easily enough. The reason appears to lie in the tendency for formation of large clones of a single sex. Sporophytes are produced from June into July; I have seen mature and immature ones as late as July 12 (*RMS 42398*), suggesting that they may mature until almost August. In the same area, sporophytes were seen as early as June 20.

Cell size in *B. trilobata* is strongly variable with vigor. For example, in the small plants, 1–1.5 mm wide, referred by Müller to "var. *depauperata*," the cells are only 20–25 μ; in medium-sized plants, 2–2.5 mm wide, from North Carolina (*RMS 40830a*), median cells are only 20–25 × 25–30 μ; in robust plants from Big Creek, Ga., which are 5–5.5 mm wide, median cells are 32–38 × 32–42 μ. There are also associated differences in leaf shape. On the smaller plants the leaves are more rectangulate, with the antical and postical bases hardly dilated; the leaves are then nearly straight and hardly falcate, spread at an angle of 75–85°. In the robust phases the leaves are strongly cordate and ampliate basally and ovate-falcate, with the distal half of the leaf spreading at an angle of 110–130°, occasionally even more. Such differences in leaf form and cell size appear to be entirely environmentally induced and hence taxonomically meaningless.

BAZZANIA DENUDATA (*Torrey*) Trev.

[Figs. 85:4, 100:2–8, 101:7–9, 102:1–3]

Herpetium deflexum var. *implexum* Nees, Naturg. Eur. Leberm. 3 : 59, 1838.
Mastigobryum denudatum Torrey, *in* G., L., & N., Syn. Hep., p. 216, 1845; Lindenberg & Gottsche, Spec. Hep. *Mastigobryum*, p. 7, pl. 1, figs. 1–4, 1851.
Jungermannia denudata Torrey, *in* G., L., & N., Syn. Hep., p. 216, 1845 (in synonymy).
Mastigobryum ambiguum Lindenb., *ibid.*, 217, 1845 (*p.p.*).
Bazzania denudata Trev., Mem. R. Ist. Lomb., Ser. 3, Cl. Sci., 4 : 414, 1877; Evans, Rhodora 25 : 89, 1923; Fulford, Amer. Midl. Nat. 17 : 412, figs. 9–10, 1936; K. Müller, Rabenh. Krypt. –Fl., ed. 3, 6 : 1157, fig. 444, 1956.
Pleuroschisma implexum Meyl., Hép. Suisse, p. 240, 1924.
Pleuroschisma tricrenatum var. *implexa* K. Müll., Rabenh. Krypt. –Fl. 6(2): 270, fig. 80, 1913 (exclusive of synonymy; *p.p.*?).
Mastigobryum ovifolium Steph., Spec. Hep. 3 : 464, 1908 (*fide* Hattori, 1957, p. 82).[56]
Mastigobryum vastifolium Steph., *ibid.* 3 : 503, 1908 (*fide* Hattori, 1957).
Bazzania ovifolia Hatt., Jour. Jap. Bot. 19 : 347, 1943.
Bazzania vastifolia Hatt., *ibid.* 19 : 348, 1943.
Bazzania ovifolia var. *vastifolia* Hatt., Contrib. Fl. Hep. Shikoku, Kochi, p. 24, 1953.
Bazzania denudata subsp. *ovifolia* Hatt., Jour. Hattori Bot. Lab. no. 18 : 82, 1957.

In *prostrate mats*, the arching stems at most procumbent, never suberect, smaller and more delicate than those of *B. trilobata*, *light or pure green to yellowish green*, in xeric sites occasionally dull to somewhat olive-green, *never with brownish*

---

[56] I am not fully convinced that the Japanese plants (*B. ovifolia*) are identical with *B. denudata*.

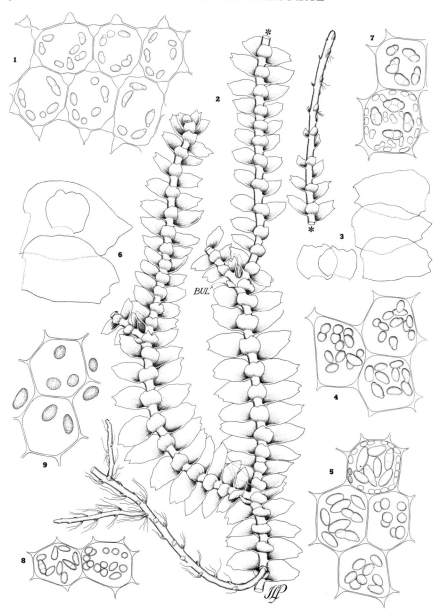

Fig. 100. *Bazzania*. (1) *Bazzania trilobata;* (2–8) *B. denudata;* (9) *B. nudicaulis*.
B. trilobata: (1) Median cells (×455). *B. denudata:* (2) Plant of the freely stoloniferous,
creeping, small form with rare caducous leaves, with terminal flagellum (at asterisk)
and three *Frullania*-type branches; at BUL′ a primary branch underleaf (×16).
(3) Leaves and underleaves (×30). (4) Median cells of phase with 9–14(16) oil-bodies
per cell (×575). (5) Median cells with oil-bodies and, upper cell, chloroplasts, of large
phase in Fig. 101:7–9 (×490). (6) Leaves and an underleaf (×33). (7) Median cells
of phase with larger, fewer, only (3)4–6(7–9) oil-bodies per cell (×555). (8) Median
cells with oil-bodies (×385). *B. nudicaulis:* (9) Median cells with oil-bodies (×600).

pigmentation. Shoots *(1.0)1.5–2(2.2) mm wide* × 1–2.5 cm long, sparingly branched, the branches lateral and (on weak shoots) *often monopodial,* but usually more or less pseudodichotomous and widely furcate; occasionally with isolated branches arising postically, and with the postical flagella (which vary from rare to frequent) transformed to leafy shoots; rarely (fo. *flagellifera*) with the leading or lateral axes attenuate and becoming flagelliferous. Stems slender and rather wiry, (110)125–190(200) *μ* in diam., 7–10 cells in diam.; hyaloderm layer poorly developed, the cortical and medullary cells similar in diam., ca. 20–25 *μ*; *epidermal cells 2–3× as long as broad* (ca. 20–25 *μ* wide × 36–58 *μ* long); interior cells ca. 170 *μ* long × 18–23 (25) *μ* wide, the thick longitudinal walls locally pitted, the end walls thin. Rhizoids of leafy shoots few, at under-leaf bases, common on flagella. Leaves distant to weakly imbricate, wide-spreading (80–90° usually), *horizontally spreading and nearly plane,* at most weakly convex, *usually freely caducous,* variable in shape, usually oblong-ovate to ovate-lingulate to short-ovate, from 320–375 *μ* wide × 550–620 *μ* long up to 400–600 *μ* wide × 600–900 *μ* long, rarely longer, *slightly to moderately asymmetric, not or hardly falcate;* antical margin moderately arched from a *weakly dilated and hardly rounded base* (the leaves antically crossing only 0.3–0.5 of the stem), postical margin straight or sometimes vaguely dilated near the base; *leaf apices very variable,* usually ranging from merely acute or obtusely pointed to bidentate to (sporadically) sharply tridentate, but *never with the majority of leaves tridentate,* the apices often merely broad and truncate and vaguely lobulate-crenate to rounded, *never deflexed;* leaf lobes or *teeth usually blunt,* occasionally acute, the antical sometimes slightly larger; *sinuses shallow, rounded to obtuse usually.* Cells rather variable in size: the marginal 20–23 *μ* (tangentially measured); the subapical 23–27 × 27–30 *μ*; the median 25–28(31) × 27–34(36) *μ*; the basal increasing to 28–34 × 35–50 *μ*, not forming a vitta-like region; upper cells largely subquadrate and nearly isodiametric, with distinct to conspicuous but *not bulging trigones;* cuticle smooth or nearly so. Oil-bodies variable, averaging *(3)5–6(7–10) per cell* in some leaves, then usually larger, 5 × 7.5–10 *μ* to 6 × 10 *μ*, and *becoming 2–4(5)-segmented* (at least with age), or else more numerous, *to 9–14(16) per median cell,* then homogeneous or faintly 2–3-segmented, from 3 × 5–6 *μ* to 4 × 5–8 *μ*. *Underleaves remote to barely contiguous,* somewhat spreading to almost squarrose, broadly orbicular-quadrate to trans-versely rounded-rectangulate, *averaging wider than long,* ca. (240)280–450(500) *μ* wide × (180)210–400(450) *μ* long, not auriculate at base, the lateral flanks evenly rounded to nearly straight, occasionally sinuate to obscurely angulate-dentate, the truncate apex variable, from subentire to emarginate to shallowly 3–4-lobulate, the sinuses vestigial to descending at most 0.2–0.25 the underleaf length, *the obscure lobes rounded or blunt.* Asexual reproduction usually present (lacking only in plants from very moist sites), by means of *caducous leaves and underleaves.*

Almost uniformly sterile. Androecia unknown. Gynoecia rather rare; bracts (of undeveloped gynoecia) to 400–550 *μ* long, ca. 450 *μ* broad, the apex

[1, *RMS 45153d,* Thompson R., N.C.; 2–4, *RMS 40830,* Chattooga R., N.C.; 5–6, *RMS 43012,* from plants mixed with *Metacalypogeia schusterana,* Cape Breton, N.S.; see Fig. 101; 7, *RMS 40830;* 8, *RMS 24412,* Cornell Mt., N.Y.; 9, *RMS 36954,* Roan Mt., N.C.]

rounded and entire to dentate or short-ciliate, the margins crenulate-dentate, bearing slime papillae; bracteole similar. Perianth unknown.

Type. "Near New York" (*Torrey*).[57]

Distribution

Bazzania denudata s. lat. is nearly circumpolar in range, although to a large extent restricted to the older portions of the continents, thus generally lacking in the interior. While not oceanic in range, it must be considered at least suboceanic. The species is a polymorphous one; it is possible that the European plant ("*Pleuroschisma implexa*" of Meylan), which has been lately considered a synonym of *B. denudata* (Müller, 1951–58), should be segregated. In eastern Asia it is reported from Japan [subsp. *ovifolia* (Steph.) Hatt.; see Hattori, 1957, p. 82], where common in the north, "less frequent towards the south."

The European plant [which may be an artificial ensemble of plants belonging in part to *B. tricrenata*; at least the British plant is referred by Jones (1958a), to this species] found in central Europe: Switzerland, Austria; Bavaria, the Allgäu and Black Forest, Westphalia, in Germany; also in Czechoslovakia (Bohemia, the Tatra Mts.), the French Pyrenees, and England.

In North America found in the West from Alaska to British Columbia, Washington, and Alberta. Eastward found as follows:

GREENLAND. *Fide* Macoun (1902); this report surely erroneous.   QUEBEC. Montagne du Collège de Ste-Anne (Lepage, 1944–45).

MIQUELON I. (Lepage, 1944–1945).   ONTARIO. *Fide* Macoun (1902).   NEW-FOUNDLAND. Channel and Placentia Bay (Evans, 1923e); Gaultois, Pushthrough, Hare Bay, and Ramea, all on S. coast (Buch & Tuomikoski, 1955), Bay Bull's Arm (*Waghorne 1890*)!   NOVA SCOTIA. Louisburg, Port Clyde, Purcell's Cove, Halifax Harbor; near Indian Brook, mountains N. of Barrasois R., Barrasois R. valley, and Cape Dauphin, all on Cape Breton (Evans, 1923e); Halifax (Fulford, 1936a), Cap Rouge, Cape Breton with *Metacalypogeia schusterana* (*RMS 43011*); Barrasois Valley, Cape Breton (*Nichols 1322!*); Indian Brook, Cape Breton (*Nichols, Aug. 1909!*).

MAINE. Mt. Katahdin (*RMS*); Greenville, Round Mountain L., Franklin Co.; Jordan Mt., Mt. Desert I. (Evans, 1923e).   NEW HAMPSHIRE. Base of Mt. Washington; Flume, Gorham, Franconia, Thompson's Falls, White Mts.; Shelburne; Mt. Willard; Mt. Prospect, Plymouth; Waterville; Ice Gulch, Randolph; Mt. Madison (Evans, 1923e); Franconia Mt. (*T. P. James 1851!*); Gorham (*James!*); White Mts. (*James 1853!*).   VERMONT. Boulton; Lake Dunmore; Salisbury; Willoughby; Granville Notch; Downer's Glen, Manchester; Brandon (Evans, 1923e); Mt. Mansfield, near Stowe (*RMS*); Haystack Mt., Windham Co.!   MASSACHUSETTS. Plainfield (part of the type material of *M. denudatum*; from the Hooker Herbarium, now in the

---

[57] Evans (1923e, p. 90), however, points out that "part of the original material of *M. denudatum* from the Hooker Herbarium, now in the Mitten Herbarium [NYBG]," is from Plainfield, Mass., while another specimen from Middlefield, Mass. (*Emmons 120*) [NYBG], is labeled "*J. denudata* sp. nov." Either specimen could be regarded as a lectotype.

Mitten Herbarium); Middlefield (Emmons, 1822; NYBG, labeled "*J. denudata*, sp. nov."); Mt. Greylock (*Andrews;* cited *in* Andrews, 1904, as *B. triangularis*); Everett Brook, Sheffield; Alandar, Berkshire Co. (Evans, 1923e); Chesterfield Gorge, near W. Chesterfield, Hampshire Co. (*RMS 41306*). CONNECTICUT. Salisbury, Beacon Falls,[58] Naugatuck, Redding (cited as *B. tricrenata in* Evans & Nichols, 1908; see Evans, 1923e).

NEW YORK. Ca. 1 mi N. of Olean, Rock City, Cattaraugus Co., and Martiny Rocks, W. branch of Four Mile Rd., Allegany, Cattaraugus Co. (Schuster, 1949a); Catskill Mts. (*Peck!*; cited by Peck, 1866, p. 70, as *Mastigobryum deflexum*); Rocky Falls, North Elba, Essex Co. (Peck, 1899, as *B. deflexa*); Little Falls; Clareyville; Chapel Pond Brook, Adirondack Mts.; Undercliff, Essex Co.; Little Moose L., Herkimer Co. (cited by Haynes, 1906a, as *B. triangularis*, and distributed as *B. tricrenata*; corrected label in Amer. Hep. no. 39). East R. Falls above L. Golden; L. Mohonk (all cited in Evans, 1923e); Cornell, Wittenberg, and Slide Mts., Catskill Mts. (*RMS*). PENNSYLVANIA. Meadville! VIRGINIA. Near summit of White Top Mt. and Mt. Rogers (Schuster & Patterson, 1957); Cascades, near Mountain L., Salt Pond Mt., ca. 3000 ft (*RMS*). WEST VIRGINIA. Tibbs Run, Monongalia Co. (Evans, 1923e); cited as *B. deflexa* by Evans, 1892a); road to Cheat View; Spruce, Pocahontas Co. (Evans, 1923e). NORTH CAROLINA. Grandfather Mt. (cited by Andrews, 1914, as *B. tricrenata*); Chattooga R., about 1 mi below Narrows, Macon Co. (*RMS 40830*; fo. *flagellifera*, with *Bryoxiphium norvegicum*); Roan Mt., Mitchell Co. (*RMS*); Mt. Mitchell and Mt. Craig, Black Mts., Yancey Co. (*RMS 24827,*); Andrews Bald near Clingmans Dome, Smoky Mt. Ntl. Park, Swain Co. (*RMS*); Dry Falls and below Cullasaja Falls, Cullasaja Gorge, NW. of Highlands, Macon Co. (*RMS*); Mt. Mingus (Fulford, 1936a); above Looking Glass Falls, N. of Brevard (*RMS 61251a*); Grandfather Mt., Caldwell Co. (*RMS 30180, 30185*); Linville Gorge, Burke Co. (*RMS 28887a*); Craggy Gardens, Craggy Mts., Blue Ridge Parkway (*RMS 24765b, p.p.*); Avery, Durham, Haywood, Polk, Swain, Watauga Cos. (Blomquist, 1936). SOUTH CAROLINA. Chattooga R., S. of Ellicott's Rock, Oconee Co. (*RMS*). GEORGIA. Big Cr., 0.2–0.5 mi below High Falls, N. edge Rabun Co. (*RMS*). TENNESSEE. Roan Mt., Carter Co. (*RMS*); Mt. LeConte (Fulford, 1936a); Mt. Collins (Fulford, 1936a); Carter, Cocke, Sevier Cos. (Sharp, 1939).

Westward extending as follows: ONTARIO. *Fide* Macoun (1902). OHIO. Kunkle's Hollow, Hocking Co. (*M.S. Taylor*); Athens, Fairfield, Hocking Cos. (Miller, 1964). KENTUCKY. Powell and McCreary Cos. (Fulford, 1936a).

Bazzania denudata is the most widespread species of the genus in North America, except for *B. trilobata*. Although it is always montane, it descends in the Southern Appalachians to as low as 1800–2000 feet, following the Escarpment gorges down. It is, therefore, not nearly as limited to high altitudes (and latitudes) as *B. tricrenata*.

Ecology

Bazzania denudata occurs in two general types of habitats—on the bases of trees, rarely to a height of 8–12 feet or more, particularly in humid and sheltered areas, such as deep gorges, and on sheltered, humid, or moist rocks.

The species occurs rather generally on trees in the Southern Appalachians, particularly often on the bark of *Tsuga canadensis* (where it may

---

[58] Evans (1912, p. 20, fig. 27) illustrated a plant from this collection as *B. tricrenata* but later corrected this determination to *B. denudata* (1923e, p. 90); in spite of this, his figure is redrawn and given under *B. tricrenata* in Frye & Clark (*loc. cit.*, p. 670, fig. 3).

be closely associated with *Bazzania trilobata, Anastrophyllum michauxii, A. hellerianum, Radula tenax, Lejeunea ulicina, Harpalejeunea ovata*). At higher elevations, as in the Black Mts., often at tree bases; then sometimes associated with *B. tricrenata*.

The species is more widespread in the Southern Appalachians on the vertical faces of rocks and in damp, shaded, and sheltered recesses of rocks. Here it may occur with *Bazzania trilobata* and *Bryoxiphium norvegicum* (Chattooga R., N.C.), with *Bazzania tricrenata* and *nudicaulis* (Black Mts., N.C.), and with *Herberta adunca tenuis, Plagiochila austini, P. sullivantii*, etc. (Dry Falls, N.C.).

Northward, on "mossy cliffs," "boulders," and humus-covered rocks, associated with *B. trilobata, Dicranum fuscescens, D. scoparium, Dolichotheca turfacea, Lepidozia reptans, Lophozia ventricosa* var. *silvicola, Diplophyllum albicans, Isopterygium elegans, Geocalyx graveolans, Mnium hornum, Mylia taylori*, and *Tetraphis geniculata* (Buch and Tuomikoski, 1955).

I have also collected this species from dry erratic boulders, and damp to dry ledges and cliffs, and on dry ledges and rock walls, in the Catskill Mts., N.Y., associated with some of the above species (*L. v.* var. *silvicola, Lepidozia reptans, Mylia taylori, Bazzania trilobata*) but also with *B. tricrenata, Lophozia attenuata, Anastrophyllum michauxii* as well as *A. minutum, Blepharostoma trichophyllum, Cephalozia media*, and *Jamesoniella autumnalis*. In western New York over damp quartz-conglomerate boulders, on the vertical faces, associated with many of the preceding species but also with *Ptilidium pulcherrimum, Tritomaria exsecta*, and *T. exsectiformis* (Schuster, 1949a).

As is characteristic of the other species of the genus, *B. denudata* is absent or virtually absent from calcareous regions.

Variation and Differentiation

The taxonomic history of *B. denudata* has been checkered and confused. For many decades *B. denudata* was considered identical with *B. tricrenata* ("*Mastigobryum deflexum*"), as, e.g., by Austin (1873; Hep. Bor.-Amer. Exsic. no. 80) and Müller (1905–16) until Evans (1923e) carefully studied the species and portrayed its history in detail.

Bazzania denudata differs from *B. tricrenata* in its inability to develop secondary pigmentation; in the normal presence of caducous leaves; in the more numerous oil-bodies per cell; in the hardly falcate, nearly plane leaves, whose apices are not deflexed; in the more nearly prostrate growth; and in the widely divergent branches. In general the two species are separable at a glance, even in the field. Even the slender phases of *B. tricrenata*, from the Southern Appalachians, are readily separable by their convex leaves with deflexed apices, and by the darker coloration and tendency for the surface to be subnitid.[59]

This polymorphous species is usually readily recognized, even in the field, because of the smaller size (rarely exceeding 2 mm wide) as compared with *B. trilobata*, linked with a generally light to pure green color that suggests *B. trilobata*, and nearly flat, often somewhat lingulate leaves. In addition there is usually, although hardly invariably, development of caducous leaves, which are of exceptional occurrence in *B. trilobata*. Under very mesic, humid conditions, *B. denudata* is often devoid of caducous leaves; it may then also attain its maximal size, of about 2–2.2 mm, and may develop the optimal dentition, with a high incidence of tridentate leaves. Such plants may be difficult to separate from the smaller phases of *B. trilobata* from which they differ in the laxer leaves, usually contiguous to slightly imbricate, with the antical bases hardly ampliate; as a consequence, the stem is usually at least locally exposed in antical aspect.

Under xeric conditions, *B. denudata* is typically very much denuded, with almost all older leaves caducous, and large sections of the wiry, gray- to olive-green stems exposed. Such plants, in extreme cases, may be mistaken by the beginner for *B. nudicaulis*, from which they differ in the ± toothed, wider than long underleaves, the nearly plane lateral leaves, the larger leaf cells, and the more numerous oil-bodies. I have, e.g., collected a phase of *B. denudata* from a dry, erratic boulder in open deciduous woods (Slide Mt., N.Y.), in which the plants are virtually devoid of leaves, with branches brownish or blackish, older leaves (where present) olive-green and largely acute. By contrast, in the same area, the species may occur on damp, shaded rock walls, where it totally lacks caducous leaves, even with age.

Bazzania denudata is tremendously variable with differing environmental conditions and hence is not at all times easily recognized. On damp, shaded rocks it may occasionally show development of flagella from the apices of ordinary, leafy, lateral shoots; our other species apparently lack this ability to form terminal flagella. The leaves vary considerably in shape, ranging from obliquely tridentate on optimally developed phases to variably bidentate or sinuous to merely acute on weakly developed manifestations. On weak phases the leaves may be remote to contiguous, narrowly ovate to virtually lingulate (ca. 320–375 μ wide × 550–620 μ long). Such plants correspond well with the

---

[59] European writers appear to have confused the boundaries of these two species, Pearson's "*Bazzania triangularis*" appearing to represent a series of plants which sometimes possess, sometimes lack, caducous leaves. These plants are variously cited as synonymous with *B. denudata* (Müller, 1951–58; *fide* Jones, 1958a), a separate species, *B. implexa* (Meylan, 1924), a variety of *Bazzania tricrenata*, var. *implexa* (Müller, 1905–16), a form of *B. tricrenata* (Jones, 1958a). This leaves the range of *B. denudata* in Europe quite uncertain, if, indeed, it occurs there at all. Fulford (1936a) follows Evans (1923e) in leaving the placement of these European collections an open question.

European plants assigned to *B. denudata* in Müller (1951–58, p. 1151), according to whom the leaves of *B. denudata* are tongue-shaped and twice as long as broad, with the apex merely bidentate. The plant portrayed and described by Müller, however, fails to correspond closely to typical American plants. For example, he emphasizes, as a species criterion, the smaller cells ("in der Blattmitte 20–25 μ"; "Randzellen 16 μ") supposedly separating the species from *B. tricrenata*. By contrast, Evans (1923e) gives the median cells as 25 × 27 μ, and Frye and Clark as 31 × 36 μ. There is thus no consistent difference in cell size between this species and *B. tricrenata*. Müller also states that the oil-bodies in *B. denudata* occur "3–4 je Zelle . . . aus einzelnen Kügelchen zusammengesetzt." In American plants, however, they are much more numerous (usually a minimum of 4–6 or even 9–16 per cell), only isolated, small cells having as few as 3–4 oil-bodies.[60] Furthermore, the oil-bodies are either homogeneous or coarsely 2–4-segmented. In oil-body form and variation, *B. denudata* closely parallels *B. trilobata*, although the former usually possesses more numerous oil-bodies.

The identification of *B. denudata*, particularly its separation from *B. tricrenata* and *B. nudicaulis*, is usually easily accomplished, when living plants are at hand, on the basis of the usual 6–12 oil-bodies per median cell; in *B. tricrenata* there are 2–5(6) oil-bodies, and in *B. nudicaulis* only (1)2–4 oil-bodies, per cell. The oil-body number in *B. denudata* is thus the highest among our eastern North American species of *Bazzania*, since *B. trilobata* possesses only (3)4–6(7–8) oil-bodies per cell.

Branching in *B. denudata* is distinctive; the lateral branches are often mono-podial rather than furcate—i.e., the branch issues at a wide angle from an axis that retains its dominance. In the often loosely and remotely monopodial branching the species differs at once from *B. trilobata*, in which branching is clearly furcate and pseudodichotomous.

Family CALYPOGEIACEAE *H. W. Arn.*

Calypogeiaceae H. W. Arn. *in* O. R. Holmberg, Skand. Fl., 1928.

Medium-sized (usually 0.8–3.5 mm wide), *sparsely branched; vegetative branches usually postical-intercalary* (often terminal and of the *Frullania* type in *Metacalypogeia*), *of indeterminate length*, the sexual branches always *short and of determinate length*, generally bearing only reduced leaves. Stems soft-textured, *without a sharply delimited cortex; vegetative stems uniform,*

---

[60] The few oil-bodies in the European plant suggest that a taxon nearer *B. tricrenata* is at hand, which is also the conclusion advanced by Jones (1958a). Such a conclusion is difficult to reconcile to the claim made by Müller (p. 1159) that plants of *B. tricrenata* and "*B. denudata*" from the Feldberg retained their distinctive features under culture. *If*, then, central European "*B. tricrenata* var. *implexa*" or "*B. denudata*" is not identical with American *B. denudata*, and if Müller is correct that it remains distinct from *B. tricrenata* in culture, it will be necessary to follow Meylan (1924), an otherwise very conservative worker, who separated the plant as *B. implexa*. This plant apparently is not identical with the British plant alluded to by Jones (1958a).

normally lacking differentiation into leafy shoots and postical leafless flagella or stolons; stems essentially simple in morphology, undifferentiated; the cortical cells often slightly shorter than the medullary, but little or no more thick-walled, and of approximately the same diam. (thus with virtually no dorsiventral or cortical-medullary differentiation). Leaves inserted by a *long, incubously oriented, nearly straight,* scarcely arched *line* that is so strongly oblique as to appear nearly parallel with the stem; leaf essentially ovate to ellipsoidal, rarely oblong, averaging between 0.7 and 1.5(2.0) × as long as wide, the *apex usually entire or very briefly bidentate* (never deeply lobed; margins always edentate); leaf orientation strongly oblique, the virtually flat leaves incubously shingled. Underleaves *large to very large,* varying from *bilobed* (with lateral, outer margins often with a sharp to obtuse lateral tooth or lobe) *to suborbicular or ovate and entire* or retuse-emarginate; underleaf bases medially 2–3 cell layers thick, *more or less cushioned at juncture with stem,* formed of small, nearly isodiametric cells, the inferior layer of which actually or potentially forms rhizoid-initials; rhizoids absent elsewhere on stem (except sometimes in *Metacalypogeia*). Leaf and underleaf cells not or slightly, rarely moderately, collenchymatous, usually *large* (not less than 25–35 μ in leaf middle) *and thin-walled* (rarely with discrete or even bulging trigones), *normally unable to develop any secondary pigmentation;* chloroplasts rather small (ca. 4 μ); oil-bodies mostly 3–10 per cell, distinctly formed of protruding, coarse spherules that are colorless or blue, less often finely granular-segmented and opaque.

Monoecious or dioecious: *usually monoecious* (auto- and paroecious). ♂ *Branches short, postical, compact,* ceasing their growth with maturation of the androecium; androecium spicate, formed of several to many pairs of very small concave and erect to suberect 2–3-lobed leaves, each of which usually bears a single (sometimes 2–3) antheridia. Antheridia obovate to ellipsoidal, the stalk 1–2-seriate. Archegonial branches short, merely bearing reduced vestigial leaves or bracts (somewhat elongate in *Metacalypogeia schusterana,* then with larger leaves); stem tissue basad to and beneath the several archegonia growing downward to form a *fleshy, thick, rhizoidous perigynium,* near whose base the archegonia eventually come to be situated; a distinct calyptra retained. Sporophyte with seta becoming exceedingly elongate, of *numerous rows,* epidermal cells *little larger than internal,* in surface view rectangular, their ends irregular in position (never with all adjoining cells ending at identical points, or with alternating cells of each tier ending at identical points). Capsule rarely ovoid (*Metacalypogeia*), usually cylindrical and elongate, splitting to base into 4 *linear to lingulate, parallel-sided valves*. Capsule wall *bistratose*, the outer layer with 8–16 cell rows; longitudinal radial walls of adjoining cells alternately nearly evenly or irregularly thick-walled (in the latter case with nodular thickenings) and alternately quite thin-walled (except *Metacalypogeia*); inner cell layer with semiannular, U-shaped bands of thickenings, the inner tangential walls with complete bands. Spores 9–16 μ; elaters free, diam. 7–12 μ, 2(3)-spiral.

Asexual reproduction normally obvious, by means of 1–2-celled ellipsoidal, thin-walled greenish gemmae produced in fascicles from the margins and abaxial leaf surface of the reduced-leaved ascending extensions of normally leaved shoots.

Genera. Calypogeia (cosmopolitan); *Metacalypogeia* (four species; Japan, Korea to Hawaii; Alaska and Greenland to Nova Scotia; presumably more widespread).

The Calypogeiaceae are a nearly cosmopolitan family, represented in North America by 10 or 11 species of *Calypogeia* and by a single, evidently exceptionally rare species of *Metacalypogeia* (Schuster, 1968a). Although the group includes only a limited number of regional species, these are exceptionally technical because of a combination of several factors: relative uniformity in organization and hence limited intrinsic morphological speciation, coupled with a labile and often drastic response to environmental differences; the very ephemeral sporophytes, often produced rather infrequently, so that sporophyte criteria are rarely available; the evident existence of 9- and 18-chromosome races (Tatuno, 1941; see p. 125); and the apparently extensive range of some species, such as *Calypogeia integristipula* and *C. sphagnicola* (Schuster, 1968a), with associated geographically related polymorphism. Since any clear comprehension of the taxa depends on an understanding of the responses of the various species to varying environmental conditions—phenocopy production being frequent—studies based predominantly on herbarium specimens are nearly useless. Indeed, all treatments before that of Buch (1936a) are best neglected, and some published subsequently are of limited value, in part because of neglect of criteria derived from cytology and from sporophyte anatomy. These last criteria *must* be taken into account in any thorough study of the group. The problems posed by—and the limitations of—herbarium taxonomy in the Calypogeiaceae have been discussed (Schuster, 1968a); in that treatment the malleability of various species (i.a., *Calypogeia integristipula, C. neesiana, C. muelleriana, C. fissa, C. peruviana*) is presented in greater detail than is possible here.

The difficulties of the group are such that the two chief students of the genus *Calypogeia* in Europe, Buch (1936a, 1942) and Müller (1947a, 1951–58), were never able to agree on the identity and/or disposition of several taxa: e.g., *Calypogeia neesiana* var. *laxa*, *C. fissa*, *C. trichomanis* vs. *muelleriana*, *C. integristipula*. The more recent review of the European species by Bischler (1957) adds little new; the treatment of the North American taxa by Frye and Clark (1937–47) is perfunctory and obsolete (they report only 6 of the 10–11 North American species). The only recent detailed monograph of the North American taxa is that of Schuster (1968a), on which the following abridged treatment is closely patterned.

The Calypogeiaceae are typically terrestrial plants (a very few species, in the wet tropics, become corticolous) which are almost always mesophytic or hygrophytic; true xerophytes are very rare, although several taxa may produce xeromorphic, small-leaved modifications (mod. *parvifolia-densifolia*). Most species occur on acid or neutral substrates, although two (*Metacalypogeia schusterana*, *Calypogeia integristipula*) at least tolerate somewhat calcareous substrates; other taxa (e.g., *Calypogeia sphagnicola*, *C. neesiana*, *C. suecica*) are strictly limited to acid sites—peat, decaying logs, humus. Many taxa (i.a., *Calypogeia sullivantii*, *C. fissa*, *C. muelleriana*) are commonly confined to raw inorganic substrates, such as damp rocks or loamy, clayey, or silty soils. The species appear to release spores in spring, although *Metacalypogeia schusterana* may be autumn-"fruiting." Most species (all of ours exc. *C. suecica*) are bisexual, and sporophytes are sometimes rather frequent. However, sporophytes are exceedingly ephemeral, and since there is no externally evident perianth (the marsupium being hidden *in situ*), the general impression is one of predominant sterility.

Affinity and Circumscription

A close relationship of the Calypogeiaceae to other families hardly exists, the group being one of a minority of sharply definable families in the Jungermanniales. It can best be regarded as a very specialized offshoot from a primitive Lepidozioid type. However, the relatively unspecialized seta of the Calypogeiaceae (Fig. 107:5; as contrasted to the more specialized one of the Lepidoziaceae: see p. 13) probably prohibits our deriving the group directly from any Lepidoziaceae. Although perhaps retaining a more primitive seta anatomy than the Lepidoziaceae, the Calypogeiaceae are more specialized in almost all their other criteria: (*a*) the capsule has become cylindrical, with spirally twisted valves (exc. *Metacalypogeia*); (*b*) the capsule wall has become uniformly 2-stratose; (*c*) the perianth and the large, isophyllous bracts are lost, being replaced by a few small scales and a pendent, rhizoidous marsupium, respectively; (*d*) lobing of the leaves is greatly reduced—at most two small teeth of the leaf apices are retained, but in many cases even these are lost; (*e*) lobing of the underleaves undergoes reduction or obsolescence—at most two distinct lobes are retained, with sometimes a lateral tooth of each lobe; (*f*) gemmae are frequently present on erect attenuated branches; (*g*) branching modes are restricted. Most recent treatments implicitly state that only postical-intercalary branching occurs in the Calypogeiaceae (in contrast to the fantastic diversity of branching modes retained in all but the most advanced Lepidoziaceae); see, e.g., Müller (1951–58) and Bischler (1963). However, as pointed out recently (Schuster, 1968a), in the genus *Metacalypogeia* furcate, *Bazzania*-like terminal branching is found, and in at least three species of *Calypogeia* (the regional *C. muelleriana* and *C. sphagnicola*; the antipodal *C. tasmanica* Rodway, which may be identical with *C. sphagnicola*; see Schuster, 1963, 1968a) sporadic to frequent terminal, *Frullania*-type branching occurs.

The frequent occurrence of pseudodichotomous, *Bazzania*-like terminal branching in *Metacalypogeia schusterana*, together with the bidentate leaves and 2–4-lobed underleaves of that species, and—especially—the rather leafy gynoecial branches are strongly suggestive. This species lies closer to the point of origin of the Calypogeiaceae than any other, and its *Bazzania*-like facies is perhaps not accidental. However, even this species lacks the microphyllous, geotropic flagella or stolons that are such a dominant feature of almost all Lepidoziaceae. (Such stolons have been seen, as an exceptional feature, in *Calypogeia neesiana*; see Fig. 130:1. In this I have also once seen an *Acromastigum*-type branch [*RMS 68–315*]).

Key to Genera of Calypogeiaceae

1. Yellow to deep green, *chlorophyllose, opaque*; cells with conspicuous trigones, with *granular*, opaque oil-bodies; leaves entire to bidentate, often pointed or bidentate, *Bazzania*-like. Marsupium appearing stipitate (the branch with at least 1–2 pairs of leaves), obovoid to obpyriform, short and stout, with few rhizoids, developed to a point above that of the sexual branch proper (the stalklike ventral branch that bears it appearing inserted *on the side* of the marsupium). Capsule ovoid, only about 1.5–3× as high as in diam. Rhizoids in part scattered. With lateral vegetative branches present, the shoots pseudodichotomously furcate. No asexual reproduction. . . . . *Metacalypogeia* (p. 103)
1. Pale to grayish to bluish green, usually *translucent and delicate, little chlorophyllose*; cells mostly ± leptodermous, with coarse-segmented, *botryoidal* oil-bodies; leaves (in ours) usually ovate or dimidiate-ovate to ovate-triangular, not *Bazzania*-like. Marsupia appearing sessile, from exceedingly abbreviated branches that lack leaves or underleaves, cylindrical and terete, densely rhizoidous, subterranean, and developed only at and below the level of the short axis from which it originates ("stalk" inserted at side of summit

of marsupium). Capsule cylindrical, with twisted valves. Rhizoids strictly confined to underleaf bases. Branches all (normally) postical-intercalary. Asexual reproduction by means of gemmae. . . . *Calypogeia* (p. 112)

METACALYPOGEIA (Hatt.) *Inoue*

Calypogeia subg. *Metacalypogeia* Hatt., Jour. Hattori Bot. Lab. no. 18:83, 1957.
Metacalypogeia Inoue, *ibid.* no. 21:231, 1959.[1]

Plants relatively firm, yellow-green to dark green, at times brownish-green, closely decumbent or prostrate. Stem 0.5–3 cm long × 150–250 μ in diam., (1.0)1.4–2.5 mm wide with leaves, nearly simple or sparingly branched, the branches partly postical-intercalary (in at least subg. *Eocalypogeia* the branches in part lateral-terminal); without ventral flagella. Leaves contiguous to moderately imbricate, incubous, ovate-triangular to ovate-falcate, entire-margined but *the apex acute to bidentate*, widely spreading, *often with deflexed tips*. Underleaves remote, transversely to arcuately inserted, entire and unlobed to bilobed at apex, as wide as or wider than long. *Rhizoids abundant* at underleaf bases (and in *Eocalypogeia* also scattered over ventral stem surface). *Cells opaque* because of the *large and granular-botryoidal oil-bodies* and numerous chloroplasts, rather large (median 25–30 × 28–40[50] μ); the cell walls thin or slightly thickened, *trigones conspicuous and often somewhat bulging; the walls often brownish*; cuticle smooth or verruculose; oil-bodies from 2 to 30 in each cell, ovoid to ellipsoidal or rarely spherical, 3–6 μ to 4–6 × 4.5–8 μ up to 6–6.5 × 16–20 μ, *finely granular-segmented or botryoidal-granular, gray-brown and opaque*. Asexual reproduction lacking.

Autoecious or dioecious. Androecia small, from axils of underleaves; bracts in 2–7 pairs; monandrous. Marsupium obovoid and chloro-phyllose, with few or no rhizoids (*Eocalypogeia*) and then on a distinct, more or less leafy, postical sexual branch, or (subg. *Metacalypogeia*) cylindrical, rhizoidous, becoming yellowish brown, and essentially sessile (as in *Calypogeia*). Mature sporophyte known only in subg. *Metacalypogeia* (see below); capsule *ellipsoidal, with nearly straight valves*.

Type. Metacalypogeia cordifolia (Steph.) Inoue (= *Calypogeia cordifolia* Steph., Spec. Hep. 3:393, 1908).

Metacalypogeia includes two strongly divergent elements, one constituted by the Japanese-Korean-Hawaiian taxa (subg. *Metacalypogeia*) and the other by the North American species, *M. schusterana* (subg. *Eocalypogeia*); the latter may possibly deserve generic rank.

The genus is more easily confused with *Bazzania* than with *Calypogeia*. In fact, the Japanese *M. montana* was originally described by Horikawa

[1] Although this issue of the journal is dated October, 1959, my copy was received in February, 1960; the date of issue is uncertain.

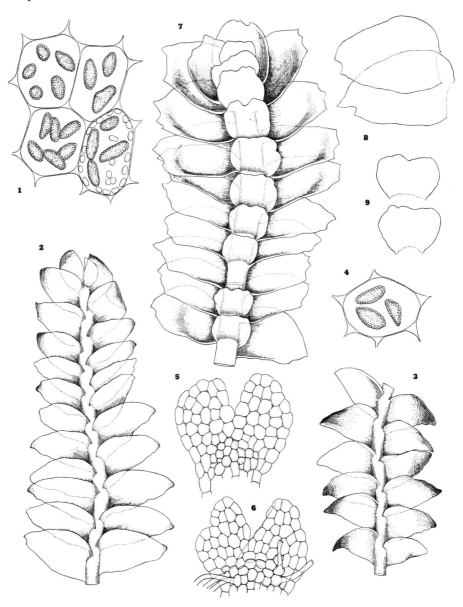

Fig. 101. *Metacalypogeia schusterana* (1–6) and *Bazzania denudata* (7–9). (1) Median cells with oil-bodies and, lower right, chloroplasts (×460). (2) Shoot, dorsal aspect (×26). (3) Same, from mod. *angustifolia* (×22). (4) Cell with oil-bodies (×690). (5–6) Medium-sized underleaves (×122). (7) Shoot-sector, ventral view (×26). (8–9) Two leaves and two underleaves (×37). [All from near Presque Isle, Cape Breton, N.S.; 1–6, *RMS 43011a;* 7–9, *RMS 43012.*]

as a *Bazzania*. The genus, of course, differs from *Bazzania* in the lack of a perianth, in the different stem anatomy, in the lack of ventral flagella, in the different oil-bodies, and in the seta anatomy.

Inoue (1959c) elevated *Metacalypogeia* to generic rank on the basis of the less cylindrical, more oblong capsule of *Metacalypogeia* vs. *Calypogeia,* with the valves of the capsule "not twisted," [2] and the peculiar ornamentation of the epidermal cells of the capsule wall.

The inner cells of the capsule bear tangential bands extended usually completely across the inner tangential walls, much as in other Calypogeiaceae. However, the form and ornamentation of the epidermal cells are, to my knowledge, unique. First of all, the cells are *transversely* rather than longitudinally elongated, averaging more than twice as wide as long. Second, both faces of the *alternating* longitudinal (shorter) radial walls bear one, rarely two, strongly nodular thickenings, and the adjoining portions of *all* transverse (longer) walls bear similar thickenings. Both faces of the alternating longitudinal walls, and the adjoining parts of the transverse radial walls, lack thickenings. Thus a nearly unique capsule-wall ornamentation results. As a consequence of the transverse orientation of the long axis of the cells, the majority of the nodular thickenings are on these transverse walls. The epidermal cells typically lie in eight rows; since the adjoining walls of alternate cell walls lack thickenings (as do the walls forming the valve margins), a 1:1 pattern of alternation between ornamented and unadorned walls results (see discussion under *Calypogeia*). Also, on superficial examination, each valve, as a consequence, appears to bear four lines formed by the thickenings on and adjacent to the longitudinal walls.

Key to Subgenera of Metacalypogeia

a. Underleaves and (often) leaves 2-lobed or 2-dentate; rhizoids not confined to underleaf bases; cuticle ± smooth; capsule only 1.5–2 × as long as in diam.; oil-bodies only 2–6 per cell; perigynium stipitate, at apex of a distinct ventral branch that bears 2-several pairs of more or less reduced leaves; autoecious. (Plants with a *Bazzania*-like facies; with frequent terminal, pseudodichotomous branches as in *Bazzania*.) . . . .
. . . . subg. *Eocalypogeia*
a. Leaves and underleaves normally entire, the leaves rarely minutely bidentate; rhizoids only at underleaf bases (as in *Calypogeia*); cuticle ± verruculose; capsule 2.5–3 × as long as in diam., the valves weakly twisted; oil-bodies 8–20 per cell; perigynium cylindrical, densely rhizoidous, appearing sessile laterally on leading stem (the ventral branch bearing it extremely short, without leaves, as in *Calypogeia*); dioecious. (Plants firm but without a *Bazzania*-like facies.) . . . . subg. *Metacalypogeia*

The subg. *Metacalypogeia* does not occur regionally; it includes three species, of which all occur in Japan, one each in Korea and China (Inoue, 1959c,

---

[2] Inoue's figs. 1g and 1m, however, show slightly twisted capsule valves, suggesting that the differences between the generic types of *Metacalypogeia* and *Calypogeia* proper are of degree only.

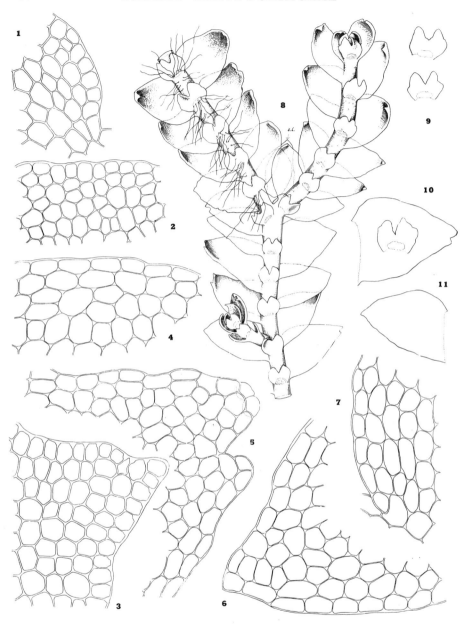

Fig. 102. *Bazzania denudata* (1–3) and *Metacalypogeia schusterana* (4–11). (1) Cells of antical leaf base (×ca. 170). (2) Cells along middle of dorsal leaf margin (×ca. 170). (3) Cells of apex of bilobed leaf (×ca. 170). (4) Cells along middle of dorsal leaf margin (×ca. 170). (5) Cells of apex of leaf in Fig. 10 (×ca. 170). (6) Cells of apex of edentate leaf (×ca. 170). (7) Cells of base of antical margin (×ca. 170). (8) Shoot, postical aspect (only apex of main stem with rhizoids drawn in; on right-hand branch, only rhizoid-initials scattered on internodes drawn in; at s.l. the supporting stem half leaf associated with the *Frullania*-type branch at right; below a ventral-intercalary

1963a; Hattori, Hong, and Inoue, 1962), and one, *M. montana* (Horik.) Inoue, has been reported from Hawaii (Miller, 1963). Inoue (1963a) has shown that this last is identical to *M. remotifolia* (Herz.) Inoue.

Subgenus *EOCALYPOGEIA* Schust., subg. n.

Subgenus subg. Metacalypogeiae simile, differens, autem, ut rhizoidea partim sparsa; folia angusta 0–2-dentata; marsupia brevia crassaque, lateraliter affixa; rami gynoeciales ± elongati, aliquot paria bractearum foliiformium plerumque distincta praebentes.

Yellow-green to deep green, densely chlorophyllose and opaque, creeping and closely prostrate, with slender but somewhat fleshy stems. Shoots subsimple to infrequently branched, the *vegetative branches often lateral and of the Frullania type* (occasionally with intercalary-postical, vegetative branches from leading stems of fertile plants); *sexual branches all postical, abbreviated; ventral flagella lacking. Rhizoids numerous*, often pale brownish, long, in dense masses from underleaf bases and also *some from scattered stem cells elsewhere* on the ventral merophytes. Leaves *incubous*, slightly to moderately imbricate, ovate-triangular to lanceolate-triangular, *acute and entire at apex or bidentate*. Underleaves moderately large, bilobed, nearly transversely inserted. Cells moderately large but firm, thin-walled, with pronounced trigones; cuticle ± roughened; *oil-bodies fine-segmented (botryoidal), grayish or brownish and opaque, large* (from $4.5–6 \times 4.5–8 \mu$ up to $6–6.5 \times 16–20 \mu$), 2–5(6) per cell. Asexual reproduction evidently lacking.

Autoecious or ?paroecious. Androecia as in *Calypogeia:* small, compactly spicate postical branches, not proliferating apically; bracts usually in 2–4 pairs, erect and ± concave, bi- or trilobed, monandrous. *Gynoecia on relatively elongate postical branches (the perigynia appearing stipitate)*, the gynoecial branch with 2–3 pairs of small, reduced leaves and (distally) *often with 1–2 pairs of incubous bracts that are rather large* (in extreme cases as large or larger than vegetative leaves), at other times with leaves reduced in size and scalelike; gynoecial branch postically *with distinct underleaves*, similar to those of the stem, although at times somewhat smaller. *Perigynium massive and fleshy*, bright green, appearing attached laterally rather than at or near its summit (i.e., with stem tissues forming the perigynium elaborated to a level *above* the point of insertion of the stalklike gynoecial axis), the *perigynium broadened above, somewhat conoidally narrowed below, not cylindrical, sparingly or hardly rhizoidous;* apex of perigynium with several small scalelike bracts and, usually, with at least one pair of large, winglike or leaflike, laterally patent leaves (which hide the perigynium from surface view). *Capsule ovoid (not more than twice as long as wide).*

Type. Metacalypogeia schusterana Hatt. & Mizut.

I regard as the primary characters, separating *Metacalypogeia* subg. *Eocalypogeia* from both *Metacalypogeia* subg. *Metacalypogeia* and *Calypogeia*, the following: (*a*) commonly with pseudodichotomous lateral branching, of the *Frullania* type—the plants thus at times with a *Bazzania*-like facies; (*b*) oil-bodies granular-botryoidal; (*c*) rhizoids, at least in part, scattered

---

branch (×22). (9) Medium-sized underleaves (×34). (10) Leaf and associated large underleaf (×34). 11. Smaller, unlobed leaf whose apex is shown in fig. 6 (×34). [1–3, *RMS 43012;* 4–12, type, *RMS 43011a*, Cape Breton.]

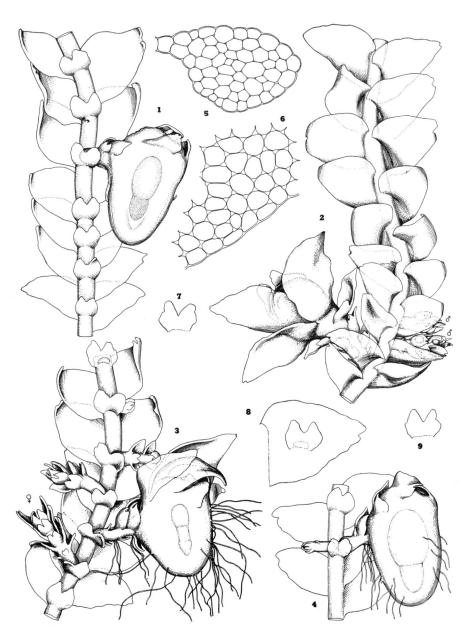

Fig. 103. *Metacalypogeia schusterana*. (1) Shoot, ventral aspect, with marsupium-bearing branch (×23). (2) Autoecious plant, dorsal view, with marsupium-bearing branch at left, with maximal development of bracts, and two androecia at right (×23). (3) Sector of plant, postical view, with marsupium-bearing branch and young gynoecial branches (×23). (4) Shoot sector, with two ventral-intercalary branches from one underleaf axil, one marsupium-bearing (×23). (5) Stem cross section (×ca. 170). (6) Cells along leaf margin (×ca. 170). (7–9) Three underleaves and a leaf (×36). [All from type, *RMS 43011a*, Cape Breton.]

over the stem and not confined to underleaf bases; (*d*) gynoecia on relatively elongated postical branches, bearing at least two pairs of reduced lateral leaves and a pair of underleaves, the perigynia appearing longly stipitate, and thus not lying under the parent axis and hidden; (*e*) gynoecial branch often with leaflike bracts near the apex; (*f*) perigynium rather short, not cylindrical, only slightly rhizoidous, obovoid in profile and widest near the summit; (*g*) capsule ovoid and only about twice as long as wide.

These differences suggest that *Eocalypogeia* should possibly be regarded as a discrete genus. Elevation to generic status must await discovery of the mature sporophyte. If *Eocalypogeia* possesses a capsule-wall anatomy differing from the unique capsule wall of *Metacalypogeia* subg. *Metacalypogeia*, it will have to be considered as generically distinct. The relatively elongated and relatively primitive form of the gynoecial branches is, in itself, almost enough to warrant the segregation of *Eocalypogeia* into an independent genus.

On the basis of the deep color and leaf form, our single species of *Eocalypogeia* is most likely to be mistaken for a *Bazzania*, when found sterile; this is particularly true when the terminal-lateral branches are clearly evident.

METACALYPOGEIA SCHUSTERANA Hatt. & Mizut.

[Figs. 101:1-6, 102:4-11, 103]

Metacalypogeia schusterana Hatt. & Mizut., Misc. Bryol. et Lichen. 4(8):121, fig. 1, 1967.

Opaque, yellow-green, weakly glistening when dry, creeping as individual plants or in thin patches, rather small; shoots to 5–12 mm long × (1.0)1.4–1.6 mm wide when mature. Stems ca. (120)150–175(200) μ in diam., yellow-green; *rhizoids relatively abundant*, not only arising from underleaf bases but also *scattered along the ventral merophytes*, some arising almost axillary in the underleaves; branching rare and diffuse, *lateral and pseudodichotomous;* with sporadic ventral branches (almost all sexual), but *without flagella.* Leaves somewhat asymmetrically, *narrowly ovate-lanceolate to ovate-triangular*, sometimes weakly falcate, *their apices rather strongly deflexed* in most cases, strongly narrowed to the tips, which are *bluntly acute, acute, or relatively infrequently asymmetrically bidentate*, varying from 450 μ wide × 625 μ long up to 600–620 μ wide × 740–760 μ long; *leaves slightly to moderately imbricate*, but in antical aspect the stem not hidden, ± *extensively exposed*, because of the *slight dilation* of the antical leaf base and relatively narrow leaf shape, with the antical margin only moderately arched. *Underleaves distant*, quite small (width 1.0–1.5 × the stem width usually), varying from 250–265 μ wide × 205–210 μ long up to 270–285(310–330) μ wide × 200–230 μ long, almost *constantly bifid*, with usually an acute, V-shaped sinus descending 0.4–0.6 the underleaf length, the lobes bluntly triangular or rounded at the apex, often but not always with a lateral rounded or blunt angulation or tooth, the *margin otherwise sinuous* to irregularly wavy, but without additional teeth.

Cells large, thin-walled, mostly *with salient*, weakly (rarely sharply) bulging trigones (in mod. *parvifolia* occasionally $\pm$ entirely leptodermous), the apical varying from 22 to 28 μ, the median from 23–25 $\times$ 25–30 μ (mod. *parvifolia*) to 25–30(32) $\times$ (28)35–40(50) μ; cuticle faintly to distinctly verruculose-striolate, at least in basal portions of leaves; *oil-bodies brownish and opaque, distinctly botryoidal*, in normal leaves 2–5(6) *per cell* (mod. *parvifolia*, to 6–12 per cell), ovoid to ellipsoid, 5–6 $\times$ 8 to 5 $\times$ 11–15 μ to 6–6.5 $\times$ 16–18(20) μ, *very conspicuous*. Marginal cells below and near the leaf tip tending to be *strongly elongated tangentially*, although isolated cells nearly isodiametric, *forming locally a weakly defined border*, measuring (middle of postical margin) ca. 14–16 μ wide $\times$ 26–45 μ long, locally averaging up to 2–3 $\times$ as long as wide (locally $\pm$ isodiametric and 18–24 $\times$ 25–30 μ); cells of antical base and in margin also $\pm$ strongly elongated, ca. 18–22 $\times$ 25–45 μ. Asexual reproduction evidently *absent*, the *leaves persistent*.

Autoecious (sometimes paroecious?). Androecia small, sessile, spikelike, postical-intercalary branches only 300–310 μ wide $\times$ ca. 500–550 μ long, with 2–3(4) pairs of small, erect, basally concave bracts, inserted nearly transversely, at the apex 2–3-dentate, the lowest tooth sometimes lobelike, the dorsal two often small and approximated; *monandrous*. Gynoecia on distinct, if short, postical-intercalary branches, which bear 2–3 pairs of reduced, often nearly scalelike leaves (some of which so closely resemble ♂ bracts that it is probable some inflorescences are paroecious); perigynium obovoid to obpyriform, green, very fleshy, ca. 700–750 μ in diam. $\times$ 990–1250 μ high, bearing usually few or almost no rhizoids, except at juncture with branch that gives rise to it, *attached laterally*, the broadened crown bearing 1–2 pairs of bracts that may be small and irregular in shape, scalelike or often leaflike (then to 990–1050 μ long $\times$ 550 μ wide), ovate-lanceolate and subfalcate, arched in a hoodlike fashion over the perigynium (perigynium with bracts to 1650–1700 μ wide); under-leaves few, the apical one with age typically splitting into two widely separated scales. Capsule (immature, included) 400 μ in diam. $\times$ 550–600 μ high. Mature sporophyte unknown.

Type. Small ravine emptying into Gulf of St. Lawrence, near the northern end of Presque Isle, south of Cap Rouge, Cape Breton I., Nova Scotia (*RMS 43011a*). Type in herbarium of author and of S. Hattori.

Distribution

Known only from the type, Newfoundland, west Greenland and Alaska.

WEST GREENLAND. Marrait, Kangigdlit, below Anana Mamagissa, Nugssuaq Pen., 70°30′ N., 54°10′ W. (*RMS & KD 726*).   NOVA SCOTIA. Over thin, damp soil on weakly basic, slaty rocks, small ravine emptying into Gulf of St. Lawrence at the N. end of Presque Isle, S. of Cap Rouge, Cape Breton I. (*RMS 43011a, 68–961*).   NEW-FOUNDLAND. N. of Daniels Harbour, N. Pen., crevassed limestone (*RMS 68–1455*). ALASKA. Driftwood Camp, near headwaters of Utukok R., N. slope of De Long Mts., Brooks Range, ca. 68°53′N., 161°10′W. (*Steere 16401*).

Ecology

The type material was collected from thin soil over basic rocks and in part from bare rock itself, on the steep ravine sides above the zone of submergence. Associated with the lime-tolerant *Lophozia gillmani* and *Conocephalum conicum* and with *Fissidens*.

The Greenland plant came from a calcareous, wet fen traversed by small rills, associated with *Eriophorum*, *Lophozia gillmani*, *L. rutheana*, *Odontoschisma macounii*, *Scapania gymnostomophila*, and *Blepharostoma trichophyllum brevirete*. The last two taxa occur admixed with the Newfoundland plants, which occurred in deep crevices of coastal limestone ledges.

Differentiation

Sterile shoots of this rare and interesting species superficially have a facies intermediate between those of the two widespread species *Bazzania tricrenata* and *B. denudata*. They possess a sharp tendency for the leaf apices to be deflexed as in *B. tricrenata* and *B. triangularis*, thus differentiating them from *B. denudata*. The species differs further from *B. denudata* in that the leaves are either sharply asymmetrically bidentate at the tip or are merely acute—with tridentate leaves absent or very rare and sporadic in occurrence. Furthermore, the underleaves generally have the lateral margins entire and little bulging, while the apices are usually 2-lobed.

The species, although obviously affiliated with *Calypogeia*, is hardly likely to be mistaken for that genus. The deeper color and more opaque appearance and the *Bazzania*-like leaves, as well as the large, botryoidal oil-bodies, are distinctive; also diagnostic in separating *Metacalypogeia schusterana* from *Calypogeia* is the branching pattern. Lateral, *Frullania*-type branches such as are common in the former occur only as a rare exception in *Calypogeia*. The stipitate perigynia, and their obovoid to obpyriform shape, are also very different from those of *Calypogeia*. In association with the fact that the perigynia are not hidden by the main stem and its leaves, the perigynia are densely chlorophyllose. Since the capsule is considerably less elongated, the perigynia are much less deep and not nearly as subterranean; as a consequence they bear few rhizoids.

The Alaskan plant, studied *after* the discussion of this genus was set in type, deviates slightly from the diagnosis. It has, more often, merely narrowly acute leaf apices, and when bidentate leaves occur, these are often more deeply bidentate. Underleaves are very similar to those of the type. Even more impressive is the similarly frequent terminal branching and the frequent development (with age) of scattered rhizoids. Branches are usually sparing— one terminal branch per plant is most generally observed, and then growth of the branch may somewhat displace the main axis to one side (as is shown in Fig. 102:8), but often the main axis does not lose its dominance, and branching remains strictly monopodial. In a very few cases I have observed a terminal branch from adjoining but opposed lateral merophytes, so that terminal branches occur very closely juxtaposed; in such cases the branches are con-siderably less vigorous than the main axis and do not displace this at all. Although, in many "lower" Jungermanniales (e.g., *Chaetocolea palmata*, also *Blepharostoma trichophyllum*; see Vol. I, Fig. 36:2) terminal branching of one side of the axis may then be of the *Microlepidozia* type, this is apparently never so in *M. schusterana*.

M. schusterana is allied only to *M. quelpartensis* Hatt. & Inoue (see Hattori and Mizutani, 1967) of Korea and Japan, which has an obscurely toothed antical leaf margin and more ovate leaves.

CALYPOGEIA Raddi emend. Nees

Calypogeia Raddi, Mem. Soc. Ital. Sci. Modena 18:31 (p. 42 of reprint publ. in 1820), 1818, in part (section B only); Nees, Naturg. Eur. Leberm. 3:7, 1838.
Cincinnulus Dumort., Comm. Bot., p. 113, 1822.
Kantius S. F. Gray, Nat. Arr. Brit. Pl. 1:706, 1821.
Kantia Lindb., Acta Soc. Sci. Fennica 10:506, 1875; Schiffner, *in* Engler & Prantl, Nat. Pflanzenfam. 1(3):100, 1893.
Parentia Leman, *in* Dict. Sci. Nat. 37:536, 1825.[3]

Rather *subhyaline, pale green to whitish green,* to grayish or ± deeply bluish green, rarely (exotic taxa) feebly brownish, covered with a water-repelling waxy cuticle, dull to glistening; normally *little chlorophyllose;* walls without secondary pigmentation. Plants creeping and prostrate, often closely adnate, with slender yet ± *fleshy, soft stems;* stems with *little or no cortical differentiation;* shoots medium-sized (0.8–4.5 mm wide usually), *subsimple or sparingly, irregularly branched.* Branches normally all *postical-intercalary* (exceptionally some lateral-terminal in origin); ventral *stolons or flagella normally lacking.* Rhizoids abundant, confined to sharply de-limited areas at *underleaf bases.* Leaves planodistichous, incubous, ± horizontally oriented, remote, barely contiguous or moderately imbricate, laterally patent, ovate to ovate-elliptical to ovate-triangular, flat or slightly convex, *rounded and entire at apex or ± minutely bidentate.* Under-leaves large to moderately large, *varying from bifid* (and then sometimes with a sharp tooth or small lobe on either side, approaching bisbifid) *to retuse to unlobed.* Cells usually ± *thin-walled and with trigones minute to small* (rarely conspicuous), usually ± delicate and *relatively large* (mostly 30–45 μ wide × 40–60 μ long in leaf middle), ± pellucid, smooth or finely verruculose; oil-bodies several (2–13) per cell, finely *botryoidal to coarsely segmented,* glistening, colorless or blue. Asexual reproduction by *1-2-celled gemmae* from the uppermost, reduced leaves and underleaves of *erect, attenuate shoot apices.*

Autoecious or paroecious, less often dioecious.   ♂ Branches small,

---

[3] Leman had the habit of giving new names to pre-existing, published genera, when he discovered the valid generic name too late to insert it in the proper volume of the Dict. Sci. Nat.!

spicate, postical-intercalary, of 2–8 pairs of bracts; bracts imbricate, 2–3-lobed at apex, concave, 1–2(3)-androus; bracteoles small, flat. Gynoecia subsessile, at apices of *very short, leafless,* postical-intercalary branches, each developing a *long, cylindrical, pale, pendulous, subterranean rhizoidous marsupium, appearing virtually sessile,* crowned at apex by several ephemeral or persistent *small, scalelike bracts.* Capsule *cylindrical; valves linear, spirally twisted.* Epidermal cells oblong, variable, with thickenings of some (almost *never all*) longitudinal walls; transverse walls ± hyaline; secondary longitudinal walls either with nodular thickenings or with bandlike or sheetlike thickenings which may extend slightly on to tangential walls. Inner cell layer with numerous semiannular bands. Elaters subequal to spores in diam., 2(3)-spiral; spores 8–14(17.5) μ in diam. Chromosome no.: $n = 9(8 + x/y$ or $7 + H + h)$ or 18 (Lorbeer 1934; Tatuno 1941; Inoue 1966c).

Type. Calypogeia fissa Raddi (*Mnium fissum* L., *p.p.*).

The genus *Calypogeia* has had a checkered nomenclatural history: see Levier (1902) and Evans (1907). Raddi (1818) established it to include two sections: A (Examphigastriae) for species without underleaves, and B for *Calypogeia fissa* (*Mnium fissum* L. in part) and several other plants, including *Jungermannia trichomanis* and *J. calypogea* Raddi, which were quoted as synonyms.

Three years later S. F. Gray (1821) established *Kantius,* with the single species *K. trichomanis* (with *Mnium trichomanis* and *M. fissum* cited as synonyms). The following year Dumortier (1822) proposed *Cincinnulus,* for *J. trichomanis. Cincinnulus* and *Kantius* are thus based on the same element, the *Calypogeia fissatrichomanis* element, as sectio B of *Calypogeia* of Raddi. *Calypogeia* was also accepted by Corda (1828, p. 653) to include *G. trichomanis,* and by Nees von Esenbeck (1836, *in* 1833–38, p. 405; 1838, p. 7).

Nees, recognizing that the two sections of Raddi were not congeneric, proposed the genus *Gongylanthus* for succubous-leaved plants, including in it *Calypogeia ericetorum* of Raddi. *Calypogeia* was restricted to the incubous-leaved second section. Thus, Nees was the first to delimit *Calypogeia* in its modern sense.

Lindberg (1875), however, attempted to restore *Calypogeia* for Raddi's section A, equivalent to *Gongylanthus* Nees, and used for section B the next oldest name, *Kantia.* In this Lindberg found some following, particularly among Scandinavian botanists. Other workers adopted *Calypogeia* in Lindberg's sense but used *Cincinnulus* for Raddi's section B (*Calypogeia* in the modern sense). Levier (*loc. cit.*), however, reasoned that, since Raddi quotes *Jungermannia calypogea* as a synonym of *Calypogeia fissa,* the latter species gave the genus not only its name but also its basic characters. Evans (1907) accepted this reasoning and so have most modern students.

Distribution

Calypogeia, with about 90 described species, is almost worldwide in range, but is absent from nearly all desert and semidesert regions, from the high Arctic (the northernmost report of the genus and family is that of *C. sphagnicola* from West Greenland at 77°30′ N.; see Schuster, 1968a) and from much of the cooler or cold Antipodes. It is absent from much of Australasia [but there is one species, *C. tasmanica* Rodway (1916) in Tasmania and New Zealand] and is rare in the southern tip of South America (one species, evidently identical with *C. tasmanica*; see Schuster, 1968a; the revision of the South American taxa by Bischler, 1963, fails to include this). The report of *C. tasmanica* from Tierra del Fuego at ca. 50° S. (Schuster, 1968a) appears to be the southernmost for the genus and family.

Differentiation

Recognition of *Calypogeia* offers few if any difficulties. Distinctive are the rather translucent appearance; the often whitish to light to gray-green color (coupled with a general lack of secondary wall pigments); the conspicuously incubous leaves, entire or merely feebly bidentate at the apex, which are usually nearly flat and are obliquely to widely laterally patent—the plants being flat and planodistichous in aspect; the simple or subsimple shoots, with sparing branching, almost always postical in origin; the frequent occurrence of erect, microphyllous, attenuate, gemmiparous shoot apices; the prostrate and often rather closely adnate growth pattern. On the basis of the simple, incubous leaves this genus is subject to confusion possibly with *Bazzania* and *Metacalypogeia*, among our genera. The general lack of terminal branching and of slender ventral flagella, and the much more subhyaline, delicate appearance, separate all our species from *Bazzania*. The distinction of *Metacalypogeia* from *Calypogeia* has already been discussed (p. 111).

Taxonomy

Calypogeia has been divided into three (Bischler, 1963) or four subgenera (Schuster, 1968a), of which two [*Caracoma* Bischler and *Mnioloma* (Herz.) Bischler; see Bischler, 1963] are known to date only from the neotropics. In these two subgenera the leaf- and underleaf-margins are, or tend to be differentiated—usually as a border formed by strongly elongated cells that are oblique and overlap for much of their length (*Mnioloma*), or by cells that tend to be elongated at right angles to the margin and are often produced as crenulations (*Caracoma*). In both of these subgenera leaves (and usually underleaves) are unlobed, and leaves are most often oblong with nearly parallel sides and a rounded-truncate apex (produced as a beaklike point in *Mnioloma*). Both *Mnioloma* and *Caracoma* consist, basically, of firm taxa, often with pigmented, firm, sometimes distinctly collenchymatous cells; in this respect they stand in rather sharp contrast to the species of our regionally represented subgenera.

The subgenera, *Asperifolia* and *Calypogeia s. str.*, include, respectively, one and eight regional species. *Asperifolia*, with one species is no taxonomic problem; the group becomes taxonomically complex in the neotropics, in which many species are found. *Calypogeia s. str.*, however, is one of the most technical and critical of the regional groups (see, e.g., Schuster, 1968a; see also p. 124). There is no space, within the present confines, to go over this ground in detail.

Key to Regional Subgenera of Calypogeia

a. Capsule valves with epidermal cells in basically 8 rows (rarely 1–2[3] of these longitudinally divided), whose width averages twice that of cells of the inner layer; alternating longitudinal walls hyaline and thickened, or all thickened. Stems elliptical to subterete, in section, dorsally formed of 6–8(9–14) or more rows of cells, ventrally formed of 8–10 cell rows or more, 7–9(10) cells high, lateral margins (at point of attachment of leaf) formed by 2–3 cell rows; cuticle (in our species) smooth; underleaves bilobed to entire, lobes (if any) broad and ± obtuse to acute, unarmed laterally or with a blunt tooth; leaf apices entire or faintly bidentate; oil-bodies linear or ellipsoidal, coarsely segmented (segments 1.5–4 μ). . . . . subg. *Calypogeia*
a. Capsule valves with epidermal cells, like inner cells, in basically 16 rows (very rarely with 12–14 rows through suppression of 1–several longitudinal walls); longitudinal walls of the valve midline and the walls situated midway between midline and margins of valves hyaline and unthickened, the other longitudinal walls with thickenings. Stems flattened, usually of 4–6 cell rows dorsally, 4–6 cell rows ventrally, only 5–6 cells high, bounded on each side by a single row of large, hyaline cells (to which leaves are attached); cuticle (in ours) verruculose; underleaves bisbifid, bifid to within 1(2) cells of rhizoid-initial field, the acuminate lobes armed externally with a sharp tooth; leaf apices sharply bidentate; oil-bodies spherical to broadly ovoid, 4–5 (rarely 4 × 7–10) μ, of minute spherules (ca. 1 μ). . . subg. *Asperifolia*

Subgenus *ASPERIFOLIA* (*Warnst.*) *K. Müll.*

Calypogeia sect. *Asperifolia* Warnst., Bryol. Zeitschr. 1:111, 1917.
Calypogeia subg. *Asperifolia* K. Müll., Rabenh. Krypt.–Fl. ed. 3, 6:1164, 1956.

Relatively delicate, mostly under 2 mm wide, with almost horizontally inserted, weakly imbricate or contiguous leaves; branching sparing, postical-intercalary. Stem *flattened, much broader than high,* the *cortical cells dorsally and ventrally in usually only 4–6 rows,* the stem laterally with a single row of *very large cells* on each side, forming the point of union with the leaves; medullary cells few in number, relatively small, much smaller than lateral cortical cells. Leaves ovate-elliptical, remote to feebly imbricate, inserted on a feebly arcuate or

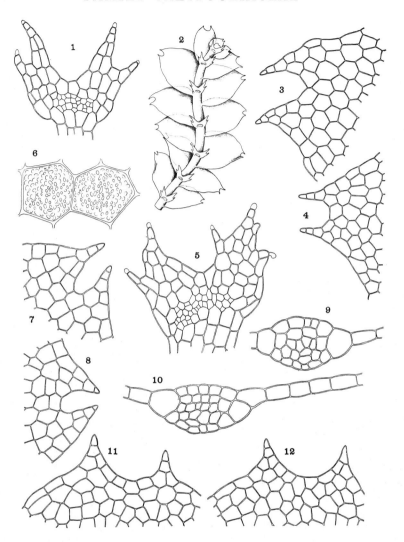

Fig. 104. *Calypogeia sullivantii* (1–10) and *C. arguta* (11–12). (1) Typical underleaf (×98). (2) Shoot, postical aspect (×12.5). (3–4) Leaf tips (×90). (5) Large, unusually dentate underleaf (×98). (6) Two cells, showing surface papillae (×ca. 320). (7–8) Leaf apices (×110). (9–10) Stem cross sections (×115). (11–12) Leaf apices (×110). [1–5, *RMS 28221;* 6–10, *RMS 29389,* all North Carolina; 11–12, from Schiffner's Hep. Eur. Exsic. no. 1444.]

nearly straight line, flat or nearly so, at the apex always *sharply bidentate.* Cells large, *usually sharply asperulate or papillate*; oil-bodies *finely botryoidal, many-segmented*, several per cell. *Underleaves few-celled and rather small:* typically *remote, bisbifid, Lophocolea-like,* divided to within 1–3 cells of rhizoid-initial field into two main lobes which are *acute to short-acuminate, each armed laterally with a sharp tooth or secondary lobe.* Gemmae frequent.

Auto- or dioecious. Capsules with epidermal cells *in ca. 16 rows* ($\pm$ 1–2), each of the 4 rows of primary cells (whose walls lack pigmented thickenings) subdivided by *3 secondary, longitudinal walls bearing nodular thickenings* (thus with a regular 3:1 alternation of pigmented and hyaline longitudinal walls); inner cells *equal in number* to epidermal cells, with semiannular bands. Elaters bispiral.

Type. Calypogeia arguta Mont. & Nees.

The group *Asperifolia* was proposed by Warnstorf (1917) without indication of status; Müller (1951–58) referred to it as a subgenus, and it is tentatively accepted here with this status. The group roughly corresponds to "subgroup 5" of subg. *Calypogeia* in Bischler (1963).

I have adopted *Asperifolia* as a subgenus, following my recent treatment (Schuster, 1968a), because I was able to demonstrate that all of the dozen true *Calypogeia* species studied so far—isolated exceptions aside—showed epidermal capsule cells formed by *single, pigmented, secondary walls* subdividing each of the primary cells that lie in four longitudinal series. This distinction seems infinitely more definitive than the often vague and quantitative differences used by Bischler to separate her "subgroups" of *Calypogeia*. *If* the distinction maintains itself and shows reasonable correlation with a papillose cuticle and bidentate leaves, Warnstorf's group *Asperifolia* may well be accepted as a subgenus. In the absence of any sporophytic data of value in any recent work on nonholarctic Calypogeiaceae, this group must remain of hypothetical validity. In any case, the *Calypogeia arguta-sullivantii* complex represents an isolated group, without close relatives among any other holarctic Calypogeiaceae.

Key to Species of Subgenus Asperifolia

1. Dioecious; leaves with apical teeth usually clearly divergent to divaricate, the sinus between them relatively broad and lunate; stem with medullary cells in up to 4–5 layers, with up to 9 dorsal cortical cells and 9–10 (10–13) ventral cortical cells. European; adventive in North America in greenhouses . . . . *C. arguta* (p. 123)
1. Monoecious (? sometimes dioecious); leaves with apical teeth subparallel to slightly convergent, the sinus between them acute to very narrowly lunate, the lobes approximate; stem with medullary cells in 3–4 layers, with 4–6 cortical cells dorsally and ventrally . . . . *C. sullivantii* (p. 117)

CALYPOGEIA SULLIVANTII Aust.
[Figs. 104:1–10, 105–106]

Calypogeia sullivantii Aust., Hep. Bor.–Amer. Exsic. no. 74b, 1873; Austin, Bull. Torrey Bot. Club 6:18, 1875; Evans, Rhodora 9:67, pl. 73, figs. 1–8, 1907.
Kantia sullivantii Underw., Bot. Gaz. 14:196, 1889.
Calypogeia arguta Sharp (nec. Nees & Mont.), Amer. Midl. Nat. 21(2):279, 1939.
Calypogeia arguta var. *sullivantii* Frye & Clark, Univ. Wash. Publ. Biol. 6(4):687, 1946.

On mineral substrates: scattered or in loose mats, pure deep green (in shade) to, rarely, yellowish green (in partial sun), dull. Shoots *small, 1.2–1.8(2.1) mm wide*, prostrate, simple or sporadically branched; branches all postical-intercalary; stems laterally *with a large-celled, hyaline cortex* and a smaller-celled medulla (formed of 3 layers of cells, 4–6 cell rows wide); *cortical cells usually in 10–13 rows, 4–5 antical, 4–6 postical, one lateral on each side; lateral cortical cells large,* 45 μ, in diam. $\times$ 45–70 μ long. Leaves small near shoot base, increasing in size upward, *distant to slightly imbricate* (mod. *densifolia* never produced in nature), spreading at an angle of 70–90°, flat; leaves ovate to broadly ovate, ca. 1.1–1.3 $\times$ as long as wide, 0.75–1.2 mm long $\times$ 0.6–0.85 mm wide; postical margin usually more strongly curved than antical, very short-decurrent on stem; leaves gradually tapering, *uniformly bidentate*; *teeth sharp*, 3–4(5) cells long, 2–4 cells broad at base, *tipped by a uniseriate row of 2–3 cells, the rather narrow sinus between subacute to obtuse to lunate, the teeth usually subparallel.* Cells leptodermous, *delicately verruculose*, never with discrete trigones; marginal apical cells 45–50 μ, subisodiametric to slightly tangentially elongate; median cells ca. 40 $\times$ 60 μ; *oil-bodies 6–12(13) per cell*, spherical and 5–7.5 μ to ovoid or ellipsoid and 6 $\times$ 8 to 6.5 $\times$ 10–12 μ, of $\pm$ protuberant, *small (1–1.3 μ) spherules;* chloroplasts 4.5–6 μ. Underleaves remote, small, ca. 250–300(400–420) μ wide $\times$ 200–240(275) μ long, obtrapezoidal, usually *bifid to within 1 cell of the rhizoid-initial field*; sinus usually narrowly lunate; lobes divaricate, acuminate to acute; each lobe 2–3, sometimes 4–5, cells wide at base, 3–6 cells long; lateral underleaf margins *uniformly armed with a small, sharp tooth at lobe bases*, 1–2(3) cells long; rhizoid-initial region transverse, usually 3–4 cells high. Asexual reproduction frequent by erect, gemmiparous shoots; gemmae spherical to ovoid, pale green.

Autoecious. Antheridial branches reduced, budlike, usually of two pairs of monandrous bracts; bracts oblong, ca. 150–240 μ long, trilobed up to ca. the middle, middle lobe widest (to 5 cells wide $\times$ 4–5 cells long), lateral lobes toothlike, 1–2 cells wide at base, 3 cells long; lobes each terminated by 2 superimposed, single elongated cells. Antheridia obovate to broadly ellipsoidal, ca. 60 $\times$ 75 μ, on a short, uniseriate stalk; antheridial jacket formed of a group (4?) of basal cells, an apical cap of usually 4 smaller cells, between which lie 2 tiers of rectangulate cells. Gynoecia initially budlike, with at most 2–3 pairs of scalelike, *closely imbricate bracts that are bilobed or trifid, often unequally so, with acute to acuminate lobes*, often armed on one (or both) external margins with a cilium-like tooth. Capsules with linear valves 260–280 μ wide $\times$ ca. 2000 μ long. Interior cells linear, (13)16–18 μ wide $\times$ 140–160 μ long, mostly in 16 longitudinal rows (1–3 may be subdivided to give 18–21 cell rows); external tangential walls with complete reddish transverse bands, extending across both radial walls (but not, or only occasionally, extended onto the interior tangential walls: thus with semiannular thickenings). *Epidermal cells in ca. 16 longitudinal rows* (occasionally, with suppression of 1–2 longitudinal walls, in 13–15 rows), the cells averaging 18 (20–21) μ wide $\times$ 70–100 μ long, narrowly rectangulate; secondary longitudinal walls with $\pm$ *distant nodular thickenings of both faces* (only the primary walls lying in the midline of the valve, and those lying midway between axis and lateral valve margins bear no or only scattered, faint thickenings; thus three longitudinal lines per valve are essentially unthickened). Spores reddish brown, faintly granulose, 14–16(17–18) μ. Elaters bispiral, 8–10 μ in diam., 180–300 μ long; spirals reddish brown, ca. 3 μ.

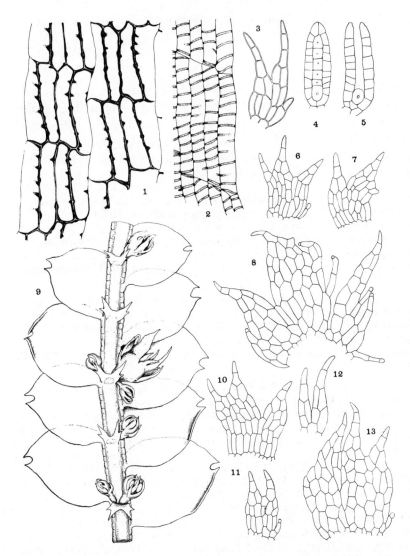

FIG. 105. *Calypogeia sullivantii.* (1) Epidermal cells of capsule wall ($\times$220). (2) Inner cells, capsule wall, aspect of free tangential walls ($\times$220). (3) Large ♂ bracteole ($\times$190). (4–5) Archegonia ($\times$220). (6–7) Smaller ♂ bracts ($\times$100). (8) ♀ Bract, and at right, bracteole ($\times$100). (9) Autoecious shoot with six androecia and a young gynoecium ($\times$31). (10–11) Large ♂ bract and associated bracteole ($\times$100). (12) Small ♀ bracteole ($\times$100). (13) Outer, small ♀ bract ($\times$100). [1–2, *RMS 28255*, Hanging Rock State Park, N.C.; 3–13, *RMS 29389*, Johns Cr., near Marion, McDowell Co., N.C.]

Type. Alabama: Mobile, 1845, *Sullivant s.n.*, Austin, Hep. Bor.-Amer. Exsic., no. 74B, 1873.

Distribution

Nova Scotia and Maine southward to Florida, westward to Ohio and Kentucky, to Arkansas and Mississippi; also reputedly in the West Indies. Essentially oceanic and Appalachian in range, although in the Appalachians it is confined to low elevations (largely to the Escarpment region). A sibling derivative of the European *C. arguta*, from which it is doubtfully specifically distinct.

NOVA SCOTIA. 2 mi S. of Lower Argyle, Yarmouth Co. (*RMS 43061a*); L. Anis, Yarmouth Co.; Fergusons Cove and Upper Sackville, Halifax Co.; Mt. Uniacke (Brown, 1936a). MAINE. Mt. Desert I. (Lorenz, 1924). NEW HAMPSHIRE. Passaconaway (Evans, 1914). MASSACHUSETTS. Newton; Worcester; Leicester; Guilder Pond, Mt. Everett, 2060 ft, Berkshire Co. (*RMS 68–167;* traces only). RHODE I. Middleton. CONNECTICUT. East Haven, Milford, Woodbridge, in New Haven Co.; Waterford, New London Co. (Evans & Nichols, 1908; etc.)

NEW YORK. Fishers I. (Evans, 1926); Cold Spring Harbor, Long I. (Burnham & Latham, 1917). NEW JERSEY. Atsion, "on sandy banks of a pond" (Underwood & Cook, Hep. Amer. no. 156); Delaware Water Gap (Austin, Hep. Bor.-Amer. Exsic., no. 74b, 1873). DISTRICT OF COLUMBIA. OHIO. Hocking Co. (see Conklin, 1926). ARKANSAS. See Evans (1907b); Saline R., SE. of Mena, Ouachita Mts., Polk Co. (*Anderson 11458!*).

WEST VIRGINIA. Brooke, Ohio, Preston Cos. (Ammons, 1940). KENTUCKY. Laurel, Powell Cos. (Fulford, 1934, 1936). TENNESSEE. Blount, Grainger, Morgan Cos. (Sharp, 1939; as *C. arguta*). VIRGINIA. NE. of Cypress Chapel, W. edge of Dismal Swamp, Nansemond Co. (*RMS 34502*); Giles, Botetourt, Chesterfield, Henrico, Spottsylvania Cos. (Patterson, 1950, etc.); Cascades, Little Stony Cr., near Mountain L. (*RMS 40204c*). NORTH CAROLINA. Johns Cr., 6 mi N. of Marion, McDowell Co. (*RMS 28810c, 29388, 29389*); New Hope Cr., near Durham (*RMS 28524*) and Laurel Hill on New Hope Cr., SW. of Durham (*RMS 28471*), Orange Co.; Cascades, near Hanging Rock State Park, Stokes Co. (*RMS 28255, c. caps; 28274*); Eno R., at Laurel Hill, near Durham, Durham Co. (*RMS*); Cumberland, Forsyth, Moore Cos. (Blomquist, 1936); Southern Pines (*Haynes, 1905*). SOUTH CAROLINA. Below Tallulah Falls, Rabun Co. (*RMS 34384, 34384a, p.p.*); Thomasville, Thomas Co. (Brown, 1924). MISSISSIPPI. Tishomingo State Park, Tishomingo Co. (*RMS 19738, M-569*); Wyatt Hills, George Co. (*RMS 27737*); near Van Cleave, Jackson Co. (*RMS 19204*). ALABAMA. Mobile (*Sullivant, 1845*). FLORIDA. Torreya State Park, Liberty Co. (*RMS 44042a*); Seminole Co. (Redfearn, 1952); Sanford (*Rapp*; Haynes, 1915).

Also reported (Wagner, 1946) from INDIANA: Pinhook Bog, Laporte Co., *Richards, 1940;* this report should be verified. The habitat cited and the stated occurrence "with *Sphagnum*" almost preclude *C. sullivantii*. OHIO. Hocking Co. (Miller, 1964).

Ecology

Usually inhabiting deeply shaded sites, on damp or moist soil, more rarely over rock faces. Frequent in the Southern Appalachian Escarpment, on steep banks along small streams, in deep shade of *Rhododendron maximum* and *Tsuga*, with *Calypogeia muelleriana* fo., *C. peruviana*, *Microlepidozia sylvatica*, *Diplophyllum apiculatum*; also on acid, moist, peaty soil, along shaded brooks, associated with *Nardia lescurii* and *Diplophyllum apiculatum*. On moist loamy or clayey banks often with *D. apiculatum*, occasionally *Calypogeia muelleriana*, southward also with *Cephalozia macrostachya*, *Odontoschisma prostratum*. The plant also frequently occurs, southward, on thin soil over sandstone boulders, along cascading brooks with *Pellia epiphylla*, *Cephalozia macrostachya*, and *Solenostoma crenuliformis*, or on moist, alluvial soil amidst and over rocks, close to the water edge, with *Calypogeia fissa*, *Solenostoma crenulatum*, *Scapania nemorosa*. Also often on damp sandstone; here associated with some of the preceding taxa (*Scapania nemorosa*, *Calypogeia fissa s. lat.*) and others (*Riccardia multifida s. lat.*, *Reboulia hemisphaerica*, *Fossombronia brasiliensis*, *Cephaloziella hyalina*).

Calypogeia sullivantii is largely a mesophytic and sciaphytic species; it rarely occurs in sites with more than a maximum illumination of 65–100 foot candles and usually is found on the steep sides of banks or of ledges, where incident light is oblique and diffuse. Occurrence adjacent to springs or along small tumbling streams or rivulets is rather constant and appears to be correlated with a slight ability to withstand desiccation or even a high atmospheric saturation deficit.

At the northernmost edge of its range (Nova Scotia) the species occurs on wet, sandy-loamy soil of roadside ditches, associated with *Nardia insecta*, *Cephalozia bicuspidata*, *Cladopodiella francisci*, and *Calypogeia neesiana*.

Differentiation

A primary differential character between *C. arguta* and *C. sullivantii* is the form of the leaf apex, *C. arguta* supposedly differing from *C. sullivantii* "mainly by the longer and more divergent teeth of the leaves, separated by a broader sinus" (Evans, 1907b). In *C. arguta*, the apical teeth of the leaves are 3–4 cells long (in *C. sullivantii* mostly 2–3 cells long, but, *fide* Evans, also 3–4 cells long); in *C. arguta*, the sinus between the teeth is rounded to lunate (Fig. 104:11, 12) and the lobes are more or less divaricate (in *C. sullivantii*, the teeth may be closely juxtaposed, separated by a narrow sinus, and lie subparallel or are slightly connivent; Fig. 105:3–4, 7–8).

Frye and Clark (1937–47, p. 687) attempted to draw another distinction between the two taxa on the basis of the underleaves. In *C. arguta* these, admittedly, usually have lobes 3–4 cells long and 2 cells wide (at least for the basal 1–2 tiers of cells), while the lateral teeth are 2 cells wide at the base terminating in a single elongate cell. By contrast, in *C. sullivantii*, the lobes are often 2.5–3 cells long (from the insertion of the lateral teeth) and uniseriate throughout, and the lateral teeth are constituted by a single cell. These

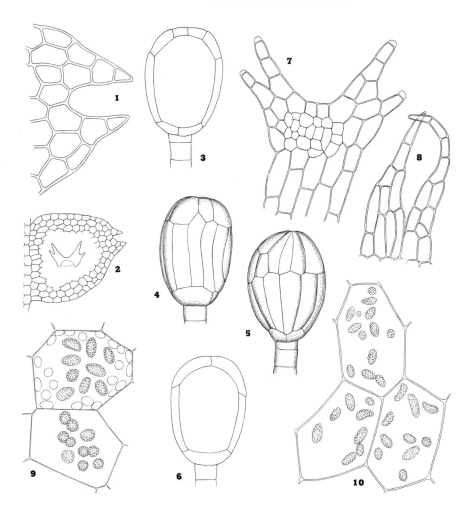

Fig. 106. *Calypogeia sullivantii.* (1) Leaf apex (×200). (2) Leaf and under-leaf (×48). (3) Antheridium, optical section (×350). (4) Same, surface view (×350). (5) A second antheridium (×350). (6) Antheridium in fig. 4, optical section, turned 90° from fig. 3 (×350). (7) Underleaf, with rhizoid-initials (×205). (8) ♂ Bracteole (×160). (9) Cells with oil-bodies and, upper cell, chloroplasts (×470). (10) Cells with oil-bodies (×405). [1–2, 7, *RMS 27737*, Wyatt Hills, Miss.; 3–6, 8, *RMS 29389;* 9, *RMS 37675*, Estatoe R., S.C.; 10, *RMS 43061a*, Nova Scotia.]

differences in the underleaves are inconstant. As is evident (Fig. 104:1, 5), underleaves of *C. sullivantii* may be much larger than described by Evans (1907b) or Frye and Clark, up to 400–420 μ wide $\times$ 285 μ long; the lobes may be (measured from base of lateral teeth) 4 cells long and 2 cells wide for the basal 2 cells. Occasional underleaves from mature shoots of the species (2–2.1 mm. wide) are more complex than any I have found in *C. arguta*. On such shoots, also, no difference *in size* of the apical teeth of the leaves exists, when compared with those of *C. arguta*. As is evident from Fig. 104:5, the apical teeth may be 3–4 or even 5 cells long $\times$ 2–3 or even 4 cells broad at the base.

Therefore, the sole remaining vegetative difference between the two taxa appears to be derived from the *direction* of the apical teeth of the leaves. This, by itself, seems insufficient to warrant retention of *C. sullivantii* as a species. However, the plants from North Carolina (*RMS 29389*) are monoecious, while *C. arguta* is always dioecious. Evans (1907b, p. 68) and Frye and Clark state that *C. sullivantii* is dioecious. Should *C. sullivantii* prove to be consistently autoecious, that would warrant its separation from the European *C. arguta*. In the autoecious inflorescences of the North Carolina plant, *C. sullivantii* closely approaches the Hawaiian *C. bifurca* (Aust.) Evans.

Another difference between *C. sullivantii* and *C. arguta* may lie in the oil-bodies. Those of *C. arguta* occur 4–10 per cell (Müller, 1947a, p. 417) and are rather small (4–5 μ with occasional ones 4 $\times$ 7–10 μ), formed of only 8–10 globules each 1 μ in diam. In *C. sullivantii*, median cells contain 6–12(13) oil-bodies, which vary from 4–5 $\times$ 5–7.5 μ when subspherical to 6 $\times$ 6 μ up to 4–7 $\times$ 8–12 μ; they are formed of more numerous (ca. 12–25) globules 1–1.5 μ in diam. (Fig. 106:9–10).

Calypogeia arguta was reported from a greenhouse in Philadelphia, Pa. (Underwood, 1892b, p. 300; Evans, 1907b, p. 69) and was cited by Sharp (1939) from Tennessee; the latter report is based upon the erroneous consolidation of *C. sullivantii* and *C. arguta*. As far as known, the single report from Pennsylvania aside, *C. arguta* has not been again introduced into this country and has evidently not become established. It is not further treated here.

Subgenus *CALYPOGEIA* Raddi

Usually more vigorous than subg. *Asperifolia*, (1)2–4.5 mm wide. Stem *biconvex, little flattened, of numerous cell rows* (dorsally, between leaf insertion, *usually 7–14 cell rows occur*), the leaves inserted laterally on 2–3 cell rows *that are not distinctly enlarged*. Branching typically all postical-intercalary (but species of the *C. sphagnicola* complex also with terminal, *Frullania*-type branching). Leaves variable in form: usually ovate to dimidiate-ovate to ovate-cordate, 0.85–1.4 $\times$ as long as wide, usually clearly widest in basal 0.1–0.25, never oblong or lingulate, *apex narrowly rounded, pointed, or closely bidentate; leaves never margined by radially or obliquely oriented, differentiated cells.* Underleaves variable, unlobed to bifid, *never bisbifid, formed of usually numerous (60–250) cells*, not distinctly margined; lobes (if any) *rounded to blunt to subacute, broad.* Cells usually (always in our species) *smooth, normally delicate* and with small, vestigial or no trigones; oil-bodies formed of *few (usually 2–18) coarse segments, botryoidal.*

Capsules with epidermal cells *in 8 (sometimes ± 1–3) rows*, basically with 3 continuous lines formed by ± unthickened, often unpigmented primary, longitudinal walls (situated at valve midline and midway between midline and valve margins) *alternating with which are interpolated four continuous lines formed by secondary walls bearing pigmented thickenings*, a 1:1 alternation of thickened and unthickened walls resulting (minor secondary subdivision excepted). Inner cells usually *ca. twice as many (and half as wide) as the epidermal cells*, with semi-annular bands. Elaters 2(–3)-spiral.

Type. Calypogeia fissa Raddi.

Subgenus *Calypogeia* is here defined differently from in Bischler (1963). Primary emphasis in a subdivision of *Calypogeia* is based on differences in capsule-wall anatomy, not alluded to in Bischler's account. Subgenus *Calypogeia*, as here defined, includes basically species placed by Bischler in her subgroups 1–3. I cannot ascertain from Bischler's account whether any of the taxa she places in her subgroup 4 are referable to subg. *Calypogeia.*

None of our species of subg. *Calypogeia* possesses a roughened cuticle; none has a reduced axial anatomy; none has bisbifid underleaves, although *C. fissa* and *C. peruviana* may show an approach to such under-leaves. Also, in all of the species oil-bodies are rather coarsely botryoidal—ideally, as in *C. peruviana* and *C. trichomanis* (where they are indigo-blue), they almost perfectly approach the "Trauben typus" of Müller (1939).

The separation of the species of this subgenus is treated in the conspectus (p. 127) and in the keys that follow this classification (pp. 128, 130).

Classification

The taxonomy of the holarctic species of subg. *Calypogeia* remains subjective. Buch (1936a), using simple culture techniques, prepared a stimulating account; he was followed by Müller (1947a, 1951–58), who sharply disagreed with Buch in certain particulars: (*a*) *C. fissa sensu* Buch was stated by Müller not to represent *C. fissa* at all; (*b*) *C. meylanii* Buch was regarded by Müller first (1947a) as a synonym of *C. neesiana* and then (1951–58), following Schuster (1949a), as a variety of *C. neesiana; (c) C. trichomanis sensu* Buch was considered by Müller to represent *C. muelleriana.* Buch (1942, 1948) defended the distinction between *C. meylanii* and *C. neesiana*, but seemingly agreed that his concepts of *C. fissa* and *C. trichomanis* were evidently based on *C. muelleriana* and that he had not studied *C. trichomanis sensu* Müller at all. Other accounts needlessly confusing *C. muelleriana* and the *C. meylanii-neesiana* complex appeared. Although the European species of the group were reviewed by Bischler (1957) and the English species by Paton (1962), these papers added little new information. A detailed review attempting a complete re-evaluation and introducing new data is that of Schuster (1968a); in this there is a detailed evaluation of taxonomic criteria and their relative "values" and degree of variability. In this monograph I point out (as I had earlier; Schuster, 1953) that the delimitation of certain taxa in European works did not seem to "hold" for American phenotypes, and that the European classification (and delimitation of taxa) could not be adopted *in toto*. Some of the taxonomic problems with the American species of *Calypogeia* are inherent in the fact that these species appear to be undergoing evolution along lines different from the European populations. Hence, in attempting to adapt European treatments to the American material, particularly with respect to *C. fissa*, *C. meylanii* (= *C. integristipula*) and *C. neesiana*, the student soon runs into difficulties.

American species of subg. *Calypogeia* appear to be divisible into five small complexes, separable by the subjoined conspectus. Between each of these small units there is no *actual* intergradation. For example, I have collected *C. neesiana* and *C. suecica* intermingled, and they are abundantly distinct; similarly, *C. peruviana*, *C. neesiana*, and *C. muelleriana* were collected admixed and are obviously distinct, even under the dissecting microscope (Fig. 126). Superficial transitions between members of these groups may occur, leading to confusion. Thus, *C. peruviana* and *C. fissa* have been confused in North America; *C. neesiana* and *C. muelleriana* have been widely confused, even though texture alone allows a certain separation of *all* phenotypes (see Schuster, 1968a). *Within* the more polymorphic of these complexes, however, there is so wide a latitude of variability (intrinsic and environmentally induced) that perfectly valid distinctions between species may be masked. In other cases [e.g., *C. fissa* vs. *muelleriana* and *C. neesiana* vs. *integristipula* (= *meylanii*)] we are still not sure that the intergradation is entirely due to this "masking" effect, and suspect it may actually be genetic. In other words, our species concept may be too narrow; or, alternatively, as I suggested (Schuster, 1968a) for the *C. neesiana-meylanii* (= *integristipula*) complex, hybridization may occur—or may have once occurred—leading to populations with intermediate characteristics.

Müller (1947a) used chromosome number as a major species criterion; it surely is of trivial importance in this subgenus, as is clear from the fact that both 9 and 18 chromosome races are now known in *C. trichomanis*, *C. muelleriana* (18 in European plants, *teste* Müller, 1951–58, p. 1180; 9 in American subsp. *blomquistii*) and *C. neesiana*. Sexuality, spore size, and capsule-wall anatomy appear more significant taxonomically.

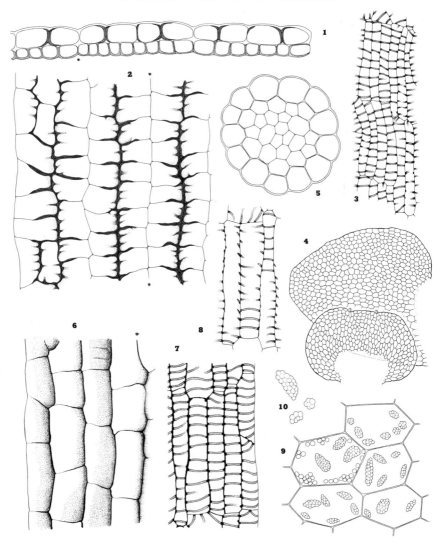

FIG. 107. *Calypogeia:* general plate. (1–4) *Calypogeia integristipula;* (5) *C. sphagnicola;* (6–10) *C. peruviana.* (1) Capsule wall cross section, valve margin at right, valve margins at asterisk (×210). (2) Epidermal cells of capsule wall, valve margin at left, valve midline at asterisks; note unusually short cells (×210). (3) Inner cells of capsule wall (×210). (4) Typical leaf and below it an underleaf, rhizoid-initial area not drawn in (×36). (5) Seta cross section (×114). (6) Epidermal cells, capsule wall, valve margin at left, valve midline at asterisk; note, in contrast to *C. integristipula,* the "normally" elongate cell shape (×210). (7) Inner cells, surface view, valve margin at right (×210). (8) Three inner cells from fig. 7, but level of focus on internal tangential walls (×210). (9) Median cells with oil-bodies and, upper left, chloroplasts (×315). (10) Individual oil-bodies (×630). [1–4, *RMS 38942,* Kalkaska Co., Mich.; 5, *RMS & KD 66054a,* Sondrestrom Fjord, Greenland; 6–8, *RMS 28810,* North Carolina; 9–10, *RMS, Aug. 1962,* Whitewater R., S.C.]

Conspectus of Species Complexes within Subgenus Calypogeia

a. Chromosome no. = 18 (exc. *C. muelleriana* subsp. *blomquistii* and races of *C. neesiana*); cells medium-sized to large (30–45 μ wide × 40–60 μ long or more in leaf middle); spores 10–17.5 μ; monoecious . . . . *b.*

    b. Capsule with epidermal cell thickenings (of at least alternating longitudinal cell walls) on *both* adaxial and abaxial faces; oil-bodies colorless (except in the broad-leaved *C. trichomanis*); underleaves entire or barely to greatly divided, but usually only to within 2–5 cells of rhizoid-initial region . . . . *c.*

        c. Underleaves subentire or emarginate, length along midline usually (6)8–12 cells; underleaf cells 20–25 μ wide medially; leaf cells only ca. 30–35 μ wide medially; texture of leaves (except marginal cells) dull, feebly glaucous; capsule with epidermal cells with well-developed nodular thickenings; spores 10–12 μ. (Only with intercalary branching) . . . . *C. neesiana* complex[4]
                                      (*C. integristipula, C. neesiana*)

        c. Underleaves bilobed (to within 2–6 cells of base usually); underleaves with median cells usually 30–33 μ wide; leaves weakly but distinctly nitid; capsules variable but often with thickenings in nearly evenly thick sheets; spores usually 12–16(17.5) μ . . . . *d.*

           d. Branching frequently terminal, furcate, of the *Frullania* type; cells often with ± distinct trigones, frequently smaller, 30–35 μ wide medially . . . . *C. sphagnicola* complex
                                      (*C. sphagnicola*)[5]

           d. Branching usually only postical-intercalary; cells without distinct trigones, usually larger, (32)35–45 μ wide medially . . . .
       . . . . *C. fissa* complex
                 (*C. fissa, C. muelleriana, C. trichomanis*)

    b. Capsule with epidermal cell thickenings (of usually all rows) confined to *adaxial* portions of radial and adjoining tangential walls; oil-bodies deep blue, of fine segments, up to 7 × 18 μ; leaves bidentate; underleaves divided to within 1–2(3) cells of rhizoid-initial region; cell size as in *C. fissa* complex . . . . *C. peruviana* complex
                                  (*C. peruviana*)

a. Chromosome no. = 9; cells small, usually collenchymatous, 25–30 μ wide in leaf middle; spores 8–11 μ; capsule wall as in *C. peruviana* complex; dioecious . . . . *C. suecica* complex
                                  (*C. suecica*)

Within these complexes, *C. sphagnicola* shows major deviations. It has the small cells of the *Neesiana* complex, but the leaf and underleaf forms of the *Fissa* complex. It also diverges in producing rather regularly, if sparingly, *Frullania*-type branches.

---

[4] I use the term "complex," which has no formal hierarchial standing, rather than "section," since I do not believe that the groups should have this status.

[5] *Calypogeia tasmanica* Rodway (of Tasmania; see Schuster, 1963) also has terminal branching; it (or a plant essentially indistinguishable) occurs in Tierra del Fuego, Argentina (Valle Carbajal, *RMS*); both may be identical to *C. sphagnicola*. The only other instance of terminal branching seen is in *C. neesiana*, in which I have seen one *Acromastigum*-type branch.

Our species of subg. *Calypogeia* all appear to fall within *Calypogeia* "subgroup 1" of Bischler (1963, pp. 56–57). They share with neotropical species of that group the following complex of criteria: underleaves feebly to distinctly decurrent (except in *C. neesiana-integristipula* complex), usually 2–3× the stem width; stems somewhat fleshy, 200 μ in diam. or more; leaf insertion feebly arched, toward antical end at least, the antical leaf base never auriculate; leaf apices pointed, rounded, or bidentate, the leaf form basically ovate to dimidiate-ovate, never oblong, the leaf length usually 0.85–1.4× the width; leaves feebly to distinctly, never closely imbricate, weakly convex at best, at postical base usually little or not decurrent; leaf margins usually not differentiated (exc. *C. neesiana*); leaf cells always smooth, medium or medium-large, 25–48 μ wide in leaf middle × 40–60(70) μ long; trigones absent or very small (exc. *C. suecica*); cortical cells of stem mostly 45–80 μ long; plants light to yellowish to rather deep green, rarely very whitish green.

As I have here redefined "subgroup 1," which constitutes the sectio Trichomanoides, I would eliminate from it several species assigned to it by Bischler, i.a., *C. muscicola* and *C. lophocoleoides*, of which at least the latter has a roughened cuticle. The species of this section are, perhaps, the most technical and polymorphous of the entire genus. As is developed elsewhere, unresolved taxonomic affinities occur between several North American taxa and taxa occurring in Latin America. Thus, *C. fissa* subsp. *neogaea* appears immediately allied to *C. subintegra* (G., L., & N.) Bischler, especially its var. *dussiana* (Steph.) Bischler, and *C. muelleriana* appears to be very close to *C. andicola* Bischler. As long as the Latin American taxa are known only from sterile—and dead—material, any proper evaluation of the *degree* of affinity remains impossible.

Key to Species of Subgenus Calypogeia (Living Plants)

1. Underleaves deeply bifid; midline of the underleaf, from sinus to rhizoid-initial region, *formed by* (*1*)*2–5, rarely 6–7 cells;* distal marginal cells of leaves never elongate; underleaves more or less decurrent, narrowed basally; plants when dry nitid or subnitid . . . . 2.
   2. Oil-bodies colorless . . . . 3.
      3. Cells small: less than 30(32) μ in leaf tips, 25–35 × 30–45 μ in leaf middle; oil-bodies formed of 2–4 (5–7) segments; only on organic substrates; small plants, 0.8–1.8 mm wide . . . . 4.
         4. On *Sphagnum;* cells 30–35 × 40–45 μ in leaf middle; underleaves less than 2× stem width, remote; monoecious; with some terminal, *Frullania*-type branches . . . . *C. sphagnicola* (p. 133)
         4. On decaying wood; cells 25 × 30 μ in leaf middle; underleaves 2–3.5× stem width, contiguous or imbricate; dioecious; with only intercalary-postical branching . . . . . *C. suecica* (p. 143)
      3. Cells large: median (32)36–45(48) × 45–60–75 μ or larger; oil-bodies (in some cells at least) formed of numerous segments; larger plants, 1.8–4 mm wide when mature; no terminal branching. . . 5.
         5. Leaves entire, blunt or narrowly rounded at tips; underleaves

orbicular (or sometimes elongate), less than 1.35× as wide as long, with lobes not strongly divergent; underleaves divided to within (3)4–5(6–7) cells of rhizoid-initial area, their lateral margins evenly rounded, only occasionally angulate or dentate. (Plants robust: to 2.5–4 mm wide) . . . . *C. muelleriana*. . . 6.

6. Leaves of robust mature shoots obliquely ovate, usually nearly or quite as wide as long; underleaves orbicular to transversely oval, the lobes not drawn out; leaf margins and underleaf margins not, or rarely, obscurely crenulate . . . .
. . . . *C. muelleriana* subsp. *muelleriana* (p. 174)

6. Leaves of mature shoots narrowly ovate, longer than wide, spreading almost at right angles to stem; underleaves polymorphic, some with lobes elongate and acute, drawn out: the underleaves becoming strongly elongated; leaf and underleaf margins normally strongly crenulate . . . .
. . . . *C. muelleriana* subsp. *blomquistii* (p. 187)

5. Leaves frequently bidentate at apex (or sharp-pointed); underleaves strongly transverse, only 1.2–1.5× as wide as stem, 1.5–2× as wide as long, bifid to within (1)2–3 cells of rhizoid-initial region, with spreading lobes which are usually armed (on their external margins) by an angulation or tooth. (Leaves usually narrowly ovate; margins never crenulate; plants only 1.5–2.5 mm wide) .
. . . . *C. fissa s. lat.* (p. 164)

2. Oil-bodies deep blue, conspicuous . . . . 7.

7. Leaves broadly ovate-cordate, as wide as or wider than long, entire at apices; underleaves bilobed to within (3)4–5 cells of rhizoid-initial field, with lateral margins almost invariably entire or with a low angulation. Northern . . . . *C. trichomanis* (p. 149)

7. Leaves ± narrowly ovate, usually longer than broad, frequently or usually bidentate at apex; underleaves deeply bilobed, to within 2–3 cells of rhizoid-initial field, the lateral margins usually armed on one or both sides with an acute tooth. Southeastern (north to North Carolina) . . . . *C. peruviana* (p. 156)

1. Underleaves entire or shallowly notched (the lobes then broad, rounded); midline of underleaf, from rhizoid-initial region to sinus *formed by a row of* (6)7–10(11–12) cells; marginal cells frequently more or less elongate, forming a border (at least on some leaves and underleaves); underleaves, at least on mature shoots, orbicular, swiftly rounded and narrowed at base, virtually nondecurrent; plants when dry ± dull, marginal cells excepted; oil-bodies colorless . . . . 8.

8. Border of elongate cells vestigial or almost absent; underleaves usually 1.2–1.6× as wide as long, almost invariably entire or shallowly, *broadly* retuse or emarginate; rhizoid-initial area conspicuous, broadly ellipsoidal; leaves usually evenly, narrowly rounded distally, obliquely, narrowly ovate, widest basally . . . . *C. integristipula* (p. 204)

8. Border of elongate cells well developed (on some or many leaves); underleaves (1.0)1.1–1.2× as wide as long, typically with a shallow but

sharp V-shaped sinus; rhizoid-initial area inconspicuous; leaves ovate, often broadly so, with apex usually ± truncate, widest some distance above base . . . . *C. neesiana* (p. 193)

Supplementary Key to Species of Subgenus Calypogeia

1. Leaf apices of mature plants with marginal cells 25–31(32) μ or less; median cells 35 × 40 μ or less, often with discrete to slightly bulging trigones; underleaves deeply bilobed, with acute to subacute lobes, often armed laterally; leaves entire, when well developed dimidiate-ovate to broadly ovate, nearly or quite as wide as long; oil-bodies formed almost exclusively of 2–5 segments, the largest only 4 × 5–7(8) μ, colorless. Plants small: 0.5–1.8(2.0) mm wide; only on organic substrates. . . . . 2.
   2. Underleaves contiguous to imbricate, over 2× the stem width, broader than long, usually unidentate laterally; cells of leaf middle averaging 25 × 30 μ; dioecious; capsule with epidermal cells without nodular thickenings; on decaying wood; branching all postical-intercalary . .
   . . . . *C. suecica* (p. 143)
   2. Underleaves remote, usually little wider than stem, scarcely or no broader than long, most frequently entire laterally; cells of leaf middle averaging 30–35 × 40 μ; monoecious; epidermal cells of capsule with nodular thickenings; over peat; branching partly lateral, terminal . . . . . .
   . . . . *C. sphagnicola* (p. 133)
1. Marginal cells of leaf apices (28)30–45 μ or longer (measured on margins); median cells 32–45 μ wide; without readily evident trigones; oil-bodies (in well-developed plants at least) larger, mostly of 4–35 globules in surface view, the largest at least 4–5 × 9 μ. Plants more robust, 1.8–4 mm wide. 3.
   3. Underleaves orbicular or transversely oval, entire or retuse at apex (or 0.1–0.25 emarginate), their length medially (6–7)8–10(11–12) cells (from rhizoid-initial field to tip); leaves with marginal cells of apex locally or consistently quite elongate parallel to margin (in 1–2 rows), at least on some leaves; leaves dull when dry (marginal cells excepted); lateral leaves ovate (except occasionally on ± xeromorphic plants with leaves wider than long), rounded to narrowly truncate at apex; leaf cells smaller, the subapical often only 23–34 μ wide × 33–38 μ long, the median typically 33–38(42) μ wide × 34–40(48) μ long; oil-bodies colorless. Epidermal cells of capsule wall with nodular thickenings on alternating longitudinal walls; spores 10–12(14) μ . . . . 4.
   4. Marginal cells (often only locally) elongate (at least on smaller leaves), averaging 42–48(50–60) μ long × 22–28 μ wide, forming a ± distinct border usually one cell row wide; leaves frequently truncate to truncate-retuse at tip, broadly ovate and widest somewhat above base; oil-bodies usually absent from some, many, or all median cells, usually few-segmented, mostly of 2–6(8) coarse globules; underleaves typically orbicular, emarginate to shallowly bilobed, 1.1–1.25× as wide as long, with a strongly transverse rhizoid-initial field at the base, 3–4× as wide as high; plants small, whitish green, 1.5–2.1 mm wide . . . . *C. neesiana* (p. 193)

4. Marginal cells (of smaller leaves at least) little elongate, averaging 32–40 × 28–36 μ, forming at best an obscure and much-interrupted border 1 cell wide, many of the cells isodiametric; leaves obliquely ovate, widest basally, typically narrowly rounded distally, usually elongate; oil-bodies present in all leaf cells, from 4 × 8 to 5 × 13 μ, rarely larger, the larger of 12–18 fine globules; underleaves typically orbicular-reniform, rounded to truncate distally, 1.25–1.6× as wide as long, with a broader rhizoid-initial field, 1.5–2× as wide as high; plants grayish green, to 2.5–3.2 mm wide . . *C. integristipula* (p. 204)

3. Underleaves consistently bilobed, the sinus narrow to broad, usually descending at least 0.3–0.4 the underleaf length, the underleaf medially (to sinus) 2–5, rarely 6, cells in length; marginal cells of leaf apex ± isodiametric; leaves ± nitid when dry; leaves broadly ovate and heart-shaped, as wide as or wider than long, or narrowly ovate; cells of leaf middle normally larger, mostly (32–36)40–48 × 48–60(92) μ, always with oil-bodies. Epidermal cells of capsule wall usually without sharply defined nodular thickenings; spores averaging 12–16(17.5) μ. . . . 5.

5. Cells with oil-bodies colorless; plants (when living) with shoot apices greenish to whitish or grayish green . . . . 6.

6. Underleaves typically ovate to suborbicular, to 1.4× as wide as long, usually shallowly divided (with sinus descending to ⅓, rarely ⅗, the underleaf length), the lateral margins typically evenly rounded, unarmed; underleaf along midline 4–5(6–7) cells long; leaves typically broadly ovate and somewhat cordate-heart shaped, averaging as wide as or wider than long, at apex narrowly rounded or obtuse, rarely subacute . . . *C. muelleriana* (p. 174) . . . 7.

7. Leaf margins smooth (the marginal cells with straight walls); leaves (if narrowly ovate) either acute or apiculate, or bidentulate at apex; underleaves usually distinctly wider than long, with short, broad, often obtuse lobes, frequently armed laterally; plants with median cells averaging (32)36–45(48) × (45)55–65 μ long; chromosome no. = 18 . . . .
. . . .*C. muelleriana* subsp. *muelleriana* (p. 174)

7. Leaf margins ± crenulate-sinuate (the marginal cells each concave medially along their outer edges); leaves narrowly ovate (usually rounded or blunt distally); underleaves with lobes erect, parallel, often drawn out and greatly elongated, crenulate-sinuate on margins like lateral leaves, their sides unarmed; mature plants with strongly narrowed and elongated median cells, usually averaging 32–37(42) μ wide × 76–90 μ long; chromosome no. = 9 . . . .
. . . .*C. muelleriana* subsp. *blomquistii* (p. 187)

6. Underleaves strongly transverse, typically 1.3–1.8× as wide as long, divided almost to base by a broad, often angular sinus, the lobes subacute to obtuse (never rounded), the sides typically armed with a humplike angulation or tooth; underleaf (1)2–3 cells long on midline; leaves typically narrowly ovate, averaging longer than

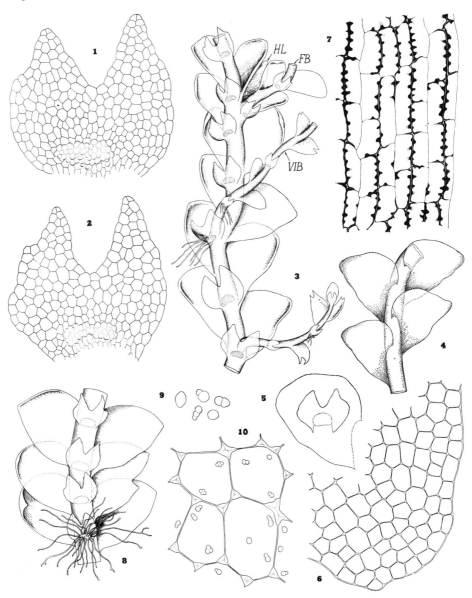

FIG. 108. *Calypogeia sphagnicola* (for seta cross section see Fig. 107:5). (1–2) Under-leaves (×89). (3) Shoot, ventral view, with two ventral-intercalary branches (VIB) and, above, a terminal, *Frullania*-type branch (FB) with associated stem half leaf (HL) (×24). (4) Shoot sector, antical view (×24). (5) Leaf and underleaf (×33). (6) Leaf apex (×190). (7) Epidermal cells, capsule wall, valve margin at right (×230). (8) Sector of mod. *densifolia*, ventral view (×34). (9) Oil-bodies (×740). (10) Median cells with oil-bodies (×560). [1–7, *RMS 45603*, Kânâk, N.W. Greenland; 8–10, after Schuster, 1953 (8, *RMS 12233b*, Belle Rose I., Cook Co., Minn.; 9–10, Grand Portage, Minn., *RMS*.)]

wide, at apex either acute or apiculate (rarely obtuse) or $\pm$ bidentate . . . .*C. fissa* (p. 164)

5. Cells with oil-bodies blue; plants (when living) appearing azure blue at apices (even to naked eye), when dead often with cell walls or the cytoplasmic layer adjacent to them bluish-pigmented; oil-bodies often very large and up to $5-9 \times 15-18 \mu$ . . . . 8.

8. Leaves typically broadly ovate, somewhat cordate and heart-shaped, averaging as broad as or broader than long, the apex subacute to narrowly rounded to retuse; underleaves little or scarcely wider than stem, varying from as long as to slightly longer than wide, bilobed by an acute and narrow sinus into 2 acute to subacute lobes; lateral margins of underleaves usually rounded and entire; oil-bodies often of only 4–10 coarse segments (surface view), rarely more than $5-6 \times 8-12 \mu$ . . *C. trichomanis* (p. 149)

8. Leaves ovate to narrowly ovate, usually longer than wide, the apex frequently to consistently bidentate; most underleaves much wider than stem, typically broader than long, bilobed by an open sinus to within 2–3 cells of rhizoid-initial field, the lobes usually divergent, often obtuse; lateral margins of underleaves frequently to almost consistently angulate or armed with a tooth; oil-bodies variable in size, but the largest up to 6×8 to $9 \times 18 \mu$, 3–5 rows of globules wide medially (the globules only $1-1.5 \mu$) . . . . *C. peruviana* (p. 156)

CALYPOGEIA SPHAGNICOLA (*Arn. & Perss.*)
Warnst. & Loeske
[Figs. 107:5, 108–111]

Calypogeia trichomanis var. *tenuis* Aust., Hep. Bor.-Amer. Exsic. no. 74, 1873.
Kantia trichomanis var. *tenuis* Underw., *in* A. Gray, Manual of Bot. 6 ed:713, 1889.
Kantia sphagnicola Arn. & Perss., Rev. Bryol. 29:26, figs. 1–8, 1902.
Calypogeia sphagnicola Warnst. & Loeske, Verh. Bot. Ver. Brandenburg 47:320, 1905.
Cincinnulus trichomanis var. *sphagnicola* Meyl., Bull. Herb. Boissier, Ser. 2, 6:499, 1906.
Calypogeia paludosa Warnst., Krypt.-Fl. Mark Brandenburg 2:1117, 1906; Evans, Rhodora 17:119, 1915.
Calypogeia tenuis (Aust.) Evans, Rhodora 9:69, pl. 73, figs. 9–14, 1907.
Calypogeia trichomanis var. *sphagnicola* Meyl., Rev. Bryol. 36:53, 1909.

Creeping on dead or living plants of *Sphagnum, usually as isolated plants* or in small patches, *pale or whitish green, pellucid, nitid when dry. Shoots small,* (0.5)1.5–1.8(2.0) mm wide $\times$ 0.5–2(3) cm long, sparingly branched; *branching partly terminal, Frullania type* (and narrowly furcate), partly postical-intercalary. Leaves nearly flat to *slightly convex,* at an angle of 40–60°, *distant to approximate,* obliquely ovate to *ovate-triangular,* widest basally, on mature shoots rarely almost as wide as long (ca. 650 μ wide $\times$ 750 μ long to 600–700 μ wide $\times$ 850–1000 μ long), on mod. *laxifolia* often more elongated, *the apex typically subacute and entire.* Cells of leaf apices near and on margins isodiametric or nearly so, ca. *23–27 μ, more rarely to 32 μ,* the marginal cells not forming a border; *median cells 22–35 $\mu \times$ 26–42 μ* (mostly averaging 30 $\times$ 38 μ); *cells thin-walled and with small or vestigial trigones to* (most often) *discrete or feebly bulging trigones*

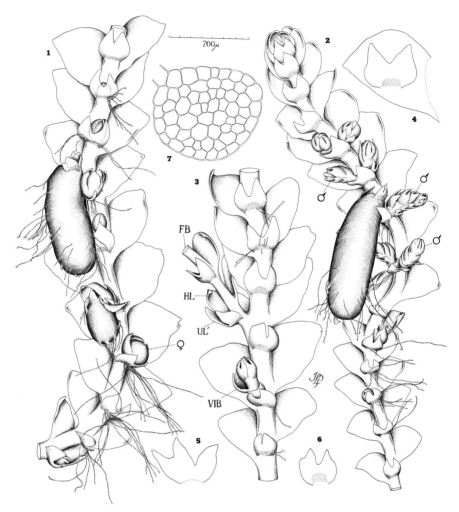

Fig. 109. *Calypogeia sphagnicola*. (1) Shoot, postical aspect, with semimature marsupium and seven gynoecial branches of varying ages (×27). (2) Same, but with mature marsupium and several ♂ branches and juvenile ♀ branches (×27). (3) Shoot sector, ventral aspect, showing a terminal, *Frullania*-type branch at FB, the stem half-leaf at HL, the primary, modified branch underleaf at UL', and a ventral-intercalary branch at VIB (×31.5). (4) Leaf and underleaf (×31.5). (5–6) Underleaves, one abnormal and trifid (×31.5). (7) Stem cross section (×200). [All from *RMS & KD 66–061*, Sondrestrom Fjord, W. Greenland; 3–6 to scale at top of figure.]

(with age the cell walls sometimes evenly thick-walled); cells usually *with 2–3(5) colorless oil-bodies*, these irregular in shape, *largely formed from 1–4(5–6) globules or segments* 1.5–2 μ in size, mostly less than 4 × 7 μ, rarely to 4–5 × 9 μ. *Underleaves distant* to (more rarely) approximate, suborbicular *to slightly longer than wide*, varying from ca. 1.2–1.8× as wide as the stems, *uniformly bilobed* ⅗–¾ their length (to the rhizoid-initial field), *the lobes acute or subacute*, narrowly triangular or ovate-triangular, the sinus usually acute or subacute, rarely narrowly rounded at base; lateral margins frequently to infrequently armed (on one side; *very rarely* on both sides) with a small tooth or angulation; rhizoid-initial area ca. twice as wide as high, ellipsoidal. Asexual reproduction by gemmae almost invariably present (except in hygric phases); *gemmiparous shoots often abundant, filiform, slender, erect;* gemmae spherical to ovoid or broad-ellipsoidal, mostly 2-celled when mature, 15–20 μ long.

Monoecious (and rarely paroecious?). ♂ Bracts in 4–5, rarely to 8–11, pairs, forming a compact, small postical spike; bracts bilobed, occasionally with a small antical tooth, acutely lobed for ca. ½ their length. Capsules with valves 6–7× as long as broad (averaging 285–295 × 1375–1400 μ); epidermal cell layer with *cells in 16 rows*, the secondary longitudinal walls (on both faces) irregularly thickened, *with discrete nodular thickenings;* the alternating (primary) longitudinal walls (between those provided with thickenings) either lacking thickenings or having them poorly developed; inner cell layer with semiannular thickenings, the cells averaging 11–13 μ wide. Spores 10–12 μ in diam. Chromosome no. = 18.

Type. Sweden (Mora, Prov. Delarne, *J. Persson*!).

Distribution

Widely distributed in northern and central Europe and principally in northern North America. The species is restricted to subarctic and subalpine situations, to the Coniferous Forest Panclimax of North America and Europe, but extends north to 77°30′ N. in west Greenland. In North America found from Quebec and Ontario south to Connecticut, central and western New York, northern New Jersey, West Virginia, and Virginia, westward to Minnesota, Alberta and British Columbia, and Alaska (Persson, 1946; Persson and Weber, 1958). In New York, Connecticut, and New Jersey south to Virginia found in spruce or spruce-tamarack bogs occurring as disjuncts in deciduous forest. However, I know of no authentic sites in purely deciduous forest.[6]

Reported by Allorge & Allorge(1948) from the Azores. Recently found in Hokkaido, Japan (Inoue, pers. comm.).

I can find no real differences between the plant from Tasmania (*Calypogeia tasmanica* Rodway; Rodway, 1916) and *C. sphagnicola.* I have collected this

[6] Reports by Blomquist (1936) from North Carolina represent misdeterminations.

plant on Mt. Rufus, Tasmania, from among *Sphagnum* and, again from among *Sphagnum*, in the Valle Carbajal, Tierra del Fuego, Argentina (*RMS, February 1961*). Like *Ptilidium ciliare, C. sphagnicola* appears to be bipolar in range, with the antipodal stations infrequent and scattered. A recent station is from Dunedin, New Zealand.

NW. GREENLAND. Inglefield Bay, on *Sphagnum* near Kânâk, 77°30′ N. (*RMS 45603, 45606*; northernmost report of genus and family!). W. GREENLAND. Ikertoq Fjord, 66°45′ N. (*J. Vahl 192, Aug. 1832!*); Tupilak I., Egedesminde, 68°42′ N., 52°55′ W. (*Holmen 16999!*); Ameralik Fjord, 64°03′ N., and Igdlorsuit, Ameralik Fjord, 64°10′ N. (*J. Vahl 130!, Aug. 1830 and Aug. 1831!*; publ. by Lange & Jensen, 1887, as *C. trichomanis*); SE. of Jacobshavn, Disko Bay, 69°12′ N., 51°05′ W. (*RMS & KD 66-224, 66-225a; 66-228,* fo. *paludosa*); Sondrestrom Fjord, near shore of L. Ferguson, on *Sphagnum,* 66°5′ N. (*RMS & KD 66-054, 66-055, 66-056, 66-062; 66-061,* with marsupia); Tupilak I., W. of Egedesminde, 68°42′ N. (*RMS & KD 66-104a, 66-107*); E. end of Agpat Ø, neck E. of Umiasugssup ilua, 69°52′ N., 51°38′ W. (*RMS & KD 66-1422*); Lyngmarksbugt, ca. 25-125 m, 69°15′ N., 53°30′ W. (*RMS & KD 66-561*) and Blaesedalen, Godhavn, Disko I. (*RMS & KD 66-1570*). S. GREENLAND. Narssaq, 60°55′ N., 46°02′ W. (*J. Vahl 240, Sept. 1829!*; published as "*C. trichomanis*" by Lange & Jensen, 1887); Tasermiut Fjord, 60°05′ N. (*J. Vahl. Aug. 1829!*; det. by C. Jensen as *C. trichomanis*); Narssalik, 61°04′ N., 49°19′ W. (*Damsholt, July 26, 1963!*; probably!). QUEBEC. Shigawake, Gaspé Co. (*Kucyniak & Ray 56060*). ONTARIO: W. of Raith, Thunder Bay (Williams & Cain, 1959); Cochrane Distr., in black spruce bog. NOVA SCOTIA. French Mt., Cape Breton; L. Charlotte, Halifax Co. (*RMS*). NEW-FOUNDLAND. N. of Robinson's and SW. of St. George's, W. Nfld. (*RMS*).

MAINE. Mt. Desert I. (*Lorenz, 1924*); Matinicus I.; foot of Mt. Katahdin (*RMS*); Schoodic L., Piscataquis Co. (Evans, 1907b). NEW HAMPSHIRE. Mt. Monadnock; Eagle L., Mt. Lafayette, and Lonesome L., Franconia Mts. (Lorenz, 1908c). VERMONT. Rochester; Willoughby (*Lorenz*). MASSACHUSETTS. Hawley Bog, W. Hawley, Franklin Co. (*RMS 43953a*); Reading. RHODE I. Hopkinton, with *Cephaloziella elachista;* see also Evans (1913). CONNECTICUT. In tuft of *Dicranum bergeri*, New Milford (Evans, 1907b, p. 65); Woodbury (Evans, 1907b, p. 69; type of *C. tenuis*).

NEW YORK. Bergen Swamp, Genesee Co.; Junius Bog, Seneca Co.; Cattaraugus Co. (Schuster, 1949a); L. George, N. of Glen L. (*Burnham*) [also (almost surely erroneously) reported from Cold Spring Harbor, Long I., by B. Andrews, 1931]. NEW JERSEY. Closter (Austin, Hep. Bor.-Amer. no. 74, 1873!; type of "var. *tenuis*"). MICHIGAN. Mud L., Cheboygan Co. (*Cain 3917*); bog bordering Little L. Sixteen, Cheboygan Co. (*Steere*). MINNESOTA. Big Susie I., Long I., Porcupine I., Belle Rose I., Lucille I., all in Susie Isls., Cook Co.; Grand Portage and Pigeon R., Cook Co; near L. George, Hubbard Co.; near Jaynes, Itasca Co.; Black Bay, Rainy L., and SE. of Ericsburg, Koochiching Co.; Kerrick, Pine Co.; N. of Orr and SW. of Gheen, St. Louis Co. (Schuster, 1953). WEST VIRGINIA. Cranberry Glades, Pocahontas Co., 3500-3600 ft (*RMS 61211, p.p., 61213; RMS 61208 =* fo. *bidenticulata, 61210*). Also reported from Bald Knob in Pocahontas Co. (*Gray; fide* Frye & Clark, 1937-47). VIRGINIA. Cranberry Bog, near Mountain L., Salt Pond Mt., Giles Co. (*RMS 61279*).

Ecology

Calypogeia sphagnicola, as the name implies, is nearly restricted to peat bogs and to *Sphagnum*-capped crests of cliffs. I (Schuster, 1953) have found the species frequent in *Sphagnum* bogs, "associated with *Mylia anomala, Cephaloziella elachista, Cephalozia pleniceps* and *connivens*, and occasionally with *Lophozia marchica* and *Cladopodiella fluitans* .... It also occurs over peat on sphagnous crests of cliffs ... pH 3.7–3.9, with *Mylia, Cephalozia loitlesbergeri, Odontoschisma denudatum, Lophozia ventricosa*."

The occurrence of *C. sphagnicola* in bogs, as a typical member of the *Mylia-Cladopodiella* Associule (see Schuster, 1957), or the *Mylia-Odonto-schisma* facies of that community of peat-covered ledges, is diagnostic of this species. In bogs usually confined to the heath stage (*Chamaedaphne* associes), often with *Ledum groenlandicum, Kalmia polifolia, Vaccinium oxycoccus, Chamaedaphne calyculata*, and *Andromeda polifolia*; on peat-capped moist ledges *Ledum* (more rarely also the *Andromeda*), *Vaccinium vitis-idaea*, and other species of *Vaccinium* (*oxycoccus, macrocarpon*) are frequently associated.

In the isolated sites in the high Arctic, as at Kânâk (*RMS 45603, 45606*), the species occurs in typical form among *Sphagnum*, together with *Lophozia opacifolia, L. alpestris s. lat.*, and *Cephaloziella* sp., together with *Cassiope tetragona, Vaccinium uliginosum* var. *microphyllum*, and *Dryas chamissonis*. In less extreme latitudes, as at Ikertoq Fjord (*Vahl 192*), the species occurs with *Mylia anomala* and *Lophozia binsteadii*. In the arctic stations and the few stations from the Susie Isls., Minn., *C. sphagnicola* often occurs in areas juxtaposed to *Empetrum* (usually *E. hermaphroditicum*). It is noteworthy that in Tierra del Fuego *C. sphagnicola* (or a plant I cannot separate from it) occurs with *Empetrum rubrum* in bogs.

Variation

Correlated with a relatively narrow environmental range frequented by *C. sphagnicola*, limited variability is found. However, depending on the amount of light and moisture available, considerable phenotypic variation occurs. This can be easily studied in bogs, when populations from different levels above the water table are accessible. The range in variation in cell size has already been commented upon (Schuster, 1953). Correlated with this, in going from hygric extremes to the more xeric conditions found at the exposed crests of peat hummocks, there is a gradual development and increase in size of the trigones; under extreme conditions, these may be bulging (Fig. 108:10).

With hygric conditions, plants develop into a lax-leaved phase (mod. *leptoderma-laxifolia*) with distant or approximate leaves, narrower than long (mod. *angustifolia*), which are strongly decurrent at base (Figs. 108:4, 110:9); such plants typically also have small underleaves, barely wider than the stems (mod. *parvistipula*).

With xeric conditions, leaves become moderately dense and broader (mod. *meso-vel-pachyderma-densifolia-latifolia-densistipula*), smaller (mod. *par-vifolia*), strongly ovate in shape, fully as broad as long, and not strongly

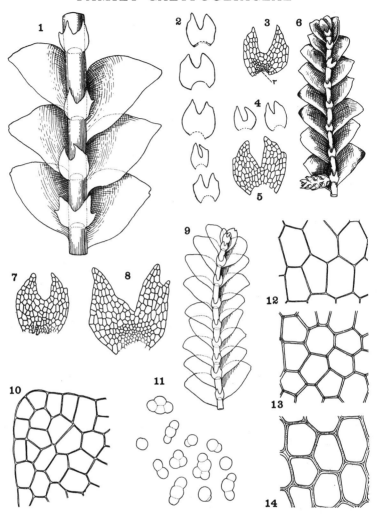

FIG. 110. *Calypogeia sphagnicola.* (1) Shoot sector, ventral aspect (×ca. 28). (2, 4) Underleaves (×28). (3, 5, 7–8) Underleaves (×ca. 58), the rhizoid-initial area at *r*. (6) Shoot with androecium, ventral view (×ca. 12). (9) Shoot, ventral view (×12). (10) Leaf apex cells (×ca. 245). (11) Oil-bodies (×1700). (12–14) Median cells, from young to old leaves, showing progressive thickening of wall in some phases (×290). [1, 11, from Belle Rose I., Minn., *RMS;* 2–10, Junius Bog, near Geneva, N.Y., *RMS;* 12–14, Bergen Swamp, Genesee Co., N.Y., *RMS.*]

decurrent (Fig. 108:8). Such phases also have nearly approximated underleaves which become broader (up to 1.5–1.8× as wide as the stem). Unlike *C. suecica, C. sphagnicola* does not appear to produce an extreme dense-leaved phase in nature. Buch (1936a) emphasized that such phases, with imbricate underleaves, are produced only in culture.

The xeric extreme has been considered to represent "typical" *C. sphagnicola*. With age, cells of this phase may become equally thick-walled (such a plant was the basis for the illustration in Macvicar, 1926, p. 319). By contrast, Evans (1907b) attempted to maintain a separate *C. tenuis* based on the more hygric extreme; this leptodermous plant, always gemmae-free, may be more branched— and always shows some terminal, *Frullania*-type branches.

Xeric phases occurring over exposed peat on cliffs tend to be smaller (often less than 1 mm wide) and produce enormous numbers of extremely slender, erect gemmiparous shoots.[7] By contrast, laxer bog forms are often virtually devoid of gemmae. It is possible that the greater intermittency of moisture conditions on *Sphagnum* over cliff crests stimulates formation of gemmiparous shoots.

Considerable variations of *C. sphagnicola* occur at the northern edge of its range, in Greenland, and at the southern edges of the American range, in the Southern Appalachians.

In Greenland, plants from Ikertoq Fjord and Tasermiut Fjord, e.g., are thoroughly "normal," and possess sporadic to frequent terminal branches (as occur also in the type of *C. sphagnicola* from Mora, Sweden), are small-celled [apical cells from 22–26 to 23–31 μ; median cells 24–32(34) $\times$ 25–38(48) μ], average only 900–1530 μ wide, and have obliquely ovate leaves ca. 750 μ long $\times$ 650 μ wide. Other Greenland plants, which appear to correspond nearly to the "*tenuis* extreme," may be larger, with larger cells. For example, the Tupilak I. material has broader, often notched leaves and broad underleaves frequently provided with lateral angulations or teeth; it approaches *C. muelleriana* in facies and in size (shoots to 2.3 mm wide) and cells are unusually large (apical 26–36 μ; median 34–42 $\times$ 48–60 μ). Yet this plant often develops terminal branches—sometimes 2–3 per plant!

In the Southern Appalachians two extremes occur that may represent polyploids of the species. The three most evident phenotypes are separable as follows.

Key to Forms of Calypogeia sphagnicola

1. Small or very small, ca. 0.5–1.6(1.8–2) mm wide, with leaf apices bluntly pointed to obtuse; marginal cells of leaf apices 25–32 μ; median cells up to 30–35 $\times$ 38–42 μ; oil-bodies to 4 $\times$ 5–7 μ, 2–4(6)-segmented, 2–5 per cell; often with conspicuous trigones; gemmiparous, erect shoots usually common. Northern (south to New York, Minnesota, and Wisconsin) . . . . Typical *C. sphagnicola*
1. More robust, the mature ones 1.5–2.05 to 2.0–2.9 mm wide; marginal cells of leaf apices 26–32 to 30–35 μ on an average; median cells 32–38 $\times$ 45–55

---

[7] Discovery of most populations growing on living *Sphagnum*, over shaded to sunny cliff crests, resulted from noting the filiform and rigid, erect, yellow-green-tipped gemmiparous shoots (the leafy plant being inconspicuous and growing appressed to *Sphagnum*). Macvicar (1926, p. 315) attempts to separate *C. suecica*, with "gemmiform stems numerous and conspicuous," from *C. sphagnicola*, supposedly with "gemmiform stems not conspicuous." This distinction does not hold.

μ or larger; oil-bodies to 4–5 × 9–11 μ, the larger of 6–12(16) segments, 4–8 to 7–12 per median cell; no distinct trigones; never with gemmae. 2.

　2. Leaves with apices often to usually bidentulate on mature shoots; apical cells 26–32 μ, median cells (28)32–38 × 45–55 μ; oil-bodies 4–8 per cell . . . . fo. *bidenticulata*, fo. n.

　2. Leaves with apices bluntly acute to acute, never bidentate; apical cells (27)30–35(38) μ, median cells 40–52 × 55–85(100) μ; oil-bodies 7–12 per median cell . . . . fo. *paludosa* (Warnst.) Schust.

CALYPOGEIA SPHAGNICOLA fo. PALUDOSA
(*Warnst.*) *Schust., comb. n.*
[Fig. 111:1–8]

Calypogeia paludosa Warnst., Krypt.–Fl. Mark Brandenburg 2:1117, 1906.

Often unusually vigorous, the larger (2.0)2.5–2.9 mm wide, very sparingly branched (only postical-intercalary branches seen). Leaves asymmetrically ovate to dimidiate-ovate, often long-decurrent postically, ca. 1050 μ wide × 1200 μ long, the apices bluntly pointed to narrowly rounded or obtuse, *always entire. Cells large:* apical cells (27)30–35(38) μ; median cells ca. 40–52 × 55–85(100) μ, delicate, polygonal to oblong-polygonal, with feebly indicated trigones and thin walls; oil-bodies *more* numerous (7–12 per cell in median cells), *larger* (to 4–5 × 8–11 μ), formed in part *of more numerous segments* (to 7–12 in surface view). Underleaves not squarrose, deeply bifid, seemingly never with angulate or dentate margins, to ca. 550–575 μ wide × 570–600 μ long, the lobes 8–9 cells wide at base.

The preceding diagnosis is fashioned from the West Virginia plant. I have reluctantly adopted Warnstorf's epithet for this phenotype, rather than base a new one on it; however, I have not seen Warnstorf's type.

Distribution

WEST VIRGINIA Cranberry Glades near Mill Point., ca. 3500 ft, Pocahontas Co. (*RMS 61210;* admixed with *Polytrichum, Sphagnum,* etc.).

Some of the Greenland plants (see p. 139) also appear referable to the *paludosa* extreme, but differ in producing terminal branches.

This variant is placed with some reluctance under *C. sphagnicola,* with which it shares the entire facies: the dimidiate-ovate, long-decurrent, oblique leaves, always at least feebly longer than wide, often remote to barely contiguous; the deeply bifid underleaves averaging as long as to slightly longer than wide; the occurrence over *Sphagnum* in open bogs. Dead plants, except for size (to 2.5–2.9 mm wide), are separable from "normal" var. *sphagnicola* principally by the much larger leaf cells.

The large leaf cells and the number, segmentation, and size of the oil-bodies, together with the bifid underleaves, suggest *C. muelleriana.*

Evans (1915) discusses *C. paludosa* and, although recognizing it as distinct, admits that its acceptance "rests on a rather insecure basis." He follows Schiffner (Krit. Bemerk. 13:8–9, 1914) in giving it the rank of a species, partly on the basis of the leptodermous and larger cells, the more vigorous size, and the less deeply bifid leaves with broader lobes. These are all criteria that suggest *C. muelleriana* (*C. trichomanis* auct., in large part). Indeed, Schiffner's description of the capsule of *C. paludosa* as having 8 epidermal cell rows lacking nodular thickenings appears to prove conclusively that he confused *C. muelleriana* (and/or *trichomanis*) with the *C. sphagnicola-paludosa* complex.

CALYPOGEIA SPHAGNICOLA fo. BIDENTICULATA, fo. n.
[Fig. 111:9–15]

Forma satis magna, usque ad 1.8 mm lat.; ramificatione partim terminali; folia apices acutos ad, saepius, minute bidentatos habentia; amphigastria parva, saepe squarrosa, profunde bifida.

Moderately vigorous, the mature shoots 1.5–1.8(2.05) mm wide, sparingly branched, the branches in part terminal, of the *Frullania* type, in part postical-intercalary. Leaves ovate to dimidiate-ovate, the postical margins from nearly straight to moderately convex, but less so than the strongly arched antical margin, which is subcordately dilated at antical base; leaves from 660 μ wide × 780 μ to 880–990 μ wide × 1050–1100 μ long, always slightly to distinctly longer than wide, at the *apices sharp-pointed to (more usually) minutely bidentate*. Apical cells from 26 to 30(32) μ, here and there a few marginal apical cells elongated and ca. 20–25 × 40–48 μ; median cells ca. (28)32–38 × 45–55 μ; cells leptodermous, with minute trigones; oil-bodies in all leaf and under-leaf cells, in median leaf cells ca. 4–8 per cell, variable, some ca. 4–4.5 × 5.5–7 μ and few (3–7)-segmented, but many larger, linear to irregularly linear-fusiform or ellipsoidal, ca. *3.5–4.5 × 9 μ, formed of 7–11(16) segments in surface view*. Underleaves *small, often squarrose, deeply bifid*, only 1.2–1.5× as wide as stem.

Type. West Virginia: Cranberry Glades near Mill Point, ca. 3500 ft, Pocahontas Co. (*RMS 61208;* admixed with *Cephalozia connivens* and *C. pleniceps*, all growing over *Sphagnum*).

This variant of *C. sphagnicola* seems distinct to me in that mature, normal, weakly imbricate-leaved shoots bear almost consistently bidenticulate, less often pointed, leaves, reminiscent in their apices of *C. fissa* and its var. *neogaea*. On weaker stems the apices may be sharply and consistently bidentate. Equally distinctive seem to be the more copiously segmented oil-bodies, many of which are formed of 8–12 or more segments.

The fo. *bidenticulata* may be a mere extreme of what Evans (1907b, p. 69) described as *C. tenuis*. Evans stressed the tendency for the leaves to be bidentate or bilobed, and for the cells to be larger than in *C. sphagnicola*. However, Evans' *C. tenuis* was somewhat more broadly delimited and seems to represent merely the leptodermous bog phases of *C. sphagnicola s. lat.* For that reason, and because I think the present plant is at best a minor genotype, it is described as a forma, under a separate name.

Differentiation

Calypogeia sphagnicola is a relatively well-defined species, with a sharply circumscribed physiological range, confined to acid sites (pH usually

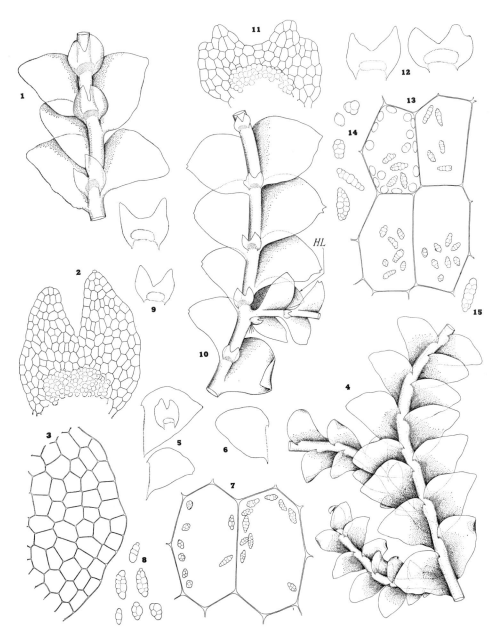

FIG. 111. *Calypogeia sphagnicola*. (1–8, all fo. *paludosa;* 9–15, all fo. *bidenticulata*). (1) Shoot sector, postical aspect (×18). (2) Underleaf (×64). (3) Cells of leaf apex (×145). (4) Plant, antical aspect, with two ventral-intercalary branches (×14). (5–6) Leaves and an underleaf (×14). (7) Median cells with oil-bodies (×350). (8) Individual oil-bodies (×725). (9, 12) Underleaves (×42). (10) Shoot sector, ventral view with *Frullania*-type terminal branch, at HL the narrow dorsal half leaf (×18). (11) Underleaf (×87). (13) Median cells with oil-bodies and, upper left, chloroplasts (×530). (14–15) Individual oil-bodies (×950). [All from Cranberry Glades, W. Va.; 1–8, *RMS 61210;* 9–15, *RMS 61208*.]

3.7-4.2); see Schuster (1953, 1957). Correlated with the narrow tolerances, the rigidly circumscribed distribution and ecological range differentiate the species from all others of its genus—as do the morphological
characters.

Morphological differences of note are (*a*) the small—but, as in other
species, variable—cell size and well-marked ability to develop discrete,
even weakly bulging trigones; (*b*) the small overall size of the plants; (*c*)
the deeply bilobed underleaves with acute to subacute, not divaricate,
lobes; (*d*) the essentially broadly ovate-triangular leaves; (*e*) the usually
few (2-4-5 per cell), small, few-segmented oil-bodies; (*f*) the frequent—
probably regular—presence of terminal, *Frullania*-type branches. Deviations from this ensemble of characters are frequent and have been
previously noted.

All but the last of these characters are shared with *C. suecica*, to which *C.
sphagnicola* is probably only distantly related because of (*a*) the fundamental
habitat (and consequent physiological) differences; (*b*) the monoecious inflorescences; (*c*) the presence of distinct nodular thickenings of *both* faces of
secondary longitudinal walls of the epidermal cells of the capsule; and (*d*)
the haploid chromosome no. of 18 (9 in *C. suecica*). Since *C. sphagnicola* is only
infrequently fertile and since a chromosome count is not readily made, the
first-cited difference is the most important one, in practice, in separating the
two taxa. However, additional differences are usually evident, among them
(*a*) the slightly larger cells of *C. sphagnicola*, always over 25 × 30 μ in leaf
middle; (*b*) the apparently constant inability to develop an extreme dense-
leaved phase, with imbricate underleaves (mod. *densifolia*, with subimbricate
to imbricate underleaves, is commonly produced in *C. suecica*); (*c*) the underleaves rarely over 1.8× as wide as the stem and usually not wider than long (in
C. suecica usually 2-2.5× as wide as the stem, usually somewhat broader than
long).

The largest extremes of the lax-leaved, leptodermous mod. *laxifolia-leptoderma*,
found in wet bogs, may approach some extremes of *C. muelleriana*. Such phases
of *C. sphagnicola* may have larger cells than normal, approaching those of *C.
muelleriana*. However, such bog phases rather freely produce terminal, *Frullania*-
type branches, which I have rarely seen in any phase of *C. muelleriana*.

CALYPOGEIA SUECICA (Arn. & Perss.) K. Müll.

[Fig. 112]

Calypogeia trichomanis auct. (*p.p.*)
Kantia suecica Arn. & Perss., Rev. Bryol. 29:29, figs. 1–6, 1902.
Cincinnulus suecicus K. Müll., Beih. Bot. Centralbl. 13:99, 1902.
Calypogeia suecica K. Müll., *ibid.* 17:224, 1904.
Cincinnulus trichomanis var. *suecicus* Meyl., Bull. Herb. Boissier, Ser. 2, 6:499, 1906.
Calypogeia trichomanis var. *suecica* Casares-Gil, Fl. Ibérica, Hepáticas, p. 575, 1919.

In ± *shiny*, pellucid, pale yellowish green or pale green patches (with age,
pale brownish), exclusively *on decaying, often peaty logs*. Shoots *small, usually*

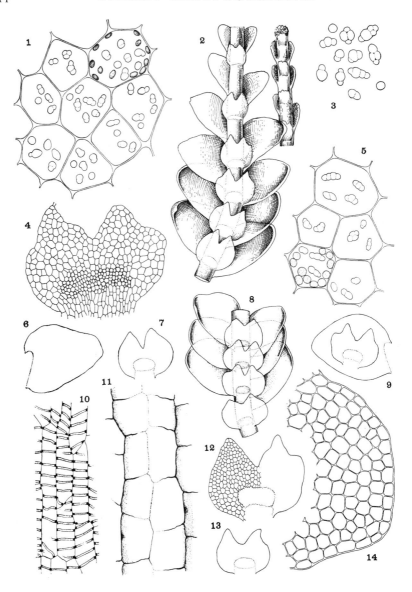

Fig. 112. *Calypogeia suecica.* (1) Median cells with oil-bodies, in upper right also chloro-plasts (×ca. 470). (2) Gemmiparous shoot, apex at right (×ca. 24). (3) Oil-bodies (×755). (4) Underleaf (×64). (5) Cells with oil-bodies and lower left, chloroplasts (×462). (6–7) Leaf and underleaf (×27). (8) Shoot sector, of the normal mod. *densifolia* (×25). (9) Leaf and underleaf (×27). (10) Inner cells, capsule wall (×285). (11) Epidermal cells of capsule wall, valve midline towards the left (×285). (12) Under-leaf (×31). (13) Underleaf (×27). (14) Leaf apex (×135). [1–3, Amygdaloid I., Isle Royale, Mich., *RMS;* 4–8, 13–14, *RMS 24000*, Mt. LeConte, Tenn.; 9, 12, *RMS 24826*, N.C.; 10–11, *Wilson 10*, Mt. Marcy, N.Y.]

(0.9)1.2–1.8, rarely to 2–2.5, mm wide, rather frequently with postical branches, *without terminal branching*; stems relatively stout, 200–250 μ wide. Leaves usually *strongly imbricate, somewhat convex*, typically with decurved apices, obliquely ovate to *broadly ovate-cordate*, 0.9–1.1 × as wide as long, varying from ca. 750 μ long × 720 μ wide to ca. 1000 μ long × 1100 μ wide, *narrowly to broadly rounded* to narrowly truncate *at apex*, sporadically obscurely bidentate, maximal width somewhat above the base, barely decurrent ventrally. Cells ± collenchymatous, frequently with *slightly bulging trigones* and rounded lumina, *nearly isodiametric throughout leaf*, no or little larger below middle than near apex; apical cells of leaves small, usually quadrate and isodiametric, (23)25–28 μ (more rarely a few 30–32 μ long × 25–28 μ wide); *median cells and subapical cells 25–28* to (mod. *leptoderma*) *30–35 μ, virtually isodiametric;* cells at leaf base 26–32 × 35–45(54) μ. *Oil-bodies colorless*, 2–7 per cell, 4 × 5.5 to 5.4 × 9 μ, a few to 6 × 8 or 4 × 10 μ, *usually of 2–4(5) coarse, globular segments 2–3.5* (occasionally 4) μ in diam.; chloroplasts ca. 4 μ. *Underleaves very large*, approximate to *distinctly imbricate*, broadly, *transversely ovate to suborbicular* (*1.3–1.6×* as wide as long), 2–3(3.5)× as wide as stem [from (310)440–500 μ long × (500)560–580 μ wide to 525 μ long × 675 μ wide], *divided ⅖–⅗ their length (to within 4 cells of rhizoid-initial region)*; lobes acute to obtuse, *often with an obtuse angulation* or tooth on one or both outer margins, sinus acute to rectangular; rhizoid-initial field broadly elliptical, 2–2.5× as wide as high, conspicuously, abundantly rhizogenous. Median underleaf cells barely elongate, 20–30 μ wide × 23–30 μ long. Asexual reproduction by gemmiparous shoots sometimes rare, usually present only when the species occurs as a pioneer in insolated sites; gemmae 2-celled, 15–20 μ, thin-celled, spherical to ovoid.

Dioecious, but frequently producing capsules. Valves 1300–1600 μ long × 160–240 μ wide (7–8× as long as wide). Epidermal cells *in usually 8 rows* (cells 20–30 μ wide × 40–52 μ long); cells typically with alternating longitudinal, radial walls (= secondary walls) with an *even or nearly even layer of reddish secondary thickening* (without or only rarely with irregularities in thickening), the reddish deposit usually thin, *nearly confined to adaxial radial faces of cells* (extending for a short distance onto the outer tangential walls *as a thin layer of reddish deposit*); intervening (= primary) longitudinal radial walls unthickened, pellucid (and difficult to demonstrate), or occasionally with a slight layer of reddish deposit; transverse radial walls with a thin layer of reddish thickening, becoming colorless usually as the longitudinal colorless radial walls are approached. Inner layer in 15–16 cell rows with narrow cells, the individual cells 11–14 μ wide × 60–80 μ long, on the external (inner) wall with numerous complete tangential bands, extending over the radial walls and continued as spurs onto the internal tangential walls (but never complete there). Spores (8)9–11 μ, minutely verruculose, pale brownish. Chromosome no., $n = 9$.

Type. Hede, Prov. Herjedalen, Sweden (*J. Persson, Aug. 27, 1899*).

Distribution

Frequent in northern and central Europe (see, e.g., Müller, 1951–58); disjunctly occurring on the Azores (Allorge & Allorge, 1950).

In North America almost restricted to the boreal spruce-fir forests.
Unlike *C. sphagnicola*, appearing absent from the small disjunct outliers
south of that region, consequently a good index species of the subarctic-
subalpine region, except for anomalous occurrences in the Southern
Appalachian Escarpment (South Carolina, Georgia). As Buch (1936a)
points out, *C. suecica* usually occurs on decorticated spruce and fir logs;
in the normal habitat of this species these are the only common species of
trees present. Frequent from Quebec to British Columbia, southward in
the East to New York and Connecticut, with disjunct stations in the
spruce-fir forested summits of the Southern Appalachians (in Virginia,
North Carolina, and Tennessee), where locally common in the dense
Fraser fir-spruce forests of the high summits (above 5500 feet). Southward in the Midwest to Michigan, Wisconsin, and Minnesota, and in the
West to California.

NEWFOUNDLAND. Placentia Bay (*Waghorne 40, 41, 1891!*; among *Lophozia incisa,
Lepidozia reptans, Cephalozia leucantha, Blepharostoma*); Spread Eagle (*Waghorne 227!*;
among *Lophozia incisa*); Hogans Pond, Aquaforte, and Biscay Bay, in Avalon Distr.;
Cape Ray, Piccadilly, Steady Brook, Benoit Brook, all in W. Nfld.; between Cow Head
and Stanford R., and at Cow Head, N. Pen. (Buch & Tuomikoski, 1955).   QUEBEC.
Rivière Ste-Anne-des-Monts (Lepage, 1944–45); near St. Jovite, Montcalm Co.
(Crum & Williams, 1960); N. of Malbaie R., 10–12 mi NW. of Bridgeville, E. end of
Gaspé Pen. (*RMS 43056c*).   ONTARIO. Durham and Halton (Williams & Cain, 1959);
near Millbrook, Durham Co. (*Williams 959!*).   NOVA SCOTIA. Ingonish, Cape Breton
I. (Brown, 1936a); Benjy's L., French Mt., Cape Breton (*RMS*).
MAINE. Mt. Desert I. (Lorenz, 1924); Mt. Katahdin (Evans, 1907b; *RMS*).   NEW
HAMPSHIRE. Wildwood Path, Franconia Mts. (Lorenz, 1908). VERMONT. Willoughby
(*Lorenz*); Long Trail below Taft Lodge, Mt. Mansfield, 3200–3300 ft (*RMS 43823*);
Haselton Trail, 2800–3200 ft, Mt. Mansfield (*RMS 43802, 43844*).   CONNECTICUT.
Stafford, Tolland Co. (Evans, 1907b).   NEW YORK: Catskill Mts.; Washington Co.
(*Burnham*).   MICHIGAN. Kauffman (1915), according to Frye & Clark (1937–47);
this report probably incorrect; edge of Rapid R., 6–7 mi NNE. of Kalkaska, Kalkaska
Co. (*RMS 38966*).   MINNESOTA. Big Susie I., Susie Isls., Cook Co. (*RMS 12013b*);
Lutsen, Beaver Dam, Arrowhead R., and Hungry Jack L. Trail (Conklin, 1942;
Schuster, 1953).

VIRGINIA. Buffalo Cr., Rockbridge Co., along main highway (*Carroll 304!*); in spruce-
fir zone, ca. 5000 ft, Mt. Rogers, Grayson, and Smyth Cos. (*RMS & Patterson 38040*).
[The report of the species from Mountain L. area by Sharp (1944) is incorrect; the
juvenile, fragmentary specimen at the Mountain L. Biological Station is indeterminable
but cannot be this species, since the plants are growing on soil. However, I repeatedly
collected the species just above The Cascades, Salt Pond Mt., Giles Co., near Mountain
L. (*RMS 40385a, 61981*), in an abundantly gemmiparous, typical phase, on decaying
logs.]   NORTH CAROLINA. Black Mts., below summit of Mt. Mitchell, Yancey Co.
(*RMS 24826*); Roan Mt., Mitchell Co., 6000 ft (*RMS 40318*); Andrews Bald, 1.5 mi
S. of Clingman Dome, ca. 5800 ft, Swain Co. (*RMS 36612c*).   TENNESSEE. Roan Mt.,
on exposed N.-facing cliffs of Roan High Bluff, 6100–6200 ft, Carter Co. (*RMS 61282*);
above and near Myrtle Pt., Mt. LeConte, Sevier Co. (*RMS 45322, 45326*); Mt. Kephart,
Smoky Mts. Natl. Park, Sevier Co. (*RMS*); along gorge emptying into Middle Prong of

Little R., at ca. 1600 ft below Tremont, Blount Co. (*RMS 34644*).   OHIO. Champaign Co. (Miller, 1964).

With wholly disjunct peripheral stations at low elevations in the Southern Appalachian Escarpment gorges: SOUTH CAROLINA. Moist decaying log in ravine of Estatoe R., Pickens Co., ca. 1200 ft (*RMS 37662*, p.p.).   GEORGIA. Big Cr., ca. 0.3–0.4 mi below High Falls, ca. 1900 ft, N. edge of Rabun Co., S. of Highlands, N.C. (*RMS 40719*).

The Georgia and South Carolina stations in the *Hymenophyllum-Trichomanis* "Old Tropical" zone of the Appalachian Escarpment,[8] in *Leucothöe editorum-Rhododendron maximum-Tsuga canadensis* shaded ravines. Associated Hepaticae were *Nowellia, Cephalozia catenulata* and *media, Harpanthus scutatus, Scapania nemorosa* and the moss *Tetraphis pellucida*, all species associated also in more northern habitats.

Ecology

An obligate xylicole; on decorticated logs it occurs relatively rarely as a strict pioneer (then occasionally with frequent rigid, stiffly ascending or erect gemmiparous shoots), but more often after previous colonization by other xylicolous Hepaticae of the *Nowellia-Jamesoniella* Associule (see, e.g. Schuster, 1957). Most commonly associated are *Geocalyx graveolans, Harpanthus scutatus, Cephalozia media* and *catenulata, Riccardia latifrons* and *palmata, Geocalyx graveolans, Lepidozia reptans, Blepharostoma trichophyllum, Anastrophyllum michauxii*, oftener *A. hellerianum, Nowellia curvifolia, Scapania umbrosa* (more rarely *apiculata*), *Lophozia porphyroleuca, L. incisa*, and *L. ascendens*. More rarely *Cephalozia bicuspidata, Odontoschisma denudatum, Bazzania trilobata, Lophozia attenuata*, and the moss *Tetraphis pellucida* are consociated. The species is strongly oxylophytic (pH 4.6 or lower) and most often sciaphytic.

Variation

Calypogeia suecica undergoes considerable variation, concurrent with differences in (*a*) light intensity and (*b*) amount of available moisture. As Buch (1936a) has pointed out, the plant in nature occurs usually as a mod. *densifolia-densistipula* (because of the subxeric sites the species usually frequents), with imbricate leaves and often imbricate underleaves. With relatively intense light, the species may still develop leaves of maximal size (ca. 1 mm long), fully as broad as long. Under xeric (or perhaps also nutritionally difficult) conditions, a mod. *parvifolia-angustifolia-densifolia* is produced, with ovate leaves, narrower than long, and closely imbricate underleaves; such plants may be only 1–1.3 mm wide. Under natural conditions the mod. *laxifolia-parvifolia* appears to be rare or absent; however, Buch produced it in the laboratory.

---

[8] In both gorges the extremely local and (regionally) rare *Hymenophyllum tunbridgense* occurs (or occurred).

Concurrently with the usually subxeric environment, *C. suecica* frequently occurs as a mod. *meso-vel pachyderma*, with somewhat thick-walled cells provided with distinct, sometimes bulging trigones. Such modifications are not developed in our other species (with the exception of the mod. *densifolia-parvifolia* of *C. sphagnicola*). The xeric phase has smaller cells (ca. 25–28 μ) than more mesic extremes. Meylan (Rev. Bryol., 1908, p. 74) has described the mesic extreme as fo. *erecta*; this has cells 30 × 35 μ medially and vestigial trigones.

If *C. suecica* did not occur almost invariably as a mod. *densifolia-densistipula* and *C. sphagnicola* as a mod. *laxifolia-laxistipula*, the two could be easily confused. Buch (1936a) has shown that corresponding modifications of the two species are closely similar.

The changes in leaf form in *C. suecica* are at times difficult to comprehend. Usually the lax forms of each species of *Calypogeia* have narrow leaves; dense-leaved forms develop broader, overlapping leaves. However, we occasionally find in *C. suecica* that the smallest (only 1.15 mm wide) shoots of mod. *densifolia-densistipula*, from somewhat xeric sites, may have decidedly narrow leaves. The leaves may undergo allometric growth, with the smallest modifications tending to produce narrow leaves, even where growing under xeric conditions, in relatively direct light. Whatever the exact interpretation may be, it is a fact that the leaves of this species undergo very wide changes in shape.

The capsule is usually relatively small (length of valves 1300–1600 μ)[9] and appears quite uniform in structure, judging from a series of capsules from diverse collections. The inner cell layer always has numerous transverse, complete bands of the superficial wall (inner wall), which extend across both adjacent radial walls and, for some distance, as tapering spurs across the internal tangential wall (e.g., the wall adjoining the epidermal cells). The inner tangential extensions, though spurlike and incomplete, are unusually well developed (compared to those in *C. neesiana*, *C. trichomanis*, and other species). The epidermal cell layer has short-rectangular cells, with alternating walls, in complete longitudinal rows, thickened and reddish brown. The secondary layer of pigmented deposit is largely limited to the adaxial faces of the radial walls (with reference to the axis of the valve); it is sparsely developed and thin and often extends for a rather short distance across the adjacent external tangential wall (Fig. 112:11), resulting in a band of reddish pigment along the adaxial margins of most of the tangential cell walls. Müller (1913, *in* 1905–16, p. 234) incorrectly reported that the epidermal cells had weakly nodular thickenings; this inaccurate description was taken over by Frye and Clark but later corrected by Müller (1947a).

Buch (1936a) correctly stressed that gemmiparous shoots were produced

---

[9] Frye and Clark state 2–3 mm. This figure is certainly erroneous. Since these authors also incorrectly describe the epidermal cells as having the "longitudinal walls weakly nodular," it appears that they confuse this species with other, unrelated ones.

only rarely in this species. However, under the restricted conditions (usually pioneer occurrences), when the correct stimuli are present to initiate their development, gemmae may be frequent. It is misleading, however, to describe these shoots as "numerous, conspicuous," as in Frye and Clark (1937–47). The populations at low elevations in the Southern Appalachians (Giles Co., Va.; Estatoe R., S.C.; Big Creek, Ga.) all, however, are abundantly gemmiparous. These plants are, in other respects, quite typical. Thus, in the Georgia plants the median leaf cells are only 25–30 μ and are very distinctly collenchymatous, and the plants are clearly dioecious (only androecia were seen).

Differentiation

Calypogeia suecica is a local species, of restricted distribution, apparently almost confined to the spruce-fir zone. The small size (rarely over 1.5–1.8 mm wide) and the rather small cells suggest *C. sphagnicola*. With this species *C. suecica* also shares the inability to undergo ecesis on mineral substrates, few-segmented oil-bodies, and a distinct tendency to develop collenchymatous cells, leading to potential confusion. However, a number of important differences suggest that these similarities are more superficial than indicative of close phylogenetic relationship: *C. suecica* has a chromosome no. of 9 (18 in *C. sphagnicola*); it lacks nodular thickenings of the epidermal cells of the capsule; it is dioecious; it usually has distinctly smaller cells than *C. sphagnicola*; its underleaves are usually fully twice (often 3 ×, rarely to 3.5 ×) as wide as the stem. In *C. sphagnicola* the underleaves vary from slightly wider to at most 1.8–2 × as wide as the stem.

The chromosome no. of 9, together with the small collenchymatous cells and the dioecious inflorescence, suggests that *C. suecica* is one of the most discrete of our species of the genus. The type and localization of secondary thickenings of the epidermal capsule-wall cells also suggest that *C. suecica* has no near allies; a similar capsule-wall anatomy occurs only in the unrelated *C. peruviana* complex.

I have spent many hours searching without success for terminal branches in *C. suecica*. All branches, sterile and fertile, appear to be uniformly postical-intercalary. In this respect, *C. suecica* is strikingly distinct from *C. sphagnicola* and more closely approaches our other species of the genus.

CALYPOGEIA TRICHOMANIS (L.) Corda emend.
K. Müll. (1947).
[Figs. 113, 114]

Mnium trichomanis L., Spec. Pl., p. 1114, 1753 (in part?).
Jungermannia trichomanis Dicks., Pl. Crypt. Brit., Fasc. 3, p. 10, pl. 8, fig. 5, 1793; Hooker, Brit. Jungerm., p. 79, 1816.[10]
Kantius trichomanis Gray, Nat. Arr. Brit. Pl. 1:706, 1821.[10]

[10] It is improbable that Dickson or Gray or Hooker had the present species.

Cincinnulus trichomanis Dumort., Comm. Bot., p. 113, 1822 (? in part).
Calypogeia trichomanis Corda, *in* Opiz, Beiträge, p. 653, 1829 (? in part); K. Müller, *in* Rabenh.
    Krypt.-Fl. 6(2):247, fig. 72, 1912 (in part, excl. fo. *Mülleriana*).
Calypogeia fissa var. *integrifolia* Raddi, Mem. Soc. Ital. Sci. Modena 18:44, 1820 (of reprint).
Calypogeia paludosa Schiffn. (nec Warnst.), in part, Krit. Bemerk. über die Eur. Leberm., Ser.
    XIII, 1914.
Calypogeia trichomanioides Warnst., Bryol. Zeitschr. 1:102, 1917, *fide* K. Müller (1947).
Calypogeia variabilis Warnst., *ibid.* 1:111, 1917, *fide* K. Müller (1947).

   Bluish to grayish green, in dark, flat patches. Shoots (1.8)2–2.5, occasionally
2.6–3.0 mm wide × 1–2.5 cm long, *the apices (when living) strongly turquoise blue;*
stems 250–300(330–400) μ in diam., usually simple, without terminal branching.
Leaves laterally spreading, at an angle of 45–55(65)° with the stem, *typically
heart-shaped, very broadly cordate-ovate, wider than long* [*ca.* 920–950(1000) μ long ×
1000–1150 μ wide up to 1150–1250 μ long × 1550–1750 μ wide], the width-
length ratio varying from *1.1: to 1.45:1;* leaves nondecurrent, the antical
base dilated, *often extending across and beyond the stem,* the maximal width somewhat
above the base; *leaf apices entire, subacute to narrowly rounded,* only on juvenile
shoots (or lowermost leaves of mature shoots) occasionally bidentate; marginal
cells never crenulate. Marginal cells little or no longer than wide, 22–35 μ,
not forming a border; subapical cells (23)25–34(36) μ wide × 30–38 μ long;
median cells (28)30–38 μ wide × 36–45 μ long; basal cells 29–38 μ wide ×
42–45 μ.[11] *Oil-bodies deep blue,* 2–5(7) per average median cell, segmented; oil-
bodies from 2–4-segmented and 5–6 μ long to 4–5 × 9 to 7 × 10 μ and of
5–8(9–10) segments;[12] the segments 2–3 μ in size; oil-bodies in all leaf, underleaf, and cortical stem cells, and in cells of seta. *Underleaves 1.8–2.2× as wide
as stem, subrotundate,* barely longer than wide (ca. 580 μ long × 560 μ wide)
to 1.0–1.2× as wide as long (460–510 μ long × 480–600 μ wide, to 520–570 μ
long × 580–670 μ wide); underleaves divided (to ca. 0.5–0.65 their length to
rhizoid-initial field) by an acute or subacute sinus, the lobes broadly *lanceolate
to triangular, acute or subacute,* the tips rarely narrowly rounded or obtuse; flanks
of underleaves *evenly rounded,* rarely with a broad, vague angulation or obtuse
tooth; underleaf bases shortly decurrent to a point a little below the broadly
ellipsoidal rhizoid-initial field; *underleaf length, from sinus to rhizoid-initial field,
4–5 cells.* Median cells of underleaves 22–27 × 34–44 μ,[13] up to 30 × 50–65 μ.
   Monoecious. Capsule valves ca. 340–350 μ wide, length ca. 7–9× their
width; epidermal layer with *both* faces of alternating longitudinal (radial) walls
more or less strongly thickened, reddish,[14] the intervening (= primary)

----

[11] These are from *RMS 17019.*
[12] According to Müller (1947a, p. 416), to 4–6 × 12–14 μ, formed of 5–10 segments up to
3.5 μ in diam.
[13] In the plants of *RMS 17019;* these plants also with small leaf cells.
[14] Schiffner (1914, *in* 1901–43), Müller (1947a) and Meylan (1924) state that these thickenings
form uniformly thick wall layers. In North American collections (*RMS 17019*), alternating
(secondary) longitudinal walls may be irregularly thickened on their radial faces. These
irregularities in the thickness of the reddish wall layer are evident, on superficial view, as "nodular" thickenings, which may extend for variable distances across the adjacent tangential walls
(Fig. 113:9), gradually thinning out and disappearing. Furthermore, the tangential walls,
adjacent to the radial walls that bear these thickenings, may have more or less well-developed,
sheetlike extensions of the pigmented layers of radial wall deposit (differing in thickness and,
as a consequence, in depth of pigmentation). A number of capsules have been examined, and
in all cases the pattern of localized thickenings indicated in Fig. 113:9 was found.

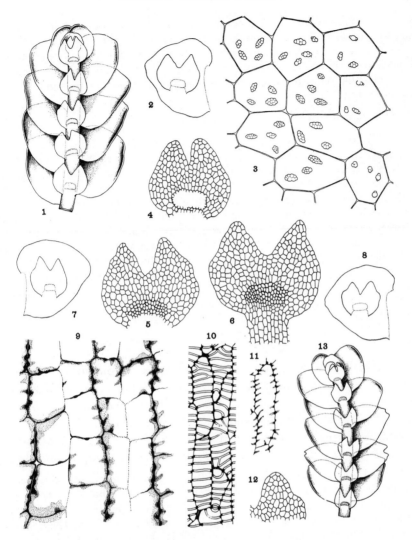

FIG. 113. *Calypogeia trichomanis*. (1) Shoot apex, ventral aspect (×12.5). (2) Leaf and underleaf (×18). (3) Median cells with oil-bodies (×375). (4–6) Underleaves, in fig. 6 with some ventral cortical cells drawn in (×45). (7–8) Leaves and, within, underleaves (×18). (9) Epidermal cells of capsule wall (×270). (10) Inner cells of capsule wall, free tangential wall surface (×270). (11) One cell from fig. 10, focus at internal tangential wall (×270). (12) Lobe of underleaf (×45). (13) Small shoot, showing occasional bidentate leaves (×12.5). [All from *RMS*, Mt. Katahdin, Me.]

longitudinal walls colorless and devoid of thickenings; epidermal cells ca. 30–45 μ wide × 36–60 μ long. Inner cell layer, on superficial view, with numerous reddish, complete semiannular bands extending across radial (vertical) walls and either ending on the inner edges of these walls or extending for short

distances across the interior tangential walls; the interior tangential walls consequently with nodular thickenings, often produced as short, tapering tangential
spurs; the inner cell layer with cells ca. 18–22 μ × 80–116 μ. Spores 14–17.5
μ, pale reddish brown, faintly granulose-papillose. Elaters bispiral, 9–10 μ
in diam., the reddish brown spirals 3 μ wide. Chromosome no.: $n = 18$
(Lorbeer, 1934) or 9 (Inoue, 1966e, for Japanese plants).

Type. European. Probably no longer extant.

It is doubtful whether Linnaeus had the present plant in hand when he
described *Mnium trichomanis*. This opinion is buttressed by (*a*) the relative
rarity of *C. trichomanis*, as restricted by K. Müller, in the north of Europe and
(*b*) the much greater frequency of *C. muelleriana*, particularly in northern Europe.
Therefore it would have been wiser, when *C. trichomanis s. lat.* was "split," to
have restricted the name *trichomanis* to the plant with colorless oil-bodies, now
called *C. muelleriana*, rather than to the plant with blue oil-bodies. Buch (1936a),
in his careful study of the genus in Finland, also used "*trichomanis*" for the plant
with colorless oil-bodies. As a consequence, adoption of the name *trichomanis*
for the relatively rare plant, of restricted distribution, with bluish oil-bodies
appears disadvantageous. However, since this change has been widely accepted
in Europe, it is followed here. Any further nomenclatural emendations would
result in chaos.

Distribution

The North American distribution almost totally undefined. The plant
has been reported from central New York (Schuster, 1949a), but only
part of those reports are correct; the rest refer to *C. muelleriana*. The only
recent reports that appear reliable are those of Persson (1952, pp. 6–7).
In addition, I have found typical material of the species in Maine and
Vermont. The following cited distribution is certainly quite incomplete.

On the basis of extensive field work, I can state, however, that the species is
completely—or nearly—absent from many loci from which it has been repeatedly reported, as, e.g., by Frye and Clark;[15] Conklin, 1929, 1942; Schuster,
1949a; Clebsch, 1954; Carroll, 1945; Plitt, 1908; Redfearn, 1952; Fabius,
1950; Blomquist, 1936; Underwood, 1892; Brown, 1936a; Little, 1936;
Haynes, 1915; Fulford, 1936; Gier, 1955; Ammons, 1940; Lehnert, 1886;
Evans, 1907b; Evans & Nichols, 1908; Sharp, 1939; and Patterson, 1949.
Among these loci are Labrador, Miquelon I., Nova Scotia, Michigan, North
Carolina, Pennsylvania, Massachusetts, Quebec, Ontario, Indiana, Illinois
(see, e.g., Arzeni, 1947), Wisconsin, Minnesota, New York, Missouri, Virginia,

---

[15] Frye and Clark (1937–47) report the species from Labrador south to Florida, west to
Alaska and California, citing it from 34 states and provinces, from Mexico, the West Indies
Asia, Europe, and Bermuda. Their description is vague and almost certainly not based on the
present plant. Of their figures, only those taken from Müller (figs. 1–2) are *C. trichomanis*;
fig. 5, from Kurz & Little (1933), cannot be *C. trichomanis*, and the old illustration taken from
Hooker (1816) is probably not this species. As is evident, neither the cited distribution, the
figures, nor the description appears to apply to *C. trichomanis* and probably are based largely on
C. muelleriana.

West Virginia, District of Columbia, Kentucky, Tennessee, Florida, Georgia Kansas (see, e.g., MacGregor, 1955), Oklahoma, Missouri. Extensive collecting in the upper Midwest and the Coastal Plain of the Southeast has totally failed to reveal the species there. [The Greenland reports of Lange and Jensen (1887) all represent *C. sphagnicola*.]

QUEBEC. Larch R., 18 mi above the fork (*Lepage 13736*; *fide* Persson, 1952, p. 6).

NOVA SCOTIA. Paquette L. near Neils Harbour (*RMS 68–657*) and Benjy's L. track, French Mt., Cape Breton (*RMS 68–827*).

MAINE. Mt. Katahdin, Harvey Ridge, on "Four Fools Trail" into Northwest Basin, ca. 3200 ft (*RMS 17019*). VERMONT. Mt. Mansfield, *ca.* 4000 ft (*RMS*). NEW YORK. South of Dryden, Tompkins Co. (*RMS 1651a*; basis for pl. 65, figs. 9, 12, *in* Schuster, 1953). [SOUTH CAROLINA. West branch of Estatoe R. ravine, Pickens Co. (*RMS 37663*). The South Carolina plants hardly typical, individually discussed subsequently.]

A "*Calypogeia trichomanis* var. *aquatica* Ingham" (see Rev. Bryol. 1906), was reported from a bog at Shad Bay Rd., Nova Scotia (Brown, 1946; based on a determination by Brinkman). This report is quite ambiguous.

Ecology

Very poorly known. Müller (1947a) speaks of both swamp and terrestrial forms of the species, giving it wide ecological tolerances. In my limited experience, it is rare on humus and on decaying shaded logs in the northeastern United States; occurring, e.g., with *Jungermannia lanceolata* and *Blepharostoma trichophyllum* on shaded, decaying logs in spruce-fir forest. Frequent in Alaska (*fide* Persson), appearing to occur under a variety of conditions, to judge from associated plants. (Cryptogams: *Hookeria lucens, Conocephalum, Sphagnum girgensohnii, S. nemoreum, Aulacomnium palustre, Hypnum subimponens.* Vascular plants: *Coptis asplenifolia, Viola langsdorfii, Trisetum spicatum, Pinguicula vulgaris, Geum pentapetalum, Agrostis borealis, Drosera rotundifolia, Tofieldia* sp., *Oxycoccus microcarpus, Andromeda polifolia.*) A number of these species clearly indicate acid, peaty conditions; others, very moist conditions.

In South Carolina, found as an atypical phase, showing many of the features of *C. peruviana*, in the "typical" habitat of the latter species: on humus and moist decaying logs in a shaded ravine, associated with *Leucothoë editorum* and *Rhododendron maximum*, together with *Cephalozia catenulata, C. media*, and *Nowellia*.

Variation

Study of the limited European and American material definitely assignable to this species does not give a well-rounded picture of the pattern of variability. However, the following points are noteworthy. (1) *A* pronounced mod. *angustifolia* does not occur in nature; i.e., all plants found in nature have characteristic, broadly ovate-cordate leaves, often rather broadly rounded distally. (2) With subxeric conditions, the underleaves may become rather large (up to 675 μ wide) and approach those of *C. muelleriana* in size; they may actually become contiguous. (3) The blue oil-body color becomes less intense in swamp forms; *fide* Müller (1947a). (4) There is a more sharply developed tendency in this species than in *C. muelleriana* for the development of occasional bidentate leaves on juvenile shoots (Fig. 113:13).

In addition to the preceding variations, there appears to be a wider range in cell size than Müller (1947a) admits. Supposedly, *C. trichomanis* has large cells, 40 × 50–60–90 μ (Müller, 1947a); however, in Maine material they are only (28)30–38 × 36–45 μ and in a specimen from the Schwarzwald, Germany (K. Müller), 32–42(45) × 42–55(65) μ. Müller also states that the underleaf middle possesses cells ca. 30 μ wide and 50–80 μ long; however, in Maine plants these cells are only 22–27 × 34–44 μ, and in Schwarzwald material, 22–28 × 40–50 μ. Müller also indicates in his key that *C. trichomanis* has underleaves scarcely wider than the stem. Yet, in Maine material, robust plants (mod. *densifolia*) have underleaves 1.8–2.2× as wide as the stem; in German plants, underleaves are up to 650 μ wide on a stem 400 μ in diam. Clearly, a wider variability is found in underleaf size also.

In addition to such gametophytic variation there is considerable variation in the sporophyte. The capsule wall is stated to possess epidermal cells with alternate, longitudinal walls bearing even layers of thickening (Müller, 1947a, fig. 5b). Maine plants (Fig. 113:9) have, on alternating longitudinal walls, very irregularly developed thickenings of the radial walls; thus the radial walls, seen end-on, have a sinuate-undulate appearance, and irregular "nodular" thickenings are present. The thickenings of the radial walls are ± extended to the tangential walls, being evident in surface view as a reddish stain (as contrasted to the hyaline, unthickened remainder of the tangential walls). Müller considers the nature of the capsule wall as one of the major distinctions between species; this variability thus appears unfortunate. However, all capsules checked have been found to agree in the presence of such nodular thickenings.

In addition to the preceding variability, we find gametophytic variation tending to obscure the otherwise sharp differences between *C. trichomanis s. str.* and the *C. peruviana-heterophylla* complex. Plants collected in the Blue Ridge Escarpment in the gorge of the Estatoe R. (*RMS 37663*) are difficult to assign to either of these taxa, although perhaps they are closer to the latter. They agree with *C. trichomanis s. str.* in the broadly ovate-cordate leaves, ranging from 1800 to 2050 μ long × 2000 to 2200 μ broad (from 1.07 to 1.12× as broad as long); in the smaller, deep blue oil-bodies, ranging from 6 μ to 5 × 8 μ up to 6–7 × 8–12 μ, usually formed of only 4–12 globular segments ranging from 2 to 4 μ in size; in the generally laterally unarmed or sporadically uniangulate flanks of the underleaves; in the occasionally shallower sinus of the underleaves, sometimes descending to merely within 4–6 cells of the rhizoid-initial field. However, the plants closely approach the *C. peruviana-heterophylla* complex (and depart from the usual concept of *C. trichomanis*) in the most often obscurely bidentulate leaf apices, in the more frequently angulate lateral underleaf margins, and in the often more deeply descending sinus of the underleaves.

Plants showing such intermediate features are perhaps identifiable only on the basis of sporophytic criteria.

Differentiation

Calypogeia trichomanis, a rare species of sporadic distribution in North America, appearing to be largely confined to the northern portion of the

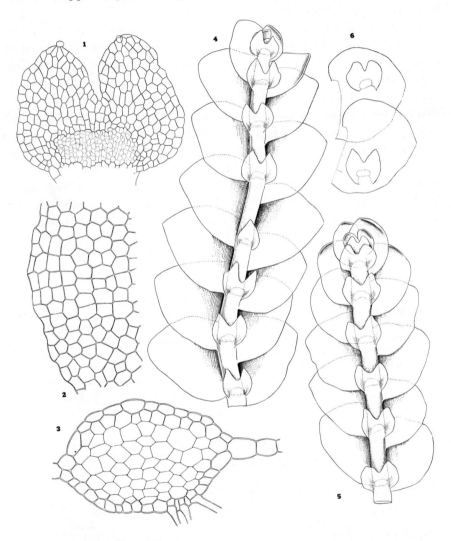

FIG. 114. *Calypogeia trichomanis*. (1) Underleaf (×ca. 72). (2) Cells of leaf apex (×ca. 140). (3) Stem cross section (×ca. 125). (4) Shoot, postical aspect, with one atypical unlobed underleaf (×12.5). (5) Shoot, of mod. *latifolia*, ventral aspect (×12.5). (6) Two leaves and adjoining underleaves of mod. *latifolia* (×ca. 12.5). [All from *RMS*, Mt. Mansfield, Vt.]

United States and to Canada and Alaska, is separable by the bluish color of the oil-bodies (giving the living plant a rather deep, bluish gray or bluish green color, with azure-blue shoot apices) from our other species, except *C. peruviana*. The broadly ovate to cordate leaves and the simply bilobed, laterally rounded underleaves, as well as the usually almost consistently entire leaf apices, serve to generally differentiate *C. trichomanis* from the tropical and subtropical *C. peruviana*.[16]

When only dead plants are available for study, separation of *C. tricho-manis* is difficult. The broadly ovate-cordate leaves, usually rounded or obtuse at the apex, and the form of the bilobed underleaves suggest phases of *C. muelleriana* (with which *C. trichomanis* was once consistently confused). Other than oil-body color, differences which serve to distinguish these two species are slight and are at times masked by the considerable varia-bility of both taxa. However, *C. trichomanis* does not appear able to readily produce the mod. *angustifolia* and only rarely produces the mod. *megastipula*. By contrast, in *C. muelleriana* the mod. *angustifolia* is frequently produced, while the mod. *latifolia* is almost always also a mod. *megastipula*. Whether these differences in behavior are constant is unknown, but they suggest that distinct physiological responses exist. The features separating the two species are summarized as following:

C. trichomanis	*C. muelleriana*
Oil-bodies deep blue.	Oil-bodies colorless.
Leaves with width-length ratio typically 1.2–1.4:1.	Leaves with width-length ratio typically 1.0–1.1:1, frequently < 0.85–0.9:1.
Underleaves usually 1.5–1.8× as wide as stem, with lobes subacute, triangular to lanceolate; lateral margins not or rarely obscurely angulate.	Underleaves usually ca. 2.2–2.5× as wide as stem (exc. in mod. *angustifolia*), usually with obtuse, often broadly triangular lobes; lateral margins fre-quently angulate or dentate.

CALYPOGEIA PERUVIANA Nees & Mont.
[Figs. 107:6–10, 115–116]

Calypogeia peruviana Nees & Mont., *in* Montagne, Ann. Sci. Nat. Paris, Ser. 2, 9:47, 1838; Nees, Naturg. Eur. Leberm. 3:26, 1838; Bischler, Candollea 18:61 [1962] 1963.
Kantia peruviana Trev., Mem. R. Ist. Lombardo, Ser. 3, Cl. Sci., 4:425, 1877.
Kantia portoricensis Steph., Hedwigia 27:280, 1888.
Calypogeia portoricensis (Steph.) Evans, The Bryologist 10:30, 1907.
Calypogeia dussiana Steph., Spec. Hep. 3:404, 1908 (*p.p.*).
Calypogeia fissa auct. Amer. (incl. Evans, 1907), in part, nec Raddi.

---

[16] Except for transitional specimens from the Southern Escarpment of the Blue Ridge, dis-cussed above.

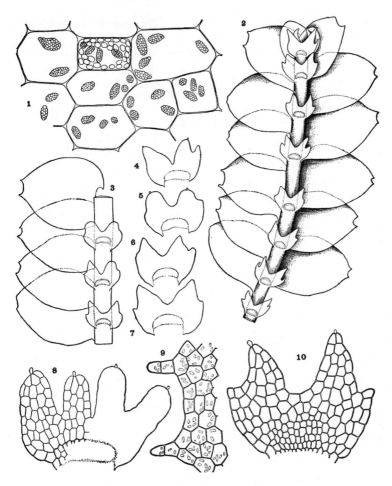

FIG. 115. *Calypogeia peruviana* (for capsule wall details and cells, see Fig. 107:6–10).
(1) Median cells with oil-bodies, one cell also with chloroplasts (×330). (2) Shoot,
ventral aspect (×20). (3) Sector of shoot with leaves and underleaves (×16). (4–7)
Underleaves of normal form (×26). (8) Abnormal underleaf (×60). (9) Leaf apex,
oil-bodies indicated (×115). (10) Normal underleaf, with slime papillae indicated
(×80). [All drawn from *RMS 19215*, 6–7 mi NE. of Van Cleave, Miss.]

In flat, depressed mats, *deep grayish green to dark green* throughout, with a
slightly bluish gray cast (due to the bluish oil-bodies), living *shoot tips azure blue*.
Shoots 2–2.4(3.5) mm wide; branches infrequent or sporadic, postical-
intercalary. Stems 200–320 µ wide; ventral cortical stem cells averaging 25 µ
wide (extremes: 20–34 µ) × 60–80 µ long. Leaves spreading at an angle of
75–90°, nearly flat, narrowly to broadly ovate, usually widest distinctly above
the base, nondecurrent, normally *slightly to very distinctly longer than wide*,
averaging 900–950(1350) µ long × 750–880(1200) µ wide, apices usually
slightly but sharply bidentate, the teeth ± widely separated, often formed by two

single cells in a row, sometimes $\pm$ blunt. Underleaves large, remote, (1.6)1.8–2.5× the stem width, averaging 480–710 μ wide × 375–500 μ long, ca. 1.2–1.3× as wide as long, *laterally slightly to distinctly decurrent*, bilobed for about ⅔ their length (with rhizoid-initial area considered part of stem), the lobes triangular, narrow, strongly divergent, separated by a rounded to obtuse, *broad sinus* descending *to within 2–3 cells* of rhizoid-initial field; *lobes armed on outer sides in almost all cases* with a broad triangular to sharp tooth. Cells large, in leaf apex 30–36 × 32–42 μ, along margins of leaf apices 36–46 μ long, the marginal cells not or scarcely elongate; median cells 32–40(45) μ wide × 50–60(70) μ long; cell walls thin, trigones absent or minute; chloroplasts 4.5–5 μ; *oil-bodies deeply indigo tinted*, of numerous *relatively small globules* (1–1.5 μ in diam.) which distinctly protrude, large (6 × 10–15 μ to 7.5–8 × 14–19 μ, occasional ones circular and only 5–7 μ); oil-bodies in all cortical, underleaf, and leaf cells, in leaf middle mostly 2–5(7–9) per cell. Underleaf cells in lobes 32–36 μ wide × 48–54 μ long.

Autoecious. Capsules deep purplish red; valves ca. 280–300 to 320–430 μ wide × 1700–1800 to 2500 μ long (ca. 1:6–8 ratio). Epidermal wall of valves ca. 8–9(10) cells wide, the cells 32–40 to 40–42 μ wide × (56)64–76 μ long; epidermal cells with *all longitudinal radial walls slightly and* $\pm$ *evenly thickened* by a layer of deep reddish deposit, apparently *laid down chiefly on the adaxial face* (with the midline of the valve considered the axis), not at all produced as nodular extensions, but at least locally extending over the tangential exterior faces of the cells for a variable distance (adaxial ¼–½ of the outer tangential walls thus more deeply reddish-pigmented than abaxial portions). Inner layer of much narrower cells, 15–19 cells across valve, averaging 0.5 the width of epidermal cells, ca. 15–19 μ wide on an average × 120–175 μ long; radial longitudinal cell walls somewhat to considerably thickened and bearing numerous vertical bands extending across 0.2–0.5 of inner tangential walls, gradually tapering and becoming indistinct (almost never with complete bands of the inner tangential walls). Spores 11–14 to 14–16 μ, reddish brown, nearly smooth. Elaters ca. 10–12 × 160 μ, 2(3)-spiral; spirals 3.5 μ wide.

Type. Peru: between Chupe and Yanacachu, *D'Orbigny, s. n.* and *s.d.* (G).

Distribution

As broadly delimited by Bischler (1963), this species extends from Brazil and Peru to Columbia and Venezuela, north to the West Indies (Jamaica, Puerto Rico) Guatemala and Mexico (Bischler, 1963) and into the southeastern United States, as follows:

NORTH CAROLINA. Along Rte. 70, 3–4.5 mi NW. of Durham, Durham Co. (*RMS 32210, 32201a, 29711*; northernmost report and only one from Piedmont); Whitewater R. gorge, Jackson Co. (*RMS 31914*); below the Narrows, Chattooga R., SE. of Highlands, Jackson Co. (*RMS 43480*); near Cape Fear R., 10 mi from Wilmington (*RMS 28814*); John's Cr., ca. 6 mi N. of Marion, McDowell Co. (*RMS 28810*); W. side of Chattooga R., above jct. with Ammons Branch, SE. of Highlands, Macon Co. (*RMS*

40833); Chattooga R., ca. 1 mi below Narrows, Macon Co. (*RMS 39417a, p.p.*); Thompson R., 0.1–0.2 mi N. of South Carolina state line, NE. of Jocassee, S.C. (*RMS 45170*). SOUTH CAROLINA. Table Rock State Park, Pickens Co. (*RMS 44822*); Whitewater R. gorge, below Lower Falls, Oconee Co. (*RMS 25100, 40961, 40940*); gorge of Estatoe R., Pickens Co. (*RMS 35620, 37650a, 37650*); W. branch of Estatoe R., in gorge, Pickens Co. (*RMS 37660a, 37687, 37680*). MISSISSIPPI. 6–7 mi N. of Van Cleave, springy slope near Pascagoula R., Jackson Co. (*RMS 19202*). GEORGIA. Ca. 0.2–0.3 and 0.5–0.8 mi below High Falls of Big Cr., N. Rabun Co., S. of Highlands, N.C. (*RMS 41621, 41790, 40694, c. C. muelleriana blomquistii; 40032, c. C. neesiana; 4073a*). FLORIDA. Torreya State Park, Liberty Co. (*RMS*); headwaters of St. John R., at Juniper Springs, Ocala Natl. Forest, Marion Co. (*RMS 29561*); Alexander Springs, Ocala Natl. Forest, Lake Co. (*RMS 33435, 33435a*); on decayed wood in swamp, Sanford, Seminole Co. (*S. Rapp 93 = 6811, Mar. 1923*). Previously reported from the United States only by Schuster (1968a).

Some of the published reports of *C. fissa* from Florida (Redfearn, 1952; McFarlin, 1940; see Frye and Clark, 1937–47), Louisiana (Frye and Clark, 1937–47), Georgia (Carroll, 1945), Mississippi (Frye and Clark, 1937–47), and South Carolina (Frye and Clark, 1937–47) may represent errors in determination for the present species. The preceding records are all quite unambiguous, being based on the study of living plants; the indigo-blue oil-bodies have been observed in all the collections except that of Rapp.

Ecology

A widespread neotropical species found in North America in the Escarpment area along the eastern and southern edge of the Blue Ridge Mts. (at elevations below 2500 feet), and locally on the Coastal Plain (Mississippi to western Florida to North Carolina) and in the elevated "Oligocene I." area of central Florida. Distribution is restricted to areas of deep shade, perpetual moisture, and low saturation deficit. Consequently, in the mountains this species occurs along shaded streams and in deep gorges, while in coastal regions it is found in deep swamps or hammock forests (which are largely or partly evergreen). In each of these areas a different phenotype occurs (p. 161).

In both coastal and upland areas, however, the species is restricted to sites deeply shaded by evergreens (*Rhododendron, Kalmia, Leucothoë editorum, Tsuga* in the mountains; *Leucothoë*, evergreen *Quercus* spp., *Sabal palmetto, Myrica cerifera, Illicium floridanum*, etc., in the Coastal Plain), growing only in niches closely adjacent to running or standing water.

In the Escarpment region the species repeatedly grows associated with the rare and extremely localized *Shortia galacifolia* (known only from a handful of sites), at times forming a matrix in which the *Shortia* grows. This is probably not accidental, but suggests a similar restriction to "Tertiary habitats" for both taxa. Other associates in the Escarpment are the filmy fern, *Trichomanis petersii*, and the Hepaticae *Pellia appalachiana, Radula sullivantii, Pallavicinia lyellii, Diplophyllum apiculatum, Odontoschisma prostratum*, and occasionally *Calypogeia sullivantii* and *C. muelleriana* fo. on bare soil, or (over moist rocks) *Riccardia*

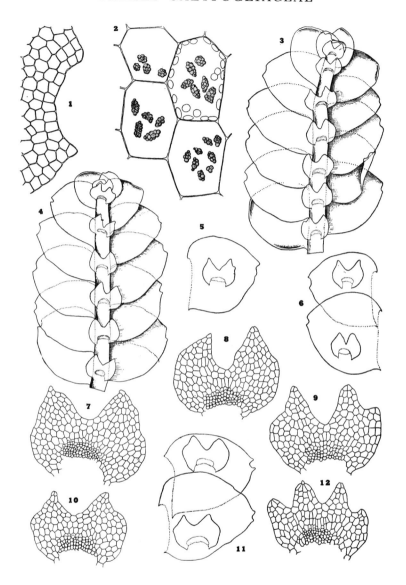

Fig. 116. *Calypogeia peruviana*, the *C. trichomanis*-like mod. *latifolia*. (1) Lea.
apex (×87). (2) Cells with oil-bodies and, upper right, chloroplasts (×385).
(3–4) Ventral aspects of shoots (×11). (5–6) Leaves and, drawn within, adjoining underleaves (×11). (7–8) *C. trichomanis*-like underleaves (×41).
(9–10, 12) *C. fissa*-like underleaves (×41). (11) Two leaves, and adjoining
underleaves (×19). [1–8, *RMS 37663*; 9–12, *RMS 37687*, all from Estatoe
R., S.C.]

multifida and *Dumortiera hirsuta*. Occasionally, also, the species may occur admixed with *Calypogeia neesiana*.

By contrast, the springy banks or swamps to which the species is confined on the Coastal Plain are characterized by the following associates: *Telaranea nematodes, Odontoschisma prostratum, Lophocolea martiana, Cephalozia connivens, Pallavicinia lyellii, Riccardia multifida*, and *Aneura pinguis*. Here the *C. peruviana* often occurs amongst mosses over tree bases and over exposed roots, occasionally on decaying logs.

Variation

Regional material comprises two different populations, the montane material differing in several respects from the Coastal Plain material (Mississippi and Florida to North Carolina). In Appalachian populations underleaves are usually unarmed laterally and are often as long as wide, or even longer than wide (particularly in South Carolina specimens). Furthermore, underleaves are rarely appreciably over 1.8 × as wide as the stem, and the underleaf lobes, when laterally armed, generally possess small, sharp, usually 1–3-celled teeth. In underleaf form, these plants therefore differ in aspect from *C. fissa* but may approach *C. trichomanis*. Furthermore, the Appalachian material has more variable leaves whose apices are usually only obscurely bidentate. The apices are either truncate and shallowly crescentically emarginate, with the teeth on either side of the emargination usually low and obtuse or, more rarely, minutely bidentulate (Fig. 116).

By contrast, coastal populations show a considerable superficial resemblance to *C. fissa*. Underleaves are broader, strongly transverse, sharply bilobed, with each lobe usually armed externally with a supplementary lobe or salient tooth. Lateral leaves, although often nearly identical in dimension and shape with those of the montane phenotype, almost invariably bear a pair of sharp apical teeth (as is the case with *C. fissa* subsp. *fissa*).

Although these distinctions are often striking, they do not, I think, preclude consideration of the two populations as a single species. Should further collections show that these differences are constant, possible varietal distinction may be indicated.

Differentiation

The gametophyte of *C. peruviana* shows similarities to both *C. fissa* and *C. trichomanis*. It agrees with the first in the often 4-lobed underleaves, but with each lobe acute or subacute and with the two larger, median lobes often widely divergent, and in lateral leaves that are mostly narrowly ovate, longer than wide, normally slightly bidentate at the tip. It differs conspicuously from *C. fissa* in the large, intensely bluish oil-bodies, formed of numerous smaller spherules, only 1–1.5 μ in diam.[17] In the bluish oil-bodies the species is similar to *C. trichomanis*. However, the usually more narrow, bidentate lateral leaves and the commonly short, often wider than long, 4-dentate underleaves *normally* separate it from the latter species.

Several minor differences between *C. peruviana* and *C. fissa* deserve emphasis. (1) In the former, the sinus between the two apical teeth is usually rather wide, often broadly crescentic; in *C. fissa s. lat.* the apical teeth—when present—are much less regularly developed, but are closely approximated, with a narrow, cutlike sinus between them. (2) In the field, *C. peruviana* stands out because of the relatively deep bluish gray-green coloration, with the shoot apices a beautiful azure blue in color; with drying and death, the oil-bodies disintegrate, but the plant nevertheless retains the dark green color.[18] In *C. fissa* living plants are pale green, with shoot tips whitish green usually; on death, the pale green color persists.

The azure-blue coloration of the shoot apices in *C. peruviana* is due to the fact that the oil-bodies attain their mature dimensions while the cells are still small. Consequently, the oil-bodies are crowded in the meristematic cells of the shoot apices, resulting in a more intense pigmentation (not masked by chloroplasts).

The capsule-wall anatomy of *C. peruviana* is extremely distinctive. A comparable epidermal cell layer occurs only in the otherwise very different *C. suecica*. The presence of a pigmented sheath of secondary deposit on *all*—or virtually all—longitudinal walls is very unusual and separates the species not only from *C. fissa* and *C. trichomanis* but also from almost all other species of *Calypogeia* known to me. The valves are short (length-breadth ratio ca. 6–8 : 1); the pigmentation unusually deep. The epidermal cell layer is somewhat similar to that of *C. muelleriana* in that the longitudinal walls lack nodular thickenings. Instead, *all* longitudinal radial walls (except in isolated cases) are pigmented and more or less strongly thick-walled; transverse walls of the valves are less or

---

[17] The oil-bodies disintegrate when the plants die (or shortly thereafter), the pigmentation usually disappearing. However, in the field, the azure pigmentation is obvious to the naked eye, the intensely blue color of the shoot apices being readily evident. The blue oil-bodies sufficiently mask the chlorophyll pigmentation of mature cells so that the plants appear to be a deep or bluish green, rather than the whitish or slightly yellowish green of the other species.

[18] In South Carolina plants, after several weeks of drying, moistening resulted in a breakdown of the oil-bodies with release of the indigo-blue pigment, staining the entire plant bluish green. In North Carolina plants, drying resulted in a breakdown of the oil-bodies but no bluish stain developed. However, remnants of the oil-bodies formed a brownish mass (or masses) in each cell, collectively darkening the plant.

not thick-walled. The deep, purplish red layer of thickening on the radial walls is largely confined to the *adaxial* cell-wall surfaces, extending onto the tangential walls for varying distances. The tangential extensions of these pigmented sheets of thickenings are evident because, in surface view, adaxial portions of the tangential walls are more deeply pigmented than abaxial portions; in fact, it is sometimes difficult to perceive where radial and tangential walls meet, since the layer of thickening of the radial walls may extend, unbroken, onto the tangential walls, thinning out gradually without a sharp break. On walls lying in the longitudinal axis of the valve and sometimes elsewhere, the tangential layers of thickening are often not, or are only barely perceptibly, developed (Fig. 107:6).

Affinities

I am assuming that the synonymy proposed by Bischler (1963) for *C. peruviana* is correct; I have not attempted to give it an independent scrutiny. I had originally assigned the North American material to *C. portoricensis* (Steph.) Evans and have been able to verify in the field (Luquillo Mts., Puerto Rico) that *C. portoricensis* has blue oil-bodies. A demonstration that *C. peruviana* has blue oil-bodies and is, in actuality, identical with *C. portoricensis* is still outstanding, the revision of Bischler being based on herbarium specimens only.

Calypogeia peruviana (including *C. portoricensis*) is allied most immediately to *C. heterophylla*, another neotropical species. Bischler (1963, p. 59) separates the two as follows:

C. peruviana: Stems not fleshy, 140–221 μ wide; lateral [*sic!*] branches numerous; leaves usually slightly elongated; leaf cells not distinctly smaller towards top than at the base of the leaves, the trigones too are of nearly equal size from top to base; underleaves small, 280–400 μ wide; underleaf cells small, in the centre . . . 50 μ long.

C. heterophylla: Stems usually fleshy, from 200–315 μ wide; lateral [*sic!*] branches numerous or rare; leaves elongated or as long as wide; leaf cells always distinctly smaller towards the top than at the base of the leaves; the trigones are more conspicuous at the top than at the base; underleaves large, 450–665 μ wide; underleaf cells medium sized or small.

Using these criteria, we find that the placement of the North American material remains ambiguous: in these plants the stems average about 200–225 μ wide; branches are infrequent; leaves are usually slightly to obviously elongated; underleaves average 480–710 μ wide; trigones are vestigial or minute everywhere. The type—and all material properly referred to *C. peruviana* (incl. *C. portoricensis*)—possesses rather smaller underleaves than are typical of *C. heterophylla* and has the lateral leaves relatively conspicuously decurrent at the postical base; in *C. heterophylla* the underleaves are larger and less remote and the lateral leaves are essentially nondecurrent. Also, Appalachian phenotypes of the North American plants disagree with *C. peruviana* in often having subentire or entire leaf apices—in which they find a parallel in many phenotypes assigned to *C. heterophylla*.

Bischler (*loc. cit.*, pp. 61, 64) draws a distinction between *C. peruviana* with "leaf apex bilobed or bidentate with narrow, pointed lobes" and *C. heterophylla* with "leaf apex very variable; on the same stem can be observed entire, rounded, pointed or apiculate leaves, and bilobed ones. But in the majority they are bilobed or bidentate, with usually narrow, pointed lobes, which occasionally are wider, triangular and rounded." I have already (p. 161) noted that North American plants *seem* to represent two phenotypes: a Coastal Plain one usually with sharply bidentate leaf apices, and a montane one with much variation in leaf apices. This last phenotype complex includes plants with wider leaves, which may be freely bidentate with rather broad and low teeth (Fig. 116:3–6, 11), and perhaps should also be circumscribed to include phenotypes with broad, predominantly entire leaves (*RMS 37663*), discussed under *C. trichomanis* (p. 154). In any case, the polymorphism in leaf shape and leaf apex form *in North American plants* is suggestive.

On these bases, collectively, the American material possibly might be better referred to *C. heterophylla* than to *C. peruviana*. However, North American plants are retained in *C. peruviana* for two reasons. (1) I have collected *C. peruviana* (= *C. portoricensis*) in the Luquillo Mts., Puerto Rico; it is a blue-green plant in the field, with the azure-blue shoot tips of living "*C. peruviana*" from the United States. (2) *C. heterophylla sensu* Bischler (*loc. cit.*) is very probably an arbitrarily—and unnaturally—delimited taxon. The description of var. *heterophylla* as "usually dark green" suggests both Puerto Rican *C. portoricensis* (= *C. peruviana fide* Bischler) and North American "*C. peruviana*." By contrast, both "var. *heterophylla* fo. *abnormis*" and "var. *heterophylla* var. *subrotunda*" are diagnosed (*loc. cit.*, p. 60) as "light or yellowish green." If these observations are correct, the latter two taxa are presumably based on plants with colorless oil-bodies (which are always pale to light yellowish green in death). Since I can find a definite point of contact, as regards oil-body color, with *C. portoricensis* (= *C. peruviana*), but not with *C. heterophylla*, it seems best to retain our plants under *C. peruviana*. Furthermore, the still poorly understood neotropical taxa must be revised on the basis of sporophyte and cytological criteria before a meaningful comparison becomes possible.

CALYPOGEIA FISSA (*L.*) *Raddi*

[Figs. 117–119]

Mnium fissum L., *p.p.*, Spec. Pl., p. 1114, 1753.
Jungermannia fissa Scopoli, Fl. Carniolica, ed. 2, 2:348, 1772.
Jungermannia calypogea Raddi, Atti Accad. Sci. Siena 9:236, 1808.
Jungermannia sprengelii Mart., *p.p.*, Fl. Crypt. Erlang., p. 133, 1817.
Calypogeia fissa Raddi, Mem. Soc. Ital. Sci. Modena 18:44, 1818 [1820].
Cincinnulus sprengelii Dumort., Syll. Jungerm. Eur., p. 73, 1831.
Calypogeia trichomanis var. *fissa* Not., Mem. Reale Accad. Sci. Torino II, 18:467, 1859.

Kantia trichomanis var. *fissa* Lindb., Acta Soc. Sci. Fennica 10:508, 1875.
Kantia fissa Lindb., Hep. Utveckl., p. 20, 1877.
Kantia calypogea Lindb., *ibid.*, p. 20, 1877.
Kantia sprengelii (Mart.) Trev., Mem. R. Ist. Lombardo, Ser. 3, Cl. Sci., 4:425 (reprint p. 43), 1877.
Calypogeia trichomanis fo. *fissa* Bernet, Cat. Hep. Suisse, p. 108, 1888.
Kantia trichomanis var. *sprengellii* Howe, Erythrea 4:50, 1896.
Cincinnulus calypogea K. Müll., Beih. Bot. Centralbl. 13:98, 1902 and Mitt. Badischen Bot. Ver. (182/183):284, 1902.

Widespread *on chiefly inorganic sites* in *pale, pellucid green* to grayish green patches, $\pm$ *shiny when dry*, usually *somewhat less robust* than *C. muelleriana*, usually 1.9–2.5, rarely 2.8, mm wide at maturity. Sparingly branched; branches postical-intercalary. Leaves mostly 900–1000 μ long, to a maximum of 1.2–1.5 mm, usually *distinctly narrower than long and narrowly ovate in shape* (to 1.15 mm wide $\times$ 1.5 mm long), widest at or a little above the base, slightly convex, blunt to *acute, sometimes apiculate at the apex*, and (at least on some shoots) *shortly and narrowly bidentate* (teeth juxtaposed) on many or most leaves.[19] Cells of leaf apex at margins 30–40 μ, in leaf middle ca. 32–40 $\times$ 48–60(68) μ, thin-walled, virtually devoid of trigones. Oil-bodies (4)4.5–5 $\times$ 7–10 μ, formed of segments 2–3 μ in diam., *of 5–14 segments oriented in 2–3 rows in most cases*, pellucid and *colorless*. Underleaves *remote, strongly transverse and wider than long*, 1.5–2.5$\times$ as wide as stem, ca. 300–325(450) μ long $\times$ 550–570(620–700) μ wide, bilobed 0.75–0.8 their length (to the rhizoid-initial field); sinus open and broad, rounded below, *descending to within 1–2 cells of rhizoid-initial field;* lobes obtuse to narrowly rounded, $\pm$ *strongly divergent, armed on their outer sides*, in most cases with an obtuse to subacute angulation or broad tooth; rhizoid-initial region narrowly elliptical, usually at least 3$\times$ as wide as high. *Frequently with gemmae.*

Autoecious or paroecious. Epidermal cells of capsule with adjoining faces of longitudinal, alternate (secondary) cell walls $\pm$ evenly thick-walled (or, subsp. *neogaea*, with *nodular thickenings*), primary cell walls thin and colorless, alternating with the thickened walls. Inner cell layer formed of cells averaging 14–20 μ wide. Valve length ca. 9–13$\times$ the width. Spores 12–16 μ. Chromosome no.: $n = 18$ (Lorbeer, 1934).

Type. European, *Calypogeia fissa* is based on *Mnium fissum* L. *Calypogeia fissa*, as a recognizable entity, dates to 1818 (Raddi).

Linneaeus (1753, p. 1114) based *Mnium fissum* on two plants, one described as "*Mnium trichomanis facie, foliolis bifidis,*" based on Dillenius (Hist. Musc. 237, pl. 31, fig. 6, 1741), and the other as "*Jungermannia terrestris repens, foliis ex rotunditae acuminatis bifidis: apertura pene visibili*" (Micheli, Nov. Plant. Gen. 8, pl. 5, fig. 14, 1729). Lindberg studied Dillenius' material, issuing three reports (1874, 1877, 1883) based on his examinations. Evans (1907) emphasized that these reports were "at variance with one another." Under specimens labeled "*Mnium trichomanis facie, foliolis bifidis*" Lindberg found two species, which he

---

[19] Based on the frequent mod. *laxifolia* in which leaves are not or little imbricate. In the rare mod. *densifolia*, especially in plants from rather xeric sites (mod. *densifolia-parvifolia*; Fig. 119), leaves are only 900–950 μ long but 1150–1200 μ wide, are broadly ovate to nearly heart-shaped, and are virtually identical with those of *C. muelleriana* in form.

called *Kantia calypogea* (Raddi) Lindb. and *K. fissa* (L.) Lindb. According to Lindberg's last study (1883), Dillenius' figure is based on the former species. One must keep in mind, however, that *K. calypogea* Lindb. is the same as *C. fissa* (*sensu* Raddi) and *K. fissa* Lindb. is *C. arguta*. Hence it could be concluded that, *Mnium fissum* L. being an aggregate species, the name should be restricted to the plant that served for the illustration in Dillenius, subsequently cited by Linnaeus. Lindberg (1883) evidently chose to follow this course and used the name *fissa* for what is now called *C. arguta* Nees & Mont. Evans (1907) believed that Lindberg's course was "difficult to defend," Raddi (1818) having properly restricted the name *C. fissa* to the other component part of the aggregate *Mnium fissum*—i.e., to the plant called by Lindberg *Kantia calypogea*.

Distribution

Chiefly "Atlantic" in distribution, in both North America and Europe. In North America extending northward into the Northern Hardwoods (hemlock-yellow birch-beech-maple) Forest, rarely into the Coniferous Forest, and becoming common only in the region of Mixed Mesophytic and Oak-Hickory Forests; southward common in the Beech-Magnolia Forests, as well as in the subtropical Magnolia-*Persea* broadleaved evergreen forests of the coastal regions. Reports from the interior (Iowa, etc.) are all of extremely dubious validity and need to be carefully checked to determine whether they refer to *C. fissa* or possibly to the more widespread *C. muelleriana*.

Calypogeia fissa has been very widely cited from North America; see, e.g., reports from as far north as the Delta region of the Mackenzie R. (Steere, 1958a), from Ontario (Williams and Cain, 1959), to Florida (Redfearn, 1952). These reports of the suboceanic and Atlantic *C. fissa* are all to be regarded as exceedingly dubious. Older reports from, i.a., Nova Scotia (Brown, 1936a), North Carolina (Blomquist, 1936), New England (see, i.a., Evans and Nichols, 1908), Michigan (see Steere, 1947a), West Virginia (Ammons, 1940), Tennessee (Sharp, 1939), Iowa (Conard, 1945), and Virginia (Patterson, 1949), cannot be accepted as authentic; doubts about the species occurring in as continental areas as Iowa were expressed already in Schuster (1953). Insofar as any of these reports belong to *C. fissa s. lat.* they all appear to pertain to subsp. *neogaea* (see p. 173).

Ecology

In the Southeast and Central Atlantic states, northward into southern New England, our genotypes of this variable species usually occur on inorganic substrates: most commonly on roadside banks and damp to moist soil of ditches, often with *Solenostoma crenulatum* and *Scapania nemorosa*, occasionally *Cephalozia bicuspidata* and *Lophozia capitata*. Sometimes on moist clayey soil after some humus deposition by decay of other pioneer taxa of such sites (*Diplophyllum apiculatum*, *Diphyscium*, *Atrichum*, *Scapania nemorosa*, etc.). Also able to invade (possibly from soil-filled crevices) adjacent rock faces. Occasionally found under more intermittent moisture conditions (as under *Tsuga*) in shade, on dry banks, with *Leucobryum glaucum*, *Calypogeia neesiana*, *Lophozia capitata*, *Nardia insecta*, *Scapania nemorosa*, *Diplophyllum apiculatum*, etc. When in such xeric habitats, robust plants may form broad-leaved modifications (mod. *densifolia-latifolia*) which approach *C. muelleriana*.

In the Southeast, particularly the southeastern Coastal Plain, *C. fissa s. lat.* is a characteristic and common plant, occurring over clay or loam banks (then usually associated with *Scapania nemorosa*, less often with *Cephaloziella hyalina*, *Odontoschisma prostratum*), or over friable, moisture-carrying sandstone. On sandstone bluffs, southward, *C. fissa* may occur with *Riccardia multifida s. lat.*, *Scapania nemorosa s. lat.*, *Reboulia hemisphaerica*, *Fossombronia brasiliensis*, *Calypogeia sullivantii*, and *Cephaloziella hyalina*.

Buch (1936a) cites *C. fissa* as usually over *Sphagnum* peat or in *Sphagnum*. Müller (1947a) points out that Buch probably did not have true *C. fissa*, since the latter supposedly never occurs on such sites. Buch (1948) now also agrees that his material was probably not *C. fissa*. However, *RMS 43818* (Mt. Mansfield, Vt., 3800 feet) is from a wet, springy, peaty slope, associated with *Sphagnum*.

Variation

Typical *C. fissa* is easily separated by its narrow leaves, which show a strong but hardly constant tendency to be minutely bilobed and by the strongly transverse, wider than long underleaves, deeply bilobed by a broad sinus and generally armed on one or both external margins with an obtuse tooth.

However, like other species of this section of the genus, *C. fissa* may produce, under conditions of greater than normal insolation and intermittency of moisture, dense-leaved, more broad-leaved forms, a mod. *densifolia-latifolia-integrifolia*; in this the leaf width averages greater than the length (ca. 900–950 μ long × 1150–1200 μ wide). Such robust plants are similar to *C. muelleriana*, differing principally in their more deeply bilobed underleaves and the frequent—if not uniform—presence of lateral teeth or angulations of the underleaves.

Calypogeia fissa, with its pattern of variation, therefore, can be readily confused with *C. muelleriana*. Numerous statements—recent ones included—that there is intergradation between *C. fissa* and *C. "trichomanis"* are based on confusion of the latter with *C. muelleriana*.

In Europe, *C. fissa* is unquestionably much more distinct than it is in the southeastern United States. All English specimens of *C. fissa* seen show a remarkably high incidence of bidentate, narrow leaves and have the remote, broad and short, deeply bilobed, laterally dentate underleaves typical of the species. In my experience, such plants are very rare or absent[20] in North America.

Macvicar (1926) calls attention to the apical teeth of the leaves, describing the leaf apex as "narrowed and bidentate, the sinus narrow, obtuse to subacute, the teeth small and variable, being triangular-ovate and 3–5 cells broad at base to nearly obsolete." This statement admirably fits European material I have studied, and the leaves may be uniformly bidentate, with teeth up to 7–8 cells wide at base and 5–7 cells long, the sinus between acute. Such phenotypes do not occur in North America; indeed I am convinced that American *C. fissa* is a variable and poorly definable taxon, freely intergrading with *C. muelleriana* in most respects. Admittedly, some collections are at hand, referred to *C. fissa*, which in their gametophytic characters actually approach European or "typical" *fissa*. These may, in part, represent juvenile or impoverished plants, which frequently show reversion and tend to develop bidentulate leaves.

Gametophytic intergradation between *C. muelleriana* and *C. fissa* in North America thus does not have an exact counterpart in Europe. K. Müller (1947a, p. 426) stresses the lack of transitional phases between the two species but admits that Joergensen (1934) claims the two are connected by transitional forms. Joergensen would refer to *C. fissa* also *C. trichomanis* var. *intermedia* C. Jensen (1915). Müller, who studied an original specimen of this taxon (Bornholm I.), found it to bear narrow, pointed leaves that are not bidentate, and underleaves that are unarmed on their lateral margins. Müller states that Joergensen's disposition of this plant under *C. fissa* "erscheint mir wahrscheinlich richtig," and believes the drawing of it in Jensen (*loc. cit.*, p. 229, fig. 9) belongs to *C. fissa*. In my opinion, this material falls into the category of a transitional phase between *C. muelleriana* and *C. fissa*. Similar narrow-leaved phases from North Carolina were determined by Müller as *C. muelleriana*.

In spite of this apparent intergradation, differences in distribution and ecology distinguish the two taxa; whether these are sufficient to separate species remains questionable. Should repeated and complete intergradation between the two be eventually demonstrated, it may prove necessary to relegate *C. muelleriana* to the status of a variety or subspecies, *C. fissa* var. (or subsp.) *muelleriana*. Such a treatment would have obvious advantages for American students, since lesser— and less constant—differences exist between the two taxa in the New World than in Europe. In fact, the disposition of most plants which have been called *C. "fissa"* in North America presents many problems.

Plants which have been called *C. fissa* in North America, with entire lateral leaves but a high percentage of *"fissa"* underleaves, appear to differ from typical *C. fissa* in sporophytic characters as well as in leaf form.

---

[20] To the worker in North America it is, at first glance, striking to find contemporary European hepaticologists in virtual agreement regarding the distinctness of *C. fissa*. The narrow-leaved forms of *C. muelleriana*, which are very abundant in the Southeast, are a particular source of trouble to the American worker; these appear midway between *C. fissa s. str.* and typical *C. muelleriana*. By contrast, in Europe such transitional forms are either rare or absent.

"Typical" *C. fissa* is thus either absent or exceedingly rare in North America; only some Texan plants approach European material. For this reason, *C. fissa* is divided into two subspecies, one restricted to America, the other found in oceanic Europe and possibly represented in North America by similar phenotypes in Texas (and perhaps elsewhere). The preceding species description is based largely on the "typical" *C. fissa* subsp. *fissa*. The following key serves to separate these two types.

Key to Subspecies of Calypogeia fissa

1. Leaves (or most of them) sharply bidentate; capsule valves 9–13× as long as wide; epidermal cells with alternating longitudinal walls lacking nodular thickenings; spores 14–16 μ. Europe . . . . *C. fissa* subsp. *fissa*
1. Leaves usually acute, only rarely slightly bidentulate; capsule valves 7–9(10)× as long as wide; epidermal cells with alternating longitudinal walls with discrete nodular thickenings; spores 12.5–14 μ. North America. . . . . *C. fissa* subsp. *neogaea* Schust.

"Typical" *C. fissa* is at once separable from all other nearctic taxa, except *C. peruviana*, by the predominantly bidentate leaves of mature plants (Fig. 117:1–3, 7–8, 10). On this account, plants of the latter species from the southeastern United States have been confused with *C. fissa*. However, in *C. fissa* subsp. *fissa* the apical teeth of the leaves are usually juxtaposed, being separated by a narrow incision, whereas in *C. peruviana* the teeth are normally separated by a shallow but relatively broad sinus. The blue oil-bodies of *C. peruviana*, however, represent the most important criterion separating this species from *C. fissa* (p. 162).

CALYPOGEIA FISSA subsp. NEOGAEA Schust., subsp. n.

[Figs. 117–119]

Subspecies subsp. fissae similis, folia, autem, ad cacumina plerumque solummodo acuta; valvae capsularum 7-9(10) plo longiores quam latae; cellulae epidermales membranae capsularum incrassationibus noduliformibus localibus praeditae.

Whitish green, in flat patches, shiny when dry. Shoots (1.9)2.2–2.9(3.0) mm wide, subsimple. Stems (216)230–300 μ wide; ventral cortical cells ca. 20–23(26) × (48)55–75(90) μ long. Leaves nearly flat, *remote to contiguous* usually, *typically narrowly ovate* (and ca. 1275 μ long × 1100 μ wide), but particularly on robust shoots sometimes broadly ovate [and (990–1100)1300–1350 μ long × (1100–1320)1400 μ wide], the *apices almost invariably entire*, from *subacute or blunt* to, infrequently, narrowly rounded, *rarely subapiculate*. Apical cells (on and within margins) ca. (27–30)32–40 μ, averaging virtually isodiametric; median cells 36–40 μ wide × (38)42–56(64) μ long; oil-bodies colorless, of coarse segments (1.5 μ large or more), up to 5 × 14 μ. Underleaves characteristically and *uniformly quite distant* (*even in dense-leaved phases*), averaging 1.5–1.8× as wide as stem, ca. 445–475 μ wide × 305–370 μ long to 530–600 μ wide × 340–375 μ long, *strongly transverse* and ca. 1.4–1.8× as wide as long,

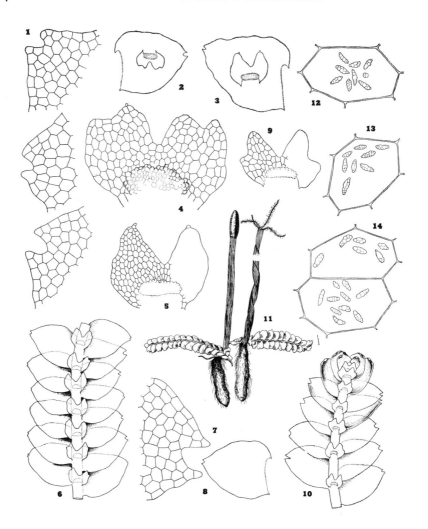

FIG. 117. *Calypogeia fissa. s. lat.* (1–8, subsp. *neogaea*, of the extreme most approaching subsp. *fissa;* 9–14, subsp. *fissa*). (1, 7) Four leaf tips (×77). (2–3, 8) Leaves and under-leaves (×14.5). (4) "*Fissa*"-type underleaf (×77). (5) "*Muelleriana*"-type underleaf (×36). (6) Postical aspect of shoot (×9.2). (9) Extreme form of underleaf (×44). (10) Shoot apex (×12). (11) Fertile plants, habitat sketch (×ca. 4.5). (12) Cell with oil-bodies (×385). (13) Cell with oil-bodies (×372). (14) Cells with oil-bodies (×320). [1–8, *Whitehouse 22536*, from Texan plant; 9–10, Krypt. Exsic. Vindob. no. 2582, from Austrian plant; 11, after Müller, 1905–16; 12–14, from English plants, leg. E. W. Jones (12–13 from phenotype with strongly bidentate leaf tips, from Bagley Woods: 14, from a typical plant from Lyndhurst).]

bifid to within 1–2 or 3 cells of rhizoid-initial field by a subacute to rectangular sinus, which is crescentic at base, less often with an open, wide crescentic sinus; lateral margins of underleaves (on one or both sides) *most often with an obtuse angulation or broad tooth*, rarely with a sharply marked, acute tooth, originating *above the middle.*

Autoecious. Capsules with valves linear, ca. 2000 μ long × 240 μ wide to 2450 μ long × 320 μ wide to 2600 μ long × 300 μ wide (varying mostly *from 7.5–9× as long as wide*). Epidermal cells fundamentally in 8 rows (and averaging 32–40 μ wide), but rarely with 1–2 supernumerary longitudinal walls (and then 9–10 cells wide); secondary longitudinal radial walls thickened *on both faces,* the thickenings irregular and in the form of more or less coalescent vertical bands, *in profile thus appearing strongly nodular to sinuous-nodular* (intervening areas of radial walls more or less thickened, sometimes so strongly that the "nodular" thickenings become less evident, the radial walls then sinuate in profile); thickenings of the radial walls ± extending across outer tangential faces in the form of short to rather long yellowish brown bands or tapering spurs; these spurs or bands never complete, often barely perceptible and coalescent as a narrow, irregular, border of tangential thickenings, adjacent to the radial walls. Inner cells fundamentally in 16 rows (and then 18–20 μ wide), but occasionally some longitudinal supplementary walls occurring, and then in 17–19 rows (and only 15–18 μ wide); length of inner cells ca. 72–85 μ; free face of inner cell layer with numerous transverse bands, extending down adjacent radial walls (but not, or only as, short spurs, across interior tangential faces). Spores (12.5)13–14, rarely 14.5, μ, brownish, finely verruculose.

Type. Loamy banks adjacent to intermittent small springs, oak-hickory forest, Tombigbee State Park, Lee Co., Miss., March 1953, *RMS 26499*. Diagnosis drawn from type and from plants from Liberty Co., Fla. (*RMS 44042a*).

Distribution

Confined to eastern North America, ranging from Texas and Arkansas to northern Florida northward to New York and Vermont, and questionably into Ontario; common in the Coastal Plain and Piedmont of the Southeast, becoming sporadically distributed northward. Ellucidation of the range complicated by the frequency of intermediate phenotypes northward and in the mountains southward.

FLORIDA. Bluff over Apalachicola R., Torreya State Park, Liberty Co. (*RMS 44042a*; typical).; ravine cut into Alum Bluff, Apalachicola R., Liberty Co. (*RMS 33528a*).   MISSISSIPPI. Tombigbee State Park, near Tupelo, Lee Co. (*RMS 26499*; type); Tishomingo State Park, Tishomingo Co. (*RMS 19741*).   ARKANSAS. Blanchard Springs, N. of Allison, Ozark Mts., Stone Co. (*Anderson 11694*); 3 mi SE. of Huntsville, Madison Co. (*Anderson 12203*); Howard's Bluff, E. of Springdale (*Anderson 12217*).   TEXAS. Alluvial soil of small creek, SE. of Zavalla, Angelina Natl. Forest, Angelina Co. (*Whitehouse 22536*); 5 mi SE. of Palestine, Anderson Co. (*Whitehouse 22381*; trace).

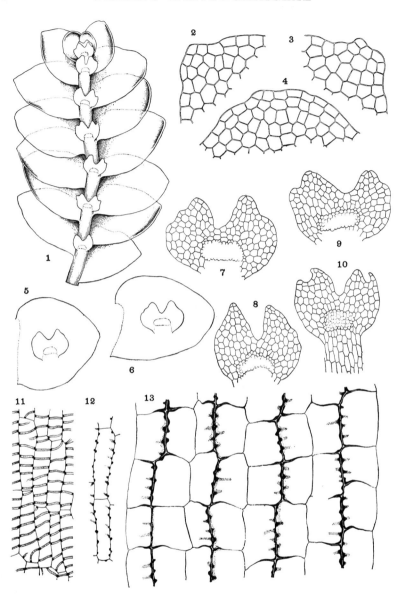

Fig. 118. *Calypogeia fissa* subsp. *neogaea*. (1) Shoot, ventral view (×16). (2–4) Leaf apices (×104). (5–6) Leaves and, drawn within, adjoining underleaves, from two separate plants; fig. 5 from mod. *latifolia* (×19). (7–8) Underleaves from plant in fig. 6 (×48.5). (9–10) Underleaves from plant similar to that in fig. 1 (×48.5). (11) Capsule wall, inner cell, free tangential walls (×225). (12) Two cells from fig. 11, but focus at internal tangential wall (×225). (13) Section across middle of capsule valve, epidermal cells (×225). [1–5, 9–13, *RMS 26499*, Tombigbee State Park, Miss.; 6–8, *RMS 27580*, Wyatt Hills, Stone Co., Miss.]

GEORGIA. Brasstown Bald, Towns Co., 4000 ft (*RMS 34340, 34342*). NORTH CAROLINA. Along Rte. 70, 3–4 mi NW. of Durham, Durham Co. (*RMS 32210a, 32210b*; colorless oil-bodies, with *C. peruviana* with blue oil-bodies admixed); bank of NE. Cape Fear R., ca. 10 mi N. of Wilmington, on Rte. 437, Pender Co. (*RMS 31796a*; isotype); on Rte. 421, ca. 6 mi NW. of Longcreek (*RMS 31833*; isotype); above Soco Falls, Soco Gap, E. of Cherokee (*RMS 24800*; a little doubtful). VIRGINIA. Mt. Rogers, Grayson Co. (*RMS 38020a*); White Top Mt., Smyth Co., 5400–5500 ft (*RMS 38098*, among *Solenostoma obscurum*). WEST VIRGINIA. Cranberry Glades, Pocahontas Co., 3500–3600 ft (*RMS 61206, 61200a*; base of red spruce, among *C. neesiana*).[21] MARYLAND. Beaverdam Rd., 2.25 mi SE. of Beltsville, Prince George Co. (*Hermann 13716*; typical).

NEW YORK. 2 mi SW. of Long Lake Village, 1700 ft, Hamilton Co. (*Hermann 14715*); Chittenango Falls, Madison Co. (*RMS 1394a*). VERMONT. Mt. Mansfield, 3300–3500 ft, near Taft Lodge on Long Trail (*RMS 43818*).

Perhaps extending westward as far as OHIO: Ashtabula, Champaign, Cuyahoga, Logan, and Scioto Cos. (Miller, 1964, as *C. fissa*), but these reports all need verification.

Variation and Differentiation

Under the infraspecific name *neogaea* I have assembled a wide range of forms, superficially lying in some respects between *C. fissa* and *C. muelleriana*. Whether all these belong to the same taxon as the type, from Mississippi, is impossible to answer without knowledge of the sporophytic characters of all these phases. However, they form, in general, an easily recognized series of phenotypes on the basis of the remote, broad and deeply bifid underleaves, coupled with the narrowly ovate lateral leaves.

The subsp. *neogaea* differs from typical *fissa* in the acute or blunt apices of the leaves, only very exceptionally bidentate on isolated (atypical) leaves of mature plants. Also, the underleaves tend to have the lateral teeth or "shoulders" situated higher up on the lateral margins. The leaf form of subsp. *neogaea* closely approaches that of mod. *angustifolia-parvistipula-laxifolia-laxistipula-fissistipula* of *C. muelleriana*. From this, *C. fissa neogaea* usually differs in the smaller underleaves (usually 1.4–1.8 × as wide as stem), divided to within 1–2(3) cells of the rhizoid-initial region, with the usually somewhat divergent lobes acute or subacute, more rarely rounded, and armed on one or both external margins (in most cases) with an obtuse to acute tooth or angulation. This difference would be insufficient to separate a discrete taxon, but the capsule wall and spores of typical *C. fissa neogaea* exhibit certain discrepancies that separate it from both *C. fissa* subsp. *fissa* and *C. muelleriana*. In the latter two taxa the spores average 14–16 μ in diam., occasionally even larger, and the longitudinal radial walls of the epidermal cells of the capsule show even deposits of

---

[21] Hardly typical, a large mod. *latifolia*, with the facies almost of *C. muelleriana* but the leaves bluntly to sharply pointed or else feebly bidentate; the much broader than long underleaves bifid to within 2–3(4) cells of the base, and the lobes usually armed laterally.

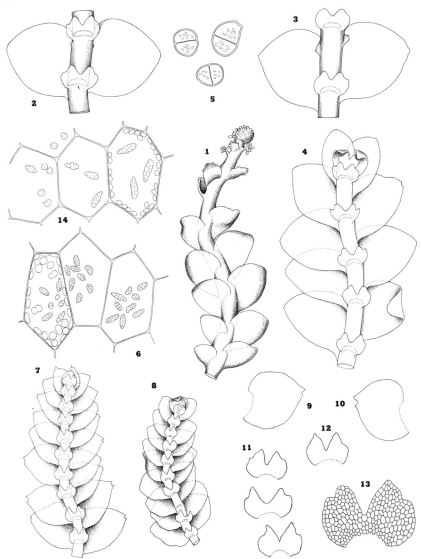

FIG. 119. *Calypogeia fissa s. lat.* (1–13, subsp. *neogaea;* 14, subsp. *fissa*). (1) Gemmiparous shoot apex in dorsal view (×ca. 12). (2–3) Shoot sectors, ventral aspect (×19). (4) Shoot, ventral aspect (×19). (5) Gemmae (×320). (6) Cells with oil-bodies and, at left, chloroplasts (×383). (7) Shoot, extreme with leaves showing maximal tendency towards bidentate condition (×10.4). (8) Shoot, ventral aspect, mod. *integrifolia* (×ca. 7). (9–10) Leaves, mod. *latifolia* (×ca. 15). (11–12) Underleaves (×ca. 15). (13) Underleaf (×ca. 30). (14) Cells with oil-bodies and, at right, chloroplasts, of mod. *integrifolia* but with underleaves of typical subsp. *fissa* (×410). [1–6, *RMS 24233*, Forest of Dean Lake, N.Y.; 7, Briggs Gully, Honeoye Lake, N.Y., *RMS;* 8–13, Dryden, Tompkins Co., *RMS*; 14, Lodge Hill, New Forest, England, *E. W. Jones*.]

thickening, except on walls that are altogether hyaline. In *C. fissa neogaea* the epidermal cells exhibit regular alternation of hyaline walls and walls bearing discrete nodular thickenings (Fig. 118:13), as in the *C. integristi-pula-neesiana* complex; furthermore, the spores are only 12.5–14 μ in diam. Also, *C. fissa* subsp. *fissa* has strongly elongated capsule valves, 9–13 × as long as wide (Müller, 1947a). In *C. fissa neogaea* the range is from 7 to 9 ×, rarely 10 ×, as long as wide (as in *C. muelleriana*).

Recognition of the subsp. *neogaea* involves two questions. (1) Are the sporophytic differences cited sufficiently constant? This question can be answered only by study of a larger series of collections. (2) How can the taxon be separated from the narrow-leaved phases of *C. muelleriana*? This problem will remain nearly insoluble so long as only sterile material of the mod. *parvistipula-fissistipula* of *C. muelleriana* is at hand. However, in *C. fissa* (including the nearctic subspecies) many or most underleaves are bifid to within 1–2(3) cells of the rhizoid-initial field and exhibit a marked tendency for development of accessory "shoulders," lobes, or teeth of all or most lateral margins.

Thus *C. fissa* subsp. *neogaea* occupies the problematical midground between European *C. fissa*—a plant seemingly absent from North America—and typical *C. muelleriana*. It shares with the latter the frequently large size, the often broad and heart-shaped leaves, and the often largely or almost exclusively entire leaf apices. Thus, viewed superficially, the plants can easily be dismissed as a mere form of *C. muelleriana*. For some years I regarded them as such—and, at times, am still tempted to so dismiss them. Yet, on critical examination, there are two salient differences: the leaf apices "tend" to be pointed, either acutely or bluntly, and often leaf tips are bidentate or retuse on mature leaves; also, the underleaves are more deeply bifid, in large part to within 2 or at most 3 cells of the rhizoid-initial field. Furthermore, at least a considerable minority of the tremendously variable underleaves show a broad angulation or shoulder, varying to a blunt tooth. In this last case, the underleaves become very broad, much broader than long. This combination of features seems to me to preclude placing these plants within *C. muelleriana*, no matter how broadly we would like to circumscribe this species. If it is insisted that the plants be so placed, we must reduce *C. fissa* to the status of a subspecies or variety of a portmanteau species including *C. muelleriana* as well.

CALYPOGEIA MUELLERIANA (*Schiffn.*) K. *Müll.*
[Figs. 120–125]

This highly technical and widespread *typus polymorphus* seems to show a rather different pattern of variability in North America from that in Europe. For this reason it is here divided into two subspecies, of which the widespread subsp. *muelleriana* displays a spectrum of variation nearly analogous to that found in Europe, whereas an Appalachian-Ozarkian series of phenotypes, distinguished as subsp. *blomquistii*, shows no parallel to anything reported from Europe.

The two extremes are distinguished in the keys (pp. 129, 131); the synonymy is cited under the extremes. Although subsp. *muelleriana* remains extraordinarily polymorphic, genetic variation, if any, is so totally masked by phenotypic, putatively environmentally induced differences that recognition of variants or subvariants (formas) seems, at this point, to be untenable.

As specifically noted elsewhere, *C. muelleriana* appears to occur in two chromosome phases, a 9- and an 18-chromosome race; only the 18-chromosome race has been reported from Europe. In this the species (*s. lat.*) shows parallelism to *C. neesiana* (see, e.g., Tatuno, 1941).

CALYPOGEIA MUELLERIANA (*Schiffn.*) *K. Müll. subsp. MUELLERIANA*

[Figs. 120–124]

Calypogeia trichomanis var. *erecta* K. Müll., Mitt. Badischen Bot. Ver. 160:94, 1899.
Kantia Mülleriana Schiffn., Lotos 1900 (7):344 (reprint p. 23), 1900 (*p.p.*)
Kantia Mülleriana var. *erecta* Schiffn., Lotos 1900(7):346 (reprint p. 25), 1900.
Calypogeia trichomanis auct. (*p. max. p.*)
Calypogeia Mülleriana K. Müll., Beih. Bot. Centralbl. 10:217, 1901.
Cincinnulus trichomanis var. *mülleriana* Meyl., Bull. Herb. Boissier, II Ser., 6:499, 1906.
Calypogeia trichomanis fo. *Mülleriana* K. Müll., Rabenh. Krypt.–Fl. 6(2):251, 1913.
Calypogeia neesiana var. *laxa* Meyl. ex K. Müll., *ibid.* 6(2):239, fig. 69b-e, 1913.

Variable in size, mod. *parvifolia* from (1.5)1.8–2.5 mm wide, more often robust, *to 2.7–3.5 mm wide* × 2–5 cm long, light yellowish green to whitish green, ± *nitid when dry*. Branching sparing, normally *postical-intercalary*.[22] Stem 280–325(360) μ thick, ventral cortical cells ca. 23–27 μ wide. Leaves typically *broadly obliquely ovate*, widest barely above base, contiguous to imbricate, in mod. *parvifolia-densifolia* ca. 800–850 μ long × 900–1050 μ wide to 950–1220 μ long × 1050–1100 μ wide; on mature shoots (of all but mod. *laxifolia-parvifolia*) the *leaf virtually as wide as or wider than long*, ca. 1.0–1.1× as wide as long, ca. 2 mm long and wide; leaf apices *obtusely pointed to narrowly rounded*, very rarely isolated leaves (mod. *parvifolia-laxifolia*) bidentate. Cells of leaf apices subisodiametric, 28–35 × 35–45 μ in leaf apex, in leaf middle usually (35)40–45(48) × (40–48)50–60(70) μ, near leaf base little larger and to 40–50 × 75–100 μ, thin-walled and with minute or no trigones, rarely with somewhat thick-walled cells and small trigones; *leaf margins not crenulate*. Oil-bodies

___

[22] At a date too late to permit re-examination of all material here attributed to *C. muelleriana s. lat.*, I discovered that in the Greenland specimens (leg. Damsholt) numerous terminal, *Frullania*-type branches occur (Fig. 120:1–3). Since this is perhaps the most typical collection of subsp. *muelleriana* seen from the Western Hemisphere, the general absence in our material of such branches may be significant. Similar phenotypes with terminal branches occur south to Nova Scotia and Newfoundland (e.g., *RMS 68-1650a*).

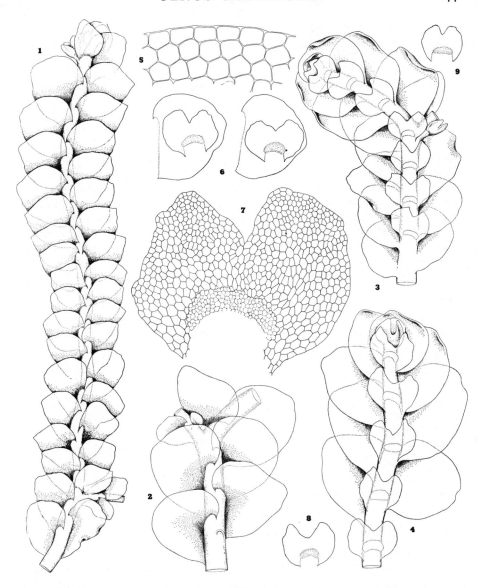

FIG. 120. *Calypogeia muelleriana* (typical, Greenland phenotype). (1) Shoot, with two *Frullania*-type branches, dorsal view (×11.4). (2) Shoot sector, dorsal view, with *Frullania*-type branch (×14.5). (3) Shoot sector, ventral view, with *Frullania*-type branch (×16). (4) Shoot sector, ventral view (×16). (5) Cells, leaf apex (×220). (6) Two leaves and, drawn within, adjoining underleaves (×14). (7) Underleaf (×60). (8–9) Underleaves (×14). [All leg. Damsholt, Unartoq I., at margin of warm spring, 60°30′N., 45°20′W., south Greenland.]

colorless, 4–8(8–13) per cell,[23] from 4–4.5 × 6–9 μ to 5 × 13 μ, rarely larger, relatively few-segmented (often of 3–5 coarse segments, each segment ca. 3.5–5 μ in size; at other times of 2 rows of segments, each ca. 1.6–2.8 μ in size, oil-bodies then 4–14 segmented); chloroplasts ca. 4 μ in diam. *Underleaves* usually (2.0)2.2–2.5(2.75)× as wide as stem, *distant to approximate*, rarely subimbricate, broadly suborbicular, ca. 1.1–1.6(1.75)× as wide as long, from 650 μ wide × 600 μ long to 660–690 μ wide × 400 μ long (forms with un- usually wide underleaves) to a maximum of 1050 μ wide × 850 μ long, from (rarely) shallowly bilobed to bilobed *for* ⅓–½ *their length* by (rarely) a shallow, acute and closed sinus or (commonly) a broader, obtuse or rounded sinus, the lobes *usually obtuse or occasionally* ± *rounded*; median underleaf cells *relatively large*, ca. 28–38(40) μ wide × 45–65(68–80) μ long, with oil-bodies present; *midline of the underleaf with (3)4–6 cells from sinus to rhizoid-initial field;* lateral underleaf margins arched, *normally unarmed* or with merely an occasional indication of a low angulation on one side of the outer margin, clearly *decurrent at base*; rhizoid-initial field usually rather strongly transverse and 2.5–3× as wide as high, ± ellipsoidal.

Autoecious and paroecious. Capsule valves narrowly lingulate, their length 7–9× their width; epidermal cells with adjoining surfaces *of alternating longi- tudinal cell walls* with *more or less equally thick sheets* of pigmented secondary thickenings; inner cell layer with transverse tangential bands, the cells averaging 11–13 μ wide. Spores 12–16 μ in diam. Chromosome no. = 18.

Type. Baden, Germany, *K. Müller.* (Topotypic material examined, *leg.* K. Müller.)

Distribution

Widespread throughout nontropical portions of eastern North America, northward into the ecotone between the Coniferous Forest and Tundra, occasionally even in true Tundra. In North America *C. muelleriana* was uniformly confused with *C. trichomanis* until the differentiating features were emphasized (Müller, 1947a; Schuster, 1949a, 1953, 1968a). Most prior reports of *C. trichomanis* from North America belong here, a few possibly to *C. fissa.*

W. GREENLAND. Disko I., Godhavn, below Lyngmarksfjeld, 69°15′ N., 53°30′ W. (*RMS & KD 66–172, 66–161, 66–176*).   S. GREENLAND. Unartoq I., at the margin

[23] The oil-bodies of this taxon appear to be highly polymorphic as regards number of segments forming each oil-body and their consequent size. Nearctic material has oil-bodies as described above and illustrated (Fig. 121:1,14;122:1,13). However, Buch (1936) states that in Finnish material they occur 5–12 per leaf cell and are of 10–25 small spherules, oriented in 2–4 rows. (Buch used the name *Calypogeia trichomanis* for this material; later, 1948, corrected by him to *C. muelleriana.*) Müller (1947a, p. 416, fig. 4d) illustrates and describes the oil-bodies of German material in terms nearly corresponding to our nearctic material; he states that the 3–8 oil- bodies per cell are 4 × 5–7 μ, occasionally 5 × 9–12 μ or even to 7 × 16 μ; they are formed of 3–14 spherules ca. 2–3 μ in diam. His illustrations correspond to those prepared from the material of *RMS 24028.* Polymorphism in oil-body form is conceivably due to the expression of a single gene-complex or may be largely or entirely environmentally induced.

of a warm spring, 60°30′ N., 45°20′ W. (*Damsholt, July 11, 1963!*; material in large mats above water at warm spring margin; quite typical, with more "typical *C. muelleriana* underleaves" than in other American phenotypes seen, with an open sinus, often blunt lobes, and smoothly curved lateral contours; *Frullania*-type lateral branches commonly developed!). QUEBEC. S. side of Great Whale R. (*Marr 66le*; Schuster, 1951); near St. Jovite, Terrebonne Co. (Crum & Williams, 1960). NEWFOUNDLAND. Widespread in Avalon Distr., on S. coast, W. Nfld., N. Pen., and in central and NE. Nfld. (Buch & Tuomikoski, 1955). ONTARIO. Manitoulin I. (Williams & Cain, 1959); Coopers Falls, Ontario Co. (*Cain 4262*). NOVA SCOTIA. 2 mi S. of Lower Argyle, Yarmouth Co. (*RMS 43064a*); Lake on Cabot Trail near summit of French Mt., ca. 1200 ft (*RMS 43038*).

VERMONT. NE. of summit of Mt. Mansfield, ca. 3200–3400 ft, along Long Trail (*RMS 43819*; mod. *densifolia*). MASSACHUSETTS. Breckenridge Rd., Hadley, Hampshire Co. (*RMS 43903a*; mod. *parvifolia*); Chesterfield Gorge near W. Chesterfield (*RMS 41307*); edge of Hawley Cranberry Bog, W. Hawley, Franklin Co., 1800 ft (*RMS 43948, 43950*); Guilder Pond, 2060 ft, Mt. Everett (*RMS 68-166*).

NORTH CAROLINA. Middle Creek, ca. 3400 ft, Black Mts., Yancey Co. (*Schofield 9539*); Highlands Biological Station, Highlands, Macon Co. (*RMS 44589*); swamp N. of Chowan R., ca. 1 mi N. of Winton, Gates Co. (*RMS 34579*); near summit of Richland Balsam, E. of Rich Mt., ca. 6400 ft, Haywood Co. (*RMS 39665b*; mod. *parvifolia-fissistipula-angustifolia*). TENNESSEE. Mt. LeConte, Smoky Mt. Natl. Park, Sevier Co. (*Sharp 5613*). SOUTH CAROLINA. Gorge of Thompson R., 2–2.5 mi above jct. with Whitewater R., Oconee Co. (*RMS 41008*). GEORGIA. Tallulah Falls, Rabun Co. (*RMS 34364a, 34364, 34384, 34355, 34358*); Brasstown Bald, Towns Co. (*RMS 34334*); N. slope of Rabun Bald, Rabun Co. (*RMS 49690a*). KANSAS. Woodson Co. (McGregor, 1955; report needing verification!) OHIO. Cuyahoga, Hocking Cos. (Miller, 1964).

All preceding personal reports have been derived from living plants; hence the colorless oil-bodies could be verified.

The reports from Georgia, South Carolina, North Carolina, and Tennessee are, in part, of plants which are not readily separable from subsp. *blomquistii*, and their placement involves a subjective element. Such plants lack terminal branches. By contrast, the northeastern phases (from Nova Scotia, Newfoundland, Greenland) have terminal branches and are very different.

Ecology

A "tolerant" and widespread species, inhabiting a wide diversity of sites which have in common two features: a reasonably constant (not necessarily high) moisture supply, and the absence of much direct sunlight. Often on thin layers of soil or humus on steep slopes or in ravines, over shaded (often damp) rocks; from there growing out over bare rock surfaces (though rarely a pioneer saxicole). Also widespread over peat and peaty soil (though rarely over and among living *Sphagnum*, then as a deviant mod. *parvifolia-densifolia*, with *Scapania nemorosa, RMS 25510*). On damp rocks it may occur with other sciaphilous meso-hygrophytic species, such as *Trichocolea tomentella, Blepharostoma trichophyllum, Lepidozia reptans, Microlepidozia sylvatica*, and *Calypogeia fissa*, and may extend to humus-covered ledges under more xeric conditions (then often with *Bazzania trilobata, Diplophyllum apiculatum*). Occasionally frequent on moist soil of shaded road cuts with *Calypogeia sullivantii*, *Diplophyllum apiculatum*, *Scapania nemorosa*, *Pogonatum*, *Atrichum*, etc., or on clayey soil along banks or near streams. The species is rare on decaying logs; it is occasional on moist, acid, sandy-peaty soil at lake edges.

Northward *C. muelleriana* often occurs over loamy soil, particularly along old logging roads or paths, often where subject to some disturbance, as well as to intermittently rather dry conditions (particularly during periods of summer drought); then frequently found as a mod. *parvifolia-densifolia* varying from an *angustifolia* to a *latifolia* extreme, ranging from merely 1 to 1.5 mm wide. Such phenotypes are hardly recognizable, since the cells are reduced in size (apical often only 25–27 μ; median often only 25–28 or 25–30 × 35–40 μ), and the underleaves are atypical in shape, with the sinus descending to within 2–3 cells of the rhizoid-initial region. They may occur with *Cephalozia bicuspidata*, *Scapania nemorosa*, *Lophozia bicrenata*, and mosses (i.a., *Diphyscium foliosum*).

Variation

Associated with wide distribution under diverse ecological conditions (see above), this abundant species produces a wide range of phenotypes. Although the majority of these are probably environmentally induced, I believe that there is much minor genetic diversity. Extreme phases of this species become difficult to separate, on the one hand, from the *C. neesiana-integristipula* complex and, on the other, from *C. fissa*.

The confusion between *C. muelleriana* and the *C. neesiana-integristipula* complex is documented by the differing opinions expressed regarding the position of plants such as *C. neesiana* var. *laxa* by European authorities on the group (K. Müller, 1940, 1947a; H. Buch, 1942, 1948). Buchloh (1952) has reviewed the complex, supporting Müller's concepts.

Buch (1948) discussed the separation of *C. muelleriana* from *C. meylanii*, after Müller (1947a) attempted to reduce *C. meylanii* (= *C. integristipula*) to a synonym of *C. muelleriana*, basing this reduction on the fact that the first synonym cited under *C. meylanii* (*in* Buch, 1936a, p. 201) is *C. neesiana* var. *laxa* Meyl. ex K. Müll. (1913, *in* 1905–16, p. 239). However, Buch points out that the type of this variant cannot typify *C. meylanii*, since the type of *C. meylanii* is Verdoorn's Hep. Sel. et Crit. no. 313. Müller (1947a) insisted that *C. neesiana* var. *laxa* represents a form of *C. muelleriana*; Buch (1948) reiterated his opinion that it is identical with lax-leaved forms of *C. meylanii*, as supposedly evident from the elongate, nearly ovate leaves, and indicated that Müller (1947a) erroneously referred lax-leaved modifications of *C. meylanii* to *C. muelleriana*. As evident from the discussion under *C. integristipula* (= *C. meylanii*, p. 209), bog-inhabiting, lax-leaved forms of *C. meylanii* (= *C. integristipula*) admittedly often have leaf cells which are so large that these plants tend to key to *C. muelleriana* in Müller's key (1947a); such phenotypes also bear such large [to 5–6 × 12–15(18) μ] oil-bodies that on this basis they also would go into *C. muelleriana* (as defined by

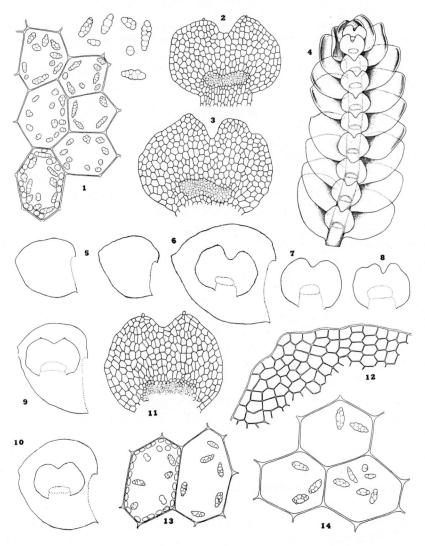

FIG. 121. *Calypogeia muelleriana.* (1) Cells of leaf middle with oil-bodies and, lower left cell, chloroplasts (×370); above, individual oil-bodies (×665). (2–3) Under-leaves (×40). (4) Shoot, ventral aspect (×12.2). (5) Leaves (×11.5). (6–8) Leaf and three underleaves (×20). (9–10) Leaves of mod. *latifolia*, with adjoining underleaves (×16). (11) Underleaf (×44). (12) Cells of leaf apex (×92). (13) Two median cells with oil-bodies and, at left, chloroplasts (×375). (14) Cells with oil-bodies (×345). [1–8, *RMS 28541*, North Carolina; 9–12, a mod. *densifolia-parvifolia*, approaching *C. neesiana*, *N.E.N. Bremekamp 2388*, from Holland; 13, Lyndhurst, England, leg. E. W. Jones; 14, New Forest, England, leg. E. W. Jones.]

Müller).[24] However, the type (and illustrations in Müller, 1905–16, fig. 69) of *C. neesiana* var. *laxa* appears to represent *C. muelleriana* (as also indicated by Buchloh, 1952). Under shaded and moist conditions, both species are evidently capable of producing a lax modification (mod. *angustifolia-laxifolia*), which must be separated on the basis of the subentire underleaves, rounded at base, of *C. meylanii* (= *C. integristipula*) vs. the evidently bilobed underleaves of *C. muelleriana* (which are clearly if weakly decurrent at base) and in particular *by texture* [*dull* in *C. meylanii* (= *C. integristipula*); *nitid* in *C. muelleriana*]. The critical problem of the distinction of *C. muelleriana* and the *C. integristipula-neesiana* complex is treated on pp. 201 and 212.

The pattern of variation in *C. muelleriana* starts, essentially, from broad-leaved forms, with short internodes and large and contiguous transverse underleaves. Such plants approach *C. neesiana s. str.*, which may also produce, if more rarely, xeromorphic forms with broadly ovate leaves, as broad as long, short internodes, and large, contiguous underleaves. Such plants, of the mod. *latifolia-densifolia-latistipula-densistipula*, may be so similar (particularly in the Southern Appalachians) that one may errone-ously conclude a single species to be at hand. [Meylan (*in* Rev. Bryol. 1910, p. 79) considered such broad-leaved forms ("var. *repanda*") of *C. neesiana* to represent *C. muelleriana* (then called *C. trichomanis*), describing the mod. *densifolia* as *C. trichomanis* var. *compacta* Meylan.] That *C. neesiana* and *C. muelleriana* are abundantly distinct species is readily evident from Buch's comparative cultures (1936a) (although he used the name *C. trichomanis* for *C. muelleriana*) and is equally clear when intimately inter-grown plants of the two species are studied from field material (p. 204 and Fig. 126).

Xeromorphic types, with increased shade and moisture, gradually undergo modification to hygromorphic shade forms which are often taxonomically very confusing. They bear much narrower leaves (1.2–1.4 × as long as wide), often rather broadly rounded distally, that are sub-contiguous (internodes long) to approximate, and possess smaller under-leaves, often only 1.8–2.2 × as wide as the stem. These underleaves, furthermore, may become more or less elongate, quite unlike those in any of our other species. Such lax forms (mod. *laxifolia-angustifolia-angusti-stipula*) are often impossible to distinguish sharply from the subsp.

---

[24] Again it is illuminating to consider Ellwein's (1926) work on *C. trichomanis*, which demon-strated great variation in cell size resulting from nutritional differences. Furthermore, the associated data and illustrations given show that even *C. muelleriana* may have the median cells small [only 34–40 × 40–48 μ in *RMS 25510* and as small as 25–30 × 35–40 μ in *RMS 49741*; thus clearly smaller than in lax forms of *C. meylanii* (= *C. integristipula*), in which the cells may be 38–42 μ wide × 48–58 (63–68) μ long]. The extreme variability of cell size, leading to confusion of the small-celled xeromorphic mod. *parvifolia-densifolia* of *C. muelleriana* with the large-celled, lax-leaved mod. *laxifolia-megafolia* of *C. meylanii* (= *C. integristipula*), is emphasized in Schuster (1949a, pp. 536–537).

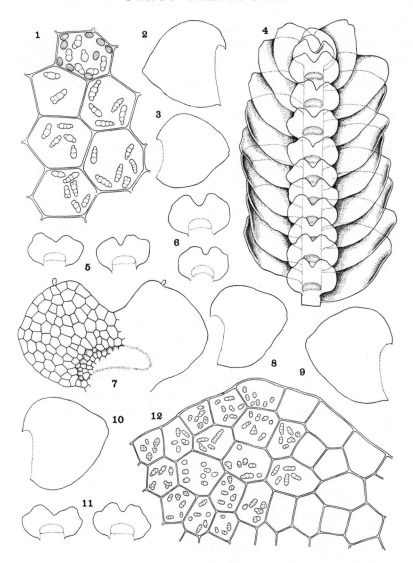

FIG. 122. *Calypogeia muelleriana* (mod. *densifolia-latifolia-densistipula-latistipula*). (1) Cells of leaf middle with oil-bodies and, upper cell, chloroplasts (×358). (2) Broad leaf (×20). (3) Leaf (×20). (4) Normal shoot, the bottom six pairs of leaves matured in the field, the upper three pairs developed in culture, with diffuse light, 100% RH (×18). (5–6) Underleaves (×16.5). (7) Underleaf (×78). (8–10) Leaves (×16.5). (11) Underleaves (×20). (12) Cells of leaf apex, oil-bodies indicated in most cells (×290). [All from *RMS 25510*, Linville Falls, N.C.]

blomquistii (which differs principally in the strong crenulation of leaf and underleaf margins). On the other hand, lax-leaved, narrow-leaved forms often approach *C. integristipula* in leaf form, although the underleaves and texture when dry are usually quite different, allowing a ready separation of the two.

Finally, certain ± lax-leaved phenotypes, with rounded or blunt leaf apices and narrow leaves, often have underleaves that are ± transverse and occasionally bear an obtuse lateral tooth. Such plants are often virtually impossible to separate from *C. fissa* and (in North America, at least) appear to grade into the latter; see p. 174.

The pattern of variation is perplexing, even after a careful analysis of critical specimens. Schuster (1968a) gives a critical analysis of variability in a selected series of phenotypes, illustrating the extremes found within *C. muelleriana*.

Differentiation

As one of the most widely dispersed and tolerant species of the genus, *C. muelleriana* shows a correspondingly wide range of environmental modification (and presumably a considerable spectrum of genetic variation), leading to potential confusion with *C. integristipula* and *neesiana*, on the one hand, and with *C. fissa*, on the other; see pp. 199–204.

The most important diagnostic features of the species are the typically broadly ovate to broadly elliptical leaves (length-width ratio ca. 1 : 1.0–1.2), usually widest slightly above the base; leaf apices narrowly rounded usually—virtually never pointed; leaf margins of isodiametric cells never interrupted by a series of prominently elongated cells; median leaf cells averaging ca. 38–45×48–60μ (except in xeric extremes); oil-bodies colorless; underleaves at least twice as wide as the stem, incised-emarginate to (more frequently) bilobed ⅓ or more, the length of the underleaf from sinus to rhizoid-initial field usually 4–6 cells; underleaves somewhat narrowed basally and short-decurrent on both sides; median cells of underleaves ca. 28–40×45–$65(80) \mu$.

Typically, *C. muelleriana* differs from *C. fissa* in having (*a*) broader leaves, nearly or quite as wide as long normally; (*b*) usually more shallowly bilobed underleaves (often divided by a sharper, narrower sinus), whose lateral margins are only infrequently angulate, normally evenly rounded; (*c*) almost uniformly entire lateral leaves, even of relatively "weak" shoots (in *C. fissa* with "weak" shoots as well as some normal, robust plants often bidentate-leaved). No single character affords a constant, sharp separation between these two species, and distressingly "intermediate" populations (possibly phenocopies) are frequent. Correct identification of these may at times be possible only by

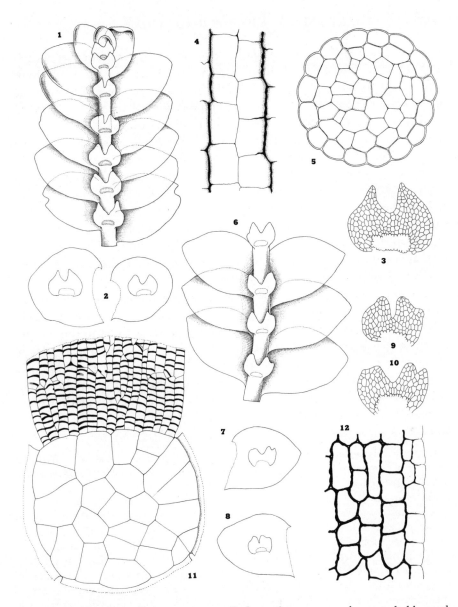

FIG. 123. *Calypogeia muelleriana*, trans. to *C. fissa* subsp. *neogaea*. (1–5, probably mod. *angustifolia-densifolia* of *C. muelleriana*; 6–12 probably mod. *integrifolia* of *C. fissa* subsp. *neogaea*). (1) Small shoot with narrower than normal leaves (×19). (2) Two leaves of the broad-leaved, more mature extreme intermixed, with adjacent underleaves (×15). (3) Underleaf (×38). (4) Epidermal cells of capsule wall (×188). (5) Seta cross section (×ca. 110). (6) Shoot of mod. *angustifolia* (×21). (7–8) Leaves and adjoining underleaves (×15). (9–10) Underleaves (×38). (11) Base of capsule, showing uppermost cells of seta and inner face of one valve base (×155). (12) Epidermal cells from valve middle, at right valve margin (×155). [1–5, *RMS 24750*, Soco Falls, N.C., from plants growing admixed with *C. neesiana* portrayed in Fig. 130:2–7; 6–12, *RMS 1394a*, Chittenango Falls, near Cazenovia, N.Y.]

culturing the plants, but can usually be achieved by using a constellation of characters. Possible hybridization occurs between these taxa; perhaps only a single polymorphic species should be recognized here, although the obvious differences between "typical" populations of the two make such a conclusion hard to reconcile with present observations.

Typically, *C. muelleriana* differs from *C. integristipula* as follows. (1) The broader, nearly heart-shaped leaves average, except in lax forms, nearly as wide as or wider than long and are more nitid in texture. (2) There is a tendency (when growing under identical conditions) to produce larger cells of the leaf and underleaf middle. (3) Underleaves tend to be less suddenly, less strongly constricted basally and more or less decurrent; underleaf tips are obtuse to occasionally rounded, with the lobes separated by a sinus varying from acute to broad and crescentic; the sinus descending usually ⅓ or deeper. (4) Consequent upon their larger cell size and deeper sinus, the underleaves are formed medially of only (3)4–5 or rarely 6–7 cells per longitudinal row. (5) Apical leaf cells are isodiametric, only rarely are short rows of 2–3 cells weakly elongated parallel to the margin (in *C. integristipula* "foreshadowing" of the *C. neesiana* marginal border of elongate peripheral leaf cells is often found on at least some leaves). No single characteristic cited above—except leaf texture—is completely constant, and almost all are strongly subject to environmental modification. Hence, it must be stressed, a constellation of characters must often be used to effect a separation.

Confusion between *C. muelleriana* and *C. integristipula* can usually be avoided because of the narrow leaves and the entire to retuse underleaves of *C. integristipula*, contrasted to the broader leaves and clearly bilobed underleaves of *C. muelleriana*. In both of these characters, *C. muelleriana* is more likely to be confused with *C. neesiana s. str.*, since the latter (especially the fo. *repanda*) may have broader leaves, as wide as or wider than long, and usually has shallowly incised or at least emarginate underleaves. Indeed, in the Southern Appalachian region (from which *C. integristipula* appears absent) difficulties arise in separating *C. neesiana* from *C. muelleriana*. Broad-leaved forms of *C. neesiana* can usually be separated from *C. muelleriana* on the basis of the (*a*) distinct margin of some or most leaves, formed by elongate, narrow cells; (*b*) more rounded bases of the underleaves; (*c*) dull texture of dry plants. In spite of the occasional occurrences of phenocopies that are difficult to place, the distinctness of these two species is readily apparent when they occur intermingled. Under such conditions, the *C. muelleriana* looks paler green (less bluish gray), is distinctly shiny when dry—rather than dull-textured, and is somewhat larger.

Calypogeia muelleriana differs from *C. trichomanis* at once in the colorless oil-bodies. When the plants are dead, differentiation of the two species is usually difficult (p. 156). However, in *C. muelleriana*, we find that (*a*) the

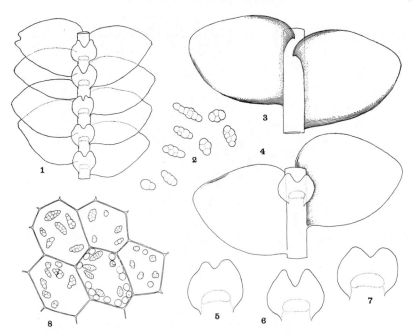

FIG. 124. *Calypogeia muelleriana*, trans. to *C. fissa* subsp. *neogaea*, probably mod. *angustifolia* of *C. muelleriana*. (1) Shoot sector, ventral aspect (×17). (2) Oil-bodies (×730). (3–4) Shoot sectors, antical and postical aspects (×17). (5–7) Underleaves (×36). (8) Cells with oil-bodies and, in one cell, chloroplasts (×390). [1, 2, 8, *RMS 24028*, Alum Cave, Mt. LeConte, Tenn.; 3–7, *RMS 19204*, Van Cleave, Miss.]

sinus of the underleaves is often relatively shallow; (*b*) the lobes of the underleaves are more often obtuse or even rounded; (*c*) the leaves are less heart-shaped. These differences are neither sufficient nor constant enough to effect a certain separation between the two taxa. Hence many herbarium specimens of this complex, for which cytological data are unavailable, are taxonomically useless.

CALYPOGEIA MUELLERIANA subsp. BLOMQUISTII
Schust., subsp. n.
[Figs. 125, 126]

Folia normaliter anguste ovata, plerumque ± acuminata, in marginibus crenulata; amphigastria margines plerumque crenulates, lobis saepe elongatis, habentia.

Robust, pale (somewhat pellucid) green, *glistening when dry*, forming flat, thin patches in areas with exceedingly diffuse light (often <20 foot candles). Shoots when mature 2.8–3.8(4) mm wide × 1.5–2(2.5) cm long, prostrate, only loosely adnate; occasional shoots become erect and small-leaved, gemmiparous

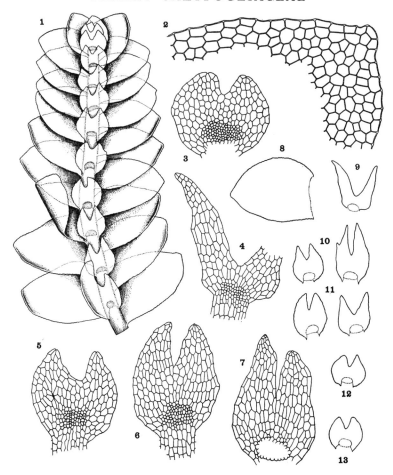

FIG. 125. *Calypogeia muelleriana* subsp. *blomquistii*. (1) Mature shoot, ventral aspect (×13.5). (2) Leaf apex cells (×88). (3–7) Underleaves of divergent types (×43). (8) Leaf (×15). (9–13) Underleaves of divergent types (×15). [All from type, Linville Gorge, N.C., *RMS*.]

at apex. Stems ca. 240–300 μ in diam.; ventral cortical cells ca. 20–25 × 60–100 μ, relatively thin-walled. Lateral leaves flat, *consistently narrowly triangular-ovate*, only slightly obliquely so, *spreading at an angle of 75–90°*, widest at the base or barely above it, averaging ca. 1100–1150 μ wide × 1550–1600 μ long, on robust shoots to 1450 μ wide × 1850–1900 μ long; *typically ca. 1.3–1.45× as long as wide*; leaf apices normally narrowly to somewhat broadly rounded, never truncate or bidentate on mature plants, occasional leaves obtuse to subacute. Marginal cells at leaf apex ca. 36–44 μ (measured tangentially), within margins 30–35 μ and ± isodiametric; median cells strongly elongated on mature leaves, (30)32–39(40–42) μ wide × (65)72–85(90–100) μ long, often averaging twice as long as wide; *marginal cells crenulate, the radial walls thickened toward their external or marginal ends, appearing to protrude*, the tangential marginal walls concave between them. Typically 6–12 *colorless oil-bodies* per cell; oil-bodies mostly irregularly fusiform to ellipsoidal, 5 × 8 to 5 × 10–12 *μ*, a few 6 × 12 *μ*, the larger 8–15-segmented, usually in 2, locally in 3, rows medially; segments ca. 1.5–2 *μ* in diam., occasional ones larger; chloroplasts unusually large: to 5–7 *μ* in longer diam. *Underleaves with extreme intraclonal variation*, on shoots with short internodes usually of the *muelleriana* type, e.g., ca. 720 *μ* wide × 560 *μ* long, *divided to* ⅓, *with obtuse to rounded lobes separated by a broadly U-shaped or broadly V-shaped sinus;* on shoots with longer internodes some or *many underleaves becoming elongated, often strongly so*, varying from 550 *μ* wide × 650 *μ* long up to 550 *μ* wide × 950 *μ* long, occasionally 700 *μ* wide × 1250 *μ* long; the more elongate underleaves most often bilobed fully ½ their length, occasionally bifid nearly to base or merely shortly emarginate and ovate-lanceolate in shape; *underleaf margins crenulate or crenulate-sinuate like leaf margins;* rhizoid-initial region transversely ellipsoidal, ca. twice as wide as high; cells of underleaves averaging 24–28 *μ* wide × 42–52 *μ* long in the middle (between sinus and rhizoid-initial field); underleaves with *lateral margins evenly curved*, exceedingly rarely armed with an obtuse angulation or small tooth. Chromosome no. $n = 9$.

Type. North Carolina: shaded rock walls, east side of Linville R. gorge, ca. 1.5–2 miles below the falls, *RMS 28940*; type in herbaria of author and of Duke University. *Cotype.* Same data, with *Hookeria acutifolia, RMS 28944.*

Distribution

Very largely restricted to the Southern Appalachians and the Ouachitas. A relatively rare and local taxon, its distribution still imperfectly known, partly because so little material in the group can be determined with certainty.

NORTH CAROLINA. Linville Gorge, Burke Co. (*RMS 28940, 28944*, type material; *RMS 28964, 29479, 29478?*); Neddie Cr., near jct. with E. fork of Tuckaseegee R., Rte. 281, Jackson Co. (*RMS 29449, 29450, 36672*); Bridal Veil Falls, Cullasaja Gorge, NW. of Highlands, Macon Co. (*RMS 29510*); E. bank of Horsepasture R., near Rainbow Falls, ca. 3000 ft, Transylvania Co. (*RMS 30018, 30019*); Soco Falls, Jackson Co. (*RMS 36672, 36678*); Clingmans Dome, near Andrews Bald, ca. 6200 ft, Swain Co. (*RMS 36656*); Chattooga R., 1 mi below Narrows, Jackson Co. (*RMS 39455, 39457*); Glenn Falls, 3500 ft, near Highlands, Macon Co. (*RMS 40610*); Bubbling Spring, 0.5 mi N. of Beech Gap, near Blue Ridge Parkway, Haywood Co., 5100–5200 ft (*RMS 39370, p.p.*); gorge of Crow Cr., below High Falls, near Cullasaja R., between Highlands and Franklin, Macon Co. (*RMS 40871, c. caps.*; *40868a*, with *C. neesiana*); W. Branch from Fork Ridge of W. Fork of Pigeon R., near jct. with Flat Laurel Cr., 4500–4600 ft, S. of Sunburst (*RMS 39380*). TENNESSEE. Gorge above Middle Prong of Little R., ca. 1600 ft. below Tremont, Blount Co., Smoky Mt. Natl. Park (*RMS & Sharp 34642*). GEORGIA. Brasstown Bald, ca. 4500 ft, Towns Co. (*RMS 34316*); below Tallulah Falls, Rabun Co. (*RMS 34358*); gorge of Big Cr., 5 mi S. of Highlands, N.C., Rabun Co., ca. 2000 ft (*RMS 40758*, autoecious; *40694a*, with *C. peruviana*; *40718, 40754, 40755,*

40056).   SOUTH CAROLINA. Whitewater R. Gorge, 0.3–0.4 mi below Lower Falls, above Jocassee, Oconee Co. (*RMS 40961a*).   VIRGINIA. The Cascades, Little Stony Cr., near Mountain L., Giles Co. (*RMS 40242, 40243, 40384a, 40251*).   ARKANSAS. On dripping cliff, densely wooded cove, Saline R., SE. of Mena, Ouachita Mts., Polk Co. (*Anderson 11459*).   Rarely northward: MASSACHUSETTS. Green R. gorge near Colrain (*RMS 68–121*).

Ecology

Found in deeply shaded and humid sites, usually on moist ledges and cliffs, often in deeply wooded ravines or on ledges shaded by dense stands of *Leucothoë*. Associated in the Appalachians are such sciaphilous mosses as *Hookeria acutifolia* and *Dicranodontium denudatum* and such Hepaticae as *Calypogeia sullivantii*, *Pellia epiphylla*, and *Solenostoma hyalinum*, as well as tolerant species like *Diplophyllum apiculatum* and *Marsupella emarginata*.

The ecology of *C. muelleriana* subsp. *blomquistii* closely matches that of typical *C. muelleriana* in the Southern Appalachians. This point is very significant, since in the circumscription of this new taxon it must be emphasized that a narrow-leaved shade form of typical *C. muelleriana* is seemingly not involved. For example, *C. muelleriana* has been found repeatedly in shaded, moist sites, bearing "normal," broad, heart-shaped leaves and "normal," transversely subrotundate underleaves, the margins of neither showing obvious crenulations.

Variation and Differentiation

Although in the field easily overlooked for *C. fissa* subsp. *neogaea* (because of the narrowly ovate leaves with entire apices, and the pale, rather whitish green color), *C. muelleriana* subsp. *blomquistii* appears distinct from other nearctic taxa in the following respects. (1) Leaves are typically narrow, narrowly to rather broadly rounded at the tip, only exceptionally subacute. (2) There is a tendency toward a distinct crenulation of the leaf (and underleaf) margins—the marginal cells appearing to have the radial walls pushed out, with the outer tangential walls concave between. Not all leaves show this characteristic equally well, but it is always readily evident. (3) Leaf cells are typically strongly elongated medially, and in well-developed plants often average more than twice as long as wide (often ca. 32–39 μ wide × 76–90 μ long). (4) Underleaves, in a genus notorious for wide variation in underleaf shape, are extraordinarily poly-morphic, with a range of variation quite foreign to typical *C. muelleriana* *and to our other species of the genus.*

Underleaves are perhaps most frequently suborbicular to slightly broader than or slightly longer than wide; they range from 1.8–2.2 × as wide as the stem on well-developed plants, averaging ca. 720 μ wide × 560 μ long. Even when suborbicular, underleaves are rounded on their external margins and are almost identical with those of *C. muelleriana*, except for the typically distinctive

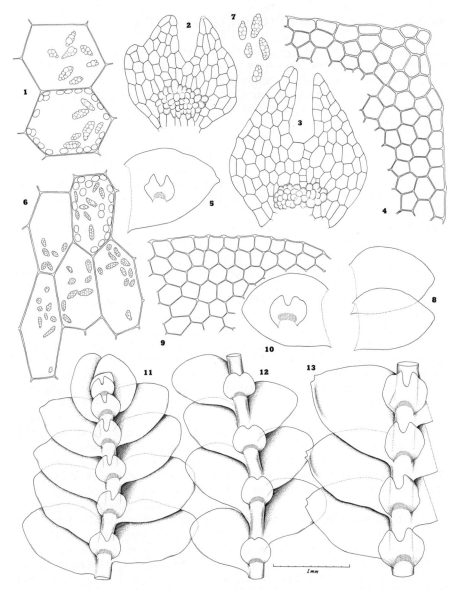

FIG. 126. *Calypogeia muelleriana blomquistii* (1–11); *C. neesiana* (12) and *C. peruviana* (13).
(1) Median cells with oil-bodies and, lower cell, chloroplasts (×430). (2–3) Underleaves
(×105). (4) Cells of leaf apex, showing the typical crenulation (×185). (5) Leaf and
underleaf (×26). (6) Median cells with oil-bodies (×315). (7) Individual oil-bodies
(×590). (8) Two leaves (×18.5). (9) Marginal cells along antical margin just below
apex (×133). (10) Leaf and underleaf (×18.5). (11) Typical shoot apex (×21.5).
(12) Shoot sector (×21.5). (13) Shoot sector of phase with underleaves laterally almost
uniformly unarmed, but typical in the blue oil-bodies (×21.5). [1, *RMS 39354*,
Haywood Co., N.C.; 2–5, *RMS 39480a*, Chattooga R., N.C., a phenotype with chromosome no. *n* = 9 (det. by Dr. V. Bryan); 6–8, from type, *RMS 28940*, Linville Gorge,
N.C.; 9–13, drawn to lower scale, all from *RMS 39417a*, Chattooga R., N.C., with
C. muelleriana blomquistii, *C. neesiana* and *C. peruviana* growing intimately interwoven.]

crenulate-sinuate margins. In general, they are divided for 0.45–0.55, often 0.6, their length, with the lobes erect, subparallel, or even slightly convergent; the sinus is only occasionally V-shaped but often narrowly U-shaped, so that its base, even if narrow, is almost always distinctly rounded. In addition, some underleaves vary greatly in form, all showing distinct elongation. The least modified type of elongate underleaf is one with the drawn-out leaf lobes becoming hornlike (the underleaf ca. 550 μ wide × 650 μ long, divided for 320 μ); with elongation of the underleaf, the cells become elongated. In extreme cases, the lobes become narrowly lanceolate, with the underleaf 510–550 μ wide × 840–950 μ long, divided for 420–500 μ. Underleaves of the latter type and transitions between them and previous types are frequent. In addition to the foregoing, extreme underleaf types, derived from the elongate type previously discussed, are produced: (a) underleaf 520 μ wide × 820 μ long, divided nearly to base into two slightly divergent, linear-lanceolate lobes; (b) underleaf 700 μ wide × 1250 μ long, extremely asymmetrical; (c) under-leaf essentially narrowly ovate in shape, 560 μ wide × 950 μ long, shallowly bilobed (100 μ) at apex into two small triangular lobes. Development of the elongate underleaves appears largely (but not totally) correlated with environmental conditions that stimulate the production of long internodes (mod. *laxifolia*); inversely, production of transverse, broad, orbicular underleaves of the *C. muelleriana* type is correlated most often with development of short internodes (mod. *densifolia*) and also often with smaller leaves (mod. *parvifolia*). Thus, if by chance a plant of the mod. *densifolia-parvifolia* is examined, and no others, the student, on superficial examination, may conclude that a narrow-leaved form of *C. muelleriana* or the frequent southeastern form of *C. fissa* (subsp. *neogaea*) is at hand. However, neither of the latter taxa shows the characteristic crenulation of leaf and underleaf margins of subsp. *blomquistii*, or the tendency for elongation of the underleaves, with their lobes hornlike and drawn out.

If a single plant is studied, it becomes evident that the cell size is to a large degree a function of leaf shape. In the lower, broader, and shorter leaves (mod. *latifolia-parvifolia-densifolia*) the cells may be of nearly normal width (34–39 μ average) and average only 50 μ long. Near the middle of the shoot, where the plant has attained its characteristic form and dimensions, the leaves are narrower and more distant (mod. *angustifolia-laxifolia-megafolia*); here the cells average 30–32 μ (occasionally 35 μ) wide × 65–85 μ long (with occasional short cell rows even 85–100 μ long per cell). One need not emphasize that the latter measurements, from mature leaves, differ considerably from those of the *C. fissa-muelleriana* complex (where the basic width averages 36–40 μ up to 38–42 μ; the length averaging 50–60 μ, rarely longer; consequently, the median cells in mature leaves never average as much as twice as long as wide). In the inferior cell width (usually ranging between 32 and 38 μ), *C. muelleriana blomquistii* agrees more closely with the otherwise quite unrelated *C. integristipula-neesiana* complex. In the latter, however, the median cells rarely average more than 40–48 μ long.

On study of a series of specimens, *C. muelleriana blomquistii* acquires a level of variability approaching that of *C. muelleriana*. Extreme plants, as in other species of this genus, are not always easily recognizable, although

subsp. *blomquistii* is usually readily differentiated from all other phases of
C. muelleriana (and from *C. fissa*) by the following combination of features
(some of which are, admittedly, mere tendencies): leaves elongated,
usually markedly so; underleaves with lateral margins never distinctly,
armed, the sinuses often U-shaped and broad, the lobes erect or even
connivent, often drawn out and hornlike; underleaves frequently tending
to be clearly longer than broad, and then rarely more than 1.2–1.6× as
broad as the stem; leaf and underleaf margins tending to be distinctly
crenulated, under "normal" conditions quite strongly so.

Initially I was of the opinion that subsp. *blomquistii* represented a fully
autonomous species (and some herbarium material labeled *C. blomquistii* has
been distributed); the chromosome number of $n = 9$ seems to further support
such a treatment, since Müller (1947a, 1951–58) reports $n = 18$ for this species.
However, there is only very imperfect linkage between crenulation + elongated
leaves + elongated underleaves, and in the years since the type plants (which
remain the most typically developed plants seen) were studied I have repeatedly
collected material that is seemingly transitional. Collections with the broader
leaves and broader underleaves of subsp. *muelleriana*, but with the strong
crenulation of both leaves and underleaves of subsp. *blomquistii*, can be found
in both North Carolina and Tennessee when carefully searched for. Their
disposition may hinge on obtaining a correct chromosome count.

CALYPOGEIA NEESIANA (Mass. & Carest.)
K. Müll. emend. Buch
[Figs. 126:12; 127–130; 131:1–4, 6–11]

Kantia trichomanis var. *neesiana* Mass. & Carest., Nuovo Giorn. Bot. Ital. 12:351, 1880.
Calypogeia trichomanis var. *neesiana* K. Müll., Beih. Bot. Centralbl. 10:217, 1901.
Calypogeia neesiana K. Müll. ex Loeske, Verh. Bot. Ver. Brandenburg 47:320, 1905; K. Müller,
   in Rabenh. Krypt.–Fl. 6(2):236, 1913 (in larger part, exc. var. *laxa*).

Smaller than *C. integristipula*, usually 1.5–2.1(2.5–2.6) mm wide, *dull*, whitish
green to gray green, subglaucous except marginal leaf cells; sparingly branched,
branches postical-intercalary. Leaves approximate to imbricate, *usually slightly
longer than wide* and ovate-ellipsoidal (width 0.85–0.95 length), infrequently up
to 1.1–1.2× as wide as long and broadly ovate-oval; *maximum width above the
somewhat narrowed base*; typically 930–950 μ wide × 1000 μ long or more, to a
maximum of 1.3–1.5 mm long,[25] apex usually truncate-retuse, more rarely
narrowly rounded. Median and basal cell size as in *C. integristipula*; *marginal
cells of distal portions of leaf normally strongly tangentially elongate in one* (rarely partly
in two) ± *interrupted row*, the cells 22–30 × 40–58(60–70) μ long; cells of leaf
apex (immediately within elongate border cells) ± isodiametric, ca. 25–32(36)
μ;[26] median cells ca. 33–38 × 34–40(45) μ long; near base ± elongate,

---

[25] In xeric modifications, leaves may be 930–950 μ wide × only 800–860 μ long to 1050 μ
wide × 950 μ long.
[26] In xeric extremes occasionally only 21–24 μ wide × 33–38 μ long.

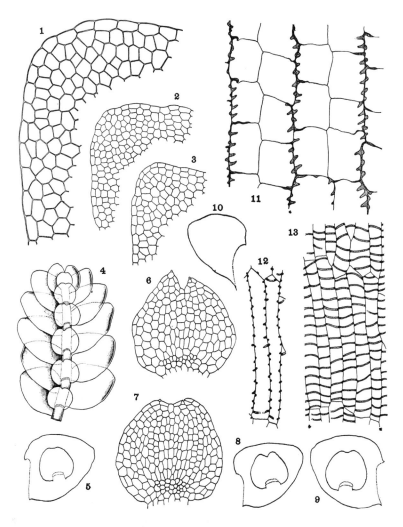

FIG. 127. *Calypogeia neesiana*. (1) Leaf lobe apex, with almost optimal development of border (×115). (2–3) Leaf lobe apices, with border obscure (×58). (4) Shoot, postical aspect (×13). (5) Leaf and underleaf (×22). (6–7) Underleaves, note slime papillae *on lobe apices* and the slight rhizoid-initial areas (×58). (8–9) Leaves and underleaves (×22). (10) Leaf (×ca. 20). (11) Epidermal cells of capsule wall; note shape of cells and form of thickenings; compare with Fig. 107:2 of *C. integristipula* (×277). (12) Inner cells, capsule wall, focus on internal tangential walls (×277). (13) Inner cells, capsule wall, focus on free tangential wall (×277). [1–9, *RMS 15998a*, Mt. Katahdin, Me.; 10–13, *Blomquist 2481*, Hendersonville, N.C.]

usually 31–40 × 52–60 μ. *Oil-bodies colorless*, typically in 1–3(4) marginal cell rows and in all or most basal cells, but *absent in some or all* (rarely only in isolated) *median cells*; oil-body usually formed of 2–8(10) *relatively large globules*, 3–4.5 (rarely 4 × 6) μ in size; oil-bodies from 4 × 7 to (more commonly) 4.5 × 9 to 4–4.5 × 10–11(13) μ, mostly 3–6 per cell. *Underleaves orbicular, averaging as long as wide* (550–650 μ wide × long), occasionally clearly wider than long (*to 1.1–1.25× as wide as long*) and then 700–780 μ wide × 510–600 μ long; median portion with (7)8–9(10–12) *cells per longitudinal row* from rhizoid-initial field to apex; median cells (20)23–27 μ wide × 30–40(52–75) μ long; *under-leaf width 1.9–2.5(3.0)× stem width*; apex *normally retuse to shallowly emarginate, occasionally* (less robust plants) *incised to 0.2–0.25 the length, then provided with obtuse, rounded, short lobes.* Underleaves usually with *oil-bodies all or nearly all absent*; rhizoid-initial field usually *small, strongly transverse*, its width ca. 3–4× its height, *often merely a linear transverse line 2–3 cells high*; underleaf bases rounded, not or very slightly decurrent. Gemmae frequent.

Apparently both autoecious and paroecious. Capsule valves reddish, 250–320 μ wide, ca. 6–9× *as long as wide*; epidermal layer with cells 30–45 μ wide × 40–50(55–75) μ long, with transverse and primary longitudinal walls delicate, devoid of thickenings, whereas both faces of alternating (secondary) longitudinal walls bear distinct (rarely obscure) *irregular, nodular, brownish thickenings.* Inner cell layer with narrow cells *in* (20)22–28 *rows*, (11)12–14(15) μ wide × 150–200 μ long; free tangential faces with numerous complete transverse bands extending across the radial walls to the interior tangential faces. Spores 10–14(16) μ.[27] Elaters ca. 8 μ in diam., bispiral; spirals reddish, 2.5 μ wide.

Type. Italy; *Massalongo* and *Carestia.*

Distribution

Except in the Southern Appalachians, an infrequent species in eastern North America. In the northern United States much less frequent than the more ubiquitous *C. integristipula*, but in the Southern Appalachians replacing the latter (which is not known south of southern New York, New Jersey, and Pennsylvania; in these regions it is montane or sub-montane). Existing reports of *C. neesiana*[28] indicate a range from the ecotone between the Arctic and Subarctic, in northern Quebec, southward to the Northern Hardwoods Forest of New England, and then reappearing in the Oak-Hickory and Mixed Mesophytic Forest Regions of the Southern Appalachians (in North Carolina and Tennessee to northern South Carolina and Georgia).

[27] Stated by Müller to be 10–12 μ, but in *Blomquist 2481* up to 15–16 μ and in *RMS 36959* largely 13–14 μ.

[28] See, e.g., those from Quebec (Lepage, 1944–45; Kucyniak, 1947), Georgia (Carroll, 1945), Kentucky (Fulford, 1936), Michigan (various reports by Steere and others), Minnesota (Conklin, 1942), Ontario ("widely distributed and very common" according to Cain & Fulford, 1948, who, however, confuse *C. neesiana* with *C. meylanii*), and Wisconsin (Conklin, 1929). The bulk of these reports cannot be accepted, and all need verification.

In spite of the sharp distinction made by Buch some 30 years ago between *C. "meylanii"* (= *C. integristipula*) and *C. neesiana*, no such distinction has been made in the more recent literature for North American material, except in the papers by Schuster (1949a, 1953) and Persson (1952). Statements about the range of *C. neesiana s str.* in other recent papers are unreliable.

W. GREENLAND. Sequineqarajutoq at Orpigsoq, 68°42' N., 50°55' W. (*L. Harmsen 18!*)[29] Aug. 12, 1932); Sondrestrom Fjord, near shore of L. Ferguson, near head of fjord, in *Salix*-moss tundra, 66°4' N. (*RMS & KD 66–078*). NORTHWEST TERRITORY. E. coast of Hudson Bay, near Haig Inlet, Flaherty I. (*Marr M669*). NEWFOUNDLAND. Hogan's Pond, Waterford Bridge, near St. Johns, and South Side Hill, near St. Johns, and Biscay Bay, all Avalon Distr.; Pushthrough, Hare Bay, and Grand Bruit, S. coast; Stephenville Crossing, W. Nfld.; Hampden and Kitty's Brook, central and NE. Nfld. (Buch & Tuomikoski, 1955). QUEBEC. St. Arsène (*Lepage 2701*); Shigawake, Gaspé Co. (*Kucyniak & Raymond 56060*). ONTARIO. Manitoulin I., bog on Rte. 68, S. of Sanfield (*Hermann 15565*); Norway L., Algonquin Park (*Cain 4264*); Kokoko Bay, L. Timagami (*Cain 2434*); bog. SW. corner of Grass L., 1 mi N. of Silver Islet, Sibley Pen., Thunder Bay Distr. (*Garton 4410; trans. ad C. integristipula*).

MAINE. Mt. Katahdin, high on Garfield Ridge, above treeline (*RMS*); also reported from Mt. Desert I. (Lorenz, 1924). VERMONT. Mt. Mansfield, Long Trail, above treeline, 4000 ft (*RMS 43809*); Mt. Mansfield, Haselton Trail, ca. 2900–3200 ft (*RMS 43837*). MASSACHUSETTS. Upper end of Amethyst Brook, just W. of town of Pelham, Hampshire Co. (*RMS. 43990b,c*); Hawley Cranberry Bog, W. Hawley, Franklin Co., 1800 ft (*RMS 43961, 43962*); Guilder Pond, Mt. Everett, Berkshire Co. (*RMS 68–164*); *Chamaecyparis* swamp, S. Wellfleet, Cape Cod (*RMS 68–315*). NEW YORK. Forest of Dean L., Ft. Montgomery (*RMS 24238, 24241*); Plymouth Bog, Chenango Co. (Schuster, 1949a; as *C. muelleriana*); also cited by Burnham from L. George, and by others elsewhere; these reports need verification. NEW HAMPSHIRE. SW. of Mt. Pleasant, White Mts. (*RMS 68–2132*).

VIRGINIA. Salt Pond Mt., at Cascades, Giles Co. (*R. P. Carroll 275*; on rotten wood); above and near The Cascades, Little Stony Cr., near Mountain L., Giles Co. (*RMS 61300a, 40212, 40243a; Patterson 2727*); White Top Mt., Smyth Co., 5400–5500 ft (*RMS 38066, 38075, 38062, 38071*); Mt. Rogers, Grayson Co., 5200 ft (*RMS 38020*). NORTH CAROLINA. Laurel Park, Hendersonville, Henderson Co. (*Blomquist 2481*); Glade Valley, Alleghany Co. (*P. O. Schallert 2179*); Mt. Sterling, Haywood Co. (*Blomquist 21932*); Whitewater R. Falls, Transylvania Co. (*Blomquist 11166*); near Glenville, Jackson Co. (*Anderson 639*); artificial lake SW. of Linville, Avery Co. (*Blomquist 10764*); base of Shortoff Mt., 3 mi W. of Highlands, hemlock-hardwood cove, 3500 ft (*Anderson 10668*); Soco Falls, below Soco Gap, W. of Maggie (*RMS 24750; mixed with C. muelleriana; RMS 24761*); Roan Mt., near SW. summit, at ca. 6000 ft, Mitchell Co. (*RMS 36959*, typical); Bubbling Spring, 0.5 mi N. of Beech Gap, N. of Blue Ridge Parkway, Haywood Co., 5100–5200 ft (*RMS 39370*); Chattooga R., 1 mi below The Narrows, Macon Co. (*RMS 39417a, p.p.*); Wayah Bald, 0.5 mi N. of summit,

[29] The Harmsen specimen is somewhat of a puzzle. It is a small, xeromorphic phenotype, to 2.2 mm wide, with free gemmae formation. Leaves are rather typical: many are truncate or truncate-retuse at the apex, but the elongate border is developed only locally and infrequently; many plants almost lack it and approach the much larger *C. integristipula*. Underleaves, as more nearly typical of *C. integristipula*, are not notched, even though the length from rhizoid-initial field to apex is usually 9–10 cells; the underleaves, however, are usually little or no wider than long, typical of *C. neesiana*. The majority of characters (size, dull texture, free gemmae formation, notched and relatively narrow, never reniform underleaves, and locally perceptibly bordered leaves) speak for *C. neesiana*.

8 mi W. of Franklin, Macon Co., 4900 ft (*RMS 39251*; mod. *latifolia* = "var. *repanda*");
Crow Cr., between Falls and Cullasaja R., between Franklin and Highlands, Macon Co.
(*RMS 40868*; with *C. muelleriana blomquistii*); near summit of Roan Mt., near Roan High
Bluff, 6000 ft, Mitchell Co. (*RMS 40318a*; with *C. suecica*); above Glenn Falls, near
Highlands, Macon Co. (*RMS 40606*).   WEST VIRGINA. Cranberry Glades, W. of Mill
Pt., Pocahontas Co., 3500–3600 ft [*RMS 61204, 61215, 61226*, all typical; *61200,
61206, p.p. ± trans. ad C. integristipula*, (which see)].   TENNESSEE. Mt. LeConte, on
Alum Cave Trail, Sevier Co. (*RMS 24176*); Roan Mt., SW. slope, at ca. 6100 ft, near
North Carolina border in Carter Co. (*RMS 36998; 37971c*).   GEORGIA. High Falls
of Big Cr., S. of Highlands, N.C., in N. edge of Rabun Co., ca. 1900–2000 ft (*RMS
40754*); N. slope of Rabun Bald, 4000 ft, Rabun Co. (*RMS 40690b*).

Ecology

Buch (1936a) stressed that *C. neesiana* occurs—"at least in Fenno-
Scandia"—mostly on partly destroyed *Sphagnum* patches or *Sphagnum* peat
in small groups or often individually amidst *Cephalozia media, C. bicuspidata,*
and *C. loitlesbergeri,* rarely on nearly undestroyed *Sphagnum* polsters, on
the same sites as *C. sphagnicola* (and often consociated with the latter) but
also on peaty wood in *Sphagnum* bogs, the *C. neesiana* usually occurring on
more acid substrates than *C. meylanii* (= *C. integristipula*)—a fact clear
from the abundant occurrence of the latter on clayey, residual soils in
subcalcareous regions, and over neutral or even subcalcareous shale or
sandstone. However, *C. integristipula* may also occur on peat in marl bogs,
in sites where no *C. neesiana* is present. *Calypogeia neesiana* often occurs in
areas of acidic igneous rocks, but usually no *C. integristipula* is found there.

Northward *C. neesiana* occurs only rarely under conditions similar to
those described as typical in Fenno-Scandia (Buch, 1936a), as on Mt.
Katahdin, Me., in a montane *Sphagnum-Vaccinium* heath, with *Empetrum
hermaphroditicum* and such associated hepatics as *Calypogeia sphagnicola,
Cephalozia loitlesbergeri, Mylia anomala.* The species often occurs under
Tsuga on acid, loamy-sandy, sometimes rather dry banks, together with
Calypogeia fissa s. lat., Nardia insecta, Cephalozia bicuspidata, Scapania nemorosa,
and the calciphobous moss, *Leucobryum glaucum.* Somewhat analogous
modes of occurrence are on sandy-loamy soil in moist, acid ditches, mixed
with some of the preceding species (*Cephalozia bicuspidata, Nardia insecta*)
as well as *Cladopodiella francisci, Calypogeia sullivantii,* and others.

The restriction to acid sites is also constant in the Southern Appalachians. Associated on moist black humus may be *Odontoschisma prostratum,
Microlepidozia sylvatica,* and *Cephalozia media,* particularly when the *Calypogeia* occurs on shaded, humus-covered banks (as under *Kalmia* or
Rhododendron). However, the species may also occur on damp, shaded,
much decayed logs, with *Tetraphis pellucida.* Thus, *Calypogeia neesiana* is
always a species of acid, organic substrates; whereas, although the very

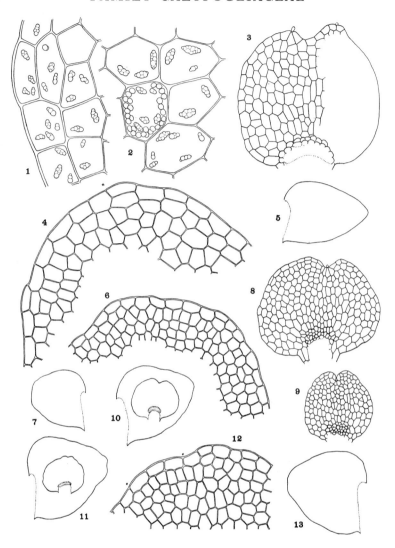

FIG. 128. *Calypogeia neesiana.* (1) Cells of leaf apex with oil-bodies drawn in where present (×285). (2) Median cells, same leaf, with well-developed oil-bodies (×310). (3) Underleaf, note slime papillae terminal on lobes (×63). (4) Leaf apex, with optimal border development (×135). (5) Leaf, mod. *angustifolia* (×14). (6, 12) Cells of leaf apices (×90). (7, 13) Leaves (×16). (8–9) Underleaves, of vigorous and weak small shoots, respectively (×40). (10–11) Leaves and underleaves, of robust shoot (×16). [1–5, *RMS 24176*, Mt. LeConte, Tenn., an atypical mod. *angustifolia* with cells unusually large (36–42 × 42–55 μ in leaf middle) and with oil-bodies in all leaf cells; 6–13, *Blomquist 2481*, Laurel Park, Hendersonville, N.C., a mod. *latifolia*.]

similar *C. integristipula* may also be found on such substrates, it is more common on clayey or loamy soils, or as a pioneer on damp rocks—i.e., as a pioneer on inorganic substrates. These differences in ecology, though not resulting in a mutually exclusive pattern of occurrence, appear significant in weighing the taxonomic merit of *C. integristipula*.

Calypogeia neesiana occurs from hygric to semixeric sites. The most extreme habitat from which I have seen the species is from bark at the base of a tree, in a hemlock-hardwoods cove, associated with the moss *Brotherella tenuirostris* and the xeric *Frullania asagrayana*. The *Calypogeia* occurred here as a broad-leaved, dense-leaved, xeromorphic extreme.

The ecological "preferences" of *C. neesiana* evidently broadly overlap those of *C. muelleriana*, as is evident from collections studied (from New York, North Carolina, and England) where the two taxa occurred intermingled. The implications of the fact that *C. neesiana* and *C. integristipula* never appear to occur intermingled need no emphasis.

Variation

Normally developed, large-leaved forms of *C. neesiana* *s. str.* show variation chiefly in the direction of *C. integristipula*, as is clear from the discussion under the latter (p. 210). In addition, phases disturbingly similar or parallel in form to *C. muelleriana* are often produced, particularly in exposed sites.

Buch's (1936a) culture experiments suggested that the pattern of variation of *C. neesiana* differs in some respects from that of *C. integristipula*, as follows. (1) The mod. *laxifolia* is produced in *C. neesiana* with stronger insolation and drier atmospheric conditions than the similar modification in *C. integristipula* (and in *C. muelleriana*); this modification develops long internodes under conditions in which the other species still show "normal" internodes. (2) Lateral leaves attain their maximal length (ca. 1.5 mm) in stronger light than in the other taxa; conversely, with lowered insolation, *C. neesiana* develops the mod. *parvifolia* under a light intensity in which *C. muelleriana* still develops normal-sized leaves.

Mature plants of *C. neesiana* rarely have leaves as narrow as those commonly produced by *C. integristipula*, but the frequent xeric phases of *C. neesiana* (mod. *parvifolia-densifolia-latifolia*) produce leaves which are as wide as or wider than long. Such broad-leaved modifications often are difficult to differentiate from small phases of *C. muelleriana* if the very different leaf textures are overlooked.

Differentiation

Calypogeia neesiana represents a specialized species in the development of a generally discrete border of tangentially elongated marginal leaf cells

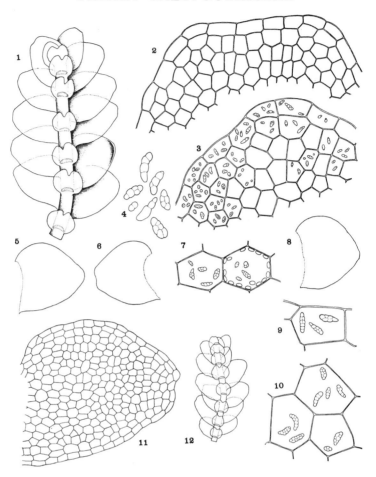

Fig. 129. *Calypogeia neesiana.* (1) Shoot, ventral aspect (×14). (2) Cells, leaf apex, oil-bodies omitted (×155). (3) Cells, leaf apex, with oil-bodies drawn in those cells where actually present (×180). (4) Oil-bodies (×640). (5–6) Leaves (×13). (7) Cells of leaf middle with oil-bodies present (×260). (8) Leaf, mod. *latifolia* (×16). (9) Cell of leaf margin from leaf in fig. 8 (×375). (10) Cells of leaf middle, from leaf in fig. 8, with oil-bodies in some cells but mostly without them (×350). (11) Leaf of mod. *parvifolia* (×59). (12) Plant, ventral aspect (×10). [1–10, *RMS 24238*, Forest of Dean Lake, N.Y.; 11–12, from a plant from Finland, leg. Buch.]

(a border already "hinted at" in many forms of the nearly related *C. integristipula*) and in the frequent suppression of oil-bodies in median leaf cells. The species shares other "specialized" characters with *C. integristipula*: the orbicular, usually essentially undivided or only notched underleaves and the dull texture of the gray-green leaves (only the marginal cells are nitid).

Typically, *C. neesiana* differs from the nearly related *C. integristipula* in the following characters: (*a*) smaller size; (*b*) more strongly bordered apical region of the leaf; (*c*) more coarsely and fewer segmented oil-bodies, which show a tendency to be suppressed in some or most median leaf cells and underleaf cells; (*d*) slight to distinct constriction at the leaf base, resulting in an obliquely, broadly elliptical-ovate to heart-shaped leaf form, whose greatest width occurs some distance above the base; (*e*) narrower, more sharply transverse rhizoid-initial field; (*f*) less broad underleaves that show a tendency to develop a distinct, often sharp emargination. Not one of these characters by itself affords a constant separation from *C. integristipula.* Therefore, discrimination depends on a constellation of characters, rather than on any single differential trait.

The distinction between *C. neesiana s. str.* and *C. integristipula* is admittedly often so slight that the latter segregate has not been recognized by some workers (Müller, 1947a, 1951–58) or has been reduced to the status of a variety of the former (Schuster, 1949a), as *C. neesiana* var. *meylanii*. However, in addition to the admittedly limited differences between the two, study of a long series has revealed not only a different pattern of modification but also a different pattern of distribution. If *C. integristipula* is a mere modification of *C. neesiana* (as Müller, 1947a, insists), one would expect it to recur throughout the range of the latter. However, study of thousands of individual plants of *C. integristipula* from Minnesota revealed no specimens clearly referable to *C. neesiana*; study of hundreds of plants, from several sites, of *C. neesiana* from the Southern Appalachians revealed no plant that could conceivably be considered *C. integristipula.* Under such conditions, two discrete entities must be at hand, even though final judgment as to the status of the taxa is withheld. For the time being, the above observations plus the culture work of Buch remain the deciding factors in maintaining the two entities as specifically distinct.

Forms of *C. neesiana* with emarginate underleaves may also develop broad leaves ("var. *repanda*"), averaging wider than long; these forms may be confused with *C. muelleriana,* from which they are superficially indistinguishable. The presence, on most mature leaves, of a distinct border of elongate cells, and the fewer-segmented oil-bodies, appears to afford a "safe" separation between the two species, as does texture. Such phenocopies appear to be formed only by the mod. *densifolia-latifolia-fissistipula* of *C. neesiana.*

Development of such broad-leaved forms superficially identical with *C. muelleriana* raises problems, at least of a practical nature, as regards separation of the two species.[30] Buch (1948) points out that large-leaved plants of *C. neesiana* often fail to develop the sharply discrete, elongate-celled border of marginal leaf cells (such plants, on culture, however, readily develop marginate

---

[30] Müller (1913, *in* 1905–16) describes such a broad-leaved form of *C. neesiana* as the var. *repanda.* He also emphasizes (1947a, p. 413) that the narrow leaf form of *C. neesiana* becomes useless as a diagnostic character with the var. *repanda,* in which the leaves often are not longer than wide.

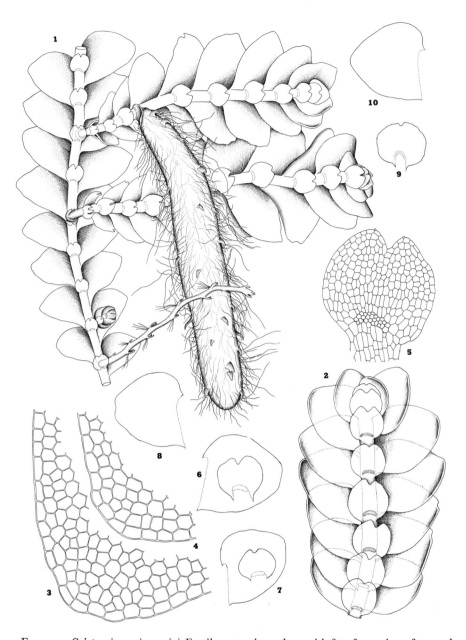

FIG. 130. *Calypogeia neesiana*. (1) Fertile autoecious plant with free formation of ventral-intercalary stolons, with relatively small underleaves (mod. *parvistipula*) (×14.5). (2) Shoot of mod. *densifolia-densistipula* (×21). (3–4) Leaf apices (×112). (5) Underleaf; note small rhizoid-initial area and slime papillae *at apices of lobes;* compare with Fig. 107:4, of *C. integristipula*, in which slime papillae are inserted in indentations (×54). (6–10) Leaves and underleaves (×21). [1, *RMS 61300a*, Mountain Lake, Va.; 2–7, *RMS 24750*, Soco Falls, N.C., from plants growing intermingled with *C. muelleriana* fo. shown in Fig. 123:1–5; 8–10, *Blomquist 2481*, also shown in Fig. 127.]

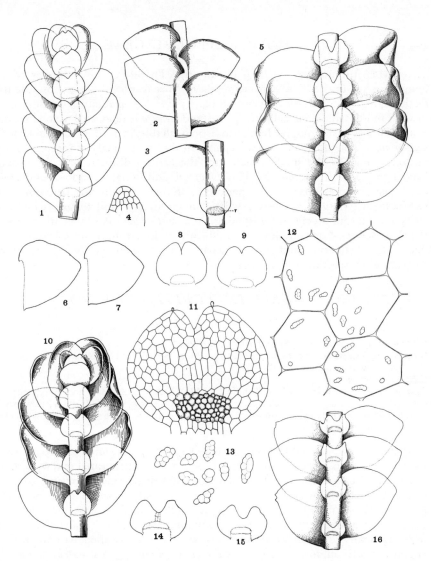

Fig. 131. *Calypogeia integristipula-neesiana-muelleriana* "intergrades." (1–4, 6–11, *C. neesiana* forms; 5, 12–16, ? mod. *latifolia* of *C. muelleriana*). (1) Shoot, postical aspect, (×14). (2) Shoot sector, antical aspect (×ca. 15). (3) leaf and underleaf, *in situ* (×ca. 15). (4) Underleaf lobe (×ca. 30). (5) Shoot sector, ventral aspect, of superficial transition between *C. neesiana* and *C. muelleriana* (×20). (6–7) Lateral leaves (×14.5). (8–9) Underleaves, of a typical *C. neesiana* (×25). (10) Shoot, ventral aspect, typical *C. neesiana* (×17). (11) Underleaf; note position of slime papillae at lobe tips (×65). (12) Median cells (×515). (13) Oil-bodies (×750). (14–15) Underleaves (×30). (16) Shoot sector, postical aspect (×15). [1, from Feldsee, Baden, Germany, leg. K. Müller, as *C. muelleriana* from the "locus typicus," but in my opinion, *C. neesiana* s. str.; 2–4, Headwaters Swamp, Tioga Co., N.Y., *RMS*; 5, Allegany State Park, N.Y., *RMS*; 6–11, Plymouth Bog, Chenango Co., N.Y., erroneously illustrated in Schuster, 1953 as *C. muelleriana* and reported as this in Schuster, 1949a; 12–13, Dreamers Rock, Manitoulin I., Ont., *Cain*; 14–16, *RMS 25627*, Taughannock Gorge, Tompkins Co., N.Y.]

leaves when the mod. *parvifolia* is produced). Therefore, the character of the marginate vs. immarginate leaves tends to "break down" on robust plants; furthermore, as noted above, nearctic plants of *C. neesiana* may show oil-bodies in all or most median leaf cells, while the mod. *densifolia* tends to develop broadly ovate, almost heart-shaped leaves indistinguishable in form from those of *C. muelleriana*. Therefore, one must depend on finding strongly marginate leaves on some or all of the smaller plants for the certain distinction of the broad-leaved forms of *C. neesiana*.[31] However, an additional criterion, uniformly overlooked by previous workers, permits *in all cases* a distinction between these two taxa: *C. neesiana* has leaves with a diagnostically dull texture when dry (only the marginal cells are glistening); *C. muelleriana*, like other members of the *C. trichomanis-fissa* complex, has uniformly, if not strongly, nitid leaves when dry. This distinction and others clearly emerge when the two species grow intermingled (see Fig. 126:11–12).

CALYPOGEIA INTEGRISTIPULA Steph.

[Figs. 107:1–4; 132–134]

Calypogeia integristipula Steph., Spec. Hep. 3:394, 1908.
Calypogeia meylanii Buch, Ann. Bryol. 7:161, 1934; Mem. Soc. F. et Fl. Fennica 11(1934–35):201, 1936 (new synonymy).
Calypogeia neesiana Auct., *p.p.*
Calypogeia muelleriana Jovet-Ast [nec (Schiffn.) K. Müll.], Bull. Soc. Bot. France 91:37, 1944.
Calypogeia neesiana var. *meylanii* Schust., Amer. Midl. Nat. 42:535, 1949; Buchloh, Rev. Bryol. et Lichén. 21:267, 1952; Müller, Rabenh. Krypt.-Fl. ed. 3, 6:1183, 1956; Bischler, Candollea 16:63, 1957.

In flat, dull (marginal leaf cells excepted), *pale green to grayish green* patches or mats. Plants subsimple, rather large, to *2.5–3 mm wide*; branches few, *postical-intercalary*. Leaves distant or feebly contiguo-imbricate (mod. *laxifolia*) to, very often, distinctly imbricate (mod. *densifolia*), normally *narrowly ovate and clearly longer than wide*, usually (0.7)0.75–0.85 as wide as long (rarely, mod. *latifolia*, 0.95–1.15× as wide as long), 1.3–1.75 mm long, *widest at base and gradually tapering to* (typically) *a narrowly rounded, entire apex*. Cells thin-walled, without distinct trigones, nearly uniform, along apex the *marginal cells ± isodiametric, occasionally interrupted isolated or short rows of more or less elongate cells* that may form a vestigial border (in other leaves, apical cells approaching isodiametric, and 29–36 μ); median leaf cells somewhat rectangularly elongate, usually ca. (28)34–38(45) μ wide × (30)38–45(64) μ long, near base mostly 34–45 μ wide × 45–64 μ long; *oil-bodies in all leaf cells*, usually formed of 8–14 globules of ca. 1.8–3.6 μ diam., the oil-body varying from 4–5 × 7.5 μ to 5.5 × 11 μ (occasional ones to 5.5 × 12.5–16, rarely 18, μ; in some or many cells the oil-bodies only 1–3-[occasionally 4–7-] segmented, and then only ca. 3 × 5 to 4 × 6.5–7.5 μ);

---

[31] The sporophytic differences perhaps may prove to be more constant. *Calypogeia neesiana* and *C. integristipula* show weak nodular thickenings of alternating longitudinal walls of epidermal cells of the capsule; *C. muelleriana* lacks these. However, Müller (1947a) notes that he once found such thickenings in a patch of *C. muelleriana* plants. Perhaps these actually represented *C. integristipula*, since, as Buch indicates, Müller appears to have confused lax-leaved forms of *C. integristipula* with *C. muelleriana*.

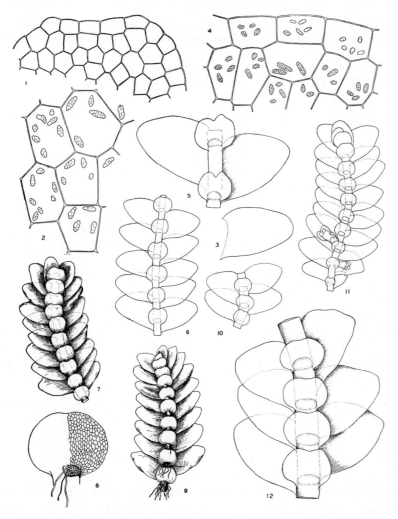

FIG. 132. *Calypogeia integristipula.* (1) Cells of unusually truncate leaf apex (×128).
(2) Median cells with oil-bodies (×313). (3) Normal leaf (×12.6). (4) Cells of
leaf apex, with oil-bodies (×340). (5) Sector of small, impoverished shoot
(×18.6). (6) Sector of more robust plant (×8.6). (7) Mature shoot, mod. *densifolia-
densistipula* (×9). (8) Underleaf from fig. 7 (×ca. 30). (9) Mature shoot, ventral
view (×9). (10) Shoot sector (×7.6). (11) Shoot sector, ventral view, with two
ventral-intercalary branches (×7.6). (12) Shoot sector of mod. *latifolia* (×19). [1–3,
RMS 24280, Mt. Wittenberg, Catskill Mts., N.Y.; 4, *RMS 24300*, same loc.; 5–6,
RMS 1676, Ringwood Swamp, Tompkins Co., N.Y.; 7–9, Headwaters Swamp,
Tioga Co., N.Y., *RMS;* 10–12, *RMS 1680a* and *1002a*, Bergen Swamp, Genesee Co.,
N.Y., from Schuster, 1953.]

oil-bodies mostly 4–8 per cell in the leaf middle. Underleaves *broadly orbicular to orbicular-reniform*, usually somewhat to *strongly transverse, 1.3–1.6× as wide as long*; 2–3× as wide as stem, ca. 510–600 μ long × 730–900 μ wide to a maximum of 620 μ long × 1000 μ wide; underleaf apex *entire or retuse* (the vaguely indicated lobes then *broadly rounded*), at base rounded, *not or imperceptibly decurrent on each side*; underleaves with oil-bodies *in all or most cells*, usually to 7.5 μ long (often occurring in cells with no chloroplasts), usually 1–5 per cell, averaging smaller than in the cells of adjacent leaves; rhizoid-initial region at underleaf base *prominent, densely rhizoidous, usually broadly transversely oval*, less than twice as wide as high; median underleaf length (from rhizoid-initial area to apex) in series of *from (6)7 to 10(12) cells*. *Asexual reproduction* by gemmae on erect, reduced-leaved shoots *frequent*.

Autoecious (? sometimes paroecious). Capsule valves (220)250–330 μ wide × (1100)1400–1850 μ long (4.5–7.5× as long as wide),[32] spirally twisted. Epidermal cells basically in 8 rows (ca. 25–40 μ wide), but occasionally several of these undergoing supernumerary division, the valves then 10–12 cell rows wide (cells then ca. 20–25 μ wide); epidermal cells to 32–70 μ long; alternating (secondary) longitudinal walls $\pm$ thick and reddish brown, without supernumerary divisions, the thickened walls weakly to vigorously produced as scattered nodular thickenings $\pm$ extended onto the outer tangential walls.[33] Cells of inner layer in 16 "primary" rows, but usually freely subdivided longitudinally (thus *with usually 22 or more, up to 28–32, rows*) averaging ca. 8–(10)–12 μ wide × (75)100–120 μ long; outer tangential faces with numerous transverse bands, extending across both adjacent radial walls. Cells at base of capsule irregular in shape, all or many bearing numerous discrete tangential bands (as on free wall of inner cell layer of valves).[34] Spores pale reddish brown, 10–12 μ, faintly verruculose; elaters ca. 7–8 μ × 130–150 μ, bispiral, the reddish spirals 2–3 μ broad.

Type. Lectotype of *C. integristipula* from Saxony, Unterwalden Grund (*Stephani, s.n., July 1888; G*). The type material of *C. meylanii* from Finland (*leg.* H. Buch); distributed *in* Verdoorn, Hep. Select. et Crit. no. 113. The lectotype designated by Bischler (1957) is a vigorous plant, identical in every respect with *C. meylanii*; there can be no doubt about their identity. I compared the types (Geneva) in 1963.

Distribution

A widespread and evidently holarctic species, ranging rarely into the southern portion of the Tundra, southward through much of the Spruce-Fir Biome. Unlike *C. neesiana*, which seems to have slightly "oceanic

---

[32] Frequently the valves are unusually short in this species, rarely as little as 5× as long as wide; in *Lepage 1680* some valves are 290 μ wide × only 1200 μ long (barely over 4× as long as wide). The longest valves were 240 μ wide × 1850 μ long—thus almost 8× as long as wide. This range is below that for *C. neesiana*: 6 to 9× as long as wide.

[33] Nodular thickenings are rarely faint and have either no or only a very delicate extension onto the external radial walls.

[34] These thickenings occur in all capsules of all capsule-bearing collections examined (ca. 20 capsules). In the related *C. neesiana* the few capsules seen show no such thickenings, agreeing in this respect with the other species of the genus studied.

leanings," *C. integristipula* occurs indifferently in both oceanic and sub-oceanic regions, and in the most continental sections of North America.

The type was from Germany, and the species is evidently widespread there; Stephani also cites Japan as an original locality. [Since the lectotype of *C. integristipula* was designated by Bischler (1957) from German material, *C. neesiana* var. *japonica* Hatt. does not automatically fall into synonymy, but unfortunately *C. meylanii* does. The present taxon has been reported almost exclusively under the name *C. meylanii* or *C. neesiana* var. *meylanii*.] The species appears equally widespread in northern Europe, ranging south to northern Italy. The European range is not elucidated, since Müller (1951–58) lumped *C. integristipula* (as *C. neesiana* var. *meylanii*) with *C. neesiana*; it is also obvious from the drawing of "*C. neesiana*" by Bischler (1957), which represents typical *C. integristipula*, that she confused the two taxa.

Widespread in the northern portion of the Deciduous Forest Region (Northern Hardwoods Forest or Lake Forest) and through the Coniferous Forest (*Abies-Picea*) Region. Southward in eastern North America from Ontario to Minnesota, New York, and Pennsylvania. The distribution of *C. integristipula* cannot be adequately defined at present, since until recently (Buch, 1936a) it was confused with *C. neesiana*—in this country as late as 1948 (Cain and Fulford). Selected stations follow.

S. GREENLAND. Qasigialik Fjord, 61°31′ N., 49°04′ W. [*Damsholt, July 27–28, 1963* (two collections, one mixed with *Blepharostoma trichophyllum* subsp. *trichophyllum* and *Lophozia hatcheri*)]; Ivnarssuangûp nûna, 60°49′ N., 47°22′ W. (*Damsholt, July 29, 1963!*). NEWFOUNDLAND. Channel (*Howe & Lang 875, p.p.*, among *Diplophyllum albicans*! NY; typical); widespread in Avalon Distr. on S. coast in W. Nfld., on N. Pen., and in central and NE. Nfld. (Buch & Tuomikoski; 25 collections). QUEBEC. Near St. Jovite, Terrebonne Co. (Crum & Williams, 1960); Causapscal, Gaspé (Lepage, 1944–45). ONTARIO. Smith L., Algonquin Park (Cain & Fulford, 1948); Durham, Halton and York (Williams & Cain, 1959); Cold Creek Swamp, Nobleton, York Co. (*Cain 4261*); Algonquin Park (*Cain 4214*).

MAINE. Saddle Trail, ca. 3500 ft, Mt. Katahdin (*RMS 33018*; xeromorphic phenotype, superficially approaching *C. neesiana* but with the very broad, subreniform underleaves of *C. integristipula* and dense rhizoid-initial areas of that species). MASSACHUSETTS. Hawley Bog, W. Hawley (*RMS;* many collections). NEW YORK. Centreville (Underwood & Cook, Hep. Amer. no. 15, as *Kantia trichomanis*, "on ground in cedar swamps"). MICHIGAN. Along Rapid R., 6–7 mi NNE. of Kalkaska, in swampy cedar woods, Kalkaska Co. (*RMS 38942, 38963; c. caps.*); shore of L. Lilly, near Ft. Wilkins, NE. of Copper Harbor, Keweenaw Co. (*RMS 39162a,b*). MINNESOTA. Cedar Cr. Bog, Anoka-Isanti Cos.; Itasca Park, Clearwater Co.; Susie Isls., Cook Co.; Big Bay at Hoveland, Cook Co.; Grand Portage and Cascade R., Cook Co.; W. of Togo, Itasca Co.; SE. of Ericsburg, Koochiching Co.; Angle Inlet, Lake of the Woods Co. (Schuster, 1953). Also MASSACHUSETTS. Green R. Gorge, Colrain (*RMS 68–102b*); Mt. Everett, Berkshire Co. (*RMS 68–134*).

Also reported from KANSAS (Leavenworth Co.: McGregor, 1955; the only specimen from Kansas seen, submitted by McGregor, is *Leucolejeunea clypeata*!) and from the Delta region of the Mackenzie R., at Eskimo Lake (Steere, 1958a). Extending west to the Yukon and Alaska (Persson, 1952).

Many or most reports of "*C. neesiana*" from Wisconsin (Oneida, Douglas, Superior, Barron, Grant, and Bayfield Cos.; see Conklin, 1929) must represent "*C. meylanii*";

similarly, all or most reports from Michigan (by Steere, Darlington, etc.) must represent "*C. meylanii.*" To date typical *C. neesiana* has not been seen from the continental portions of North America nor have I seen *C. integristipula* in the Southern Appalachians.[35]

Ecology

A species of widespread occurrence, in part because of the wide latitude of environmental conditions tolerated. On purely mineral soils (clay and loam) as well as purely organic substrates (peat and decaying logs); often a pioneer on shaded rock walls in moist places. Generally definitely mesophytic, often hygrophytic; rarely found under strongly intermittently moist conditions. Frequently a species of strongly shaded sites, e.g., lining peaty holes in boggy areas or in shaded ditches.

I have treated the ecology of this taxon in detail (Schuster, 1953, p. 546) and find (with Buch, 1936a) that it is less acidiphilic than *C. neesiana.* It is a common species of

> . . . coniferous bogs, chiefly of *Thuya*, or Cedar Swamps, in the densely shaded, wet ground layer of which it may be a very abundant component. . . . The species . . . is decidedly lime-tolerating . . . . In marly *Thuya* swamps it occurs often with *Riccardia multifida* and *latifrons*, as well as *Geocalyx, Moerckia flotowiana* or *hibernica*, and occasionally *Lepidozia reptans*. Though locally a species of slightly calcareous regions, it seems to rarely invade bare calcareous or sub-calcareous rock surfaces . . . but is most often a species invading after the marl-containing substrate is carpeted over by humus. This peculiar combination of underlying calcareous conditions coupled with rather high content of humic acids seems especially favorable for this species. In very marly areas it may be restricted to wet decayed stumps and humus, together with *Geocalyx, Blepharostoma, Cephalozia media*, etc. In more acid bogs, it often persists, occasionally associated with *Lophozia ventricosa, Calypogeia sphagnicola, Mylia anomala* . . . . (Schuster, *loc. cit.*).

However, the species is by no means confined to regions with calcareous outcrops. It may be frequent, e.g., on damp, shaded rock walls and ledges in the mountains (at above 3500 feet elevation), where sometimes a pioneer over bare rocks, but more often a secondary species, sometimes with such index species of noncalcareous sites as *Mylia taylori* and *Bazzania tricrenata* (as in the Catskill Mts., N.Y.; *RMS 24280*).

Calypogeia integristipula may be very frequent in peat bogs; however, in my experience, it totally fails to occur in such bogs when these are not underlain by calcareous drift or rocks. This difference in ecological behavior suggests that *C. integristipula* is at least ecotypically distinct from *C. neesiana.*

---

[35] However, I have seen several specimens from Cranberry Glades, Pocahontas Co., W. Va. (*RMS 61200, 61200a, 61206, p.p.*), which, although probably closer to *C. neesiana*, show a high incidence of *C. meylanii* (*integristipula*) characters.

Variation

Calypogeia integristipula exhibits variation toward both *C. muelleriana* and *C. neesiana s. str.*; this has resulted in protracted controversy among European workers (Buch, Müller, Buchloh, Jovet-Ast, etc.) as to the status of *C. integristipula*.

This variation embraces the following elements: (*a*) degree of development of a leaf border of elongate marginal cells; (*b*) leaf shape; (*c*) oil-body size and form; (*d*) underleaf form; (*e*) cell size. Of these variations, all but the first may cause difficulty in the separation from *C. muelleriana*.

The commonest form of *C. integristipula* is a dense-leaved phenotype with short internodes (mod. *densifolia-densistipula*), imbricate leaves, and approximate to imbricate underleaves. The leaves show their typical, elongate form, and underleaves are characteristically transversely reniform-suborbicular (without more than a trace of emargination; more commonly quite entire). This type is illustrated in Fig. 132:7, 9. From this "typical" phase, variation is in two directions: (*a*) formation of lax-leaved phenotypes with barely imbricate, narrow leaves and distant underleaves; (*b*) formation of xeromorphic, dense-leaved extremes, with closely imbricate leaves and underleaves, the leaves becoming as wide as or wider than long.

Lax-leaved phenotypes *always* possess narrow leaves (Fig. 132:5, 6) and remote underleaves. The latter tend to be somewhat more distinctly emarginate (Fig. 132:5); but only under extreme conditions, with formation of small, juvenile shoots (as illustrated in Buch, 1936a, fig. II:1–2) do the underleaves become sufficiently incised to suggest other species (such as *C. muelleriana*). Lax-leaved phenotypes, because of their narrow leaves and frequently emarginate underleaves, have given rise to confusion with *C. muelleriana*. The lax-leaved, narrow-leaved phase of the latter (known for some time as *C. neesiana* var. *laxa* Meyl.) was considered by Buch (1936a, 1948) to represent *C. integristipula;* both Müller (1947a) and Buchloh (1952) claim that var. *laxa* Meyl. belongs to *C. muelleriana*.

Admittedly, *C. integristipula* can produce phases that are distressingly similar to the lax-leaved mod. *angustifolia-laxifolia* of *C. muelleriana*. For instance, Fig. 132:5 represents such a lax form of *C. integristipula* (growing with the more typical plant in Fig. 132:6); such plants approach the type material of *C. neesiana* var. *laxa* Meyl., as figured in Müller (1913, *in* 1905–16, fig. 69b). Consequently, confusion between the two species may result *unless* the texture of the leaves, the form of underleaf bases, and the sizes of cells in leaves and underleaves are used to discriminate between them.

By contrast, xeromorphic phases of *C. integristipula* also cause difficulties. I have collected plants of *C. integristipula* which represent a mod. *densifolia-latifolia;* in these plants the leaves may be as broad as long (ratio of width to length, in microns, as follows: 1350:1360; 1425:1380; 1395:1360; 1275:1240; 1255:1200; 1425:1250, e.g., from 0.96 to 1.2:1). Such broad-leaved forms approach "normal" *C. muelleriana*, particularly since at times they produce larger than normal cells. Separation of such plants from *C. muelleriana* hinges on the fact that the broad, shallow underleaf sinus never becomes deep and sharp (typical of *muelleriana*); the lateral leaves of most smaller shoots are clearly more narrowly ovate; the underleaves are smaller celled. Only extreme xeromorphic phases (mod. *latifolia-densifolia-densistipula*) of *C. integristipula* ever approach *C. muelleriana* in the form of the lateral leaves.

It is thus evident that both dense- and lax-leaved phenocopies (e.g., mod. *densifolia-latifolia* and *laxifolia-angustifolia*) "approaching" *C. muelleriana* occur. This variation, superficially bridging the differences between *C. integristipula* and *C. muelleriana*, is due to overlapping under extreme ecological conditions, i.e., due to production of phenocopies, as the crucial and constant differences in cuticle will attest (p. 212).

I do not have at hand sufficient material of either species with sporophytes to be able to add to the distinctions which Müller attempted to draw between them on that basis (see diagnoses and keys). [It may not be out of place to point out that Müller (*loc. cit.*, p. 448) once found nodular thickenings in a capsule associated with *C. muelleriana* (*muelleriana* supposedly lacks nodular thickenings of epidermal cells; the *integristipula-neesiana* complex has them). Possibly, therefore, the two complexes are not sharply discrete even in this respect.]

Differentiation

Calypogeia integristipula remains a doubtful species, most closely related to *C. neesiana* as, e.g., in color and dull texture of the leaves. Typically it differs from the latter in its (*a*) larger size; (*b*) obliquely ovate leaves, widest basally, narrowly and evenly rounded distally; (*c*) oil-bodies, when developed to maximum degree, formed of 8–14 or rarely more oil-globules, and present in all leaf cells, up to 5.5×12–$15(18)$ μ in size; (*d*) leaf tips with marginal cells only sporadically tangentially elongated, rarely 2–4 elongated cells juxtaposed—but then interrupted by isodiametric cells—thus not forming a continuous border; (*e*) rhizoid-initial region usually large, less than $2.5 \times$ as wide as high; (*f*) underleaves rarely distinctly emarginate, usually much wider (1.2–1.5:1) than long, often orbicular-subreniform. None of these criteria is constant, except perhaps

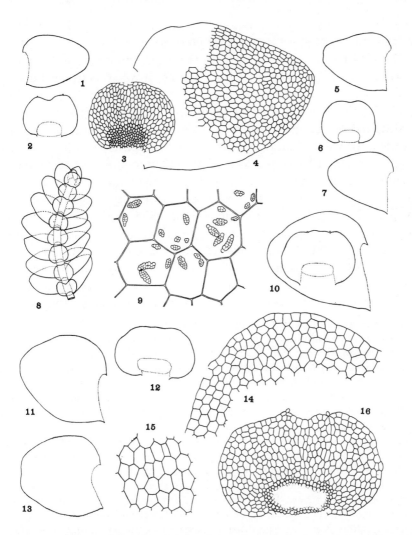

FIG. 133. *Calypogeia integristipula*. (1) Leaf (×12). (2) Underleaf (×16.5). (3–4) Underleaf, showing extensive rhizoid-initial area, and leaf (×39.5). (5–6) Leaf (×12) and adjoining underleaf (×16.5). (7) Leaf (×13.8). (8) Shoot, ventral aspect (×8). (9) Median leaf cells (×ca. 320). (10) Leaf and underleaf; note slime papillae situated in underleaf notches (×21). (11–13) Two leaves and an underleaf (×21). (14) Leaf apex (×73). (15) Median cells (×90). (16) Underleaf; note right lobe with hyaline papillae in the "normal" position — in a shallow notch of the vestigial lobe; left lobe atypical, with terminal slime papilla (×45). [1–2, 5–7, 9, from Cascade R., Minn., *RMS;* 3–4, *RMS 1674a*, Zealand Falls, White Mts., N.H.; 8, Headwaters Swamp, Tioga Co., N.Y., *RMS;* 10–16, *RMS H4011*, Bergen Swamp, Genesee Co., N.Y., an xeromorphic phase over peat; figs. 1–9 from Schuster, 1953.]

the maximum size of the individual plant, but by use of an ensemble of characters identification is generally possible.

On the other hand, confusion with the more remotely allied *C. muelleriana* is also possible (as evident from the controversy between Buch and Müller as to the disposition of material of these two species; see Buch, 1948). Typically, the following characters of *C. integristipula* will serve to separate it from *C. muelleriana:* (*a*) narrowly ovate lateral leaves, rarely as wide as or wider than long; (*b*) subentire to barely emarginate underleaves that are little or not decurrent, with smoothly curved flanks; (*c*) generally smaller median cells of both leaves and underleaves; (*d*) occasional presence of traces of a border formed by elongate cells in the leaf tips; (*e*) dull texture (marginal cells, which are nitid, omitted) and gray-green color. Again, the amplitude of variation of *C. integristipula* is so great that each individual characteristic (except perhaps texture, color, and the depth of sinus of the underleaves) tends to overlap the amplitude of variation of *C. muelleriana*. Therefore, in extreme cases, a constellation of characters must be used to effect a separation between the two taxa.

In the final analysis, the characters separating *C. integristipula* and *C. muelleriana* are more valid than those differentiating the former from *C. neesiana*. For that reason *C. integristipula* was not accepted at all by Müller (1947a), who considered it a synonym of *C. neesiana*. Schuster (1949a), Buchloh (1952), Bischler (1957), and Jones (1958a) accepted it as a variety of *C. neesiana*. However, on further study, I (Schuster, 1953) suggested that combining *C. "meylanii"* and *C. neesiana* was premature, since final data on the biological isolation of the two taxa were not available.

In the intervening time, intensive study of "typical" *C. neesiana* has convinced me that two taxa are at hand. The argument for separation of the two involves the following, partly nonmorphological considerations:

1. *Distribution.* In the East, *C. integristipula* does not occur south of southern Minnesota and northern Pennsylvania; it apparently has an east-west distribution ranging at least as far westward as southeastern Manitoba and eastward to eastern New York. By contrast, *C. neesiana* occurs sporadically in the Far West (California to Alaska), appears absent in the midcontinental portion—at least in the United States—and recurs in eastern North America; there it has an essentially Appalachian (north-south) distribution, ranging from Quebec to Tennessee, North Carolina, and the northern, montane edges of South Carolina and Georgia.

2. *Regional variability.* In the Southern Appalachians *C. neesiana* is often difficult to separate from *C. muelleriana*. However, few of the plants show any close approximation to *C. integristipula*. In fact, the characters of southern *C. neesiana* that make separation from *C. muelleriana* very difficult (almost uniformly broadly ovate leaves; almost uniformly emarginate underleaves) *are exactly those that do not normally occur in C. integristipula.*

Hence, *C. integristipula* cannot be a mere phenotypic expression of *C. neesiana*,

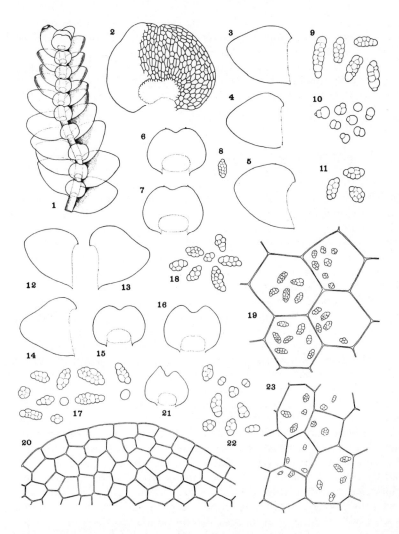

Fig. 134. *Calypogeia integristipula.* (1) Shoot, of phenotype with large cells shown in Fig. 133:3–4 (×9.5). (2) Underleaf (×39). (3–4) Normal leaves and (5) an unusually broad leaf (×12). (6–7) Underleaves, with hyaline papillae (×12). (8) Oil-body from underleaf middle of fig. 6, smaller than normal in leaf cells (×665). (9–11) Oil-bodies from three different median cells from leaf in fig. 4 (×610). (12–14) Leaves (×10.5). (15–16) Underleaves (×14). (17) Oil-bodies from leaves of figs. 12–14 (×645). (18) Oil-bodies from Fig. 19 (×610). (19) Median cells of phenotype with large oil-bodies (×275). (20) Cells of leaf apex (×103). (21) Unusually deeply emarginate underleaf (×14). (22) Oil-bodies of small oil-body phenotype (×565). (23) Median cells, small oil-body phenotype (×275). [1–2, *RMS 1674a*, Zealand Falls, N.H.; 3–11, *RMS 11695*, Grand Portage Bog, Minn., a large-celled phenotype, similar to *C. muelleriana* in cell size; 12–23, *RMS 11410*, Cascade R., Minn., phenotype with unusually large oil-bodies.]

since, if it were, *C. integristipula*-like plants would occur in the Southern Appalachians. Inversely, *C. integristipula* is abundant in central and western New York (Schuster, 1949a) and northern Minnesota (Schuster, 1953), *yet* eight years of collecting in these areas has failed to reveal clear-cut intergradation to *C. neesiana*. These conclusions appear diametrically opposed to those of Buchloh (1952), who states: "Eine Untersuchung reichlichen Standort- und Kulturmaterials ergab, dass Übergänge zwischen *C. neesiana* emend. Buch und *C. meylanii* Buch *eindeutig vorhanden sind* . . . ." However, Buchloh had studied the complex in an area where both *C. "meylanii"* (=*C. integristipula*) and *C. neesiana* occurred and where, conceivably, hybridization can occur—or once could have occurred. Also, with the great amplitude in variation of the two taxa, the impression that the two intergrade could scarcely be avoided.

Buchloh's conclusion that *C. integristipula* was not distinct from *C. neesiana* was based partly on his ability to show that *some* of the differential characters stressed by Buch failed to occur *consistently*. He stressed, e.g., that typical *C. neesiana* may have oil-bodies in all median leaf cells, exactly as is the case in *C. integristipula*. I (Schuster, 1953) independently demonstrated that nearctic *C. neesiana* exhibits wide variation in this regard, stating, "Some lateral leaves show no oil-bodies in most median cells; others show them present in all but isolated cells; others appear to show oil-bodies in every living cell examined." Admitting that this *one criterion* affords no separation between *C. integristipula* and *C. neesiana* does not necessarily prove that the two are conspecific.

3. *Amplitude and direction of variation.* Although difficult to express in concrete terms, differences in leaf shape and underleaf form, as well as size, occur, which suggest two species are at hand. Typical *C. neesiana* often produces broad-leaved modifications (var. *repanda* K. Müll.) with smaller cells, dense leaves and underleaves, and rather deeply bilobed underleaves, e.g., a mod. *parvifolia-latifolia-densifolia-densistipula-fissistipula*. This phenotype is usually only ca. 1.5–2 mm wide and typically shows very clearly the border of elongated cells. Identical modifications have not been seen in *C. integristipula*, although admittedly plants occur with leaves as wide as or wider than long and shallowly—but broadly—notched underleaves. Such phenotypes, however, lack the border of elongated marginal cells characteristic of *C. neesiana* s. str. and are larger (2.2–3.2 mm wide); they also show very strongly transverse underleaves (1.5× as wide as long) and larger rhizoid-initial areas.

4. *Morphological differences.* Although, as has been stressed, the morphological differences between *C. neesiana* and *C. integristipula* are slight, and individually "break down," collectively they appear to afford a constant means of separation. Six differences between the two species have been cited above. Buchloh (1952) indicated that the morphological differences stressed by Buch as separating *C. integristipula* from *C. neesiana* do not hold; he stated that the lateral leaves and underleaves of *C. integristipula* and *C. neesiana* "zeigen die selbe Formenamplitude." Although this may be the case in Europe, it is not so in North America, where large geographical regions exist in which one or the other taxon occurs—without obvious intergradients.

I must confess to a frequent difficulty in morphologically separating *C. neesiana* from *C. muelleriana* (when the marginal border of elongate cells is overlooked) *if the crucial differences in texture of the plants are neglected*; I experience this difficulty principally in separating material from areas (e.g., New England) where their distribution overlaps and where (present or past) hybridization may be a factor.

All previously published discussions (and that on the preceding pages) regarding the validity of *C. integristipula* (as distinct from *C. neesiana*) have centered on characters of the gametophyte. Sporophytes of typical examples of both taxa admittedly show considerable similarities, among them the width of valves (ca. 250–330 μ); the secondary longitudinal (usually alternating) walls of epidermal cells with distinct nodular thickenings; the inner cell layer formed by up to 28 cell rows, e.g., with widespread supernumerary longitudinal divisions of the inner cells; the spores usually 10–12 μ in diam.; the elaters 7–8 μ in diam., with the spirals 2–2.5 μ wide. Similarities in elater diam., spore size, and form of thickenings of the epidermal cell walls are also suggestive of a close relationship.

However, several differences are suggested by the material available, among them the following two. (1) *C. integristipula* has the cells at the summit of the seta, facing the spore cavity of the capsule, provided with numerous tangential bands, much like the tangential faces of the inner cell layer of the capsule proper;[36] such thickenings are absent in examined material of *C. neesiana*, as well as in our other species. (2) *C. integristipula* shows a strong tendency for occasional accessory longitudinal division of the epidermal cells, with the supernumerary walls usually thick-walled and provided with nodular thickenings; consequently, as many as 10–12, occasionally 13, cells occur across the middle of each external valve face. In *C. neesiana s. str.* there seem to be almost invariably 8 rows of epidermal cells, with a very regular alternation of thickened (secondary) walls and thin (primary) longitudinal walls. The significance (and constancy) of these differences must be verified from further collections before much weight can be attributed to them.[37]

[36] The capsule-bearing plants studied were those of *Cain 3425*. All capsules showed the same ornamentation of the cells at the foot.

[37] Grolle (1955, p. 93) states that he considers *C. integristipula* "gut unterscheidbar und sah weder Übergänge zu *C. neesiana* noch zu *C. muelleriana*" and that *C. integristipula* differs from *C. neesiana in lacking nodular thickenings of the epidermal cells*. In conjunction with the diagnosis of *C. integristipula* I have already pointed out that, in some instances, the nodular thickenings may be faint. However, I have again studied many capsules of typical, abundantly fruiting plants of *C. integristipula* collected in Michigan in 1957 (*RMS 38942*). All these clearly show that nodular thickenings are present (see Fig. 107:2, p. 126).

SUBORDER JUNGERMANNIINAE
Schust.

Suborder Jungermanniinae Schust., Amer. Midl. Nat. 49(2):302, 1953; Schuster, The Bryologist 61(1):6, 1958; Schuster, Jour. Hattori Bot. Lab. no. 26: 193, 230, 1963.

Gametophyte growth by means of a tetrahedral apical cell. Plants varied: minute to robust, with sparse to frequent, but always *irregular branching*; the branches primitively *both* axillary-intercalary (lateral or postical, or both) and lateral-terminal (terminal branches *only* of *Frullania* type, almost *never infra-axillary*), the shoot systems irregular, from 3- to 2-dimensional, never pinnate to plumose. *Usually anisophyllous or distichous.* Leaves with insertion varying from virtually transverse (at least feebly succubous in part of the insertion) to very oblique and succubous, *never incubous*; leaves in a few primitive members (Lophoziaceae) retaining the ancient atavistic tendency toward formation of secondary lobes (then 3–4-lobed; *never bisbifid*) but usually only bifid varying to unlobed; leaves flat to convex to adaxially concave to complicate-bilobed, if the latter, *dorsal lobe always smaller than ventral*, never the reverse; normally *without water-sac elaboration*. Ventral merophytes rarely broad and with large underleaves (0.5–0.85 the size of lateral leaves), usually reduced and with small underleaves of different form from lateral leaves, often underleaves vestigial or reduced to slime papillae; rhizoids usually frequent, normally scattered (rather infrequently only at underleaf bases). *Asexual reproduction usually present:* by 1–2–few-celled gemmae, less often by fragmenting or caducous leaves (or perianths); *never by discoid gemmae or brood-bodies*, never by cladia.

Androecia variable: bracts with or without paraphyses; antheridial stalk usually 1–2–(4)-seriate; androecial bracteoles, if retained, *without* antheridia. *Archegonia usually (5–9) 10–25 per gynoecium.* Gynoecia variable in position; bracts in cycles of 3 varying to cycles of 2; perianth primitively present, very variable (but almost never with a sharp beak), often reduced or lost, and then replaced by a *Calypogeia-* or *Tylimanthus-*type marsupium or an *Isotachis-*type perigynium; a true *coelocaule never formed*. Sporophyte with seta *usually elongated*, variable, never articulate; capsule spherical to ellipsoidal to cylindrical, usually with straight or feebly twisted valves, *dehiscent to base*. Capsule wall 2–8(10)-stratose, anatomy diverse (but not, or very rarely, with epidermal cells bearing thickenings confined to angles); inner cell layers with radial to U-shaped thickenings that are usually nonanastomosing. Spores $1-2(3)\times$ elaters in diam.; *elaters all free*, $1-2(3-4)$-spiral. Spores without intracapsular germination; sporeling forming a filament (*Cephalozia*-type) or a cell body (*Nardia*-type) protonema, the *spore exine not distending to cover sporeling* (exc. sometimes in *Plagiochila*); *protonema at no stage thallose.*

The suborder Jungermanniinae is the most difficult of the groups of Jungermanniales to circumscribe, perhaps because it is "still heterogeneous" (Schuster, 1963, p. 231). It is defined here to include the Cephaloziaceae *s. lat.*—a family having certain affinities, possibly superficial, to the Lepidoziinae but differing from them in two crucial criteria: the scattered rhizoids and the seta anatomy (Schuster, 1965b).

In essence, the basis for inclusion of a group in the Jungermanniinae has been negative: absence of positive similarity to one of the seven other suborders recognized. Nevertheless, the series of families grouped here in the Jungermanniinae (see Vol. I, key, p. 671; table, p. 696) have the preceding ensemble of criteria in common; they represent a difficultly divisible complex. The group is characterized, in my opinion, chiefly by *tendencies*: it would be too much to expect such a large and varied assemblage to be sharply definable.

Among such tendencies are the following: (1) Development in most families of 1–2-celled gemmae as the dominant mode of asexual reproduction (note the occurrence of this in such diverse families as Cephaloziaceae, Odontoschismaceae, Gyrothyraceae, Lophoziaceae, Geocalycaceae, and Jungermanniaceae; absent only in a few families, e.g., Gymnomitriaceae and Plagiochilaceae). (2) General retention of *both* lateral and postical branching; groups in which lateral branches *never* occur do not seem to have evolved; groups in which all trace of postical branching is lost seem to be very rare. Even in families with predominantly postical branching, such as the Cephaloziaceae, terminal branching is common in certain primitive taxa, e.g., *Cephalozia bicuspidata*, the genus *Pleuroclada*, the antipodal genus *Metahygrobiella* Schust. (3) General absence of experimentation with development of water sacs, of either dorsal or ventral lobes; minor exceptions are the local genus *Nowellia* and the isolated family Delavayellaceae (Schuster, 1961d). (4) Although varied perigynoecial modifications occur, involving both receptacular and infrareceptacular axial tissue and leaf tissue, the Jungermanniinae never appear to have evolved coelocaules, *s. str.*[1]

[1] The sole possible exception is the family Schistochilaceae, traditionally intercalated near the Scapaniaceae, which I retained (Schuster, 1963, p. 230) in Jungermanniinae "only tentatively." My investigations (in part reported in Schuster, 1963, p. 230; see also Schuster, 1964f) have shown that the coelocaule-developing Schistochilaceae are probably closer to the Perssoniellinae Schust. (Schuster, 1963, p. 229) in the long, 4–7-seriate antheridial stalk;

The circumscription of the Jungermanniinae has already been touched upon (Vol. I, p. 650). Phylogenetic lines, where they can be drawn, will be discussed in the consideration of the individual families.

Family LOPHOZIACEAE *Cavers*

Lophoziaceae Cavers, 1910–11; Müller, 1951–58; Schuster, 1951, 1953.

Small to very large (shoots mostly 0.5–5 mm wide × 1–12 cm long), terrestrial, occurring on rocks or organic substrates, more rarely on mineral soils or corticolous, never aquatic. *Anisophyllous or distichous,* ± chlorophyllose, often with brownish to reddish pigmentation, usually relatively firm in texture. Stems erect to prostrate; *cortex opaque, never forming a hyaloderm,* usually ± strongly differentiated in 1 stratum, rarely 2 strata (differing in their smaller, narrower cells from the medullary cells); dimorphism of stems absent, usually *without postical stolons or flagella*; *branching chiefly axillary,* from the lower halves of the axils of normal leaves,[2] but *occasionally to often terminal,* then with the branch *from the ventral half of a segment, replacing the ventral half of a leaf*; more rarely with branching ventral-intercalary. *Rhizoids scattered,* often in a dense postical mat, a rhizoid-initial field never developing. Leaves *alternate,* fundamentally succubously inserted and oriented, varying from *nearly transverse* in insertion (line of insertion slightly acroscopically arched and at least slightly succubous) *to succubously inserted* and oriented, *the insertion line extended to stem midline dorsally* (but, unlike in Gymnomitriaceae, lateral merophytes not interlocking dorsally); leaves varying from 2–4-lobed (sinus rarely obsolete, as in some species of *Anastrophyllum* and in *Hattoria*), primitively in 3 unequal rows (the postical stem sectors then up to ca. 0.25 the perimeter of stem) but *usually with the ventral row of leaves strongly reduced, often suppressed* or reduced to stalked or unstalked slime papillae (ventral merophytes then reduced to a width of 1–2 cells). *Lateral leaves and underleaves with a strong tendency toward development of supernumerary lobes*

---

ellipsoidal capsule, with lack of defined nodular thickenings of the epidermal cells; complicate-bilobed leaves with a tendency toward incubous insertion and/or orientation; lack of asexual reproduction; identical form of the coelocaule; lack of postical branching of any sort. In any event, since the group Schistochilaceae does not occur regionally, the problem of its disposition can be left unresolved here.

[2] Except in *Andrewsianthus* Schust. (Schuster, 1961b), where often or usually intercalary from dorsal end of axil. In a few cases (e.g., *Lophozia gillmani, Tritomaria polita*) we also find basiscopic terminal branching, of the *Radula* type. See Fig. 226:2.

(and teeth or cilia of one or both bases). Cells *usually collenchymatous*, moderate to large in size (18–50 μ usually), often producing bulging trigones under conditions of moderate to intense transpiration; *oil-bodies present*, typically *several (2–8) or many* (12–50) per cell, medium-sized to large (3–9 μ long, rarely larger), *formed of numerous fine, subequal globules*, thus appearing finely papillose or nearly smooth, *never coarsely segmented*. *Vegetative reproduction usually present*, normally by means of 1–2-celled *gemmae produced on the leaf lobes in branching catenate fascicles*, rarely by caducous perianths, never by caducous leaves or shoots.

Dioecious or paroecious, *rarely autoecious*. *Archegonia numerous* (exc. in *Gerhildiella*); archegonia (and perianth) always *terminal on main or long, normal lateral shoots*;[3] androecia at first terminal on main or long lateral shoots, becoming intercalary by subsequent vegetative apical growth.[4] Gynoecia anisophyllous. Perianth present and generally free from bracts (in *Mesoptychia* both a perigynium and perianth present); perianth typically *cylindrical and inflated*, terete at least below and *without any indication of angles*, with apex more or less gradually narrowed and obtusely 4–5(9–10)-plicate, occasionally smooth and devoid of plicae. Androecia with *bracts subequal in size, form, and texture to normal leaves*, similar to them but more or less saccate or ventricose at base, and often with an additional antical basal lobe or tooth; bracts more transversely inserted than leaves, each with *1–4 short-ovoid and rather short-stalked* antheridia; paraphyses present or lacking; bracteoles, if present, *lacking antheridia*; antheridial *stalk variable: biseriate or uniseriate*. Sporophyte with *seta usually unspecialized*: of numerous rows of cells, but reduced to (2–3) 4 inner, (6–7) 8 outer rows in some taxa. Capsule *short-ovoid*, rarely spherical, with wall generally of 3–5 layers of cells (in *Gymnocolea* and *Isopaches* becoming 2-stratose), the inner layer(s) usually with semiannular (U-shaped) thickenings, the outer with radial (nodular) thickenings only,[5] developed

---

[3] "*Sphenolobus*" *flagellaris* (Hatt.) Grolle has gynoecia on short, ventral-intercalary branches (Grolle, 1963a). Grolle admits that sporophytic criteria will be needed to establish the position of this plant in its correct genus and family. *If* it belongs to the Lophoziaceae, it forms an exception in its abbreviated gynoecial branches. This plant also has isophyllous ♀ bracts. As Kitagawa (1965a) has shown, it properly belongs in the genus *Iwatsukia*, which he refers to the Lepidoziaceae but which belongs to the Cephaloziaceae (Schuster, 1968c).

[4] Frye and Clark (1937–47, pp. 342–343) attempt a separation between taxa on the basis of whether androecia are "terminal, just below the tip, or farther down the shoot," not realizing that the apical cell is never used up in formation of androecia and hence androecia are not really terminal in Lophoziaceae. An androecium is terminal when first developed; it later becomes intercalary. In many species 2–3 androecia may be evident on the same shoot, the older ones residual from preceding seasons.

[5] In reduced sporophytes, as of *Gymnocolea* and *Isopaches*, the inner cell layer has reduced (*Isopaches, Protomarsupella*) or, rarely, virtually lacks tangential bands (*Gymnocolea*); in other genera median portions of the tangential bands are occasionally ± obsolete.

on essentially all transverse and longitudinal walls; capsules divided to base, the *valves lanceolate, not spirally twisted.* Spore-elater (diam.) ratio always approximately $2:1$;[6] spores small, $\pm$ papillose. Elaters usually narrowly *bispiral, free, with tapering but not thickened ends.* Chromosome no.: $n = 9$ ($7 + H + h$, *fide* Tatuno, 1941) or *18* (Heitz, 1927; only *Tritomaria exsectiformis*).

The Lophoziaceae, as restricted above, include only perianth-bearing taxa (perigynia or marsupia are lacking, except in *Mesoptychia*, in which, however, a perianth is preserved). The group is in some respects primitive (i.a., in the variability of leaf-lobe number; in the unspecialized form and position of gynoecia and androecia; in the well-preserved ventral merophytes and underleaves of some taxa, such as *Chandonanthus* and some species of *Orthocaulis*). In contrast are certain derivative traits (i.a., the scattered rhizoids; widespread reproduction by gemmae; pronounced anisophylly, characterizing even the gynoecium and androecium). In the combination of large, bifid underleaves and succubous, quadrifid lateral leaves, certain taxa, such as *Chandonanthus* subg. *Tetralophozia* and *Lophozia* subg. *Orthocaulis* (*L. quadriloba*), seem primitive and show a superficial approach to such Blepharostomaceae as the antipodal *Temnoma* Mitt. The scattered rhizoids, among other characters, suggest that these similarities represent a mere common retention of certain primitive traits and should not be overemphasized. Thus the relationships with—and derivation of—the Lophoziaceae from "older" families are obscure. In any event, the relatively slight anisophylly of the most primitive Lophoziaceae suggests that the group as a whole occupies a "low" phylogenetic position.

The Lophoziaceae are best developed in the Northern Hemisphere, where the following genera and subgenera are found: *Lophozia* (including *Isopaches, Lophozia s. str., Barbilophozia, Orthocaulis, Leiocolea, Hattoriella*), *Tritomaria, Saccobasis, Anastrophyllum* (several subgenera, incl. *Sphenolobus, Acantholobus, Eremonotus s. str.*), *Anastrepta, Gymnocolea, Chandonanthus*, and *Mesoptychia*. Southern Hemisphere taxa were once believed to be few, *Chandonanthus* subg. *Chandonanthus* excepted. However, a whole flora of interesting stenotypic taxa has been discovered, chiefly in the cooler portions of the Antipodes (Schuster, 1961b, 1963, 1965), among them *Andrewsianthus* Schust., *Protomarsupella* Schust., *Roivainenia* Perss. (Persson and Grolle, 1961), *Cephalolobus* Schust. Also, isolated taxa have been found in Japan (*Hattoria* Schust.; see Schuster, 1961b), and an exceedingly reduced genus, *Gerhildiella*, has been described from the Himalaya (Grolle,

---

[6] Up to $3.5:1$ in *Andrewsianthus* Schust.

1966d). In addition, *Lophozia s. str.*, *Anastrophyllum*, and *Barbilophozia* are represented by a few antipodal species, and the stenotypic genus *Cuspidatula* Steph. is restricted to the Antipodes.

Several other genera have been assigned in the past to the Lophoziaceae, i.a., *Acrobolbus* Nees (still placed here by Müller, 1951–58, but surely related much more closely to *Tylimanthus* Mitt., in a group Acrobolbaceae; see Vol. III); *Notoscyphus* Mitt., *Stephaniella* Jack, and *Symphyomitra* Spr. have sometimes been placed in the Lophozioid complex, incorrectly so. Müller (1951–58, p. 197) placed *Temnoma* Mitt. in the Lophoziaceae; as Schuster (1959, 1961a) has shown, this genus belongs in the Blepharostomaceae. Two stenotypic, isolated genera, producing flagelliform branches or stolons, belong in the Lophoziaceae but are still too poorly known to make a definitive assignment possible: *Stenorrhipis* Herz. (1950) and *Andrewsianthus* Schust. (1961b; see also Grolle, 1963b). *Andrewsianthus* and *Cephalolobus* appear to be closely allied (Schuster, 1965), and, through the latter, related to the *Anastrophyllum-Eremonotus complex*.

With the exception of several rather questionably assigned genera, the Lophoziaceae are predominantly developed in the cooler and cold portions of the Northern Hemisphere, ranging to the highest portions of the Arctic, and, to a somewhat lesser extent, in the cooler portions of the Antipodes.[7] However, *Chandonanthus* and *Anastrophyllum*, isolated species aside, are found primarily in montane tropical and temperate rain forests, although *Chandonanthus* subgenus *Tetralophozia* is holarctic. The various taxa are found predominantly on mineral substrates or on organic soils, less often as xylicoles or corticoles. The majority of species are oxylophytes, excepting species of *Leiocolea*. The group, therefore, is rather primitive in its ecology. The majority of species regularly reproduce sexually (hence reproductive characters are readily available for taxonomic use), even though many taxa (excl. *Anastrophyllum* and *Chandonanthus*) regularly produce gemmae.

Relationships

The family Lophoziaceae is delimited here essentially as in Joergensen (1934) and Müller (1939–40), certain marsupia-bearing genera excluded. The family, so circumscribed, includes a large number of the genera placed in the Jungermanniaceae or "Epigonianthaceae" (of Buch, Evans, and Verdoorn, 1938; Evans, 1939; Verdoorn, 1932a). The remaining genera are now placed in the Jungermanniaceae, Arnelliaceae, and Gyrothyraceae, with isolated genera transferred to the Acrobolbaceae and

---

[7] A few species, i.a., *Lophozia excisa* and *L. hatcheri*, are actually bipolar, occurring in both regions.

Plagiochilaceae. As circumscribed, the Lophoziaceae essentially include genera in which the early ontogenetic lobing of the leaves persists into the mature state, and in which a more or less terete (never strongly compressed) perianth is found, and oil-bodies formed of numerous minute spherules surrounded by a relatively firm membrane (i.e., not formed of several to many distinct segments), while the capsule wall remains 3–5-cell-layered (with few exceptions) and rhizoids are always freely scattered. I place much emphasis on the last criterion; for this reason I cannot follow Inoue (1966), who would place *Syzigiella*, with rhizoids in fascicles, into the Lophoziaceae.

The Lophoziaceae, thus circumscribed, largely include relatively unspecialized genera. The less advanced groups (*Lophozia* subg. *Orthocaulis*, *Chandonanthus*) are characterized by erect or suberect growth and an approach to the ancestral isophyllous condition, with large underleaves and with retention of ventral stem sectors including about ¼ the circumference of the stem. Associated with the unspecialized stem structure is retention of a primitive tendency to form deeply lobed leaves; there is further retention, in "low" taxa such as *Orthocaulis* and *Chandonanthus*, of an apparently primitive lack of stability in leaf-lobe number: the 2 primary leaf lobes exhibit a well-defined and apparently atavistic tendency to undergo secondary division (resulting in a maximum of 4 lobes); furthermore, the leaf bases in these groups tend to be ciliate. In these characteristics the lower Lophoziaceae, in some measure, approach other "low" families of Jungermanniales, such as the Ptilidiaceae and the Blepharostomaceae.

The Lophoziaceae, as a group, differ from other associated groups (Jungermanniaceae, Gyrothyraceae, Plagiochilaceae) essentially in lacking the innovations or specializations which the latter exhibit. Although perhaps not directly ancestral to these families, the Lophoziaceae possess features that warrant derivation of some of these families (as well as the Scapaniaceae and possibly the Gymnomitriaceae), from near them. The limits of the family can perhaps be best drawn from a comparison with these relatively specialized groups as follows.

The Jungermanniaceae represent an advance over the Lophoziaceae in having typically acquired entire lateral leaves (with suppression or reduction of the underleaves).[8] Furthermore, the Jungermanniaceae have the capsule wall generally reduced to the bistratose condition (except for the "old" and somewhat

[8] Note, however, that a very few presumed Lophoziaceae, such as the Japanese *Hattoria* Schust. (Schuster, 1961), have independently evolved entire or subentire leaves. The position of *Hattoria* in the Lophoziaceae is not above suspicion; Inoue would put it in a subfam. Jamesonielloideae, of the Lophoziaceae (see also pp. 710, 810).

annectant genus *Jamesoniella*). They have conspicuously "experimented" with perigynium or marsupium development, although they retain the 2:1 spore-elater ratio, as well as the perianth form of the Lophoziaceae, and may be regarded as on the next higher developmental level above the latter family. The Jungermanniaceae may well be considered as merely a specialized offshoot or subfamily of the Lophoziaceae.[9]

In the Plagiochilaceae, the leaves in various species have acquired the ability to develop marginal teeth or lobes (though species with sharply 2-lobed leaves are rare); the spore-elater ratio remains 2:1; the capsule wall (as in most Lophoziaceae) remains 3–7-stratose. Unlike in preceding groups, the perianth of the Plagiochilaceae acquires a characteristically laterally compressed form. The group differs fundamentally from the Lophoziaceae in patterns of asexual reproduction.[10]

The Scapaniaceae may have been derived from *Anastrophyllum*- or *Eremonotus*-like ancestral types in the Lophoziaceae. Like the latter family, the Scapaniaceae retain a 3–7-stratose capsule wall, bilobed leaves (with the dorsal lobe more or less smaller than the ventral), generally a 2:1 spore-elater ratio, and oil-bodies of the *Lophozia* type (i.e., formed of numerous little or moderately distinct, scarcely protuberant globules). The Scapaniaceae represent an advance over the Lophoziaceae in having acquired a complicate type of leaf, with the smaller dorsal lobe folded over the ventral, and generally separated by a discrete keel. Furthermore, terminal branching has been lost (the primitive *Blepharidophyllum* aside) and the perianths have become dorsiventrally compressed to a greater or lesser degree (unlike in previous groups). Though superficially the Scapaniaceae and the Lophoziaceae are very different in appearance, the fundamental characters separating the two are relatively weak, and the Scapaniaceae are not a sharply isolated family. The similar mode of formation of the fasciculate gemmae indicates a common origin.[11] Suggestive, also, is the common retention of an ability to develop secondary division of the lobes on sporadic leaves and the rare and sporadic—but unquestionable—development of basiscopic, terminal, *Radula*-type branching.

In evaluating the affinities of the Lophoziaceae to the foregoing families, it should be noted that branching modes do not afford a sharp separation.

---

[9] See in this connection the discussion under *Mesoptychia*, p. 802. However, derivation of the Jungermanniaceae directly from the Lophoziaceae is difficult. The former differs from the latter in almost uniformly lacking gemmae (occurring, rarely, in, e.g., *Jungermannia lanceolata*, and here of a different form from those found in Lophoziaceae). The Lophoziaceae almost uniformly are able to produce polygonal, characteristic gemmae (smooth in a few species of *Massula*; absent in *Chandonanthus*).

[10] Inoue (1966) has transferred the genus *Syzigiella*, previously generally included in the Plagiochilaceae (Müller, 1951–58; Schuster, 1959–60) to the Lophoziaceae. *If* this proposal is justified, the laterally compressed perianth of the Plagiochilaceae (which sporadically recurs in *Syzigiella*) will no longer adequately delimit this group. I am exceedingly sceptical of this proposal: both the opposed leaves and the fasciculate rhizoids isolate *Syzigiella* from *all* Lophoziaceae.

[11] The subg. *Jensenia* of *Scapania* forms, superficially, almost a total morphological transition to the Lophoziaceae; its species lack the sharp keel of the Scapanioid leaf and often have merely strongly concave leaves.

Although *both* lateral and postical-intercalary branching occurs in Lophoziaceae, lateral-terminal, *Frullania*-type branching is widespread in all but a few derivative genera, such as *Andrewsianthus*.[12] Such terminal branches also occur in many Jungermanniaceae. They are of widespread occurrence in Plagiochilaceae. In the Scapaniaceae, almost all of which have lost the potentiality to develop terminal branches, the antipodal *Blepharidophyllum* Ångstr. possesses terminal, *Frullania*-type branching (Schuster, 1963). In at least one species of *Blepharidophyllum*, fasciculate, Lophozioid gemmae and large underleaves (within the gemmiparous shoot sectors) still persist. However, a definite difference in branching *does* seem to separate the Lophoziaceae from the Gymnomitriaceae, the latter group appearing to generally have lost the ability to form terminal branches.

The Gymnomitriaceae appear likely to be of Lophozioid derivation. They lack the *Frullania*-type branching found at least sporadically in the preceding families. As in the Lophoziaceae, the leaves remain generally bilobed (3–4-lobed leaves are never developed) and range from somewhat succubous to virtually transverse. The leaves have secondarily become nearly equally bifid, with the sinus quite obsolete in a few cases. As in some Lophozioid genera, the ventral merophytes are uniformly narrow and generally lack the ability to develop perceptible underleaves. Unlike the Lophoziaceae, the Gymnomitriaceae develop extreme perianth reduction and, often, associated perigynium or marsupium development. If only holarctic taxa are studied, one could conclude that there is little or no connection between the two families.[13]

From the above outline it is evident that the Lophoziaceae represent a highly synthetic group: they retain characters found in a relatively large ensemble of families and appear to occupy a "low" position near the point from which these various families diverge. The Lophoziaceae, in their "strategic" position, are therefore of considerable interest to the phylogenist.

---

[12] Terminal branching in the Lophoziaceae often involves only slight or no modification of the subtending leaf. In *Gymnocolea*, e.g., the terminal branches may be subtended by perfectly normal, bilobed leaves, although others are subtended by lanceolate or ovate, undivided leaves. In *Lophozia* subg. *Barbilophozia* the subtending leaf is usually bilobed and smaller compared with the quadrilobed normal leaves (Fig. 160:3). Even the more specialized genera with loosely complicate leaves possess at least sporadic terminal branching, e.g. *Tritomaria exsecta*, and *Anastrophyllum* spp. (in which the branch may be subtended by an atypical, nearly symmetrically bilobed leaf).

[13] However, in the two stenotypic New Zealand genera *Protomarsupella* and *Acrolophozia* (Schuster, 1963, 1965) we find a Marsupelloid facies, occasioned by the pectinate, transversely oriented leaves, linked together with a lack of underleaves and of terminal branching—but a "free" Lophozioid perianth. It is possible, as is developed in the section on the Gymnomitriaceae, that these genera serve as contact points between the Lophoziaceae and Gymnomitriaceae. *Acrolophozia* seems closer to the Gymnomitriaceae but is transitional to the Lophoziaceae in sporadically developing trifid leaves. The Gymnomitriaceae are advanced in their dorsally "interlocking" merophytes, a feature already clearly developed in *Acrolophozia*.

Classification

The Lophoziaceae represent perhaps the most polymorphic and difficult of the holarctic families of Hepaticae. Transitions, at least of a superficial nature, occur between most so-called genera, and not infrequently between so-called species. As a consequence generic and species concepts in the group remain fluid. A tendency has developed to divide the "old" genus *Lophozia* into numerous generic groups. The basis for such subdivision has gradually varied. Loeske and Schiffner used the number of leaf lobes; Müller (1910, *in* 1905–16) introduced as a basis the perianth form and later (1939) the oil-body form; Buch (1933a) extensively used stem anatomy; Schuster (1961b) first emphasized branching. Whatever criteria were selected, the result has been a gradual fragmentation of the portmanteau genus *Lophozia* into smaller, more homogeneous genera. With the use of different criteria, diverse systems of classification have evolved, some at odds with others in significant respects.

The subjective nature of the intrafamiliar classification is due in part to the considerable polymorphism in form of the leaves and underleaves. Within the single genus *Lophozia* there may be variation in the width of the ventral merophytes and consequently in the degree to which underleaves are suppressed (cf. *Orthocaulis*, *Barbilophozia*, *Leiocolea*). Similarly, leaves range from bilobed to trilobed or even quadrilobed within a single genus (*Chandonanthus*; also *Lophozia* subg. *Orthocaulis*; with nutritional differences also within *Barbilophozia*), or may vary from bilobed to entire (*Anastrophyllum*). Polymorphism may be so pronounced within one species (cf. *Lophozia quadriloba*) that criteria which are valuable in distinguishing genera in other families become invalid in separating species.

As a consequence, an impressive accumulation of literature (Müller, 1910 *in* 1905–16, 1939, 1942, 1947, 1954 *in* 1951–58; Buch, 1933a, 1942; Evans, 1935; Meylan, 1939; Andrews, 1948; Schuster, 1949a, 1951a, 1961b; Inoue, 1957) deals with the Lophozioid genus and its circumscription. The problem as it pertains to the holarctic taxa has been recently reviewed (Schuster, 1951a, 1961b); I here follow this classification with some deviations. For a thorough analysis of the problem, a critique of previous systems, and details of the basis for the classification used in this text, the student is referred to the papers cited above.

Possible evolutionary patterns within the family have been schematically presented (Schuster, 1951a) as follows. The ancestral" prototype" for the Lophoziaceae approached the Herbertoid-Temnomoid organization in (1) presence of several oil-bodies per cell; (2) a pluristratose capsule wall; (3) retention of both terminal and intercalary branching; (4) virtually isophyllous and erect organization, with the lateral leaves almost transversely (but, unlike *Herberta*, probably very slightly succubously) inserted; leaves and underleaves deeply lobed, with a tendency toward development

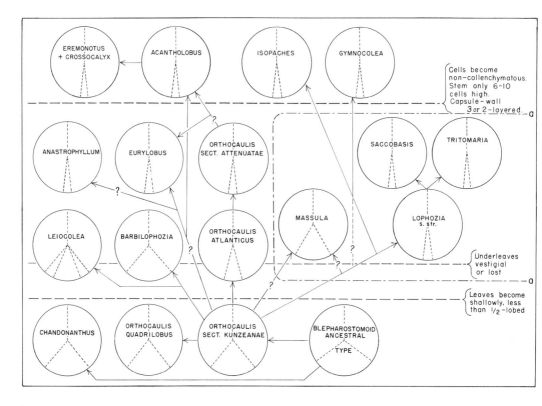

FIG. 135. Possible interrelationships of regional taxa of Lophoziaceae; the degree of reduction of ventral merophytes is schematically portrayed; for details see text.

of basal cilia or teeth; (5) collenchymatous cells; (6) terminal perianth. From such a prototype, evolution proceeded by (*a*) reduction in size and complexity of both sporophyte and gametophyte; (*b*) progressive development of dorsiventrality with corresponding reduction of ventral merophytes; (*c*) correlated with (*b*), reduction and gradual elimination of underleaves; (*d*) with reduction in size, the axis of the gametophyte undergoing progressive reduction; (*e*) with reduction in size, the axis of the sporophyte undergoing similar reduction (to a minimal condition consisting of 2 concentric cell layers, the inner of 2-4, outer of 6-8, rows of cells); (*f*) with reduction in size, the capsule wall gradually reduced from a 5-stratose to a 2-stratose condition; (*g*) development of asexual reproduction by means of gemmae.

The general pattern of reduction appears to have become established several times, independently, leading to parallelisms in evolution that have not simplified understanding of the group. These parallelisms are indicated in Fig. 135, in which the degree of preservation of the ventral merophytes is indicated schematically for each group.[14]

This construction is based on the same assumption used by Buch (1933a, p. 296) and Schuster (1951a), namely, that the oldest acrogynous Hepaticae had 3 equal rows of bilobed leaves. The predominance of a 3-sided apical cell and corresponding development of 3 rows of stem segments, together with the constantly bilobed form of the embryonic leaves, warrant such a supposition. Although neither the Herbertaceae nor Blepharostomaceae can be looked upon as ancestral to the Lophoziaceae, they suggest the probable form of the archetype; in further characterizing such an archetype, the following conclusions are probably valid. (1) *The leaf cells were collenchymatous* (this appears to be the primitive condition throughout the acrogynous liverworts). (2) *The sporophyte was relatively unreduced*, and had the "general type" of seta (Douin, 1908) and a multistratose capsule wall. (3) *The antheridial stalk consisted of at least 2 rows of cells.* (4) *The leaf-lobe number was not strictly fixed* but was determined in part by environmental conditions. In this I differ from Buch. I think that the ability to develop supplementary leaf lobes, which recurs so often in various "low" families, such as Ptilidiaceae, Lepicoleaceae, Lepidoziaceae, Herbertaceae, Vetaformaceae, Isotachaceae, and Trichocoleaceae, must be regarded as a primitive characteristic. In Lophoziaceae retention of wide ventral stem sectors, a primitive character, is correlated with the ability to develop accessory leaf lobes as, e.g., in *Orthocaulis, Chandonanthus, Protolophozia, Barbilophozia*, and *Massula*—only in *Leiocolea*[15] is there an exception to the rule. If this assumption is accepted (and it must be if *Orthocaulis* is taken as the most synthetic and hence archaic type), derivation of various taxa with 2-lobed leaves, as well as of 3–4-lobed types, is relatively uncomplicated. (5) *Leaf insertion was originally oblique*, and though not transverse, was probably only a little succubous (such a lateral leaf insertion appears to have evolved in almost all lower acrogynous liverworts, except for incubous-leaved forms). The nearly horizontal or extremely oblique leaf insertions found in *Barbilophozia, Leiocolea*, and *Massula* are readily derived from such a type, while the *Anastrophyllum* and *Tritomaria* types are also readily derived from it, and are associated with less prostrate growth. (6) *The oilbodies were moderate in size* (4–9 μ) and moderately numerous per cell (ca. 4–15). Deviations from this type are always accompanied by specialization in other features (note *Massula* and *Leiocolea*).

With these assumptions, the relationships expressed in Fig. 136 become intelligible. It is presumed in this figure that other parallelisms in development occurred concurrently with increase in dorsiventrality, resulting in the evolution of diverse groups on identical "developmental levels." This

[14] Limited to regionally represented taxa; if the "exotic" genera were to be included, a considerably more complex table would be necessary. The lines connecting the various groups should not be interpreted too literally. For example, in Fig. 135 *Massula, Lophozia*, and *Leiocolea* are to be understood as derived from *L. (Orthocaulis) kunzeana*-like ancestral types, not from *L. (O.) kunzeana*.

[15] For ease of discussion generic and subgeneric names are used interchangeably in this presentation.

parallelism has led to difficulty in classification of the complex. The parallel steps may be listed as follows (the numbers corresponding to those in Fig. 136).[16]

1. Original isophylly gives way to distinct anisophylly, but large underleaves are retained; no retention of antheridia in axils of underleaves; leaves acquire a perceptibly succubous line of insertion; the stem retains a suberect type of growth and lacks differentiation of the medulla.

2. Ventral stem sectors remain wide (6 or more cells in width), but lose the ability to develop underleaves on sterile shoots; the plant becomes completely bilateral; the ability of the leaves to develop basal cilia is also lost.

3. Ventral stem sectors become reduced in width (usually merely 1–2 cells wide; locally, especially where endogenous buds are laid on, secondarily wide); there are never discrete underleaves.

4. The stem, with increasing dorsiventrality, becomes dorsiventrally differentiated, with the medulla developing a small-celled ventral band of mycorrhizal cells.

5. The upper portion of the leaf insertion loses any tendency to be inserted in a line curving toward the stem base (i.e., it becomes transverse or even curves toward the stem apex).

6. The dorsal half of the leaf insertion becomes curved toward the stem apex.

The various specializations that cut off the individual groups or lines may be briefly summarized as follows, the letters corresponding to those in Fig. 136.

a. The leaves remain deeply and often 3–4-lobed; the perianth becomes deeply pluriplicate to base. The ability to form gemmae is lacking.

b. The leaves become nearly horizontally inserted; cilia of the leaf bases are formed by elongate cells; the leaves become constantly 4-lobed.

c. The leaves become nearly horizontally inserted; cilia of the leaf base are generally lost; leaves become regularly 2-lobed; leaf cells tend to become large; terminal branching usually lost; oil-bodies become large and few; the perianth becomes beaked.

d. Leaves become usually regularly 2–3- or 2-lobed; underleaves are lost (exc. *L. obtusa*); leaves become more oblique in insertion (generally?); terminal branching retained.

e. Both leaf and stem cells become noncollenchymatous; stem cells become narrow and elongate (usually?); oil-bodies become minute and numerous; leaves retain the ability to sporadically form 3–4(5) lobes; ventral stems sectors stay broad (6 cells or more).

f. Ventral stem sectors are reduced to a width of 1–2 cells; stem cells do not become very narrow and elongate; oil-bodies remain moderate in size and number; leaves lose ability to develop other than an occasional third lobe.

g. Cells become noncollenchymatous, equally thick-walled; stems become only 8–12 cells high; medulla remains uniform.

[16] Again, in order to maintain optimal clarity and simplicity, I have left out all "exotic" genera.

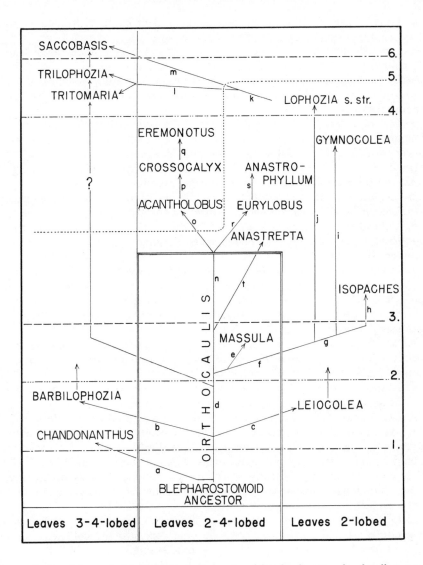

FɪG. 136. Possible evolutionary patterns in regional Lophoziaceae; for details see text.

h. Perianth remains plicate, persistent; gemmae commonly formed; leaves retain acute lobes; ♂ and ♀ bracts remain dentate.

i. Perianth becomes inflated, caducous; gemmae absent or rare; leaves become obtuse-lobed; lobes of ♂ and ♀ bracts become entire.

j. Stems develop a dorsiventrally differentiated medulla but normally remain 15–20 cells high; cells remain strongly collenchymatous, developing bulging trigones.

k. Leaves become rather regularly 3-lobed; dorsal cortical stem cells become elongate; dorsal half of leaf insertion becomes directed more toward stem apex (never toward base); basal leaf cells become elongate.

l. Leaves remain ± asymmetric; cortical cells remain 2–3.5× as long as wide; dorsal end of leaf insertion remains transverse or nearly so.

m. Leaves become symmetrical; cortical cells become extremely elongate; dorsal sector of leaf insertion becomes directed toward stem apex.

n. Ventral stem sectors are reduced to width of 1–2 cells [already occurring in *Lophozia (O.) attenuata*]; leaf lobes fixed at 2; dorsal half of leaf becomes more transversely inserted.

o. Dorsal half of leaf becomes completely transverse in insertion; cells become noncollenchymatous, nearly equally thick-walled; capsule wall becomes 2- or 3-stratose; dorsal portion of leaf not expanded, the base extending only to stem midline; gemmiparous shoots become erect, slender, with appressed leaves.[17]

p. Seta becomes reduced to 4 inner, 8–9(10) outer, rows of cells; gemmiparous shoots become extremely specialized; perianth acquires a distinct, deep antical sulcus; stems become merely 6–7 cells high.

q. Ability to produce gemmae is lost; cells become extremely small; stolons develop; only lateral-intercalary branching preserved.

r. Dorsal end of line of insertion curves down and runs down the dorsal edges of the lateral stem sectors, hence is not entirely transverse; cells remain collenchymatous; capsule walls remain 3–4-stratose; dorsal portion of leaf expanded, extending across stem; gemmiparous shoots do not become modified (ability to develop gemmae is, generally, lost); leaves remain relatively broad (much wider than long); leaf lobes retain strong tendency to be incurved (as in *Orthocaulis*).[18]

s. Leaves become widest near base, as long as or longer than wide; ability to produce 3-lobed leaves lost; cells become very strongly collenchymatous, those of base become elongate; leaves become strongly canaliculate.

---

[17] Whether this indicates any phyletic connection to such species as *Lophozia (Orthocaulis) attenuata*, *L. binsteadii*, and *L. atlantica*, in which reduction of the ventral stem sectors and development of more or less slender, erect gemmiparous shoots have already occurred, is an open matter. The correlation is interesting, at any rate, and indicates that there may be decided convergence here. It should be noted that the "lower" species of *Orthocaulis* do not have specialized gemmiparous shoots, and also retain underleaves and wide ventral stem sectors.

[18] The connection between *Orthocaulis* species (such as *binsteadii* and *atlantica*) and *Anastrophyllum (Eurylobus) saxicolus* seems significant. The leaf insertion is strikingly similar; the ventral stem vectors are equally reduced, leaves are broad and have incurved lobes. This similarity is emphasized by the fact that *A. saxicolus* still retains the ability to develop occasional 3-lobed leaves on sterile shoots (Fig. 239:1), much as is the case in the two species of *Orthocaulis* mentioned. There thus seems to be obvious convergence toward the more specialized species of *Orthocaulis*, on the one hand, and the less specialized species of *Anastrophyllum*, on the other.

An attempt is made to indicate in Fig. 136 the continuum often existing among the various groups. *Barbilophozia* is contiguous to *Orthocaulis* (through *L. barbata*); *Orthocaulis* almost approaches *Chandonanthus* (through *L. quadrilobus*); *Massula* appears to grade into *Barbilophozia* (through *L. obtusa* and *L. barbata*; also through *Protolophozia*); *Orthocaulis* appears to approach *Anastrophyllum* (through *L. binsteadii* and *atlantica*). The subg. *Lophozia* appears to approach *Massula* (through *L. excisa*) and *Tritomaria* (through *L. longidens* and *T. scitula*). These seeming or real transitions make a clear-cut circumscription of genera very difficult and hence prompt conservative treatments, such as those of Andrews (1948, 1957), in which all of the American species are placed in the single genus *Lophozia*.

I have suggested (Schuster, 1951a) for the nearctic members of the family two possible classifications, one more conservative than the other. In this work, the more conservative classification is adopted; it is taken almost directly from Schuster (1951a), with minor changes. The two systems are outlined as follows:

"CONSERVATIVE"	"RADICAL"
1. *Lophozia*	1. *Barbilophozia*
a. Subg. *Orthocaulis*	*a.* Subg. *Orthocaulis*
b. ,, *Barbilophozia*	*b.* ,, *Barbilophozia*
c. ,, *Leiocolea*	2. *Leiocolea*
	3. *Lophozia*
d. ,, *Massula*	*a.* Subg. *Massula*
e. ,, *Isopaches*	*b.* ,, *Isopaches*
f. ,, *Lophozia*	*c.* ,, *Lophozia*
2. *Tritomaria*	4. *Tritomaria*
a. Subg. *Tritomaria*	*a.* Subg. *Tritomaria*
	b. ,, *Trilophozia*
b. ,, *Saccobasis*	5. *Saccobasis*
3. *Anastrophyllum*	6. *Anastrophyllum*
a. Subg. *Eurylobus*	*a.* Subg. *Eurylobus*
b. ,, *Anastrophyllum*	*b.* ,, *Anastrophyllum*
	7. *Eremonotus*
c. ,, *Sphenolobus*[*Acantholobus*]	*a.* Subg. *Sphenolobus*
d. ,, *Crossocalyx*	*b.* ,, *Crossocalyx*
e. ,, *Eremonotus*	*c.* ,, *Eremonotus*
4. *Chandonanthus*	8. *Chandonanthus*
5. *Gymnocolea*	9. *Gymnocolea*
6. *Mesoptychia*	10. *Mesoptychia*

In spite of criticisms by, i.a., Andrews (1948) and Schuster (1951a) of the division of the *Lophozia-Sphenolobus* complex into numerous genera, as in Buch (1933a), a generic fragmentation has been adopted to a large extent by European authors, i.a., Müller (1954, *in* 1951–58), who previously (1939, 1942) criticized the attempt by Buch to subdivide the complex. S. Arnell (1956) accepted Buch's genera as originally proposed and, indeed, added the "genus" *Obtusifolium* for a single, rather isolated species, *Lophozia obtusa*.

A striking example of the problems involved—which have not been faced by European authors who favor fragmentation of the Lophozioid genera—is offered by *Anastrophyllum mayebarae* Hatt. (see p. 714). Even more impressive evidence of confusion is found in the case of *Lophozia diversiloba* Hatt., which was later shifted by its author to *Acrobolbus*; in a letter to Inoue, Hattori then stated that this species should go into "*Dilophozia*." Inoue (1957) stated that it possesses the collenchymatous cells of "*Dilophozia*" (genus *Lophozia s. str.* of Buch, minus *Massula*), the large and few oil-bodies of *Leiocolea* (considered a discrete genus by current European authors), a 2-stratose capsule wall as in *Isopaches* (variously delimited by Buch, Arnell, and Müller, but now accepted as a genus by all), and paraphyllia of the perigonal bracts, much as in *Leiocolea rutheana* and in *Orthocaulis* (considered a valid genus by Buch, as well as by Arnell; a synonym of *Barbilophozia* by Müller). The lack of gemmae, the stem anatomy, the oil-bodies, the papillose cuticle, and the collenchymatous cells suggest *Leiocolea*; the lack of underleaves and the somewhat plicate perianth suggest "*Dilophozia*." To Inoue, the bistratose capsule wall suggests an affinity to *Isopaches*. Thus we have in a single species linkage of characters assigned to four "genera." If these so-called genera of Lophoziaceae are adopted, *L. diversiloba* possibly belongs in a genus of its own. Inoue, accepting my relatively conservative treatment (Schuster, 1951a), placed the species in *Lophozia* in a subg. *Hattoriella*—but later (Inoue, 1961) elevated this group to generic rank.[19]

Lophozia diversiloba and *Anastrophyllum mayebarae* illustrate a point I emphasized in 1951a: as soon as exotic species, i.e., those not found in Europe, are studied, the system proposed by Buch breaks down. *Lophozia herzogiana* of New Zealand is equally illuminating (Grolle, 1962a); see p. 257. These species clearly illustrate the dangers inherent in proposing a generic system based only on species from the relatively impoverished European flora, and, indeed, from any area that has undergone recent glaciation.[20] It is not even necessary to resort to "exotic" species to demonstrate the subjectivity of the concept of genus in the Lophoziaceae. "*Orthocaulis*" *cavifolius* Buch & Arn., described in 1951, was transferred three years later to *Sphenolobus* by Müller! Two final pertinent comments may be made here: (*a*) Buch, Arnell, and Müller mastered the European hepatic flora to a point unsurpassed by other recent workers, so that the differences of opinion are not due to a lack of knowledge, per se; (*b*) even this writer, with a more conservative generic concept, places *Orthocaulis* and "*Sphenolobus*" into different genera!

---

[19] Kitagawa (1962b) has recently refused to recognize *Hattoriella* as a distinct genus; Inoue (pers. comm.) informs me that he now, also, no longer would maintain it! Kitagawa (1965, p. 271) now places *Hattoriella* as a synonym of *Lophozia* subg. *Leiocolea*. If a single species is subject to such varied placements, perhaps our generic lines are drawn too narrowly.

[20] See also in this connection the discussion of the "*Sphenolobus*"-*Anastrophyllum* problem, p. 706.

Before any worker dealing with the relatively limited holarctic flora adopts the "radical" systems propounded by Buch and taken over by Arnell, and to a lesser extent by Müller, he should, in addition, study the following species, *none* of which "fits" into the generic system of Buch: *Lophozia perssoniana* Miller (1963) of Hawaii; *"Sphenolobus" flagellaris* (Hatt.) Grolle of Japan (see Grolle, 1963a); *Hattoria yakushimensis* (Horik.) Schust. (olim *Anastrophyllum yakushimense*) of Japan. I would also suggest that *Lophozia (Orthocaulis) hamatiloba* Grolle of New Guinea (see Grolle, 1965, p. 50, fig. 4), a species that hardly fits *Orthocaulis* as presently construed, be examined. Relevant too is the fact that an apparently simple mutation in "normal" distichous, underleaf-free *Anastrophyllum saxicolus* can result in production of tristichous, isophyllous plants (see p. 739)! Finally, in conjunction with any decision of whether or not to adopt a generic division between the *Eremonotus-"Sphenolobus"-Crossocalyx* complex and *Anastrophyllum*, one should examine *Anastrophyllum sphenoloboides* Schust. of Greenland (see p. 741) and *A. novazelandiae* Schust. of New Zealand (see Schuster, 1965, p. 282, fig. 4). Indeed, when tropical and antipodal taxa are studied, generic concepts so laboriously built up for holarctic taxa collapse. These observations are relevant since, discussing my "conservative" system of 1951, Jones (1958a) adopted only portions of it because he felt that I had placed too much emphasis on the existence of one or two intermediate species. Jones, in effect, stated that adoption of such a conservative system hinged on existence of a broad spectrum of intermediates; that such exist can scarcely be denied today. Hence, after much hesitation, I continue to utilize a system closely patterned on the conservative system propounded in 1951—and have, indeed, become even more conservative in now reducing *Saccobasis* to a subgenus of *Tritomaria*.

Conspectus of Genera and Subgenera of Lophoziaceae[21]

1. Perianth well developed, never with a marsupium at base, lying in the axis of the stem; leaves 2–4-lobed for 0.15–0.9 their length, the lobes conspicuous.
. . . . 2
    2. Leaves lobed 0.15–0.5 (occasionally to 0.65–0.7) their length into 2–4 lobes, lobes not channeled, without strongly reflexed margins; perianth obtusely 4–5-plicate terminally (one or more plicae occasionally approaching the base) or smooth; often with gemmae; ventral stem sectors 1–12 cells wide usually, the ventral merophytes occupying less than 0.25 the stem periphery (usually less than 0.15); leaves obliquely inserted: ventral ⅔ of line of insertion at an angle of 15–60° to stem. . . . . . 3.
    3. Leaves ± obliquely to horizontally oriented, the leaf *insertion oblique throughout*, that of dorsal end clearly directed toward stem base (ventral half also oblique in insertion). [Ventral stem sectors often 4–12 cells wide, developing underleaves; leaves 2–4-lobed; if more than 2-lobed never with dorsiventral differentiation of stem medulla; leaves rarely

---

[21] The complexity of the group is such that the present key is of necessity technical; a simpler artificial key begins on p. 239.

canaliculate or complicate-canaliculate: dorsal lobe rarely lying over
ventral, lobes lying generally in ± same plane][22] . . . . 4.

4. Leaves generally with *acute or subacute lobes*;[23] leaf cells rarely
equally thick-walled and capsule wall generally 3–5-stratose;[24]
inner layer of capsule wall with semiannular, rarely only radial
thickenings; perianth plicate (if smooth, beaked), *persistent*;
perichaetial bracts larger than leaves, generally different in shape
(usually with more lobes and/or with marginal teeth); rhizoids in
a dense ventral mat; gemmae generally common; stem 8–22 cells
high[25] . . . . *Lophozia* Dumort. *s. lat.* . . . *a.*

 a. Leaves *all or most 3–4-lobed* (on well-developed shoots), on postical
  bases often with 1–5 cilia; ventral stem sectors well developed,
  4–12 cells wide, usually producing distinct underleaves;[26]
  perianth plicate; medulla of stem uniform, lacking a ventral
  mycorrhizal band; capsule wall 3–4-stratose, the inner layer
  with semiannular thickenings; branches normally all lateral-
  terminal, exceptional ones lateral-intercalary. (Oil-bodies 2–8
  per cell, usually moderate in size, ca. 4–6 × 5–9 μ) . . . *b.*

 b. Leaves *generally ± bluntly 3-lobed, with dorsal lobe largest* (occasionally only 2-lobed; in *L. quadriloba* mostly 4-lobed); *plants
  suberect or ascending in growth*, with an arched line of insertion at
  about a 45–60° angle to axis, curving toward stem apex; postical leaf base with *cilia of isodiametric cells*, when present; leaves
  concave, often with incurved lobes, occasionally canaliculate .
  . . . . subg. *Orthocaulis* (Buch) Schust.

 b. Leaves generally all or *mostly 4-lobed*; lobes ± acute or mucronate; plants *horizontal in growth*, leaves with a nearly straight
  line of insertion at about a 25–30° angle to axis; postical leaf
  base with cilia, when present, *of elongate cells*; leaves nearly flat
  or somewhat dorsally convex (rarely concave), widely spreading,
  the lobes rarely distinctly incurved, the leaves never canaliculate
  . . . . subg. *Barbilophozia* Loeske

 a. Leaves *normally 2-lobed* (occasionally 3-lobed), the *postical bases
  lacking cilia*;[27] ventral stem sectors usually unable to produce
  underleaves (or perianth smooth and beaked) . . . . . . *c.*

 c. Perianth smooth, terete, *abruptly contracted into a beak*; *oil-bodies
  large*, 4–7 × 8–15 μ, mostly 2–5(6–9) *per cell*; cuticle usually very
  distinctly papillose; ventral stem sectors well preserved, 4 or
  more cells wide, *usually with distinct underleaves*; medulla of stem

---

[22] In a few species of *Lophozia* subg. *Orthocaulis* (*L. atlantica, kunzeana*, etc.) the leaves may be
strongly concave and more or less complicate. These species differ from all those falling under
the second *3* (p. 236) in having some or many leaves trilobed, with the *dorsal* lobe the largest.

[23] Only in *L. obtusa* are the lobes generally rounded or obtuse at the apex.

[24] In *Isopaches*, however, the cells are equally thick-walled and the capsule wall is bistratose.

[25] In *Isopaches* only 8–10 cells high; in other groups usually 10–22 cells high.

[26] In a few species of *Orthocaulis* reduced to a width of 2 cells and unable to develop underleaves.

[27] In *L. rutheana* occasionally with a single vestigial basal cilium.

unmodified; capsule wall 2–4-stratose; cells of leaves *usually very large* (35–60 μ), and generally strongly collenchymatous; branches few, normally lateral-intercalary . . . .
. . . . subg. *Leiocolea* K. Müll.

c. Perianth *distinctly plicate, gradually narrowed toward apex, never beaked*; oil-bodies smaller, 2–4 × 3–8 μ or less in size, *mostly 8–50 per cell*; cuticle smooth or very finely papillose; ventral stem sectors only 1–3 cells wide (or cells noncollenchymatous and oil-bodies 18–60 per cell); branches few to frequent, normally lateral-terminal . . . . *d.*

d. Stem with medulla homogeneous: not dorsiventrally differentiated, *free of mycorrhizae* or the ventral 1–3 layers of stem cells infested, the ventral medullary cells never distinctly smaller than the dorsal medullary cells; *leaf cells noncollenchymatous*: either thin-walled or equally thick-walled; leaves as wide as or wider than long; ♂ and ♀ *bracts generally more or less strongly dentate*
. . . . *e.*

e. Cells ± *thin-walled, relatively large* and usually 30–60 μ, *with numerous* (18–60) minute (ca. 2–5 μ) *highly refractive*, nearly homogeneous *oil-bodies per cell*; stems ca. 12 cells high, with the thin-walled cortical cells very similar to medullary cells; mycorrhizae usually absent or confined to the ventral cortical cells; *ventral stem sectors very wide* (usually 8–12 cells or more), occasionally producing underleaves; leaves very obliquely, nearly horizontally inserted, generally distinctly wider than long, often developing supplementary lobes or marginal teeth; capsule wall 3–5-stratose, the inner layer with semiannular thickenings
. . . . subg. *Massula* K. Müll.

e. Cells *thick-walled and small*, 20–30 μ, *with few* (5–15) *relatively large* (4–9 μ), *less refractive*, granular-botryoidal oil-bodies per cell, *the marginal cells usually lacking oil-bodies*; stems merely 7–10 cells high, the cortical cells ± strongly thick-walled; mycorrhizae usually present in the 1–3 ventral stem layers; ventral stem sectors very narrow (2–4 cells wide at most), lacking underleaves; leaves inserted at an oblique angle of about 45–50°; leaves subcircular, never developing supplementary lobes or teeth (except when gemmiparous); capsule wall bistratose, both inner and outer layers with radial (nodular) or incompletely semiannular thickenings . . . subg. *Isopaches* (Buch) Schust.

d. *Medulla distinctly dorsiventrally differentiated* into a dorsal band of large, hyaline cells and a ventral region of smaller and shorter cells, ± destroyed by mycorrhizal activity, at maturity 10–22 cells high or more; leaf cells collenchymatous, rarely scarcely so, never with equally thickened walls; leaves longer than wide to somewhat wider than long; ♂ and ♀ bracts with lobes not denticulate (except in *L. excisa*); oil-bodies 8–24 per cell, 3–4 × 5–8 μ, distinctly botryoidal or granular; ventral stem

sectors obsolete, usually 1–3 cells wide; capsule wall 3–5-stra-
tose, the inner layer with semiannular, the outer with nodular,
thickenings . . . . subg. *Lophozia* Dumort. emend. Schust.

4. Leaves uniformly bilobed, *with obtuse to blunt lobe apices*; leaf cells
noncollenchymatous, ± *equally thick-walled*, brownish; capsule wall
2-stratose; perianth smooth, inflated, not beaked at mouth, *usually
caducous*; ♀ bracts identical in size and form to leaves; rhizoids
very few, scattered; gemmae rare or none; stem 6–8(9) cells high,
nonmycorrhizal, with medulla undifferentiated: ventral stem
sectors narrow, 1–3 cells wide, lacking underleaves or with minute
ones on fertile shoots . . . . *Gymnocolea* Dumort.

3. Leaves vertically oriented, the *leaf insertion oblique ventrally, but ± transverse dorsally*, or directed toward the stem apex;[28] ventral stem sectors
merely 1–2 cells wide, unable to develop underleaves on vegetative
shoots; leaves 2-lobed (and medulla uniform) or 3-lobed (and medulla
dorsiventrally differentiated), generally more or less concave-canaliculate to complicate-canaliculate, the dorsal lobe usually lying over the
ventral . . . . 5.

5. *Leaves uniformly bilobed*, with dorsal lobe little smaller than ventral;
stem medulla homogeneous, lacking mycorrhizae or with them
confined to the peripheral 1–3 stem layers; plants erect or suberect
in growth, with rhizoids sparse; capsule wall 2–4-stratose; cortical
stem cells relatively short and mostly 1–2.5× as long as wide,
slightly wider than basal leaf cells . . . .
. . . . *Anastrophyllum* (Spr.) Schiffn.   a.

a. Cells of leaf middle *more or less equally thick-walled*, able to produce
only small or minute trigones; *basal cells not elongated*; normally
developed leaves with dorsal half of insertion *strictly transverse
and not decurrent*; leaves widest ± medially, dorsal base of leaf
not expanded, *extending merely to midline of stem*; intercalary
branching lateral, from leaf axils. Plants able to develop gemmae on suberect or erect specialized shoots with erect to appressed, more or less sheathing, reduced leaves; small species,
with stem merely 5–10(11) cells high in section; antheridial stalk
of a single row of cells; perianth slightly to distinctly flattened
above, oval in section, with a deep dorsal groove (at least when
immature). Capsule wall normally 2- or 2–3-stratose . . .
. . . . b.

---

[28] This does not strictly hold for *Eurylobus* and *Anastrophyllum*. Here the dorsal half of the
insertion may show an approach to the transverse, but the line of insertion is arched and distinctly decurrent *along the dorsal midline* (i.e., along the inner edges of the lateral merophytes of
the stem). Species of these taxa, however, differ from all groups under the first *3* (p. 233)
in the strongly concave leaves, either ± canaliculate or cupped, with the lobes often incurved.
They differ further from obliquely leaved taxa in the dilated dorsal base of the leaf extending
far beyond the stem midline. The leaf in oblique-leaved groups is oblique throughout and not
merely decurrent along the midline where the lateral merophytes meet: a condition fundamentally unlike that in *Eurylobus* and *Anastrophyllum* (which, however, is scarcely evident from
examination of lateral views of the stem).

b. Seta in section of many rows of cells; stem 10–11 cells high in section; gemmae largely 2–(3–4)-celled; ♀ bracts with entire lobes; leaves below inflorescences never with a basal tooth or lobe; lateral-terminal branching frequent . . . . subg. *Acantholobus* Schust.

b. Seta in section merely of 8(9–10) peripheral and 4 inner rows of cells; stem only 6–7 cells high; gemmae largely or entirely 1-celled; ♀ bracts with ± dentate lobes; leaves below inflorescences generally with a basal tooth or lobe; only intercalary branching . . subg. *Crossocalyx* Meyl. (= *Sphenolobus* Lindb.)

a. Cells with *distinct to bulging trigones*; *basal cells* (at least near leaf midline) usually *elongated*; leaves with dorsal part of insertion usually not wholly transverse, ± *distinctly decurrent* along dorsal stem midline (or basal cells ± sinuously thickened, elongated); leaves widest near or toward base, the *dorsal base of leaf expanded, extending beyond midline of stem*. Plants with gemmae (when developed) on normal shoots with ± normal, wide-spreading leaves; usually larger species, with stems (10)12–16 cells high or more; antheridial stalk usually biseriate; perianth terete, pluriplicate and lacking a discrete dorsal longitudinal groove; seta of many cell rows. Capsule wall 3- or 3–4(5)-stratose . . *c.*

c. Leaves much broader than long, cupped and handlike; leaf cells unable to produce large trigones; basal cells hardly elongated; ♀ bracts erect, sheathing perianth, the lobes more or less divided and dentate; dorsal leaf lobes more or less incurved. No gemmae . subg. *Eurylobus* Schust. [*A. saxicolus*]

c. Leaves as long as or (usually) distinctly longer than wide, not strongly cupped, but normally canaliculate, widest at base; leaf cells able to develop large, bulging, often yellowish to reddish trigones; ♀ bracts usually spreading away from perianth, the lobes not subdivided, entire (our taxa);[29] dorsal leaf lobes spreading or squarrose (in our species), or lying loosely over ventral leaf lobes . . . . *d.*

d. Leaves *not* clearly decurrent dorsally, inserted almost transversely (as in *Eremonotus*; but basal cells elongated, collenchymatous); only with lateral-terminal branching, as far as known; gemmae absent. Small plants, with reddish pigments and cylindrical perianths . . . subg. *Schizophyllum* Schust. [*A. sphenoloboides*]

d. Leaves *clearly decurrent* along dorsal midline;[30] usually with postical-intercalary branching developed (although lateral-terminal branches may occur, often almost to exclusion of postical branching) . . . . . subg. *Anastrophyllum* (Spr.) Schiffn. *e.*

e. Gemmae present; leaves hardly longer than broad; basal leaf cells little elongated, lacking intermediate thickenings; perianth

[29] Not so in many extraterritorial species.
[30] The decurrence along the antical midline of the axis is always distinct on mature plants; on weak, especially gemmiparous shoots, it may hardly be discernible.

obpyriform; bracts simply bifid, spreading . . . .
. . . . sect. *Isolobus* Schust. [*A. michauxii* only]

e. Gemmae consistently absent; leaves elongated, usually conspicuously so; basal leaf cells ± strongly elongated, often trabe-
culate, or with confluent coarse trigones and intermediate
thickenings, the lumina usually sinuous; perianth ovoid to
ovoid-cylindrical; bracts variable . . . .
. . sect. *Assimiles* Schust. [*A. assimile*, etc.]

5. *Leaves mostly 3-lobed* (exc. gemmiparous leaves) ;[31] stem with medulla
more or less dorsiventrally differentiated, the ventral 4–5 or more
medullary cell layers small-celled, becoming mycorrhizal and
brownish, but cortical cells free or nearly free of mycorrhizae;
plants decumbent to ascending in growth (exc. *Saccobasis*), with
rhizoids numerous, usually forming a dense ventral mat; capsule
wall 3–5-stratose; cortical stem cells ± strongly elongate, mostly
2.5–8 × as long as wide, averaging narrower than basal leaf cells;
basal leaf cells 1.5–4 × as long as wide; median leaf cells usually
obviously elongate . . . . *Tritomaria* Schiffn.   a.

   a. Leaves with dorsal end of line of insertion transverse; ventral
end of line of insertion oblique to transverse, but never decurrent;
dorsal end of line of insertion situated nearest stem base; leaves
± asymmetric, rarely nearly symmetric, with acute or apiculate
lobes; curvature of the shoot apex marked; rhizoids forming a
very dense ventral mat; stem with cortical cells mostly 2–3.5 ×
as long as wide . . . . subg. *Tritomaria* Schiffn.

   a. Dorsal portion of line of insertion running obliquely toward stem
apex (except for the slightly decurrent innermost edge); ventral
line obliquely directed toward stem apex (together with dorsal
line cutting off a loop or basal pocket for the leaf), but abruptly
decurrent, the entire, complex line of insertion thus strongly
undulate; leaves symmetric, shallowly trilobed at the truncate
apex, with obtuse lobes; nearly erect in growth, the shoot apex
lacking autonomous dorsal curvature; rhizoids scattered, long,
rather sparse; stem with cortical cells mostly 4–8 × as long as
wide, much longer than basal leaf cells . . . .
. . . . subg. *Saccobasis* (Buch) Schust.

2. *Leaves lobed 0.7–0.9 their length into 3–4* (on some leaves 2) *lobes*, which are
channeled and with *reflexed margins and sinuses;* perianth deeply sharply 7–9-
plicate to near base; no asexual reproduction; ventral stem sectors 10–16
cells wide or more, the ventral merophytes occupying 0.25 the stem periphery; leaf insertion (in ours) subtransverse; underleaves very large,
bifid with ciliate margins . . . . *Chandonanthus* Mitt.

1. *Perigynium* (marsupium) *present*, terminal, at right angles to the stem; perianth present, about as high as perigynium; leaves nearly horizontally
inserted, *subentire* (bilobed 0.05–0.08), *much wider than long;* large, often
purple underleaves present on sterile shoots, bilobed, ciliate . . . .
. . . . *Mesoptychia* Lindb.

---

[31] These often asymmetrically bifid with much smaller dorsal lobe.

Artificial Key to Genera and Subgenera of Lophoziaceae (of North America)

1. Leaves unable to produce marginal rhizoids; a perianth present . . . 2.
1. Some leaves producing marginal rhizoids; no perianth but a distinct rhizoidous perigynium at right angles to the axis; stem only 6–7 cells in height. Southern Appalachians . . . .
. . . . See *Acrobolbus* (Acrobolbaceae) (see Vol. III)
2. Ventral stem sectors (merophytes) broad, at least 4–8 cells, often 10–16 cells, wide, usually *able to develop distinct underleaves*; leaves often in part or entirely 3–4-lobed; stem with medulla uniform, lacking a small-celled ventral band . . . . 3.
2. Ventral stem sectors narrow, mostly 1–2 (rarely 3–4) cells wide (the postical leaf bases thus approaching the imaginary postical midline of the stem), *never able to produce underleaves*; leaves 2- or at most 3-lobed . . . . . 8.
3. Leaves 2–4-lobed for ⅙–⅘ their length; leaf margins never strongly reflexed, the leaves concave or flat, rarely slightly convex; with a perianth (but never a marsupium) . . . . 4.
3. Leaves subentire, divided by a shallow sinus to ca. ¹⁄₁₀–⅛ their length into two broad, asymmetrical lobes; leaf margins more or less reflexed, the adaxial leaf surface at least in part convex; with large, bifid, and ciliate underleaves; perigynium (marsupium) and perianth both present . . .
. . . . *Mesoptychia* (p. 799)
4. Leaves all (or largely) 2-lobed, occasional ones irregularly lobed or with dentate margins, their insertion strongly oblique, horizontally spreading; oil-bodies specialized (either minute and homogeneous and 25–60 per cell, or very large and only 2–6 per cell); leaf cells generally large and 25–50 μ in leaf lobes . . . . 5.
4. Leaves all or mostly 3–4-lobed (in *L. kunzeana* largely 2-lobed; if 2-lobed, concave and with erect or incurved lobes);[32] if leaves subhorizontally inserted, they are almost uniformly 4-lobed; oil-bodies moderate in size (ca. 4 × 8–10 μ), finely botryoidal, 5–20 per cell; *usually with bifid or ciliate underleaves*; cells in leaf lobes mostly 13–25 μ, with small to bulging trigones . . . . 6.
5. Cells of leaves ± collenchymatous, with distinct trigones and a papillose cuticle (exc. *L. badensis*); oil-bodies ± granular, large, 2–6 per cell; usually with underleaves; perianth smooth, beaked; cortical stem cells generally ± thick-walled; stems little flattened; leaves uniformly 2-lobed; branches usually intercalary . . . . *Lophozia* subg. *Leiocolea* (p. 360)
5. Cells of leaves thin-walled, lacking discrete trigones, the cuticle normally smooth; oil-bodies minute, homogeneous, highly refractive, 2–4 μ usually, generally 25–60 per cell; underleaves normally absent (occasional in *L. obtusa*); perianth plicate, not beaked; cortical stem cells thin-walled, like medullary; leaves 2–3–4-lobed in some species or with scattered marginal teeth; branches usually terminal . . . . *Lophozia* subg. *Massula* (p. 419)
6. Leaves subhorizontal, inserted at an angle of 15–30° to axis, widely spreading, flat, varying to slightly concave or convex, 4-lobed ¼–½ their length;

[32] *Lophozia quadriloba* produces a rare, *Cephaloziella*-like extreme (fo. *paradoxa*) with almost exclusively bifid leaves, divided to 0.65–0.75 their length. The vertical or subvertical leaf orientation, and the exclusively intercalary branching are diagnostic.

postical leaf bases with cilia (when present) *of elongate cells* . . . . . .
. . . . *Lophozia* subg. *Barbilophozia* (p. 332)

6. Leaves concave, with erect or incurved lobes, not spreading, inserted at an angle of 50–80° to axis, predominantly 3-lobed (or partly 2-lobed) to 4-lobed; postical leaf bases with cilia or teeth (when present) *of isodiametric cells* . . . . 7.

7. Leaves divided ¼–⅔, the lobes with margins not strongly reflexed, not channeled; leaves (2)3-lobed (exc. *L. quadriloba*) . . . .
. . . . *Lophozia* subg. *Orthocaulis* (p. 261)

7. Leaves divided ⅘ into 4 (rarely 2–3) lobes, the narrow lobes with reflexed margins and sinuses, lobes thus channeled and adaxially convex . . . .
. . . . *Chandonanthus* (p. 241)

8. Leaves all 2-lobed (only very occasional ones with ventral lobe divided, thus with a larger dorsal lobe) . . . . 9.

8. Leaves all or mostly 3-lobed, concave and more or less canaliculate or even complicate-concave . . . . 18.

9. Dorsal end of leaf insertion oblique or decurrent (or trigones conspicuous)
. . . . 10.

9. Dorsal half of insertion transverse; antical leaf base not extended beyond stem middle; perianth with a deep dorsal groove, often extending to near base on immature perianths; leaf cells small (10–15 μ wide), mostly thick-walled; gemmae on erect, filiform shoots with $\pm$ reduced leaves; stem only 6–11 cells high . . . . 17.

10. Leaf cells equally (often strongly) thick-walled, sometimes appearing guttulate; stems merely 6–8(10) cells high, with medulla never dorsiventrally differentiated; leaves $\pm$ circular, mostly as wide as long . . . 11.

10. Leaf cells $\pm$ thin-walled but with $\pm$ distinct trigones; stems (9)10–22 cells high . . . . 12.

11. Leaf lobes blunt to rounded; gemmae absent, but generally with caducous perianths; rhizoids few; usually $\pm$ deep brown . . *Gymnocolea* (p. 783)

11. Leaf lobes acute; gemmae usually present, perianths not caducous, plicate; rhizoids numerous; $\pm$ green . . . . . *Lophozia* subg. *Isopaches* (p. 477)

12. Leaves spreading, usually little concave, rarely canaliculate, the lobes $\pm$ spreading (the entire distal half of the leaf in $\pm$ the same plane); stems with medulla dorsiventrally differentiated, a ventral region of small cells mycorrhizal; decumbent in growth usually . . . .
. . . . *Lophozia* subg. *Lophozia* (p. 490)

12. Leaves strongly concave or concave-canaliculate, the ventral lobe often $\pm$ incurved; medulla uniform, nonmycorrhizal, lacking ventral band of small cells; erect or suberect in growth . . . . 13.

13. Leaves subhemispherically concave, conspicuously broader than long; median and basal leaf cells with trigones, but these neither sharply defined nor strongly bulging; basal cells not elongated . . . . 14.

13. Leaves as long as to much longer than wide, often canaliculate but never cupped; trigones always strongly developed, sometimes confluent. . . 15.

14. Gemmae occasional; leaves subsymmetrically 0.2–0.35 bilobed, so strongly cupped that the leaf tears in flattening . . . .
. . . . *Lophozia* subg. *Orthocaulis* (p. 261)

14. Gemmae lacking; leaves asymmetrically 0.35–0.5 bilobed, both larger ventral and smaller dorsal lobe incurved, the plant with a *Scapania*-like facies, the leaves flattening without tearing . . . .
. . . . *Anastrophyllum* subg. *Eurylobus* (p. 733)

15. Basal leaf cells not elongated, never with sinuous walls and confluent thickenings; with gemmae; leaves hardly longer than broad, the ventral lobe of mature leaves incurved. Non-arctic . . . .
. . . . *Anastrophyllum* subg. *Anastrophyllum* (p. 716)

15. Basal leaf cells ± strongly elongated (length-width ratio at least partly 2–3:1), with strong, often confluent trigones, occasional intermediate thickenings, the walls ± sinuous; never with gemmae; leaves typically more than 1.2× as long as wide, both lobes spreading. Our species arctic . . . . 16.

16. Leaves not or very feebly decurrent dorsally, dorsal half of insertion transverse; only with lateral terminal branching (as far as known); small plants developing reddish pigments . . . .
. . . . *Anastrophyllum* subg. *Schizophyllum* (p. 739)

16. Leaves clearly decurrent dorsally and obliquely inserted; usually with postical-intercalary branching; larger plants, ours fuscous . . . .
. . . . *Anastrophyllum* subg. *Anastrophyllum* (p. 716)

17. Plants minute, 4–8 mm high; stems only 6–7 cells high; gemmae 1-celled, about 10 × 10 μ. On bark or wood . . . .
. . . . *Anastrophyllum* subg. *Crossocalyx* (p. 769)

17. Plants larger, 1–3 cm long; stems (9)10–11 cells high; gemmae largely 1–2(3–4)-celled, 10 × 15 μ or larger. On rocks . . . .
. . . . *Anastrophyllum* subg. *Acantholobus* (p. 755)

18. Dorsal cortical stem cells 4–8× as long as wide; dorsal half of leaf with line of insertion arched toward stem apex; leaves equally 3-lobed for ⅒–⅙ their length. . . . . *Tritomaria* subg. *Saccobasis* (p. 693)

18. Dorsal cortical stem cells 1.5–3.5(–4)× as long as wide; dorsal half of line of insertion of leaves transverse or decurrent toward stem base; leaves mostly decidedly unequally 3-lobed for ¼–½ their length . . . . . 19.

19. Leaves (when 3-lobed) with dorsal lobe ± shorter and smaller; stem with a distinct ventral medullary region of small, brownish cells; rhizoids forming a very dense ventral mat; dorsal end of leaf insertion transverse.
. . . . *Tritomaria* subg. *Tritomaria* (p. 642)

19. Leaves (when 3-lobed) with dorsal lobe obviously larger and longer; stem with medulla homogeneous, free of mycorrhizae and not distinctly dorsiventrally differentiated; rhizoids usually sparse; dorsal portion of line of insertion obviously oblique . . . . . *Lophozia* subg. *Orthocaulis* (p. 261)

CHANDONANTHUS Mitt.

[Figs. 137, 138]

Chandonanthus Mitt., *in* J. D. Hooker, Hdb. New Zealand Fl. 2:750, 1867, *p.p.*; Lindberg, Musci Scand., p. 5, 1879; K. Müller, Rabenh. Krypt.–Fl. 6(2):308, 1914 [ed. 3, 6(1):619, 1954]; Macvicar, Studs. Hdb. Brit. Hep. ed. 2:342, 1926; Schuster, Jour. Hattori Bot. Lab. no. 23[1960]:204, 1961; Kitagawa, *ibid.*, no. 28:254, 1965.
Blepharostoma Lindb., Musci Asiae Bor., p. 28, 1889, *p.p.*; not of Dumort., 1835.

Temnoma Howe, Bull. N. Y. Bot. Gard. 2:104, 1901; Frye & Clark, Univ. Wash. Publ. Biol. 6(2):413, 1945; Schuster, Amer. Midl. Nat. 45(1):75, 1951 (not of Mitten, 1864, *teste* K. Müller, 1954, *in* 1951–58).

Robust, yellowish or reddish brown to deep brown, *rigid when dry*, simple or sparingly branched, laxly caespitose or forming intertwined patches, erect or procumbent (in epiphytic taxa sometimes loosely pendulous); branches lateral-intercalary and ventral-intercalary, and/or lateral-terminal and of the *Frullania* type. Stem with moderate to strong differentiation into a small-celled, thick-walled cortex and firm but more leptodermous, hyaline-walled medulla. Rhizoids few, *scattered* over ventral face of stem. Leaves imbricate, either *almost transversely inserted* (our species) or *slightly to quite distinctly succubous*, semiamplexicaul, *deeply 2–3–4-lobed* (commonly to within 0.1–0.2 of the base; on weak or juvenile stems rarely largely 2-lobed), *the lobes canaliculate*, their margins ± recurved, often falcate, entire or more *usually armed at base* (at least at postical base) *with teeth or laciniae;* lobes equal (or, if unequal, the dorsal one[s] largest). *Underleaves large*, arising from broad ventral merophytes, nearly or quite similar in length to lateral leaves, deeply lobed like these and armed at base with teeth or cilia, but *merely bilobed*. Cells small, with obscure to distinct (in subg. *Chandonanthus*, salient) trigones or with somewhat thickened walls; oil-bodies botryoidal, of delicate or coarse spherules, several (usually 2–6) per cell. *Without asexual reproduction.*

Dioecious. Androecia intercalary, bracts similar to leaves (but less deeply divided; somewhat concave at base); antheridia 2–3, among laminar paraphyses; underleaves without antheridia. Gynoecia with single subfloral innovations; bracts and bracteoles nearly identical with leaves and underleaves, less deeply lobed; perianths large, inflated, longly emergent, ovoid-cylindrical, *strongly and deeply pluriplicate, essentially to the base*, the mouth *slightly contracted* and ciliate.

Type. Chandonanthus squarrosus (Hook.) Mitt., *in* J. D. Hooker, Hdb. New Zealand Fl., 2:753, 1867; see Fig. 137:12–14.

Chandonanthus is largely antipodal in distribution. Of 11 species listed by Stephani (1909, 1922, *in* 1898–1924), *C. pusillus* from Japan is regarded as a synonym of *C. setiformis* (Hattori, 1952a), although both Schuster (1961a) and Kitagawa (1965) regard it as distinct. Two species (*C. birmensis* and *C. hirtellus*) are widespread from Indo-Malaya northward to the Himalayas and Japan and westward to Madagascar; *C. hirtellus* also occurs in Polynesia, E. and W. Africa, Australia and British Columbia. Other taxa are almost all highly restricted in range and occur in antipodal regions, northward as far as the equatorial regions of Africa. Szweykowski (1956) gives a distribution map of all the species. Only the following species and two close allies, whose position in *Chandonanthus* has been challenged (Howe, 1901a), occur exclusively in the Northern Hemisphere, where they are confined to a circumpolar band radiating south from the imperfectly glaciated regions peripheral to the polar sea.

Chandonanthus has been generally placed in the "Ptilidiaceae" *s. lat.* Buch (1933a), employing the name *Temnoma* and considering only the circumpolar "*T. setiforme*," pointed out that the genus was more closely allied to the Lophoziaceae, specifically to such species as *Lophozia* (*Orthocaulis*) *quadriloba*. This disposition has been generally accepted in recent years.[33] It is warranted on the basis of the succubous leaf insertion and the scattered position of the rhizoids (Fig. 138: 10–11), peculiarities previously pointed out for the genus (Schuster, 1961a). Leaf insertion varies from very slightly succubous (5–15° from transverse), as in *C. setiformis* (Fig. 138:7, 10) to quite strongly succubous (40–45° from transverse) as in the widespread *C. birmensis*. Leaves vary greatly in the genus as it is now circumscribed. In the generitype they are equally bilobed to the middle as may also be the case in weak phases of *C. setiformis*. In normal *C. birmensis* the leaves are strongly unequally trifid to near the base, with the antical lobe by far the largest. This is also the case in the frequent instances when *C. setiformis* produces trifid leaves (Fig. 138:7), but leaves of *C. setiformis* and several other species are, when well developed, subequally quadrifid.

Mitten (1867) established both the genus *Temnoma* and the genus *Chandonanthus*. Our circumpolar species is clearly referable to the latter (diagnosed by Mitten as having the "perianth tubular, many-plicate; mouth open; stems erect or ascending"), rather than to *Temnoma* (diagnosed by Mitten as having the "perianth above 3-gonous, truncate"). Consequently the practice in America, dating to Howe (1901a), of referring *C. setiformis* to *Temnoma* is untenable. Although Stephani synonymized Mitten's genus *Temnoma* with *Blepharostoma*, the former represents a well-defined genus, only distantly allied to *Blepharostoma*, belonging in the Herbertinae (see Schuster, 1966b).

Chandonanthus, ecologically considered, consists of several heterogeneous elements. The generitype, *C. squarrosus* is an abundant epiphyte in New Zealand, often festooning twigs and branches; *C. hirtellus* and *C. birmensis* may occur on rocks but are also common on humus or logs or at the bases of trees. By contrast, *C. setiformis* and *C. pusillus* are found on rocks or on thin soil over rocks, chiefly at high altitudes or latitudes, usually well

---

[33] Ironically, although Buch is correct in placing "*Temnoma*" *setiforme* in the Lophoziaceae, his attempt to place *Temnoma* in this family is indefensible. Stem anatomy, the restricted rhizoid-initials, and the trigonous perianth of *Temnoma s. str.*, as typified by *T. pulchellum*, preclude *Temnoma* from being placed in the Lophoziaceae (see Schuster, 1961a, 1966b).

above the tree line. None of the species appears to occur in calcareous regions. The genus is, ecologically, rather primitive.

Chandonanthus also shows some primitive morphological features: (*a*) branching is polymorphous, with both lateral- and ventral-intercalary branching, as well as lateral-terminal, *Frullania*-type branches produced; (*b*) asexual reproduction is lacking; (*c*) all taxa are dioecious; (*d*) the capsule wall is massive; (*e*) leaves show much polymorphism as regards insertion and number of lobes. In the last respect, *Chandonanthus* shows an approach to *Lophozia* subg. *Orthocaulis*. From this (and other genera of Lophoziaceae) it is at once distinct in (1) the deeply and longly pluriplicate perianth; (2) the very deeply lobed leaves. It is probable that the first is an advanced, the second a primitive, feature. Development of conspicuous cilia of the leaf bases and the extreme development of collenchyma (in subg. *Chandonanthus*) appear to be equally advanced features. Thus, the genus is a peculiar mixture of old and new features, which isolate it from all other genera of Lophoziaceae.

According to Andrews (1948), the genus *Chandonanthus sensu* Stephani represents "an unnatural aggregation of forms." It has been divided (Schuster, 1961a, pp. 204–205) into two subgenera, of which subg. *Chandonanthus* (with sectios Squarrosi Schust., Piliferi Schust., Hirtelli Schust.) is chiefly antipodal in range, extending northward to the Philippines, Japan, and the Himalaya, with *C. hirtellus* (Web.) Mitt. as a disjunct in British Columbia (Schofield, 1968). The other subgenus, *Tetralophozia* Schust., includes three holarctic species (separated by key in Schuster, 1961a): *C. setiformis*, *C. pusillus* [Japan, Formosa], and *C. filiformis* [China-Himalaya], and a fourth species *C. quadrifidus* Steph., from Africa (Kitagawa, 1965).

These subgenera are separable by the following key:

a. Leaves of mature shoots typically bifid or asymmetrically trifid, rarely quadrifid, obliquely inserted, antically usually $\pm$ secund or dorsally squarrose; cells with coarse, nodose trigones, the intervening walls $\pm$ thin (Fig. 137); shoot apices somewhat arched away from substrate; stem smooth or often paraphyllose. Tropical-antipodal . . . . subg. *Chandonanthus*
a. Leaves when mature symmetrically quadrifid, vertically oriented, subtransversely inserted, not secund or squarrose, the leaf lobes erect or suberect; cells small and firm, walls $\pm$ thickened, trigones obscure to ill defined usually (Fig. 138:1); shoot apices straight; stem smooth. Arctic and alpine; not in Antipodes . . . . subg. *Tetralophozia*

CHANDONANTHUS subg. TETRALOPHOZIA Schust.

Chandonanthus subg. *Tetralophozia* Schust., Jour. Hattori Bot. Lab. no. 23 [1960]:206, 1961; Kitagawa, *ibid.*, no. 28:259, 1965.

Filiform, suberect to loosely procumbent, forming large, loose, yellow-brown to reddish brown carpets. *Stems with apices straight,* not autonomously arched,

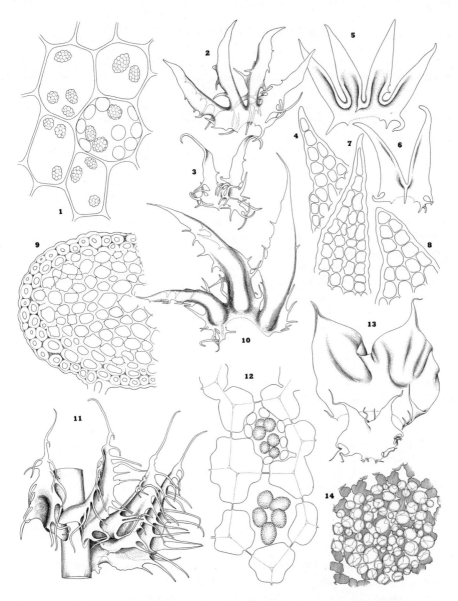

FIG. 137. *Chandonanthus:* intrageneric variations. (1–8, subg. *Tetralophozia;* 9–14, subg. *Chandonanthus*). (1, 8) *C. setiformis:* 1, median leaf cells with oil-bodies and, middle cell, chloroplasts (×800); 8, apex of leaf lobe (×295). (2–4) *C. filiformis:* 2, leaf (×50); 3, underleaf (×50); 4, leaf lobe apex (×295). (5–7) *C. pusillus:* 5, leaf (×90); 6, underleaf (×90); 7, leaf lobe apex (×295). (9–10) *C. hirtellus:* 9, stem cross section (×190), 10, leaf, adaxial view (×36); (11) *C. piliferus,* leaf, lateral view, *in situ* (×40). (12–14) *C. squarrosus,* the generitype: 12, median cells with oil-bodies (×925); 13, leaf and underleaf (×12); 14, median cells (×305). [1, *RMS & KD 66-005,* Tupilak I., W. Greenland; 2–4, China (*Delavay;* type); 5–7, Mt. Komadagake, Kai Prov., Japan (*Tamura, 1903,* type); 8, *RMS 1301,* Mt. Washington, N.H.; 9–10, *Hodgson No. 2937,* Malaya; 11, from type, Farlow Herb.; 12, *RMS 49747b,* Haast Pass, N.Z.; 13–14, Dusky Sd., N.Z. (*Menzies;* type).]

simple or with a *very few lateral- and/or postical-intercalary branches*, occasionally with *Frullania*-type terminal branches; stems rigid, firm, (3)5–14 cm long, with 2(3) rows of *thick-walled, yellowish to brownish cortical cells*, which are short-rectangular (ca. 2–4× as long as broad), gradually grading into the colorless medullary cells, which are collenchymatous and 1–3(4)× the diam. of the cortical cells. Rhizoids rather few, colorless, *scattered over the ventral merophytes*. Leaves usually closely imbricate, semiamplexicaul, the *shoots terete. Leaves transversely oriented*, the entire base strongly spreading, *but the lobes erect, fingerlike*, the leaves thus nearly handlike; insertion 10–15° (20°) succubous, hardly to slightly acroscopically arched; leaves on mature shoots *symmetrically 4-lobed to within 0.15–0.25 of base; lobes subequal*, ovate-triangular to lanceolate, their margins and especially the sinuses *strongly reflexed*, the lobes thus channeled and adaxially convex, near their bases commonly with 1-several sharp teeth or tapering cilia, the larger ± reflexed or hamate, the outer leaf margins near base similarly, usually more strongly, armed (on weak phases entire!). Underleaves large, transverse, 0.6–0.8 the length and ca. *0.3–0.4 the area of the lateral leaves, bifid* for 0.55–0.85 their length, near base (and often near bases of lobes) with a few cilia or teeth. Cells small, 10–15 μ apically, in middle 10–13 or 13–18 μ wide × 12–18 or 18–23 μ long, ± *thick-walled, the lumen rounded*, without sharply defined trigones or the large but ill-defined trigones confluent; oil-bodies 2–4 per cell, spherical to ovoid, 3 μ to 3 × 4 μ, formed of distinct spherules. Asexual reproduction lacking.

Dioecious. Andreocia becoming intercalary; bracts less deeply lobed than leaves. Gynoecia terminal on main axes, but often appearing lateral because of development of subfloral innovations. Bracts larger than leaves, the lobes narrower, sharply pointed to acuminate, coarsely dentate below apex and near base, otherwise similar to leaves. Perianth to ⅓–½ emergent, firm and rather short, ovoid-conoidal, gradually contracted to a rather narrow mouth, *deeply pluriplicate to near the base*; cilia of mouth scattered, of short, hardly elongated thick-walled cells. Capsule ovoid; wall pluristratose. Spores reddish brown, 14–15 μ; elaters 8–9 μ in diam.

Type. Jungermannia setiformis Ehrh., Beitr. z. Naturk. 3:80, 1788.

In regard to the subg. *Tetralophozia*, one is tempted to agree with Kitagawa (1965), who regards it as "a very natural taxon," that "is so different from *Chandonanthus* that it may be possible to raise it even to a generic rank." Only *C. piliferus*, with asymmetrically quadrifid leaves that are almost transversely inserted, seems to form a transition—and, in my opinion, renders a full generic cleavage inadvisable (see Fig. 137:11).

When *Tetralophozia* was originally described, only lateral-intercalary branches had been seen; with study of more abundant materials, I have also found ventral-intercalary and *Frullania*-type lateral-terminal branches—two branching types which occur also in subg. *Chandonanthus*.

CHANDONANTHUS SETIFORMIS (*Ehrh.*) *Lindb.*
[Figs. 137:1,8, 138]

Jungermannia setiformis Ehrh., Beitr. Naturk. 3:80, 1788; Hooker, Brit. Jungerm., pl. 20, 1816.
Jungermannia concatenata Smith, *in* Linnaeus, Fl. Lapp. ed. 2, p. 343, 1792.
Anthelia filum Dumort., Rec. d'Obs., p. 18, 1835.
Anthelia setiformis Dumort., *ibid.*, p. 18, 1835; *ibid.*, Hep. Eur., p. 98, 1874.
Chandonanthus setiformis Lindb., Musci Scand., p. 5, 1879; K. Müller, Rabenh. Krypt.–Fl.
    6(2):310, fig. 93, 1914 [ed. 3, 6(1):621, fig. 185, 1954]; Macvicar, Studs. Hdb. Brit. Hep.
    ed. 2:342, figs. 1–4, 1926; Meylan, Beitr. Krypt.–Fl. Schweiz 6(1):247, fig. 171, 1924.
Temnoma setiforme Howe, Bull. N. Y. Bot. Gard. 2:104, 1901; Schuster, Amer. Midl. Nat.
    45(1):75, pl. 4, 1951.

Usually in *pure, deep, loose, interwoven masses* or tufts, spongy when wet, rarely scattered among other bryophytes, *shiny, deep olive-green to yellowish brown, robust,* mature shoots *virtually terete,* (550–650)700–1100 μ wide × 2–6(12) cm long, *filiform.* Stems usually arching and ascending to *erect, rigid, brittle when dry,* almost unbranched (the few branches lateral intercalary and/or postical, occasionally also terminal-lateral), 200–250 μ in diam. Stem cross section of cells of nearly equal diam., the cortical thick-walled. *Rhizoids very few or almost none,* nearly confined to shoot bases. Leaves inserted by a slightly acroscopically arched, *almost transverse* line, at ca. 75–80° with longitudinal axis of stem, semi-amplicaul, slightly to rather densely imbricate, not decurrent, handlike, obtrapezoidal-reniform, ca. 1.3–1.6× as wide as long, *3–4-lobed for 0.7–0.85 their length, the short, undivided portion widely spreading, the lobes erect from this spreading base; lobes broadly ovate-lanceolate to oblong-lanceolate,* ca. 250–300(400) μ wide × 450–550 μ long, averaging less than 2× as long as wide, erect, acute or blunt at apex (terminated by 1 or rarely 2 single, nonelongated cells), *strongly channeled, with the margins and sinuses between them broadly reflexed;* margins of leaves entire above, but both *antical and postical bases, and bases of lobes* (in the reflexed sinuses) *coarsely dentate,* teeth polymorphous, usually broad-based, the longer often falcate and reflexed; sinuses sharp, usually acute (but when strongly gibbous sometimes rounded). Cells of leaf apices and tips ca. 10–15(18) μ, of leaf-lobe middle 12–15(18–20) × (15)18–23(25) μ, scarcely larger near base, *walls somewhat thickened, never sigmoid and undulate, obscuring the small or rather large trigones; lumen ± rounded;* cuticle thick, *smooth* or faintly papillose; oil-bodies 2–4 per cell, formed of small spherules, spherical to ovoid, 3–4 μ to 3–4(5) × 4–6(7) μ. *Underleaves extremely large,* ca. 0.65–0.8 the length of leaves, oblong to ovate, with marginal cilia 300–350(450) μ wide × 320–460(500) μ long, bifid for 0.6–0.8 their length, the sinus acute (often with a reflexed, narrowly rounded base), *their basess pinose-dentate to ciliate,* the lobes often falcate, acute to acuminate.

Almost always sterile. Androecia intercalary; bracts like leaves, less deeply divided, somewhat more concave at base; 2–3-androus, the antheridia among paraphyses. ♂ Bracteole large, like underleaves, without antheridia. Gynoecia rarely produced, with solitary subfloral innovations. Bracts like leaves but larger, slightly less deeply 4–5-lobed, the lobes narrow and lanceolate, longly acute to acuminate, both margins and bases of lobes spinose-ciliate to sharply

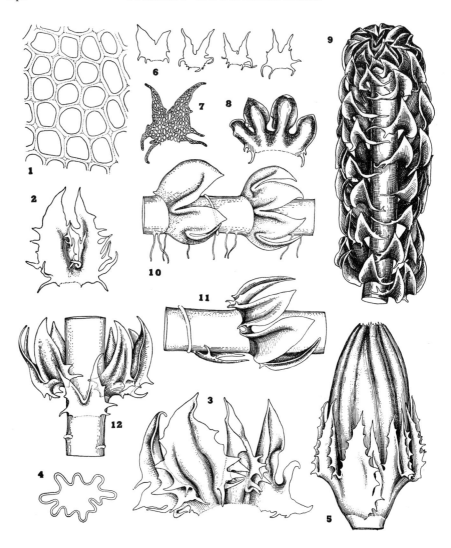

FIG. 138. *Chandonanthus* (*Tetralophozia*) *setiformis*. (1) Median cells (×350). (2) Under-leaf (×60). (3) Mature leaf (×60). (4) Cross section through perianth (×25). (5) Somewhat immature perianth and bracts (×25). (6) Underleaves (×29). (7) Under-leaf (×50). (8) Leaf, of simple type, adaxial aspect (×29). (9) Shoot apex, antical aspect (×29). (10) Shoot sector, lateral aspect, with two leaves and scattered rhizoids (×44). (11) Same, with leaf, underleaf, and leaf insertion (×48). (12) Shoot sector, ventral aspect (×48). [1–5, from Schiffner's Hep. Eur. Exsic. no. 454; 6–12, from *RMS, 1944*, Mt. Washington, N.H.]

dentate. Bracteole free, bilobed ca. 0.5 its length, at base armed like bracts, the acuminate lobes ciliate-dentate at base. Perianth oblong to oblong-ovoid, 0.4–0.6 emergent, *deeply and conspicuously 7–10-plicate almost to base*, contracted to mouth, the mouth lobulate and with cilia; cilia 3–6 cells long, with occasional 1–2-celled teeth interspersed. Elaters 8–9 μ in diam.; spores 13–15 μ.

Type. Harz Mts., Germany (*Ehrhart*).

Distribution

A widely dispersed arctic-alpine species, imperfectly circumpolar in range. A characteristic and frequent species of relatively dry and rocky, acidic sites throughout all but the northernmost edge of the Tundra. Müller (1951–58, p. 622) regards it as a relict of earlier geological ages; it appears to range southward from the imperfectly glaciated areas at the northern periphery of the Northern Hemisphere. Like almost all such "old" species, the plant is a pronounced oxylophyte, has unusually small cells, is dioecious, and extremely rarely bears capsules (these have been found only once or twice). In Europe widespread in the north, from Spitzbergen (there to a northern limit of 80°40′ N.) and throughout northern Scandinavia (Norway, Sweden, Finland), southward to the higher alpine peaks of Scotland, recurring in various alpine (and rarely subalpine) sites in Germany (Harz, Iser Mts.; Thüringer Forest; Black Forest; near Bonn, etc.), Austria, France (the Ardennes), the Tatra Mts., and (*fide* Nees) the Carpathians; also reputedly in the Italian Tyrol. Extending, in in the east, into Asia (Siberia, north to Novaya Zemlya, 72° N., on the Arctic Sea). Also in Iceland. Reported from the Sikkim-Himalaya (Szweykowski, 1956, fig. 2) and Japan (Hattori, 1948, 1952), but these reports are based on other, related taxa (Schuster, 1961a; Kitagawa, 1965).

Transcontinental at high latitudes, in the West occurring from Alaska (Sherrard, 1957; Evans, 1900; Persson, 1952; Persson and Weber, 1958) and the Yukon southward to British Columbia (but surprisingly not known from alpine situations in the Rocky Mts. to the south). In our area as follows:

NW. GREENLAND. Widespread in Inglefield Bay: Siorapaluk; Kânâk (*RMS 45645, 45625, 45626, 46152*, etc.); Kekertat, near Heilprin Gletscher (*RMS 46093, 46029*); Kangerdlugssuak (*RMS 45778, 45962, 45961a,b*), from 77°22′ to 77°40′ N. E. GREENLAND. Scoresby Sund (*Hagerup, 1924!*); Danmarks Ø, Scoresby Sund (*Hartz, 1892!*); Vahl's Fjord, 66°22′ N. (Jensen, 1897). W. GREENLAND. Ritenbenk, Jakobshavn, Godhavn, Tasiusak, Autlatsivik Fjord; Sarfanguak, Praestefjeldet near Holstensborg; Alangua, Maneetsok; Ivigtut; Pakitsok, Ameralik; Tunugdliarfik (all from Lange & Jensen, 1887). Also seen are two specimens, presumably from W. Greenland, without loc., collected by the "Kane Arctic Exped." and the "Peary Exped., 1896";

the first specimen may be the one attributed to Kane, from Fiskefjord, by Lange & Jensen (1887, p. 100). Godhavn, on Disko I.; Tupilak I., near Egedesminde; S. end of Arveprinsens I., SE. of Jacobshavn, Disko Bay, and elsewhere in W. Greenland (*RMS 66-243, 66-194, 66-202, 66-005, 66-001*).   General in s. GREENLAND.

ELLESMERE I. Bedford Pim I. (Bryhn, 1906).   DEVON I. REGION. Philpots I. (Dickie, 1871)   BAFFIN I. Pond Inlet, N. Baffin (Polunin, 1947); Clyde, Exeter Sd., and Pangnirtung, all in C. Baffin (Harmsen & Seidenfaden, 1932; Polunin, 1947); Dorset, S. Baffin (Polunin, 1947); Lake Harbour (*Polunin 51*!, *52*!).   MELVILLE PENINSULA. Repulse Bay (Steere, 1941).   NORTHWEST TERRITORY. Tukarak I., Belcher Isls., "on arkose" (Schuster, 1951).   LABRADOR. Hebron (*Polunin 1052b-1*!); Venison Pickle (*Waghorne 9*!); Chidley Pen.; W. side of Knob L., SE. of Slimy L. (*Harper 3353*!).   HUDSON BAY REGION. Southampton I. (Steere, 1941).

ONTARIO. Ouimet Canyon, 10 mi N. of Dorion Station, Thunder Bay (Williams & Cain, 1959).   QUEBEC. Wakeham Bay (*Polunin, 1947*!); The Narrows, Cairn I., Richmond Gulf (Schuster, 1951); near mouth of Great Whale R. (Schuster, 1951); Mont Albert (*Macoun 1892*!, with *Anastrophyllum saxicolus*) and Mont la Table, Gaspé (*Lepage 3920*!; *Collins 4472-B*!, *4096*!); Port Harrison (*Taylor 75*!).   NEWFOUND-LAND. Big Bay, Rencontre W., S. coast; Kitty's Brook and Gaff Topsail, C. and NE. Nfdl. (Buch & Tuomikoski, 1955).

MAINE. Pomola, the "Knife Edge," Baxter Peak, Hamlin Ridge, summit of Cathedral Trail, and elsewhere on Mt. Katahdin (*RMS 15806, 15964, 15930, 17010, 17004, 15814, 17001, 32955a*, etc.; Lorenz, 1917); W. peak, Mt. Bigelow, Dead R. (*Collins, 1896*!); Augusta (*Scribner, 1875*!).   NEW HAMPSHIRE. Mt. Washington (*July 1851, Herb. James*!; *RMS*; Underwood & Cook, 1889); "on top of Mt. Washington, White Hills," 1843 (*s. coll.*!); Mt. Lafayette (*Evans, 1908*!; *RMS*); Mt. Clay (*Evans, 1890*!); Mt. Monroe (*RMS*); Mt. Adams (*Evans, 1917*!); Kings Ravine, Coos Co. (*Evans, 1917*!); "White Mts." (Oakes; Austin, Hep. Bor.-Amer. Exsic. no. 49).   VERMONT. Mt. Mansfield (Evans, 1913; *RMS 43869, 43812*, etc.); Smuggler's Notch, Mt. Mansfield (*Dutton 1641*!).   NEW YORK. Top of Mt. MacIntyre, Adirondack Mts. (*Peck*! NY); ice caves, E. shore of Lower Ausable L. near S. end, Keene, Essex Co., 2000 ft (N. G. Miller, 1966).

An abundant and characteristic species, above treeline, in the highest New England mountains, often occurring in fine, pure patches. The Ontario report is of a disjunct population, but from an area where many other arctic-alpine taxa are known (Schuster, 1958b).

Ecology

Apparently lacking or very rare in regions of basic rock outcrops, but abundant throughout most of the Arctic wherever rather dry, exposed rock faces and ledges occur, very often on scree slopes, and common in similar sites in alpine situations further southward. Although often on rather dry granitic rocks, the species rarely occurs under as xeric and exposed conditions as are frequented by our species of *Gymnomitrion*. In the mountains, above treeline, most frequently in shaded or sheltered recesses, or near the bases of large boulders or ledges, where a little soil— and moisture—are usually present. Here often in large, pure mats, at other times slightly admixed with various mosses (*Rhacomitrium, Andreaea*) more rarely with *Gymnomitrion concinnatum* or *G. corallioides*, occasionally with *Lophozia atlantica*, *L. alpestris*, *L. attenuata*, *L. kunzeana*, *Anastrophyllum saxicolus* and *A. minutum*. Less frequently with *Lophozia ventricosa*, *L. longidens*, *Mylia taylori*, *Ptilidium ciliare*, *Tritomaria quinquedentata*.

In the high Arctic Tundra a typical component of the rocky fell vegetation, associated in arid sites, often between boulders, with *Rhacomitrium lanuginosum* and *Polytrichum* and, occasionally, *Lophozia hatcheri*.

In the Tundra-Taiga border, on the eastern coast of Hudson Bay, often in somewhat moister situations, associated on damp rock walls with various lichens, *Lophozia longidens*, *L. ventricosa*, and *Ptilidium ciliare*, or on margins of pools in granite, with *P. ciliare*, *Cephaloziella arctica*, *Calypogeia*, *Lophozia bicrenata*, *L. wenzelii*, *Mylia taylori*, *Cephalozia bicuspidata*, and *Gymnocolea inflata* (Schuster, 1951). Occurrences of this type, where the species is admixed with a variety of others characteristic of moist Tundra, are somewhat exceptional. However, Buch and Tuomikoski report the species, under somewhat similar conditions, from Newfoundland, associated with *Jamesoniella*, *Lophozia longidens*, *L. ventricosa silvicola*, and *Ptilidium pulcherrimum*.

Differentiation and Variation

A strongly isolated species which can hardly be confused with any other. The deeply, usually 4-lobed leaves, somewhat like a cupped hand in shape, whose diagnostically sulcate lobes are erect and imbricate from a spreading base, are distinctive. The leaves half embrace the stem and together with the large underleaves, form a complete ring of erect, slender lobes girdling the stem. As a consequence, the shoots have a characteristically terete appearance. Confusion is barely possible between small phases of *C. setiformis* (with nearly entire leaf bases) and *Lophozia quadriloba*. However, leaves in the latter species are less deeply lobed, never have as strongly canaliculate lobes, and have somewhat spreading, rather than erect, lobes, thus lending an entirely different facies to the plant.

Chandonanthus setiformis produces small phases, or modifications, which are at times very different from the mature plant. Hooker (1816, pl. 20, figs. 1, 3–4) described such a phase as the var. *alpina* [= *Chandonanthus setiformis* var. *alpina* (Hook.) Kaal., Nyt. Mag. Naturvid. 33:227, 1893; = *Temnoma setiforme* var. *alpina* (Hook.) Frye & Clark, Univ. Wash. Publ. Biol. 6(3):415, (1944) 1945]. In this plant the leaves are simple, 3–4-lobed, with occasional leaves even merely 2-lobed; leaf bases are subentire, while the sinuses are always edentate. Such plants, in my opinion, are merely juvenile (or impoverished) manifestations of the species, in which the "species characteristics" (such as basal teeth and cilia) are much less marked. The "var. *alpina*" occurs very commonly among perfectly "normal" *C. setiformis* in various localities in New England. For instance, the plant from which the stem sector in Fig. 138:10 was taken is clearly referable to "var. *alpina*," yet this plant grew in the same mat with plants from which Fig. 138:9,11 were drawn, which represent "normal" *C. setiformis*. At times whole mats can be found in which all plants are similar to "var. *alpina*." Plants of such mats are usually growing under nutritionally difficult conditions and represent a mere juvenile modification (mod. *parvifolia-integrifolia*), rather than a distinct variety.

Schiffner (Lotos 60:47–48, 1912) found that some of the plants assigned to "var. *alpina*" have acute lobes, like the "species," whereas others have obtuse lobes. He divided "var. *alpina*" into two varieties, "var. *obtusiloba* Schiffn.," with obtuse lobes, and "var. *subintegerrima* Schiffn.," with acute lobes. I regard the latter as a synonym of "var. *alpina*"; "var. *obtusiloba*" is only an extremely reduced phase, in which the leaf lobes do not possess the usual sharp apex, or a mod. *parvifolia-integrifolia-obtusiloba*. Taxonomic recognition of these phases, therefore, appears hardly necessary. As has been stressed (Schuster, 1951a, p. 75), such small phases of *C. setiformis* "generally have the leaves less deeply, and often merely ⅔, divided. Such small forms, which lack all or most of the marginal cilia or teeth of the lobes, are strikingly similar to well-developed forms of *Lophozia* (*O.*) *quadriloba*, which have the leaves divided to a maximum of ⅔." Small, brown, rigid forms may also occur in exposed sites (as amidst *Gymnomitrion* on dry rock walls); such phenotypes may lack cilia of the leaf bases and have reduced bifid underleaves and bifid lateral leaves. Such phases may be very similar to *Lophozia kunzeana*.

Relationships

Chandonanthus setiformis is allied only to the Japanese-Formosan *C. pusillus*, which has sharper and narrower leaf lobes and coarser trigones (see Schuster, 1961a; Kitagawa, 1965), the Himalayan-Chinese *C. filiformis* (a key separating the three taxa is in Schuster, 1961a, pp. 209–210; see also Fig. 137: 1–8), and *C. quadrifidus* of Africa.

LOPHOZIA Dumort. emend. K. Müll. (*1910*)[1]

Jungermannia sect. *Lophozia* Dumort., Syll. Jungerm. Eur., p. 53, 1831.
Lophozia Dumort., Rec. d'Obs., p. 17, 1835.
Hattoriella Inoue, Jour. Hattori Bot. Lab. no. 23[1960]:37, 1961.

Gametophyte variable, with *leaves 2–4-lobed* (if 3-lobed, dorsal lobe obviously larger), *fundamentally succubous*, inserted obliquely ventrally and *obliquely* to almost (*but never quite*) transversely *dorsally*, the leaf plane varying from nearly transverse to the axis (*L. quadriloba*) to nearly horizontal (*Barbilophozia*), but never with dorsal half of leaf folded over the ventral (the *leaf never distinctly canaliculate-complicate*).[2] Stem usually 8–15–20 cells high, either suberect or decumbent. Branching sparing: in part intercalary (the intercalary branches *never* from dorsal end of leaf axils); frequently with ± furcate, terminal-lateral branching; flagella or stolons

---

[1] A full synonymy of the taxa accepted here as subgenera is given under each individual subgenus.
[2] Occasional species of *Orthocaulis* develop xeromorphic modifications with more or less complicate leaves, such as *L. kunzeana* mod. *plicata*.

absent.[3] Ventral stem sectors primitively about ⅕ the width of the circumference, but often reduced to 1–2 cells wide, the *underleaves varying from large* (and then lobed and ciliate) *to absent*. Frequently with asexual reproduction by fasciculate *1–2-celled, usually angulate gemmae*.

Dioecious or paroecious, rarely autoecious. Perichaetial bracts usually larger than leaves, sheathing perianth at base; bracteole present, smaller than bracts, *usually fused with one* (*rarely both*) *bract(s)*. Perianth well developed, exserted when mature, cylindrical, plicate in distal 0.2–0.5 and gradually narrowed distally or smoothly terete and suddenly narrowed to a small beak; *perianth free from bracts, not subtended by a perigynium*. Capsule ovoid, wall *normally* (*2*) *3–5-stratose*; inner layer with semiannular, U-shaped thickenings (radial bands, connected by complete tangential bands of the free faces of the cells; these latter bands sometimes incomplete, resulting in ± nodular thickenings); epidermal cell layer with cells larger, with coarser nodular thickenings (radial bands, slightly or not extended over the free tangential faces, occasionally extended as spurlike extensions); seta elongate, of numerous rows of cells. Spores ca. 10–18 μ; elaters 6–9 μ in diam., bispiral. Androecia spicate, becoming intercalary on main shoots, the bracts similar to leaves in size, with base ± ventricose; antheridia usually 2 or more,[4] often accompanied by several paraphyllia.

Type. Lophozia ventricosa (Dicks.) Dumort.

The above diagnosis is at some points disturbingly negative in nature; at others it is of necessity so vague that it scarcely allows a separation of the genus from related genera. In fact, because of the very varied types found in *Lophozia*, it is almost impossible to prepare an accurate description of the genus that will serve effectively to separate it from less synthetic, related groups, such as *Tritomaria*, *Anastrophyllum*, *Chandonanthus*, and *Gymnocolea*. It is obvious from Fig. 136 that *Lophozia s. lat.* occupies a central position from which there has been divergent evolution in a number of directions.

These evolutionary "lanes" lead to groups very different in facies, and therefore an attempt has been made to segregate them as genera. Buch (1933a, 1942) attempted a separation of generic nature along the lines *a–a* (in Fig. 135) on the basis of supposed differences in stem structure. He considered this separation fundamental. As was shown (Schuster, 1951a), groups with identical stem anatomy occur on both sides of this line, invalidating his distinction. All other proposed generic segregates from *Lophozia s. lat.* are of dubious value.

---

[3] Except in the South American subg. *Hypolophozia* Schust. (Schuster, 1965). This group may merit treatment as a separate genus.

[4] Rarely [e.g., subg. *Anomacaulis* (see p. 263)] with monandrous bracts. [I am now convinced *Anomocaulis* must be segregated as a wholly distinct genus; see Schuster, 1969a.]

Lophozia, as broadly limited here, is likely to be confused with a variety of other genera bearing lobed leaves, among them *Lophocolea*, *Marsupella*, and *Geocalyx*. From these it differs, collectively, in (*a*) the tendency toward production of fasciculate, often pigmented gemmae; these are absent in only a few species; (*b*) the relatively irregular, polymorphic, and ciliate underleaves, in those species bearing underleaves; (*c*) the presence of free, elongate, terete perianths, usually plicate distally; (*d*) the terminal position of inflorescences, on main shoots (the androecia always becoming intercalary because of apical proliferation). Confusion is more probable with members of same family, particularly *Anastrophyllum*, *Gymnocolea*, and *Tritomaria*. Generic differences are emphasized in the keys (pp. 233–241).

The species of the genus *Lophozia*, in the broad sense, range most frequently from the Tundra into the Taiga; the arctic taxa are particularly troublesome, and the genus in the Arctic seems to be in an active state of evolution. Only a minority of taxa extends southward into the region of deciduous forests, and only a single species extends into the subtropical region of central Florida. In the tropics proper the genus is scarcely represented, and only a few taxa occur in the Antipodes.

In the north the genus is represented by many species, often extremely difficult to distinguish. Indeed, without very careful study of the living plant and of the stem anatomy, it is often difficult to place plants in the correct subgenus. There is perhaps a good argument to be made in favor of recognizing three genera: *Barbilophozia*, with two subgenera, *Barbilophozia* and *Orthocaulis*; *Leiocolea*; and *Lophozia s. str.*, with three subgenera, *Massula*, *Isopaches*, and *Lophozia*.[5] The first of these "genera" includes species with predominantly 3–4-lobed leaves, often ciliate at base, although several taxa have predominantly bilobed leaves; the species usually—but not always—retain underleaves. The second "genus" (*Leiocolea*) differs in the smooth, beaked perianths, few oil-bodies per cell, and homogeneous stem structure. *Lophozia s. str.* would then be restricted to those types with uniformly 2-lobed leaves, plicate and nonrostellate perianths, and cells with rather numerous (usually 8–20 or more) oil-bodies. However, when such competent specialists as Müller (1942, 1951–58), Buch (1933a), and Meylan (1939) disagree as to the generic disposition of *Lophozia obtusa* (among the genera postulated above), it appears futile to attempt to maintain generic distinctions.[6] As is emphasized (Schuster, 1951a), excellent

[5] Müller (1954, *in* 1951–58), after vigorously criticising Buch for adopting such a narrow generic concept, later acknowledged *Barbilophozia* (incl. *Orthocaulis*), *Leiocolea*, and *Isopaches* as genera, distinct from *Lophozia*. The study by Inoue (1957) of the Japanese *L. diversiloba* has rendered futile any attempt to separate *Leiocolea* and *Isopaches* from *Lophozia* as other than sections or subgenera (see p. 232). I also show (p. 363) that *L.* (*Leiocolea*) *badensis* phenotypes may have strongly plicate perianths.

[6] Also, as noted elsewhere, as soon as the divergent exotic taxa are considered, the preceding distinctions become invalid. For example, where would subg. *Protolophozia* Schust. be placed?

arguments can be adduced in favor of a conservative treatment of these taxa, which are here considered to represent subgenera. The separation of these subgenera may be effected by use of the keys (pp. 233–241) and the associated table (Table 1).

Extraterritorial Taxa

In addition to the subgenera I recognize for North America, and the questionably distinct subg. *Hattoriella* Inoue from Japan, there are other isolated elements in *Lophozia s. lat.* which I believe deserve subgeneric or at least sectional recognition.[7]

1. Subg. *HYPOLOPHOZIA* Schust., Rev. Bryol. et Lichén. 34:276, 1966.

Type. Lophozia subinflata (Spr.) Schust. = *Jungermannia* (*Lophozia*) *subinflata* Spruce, Trans. Proc. Bot. Soc. Edinb. 15:513, 1885. The type is the first species cited in that convenient receptacle, established on a generic level, *Sphenolobus*! Also belonging here is "*Sphenolobus*" *argentinus* Steph. = *Lophozia argentina* (Steph.) Schust.

This subgenus is discussed *in extenso* in Schuster (1966i). It differs from other subgenera of *Lophozia* with bilobed leaves as follows: (*a*) postical-intercalary flagella are frequent to common; (*b*) the capsule wall is 2-stratose, with the epidermal cells slightly higher than the inner; both bear vertical thickening bands, or "nodular" thickenings, as in *Isopaches;* (*c*) gemmae are lacking; (*d*) leafy branching is of the *Frullania* type, rarely postical-intercalary. *Hypolophozia* approaches subg. *Lophozia* in the form of the bracts, in the distally 5–8-plicate perianth, in the cells and leaves. It differs from subg. *Lophozia* in, i.a., capsule-wall anatomy, the ventral stolons, and the stem (only 8–10 cells high, lacking dorsiventral differentiation of the medulla, but with two layers of cortical cells forming a weakly differentiated, yellowish brown cortex). Both species referred here are South American.

2. Subg. *PROTOLOPHOZIA* Schust., subg. n.

Pallidevirens vel albescens, saepe rubro-violaceo tincta; caulis irregulariter pauciramosus; cellulae corticales leptodermaceae, rectangulatae; sectione

---

[7] It is exactly the *existence of such elements*—becoming more and more numerous as the Lophoziaceae are investigated more carefully—that makes it difficult to follow the narrow generic concepts of Buch (1933a), Müller (1951–58), and Arnell (1956), as I have mentioned at several points; see also the discussion (p. 262) under subg. *Orthocaulis*. Before the worker whose experience has been solely with the limited holarctic flora decides on the merits of "genera" such as *Obtusifolium* S. Arn. and *Isopaches* Buch, he should study a suite of exotic taxa. Thus Grolle (1960e), after once accepting *Barbilophozia* as a distinct genus, now (1962a) considers *Lophozia herzogiana* as not fitting into any of the subgenera of *Lophozia* Dumort. *s. lat.*, but standing between "Subgenus *Massula* K. Müller und dem Subgenus *Barbilophozia* (Loeske) K. Müller *sensu amplo*" (including the "genus" *Orthocaulis* Buch). See in this connection also the discussion (Schuster, 1965) of "*Sphenolobus*" *leucorhizus* (Mitt.) Steph., a species which fails to fit well into existing genera or subgenera. This species once again illustrates the problems confronting the student *as soon as he tries to apply "European" generic concepts to non-European taxa*!

TABLE 1

Subgeneric Criteria in *Lophozia**

	Orthocaulis	*Barbilophozia*	*Massula*	*Leiocolea*	*Isopaches*	*Lophozia*
Leaf lobes	(2)3(4)	4	2(3–4)	2	2	2
Leaf insertion	Acroscopically arched, little succubous		Scarcely arched, strongly succubous			
Median cells usually	<30 × 35 μ Strong trigones	<30 μ Weak trigones	30–60 μ No trigones	30–50 μ Strong trigones	<30 μ Thick-walled	<35 μ Weak to strong trigones
Oil-bodies	Papillose	Papillose	Smooth	Papillose	Papillose	Papillose
Number per cell	3–6(8–10)	3–6(12)	20–50	2–6	6–15(20)	6–20(25)
Size	5–9 μ long	4–9 μ long	1.5–3(4–5) μ	4–7 × 8–15 μ	4–9 μ	4–9 μ
Cuticle usually	Verruculose	Verruculose	Smooth	Verruculose	Smooth	Smooth
Cortex cells with walls	Thick	Thick	Thin	Thin–thick	Thick	Thick
Medulla	Homogeneous	Homogeneous	Homogeneous	Homogeneous	Homogeneous	Dorsiventrally differentiated
Stem height in cells	11–15 or more	11–15 or more	11–15 or more	9–15	7–10	10–22
Cortex cell width:length	1:2–3	1:2–3	1:3–8	1:2–3	1:2–3	1:2–3
Leaf width: length	0.9–1.6:1	1.2–1.8:1	1.0–1.6:1	0.9–1.2:1	0.9–1.1:1	0.7–1.2:1
Underleaves	Large > small > absent	Large > small	Small > absent	Large > absent	Absent	Absent
Ventral merophyte width	8–2 cells	12–6 cells	12–6 cells	12–4 cells	Ca. 2 cells	Ca. 2 cells
Leaf bases	Usually ciliate	Usually ciliate	Entire	Entire	Entire	Entire
Gemmae	+	+(–)	+	–(+)	+	+
Perianth	Plicate	Plicate	Plicate	Beaked, smooth	Plicate	Plicate
Growth	Suberect	±Prostrate	Prostrate	Prostrate	Prostrate	Prostrate > Suberect

* Isolated exceptions are of necessity disregarded in this table.

transversali cellulae subaequimagnae, pallidae, vix incrassatae; ramificationes intercalares ventrales. Folia caulina oblique patula, parum oblique inserta, profunda (0.25–0.75) 2–3–4-lobata; lobis lanceolatis, divergentibus. Cellulae pallidae, parietibus tenuibus, trigonis nullis vel minutis, cuticula laevi vel ad basin striatula; medianae 21–28 × 25–36 μ vel (35)40–70 μ. Amphigastria magna, foliis suis 0.25–0.5 solum minora, lanceolata vel bilobata. Gemmulis (1)2-cellularibus, angulatis. Involucra feminina ± crispata; folia involucralia libera, usque ad ⅓ quadriloba; amphigastrium involucrale bilobatum. Perianthia tubulosa, exserta, ad orem 4–5-plicata, ore contracto, crenulato vel dentato. Sporogonia exserta, valvulis (3–)4-stratosis; stratum intimum annulatim incrassatum. Typus: *Lophozia herzogiana* Hodgs. & Grolle, Rev. Bryol. et Lichén. 31:152, cum fig., 1962; from New Zealand.

To judge from the extremely brief diagnosis, *Lophozia perssoniana* Miller of Hawaii [Ark. Bot. 5(2):508, fig. 9, 1963] must also belong here.[8] Indeed, the two seem strikingly similar and may perhaps even prove identical; Miller's plant is so imperfectly described that it is impossible to arrive at any certain conclusion.

In addition, two other taxa appear to belong here, *L. crispata* Schust., sp. n., from Tierra del Fuego and, less surely so, *L. multicuspidata* (Hook. f. & Tayl.) Grolle of Campbell I., New Zealand. These taxa share the relatively soft texture, the stem without a defined, thick-walled cortex or a cortex of smaller cells, and the flaccid and polymorphous leaves, in part at least 3-, rarely also 4-, lobed on sterile shoots. The following (highly tentative) key will separate the taxa:

1. Leaves (on larger shoots, then many 3–4-lobed) as wide as or wider than long, divided for 0.4–0.75 their length; with conspicuous underleaves of all but the most reduces axes . . . . 2.
   2. Leaf lobes sharply acute in all cases; sinuses not or obscurely gibbose; median cells under 50 μ on an average. Antipodal . . . . 3.
      3. Dioecious; gemmae on slender, microphyllous, *L. attenuata*-like innovations; cells large, median ca. (35)40–50 μ; leaves 2–3-lobed for 0.5–0.75 their length; underleaves broad, bifid to middle. New Zealand . . . . *L. herzogiana* Hodgs. & Grolle
      3. Autoecious; no gemmae; cells smaller, median 21–28 × 25–36 μ; leaves 2–3–4-lobed for 0.4–0.65 their length; underleaves narrow, bifid 0.6–0.7 or lanceolate and unlobed. Tierra del Fuego . . . . *L. crispata* Schust., sp. n.[9]

[8] The selection of this species epithet is unfortunate because of potential confusion with the earlier *L. perssonii* Buch & Arn.

[9] Plants light green, whitish below, purplish brown at times above, suberect in growth to procumbent; freely branched by lateral-intercalary branches (from ventral half or ventral angle of leaf insertion), rarely by ventral-intercalary branches. Leaves crispate-undulate, 2–3–4–lobed, divided 0.4–0.65, sinuses often angular or rounded, lobes sharply acute; underleaves narrow, oblong to lanceolate, bifid ca. 0.5–0.7 or undivided. No gemmae. Median cells thin-walled with small trigones, 21–28 × 25–36 μ. Autoecious. Androecia spicate, usually on short, lateral-intercalary branches.

Type. Argentina: Tierra del Fuego, Rio Harubre Valley, 2–3 km south of Paso Garibaldi (*RMS 59417b*). For figure see Schuster (1969b).

2. Leaf lobes acute to blunt to rounded; sinuses often rounded and sharply recurved; median cells 50–70 μ; mature leaves 3–4-lobed. Hawaii . . . . *L. perssoniana* Miller

1. Leaves mostly oblong, 2–3(4)-lobed and all longer than broad, divided for 0.25–0.4 their length into sharp lobes; underleaves rudimentary or lacking? Campbell I., New Zealand . . *L. multicuspidata* (Hook. f. & Tayl.) Grolle.

Subgenus *Protolophozia* bears affinities to both subg. *Massula* (in, i.a., the polymorphous leaves; the large, leptodermous leaf cells with essentially smooth cuticle; the soft stem, without a defined cortex or a dorsiventral medulla; the often quadrifid bracts) and subg. *Barbilophozia* (the large underleaves; the 3- or even 4-lobed leaves). Grolle noted that his species would probably prove to represent a separate subgenus. With the discovery of additional species in the same complex, it seems wisest to place them into a subgenus of their own. In my opinion, the axial anatomy, the occasionally vinaceous pigmentation, the soft texture, the large and leptodermous cells, the form of the bracts, the perianth, and particularly the polymorphous leaves, with 2–3–4-triangular lobes, suggest that the subgenus is closest to subg. *Massula*. From this it differs primarily in the large and sometimes multiform underleaves, which are up to 0.5 the size of the lateral leaves.

In addition to the six regional subgenera represented and the two treated above, there are three stenotypic subgenera, *Hattoriella* from Japan (*L. diversiloba* Hatt., type) and *Xenolophozia* from South Africa (*L. capensis* S. Arn., type) and *Anomacaulis* of New Guinea (*L. hamatiloba* Grolle, type). The following conspectus separates all these groups:[10]

1. Leaves typically (2)3–4-lobed; ventral merophytes remaining broad, usually developing underleaves . . . . 2.
  2. Cells ± collenchymatous, firm, small (under 25–30 μ in leaf middle); terminal branches usually present; leaves never with vinaceous pigmentation; leaves usually similar on any one shoot, with limited polymorphism only. (Often with gemmae) . . . . 3.
    3. Plants with leaves subtransversely oriented, 2–4-lobed, the lobes never mucronate, postical margin never with cilia of slender, narrow cells . . . . *Orthocaulis*
    3. Plants with leaves nearly horizontally inserted and laterally ± patent, 4-lobed (except on juvenile stems), lobes often cuspidate or mucronate, postical margin usually armed with cilia of elongate, narrow cells . . . . *Barbilophozia*
  2. Cells thin and delicate, without trigones or with minute ones, large (to 30–60 μ in leaf middle); mostly with lateral-intercalary branches; usually with vinaceous pigmentation, much as in *Massula*; leaves strongly polymorphous, 2–4-lobed, often on a single shoot, frequently crispate or undulate. Pacific (Hawaii, New Zealand), and Tierra del Fuego . . . . *Protolophozia*, subg. n.

[10] For the separation of the regional subgenera assigned to *Lophozia s. lat.*, see the keys on pp. 233–241.

1. Leaves 2-lobed (in *Massula* with leaves in part 3–4-lobed; in this without underleaves) . . . . 4.
    4. Perianth abruptly contracted to the small mouth, ± sharply beaked, terete and eplicate below; usually with distinct, ciliate underleaves; no medullary differentiation of stem; cortical cells ± large, ± thin-walled; gemmae rare, smooth . . . . *Leiocolea*
    4. Perianth (at least in distal 0.25) conspicuously plicate, not rostrate [or *Anomacaulis*, beak externally ciliate];[11] almost always with gemmae . . 5.
        5. Seta with 16–24 or more epidermal cell rows, 6 or more cells in diam.; dioecious or paroecious; underleaves normally absent; bracts (where known) free from bracteole, or only one bract united with bracteole . . . . 6.
            6. Cells collenchymatous: with often coarse, bulging trigones or else becoming equally thick-walled (lumen then guttulate); oil-bodies (unknown in *Anomacaulis*) fewer (either 4–8 or 8–24 per cell), granular-segmented or botryoidal . . . . 7.
                7. Capsule wall 3–5-stratose; leaf cells thin-walled, with distinct to coarse trigones; stem with strong dorsiventral differentiation of medulla, the ventral layers of small, brownish cells; ♂ and ♀ bracts normally edentate . . . . *Lophozia s. str.*
                7. Capsule wall 2(3)-stratose (*Anomacaulis?*); stem without distinct dorsiventral differentiation of the medulla; ♂ and ♀ bracts often sharply dentate, as well as lobed . . . . 8.
                    8. Postical intercalary flagella present; both epidermal and inner cell layers of capsule with nodular thickenings; no gemmae; stem 8–10 cells high only; cells thin-walled, with large trigones. South America . . . . *Hypolophozia*
                    8. Flagella (and postical branching) lacking . . . . . . 9.
                        9. Leaves subsymmetrically ovate or orbicular or quadrate-orbicular, without shingled and very conspicuously ampliate ventral leaf bases; leaves 0.2 or more bilobed, the lobes never longly apiculate, never dentate (in absence of gemmae), not sharply, hamately incurved. Northern Hemisphere . . . . 10.
                          10. Leaf cells becoming equally thick-walled; oil-bodies smaller, (3–4)5–15 per cell; ♂ bracts bilobed, symmetric; rusty to red angulate-stellate gemmae abundant. Europe, North America . . . . *Isopaches*
                          10. Leaf cells (and usually oil-bodies) as in *Leiocolea*: with coarse trigones and with few (4–5) very large, conspicuous oil-bodies; ♂ bracts 3–5-lobed and toothed, asymmetric; gemmae absent. Asia . . . *Hattoriella*[12]

---

[11] In *Anomacaulis*, which differs from *Leiocolea* in the rigid, collenchymatous axis with a thick-walled cortex, in the broadly ovate-trigonous, shallowly bilobed leaves with ampliate postical bases, in the lack of underleaves, the perianth apex is uniquely bristly externally.

[12] *Hattoriella* is placed by Kitagawa (1965) as a synonym of subg. *Leiocolea*, probably correctly so. The type species, with the distal part of the perianth plicate, suggests that perhaps a separate

9. Leaves asymmetrically ovate-trigonous, the postical bases strongly ampliate, shingled to hide the stem; leaves less than 0.12 bilobed, the longly apiculate lobes often ± denticulate, hamately incurved. [Cells with coarse, nodose trigones, the basal cells elongated; plants vigorous, to 5 cm long, intensely vinaceous above; stem with a rigid cortex of thick-walled cells and rigid, collenchymatous medulla.] New Guinea . . . .
. . . . *Anomacaulis*, subg. n. (see p. 263)

6. Cells thin-walled (or becoming slightly, equally thick-walled; rarely with conspicuous trigones); oil-bodies numerous, 20–50 per cell, usually minute (2–4 μ, rarely 5 μ), usually ± homogeneous and glistening; capsule wall 3–5-stratose; no dorsiventral medullary differentiation, and cortex of thin-walled cells. . . . .
. . . . *Massula*[13]

5. Seta with 8 extremely large, pellucid epidermal cell rows surrounding 4 minute inner ones; no dorsiventral medullary differentiation; distinct underleaves; with gemmae; bracteole broadly connate with *both* bracts; autoecious . . . . *Xenolophozia*, subg. n.[14]

It is possible that, with study of the remaining and poorly known exotic species, additional sectional or subgeneric groupings may have to be established. The above conspectus of subgenera regards as valid all proposed groups, except the monotypic *Obtusifolium* (proposed as a subgenus of *Barbilophozia* by Buch, 1942, to include *L. obtusa;* elevated to generic rank by Arnell, 1956) and *Barbicaulis* and *Barbifolium*, established by Buch (1942) as so-called subgenera of *Barbilophozia s. str.* These groups I regard as of sectional rank.

There are still taxa which fail to fit well into the above grouping. For example, *Lophozia anomala* Schust. (Schuster, 1965) and *L. leucorhiza* (Mitt.) Schust., although both referred to *Orthocaulis*, fail to correspond well to that subgeneric concept and should perhaps be placed in another subgenus or section (Schuster, 1965). With study of such antipodal taxa, the generic and subgeneric grouping proposed by Buch (1933a), and all arrangements based on this classification, become essentially untenable.

subgenus is involved; however, even in *Leiocolea* (e.g., *L. badensis*) the distal 1/4 of the perianth may be quite plicate.

[13] Kitagawa (1965, p. 289) has proposed that *Lophozia cornuta* (Steph.) Hatt., Bull. Tokyo Sci. Mus. 11:35, 1944, be segregated as a separate and monotypic subgenus, *Schistochilopsis;* I would regard this as a well-defined section of *Massula;* see p. 422.

[14] *Lophozia capensis* S. Arn., type (Svensk Bot. Tidskr. 47:112, figs. 3–4, 1953). I cite the abbreviated species diagnosis to validate the subgenus. Autoica, minor, rufo-brunnea–viridis; caulis ad 4 mm longus; folia integerrima–dentata, concava, ad 0.5 biloba; folia floralia connatas; perianthia ovato-oblonga, plicis 5, apice crenulata. Known only from Cape Province, South Africa. I have seen *Arnell 1181*, one of four specimens cited in the original diagnosis.

Subgenus *ORTHOCAULIS* (*Buch*) *Schust.*

[Fig. 139]

Orthocaulis Buch, Mem. Soc. F. et Fl. Fennica 8:293, 1933.
Lophozia subg. *Orthocaulis* Schust., Amer. Midl. Nat. 45(1):40, 1951.
Barbilophozia K. Müll., Rabenh. Krypt.–Fl. ed. 3, 6:622, 1954; Grolle, Nova Hedwigia
   2(4):555, 1960 (*p.p.*).
Barbilophozia subg. *Orthocaulis* Buch, Mem. Soc. F. et Fl. Fennica 17:289, 1942; Kitagawa
   Jour. Hattori Bot. Lab. no. 28:268, 1965.

Green to brownish pigmented, *never with clear reddish pigmentation*, robust
(largely 0.8–2.5 mm wide), *suberect to strongly ascending*, often stiffly so, sparsely
branched; stems rather rigid, flexuous, firm, 11–16 cells high usually, showing
no pronounced dorsiventral differentiation, the 1–2 postical cell layers occasionally
weakly mycorrhizal (but never with a discrete medullary mycorrhizal band);
cortical cells subequal in width to basal leaf cells, short rectangular (usually
2–3× as long as wide). Rhizoids usually rather sparse. *Leaves polymorphous*,
varying about a *fundamentally trilobed* type: *typically with a large, longer dorsal lobe
and 2 equal, smaller postical lobes* (but sometimes the postical lobes united, then
subequally bilobed; sometimes with dorsal lobe also bilobed, then subequally
4-lobed); leaf insertion slightly (10–20°) to moderately (25–45°) oblique, the
line of insertion *perceptibly acroscopically arched*, leaves fundamentally only slightly
succubously oriented, *often vertical or subvertical, often rather strongly concave*, sometimes nearly "cupped," or handlike, with the *lobes strongly incurved* or at least
erect from a spreading base; *leaves* usually quite *distinctly broader than long* (width-
length ratio ca. 1.1–1.6:1); postical leaf bases (at least of some leaves) often
with 1–several short cilia, formed of isodiametric cells, occasionally reduced to 1–
several sessile or stalked slime papillae. Ventral merophytes normally 4–10 cells
broad and bearing *distinct, usually bifid underleaves* (often bearing 1–several basal
cilia terminated in slime papillae), but sometimes reduced to a width of 2 cells
(and unable to develop underleaves). Cells thin- to slightly thick-walled, *more
or less strongly collenchymatous, firm and opaque, rather small* (13–17 μ in leaf lobes)
and opaque; *cuticle verruculose;* oil-bodies rather large (frequently up to
5–7 × 9 μ), *usually 3–6(8–9) per cell.* Gemmae absent (*L. floerkei*, *L. quadriloba*)
or freely formed, either on leaves of normal size (*L. kunzeana*) or on reduced,
erect-appressed, imbricate leaves (most species).

Dioecious. Perianth large, *plicate*, somewhat *narrowed to the weakly dentate
apex.* Bracts usually 3–5-lobed, often with supplementary smaller lobes and
teeth, more or less connate with the bracteole; bracteole 2–4-lobed, the lobes
often laciniate, often with supplementary teeth. ♂ Bracts with 2 or more
antheridia, often *with paraphyses.*

Lectotype. Lophozia (*O.*) *kunzeana*, by present designation. No type was
designated originally.

The subgenus is exactly equivalent to the genus *Orthocaulis* of Buch (1933a);
its chief distinguishing characters have been stressed by Buch. Buch emphasized
that the mode of origin of the leaf lobes appears to be different in *Orthocaulis* and
Barbilophozia. In *Orthocaulis* the basic leaf-lobe number is 2 (as in most material
of *L. kunzeana*), with a well-developed tendency for the ventral lobe to subdivide
(resulting in a trilobed leaf with 2 small ventral lobes, contrasted to a larger

dorsal lobe); occasionally, as in *L.* (*O.*) *quadriloba* the dorsal lobe may also be subdivided, resulting in a symmetrically 4-lobed leaf. In *Barbilophozia*, on the other hand, the origin of the dorsal and ventral outer lobes appears to be from the basal cilia of the leaves; *fide* Buch.[15]

Subg. *Orthocaulis* is extremely synthetic in nature and comes close to the hypothetical common ancestral form from which the entire Lophozoid complex may have originated; see Figs. 135, 136, pp. 226 and 229. The approach toward large underleaves and wide ventral stem sectors, lack of stability in the development of leaf lobes, suberect or ascending growth, generalized form of the oil-bodies, unspecialized stem, and lack of obviously derivative features all indicate a "low" position for *Orthocaulis*. Through *L. obtusa* the subgenus appears to be connected to the *Massula-Isopaches-Lophozia* complex. Through *L. barbata* and *floerkei* it appears related to *Barbilophozia*; through *L. rutheana* and *heterocolpa*, to *Leiocolea*. Through *L. quadrilobus* the subgenus is possibly remotely connected with *Chandonanthus*. Through *L. binsteadii* and *atlantica* it may be related to the *Anastrophyllum* complex (probably most directly to *A. saxicolus*) and through *L. cavifolia* to *Acantholobus* (see p. 332). These annectant species are discussed more fully under their respective groups.

The tendency toward development of 3-lobed leaves may result in confusion with *Tritomaria* subg. *Saccobasis* and subg. *Tritomaria*. However, the former is very distinct from *Orthocaulis* in the equally trilobed leaves, the mode of leaf insertion, and the very narrow, elongate cortical cells. *Tritomaria* subg. *Tritomaria* has a fundamentally different leaf, with the postical lobes largest (and postical leaf margins longer), whereas the antical lobe is shorter, often giving the appearance of a mere tooth at the antical margin of a bilobed leaf.

The species of subg. *Orthocaulis*, with few exceptions, are all holarctic. The only exceptions are: (1) *Lophozia anomala* Schust. (Schuster, 1965) [*Orthocaulis longiflorus* Herz., Rev. Bryol. et Lichén. 23(1–2):32, fig. 2, 1954] is from Andean Chile; (2) *L. floerkei* has been reported from the Andes (see Grolle, 1960e), and (3) "*Sphenolobus*" *leucorhizus* (Mitt.) Steph. has been placed in *Lophozia* (Schuster, 1965); it was described from Kerguelen I. and seemingly belongs in *Orthocaulis*.

The holarctic species have served to establish the basic perimeters of subg. *Orthocaulis*. Two Southern Hemisphere taxa that belong in the "vicinity" of *Orthocaulis* serve to demolish these perimeters, *if* we retain them in *Orthocaulis*. The first is *Lophozia badia* (G.) Steph.[16] This species has a nearly straight, quite oblique (ca. 45–50° succubous) insertion which is not acroscopically arched— unlike *Orthocaulis*. It has consistently bilobed leaves, which are suberect, sub-vertical, and rather concave, with the rounded or blunt lobes incurved. The leaves are reminiscent of a hypothetical cross between *L. obtusa* and *L. kunzeana*. The collenchymatous cells have coarse, high, close, cuticular papillae, as, e.g., in *L. quadriloba*. Unlobed to bifid underleaves, bearing several basal teeth or

---

[15] This is readily apparent from studying poorly developed plants of *L. hatcheri* and *lycopodioides*. However, it is not at all evident in *L. barbata*, in my experience.

[16] I have not seen the type; my discussion is based on a specimen from South Georgia (*Skottsberg 21*, G; det. Stephani).

cilia, are present throughout. The stem has a poorly marked cortex and no intramedullary differentiation.

Cuticular papillae, cell type, leaf form (but not insertion), and stem anatomy suggest subg. *Orthocaulis*. *Lophozia badia* is quite distinct from "normal" species in the more succubous and more nearly straight leaf insertion—one of the criteria Buch emphasized in separating *Barbilophozia* from *Orthocaulis*. Yet, in spite of the insertion, *L. badia* seems closer to *Orthocaulis* than to *Barbilophozia*. Obviously, another antipodal type is at hand that fails to fit clearly into any of the Northern Hemisphere "pigeonholes." If it is placed into *Orthocaulis*, as I would suggest, that group loses some of its homogeneity. The alternative—to propose a new subgenus—is impractical; if this is done for every deviant species, we shall be lost in subgenera.

The second deviant Southern Hemisphere taxon is *L. (O.) hamatiloba* Grolle (Grolle, 1965, p. 50, fig. 4) from New Guinea. This has very shallowly bilobed, asymmetric leaves with unusually ampliate ventral bases that shingle each other to hide the stem ventrally; leaf lobes are longly apiculate, sometimes sparingly denticulate, and strongly incurved. The leaf form, especially the broadly ampliate ventral bases, and the intensely vinaceous color of the plants preclude this taxon's being placed in *Orthocaulis* or, seemingly, in any other subgenus of *Lophozia*. I suggest that it should go into a subgenus of its own, thus making possible some remaining delimitation of *Orthocaulis*. For it the subg. *Anomacaulis* Schust., subg. n., is proposed, with the following portions of the original diagnosis of Grolle cited to delimit the group [see also footnote, p. 253]:

Anomacaulis Schust., subg. n.

Dioica; robusta, infra flavo-pallida, superne optime purpurata; caulis repens; series merophytorum ventralium angustissima; sectione transversali ca. 12 cellulas altus, medulla cellulis majoribus pallidis, trigonis magnis; rami intercalares-laterales et terminales-laterales; folia conferta, oblique succubatim inserta, homomalla, lamina satis concava, ambitu late ovato-trigona, apice brevissime bilobata, sinu lunulato; lobis apiculatis, hamatim incurvis, conniventibus, marginibus integerrimis vel interdum superne sparsim denticulatis; margo basi ventrali valde ampliatus; amphigastria nulla; antheridia solitaria.

Our species of *Orthocaulis* have a distribution ranging from the Tundra (in which most of them are found) into the Coniferous Forest Biome, in which usually only the widespread *Lophozia attenuata* occurs. This latter species occurs southward to North Carolina and Tennessee, at medium and high elevations in the mountains; otherwise the species of the genus are restricted in the United States to a few high montane peaks and to isolated sites in the Great Lakes region. In the arctic-alpine region they are often abundant and conspicuous.

Key to Species of Orthocaulis

1. Underleaves large, obvious, usually bilobed and with 1–several cilia; leaves with postical bases occasionally or usually with 1–several cilia; gemmae normally absent (exc. *L. kunzeana;* here gemmiparous plants with leaves not reduced); ventral stem sectors 6–16 cells wide . . . sect. Orthocaulis 2.

FIG. 139. *Lophozia* subg. *Orthocaulis*. (1–4) *L. quadriloba* var. *collenchymatica:* 1, two leaves (×37.5); 2, 3, dorsal and ventral lobe apices, respectively (×160); 4, median cells (×215). (5–9) *L. kunzeana* fo. *acuta:* 5, 6, underleaf and leaves (×42.5); 7–9, leaf lobe apices (×160). (10–15) *L. attenuata*, alpine extreme with purplish gemmae, and, in part, coarse trigones, approaching *L. binsteadii:* 10, gemmae (×360); 11, median cells of collenchymatous extreme with yellowish trigones (×635); 12, median cells with oil-bodies (×525); 13, two leaves of the mod. *latifolia* (×28); 14–15, leaves *in situ*, lateral aspect, showing rhizoid bases (×26). (16–18) *L. binsteadii*, mod. *latifolia:*

2. Leaves (mature shoots!) predominantly 4-lobed for 0.5–0.6 their length, *lobes ± narrowly triangular, sharp*, with the margins (especially at sinus bases) distinctly reflexed; usually ± blackish or "inky" black; cells of lobe margins 12–15 μ, of leaf middle 17–19(23) × 18–25(29) μ; trigones small; cuticle with coarse papillae . . . . L. (O.) *quadriloba* (p. 267)

2. Leaves predominantly 2(3)- to 3(4)-lobed, lobes broadly triangular, acute, blunt, or rounded; sinuses descending 0.25–0.5; lobe margins (and often sinus bases) little reflexed; green to brown, occasionally fuscous; cells of lobe margins 14–19 μ or larger . . . . 3.

   3. Leaves predominantly 2-lobed, exceptionally (and sporadically) 3-lobed, sinuses descending to 0.5; lobes quite blunt to broadly rounded; gemmae frequently present; at most 1 minute postical cilium. [Marginal cells 17–18.5 μ; median cells 20–26 × 22–28 μ; cuticular papillae numerous but not coarse.] . . . . L. (O.) *kunzeana* (p. 288)

   3. Leaves (2)3(4)-lobed, sinuses variable, mostly descending 0.25–0.45; lobes mostly subacute to acute; never with gemmae; mature leaves with 1–2(3) coarse laciniae or teeth of postical base . . . . . . 4.

      4. Leaves broad, in part obtrapezoidal, variable: 2–4-lobed, even on a single plant, divided for 0.3–0.4(0.5) their length; lobes often only subacute or even bluntish; postical (less often so antical) margin at base with 0–1–2(3) coarse teeth, isolated ones verging on laciniae. Cells large: marginal cells in lobes (18.5–21)22–25 μ; median cells (23)25–29 × 26–32 μ, each with usually 7–15 oil-bodies; cuticle coarsely papillose. High arctic . . . .
. . . . L. (O.) *hyperborea* (p. 282)

      4. Leaves subquadrate in outline, almost invariably 3-lobed, with weakly arched, almost parallel lateral margins, divided 0.25–0.35 into 3 sharply acute, nearly equal lobes; postical (but never antical) margins with 2–3(4–5) slender cilia. Cells small: marginal cells in lobes 14–19 μ; median cells 16–19 × 20–25 μ, each with 2–6 oil-bodies; cuticle faintly papillose. Subarctic-antipodal . . . .
. . . . L. (O.) *floerkei* (p. 300)

1. Underleaves absent or mere linear to lanceolate, undivided, obscure vestiges, or in the form of 1–3 linear cilia; postical leaf bases never ciliate; plants usually gemmiparous; leaves (with rare, sporadic exceptions) 2- or 3-lobed for 0.25–0.5 their length; ventral stem sectors reduced to a usual width of 2–4 cells . . . . sect. Attenuatae 5.

5. Marginal cells of free lobes (17–18)19–24(–27) μ, median 19–23 × 23–28 μ or larger; median cells with distinct to very coarse trigones, usually slightly to strongly bulging, the walls almost invariably brownish; gemmae purplish, on the margins of ± normal leaves (or on somewhat

---

16, lobe apex (×215); 17, leaves (×31); 18, median cells of mod. *mesoderma-viridis* (×810). [1–4, *RMS & KD 66-075*, Sondrestrom Fjord, W. Greenland; 5–9, *RMS & KD 66-022*, Tupilak I., W. Greenland; 10–15, Mt. Katahdin, Me. (10–11, *RMS 32998*, 12–15, *RMS, 1949*); 16–18, *RMS 45951* and *46051*, Siorapaluk, N.W. Greenland.]

reduced, dense leaves; however, very rarely with gemmiparous shoots strongly and slenderly flagelliform). Arctic-alpine . . . . 6.

6. Leaves of mature shoots predominantly 3-lobed, with lobes spreading to moderately incurved, but leaves never cupped; plants not with facies of *Anastrophyllum minutum* . . . . 7.

  7. Leaves spreading to erect-spreading, divided 0.25–0.35 their length into 3 (occasionally 2) incurved, *broad* lobes; leaves averaging *1.3–1.5×* *as wide as long* (when mature and 3-lobed), the *lobes broader than long;* underleaves of apices of mature shoots occasional to frequent, of 1–3(4) cilia usually; cells with trigones not or rarely confluent; green to ± fuscous brown. . *L. (O.) atlantica* (p. 324)

  7. Leaves erect-spreading or suberect, strongly concave or cupped, divided 0.35–0.5 their length into 3 *ovate-lanceolate, sharp lobes;* leaves averaging *0.9–1.2× as wide as long*, the lobes usually *longer than broad;* underleaves absent; cells with trigones often confluent, very coarse; warm brown . . . . . *L. (O.) binsteadii* (p. 315)

6. Leaves all or almost all 2-lobed, hemispherical, with incurved lobes (so strongly concave that leaves cannot be flattened without tearing); with facies of *A. minutum.* (Leaves much broader than long, with very broad and obtuse to bluntly pointed lobes.) . . . .
. . . . *L. (O.) cavifolia* (p. 330)

5. Marginal cells of free lobes 12.5–17 μ, median 16–20 × 17–25 μ; cells slightly thick-walled, with ± small, concave-sided trigones, walls rarely strongly brownish-pigmented; gemmae greenish to yellow-brown (in alpine forms rarely purplish), on reduced leaves of strict, attenuate, rigid, filiform axes; leaves usually 0.9–1.2 × as wide as long, almost uniformly trilobate for 0.25–0.35 their length, lobes not or rarely incurved, usually as long as broad, pointed. Non-arctic . . . . . *L. (O.) attenuata* (p. 307)

Sectio ORTHOCAULIS

Lophozia subg. *Orthocaulis* sect. Kunzeanae Schust., Amer. Midl. Nat. 45(1):42, 1951.

Ventral merophytes well preserved, broad, producing underleaves throughout nongemmiparous regions, usually bifid and often with teeth or cilia toward base; leaves at ventral bases often ciliate; gemmae lacking or (*L. kunzeana*) produced on shoots with unreduced, unmodified leaves.

Type. Lophozia kunzeana.

This section includes, in my opinion, the most plastic[17] and least specialized extant members of the Lophoziaceae (I would regard *Chandonanthus* as already more specialized). Associated with this, the branching patterns are relatively plastic. Even though some taxa seem to have exclusively *Frullania*-type terminal branches (e.g., *L. kunzeana*; Fig. 150), others show much internal variation. For example, in *L. quadriloba* var. *quadriloba*

[17] To demonstrate graphically the extraordinary plasticity in the group I have treated *L. quadriloba* in considerable detail. Compare Figs. 139:1–4, 140–144, 145:1–6, 146:1–4.

branching is apparently nearly always of the *Frullania* type, with the branch often arising some distance above the supporting "half-leaf"; the latter is either ovate and unlobed or bifid, depending on whether the parent axis bears mostly 2–3 or 3–4-lobed leaves. In var. "*heterophylla*" similar branching occurs, but I have seen occasional lateral-intercalary branches from the ventral angle of the leaf. By contrast, the fo. *cephaloziel-loides* normally bears exclusively ventral- and lateral-intercalary branches!

LOPHOZIA (ORTHOCAULIS) QUADRILOBA
(*Lindb.*) *Evans*
[Figs. 139:1–4; 140–144; 145:1–6; 146:1–4]

Jungermannia quadriloba Lindb., Meddel. Soc. F. et Fl. Fennica 9:162, 1883; Arnell & Lindberg, Kgl. Sv. Vetensk.–Akad. Handl. 23(5):55, 1888.
Lophozia quadriloba Evans, Proc. Wash. Acad. Sci. 2:304, 1900.
Sphenolobus quadrilobus Steph., Spec. Hep. 2:168, 1902.
Lophozia quadriloba var. *heterophylla* Bryhn & Kaal., Rept. 2nd Norwegian Arctic Exped. in the "Fram" 1898–1902, 2(11):39, 1906.
Barbilophozia quadriloba Loeske, Hedwigia 49:13, 1909; K. Müller, Rabenh. Krypt.–Fl. ed. 3, 6(1):625, fig. 186, 1954.
Lophozia (subg. *Barbilophozia*) *quadriloba* K. Müll., Rabenh. Krypt.–Fl. 6(1):640, fig. 303, 1910; Macvicar, Studs. Hdb. Brit. Hep. ed. 2:203, figs. 1–4, 1926.
Orthocaulis quadrilobus Buch, Mem. Soc. F. et Fl. Fennica 8(1932):294, fig. 2 (12), 1933; Frye & Clark, Univ. Wash. Publ. Biol. 6(3):411, (1944) 1945.
Lophozia (subg. *Orthocaulis*) *quadrilobus* Schust., Amer. Midl. Nat. 45(1):42, 1951; Schuster, *ibid.* 49(2):321, pl. 4, 1953.

Slender, caespitose, in loose tufts or in dense patches, *the shoots erect*, often densely crowded, *dull olive-green* to brownish, in sun forms plants *often blackish or a bluish green* in drying; *slender, 800–1250(1750) μ wide, appearing terete*, (0.8)1.5–4 cm high, simple or occasionally with a terminal branch (replacing the lower 2 lobes of the 4-lobed leaf). Stems almost terete, rigid and stiff, nearly or quite straight, 220–275(360) μ in diam.; dorsal cortical cells (13)14–16(17) μ wide × 35–48 μ long, rectangulate (3–4.5× as long as wide), slightly to somewhat thick-walled, their diam. only slightly less than that of adjacent medullary cells; ventral 1–2 rows of cells mycorrhizal with age. Rhizoids frequent, scarce above, colorless. Leaves inserted by a distinctly acroscopically arched line, *only slightly oblique* (overall insertion at ca. 15–25° [>35°] from transverse), short decurrent antically and postically, *distant to contiguous*, imbricate only near shoot apices, broadly reniform-obtrapezoidal in outline, semi-amplexicaul, *typically with the undivided base distinctly spreading, but the lobes erect or suberect, the leaves thus distinctly cupped, almost handlike*; well-developed leaves from 875–975 μ wide × 600–675 μ long up to 1000–1500 μ wide × 820–900 μ long, (*2–3*)*4-lobed for 0.6 their length*, but sometimes divided for only 0.5 their length; lobes (on 4-lobed leaves) subequal or (on 3-lobed leaves) the dorsal considerably larger; *lobes narrow, ovate-triangular to ovate-lanceolate and pointed*, their apices often incurved, *their margins usually distinctly revolute, at the sinus base so strongly reflexed that it is rounded*, sinuses closed and acute below (but often narrowly rounded, where broadly reflexed), flaring and becoming broad

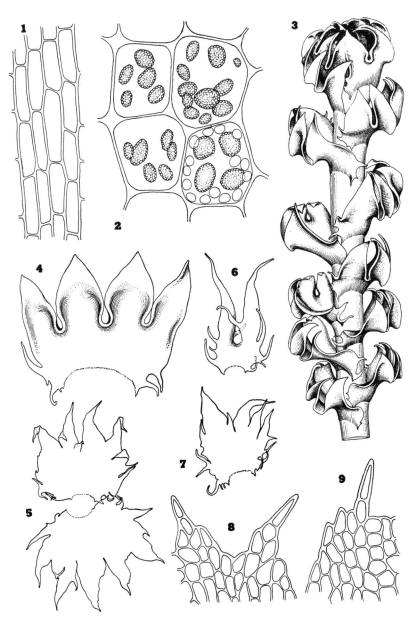

Fig. 140. *Lophozia* (*Orthocaulis*) *quadriloba*. (1) Dorsal cortical cells (×340). (2) Median cells with oil-bodies and, lower right cell, chloroplasts (×870). (3) Robust sterile shoot (×28). (4) Leaf (×50). (5) ♀ Bracts (×20). (6) Underleaf (×50). (7) ♀ Bracteole (×20). (8–9) Perianth mouth sectors (×250). [All from *RMS 35108a*, Mt. Pullen near Alert, Ellesmere I.]

distally; *postical base with usually 1–3 cilia or teeth, antical base often similarly, but usually less strongly, ciliate. Cells small*, rounded-polygonal, the walls thin to somewhat thickened, the trigones small to moderate, rarely weakly bulging, the cell lumen usually angular to rounded; marginal cells of lobes (12)13–15(18) μ, median cells of lobes 13–17 × 17–20(23) μ, cells of leaf middle (17)18–19(23) × (18)20–27(29) μ; basal cells hardly larger than median, (17)18–23 μ wide × 23–35 μ long, averaging somewhat broader than dorsal cortical cells; *cuticle coarsely and distinctly verrucose*, in basal half of leaf becoming striolate-verruculose;[18] oil-bodies in median cells usually 4–9 per cell, distinctly papillose-segmented, very variable in size, mostly ovoid or ellipsoidal, from 3.5 × 4.5–5 μ to 5–6 × 6–7 μ, a few to 6.5 × 9 μ or 5 × 11 μ; chloroplasts ca. 3.5–4 μ. *Underleaves very large*, from 600 μ long × 400 μ wide to 850 μ long × 600–700 μ wide, bifid for 0.75–0.9 their length, their bases (and bases of lobes) *pluriciliate*, the lobes linear-lanceolate, acuminate. *Gemmae absent.*[19]

Dioecious, usually sterile. Androecia becoming intercalary, longly spicate, the ♂ bracts in up to 10–20 pairs, imbricate, similar to leaves but 3–4-lobed only for 0.3–0.4 their length, concave at base, pluriciliate at both antical and postical bases; antheridia 2–4, situated among prominent, filiform to lanceolate paraphyses. Gynoecia with bracts usually barely or slightly larger than leaves, 1100–1500 (1600–1700) μ wide × 1000–1250 μ long, *4–5-lobed to the middle*, the bases of the lobes commonly with 1–2 cilia or laciniae, the undivided lateral margins of the bracts with several cilia or laciniae (some of which may be large enough to simulate small lobes); lobes acute to acuminate (rarely blunt), entire, or occasionally with a few vague teeth, with margins recurved, especially at base of sinus. Bracteole very large, 900 μ wide × 1000–1100 μ long, bilobed to middle, the reflexed base of sinus often with 1–2 teeth, the undivided base laciniate-lobed to ciliate (like bracts). Perianth oblong-obovoid, to 0.8 emergent, plicate, rounded and narrowed to mouth in distal fourth; mouth irregularly lobulate-dentate, the lobes faintly serrate, ending in 2–3 superimposed cells (the distal cells acuminate, ca. 12–14 × 25–35 μ). Sporophyte unknown.[20]

Type. European.

Distribution.

Evidently incompletely holarctic at high latitudes, extending from the northernmost land points (83°6′ N., at least) southward through much of the Tundra, chiefly where calcareous and subcalcareous conditions occur. Often nearly ubiquitous at very high latitudes in calcareous areas where the ground is snow-free for less than 3 months per year.

[18] According to Macvicar (1926), this character serves to separate *L. quadriloba* from *L. floerkei.*

[19] I have searched in vain for gemmae in this species. If ever present, they must be very rarely produced, although they have been described as polygonal to tetrahedral, yellowish, thick-walled, 18–27 μ long. H. W. Arnell appears to have been the only one who has seen them.

[20] In view of the fact that *L. quadriloba* is extraordinarily variable, I have drawn the above diagnosis to include, basically, only var. *quadriloba;* the deviant variations are treated on pp. 276–282.

In Europe both arctic and alpine, occurring from Spitsbergen and northernmost Scandinavia (Lapland) south to southern Norway, Sweden, and Finland; eastward to the Soviet Union; recurring in the Alps of Switzerland (Meylan, 1924), the Allgäu, Steiermark, and the Salzburg District of the Austrian Alps; also in the Italian Alps in Savoy and in the High Tatra Mts. Also in Scotland. Eastward extending into Asia (Siberia: Novaya Zemlya, the Yenisey R., and eastern Siberia); absent from Japan. Arctic-alpine in Europe but absent from the alpine Tundra in North America.

In North America transcontinental, ranging in the West from Alaska to Alberta and British Columbia. Eastward as follows:[21]

E. GREENLAND. Danmarks Ø, Gaaseland, Runde-Fjeld, all in Scoresby Sund (*Jensen*); Nyhavn Hills and Labben Hills, Mesters Vig (*Raup 716, 718*!); Cap Ravn (*Böcher 955*; var. *heterophylla*!); Nordostbugt; Hurry Inlet; Scoresby Sund (Jensen, 1900, 1906a; Harmsen, 1933); Dove Bugt (76°40′ N., among *Ditrichum flexicaule, Stereodon bambergeri;* Jensen, 1910); Zackenberg (*Holmen 33582;* a vigorous, typical form, deviating only in the usually blunt to narrowly rounded leaf tips). N. GREENLAND. Low Pt. (83°6′ N.); Cape Benet (83°2′ N.); Lemming Fjord (82°53′ N.); Somerdalen (82°29′ N.; *Thorild Wulff*!); Cape May (82°27′ N.), *teste* Hesselbo (1923a); Herlufsholm Strand, E. coast, 82°40′ N. (*Holmen 7449*!, with *O.h. paradoxa; Holmen 7623*!, var. *glareosa*). NW. GREEN-LAND. Nucluet, Booth Sd. (*RMS 46201*, etc.); Kânâk, Red Cliffs Pen., Inglefield Bay (*RMS 45556a*); Igdlorssuit, Bowdoin Bugt, Inglefield Bay (*RMS 45779d*); Dundas, North Star Bay, Wolstenholme Fjord (*RMS 46266*). W. GREENLAND. Sondrestrom Fjord, near shore of L. Ferguson, in *Salix*-moss tundra (*RMS & KD 66-063a, 66-079*). Disko I.: Fortune Bay, 69°16′ N., 53°50′ W. (*RMS & KD 66-135a, 66-141, 66-143; 66-207*, fo. *glareosa trans. ad* var. *collenchymatica*). Godhavn: mouth of Blaesedalen, near Lyngmarksfjeld (*RMS & KD 66-281c*); below Lyngmarksfjeld, 69°15′ N., 53°30′ W. (*RMS & KD 66-157, 66-165, 66-170*); slopes below Skarvefjeld, NE. of Godhavn, ca. 150–200 m (*RMS & KD 66-232*); below Skarvefjeld, E. of Godhavn, basalt cliffs near shore (*RMS & KD 66-320*). Eqe, near Eqip Sermia (Glacier), 69°46′ N., 50°10′ W.: ca. 1 mi from Inland Ice, 350–400 m alt. (*RMS & KD 66-291, 66-292, 66-296, 66-340*, etc.); basalt dike (*RMS & KD 66-288b, 66-333*). Mouth of Simiutap Kûa Valley, Umîarfik Fjord, 71°59′ N., 54°35–40′ W. (*RMS & KD 66-914h, 66-916b, 66-917*); Magdlak, Alfred Wegners Halvö, 71°06′ N., 51°40′ W. (*RMS & KD 66-1350*, var. *collenchymatica*); Kangerdlugssuakavsak, head of Kangerdluarssuk Fjord, 71°21′ N., 51°40′ W. (*RMS & KD 66-1288*); head of Marmorilik Fjord, N. of Akuliarusikavsak 71°05′ N., 51°12′ W. (*RMS & KD 66-1206*).

ELLESMERE I. Mt. Pullen, 5 mi S. of Alert, NE. Ellesmere, 82°26′ N. (*RMS 35108a, 35101, 35122b*, etc.); The Dean, S. of Alert (*RMS 35177, 35188*, etc.); S. end of Hilgard Bay, 12–15 mi SW. of Alert (*RMS 35231c, 35235*, etc.); between Mt. Olga and Mt. Erica, ca. 5 mi SW. of end of Hilgard Bay, 850–900 ft (*RMS 35520a, 35548a*); E. edge of U.S. Range, 9–10 mi W. of Mt. Olga (*RMS 35598, 35600*); Lincoln Sea, 1.5–2 mi W. of Cape Belknap, Alert (*RMS 35635, 35752*); Twin Valley Glacier, in Hayes Sd. region; Fram, Harbour, and Goose Fiords on S. coast (Bryhn, 1906; typical material); ("var. *heterophylla*" more widespread, occurring in Beitstad Fiord, Lastraea Valley, Cape Rutherford, Bedford Pim I., and Fram Harbor in Hayes Sd. region; Fram, Harbour, Goose Fiords on S. coast; Lands End and Reindeer Cove on W. coast; *fide* Bryhn 1906). DEVON I. REGION. North Kent I. (Bryhn, 1906; "var. *heterophylla*"). BAF-FIN I. Arctic Bay, N. Baffin (Polunin, 1947). SOUTHAMPTON I. *Fide* Steere (1941).

[21] Distribution basically that of typical *L. quadriloba* and of minor small forms often called "var. *heterophylla*." The collections clearly belonging to extreme forms, referred to recognizable forms and varieties, are treated separately (pp. 276–282).

NEWFOUNDLAND. River of Ponds Prov. Pk. (*RMS 68–1480*).   MICHIGAN. Copper Harbor, and at the "Devils Washtub" near Eagle Harbor, Keweenaw Co., 47°30′N. (*RMS, Sept. 9, 1949;* see Schuster, 1953; plate; southernmost known station of the species!).   ONTARIO. Little Cove, Bruce Pen. (Williams, 1968).

Ecology

Lophozia quadriloba is an abundant and characteristic species of slopes below persistent snow banks in basaltic or calcareous districts of the high Arctic (i.a., on Ellesmere I.); occurring particularly commonly on thin soil or humus over calcareous and subcalcareous rocks. It often forms extensive pure patches, generally tinged a diagnostic fuscous or even almost bluish green-black, particularly when nearly dry; at other times it occurs as straggling individuals amongst caespitose mosses. Associated Hepaticae at the northern extremity of its occurrence include *Tritomaria quinquedentata, T. heterophylla, Anthelia juratzkana, Cephaloziella arctica, Arnellia fennica, Lophozia heterocolpa, L. gillmani, L. hyperborea, Scapania gymnostomophila,* and *Odontoschisma macounii.* At the southern extremity of its range it is found either in the spray-moistened zone along Lake Superior, with *Lophozia gillmani* and *Scapania cuspiduligera,* or above this zone on drier rocks, with *Lophozia hatcheri, Tritomaria quinquedentata, Scapania mucronata,* and *Cephaloziella arctica,* in both cases over basaltic rocks. The ecology of this species is discussed in Schuster (1953) and Schuster et al. (1959).

The occurrence of the species with a whole series of "calciphytes" and of lime-tolerant species is notable. Müller (1905–16), however, notes that the species is found both over basic rocks and over acidic rocks.

Even in the Arctic, where *L. quadriloba* is common, in areas with acidic rocks it occurs rarely and then mostly as poorly developed and persistently juvenile manifestations. It is, indeed, present almost everywhere except in the most acid sites (as, e.g., between *Sphagnum,*) at least as isolated traces.

Differentiation

Associated with its extraordinarily varied series of habitats, *Lophozia quadriloba* occurs, in the Arctic, in an equally extraordinary series of phenotypes. As stated above, it is usually present in most sites that are not too acid—even when clearly growing under severely submarginal conditions. Then *L. quadriloba* occurs in one or another of countless juvenile phases that hardly "key out," no matter how flexible a key is constructed. Many or most of these small forms, with 2- or 2(3)-lobed leaves that generally lack gibbose sinuses and teeth or cilia of the leaf bases, belong in or near "var. *glareosa.*" Yet this entity (and "var. *heterophylla*"—which I think can hardly be separated) is chiefly a convenient receptacle in which all "underdeveloped" phenotypes are placed (see p. 276).

In the field at once characterized, when well developed, by the color. The plants, when in patches, give almost the impression of being fuscous or even bluish black, much as do some forms of *Gymnocolea inflata*. In other cases, plants are dull olive or brownish-green, never with a warmer brown cast. This separates the species from our other species of *Orthocaulis* in the

Fig. 141. *Lophozia* (*Orthocaulis*) *quadriloba*. (1) Apex of shoot with two *Frullania*-type terminal branches (×18). (2–3) Underleaves (×29). (4–7) Leaves, showing variation in lobe number; leaf in fig. 4 a "half-leaf" associated with a branch (×21). (8) Paraphyses from ♂ bract (×167). (9) ♂ Bract, *in situ* (×34). (10) Large paraphyllium from fig. 9 (×ca. 65). (11) Leaf, not flattened (×16). [1,4–7, from "Devil's Washtub," Keweenaw Pen., Mich., *RMS, 1949;* 2–3, 8–10, from Norwegian plants (Sör-Tröndelag, Opland), leg. Buch; after Schuster, 1953.]

field. The very deeply 4-parted leaves are characteristic. They strongly suggest a slender phase of *Chandonanthus setiformis*, but the latter is a warmer brown in color and distinctly nitid when dry. The similarity to *Chandonanthus* is increased by the cupped leaf form, with the undivided leaf base somewhat spreading or erect-spreading, but the lobes erect with their apices commonly incurved. The reflexed lobe margins, the strongly gibbous sinuses, and the presence commonly of 1-several cilia or teeth of the antical leaf base further enhance this resemblance to *Chandonanthus*, as do the large underleaves. However, the leaves of *L. quadriloba* are distinctly less deeply divided than those of *Chandonanthus setiformis*, and the sinuses between the lobes are never dentate. Furthermore, *L. quadriloba* usually has much laxer leaves.

When well developed, *L. quadriloba* cannot be confused with any other species of *Orthocaulis* (or other Lophoziaceae except for *Chandonanthus setiformis*). It differs from the other species of *Orthocaulis* in (*a*) the more deeply parted leaves; (*b*) the usual presence of 4 leaf lobes, the leaves less commonly being trilobate, with the antical lobe the largest; (*c*) the sporadic to common occurrence of basal and subbasal teeth or cilia of the antical leaf margin; (*d*) the lack of gemmae; (*e*) the peculiar coloration. Unfortunately, slender phases of *L. quadriloba* ("var. *heterophylla*") are very common in which many or the majority of leaves are only 2–3-lobed. Although described by Bryhn (1906) as a distinct variety, this phase of the species may be a mere modification occurring under difficult nutritional conditions. Much as in *Chandonanthus setiformis*, the weaker, more or less persistently "juvenile" shoots show reduction in lobe number, more shallowly incised leaves (often only for 0.5, occasionally only 0.45, the leaf length), and reduction or obsolescence of the teeth of the leaf bases. Such weak phases are subject to confusion with both *L. kunzeana* and *L. floerkei*. They agree with the former in the rather deeply lobed leaves, with the latter in the high incidence of trilobed leaves (and of occasional cilia, at least of the postical leaf base). The more sharply reflexed and commonly gibbous leaf sinus, the frequent occurrence of cilia of the postical leaf base, the frequent presence of trilobate leaves, and the usually acute or subacute leaf lobes separate slender phases of *L. quadriloba* from *L. kunzeana*. The separation from *L. floerkei* is more difficult, as has been emphasized in Schuster (1951). However, all phases of *L. quadriloba* have more deeply lobed leaves and a coarsely papillose cuticle. Other differences are cited under *L. floerkei*.

Lophozia quadriloba often occurs in the high Arctic with *L. hyperborea* (the two, e.g., are found admixed in the collection of Thorild Wulff from Somerdalen; see Fig. 145) and has been much confused with this species, particularly since both taxa show strongly variable leaf-lobe numbers. It differs from all forms of *L. hyperborea*, however, in (*a*) the narrower and usually more acute leaf lobes; (*b*) the smaller, less collenchymatous leaf cells; (*c*) typically in the deeper color, although transitions in this respect occur. Since these two taxa occur together (as in the Somerdalen collections and as in northern Ellesmere I.; see Schuster et al., 1959, p. 22) and maintain their differences, they obviously belong to two different species.

Variation

Lophozia quadriloba and L. hyperborea, together, constitute one of the critical and polymorphous complexes among Northern Hemisphere taxa, and their variability is best treated jointly. The following treatment, even though seemingly complex, is still hardly adequate to encompass the variations which occur—some of which will surely prove environmentally induced (?var. *glareosa*), whereas others seem genetic in origin. For the time being, it appears most rational to recognize these variants, of still unresolved status, taxonomically. For example, some collections seen have obtuse to narrowly rounded leaf-lobe apices, whereas "normal" L. quadriloba possesses sharp lobe tips—in fo. *cephalozielloides* even ending in up to 3–4 superimposed single cells. What is the taxonomic significance of

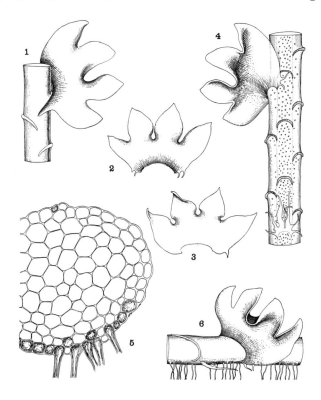

Fig. 142. *Lophozia (Orthocaulis) quadriloba.* (1) Leaf, *in situ*, on stem (×26). (2–3) Leaves (×20). (4) Leaf, underleaf, *in situ*, on stem, showing insertions, ventral mero-phyte boundary, and rhizoid-initial bases (×26). (5) Stem cross section, with mycor-rhizal infestation limited to cortex (×165). (6) Leaf and underleaf, *in situ*, showing the acroscopically arched insertion (×26). [1–2, 4, 6, Norwegian plants (Sör-Tröndelag, leg. Buch); 3, 5 "Devil's Washtub," Keweenaw Co., Mich. *RMS.*]

these variations? Are they environmentally induced or not? These are questions which only prolonged cultural studies can hope to answer.

The following key crystallizes the major patterns of variation:

Key to Variants of L. quadriloba and L. hyperborea

1. Cells on lobe margins 12–17(18–20) μ; median cells with (3)4–6(7–9) oil-bodies. [Leaves mostly with lobes narrow, much longer than broad, the sinuses usually reflexed; cuticle coarsely, usually contiguously papillose.] . . . . *L. quadriloba*   2.

  2. Cells with trigones absent or obscure, the cell walls usually equally thickened; leaf lobes normally sharply pointed; cells smaller, on leaf margins mostly 12–16(17) μ, in leaf middle 17–20(23) × 18–25 μ. . 3.

    3. Plants larger, often inky black or fuscous-black, leaves mostly (3)4-lobed, with strongly recurved margins and sinuses; postical (and occasionally antical) leaf bases with 1–2 cilia or teeth; underleaves large, bifid, usually ± ciliate . . . . var. *quadriloba*

    3. Plants small, often under 0.6 mm in diam., fuscous, with leaves mostly 2–3-lobed, margins usually unarmed and little or not recurved; under-leaves small, lanceolate or bifid, with 0–1(2) cilia . . . . var. *glareosa* (incl. var. *heterophylla*)   4.

      4. Black or fuscous; plants to 500–600 μ wide, often rather dense-leaved; branching mostly terminal-lateral; leaves with lobes often sharp-pointed but points 1–2-celled; sinuses not sharply V-shaped usually . . . . fo. *glareosa*

      4. Green to pale brown; plants 280–400 μ wide usually, remote-leaved, often with ventral- and lateral-intercalary branches; leaves in large part 2- (less often 3-, almost never 4-)lobed; lobes 4–8 cells wide usually, short-acuminate with mostly 2–4 superimposed cells at apex; sinuses narrow and very acute, to 0.75 the leaf length on smaller leaves . . . . fo. *cephalozielloides*

  2. Cells with trigones large, often rather bulging, walls between thin; most leaf lobes blunt to narrowly rounded; cells larger, the marginal cells of lobes 16–19 up to 18–20 μ, in leaf middle mostly 20–24 μ wide . . . . var. *collenchymatica*

1. Cells on lobe margins (18.5–20)21–25(26) μ, in leaf middle (21)22–26(28) × 24–32(36) μ. Leaves 2–3(4)-lobed. [Small, often under 0.7 mm wide, brownish to fuscous as in var. *glareosa*, with lateral leaf margins unarmed or with 1–2(3) small, short, often blunt cilia or teeth; underleaves small, often lanceolate, as in var. *glareosa*.] High arctic only (north of 70°) . . . . *L. hyperborea*   5.

  5. Leaf lobes broadly triangular, usually nearly as broad or broader than long, sharp to rounded at apices; sinuses descending 0.35–0.45(0.5) leaf length; underleaves lanceolate to bifid, the lanceolate lobes ± ciliate; trigones small, cuticle with at least locally very coarse papillae (as in *L. quadriloba*); oil-bodies (4–6)7–15 per cell . . . . *L. hyperborea* var. *hyperborea*

5. Leaf lobes usually narrowly triangular, longer than wide, acute to bluntly pointed; sinuses descending 0.5–0.65 leaf length; underleaves minute, lanceolate to bifid, with 0–2–3 small teeth or minute cilia; trigones coarse, often bulging, the cuticle moderately papillose to nearly smooth; oil-bodies usually 2–5 per cell . . . . *L. hyperborea* fo. *paradoxa*

LOPHOZIA QUADRILOBA var. GLAREOSA (*Joerg.*) *Joerg.*

Jungermannia quadriloba var. *glareosa* Joerg., Bergens Mus. Aarbok 18:0.p, 1895 (ref. not seen).
Lophozia quadriloba var. *glareosa* Joerg., Bergens Mus. Skrifter no. 16:142, 1934.
Lophozia quadriloba var. *heterophylla* Bryhn & Kaal., Rept. 2nd Norwegian Arctic Exped. in the "Fram" 1898–1902, 2(11):39, 1906.

Small, usually under 600 μ wide × 15 mm long, with shoots often rather rigid, often filiform, blackish brown to fuscous green but without the "inky" cast of well-developed *L. quadriloba*. Leaves 2–3(4)-lobed usually for 0.35–055 with ± convex outer margins which are never armed; lobes acute to sharply apiculate. Underleaves much reduced, lanceolate or narrowly lingulate-bifid, with only 0–1(2) small teeth or cilia towards base.

Type. ?

Distribution

Found evidently throughout much of the range of the typical species; some literature reports, perhaps only questionable here, have been cited (as "var. *heterophylla*") under the species proper. The following are representative of my concept of var. *glareosa*:

E. GREENLAND. Cap Ravn, Urteli, 68°31′ N., ca. 28°15′ W. (*Böcher 955, p.p.*, with *Cephaloziella arctica, Preissia quadrata*, etc.; reported as *L. quadriloba* by Harmsen, 1933. The plants have the leaf cells very slightly larger than normal: marginal cells of lobes average 15–17 μ, median cells average 18–21 μ wide. In other features—the predominantly 2–3-fid leaves, the occasional sporadic occurrence of single 4-fid leaves, and the sharp lobes and sharp sinuses, the plants are typical var. *heterophylla*); Fleming Inlet, 70°40′ N., ca. 23° W. (*N. Hartz or Kruuse, Aug. 26, 1900;* cited by Jensen, 1906a, as *L. elongata*). N. GREENLAND. Herlufsholm Strand, E. coast of Peary Land, 82°40′ N., ca. 21° W. (*Holmen 7741;* also *7623, p.p.;* mixed with *L. heterocolpa* var. *harpanthoides* and *Blepharostoma trichophyllum*); Station Nord, 81°36′ N., 17° W. [*Troelsen, Aug. 22, 1954;* det. and publ. as *Orthocaulis floerkei* by Arnell (1960)]. W. GREENLAND. Ikorfat, Nugssuaq Pen., 70°45′ N., 53°07, W. (*Holmen 62-1574*); Ritenbenk, Agpat, 69°46′ N., 51°20′ W. (*RMS & KD 66-260*, ± var. *glareosa; 66-264a*); Eqe, near Eqip Sermia (Glacier), ca. 1 mi from Inland Ice, 350–450 m, 69°46′ N., 50°10′ W. (*RMS & KD 66-339b*, with fo. *cephalozielloides*); E. side of Hollaenderbugt, N.W. Nugssuaq Pen., slopes below Qârssua, 60°47′ N., 53°47′ W. (*RMS & KD 66-1505c*).

The above plants appear to be identical with *L. quadriloba* var. *glareosa* (Joergensen, 1895), which Joergensen (1934, p. 142) compares to both *L. kunzeana* and *L. quadriloba* var. *heterophylla*. Joergensen (1934) states: "A very characteristic and almost unrecognizable form, which easily can be confused with a small *L. kunzeana*, is var. *glareosa* Joerg., 1895."

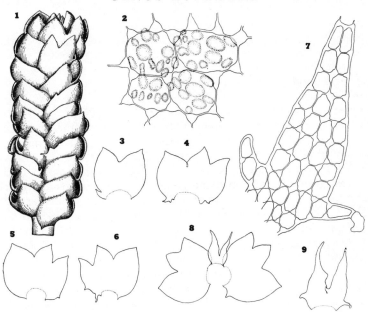

Fig. 143. *Lophozia* (*Orthocaulis*) *quadriloba*, trans. ad *L.* (*O.*) *floerkei*. (1) Sterile shoot apex (×27). (2) Median cells, with cuticular papillae (×518). (3–6) Leaves (×27). (7) Lobe of underleaf in fig. 9 (×265). (8) Two leaves and juxtaposed underleaf (×27). (9) Underleaf (×57). [All from Marr, 1939, from Tukarak I., Belcher Isls., NWT.; reported as *L. floerkei* by Schuster, 1951; close to var. *glareosa*.]

This is described as "blackish brown in color, more threadlike (filiform) than the main species (only 0.6 mm broad). Leaves usually 3-lobed, partly 2- and 4-lobed, with little or not recurved margins in the leaf lobes, lobes often sharp or shortly apiculate."

Joergensen states, "A similar reduced arctic form (julaceous because of the leaves closely inserted together, the leaves often 2-lobed and often with recurved leaf margins) is var. *heterophylla* . . . ." I consider both of these as, basically, impoverished extremes of difficult sites, only questionably based on genetic differences, and therefore treat them here together.[22]

The plants of *Holmen 7623* were subjected to repeated study because they form part of the basis for the report of *Lophozia kunzeana* from Peary Land in Arnell, 1960, p. 9. The plants are small, rather fuscous green, with largely 2–3-lobed leaves, although I have seen a very few with 4-lobed leaves; sinuses descend 0.45–0.6 and on the best-developed plants show a slight *L. quadriloba*-like gibbosity at the sinus bases; marginal cells of the lobes average 13.5–15.5 μ; trigones are small or very small and concave-sided. Underleaves are present

---

[22] My primary reason for continuing to give the small phenotypes with 2–3(4)-lobed leaves continued taxonomic recognition lies in the fact that, to date, they have most often been confused with other taxa, e.g., *L. floerkei* (Arnell, 1960), *L. kunzeana* (Arnell, 1960), or *L. elongata* (Jensen, 1906a); see footnote, p. 291.

throughout, small, lanceolate or deeply bifid, without conspicuous cilia. The plants from Nord (*Troelsen*), which were published by Arnell as *Orthocaulis floerkei*, are very similar, with marginal cells of the lobes mostly 13.5–16 μ in diam., and the lobes narrow, longer than broad; they represent an inky black phase very common in polar regions.

The plants from Fleming Inlet, reported by Jensen (1906a) as *L. elongata*, do not agree with the type, from which they differ as follows. (1) Underleaves are better developed, ovate-lanceolate to lanceolate and either unlobed (then ca. 90 × 180–200 μ) or bifid (then 100–120 × 225–275 μ), the margins near the base occasionally with a slime papilla, but not ciliate. (2) Leaves are variably 2-lobed, with rare 4-fid leaves; sinuses descend 0.35–0.65 the leaf length. Leaves range from bilobed and subquadrate and ca. 380–500 μ in length and width, to asymmetrically trifid and *Orthocaulis*-like, then to 620 μ wide × 400 μ long. The succubously inserted leaves are rather subvertical at times, often *Orthocaulis*-like in insertion and orientation. (3) Cells have very to moderately small trigones, are relatively small (marginal cells of lobes chiefly 14–15 μ), and range from nearly smooth to relatively weakly papillose-striolate; median cells average ca. (14)15–19(21) × 18–23 μ. (4) The stem has a rather rigid cortex which is brownish, moderately thick-walled on old stems, the cells 14–16 μ in diam. and little smaller than the hyaline and nonmycorrhizal medullary cells. When these features are compared to the diagnosis and figures (pp. 431–436) of the type of *L. elongata*, it is evident that the Fleming Inlet plants cannot belong there.

LOPHOZIA QUADRILOBA var. GLAREOSA fo. CEPHALOZIELLOIDES Schust., fo. n.

[Fig. 144]

Ramificatio variabilis; ventrali- et laterali-intercalaris et typi Frullaniae; planta minuta reductaque, folia profunde bifida, lobis tenuiter lanceolatis, ad basim 5–6 cellulis latis, plerumque habens; bases foliorum non armatae. Amphigastria plerumque lanceolati-subulata.

Small, slender, olive-green to light dull brownish, creeping; ventral surface copiously producing very long, thick-walled rhizoids, 7.5–12 μ in diam. Shoots with leaves only ca. 280–380 μ wide × 2–6(8) mm long, simple or rarely branched (branches seen both *ventral-intercalary and lateral-intercalary*, at almost right angles to parent axis, and terminal, *Frullania* type). Stems terete, slender, ca. 75–110 μ in diam., of ca. 10–11 to 17–18 rows of rather thick-walled cortical cells, striolate on their convex, free faces, each 14–16 μ in radial diam., (12)14–18 × (22)25–48(58) μ, ± oblong, slightly larger to subequal in diam. to the few (usually 9–20, in 3–4[5] tiers) rows of thick-walled medullary cells; medullary cells 12–18 μ in diam.; no mycorrhizal medullary infestation, but external hyphae over rhizoids (and often spirally curled over them) common. Leaves remote except at shoot tips, vertical (the transverse insertion antically attaining stem midline), suberect to erect-spreading and usually somewhat concave, the leaf tips occasionally weakly or distinctly incurved; leaves oblong-ovate, ca. (140)195–210(220) μ wide × (180)270–320(335) μ long, bifid (rarely trifid) for 0.55–0.75 their length; when trifid widest distally, to 365 μ, the lobes erect,

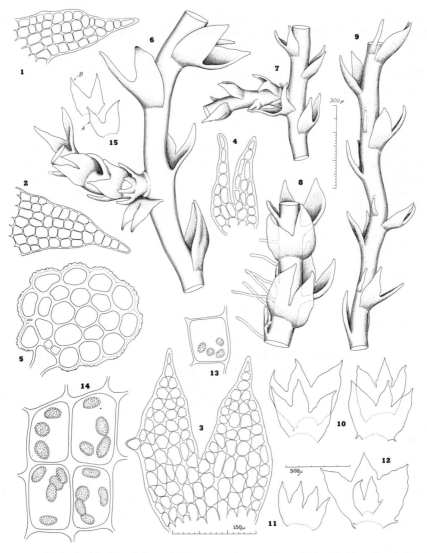

FIG. 144. *Lophozia* (*Orthocaulis*) *quadriloba* var. *glareosa*, fo. *cephalozielloides*. (1–2) Leaf lobes, fig. 1 from lobe A in fig. 15; fig. 2 from lobe B in fig. 15 (×185). (3) Leaf (to scale below; ×185). (4) Underleaf (×185). (5) Stem cross section, of small stem (×425). (6–7) Shoot sectors, dorsal view, with lateral-intercalary branches (×70). (8) Shoot sector, lateral view (×70). (9) Shoot sector extreme, ventral view (×70). (10) Four leaves of large extreme, grading into fo. *glareosa* (×35). (11) Unusual and rare, quadrilobed leaf (×35). (12) Large leaf and underleaf (×35). (13) Cell from lobe margins (×700). (14) Median cells (×820). (15) Two leaves of "normal" size (×35). [All drawn from *RMS & KD 66-326*, Eqe, W. Greenland; type. Figs. 1–4, to scale below fig. 3; 6–9, to scale left of fig. 9; 10–12, 15, to scale adjoining fig. 12.]

on trifid leaves sometimes the larger dorsal ones divergent, slenderly lanceolate, (4)5–6(8) cells wide at base, longly acute to acuminate, ending in 1–2-3(4) in part elongated single cells; lateral margins feebly arched, entire or, on larger leaves, occasionally with a small, sharp, 1–2(3)-celled tooth (sometimes from a 2-celled base) at or near base or from near lobe base; sinuses V-shaped, acute. Cells at lobe bases ca. (14)15–21 × 16–26(28) μ, firm, somewhat thick-walled with rounded corners or rather distinct, concave-sided to weakly convex trigones; cuticle faintly papillose-striolate above, striolate toward base; oil-bodies (2)3–5 per cell in lobes, moderate-sized or small, colorless, distinctly granulate-botryoidal, ca. 3 × 3–5 to 4–5 × 6–6.5(8), rarely 3.5 × 9 μ. Underleaves, except on smallest shoots, to 200 × 60 μ large, lanceolate-subulate to lanceolate-lingulate, 2–3(4) cells wide at base usually, often with a sharp basal tooth of one or both sides, the subulus terminated in 4–6 isodiametric to somewhat elongated single cells, initially faintly arched away from axis, the apex often curved toward axis; underleaves occasionally bifid to near base. No gemmae. Otherwise unknown.

Type. Eqe, near Eqip Sermia, west Greenland, 69°46′ N., 50°10′ W. (July 20, 1966; *RMS & KD 66–339b*), over thin soil around moist vertical basalt dike; associated with *Peltolepis grandis, Cephalozia pleniceps, Blepharostoma trichophyllum brevirete, Lophozia heterocolpa, Tritomaria quinquedentata, Anastrophyllum minutum;* ca. 450 m, within 1 mile of Inland Ice.

Smaller plants superficially resemble a large, large-celled *Cephaloziella*, which they approach in the distinct underleaves; the vertical, transverse, remote, deeply bifid lateral leaves; the rather small—but for *Cephaloziella* quite large—and often rather evenly thick-walled cells; the roughened cuticle; the oil-bodies, several in each cell; the intercalary branching; the scattered rhizoids; the tendency, especially marked in *Cephaloziella* sect. *Schizophyllum*, to have a lateral tooth of some leaves. The underleaves, however, are impossible to reconcile with *Cephaloziella*; they are unusually long—often 0.5–0.65 the leaf length, are often setaceous for much of their length or formed of two setaceous segments, and are arched, with their tips juxtaposed to the axis, the median part usually arched away from the stem.

The size and facies of most plants may remind one of *Cephalozia leucantha*, even though the rather brownish color and presence of oil-bodies (as well as the striate cuticle of stem and leaf bases) eliminate this species—and genus.

The fo. *cephalozielloides* has been treated in detail because, even with careful study, it can lead the student far astray. The predominance of lateral- and ventral-intercalary branches and the very deeply bifid leaves with narrow lobes are quite uncharacteristic of any other regional taxon of Lophoziaceae. I eventually found intermingled a few axes approaching nearly typical var. *glareosa* and placed the plant under this. However, in studying numerous other juvenile phenotypes of *L. quadriloba*, I have not, again, met with a plant as reduced as this one.

LOPHOZIA QUADRILOBA var. COLLENCHYMATICA
Schust., var. n.

[Fig. 139:1–4]

Varietas var. quadrilobae similis, cacumina foliorum, autem, saepissime obtusa aut anguste rotundata; cellulae saepe paululo maiores, cellulis marginalibus 16–20 μ, cellulae valde collenchymatosae, trigonis ± inflatis.

Fuscous, erect, rather rigid and brittle. Leaves 2–3–4-lobed, rather dense, divided variously 0.3–0.45 to 0.45–0.55, lobes usually much longer than broad, their tips *mostly obtuse or narrowly rounded;* sinuses ± strongly to feebly recurved or gibbous. *Cells often larger* than in var. *quadriloba:* marginal cells 16–19 grading to 18–20 μ; median cells 20–24 × 25–36 μ, thin-walled, with sharply defined, often nodose trigones; cuticle coarsely, contiguously papillose; oil-bodies 2–6 per cell. Underleaves bifid to near base, with 2–3 small basal cilia.

Type. Head of Sondrestrom Fjord, west Greenland, near Lake Fergu-son, *RMS & KD 66–075,* with *Scapania brevicaulis, Anastrophyllum minutum.*

The type plants have very variable leaves, often unarmed with basal teeth or cilia, but some have cilia of the postical bases, and a few show a tooth of the antical base as well. Cell size varies somewhat with vigor but always ranges well over that of "normal" *L. quadriloba* (in which cells range mostly from 12–15 μ on the margins).

Distribution

N. GREENLAND. Cape Benet, 83°02′ N. (*Thorild Wulff, June 5, 1917;* grading to var. *heterophylla*); Low Pt., 83°06′ N. (*Th. Wulff, July 10-13, 1917;* mixed with *L. h. paradoxa*). N.W. GREENLAND. Kangerdluarssuk (*RMS 45779c*); Kânâk (*RMS 45556a*). W. GREENLAND. Tupilak I., Egedesminde, 68°42′ N., 52°55′ W. (*Holmen, June 30, 1956;* forma with shallower than normal lobed leaves); Sondrestrom Fjord, near shore of L. Ferguson, in *Salix*-moss tundra (*RMS & KD 66–075*); Fortune Bay, Disko I., 69°16′ N., 53°50′ W. (*RMS & KD 66-139*); Magdlak, Alfred Wegeners Halvö, 71°06′ N., 54°40′ W. (*RMS & KD 66-1350*). [Approaching (but not identical) are plants shown in Fig. 143.]

Although leaves on mature shoots are mostly 3–4-lobed, on weak shoots a high incidence of (2)3-lobed ones occurs. Plants are brownish in aspect, and have the sinuses mostly descending only 0.3–0.4, rarely 0.45, the leaf length. Leaves spread rather stiffly, with the lobes often hardly erect or suberect, so that (together with the shallower than normal sinuses) a *L. floerkei*-like aspect obtains. The cells, however, are coarsely papillose; marginal cells are ca. 16–19 μ; cells tend to be relatively thin-walled with large, conspicuous trigones. The slight development of the reflexed sinus bases is unusual, and also the normally blunt or even narrowly rounded leaf lobes. Associated with the shallower sinuses, lobes are generally rather broader. In most of the deviant features an approach to *L. floerkei* is evident—yet the high incidence of 4-lobed leaves, the coarsely papillose cuticle, and the occasionally more deeply divided leaves of some shoots clearly place these plants in *L. quadriloba*.

LOPHOZIA (ORTHOCAULIS) HYPERBOREA (Schust.) Schust., stat. n.

[Figs. 145:7-10; 146:5-15; 147]

Lophozia (Orthocaulis) floerkei var. *hyperborea* Schust., in Schuster et al., Natl. Mus. Canada Bull. 164:21, 1959.

Erect to loosely procumbent, with facies of *L. atlantica* or of a dense-leaved *L. kunzeana*, sometimes dense-leaved and almost julaceous, brown to brownish black, unbranched or rarely branched; shoots mostly 800–1050(1100) μ wide × 1–2 cm long. Leaves subvertical to vertical, subcontiguous to moderately imbricate, subpectinate to pectinately oriented, rather *strongly concave to concave-folded, lobes usually incurved; leaves unusually broad, very variable, 2–3(4)-lobed*, mostly transversely oblong-rounded to subobtrapezoidal, usually widest medially, when bilobed from 860 μ wide × 600 μ long to 1030 μ wide × 750 μ long (1.1–1.3× as long as wide), when 3(4)-lobed, 920–1000 μ wide × 650–700 μ long *(to 1.3–1.5× as wide as long), divided for (0.3)0.35–0.5 their length*, the V-shaped sinuses usually slightly reflexed at base but *lobes never canaliculate; lobes broadly triangular to ovate-triangular, normally as wide as or wider than long, the tips blunt to ± acute* (but often ending in 2 superposed cells which are normally wider than long); postical leaf bases with 1–3 blunt to sharp short teeth and/or cilia, less often also antical bases. *Underleaves large*, contiguous to imbricate, almost always bifid 0.7–0.85 their length, basally with several short teeth and, occasionally, 1–2 cilia or laciniae, the *lanceolate-acuminate lobes* near base usually pauciciliate to ciliate-dentate. *Large-celled: marginal cells* in lobes (18.5–21)22–25(27) μ; median cells of lobes variable, 20–25(26) × 20–27 μ up to 22–25(27) × 22–26(28) μ; cells of middle of disc (23)25–29 × (25)26–30(32) μ; *cuticle closely and coarsely papillose-verrucose* with elliptical papillae that range up to 4–5 × 8 to 6–8 × 10–12 μ; *oil-bodies (4–6)7–15 per median cell*, feebly botryoidal, from 4.5–6.5 μ when spherical to 4.5–7 × 6–8(9) when ellipsoidal. *Gemmae never present.*

Dioecious. Androecia eventually intercalary, of 3–5 (rarely more) pairs of somewhat antically inclined, rather strongly ventricose, suberect to obliquely patent, 2(3)-lobed bracts. Gynoecia with bracts leaflike, 2–3-lobed, with *erect, blunt to subacute lobes*, 860–1030 μ wide × 600–750 μ long, sinus bases sharply acute, usually reflexed; margins sinuous to entire, with only occasionally a blunt to acute subbasal tooth. Perianth ovoid, feebly plicate near the contracted, lobulate-ciliate and ciliate-dentate mouth, with a deep longitudinal antical sulcus, ca. 940 μ in diam. × 1350–1500 μ long; perianth mouth with shallow, irregular lobes ending usually in cilia formed of 2(3) superposed, somewhat elongated cells.

Type. Mt. Pullen, near Alert, ca. 82°25′ N., northeast coast of Ellesmere I., on shale substrate (*RMS 35101*); fragments of the type in herb. Grolle and Ottawa.

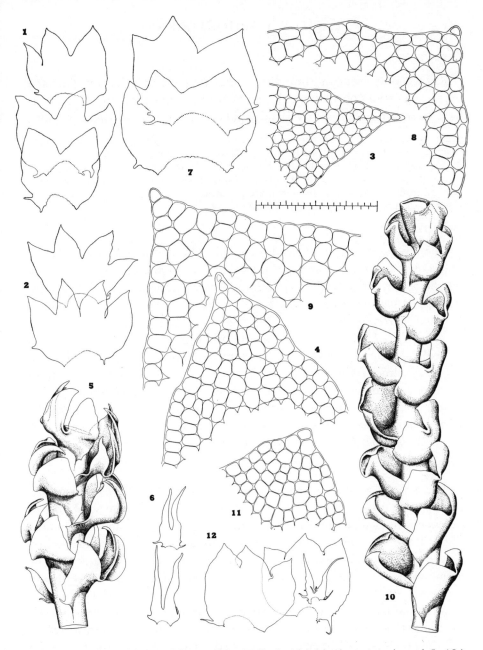

FIG. 145. *Lophozia* (*Orthocaulis*) *quadriloba* (1–6), *L.* (*O.*) *hyperborea* (7–10), and *L.* (*O.*) *floerkei* (11–12). (1–2) Leaves, showing variation (×36). (3) Leaf lobe apex (×165). (4) Leaf lobe apex (×220). (5) Shoot sector, dorsal view (×32). (6) Two underleaves (×36). (7) Two leaves (×36). (8) Leaf lobe apex (×165). (9) Leaf lobe apex (×220). (10) ♂ Shoot apex (×32). (11) Leaf lobe apex (×220). (12) Two leaves and an under-leaf (×20). [1–10, from intermingled plants, from Somerdalen, N. Greenland, *Thorild Wulff, July 13, 1917;* 11–12, *RMS 32968,* Mt. Katahdin, Me. Figs. 3, 8, drawn to top scale = 200 μ; 4, 9, 11, to bottom scale = 150 μ.]

Distribution

Known only from the imperfectly and perhaps only locally glaciated northern portions of Ellesmere I. and Peary Land (Greenland), and northwest Greenland. The Peary Land plants referred to fo. *paradoxa* (p. 286).

ELLESMERE ISLAND. Same general locality as type (*RMS 35366, 35366a, 35121, 35132; cotypes*); The Dean, just E. of Mt. Pullen (*RMS 35597; cotype*); moist snow-fed scree slope, 2000–2400 ft, E. edge of U.S. Range, ca. 9–10 mi due W. of Mt. Olga (*RMS 35572, 35576, 35580; cotypes*). Plants of *RMS 35118a*, from Mt. Pullen, originally referred to *L. quadriloba* var. *grandiretis*, are probably an extreme phase of *L. hyperborea*.
NW. GREENLAND. Nucluet (*RMS 46206*).

Ecology

The type and cotype material occurred in limited irrigated sites, below "hanging" snow banks and firn-ice "cliffs," in high arctic desert regions, where there are less than 4–6 in. of precipitation. Associated were *Tritomaria quinquedentata* and its var. *grandiretis, Anastrophyllum minutum, Scapania praetervisa* var. *polaris, Lophozia excisa, L. kunzeana* fo. *rotundiloba, L. heterocolpa, L. alpestris* subsp. *polaris*, and *Cephaloziella arctica*, as well as typical *Lophozia quadriloba*. The last-named also occurred with the Thorild Wulff specimen (cited below as fo. *paradoxa*). Associated mosses were (Schuster et al., 1959) principally *Hypnum bambergeri, H. revolutum, Encalypta procera, Tomenthypnum nitens, Myurella julacea, Barbula icmadophila, Distichium capillaceum, Ditrichum flexicaule, Timmia austriaca, Tortula ruralis, Grimmia apocarpa, Rhacomitrium lanuginosum*, and *R. canescens*.

Differentiation

This species appears to be a high arctic derivative of *L. (O.) floerkei*, from which it differs in (*a*) the less quadrate, more transverse-oblong to obtrapezoidal, generally much broader leaves; (*b*) the more polymorphous leaves—a single stem bearing usually a mixture of 2- and 3-lobed leaves with, sporadically, a 4-lobed leaf present; (*c*) the less apiculate or acute leaf lobes; as Fig. 145:8, 9, 11 shows, this distinction is not always realized; (*d*) the shorter, coarser, usually fewer lacinia and cilia of the postical leaf base, but with a tendency (absent in all *L. floerkei* I have seen) for the antical leaf base to develop 1–2 teeth; (*e*) the deep brown color; (*f*) the much larger leaf cells; for these see diagnosis and figures; (*g*) the generally more coarsely papillose cuticle; (*h*) the more numerous oil-bodies, usually 7–15 per cell, only sporadically a cell with as few as 4–6. Although I had originally assigned the plants to *L. floerkei* as a high arctic variety, I now believe that treatment as a distinct species is warranted.

Lophozia hyperborea also shows some conspicuous similarities to *L. quadriloba*, a species which is so variable in the high Arctic as to almost defy

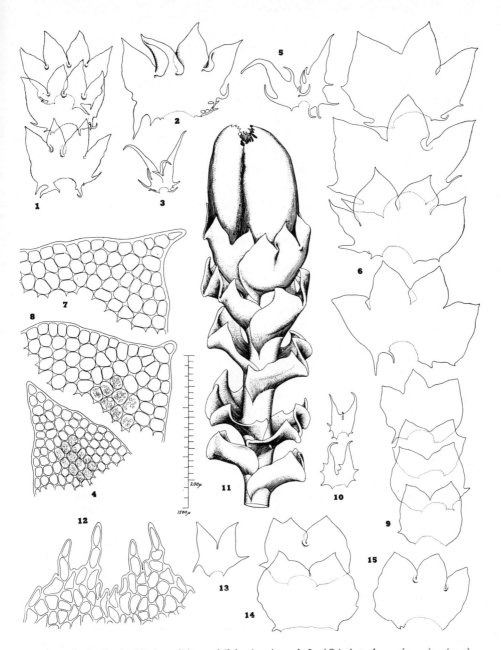

FIG. 146. *Lophozia (Orthocaulis) quadriloba* (1–4) and *L. (O.) hyperborea* (5–15). (1–2) Mature leaves of vigorous plant (×26.5). (3) Underleaf of optimal size (×26.5). (4) Leaf lobe, cuticular papillae drawn in locally (×165). (5) Large underleaf (×26.5). (6) Optimally developed leaves (×26.5). (7–8) Leaf lobe apices, locally with cuticular papillae drawn in (×165). (9–10) Normal-sized leaves and underleaves (×26.5). (11) Perianth-bearing shoot (×26.5). (12) Perianth mouth cells (×165). (13) Subfloral underleaf (×26.5). (14) Subfloral bract (above) and ♀ bract (below) (×26.5). (15) ♀ Bract (×26.5): [1–4, *RMS 35108a*, Mt. Pullen, Ellesmere I., of optimally developed typical *L. quadriloba;* 5–7, *RMS 35118a*, same microlocality as figs. 1–4; 8–15, *RMS 35101*, type of *L. hyperborea*. Figs. 1–3, 5–6, 9–11, 13–15 all to scale at left; 4, 7–8, 12 to scale at right].

understanding. Indeed, for some time I speculated with the idea that *L. hyperborea* was a larger-celled, high arctic subspecies of *L. quadriloba*. The coarsely papillose cuticle of both taxa suggests this possibility. However, in the very sparing plants of Thorild Wulff from Somerdalen (Fig. 145), material of *both* taxa is admixed. *L. hyperborea*, when thus studied juxtaposed to *L. quadriloba*, is obviously distinct in (*a*) the much larger leaf cells; (*b*) the broader and usually blunter leaf lobes; (*c*) the rather shallower and more often obtuse sinuses. Since I have also collected *L. hyperborea* and *L. quadriloba* admixed in Ellesmere I. (see Fig. 146; see also Schuster et al., 1959), it seems self-evident that two distinct species must be at hand.

On the basis of the often bilobed leaves and blunt lobes, and the large underleaves, confusion with *L. kunzeana* is possible. Leaf tips, however, are usually sharper in *L. hyperborea*—often ending in two superposed cells, the cilia of the leaf bases are better developed, gemmae are lacking, and leaf cells are consistently larger. Since cotype material of *L. hyperborea* was found admixed with *L. kunzeana* (see Table I in Schuster et al., 1959, where the differences in cell size between the intermingled plants are contrasted), the two taxa are obviously distinct. *Lophozia hyperborea* has, also, a conspicuously coarser papillose cuticle.

LOPHOZIA HYPERBOREA fo. PARADOXA
Schust., fo. n.

Varietas parva, ad 12–15 mm long.; cellulas magnas velut in var. *hyperborea* (cellulis marginalibus 21–25 μ, mediis 22–24 × 24–32 μ), solummodo 2–5(6) guttas olei, autem, omni in cellula habens; cellulae trigona grossa, saepe inflata praebentes; cuticula parum papillosa, aut fere omninove levis.

Small (typically under 700 μ wide × 12–15 mm long), fuscous to deeply brown-fuscous, with subcontiguous, stiffly patent leaves. Leaves 2–3-lobed, often mostly 2-lobed, the 2-lobed ones subquadrate to broadly quadrate (380 × 380 μ to 680 × 550 μ, then wider than long), the 3-lobed ones wider than long (600–700 × 450–550 μ up to 830 × 640 μ) and obtrapezoidal, all leaves divided 0.5–0.6(0.65) their length, the sinuses sharp, often a little recurved and narrowly gibbous, but lobe margins flat or essentially so; lobes acute to subacute, occasionally blunt, longer than wide; postical margin (rarely antical) with 0–1–2 small teeth (never cilia) that are often blunt or only subacute. Cells with brown walls and quite obvious, often bulging trigones; cuticle smooth to feebly papillose-striolate; marginal cells (19)21–25(26) μ; median cells (29)22–24(25) × 24–32(36) μ, very variable in size from cell to cell; oil-bodies in median cells and in lobes 2–5(6) per cell, variable in size, spherical and 3.5–5.5(7.5) μ to ellipsoidal and 4–4.8 × 7–8 μ. Underleaves small or minute, appressed, from lanceolate (ca. 160 × 180–200 to 200 × 300 μ, with 2–4 short cilia or teeth) to bifid and eciliate (then 160 × 180 μ or larger).

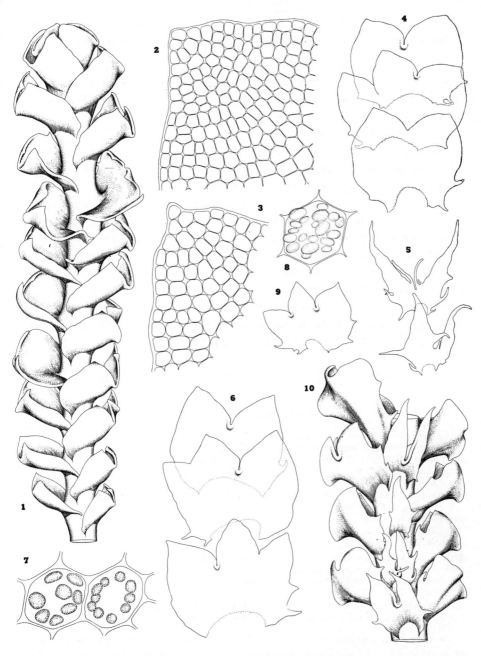

FIG. 147. *Lophozia (Orthocaulis) hyperborea.* (1) Well-developed shoot (×32). (2) Lobe apex cells from lowest leaf in fig. 4 (×164). (3) Lobe apex cells from uppermost leaf in fig. 4 (×164). (4, 6) Leaves (×39). (5) Underleaves (×39). (7) Two cells with oil-bodies (×508). (8) Cell showing cuticular papillae (×508). (9) Leaf with optimal development of teeth of bases (×ca. 25). (10) Robust shoot-sector, ventral aspect (×42). [All from *RMS 35101*, type.]

Type. North Greenland: Herlufsholm Strand, eastern coast of Peary Land, 82°40′ N., ca. 21° W. (*Holmen 7449, p.p.;* Copenhagen!).

Additional localities: North Greenland: Low Point, 83°06′ N. (*Thorild Wulff, July 10–13, 1917!* mixed with *L. quadriloba*); Somerdalen, northern coast of Wulff's Land, 82°29′ N. (*Th. Wulff, July 13, 1917,* with *L. quadriloba, L. heterocolpa* var. *harpanthoides, Arnellia fennica,* etc.).

No other collections are known.

The type occurred admixed with *Arnellia fennica,* traces of *Blepharostoma trichophyllum* var. *brevirete, Lophozia heterocolpa* var. *harpanthoides,* and ?*Lophozia badensis,* in a dry, stony fell-field.

This paradoxical variant (hence the name) has the large cells of var. *hyperborea,* yet has on an average only ⅓ as many oil-bodies per cell. (Fortunately the type still had many leaves with well-preserved oil-bodies.) It also differs in the coarse trigones—an unusual feature in *L. quadriloba s. lat.*—and is unique in having an almost smooth cuticle. Only in the median and basal cells have I been able to perceive weak striolations. The leaves are, as in arctic phases of *L. quadriloba,* extraordinarily variable, with lobes varying from blunt to quite sharp. The Wulff specimens (Fig. 145:7–10) less surely belong to this form.

LOPHOZIA (ORTHOCAULIS) KUNZEANA (Hüben.)
Evans
[Figs. 139:5–9; 148–150]

Jungermannia kunzeana Hüben., Hep. Germ., p. 115, 1834.
Jungermannia barbata β. floerkei var. *obtusata* Nees, Naturg. Eur. Leberm. 2:170, 1836.
Jungermannia plicata Hartm., Skand. Fl. ed. 3(2):329, 1838.
Jungermannia colpodes Tayl., London Jour. Bot. 5:280, 1846.
Jungermannia plicata var. *kunzeana* Hartm., Skand. Fl. ed. 10, 2:137, 1871.
Jungermannia kunzei Lindb., Musci Scand., p. 8, 1879 [not of Lehm. & Lindenb., *in* Lehmann, Stirp. Pugil. 6:50, 1834].
Jungermannia kunzei var. *plicata* Lindb., *ibid.,* p. 8, 1879; Lindberg & Arnell, Kgl. Sv. Vetensk. –Akad. Handl. 23(5):55, 1889.
Lophozia kunzeana Evans, Proc. Washington Acad. Sci. 2:305, 1900.
Jungermannia minuta var. *lignicola* Velenovsky, Jatrovky Čseká 1:6, 1901.
Lophozia colpodes Macoun, Cat. Canad. Pls. 7:19, 1902.
Sphenolobus kunzeanus Steph., Bull. Herb. Boissier, Ser. 2, 2:168, 1902; also Spec. Hep. 2:160, 1902.
Orthocaulis kunzeanus Buch, Mem. Soc. F. et Fl. Fennica 8(1932):294, 1933.
Barbilophozia kunzeana (Hüben.) K. Müll. *in* Rabenh. Krypt.–Fl. ed. 3,6:626, 1954.

As scattered, creeping to erect or suberect stems among sphagna or mosses, or forming loosely tufted patches, characteristically yellow brown (more rarely green) to chestnut bown. Shoots 2–5 cm long, rarely to 8 cm long, mostly 0.6–0.8 mm wide, but on fertile shoots to 1.2–1.4 mm wide; stems erect or ascending, rigid, mostly yellow to dark brown, diffusely or virtually not at all branched, but often innovating below the perianth. Cortical stem cells elongate; stem with cortex of 2–3 layers of cells smaller in cross section than the interior cells; medullary cells thin-walled, with no or small trigones. Rhizoids rather abundant, long, colorless. Leaves mostly approximate to subimbricate, occasionally distant, slightly succubous, the upper nearly transversely oriented, scarcely decurrent at dorsal base; lower leaves horizontally spreading, the upper erect-spreading, appearing nearly erect because of the incurved lobes (the basal portion of the leaf mostly distinctly spreading); leaves subquadrate to subcircular, most of them distinctly wider than long, on sterile shoots mostly 850–1000 μ wide and 675–700 μ long, but on robust or fertile shoots about 1200 × 1000 μ; at base semiamplexicaul, and very shortly sheathing, divided ⅓–½ (usually about ⅖), *into 2, rarely 3, broadly ovate to triangular lobes* that are *obtuse or rounded* at apex and distinctly incurved, the leaf thus appearing concave or somewhat conduplicate; sinuses narrow and very acute at base, suddenly flaring outward, *usually reflexed and gibbous; postical leaf margin normally unarmed at base,* but occasionally with a small cilium or tooth. Cells of leaf apices and margins nearly isodiametric, about 14–18 μ, cells of leaf middle 17–20 × 21–26 μ, the lumen mostly rounded, and the trigones small to moderate, but frequently somewhat bulging and yellowish; walls rather thin or a little thickened, especially near leaf margins; *cuticle striate-verruculose;* oil-bodies (1)2–4(5) per cell, spherical and 3.8–5.4 μ to short-ovoid and 4.2 × 7.2–7.5 μ, formed of fine spherules, appearing grayish and papillose. *Amphigastria present throughout,* mostly erect-spreading, the apices again incurved, frequently bifid nearly to base and ca. 300–320 μ wide and long, but occasionally simple and lanceolate, the lobes long, lanceolate or lanceolate-acuminate, entire, or occasionally with a basal tooth (rarely 2–3) on each or one side. *Gemmae not common,* in catenations on the apices of the *unmodified upper leaves,* which become lacerate-erose; gemmae ovoid to irregularly 3–5 angulate in profile, 1–2-celled, 12–18 × 18–23 μ, *yellow-green to brown or orange-brown.*

Dioecious. ♀ Plants normally larger than sterile plants, the leaves beneath the bracts crowded, forming a crispate-undulate head. Bracts 1000–1200 μ long × 1350–1450(1500) μ wide or more, erect or erect-spreading, 3–4(5–6)-lobed for ¼–⅓ their length, the basal margins with a tooth on one or both sides; the lobes variable, ovate to broadly triangular, obtuse to apiculate at apex, the *sinuses narrow, acute, gibbous;* subinvolucral leaves similar, gradually diminishing in size; bracteole (and underleaves directly beneath it) very large, mostly ovate-quadrate, occasionally lanceolate, usually bilobed about ⅓–½, and on the base, on each side, with a tooth or spine, ca. 1050 μ long and wide. Perianth oblong-ovoid, ¾–⅘ exserted, the upper ¼–⅓ plicate, gradually contracted to mouth apically; mouth dentate-serrate, the teeth ending in cilia 2–4(6) cells long, formed of cells 12–13 μ wide × 18–23 μ long. Seta about 5 mm long; in cross section the cells nearly uniform. Capsule ovoid-elliptical, deep reddish brown, wall 3–4-stratose; epidermal cells about 3 times as large as in inner layer, with nodular thickenings. Elaters 10 μ in diam., thick, closely bispiral, vinaceous red. Spores pale reddish, minutely granulose, 10–14 μ. Androecia rather compact, terminal at first, later intercalary (2 or 3 androecia, separated by regions of sterile leaves, often evident on the same

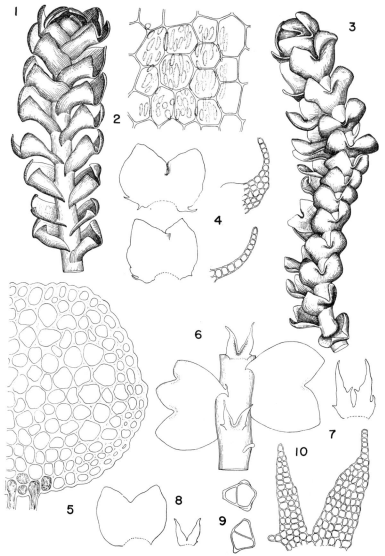

FIG. 148. *Lophozia* (*Orthocaulis*) *kunzeana*. (1) Dorsal aspect of mod. *densifolia*, the "fo. *plicata*" (×27.5). (2) Median cells, cuticular papillae drawn in (×332). (3) Lateral aspect of shoot (×29). (4) Two leaves (×21); at right the basal cilia more highly magnified. (5) Stem cross section (×196). (6) Shoot sector, ventral aspect, of the green mod. *viridis-laxifolia* (×23). (7) Underleaf (×42). (8) Leaf and underleaf (×23). (9) Gemmae (×460). (10) Underleaf (×94). [1, 3–5, 7–8, 10, Big Susie I., Cook Co., Minn., *RMS;* 2, Grand Marais, Minn., *RMS;* 6, Belle Rose I., Minn., *RMS;* 9, Little Susie I., Minn., *RMS;* all from Schuster, 1953.]

shoot), of 6–10 pairs of bracts usually; bracts generally slightly shorter than sterile leaves of same shoot, more erect and usually more closely imbricate, gibbous and saccate at base, with the lobes erect (the antical usually more or less incurved); apices of bracts very variable: some predominantly 2-lobed, with or without an antical, basal tooth, and little or no wider than long, others predominantly 3- or 4-lobed, from about 720 to 860 μ wide, or more, and about 1.3–1.5 × as wide as long (the antical lobe on such bracts either small and toothlike, or large and lobelike); sinus descending $\frac{2}{5}$–$\frac{3}{5}$ the length of the bracts; lobes narrowly to (usually) broadly triangular, acute to subobtuse at apices. Antheridia ovate, about 0.14 mm long, on a stalk about 50–60 μ long, normally 2–3 per bract, shielded mostly by 3–5 paraphyses; paraphyses linear to lanceolate, up to about 0.3 mm long. $\male$ Plants usually somewhat more robust than sterile shoots, with underleaves very distinct for their entire length, and with the sterile leaves (between the androecia), in material seen, predominantly 3–4 lobed.[23]

Type. Harz Mts., Germany (*Hübener*).

Distribution

An arctic-alpine species, circumboreal in the Tundra; with isolated, more or less disjunct extensions southward into the Coniferous Forest (in the Lake Superior, Michigan, and Huron regions). Widespread in arctic and alpine situations in Europe (from the montane northern portions of Italy and from Switzerland, France, and Germany, to the Tatra region of Czechoslovakia, to Scotland, Norway, Sweden, and Finland), eastward to Siberia, where widespread (*fide* Arnell and Kaalaas). Apparently unknown from Japan, but in North America from Alaska (Yakutat Bay) to the Yukon, British Columbia, Alberta and southward to Colorado. Recurring eastward as follows:

N.W. GREENLAND. Dundas, North Star Bay, Wolstenholme Sd. (*RMS 46251a*); Siorapaluk, Robertson Bay, 77°48′ N. (*M.P. Porsild, Aug. 15, 1943!*).[24]   E. GREENLAND. Cape Dalton, Tunok (65°53′ N.), and Tasiusak (Jensen, 1906a); d'Aunay Bay; Cape Daussy; Ingermikertorajik (65°54′ N.); Danmarks Ø and Hekla Havn in Scoresby Sund (Jensen, 1897); Geographical Society I., 72°44′ N., 22°30′ W. (*Holmen, 18.101; wet Vaccinium uliginosum* heath, with *L. rutheana, Gymnocolea inflata, Blepharostoma tr. brevirete*). W. GREENLAND. Ameralik (*Vahl, 1830!*, mixed with *L. lycopodioides* and *L. hatcheri*,

[23] Such plants are likely to be mistaken for forms of *L.* (*Orthocaulis*) *floerkei*, which normally has the leaves more than 2-lobed. The more deeply divided leaves, and slightly or not ciliate amphigastria, will separate *L. kunzeana* from the latter. $\male$ Plants of *L. kunzeana*, furthermore, often bear abundant, golden green to brownish green gemmae, in dense masses at the apices of the shoots.

[24] *Lophozia kunzeana* was also reported from Bjørling Ø, Cary Isls., in northwest Greenland (Harmsen & Seidenfaden, 1932); the two collections involved represent an indeterminable *Scapania*, and *L. alpestris.* Reports of *L. kunzeana* from Peary Land, Herlufsholm Strand (*Holmen 7623, 7732, 7742!*), cited by Arnell (1960), represent *L. quadriloba* var. *glareosa.* The other Peary Land collection, from G. B. Schley Fjord (*Troelsen 8762!*), also reported by Arnell as *L. kunzeana*, is a small phase of *L. quadriloba* var. *glareosa*, with 2–3-lobed leaves.

reported by Lange & Jensen, 1887, as *J. lycopodioides* var. *floerkei*; *Vahl 78, 79*!, reported by Lange & Jensen, 1887, as *J. attenuata*); Tasermiut Fjord (reported by Lange & Jensen, 1887, as *J. lycopodioides floerkei*); Frederikshaab (*Vahl, June 1829*!, mixed with *L. atlantica*; reported by Lange & Jensen, 1887, as *J. attenuata*); Sondrestrom Fjord, near shore of L. Ferguson, 66°5′ N. (*RMS & KD 66-057, 66-090, 66-068, 66-099; 66-051a*, fo. *wenzelioides; 66-093*, fo. *acuta*); Ritenbenk, Agpat, 69°46′ N., 51°20′ W. (*RMS & KD 66-254, 66-273, 66-270a*); SE. of Jacobshavn, Disko Bay, 69°12′ N., 51°05′ W. (*RMS & KD 66-221, 66-222, 66-220, 66-218*); Fortune Bay, Disko I., 69°16′ N., 53°50′ W. (*RMS & KD 66-215*); Godhavn, Disko I.: below Lyngmarksfjeld, 69°15′ N., 53°30′ W. (*RMS & KD 66-191*), Lyngmarksbugt, ca. 25-125 m (*RMS & KD 66-343*). Tupilak I., W. of Egedesminde, 68°42′ N. (*RMS & KD 66-104, 66-112, 66-111a, 66-107*, etc.; *66-022*, fo. *acuta*).

ELLESMERE I.   Mt. Pullen, S. of Alert, ca. 82°24–26′ N. (*RMS 35124a; RMS 35366, 35366a*, fo. *rotundiloba;* northernmost reports!).   N. BAFFIN I.   Pond Inlet (Polunin, 1947).   NORTHWEST TERRITORY.   Tukarak I., Belcher Isls. (Wynne & Steere, 1943). QUEBEC.   Cairn I., Richmond Gulf; Mouth of Great Whale R.; Runway Bog, Ft. Chimo, Ungava Bay (see Schuster, 1951); L. Mistassini; Rigaud; La Tuque; Mt. Albert (Lepage, 1944–45).   ONTARIO.   Manitoulin I., L. Huron(*RMS*); Humboldt I., L. Nipigon; peat bog on Nipigon R.; Sudbury Jct.; Algonquin Park; Thunder Bay, L. Superior (see Macoun, 1902; Cain & Fulford, 1948); L. Nipigon (*Macoun, July 17, 1884*!).   NOVA SCOTIA.   Tiverton, Digby Co. (*Nichols*).

MAINE.   Mt. Desert (*Lorenz*); Mt. Katahdin (*RMS*).   NEW HAMPSHIRE.   L. of the Clouds, Mt. Washington, and Mt. Monroe (*RMS*); Mt. Lafayette (*Evans et al., 1908*!); Waterville, Mt. Osceola, headwall of Split Cliff Ravine (*Lorenz 667*!).   NEW YORK. Indian Pass, Adirondack Mts. (Peck, 1899; N. G. Miller, 1966); Cascades, NW. side of Cascade Mt. between Upper and Lower Cascade Ls., SE. of L. Placid, Essex Co., 2400–3200 ft (N. G. Miller, 1966); Mt. Marcy summit (N. G. Miller, 1966); Wilmington Notch, Essex Co. (N. G. Miller, 1966).

MICHIGAN.   Keweenaw Co. (Steere, 1937); Huron Mts. (*Nichols*); Deer L., W. of Munising, Baraga Co. (*RMS*).   MINNESOTA.   Grand Marais, Cook Co.; Belle Rose, Little Susie, Porcupine, and Big Susie Isls., near Grand Portage, Cook Co. (Schuster, 1953).

Ecology

Lophozia kunzeana is widespread under arctic conditions, chiefly on peaty ledges, in peaty places on rocky shelves near cold lakes or streams, amidst *Sphagnum* around tarns and rock pools, often in rather strongly insolated sites; in the Far North a typical heath species. In my experience, *L. kunzeana* is particularly frequent as a secondary species, over accumulated soil, or in shallow soil-filled crevices, in damp to wet depressions in exposed, sunny rocks, or even bordering sunny acid rock pools, at points where seepage keeps the vegetation moist. Here it may be abundant with *Polytrichum* and *Sphagnum*, various Ericaceae (*Ledum, Cassiope hypnoides, Phyllodoce caerulea, Andromeda glaucophylla, Vaccinium macrocarpon, V. vitis-idaea*), *Drosera rotundifolia, Empetrum nigrum, E. atropurpureum*, and other plants.

Although usually a secondary species, *L. kunzeana* may be strictly pioneer in seepage-moistened crevices in granite, or on arkose, less frequently over basaltic rocks (trap rock and conglomerate, etc.) of less acidity; rarely on dolomitic rocks (as at Lake Mistassini). Most commonly associated under such conditions are *Gymnocolea inflata*, *Lophozia ventricosa* and *wenzelii*, *Cephaloziella rubella* and *hampeana*, *Ptilidium ciliare*, *Mylia taylori*, *Cephalozia bicuspidata*, *leucantha*, and *loitlesbergeri*; more rarely, and only in the Far North, *Lophozia hatcheri* and *atlantica*, *Chandonanthus setiformis*, and *Pleuroclada albescens*. In bogs the species often occurs with *Mylia anomala*, while over moist, shaded granitic rocks it is frequently consociated with *M. taylori*.

In the Far North also frequent as a helophyte in bogs, particularly in bog holes (often with *Gymnocolea inflata*, *Mylia anomala*, etc.). Although reported from decaying wood, the species undergoes ecesis on this only exceptionally (and then on tree trunks lying in bogs). Under strictly arctic conditions it is a common heath or moor species, coming in under rather xeric conditions (in lichen heaths, with *Dicranum fuscescens*, or in *Salix* heaths) or under hygric conditions in moors (then often with *Aulacomnium palustre*).

Differentiation

In spite of having predominantly bilobed leaves, *Lophozia kunzeana* bears a much closer relationship to *Orthocaulis* than to such bilobed-leaved groups as subg. *Leiocolea* or *Lophozia*. The rather subtransversely inserted and oriented, deeply bilobed leaves, together with the presence of distinct, often bifid underleaves, generally separate the species from the latter two subgenera.

The large underleaves of *L. kunzeana* separate it from sectio Attenuatae of *Orthocaulis* (as well as from such unrelated but superficially similar types as *Anastrophyllum michauxii* and *A. saxicolus*). Consequently, confusion should arise only between *L. kunzeana* and the two closely allied species, *L. floerkei* and *L. quadriloba*. The predominant development of bilobed leaves separates *L. kunzeana* from normal forms of both these species. However, slender phases of *L. floerkei* and *L. quadriloba* may have a high incidence of 2-lobed leaves, simulating *L. kunzeana*. Normal shoots of *L. kunzeana* bear leaves lacking basal cilia, while the other two species usually have 1 or several basal postical cilia of the leaves. This character, unfortunately, is also subject to much variation (slender phases of *L. floerkei* and *L. quadriloba* bearing largely eciliate leaves).

The bright green to yellow-brown color of *L. kunzeana* is usually distinctive. In *L. quadriloba* the color is usually drab or inky olive-green to fuscous. *Lophozia kunzeana* has rather deeply lobed leaves (the sinus frequently extending for 0.5 the leaf length); this characteristic is usually sufficient to separate it from the more shallowly lobed *L. floerkei*.

The brown, xeromorphic phase with complicate, canaliculate leaves ["fo.

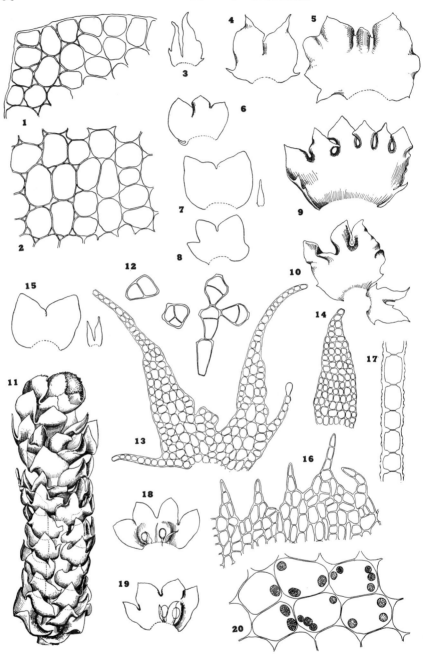

FIG. 149. *Lophozia* (*Orthocaulis*) *kunzeana*. (1) Cells of lobe apex (×325). (2) Median cells (×325). (3) Large underleaf, from fertile shoot (×37). (4–5) ♀ Bracteole and bract (×23). (6) Leaf from below androecium (×23). (7–8) Leaves and an underleaf (×23). (9) ♀ Bract, closely connate with bracteole, on left (×23). (10) Sub-involucral bract and bracteole, the bracteole almost free (×23). (11) Androecial shoot, gemmiparous at apex (×17). (12) Gemmae (×400). (13) Large underleaf

plicata" (Hartm.) Lindb.; Fig. 148:1] has the leaves so nearly oriented as in the *Anastrophyllum-Sphenolobus* complex that some authors (e.g., Stephani, 1902) actually placed *L. kunzeana* in "*Sphenolobus.*" The presence of large discrete underleaves at once separates the species from this complex.

Variation

Lophozia kunzeana undergoes considerable variation in color, size, and orientation of the leaves. Palustral shade forms are often found creeping amidst *Sphagnum* (then disturbingly similar to forms of *Cladopodiella fluitans*); they bear distant leaves, and the entire plant is green and strongly elongated. This represents the helophytic shade form, mod. *laxifolia-viridis-parvistipula*; in this form, leaves are almost constantly bilobed, and, associated with less robustness, underleaves may be small and merely lanceolate. Such plants have somewhat concave leaves, usually with ± inflexed lobes. The leaves, as a result of the attenuation of the stem, show the somewhat succubous insertion and orientation to an enhanced degree.

The "normal" phase of the species occurs on ± shaded rocks or over rather dry peat. There is then a tendency toward yellow-brown to golden-brown pigmentation; the leaves are denser, less obviously oblique, and occasionally 3-lobed on robust stems; the underleaves are well developed. This phase is the commonest one met; it represents a mod. *mesoderma-colorata-megastipula*. The plants show concave leaves, but the lobes are not sharply bent over each other.

Under xeric conditions, particularly when strongly insolated, a brown plant occurs, frequently forming pure mats, with the stems often erect. These plants represent an xeromorphic type, with strongly collenchymatous cells and yellowish walls; the leaves are dense and virtually uniformly bilobed, with the two halves somewhat complicate, strongly canaliculate, slightly oblique. This plant represents the extreme sun form, mod. *colorata-pachyderma-densifolia-plicata*. It was described as a distinct species, *Jungermannia plicata*, but represents merely a modification of strikingly different appearance, suggesting *Anastrophyllum* rather than *Lophozia*.

The underleaves are mostly small, most often merely lanceolate, occasionally bifid. On the most robust, sterile shoots and on most fertile shoots they are bifid nearly to base, with each half lanceolate or lanceolate-acuminate to subulate; the basal portion of each lobe bears 1 or 2 cilia (rarely 3), usually on

---

(×100). (14) Lanceolate unlobed underleaf (×ca. 100). (15) Leaf and underleaf (×23). (16) Perianth mouth cells (×165). (17) Leaf cross section (×270). (18–19) ♂ Bracts (×26). (20) Median cells, with oil-bodies (×480). [1–5, 7–10, 14–15, 17, *RMS 11711*, Big Susie I., Minn.; 6, 11–13, 18–19, Little Susie I., Minn., *RMS, Sept. 7, 1947*; 16, *RMS 13030*, Big Susie I., Minn.; 20, Manitoulin I., Ontario, *RMS, Sept., 1949*.]

the other side of each lobe, but occasionally (Fig. 148:7) with a tooth on the inner side as well. These teeth are of isodiametric cells, like those of the leaf bases.

In my experience the leaves are simply bilobed in about 98% of all cases, even on fertile shoots. The presence of a basal cilium of the postical base, as described in the literature, is difficult to demonstrate. In about 100 leaves from sterile shoots examined, I have found no case in which the basal postical cilium occurred. On fertile shoots, the leaves occurring below the subinvolucral bracts occasionally or frequently have such a basal cilium, and a very few leaves (Fig. 148:4) have 1–2 small cilia or teeth on the antical margin as well. It has been generally considered that the leaves associated with the perianth retain the generalized or archaic features of the species more than the purely vegetative leaves; the presence of postical, as well as antical, cilia would seem to connect *L. kunzeana* with *L. quadriloba*. In fact, the perichaetial bracts and bracteole of *kunzeana* often bear considerable resemblance to normal leaves and underleaves of *quadriloba*.

Although the tendency for the development of trilobed leaves is poorly marked in this species, they occasionally occur, especially on robust shoots, and sometimes also on relatively slender shoots, where they may even be weakly quadrilobate (Fig. 149:8).

Arctic Phenotypes

Lophozia kunzeana occurs in the Arctic in four usually readily distinguishable but intergrading extremes, separable by the following key:

Key to Variations of L. kunzeana

1. Leaves 0.35–0.55 bilobed or sporadically trilobed; sinuses ± gibbous . . 2.
   2. Leaf lobes blunt to rounded; underleaves large, mostly bifid, often
      ciliate toward base . . . . 3.
      3. Leaf lobes blunt to obtusely pointed, never decolorate on margins. .
      . . . . fo. *kunzeana*
      3. Leaf lobes broadly rounded, isolated exceptions aside, often decolorate
      on margins; usually dense-leaved . . . . fo. *rotundiloba*
   2. Leaf lobes, with isolated exceptions, long and sharp, the lobe length
      usually much greater than the width, tips ending in 1 or often 2 super-
      posed cells; underleaves small, locally vestigial, of 1 or 2 small, lanceolate
      segments . . . . fo. *acuta*, fo. n.
1. Leaves 0.25–0.3 bilobed or 3–4-lobed; bilobed leaves with incurved lobes,
   the leaf saucer-shaped, sinuses usually not reflexed; lobes usually blunt;
   underleaves small, often unlobed . . . . fo. *wenzelioides*, fo. n.

To what extent these forms are environmentally induced or genetic in nature remains to be demonstrated by experimental work. I assume that they have a genetic basis, since, near the southern periphery of its range,

L. kunzeana is relatively stenotypic, occurring in a small series of phenotypes only.

LOPHOZIA KUNZEANA fo. ROTUNDILOBA
(*Schust.*) *Schust.*, *comb n.*
[Fig. 150:1–8]

Lophozia (*O.*) *kunzeana* var. *rotundiloba* Schust., *in* Schuster et al., Natl. Mus. Canada Bull.
    164:24, 1959.

Plants rather dense-leaved, with concave, 2(3)-lobed leaves for 0.5–0.65 their
length, smooth-margined, except for an occasional solitary tooth of postical
base, the *lobes bluntly rounded to typically very broadly rounded, often becoming decolorate*
at the margins. Marginal cells 17–18.5 μ, often subquadrate; median cells of
lobes (18)20–22 × 21–25 μ; cells of middle below sinus 20–25(26) ×
22–26(28) μ; cuticle, at least locally, closely but not very coarsely, densely
papillose; papillae 2–3 × 2–4(5) μ; underleaves predominantly lanceolate
with 1–2 basal cilia to, more rarely, bifid and usually eciliate. Gemmae rare,
purplish.
  Dioecious. Perianth and bracts as in *L. kunzeana*. Spores larger than in
L. kunzeana, 14–16 μ; elaters 6–8 μ broad.

  Type. Moist humus over low scarp, N. of Mt. Pullen, near Alert,
Ellesmere I., 82°25–26'N. (mixed with *Lophozia heterocolpa*, *L. quadriloba*,
and *Scapania praetervisa*, *RMS 35366*).

Distribution

Known, to date, only from Ellesmere I. (type) and from ALASKA: Old Johl
Lake, S. slope of Brooks Range (*O. Mårtensson, July 15, 1961*).

Ecology

The type occurred over thin humus on a low scarp, where irrigated by
persistent snow, admixed with *Lophozia heterocolpa*, *Scapania praetervisa* var.
polaris, *Anastrophyllum minutum* and the weakly similar *L. hyperborea* (under
which the ecology is more adequately discussed).
  I initially described this plant as a variety of *L. kunzeana*, to which it bears
the most immediate affinity in the deeply 2–3-lobed leaves, underleaf form,
less coarsely papillose cuticle (as compared with *L. quadriloba*), lack of cilia of
the postical leaf base, and its light green to warm brown coloration. Arnell
(*in litt.*, Jan. 16, 1964) has expressed the opinion that it "deserves the rank of
species . . ." The broadly rounded lobes of the leaves are distinctive, and easily
separate the plant from *L. kunzeana*. Also, I have never seen purplish gemmae
in *L. kunzeana*, which has brown gemmae, even in sun forms. The spores in fo.
rotundiloba are also somewhat larger in diameter. Underleaves, as contrasted
with *L. kunzeana*, are most often unlobed, but bear 1–2 small basal teeth.
  The status of this plant must remain an open matter until adequate additional
collections are at hand.

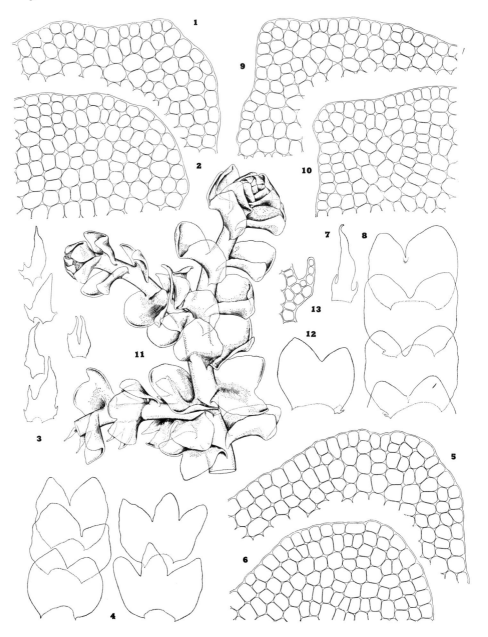

FIG. 150. *Lophozia* (*Orthocaulis*) *kunzeana* (9–13) and *L.* (*O.*) *kunzeana* fo. *rotundiloba*
(1–8). (1–2) Leaf lobe apices (×205). (3) Underleaves (×35). (4) 2- and 3-lobed
leaves (×35). (5–6) Leaf lobe apices (×205). (7) Underleaf (×35). (8) Three leaves
and apex of a fourth (×35). (9–10) Two leaf lobe apices (×205). (11) Gemmiparous
shoot, with two terminal branches (unlobed stem leaves associated with branch, or

LOPHOZIA KUNZEANA fo. *ACUTA*, fo. n.

[Fig. 139:5–9]

Forma plerumque dense foliacea, lobis foliorum anguste ovato-triangularibus, multo longioribus quam lati, plerumque acuminatis et in 1–2 singulis cellulis terminantibus

Leaves typically dense, with imbricate, erect lobes from a spreading base, all bifid for 0.35–0.65 their length; lobes narrowly ovate-triangular, often longer than wide, usually sharp-pointed and ending in 1 or often 2 superposed cells, exceptionally blunt. Gemmae rare, tawny. Underleaves ± small, less than 0.3 leaf length usually, of 1 or 2 linear to lanceolate segments, usually cilium-free.

Type. Tupilak I., west of Egedesminde, west Greenland (*RMS & KD 66–022*). *Paratype.* Head of Sondrestrom Fjord, Lake Ferguson, west Greenland (*RMS & KD 66–093*).

The type specimen gives one, initially, the impression of a greenish *L. binsteadii*, which it approaches in the sharp leaf lobes and the cupped, hollowed, and usually rather dense leaves. However, the almost consistently bifid leaves and the small yet distinct underleaves prohibit confusion with this species. I have seen no other collections of such sharp-lobed *L. kunzeana*.

LOPHOZIA KUNZEANA fo. *WENZELIOIDES*, fo. n.

Forma laxe foliacea; folia vadose (0.25–0.3) bilobata, lobis obtusis ad rotundatos, incurvatis; bases sinus non aut rare gibbosae.

Rather lax, yellow-green to green, with predominantly shallowly (0.25–0.3) bilobed, concave leaves; lobes incurved, broad, blunt or narrowly rounded, varying locally to subacute. Isolated leaves, particularly on lower parts of stems 3- or rarely even 4-lobed, then with lobes spreading. Sinus bases not or rarely gibbous. Underleaves variable: often reduced to an acuminate, linear-lanceolate lamella, less often bifid, locally subobsolete.

Type. West Greenland: near Lake Ferguson, head of Sondrestrom Fjord (*RMS & KD 66–051a*).

This plant differs from all other phases of *L. kunzeana* I have seen in the almost constantly shallowly bilobed, cupped leaves, which remind one of an obtuse-lobed *L. wenzelii* (hence the name). It does not readily key to *L. kunzeana* because of the shallow sinuses, but will not fit into any other species of *Orthocaulis* on account of the almost constantly bilobed leaves.

---

half leaves) (×23). (12) Leaf (with postical base cilium indicated by stippled line) (×22). (13) Postical leaf base cilium of fig. 12 (×41). [1–4, from Alaskan material, leg. Mårtensson; 5–8, from type, Alert, Ellesmere I., *RMS;* 9–13, *RMS 14868*, Minn.]

LOPHOZIA (ORTHOCAULIS) FLOERKEI (Web. & Mohr) Schiffn.

[Figs. 145:11–12, 151]

Jungermannia floerkii Web. & Mohr, Bot. Taschenb., p. 410, 1807.
Jungermannia naumanni Mart., Fl. Crypt. Erlang., p. 143, 1817.
Jungermannia barbata var. *floerkii* Nees, Naturg. Eur. Leberm, 2:168, 1836.
Jungermannia barbata var. *naumanniana* Nees, *ibid.* 2:170, 1836
Jungermannia barbata var. *naumanni* Dumort., Hep. Eur., p. 73, 1874.
Jungermannia lycopodioides var. *floerkei* Lindb., Musci Scand., p. 7, 1879.
Jungermannia floerkii var. *alpina* Pears., List Canad. Hep., p. 22, 1890.
Lophozia floerkei Schiffn., Engler & Prantl, Nat. Pflanzenfam. 1(3):85, 1893.
Barbilophozia floerkei Loeske, Abh. Bot. Ver. Brandenburg 49:37, 1907; Grolle, Nova Hedwigia 2(4):559, 1960.
Orthocaulis floerkii Buch, Mem. Soc. F. et Fl. Fennica 8 (1932):294, 1933.

In loose to (more rarely) compact tufts, pale or yellowish green to dark green or *(rarely) brown*, often forming extended masses, frequently very diffuse, the plants scattered among pleurocarpous mosses. Stems mostly 2–5 (rarely to 10) cm long, the *shoots 1.8–3, occasionally 4–5, mm wide, erect or suberect*, rather rigid, green to brown, very sparingly branched (the branches suberect or erect); stems mostly 330–450 μ wide, rather densely covered postically with short rhizoids. Leaves varying from rather close to imbricate, the internodes usually short, inserted on a slightly acroscopically arcuate line *at ca. a 45–65° angle with the axis, the leaves appearing only slightly oblique*, especially above; *leaves stiffly spreading to subsquarrose* (suberect only in xeromorphic extremes), rather strongly *crispate, quadrate-rotund to somewhat transverse* (usually 1.1–1.3× as wide as long; in xeromorphic extremes rarely 1.3–1.5× as wide as long), usually ca. 950–1050 μ long × 1250–1350 μ wide, to 1000–1100 μ long × 1400–1500 μ wide, almost *uniformly trilobate, the lobes frequently subequal*, broadly ovate-triangular, *short, obtusely pointed*, the lobes usually incurved, more or less wavy; sinuses rectangular to obtuse, *strongly gibbous, descending usually* ¼–⅓ *the leaf length; postical and antical margins subequal*, the antical short-decurrent, the postical margins with *normally 1–3, occasionally 4–6, subbasal cilia; cilia relatively short, at base often 2–3(4–5) cells wide, formed of isodiametric cells.* Cells with distinct to rather strongly bulging trigones, the *cuticle faintly verruculose* (distally obsoletely so; basally somewhat striolate; in the rare xeromorphic phases more distinctly verrucose); marginal cells of lobes (14)16–17(19) μ, the cells of the free lobes scarcely larger; median cells ranging from (16)18–19× 20–25 μ, occasionally to 20 × 19–25 μ; cells each with 2–6, mostly 3–5, relatively large oil-bodies, varying from spherical and 4–5 μ to ovoid or ellipsoid and 4–5.5 × 6 to 5 × 10 μ, appearing moderately papillose (formed of discrete spherules); chloroplasts smaller, ca. 2.5–3.5 μ long. *Underleaves large, present throughout* (occasionally obscured by the rhizoid mat), normally *bifid nearly to base*, mostly 250–285 μ wide (exclusive of cilia) × 550–650 μ long, the lanceolate-acuminate lobes ending in a cilium and usually bearing 1-several cilia near base. *Gemmae* absent (*fide* Müller, and according to my observations) or "*very rare*, on the leaf lobes at the apex of branches, oblong to 3–4-angled, greenish-white to reddish, usually 1-celled" (*fide* Macvicar).

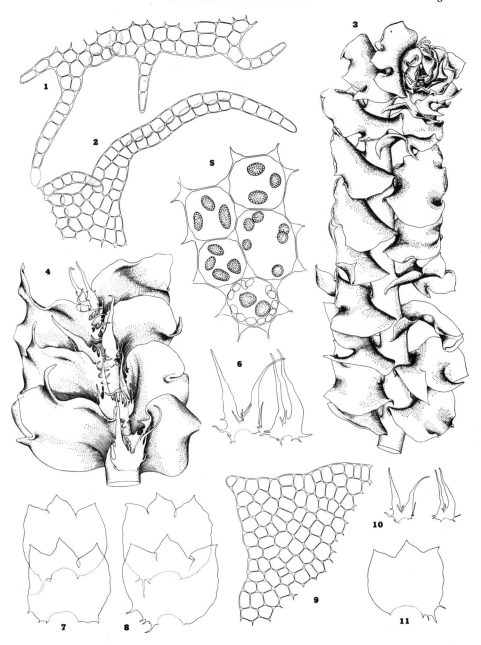

FIG. 151. *Lophozia* (*Orthocaulis*) *floerkei*. (1–2) Cilia from postical leaf base (×215). (3) Mature shoot, antical view (×19). (4) Mature shoot sector, postical aspect (×25). (5) Median cells with oil-bodies and, lower cell, chloroplasts (×660). (6) Underleaves (×33). (7–8) Leaves (×20). (9) Cells of leaf lobe apex (×220). (10–11) Two underleaves and a leaf (×20). [All from *RMS 32968*, Mt. Katahdin, Me.]

Dioecious. ♂ Plants often admixed with ♀. Androecia becoming inter-calary; the bracts in 4–7 or more pairs, imbricate, concave, ventricose basally, with usually an antical tooth, otherwise like leaves (but more transverse, usually smaller than normal leaves); antheridia 1–3, among paraphyses, oval-globose. ♀ Bracts somewhat sheathing perianth, transversely quadrate-rotund, erect, plicate and crispate above, with undulate margins, shallowly divided into 3–4 (rarely 5–7) short, triangular, obtuse to cuspidate lobes, at both antical and postical bases usually with 1-several cilia or cilia-like teeth; bracteole more than half the bracts in size, virtually as long, quadrate to ⅓ bilobed, the lobes often cuspidate, the margins basally ciliate as on the bracts. Perianth ovate-cylindrical, narrowed distally, *deeply plicate to the middle or deeper, the mouth dentate with projecting cells or ciliolate with teeth 1 cell long.* Capsule ovoid, the wall 4-stratose; epidermal layer of very large cells, the 3 inner layers of much smaller cells; all inner cell layers with semiannular thickenings. Spores brownish, ca. 12 μ; elaters ca. 8 μ in diam., bispiral.

Type. Austria: Salzburg, Zillerthal (*Flörke, 1798*); W.

Distribution

A species with a strongly disjunct range, involving a single known station in the Cordillera of Peru (at 4600 m; see Grolle, 1960e), and a scattered, essentially subarctic range in the Northern Hemisphere. Found in the montane portions of western and central Europe (Switzerland, Germany, Austria, Czechoslovakia) south at high elevations as far as the Pyrenees (northern Spain), Yugoslavia (Montenegro), Bulgaria, Rumania, north-ward to northern Germany (where at times at low elevations), Belgium, Holland, Great Britain, and Ireland; more ubiquitous in Scandinavia (The Faroes, Denmark, Norway, Sweden, Finland), and eastward barely to Russia. Reported from the Caucasus and Siberia (Yenisey R. and eastern Siberia); Grolle (1960e) has shown that the Siberian and Cau-casian plants belong elsewhere.[25] In North America found from the Aleutian Isls. and continental Alaska! southward (in the west) to the Yukon!, Alberta, British Columbia!, and Washington, and at high eleva-tions in the Rocky Mts. to Montana!, Colorado!, and New Mexico. In our area rare, distributed as follows:[26]

[25] The reports from Jan Mayen I. (Jensen, 1900), from Iceland (Hesselbo, 1918), from Spitsbergen, and from Greenland (Lange & Jensen, 1887, as *J. lycopodioides* var. *floerkii;* Jensen, 1900, 1906a) all represent misdeterminations (Grolle, 1960e). Two specimens from Scoresby Sund in the NYBG (*N. Hartz, June 1892, July 1892,* respectively), determined by Jensen as *Lophozia floerkei,* are *L. hatcheri,* of the form with deeply divided leaves, approaching "var. *palmatifolia* Meylan." The report (Arnell, 1960) of *L. floerkei* from North Greenland (Station Nord) is based on *L. quadriloba* var. *glareosa.* Just reported by Kitagawa (1967a, p. 39) from Toyama Pref., Honshu, Japan.

[26] Clarification of the distribution of *L. floerkei* is greatly complicated by the fact that for a long time *L. hatcheri* was confused with this species. For example, the several collections from South Georgia and Louis Phillipe Land, in the western Antarctic (determined as *L. floerkei* by Stephani;

ELLESMERE ISLAND. Lastraea Valley, Twin Glacier valley, and Fram Harbor, all in Hayes Sd. region (Polunin, 1947). These reports, based on Bryhn (1906), probably all represent errors in determination.   QUEBEC. Island near mouth of Seal R., Cape Jones; Ft. Chimo Air Base, Ungava Bay; Great Whale R.; Montreal; Mont-la-Table (see Schuster, 1951; Lepage, 1944–45); Old Factory, James Bay; Roggan R., Ungava (Lepage, 1953; det. Arnell).   NORTHWEST TERRITORIES. Tukarak I., Belcher Isls., Hudson Bay (see Schuster, 1951); Paint Hills Isls., James Bay (Lepage, 1953; det. Arnell).   NEWFOUNDLAND. Steady Brook, W. Nfld. (Buch & Tuomikoski, 1955). NOVA SCOTIA. Track to Benjy's L., Cape Breton (*RMS*).

MAINE. Mt. Katahdin, along Saddle Trail, ca. 3500 ft (*RMS 32968*).   NEW HAMPSHIRE. Mt. Washington: Tuckerman's Ravine (*Evans, 1902!*), L. of the Clouds (*Evans, 1917!*; *RMS 68–2080*), SW. of Mt. Pleasant and SW. of Mt. Adams, White Mts. (*RMS 68–2121, 68–2235*); S. Twin Mt. (*RMS*).   VERMONT. Mt. Mansfield, near summit, 4200 ft (*Lorenz, 1917!*).

The species, except in areas approaching the Tundra, occurs at over 3200 feet in eastern North America, but in the higher mountains (Europe; Rocky Mts.) it may go up to 10,000 feet (as on Pike's Peak, *Holzinger, 1896!*). The plant appears to be much rarer and more sporadically distributed in eastern North America than in the West or in Europe. In North America the species is lacking from midcontinental portions. A map showing the entire range is given in Grolle (1960e).

Ecology

Lophozia floerkei, at least in our area, has much the same ecological tolerance and distribution as *L. lycopodioides*. It occurs usually in the uppermost portions of the Spruce-Fir Zone (often in the scrubby extensions near the summits of the mountains, up to elevations, locally, of 3500–5500 feet or more), where the plants occur scattered or in diffuse patches under spruce and fir, usually on rocky slopes or over boulders covered with needle humus, and in rather deep and permanent shade. The species is one of the very few Hepaticae persisting in the undisturbed climax forest, together with *Lophozia lycopodioides* (rarely also *L. barbata*) and *Bazzania trilobata*; it is frequently found with both these species, as well as such climax mosses as *Dicranum*, *Hylocomium splendens*, *Ptilium crista-castrensis*, and more rarely *Leucobryum* (or in wet places, *Sphagnum*). The survival of the plant apparently depends on (*a*) the ability to grow on a substrate of

cited also by Gottsche, 1890, Stephani, 1905a, 1911, and Steere, 1961) all represent *L. hatcheri*.

Grolle (1960e) has shown that prior reports from Greenland, Spitsbergen, Iceland, Novaya Zemlya, Bären I., Archangel, Siberia, the Caucasus, etc., refer to other species. There is an as yet unclarified report (based on a Stephani determination; see Stephani, 1902, *in* 1898–1924) from the Azores. All actual collections from Greenland that I have seen, based on published reports of *L. floerkei* (Lange & Jensen, 1887; Arnell, 1960) belong elsewhere, as follows: Nord, northeast Greenland (*Troelsen*) = *L. quadriloba* var. *glareosa*; Tunugdliarfik, south Greenland = *L. hatcheri*; Holstensborg, west Greenland = *L. hatcheri*; Amerilak, west Greenland (*Vahl, 1830*) = *L. kunzeana* + *L. lycopodioides* + *L. hatcheri*; Kobbefjord (*Warming & Holm, 1884* = *Tritomaria* subg. *Saccobasis*, probably *polita*); Tasermiut Fjord = *L. kunzeana*.

spruce and fir needles, at a low pH; (*b*) the ability to grow virtually erect and at a rather rapid rate. In our area almost exclusively over acidic rocks, although it has been reported (in Europe) from areas with basic rocks (where also a climax species, long after accumulation of humus has masked the basic strata beneath).

Northward the species descends to sea level and becomes a Tundra plant. There it may occur over moist granite, with such associates as *Anastrophyllum minutum, Lophozia wenzelii,* and *Scapania irrigua.* However, *L. floerkei,* like *L. lycopodioides,* is essentially confined to the Hudsonian zone of spruce and fir forests when it occurs south of the Tundra region, disappearing before the barren summits of the mountains are reached.

Differentiation

Although once much confused with *L. lycopodioides, L. hatcheri,* and *L. barbata* (as well as *Tritomaria quinquedentata*), the plant is very distinct from all of these. *Lophozia floerkei* differs from the latter two in the large under-leaves of sterile shoots and the cilia of the postical leaf margins, as well as in the form of the leaves. The ciliate postical leaf margin and large under-leaves are shared with *L. lycopodioides* and *hatcheri,* but *L. floerkei* differs from these two species in the more subtransversely inserted and oriented leaves, the more bluntly tipped leaf lobes, and the form of the cilia of the postical leaf margins. These cilia are formed of isodiametric cells (in the other two taxa of prominently elongated cells). Leaves in *L. floerkei* are also subquadrate; in the other two species, obtrapezoidal and much broadened toward the apex. Furthermore, *L. floerkei* habitually lacks gemmae (although there are a few—surely erroneous—reports of gemmae), while both *L. hatcheri* and *L. lycopodioides* produce them freely under xeric conditions.[27]

Lophozia floerkei has cells with oil-bodies much as in the related species of the subgenera *Orthocaulis* and *Barbilophozia.* They occur 2–6 per cell (most often 3–4), are rather large for the cell size, ranging from 4–5 μ and spherical up to 4–5.5 × 6 μ, occasionally to 5 × 10 μ; they are papillose in appearance, as in other related species, being formed of rather discrete, projecting globules.

Variation

The "normal" form of the species is by far the commonest; it is charac-terized by occurrence under mesic conditions; the plants show moderately

---

[27] The habitual similarity between erect, xeromorphic phases of *L. hatcheri* and *L. floerkei* is admittedly great; both produce forms with the postical cilia nearly or quite obsolete, and with more or less obtuse leaf lobes. However, almost all phases of *L. hatcheri* produce at least occasional apiculi of the leaf lobes, formed by elongate, large cells, and occasional contorted cilia of the postical leaf bases, formed of prominently elongated cells; such cilia and apiculi are quite lacking in *L. floerkei.*

short internodes, with the characteristically stiffly spreading leaves not imbricate in appearance, the lobes not strongly incurved, and the leaves quite obviously crispate and fluted (especially above). Such plants usually average 2–3 mm wide and 2–5 cm high; they are generally green in color or show slight brownish pigmentation.

Under unusually moist conditions, in deep shade, a lax form occurs, 5–10 cm high, forming swelling pale yellow-green tufts. The plants are laxly leafy, with the leaves almost flat, stiffly standing away from the stem or squarrose, rarely at all crispate; this represents the mod. *laxifolia-leptoderma-viridis* and has been described by Nees (Naturg. Eur. Leberm. 2:170, 1836) as fo. *naumanniana*.

By contrast, the xeric phase attributed to this species (dealt with in some detail by Schuster, 1951, pp. 4–5), may be "merely 5–8 mm high, appear almost terete, because of the densely imbricate, strongly concave, erect leaves . . . ." This form represents a mod. *densifolia-pachyderma-parvifolia*; it was described by Nees (*loc. cit.*, p. 168) as fo. *densifolia*. A plant of this type has been described by Schuster (1951, pp. 4–5) from Quebec and the Northwest Territories and is here illustrated (Fig. 143:1–9). This plant differs from the "normal" forms of *L. floerkei* in a considerable number of features, among them (*a*) reduction of cilia of the leaves and of the underleaves; (*b*) shorter, wider leaves, averaging 1.3–1.5× as wide as long, somewhat more deeply divided (0.3–0.4–0.5) with the lobes rather unequal; (*c*) coarsely papillose cuticle, with only 3–4 large papillae per cell; (*d*) occurrence under basic or subbasic conditions. The plants are found associated with *Anthelia juratzkana*, *Tritomaria scitula*, *Scapania curta*, *S. calcicola*, *Blepharostoma*, *Lophozia hatcheri*, *L. heterocolpa*, *L. bantriensis*, *Cephalozia pleniceps*, *Tritomaria polita*, *Preissia*, *Odontoschisma macounii*, *Plagiochila asplenioides*; all these species either tolerate calcareous conditions or are obligatory calciphytes. For some time it was believed they might be referred to the lime-tolerating *L. quadriloba*, a closely allied plant, especially since they have a coarsely papillose cuticle (normal for *L. quadriloba*), the sinuses of the leaves descending 0.4 or even 0.5 the leaf length, broader leaves, and an olive-green, dull color. However, the 2–3-lobed leaves have the fundamental form of those of *L. floerkei*; in particular, the leaf lobes are broader and much shorter, with less reflexed margins than in *L. quadriloba*.[28] In spite of this, the position of such plants remains ambiguous.

[28] Because of the virtually eciliate leaf bases and underleaves, and the somewhat smaller than normal underleaves, these plants would key to *L. atlantica* in Frye & Clark (1937–47, pp. 401–402), as a result of the ill-advised selection of key characters in this work. Macvicar (1926, p. 200) also speaks of "a small form of *L. Floerkii* with only a few subulate underleaves, which is much like *L. atlantica*, but even in this form the underleaves are rather more in evidence . . . and the cells are rather larger." As I have pointed out (Schuster, 1951, p. 4), the inverse relationship holds, *as regards the marginal cells*. In all forms of *L. floerkei* marginal cells average 16-17, rarely 18, μ; in *L. atlantica* they average 19-25 μ. In *L. atlantica* the largest underleaves average only 0.2–0.25 the length of the leaves and are formed of only 1-2 (or rarely 3-4) ciliary divisions; in even the small forms of *L. floerkei* underleaves are half as long as the leaves and are formed of 2 lanceolate lobes, at least in part. In *L. atlantica* the ventral stem sectors are only 4-5 cells broad; in *L. floerkei* they are at least 6-8 or even 10 or more cells broad, "only on occasional shoots 5 cells broad."

Sectio *ATTENUATAE* Schust.

[Figs. 139:10–18; 152]

Lophozia subg. *Orthocaulis* sect. Attenuatae Schust., Amer. Midl. Nat. 45(1):43, 1951.

Folia plerumque 3-lobata, basibus posticis saepius sine ciliis dentibusve; amphigastria obsoleta aut nulla; gemmae saepius praesentes, surculis qui gemmas ferunt saepe flagelliformibus aut digitiformibus.

Leaves *almost uniformly (2)3-lobed* on mature shoots, their *postical bases without cilia* (rarely with a short-stalked slime papilla). Stem with ventral merophytes reduced, on nongemmiparous shoots, merely 2–5 cells wide, with *underleaves obsolete* (reduced to lanceolate or linear vestiges, or to a few uniseriate cilia) *or totally lacking*. Gemmae development sometimes more or less inhibiting growth of the leaves, which become *reduced, dentate, erect or erect-appressed*, the shoots bearing them consequently more or less *flagelliform with prolonged gemmae development*.

Type. Lophozia attenuata (Mart.) Dumort.

This section includes plants perceptibly more specialized (in all of the criteria cited above) than sectio Orthocaulis. The stem leaves usually have acquired a fixed number, although in *L. atlantica*, the least derivative taxon, sporadic variation occurs and 2- or 4-lobed leaves are found.

Branching in the Attenuatae is, in my experience, usually lateral-axillary, with the branches arising from the ventral halves of the leaf axils. I have seen only such branching in *L. binsteadii*. The loss of ability to develop terminal branches, at least as a "normal" phenomenon, is a decidedly specialized feature. (*L. attenuata*, however, occasionally is frequently terminally branched.)

The sectio Attenuatae is sharply isolated from the sectio Orthocaulis by the preceding characters. The regional species of the section, however, are exceedingly close to each other except for *L. cavifolia*. Fundamentally, two taxa are involved. One is a lowland plant of the spruce-fir and upper margins of the deciduous forests (*L. attenuata*), in which the cells are relatively small and never strongly collenchymatous; this plant will probably prove haploid. In the arctic and alpine Tundra it is replaced by strongly pigmented plants with consistently larger cells that develop coarse trigones even under relatively humid conditions. They have been separated into either two (*L. atlantica*, *L. binsteadii*) or three species (*L. herjedalica* is separated by Schiffner, *in* Arnell, 1906, from *L. binsteadii*; *L. ambigua* is separated from *L. binsteadii* by Joergensen, 1934).

I find it difficult to maintain two species here and totally fail to appreciate the basis for the further segregation attempted by Schiffner and Joergensen. Fundamentally, the two arctic-alpine taxa differ from each other in their ecology, and slightly in cell size (see Table 3, p. 322). *Lophozia binsteadii* is usually a bog species; *L. atlantica* is a species of shaded to exposed rocks (only occasionally over peat in such sites). Of the two, *L. binsteadii* is exceedingly close to *L. attenuata*, probably representing an arctic-alpine derivative; it has only slightly larger cells, but better development of collenchyma. Possibly *L. binsteadii* will prove to be diploid. By contrast, *L. atlantica* is a more sharply isolated taxon, differing from *L. binsteadii* and *attenuata* in larger average cell size, in broader leaves, and in the presence of distinct underleaves of gemmiparous shoots.

LOPHOZIA (ORTHOCAULIS) ATTENUATA (*Mart.*)
Dumort.

[Figs.139: 10-15; 152: 1-11, 153, 155:17-19]

Jungermannia gracilis Schleich., Pl. Crypt. Helvetiae Exsic., Cent. 3, no. 60, 1804, *nomen nudum;* Lindberg, Musci Scand., p. 7, 1879.
Jungermannia barbata var. *minor* Hook., Brit. Jungerm., pl. 70, figs. 18–20, 1816.
Jungermannia quinquedentata var. *attenuata* Mart., Fl. Crypt. Erlang., p. 177, 1817.
Jungermannia attenuata Lindenb., Syn. Hep. Eur., p. 48, 1829.
Jungermannia attenuata var. *gracilis* Lindenb., *ibid.*, p. 48, 1829.
Jungermannia quinquedentata var. *gracilis* Hüben., Hep. Germ., p. 203, 1834.
Lophozia attenuata Dumort., Rec. d'Obs., p. 17, 1835.
Jungermannia barbata var. *attenuata* Nees, Naturg. Eur. Leberm. 2:163, 1836.
Jungermannia barbata var. *attenuata* β *gracilis* Nees, *ibid.* 2:164, 1836.
Lophozia gracilis Steph., Spec. Hep. 2:147, 1902; K. Müller, Rabenh. Krypt.–Fl.6(1): 652, fig. 306, 1910.
Lophozia trifida Steph., *ibid.* 2:157, 1902 (new synonymy).
Barbilophozia attenuata Loeske, Abh. Bot. Ver. Brandenburg 49:37, 1907.
Orthocaulis gracilis Buch, Mem. Soc. F. et Fl. Fennica 8(1932):294, 1933; Frye & Clark, Univ. Wash. Publ. Biol. 6(3):407, 1945.
Orthocaulis attenuatus Evans, *in* Buch, Evans, & Verdoorn, Ann. Bryol. 10(1937):4, 1938.
Lophozia (Orthocaulis) gracilis Schust., Amer. Midl. Nat. 42(3):564, pl. 7, figs. 8–11, 1949.
Lophozia (Orthocaulis) attenuata Schust., *ibid.* 45(1):43, pl. 2, 1951; Schuster, *ibid.* 49(2):324, pl. 3, 1953.
Barbilophozia gracilis K. Müll., Rabenh. Krypt.–Fl. ed. 3, 6:637, fig. 192, 1954; Kitagawa, Jour. Hattori Bot. Lab. no. 28:268, fig. 7, 1965.

Forming pure *green* to somewhat brownish, rather loose but intricate, often extensive mats, *rather small* (shoots *750–1400* μ, rarely to 1600–1800 μ, wide × 1–3, occasionally 4 cm long). Stems mostly 200–280(325) μ in diam., usually green, occasionally brownish, flexuous, ascending to suberect, simple or sparingly branched, *freely developing filiform, terete, elongate appressed-leaved innovations from both sterile and fertile shoot apices*; rhizoids frequent, rather short, absent only on innovations. Leaves *approximate to loosely imbricate, usually green* (or with slight fuscous-brown pigmentation), somewhat obliquely inserted and oriented, *spreading, but lobes often slightly incurved, subquadrate* and often slightly obtrapezoidal, varying from 720 μ long × 750 μ wide to 680–750 μ long × 870 μ wide, rarely to 850 μ long × 1000 μ wide (*width averaging 1.0–1.25× length* of ventral half, but on weak shoots often only 0.9 as wide as long), *normally 3-lobed* (on weak shoots occasional leaves 2-lobed, often a few leaves subequally 4-lobed) *for ¼–⅓ their length*; lobes sometimes subequal, when trilobed the postical somewhat smaller, *triangular, acute or subacute* (often terminated by 2 superposed cells); sinuses acute to rectangulate, rarely obtuse, rarely gibbous; *postical margin lacking basal cilia*. Cells quadrate-hexagonal, with walls usually slightly

FIG. 152. *Lophozia* subg. *Orthocaulis*, sect. *Attenuatae*. (1–11) *Lophozia attenuata;* (12–13) *L. atlantica;* (14–15) *L. binsteadii.* (1) Ventral leaf lobe apex (×250). (2) Leaves (×50). (3) Leaf lobe apex (×250). (4) Leaves (×50). (5) Two leaves, *in situ* (×22). (6) Shoot sector, antical aspect (×19). (7–8) Leaves on shoot sectors, lateral and antical views (×22). (9) Leaf section (×230). (10) Underleaf, abnormally present (×ca. 230). (11) Stem cross section (×165). (12) Leaves (×50). (13) Ventral leaf lobe apex (×250). (14) Leaves (×50). (15) Ventral leaf lobe apex (×250). [1–2, *RMS 43808*, from "normal" but high altitude phase of *L. attenuata*, with purplish gemmae, Mt. Mansfield,

thickened, *trigones small, generally ± concave-sided;* cuticle weakly verruculose; *marginal cells of lobes averaging 13–16 μ,* locally occasionally to 17 μ; median cells varying from 16–19 × 17–23 μ to a maximum of 18–21 × 20–26 μ; oil-bodies varying from spherical (then up to 6–10 per cell) and 4.8–7.2 μ, to ovoid or ellipsoid (then mostly 3–5 per cell) and 5–6 × 6–8(10) μ; in lobes mostly 3–4 per cell and only 3.6–4.8 to 4 × 8 μ; chloroplasts 3.5–4 μ, smaller than oil-bodies. *Underleaves absent,* or represented merely by a cilium formed of 1–3 cells, terminated by a slime papilla (the whole only 45–60 μ long × 9–12 μ wide), the *ventral merophytes normally 2 cells wide. Gemmae formation restricted to stiffly erect, exceedingly attenuate* (380–550 μ, rarely 600 μ, wide × 6–15 mm long) *slender innovations, bearing imbricate, erect-appressed,* oblong-quadrate leaves, with erose-truncate apices weakly or barely trilobate, giving rise to *pale greenish gemmae* (in sun forms often orange to fulvous; in alpine phases sometimes tardily vinaceous); gemmae 1–2-celled, polymorphous, from pyramidal to quadrate to irregularly fusiform to obtusely stellate, their walls slightly thickened, 19–20 × 22 μ to 22 × 32 μ; *gemmiparous innovations without underleaves* (or with them reduced to stalked slime papillae, the ventral merophytes remaining 1–2 cells wide).

Dioecious. Androecial plants often slender; androecia becoming intercalary, of 4–7 or more pairs of bracts; bracts imbricate, the basal half strongly concave, erect, the apical portions and lobes spreading; apices of bracts trilobate, frequently with gibbous sinuses, the antical margin above base usually with a tooth; 1–2-androus. ♀ Bracts larger than leaves, somewhat sheathing perianth but apical portions spreading, ¼–⅓ divided into 3–5 triangular lobes; lobes acute or cuspidate, frequently unequal, often wavy; bracteole ovate, irregularly 2–4-lobed, the lobes similar to those of bracts. Perianth cylindrical or oblong-cylindrical, long-exserted, plicate and contracted toward apex; mouth laciniate and armed with uniseriate, unequal cilia, *the longer to 6 cells long.* Capsule wall 3-stratose [Müller (1910, p. 654) incorrectly describes it as bistratose], the innermost with partial or complete semiannular thickenings, the outer with nodular thickenings. Spores vinaceous red-brown, (10)14–15 μ, verruculose; elaters 7–8 μ in diam., vinaceous, loosely bispiral.

Type. European.

Distribution

The most widespread and abundant species of subg. *Orthocaulis,* occurring further southward than the other species (the only species found in deciduous forests); usually montane or submontane. Circumboreal in distribution (Müller, 1954, *in* 1951–58, p. 638, regards it as circumpolar,

---

Vt.; 3–11, *RMS 11731,* Big Susie I., Susie Isls., Minn., a phase approaching *L. atlantica* in the often broader leaves, occasional small underleaves, and large trigones and cells; reported questionably as *L. atlantica* in Schuster, 1953; 12–13, *L. atlantica: A.R.A. Taylor 43,* Lake Minot, Quebec, growing with *L. binsteadii* (figs. 14–15); 14–15, *L. binsteadii;* see above.]

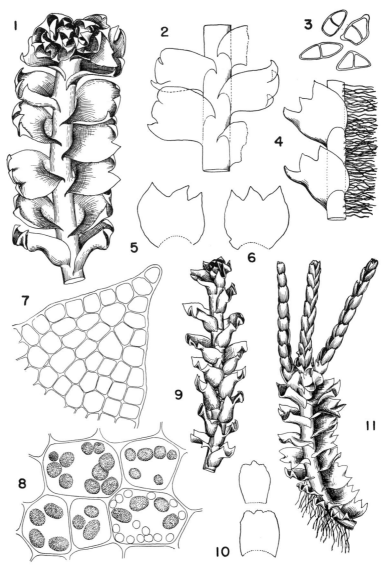

FIG. 153. *Lophozia (Orthocaulis) attenuata.* (1) Shoot, dorsal aspect (×22). (2) Shoot sector, postical aspect, rhizoids omitted (×26). (3) Gemmae (×280). (4) Shoot sector, lateral aspect (×26). (5–6) Leaves (×28). (7) Cells of lobe apex (×343). (8) Median cells of ventral lobe with oil-bodies and, lower right cells, chloroplasts (×700). (9) ♂ Shoot apex (×ca. 18). (10) Two leaves from erect-appressed leafy regions of gemmiparous shoots (×ca. 30). (11) Gemmiparous shoot-apex (×ca. 12). [1, 2, 4, Gunflint Lake, Minn., *RMS;* 3, 9–11, Blue Mt. Lake, Adirondacks, N.Y., *RMS;* 5–6, Raquette Lake, Adirondacks, N.Y., *RMS.*]

but the distribution pattern of the species does not conform to this viewpoint).

Found throughout much of Europe, particularly in the elevated portions, from the Pyrenees and the Alps of northern Italy eastward to the Tatra Mts. and northward to northern France, Belgium, England, and Scotland, Ireland, Scandinavia (Norway, Sweden, Finland); also reported from Spitsbergen and eastward in Novaya Zemlya, Siberia (but these reports possibly refer in part to *L. binsteadii* or *atlantica*); reported from the alpine region of Japan (Kitagawa, 1965; Hattori, 1952),[29] and from Taiwan (Kitagawa, 1965). Also on the Azores (Allorge & Allorge, 1948).

In North America reported from Alaska (St Paul I., *fide* Frye and Clark; Müller, *loc. cit.*, p. 639, doubts the accuracy of this report), southward to British Columbia, Washington, and in the Rocky Mts. to Montana, Wyoming, and Colorado. Widespread in our region (authorities consequently cited only for personal examinations and collections):

GREENLAND. Reported from Danmarks Ø, Gaasefjord, Runde-Fjeld, and near Cape Stewart, all near Scoresby Sund (Jensen; Müller, *loc. cit.*, correctly presumes these reports may refer to *L. binsteadii*).[30] BAFFIN I. Pangnirtung, C. Baffin, and Dorset, S. Baffin (Polunin, 1947). SOUTHAMPTON I. South Bay (Polunin, 1947). LABRADOR. "W. Deep Water Creek" (*Waghorne 74, 1891*! typical, marginal cells 13–15 μ, trigones small, leaves patulous, innovations filiform); also two collections, with coarse trigones, marginal cells 15–17 μ or 16.5–18 μ, from Cartridge Bight (*Waghorne, 24, 25, 1891*!; these grading into *L. binsteadii*, except in leaf shape). [Of the many collections of *L. attenuata* from Labrador (det. Underwood; the basis for the Underwood report, 1892) these are the only ones referable to *L. attenuata*; the rest represent *L. atlantica* or *L. binsteadii*.] NEWFOUNDLAND. Broad Cave (*Waghorne, 115, 118, 125*!; *119*!, grading into *L. binsteadii*, with the cells strongly collenchymatous, 17–18 μ on the margins!); New Harbour (*Waghorne 35*!; grading into *L. binsteadii*; strongly collenchymatous cells 16.5–19 μ on margins); Harbour Grace (*Waghorne 127*!; cells 15–17 μ on margins); Placentia Bay (*Waghorne 23*!). [Other collections, forming part of the basis of previous reports of this species from Newfoundland, represent *L. binsteadii*.] QUEBEC. Tadoussac (*Evans 73*!: a form of dry rocks, golden-brown, with purplish gemmae, partially with bulging trigones; close to *L. binsteadii*, but cell size typical for *L. attenuata*); Bic (*Evans, 134*!); Rigaud; Oka; St-Laurent des Monts; Waterloo; Mt. Orford; Mt. du Collège de Ste-Anne; I. Canuel near Rimouski; R. Ste-Anne-des-Monts; Mt. Albert (Lepage, 1944–45); Mt. Lac des Cygnes (Kucyniak, 1947); Mt.

---

[29] The type of *Lophozia trifida* Steph. from Japan: Mt. Hakusan, Prov. Kago, *leg.* Matsumura, is identical to *L. attenuata*.

[30] I have seen a specimen from west Greenland, from Ritenbenk (*Berggren, 1870*; Lund). This is poorly preserved and developed, with leaves 2(3)-lobed 0.55–0.65 their length, often folded, with sharp and narrow lobes, and with large, often bifid underleaves. A juvenile phase of *L. quadriloba* is at hand. I have also collected intensively at Ritenbenk but could not locate the species there.

Other reports (Lange & Jensen, 1887) of this species from south and west Greenland have been checked: (1) Sydostbugten ved Christianshaab (*Vahl 108*) = *L. binsteadii*; (2) Maneetsok, Kobbefjord (*Warming & Holm*) = *L. binsteadii*; (3) Ameralik Fjord(*Vahl 78*) = *L. kunzeana*; (*Vahl 79*) = *L. kunzeana* + *L. binsteadii*; (4) Frederikshaab (*Vahl; June 1829*) = *L. kunzeana* + *L. atlantica*.

Albert, Gaspé (*J. F. Collins 4216b*!).   ONTARIO. Rockcliffe; Hull; Otter Head, L. Superior; Sudbury; L. Timagami; La Cloche Mts., Sudbury Distr. (*Cain 1867*!); Ottawa (*Macoun 401, 1891*!).

MAINE. Mt. Katahdin, on Cathedral Trail above treeline, ca. 4880 ft (*RMS 32983*, with *L. atlantica* and *L. kunzeana*!); foot of Hamline Ridge, Mt. Katahdin, 3000 ft (*RMS*); Beech Mt., Mt. Desert I. (*Rand, 1891*!); Round Mountain L. (*Lorenz*). NEW HAMPSHIRE. Mt. Washington (*RMS*); Mt. Lafayette (*RMS*); Zealand Notch in White Mts. (*RMS*); "White Mts." (*Underwood & Cook, 1889*); Tuckerman's Ravine (*RMS*); Pinkham Notch (*RMS*); Crystal Cascade, Mt. Washington (*Underwood & Cook, 1889*!); Mt. Adams (*Farlow*!; *Evans, 1917*!, near Storm L.); Mt. Moosilauke, 4800 ft (*Kingman 2014*!); Mt. Madison (*Evans, 1917*!).   VERMONT. Jay Peak (*Faxon 50*!); L. Dunmore (*Farlow 510b*); Stratton (*Howe, 1899*!); Mt. Mansfield (*RMS*; *Dutton 1674B*!)   MASSACHUSETTS. Mt. Everett, Town of Mt. Washington (*Lorenz, 1915*!); Mt. Machuset; Sheffield; 1000 ft, Bare Rock Falls (*Lorenz, 1915*!).   CONNECTICUT. Salisbury, Litchfield Co.

NEW YORK. Little Moose L., Herkimer Co. (Haynes, Amer. Hep. no. 38!); Onondaga Co.; Raquette L.; Blue Mt.; Lyon Mt., near Ellenburg, Adirondacks (*RMS*); Peasleeville (*RMS*); Rock City near Olean, Cattaraugus Co. (*RMS*); Waterman Swamp, Napoli, Cattaraugus Co. (Schuster, 1949a); at summit of Thorn Mt., Palisades Interstate Park in Oak-Hickory Forest! (*RMS 33100*); Slide, Cornell, and Wittenberg Mts., Catskill Mts. (*RMS*); Mt. Marcy (*Britton, 1892*!) and Mt. Whiteface (*Britton*!), Adirondacks; Au Sable L., Adirondacks, *1902*!; Panther Mt., Little Moose L. (*Haynes 29*!). PENNSYLVANIA. Near Lock Haven, Clinton Co.   VIRGINIA. Summit of Bald Knob, Salt Pond Mt., Giles Co., 4360 ft (*Patterson*!).   NORTH CAROLINA. Flat Rock, near Grandmother Notch, vic. Linville (*RMS*); Buncombe and Transylvania Cos. (Blomquist, 1936); Mt. Craig, Black Mts., Yancey Co. (*RMS*); Devil's Courthouse, Blue Ridge Parkway (*RMS*).   TENNESSEE. Alum Cave, Mt. LeConte, Sevier Co., abundant! (*Schuster, 1952*; Sharp, 1939).

WISCONSIN. Sand I., Apostle Isls. (*RMS*; see also Conklin, 1929).   MICHIGAN. Copper Harbor, Ft. Wilkins, Keweenaw Co. (Steere, 1937; *RMS*); Sugarloaf Mt., near Marquette (*Nichols 130*!); Canjon L., Cliff R., Huron Mts.; N. of Burt L., Cheboygan Co. (*Steere, 3096*!); Gogebic, Chippewa Cos.; E. of Negaunee (*Nichols, 1935*!). MINNESOTA. Carlton, Carlton Co.; French R., Nopeming, and Oneota, St. Louis Co.; Great Palisade, L. Superior, Lake Co. (*RMS 11781, 11255*); Big Susie I., Long I., Lucille I., Belle Rose I., all Susie Isls., near Grand Portage, Cook Co. (*RMS 12104, 11900, 7182a, 7256a, 4761, 13040, 13376, 12216*); Gunflint L., Cook Co. (*RMS 13401*); Lutsen, Cook Co. (Conklin, 1942; Schuster, 1953).

Ecology

Found throughout the Spruce-Fir Forest Region, on crests and faces of boulders and ledges (with a variety of species belonging to the *Lophozia-Scapania* associule; chief among them *Lophozia ventricosa* var. *silvicola*, *L. longidens*, *Anastrophyllum michauxii*, *Tritomaria quinquedentata*, *Jamesoniella*, *Bazzania trilobata*, *Scapania nemorosa*). The species may be pioneer here but more often forms extensive mats after various mosses have come in and may then grow amidst *Leucobryum*, *Dicranum*, *Polytrichum*, and other coarse acrocarpous mosses, ocasionally with *Sphagnum*, *Lophozia ventricosa silvicola*, and *Bazzania*, often forming extensive mats. It occurs similarly on talus and rocky slopes.[31]

More rarely *L. attenuata* occurs south of the Spruce-Fir Zone, in open and rather dry Oak-Hickory Forest, either in swelling masses at the crests of ledges or on the shelves beneath them (where soil accumulates); associated then are *Bazzania trilobata*, *Scapania nemorosa*, *Leucobryum*, *Ptilidium pulcherrimum*, etc. To the south (Flat Rock, N.C., near Linville) the species also occurs at such sites and at the edges of shallow, soil-filled depressions, associated with *Selaginella tortipila*, *Gymnocolea inflata*, *Bazzania trilobata*, and *Anastrophyllum michauxii*, as well as various mosses (*Dicranum*, *Polytrichum*, *Leucobryum*, *Sphagnum*), under very acid conditions, with various Ericaceae.

In addition to occurrences on and adjacent to rocks, the plant is abundant (in the Spruce-Fir Zone) on decaying logs and stumps, often with *Bazzania trilobata*, *Dicranum*, *Anastrophyllum michauxii*, and *Tritomaria exsectiformis*, occasionally *Lophozia ascendens*, *longidens*, and *ventricosa silvicola*. On decaying logs usually not a pioneer (coming in after ecesis of the initial *Lophozia longidens-ascendens-Anastrophyllum hellerianum* facies), but persisting until almost total disintegration has set in, then competing with coarser acrocarpous mosses.

The species also penetrates well into the alpine Tundra, at least in New England, where frequent at elevations of 4000–5200 feet. It may occur here at the foot of boulders in the alpine "fels," together with *Lophozia* (*O.*) *atlantica* and *kunzeana* (as in *RMS 32983*, Mt. Katahdin). Tundra phases differ considerably from plants of the Spruce-Fir and Oak-Hickory regions in that apices of gemmiparous shoots are more or less vinaceous, mature gemmae usually being similar in color to those of *L. atlantica*.

Lophozia attenuata has, in general, a wide range under diverse edaphic conditions, with occasional occurrences on moist basaltic rocks under circumneutral conditions, with the opposite extreme found on peat (pH as low as 3.8). It occurs indifferently on organic and inorganic substrates and is found both in the deepest shade under spruce and fir, and in open sun on wet peat in alpine Tundra. Correlated with the wide tolerances is a nearly ubiquitous distribution in the Spruce-Fir Zone.

Differentiation

Generally recognizable in the field, even with the naked eye, because of the abundant development of the filiform, stiff, erect innovations, linked with nearly uniformly trilobate leaves. Gemmiparous innovations are rarely lacking (fo. *eflagellis* Schiffn.); such forms are more difficult to recognize. The absence of underleaves, the subquadrate trilobed leaves,

---

[31] The species may persist on moist ledges until a thick peat layer has formed; it then occurs with such helophytes as *Mylia anomala*, and helophytic phases of *Odontoschisma denudatum*, at a pH as low as 3.8–3.9 (Schuster, 1953, p. 324).

which are only slightly obliquely oriented, and the lack of cilia of the postical leaf bases are other important diagnostic features. The absence of postical cilia and of distinct underleaves separates the species from *L. floerkei* and *L. hatcheri*, small forms of which may in other respects be superficially similar.

One of the most important diagnostic features of *L. attenuata* is the small size of the leaf cells. On lobe margins these average usually 13–16, rarely 17, μ. Furthermore, the species appears unable to develop coarse, bulging trigones except in certain arctic phenotypes, even under conditions of extreme insolation and very intermittent moisture. For example, plants from open, dry Oak-Hickory Forest growing closely adjacent to the xeromorphic scrub, *Quercus ilicifolia*, although largely shaded, may grow amidst the xerophytic *Leucobryum*—yet show exceedingly small, concave-sided trigones. Under extreme conditions both the cell wall and the angles are thickened, and the cells become rather guttulate.

Lophozia attenuata is closely related to *L. binsteadii* and *L. atlantica*. It differs from both in two important features: (*a*) smaller cell size; (*b*) general inability to develop pronounced collenchyma of the leaf cells. Furthermore, the gemmiparous flagella are ultimately more slender, more attenuate, and usually much more freely produced in *L. attenuata* than in *L. atlantica*. Usually *L. binsteadii* is distinct because of the warmer, golden to reddish or chestnut-brown coloration, the more imbricate secund and suberect, strongly cupped leaves, and the extreme development of collenchyma. However, northward, in the area where the two species overlap, (Labrador, Nova Scotia, Newfoundland), they at times appear to grade imperceptibly into each other. See in this connection *L. binsteadii* (p. 321).

There has been some reluctance to accord *L. atlantica* specific status, next to *L. attenuata*, most recently expressed in Müller (1954, *in* 1951–58, p. 640). With study of alpine forms of *L. attenuata*, the distinctions between this and *L. atlantica* become admittedly somewhat tenuous, since alpine forms develop vinaceous gemmae, identical in color and size to those of *L. atlantica*. (Müller, Macvicar, and almost all other authors describe the gemmae of *L. attenuata* as pale green to yellowish red, or pale brownish. Müller attempts a further distinction, characterizing the gemmae of *L. attenuata* as 2-celled, those of *L. atlantica* as 1-celled; this distinction is quite invalid, mature gemmae being usually 2-celled in both species.) Alpine phases of *L. attenuata* also tend to be somewhat fuscous or brownish green or even blackish under extreme conditions, much as in *L. atlantica*. However, there can be no question about the specific distinction between the species, since I have found the two growing intimately admixed (above treeline, Mt. Katahdin, *RMS 32983*). Mature plants from this collection are contrasted in Table 2, in which the major differences are clearly apparent.

TABLE 2

*Distinctions between L. attenuata and L. atlantica**

Criterion	*L. attenuata*	*L. atlantica*
Leaf dimensions	720 μ long × 750 μ wide > 680–750 μ long × 870 μ wide	650 μ long × 920 μ wide > 800–880 μ long × 1000–1120 μ wide
Width-length ratio	1.04–1.28:1	1.25–1.41:1
Leaf lobes	3(4)	2–3
Marginal lobe cells	15–16(17) μ	(18–19)20–25 μ
Median cells	16–18 × 17–23 μ	24 × 24 to 28–32 × 28–34 μ
Gemmiparous shoots	Filiform 450–550(600) μ in diam.	Stout, fingerlike 600–1200 μ in diam.
Gemmae	1-2-celled, tardily purple 19–20 × 22 μ to 22 × 32 μ	1-2-celled, early vinaceous 23 × 23–26 μ to 21–23 × 35 μ
Oil-bodies	(3)4–6(7–9) per cell 4.5– 5 > 5 × 6 > 6 × 7–8 μ	(3–4)5–8(9) per cell 5 > 5–6 × 7–9 μ

* See also Table 3, p. 322. Distinctions between *L. binsteadii* and *L. atlantica*.

LOPHOZIA (ORTHOCAULIS) BINSTEADII (Kaal.)
Evans
[Figs. 139:16–18; 152:14–15; 154]

Jungermannia floerkei subsp. *ambigua* Joerg., Vidensk. Selsk. Forh. Kristiana 8:54, 1894.
Jungermannia binsteadii Kaal., Vidensk. Skrift. Kristiana, 1898 (9):9, 1898.
Lophozia binsteadii Evans, Ottawa Nat. 17:22, 1903; K. Müller, Rabenh. Krypt.–Fl. 6(1):655, 1910; Schuster, Amer. Midl. Nat. 45(1): 43, 1951; Schuster, Natl. Mus. Canada Bull. 122:5–6, 1951.
Jungermannia herjedalica Schiffn., apud Arnell, Bot. Notiser 1906:152, 1906.
Barbilophozia binsteadii Loeske, Hedwigia 49:13, 1909; K. Müller, Rabenh. Krypt.–Fl. ed. 3, 6:640, fig. 194, 1954.
Orthocaulis binsteadii Buch, Mem. Soc. F. et Fl. Fennica 8 (1932):294, 1933; Frye & Clark, Univ. Wash. Publ. Biol. 6 (3):403, figs. 1–6, 1945.
Lophozia ambigua Joerg., Bergens Mus. Skrifter no. 16:138, 1934.

Erect or suberect, usually crowded and caespitose, among mosses (especially *Dicranum* and *Sphagnum*) or in small patches, usually *golden brown to chestnut brown,* the *shoot apices often dull, ferruginous,* slender, *shoots usually* (0.5)0.6–1.2 mm *wide×* 1–3 cm high, *appearing terete* because of the erect leaves. Stems subsimple, rather stiff and flexuous, 150–265 μ thick, with rather free development of short rhizoids, even near shoot apices; branches occasional only, lateral-intercalary from lower halves of leaf axils. Leaves usually rather dense, often *quite imbricate,* obliquely inserted, *somewhat dorsally secund, suberect and strongly concave and with incurved lobes, when flattened subquadrate or narrowly obtrapezoidal,* when mature *3-lobed,* from (640–700)710–720 μ wide × (575–600)640–760 μ long to 800 μ wide × 840–950 μ long to 900 μ wide × 850 μ long (type; *0.85–1.05× as wide as long*) on slender plants; on more robust plants from

810 μ wide × 725 μ long to 1060–1120 μ wide × 850 μ long ["var. *herje-dalica*"; *1.1–1.2(1.3) × as wide as long*], *usually nearly symmetrical* (postical lobe rarely conspicuously larger), narrowed basally; *lobes usually subequal, longer than broad, ovate-triangular to broadly lanceolate*, usually 3, exceptionally 2, *pointed*, often to usually *tipped by 2 superposed cells; sinuses sharp and acute*, descending ⅓–½ the leaf length. Cells usually (but not always) strongly collenchymatous, at least in the lobes, with *bulging to subconfluent trigones; marginal cells of lobes averaging (16–18)19–21(22–23) μ; median cells (17)18–23 × (18)24–30 μ;* cuticle usually closely and distinctly verruculose, becoming striolate below. *Underleaves absent on sterile, nongemmiparous stems, usually absent on gemmiparous shoots as well. Gemmae rare*, on the margins of ± normal leaves or of somewhat reduced leaves (but *never on flagelliform shoots*), *vinaceous red (in masses deep purple)*, 1–2-celled, thick-walled, polygonal, *mostly (18)20–25 μ.*

Dioecious. ♂ Plants often more slender; bracts like leaves, but with an antical tooth. ♀ Bracts sheathing perianth, quadrate to obtrapezoidal in outline, 3–5 lobed for ½–⅗ their length (the outer lobes often mere spinose teeth), ca. 875–1100 μ wide × 1050–1150 μ long, the lobes lanceolate-acuminate, often sparingly and irregularly repand-dentate or with 1–several sharp to spinose teeth, *distally ending in a cilium often formed of 3–7 superposed cells*, near base often with 1–2 small teeth formed of isodiametric cells. Bracteole very large, ca. 450–600 μ wide × 800–1000 μ long, bifid for ½–⅗ its length into lanceolate lobes usually terminated like those of bracts, bearing often 1–2 sharp lateral teeth (as on bracts), free or united for ⅕–¼ with one bract. Perianth longly exserted, longly cylindrical-clavate, ca. 1000–1050 μ in diam. × 3000 μ long, contracted and deeply 4–5-plicate at the decolorate mouth; mouth lobulate with small lobes, which are ciliate-dentate, ending in cilia formed of (1–2)3–5 superimposed cells.

Type. Near Kongsvold, Dovre, Norway (*Binstead, 1892*). Schiffner (*Hep. Eur. Exsic. no. 433*) distributed material from Gudbrandsdalen, Norway (*Kaalaas, 1907*), erroneously labeled "Orig. Ex."

Distribution

An arctic-alpine species, usually of bogs, descending (in Greenland and northern Europe) to sea level, but more common at higher elevations (1000–1500 m). In Europe in the southern to northern half of Norway, in Swedish Lapland, the Sarek Mt. region of Sweden, northern Finland, northern USSR (Kola region), eastward to Siberia (Novaya Zemlya and eastern Siberia). In North America transcontinental at higher latitudes. The species is imperfectly circumpolar and essentially arctic. Unlike the related *L. atlantica*, without alpine extension southward.

In North America much more common under arctic conditions than previously assumed (almost all earlier collections were misidentified and have given *L. binsteadii* an erroneously "spotty" distribution). Found in the west from Alaska (Evans, 1919b) and Yukon (Dawson, Bonanza

Creek!) to British Columbia (soil among rocks, Simpson Pass, Monarch Mt., 8500 feet, *Brinkman, 1913*; det. G. H. Conklin!),[32] and Alberta; eastward distributed as follows:

N. GREENLAND. Neergard Elv, S. coast of Peary Land, 82°00′ N., 26° W. (*Holmen 7717*; one plant, mixed with calciphytes such as *Arnellia fennica* and *Scapania praetervisa polaris;* possibly admixed in packeting—in any event I am suspicious of the association). E. GREENLAND. Kaiser Franz Joseph Fjord; Röhss Fjord (Jensen, 1906a, p. 304); E. Greenland coast between 69°30′ N. and 65°37′ N. (Harmsen, 1933). NW. GREENLAND. North Star Bay, Inglefield Gulf, 77°30′ N. (*E. O. Hovey, 1916*, det. Evans, 1918!);[33] Siorapaluk, Robertson Bugt, 77°40′ N. (*RMS 45051, 45052b*). W. GREENLAND. Ikertoq Fjord, 66°45′ N.; with *Calypogeia sphagnicola, Mylia anomala* (*Vahl 192, Aug. 1832!*); Sydostbugten ved Christianshaab (*Vahl 108!*; reported by Lange & Jensen, 1887, as *J. attenuata*); Maneetsok, Kobbefjord (*Warming & Holm!*; reported by Lange & Jensen, 1887, as *J. gracilis*); Ameralik Fjord, 64°10′ N. (*Vahl 79!*; mixed with *L. kunzeana*; reported by Lange & Jensen, 1887, as *J. attenuata*); Oqaq, Kronprinsens Eiland, 69° N. (*Kruuse;* det. by Jensen as *J. gracilis*); Tupilak I., W. of Egedesminde, 68°42′ N. (*RMS & KD 66-112, 66-024, 66-012, 66-009b*, etc.; *Holmen 16-994*); Fortune Bay, Disko I., 69°16′ N., 53°50′ W. (*Holmen & Steere 62-038; RMS & KD 66-208*); SE. of Jacobshavn, Disko Bay, 69°12′ N., 51°05′ W. (*RMS & KD 66-235*); Lyngmarksbugt, Godhavn, ca. 25-125 m, 69°15′ N., 53°30′ W. (*RMS & KD 66-343, 66-561*); Sondrestrom Fjord, 66°5′ N., L. Ferguson near head of fjord (*RMS & KD 66-090, 66-091, 66-085*).

ELLESMERE I. Cape Rutherford, Bedford Pim I., and Fram Harbor, all in Hayes Sd. region; Goose Fiord, on S. coast (Polunin, 1947; based on Bryhn, 1906). DISTRICT OF KEEWATIN. *Fide* Steere, 1941. NORTHWEST TERRITORIES. Paint Hills Isls., James Bay (Lepage, 1953, det. Arnell). LABRADOR. Cartridge Bight (*Waghorne 32!*; typical, bearing stout gemmiparous shoots with purple gemmae; no underleaves or only stalked slime papillae; cells 23 μ on margins); Venison Pickle (*Waghorne 4, 11*; typical!). QUEBEC. Sugluk Inlet (Lepage, 1944-45); Ft. George, James Bay (Lepage, 1953; det. Arnell); tributary of Clearwater R., Richmond Gulf (Lepage, 1953; det. Arnell); island in L. Minto, 57°06′ N., 75°25′ W. (*A.R.A. Taylor 40, 43, p.p.!*). NOVA SCOTIA. Barrasois Valley, Cape Breton I. (*Nichols!*; see Brown, 1936a).

Ecology

Typical *L. binsteadii* is a taxon with a narrow environmental range and with evidently narrow physical variation as well. It is restricted, although not exclusively, to bogs or insolated peaty ground around shallow lakes, where it normally grows erect, crowded among various mosses or creeping over *Sphagnum*, usually forming warmly pigmented, golden-brown, sometimes chestnut-colored patches, the shoot apices often slightly purplish brown. Gemmae are rarely formed, are normally metallic purple in

---

[32] This material with mature leaves only ¼-⅓ bilobed, ca. 1.2-1.3× as wide as long; leaves concave and suberect as in *L. binsteadii*, however. Apparently transitional to *L. atlantica*.

[33] This material has asymmetrical, very broad leaves 1.25-1.36× as wide as long, as in *L. atlantica*; the leaves are divided for usually less than ⅓. These plants might possibly be referred to *L. atlantica*.

color, and usually only slightly inhibit normal growth of the shoot (rarely and sporadically attenuate gemmiparous shoots are formed).

Associates are *Mylia taylori*, *M. anomala*, *Cephalozia leucantha*, *C. bicuspidata*, *Calypogeia sphagnicola*, *Anastrophyllum minutum*, *Polytrichum*, several species of *Sphagnum*, *Dicranum* spp., *Cephaloziella subdentata*, *Lophozia kunzeana*, rarely *L. atlantica*, and various mosses. The association with *Calypogeia sphagnicola* is a particularly constant one. In west Greenland, where ubiquitous in sphagnous areas, usually associated with *Calypogeia sphagnicola*, *Mylia anomala*, *Cephalozia bicuspidata*, *Lophozia ventricosa s. lat.*, *Riccardia sinuata*, and *Anastrophyllum minutum*. The species may here form large, pure mats, usually quite devoid of gemmae.

Differentiation

The somewhat larger cells, generally large trigones, diagnostically concave leaves (with narrow, erect or incurved, sharp lobes), warm brown coloration, and arctic habitat are usually sufficient to separate the species from the closely allied *L. attenuata*. In the final analysis, only the ability to develop pronounced collenchyma and the larger cell size (marginal cells averaging usually 19–21 μ) serve to consistently separate the two species.

Concurrently with erect, caespitose growth in bogs, *L. binsteadii* normally has a habitually distinctive facies: leaves are erect or suberect, quite concave, with incurved lobes, and are usually quite imbricate. As a consequence, the shoots are nearly terete. The warm, golden to chestnut color of the plants (the mod. *viridis* is rare!) is also distinctive. However, both pigmentation and orientation of the leaves are associated with the specialized habitat and do not consistently separate *L. binsteadii* from the similar *L. atlantica*.

Normally *L. binsteadii* differs from *L. atlantica* in the more cupped leaves (those of *L. atlantica* are more spreading basally, with incurved lobes) and the narrower leaf form, with leaves averaging about 0.9–1.1 × as wide as long. The leaves usually have a distinctive facies, due to the narrow and deep sinuses (descending usually ⅓–½ the leaf length), and the consequently narrowly triangular lobes. However, distinctions in leaf form, depth of the sinus, and form of the lobes are subject to variation. For instance, *L. binsteadii* may have leaves 1120 μ wide × 850 μ long (1.2 × as wide as long) in the "var. *herjedalica*."

Plants collected and identified by H. W. Arnell as *L. binsteadii* (Sweden: Herjedalen, Storsjö, *July 7, 1904*, NY) have even more nearly intermediate

leaves. On robust shoots they may be from 830 μ wide × 700 μ long (1.2 ×
as wide as long) to 910 μ wide × 650 μ long (1.4× as wide as long); on such
leaves the obtuse to rectangular lobes may be quite short, the sinuses descending
only to ¼–⅓ the leaf length.[34] These plants also bear somewhat attenuate
gemmiparous shoots (exactly as in *L. atlantica* from Maine), and occasionally
have minute and vestigial underleaves on nongemmiparous shoots. Similarly,
material from North Star Bay, Greenland (determined by Evans, 1918, as *L.
binsteadii*; NY) has very unusually asymmetric leaves, with the antical lobe
smaller than the median, the median much smaller than the postical; the
plants are a large form, with leaves from 1360 μ wide × 1000 μ long to 1400 μ
wide × 1100 μ long (1.25–1.35× as wide as long, on a series of leaves);
sinuses descend only ¼–⅕, rarely ⅓, between postical and median lobe, and
ca. ¼–⅓ rarely ⅖, between antical and median lobe; the postical lobe is often
rectangular or even obtuse. I would regard such plants with broad leaves as
close to *L. atlantica*. Wherever they are assigned, they illustrate the extreme
difficulty of maintaining a distinction here between these two species.[35]

Buch (1937b, p. 126) also emphasizes the immediate relationship
between *L. atlantica* and *L. binsteadii*, and admits that (except for extreme
pachydermous typical phases of *L. binsteadii*, which differ in the almost
confluent trigones) the two are very close, "da die Blattzellgrösse und die
Form der Keimkörner tragenden Sprosse fast dieselbe ist und die Unter-
schiede in der Blattform sich bisweilen verwischen können." Buch implies
that *L. atlantica* always develops underleaves, at least on gemmiparous
shoots; such is apparently usually the case, though nongemmiparous
shoots may lack them. In *L. binsteadii* underleaves are usually absent,
even on gemmiparous shoots.

Thus *L. binsteadii* agrees closely with *L. atlantica* in most features. Buch
(1936) and, following him, Schuster (1951a, p. 43) attempted to separate them
on the basis of marginal cell size (*L. binsteadii*, 16–19 μ; *L. atlantica*, 18–27 μ).
Study of a series of specimens, from both North America and Europe, including
"topotypic" material of *L. binsteadii* (Schiffner, Hep. Eur. Exsic. no. 433),
shows no consistent difference in cell size between the two. In "topotypic"
material of *L. binsteadii*, lobes show marginal cells averaging between 20–23 μ;
in the typical material distributed by Macoun (Canad. Hep. no. 49, Bonanza

[34] Plants with similarly broad leaves (760–780 μ wide × 650 μ long) are also widespread
in west Greenland (*e.g.*, the Oqaq and Sondrestrom Fjord collections, and the plants from
North Star Bay).
[35] Frye and Clark (1937-47, pp. 401-402) place primary emphasis on separating these two
allied species on the number of leaf lobes, *L. binsteadii* supposedly showing "most commonly
2-lobed" leaves, vs. "most commonly 3-lobed" ones in *L. atlantica*. This attempt at a separation
was unfortunately accepted (Schuster, 1951a, p. 43), but study of a series of specimens has
shown that there is no basis for such a separation. Indeed, the inverse relationship appears to
hold, since *L. binsteadii* rarely has 2-lobed leaves, while such leaves are often common in *L.
atlantica*. Although Frye and Clark (*loc. cit.*, p. 403) describe the leaves of *L. binsteadii* as "usually
2-lobed," their figures show them to be, with few exceptions, 3-lobed.

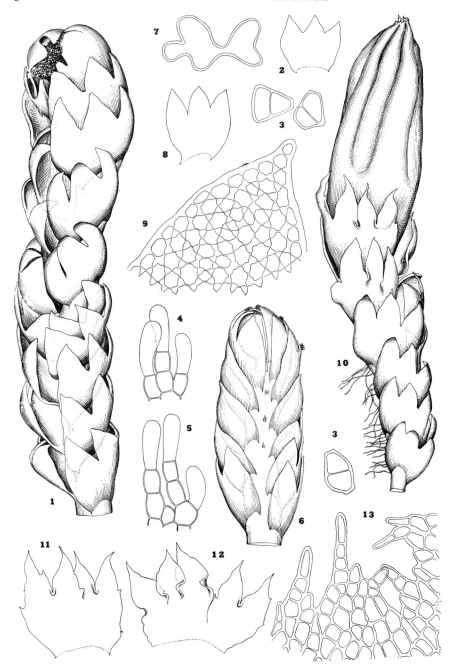

FIG. 154. *Lophozia* (*Orthocaulis*) *binsteadii.* (1) Gemmiparous shoot, dorsolateral aspect (×45). (2) Leaf from plant in fig. 1 (×25). (3) Gemmae (×440). (4–5) Underleaves from apices of gemmiparous shoots (×400). (6) Gemmiparous shoot apex, ventral aspect, with rudimentary underleaves and a few gemmae (×45). (7) Perianth cross section (×33). (8) Normal leaf (×25). (9) Leaf lobe apex (×235). (10) Fertile

Creek, Yukon) the cells average slightly smaller, 18–19.5 μ. In various specimens seen of typical *L. atlantica* the cells average mostly 19–24 μ, occasionally for short stretches 25–27 μ.[36] Müller (1954, pp. 640–41, *in* 1951–58) described the gemmae of *L. binsteadii* as 2-celled, of *L. atlantica* as 1-celled. This distinction fails to hold; mature gemmae of both are to a large extent 2-celled.

Although variation in *L. binsteadii* and *L. atlantica* is such that the two may closely approach each other, there appears to be no doubt about their distinction. Although they have not been cultured, excellent evidence is at hand that discrete taxa are involved, since the two have been found growing admixed (*Waghorne 4*, Venison Pickle, Labrador, reported by Underwood as *L. attenuata*). When growing admixed, mature plants of *L. binsteadii* differ in the more terete shoots, with handlike, concave, erect, imbricate leaves vs. leaves which are spreading, but with the lobes folded upward, in *L. atlantica*. The admixed *L. binsteadii* was a warmer brown, the *L. atlantica* more fuscous. *Lophozia binsteadii* lacked gemmae and was less robust; *L. atlantica* produced stout, fingerlike, gemmiparous shoots and was evidently more robust. In addition to these habitual differences, there are many other differences in the leaves, leaf cells, and other characters, as is also evident from Table 3, derived from study of mature plants of the two species, growing within 5 mm of each other.[37]

Variation

Lophozia binsteadii is still a poorly defined, misunderstood plant. In the ecotone region between Taiga and Tundra (as, e.g., in Newfoundland

[36] Müller (1954, p. 640, *in* 1951-58) claims that the type of *L. atlantica* has cells equal in size to those of *L. attenuata*: "other plants, by contrast, show somewhat larger cells." Although the type of *L. atlantica* has not been available for study, the concepts of Joergensen (1934), Buch (1936, 1937b), and other Scandinavian students who have critically studied the species are all based on a large-celled plant.

[37] A second collection, in which the two taxa are intimately admixed, has been seen (Lake Minto, Quebec, *Taylor 43*); from this the differences between them clearly emerge. *Lophozia binsteadii* has leaves 460–580 μ wide × 400–580 μ long, almost uniformly trifid for 0.4–0.5 their length into narrowly triangular, sharply acute lobes; the plants are rather strongly yellow-brown pigmented and have cells with strongly bulging, gibbous trigones (Fig. 152:14–15). By contrast *L. atlantica* has leaves ca. 625–635 μ wide × 435–450 μ long when trifid (some leaves are 2-, others 4-, lobed), is weakly brownish tinged, and has cells with moderate, hardly bulging trigones (Fig. 152:12–13). The differences often apparent in cell size are hardly discernible in the marginal cells of the lobes, those of *L. binsteadii* being usually 20–23 μ, sometimes averaging up to 25 μ; in admixed *L. atlantica* they are 19–23.5 μ in size! As is so often the case, *L. binsteadii* is free of gemmae while *L. atlantica* sparingly bears purplish red gemmae. These distinctions (and the similarities in cell size) are readily evident from Fig. 152:12–15.

shoot (×24). (11–12) Bract, and at right, bract with bracteole from one gynoecium (×24). (13) Perianth mouth cells (×235). [1–6, *A.R.A. Taylor 42*, Lake Minto, Quebec; 7–13, *A.R.A. Taylor 57*, Bush Lake, Quebec.]

TABLE 3

Distinctions between L. binsteadii and L. atlantica*

Criterion	L. binsteadii	L. atlantica
Leaf dimensions	710 μ wide $\times$ 640–650–760 μ long; 720 μ wide $\times$ 740 μ long	860 μ wide $\times$ 620 μ long; 915 μ wide $\times$ 700 μ long; 1100 μ wide $\times$ 770 μ long
Width-length ratio	0.93–1.1 : 1	1.3–1.4 : 1
Leaf lobes	3	2–3–4
Marginal lobe cells	(16)18–20(22) μ	20–25 μ
Median cells	18–20 $\times$ 18–23(25) μ	24–30 $\times$ 24–33 μ
Trigones	$\pm$ Confluent in lobes	Bulging but not confluent
Leaf lobes	Longer than broad, narrow, acute	As broad as or broader than long, broad, $\pm$ subobtuse
Sinuses	$\frac{1}{3}$–$\frac{1}{2}$	($\frac{1}{4}$)$\frac{1}{3}$–$\frac{2}{5}$
Underleaves	Lacking	Small and lanceolate, at least on gemmiparous plants

* See also the similar Table 2, p. 315, where admixed *L. attenuata* and *L. atlantica* are compared.

and Labrador) it appears to undergo transition to *L. attenuata*, of which it may be an arctic subspecies.

A specimen from Nova Scotia (*Nichols 319; p.p.*, among *Mylia taylori, Cephalozia leucantha;* on rocks), identified and reported as *L. attenuata*, is in the NYBG. This plant is ♂, slender, and approaches *L. attenuata* in the occasional flagelliform shoots with colorless gemmae and in the somewhat spreading leaves, with incurved lobes. However, it has cells with coarse, bulging trigones, and marginal cells average (17)19–21(23) μ. On this basis it is referred provisionally to *L. binsteadii*. Minnesota plants (which are ♀), once referred to *L. atlantica* (Schuster, 1953) because of the presence of underleaves, may be identical. Lax shade forms of the *L. binsteadii-atlantica* complex are rare and very poorly understood.

Other plants similar to the Nova Scotia material (*Nichols 319*) have been seen, i.e., with rather large cells and yellowish bulging trigones, but with slender and filiform innovations, whose apices bear colorless gemmae. These plants, intermediate in facies between *L. attenuata* and *L. binsteadii*, in most cases, have larger cells. For instance, material from Labrador [Cartridge Bight, *Waghorne 25, 1891*; reported by Underwood (1892a) as *L. attenuata*] has leaves identical with those of *L. binsteadii* fo. *herjedalica* and marginal apical cells 19–23 $\times$ 23–28 μ. Another critical collection is from Cartridge Bight (*Waghorne 24*; reported by Underwood as *L. attenuata*). Here two plants are admixed, one with narrow leaves divided $\frac{1}{4}$–$\frac{1}{3}$, averaging 730 μ wide $\times$ 700 μ long. In this the strongly collenchymatous cells, with bulging trigones and yellow-brown walls, give the distinct impression of *L. binsteadii*. The leaf form and shallow lobes, as well as the cells (marginal only 14–16 μ, rarely 17 μ; median 18–20 $\times$ 20–28 μ), approach those of *L. attenuata* (as do the attenuate gemmiparous shoots). Study of a series of specimens of the *L. attenuata-binsteadii* complex, therefore, shows intergradation in cell size, in degree of development of trigones, and to some degree in leaf form. The intergradation occurs only in transitional regions where the known ranges of the two taxa overlap. Associated with the *L. binsteadii-attenuata* intergrade of *Waghorne 24* is typical *L. atlantica* (leaves 2–3-lobed, the 3-lobed ca. 870 μ wide × 580 μ long = 1.5:1 ratio; marginal cells 20–24 μ; median cells 23–25 × 23–30 μ; gemmae on short, stout, fingerlike shoots). This collection (as well as *RMS 32983*) is critical in showing that *L. atlantica* is a sharply defined taxon, distinct from the *L. attenuata-binsteadii* complex.

The problem of a distinction between *L. attenuata* and *L. binsteadii* is equally acute in Newfoundland, where nine collections made by Waghorne (May–September 1891) were studied; all of these plants were listed as *L. attenuata* by Underwood (1892a). Five of the nine numbers (Harbour Grace no. 127; Broad Cove nos. 115, 118, 125; Placentia Bay no. 23) appear to be nearly typical *L. attenuata*. Their cells average 15–17 μ on the margins of the lobes. Two numbers are more or less intergrading but probably refer to *L. attenuata* (Broad Cove no. 119; New Harbour no. 35); in these the marginal cells of the lobes average 17–18 and 16.5–19 μ, respectively. Finally, in three numbers (New Harbour nos. 21, 17, 34) the marginal cells average 19–23 μ; the cells are also strongly collenchymatous. These plants, on the basis of leaf form and the presence of strict and filiform gemmiparous innovations, closely approach *L. attenuata* and perhaps should be placed in this species (in spite of the larger cells and coarse trigones).

A collection from Labrador (Cartridge Bight, *Waghorne 32*; reported by Underwood as *L. attenuata*), appears definitely referable to *L. binsteadii*. It does not show the *L. attenuata* features of the preceding plants (e.g., the gemmiparous shoots are fingerlike and stout and bear purplish gemmae, as in *L. atlantica*). As in *L. atlantica*, the cells are large (marginal average 21–24 μ), gemmae agree in color, and gemmiparous shoots are stout. The confluent or subconfluent, large trigones, lack of underleaves on gemmiparous and sterile shoots (occasionally with stalked slime papillae on gemmiparous shoots, e.g., with 2-celled underleaves), and leaves with deeper sinuses and narrower lobes (leaves 1.0–1.25× as wide as long; sinuses ⅖–½ the leaf length) suggest *L. binsteadii*. The only other Labrador collections seen which match "typical" European *L. binsteadii* are *Waghorne 4* and *11*, from Venison Pickle. These are small plants, terete because of the dense and erect leaves, with deeply trilobate, narrow leaves, and confluent or subconfluent trigones. As in the type plants, no gemmae are developed.

Study of this at times confusing series indicates that (*a*) *L. binsteadii* "grades towards" *L. attenuata*; (*b*) *L. binsteadii* is more polymorphic than described in manuals (such as Müller, 1951–58; Frye and Clark, 1937–47). If my interpretation is correct, a saxicolous phase exists, evidently of shaded rocks, growing with such species as *Mylia taylori* and *Cephalozia* spp. In this phenotype trigones are large but never confluent; leaves less dense, more spreading; leaf sinuses often less deep; and leaves larger, slightly

more transverse. Gemmiparous shoots may be slender and filiform (as in
L. attenuata), and gemmae colorless. If this saxicolous phase, with large
trigones and large leaf cells, is referred to *L. attenuata*, the circumscription
of the latter species becomes almost impossible, since the criteria (cell size
and type) which one can reasonably expect to be genetically sharply
fixed become meaningless.

Identification of such shade forms with patulous leaves and greenish
gemmae will certainly remain critical and (for forms in which the cell
size is somewhat deviant) may even continue to be impossible.

In addition to saxicolous phenotypes that appear intermediate between
L. attenuata and *L. binsteadii*, in the Arctic *L. binsteadii* shows a distressing
amplitude of variation in the otherwise normal bog phenotypes. For
example, in *RMS & KD 66–208* (Fortune Bay, Disko I., Greenland) many
shoots have leaves with abnormally small trigones and leaves that are
wider than normal (width ca. 860 μ × 650 μ long; thus ca. 1.33 × as
wide as long). Such plants also have marginal cells only 17–18 μ. They
occur mixed with "normal" individuals with leaves ca. 660 μ wide and
long, coarse trigones, and marginal cells 19–21 μ.

LOPHOZIA (ORTHOCAULIS) ATLANTICA (Kaal.) Schiffn.

[Figs. 152:12–13, 155:1–16]

Jungermannia atlantica Kaal., Vidensk. Skrift. Kristiana 1(9):11, 1898.
Lophozia atlantica Schiffn., Lotos 49:46 (of reprint), 1901; K. Müller, Rabenh. Krypt.–Fl.
    6(1):652, 1910; Macvicar, Studs. Hdb. Brit. Hep. ed. 2:198, cum fig., 1926.
Orthocaulis atlanticus Buch, Mem. Soc. F. et Fl. Fennica 8(1932):294, 1933; Frye & Clark,
    Univ. Wash. Publ. Biol. 6(3); 406, 1945.
Lophozia ambigua Joerg., Bergens Mus. Skrifter no. 16:138, 1934 (*p.p.*; *fide* Buch, 1938).
Lophozia (Orthocaulis) atlantica Buch, Ann. Bryol. 10(1937):126, 1938; Schuster, Amer. Midl.
    Nat. 45(1):43, pls. 3, 8:9–14, 1951; Schuster, Natl. Mus. Canada Bull. 122 (1950):5,
    1951; Schuster, Amer. Midl. Nat. 49(2):326, pls. 4, 5:9–14, 1953.
Barbilophozia atlantica K. Müll., Rabenh. Krypt.–Fl. ed. 3(6):639, fig. 193, 1954.

Green to brown-green (*the brown pigment fuscous, never warm, occasionally blackish*),
medium-sized, shoots (0.8)1.0–1.8 mm wide × 1–3 cm high, sparsely branched,
flexuous, *forming wiry, intricate, often interwoven mats on rocks or creeping over peat.*
Stem green to brown, flexuous, ascending, 220–450 μ in diam., *without small-
leaved, filiform gemmiparous innovations from tip*, but often with normal shoots
gradually narrowed into stoutly filiform gemmiparous apices. Leaves weakly
obliquely inserted, contiguous to imbricate, *stiffly spreading, but the lobes rather
strongly incurved*, thus somewhat concave, slightly dorsally connivent, appearing
somewhat complicate-concave, and *nearly transversely oriented*; leaves polymorphic,
2–4 (but usually 3-) lobed, usually 1.2–1.5× as wide as long (when 3-lobed), varying
from (625) 910 μ wide × (435) 650–670 μ long to 1020–1120 μ wide ×
680–880 μ long, to a maximum of ca. 1200 μ wide × 850 μ long, *usually*

broadly obtrapezoidal to transversely rectangulate, lobed for ¼–⅖ *their length*; lobes usually *broadly triangular, incurved* (rarely spreading), *obtuse to subacute*, most often terminated by a solitary cell; sinuses obtuse to acute, not gibbous or rarely weakly so; *postical margin without cilia at base*, rarely with a stalked slime papilla. Cells of *margins averaging* (*18*)*19–24*(*27*) μ, in the leaf middle *23–26* × *23–30* μ to 28–32 × 28–34 μ; cuticle delicately verruculose; *trigones small, concave-sided to bulging, never subconfluent*, oil-bodies (*3–4*)*5–7*(*10*) *per cell*, spherical to short-ovoid, 4.5–5.5 to 6–7.5 μ, a few to 5 × 7 to 6 × 8–9 μ, grayish, somewhat opaque, very finely granular, the spherules barely protuberant, in leaf apices mostly only 4–5 per cell. Underleaves *occasional on sterile nongemmiparous shoots*, varying from minute and filiform to narrowly lanceolate vestiges, or consisting of 1–2, rarely 3–4, short, ciliary divisions (then up to 190 μ and 10 cells long); *robust gemmiparous shoots frequently with underleaves well developed. Gemmae vinaceous* (in masses appearing metallic purple), at apices of leaves that are either slightly modified or become progressively more inhibited with gemmae formation, then *shoot apices becoming erect and stoutly flagelliform* and 600–1200 μ wide (their leaves denticulate, suberect or erect, ± concave, reduced in size); mature gemmae 1–2-celled, the 1-celled mostly 20–25 × 27–29 μ, the 2-celled from 22 × 24–26 μ up to 25–26 × 32–35 μ, isolated ones *to 24–26 × 38–40 μ.*

Dioecious, *almost invariably sterile.* Perianth cylindrical, plicate above, contracted to the shortly ciliate mouth.

Type. Malde, Norway, near Stavanger; at sea level (*Bryhn and Kaalaas, 1889*).

Distribution

Although originally considered to be an "ausgesprochene atlantische Art" (Müller, 1910, p. 652, *in* 1905–16; this concept is also inherent in the species name), the species actually appears to have a widespread although suboceanic, arctic-alpine range. In Europe now known from Switzerland (Grimsel), Brittany in France, a few stations in England, Scotland, and the Faroes; northward becoming more frequent in Norway, southern and central Sweden, and southern Finland.

In North America erroneously reported by Haynes (1925) and, following her, by Buch, Evans and Verdoorn (1938); these reports not substantiated by herbarium specimens. Apparently the first correct reports from North America are those of Lepage (1944–45), Kucyniak (1949a), and Schuster (1951, 1951a). The species has been found to be quite frequent at high elevations in New England and has been questionably reported from Minnesota (Schuster, 1953). Apparently largely confined in North America to a restricted area around Hudson Bay, the cooler portions of the Great Lakes, the highest peaks in New Hampshire and Maine, northward to Labrador and Greenland.

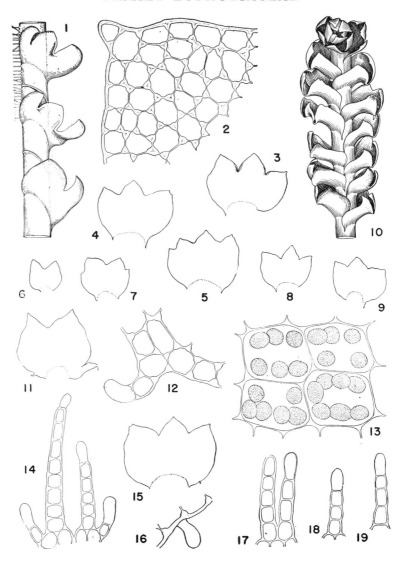

Fig. 155. *Lophozia* (*Orthocaulis*) *atlantica* (1–16) and *L. attenuata* (17–19). (1) Shoot sector, lateral aspect, rhizoids drawn in above only (×24). (2) Ventral leaf apex of leaf in fig. 3 (×350). (3–5) Leaves from a robust shoot (×24). (6–9) Leaves from less vigorous plant (×16). (10) Shoot, antical aspect (×20). (11) Large leaf (×20). (12) Cells from dorsal base of leaf in fig. 11, with stalked slime papilla (×350). (13) Median cells with oil-bodies (×650). (14) Large underleaf (×210). (15) Large leaf with sessile slime papilla (×20). (16) Slime papillae of leaf in fig. 15. (17–19) Underleaves from robust plant (×300). [1–10, Mt. Washington, N.H., *RMS*; 11–16, Great Whale R., Quebec, *Marr*; 17–19, from plants in Fig. 152:3–11, Minnesota, approaching *L. atlantica*.]

NW. GREENLAND. North Star Bay, 77°30′ N. (*Hovey, 1916;* Evans, 1918, det. as *L. binsteadii* but with the broad leaves, shallow sinuses, and broad lobes of *L. atlantica*; a doubtful specimen). W. GREENLAND. Frederikshaab (*Vahl, June 1829*!; with *L. kunzeana*; reported by Lange & Jensen, 1887, as *J. attenuata*). SW. GREENLAND. Kangerdluarssukasik, 61°55′ N., 49°18′ W. (*Damsholt 651321, p.p.*!, among *Mylia taylori* and *Blepharostoma trichophyllum* subsp. *trichophyllum*). BAFFIN I. Pond Inlet, N. Baffin I. (*Polunin, 1934,* [*B*]*695, 696* [in part]); Pangnirtung, C. Baffin I. (*Polunin, 1934,* [*B*]*510* [in part]); Dorset, S. Baffin I. (*Polunin, 1934,* [*B*]*297* [*in part*]). LABRADOR. Battle Harbor (*Waghorne 44, 1891*!; as *L. attenuata* in Underwood; *p.p.*, with *Cephalozia leucantha, Blepharostoma trichophyllum brevirete, Lophozia wenzelli*); Venison Pickle (*Waghorne 13, Aug. 16, 1891*; typical; with stout gemmiparous shoots bearing linear-lanceolate underleaves; reported by Underwood as *L. attenuata*); Cartridge Bight (*Waghorne 24, 1891*!; admixed with a plant with filiform innovations and smaller cells, probably representing the rare mod. *pachyderma-colorata* of *L. attenuata*; the *L. atlantica* with 3-lobed leaves 1.5× as wide as long; marginal cells 20–24 μ, median 23–25 × 23–30 μ); Venison Pickle (*Waghorne 4, Aug. 16, 1891*; typical, with stout gemmiparous shoots bearing underleaves; admixed with typical *L. binsteadii*; see p. 322; reported by Underwood as *L. attenuata*).

QUEBEC. Cairn I., Richmond Gulf (*Marr 630*); mouth of Great Whale R. (*Marr 655, 656, 662*); Manitounuck Sd. (*Marr 648a*); all on E. coast of Hudson Bay (see Schuster, 1951); Leaf R., 80 mi. above Ungava Bay (*Marr;* see Schuster, 1951); above L. Nemaska, Rupert R. (Lepage, 1945; det. Evans); L. Hubbard, W. shore of George R., near 55°55′ N. (Kucyniak, 1949a); Wolstenholme (Polunin, 1934, 1947); Wiachouan R., Hill Portage (Lepage, 1953; "probable identity," *fide* Arnell); L. Minto (*Taylor 43, p.p.*!). MAINE. Mt. Katahdin: near upper end of Cathedral Trail (*RMS 32983, 32978, p.p.*); upper end of Saddle Trail (*RMS 33009, 33004, 33004a, 33002, 32993*); Caribou Springs, tableland of Mt. Katahdin (*RMS*); Mt. Pomola (*RMS*); N. slope of Baxter Peak, *ca.* 4800–5000 ft (*RMS 15814*); NE. slope of Baxter Peak, Tundra, above 5000 ft (*RMS 15929*); NW. side of Baxter Peak, ca. 5000 ft (*RMS 15937*); Hamlin Ridge Trail, 4500 ft (*RMS 17013*). NEW HAMPSHIRE. Mt. Washington, near L. of the Clouds hut (*RMS 68-2091, 68-2115*).

I would now, tentatively, refer my earlier report (Schuster, 1953) of *L. atlantica* from Minnesota to *L. attenuata* (see p. 329).

Apparently common in the region to the east of Hudson Bay, north of 51°N.; to the south of this, the only reports are from the high summits of the New England mountains. The species is abundant on the summit of Mt. Katahdin, being one of the characteristic Lophoziae of the less xeric sites. It is surprising that it was not collected there earlier.

Lophozia atlantica has also been reported from Wilmington Notch, 6.5 miles southwest of Wilmington, Essex Co., N.Y. (N.G. Miller, 1966), based on a determination of Fulford. In view of the low elevation (1500–2000 feet) and the habitat ("sandy bank") the report seems highly unlikely to be correct.

Ecology

Usually occurring on exposed to shaded rocks, often deep in shaded and damp crevices between boulders at and near the summits of mountains; the plant is a characteristic species of shaded, dry bases of boulders, associated there with *Chandonanthus setiformis, Lophozia alpestris, L. attenuata, L. kunzeana,* and *Anastrophyllum michauxii*; northward also on rocks, associated with *Lophozia wenzelii, Cephalozia leucantha, Blepharostoma tricho-phyllum brevirete*, and mosses. At such sites it forms brown to blackish green, intricate mats. Equally "at home" growing in full sun over seepage-moistened *Sphagnum*, associated with *Mylia anomala, Cephalozia* spp., *Cephaloziella spinigera, Lophozia wenzelii, L. alpestris, L. kunzeana, Gymnocolea inflata*, etc., and such characteristic ericads as *Cassiope hypnoides, Phyllodoce caerulea, Vaccinium vitis-idaea*, and *Drosera rotundifolia*. On peat either in small patches or creeping over *Sphagnum*.

The species is habitually sterile but freely produces ± modified gemmiparous shoots which are usually less modified than those of *L. attenuata* and very rarely as slenderly filiform, with the leaves usually less reduced and not strongly appressed. Only Müller (1954, *in* 1951–58) describes the perianth.

Differentiation[38]

Characterized by (*a*) the arctic-alpine distribution; (*b*) development of moderately and *only gradually* attenuated gemmiparous shoots with grad-ually reduced leaves, these shoots not as suddenly flagelliform as in *L. attenuata*; (*c*) large, deeply vinaceous to purple gemmae; (*d*) large leaf cells, the marginal *averaging* 18–24 μ or even larger; (*e*) usually medium-sized to moderately bulging trigones. The plants have a distinctive facies; they usually acquire a rather deep brown pigmentation, even when rather shaded, and in sun forms become quite blackish green (by contrast, the closely related but more slender *L. binsteadii* usually is a warm, clear, chestnut brown, the apices of the shoots at times somewhat reddish brown). The leaves of *L. atlantica* are more polymorphous, even on the same plant, than those of *L. attenuata* and *L. binsteadii*. Unlike these latter species, which have almost uniformly 3-lobate leaves, *L. atlantica* shows variation from 2- to 4-lobed leaves, with the majority of leaves 3-lobed, except on exceedingly weak shoots. In the separation of *L. atlantica* (and in the preceding diagnosis) emphasis is placed on the "normal" trilobate leaves. These are usually much wider in *L. atlantica* than in the related *L. attenuata* and *L. binsteadii*, averaging 1.2–1.5 × as wide as long; see Tables 2 and 3; pp. 315 and 322.

In the separation of *L. atlantica* from the less robust *L. attenuata* the most satisfactory characters are those drawn from the leaf cells. In *L. atlantica* these are larger and about 18–24 μ along the lobe margins and in the lobes (14–17 μ in *L. attenuata*). Gemmae may occur on innovations as in *L. attenuata*, but these are shorter and usually more robust; furthermore, gemmiparous shoots, and occasionally the younger parts of sterile shoots,

[38] A useful biometric study of the species has been published by J. W. & R. D. Fitzgerald (1962).

bear minute, usually undivided underleaves (normally absent in both *L. attenuata* and *binsteadii*). The facies of the plants is also different: *L. atlantica*, like *L. binsteadii*, has the leaf lobes somewhat to obviously incurved; in *L. attenuata* the leaf lobes are more or less distinctly spreading, with at most the lobe-apices slightly incurved.

Lophozia atlantica is more closely related to *L. binsteadii* than to any other species. It differs from *L. binsteadii* in the somewhat broader leaves, with wider lobes and shallower sinuses ($\frac{1}{3}-\frac{1}{2}$ in *L. binsteadii*, $\frac{1}{4}-\frac{2}{5}$ in *L. atlantica*), and in the occasional presence of underleaves (at least on gemmiparous shoots). Whether the two species are always distinguishable is an open question, but see Table 3, p. 322.

Some material reported from New Hampshire is assigned to *L. atlantica* because of the relatively broad, shallowly lobed leaves (mostly 3-lobed, a few 4-lobed and 2-lobed), with broad, more or less obtuse to subacute lobes (lobes averaging 1.1–1.5 × as wide at base as long). Generally, *L. atlantica* is described as having the cell walls merely slightly thickened, and the trigones small to rather large (see Frye and Clark, p. 407, *in* 1937–47; Buch, 1937b, p. 126). The local material, however, had strong, bulging trigones, and the walls were distinctly brownish. Furthermore, underleaves could not be demonstrated in any of the local plants. These two characteristics would seem to indicate that *L. binsteadii* rather than *L. atlantica* is at hand: however, leaf form, upon which I place more emphasis, is typical of *L. atlantica*, and I therefore refer the local plants here.

It is of interest that Joergensen (1934) attempts to keep separate from *L. binsteadii* a *L. ambigua*. This latter (*fide* Buch, 1937b), represents a *lepto-vel-mesoderma* modification of *L. binsteadii*. Apparently *L. binsteadii* varies widely, as regards development of trigones, from a nearly leptodermous form (*ambigua* Joerg.) to a strongly pachydermous modification with bulging trigones (typical *binsteadii*). With the demonstration that *L. atlantica* may also have a mod. *colorata-pachyderma* (and I consider the White Mt. material to be such a modification), a similar range of modifications can be demonstrated for *L. atlantica*.

A nearly leptodermous extreme of a superficial intermediate between *L. attenuata* and *L. atlantica* is represented by the deviant plants reported by Schuster (1953, p. 326) from Minnesota. In these plants large to slightly bulging trigones may be present, but the leaves are green, explanate, and often unusually narrow (width barely exceeding length, in other cases actually less than length). The plants give the impression of a gemmae-free form of *L. attenuata* but differ from it in three critical features: (*a*) leaves polymorphous, from 2–4-lobed; (*b*) cells large, at leaf apices 19–22 μ; (*c*) traces of underleaves on mature shoots; see Fig. 152:3–11.

Lophozia atlantica often occurs with *L. kunzeana* and may be difficult to separate from it, particularly when the two occur over peat. However, *L. kunzeana* differs in (*a*) large, often bifid underleaves; (*b*) a much higher incidence of subcomplicate, 2-lobed leaves, whose sinus is appreciably deeper. The montane extremes of *L. kunzeana* show the same tendency as do *L. atlantica, attenuata,* and *binsteadii* to form purple gemmae under such conditions, a phenomenon that can easily lead to confusion.

Very small gemmiparous phases of *Lophozia* (*Barbilophozia*) *hatcheri*, the mod. *parvifolia-colorata*, may be markedly similar to *L. atlantica* in aspect. They differ from this species in that underleaves, even though reduced, are still more distinct, and that at least an occasional leaf bears a basal postical cilium formed of somewhat elongated cells (see p. 354 for this phenotype).

LOPHOZIA (ORTHOCAULIS) CAVIFOLIA (Buch & S. Arn.) Schust., comb. n.

[Fig. 156]

Orthocaulis cavifolius Buch & S. Arn., Mem. Soc. F. et Fl. Fennica 26(1949–50):71, figs. 1–9, 1951.
Sphenolobus cavifolius K. Müll., Rabenh. Krypt.–Fl. ed. 3, 6(1):723, fig. 234, 1954.

In patches formed of stiffly ascending to erect stems, in arctic to alpine sites. *Shoot with facies of a vigorous Anastrophyllum minutum*, 15–30 mm high × 740–1100 μ wide (with leaves), mostly brownish to reddish brown, green only in shaded sites. Stem sparingly branched, ca. 250–300 μ in diam., *13–14 cells high*, elliptical in section, brownish, nearly or quite free of rhizoids; cortical cells rather evenly thick-walled, rather short (*dorsally in 1 stratum, ventrally in 2–3-strata*), grading rather abruptly into a larger-celled medulla. Cortical cells small (superficial view), oblong, ca. (14)15–17 up to 22–26 × 26–40 to 39–50 μ, except ventrally, where much shorter, only a little more thick-walled than medullary cells; medullary cells 5–10× as long as broad, (23)25–30(32) μ in diam., collenchymatous, colorless (the outer often brownish and grading into those of cortex). Leaves remote to contiguous, transversely oriented and *almost transversely inserted*, rather strongly (60–85°) spreading and somewhat antically secund, *hemispherical*, ca. (360–510)650–750(800–900) μ long × (520)850–950(1000–1200) μ broad, *always at least 1.2–1.3× as broad as long*, 0.2–0.35 *2-lobed* (*sporadically 3-*, exceptionally even 4-lobed); sinus sharp to rectangulate, when leaf is flattened becoming sharply obtuse-angular; *lobes blunt to obtusely pointed*, usually 2–3(4)× as broad as long, *incurved* (*to the point that the leaf cannot be flattened without tearing*); *postical leaf bases eciliate*. Cells subisodiametric, *relatively large*, in lobe apices and marginally somewhat to conspicuously, nearly evenly, thick-walled with usually concave- to straight-sided, ill-defined trigones, *medially becoming thinner-walled, with conspicuous, often bulging trigones*; marginal cells *very large*, (19)*20–25*(27) μ tangentially measured, in lobes; medially gradually and *only slightly* larger, ca. 22–25(28) × (21)26–35(40) μ. *Oil-bodies small*, (4–5)*6–8*(9–10) *per cell*, somewhat fine-botryoidal, colorless, spherical and 4–5 μ to ellipsoidal and 4–4.5 × 5–6.5 μ; cuticle very delicately papillose, toward base delicately striolate. Ventral merophytes narrow, *usually not developing underleaves*; underleaves, if present, 3–12-celled, 1–2 cells broad at base, linear-subulate to linear-lanceolate, uniseriate above at least. *Gemmae reddish brown*, sparse, at lobe apices of young, essentially *unmodified leaves*, (1)2-celled, obtusely polygonal and subisodiametric, 19–25 × 22–29 μ (Buch and Arnell, 1951) to 20–24 × 21–27 μ up to 27 × 33 μ.

Dioecious. Androecia ill defined, the bracts leaflike, differing only in the somewhat more dilated basal portion and rather ventricose base. Antheridia solitary, the stalk uniseriate. Gynoecia and sporophytes unknown.

Type. Sweden (Torne Lappmark, Vassitjåkke, *S. Arnell*); type in Riksmuseum, Stockholm.

Distribution

Until recently known only from northern Sweden (Härjedalen; Torne Lappmark; Lule Lappmark) and from Norway (Nordland; *fide* Mårtensson, 1955–56, p. 30). Arnell (1960) has also reported it from one station in northernmost Greenland, but I have seen it only from western and northwest Greenland:[39]

NW. GREENLAND. Dundas, near Thule Air Base, Wolstenholme Fjord (*RMS 46264a*).
W. GREENLAND. Tupilak I., near Egedesminde, 68°43′ N; 55°55′ W. (*RMS & KD 66-037*).

Ecology

Poorly known. According to Mårtensson, "rather independent of its substratum. It grows among common mosses such as *Dicranum elongatum* and *D. fuscescens*." In northwest Greenland I have collected the plant in quantity; it grows on open slopes, in damp moss-*Cassiope* Tundra, associated with both typical *Anastrophyllum minutum* and *A. minutum* var. *grandis*. Other casual associates are *L. (Barbilophozia) barbata*, *L. hatcheri*, *Plagiochila asplenioides subarctica*, *Tritomaria quinquedentata*, *Gymnomitrion corallioides*, *Scapania perssonii*, *Ptilidium ciliare*.

Differentiation

A poorly known species. I have had opportunity to study living material intensively in the field in Greenland. The plants are almost identical in

---

[39] Plants reported by Arnell (1960) as *Orthocaulis cavifolius*, from Low Point, Peary Land, which had been reported earlier by Hesselbo (1923a, p. 272) as "*Marsupella arctica*," represent *Anastrophyllum minutum grandis*, for the following reasons: marginal cells of lobes average 15.5–18.5 μ, tangentially measured—rarely and sporadically (18)19–21(24) μ; median cells of lobes average only 18–20× (17)18–20 μ; cells of leaf middle average (18)21–24(25) × 20–27(28) μ. I could find no gemmae; all leaves are bifid; weaker plants have exactly the facies of *A. minutum*. The leaf varies from 750 μ wide × 580–620 μ long to 810 × 580 μ (width-length ratio = 1.2–1.4:1); stems, which are ca. 10–12 cells high, lack the smaller-celled cortex which typical *L.(O.) cavifolia* possesses; no trace of underleaves could be found. On these bases, the plants seem more likely to be a robust form of *A. minutum grandis*. They bear occasional branches, usually from the ventral portion of the leaf axil, but rare terminal branches and sporadic branches from the postical surface occur.

facies to robust extremes of *Anastrophyllum minutum* var. *grandis*; the Greenland plants, indeed, occur admixed with nearly typical forms of *A. minutum*. From this they differ abundantly in several characters. First, the cells are larger, the marginal cells averaging (19)20–25(26–27) μ in the lobes. Second, median cells and nonmarginal cells of lobes have (4)5–8(9–10) oil-bodies each. Third, very robust plants show at least sporadic trilobed leaves. Fourth, the stem is 13–14 cells high.

The plants differ from all but the most extreme forms of *A. minutum* var. *grandis* also in the form of the leaves. These are so strongly hemispherical —with strongly incurved lobes—that they cannot be flattened without tearing. In *A. minutum* the leaves are usually loosely conduplicate, rather than cupped, although in extremes of var. *grandis* they may be cupped. When plants of these two taxa, growing together, are carefully examined, an overwhelming impression of two distinct species is obtained. However, vigorous forms of *A. minutum* var. *grandis* may have median cells of the leaves fully as large as in *L. cavifolia*, averaging 20–23 μ wide $\times$ 21–26 μ long. However, in such cases the marginal cells remain much smaller than in *L. cavifolia*, and the 3–6 oil-bodies per cell characteristic of all forms of *A. minutum* are retained. Also, the var. *grandis* appears to be nearly consistently without gemmae. Other distinctions are brought out in the treatment of *A. minutum* (pp. 763, 769).

Müller (1954, *in* 1951–58, p. 724) states that the species, although originally placed in *Orthocaulis*, "hat aber mit dieser Gattung nichts zu tun," and that the facies and the transversely inserted, constantly bilobed leaves, the cells with thickened walls, the regularly pectinate (steplike) leaf orientation, and the 1-seriate antheridial stalks prohibit placing it in *Orthocaulis*. However, my Greenland material shows, in part, 3- (exceptionally even 4-) lobed leaves, and the median leaf cells bear very distinct trigones. Thus, if a generic distinction must be maintained, the species probably goes into *Orthocaulis* (whatever status we wish to give this group). Admittedly, my original feeling was that the plant is a mere arctic, polyploid, larger-celled variety of *Anastrophyllum minutum*. It has no close relatives in *Orthocaulis*.

Subgenus *BARBILOPHOZIA* (*Loeske*) K. Müll. emend. Buch (*1933*), Schuster (*1951*)

Lophozia, in part, of authors.
Barbilophozia Loeske, Abh. Bot. Ver. Brandenburg 49:37, 1907.
Lophozia subg. *Barbilophozia* K. Müll., Rabenh. Krypt.-Fl. 6(1):622, 1910; Schuster, Amer. Midl. Nat. 42(3):565, 1949: Schuster, *ibid.* 45(1):43, 1951; Schuster, *ibid.* 49(2):326, 1953.
Jungermannia sect. Barbatae C. Jensen, Danmarks Mosser, Bryofyter 1:110, 1915.
Jungermannia subg. *Barbilophozia* Arn., Ark. Bot. 19(10):22, 1925.
Barbilophozia subg. *Barbilophozia* Kitagawa, Jour. Hattori Bot. Lab. no. 28:262, 1965.

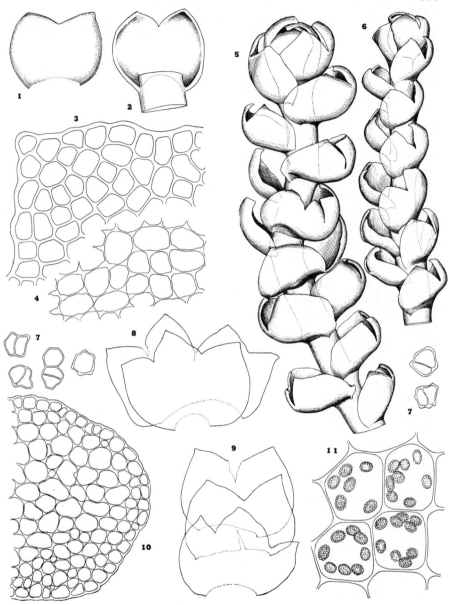

FIG. 156. *Lophozia (Orthocaulis) cavifolia.* (1) Small leaf, abaxial view, moderately flattened (×49). (2) Same, adaxial view, *in situ* (×49). (3) Cells of leaf lobe apex (×265). (4) Median cells (×265). (5) Robust shoot, with 2-3-4-lobed leaves of optimal width (×29). (6) Average-sized shoot (×29). (7) Gemmae (×300). (8) Leaves of optimal size, 3-4-lobed, so concave that they tend to tear in flattening (×35). (9) Leaves of normal size (×35). (10) Stem cross section (×205). (11) Median cells with oil-bodies (×690). [All from *RMS 46264a*, Dundas, Wolstenholme Fjord, Greenland.]

Forming large, thin, often interwoven mats, *robust* (largely 2–6 mm wide ×
3–10 cm long), *prostrate or decumbent* to weakly ascending in growth, only the
shoot apex clearly ascending in most cases; *branching sparing, lateral-terminal,
Frullania* type; no flagella. Stems rigid and firm, mostly 11–16 cells high,
cortex of ± smaller, thick-walled rectangular cells (ca. 2–3× as long as wide);
medulla *essentially homogeneous*, of large, pellucid cells, at most the postical
1–2(3) cell layers appreciably smaller, but not forming a discrete mycorrhizal
band. *Rhizoids dense*, forming a feltlike postical mat in which underleaves are
hidden. Leaves quadrate to transversely obrhombic or reniform-obtrapezoidal,
usually *distinctly to greatly broader than long* (ratio ca. 1.1–1.7:1), *normally shallowly
4-lobed* and nearly symmetrical, dorsal and postical lobes often smaller and short
(occasionally in the form of large cilia), apparently derived by elaboration from
marginal cilia; postical (more rarely to a slight degree also antical) leaf bases
with 1-several flexuous and contorted cilia, *formed of greatly elongated cells* (in
L. barbata reduced to a stalked slime papilla); *leaf insertion extremely oblique*
(usually 65–75°), very slightly or hardly curved and scarcely acroscopically
arched, the leaf somewhat decurrent on the antical stem surface, leaves usually
nearly horizontally, laterally spreading, often adaxially slightly convex, rarely with
the lobes appreciably incurved. *Ventral merophytes* usually 6–12 cells *broad* or
more, *developing bilobed underleaves* (lobes often ciliate; in *L. barbata* the underleaves much reduced, usually to 2 basally connate cilia); underleaves much
obscured by the *dense rhizoids*. Cells thin-walled, only moderately collenchymatous (trigones rarely bulging), *opaque*, with cuticle weakly papillose or
virtually smooth, *small* (ca. 22–26 μ medially); oil-bodies grayish, opaque,
small, mostly 4–5 μ and spherical and 4–5 × 6–8 μ and ovoid, mostly 4–8(12)
per cell, the included spherules small, little protruding. Gemmae very rare or
absent (except in *L. hatcheri*), angular, reddish brown.

Perianth unmodified, exserted, *plicate* and moderately *narrowed* toward a
closely dentate to ciliate mouth, the *mouth not beaked*. Capsule wall 3–4(5)-
stratose, the inner layers with semiannular bands, the outer with strong radial
(nodular) thickenings. Dioecious. Antheridial stalk biseriate; paraphyses ±
distinct, laciniiform lamella.

Type. *Jungermannia barbata* Schmid.

The group includes those species of *Lophozia* with quite succubous to
subhorizontal, typically ⅙–⅓ quadrilobed leaves; their orientation is
almost parallel to the stem, the leaves therefore lying nearly horizontally,
but the insertion is curved ventrally at the anterior end. Leaves are widest
about or above the middle and have equal lobes, or lateral lobes are
smaller, rarely reduced to teeth or cilia or wanting; dorsal and ventral
leaf margins are subequal or equal; the base of the ventral margin is
provided with 1 or more cilia consisting of very elongate cells (except
usually in *L. barbata*). Underleaves are large, bifid, and ciliate (in *L.
barbata*, small to obsolete). Stems are prostrate, at most apically ascending,

with a dorsal cortex of cells at least 1.5–2 × as long as wide; ventral cortex in 2–3 layers, grading into interior cells; interior cells 3–6 × as long as wide; in cross section cortical cells subequal in size to interior cells. Rhizoids are numerous, colorless, in feltlike mats to near stem apex.

The perianth is plicate to within half the distance to base, gradually contracted to its mouth; the sporangium wall is mostly 3–4 cells thick, the epidermal layer with nodular thickenings, the innermost layer with semiannular thickenings (Fig. 159:9–11).

Barbilophozia, as the subgenus is emended here, includes only three species, of northern distribution. It is equivalent to the genus *Barbilophozia* of Buch (1933a); the species of *Orthocaulis* are excluded. Müller (1939, 1940, 1951–58) would unite the two groups, not accepting *Orthocaulis*. The two are admittedly closely allied and could be considered as two sections of a single subgenus. However, the decumbent mode of growth, the nature of the basal cilia of the leaves, the essentially 4-lobed and more nearly longitudinally inserted leaves, with a less acroscopically arched line of insertion, suggest that *Barbilophozia* may be treated as subgenerically distinct from *Orthocaulis*.

Tritomaria quinquedentata was once included here. Prostrate, large mesophytic forms of this species, growing more or less horizontally, have broad leaves, somewhat succubously inserted (dorsally nearly transversely, however); they may resemble *Barbilophozia* much more than *Tritomaria*. Arnell (1906) regarded *T. quinquedentata* as connecting *L. lycopodioides* with *T. exsecta* and *exsectiformis*; if such an (unlikely) disposition is accepted for the species, it would suggest that *Tritomaria* is a specialized side shoot from a generalized *Barbilophozia*.

A discussion of the relationship of *Barbilophozia* to *Orthocaulis* is given under the latter; the two should be considered near the base, rather than the apex, of the Lophozioid stock.

Species of this subgenus, like those of *Orthocaulis*, are chiefly far northern, occurring in our area only in the mountains. *Lophozia barbata* is an exception, however, and extends into the Transition Zone. At least *L. barbata* and *L. lycopodioides* are lime-tolerant (the latter may be a common species in calcareous mountains like the Jura); neither they, nor *L. hatcheri*, are, however, confined to calcareous districts.

Key to Nearctic Species of Barbilophozia

1. Postical leaf margin with 1-several cilia formed of elongate cells; leaf lobes more or less distinctly mucronate-tipped; underleaves large, prominent, present throughout on sterile shoots; leaves obtrapezoidal, much wider near apex than near base . . . . [sect. Lycopodioideae]   2.

2. Leaf lobes obtuse and very broad, mostly convex-sided; antical leaf margins usually strongly convex; middle leaf lobes ± twice as wide as long (mucro omitted). Robust: usually 3.5–5 mm wide × 4–8 cm long; leaves very broadly rhomboidal, often ± asymmetrical, 2–3 mm broad, 1.5–1.8× as wide as long, almost all ca. 0.15–0.2 quadrilobed, strongly undulate-crispate; leaf tips (nongemmiparous plants) strongly mucronate with thick-walled cells from (50)60–75(100) μ long × 18–20(22) μ wide (averaging 3.5–6× as long as wide); postical leaf bases with (3)5–7(8) long, tortuous cilia formed largely of cells averaging (65)90–110 μ long (averaging 6–10× as long as wide); gemmae rarely formed . . . . *L. (B.) lycopodioides* (p. 337)

2. Leaf lobes ± rectangulate or even acute, usually straight-sided; antical leaf margin straight or slightly convex; middle leaf lobes ± rectangulate, less than 1.35× as wide as long. Medium-sized; usually 1.5–2.7 mm wide × 2–5 cm long; leaves narrow to rather broadly rhomboidal, ± symmetrical, 1.0–1.9 mm broad, 1.0–1.7× as wide as long, not or weakly crispate-undulate, ca. 0.25–0.3 tri-quadrifid; leaf tips usually weakly mucronate, largely acute or acuminate rather than mucronate, the terminal 1–2 cells mostly 28–60 × 15–20 μ (averaging 2–3× as long as wide at most); postical leaf bases with 1–5(6) short cilia formed of rather short cells at their bases, the elongate distal cells only (40)50–90(110) μ long (averaging 3–6× as long as wide); gemmae regularly developed. . . . . *L. (B.) hatcheri* (p. 345)

1. Postical leaf margin at most armed with a stalked slime papilla or unarmed; leaf lobes obtuse to merely acute; underleaves on sterile shoots vestigial, occasionally rather distinct;[40] leaves nearly quadrate (when 4-lobed) and little wider distally than at base; gemmae (rare or) absent . . . . [sect. Barbatae] *L. (B.) barbata* (p. 355)

Sectio *LYCOPODIOIDEAE* Schust.

Lophozia subg. *Barbilophozia* sect. Lycopodioideae Schust., Amer. Midl. Nat. 45(1):44, 1951.

Folia obtrapezoidea obrhombicave, ad bases ventrales ciliis, e cellulis perelongatis formatis, armata; amphigastria magna, bifida ciliataque; gemmae saepe praesentes.

Plants with obtrapezoidal to obrhombic leaves, with leaf lobes sharp, usually cuspidate or longly mucronate to ciliate-tipped; ventral leaf bases with conspicuous, tortuous cilia (rare to vestigial only in xeromorphic phenotypes) formed of strongly elongated cells. Underleaves large, usually bifid, ciliate. Rust-red gemmae occasional to abundant.

Type. *Lophozia lycopodioides* (Wallr.) Cogn.

This section corresponds to the subg. *Barbifolium* of Buch (1942), which is a *nomen nudum*.

[40] Frye and Clark (1937–47, p. 426) attempt to separate *L. barbata* from the other two species because of the supposed absence of underleaves on sterile shoots. Their fig. 11 (p. 427), which is copied from Evans (1898), however, shows a distinct underleaf on a sterile shoot.

LOPHOZIA (BARBILOPHOZIA) LYCOPODIOIDES
(Wallr.) Cogn.

[Figs. 157, 158, 159:1–4]

Jungermannia lycopodioides Wallr., Fl. Crypt. Germ. 1:76, 1831.
Jungermannia barbata var. *lycopodioides* Nees, Naturg. Eur. Leberm. 2:185, 1836.
Lophozia lycopodioides Cogn., Bull. Soc. Bot. Belg. 10:278, 1872.
Barbilophozia lycopodioides Loeske, Abh. Bot. Ver. Brandenburg 49:37, 1907.
Lophozia (Barbilophozia) lycopodioides K. Müll., Rabenh. Krypt.–Fl. 6(1):627, 1910; Schuster,
    Amer. Midl. Nat. 45(1):44, 1951.
Lophozia hatcheri var. *ciliata* K. Müll., *ibid*. 6(1):634, 1910.

In flat, depressed, loose and flocculent mats, often mixed with *Dicranum* and
other mosses, over acid soil or soil-covered rocks, *particularly over needles of spruce
and fir;* usually *rather pellucid, pale green to stramineous yellow, rarely brownish. Very
robust,* 3–5, occasionally 6–8, cm long, (3)*4–4.5* (rarely 5) mm *broad;* stems
nearly simple, ca. 500–600 μ in diam.; rhizoids forming a dense ventral mat.
Leaves *strongly imbricate*, very obliquely inserted, laterally spreading, prominently *wavy and crispate* (sinuses between the lobes reflexed to sulcate, lobes
strongly convex), *very large and much wider than long* (ca. 2200 μ wide × 1500 μ
long to 2700 μ wide × 1550 μ long, to a maximum of 2900–3000 μ wide ×
1950–2000 μ long), *strongly obrhombic* in shape, *shallowly divided by three very obtuse
sinuses* into 4 subequal and *usually convex-sided* lobes; sinuses descending 0.15–0.2,
rarely to 0.25, the leaf length; lobes very broadly, obtusely triangular, each
suddenly narrowed into a stiff, *1–3-celled mucronate tip, formed of strongly elongate,
strongly thick-walled cells* varying from 50 × 22 μ to 60–75 × 18–22 μ (postical
lobe often terminated in a longer mucroniform cilium, 3–4 or more cells long,
formed of cells 50–60 μ long × 15–20 μ wide, the terminal cells occasionally
much longer, 100–105 μ); postical leaf base *usually with 4–7* extremely long,
slender, contorted, rarely branched cilia, usually (500)600–900 μ long, formed
of mostly 5–10 *extremely narrow and elongate* thick-walled cells (the septa between
usually somewhat expanded externally), cells 10–15 μ wide × (60)*80–110 μ
long*. Underleaves large, ca. 750–900 μ long, bifid for ca. ⅘ their length, the
lanceolate lobes *bearing numerous much elongated, tortuous cilia* (usually 12–20 on
each half of the underleaf), cells of the cilia similar in size and form to those of
cilia of postical leaf bases. Cells rounded-polygonal, thin-walled, with small to
large trigones (trigones occasionally locally confluent), cells near the leaf apices,
and sometimes 1-several rows of marginal cells near the leaf apices, *decolorate
and thick-walled, guttulate,* forming a weak border; cells of leaf apices, below the
mucronate tips, subisodiametric, ca. 20 μ; median cells ca. 20 × 25–28 μ;
cells of postical base, beneath origin of cilia, ca. 23–25 × 28–32 μ; cuticle
virtually smooth; oil-bodies as in *L. hatcheri*. *Gemmae rarely developed,* reddish
brown, ca. 17–20 μ wide, polyhedral, as in *L. hatcheri;* with (rare) free development of gemmae on immature leaves, the lobes resolved in gemmae, consequently these leaves at maturity often virtually elobulate.

Dioecious; *almost habitually sterile.* Androecia intercalary, the bracts strongly
concave at base, similar to leaves, but the antical lobe folded over the stem;
dorsal pair of lobes smaller than the longer postical lobes; antheridia among

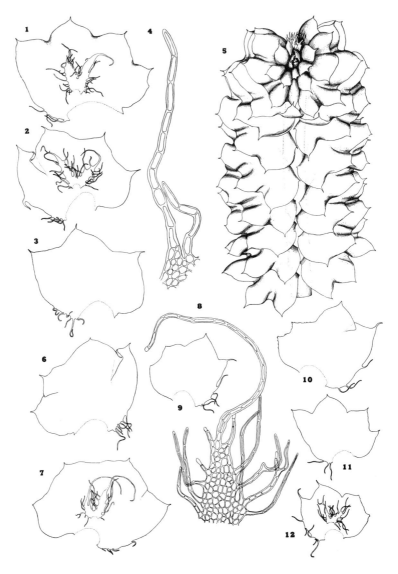

Fig. 157. *Lophozia (Barbilophozia) lycopodioides.* (1–3) Leaves and underleaves of normal size (×13.4). (4) Cilia of postical leaf base, minimally developed (×103). (5) Shoot, antical aspect (×10.5). (6–7) Leaves and underleaf (×13.4). (8) Half of underleaf (×63). (9–10) Leaves, from gemmiparous phenotype (×14.5). (11–12) Leaves and associated underleaf (×14.5). [1–5, *RMS 17029*, Mt. Katahdin, Me.; 6–7, *RMS 1089*, Mt. Washington, N.H.; 8, Feldberg, Germany, K. Müller, 1898; 9–10 and 11–12, *RMS 17322*, Amygdaloid I., Isle Royale, Mich., and *RMS 16999*, Copper Harbor, Mich., both fo. *parvifolia*, approaching *L. hatcheri.*]

numerous linear to lanceolate paraphyses. ♀ Bracts divided ⅓–½ into 4–5 irregular, unequal, undulate crispate, lobes, terminating in long, setose cilia; bracteole like bract, but divided up to ¾ into 2–4 unequal lobes terminated in similar cilia, occasionally with supplementary teeth or cilia. Perianth large, inflated, ovoid-cylindrical, 5–6 mm long, the wall pluristratose below, plicate on the distal half; mouth dentate with 1–3-celled teeth formed of elongate, pachydermous, large, sharply pointed cells, and some cilia of 4–8 cells 10–13 × 36–45 μ. Capsule wall 3-stratose; the epidermal layer of large cells with nodular (radial) thickenings, the inner layers with narrow semiannular bands. Spores brownish, 12 μ (or larger); elaters 9–10 μ in diam., contorted, bispiral, the spirals tightly wound, deep brown.

Type. Germany; exact locality not recorded.

Distribution

Almost strictly high subarctic-subalpine, with very few stations known from low elevations in the region south of the Taiga s. str.; holarctic in distribution, but (unlike the related *L. hatcheri*) unknown from antarctic regions; sporadic in all but the high Arctic.

In Europe found from the mountains of Central Europe and the Balkans (Bulgaria) and Spain (Pyrenees), northward to the Arctic, where more abundant; eastward to the Caucasus (*Brotherus; fide* Stephani), easternmost Siberia (Kamchatka; see Arnell, 1927), and Honshu, Japan (Hattori, 1948, p. 107).[41] North to Spitsbergen (Watson, 1922). In North America transcontinental, ranging from Alaska (Adak I., Yakutat Bay, to Attu I. and Bering Strait) and Yukon southward to British Columbia, Washington, Wyoming, Montana, and in the Rocky Mts. as far south as Utah, Colorado, and New Mexico (there at 3300 m in the Las Vegas range, *Cockerell*).

The species is much less frequent than *L. hatcheri* and has been much confused with it; the following reports (where not based on personal collections or marked with a !) in some cases need revision:

E. GREENLAND. Cap Ravn; Kangerdlugssuak; Scoresby Sund (doubtful; with gemmae); Tasiusak; Nenese; Gaaseland on Scoresby Sund (Jensen, 1897). W. GREENLAND. Disko I., Godhavn: below Lyngmarksfjeld (*RMS & KD 66-181*); Lyng-marksbugt, ca. 25–125 m (*RMS & KD 66-553, 66-566*). Ameralik (*Vahl, Aug. 1830;* mixed with *L. hatcheri*—at least two size classes of plants appear to be admixed—and *L. kunzeana;* reported by Lange & Jensen, 1887, as *J. lycopodioides floerkei*). BAFFIN I. Without locality, *Boas* (*fide* Stephani).

LABRADOR. Without locality, *Ahles* (*fide* Stephani). QUEBEC. Stupart Bay (Macoun, 1902); Leaf R., ca. 80 mi from Ungava Bay; Richmond Gulf, near Fishing Lake Cr.; Seal R., Cape Jones; near Montreal; Ste-Anne-de-la-Pocatière; Bic; Sacré-Coeur, côté de Rimouski; Rimouski, Mont Albert; Mt. la Table (see Schuster, 1951; Lepage, 1944–45); Ft. Chimo, Ungava Bay; Seal L. (Macoun, 1902); N. side of Malbaie R., 10–12 mi NW. of Bridgeville, Gaspé (*RMS 43502, 43511, 43512; 43513, 43522, c. caps.,*

---

[41] Kitagawa (1965, p. 264, fig. 5) treats the Japanese plant, which he describes as only 1.8–2.8 mm wide. To judge from both diagnosis and figures, he is dealing with the larger extreme of *L. hatcheri*, a species he reports from the same locality (Mt. Tsubakuro).

andr.). Ste-Anne-des-Monts R., Gaspé Co. (*Lepage 2116*).　NEWFOUNDLAND. Port au Choix, St. John I., St. Barbe Bay, Flower Cove, Cook Harbour, Ha-Ha Bay, Bartlett's R., St. Anthony, Great Harbour Deep, all on N. Pen. (Buch & Tuomikoski, 1955).　NOVA SCOTIA. Pirate's Cove; Mary Ann Falls, S. of Neils Harbour, Cape Breton (*RMS 66312, 66314, 66313a*). NEW BRUNSWICK. Campbellton (Macoun, 1902).　ONTARIO. Ottawa (Macoun, 1902); South Twin I., James Bay (Cain & Fulford, 1948).

MAINE. Mt. Katahdin, Saddle Slide Trail, 3500–4000 ft, and foot of Hamline Ridge and above Northwest Basin (*RMS 17029, 32970a, 32972, 32908*); Round Mountain L., Franklin Co. (*Lorenz*).　NEW HAMPSHIRE. Thorn Mt.; Carter Dome (*Evans, 1902!*); Waterville; Mt. Washington, on Glen Boulder-Boot Spur Trail, 3500–4000 ft and 4200 ft (*RMS 1089*, etc.); Davis Trail, 4200–5000 ft, Mt. Washington (*RMS 23106, 23115*). MICHIGAN. Keweenaw Co: Phoenix (*Taylor 500A;* det. Evans!); summit of Mt. Bohemia (*Steere;* det. Evans!); Amygdaloid I., Isle Royale (*RMS 17322*); Copper Harbor, Ft. Wilkins, (*RMS 16999*).

As Evans emphasized, the species is mostly confined to the higher summits of the mountains in New England (usually above 4000 feet) although it descends to 2300 feet in Franklin Co., Me.; occurring northward at progressively lower elevations. On that basis, the reports from Michigan, largely at elevations between 600 and 1200 feet, are striking. Study of this material shows that the specimens are intermediate between "typical" *Lophozia hatcheri* and *L. lycopodioides.* Some reports of *L. lycopodioides* cited above possibly belong to *L. hatcheri* (see Schuster, 1951, p. 6), e.g., those accredited to Macoun (1902) from Quebec, New Brunswick, and Ontario, as well as the early (1897) Jensen reports from Greenland.

Ecology

Widespread in mountainous regions and in the far north on well-drained sites (e.g., rock walls, cliffs, steep ledges, the well-drained humus derived from spruce and fir needles on mountainous slopes), often associated with rock outcrops, although usually not a pioneer on rocks. Persisting under climax conditions on shaded forest floor, often with *Pleurozium schreberi* and *Hylocomium splendens*. The species frequently occurs on acidic rocks but is present equally freely in calcareous sites (there usually, however, on acid humus). In the New England mountains a characteristic species at and near treeline (4000–5000 feet), where it occurs under stunted red spruce and fir, forming loose, rather flocculent mats, characteristically pale green, over loose and only partially disintegrated needle humus. The plant is here often rather xerophytic in occurrence, often associated with *Dicranum* and other robust mosses (*Pleurozium schreberi, Hylocomium* spp., etc.).

When occurring in mesic sites (as under spruce and fir), the species is very robust, a characteristic pale yellowish, rather pellucid, green; it is then usually without gemmae. When occurring as a pioneer on exposed rocks, or in partial sun on soil-covered rocks near the summits of the higher mountains, it sometimes forms brownish mats, occasionally abundantly gemmiferous. When the stimulus for gemmae production is particularly intense, the leaf lobes of the juvenile gemmiparous leaves are virtually resolved into gemmae, failing to develop. Under such conditions the leaves, on maturation, are strongly obliquely obrhombic, with a virtually elobate, obliquely truncate apex; such an extreme, xeric modification has been described as the "var. *obliqua*" (K. Müller, 1905–16, p. 629, fig. 300). The closely related *L. hatcheri*, under similar conditions, reacts the same way.

Variation and Differentiation

Lophozia lycopodioides can be separated from *L. hatcheri* only by considering a constellation of features. In *L. lycopodioides* the leaves are more shallowly lobed and have much broader, typically convex-sided lobes (the lobes in *L. hatcheri*, though variable, are mostly acutely angular to rectangular, with normally straight sides); the mucronate tips of the lobes arise abruptly and are longer (in *L. hatcheri* they may even be suppressed on gemmiparous leaves or those of xeromorphic extremes, and when present arise less abruptly); the antical and, to lesser extent, postical leaf margins are usually quite convex (in *L. hatcheri*, divergent from each other but hardly or not arched). These differences in aspect are ultimately more constant and valuable in separating the species than the usual measurements, which have only a secondary validity.[42]

In typically developed plants the robust size (4–5 mm wide), the somewhat pellucid yellowish green color, and the strongly mucronate-tipped *and convex-sided* lobes of the greatly undulate-crispate leaves give this magnificent species an unmistakable facies. Fortunately the robust phase of the species, quite devoid of gemmae, is by far the most common. This phase, the mod. *latifolia-megafolia-undulifolia-leptoderma-viridis*, has leaves often 2.5–3 mm broad, extraordinarily fluted between the lobes (which become as a consequence quite convex), and the setose, stiff mucros of the leaf lobes are well developed on virtually all lobes.

Lophozia lycopodioides commonly develops brown, more or less strongly xeromorphic forms that show a decrease in size (mod. *parvifolia*); a stronger tendency toward development of gemmae (mod. *gemmipara*);

---

[42] Buch (1909) emphasizes the discrepancies in ecology and ability to produce gemmae as differential features. The normally gemma-free *Lophozia lycopodioides* is admittedly usually a plant of the Conifer Forest floor, often over thinly soil-covered ledges, mixed with pleurocarpous mosses, the individual plants seemingly arising from a layer of partially decayed conifer needles. By contrast, *L. hatcheri* is found mostly over rocks in more exposed places or in sites with much more intermittent moisture. Unfortunately, it is exactly the occurrence under intermittent moisture conditions which seems to lead to gemma formation and to production of a mod. *parvifolia-angustifolia*. An analogous case is that of *Anastrophyllum michauxii*, in which robust, luxuriant phenotypes of moist, shaded logs are often gemma-free, whereas the slender form, the mod. *parvifolia*, on relatively dry rocks or on bark of trees, is abundantly gemmiparous.

an abbreviation in length and a decrease in number of the postical cilia of the leaves; and a well-marked tendency for the leaves to be less fluted and undulate and for the suppression of the mucronate teeth of some lobes, their cells becoming notably less elongate. Such plants were described as "var. *parvifolia*" (Schiffner, Österr. Bot. Zeitschr., vol. 58, p. 381, 1908).

Müller (*loc. cit.*, p. 629) correctly states that the plants of "var. *parvifolia*" are only ca. one-half as large as those of the "typical" species and are habitually confusingly similar to *L. hatcheri*. According to Müller, the variety differs from *L. hatcheri* in the "characteristic *Lycopodioides* leaf shape." However, Müller goes on to state (pp. 634–635) that transitional phases between "var. *parvifolia*" and *L. hatcheri* occur, suggesting that even this characteristic may "fail" to distinguish the two taxa. This is particularly clear when the plants, referred to "var. *parvifolia*," collected in Michigan (e.g., at low elevations) are studied. Evans determined some of these plants as *L. lycopodioides* (see Evans and Nichols, 1935; Steere, 1942a). Several other collections, made by the writer along the shores of Copper Harbor and on Isle Royale (*RMS 17322, 16999*), fall into this same category. In these plants we find the strongly asymmetrical leaf shape of *L. lycopodioides*, with broad and shallow lobes, and discrete, mucronate, sometimes setose apices of the leaf lobes; the cilia of the postical leaf bases are also predominantly long and tortuous, formed of very elongate cells. In leaf size and shape these plants approach *L. lycopodioides* (in *RMS 17322* the leaves vary from 1720 to 2200 μ wide; in *RMS 16999*, from 1600 to 2000 μ wide), but show intergradation to more or less typical *L. hatcheri*. Schiffner (1903, *in* 1901–43, p. 50) also reports such intermediates and remarks that these forms, though common in northern Europe, are lacking in central Europe. Schiffner also emphasizes that the cells of *L. hatcheri* are smaller than those of *L. lycopodioides*. Intramarginal cells, within the leaf apices, average isodiametric and ca. 19–22 μ in *L. lycopodioides*, while marginal cells below the leaf tips are often elongated tangentially in short rows, averaging 21–24 μ long, occasionally 24–28 μ. In *L. hatcheri*, the lobe cells supposedly average 15–18 μ, while marginal cells are almost uniformly isodiametric, averaging 16–19 (rarely 20) μ, measured tangentially. However, in the difficult robust forms of *L. hatcheri* (e.g., *RMS & KD 66–203*, Disko I., Greenland) subapical cells may average 17–23 μ and marginal cells 20–24 μ. Clearly no consistent distinction seems possible.

The feeble border formed of tangentially elongated cells, when discernible, separates critical material from *L. hatcheri*. Such a border is perhaps best developed in extreme xeromorphic phases, which do not always develop gemmae. Such a plant has been described (Schuster, 1951, p. 6) as bearing unusually imbricate and concave, brownish-pigmented leaves. Mature leaves have cell walls and bulging trigones brownish-pigmented, except for the cilia and leaf margins, which are decolorate and hyaline, and formed of strongly thick-walled cells (forming a discrete border). Leaves are not horizontally spreading but are rather concave, strongly imbricate, forming terete, somewhat ascending shoots, appearing superficially closely similar to *L. (Orthocaulis) floerkei*.

Particularly difficult to place are occasional gemmiparous forms of *L. lycopodioides* in which leaves become strongly asymmetrical and leaf lobes

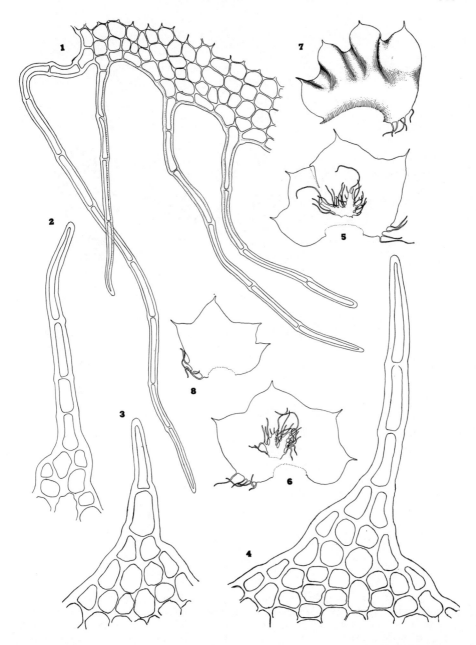

FIG. 158. *Lophozia* (*Barbilophozia*) *lycopodioides*. (1) Cilia at base of leaf (×170). (2) Cells of lobe apex (×238). (3–4) Cells of lobe apices (×ca. 280). (5–6) Leaves and underleaves, same scale (×13). (7) Leaf, not flattened (×13). (8) Leaf, of phase transitional to *L. hatcheri* (×13). [1–6, from plants from Feldberg, Baden, Germany, K. Müller, 1898; 7, *RMS 1089*, Mt. Washington, N.H.; 8, *RMS 16999*, Amygdaloid I., Mich.]

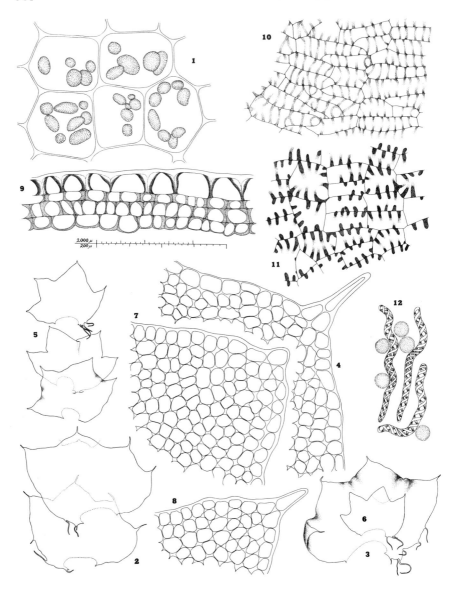

Fig. 159. *Lophozia* subg. *Barbilophozia*. (1–4) *L. lycopodioides*, (5–12) *L. hatcheri*. (1) Median cells (×650). (2) Two leaves (×13). (3) Leaf (×13). (4) Leaf lobe apex from leaf in fig. 2 (×175). (5) Three leaves (×13). (6) Leaf, drawn to scale of fig. 3 (×13). (7) Leaf lobe from leaf in fig. 6 (×175). (8) Dorsal leaf lobe (×175). (9) Capsule wall cross section (×305). (10) Inner cell of capsule wall, surface view (×305). (11) Epidermal cells, capsule wall, surface view (×305). (12) Elaters and spores (×305). [1, *RMS & KD 66-181*, Godhavn, W. Greenland; 2–8, drawn from intermixed *L. hatcheri* and *L. lycopodioides*, *RMS & KD 66-566*, Godhavn, W. Greenland; 9–12, *RMS & KD 66-241*, Jacobshavn, W. Greenland.]

obsolete, with the lobes blunt and lacking the usual apiculi, as, e.g., in
RMS 43513 from the Gaspé. Associated with this is a somewhat smaller
size (2.5–2.7 mm wide), although the leaves remain strongly fluted in most
cases and bear 3–6 tortuous cilia on the postical margins. The cilia are
variable, often strongly elongate, with terminal and subterminal cells
ranging to 13 × 95–105 μ (ratio of 1 : 7–8 width to length). In such plants
the asymmetric condition and reduction of the lobes and mucros of the
lobes are associated with free gemmae formation. Phases such as this
constitute the var. *obliqua* Müll., which I regard as a mere environmental
phase of *L. lycopodioides*. Reduction of the lobes and loss of the terminal
mucros are not accompanied by a reduction in width of the leaves, which
remain about 1.5× as wide as long. Parallel modifications also occur in
L. hatcheri, but here the gemmiparous leaves are much less transverse and
often are as long as wide.

If we omit from consideration the occasional gemmiparous "intermediate"
forms, some of which should possibly be placed as luxurious forms under *L.
hatcheri*, the separation of *L. lycopodioides* is simplified, since normal forms are
always much more vigorous than *L. hatcheri*, are provided with more prominently mucronate leaf lobes and more undulate-fluted leaves, and are almost
constantly lacking sexual organs. The differences in degree of fertility can be
observed in the field, in areas where both taxa occur (e.g., the eastern tip of the
Gaspé Peninsula). In *L. lycopodioides* occasional ♂ and ♀ patches are to be
found, yet perianths appear to be very rarely produced, and gemmae are
consistently lacking. By contrast, the equally frequent *L. hatcheri* is distinct to
the naked eye by its smaller size (the plants averaging at most two-thirds the
width of the *L. lycopodioides*, growing under identical conditions), less fluted
leaves, and abundant development of gemmae, at least on some plants in every
patch. Free gemmae development does not seem to inhibit development of
androecia and perianths, and even capsules (which mature in August and
possibly into early September). The nearly equal degree of frequency of the
two species in some areas, under apparently identical conditions, is mystifying,
considering the nearly uniform lack of reproduction—sexual and asexual—in
L. lycopodioides, contrasted to the free production of both gemmae and spores in
L. hatcheri.

Capsules are, however, occasionally produced in *L. lycopodioides*. As in *L.
hatcheri*, mature sporangia occur in mid-August in the Gaspé; they are probably
produced into early September.

LOPHOZIA (BARBILOPHOZIA) HATCHERI (*Evans*)
Steph.

[Figs. 159:5–12, 160, 161]

Jungermannia barbata var. *pusilla* Schiffn. & Schmidt, Lotos 34:25, 1886.
Jungermannia collaris Massal., Atti Soc. Veneto-Trent., Sci. Ser. 2, 2:29 [of reprint], 1895
[not of Nees, 1817].

Jungermannia hatcheri Evans, Bull. Torrey Bot. Club 25:417, 1898.
Jungermannia floerkii var. *baueriana* Schiffn., Österr. Bot. Zeitschr. 50(8):274, 1900.
Lophozia hatcheri Steph., Résult. S. Y. Belgica, Hepat., p. 4, 1901; Bull. Herb. Boissier, Ser. 2, 2:167, 1902; Spec. Hep. 2:159, 1902.
Jungermannia floerkii var. *aculeata* Loeske, Moosfl. Harzes, p. 86, 1903.
Lophozia baueriana Schiffn., Lotos 51:221, 1903 [Krit. Bemerk. 3:9].
Jungermannia baueriana Arn., Bot. Notiser, p. 145, 1906.
Barbilophozia hatcheri Loeske, Abh. Bot. Ver. Brandenburg 49:37, 1907; Frye & Clark, Univ. Wash. Publ. Biol. 6:430, 1944; Grolle, Nova Hedwigia 2(4):555, pl. 94, 1960.
Barbilophozia baueriana Loeske, Hedwigia 49:13, 1909.
Lophozia (Barbilophozia) hatcheri K. Müll., Rabenh. Krypt.–Fl. ed. 2, 6(1):631, fig. 31, 1910.

In loose, green to brown patches; *shoots (1)1.5–2(2.7) mm wide*. Stems little branched, 2–5 cm long, prostrate to ascending to ± erect, flexuous, with 1–2 dorsal cortical layers of cells somewhat equally thick-walled, 18–20 μ in diam.; medullary cells thin-walled, averaging 26–30(36) μ, becoming smaller ventrally (where with a trace of a small-celled medullary band, formed of 2–3 layers of cells only 15–18 μ in diam.) and equal in diam. to the ventral cortical cells; ventral cortical and ventral medullary layers of cells slightly thick-walled, not or rarely mycorrhizal. Rhizoids dense, rather short. Leaves *little or not undulate*, sinuses weakly reflexed, very obliquely inserted, variously oriented: in green shade forms strongly laterally spreading, horizontal and nearly flat, in brownish sun forms erecto-patent and concave, with often somewhat incurved lobes (then assuming the appearance of *Orthocaulis*), approximate to imbricate; leaf shape very variable, on slender shoots quadrate-obrhomboidal (ca. 1100–1200 μ wide × 990–1100–1225 μ long, scarcely broader than long), then most *often trilobed*, on more robust shoots becoming *broadly obrhomboidal from a narrow base* (500–600 μ wide), *usually distinctly asymmetrical*, with postical margin longer and somewhat arched, the shorter antical margin often nearly straight, 1800–1900 μ wide × 1300–1400 μ long, *almost invariably (3)4-lobed*; leaf lobed for usually 0.2–0.4 its length, the *lobes ovate-triangular* (in strongly xeromorphic, gemmiparous shoots often with lobes obsolete and very obtuse), *mostly straight-sided*, *terminated typically by rather weakly differentiated apiculi* or cusps formed usually of 1–2(3) superimposed cells averaging 28–36(45) μ long × 18–29 μ wide; sinuses obtusely angular, often appearing more acute when leaf is not flattened, and weakly to (rarely) strongly reflexed and gibbous, the lobes consequently convex; postical base of leaves *with 2–3* (more rarely 4–5) *relatively short*, usually 180–300 μ long, occasionally bifurcate cilia, normally formed of 5–9 superposed elongated thick-walled cells; *terminal cells elongated, usually (40)50–60* (rarely 75–90) μ long × 12–15 μ wide. Underleaves *large* (ca. 750–800 μ long), bifid ¾–⅘, the lobes and base ciliate with long, often somewhat contorted cilia, the narrowly lanceolate lobes ending in long cilia; apices of cilia formed largely of cells 40–80(100) μ long × 12–18 μ wide. Cells of leaf middle 20(22–23)–26 μ long × 17–(20)24 μ wide, quadrate to hexagonal, thin-walled, and with small to slightly bulging trigones; cells of apices of leaf lobes ca. 15–18 to 18–24 μ; cells typically with 2–5 (rarely 6–10) spherical to short-ovoid oil-bodies, each ca. 4–5.5 μ long to a maximum of ca. 7.5 to 5 × 9 μ, formed of fine, barely protruding spherules, grayish. Gemmae *reddish brown, abundant* (on green shade forms rendering leaf lobes slightly dentate; when abundantly produced on sun forms, resulting in obtuse and vestigial lobes),

2-celled, polygonal to pyriform, ca. 20–22 × 22–27 μ, formed not only from leaf lobes, but occasionally from cilia of underleaves.

Dioecious. Androecia becoming intercalary, *frequently developed* bracts like leaves, in 5–7 pairs, but saccate basally, rather strongly concave, imbricate, quite asymmetrical, the ventral lobe larger; antheridia 2–3(4–5) per bract. Antheridia with body ca. 190–200 × 170–175 μ; stalk 2-seriate, 60–70 × 28–30 μ. Perichaetial bracts 3–5-lobed (to a depth of 340–420 μ), similar to leaves but the lobes often terminated in ± long cilia, occasionally with lobe margins or bract bases with 1–2 supplementary cilia, as large as or little larger than leaves (to 1550–1800 μ wide × 1275–1400 μ long); bractlike bracteole usually shallowly 2–3-lobed, often with 1–2 marginal cilia, smaller than bracts (ca. 1050–1350 μ wide × 1200 μ long), united with one or both bracts for a height of up to 350 μ. Perianth ovoid-cylindrical, apical half plicate, contracted at the lobed and short-dentate mouth. Mouth with crowded short cilia or teeth of 1–4(5) little elongated cells (terminal cell excepted; this 7–9 × 14–22 μ).

Capsule wall opaque, rather thick (52–60 μ), of (3)4(5) layers. Epidermal cells ca. 17–21 μ high, about twice as high as each of the inner strata, short-oblong in surface view, ca. (14)22–25(29) μ wide × 28–40 μ long, with rather strong nodular (radial) thickenings extended as spur- or bandlike tangential extensions. Innermost layer of more irregular, narrower cells, ca. 12–17 μ wide, variable in length, with numerous complete semiannular bands. Spores 14–15.5 μ in diam., the exine finely granular-papillose. Elaters contorted, rather short, ca. (65)75–130(150) μ long × 7–9.5 μ in diam., closely bispiral.

Type. Lapataia, Patagonia, South America (*Hatcher*).

Distribution

Bipolar and essentially high subarctic-subalpine to arctic-alpine. In Antarctica S. to Hook I., Graham Coast, 65°38′ S., 65°10′ W. (*RMS*); in the Northern Hemisphere clearly holarctic, extending from Honshu, Japan (Hattori, 1948, p. 107), to northern Siberia (Novaya Zemlya) and on the Yenisey R. at Patapovskoje (68°35′ N.), according to Persson (1947). Also in western Spitsbergen (Arnell and Mårtensson, 1959). Kitagawa (1965, p. 267) reports it from Taiwan and the Himalayas. Widespread in northern and central Europe, England, westward to Jan Mayen I., Greenland, across northern North America, southward to New Hampshire, Vermont, Michigan and Minnesota in the Midwest, Wyoming and Colorado to California in the West; disjunct in North Carolina.

E. GREENLAND. Lille Pendulum Ø; Canning Land; Fleming Inlet; Nualink; Braefjord near Cape Wandel; Akiliarisek; Depot Ø; Smalsund; Sarfak Pynt; Ingmikertorajik; Tasiusarsik in Angmagsalik Fjord; Adloe; Cape Beaugaré, in hot spring (38°C), among *Aulacomnium palustre*; Cape Deichmann; also variously between 74°30′ and 65°31′ N. NW. GREENLAND. Kangerdluarssuk, Bowdoin Bay, Inglefield Bredning, 77°33′ N., 68°35′ W. (*RMS. 45817a, 45813, 45975, 45803, 45796, 45805*); Kekertat, Harward Øer near Heilprin Gletscher, 77°31′ N., 66°40′ W., E. end Inglefield Bredning

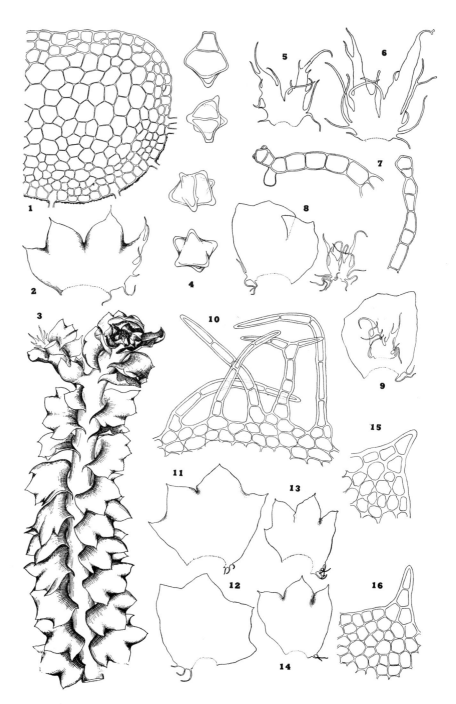

FIG. 160. *Lophozia (Barbilophozia) hatcheri*. (1) Stem cross section (×ca. 150). (2) Leaf of maximal size (×ca. 20). (3) Gemmiparous shoot, with base of *Frullania*-type branch at left (×11). (4) Gemmae (×ca. 550). (5–6) Underleaves, in fig. 6 with gemmae formation from apical cilia (×30 and 40, respectively). (7) Gemmiparous ends of cilia of underleaves (×ca. 280). (8–9) Leaves and underleaves from xeromorphic,

(*RMS 46035, 46074, 46101*); Kangerdlugssuak, S. side of Inglefield Bredning, 77°23′ N,. 67° W. (*RMS 45961b*); Kânâk, Red Cliffs Pen., 77°30′ N. (*RMS 45622, 45648, 45624*); Hackluyt I., Smith Sd., 77°24′ N., 72°31′ W.!; Bjørling I., Cary Isls., 76°43′ N., 72°22′ W. (*Seidenfaden 41*!; reported by Harmsen & Seidenfaden, 1932 as *Lophozia barbata*).   W. GREENLAND. Ubiquitous in all but very basic areas, e.g., Sondrestrom Fjord, near shore of L. Ferguson, in *Salix*-moss tundra, 66°5′ N. (*RMS & KD 66-057, 66-065, 66-086, 66-092*); Tupilak I., W. of Egedesminde, 68°42′ N. (*RMS & KD 66-015, 66-008, 66-004a, 66-018*); SE. of Jacobshavn, Disko Bay, 69°12′ N.. 51°05′ W. (*RMS & KD 66-216*, mod. *parvifolia-colorata*; *66-243, 66-241, 66-239, 66-238*); Ritenbenk, Agpat, 69°46′ N., 51°20′ W. (*RMS & KD 66-255, 66-266*); E. end of Agpat Ø, neck E. of Umiasugssup ilua, 69°52′ N., 51°38′ W. (*RMS & KD 66-1469a*). Disko I.: Godhavn, below Lyngmarks-fjeld, 69°15′ N., 53°30′ W. (*RMS & KD 66-192*); Fortune Bay, 69°16′ N., 53°50′ W. (*RMS & KD 66-203, 66-204; 66-205*, mod. *parvifolia; 66-206, 66-132*); Flakkerhuk, near Mudderbugten, E. end of Disko I., low, wet sandy-peaty plain near shore, 69°39′ N., 51°55′ W. (*RMS & KD 66-173a*); Lyngmarksbugt, ca. 25–125 m, 69°15′ N., 52°30′ W. (*RMS & KD 66-566, 66-567, 66-558*). Holstensborg (*Warming & Holm 1884*!; reported as *J. lycopodioides floerkei* by Lange & Jensen, 1887); Ameralik (*Vahl, Aug. 1830*!; apparently mixed with *L. lycopodioides*; reported as *J. l. floerkei* by Lange & Jensen, 1887).   S. GREENLAND. Tunugdliarfik (*Vahl 1017*!; reported as *J. l.* var. *floerkei* by Lange & Jensen, 1887; a gemma-free form less than 2 mm broad with a fair number of trifid leaves that is possibly a mod. *colorata-pachyderma-parvifolia* of *L. lycopodioides* and closely matches plants assigned to this taxon [?erroneously] by Kitagawa, 1965).

ELLESMERE I. Twin Glacier Valley and Fram Harbor, in Hayes Sd. region; Fram Fiord, S. coast (Polunin, 1947). The following reports of *Tritomaria quinquedentata* var. *turgida* ("*Lophozia lyoni turgida*") in Bryhn (1906) have been checked and represent *L. hatcheri*: Framshavn, 78°45′ N. (*Simmons 753, 754, 740*); Bedford Pim I., 78°40′ N. (*Simmons 1320*). BAFFIN I. Cape Dorset, Hudson Strait (Steere, 1939).

LABRADOR. Attikamagen L., NW. end of Iron Arm (*Harper 3558*!).   QUEBEC. Cairn I., Richmond Gulf; mouth of Seal R., Cape Jones; S. side of Great Whale R.; Chimo Air Base, Ungava Bay; Leaf R., 80 mi SW. of Ungava Bay (Schuster, 1951); Wakeham Bay; Ivuyivik; also from Tadoussac; Rigaud; Bic; Islet Canuel near Rimouski; Sacré-Coeur, côté de Rimouski; Mont la Table (Lepage, 1944–45); Mt. Lac des Cygnes, St. Urbain, Charlevoix Co. (Kucyniak, 1947); George R., near 57°21′ (Kucyniak, 1949a); Scoter L., 58°30′ N., 76°44′ W. (*A.R.A. Taylor 81*!, mod. *parvifolia*). NEWFOUNDLAND. Table Mt. of Cape Ray, W. Nfld.; Port au Choix, Bartlett's R., Highlands of St. John, Ha-Ha Bay, Williamsport, Great Harbour Deep, all on N. Pen.; Gaff Topsail, Mint Brook, and Gambo Pond, C. and NE. Nfld. (Buch & Tuomikoski, 1955.)   NOVA SCOTIA. L. Kedji; Maitland Falls; Mary Ann Brook, S. of Neils Harbour, Cape Breton I. (*RMS 42667a,b, 42679, 42685*).   ONTARIO. L. Penage, Sudbury Distr. (Cain & Fulford, 1948); Long Pt., L. Timagami (*Cain 4259 p.p.*!); below Cache L., Algonquin Park (Williams & Cain, 1959).

MAINE. North Basin, ca. 3000 ft, Mt. Katahdin (*RMS 32918*); Little Saddleback Mt., Franklin Co.   NEW HAMPSHIRE. Mt. Washington and vicinity (*RMS*).   VERMONT. Mt. Mansfield (*RMS*).   NEW YORK. Cascades, between Upper and Lower Cascade Ls., 6.5 mi SE. of L. Placid; Wilmington Notch, W. Branch Ausable R., 6.5 mi SW. of Wilmington, Essex, Co. (N. G. Miller, 1966).

---

strongly gemmiparous phase (×17). (10) Cilia from base of leaf in fig. 13 (×190). (11–12) Leaves from optimal-sized shoot (×17). (13–14) Leaves from "normal shoot" (×17). (15–16) Apices of median leaf lobes (×190). [1–7, *RMS 13400*, Gunflint Lake, Minn.; 8–9, *RMS 11959*, Sailboat I., Susie Isls., Minn.; 10–16, from plants from Bavaria, issued by Schiffner in Hep. Eur. Exsic. no. 106, as type of *L. baueriana.*]

MICHIGAN. Copper Harbor, and between Phoenix and Copper Harbor, Keweenaw Co. (Steere, 1937; *RMS*); Amygdaloid I., Isle Royale (*RMS*); Huron Mts. (*Nichols*); "Devils Washtub," Keweenaw Co. (*RMS*). MINNESOTA. Cook Co.: Pigeon R., N.W. of Grand Portage; Gunflint L.; Sailboat I., Susie Isls., near Pigeon Pt., L. Superior (Schuster, 1953.)

Disjunct in NORTH CAROLINA: Grandfather Mt. (*RMS*); NW. facing granitic cliff, ca. 5200 ft, Watauga Co. (*M. L. Hicks 1130!*).

Further westward also in Alberta, British Columbia, District of Keewatin, southward to Wyoming, Colorado, Montana, Washington, Oregon, and California, northwestward to the Yukon and Alaska (Persson, 1947).

Lophozia hatcheri is abundant in both the upper half of the Taiga and through all but the northernmost portions of the Tundra (neither Schuster nor Holmen was able to find the plant in northernmost Ellesmere I. or Peary Land, respectively). Many, if not most, published reports of "*L. lycopodioides*" pertain to this species, and a large number of early reports of "*L. floerkei*" also belong here; as Schuster (1958, p. 286) showed, several reports of "*Tritomaria quinquedentata* var. *turgida*" from southern Ellesmere I. (Bryhn, 1906) represent *L. hatcheri*.

Ecology

A species abundant both in the Arctic and in the northern portion of the Taiga: in the Taiga either in the lowlands or in submontane areas, often in partial shade under spruce or fir, in large green mats (frequently with *Dicranum*, *Ptilidium ciliare*, etc.). Under ecologically less advanced conditions abundant on rocks: usually on steep or vertical sides of cliffs or ledges, under conditions of intermittent moisture. It may occur on either acidic rocks (granite, arkose, etc.) or ± basic rocks (basaltic trap-rock and conglomerate), and does not appear to show a strong preference for acid conditions; however, it is rare or absent on basic sedimentary rocks. Associated on dry to damp ledges under such conditions are various other Hepaticae, among them *Ptilidium ciliare*, *Tritomaria quin-quedentata*, *Lophozia barbata*, *L. attenuata*, and *Scapania nemorosa*; the measured pH ranges from 4.5 to 6.0. Although the species may be a pioneer under such conditions, it is more frequent as a secondary species. As is evident from the frequent occurrence under spruce and fir (particularly in the mountains below the treeline, where the trees are low and form a dense mat), the species persists until the climax is attained, and is then in open competition with coarser mosses and *Ptilidium ciliare*.

In higher mountains (as in New England and New York), the species drops out and is largely replaced by the closely related *L. lycopodioides*, in treeless areas near the summits of mountains. Macvicar (1926) also notes that in England the species is subalpine and only "occasionally alpine," in contradistinction to *L. lycopodioides*, which is confined to the "upper parts" of the "higher mountains." However, *L. hatcheri* appears to be at least fully as common as *L. lycopodioides* in the Tundra of the far north, under conditions simulated at the summits of many of our New England mountains.

In all but the lower parts of the Arctic *L. hatcheri* is, indeed, much commoner than *L. lycopodioides*. It occurs here as a typical xerophyte, in arid, exposed fell-fields, often in clefts between boulders or bedrock or in crevices of ledges, where it is sheltered from the wind but where (during the growing season) there is little available moisture; in such sites, furthermore, much snow accumulates (and persists). Thus the species occurs, often as a pioneer, in quite inhospitable sites—sites from which it seems consistently lacking in the alpine region of New England. In such arid niches there is a diagnostic series of associates: *Rhacomitrium canescens* and other *Rhacomitria*, various lichens, *Ptilidium ciliare*, *Chandonanthus setiformis*, more rarely *Lophozia rubrigemma*. Under extreme conditions it may occur as a brown mod. *parvifolia*, in rock crevices, together with *Lophozia alpestris s. lat.*, *Anastrophyllum minutum*, and *Gymnomitrion corallioides*.

Differentiation

Likely to be confused with both *L. barbata* and *L. lycopodioides*. As in these, the vast majority of leaves are 4-lobed (the few to frequent 3-lobed leaves have the dorsal, undivided lobe usually distinctly larger), and the leaves are very obliquely inserted and frequently horizontally spreading or even approach a squarrose condition from the oblique base.

Lophozia hatcheri is slightly smaller than most forms of *L. barbata* and much smaller than normal *L. lycopodioides*. These size differences are bridged by intermediate forms. Under the microscope, the present species differs from *L. barbata* in the presence of cilia of the basal portion of the ventral leaf margin (unlike in *Orthocaulis* species, consisting of elongate cells); furthermore, there are distinct underleaves, often large and mostly multiciliate (in *L. barbata* the underleaves are absent, except below the ♀ inflorescence, or are minute and not conspicuously ciliate). The presence of distinct mucronate points of at least some of the leaf apices (consisting of 1–4, occasionally more, elongate cells) will separate all phases from *L. barbata*.

In the ciliate ventral leaf bases, the large ciliate underleaves, and frequent presence of mucronate points on leaf lobes, *L. hatcheri* closely corresponds to *L. lycopodioides*. In fact, differentiation between the two species is often very difficult, as pointed out by, i.a., Persson (1947); Joergensen (1934) believed they might be mere varieties or ecospecies of each other.[43] Schiffner separated a var. *parvifolia* placed under *L. lycopodioides* that is almost inseparable from *L. hatcheri*; it agrees with the latter in smaller size, presence of reddish brown gemmae, and facies. Macvicar also

---

[43] Buch (1909) states that he has never encountered intergrades between the two taxa in the field but admits to having seen transitional herbarium specimens; on that basis he considers it "sehr wahrscheinlich" that the two are environmental modifications of a single species. He claims that the presence of gemmae usually separates *Lophozia hatcheri* from *L. lycopodioides* but admits—as did Arnell (1960) before him—that *L. lycopodioides* sometimes bears gemmae.

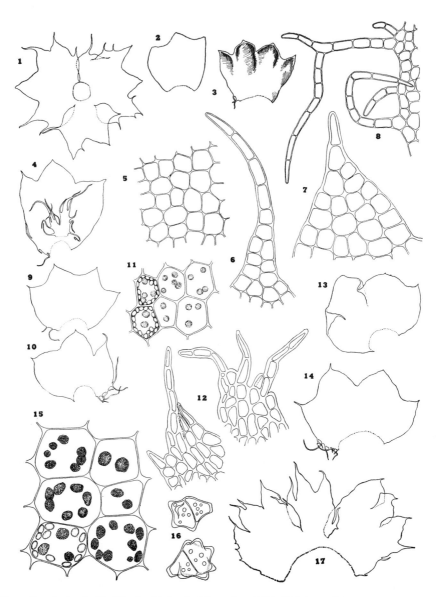

FIG. 161. *Lophozia (Barbilophozia) hatcheri* (1–11) and "*L. (B.) lycopodioides* var. *parvifolia*"
(12–17). (1) Bracts and bracteole (×15.5). (2–3) Leaves of xeromorphic shoot (×17.5).
(4) Leaf and underleaf (×19.5). (5) Median cells (×ca. 185). (6–7) Apices of lateral
and median lobes, respectively (×250). (8) Cells of postical leaf base (×155). (9–10)
Leaves (×17.5). (11) Median cells with oil-bodies and, at left, chloroplasts (×395).
(12) Lobes of perianth mouth (×210). (13–14) Leaves, from xeromorphic phase
(×15.5). (15) Median cells with oil-bodies and, lower cell, chloroplasts (×665).
(16) Gemmae (×520). (17) Bracts and bracteole (×14.5). [1–11, *RMS 13400*,
Gunflint Lake, Minn., typical *L. hatcheri*, also in Fig. 160:1–7; 12–14, 17, *RMS 16999*,
Copper Harbor, Mich., of plants transitional from *L. hatcheri* to *L. lycopodioides* (the "var.
parvifolia"); 15–16, *RMS 17322*, Amygdaloid I., Mich., same phenotype as figs. 12–14,
also drawn in Fig. 157:9–12.]

speaks of intergrading forms that cannot be placed with certainty. *Lophozia hatcheri*, in the less modified leaf shape, deeper division into lobes, less fixed and lower average lobe number (often varying on the same plant from 3 to 4), and frequent development of gemmae, is the more primitive species. When *L. lycopodioides* occurs under suboptimal conditions it apparently shows tendencies to revert: the size becomes smaller, and the leaves more lobed and less undulate. Such forms, though genetically *L. lycopodioides*, apparently cannot be placed with certainty from their vegetative characters. It would seem that in the subarctic-subalpine regions *L. lycopodioides* is especially likely to produce such reversions.

Typical material of the two species can scarcely be confused. *Lophozia hatcheri* is only 2.2–2.7 mm wide normally and 2–4(5) cm long; it has more deeply 3–4-lobed leaves (a few of which are often merely 3-lobed), with the lobes ovate-triangular or acutely triangular, *with diagnostically straight sides*. The leaf form is rather variable: from nearly quadrate-rhombiform to rhombiform-reniform, but usually considerably narrower at the base than near the apex. Gemmae are very frequent. The species is normally a pure, deep green but may become deeply brownish under xeric and insolated conditions.

Lophozia lycopodioides, on the other hand, is often 4–5 mm wide and 4–8 cm long and has 4-lobed leaves, with occasional ones 5-lobed, rarely (on poorly developed shoots) a few 3-lobed, with the broad and *convex-sided* lobes shallow, only $\frac{1}{5}$–$\frac{1}{4}$ the length of the leaf, very obtusely triangular; the apex of each lobe is more abruptly mucronate. Leaves are always more or less strongly rhomboidal or rhomboidal-reniform and at least 1.5–1.7 × as wide as long, mostly about twice as wide at apex as at base. The plants are usually pale, rather hyaline yellowish green, occasionally becoming a light, pure green. The leaves are always undulate-crispate: a condition rarely approached by those of *L. hatcheri*. The occurrence in gemmiparous material of *L. hatcheri* of leaves with very broadly triangular lobes, lacking mucronate points, gives such material the facies of a small form of *L. lycopodioides*. Occasional plants also occur, not exhibiting gemmae formation, that have a large number of such leaves; these plants must be placed with care.[44]

It is of interest that gemma formation in this species is not restricted to the leaf lobes but also occurs at the apices of the cilia and leaf lobes of the uppermost underleaves (Fig. 160:7).

Some strongly gemmiparous material (particularly from the Great Lakes region), previously referred to *L. hatcheri*, is here provisionally referred to the xeromorphic, gemmiparous phase of *L. lycopodioides* ("var. *parvifolia*").

[44] Frye and Clark (1937–47) use as a key character separating *hatcheri* from *lycopodioides* the supposed presence of an "unlobed or 2-lobed" bracteole. Fig. 161:1,17 shows that the bracteole may be 4-lobed, which they consider normal for *L. lycopodioides*; thus this is not a valid character for the separation of the two. These writers also state that the perianth mouth is "lobed and shortly toothed," contrasted to the "ciliate" perianth mouth of *lycopodioides*. This is apparently a "paper" separation, since Müller (1905–16, p. 634) describes an xerophytic form of *hatcheri* (as var. *ciliata*), in which the perianth mouth is fringed with cilia 10–20 cells long, obviously more deeply ciliate than in *lycopodioides*, with cilia "1–3 cells long."

Tendencies to automatically refer gemmiparous plants to *L. hatcheri* and nongemmiparous ones to *L. lycopodioides* appear to be unwarranted, although *L. hatcheri* more often occurs under xeric conditions, in rather insolated sites, and is then abundantly gemmiparous.

The basis for disposition of such plants is slight and rests on four differences. (1) The leaf shape in even smaller, brownish, xeromorphic forms of *L. lycopodioides* tends to be broader (average figures being 1950–2200 μ broad × 1380–1475 μ long), with often as many as 5–7 cilia of the postical base; analogous phases of *L. hatcheri* have leaves rarely over 1800 μ wide and average only 3 or 4 cilia per postical base. (2) The underleaves in even small xeromorphic phases of *L. lycopodioides* are almost totally bifid and bear numerous cilia (8–12 per lobe) formed of strongly elongated cells; in *L. hatcheri* each underleaf lobe bears 4–6, rarely more, cilia, with less elongate cells (ca. 40–80 μ long usually). (3) The median leaf lobes in *L. lycopodioides* are very broad and convex-sided, and are most often terminated by sharp, sudden, setose mucros, formed of 2–3 strongly elongated, thick-walled cells (each cell ca. 18–20 × 60–100 μ); in *L. hatcheri* the lobes usually are much narrower, normally are straight-sided and vary from merely acute to apiculate, rarely mucronate, the apex formed by 1–2 superposed cells usually only 30–36 μ long × 18–20 μ wide. (4) The postical leaf bases of *L. lycopodioides* have cilia very long, tortuous, and sinuate, formed of cells 12–15 μ wide × 90–110 μ long usually; in *L. hatcheri* there is much variation in the form of the cilia, but these are usually shorter, less flexuous, with at least the basal cells shorter, distal cells rarely averaging beyond 12–15 × 50–75 (90) μ.

Not one of the above distinctions is absolute, but I believe that the basic leaf outline offers the safest point of distinction. In particular, the very low and broad, convex-sided, median leaf lobes of *L. lycopodioides* will almost always separate it from *L. hatcheri*.

Small, xeromorphic extremes (mod. *densifolia-colorata-parvifolia*) of *L. hatcheri* may be habitually so different from "normal" phases that they are at first unrecognizable. Material from Scoter Lake, Quebec (*Taylor 81*), is characteristic of this phase: leaves vary from 700 μ wide × 550 μ long to 720–800 μ wide × 530–580 μ long, and range from 3–4-lobed; the triangular lobes are rarely provided with enlarged apical cells or apiculi and are sometimes blunt or even narrowly rounded; postical leaf bases are devoid of cilia or bear 1–2 cilia 80–125 μ long (formed of 2–3 superposed cells, each up to 45–58 μ long); underleaves are vestigial (ca. 80 × 300 μ) and formed of two slender segments 2–3 cells wide at base, which may bear a single tooth each. Such phases have the color and size of *L. alpestris* and *L. binsteadii*. They are possibly subject to confusion with *L. atlantica* because of the minute underleaves. However, the occasional cilia of the postical leaf bases, formed of elongated cells (identical to the cells of the tips of the underleaf divisions) place such extremes in *Barbilophozia*.

The very small mod. *parvifolia-colorata* of *L. hatcheri* (e.g., *RMS & KD 66–216* from Jacobshavn, Greenland) may be less than 1 mm wide, concave-leaved with incurved and often blunt lobes that almost never develop even a vestigial 1-celled mucro, and almost never possess basal cilia of the leaves. Such plants freely develop gemmae, have almost constantly trifid leaves, and may have greatly reduced underleaves formed of only a pair of tapered cilia. Such modifications have given rise to confusion with *L. floerkei* and *L. atlantica*; they differ from both in that with prolonged search one can always find leaves with at least one short basal cilium, often only 2–3 cells long, formed in part of elongated cells. The free development of gemmae also separates such plants from *L. floerkei*, and the small trigones from *L. atlantica*.

At the opposite pole are vigorous, large, green shade forms which show limited gemma development and very restricted formation of cilia of the ventral leaf bases (usually only 1–2 cilia occur per leaf base). Such plants may also show unusually small underleaves that are often simply lanceolate-acuminate, with a few cilia near the base, rather than bifid (e.g., *RMS & KD 66–238*, from Jacobshavn, Greenland).[45] Similar plants are frequent in Antarctica.

Sectio *BARBATAE* Jensen

Jungermania-Gruppe 4, *J. barbatae* Jensen, Danmarks Mosser, Bryofyter 1:110, 1915.
Lophozia subg. *Barbilophozia* sect. Barbatae Schust., Amer. Midl. Nat. 45(1):45, 1951.

Plants with subquadrate and subhorizontal leaves, the lobes acute to blunt, never cuspidate to ciliate-tipped; ventral leaf bases unarmed. Underleaves vestigial or small, hidden amidst the dense rhizoids. No gemmae.

Type. Lophozia barbata (Schmid.) Loeske.

This section corresponds with subg. *Barbicaulis* Buch (1942), a *nomen nudum*. Jensen's "Jungermaniae barbatae" includes sectio Lycopodioideae as well and hence is equivalent, taxonomically, to subg. *Barbilophozia*, as well as including most of subg. *Orthocaulis*.

LOPHOZIA (BARBILOPHOZIA) BARBATA (Schmid.) Dumort.

[Fig. 162]

Jungermannia barbata Schmid., Icones Pl. et Anal. Part., p. 187, 1747; *ibid.*, pl. 48, 1762; ed. 2, 1793; Evans, Plant World 1(7):97–102, figs. 1–15, 1898.
Lophozia barbata Dumort., Rec. d'Obs., p. 17, 1835.
Jungermannia barbata var. *schreberi* Nees, Naturg. Eur. Leberm. 2:189, 1836.
Barbilophozia barbata Loeske, Abh. Bot. Ver. Brandenburg 49:37, 1907; Frye & Clark, Univ. Wash. Publ. Biol. 6:426, 1944.
Lophozia (Barbilophozia) barbata K. Müll., Rabenh. Krypt.–Fl. ed. 2, 6(1):656, fig. 307, 1910; Schuster, Amer. Midl. Nat. 45(1):45, pl. 8, figs. 1–8, 1951; Schuster, *ibid.* 49(2):328, pl. 5, figs. 1–8, 1953.
Barbilophozia (Barbilophozia) barbata Kitagawa, Jour. Hattori Bot. Lab. no. 28:262, fig. 4, 1965.

In loose, depressed mats, relatively *dark green*, in sunny locations brownish green, rarely deep brown; shoots 2.0–5.0 mm wide. Stems 3–8 cm long, ca. 320–500 μ in diam., with cortical cells little distinct from medullary in size,

---

[45] The phenomenal variability of *Lophozia hatcheri* has been emphasized by H. W. Arnell (1906, p. 147) as the most marked feature of the species, even though he admits that separation from *L. lycopodioides* is often difficult.

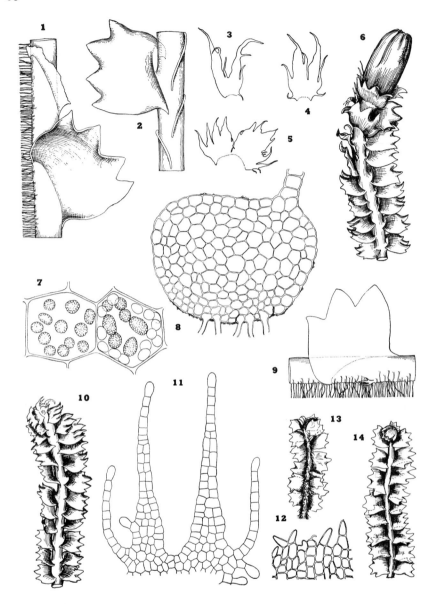

Fig. 162. *Lophozia (Barbilophozia) barbata.* (1) Shoot sector, lateral aspect (×14.5). (2) Same, dorsal aspect (×14.5). (3–4) Subinvolucral underleaves (×30). (5) Bract and bracteole (×6). (6) Perianth-bearing shoot (×ca. 6). (7) Cells, with oil-bodies and, at right, chloroplasts (×850). (8) Stem cross section (×150). (9) Leaf, in lateral aspect, with rhizoids and, almost hidden, underleaf (×14.5). (10, 13–14) Shoots, dorsal and ventral aspects (×ca. 4–5). (11) Large underleaf from sterile shoot (×155). (12) Perianth mouth cells (×ca. 100). [1–2, 9, 11, from Belle Rose I., Minn., *RMS*; 3–6, 10, 12, Peasleeville, Adirondack Mts., N.Y., *RMS*; 13–14, Delaware Water Gap, Pa., *RMS*; 7, *RMS 29601*, Nigger Mt., N.C.; 8, Gunflint Lake, Minn., *RMS*.]

isodiametric in cross section; medullary cells 20–36 μ in diam. dorsally, becoming somewhat smaller ventrally; ventral medullary cells in 2–4 tiers smaller than dorsal (ca. 14–20 μ), forming a vestigial ventral band, more thick-walled, hyaline or yellowish, becoming brownish with age, similar to ventral cortical cells; mycorrhizal infection slight, usually confined to ventral cortical cells, largely external; stems simple or slightly branched, procumbent or apically ascending, elliptic in cross section. Rhizoids colorless or pale brownish, abundant, short, in a feltlike mat on ventral side of stem. Leaves strongly succubous, very obliquely inserted, somewhat decurrent dorsally, subimbricate to imbricate, widely spreading, mostly 4-lobed (occasionally 3- or 5-lobed), obcuneate-quadrate, symmetric or nearly so, entire-margined, rarely with obsolete teeth or cilia near leaf bases; lobes subequal in size, *lying in a line almost parallel to stem*, the two lateral ones barely smaller, triangular-ovate to ovate, *obtuse to subacute at apex*, separated by obtuse to rectangulate sinuses $\frac{1}{6}$–$\frac{1}{4}$ the leaf length. Cells of leaf middle 21–24(26) × 23–25(28) μ, of lobes slightly smaller (18–23 μ), rounded-polygonal, thin-walled with trigones ± distinct, concave-sided; oil-bodies (4–5)7–9(12) per cell, mostly 4–5 μ and spherical to 4–5 × 5–7 μ, a few to 6 × 10 μ and ovoid, formed of numerous fine spherules, appearing faintly segmented or papillose; chloroplasts 4 μ. *Gemmae absent* or rare, "reddish yellow, 1–2 celled, 3–6 angulate or oblong, at apices of leaf lobes." *Underleaves small or obsolete*, subulate to lanceolate, sometimes deeply 2-lobed on sterile shoots: rarely bifid and with ciliform lobes, then to 320 μ high; toward fertile areas larger.

Dioecious. ♂ Plants in separate or same patches as ♀; perigonal bracts closely imbricate, in 5–10 pairs, the dorsal margin with 1–2 anteriorly bent teeth; antheridia 2–5, oval, among lanceolate to irregularly lobed paraphyses. ♀ Bracts similar to, and scarcely larger than, leaves, somewhat more deeply lobed, the lobes often rather spinose-dentate, more sharply acute, the sinuses often gibbous; bracteole large, 2–3-lobed and often with 1–2 laciniate teeth. Perianth strongly emergent, oblong-cylindrical, deeply 6–8-plicate in apical third, constricted apically, the mouth crenulate or dentate with projecting teeth formed by 1, more rarely 2, cells. Sporangium ovoid, the wall 4-stratose. Elaters 8 μ in diam., bispiral; spores ca. 15 μ, verrucose, yellowish brown.

Type. European; exact locality not given.

Distribution

Holarctic, in cold or cool climates, with a wide range extending from the arctic-alpine regions (where infrequent), throughout the Taiga, southward into northern edges of the Deciduous Forest Regions, where it becomes rare and of relatively sporadic occurrence (evidently as a relict of the Pleistocene glaciations). Distributed in Europe from Portugal, Italy, and the Dalmatian coastal region northward to Scandinavia, eastward into Siberia (64° N.) and northern Japan;[46] also recorded variously

---

[46] Hattori (1955a, p. 82) reports the species from Mt. Apoi, Hokkaido, from serpentine, pH 6.5–7.0.

from "Asia"; in North America found from Alaska to Greenland and southward; of North American reports, only peripheral stations are cited in detail.

To the west of our region found from Alaska southward to Washington, British Columbia, Idaho, Montana, Alberta, and at high elevations to Colorado and New Mexico. Also known from Iceland.

NW. GREENLAND. Dundas, North Star Bay (*RMS 46251*; typical! northernmost report of this species).[47] W. GREENLAND. Nakajanga Umivit, Sondrestrom Fjord, 67° N. (*Holmen 4320!*; typical!). S. GREENLAND. Ûnartoq Fjord, 60°37′ N., 45°15′ W. (*P.M. Peterson 174!*). DEVON I REGION. Philpots I. (Polunin, 1947). BAFFIN I. Pond Inlet (Polunin, 1947); without locality (Steere, 1938a). HUDSON BAY REGION. Island in NE. Hudson Bay (Polunin, 1947). NEWFOUNDLAND. Air Force station, N. of Port au Port (*RMS 68–1894*).

QUEBEC. Widespread: Seal R., Cape Jones; Wakeham Bay; Manitounuck Sd.; Cairn I., Richmond Gulf; Cape Smith; southward to Bic; Rimouski; Mt. Albert, Miquelon I., La Tuque; Montmorency R.; Mt. Shefford; Montreal, etc. (Schuster, 1951; Lepage, 1944–45; Fabius, 1950). MIQUELON I. (Delamare et al., 1888; Buch & Tuomikoski, 1955, doubt the identification; they did not find the species in nearby Newfoundland). NEW BRUNSWICK. Grand Manan; Northern Head; Woodstock. NORTHWEST TERRITORY. Tukarak I., Belcher Isls. (Schuster, 1951). LABRADOR. NOVA SCOTIA. Ship Harbour; Ellershouse; Port Mouton (Brown, 1936a); Mary Ann Brook, S. of Neils Harbour, Cape Breton (*RMS 66313*). ONTARIO. North Hastings; Chelsea near Ottawa; Algonquin Park; L. Nipigon; Thunder Bay, L. Superior; Sudbury Jct., etc. (Macoun, 1902); Manitoulin I., L. Huron (*Schuster, 1949*); Moose R., James Bay!; N. Shore, L. Penage, Sudbury (*Cain 2135!* det. M. Fulford as "*Orthocaulis kunzeanus*"!); Cedar L., 28 mi N. of Vermilion Bay, Kenora Distr. (*Cain 5606!*); Bruce Mines, Algoma Distr. (*Cain 3438!*); West Hill, Owen Sd. (*Conard 38-306!*).

MAINE. Round Mountain L., Franklin Co.; Mt. Desert (*Lorenz*); Orono (*s. coll.*); North Basin, 2800 ft, Mt. Katahdin (*RMS 32906*; among *Tritomaria quinquedentata*). NEW HAMPSHIRE. Mt. Washington (*RMS*). VERMONT. Jamaica, Windham Co. (*Dobbin & Burnham*); Birdseye Mt., Rutland Co. (*Carpenter & Burnham*); Mt. Equinox, Bennington Co. (*Dobbin & Burnham*); Mt. Mansfield; Pico Peak (*RMS, s.n.*); Willoughby Mt., E. of Willoughby L. (*RMS 46212b*). MASSACHUSETTS. Princeton; Mt. Toby, Sunderland (*RMS, s.n.*); Chesterfield Gorge, W. Chesterfield, Hampshire Co. (*RMS 41309*). CONNECTICUT. Litchfield, Hartford, New Haven, Middlesex Cos. (Evans & Nichols, 1908). NEW YORK. Little Moose L., Herkimer Co. (*Haynes*); Peasleeville, Clinton Co. (*RMS A-256, A-257*); Coy Glen and Six Mile Cr., Tompkins Co.; Middlesex, Yates Co.; Primrose Hill, Onondaga Co.; Elmira, Chemung Co.; Woodhull, and near Haskinville, Steuben Co. (Schuster, 1949a); Cattaraugus Co. (Boehner, 1943); W. Fort Ann, Sugarloaf Mt., and South Bay, Washington Co. (*Burnham*); Saratoga Battlefield, Saratoga Co. (*Burnham*); Summit of Cornell Mt., and between Wittenberg and Cornell Mts., Catskill Mts. (*RMS 24471, 15785*); Sand L., Rensselaer Co. (*Burnham*); Mt. Markham, Plainfield, Otsego Co. (*S. A. Brown*); Indian Ladder, Helderberg Mts.,

---

[47] The species is known, in Siberia, north to 64° N.; it is not reported from Spitsbergen by Arnell & Mårtensson (1959). However, there are two prior reports from northwest Greenland by Harmsen & Seidenfaden (1932) from Hackluyt I., 77°24′ N. (= *Lophozia heteromorpha*) and from Bjørling's Ø, 76°43′ N. (= *L. hatcheri*). The species has also been reported from Cap Ravn, east Greenland, based on *Böcher 817* (published in Böcher, 1933; based on det. by Harmsen); this specimen represents *Saccobasis*! All Greenland reports, except those here verified, appear based on errors in determination.

Albany Co. (*Burnham*); Ellenburg Mt., Clinton Co. (*Schuster, 1945*); Martiny Rocks, Cattaraugus Co. (*Boehner!*).   PENNSYLVANIA. Delaware Water Gap (*Schuster, 1943*).

VIRGINIA. Shenandoah Natl. Forest, 18–20 mi S. of Panorama, Skyline Drive, 3200–3600 ft (*RMS 18391*); Mountain L., Giles Co.; Skyline Drive, Mile 93, Timber Hollow Overlook, SW. of Pollock Knob, 3200 ft (*RMS 19876a*).   NORTH CAROLINA. Summit of Nigger Mt., near W. Jefferson, Ashe Co. (*Blomquist; RMS 29601*); Doughton Park, Blue Ridge Pkwy., N. Wilkes Co. (*RMS 34295*); southernmost stations east of the Rocky Mts.

INDIANA. Lawrence Co. (Wagner, 1947; a vague and unreliable report). MICHIGAN. Amygdaloid I., Isle Royale (*RMS 17321; c. caps*); Copper Harbor, Ft. Wilkins (*RMS*); Huron Mts.; Gogebic, Luce, Marquette, Chippewa, Cheboygan Cos. (*Steere*).   WISCONSIN. St. Croix Falls (*RMS*); Vilas, Oneida, Lincoln, Bayfield, Polk, Sauk, Douglas, Iron, Ashland Cos. (Conklin, 1929); Sand I., Apostle Isls. (*Schuster, 1949*).   MINNESOTA. Carlton, Chisago, Cook, Lake, Lake of the Woods, St. Louis Cos. (Schuster, 1953). Also cited from OHIO: Geauga Co. (Miller, 1964).

Ecology

Correlated with the extensive range is a wide ecological tolerance. The species varies from mesophytic to xerophytic; it occurs from deeply shaded sites to sites with direct sunlight, undergoing ecesis on a variety of substrates, although "usually associated with cliffs or rock outcrops" or "on loamy or rocky banks . . . or almost as a pioneer on thinly soil-covered rocks." The species has a wide pH tolerance: "it may occur under quite acid conditions on humus, but is more often common under more nearly circumneutral conditions, with the pH from 5.8 to 6.5. When the pH gets above 6.5 the species begins to drop out" (Schuster, 1953). The rather moderate toleration for lime was also pointed out in Schuster (1949a), where it was indicated that the species was often found with the calciphilous *Lophocolea minor*.

Lophozia barbata, in its largely epipetric occurrence (or, at least, restriction to sites where rock outcrops freely occur), closely agrees with the two other species of *Barbilophozia*. In the Coniferous Forest Region, where the species is perhaps most abundant, it is found over ledges with a large variety of other Lophoziaceae (*Lophozia ventricosa*, *L. hatcheri* and more rarely *lycopodioides*, *L. attenuata*, *Anastrophyllum saxicolus*, *Tritomaria quinquedentata*), and *Scapania* spp. (*nemorosa*, *undulata*). More rarely found in peaty depressions in basaltic rocks, associated with *Sphagnum* and *Lophozia kunzeana*; however, it is absent from true bogs. *Lophozia barbata* rarely persists long after formation of a deep humus layer and only exceptionally undergoes ecesis on decaying logs.

Rather rarely, at least in our region, invading the faces and soil-filled crevices of quite dry rocks (as at Delaware Water Gap, Pa.,); it then forms a small, impoverished phase, 1.5–2 mm wide, bearing 2–3-lobed leaves on most plants. Occasionally the species also occurs on acid, sterile rocky soils, or rocky crests of ravines where the soil is much leached, then with members of the xerophytic *Lophozia bicrenata-Cephaloziella* Associule of acid, sterile soils.

At the southern periphery of its range, in Virginia and North Carolina, *L. barbata* is confined to shaded but relatively dry, acidic rocks, usually on the vertical faces of ledges or in soil-filled crevices of such ledges. Here it may be associated with a series of Appalachian pioneer species, among them *Metzgeria crassipilis, Harpalejeunea ovata, L. (Microlejeunea) ruthii* and *Frullania tamarisci* subsp. *asagrayana.*

Variation and Differentiation

In spite of the wide geographical and ecological range, *L. barbata* shows a surprisingly narrow diversity of phenotypes. It produces no forms that are likely to lead the experienced student to errors in determination.

The "normal" (i.e., most widespread) plant is usually a pure and rather deep, opaque green because of the densely chlorophyllose cells. Such plants have small trigones and almost constantly 4-lobed leaves (mod. *viridis-leptoderma-quadriloba*). The quadrate, almost uniformly 4-lobed leaves, the robust size, and the rather deep green color are diagnostic.

From other regional species with 4-lobed leaves (*Lophozia lycopodioides, hatcheri, quadriloba*) *L. barbata* differs in the small, triangular leaf lobes, never terminated in cilia, with the sinuses never strongly reflexed, in the absence of cilia of the postical leaf bases, and in the very small underleaves, quite hidden in the feltlike, dense mat of short rhizoids of the postical stem surface. Confusion is most likely with forms of *L. hatcheri* that have merely acute, not mucronate, leaf lobes. However, *L. hatcheri* is almost constantly freely gemmiparous, while *L. barbata* is constantly gemmae-free (at least in our region).[48]

Under nutritionally difficult conditions, particularly xeric sites, the species produces small phases, deep brown in color, often only 1–1.2 mm wide; these plants show a high incidence of 2- or 3-lobed leaves. They have been dignified with the name "fo. *biloba*" by Schiffner (Lotos 59:20, 1911). Such plants could be mistaken for species of subg. *Lophozia*, were it not for the fact that more typical plants bearing some 4-lobed leaves always occur admixed.

Subgenus *LEIOCOLEA* K. Müll.
[Fig. 163]

Lophozia subg. *Leiocolea* K. Müll., Rabenh. Krypt.–Fl. 6(1):711, 1910; Macvicar, Studs. Hdb. Brit. Hep. ed. 2, p. 166, 1926; Schuster, Amer. Midl. Nat. 45(1):45, 1951; Kitagawa, Jour. Hattori Bot. Lab. no. 28:271, 1965.

[48] Müller (1910, p. 658, *in* 1905–16) describes gemmae as reddish yellow, 3–6-angular, almost stellate, 1–2-celled, and spherical in outline. I have never seen gemmae. Macvicar (1926, p. 202) also describes the gemmae, his description being almost a translation of that of Müller; furthermore, he gives a figure of a 2-celled gemma.

Leiocolea Buch, Mem. Soc. F. et Fl. Fennica 8:288, 1933; K. Müller, Rabenh. Krypt.–Fl. ed. 3, 6:688, 1954.
Lophozia subg. *Hattoriella* Inoue, Bot. Mag. Tokyo 70:360, 1957.
Hattoriella Inoue, Jour. Hattori Bot. Lab. no. 23:39, 1960.

Small (0.5–1mm wide × 8–15 mm long) to very robust (3–5 mm wide × 4–8 cm long), green- to brownish-pigmented, more rarely with reddish or dull purplish pigmentation; often ± fragrant. Stems, unless crowded, growing *nearly horizontally*, little ascending (except gemmiparous shoots of *L. heterocolpa*), but shoot apices with distinct dorsal autonomous curvature; branches few or none, normally *lateral-intercalary*, occasionally (*L. heterocolpa* at least) terminal. Stems 7–16 cells high, with *medulla entirely homogeneous*, but with the lower 1–3 layers of medullary cells often somewhat smaller than dorsal medullary cells, *not forming a ventral mycorrhizal band*; cortical cells able to develop only ± weakly thickened walls (*usually in 1 cell layer*), somewhat to scarcely smaller than the medullary cells. Rhizoids usually forming a *dense postical mat*. Ventral stem sectors broad, *mostly (4)6–12 rows of cells broad on sterile shoots, and able to develop discrete underleaves;* in a few species (*L. badensis, L. turbinata*) ventral stem sectors are only about 2–3 cells broad and unable to develop underleaves. Leaves *uniformly bilobed, very obliquely inserted* (appearing nearly horizontal laterally, scarcely concave to nearly flat, becoming dorsally convex in *L. rutheana*); *postical leaf margins unable to develop basal cilia* (except in *L. rutheana*). *Leaf cells large*, mostly 30–50 μ in leaf middle, normally strongly collenchymatous and with bulging trigones, but sometimes thin-walled and lacking trigones, generally very distinctly *papillose to striolate-papillose*; oil-bodies grayish, *very large, mostly 4–6 × 7–12 μ or larger, mostly only 2–5(6–10) per cell*, appearing somewhat opaque and formed of numerous very minute, scarcely distinct oil-globules that do not protrude (occasionally also with 1-several larger included oil-globules); chloroplasts averaging much smaller than the oil-bodies. *Gemmae absent* (except in *L. heterocolpa*, where brownish to reddish, 2-celled, ovoid and smooth, at tips of slender, erect, subisophyllous shoots with reduced leaves).

Dioecious or paroecious. Perianths slender, long, longly exserted, *tubular and terete, virtually nonplicate, swiftly constricted* near apex into a *small, narrow beak;* mouth crenulate to ciliolate. Perichaetial bracts larger than leaves, but generally *bilobed like leaves*. Sporophyte with seta of numerous rows of cells (except in *L. badensis*); capsule wall 2–3- (occasionally 3–4-) stratose, the inner wall with annular, the outer with radial (nodular), thickenings. ♂ Bracts with or without paraphyses. Antheridial stalks mostly biseriate;[49] wall of large (ca. 24–26 × 24–28 to 20 × 26–30 μ) polyhedral to quadrate cells.

Type. No type was originally proposed; I suggest *L. rutheana* as lectotype.

Leiocolea includes a limited number (about eight) of calcicolous species, purely of holarctic distribution. Except for three species reported from

---

[49] In at least *L. gillmani* the antheridial stalk varies from uniseriate (with 1–2 cells divided, thus locally biseriate), to completely biseriate, to triseriate near the distal end and biseriate near the base. Müller (1948a, p. 14), however, states "bei allen *Leiocolea*-Arten zweizellreihig."

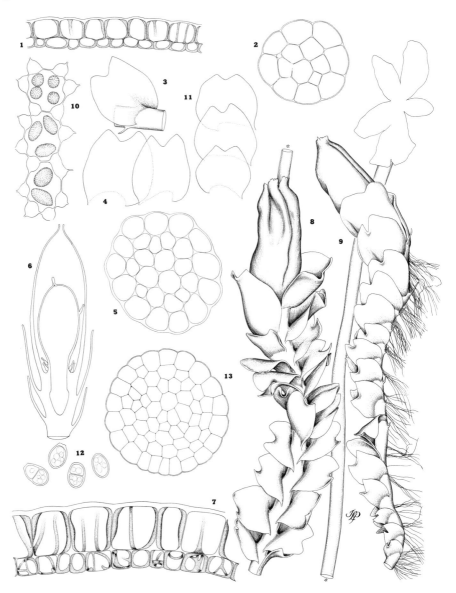

Fig. 163. *Lophozia* subg. *Leiocolea;* intrasubgeneric variation. (1–2) *L. badensis:* 1, transverse section through capsule wall ($\times300$); 2, seta cross section ($\times142$). (3–9) *L. heterocolpa* var. *heterocolpa:* 3, leaf and underleaf, *in situ* ($\times27$); 4, leaves ($\times27$); 5, seta cross section ($\times127$); 6, longisection through perianth-bearing shoot apex ($\times18$); 7, capsule wall cross section ($\times445$); 8–9, dorsal and lateral aspects of gynoecial plants, in fig. 8 with extruded sporophyte showing irregular dehiscence ($\times13.5$). (10–12) *L. heterocolpa* var. *harpanthoides:* 10, median cells with oil-bodies ($\times505$); 11, leaves ($\times27$); 12, gemmae ($\times415$). (13) *L. gillmani;* seta cross section ($\times127$). [1–2, *RMS 10987*, Letchworth State Park, Genesee Co., N.Y.; 3–9, *RMS & KD 66-311*, Blaesedalen, Disko I., Greenland; 10–12, *RMS 35792a*, Mt. Pullen, Alert, Ellesmere I.; 13, *RMS 67989*, Bear River, Conway, Mass.]

Japan and the chiefly Mediterranean *L. turbinata*, all other known taxa occur in our region.

Hattoriella Inoue (1961) includes two Japanese taxa, *H. diversiloba* and *H. mayebarae*. The latter species appears to be a synonym of *L. badensis*; the former differs from other species in *Leiocolea* in the rather more plicate distal portions of the perianth and the more elaborated and spinose-dentate ♂ bracts. It is possible (as indicated in the key, pp. 258–260) that *Hattoriella* can be retained as a distinct subgenus for only the generitype, *L. diversiloba* Hatt., which is fully illustrated by Inoue (1957). However, Kitagawa (1962b, 1965) is probably correct in simply placing *Hattoriella* into the synonymy of *Leiocolea*. A third Japanese species referred to *Leiocolea* (by Inoue, 1962a, p. 190) is *Lophozia igiana* Hatt., Jour. Jap. Bot. 31:201, fig. 67, 1956.

The subgenus as restricted above includes a closely allied group of species (Schiffner, 1904, p. 381), agreeing in the nearly horizontally inserted leaves with rather large, strongly collenchymatous, papillose cells, provided with peculiarly large oil-bodies. Stem structure is uniform throughout the group, and the perianth is extremely characteristic (never recurring in such a form elsewhere in the genus).[50] The group includes only one really divergent species group (sectio Heterocolpae), which approaches the ancestral form in the smaller leaf cells, the less horizontal leaves, and the somewhat ascending growth, as well as the ability to produce gemmae. The other somewhat isolated species is *L. badensis*, a small species with the seta reduced to a finite number of cell rows and with reduction of ventral stem sectors and suppression of underleaves. It is also unique in the thin-walled cells (with small trigones) whose cuticle is rarely distinctly papillose. Nevertheless, it would be absurd to consider such an obviously reduced, small form as separate from *Leiocolea*.[51]

It is a moot question, answerable only on a personal basis, whether to consider *Leiocolea* a subgenus of *Lophozia* or to separate it generically. If retained as a subgenus, it is certainly a sharply circumscribed and very isolated subgeneric group in *Lophozia*. According to Kitagawa (1965), *Hattoriella* (Inoue) Inoue is a simple synonym of *Leiocolea*, although the generitype supposedly departs somewhat from *Leiocolea* (in, e.g., having paraphyses in the axils of ♂ bracts and in the somewhat plicate perianth apex).[52]

[50] Deviations in perianth form occur in both *Lophozia gillmani* (see p. 379; see also Schuster et al., 1959), in which the perianth may be lobulate-incised at the mouth and nonrostrate, and in *L. heterocolpa* (p. 408), in which the perianths may become strongly laterally compressed and nonrostrate; in some phases of *L. badensis* (e.g., *RMS 68-1628*, Newfoundland) the perianth may be strongly plicate in the distal 0.2–0.4.

[51] Analogously, I do not consider the small, reduced "*Sphenolobus*" *hellerianus* as generically separated from "*Sphenolobus*" *minutus*. In both cases obvious parallel reduction in size has taken place, correlated with reduction in cell rows of the gametophyte and sporophyte axes.

[52] *Lophozia rutheana*, furthermore, typically has lanceolate, polymorphous paraphyses associated with ♂ bracts; these may occasionally be very crowded and conspicuous. Thus the presence of paraphyses in *Hattoriella* serves to unite it to, rather than separate it from, *Leiocolea*.

I have, ultimately, retained *Leiocolea* as a subgenus of *Lophozia* because most of the criteria used by Müller (1951–58, p. 688) to segregate *Leiocolea* recur in *Lophozia* subg. *Massula* and *Lophozia*: e.g., the obliquely inserted leaves recur in *L. obtusa* and *L. alpestris polaris;* distinct underleaves occur in the antipodal subg. *Protolophozia* Schust. (p. 255) and sporadically in the Greenland *L. heteromorpha,* but are lacking in two of the *Leiocolea* species; the "in geringer Zahl vorhandenen aber verhältnismässig grossen Ölkorper" tend to recur in *Lophozia ventricosa* var. *confusa* and *L. wenzelii* (pp. 584, 598); the tooth at the antical base of ♂ bracts recurs, as Müller himself states (p. 647), in sectio Excisae of subg. *Lophozia.* I find one other, but inconstant, distinction: *Leiocolea* normally has only lateral-intercalary branches; *Lophozia s. str.* normally has terminal branches.

Taxonomy

In spite of the limited number of known species, *Leiocolea* is, in northern regions, a baffling and difficult group. Determination of sexuality is often difficult since ♂ bracts are often hardly morphologically distinct; furthermore, as is noted under both *L. gillmani* and *L. rutheana*, isolated plants of these species develop perianths without evidently having first developed antheridia. In many cases, as discussed under *L. gillmani*, correct determination of sex involves finding either (*a*) plants with young gynoecia at shoot tips, with antheridia still preserved, or (*b*) subfloral innovations from beneath unfertilized perianths, on which ♂ bracts can be found.

Often *L. rutheana* (paroecious) is hardly separable from *L. bantriensis* (dioecious), and I have had to discard all but one or two reports of the latter from our region. *Lophozia gillmani* forms a distressingly difficult complex, and we find (typical form) an approach to *L. collaris;* sexuality overlooked, plants may be searched for under that species. On the other hand, *L. gillmani* fo. *orbiculata*, because of its poorly defined ♂ bracts, may be regarded as dioecious, and sought—on the basis of small size and broadly orbicular leaves with narrowed apices—under *L. badensis!*[53]

Lophozia collaris itself may be confused with *L. heterocolpa;* forms of the latter, which very rarely develop gemmiparous, erect shoots, occur under rock shelves and in very moist places and are almost impossible to separate from *L. collaris.* On the other hand, small forms of *L. collaris*, when sterile, may be difficult to separate from *L. badensis.*

It is thus clear that in many cases fragmentary specimens can hardly be definitely named. Fortunately, the complexities cited above do not always apply in temperate and subarctic regions; difficulties are encountered principally with arctic phenotypes. It should be especially noted that considerable

---

[53] To document the considerable intraspecific variability within the species of *Leiocolea*, I have devoted a special study to *Lophozia gillmani*, which is perhaps the most polymorphous and difficult of the species. There is not sufficient space to document in equal detail variation in other taxa.

variation in cell size occurs within each species, caused at least in part by differences in exposure and nutritional conditions. Also, variations in pigmentation patterns within the several species are hardly understood. For example, *L. rutheana* may be brownish-pigmented in one place, largely purplish in another. Under one set of conditions, *L. heterocolpa* var. *harpanthoides* may be chestnut brown, yet have reddish shoot tips in another site.

Key to Species of Leiocolea

1. Paroecious; gemmae absent; underleaves present. Plants usually medium-sized or large, (1.2)1.8–5 mm wide × 2–8 cm long; cells of the leaf middle mostly (28)30–36 × 35–48 μ or larger, of the leaf apices 25–30 μ or larger, with ± bulging trigones . . . . .2.
   2. Leaves dorsally convex, widest basally and wider (to 2–3 mm) than long, ventrally more or less decurrent; underleaves and bracteoles extremely large, freely ciliate (some more than half as long as leaves); plants 3–5 mm wide × 4–8 cm long; perianth short-beaked . . . .
      . . . . *L. rutheana*[54] (p. 366)
   2. Leaves flat or dorsally concave, widest in middle and orbicular or ovate, as wide or longer than wide (at most 1.8 mm wide), ventrally not decurrent; underleaves and ♀ bracteole small, inconspicuous; plants (1.2)1.6–2.8 mm wide × 1.5–4 cm long; perianth long-beaked . . . .
      . . . . *L. gillmani*[55] (p. 373)
1. Dioecious (some taxa with gemmae; some taxa lacking underleaves) . . 3.
   3. Underleaves present; cuticle distinctly papillose; leaves with generally distinct to bulging trigones; plants larger, more robust, 1–4 mm wide × 0.8–8 cm long . . . . 4.
      4. Gemmae present at the leaf apices (generally on more or less erect shoots with erect or appressed, often reduced leaves); leaf cells with large, bulging trigones (often yellowish), 25–30 μ in leaf middle or smaller; plants medium-sized, 1–2 mm wide × 8–25 mm long; usually sterile . . . . 5.
         5. Leaf sinuses acute to rectangular, ¼–⅓ the leaf length; leaf lobes ± acute; gemmae common, brownish . . . .
            . . . . *L. heterocolpa* var. *heterocolpa* (p. 398)
         5. Leaf sinuses crescentic or lunate, only ⅙–¼ the leaf length; leaf lobes blunt to rounded; gemmae rare, brown to reddish . . . .
            . . . . *L. heterocolpa* var. *harpanthoides* (p. 405)
      4. Gemmae constantly absent: plants (unless crowded) ± horizontal in growth, with mostly nearly horizontal, spreading leaves; leaf cells

[54] Frye and Clark (1937–47, p. 380) state that *Lophozia rutheana* is separated from *L. harpanthoides* by the "thin and colorless" walls of the leaf cells. However, in much material the very large, bulging trigones of *L. rutheana* are a deep golden brown; the species is often scorched in appearance.

[55] Frye and Clark (*loc. cit.*, p. 380) attempt to separate *Lophozia gillmani* from *L. collaris* because its trigones are supposedly "small" (compared to "medium large to bulging" for *L. collaris*). There is no separation between the two species in this regard, and *gillmani* often has bulging trigones. With the criteria in the above key, *L. schultzii* var. *laxa* Schiffn., described from England, would fall under *L. gillmani* and appears to be a marl-bog form of the latter species. I believe the variety to be better placed as *L. gillmani* var. *laxa* (Schiffn.) Schust. Macvicar (1926) illustrates and describes this anomalous variant.

with small or moderate (rarely strongly bulging) trigones, 25–30 ×
30–40 μ or larger in leaf middle; cortex of stem with thin hyaline
walls; leaves not or scarcely concave; plants often larger, 1–4 mm
wide × 1–8 cm long . . . . 6.

    6. Cells in leaf apices 30–35 μ, in leaf middle 35–40 μ wide;
    plants large, 3–4 mm wide × 2–8 cm long; leaves usually more
    or less dorsally convex; spores 12–15 μ; ♀ bracts with lobes
    entire . . . . *L. bantriensis* (p. 387)
    6. Cells in leaf apices ca. 25 μ, in leaf middle 25–30 μ wide;
    plants smaller, 1–2.8 mm wide × 1–4 cm long; leaves not
    convex; spores 10–12 μ; ♀ bracts usually with lobes denti-
    culate . . . . *L. collaris* (p. 392)

3. Underleaves lacking or mere stalked slime papillae; cuticle smooth or
nearly so; cells thin-walled, with slight or no trigones; plants small,
mostly 3–8 mm long, 0.6–1 mm wide; cortical stem cells large (23–30 ×
48–90 μ or larger), thin-walled, hyaline, 2–4(5)× as long as wide; cells
(24)28–35 μ in lobes, 32–38 μ wide in leaf middle or larger; gemmae
absent . . . . *L. badensis* (p. 408)

LOPHOZIA (LEIOCOLEA) RUTHEANA (Limpr.)
Howe

[Fig. 164]

Jungermannia schultzii Nees, Naturg. Eur. Leberm. 2:30, 1836 (not of Spreng., Pugil. 1:64, 1813).
Jungermannia rutheana Limpr., Jahresb. Schles. Gesell. Vaterl. Kult. 61:207 [reprint, p. 4],
    1884.
Jungermannia lophocoleoides Lindb., Meddel. Soc. F. et Fl. Fennica 14:66, 1887.
Lophozia rutheana Howe, Bull. N. Y. Bot. Gard. 2:102, 1901; Stephani, Spec. Hep. 2:132, 1901.
Lophozia schultzii Schiffn., Verh. Zool.–Bot. Gesell. Wien 54:397, 1904.
Lophozia (subg. *Leiocolea*) *schultzii* K. Müll., Rabenh. Krypt.–Fl. 6(1):713, fig. 325, 1910;
    Schuster, Amer. Midl. Nat. 49(2):333, pl. 7, 1953.
Leiocolea schultzii Joerg., Bergens Mus. Skrifter 16:161, 1934.
Leiocolea rutheana K. Müll., *in* Gams, Kl. Kryptogamen Fl. 1:40, 1940; Frye & Clark, Univ.
    Wash. Publ. Biol. 6(3):391, figs. 1–10, (1944) 1945.
Lophozia (subg. *Leiocolea*) *rutheana* Schust., Amer. Midl. Nat. 42(3):567, 1949; Schuster, *ibid.*
    45(1):47, pl. 5, 1951.

Usually scattered among mosses, sometimes in large spongy tufts, depressed
(or when among caespitose mosses, ascending), *robust, 3–4(5) mm wide × (2)4–8
cm long, aromatic*, light to deep green, with age often *reddish brown or fuscous.*[56]
Stems 300–400 μ in diam., 13–16 cells high, decumbent to suberect, simple or
sparingly branched (branches lateral-intercalary), green (the underside often
brownish); cortical cells scarcely to somewhat thick-walled, 15–18 μ in diam.,
smaller than medullary cells, which are quite leptodermous and 25–32(36) μ
in diam. Rhizoids numerous, long, pale brownish, forming a rather dense mat

[56] Plants exposed to full sunlight (e.g., in insolated wet tundra) may be blackish green to
black or purplish brown with the upper leaves (and perianths) decolorate and bleached marginally and distally.

obscuring in part the underleaves. Leaves somewhat to quite imbricate, widely
spreading, generally *distinctly adaxially convex*, very obliquely inserted and
horizontally spreading, obliquely semiamplexicaul, *broadly* ovate-reniform
to transversely rectangulate-ovate on mature stems, averaging *wider than long*
[ca. 2700–2850(3200) μ wide $\times$ 1800–1900(2200) μ long; ca. *1.4–1.55$\times$ as
wide as long* on robust shoots], somewhat decurrent antically, the *strongly ampliate
and arched postical base shortly but distinctly decurrent;* ventral base rarely with 1–3
small and short cilia; apex uniformly bilobed for $\frac{1}{5}$–$\frac{1}{4}$ the leaf length, the lobes
broadly triangular but apiculate or subacute, rarely blunt, unequal, the dorsal
smaller, the sinus crescentic to obtuse, rarely rectangular, *usually gibbous and
recurved*. *Cells very large*, those of margin distally 32–36 μ, of lobes and apex
33–36 $\times$ 35–45 μ, of leaf middle (30)35–40 $\times$ (35)42–48 μ, from roundish-
polygonal distally to subrectangulate basally, *strongly collenchymatous*, the prominent, bulging trigones separated by sharply defined, thin walls; cell walls
colorless to brownish; cuticle prominently, coarsely *striate-verruculose;* oil-bodies
generally large, ovoid to ellipsoidal, 2–5(6–7) per cell, from ca. 5–6 $\times$ 7–10 μ
to 6–8 $\times$ 10–15 μ, rarely to 9 $\times$ 17 μ, much larger than chloroplasts, finely
granulose, grayish, with age sometimes with larger included globules. *Underleaves very large*, generally 1200–1700 μ long, *usually 2–3-fid and the chief lobes
pluriciliate*. *Asexual reproduction lacking.*

Paroecious. ♂ *Bracts usually in 1–2 pairs, very similar to leaves, scarcely more concave
at base* and often lacking tooth of antical base (thus not readily recognized after
decay of antheridia); antheridia 1–3, *often among lanceolate to linear paraphyses;*
one or rarely both perichaetial bracts serving also as perigonal bracts. Antheridial stalk 2-seriate; body of large cells (each ca. 24–28 $\times$ 26–30 μ). ♀ Bracts
like leaves except for the $\pm$ sheathing and concave base, scarcely or no larger,
somewhat less obliquely inserted and oriented, the antical (and more often also
postical) margin sometimes with 1–2(3–4) small cilia at base, the sinus frequently reflexed, as is often the antical and postical margin. Perianth long-
exserted at maturity, ca. 1.5–1.6 $\times$ 5–6 mm, *longly cylindrical* or weakly clavate,
smooth or with 1–2 shallow sulci, suddenly contracted distally into a *very
short beak;* mouth fringed with short teeth formed of 1–2 elongated cells.
Capsule with wall 3–4-stratose. Elaters 8–10 μ thick $\times$ 50–100 μ long, with 2
closely coiled reddish spirals; spores (12)16–20 μ, pale brown.

Type. Germany.[57]

Distribution

A localized and rare species of imperfectly circumpolar distribution.
It is essentially arctic and high subarctic in range but is lacking in the high
Arctic. In Europe sporadically distributed, more widespread in the
north (Norway, Sweden, Lapland, Finland), southward to Denmark, the

[57] Although the name *Lophozia rutheana* is selected by Frye & Clark (1937–47, p. 391),
they cite the type collection as that of Schultz, from Neubrandenburg. However, the type
collection must be that of R. Ruthe (1872), issued in Gottsche & Rabenhorst, Hep. Eur. Exsic.
no. 583.

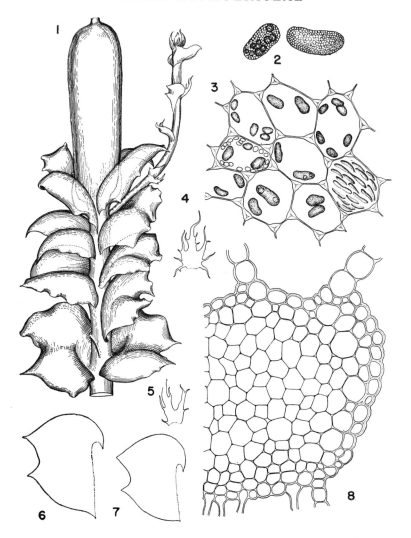

Fig. 164. *Lophozia* (*Leiocolea*) *rutheana*. (1) Perianth-bearing shoot with innovation (×7.5). (2) Oil-bodies (×940). (3) Median leaf cells with oil-bodies, cell at left with chloroplasts, at right with cuticular papillae drawn in (×300). (4–5) Underleaves (×8). (6–7) Leaves (×7.6). (8) Stem cross section (×75). [All from Grand Portage, Minn., *RMS*.]

northern portions of Germany (Westphalia; Schleswig-Holstein; Neu-brandenburg; Mecklenburg), England (Norfolk), eastward into Asia (Siberia: common on the Yenisey R. near 70° N.; surely elsewhere). Also in Iceland. In North America the distribution poorly known, occurring from the arctic shore of Alaska (Persson, 1952; Persson and Weber,

1958) to the Yukon (Evans, 1903; Persson, 1952), eastward to Alberta (Steere, 1942a) and New York.

E. GREENLAND. Mt. Zackenberg, Wollaston Foreland, 74°28′ N., 20°35′ W. (*Holmen, July 26, 1950*!; with *Cephaloziella arctica* var. *alpina, Odontoschisma macounii, Anastrophyllum minutum grandis*); Clavering Ø, Dødemandsbugten, 74°07′ N., 21° W. (*Holmen 4496*!; with *Tritomaria quinquedentata, Blepharostoma trichophyllum brevirete*); Cape Hedlund, Lyell Land, 72°50′ N., 26° W. (*Th. Sorensen, 243a*!; with *Cephalozia pleniceps*); Maria Ø, 72°58′ N., 24°50 ′W. (*Holmen 19.213*!; with *Lophozia badensis, L. pellucida*; clearly paroecious); Geographical Society I., 72°44′ N., 22° 30′W. (*Holmen 18.101, Aug. 26, 1958*; with *Lophozia kunzeana, Gymnocolea inflata, Blepharostoma trichophyllum brevirete*); Charcots Land, 71°54′ N., 29° W., 700 m, bog at lake (*Holmen 19.343*!); Gurreholm, Jameson Land, 71°14′ N., 24° 30′ W. (*Holmen 18.309*!). W. GREENLAND. Fortune Bay, Disko I., 69°16′ N., 53°50′ W. (*Steere & Holmen 62-030*! *c. per.*); Tupilak I., Egedesminde, 68°42′ N., 52°55′ W. (*Holmen 13.012*!); Magdlak, Alfred Wegeners Halvö, 71°07′ N., 51°48′ W. (*Holmen 16.841*); Simiutap Kûa, Svartenhuk Pen., 71°55′ N., 54°32′ W. (*Holmen 13.688; RMS & KD 66-917, 66-916b, 66-908, 66-914a*); Esersiutilik, head of Kangerdluarssuk Fjord, 71°17′ N., 51°32′ W. (*Holmen 13.834*); Umanak, 70°41′ N., 52°10′ W. (*Holmen 13.618*); Nugssuaq Pen.: Avfarassuaq Dal, W. of Taserssuaq, 70°25′ N., 52°26′ W. (*Holmen 13.429*!); Marrait, 70°30′ N., 54°12′ W (*Holmen, Sept. 3, 1956*!); Marrait Kangigdlit, calcareous slopes, below Anana Mamaglissa (*RMS & KD 66-727*).

MANITOBA. RCAF Base at Churchill, shore of Hudson Bay (*RMS 35004, 35009*). QUEBEC. St-Fabien, côté de Rimouski; near L. Carré, en face de l'Islet au Flacon (Lepage, 1944–45); also northern Quebec (Buch & Tuomikoski, 1955). NEWFOUNDLAND. Eddie's Cove West, Bartlett's R., and between Cow Head and St. Paul's Inlet, all N. Pen. (Buch & Tuomikoski, 1955); River of Ponds Prov. Pk., N. Pen. (*RMS 68-1561*). ONTARIO. Red Bay, Bruce Pen. (Williams, 1968); Winisk, Hudson Bay (*Sjörs*).

MINNESOTA. Grand Portage bog, Cook Co. (*RMS 11270, 11480, 11478, 11905*); Grand Marais, Cook Co.; Gooseberry R., Lake Co. (Conklin, 1942; Schuster, 1953). MICHIGAN. Eagle Harbor, Keweenaw Co.; Thuja Swamp near Munising, Alger Co.; Cecil Bay, L. Michigan, Emmet Co. (Steere, 1937, 1942a, etc.). NEW YORK. Bergen Swamp, W. of Rochester, Genesee Co. (Schuster, 1949a), at 43° N. (southernmost report of species!).

The report of the species from a marl bog at Bergen Swamp, N.Y. (Schuster, 1949a) is remarkable; it occurs here with other northern species, e.g., *Cephaloziella spinigera*.

Ecology

This magnificent species is largely or entirely confined to "wet rich fens on more or less calcareous ground" and to calcareous *Thuja* bogs and swamps, where it sometimes occurs directly on relatively exposed marly soil, sometimes creeps among *Drepanocladus* and the more robust allied moss *Scorpidium scorpioides*, and sometimes occurs in *Thuja*-shaded ground. Under the latter conditions (in Minnesota and New York) with *Moerckia hibernica, Riccardia multifida, Lophozia grandiretis*, and *Chiloscyphus pallescens*, as well as (when in sun) with *Preissia quadrata*. There also "seems to be a very general correlation between its occurrence and that of the calciphile *Potentilla fruticosa*" (Schuster, 1953).

Farther northward, as around Hudson Bay, it occurs in wet peaty Tundra overlying calcareous rocks, associated with *Odontoschisma macounii*, *Lophozia grandiretis*, *Scapania degenii*, *Aneura pinguis*, and other lime-tolerating bryophytes. With a wide toleration for extremes in insolation, occurring both in deep shade under *Thuja* and in open sun. In the latter instance it is often quite brownish-tinged or purplish brown, in extreme cases scorched in appearance. However, the species seems to always occur in more or less hygric sites. It rarely forms extensive mats, more often creeping as isolated stems among other mosses. In the Tundra it is usually found around wet holes or in depressions fringing standing water. Buch and Tuomikoski (1955) also found it with *Scapania degenii* and such mosses as *Campylium stellatum*, *Limprichtia revolvens*, *Tomenthypnum nitens*, *Loeskehypnum wickesiae*, *Fissidens adiantoides*, *Paludella squarrosa*, and *Sphagnum warnstorfianum*.

In west Greenland the species may often occur in enormous quantity in rich calcarous ground around tarns and pools, especially where fertilization by water fowl occurs; here it may occur (e.g., in *RMS & KD 66–917*, Umiarfik Fjord) with *Cryptocolea*, *Blepharostoma*, *Tritomaria scitula*, *Lophozia grandiretis*, *L. quadriloba*, and *Scapania gymnostomophila*, as well as with *L. gillmani* and *L. heterocolpa* var. *harpanthoides*. On poorly drained calcareous slopes where *Eriophorum scheuchzeri* occurs, it may be found (as at Marrait, Nugssuaq Peninsula) with *L. gillmani*, *Scapania gymnostomophila*, and the very rare *Metacalypogeia schusterana*.

Differentiation

Diagnostic, in general, for *L. rutheana* are the vigorous size; the often fuscous to scorched color, olive-green in shaded sites; very broadly inserted leaves, somewhat decurrent antically, widest toward base, laterally patent, and with decurved sides and usually a somewhat gibbous sinus; large, bifid, and ciliate underleaves; strongly collenchymatous cells and rough cuticle; consistently and conspicuously imbricate and very soft and lax leaves. The leaves, in particular, have the dorsal half very convex, with at least the distal parts of the antical margin strongly decurved, much as is normal in *Plagiochila*. Occasional leaves—the frequency varies from collection to collection—bear 1–several small, short cilia or stalked slime papillae at the ventral base. Such cilia, if found, will easily separate the species from *L. bantriensis*, in which they appear never to be formed.

Branching in *L. rutheana* is very sparing, unless the shoot tip is destroyed. All branches seen have been lateral-intercalary, from near the ventral end of the leaf axil.

In extreme cases sterile *L. rutheana* cannot be easily separated from *L. bantriensis*. However, the leaves in *L. rutheana* are typically at least 1.2 × (often to 1.35–1.4 × on vigorous forms) as broad as long; on mature shoots they have both margins strongly convex, with the antical margins, above a narrowed base, strongly convex. *Lophozia bantriensis* has much less broad and much less convex leaf margins, without a conspicuous dilation along the postical base (and without trace of postical cilia or stalked slime papillae). The differences in cell size occasionally emphasized (e.g., in Müller, 1951–58) can hardly be relied upon, the cell size in *L. rutheana* apparently being rather variable.

In spite of the paroecious inflorescences, this species very rarely produces sporophytes. In the abundant specimen from Marrait (*Holmen, Sept. 3, 1956*) numerous, often incised and malformed perianths were present, and each plant bore at least 1–3 antheridia—yet not a single case was seen where a young sporophyte was formed.

This large, handsome species undergoes only slight variation. "The large size . . . , the decidedly aromatic odor when fresh, and the broad, bilobed leaves with a distinctly decurrent postical base separate it in the field" (Schuster, 1949a). Robust extremes occasionally bear 1–several small cilia of postical leaf bases; these are consistently lacking in most of our forms, except on isolated subinvolucral leaves. The large size, with robust plants attaining a width of 4–5 mm and a length of 3–6 or even 8 cm, separates it from related species. "Confusion is most likely with *L. gillmani*, which barely approaches it in size . . . ." It differs from this and all other species of subg. *Leiocolea* in the

very large, multifid and ciliate underleaves . . . , some of which are more than half the length of the leaves. The leaves in this species are also characteristically dorsally convex and widest near base, and much wider than long on robust shoots . . . ; they may have one or several small cilia near the postical base (reminiscent of the allied subg. *Barbilophozia*). The somewhat convex leaves, with the dorsal leaf margins somewhat reflexed, are highly characteristic (Schuster, 1953).[58]

As in *L. gillmani*, "some plants appear to be devoid of antheridia, but androecia never occur on separate branches or plants" (Schuster, *loc. cit.*). In such cases, confusion with *L. bantriensis* is likely to occur. This latter species has equally large cells, may also have convex leaves, and may attain a width of 2.5–4 mm. It differs not only in the dioecious inflorescences but also in the narrower leaves, the smaller and less conspicuous underleaves, and the generally less strongly developed trigones of the cells.

[58] Although well-developed plants of *Lophozia rutheana* normally bear leaves that are much broader than long, less robust phases occur in which the leaves are merely ovate-quadrate. Such phases may also have the underleaves somewhat smaller and less copiously ciliate. I have found such plants at Churchill, Manitoba (*RMS* 35014); it is possible that "*L. schultzii* var. *laxa* Schiffn." represents such a plant, although this has been transferred (Schuster, 1951a) to *L. gillmani*. In the final analysis, the larger underleaves, convex and wavy leaves, and obsoletely beaked perianth appear to be the chief features differentiating *L. rutheana* from *L. gillmani*.

Lindberg gave this species the appropriate name of "*lophocoleoides*," in reference to the marked similarity in aspect to large species of *Lophocolea*. Both color,
the large and ciliate underleaves, and shape of the convex leaves may suggest
large phases of *Lophocolea cuspidata*, when sterile material only is at hand. The
scattered rhizoids and salient trigones, as well as markedly verruculose cuticle,
at once prohibit further confusion of *L. rutheana* with any species of *Lophocolea*.
The peculiar arctic *Mesoptychia sahlbergii* may also be mistaken for *L. rutheana*,
when sterile; indeed, the facies of the two may be quite similar. However,
M. sahlbergii tends to develop purplish underleaves and has more regularly bifid
underleaves with more regularly ciliate margins, and even more shallowly
bilobed leaves.

When mature perianths are present (as in *Holmen & Steere 62–030*), they are
firm and may be polystratose to at least the middle; toward the base they are
4–5-stratose and quite rigid, with the wall to 120–140 μ thick. A weak shoot-
calyptra is developed, the sterile archegonia being inserted on at least the basal
one-fourth of the "calyptra."

Greenland specimens are always at least somewhat brownish-tinted, often
scorched, with fuscous cell walls. Even plants from the southern edge of the
range of the species (see Schuster, 1949a) tend to show fuscous pigmentation
when growing in sun. This helps to separate the species from *L. bantriensis*,
which may be very similar but is usually a warm brown in color, much like
some forms of the much smaller *L. heterocolpa*. Frye and Clark (1937–47, p. 380)
attempt to separate *L. rutheana* from *L. harpanthoides* on the basis of cell walls that
are "colorless." Obviously this distinction is nonexistent. In addition to the
brownish pigmentation (giving the trigones a yellow-brown or yellow color by
transmitted light), extreme sun forms, such as the Cape Hedlund specimen,
from a *Carex rariflora-Eriophorum* bog, may develop intense vinaceous pigmentation of the basal parts of the leaves and of the underleaves, which may in extreme
cases suffuse almost the whole plant, more or less masking the brown color.
Such deeply pigmented forms, when dry or when seen *en masse*, look almost
black. The postical leaf bases of such plants may be exceedingly vividly pigmented, with trigones and middle lamellae a particularly deep purplish red.

Often the paroecious inflorescence is difficult to determine. In a few cases
it appears that individual plants which are only female may be produced, the
antheridia being suppressed (as in the Lyell Land specimen).

Occasionally a single population may display much variation in degree of
suppression of antheridium formation. In *Holmen 13.688*, e.g., one plant showed
a single antheridium only, near the antical base of a leaf axil, with an associated
laciniiform paraphyse; in another plant 1–3 antheridia occurred—often some
distance from the leaf axil and displaced almost to the dorsal stem midline—
associated with each of 3–6 essentially unmodified or feebly pouched, leaflike
bracts, whereas the uppermost pair of bracts (perichaetial) and 1–3 pairs below
them lacked all trace of antheridia. It is common to find the ♂ bracts exactly
like leaves, without trace of an antical pouch; then, on plants with mature
perianths, determination of sexuality is difficult. On the other hand, some phases
show ventricose bract bases (Schuster, 1953), even the ♀ bracts bearing antheridia in their axils.

LOPHOZIA (LEIOCOLEA) GILLMANI (Aust.) Schust.

[Figs. 163:13, 165–167]

Jungermannia gillmani Aust., Bull. Torrey Bot. Club 3:12, 1872.
Jungermannia kaurini Limpr., Jahresb. Schles. Gesell. Vaterl. Kult. 61:204, 1884.
Jungermannia muelleri fo. paroica Bernet, Cat. Hep. Suisse, p. 68, 1888.
Lophozia kaurini Steph., Spec. Hep. 2:130, 1901; Lorenz, The Bryologist 14:25, p.l 4, 1911.
Lophozia (subg. Leiocolea) kaurini K. Müll., Rabenh. Krypt.–Fl. 6(1):716, fig. 326, 1910;
    Macvicar, Studs. Hdb. Brit. Hep. ed. 2:174, figs. 1–5, 1926.
Leiocolea kaurini Joerg., Bergens Mus. Skrifter 16:161, 1934.
Leiocolea gillmani Evans, The Bryologist 38:83, 1935; K. Müller, Rabenh. Krypt.–Fl. ed.
    3,6(1):692, fig. 217, 1954 (? in part).
Lophozia (subg. Leiocolea) gillmani Schust., Amer. Midl. Nat. 45(1):47, pl. 6, figs. 5–12, 1951;
    Schuster, ibid. 49(2):335, pl. 8, figs. 5–12, 1953.

Fragrant, in low, prostrate patches or mats, rarely singly and scattered, pale green and rather pellucid (moist, deeply shaded localities) to deep green and rather opaque, occasionally somewhat blackish green or olive-fuscous or (very rarely) dull purplish on tips of shoots (insolated, wet sites), *medium-sized;* shoots *(1.2)1.8–2.8(3)* mm wide × 2–4 cm long, *fertile shoots sometimes to 4–5 mm wide,* creeping but with ascending tips, when crowded rarely ascending in growth, usually simple or sparingly branched below; often with gynoecial innovations. Stems ca. 300–400 μ in diam.; dorsal cortical cells 20–25(29) × (75)90–120 (140–185) μ. Rhizoids numerous, *forming a dense postical mat,* colorless to pale brownish. Leaves contiguous to moderately imbricate, usually horizontally spreading but sometimes weakly antically secund, *flat to somewhat concave* near base, ovate-rotundate to broadly oblong-ovate, *widest a short distance below middle, about as long as or slightly longer than wide* (from 1100 μ long × 1120 μ wide to 1160–1275 μ long × 1200–1220 μ wide up to 1650 μ long × 1600 μ wide), *not decurrent ventrally* and scarcely decurrent antically, bilobed for ca. 0.15–0.25 their length, the *sinus crescentic or obtuse,* less often acute, *gibbous or reflexed or not;* lobes varying from rectangulate or acute to obtuse and blunt at apex. Cells of leaf margins and apices ca. (26)27–30(33) μ, (27)30–36 × (30)36–48(50) μ *in leaf middle,* to (32)35–38 × 45–50(55) μ at leaf base, rounded-polygonal, with *somewhat to strongly bulging trigones; cuticle prominently verruculose;* oil-bodies mostly (2)3–6(9) per cell, large, subspherical to ovoid and ellipsoid, ca. 4–5 × (4.5)6.5–9 μ to 6 × 10–13 μ, a few to 7 × 10–12 μ, appearing finely granulose-papillate, much larger than chloroplasts, which are 4.0–5.5 μ long. *Under-leaves distinct but small and often nearly hidden among the rhizoids,* from subulate (and 5–10 cells long × 2–4 cells wide in basal third or less) to lanceolate, frequently terminating in a slime papilla, *often with 1–2 basal cilia of teeth* (which generally end in slime papillae) *but never pluriciliate and laciniate,* the underleaf as a whole from 180 to 350 μ high. Asexual reproduction absent.

Paroecious (but occasional gynoecia without subtending androecia). Androecia always below gynoecia, the bracts somewhat or scarcely larger than leaves, ca. 1250–1500 μ long × 1550–1700 μ wide, ovate-quadrate, in usually 2–4 (rarely more) pairs, *rather variable, often* nearly horizontally spreading and *similar to leaves, except for the ventricose, explanate or infolded antical base,* which may bear 1 (rarely 2) erect or obliquely spreading teeth, sometimes more nearly subtransversely inserted and somewhat canaliculate, sometimes concave and

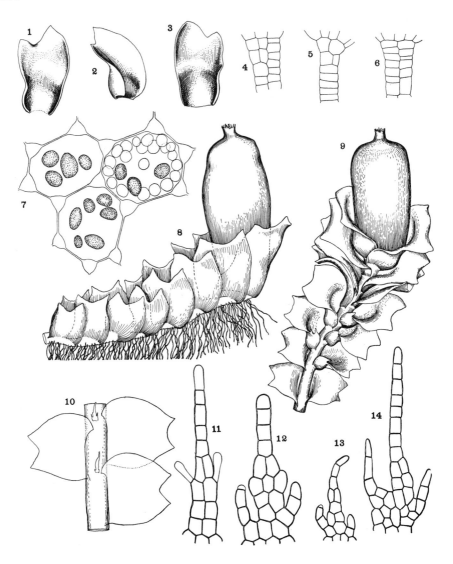

Fig. 165. *Lophozia* (*Leiocolea*) *gillmani*. (1–3) ♀ Bracts (×7.5). (4–6) Antheridial stalks (×115). (7) Median cells with oil-bodies, at right with chloroplasts (×525). (8–9) Fertile plants (×6.5). (10) Postical view of shoot sector, rhizoids omitted (×8.8). (11–12, 14) Underleaves (×138). (13) Underleaf (×115). [All from *RMS 13240a*, Temperance River, Minn.]

suberect. Antheridia (1)2–3(4) per bract, ca. 130–160 × 145–180 μ, the stalk 1–2- (usually 2-) seriate. Gynoecia with bracts like leaves, asymmetrically *bilobed 0.15–0.2*, but considerably larger, ca. (1600)1700–1900(2200) μ long × (1750)2200–2350 μ wide, rounded-ovate to broadly oblong-ovate to ovate-quadrate, virtually transversely oriented, concave and somewhat sheathing at the erect base, obliquely to widely spreading at the somewhat reflexed apices, shallowly and broadly bilobed, the sinus often gibbous or reflexed, margins sometimes undulate; *lobes blunt or rounded*. Bracteole lanceolate, *small*, with 1–few small basal teeth. Perianth terete, cylindrical-ovoid to cylindrical-clavate, up to 2.7 mm in diam. × 7.0 mm long, ca. 0.75–0.85 emergent, rather suddenly contracted at apex into a *discrete, rather elongate beak, whose mouth is shortly crenulate-setulose to ciliolate*. Elaters ca. 8 μ in diam., with 2 reddish brown spirals; spores ca. 15 μ, finely papillose or verruculose, reddish brown.

Type. Au Train Island, Alger Co., Mich., in Lake Superior (H. Gillman, 1867).

Distribution

Presumably imperfectly circumpolar, although not yet known from eastern Asia. Buch and Tuomikoski (1955) correctly note that the distribution shows "continental tendencies," since the plant is absent from Iceland, the Faroes, and Ireland and is rare in Scotland and on the west coast of Norway. It is, however, abundant on the oceanic north coast of Ellesmere I. Considered by Müller (1951–58, p. 693) to be nordic-alpine but more nearly arctic-alpine, since it extends northward at least to 82°30′ N. in Ellesmere I., where it is common. Widespread in calcareous sites in much of the northern half of the Taiga, and apparently found nearly throughout the Tundra wherever calcareous conditions occur.

In Europe scattered through much of Scandinavia (coastal Norway, Sweden, Finland), northward to Lapland; southward recurring sporadically in alpine situations in Central Europe (Germany, in the Berchtesgaden region; Savoy; Switzerland; Monte Rosa in north Italy; Bucovina) and also rarely in the Scottish highlands. Eastward extending to Siberia (Yenisey R. valley).

In North America transcontinental in the Coniferous Biome and presumably in the southern edge of the Tundra, but with a hiatus in its range between Montana and Minnesota, ranging westward from Alaska and Yukon south to British Columbia, Washington, Oregon, and eastward to Alberta and Montana. In our area as follows (selected stations):

ELLESMERE I. S. end of Parr Inlet, at Alert, NE. Ellesmere, 82°30′ N. (*RMS 35141, 35142, p.p.*); slope 2.5–3 mi W. of Cape Belknap, ca. 82°31′ N., S. of Williams I., vicinity of Alert (*RMS 25200*, fo. *heterogyna; 25203, p.p.*, etc.; northernmost station for the species). N. GREENLAND. Moist snow patch, Erlandsen Land, Brønlund Fjord, Mt. Buen, 82°12′ N., 31° W. (*Holmen 6133!*); *Cassiope tetragona* heath, Heilprin Land,

Brønlund Fjord, 82°10′ N., 31° W. (*Fristrup 251h, p.p.*). [The Holmen specimen was published by Arnell (1960) as *L. heterocolpa,* but is definitely paroecious.]   W. GREEN-LAND. Magdlak, Alfred Wegener's Pen., 71°07′ N., 51°48′ W. (*Holmen 15756, p.p.,* with *Lophozia collaris, Peltolepis, Sauteria; RMS & KD 66-1382a, 66-1350, 66-1358, 66-1364*); Fortune Bay, Disko I., 69°16′ N., 53°50′ W. (*RMS & KD 66-142a*); mouth of Simiutap Kûa Valley, Umîarfik Fjord, 71°79′ N, 54°35–40′ W. (*RMS & KD 66-916a, 66-918*); Godthaabsfjord, Itiverna, 64°22′ N., 50°25′ W. (*B. F. Nielsen 1531*!); Marrait, below Anana Mamagissa, Nugssuaq Pen., 70°30′ N., 54°10′ W. (*RMS & KD 66-726a*; with *Metacalypogeia schusterana*). [See also pp. 385 and 387].

NEWFOUNDLAND. South Branch, Piccadilly, Benoit Brook, all in W. Nfld.; Port au Choix, St. John I., Flower Cove, Cook Harbour, and Bartlett's R., all on N. Pen. (Buch & Tuomikoski, 1955).   NOVA SCOTIA. Truro (*Macoun 377*!, as *Jungermannia incisa*); Aspy Bay, Ingonish Mts. (Brown, 1936a); N. end of Presque Isle, S. of Cap Rouge, Cape Breton I., in small ravine (*RMS 43011a,* with *Metacalypogeia schusterana*).   QUEBEC. Iberville; between Baldé and La Baie des Chaleurs, Bonaventure R. (Lepage, 1944–45); The Grotto on Mt. Ste-Anne, Percé, Gaspé (*RMS 44012, 44015*).   ONTARIO. Marly spring, Manitoulin I., L. Huron (*Schuster, 1949*); E. of Nipigon, Thunder Bay Distr. (Cain & Fulford, 1948).; Little Cove and Driftwood Cove, Bruce Pen. (*Williams*).

MAINE. Round Mountain L., Franklin Co. (*Lorenz; fide* Evans, 1914).   NEW HAMPSHIRE. Beaver Falls, Colebrook, Lime Pond, Columbia (Evans, 1917).   VERMONT. Hartland (*Dutton, 1910*); Quechee Gulf, Hartford (*Lorenz, 1911*); W.-facing slope of Willoughby Mt., E. side of Willoughby L. (*RMS 46210*); just N. of Smuggler's Gap, NE. of Mt. Mansfield (*RMS 661854, 661859*).   MASSACHUSETTS. W. side of Green R. valley, N. of Greenfield (*RMS, s.n.*); Bear R. valley, E. side of the Flume, ca. 0.5 mi above jct. with Deerfield R., Conway, Franklin Co., ca. 500 ft (*RMS 67989, 661813*).

NEW YORK. Adirondack Mts.: Cascades, NW. side of Cascade Mt. between Upper and Lower Cascade Ls., 6.5 mi SE. of L. Placid, Essex Co., 2400–3200 ft; Indian Head, wet cliff face, with *Saxifraga aizoön,* cliffs along E. side of Lower Ausable L., 4 mi SW. of St. Huberts, Essex Co., 2200 ft; Rainbow Falls, N. end of Lower Ausable L., 4 mi SW. of St. Huberts, Essex Co. (N. G. Miller, 1966). Esopus Gorge, 0.5 mi N. of Atwood, Ulster Co., 280 ft; Kaaterskill Falls, 1.75 mi E. of Haines Falls, Ulster Co., 1900 ft, Catskill Mts. (N.G. Miller, 1966; *RMS, s.n., 1965*).

Westward in the Great Lakes region as follows. OHIO. West Jefferson (doubtful station! see Conklin, 1923).   MICHIGAN. Amygdaloid I., NW. of Isle Royale (*RMS, s.n.*); Tonkin Bay, Isle Royale (Conklin, 1914); Isle Royale (*Allen & Stentz 937a*): Au Train I., Alger Co. (*H. Gillman, 1867*; type); Copper Harbor, Keweenaw Pt. (*RMS, s.n.*); "Devils Washtub," SW. of Copper Harbor, near Eagle Harbor, Keweenaw Pt. (*RMS, s.n.*); Huron Mts., Marquette Co. (*Nichols*); Agate, Copper, Horseshoe Harbors, Keweenaw Pen. (Steere, 1937).   WISCONSIN. Sand I., Apostle Isls., Bayfield Co. (*RMS s.n.*); Ashland, Douglas, Bayfield, and Iron Cos. (Conklin, 1929).   MINNESOTA. Cook Co.: The Point, Grand Marais (*RMS 6570, 13791,* etc.); Belle Rose, Long, Big Susie, small island between Lucille and Susie Isls., Lucille and Little Susie Isls., NE. of Grand Portage (*RMS 12239, 8013, 13018, 4576, 13501, 13193a, 13380, 13800a,* etc.); L. Superior shore 1 mi below Temperance R. (*RMS 13240b*); Lutsen, Cascade R., and Temperance R. (*RMS 14697, 14702; Conklin 1528, 1125, 1476*); Gooseberry R., Manitou R., and Two Island R., all along L. Superior in Lake Co. (*RMS 13216a, 14157,* etc.; *Conklin 2522, 2629,* etc.).

The species has generally been regarded as rare, but I find that in much of our area this and *L. heterocolpa* are the dominant Leiocoleae; part of the implied rarity has been due to confusion of small forms with *L. collaris* and of larger forms with *L. bantriensis.*

Ecology

Like our other species of *Leiocolea*, *L. gillmani* is an obligate calciphyte,[59] consequently has a restricted distribution. It is frequent in the Great Lakes region, wherever moist basic intrusive rocks occur or calcareous sedimentaries or schists are found. *Lophozia gillmani* is more strictly "bound" to calcareous conditions than the regionally more frequent *L. heterocolpa* and consequently occurs often as either a pioneer or a near-pioneer on moist rocks. It is frequent in the "spray zone" along the Lake Superior shore line, and peripheral to waterfalls, occurring as a chasmophyte and from there spreading over the faces of ledges or cliff walls. In such habitats it is usually not a primary species, coming in often after ecesis by *Gymnostomum aeruginosum* and other calciphilous mosses, among which the *L. gillmani* may sometimes be found as isolated, creeping plants. When a mesophyte in the spray zone, on shaded ledges, *L. gillmani* is usually conspicuous by its pale, rather pellucid, and shining green appearance. Associated most frequently are *Lophozia badensis*, *L. heterocolpa*, *Scapania gymnostomophila*, *S. cuspiduligera*, *Odontoschisma macounii*, *Preissia quadrata*, *Pellia fabbroniana*, *Plagiochila asplenioides*, and *Tritomaria scitula*. The pH range, under these conditions, is from 5.5 to 7.1.

Some of the same associates occur with it in New England (*Lophozia heterocolpa*, *Plagiochila*, *Preissia* and *Scapania cuspiduligera*, at Smuggler's Notch, Vt.; *Pellia fabbroniana* in Green R., Mass.), but the species rarely occurs as low as 450 feet on cool ledges without any of these boreal associates (Bear R., Mass.; there under conditions of least elevation in eastern United States).

Less frequently around rock pools, in partial or almost full sun (as in the microclimatic "Tundra Zone" along Lake Superior; see Schuster, 1953, 1957), associated with *Scapania degenii*, *Odontoschisma macounii*, and such angiosperms as *Scirpus caespitosus*, *Primula mistassinica*, *Pinguicula vulgaris*, *Polygonum viviparum*, and also *Selaginella selaginoides*. The last three taxa and *Scapania calcicola* are associated in *RMS 68-1463* (N. of Daniels Harbour, Newfoundland). The plants, in such exposed, insolated sites, are often darker green and much more opaque.

At the northern edge of its range on calcareous clay-shale solifluction slopes below long persisting snow banks, together with *Athalamia hyalina*, *Sauteria alpina*, *Aneura pinguis*, *Lophozia collaris*, *L. heterocolpa harpanthoides*, *L. pellucida*, *Preissia quadrata*, and *Blepharostoma trichophyllum brevirete*.[60] The species under such conditions occurs in full sun, unless shaded by competing mosses, and often forms a blackish, opaque, small phase habitually similar to small phases of *L. collaris*, associated with *Salix arctica*, *Equisetum variegatum* (the latter at its

[59] Müller (1951–58, p. 693) states that the species is "occasionally also on siliceous substrata." This appears to be never the case in North America.

[60] This consociation, along the northeast coast of Ellesmere I., is so constant that, with the discovery of one of these species, the finding of the other associates is usually predictable.

northernmost known station!), *Pedicularis hirsuta, Eriophorum angustifolium triste,* etc. (Schuster, *in* Schuster et al., 1959).

Kaalaas (1893, p. 363) reports it in Norway as "on shady, moss-grown, some-what damp cliffs, by brooks and water, as well as on damp sandy soil, as a rule on . . . loose clay schist . . . it grows most often mixed with different mosses, most frequently perhaps in company with *Sphenolobus* [*Saccobasis*] *politus.*" Kaalaas also mentions that Arnell found it occasionally on rotten logs, and Lindberg and Arnell (1889, p. 45) report it from a decaying log, with *L. heterocolpa* and *Blepharostoma.* Such occurrences on organic substrates are exceptional. Buch and Tuomikoski (1955) report it as "often directly on limestone" and list as associates most of the species found in association in the Lake Superior area by Schuster (1953), as well as *Moerckia flotowiana* and *Saccobasis polita.* Apparently rare in swamps or marl bogs, as in Gaspé (associated with *Saccobasis*).

In west Greenland, frequent in calcareous areas, usually in fens or boggy ground, often around small lakes, with *Lophozia heterocolpa, L. rutheana,* and occasionally *Cryptocolea imbricata* (Umîarfik Fjord), or in ill-drained, boggy, calcareous meadows, with *Eriophorum scheuchzeri, Scapania gymnostomophila,* and the very rare *Metacalypogeia schusterana* (as at Marrait, Nugssuaq Peninsula). The association with *M. schusterana* recurs on moist rocks in Nova Scotia.

Differentiation

When fertile, and with the paroecious inflorescence evident, distinct from all other species of *Leiocolea* except *L. rutheana,* which *L. gillmani* may almost approach in size of plant and of cells. *Lophozia gillmani* differs from *L. rutheana* in the (*a*) much smaller underleaves, often inconspicuous and difficult to demonstrate, usually nearly hidden among the rhizoids, never bearing numerous ciliary divisions; (*b*) usually slightly concave or nearly flat leaves, whose lobes are not strongly deflexed, and whose sinuses are usually not reflexed or gibbous; (*c*) relatively narrower leaves, never conspicuously broader than long, which are not decurrent ventrally; (*d*) longer and more conspicuous beak of the perianth. In high arctic phases of *L. gillmani,* however, the perianth beak is often much shorter than normal.

In isolated plants of *L. gillmani* antheridia fail to develop in the axils of leaves below the perianths (Schuster, 1953, p. 335). Such individuals are almost inseparable from *L. collaris* and *L. bantriensis* (except perhaps by cell size). However, even though purely ♀ shoots of *L. gillmani* can be found, no purely androecial shoots develop. Normally 2–4 (rarely 5) pairs of leaves below the perichaetial bracts are strongly concave at base, "with the dorsal portion of the base ventricose, and the dorsal base with margin infolded to form a small pocket, often with a subbasal tooth at this point . . . when perianths are fully mature the antheridia often have decayed and dehisced (therefore the form of the bracts, rather than actual presence of antheridia, must often be used to determine the paroecious inflorescence)" (Schuster, 1953).

When plants are sterile, difficulty may be encountered in separating this species from the dioecious *L. bantriensis*. The latter has somewhat larger cells, 35–38 μ in the leaf tips, and 35–38–(40) × 38–48–(55) μ in the leaf middle, but otherwise is strikingly similar in size (ca. 2.5–3, occasionally to 4 mm wide), in the subulate underleaves with commonly 1–2 basal teeth or cilia, and the leaf shape and dimensions. If purely androecial shoots are found, these can safely be referred to *L. bantriensis*, rather than to *L. gillmani*, although ♂ bracts in the latter may be succeeded by 2–4 pairs of leaves before archegonia occur.

At times *L. gillmani* has been considered merely a paroecious variety of the dioecious *L. collaris* (Bernet, 1888), and a variety of workers have found difficulty in maintaining a sharp distinction. However, in addition to the difference in sexuality, *L. collaris* differs from *L. gillmani* in the smaller leaf cells. In American material referred to *L. collaris* (Schuster, 1951) cells are ca. 22–26 μ in the lobes, occasionally 20–27 μ, and those of the leaf middle are only 25–30 × 32–38 μ, or exceptionally even as low as 18–24 × 24–28 μ. These differences persist even when *L. gillmani* occurs in damp, fully insolated sites (as with *Aneura pinguis* in *RMS 35142*, Alert, Ellesmere I.), then as a blackish green, small manifestation (plants only 1.5 mm wide at and below the androecia, to 1.7 mm wide at ♀ bracts), in which cell walls and bulging trigones are prominently fuscous-pigmented.

Variation

In the far north *L. gillmani* almost always is found as a blackish green phase (only when growing in dense shade, straggling amidst robust mosses, is it pale and pellucid green). The plants then show a great variation in size (although clearly more robust than the often admixed *L. collaris*), and in the form of the perigonal and perichaetial bracts. The latter often serve also as perigonal bracts, containing 1–3 antheridia at base. Purely perigonal bracts (often in only 1–2 pairs) may or may not bear the antical basal tooth commonly found in robust phases (at times one of the perichaetial bracts may also bear such a tooth). Both types of bracts are often asymmetrically bilobed. The most marked discrepancy that the high arctic plants tend to exhibit, however, is in the perianths, which are often not at all sharply contracted into a beak. Indeed, in the plants of *RMS 25200* (west of Cape Belknap, northeast Ellesmere I.) there is no beak at all, and the perianths, which often appear hardly normal in form, are often irregularly incised at the mouth, the lobes varying from irregularly crenulate to ciliolate or dentate with teeth up to 3 cells long × 1–3 cells wide at the base. Such plants, with wide-mouthed, erostrate perianths, perhaps deserve recognition as a minor variant, hardly to be reconciled to environmental causes, the fo. *heterogyna*.

In leaf shape, also, *L. gillmani* undergoes considerable variation. Limpricht (1884) described a phase with obtuse lobes, while the phase with

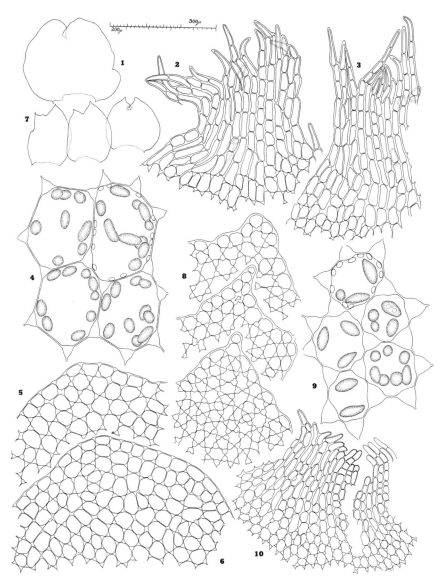

FIG. 166. *Lophozia* (*Leiocolea*) *gillmani*. (1) ♀ Bract (×11.9). (2–3) Two sectors, perianth mouth, drawn to top scale (×95). (4) Cells of lobe base, with oil-bodies and, upper right, chloroplasts (×410). (5–6) ♀ Bract lobes, drawn to bottom scale (×121). (7) Two leaves at left, a ♀ bract at right (×11.9). (8) Apices of lobes of three ♀ bracts, to bottom scale (×121). (9) Cells with oil-bodies and, top cell, chloroplasts (×450). (10) Perianth mouth cells, drawn to top scale (×95). [1–6, *RMS & KD 66-918*, var. *ciliolata*; 7–10, *RMS & KD 66-916a*, var. *gillmani* in an unusually small mod. *parvifolia-angustifolia*; both from same locality, Umîarfik Fjord, Greenland. See Table 4.]

acute leaf lobes was described as "var. *acutifolia*." This difference in leaf form[61] appears to be uncorrelated with much of the other, considerable variability of this plant. (For example, plants from Vermont have optimally acute-lobed and even apiculate leaves and bracts; plants from Massachusetts may have blunt to subacute leaves and bracts.) The wide variation in *L. gillmani*, *even in plants from a single site* (see, for example, columns 1 and 2 of Table 4, where plants from Umîarfik Fjord, Greenland, are compared), is difficult to explain; initially I believed that two paroecious taxa were involved, distinct at the species level. However, when "typical" *L. gillmani*, from near the southern periphery is compared (column 3), the distinctions to some extent disappear. However, the robust, large-celled plant with optimally ciliolate perianth mouths (column 1) evidently represents a deviant extreme, distinguished as var. *ciliolata*, as is a small extreme with broad leaves (fo. *orbiculata*).

In my considerable experience with *L. gillmani* s. *lat.*, the exposed phases of this species tend to become infuscated and do not develop a warm, brownish pigmentation; they never show traces of reddish color. High arctic phenotypes (such as described in Schuster et al., 1959, p. 27) tend to form black-green patches. By contrast, plants of *L. gillmani* from exposed sites at Umîarfik Fjord (see below) tend to become clear brown and, in extreme cases, rather intense but warm brown, the pigmentation approaching that of *L. heterocolpa* var. *harpanthoides*, with a similar but slighter tendency for some reddish pigments to develop; the shoots become rather purplish brown.

Consideration of *L. gillmani* as a rather stenotypic species has been based on widespread confusion of *L. gillmani* material with other *Leiocolea* species (particularly *L. bantriensis* and *L. collaris*).[62] A whole range of plants, all clearly paroecious, exists, showing a bewildering degree of variation in size, form of leaf cells, perianth mouth, oil-body number, and other features; these plants may occur admixed or adjacent to *L. rutheana* in the Arctic (compare columns 1, 2, and 5 in Table 4). Slighter but still striking variation occurs even in the scattered New England populations (compare columns 3 and 4). These five casually selected collections show additional variability not brought out in the table; e.g., the Vermont plants are unique, showing strikingly reflexed and gibbous sinus bases of vegetative leaves and bracts and possessing unusually slender, linear teeth of the perianth mouth that show no taper toward the apex, as well as unusually consistently sharp leaf lobes. These collections are also cited as affording proof that *L. gillmani* must be very broadly interpreted—in spite of the fact that strikingly distinct populations may inhabit juxtaposed

---

[61] Müller (1905–16, p. 719) stated it would be "natürlicher" to regard the more widely dispersed, acute-lobed phase of *L. gillmani* ("*L. kaurini*") as the type of the species. However, Limpricht had specifically regarded the blunt-lobed plant as the typical form of the species and had called the acute-lobed phase "var. *acutifolia*."

[62] There is a considerable number of reports of these two species from our region, in particular from Greenland; essentially all reports for which I have seen material represent misidentifications, and I suspect that part of the problem has been that the ♂ bracts of *Lophozia gillmani* have been overlooked, so that the plants have been erroneously placed in *L. collaris* or *L. bantriensis*.

TABLE 4

Distinctions between L. rutheana and Various Phenotypes of L. gillmani

Criteria	1 L. gillmani var. ciliolata*	2 L. gillmani
Shoot width	1.8–2.8 mm	(1.4)1.5–2.4 mm
Trigones	Medium large	Coarse
Subapical cells	(28)30–37(42) μ	25–30(32) μ
Marginal cells distally	30–35(38) μ	27–33 μ
Median cells	(30)36–45 × 45–65 μ	25–35 × 30–45

Cells along
  postical base
	19–25 × 46–72 μ
	  to 17–22 × 46–66 μ
	  (2.5–3.8:1)
		24–32 × 36–48(58)
		  μ
		  (1.5–2:1)

Oil-bodies size	7–10 per cell To 6–7 × 12–14 μ	2–7(9) per cell To 6–6.5 × 10–13 μ

Leaf
  (width × length)
	1535 × 1650 μ to
	  1745 × 1890 μ
	  (width under
	  1.0 × length)
		780 × 825 μ to
		  825–880 × 1025 μ
		  (width under
		  0.9 × length)

Sinus Length
  Shape
	0.18–0.2 leaf length
	Lunate to round-
	  angular
		0.25–0.3 leaf length
		±V-shaped

Leaf lobes	Rounded to blunt,
	  occasionally
	  subacute
		Sharp, ending in
		  1–2 single cells

Cortical cells	24–29 × 100–
	  145(185) μ
	  (4–7:1)
		30–36 × 70–95 μ
		
		  (3–4:1)

♀ Bracts
  (width × length)
	To 2000–2150 ×
	  1850 μ
	  0.15–0.2-lobed,
	  lobes round
		Ovate, 1125–1200 ×
		  1250–1400 μ
		  0.2–0.25-bilobed,
		  lobes sharp

Perianth mouth	Long; fringed with
	  short cilia (to
	  100–165 μ long),
	  of 1–3(4) narrow
	  cells
	Terminal cells to
	  14–19 × (45)54–
	  80 μ
		Long; crenulate-
		  denticulate with
		  cells free for
		  0.2–0.8(1.0) their
		  length
		Cells 17–18.5 × 38–
		  54 μ to 50–70(80)
		  μ

3 *L. gillmani* (Massachusetts)	4 *L. gillmani* (Vermont)	5 *L. rutheana*
1.8–2.4 mm Medium-sized (24)25–32 μ 27–30 μ	2.3–2.5 mm Small, concave 33–38 μ 33–38 μ	3.4–5 mm Very coarse (28)30–36(40) μ 30–36(40) μ
27–34 × 32–38 μ 19–25 × 34–62 μ	30–36 × 38–45 μ 21–25 × 47–100 μ	33–40 × 40–54 μ (25)28–38 × 34–50(58) μ
(1.8–3.2:1) (2–3)4–7(9) per cell	(ca. 2–5:1) (2)3–5 per cell	(1.2–2:1) 2–5 per cell To 7–8.5 × 18–22 μ
ca. 1000 × 1030 μ (width under 1.0 × length)	ca. 1125 × 1175– 1230 μ (width under 1.0 × length)	1840 × 1475 μ to 2000 × 1650 μ and larger (width 1.2–1.3 × length)
0.18–0.25 leaf length ±Lunate to round- angular Blunt to subacute	0.18–0.25 leaf length ±Obtusangular Sharply acute to apiculate	0.18–0.25(0.3) leaf length Obtusely angulate Acute, subacute to narrowly blunt
24–27 × 60–105 μ	20–29 × 75–115 μ	24–26 × (70)75–140 μ
(3–4:1) 1030–1230 × 1085– 1200 μ Lobes blunt to sharp to rounded Long; crenulate- setulose, cells free for 0.8–1.5 their length	(3–6:1) 1800 × 1800 μ Lobes sharply apiculate ±Long; setulose- ciliate (to 140 μ long), of 1–2 long cells	(3–6:1)
Cells 16–19 × 70– 90 μ to 19–23 × 65–70 μ	Cells 15–18 × 80–85 μ	

* The origin of the cited specimens is as follows: 1. Umîarfik Fjord, West Greenland, *RMS & KD 66–918*; 2. Same locality, *RMS & KD 66–916a*; 3. Bear R., Conway, Mass., *RMS 66–1813*; 4. Smuggler's Notch, Vt., *RMS 66–1854*; 5. Umîarfik Fjord, West Greenland, *RMS & KD 66–916*. [See Fig. 166.]

and sim ar niches in single environments (e.g., around a boggy, shallow lake at Umîarfik Fjord, west Greenland).

At Umîarfik Fjord (*RMS & KD 66–918, 66–916a*) I collected three paroecious taxa together, in an area of less than 2 sq m; the comparison in Table 4 was made from living plants (for comparison "normal" *L. gillmani* from the southern edge of its range is entered).

Key to Varieties and Forms of L. gillmani

1. Cells smaller: marginal cells of lobes ca. 24–30 μ; median cells from 25–30 × 26–35 to 25–35 × 30–45 μ; cells near postical leaf bases little elongated (1.5–2 to 1.8–3.2 : 1); oil-bodies often in large part 2–3, rarely mostly 4–7 (rarely 8 or 9), per cell; perianth mouth crenulate-denticulate to crenulate-setulose and often feebly lobed, the teeth from 0.2–0.8 to 0.8–1.5 cells long . . . . var. *gillmani*   2.
   2. Plants small or medium-sized, sterile shoots mostly 1.5–2.5 mm wide; leaves oblong-ovate, width ca. 0.9–1.0 length, flat or almost so, with feebly or hardly arched sides. ♀ Bracts ovate, length 0.95–1.2 × the width. Facies approaching that of *L. collaris* . . . . . . fo. *gillmani*
   2. Plants quite small (0.8–1.2 mm wide × 6–12 mm long); leaves mostly broadly orbicular, widest toward base, width 1.15–1.25 × the length, concave and suberect, with strongly arched sides. ♀ Bracts orbicular or nearly so, length 0.8–0.9 the width usually. Size and facies of *L. badensis*, but clearly paroecious . . . . fo. *orbiculata*, fo. n.
1. Cells larger: marginal cells of lobes 30–35(38) μ; median cells (30)36–45 × 45–65 μ; cells near postical base linear-oblong (many 2.5–5 : 1); oil-bodies 3–5 to 7–10 per median cell; perianth mouth fringed with short, crowded cilia, of 1–3(4) superimposed cells, the terminal cells narrow, 14–19 × 54–80 μ . . . . var. *ciliolata*, var. n.

LOPHOZIA GILLMANI var. *GILLMANI* fo. *ORBICULATA* Schust., fo. n.

[Fig. 167:1–3]

Planta var. gillmanio in inflorescentiis paroiciis similis, differens, autem, quod minor (ad 0.8–1.2 mm lat.), foliis late orbicularibus latissimis versus basim, bractiis femineis fere omninove orbicularibus.

Quite small, compact, often laterally compressed, only ca. 0.8–1.2(1.5) mm wide × 6–12 mm long, with aspect of *L. badensis*, but with, mostly, conspicuous warm brown pigmentation, here and there mixed with or replaced by reddish pigments, but plants never scorched in appearance. Leaves mostly 950 μ wide × 740 μ long to 900–945 μ wide × 780 μ long (*width : length 1.15–1.25 : 1*), contiguous to closely imbricate, antically assurgent, often suberect, concave, *mostly broadly orbicular and widest toward base*, with equally and *strongly arched sides*, rather strongly narrowed toward apex, ca. 0.2–0.25 bilobed, the *lobes all sharp, ending in often 2(3) single cells*, sinus usually deeply lunate to lunate-angular. Cells variable, usually with rather distinctly bulging trigones; marginal cells distally ca. 24–28 μ; subapical cells very variable, (22)25–30(32) μ on an average;

median cells also variable (23)25–30(32) × 26–35 μ; cuticle very weakly papillose-striolate. Oil-bodies mostly 2–7 per cell, variable, some small and ovoid (and ca. 4–5 × 7–8 μ), others very large, narrowly ellipsoidal, and 4.8–6 × 10–12(16) μ. Bracts suborbicular, 1200–1250 μ broad × 960–1050 μ long. Perianth with beak often reduced in length, the mouth crenulate-denticulate as in typical form; ultimate cells largely united for most of their length, fingerlike, 19–22 × 40–65 μ to 15–17 × 30–50 μ.

Type. West Greenland: east side of Hollaenderbugt, northwest Nugssuaq Peninsula, 70°47' N., 53°20' W. (*RMS & KD 66–1508a, Aug. 11, 1966*).

The type plants occurred among *Preissia quadrata, Distichium capillaceum, Lophozia pellucida, Tritomaria scitula,* and *Solenostoma pumilum s. lat.*

The small, compact, warmly brown pigmented plants of this extreme seem hardly reconcilable with the *L. gillmani* complex but resemble, rather, a vigorous *L. badensis*—with which they share the broadly orbicular leaf form and sharp lobes—or a small *L. collaris.* At first, indeed, they were taken to be the former, since no antheridia could be found in the leaves below the perianth. However, all subfloral innovations studied that were mature enough showed antheridia in the distal leaves, and young shoots, with ♂ bracts present, usually had a cluster of archegonia at their apices. *If* the sexuality of the plants is overlooked, they will, because of the cell size and trigones, be looked for under *L. collaris* or *L. badensis.*

LOPHOZIA GILLMANI var. CILIOLATA Schust.,
var. n.
[Figs. 166:1–6, 167:9–11]

Varietas var. gillmanio similis, differens, autem, eo quod cellulae maiores (mediae 36–45 × 45–65 μ), et 7–10 guttae olei omni in cellula, et os perianthii ciliis brevibus crebis, e cellulis perelongatis formatis, fimbriatum.

Relatively vigorous, 1.8–2.8 mm, or more, wide on sterile shoots. Stem with cortical cells ca. 24–29 × (75)100–145(185) μ, ca. 3–7× as long as wide. Leaves oblong-ovate, with little arched sides, to ca. 1535–1745 μ wide × 1650–1890 μ long (width slightly to clearly under 1.0 length), divided for 0.18–0.25 their length; sinuses lunate to round-angular; *lobes rounded to blunt,* varying to subacute or apiculate. *Cells relatively large:* marginal cells of lobes 30–35(38) μ, subapical cells ca. (28)30–38(42) μ, median cells ca. (30)36–45 × (38)45–65 μ, with small or medium large, not nodose trigones. Cells along postical leaf bases mostly prominently elongated, narrow-rectangulate, 17–22 × 46–66 to 19–25 × 46–100 μ (length 2.5–5× the width). Oil-bodies (2)3–5 *to 7–10 per cell,* up to 6–7 × 12–14 μ. ♀ Bracts to 2000–2150 μ wide × 1850 μ long, 0.15–0.2 bilobed, the lobes rounded or sharply apiculate. Perianth longly beaked, the beak with crowded, *short cilia* to 140–165 μ long, formed *of 1–3(4) narrow, elongated cells* [terminal ones ca. 14–19 × (45)54–85 μ].

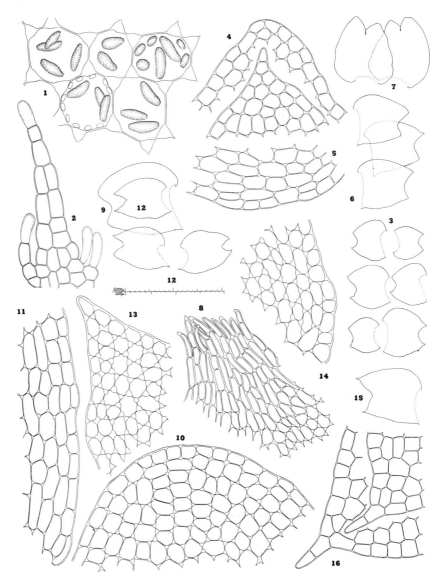

FIG. 167. *Lophozia (Leiocolea) gillmani.* (Figs. 1–8, 12–14, var. *gillmani*; 12–14, mod. *parvifolia*; 9–11, var. *ciliolata*; 15–16, a form approaching var. *ciliolata*). (1) Median cells with oil-bodies (×500). (2) Underleaf (×205). (3) Six leaves (top scale; ×13). (4) Two leaf lobe apices (bottom scale; ×132). (5) Cells of postical leaf base (bottom scale; ×132). (6) Three leaves (top scale; ×13). (7) Two ♀ bracts (top scale; ×13). (8) Perianth mouth (×95). (9) Leaf (top scale; ×13). (10) Leaf lobe apex (bottom scale; ×132). (11) Postical leaf base (bottom scale; ×132). (12) Three leaves (top scale; ×13). (13) Leaf lobe apex (bottom scale; ×132). (14) Postical leaf base (bottom scale; ×132). (15) Leaf (top scale; ×13). (16) Two leaf lobe apices (bottom

Type. Umîarfik Fjord, Svartenhuk Peninsula, west Greenland, 71°59′ N., 54°35–40′ W. (*RMS & KD 66–918*).

This plant occurred at the type station within a short distance of smaller plants with sharp-pointed to apiculate leaf and bract lobes, with smaller cells (compare Fig. 167:9 and Fig. 167:12–13), in which the perianth mouth is much less distinctly ciliolate. I was initially of the opinion that the present plant is identical to typical *L. gillmani,* but the relatively southern plants that represent the type of *L. gillmani* have smaller cells, with fewer oil-bodies (and usually coarser trigones) than var. *ciliolata.*

However, the majority of plants from Umîarfik Fjord (e.g., *RMS 66–916a*), which agree with "normal" *L. gillmani* in these characters, seem to represent an unusually small form of *L. gillmani,* with unusually sharply apiculate leaves and ♀ bract lobes. It is possible that they should be varietally divorced from var. *gillmani.* Figures 166–167 illustrates some of the taxonomic problems involved.

I tentatively place here the plant from Smuggler's Notch, Vt. (Table 4, column 4), which differs from the type chiefly in (*a*) the sharply apiculate lobes of leaves and bracts; (*b*) the gibbous, recurved sinuses, much as in *L. rutheana;* (*c*) the fewer oil-bodies per cell; and (*d*) smaller cells; see Fig. 167:15–16.

LOPHOZIA (LEIOCOLEA) BANTRIENSIS (*Hook.*)
Steph.

[Fig. 168:1–6]

Jungermannia bantriensis Hook., Brit. Jungerm., in text to pl. 41, 1816.
Jungermannia bidentata [unnamed] var., Hook., *ibid.,* Syn., p. 16, and Suppl., pl. 3, 1816.[63]
Jungermannia hornschuchiana Nees, Naturg. Eur. Leberm. 2:153, 1836.
Jungermannia subcompressa Limpr., Jahresb. Schles. Gesell. Vaterl. Kult. 61:209, 1884 [reprint p. 6].
Jungermannia muelleri var. *subcompressa* Lindb., *in* Lindberg & Arnell, Kgl. Sv. Vetensk.–Akad. Handl. 23(5):43, 1889.
Jungermannia muelleri var. *bantriensis* Kaal., Nyt Mag. Naturvid. 33:357, 1893.
Lophozia bantryensis Steph., Spec. Hep. 2:133, 1901.
Lophozia hornschuchiana Macoun, Cat. Canad. Pl. 7:18, 1902.

[63] The history of *Jungermannia bantriensis* is ambiguous at best. Hooker (in a note following the text to pl. 41) *nowhere describes this species* but merely states that *J. stipulacea* "very nearly corresponds with another new species (*J. Bantriensis MSS.*). . . which has, like the present, emarginate leaves. . . ." Then follows a statement of several differences from *J. stipulacea* but no formal diagnosis. Furthermore, on p. 16 of his "Synopsis" Hooker states that he only "hinted" at this new species (*J. bantriensis*) and now thinks it to be a mere variety of *J. bidentata* (= *Lophocolea bidentata*). Only *if* we consider the statement of differences that separate *J. bantriensis* from *J. stipulacea* as constituting a diagnosis of a new species, does this become effective publication.

scale; ×132). [1–3, *RMS 66-1508a,* Hollaenderbugt, Nugssuaq Pen., Greenland; 4–8, *RMS 67989,* Bear R., Mass.; 9–11, *RMS & KD 66-918,* Umîarfik Fjord, Greenland; 12–14, *RMS & KD 66-914,* Umîarfik Fjord, Greenland; 15–16, *RMS 66-1854,* Mt. Mansfield, Vt.]

Lophozia muelleri var. *bantryensis* Kaal., *in* Bryhn, Rept. 2nd Norwegian Arctic Exped. in the "Fram" 1898–1902, 2 (11):30, 1906.
Lophozia (subg. *Leiocolea*) *hornschuchiana* K. Müll., Rabenh. Krypt.–Fl. 6(1):723, fig. 328, 1910.
Acrobolbus septentrionalis Steph., Spec. Hep. 6:116, 1917 (new synonymy).
Lophozia bantriensis Joerg., Bergens Mus. Skrifter 16:164, 1934; Frye & Clark, Univ. Wash. Publ. Biol. 6(3):384, figs. 1–10,(1944)1945.
Lophozia (subg. *Leiocolea*) *bantriensis* Schust., Amer. Midl. Nat. 45(1):47, 1951; Schuster, Natl. Mus. Canada Bull. 122:7, 1951.

Scattered or gregarious in swelling tufts, dull or *deep green to blackish*, with a slightly greasy lustre, rather robust, (2.5–2.8) 3–4 mm wide × 2–5 (8) cm long, loosely prostrate to (with crowding) suberect. Stems brownish green, flexuous, simple or remotely furcate, with occasional lateral-intercalary branches, usually with subfloral innovations. Leaves contiguous to laxly imbricate, antically decurrent, *plane to slightly adaxially convex* usually, from widely laterally patent to antically secund, obliquely ovate to broadly oblong or subrotundate, bilobed for 0.15–0.25 their length, the sinus rectangulate to obtuse or crescentic, frequently weakly gibbous; lobes acute or obtuse, widely triangular, often slightly unequal. Cells of *leaf apices from 30–35 to 35–38 μ, of leaf middle 35–38(40) to 38–55(60) μ*, at the base to 40 × 60 μ, thin-walled and with small to coarse trigones; cuticle distinctly striolate to verruculose; oil-bodies 2–4(5–8) per cell, large,[64] pale brownish, 6–9 × 9–15 μ, fusiform to elliptical to ovoid. *Underleaves small or moderate in size, from linear-subulate to lanceolate*, rarely bifid, acuminate, usually *with 1 or 2 cilia or teeth near base*, but never pluriciliate. *Without gemmae.*

Dioecious. Androecia becoming *intercalary, loosely spicate;* bracts concave, the antical margin with an incurved tooth, otherwise similar in size and form to leaves; 1–2-androus. Gynoecia with bracts similar to leaves, erect at base, distally somewhat spreading, larger than leaves in size, *entire-margined;* bracteole subulate to lanceolate, similar to underleaves but larger. Perianth 0.6–0.75 emergent, terete and cylindrical, contracted rather swiftly but obtusely into a *short beak* with a ciliolate to shortly ciliate mouth. Capsule with a large-celled epidermal layer and two layers of small, interior cells; epidermal cells large, subquadrate, with small, nodular thickenings; inner layers of narrow, rectangulate cells with semiannular bands. Elaters 10–12 μ in diam., bispiral; spores 12–16 μ, closely papillate, reddish brown. Chromosome no. $n = 9$.

Type. Near Bantry, County Cork, Ireland (*Miss Hutchins*).

Distribution

Relatively widely dispersed in Europe, ranging from Scandinavia (Norway, Sweden, Finland; Spitsbergen) southward to the Faroes, England, and Ireland (type), through central Europe to the Alps (south Germany; Switzerland; Austria; the Vosges Mts. of France; Bucovina), southward to northern Spain (Casares Gil) and the Pyrenees; the north Italian Alps; Savoy; etc. In central Europe montane only; northward descending to sea level. Eastward extending into Asia (Siberia; Yenisey

---

[64] Frye and Clark (1937–47, p. 384) state, incorrectly, "numerous, small."

R. valley at Uskij Mys, ca. 60° N.; Arnell, 1930, *in* 1928–30, p. 16).

In North America with a characteristic, almost "Cordilleran" distribution; apparently rather widespread in the West, ranging from Alaska (Thumb Bay, *fide* Clark and Frye, 1942) to British Columbia, Alberta, Washington, Oregon, and California; also in Nevada (Clark, 1957), with an old and very dubious report from Colorado (Wolf, 1873; *fide* Austin); eastward very rare and essentially high subarctic and arctic in range, but evidently absent in northern Ellesmere I.; with "oceanic" tendencies.

E. GREENLAND. Hurry Inlet, E. coast (*Dusén, Aug. 4*!; Jensen, 1906a, p. 309, mixed with *Tritomaria quinquedentata* fo. *turgida, Blepharostoma t. brevirete*); Granatelv, 74°10′ N., Clavering Ø (Hesselbo, 1948; *not* verified). W. GREENLAND. Ikorfat, Nugssuaq Pen., 70°45′ N., 53°07′ W. (*Holmen 62-1574, p.p.*,[65] with *L. heterocolpa, L. quadriloba*); mouth of Simiutap Kûa Valley, Umîarfik Fjord, 71°59′ N., 54°35–40′ W., near pond (*RMS & KD 66-918,* ?). ELLESMERE I. Harbour Fiord, on S. coast (Bryhn, 1906; as *L. muelleri* var. *bantryensis*). NEWFOUNDLAND. St. Barbe Bay, Cook Harbour, Bartlett's R., and between Cow Head and St. Paul's Inlet, all in NW. Pen. (Buch & Tuomikoski, 1955). QUEBEC. Ft. Chimo, Ungava Bay (Schuster, 1951). ONTARIO. Sutton Ridge 54°25′ N. (Persson & Sjörs, 1960).

Ecology

Evidently similar to that of *L. collaris* and *L. gillmani*; a distinct calciphyte, like the other species of *Leiocolea*. Quebec material from "wet rocks beside stream ... among *Saccobasis, Preissia, Blepharostoma trichophyllum*" (Schuster, 1951). According to Müller (1951–58, p. 698), in large mats on wet, springy-moory sites, or on calcareous cliffs, and rarely also on siliceous rocks with a weakly basic reaction. Buch and Tuomikoski (1955) state that it occurs in Newfoundland in "more or less calcareous places, with *Fissidens* sp., *Plagiochila asplenioides, Riccardia pinguis, Bryum ventricosum, Campylium stellatum, Limprichtia revolvens, Preissia quadrata,* etc."

[65] *Holmen 62-1574* consists of few and sterile plants, which show a little of the fuscous pigmentation typical of arctic members of the *L. bantriensis-collaris* complex. They have the cells of leaf lobes and margins mostly 30-35 μ, and medially 32–36 × 35–40 μ, with rather well-developed, somewhat bulging trigones and a rather coarsely papillose cuticle. Cell size suggests *L. bantriensis*, rather than *L. collaris*, even though the plants are far inferior to "normal" *bantriensis* in size (leaves ca. 760 μ wide × 680 μ long; shoots ca. 1200–1400 μ wide). Small underleaves are present. Possibly a large-celled extreme of *L. collaris* is at hand.

The plant from Hurry Inlet, east Greenland, is typical and much more vigorous, with softer and more fragile leaves. Each of the 3 gynoecia checked lacked antheridia in subfloral leaf bases, and perianths remained juvenile (archegonia showing no fertilization). Hence a dioecious plant must be at hand. Leaf cells, although variable, averaged at least 28–35 to 30–36 μ in the leaf lobes and ca. 35–40 × 35–45 μ medially; the cuticle was conspicuously papillose-striolate. In spite of prolonged search for other collections of this species from Greenland, the Hurry Inlet station seems to be the only certain one, the Ikorfat specimen being sufficiently problematical to warrant possible exclusion.

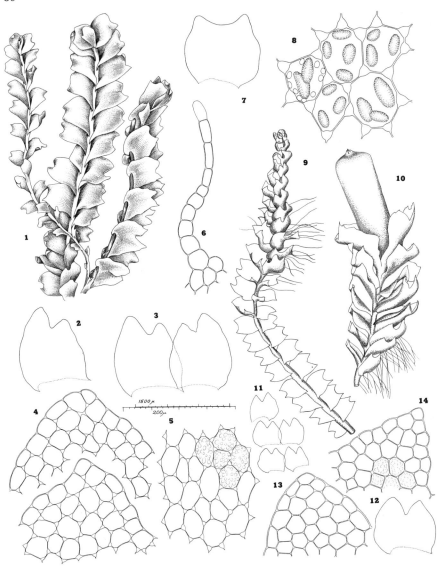

Fig. 168. *Lophozia* (*Leiocolea*) *bantriensis* (1–6) and *Lophozia* (*Leiocolea*) *collaris* (7–14). (1) Sterile shoots (×ca. 4.8). (2–3) Leaves (×17.5). (4) Leaf lobe apex cells (×156). (5) Median cells, cuticular papillae indicated in upper right (×156). (6) Underleaf (×156). (7) ♀ Bract from unfertilized gynoecium (×17.5). (8) Cells with oil-bodies and, at left, chloroplasts (×500). (9–10) ♂ and ♀ plants (×ca. 13). (11) Leaves (×17.5). (12) Leaf (×39). (13–14) Leaf lobe tips, a few cells with weak cuticular papillae drawn in (×156). [1, 9–10, after Müller, 1905–16; 2–6, from type of "*Acrobolbus septentrionalis*" Steph.; 7, *RMS 35067*, Alert, Ellesmere I.; 8, from *Wynne-Edwards 15*, Baffin I.; 11–14, fo. *pumila*, *RMS 46204*, Willoughby Mt., Vt.; 2–3, 11, all to top scale; 4–6, 13–14, all to bottom scale.]

Differentiation

This regionally very rare species, in general, is distinguished by the dioecious inflorescences, the warm brown color in exposed sites, the soft texture and often fragile leaves that tear readily, the usually convex, rather wavy leaves, the rather robust size, and large cells. The facies is often almost that of *L. rutheana*; it also bears numerous and unmistakable similarities to *L. collaris*, of which it has been considered to be a variety. It is possible that it is merely a large-celled phase of *L. collaris*; cytological study (Lorbeer, *in* Müller, 1954, pp. 698, 700) shows that both have a haploid number of 9 chromosomes and are cytologically indistinguishable. A comparative study of the capsule walls of the two species needs to be made. According to Meylan (1924), numerous "transitional" specimens occur.

The plant agrees with *L. collaris* in basic leaf shape; the small but constantly present, usually linear-subulate to lanceolate, sparingly dentate underleaves; the dioecious inflorescences; and the form of perianth. However, luxuriant phases of *L. collaris* sometimes have dentate ♀ bracts; the bract margins are never dentate in *L. bantriensis*. Of more practical importance, *L. bantriensis* has consistently larger leaf cells than *L. collaris*. In the Quebec plants, which are typical as to cell size, cells of the leaf apices averaged 35–38 μ (in other collections seen sometimes only 30–35 μ); median cells averaged 35–38 × 38–48 μ. As in *L. collaris*, trigones are always distinct and in extreme cases may be quite evidently bulging; the cuticle is always distinctly papillose or striate.

Sterile material is readily confused with the paroecious *L. gillmani*. The latter mostly has slightly smaller cells (almost intermediate between those of *L. bantriensis* and *L. collaris* in size), those of the leaf middle rarely averaging in excess of 32–35 μ wide, while apical cells average 27–33 μ. These differences in cell size are sufficiently slight so as to be unreliable. As a consequence, in the absence of other reliable stuctures allowing a separation of sterile plants, the distinction between *L. bantriensis* and *L. gillmani* must almost be based on the inflorescences and the form of the perianth (that of *L. gillmani* is typically more distinctly beaked).

For many years the name *L. hornschuchiana* was used for this species. Howe (1899, pp. 109–110) stated that the type material of *L. bantriensis* agreed with *L. hornschuchiana*. Since the anomalous mode of description of *L. bantriensis*, in a note to one of Hooker's plates, is scarcely adequate, some workers have until recently retained *L. hornschuchiana* as the correct name for this species.

The type of *Acrobolbus septentrionalis* Steph. (Estacado, Ore., *Foster, 8–7–1904*) is unquestionably *L. bantriensis*, of the form with obtuse leaf lobes. The sinuses are occasionally feebly gibbous and the leaves may be feebly falcate, because of the curvature of the usually larger ventral lobe. The underleaves are small but distinct, almost never lobed, but bear 1–2 short basal cilia ending in slime papillae. Median cells range from 35–40 × 45–55 μ and possess the large trigones and very strongly papillose cuticle of *L. bantriensis*.

LOPHOZIA (LEIOCOLEA) COLLARIS (Nees) Schust., comb. n.

[Figs. 168:7–14, 169]

Jungermannia collaris Nees, *in* Martius, Fl. Crypt. Erlang., p. XV, 1817.
Jungermannia muelleri Nees, *in* Lindenberg, Nova Acta Acad. Caes. Leop.–Carol. Nat. Cur. 14, Suppl., p. 39, 1829.
Jungermannia acuta Lindenb., *ibid.* 14, Suppl., p. 88, 1829.
Lophozia acuta Dumort., Rec. d'Obs., p. 17, 1835.
Lophozia muelleri Dumort., *ibid.*, p. 17, 1835.
Jungermannia barbata var. *collaris* Nees, Naturg. Eur. Leberm. 2:156, 182, 1836.
Jungermannia bantriensis var. *muelleri* Lindb., Acta Soc. Sci. Fennica 10:528, 1875.
Jungermannia bantriensis var. *acuta* Lindb., *ibid.* 10:528, 1875.
Jungermannia hornschuchiana var. *muelleri* Massal., Ann. Istit. Bot. Roma 3(2):162, 1888 [reprint, p. 8].
Lophozia (subg. *Leiocolea*) *muelleri* K. Müll., Rabenh. Krypt.–Fl. 6(1):719, fig. 327, 1910; Schuster, Amer. Midl. Nat. 45(1):48, 1951 (as *L. muelleri*); Schuster, *ibid.* 49(2):339, 1953.
Leiocolea muelleri Joerg., Bergens Mus. Skrifter 16:163, 1934; K. Müller, Rabenh. Krypt.–Fl. ed. 3, 6(1):699, fig. 220, 1954.

In mats (then prostrate to ascending) or scattered amongst caespitose mosses (then erect), green, in insolated sites the *leaf lobes sometimes fuscous brown to blackish*, medium-sized, *shoots 1–2(2.8) mm wide × 8–30(40) mm long*. Stems green, with age brownish beneath, 280–320 μ in diam., occasionally stouter, *11–14 cells high;* cortical cells thin-walled, or with age slightly and equally thick-walled, the dorsal and lateral 20–23(25–28) μ in diam. × (40)56–72 (85) μ long, rectangulate (*length mostly 2.5–4× the width*); ventral cortical cells often mycorrhizal with age, smaller, ca. (16)18–20 μ in diam.; medullary cells free from mycorrhizae, thin-walled and virtually without trigones, the dorsal 23–35 μ in diam., the ventral 2–5 layers gradually smaller below, (16)18–25 μ in diam.; stems simple or sparingly branched, often innovating beneath perianth. Rhizoids numerous to near shoot apex, pellucid to pale brownish. Leaves distant to weakly imbricate, *nearly flat or slightly concave*, barely decurrent antically, widely spreading laterally or sometimes antically secund, *broadly ovate to rotund-quadrate*, 575 μ long × wide to 500–540 μ wide × 450–480 μ long, to 650 μ wide × 525 μ long (♂ plants), occasionally larger on ♀ plants and then 850–950 μ long × 1350–1400 μ wide, *averaging 1.0–1.2(1.5)× as wide as long*, uniformly bilobed for ¼–⅓ *their length*, the sinus rectangulate to slightly obtuse, rarely crescentic, flat or slightly gibbous; lobes triangular, usually subequal, *acute and often with a 2-celled apiculus*, rarely narrowly obtuse. Cells of lobe margins *22–25 μ, of leaf tips 23–27 μ, of leaf middle (23)25–30 × (25)27–32 (36) μ*, at base to 30 × 40 μ; cell walls thin, *trigones small but discrete to very weakly bulging; cuticle delicately verruculose*, striolate near leaf bases and on stems; *oil-bodies 2–5(6–7) per cell, large*, grayish, finely granular-papillose in appearance, from 4.5 × 5–7 to 5 × 8–14 μ, occasionally to 4 × 10 or 6–9 × 13–21 μ, mostly ovoid or ellipsoid to fusiform; chloroplasts ca. 4 μ. *Underleaves discrete but small*, commonly 120–250(500) μ high, 3–5 cells wide at base, *lanceolate to subulate*, occasionally with 1–2(3–4) stalked slime papillae or short cilia at base, rarely bifid. *Gemmae absent.*

Dioecious. ♂ Plants slender (usually 1.0–1.2 mm wide). Androecia terminal, eventually intercalary, usually of 5–9 pairs of *closely imbricate and erect bracts, forming a rather compact spike;* bracts like leaves, sometimes somewhat smaller, but erect and strongly concave, their antical margin dilated and often bearing a large, often inflexed tooth; antheridia 1–2, sometimes amidst paraphyses. Gynoecia with erect-spreading or suberect bracts, the bracts bilobed for ca. ¼–⅙ their length, hardly to distinctly larger than leaves, rotund-ovate to oblong-ovate, lobes entire or *sometimes weakly dentate*, acute or obtuse; bracteole small, bifid, with lanceolate lobes bearing 1–few lateral teeth. Perianth 0.75–0.85 emergent, usually tubular, occasionally cylindrical-clavate, smooth, suddenly contracted at apex into a *short beak;* beak delicately ciliolate-crenulate. Capsule with epidermal cells of wall with nodular thickenings; inner cell layer with subcomplete or incomplete semiannular bands. Elaters 7–8(10) μ in diam., bispiral; spores 12–15 μ, delicately verruculose, brown. Chromosome no. $n = 9$.

Type. The type of *J. muelleri* is from Zweibrücken, Germany (Müller); that of *J. collaris* is from "Helvetia: prope Basileam (Fr. N. ab Esenb.),'' *fide* G., L., & N., Syn. Hep., p. 125, 1844.

Nomenclature

The plant long known as *L. muelleri* apparently cannot legally retain this name, since the earliest validly published designation appears to be *Jungermannia collaris* Nees, in the preface to Martius, Fl. Crypt. Erlang., p. XV, 1817. Schiffner (1900e) devoted a special paper to *J. collaris* and studied the type; he found that, unquestionably, the ♂ plant of *"Lophozia muelleri"* was at hand. Since it would appear that the first valid publication of *J. collaris* was in 1817, *not* in Nees's Naturg. Eur. Leberm., 1836, we have no alternative but to use this name. Schiffner (*loc. cit.*, p. 5 of reprint) points out that the citation of *J. collaris*, in the Martius paper "gehört hierher" on the basis of the original diagnosis. And in G., L., & N., Syn. Hep., p. 125, 1844, we find that" *Jungermannia collaris* N. ab E. in Praef. ad Mart. Fl. cr. Erl. p. XV" is cited as a synonym of "*Jungermannia barbata* var. *collaris* N. ab E.," excluding "syn. et fig. Mart.''

Distribution

A widespread and evidently common species throughout much of Europe, ranging from alpine and subalpine localities to the lowlands, from the Arctic of Spitsbergen southward from northern Scandinavia (Norway, Sweden, Finland) to Denmark, Germany, Belgium, France, Switzerland, northern Italy, northern Spain (Casares Gil, 1919) and Istria, Dalmatia, Bosnia, the Carpathians, and the Trapezunt. Also in the Tatra Mts. of

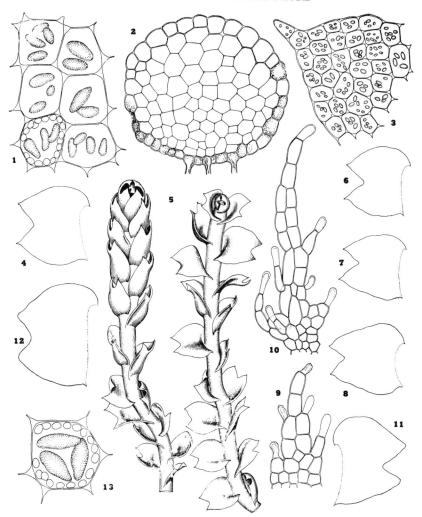

Fig. 169. *Lophozia (Leiocolea) collaris* (= *muelleri*). (1) Median cells with oil-bodies and, lower left, chloroplasts (×440). (2) Stem cross section, ventral cortical cells mycorrhizal (×154). (3) Dorsal lobe apex, ♂ plant (×200). (4) Leaf (×32). (5) Androecial and sterile shoots (×22). (6–7) Leaves (×32). (8) Leaf, ♂ shoot (×32). (9) Underleaf (×210). (10) Underleaf (×125). (11–12) Leaves from ♀ plant, that in fig. 12 an upper leaf (×20). (13) Median cell, with oil-bodies to 8 × 20 μ (×530). [1–9, *RMS 35067;* 10–13, *RMS 35070a,* both from Alert, Ellesmere I.]

Czechoslovakia, eastward to Lithuania. Infrequent in England (north to Shetland) and Ireland.

Also ranging eastward into Asia (*teste* Lindberg and Arnell, 1889). Also on Jan Mayen I.

In North America evidently rare, except perhaps in the high north; the North American distribution has been badly confused with that of *L. badensis*.[66] Cited in the Far West from British Columbia and Alberta eastward to Saskatchewan, southward to Colorado and supposedly to California (see Sutcliffe, 1941), most of these reports needing verification. In eastern North America certainly known only from:

N. GREENLAND. Graptolith Elv, N. coast of Independence Fjord, 82°14′ N. 24° W. (*Troelsen 6453*!; reported as *L. groenlandica* by Arnell, 1960).[67]   NW. GREENLAND. Kangerdlugssuak, S. side of Inglefield Bay (*RMS 45834;* probable, but possibly a totally gemma-free form of *L. heterocolpa*!).   W. GREENLAND. Magdlak, Alfred Wegeners Halvö, 71°07′ N., 51°48′ W. (*Holmen 15,756, p.p.*; amidst *L. gillmani, Peltolepis, Sauteria*); Kobbefjord (*Warming & Holm 1884*!; mixed with *Saccobasis*).[68]   ELLESMERE I. Crevices of soil polygons, calcareous foot of slope at Alert, ca. 82°30′ N. (*RMS 35067, 35070a*); Cape Belknap, 82°31′ N., 62°25′ W. (*RMS 35288*, etc.; northernmost report); near Ft. Juliane, Beitstad Fiord, Cape Viele, and Twin Glacier Valley, in Hayes Sd. region; Fram and Goose Fiords, S. coast; Reindeer Cove, W. coast (*Simmons*, ex Bryhn, 1906).   BAFFIN I. Pond Inlet (Polunin, 1947).   HUDSON BAY REGION. Tukarak I., Belcher Isls., N.W.T. (Schuster, 1951).   NEWFOUNDLAND. Limestone barrens N. of Daniels Harbour, N. Pen. (*RMS 68-1452a*).

QUEBEC. Island near mouth of Seal R., Cape Jones, E. coast of Hudson Bay (Schuster, 1951).   MIQUELON I. *Fide* Delamare et al. (1888, p. 65); but note that the more careful collections of Tuomikoski (see Buch & Tuomikoski, 1955) fail to establish the species in contiguous Newfoundland.   VERMONT. Willoughby Mt., Willoughby L. (*RMS 46204;* fo. *pumila*).   MINNESOTA. N. of Wabasha City, Wabasha Co. (*RMS 17230*);

[66] This confusion extends into Frye & Clark (1937–47, p. 387), who persist in citing old reports from Connecticut (of Evans & Nichols, 1908; see correction of this to *L. badensis* in Evans, 1910b), New York, and Quebec (see statement in Evans, 1916), all of which are unreliable or are known to be incorrect. They also cite "Examinations" from two localities in Quebec (*Macoun, 1879, 1882,*) and from Little Moose Lake, N.Y. (*Haynes, 1913*). Evans (1916) and Lepage (1944–45) state reports of the species from Quebec (Macoun, 1902) are ill founded.

Other reports, from the Rocky Mts. region and westward, need verification, especially since most reports from the East have been shown to be incorrect.

[67] The Troelsen specimen from Independence Fjord was already annotated by E. Clausen as "*Leiocolea*?" The gemma-free plants bear underleaves; leaf cells are 25–30 to 27–34 μ medially; cells of lobe margins are 24–27 μ; the cuticle is distinctly papillose. The presence of underleaves eliminates *L. badensis*; the sharp to apiculate leaf lobes and larger cells, mostly with weakly or not bulging trigones, appear to eliminate *L. heterocolpa*.

Hesselbo (1923a) had already reported a plant from I.P. Koch Fjord as "*L. muelleri.*" The plants are, however, provided with reddish angular gemmae; the broad, lax leaves, deep sinus, and large cells (24–25 μ on lobe margins) suggest *L. alpestris polaris*, but the material is too poor for absolute identification.

[68] The Magdlak specimen is of definitely dioecious plants, with underleaves, a rough cuticle, and cells of lobes ca. 25 μ. The sparing Kobbefjord plants are poorly developed, yet I cannot place them elsewhere. All gynoecia seen had failed to mature perianths and lacked all trace of antheridia from subfloral leaves and bracts; hence a dioecious population must be at hand.

bluff below the Sugarloaf, Winona, Winona Co. (*RMS 17257, 17253*, etc.); these collections referable to fo. *pumila* Nees.

Also reported, with considerable doubt, from the Lake Superior coast of Wisconsin (Conklin, 1929) and from western New York (Boehner, 1943; however I have not seen the species from this area; Schuster, 1949a). Neither of these reports is adequately established.

Ecology

In the arctic Tundra (northern Ellesmere I.), growing in moist crevices of calcareous clay-shale polygons, as well as ubiquitous on calcareous solifluction slopes, below snow banks, associated with *Arnellia fennica, Cephaloziella arctica, Solenostoma pumilum polaris, Scapania gymnostomophila, S. cuspiduligera, Sauteria alpina, Athalamia hyalina*, as well as with a series of mosses and various angiosperms (*Saxifraga oppositifolia, S. caespitosa, Festuca baffinensis, Papaver radicatum*, etc.). The species here is usually found in small quantities, growing crowded and erect, among more robust mosses; it often occurs as a slender form, only 1–1.5 mm wide, whose leaf tips may be quite fuscous-pigmented (particularly on ♂ bracts).

In the Hudson Bay area on basic soil, over basic rocks, sometimes at the edges of small ponds, associated with *Tritomaria scitula, Aneura pinguis, Solenostoma oblongifolia*(?), and *Blepharostoma trichophyllum brevirete* (Schuster, 1951).

On the bluffs along the Mississippi R., in southeastern Minnesota, *L. collaris* occurs as a relict at the edge of the Driftless Area, under very different conditions. Here on friable calcareous sandstone bluffs, forming compact, dense, often extensive patches and mats. Associated are *Preissia quadrata, Mannia rupestris*, and *M. sibirica* (Schuster, 1953), as well as, in the same general area, the equally disjunct *Athalamia hyalina* and *Asterella saccata* (Schuster, 1953). The marked "northern" aspect of this community is remarkable.

Variation and Differentiation

A variable and difficult species, differing greatly in size with dissimilar habitats. In the far north often with the growth habit and facies of a small, lax *Lophozia excisa* and *latifolia* (both of which may be essentially gemma-free). It differs from these in the (*a*) homogeneous medulla of the stem, without a mycorrhizal band; (*b*) slightly larger cells; (*c*) usually more distinct trigones and papillose cuticle; (*d*) large and few oil-bodies per cell; (*e*) presence of small but distinct underleaves. The last two characters are by far the most important.

Lophozia collaris is likely to be confused with several other species of subg. *Leiocolea*, chiefly *L. gillmani, L. bantriensis*, and *L. badensis*. It is usually smaller than *L. gillmani* and much smaller than *L. bantriensis*, but bears similar underleaves. However, all forms of *L. collaris* differ in the small median cells, which average only 25–30 μ wide in the leaf middle (at least 30–36 μ and 35–40 μ wide in *L. gillmani* and *L. bantriensis*, respectively). When fertile, the dioecious inflorescence eliminates *L. gillmani*.[69] In *L. bantriensis* the leaf cells are so much larger (30–35 μ in the leaf tips; 35–40μ in the leaf middle) that confusion with *L. collaris* is hardly possible.

However, *L. collaris*, particularly the smaller phases ("fo. *pumila* Nees"), may easily be mistaken for *L. badensis*. These may (*fide* Schuster, 1953, p. 339) be only 0.5 mm wide × 8–15 mm long, thus closely approximating smaller phases of *L. badensis*. *Lophozia collaris*, however, can be separated by the smaller cells (22–27 μ in the leaf lobes, rarely averaging as high as 30 μ; in *L. badensis* 30–35 μ in the lobes), the distinct trigones, the usually delicately verruculose cuticle, and the distinct, if small, underleaves. The underleaves may be almost obscured by the rhizoids but can almost always be readily demonstrated at apices of sterile plants.[70]

When fertile plants are available, *L. collaris* is more readily identifiable. Androecia are compactly spicate, with the erect and concave bracts closely approximated or even imbricated; in *L. badensis* androecia are loosely spicate, with usually more or less remote and spreading bracts. Often *L. collaris* has slightly dentate lobes of the perichaetial bracts, which, in all our other species of *Leiocolea*, are simply bilobed, with entire margins.

---

[69] Frye and Clark (1937–47, p. 380) separate *L. gillmani* from "*L. muelleri*" on the basis of the trigones—supposedly "small" in the former vs. "medium large to bulging" in the latter. As has been pointed out (Schuster 1951a), *L. gillmani* often has large, bulging trigones, whereas "*L. muelleri*" frequently has them small and concave-sided; there is no reliable difference between the two taxa on that basis. These authors also "separate" *L. harpanthoides* from "*L. muelleri*" (and a series of other species) on the basis of the supposedly colorless cell walls of "*L. muelleri*" vs. the "yellowish" walls of *L. harpanthoides*. In all our species of *Leiocolea* sun forms may show yellowish to brownish, or fuscous, pigmentation; no distinction on such a basis is possible.

[70] The distinction between *L. collaris* and *L. badensis* becomes hard to maintain when small sterile phases (such as *RMS 35288*, Ellesmere I.), tentatively referred to *L. collaris*, are studied. These are only 500–750 μ wide and have somewhat concave and suberect leaves, the leaf lobes fuscous to purplish-fuscous pigmented. Underleaves are reduced to 2–3 stalked slime papillae, and closely approximate those sometimes present in typical *L. badensis*. Yet, leaf cells are clearly as small as in *L. collaris* (averaging 25 μ in the leaf lobes; 23–28 × 25–32 μ in the leaf middle) and have rather distinct trigones and a ± verruculose cuticle. *If* cell size is a reliable criterion, these plants must be referred to *L. collaris*, in spite of having vestigial underleaves as in *L. badensis*.

Plants from Willoughby, Vt. (*RMS 46024*, occurring with *Athalmia hyalina*), are equally impossible to place definitely. These are delicate and laxly leafy, only 500–700 μ wide, and have nearly leptodermous, essentially smooth or obsoletely verruculose leaf cells; the ventral merophytes are 2 cells broad. Habitually they agree with *L. badensis*, to which I initially referred them. However, sterile stems consistently bear minute underleaves, usually formed of 2–4 superposed cells in a row, ending in a slime papilla, the whole structure inserted on a base 2 cells broad. Furthermore, cells of the lobes are only (22)23–25 μ, and median cells ca. 25–28 × 28–30 μ. These plants, like those above, must remain in the questionable category and appear to represent Nees's fo. *pumila*, whose status needs experimental study.

However, smaller phases of *L. collaris* lack such marginal dentition; therefore this character cannot be relied upon.

On the basis of the small cell size and small underleaves, *L. collaris* has been confused with *L. heterocolpa*. The latter, however, almost invariably has gemmae present and differs also in *usually* having yet smaller cells, only 18–24 μ in the lobes and 24–30 μ in the leaf middle. Cell size differences are not fully reliable. Sun forms of the two species are often similar except for pigmentation (*L. collaris* is fuscous, *L. heterocolpa* a warm brown). In general, *L. heterocolpa* has narrower (clearly longer than broad) and more shallowly bilobed leaves. In *L. collaris* the lobes are often apiculate, ending in a row of two superposed cells. This is rarely the case in *L. heterocolpa*.

There are considerable, and sometimes confusing, differences between the two sexes of *L. collaris*. ♂ Plants may hardly approach 1.2 mm wide and have leaves commonly only 1.0–1.2 × as wide as long; the leaf lobes are often apiculate and terminate in 2 superposed cells. The cells are commonly provided with smaller oil-bodies, rarely over 5–6 × 12 μ. ♀ Plants from the same area are up to 2.4 mm wide and have leaves which are often 1.2–1.45 × as wide as long; the leaf lobes are bluntly terminated, and the cells have oil-bodies frequently as large as 6–9 × 18–21 μ or even 9–10 × 16–18 μ. The differences in leaf shape can be due to allometric growth, leaf width increasing at a more rapid rate, with increase in general size, than leaf length. (This phenomenon is equally marked in other dioecious Lophoziaceae, such as *L. latifolia*). In the far north *L. collaris* is generally without perianths. ♂ Plants are often very slender, and the ♂ bracts (which may lack the antical basal tooth) often fail to produce mature antheridia. A plant of this type is illustrated in Fig. 169:5.

Extreme arctic phases of *L. collaris* may also be unusually compact, with leaves antically secund and suberect, almost hemispherically concave and distinctly imbricate, resulting in a nearly terete shoot. Such phases usually grow in full sun on damp, calcareous, polygon soil (associated with *Drepanocladus* and other hygrophytic mosses) and are almost black. Under the microscope the cells, although not strongly collenchymatous, have purplish black to purplish brown walls. The upper leaves of such plants (at least those of the ♀ sex) may be very broad, to 1450 μ wide × 950 μ long, with incurved lobes and a crescentic sinus. Leaf lobes of such extreme sun forms are sometimes blunt or evenly narrowly rounded.

LOPHOZIA (LEIOCOLEA) HETEROCOLPA (Thed.)
Howe
[Figs. 163:3–12, 170]

Jungermannia heterocolpos Thed., Kgl. Sv. Vetensk.-Akad. Handl., 1838:52, pl. 1, 1839.
Jungermannia muelleri var. *heterocolpos* Nees, *in* G. L. & N., Syn. Hep., p. 99, 1844.
Jungermannia muelleri fo. *attenuata-gemmipara* Jack, Leberm. Badens, p. 36, 1870.
Jungermannia wattiana Aust., Bull. Torrey Bot. Club 3:11, 1872.
Jungermannia muelleri var. *danaensis* G. ex Underw., Bot. Gaz. 13:114, pl. 6, 1888.
Jungermannia muelleri var. *maritima* Pears., List Canad. Hep. (Canad. Nat. Hist. and Geol. Surv.), p. 24, 1890 (*nomen nudum*).
Lophozia heterocolpa Howe, Mem. Torrey Bot. Club 7:108, 1899; Evans, Rhodora 14:211, 1912.

Lophozia (subg. *Leiocolea*) *heterocolpa* K. Müll., Rabenh. Krypt.–Fl. 6(1):727, fig. 329, 1910;
   Macvicar, Studs. Hdb. Brit. Hep. ed. 2: 173, 1926; Schuster, Amer. Midl. Nat. 45(1):47,
   1951; Schuster, *ibid.* 49(2):336, pl. 9, 1953.
Leiocolea heterocolpa Buch, Mem. Soc. F. et Fl. Fennica 8 (1932):284, 1933; K. Müller, Rabenh.
   Krypt.–Fl. ed. 3, 6(1):694, fig. 218, 1954.

Medium sized, forming patches or more frequently scattered among other
bryophytes, green (shade forms) but more *frequently locally golden brown to reddish
brown* in color; shoots (1.5)2.0–2.4(3.0) mm wide × 8–15 mm high, prostrate
to ascending, frequently with *shoot apices suberect or erect (especially when gemmiparous)*, simple or sparingly branched, *branches often terminal-lateral;* frequently
innovating beneath gynoecia. Stems 325–400 μ in diam.; dorsal cortical cells
moderately thick-walled, (14)15–18(20) μ wide × 40–54(60) μ long (averaging
2.5–4× as long as wide), distinctly narrower than basal leaf cells. Leaves
contiguous to imbricate, spreading at an angle of 65–85°, almost horizontal
to (xeric phases) somewhat antically secund and decurrent, *ovate to roundish
ovate to rotundate, slightly longer than wide* to slightly wider than long (ca. 1000 μ
long × 900–975 μ wide to 900–1000 μ long to 1050–1175 μ wide), simply
bilobed, the sinus descending 0.2–0.33 the leaf length, *often gibbous or reflexed*,
narrow, acute to rectangular, more rarely obtuse; lobes triangular, acute to subacute, less often obtuse or rounded. *Cells small, strongly collenchymatous*, the lumen
polygonal-stellate, thin-walled, but usually with large, bulging, often contiguous
trigones, the *trigones and walls usually strongly pigmented; cells of leaf margins and of
apices* (18)20–25 μ, of leaf middle 23–25 × 25–30(32) μ, near base becoming
23–27 × 30–38 μ; cuticle *strongly verruculose;* oil-bodies usually 2–4(5) per cell,
large, ovoid to ellipsoid to fusiform, ca. 5–6 × 8–9 μ to 6–7 × 13–15 μ, some
spherical and (4.5)5–7 μ, grayish, finely granular-papillose, sometimes largely
occluding cell lumen. *Underleaves distinct, ovate-lanceolate to lanceolate*, usually
2–4(5–7) cells wide at base, often with 1–2(3–4) short cilia or stalked slime
papillae at base but otherwise entire, rarely bifid. *Gemmae almost constantly
present, at apices of strongly ascending to erect, slightly to strongly attenuate shoots,
(1)2-celled, brownish*, from 13 × 18–21 to 15–16 × 22–24 μ, rarely to 17 × 28 μ
long, broadly to narrowly ovoid or oblong, commonly faintly constricted medially; gemmiparous shoots with oblong, shallowly bilobed, *small*, rather loosely
erect-appressed, more or less imbricate leaves and similar, large underleaves.

   Dioecious.[71] Androecia *compactly spicate,* ♂ *bracts strongly ventricose*, 2-lobed,
in 3–4(6) pairs. Antheridia 1–3 per bract; stalk biseriate, to 90–100 μ long;
body ovoid, ca. 160 × 180–200 μ. Gynoecia with bracts somewhat larger than
leaves, ca. 1550–1660 μ wide × 1350–1500 μ long, erect and (often closely)
sheathing perianth at base, with erect-spreading to somewhat squarrose obtuse
lobes, ovate-quadrate to subrotund-quadrate, 2- (rarely 3-) lobed for ⅙–⅕
their length, the lobes obtuse, the sinus often gibbous and reflexed; postical
bases often with 1–3 slime papillae, occasionally situated at apices of stalks.
Bracteole bifid, to 800 μ wide × 900 μ long, the bases and lobes with several
stalked slime papillae and short cilia or teeth. Perianth 0.5–0.6 emergent,
cylindrically oblong-ovoid, smooth and eplicate, *contracted distally to at least a
rather shortly beaked mouth;* mouth crenulate-dentate with projecting cells, the

---

[71] Frye and Clark (1937–47 p. 388) state, incorrectly, "probably bisexual."

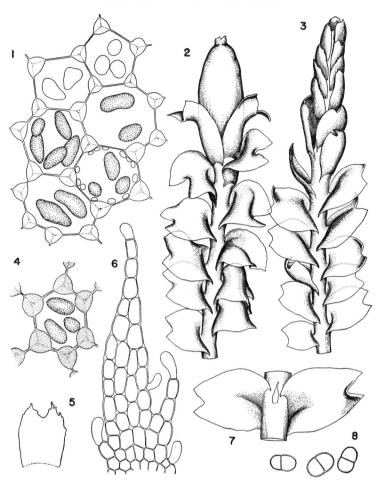

FIG. 170. *Lophozia* (*Leiocolea*) *heterocolpa*. (1) Submedian cells with oil-bodies and, lower right, chloroplasts (×545). (2) Perianth-bearing shoot (×16). (3) Gemmiparous shoot (×16). (4) Cell of extreme mod. *colorata*, approaching var. *harpanthoides* (×545). (5) Gemmiparous leaf (×30). (6) Underleaf (×ca. 280). (7) Shoot sector, postical aspect (×18). (8) Gemmae (×345). [All from *RMS 1950ia*, Porcupine I., Susie Isls., Minn.]

cells often somewhat irregular and curved, free at their tapering apices, ca. 16–18 *μ* wide × 35–46 *μ* long. Spores 11–13 *μ*; elaters 7–8 *μ* in diam.

Type. Sweden.

Distribution

A widespread imperfectly holarctic arctic-alpine species (but unreported from Japan), with isolated stations (as a relict) in subarctic-subalpine regions; locally frequent in the Great Lakes area. Distribution sharply circumscribed by restriction to somewhat calcareous sites. In Europe widespread from Spitsbergen (Arnell and Mårtensson, 1959) and Scandinavia (Norway, Sweden, Finland, north to Lapland), southward to Britain (Scottish Highlands), local in alpine situations (Switzerland, French Pyrenees, and the Auvergne Distr.; Savoy; the Tyrol; Austria; Bavaria; Baden; the Riesengebirge; the Harz Mts.; Westphalia; Czechoslovakia; Hungary), and southward to northern Spain (Prov. Huesca), the north Italian Alps, and Trieste. Also in Madeira.

Eastward extending to Siberia (Novaya Zemlya; eastern Siberia; the Yenisey R. valley). Northwest Himalaya (Rakhios Valley at 12,000 feet). Also in Iceland.

In North America transcontinental, occurring in the West from Alaska and Yukon, southward to British Columbia, Alberta, Saskatchewan, Washington, Oregon, Montana, Idaho, and California. In our area:

E. GREENLAND. Scoresby Sund: Danmarks Ø; Gaaseland, with *Astrophyllum hymenophylloides* (Jensen, 1897); Hurry Inlet (Jensen, 1900); Nordostbugt (Jensen, 1906a); Gaaselandet, W. end of Scoresby Sund, ca. 70°20′ N. (*Hartz, July 10, 1892!*).   N. GREENLAND. Low Pt., 83°06′ N. (*Th. Wulff, July 10-13, 1917!*); all other N. Greenland specimens seen belong ± definitely to var. *harpanthoides*.   NW. GREENLAND. Kekertat, Inglefield Bay (*RMS 46009*); Dundas, Wolstenholme Fjord (*RMS 46268a*).   W. GREENLAND. Ikorfat, Nugssuaq Pen., 70°45′ N., 53°07′ W. (*Holmen 1574*, among L. (?) *bantriensis*, etc., approaching var. *harpanthoides*); Sondrestrom Fjord, near shore of L. Ferguson, 66°5′ N. (*RMS & KD 66-080a*); Fortune Bay, Disko I., 69°16′ N., 53°50′ W. (*RMS & KD 66-147, 66-141, 66-135*); Magdlak, Alfred Wegeners Halvö, 71°06′ N., 51°40′ W. (*RMS & KD 66-1352, 66-1356c, 66-1358a*). Godhavn, Disko I.: basalt cliffs, near shore, below Skarvefjeld, 69°16′ N., 53°28′ W. (*RMS & KD 314a, 66-309a, 66-320b*); mouth of Blaesedalen (*RMS & KD 66-287*); slopes below Skarvefjeld, ca. 150–200 m (*RMS & KD 66-231*); Lyngmarksbugt, ca. 25–125 m (*RMS & KD 66-342, 66-569*). Mouth of Simiutap Kûa Valley, Umîarfik Fjord, 71°59′ N., 54°35-40′ W. (*RMS & KD 66-918, 66-915, 66-914b*); Eqe, near Eqip Sermia (Glacier), ca. 1 m from Inland Ice, 350–450 m, 69°46′ N., 50°10′ W. (*RMS & KD 66-339*); head of Marmorilik Fjord, N. of Akuliarusikavsak, 71°05′ N. 51°12′ W. (*RMS & KD 66-1206*).

ELLESMERE I. N. slope of The Dean, 5–6 mi SE. of Alert, and N. slope of Mt. Pullen, 5 mi S. of Alert, NE. Ellesmere, ca. 82°26′ N., 62°15′ W. (*RMS 35121, c. per.!*; *35160a*, etc.); near Ft. Juliane, Beitstad Fiord, Cape Viele, Lastraea Valley, Bedford Pim I., in Hayes Sd. region; Harbour, Musk Ox, and Goose Fiords, S. coast (Bryhn, 1906). NORTHWEST TERRITORY. Tukarak I., Belcher Isls,. Hudson Bay (Schuster, 1951). QUEBEC. Manitounuck Sd., Fishing Lake Cr. and Cairn I. at Richmond Gulf, and Island near mouth of Seal R. at Cape Jones, all on E. coast of Hudson Bay (Schuster, 1951); Ft. Chimo (Lepage, 1953); Bic (Evans, 1916); Iberville; St.-Rose du Dégelis, côté de Témiscouata; Côte de Gaspé (as *Jungermannia muelleri* var. *maritima*) (Lepage, 1944–45); E. side of Mt. Ste-Anne, above The Grotto, Percé, Gaspé (*RMS 44021*). NEWFOUNDLAND. Benoit Brook, W. Nfld.; Port Au Choix, St. John I., St. Barbe Bay, Flower Cove, Cook Harbour, Burnt Cape, Ha-Ha Bay, Bartlett's R., all in N. Pen. (Buch & Tuomikoski, 1955); Steady Brook, W. Nfld. (*RMS 68-1397*); N. of Daniels Harbour, N. Pen. (*RMS 68-1458*).

MAINE. Round Mt., Franklin Co. (*Lorenz*; Evans, 1912d, p. 211). NEW HAMPSHIRE. Beaver Falls; Colebrook; Alpine Cascade, Berlin.   VERMONT. Just N. of Smuggler's

Notch, NE. of Mt. Mansfield (*RMS 661853, p.p.*; with *Solenostoma sphaerocarpum*). NEW YORK. Essex Co.: Wilmington Notch, 6.5 mi SW. of Wilmington, 1500–2000 ft; Cascades, between Upper and Lower Cascades Ls., 6.5 mi SE. of L. Placid, 2400–3200 ft. Ulster Co.: Esopus Gorge, 0.5 mi N. of Atwood, Marbletown, 280 ft; Hillyer Ravine along Kaaterskill Cr., 2.5 mi SE. of Haines Falls, 950 ft; Kaaterskill Falls, 1.75 mi E. of Haines Falls, 1900 ft; Twilight Park, 0.75 mi SE. of Haines Falls (all N. G. Miller, 1966; also collected at Kaaterskill Falls by RMS).

ONTARIO. Tobermory, at tip of point (*Schuster, 1949*); Otter Head, NE. of L. Superior (Macoun, 1902); near Beardmore, Thunder Bay Distr. (Cain & Fulford, 1948). MICHIGAN. Amygdaloid I., Isle Royale (*RMS*); Copper Harbor, Keweenaw Pen., Keweenaw Co. (*RMS*); Port Austin, limey sandstone lake bluff; Mott I., Isle Royale (Conklin, 1914a, p. 53); mouth of Montreal R., Gogebic Co., and Phoenix, Keweenaw Co. (Evans & Nichols, 1935). WISCONSIN. Sand I., Apostle Isls., Bayfield Co. (*RMS*); Ashland, Bayfield, Douglas, and Iron Cos. (Conklin, 1929). MINNESOTA. Thompson, Carlton Co. (*Conklin 2441a*); Porcupine, Belle Rose, Sailboat, Long, Big Susie, and Little Susie Isls., Susie Isls., NE. of Grand Portage, Cook Co. (*RMS 12250b, 11752, 13042, 5726c, 14716, 7249, 7179a, etc.*); Grand Marais, at The Point (*RMS 6556, etc.*); Lutsen, Cook Co. (*Conklin 1118*); L. Superior shore 1 mi S. of Temperance R., Cook Co. (*RMS 6504, 13261, etc.*); Temperance R., Cook Co. (*RMS 14706*); Cascade R., Cook Co. (*Conklin 2352b*); Two Island R. and Knife R., Lake Co. (*Conklin 2583, 800*); Manitou R. (*RMS 14150*); Chester Cr., Fairmount Park, Condon Park, Spirit L., and French R., all near Duluth, St. Louis Co. (*Conklin 878, 702, 1646, 1566, 1582, 1604*).

Also cited from OHIO: Hocking Co. (Miller, 1964); a report requiring verification.

Ecology

Common on basic intrusive rocks and over calcareous sedimentary rocks, under Taiga conditions

associated most frequently with other lime-tolerating species such as *Tritomaria scitula* and *T. quinquedentata*, *Odontoschisma macounii*, *Scapania gymnostomophila* etc. The species is less distinctly a calciphyte than the related *L. gillmani* (or any other species of *Leiocolea*). It generally invades calcareous rocks after ecesis of pioneer species, after preparation of the substrate and deposition of considerable humus materials .... Very occasionally the species appears to be a pioneer on soil in crevices, associated with such calciphiles as *L. gillmani*, but more often the more humicolous *Blepharostoma trichophyllum*, which is associated with it under a pH range of from 6.0–4.8. When the pH is lowered (by the deposition of humus) to 5.0 or lower, *L. heterocolpa* begins to disappear. The species is almost always a distinct mesophyte and has a rather large toleration for direct sunlight (Schuster, 1953).

In addition to the preceding species, *L. heterocolpa* also occurs in the low Arctic with *Scapania degenii*, *S. calcicola*, *S. cuspiduligera*, *Preissia quadrata*, *Cephalozia pleniceps*, *Lophozia hatcheri*, *L. barbata*, *Plagiochila asplenioides*, *Tritomaria polita*, and more rarely *Lophocolea minor*, variously on slatey dolomite talus, in crevices of wet sedimentary rocks, and on muck over diabase trap rocks (Schuster, 1951).

Widespread in the high Arctic, often on mossy talus slopes below persisting snowbanks, usually over basic sedimentary rocks as a secondary species, with

Tritomaria quinquedentata, T. heterophylla, Lophozia quadriloba, Cephaloziella arctica, Blepharostoma trichophyllum brevirete, Scapania praetervisa, and *Anthelia juratzkana,* occasionally with *Arnellia fennica* and *Odontoschisma macounii.*

There are occasional reports (e.g., Evans and Nichols, 1935) of the species from decaying logs. I have seen it a single time from such a site (Copper Harbor, Mich.; Schuster, 1953); such occurrences are sporadic and depend on a saxicolous reservoir.

Reimers (1940a), Meylan (1924), and Müller (1938b) also discuss the ecology of *L. heterocolpa.* Both Müller and Reimers consider it, as does Lorenz (1911), a basophyte. Reimers emphasizes that the species, although closely allied to *L. collaris,* differs from the latter in its behavior. *Lophozia collaris* occurs earlier, successively, invading bare moist faces of calcareous rocks (in some cases gypsum), whereas *L. heterocolpa* appears later, after further colonization and often after the raw substrate has been soil- or peat-covered. According to Reimers, *L. heterocolpa* persists until such oxylophytes as *Tetraphis pellucida* and *Lophozia incisa* undergo ecesis. By contrast, Meylan (1924) regards *L. heterocolpa* as a "sous-espèce ou race humicole, calcifuge" of *L. collaris*; this is without foundation, although *L. heterocolpa* is frequently humicolous. Steere (1937) also noted that the species is usually quite secondary in occurrence and, although rarely, is also found on decaying logs.

Differentiation and Variation

Typical *L. heterocolpa* is usually easily identified. It differs from all other species of subg. *Leiocolea* in the ability to develop gemmae. The erect to stiffly ascending, somewhat attenuated, rather pale brownish, gemmiparous shoots are almost always abundantly produced. These shoots bear loosely appressed or erect-appressed reduced leaves and closely similar underleaves. Ability to develop such shoots recalls *Lophozia attenuata* (and the allied *L. atlantica*) and *Anastrophyllum hellerianum.* However, the gemmiparous shoots are stouter and usually less attenuated than in these species, and also bear smooth, ovoid-oblong, rather than angular, gemmae. *Lophozia heterocolpa* is further distinguished from our other species by the warm, brownish secondary pigmentation often present and by the distinctly smaller cell size.

Occasionally plants occur under relatively xeric conditions on banks (associated with *Odontoschisma macounii,* etc.) and then may produce gemmiparous shoots rarely or not at all. Such impoverished plants often have the leaves rather imbricate and concave, and have . . . very large and often reddish-brown or brownish trigones. Such plants may be confused with other species of *Leiocolea* (but differ in the smaller cells of the lobes, which are generally only 18–24 μ) or with *Lophozia porphyroleuca* (from which they differ in the distinctly papillose cuticle and presence of underleaves). The underleaves are lanceolate and usually quite obvious, much as in *Harpanthus scutatus.*

Such xeromorphic phases were noted by Schuster (1953) in material from northeastern Minnesota; they closely approximate the var. *harpanthoides.*

High arctic phases of *L. heterocolpa* with few or no gemmae are frequent. These usually grow crowded amidst various caespitose mosses or Hepaticae and have antically secund, sometimes almost appressed leaves. Leaves on such plants are often wider than is usual and may run from 1000 μ wide × 800 μ long up to 1100–1175 μ wide × 900–950 μ long (from 1.15 to 1.3× as wide as long); they usually have obtuse or rounded lobes. Plants of this type might be sought under *L. collaris* because of the rare occurrence of gemmae and the similar cell size. They differ from that species because at least the exposed leaf apices are a warm, fulvous brown (rather than the fuscous or blackish color of exposed phases of *L. collaris*). Usually prolonged search will demonstrate a few gemmiparous plants, even in such extreme phases of *L. heterocolpa* (as, e.g., in material from The Dean in northeast Ellesmere I.). When gemmae are totally lacking, "*L. harpanthoides*" appears to be involved; indeed, such extremes probably should all be referred to var. *harpanthoides*. Although totally satisfying proof is lacking, it is almost certain that the supposedly specifically distinct "*L. harpanthoides*" intergrades with *L. heterocolpa* and represents merely a gemmae-free extreme of strongly insolated sites. It will be noted that strongly brownish-tinged forms alluded to previously, with antically secund, imbricate, concave leaves, bear somewhat incurved leaf lobes and rarely produce gemmae. Such plants tend to be deeply castaneous brown and may have decolorate leaf margins. They also often have the apical leaves partly carmine or purplish red-tinged. Such plants appear to represent intergrades to the *L. harpanthoides* of Bryhn and Kaalaas, here reduced to a variety of *L. heterocolpa*. *Lophozia arctica* S. Arn. appears to be identical with *L. harpanthoides* Bryhn & Kaal.

At the opposite extreme from var. *harpanthoides* are mesophytic shade phenotypes, found, e.g., in damp crevices of ledges, often in the Arctic, which are rather large, virtually without gemmae, with angular leaf sinuses and sharp lobes. These are almost inseparable from *L. collaris*, and their correct identification hinges on the finding of at least sporadic, erect, gemmiparous shoots. When production of such gemmae is totally suppressed, separation from *L. collaris* is not certainly possible. Indeed, *L. collaris* seems to me to be a very rare plant in eastern North America, and it is possible that at least some of the scattered reports represent gemma-free extremes of *L. heterocolpa*.

Key to Varieties of L. heterocolpa

1. Leaves 0.2–0.33 bilobed, little or no wider than long, lobes usually acute or bluntish, sinus often angular; green to brownish, rarely with traces of reddish pigments at shoot tips; gemmae abundant on erect, tawny-brown axes, brownish . . . . var. *heterocolpa*
1. Leaves 0.15–0.2(0.25) bilobed, broader, lobes rounded or blunt, shallow sinus often crescentic or obtuse; green, with shoot tips dull to carmine reddish to purplish red pigmented, ranging to deep castaneous brown with red shoot tips; gemmae rare or lacking usually, often reddish tinged
. . . . var. *harpanthoides*

LOPHOZIA (LEIOCOLEA) HETEROCOLPA var. HARPANTHOIDES (Bryhn & Kaal.) Schust.

[Figs. 163:10–12]

Lophozia harpanthoides Bryhn & Kaal., Rept. 2nd Norwegian Arctic Exped. in the "Fram" 1898–1902, 2(11):31, 1906.
Leiocolea harpanthoides Evans, *in* Buch, Evans, & Verdoorn, Ann. Bryol. 10 (1937):4, 1938; Frye & Clark, Univ. Wash. Publ. Biol. 6(3)(1944):387, 1945.
Leiocolea arctica S. Arn., Svensk Bot. Tidskr. 44:374, figs. 1–2, 1950.
Leiocolea heterocolpos var. *arctica* Mårtensson, Bryoph. of the Torneträsk Area, I. Hepaticae:43, 1956.
Leiocolea heterocolpa var. *harpanthoides* Schust. ex Arnell, Ills. Moss Fl. Fennosc. 1:112, 1956.
Lophozia heterocolpa var. *harpanthoides* Schust., *in* Schuster et al., Canad. Natl. Mus. Bull. 164:28, 1959.

Forming castaneous to blackish brown patches, or when crowded suberect or erect; individual plants strongly yellowish to clear chestnut brown, the shoot tips often somewhat reddish. Shoots ca. 950–1300 μ wide × 8–18(25) mm long; stems ca. 180–240(280) μ in diam.; dorsal cortical cells ca. 13–16 μ wide × 25–45 μ long. Leaves clearly *imbricate, antically secund*, somewhat erect-spreading, *slightly to strongly concave and with the short lobes commonly incurved* (and then sometimes leaves almost saucer-shaped), ovate-quadrate to rotundate, from somewhat longer than wide and 800 μ wide × 1000 μ long to 725 × 725 μ or wider than long and 730–875(920) μ wide × 650–750(775) μ long, bilobed ⅙–¼, by a *crescentic or obtuse* less often rectangular or subacute sinus, not gibbous or reflexed; lobes obtuse to subrectangulate, usually *blunt to rounded at apex*. Cell walls strongly yellowish, except tips of lobes and adjacent margins often decolorate, *small*; cells of leaf tips (13)14–19 μ on the margin, 16–20(21) μ within the margins; median cells 18–23(24) μ × 18–26 μ; basal cells to 20–24 × 24–35 μ; trigones very coarse, sometimes almost confluent; oil-bodies *large*, 2–4(5–6) *per cell*, often nearly occluding lumen, subspherical and 4.5–5 × 5–6 μ to ovoid and ellipsoid and 5–6 × 8–11 μ up to 6–7 × 13–15 μ, finely granular in appearance. Underleaves lanceolate or bifid, with(partly stalked) marginal slime papillae. *Gemmae usually absent*, but rarely a few stout, erect gemmiparous shoots present (*bearing brownish purple to purple gemmae*); gemmae 1–2-celled, small, ca. 13 × 16–17 μ up to 15 × 16–19 μ. Dioecious.

Type. Fram Fiord, southeast Ellesmere I., ca. 76°20′ N., 80°55′ W. (*Simmons*).

Distribution

Evidently high arctic; known from the eastern North American Arctic and also reported (as *L. arctica*) from Swedish Lapland (67°10′–68°22′ N.) and from the USSR (Novaya Zemlya, 72°20′ N.), according to Arnell (1950a). Persson (1962, p. 8) reports it from the Bering Strait region of Alaska.

N. GREENLAND. Heilprin Land, Brønlund Fjord, 82°10′ N., 31° W. (*Holmen 8250*, moist *Anthelia* snow patch; *281e, 8714, 6749*, wet *Carex aquatilis* meadow); Lemming Fjord,

Cap Emory, 82°53′ N. (*Th. Wulff, June 16, 1917!*); Herlufsholm Strand, E. coast of Peary Land, 82°40′ N., ca. 21° W. (*Holmen 7623!*; reported by Arnell, 1960, as *L. latifolia*); Kap Glacier, S. coast of Independence Fjord, 81°48′ N., 31°45′ W. (*Holmen 7422!, 7815, p.p.*, reported by Arnell as *Leiocolea badensis*); Saxifragadal, S. coast of Independence Fjord, 81°51′ N., 31°15′ W. (*Holmen 7078!*; moist moss moor with *Cassiope tetragona*). NW. GREENLAND. Foulkefjord (Bryhn, 1906); Kangerdlugssuak, S. side of Inglefield Bay (*RMS 45831c, 45983, 45890*; the last a form with unusually large leaf cells). W. GREENLAND. Tupilak I., W. of Egedesminde, 68°42′ N. (*RMS & KD 66-034*); mouth of Simiutap Kûa Valley, near pond, Umîarfik Fjord, 71°59′ N., 54°35′ W. (*RMS & KD 66-919, 66-916b*). Godhavn, Disko I.: below Lyngmarksfjeld, 69°15′ N., 53°30′ W. (*RMS & KD 66-161, 66-173, 66-170*); basalt cliffs near shore, below Skarvefjeld (*RMS & KD 66-311*); Lyngmarksbugt, ca. 25–125 m (*RMS &KD 66-568*); mouth of Blaesedalen ved Lyngmarksfjeld (*RMS & KD 66-271*).

ELLESMERE ISLAND. Skräling I., Lastraea and Twin Glacier Valleys, Cape Rutherford, Pim I., and Fram Harbor, in Hayes Sd. region (Bryhn, 1906); Fram, Harbour, and Goose Fiords on S. coast (Bryhn, 1906); Lands End, Reindeer Cove, and Excrement Bay, W. coast (Bryhn, 1906); S. end of Hilgard Bay, 82°26′ N., 63°25′ W., NE. coast (*RMS 35250, 35256, c. per.!*); NE. slope of Mt. Pullen, 82°26′ N., 62°15′ W. (*RMS 35306a, 35307a, c. andr., gyn.*). DEVON I. REGION. North Kent I. (Bryhn, 1906). BAFFIN I. Arctic Bay, N. Baffin (Polunin, 1947).

Ecology

The material I collected in Ellesmere I. came from calcareous, weathered clay-shale and clay-sandstone slopes, over thin humus, in sites kept damp by hanging snow banks. Associated were *Arnellia fennica, Anthelia juratzkana, Lophozia quadriloba, L. pellucida, Cryptocolea imbricata, Tritomaria heterophylla* and *quinquedentata, Odontoschisma macounii, Cephaloziella arctica, Blepharostoma trichophyllum brevirete,* and *Scapania praetervisa.* Arnell (1950a) also reports the evidently synonymous *L. arctica* from "moist, calciferous earth. Collected together with *Lophozia excisa, Tritomaria quinquedentata, Blepharostoma trichophyllum, Anthelia julacea, Orthocaulis quadrilobus.*"

In northwest and northern Greenland the species has much the same ecological range and associates (e.g., *Anthelia juratzkana, Odontoschisma macounii, Cryptocolea imbricata, Tritomaria heterophylla, Blepharostoma trichophyllum brevirete, Scapania praetervisa, S. ligulifolia, Cephalozia pleniceps,* as well as *Lophozia grandiretis*).

Differentiation

Characterized by the antically connivant, usually rather closely imbricate, distinctly concave leaves (the plants thus somewhat laterally compressed in appearance). Because of the general absence of gemmae likely to be mistaken for *L. collaris* (with which it overlaps in size of leaves and of cells), but different from this species in the intense, warm chestnut brown or yellow-brown pigmentation and very coarse yellowish trigones, as well as in the somewhat smaller cell size.

The status of this plant is doubtful; I reluctantly retain it as a distinct variety, for several reasons. At some sites, in the high Arctic, *L. heterocolpa* may very rarely produce the ascending gemmiparous shoots (as, e.g., in material from The Dean, in northeast Ellesmere I.), yet nearby plants are abundantly gemmiparous, suggesting that an environmental factor may be involved. Strongly insolated plants from Hilgard Bay, referred to var. *harpanthoides*, are totally devoid of gemmae, although an especial effort was made, both in the field and under the microscope, to locate gemmiparous shoots. Abundant collections from west of Cape Belknap, along the frozen Lincoln Sea, are equally uniformly devoid of gemmae. It is possible that "*L. harpanthoides*" is a mere extreme sun form of *L. heterocolpa*.[72]

It will be noted that the cell sizes of the two plants differ slightly. In the descriptions cell sizes are based on material of typical *L. heterocolpa* from The Dean and of var. *harpanthoides* from Hilgard Bay, in adjacent localities in northeastern Ellesmere I. The cell size of var. *harpanthoides* averages perceptibly smaller than that of typical *L. heterocolpa*. Again, this may be a response to greater insolation and more intermittently moist environmental conditions. The cell size of the Hilgard Bay plants of var. *harpanthoides* closely approximates the figure (15–24 μ) assigned to it by Bryhn and Kaalaas.

Efforts made to locate gemmae in the Hilgard Bay material were without success. Yet Frye and Clark (1945, p. 387) describe them as "rare . . . subreniform, purplish," on the basis of the original description of Bryhn and Kaalaas. Such gemmae suggest that a distinct microspecies may be involved. But in *L. attenuata*, which normally has fulvous brown gemmae, alpine sun forms may have purplish gemmae. Furthermore, in one collection (*RMS 35306a*, Mt. Pullen, Ellesmere I.) prolonged search of "typical" plants of var. *harpanthoides* resulted in location of 2–3 gemmiparous plants, whose erect and imbricate apical leaves formed a stout, short, compact gemmiparous head; these plants bore normal fulvous brown "*heterocolpa*" gemmae. On the other hand, in a collection from Mt. Pullen, Ellesmere I. (*RMS 35792*), typical *L. heterocolpa* with fulvous brown gemmae masses occurred closely contiguous to "typical" *L. harpanthoides*, almost all of which lacked gemmae. A few of the "*harpanthoides*" plants, however, bore stout gemmiparous shoots, which had imbricate, concave, erect-spreading leaves, somewhat reduced (but less strongly so than similar gemmiparous leaves of "typical" *L. heterocolpa*). The upper gemmiparous leaves were purplish or reddish-tinged and bore purplish to brownish purple gemmae. These were indiscriminately 1–2-celled and averaged somewhat smaller than those of "typical" *L. heterocolpa*, ranging from 13 × 16 μ to a maximum of 15 × 19 μ. None of the gemmae were "subreniform," their shape paralleling those of *L. heterocolpa*, ranging from ovoid to broadly ellipsoidal.

The occurrence of plants with the brownish gemmae of typical *L. heterocolpa* adjacent to plants with brownish purple or purple gemmae of slightly inferior size ("*L. harpanthoides*") suggests that two distinct genotypes are probably

---

[72] Persson (1962) emphasizes that the restriction of var. *harpanthoides* to the Arctic (whereas var. *heterocolpa* is widespread under alpine and subarctic conditions) "must underline the systematical value of the taxon."

involved. It seems difficult to interpret the differences between these plants as due solely to unequal degrees of exposure.

Arnell (1950a) notes that in *L. arctica* the apical marginal cells are 20–30 *μ* and the cortical stem cells 18–24 *μ* wide; also, that the chromosome number (which could not be definitely determined) is more than 9, perhaps 14–16 (suggesting there might be 18 chromosomes). It is therefore possible that *L. arctica* is a diploid race of *L. heterocolpa harpanthoides*. A comparison of the description of *harpanthoides* (drawn from living material from Ellesmere I.) and of *L. arctica* shows a notable similarity, except for these two points. Particularly striking is the common tendency for development of reddish or purplish red pigmentation of the upper leaf apices, and the marked tendency to develop concave, antically secund, conspicuously broader than long leaves.

Lophozia heterocolpa var. *harpanthoides* may occur in dense, caespitose patches, in which case the plants are conspicuously laterally compressed. Associated with this, the perianths may be all or mostly strongly laterally compressed (as, e.g., in *RMS & KD 66–311*, Disko I., Greenland), with the mouth triangularly more or less narrowed, but not at all beaked; in extreme cases the perianth is almost Plagiochiloid and may be ± keeled dorsally. Similarly keeled and compressed perianths occur also in var. *heterocolpa* (e.g., in *RMS 68-1630*, Squires Memorial Pk., Newfoundland).

LOPHOZIA (LEIOCOLEA) BADENSIS (G. ex G. & Rabenh.) Schiffn.

[Figs. 163:1–2, 171, 172]

Jungermannia badensis G., *in* Gottsche & Rabenh., Hep. Eur. Exsic. no. 95, 1859; C. Jensen, Danmarks Mosser, Bryofyter 1:128, figs. 1–4, 1915.

Lophozia badensis Schiffn., Lotos 51:219 (p. 7 of reprint), 1903; Lorenz, The Bryologist, 14:27, pl. 5, 1911.

Lophozia gypsacea Schiffn., Verh. Zool.–Bot. Gesell. Wien 54:399, 1904 (not of Schleich., 1821).

Lophozia (subg. *Leiocolea*) *badensis* K. Müll., Rabenh. Krypt.–Fl. 6(1):730, fig. 330, 1910; Macvicar, Studs. Hdb. Brit. Hep. ed. 2:168, figs. 1–4, 1926; Schuster, Amer. Midl. Nat. 42(3):565, pl. 9, figs. 2–3, 1949; Schuster, *ibid*. 45(1):48, pl. 6, figs. 1–4, 1951; Schuster, *ibid*. 49(2):336, pl. 8, figs. 1–4, 1953.

Leiocolea badensis Joerg., Bergens Mus. Skrifter 16:166, 1934; K. Müller, Rabenh. Krypt.–Fl. ed. 3, 6(1):701, fig. 222, 1954.

Delicate, very small, in small patches or creeping among mosses (often *Gymnostomum*) mostly as isolated, inconspicuous stems, rather pellucid, yellowish green to pale green, in arctic sun forms deeper green with leaves or leaf lobes occasionally *deeply fuscous to fuscous purple*. Shoots (300–450)500–850(1000) *μ* wide × 3–10(12) mm long, prostrate and creeping, when crowded sometimes ascending, simple or sporadically furcate, with occasional postical innovations (often from beneath gynoecia). *Stems slender and translucent*, ca. 100–140 *μ* in diam. *7–9 cells in diam.*, the cortical layer of *thin-walled* rectangulate cells, 2–4(6) × as long as wide, *hyaline*, striolate, ca. (20)23–35 × (45)60–90(140) *μ*. Rhizoids frequent to sparse, elongate, colorless, brownish with age. Leaves slightly decurrent antically, *distant or contiguous, usually nearly or quite flat*, horizontally and widely spreading or occasionally (compact arctic forms!) antically secund and suberect (then ± concave), *uniformly bilobed*, oblong-ovate to broadly ovate-quadrate to *broadly orbicular*, averaging from 300–325 μ wide × 300–400 μ long (♂ plants), to (590)800–1000 μ wide × (410)550–600 μ long (♀ plants), *nearly as wide as to conspicuously wider than long* except on juvenile or ♂ shoots, with antical and postical *margins moderately to conspicuously arched, convergent distally;* lobes subequal, the postical sometimes slightly larger, erect to weakly connivent, acute (and terminated by 1 or 2 superposed cells) to blunt; *sinus* descending 0.2–0.3 the leaf length, usually acute or ± broadly rectangulate, rarely obtuse. Cells *extremely* leptodermous, trigones virtually absent, 30–35(38) μ on margins and in apices of lobes, ca. *(30)32–36(38)* × *(32)35–45 μ in leaf middle*, sharply polygonal; *cuticle usually smooth* or obscurely verruculose-striolate; *oil-bodies only moderately large*, subspherical and *4–6* × *4–8 μ* to usually ovoid or ellipsoid, ca. 5 × 8 to 6 × 9–12 μ, a few to 6–7 × 14 μ, *2–5(6–7) per cell* usually, much larger than chloroplasts, which are ca. 3–5 μ. *Underleaves absent*, but sometimes distinct as stalked slime papillae, rarely (arctic forms) vestigial, few-celled. *Asexual reproduction absent.*

Dioecious, but often fertile. *Androecia loosely spicate*, eventually intercalary; *bracts in 4–8 pairs or more, suberect* with frequently erect or incurved lobes, *contiguous or approximate*, concave at base, asymmetrically bilobed and usually with a third, sometimes incurved, antical tooth or lobe; 1–2-androus. Gynoecia with bracts somewhat concave, erect or with at most lobes widely spreading, similar to leaves but somewhat larger, 900 μ wide × 600 μ long or more, bilobed for 0.2–0.35 their length (and occasionally with a smaller third lobe), but otherwise *entire-margined*, the lobes acute or subacute, the sinus acute to obtuse; *bracteole usually absent.* Perianth 0.6–0.8 emergent, cylindrical to cylindrical-clavate, *smooth*, rather suddenly contracted at apex into the *shortly beaked mouth*, often very obscurely plicate near apex; mouth crenulate with elongated cells. *Seta of 8–12 epidermal, 4–8 inner, cell rows.* Capsule short-ovoid; wall 2–3-stratose. Epidermal cells rather short rectangular, ca. 18(20–23)27 μ wide × 25–48 μ long, with usually *3–5 nodular thickenings along each longitudinal wall*, 0–1 thickening per transverse wall. Inner layer with cells averaging somewhat more narrowly rectangular, often less regular in shape, with usually 6–8 narrow but sharply developed semiannular bands, going completely across the inner tangential face. Elaters 7–8 μ in diam., rather tightly bispiral; spirals ca. 2.0–2.5 μ wide, reddish brown. Spores (10)12–15(16) μ, finely papillate or verruculose, brown.

Type. On the Bodensee, Baden, Germany (*Jack*).

Distribution

Rather widely distributed, with the distinction of occurring farther southward than any other species of *Leiocolea* (36° N. in Tennessee), excepting only the Mediterranean *L. turbinata*, yet occurring northward to the polar basin (82°40′ N.)! Probably holarctic, occurring in much of Europe, but rare in some of the northernmost districts (but reported from Spitsbergen; Arnell and Mårtensson, 1959); found from Norway,

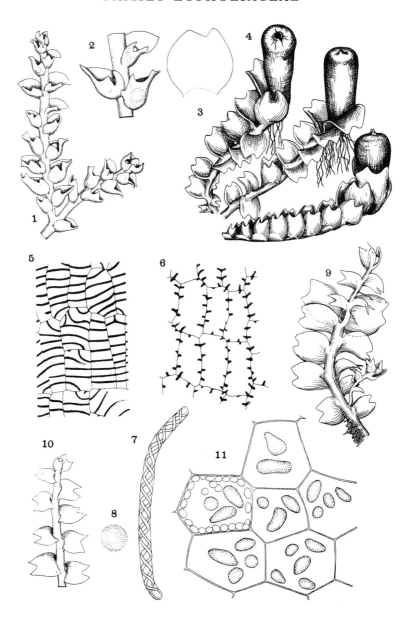

FIG. 171. *Lophozia (Leiocolea) badensis.* (1) Androecial shoot (×ca. 20). (2) Sector of androecium (×ca. 35). (3) Leaf (×36). (4) Group of perianth-bearing plants (×ca. 15). (5) Inner cells, capsule wall (×308). (6) Epidermal cells, capsule wall (×308). (7–8) Elater and spore (×470). (9) Postical view, shoot, showing innovation and lower down a lateral-intercalary branch (×ca. 24). (10) Typical, slight, sterile shoot (×ca. 15). (11) Cells with oil-bodies, at upper left also with chloroplasts (×505). [1–2 4, 9–10, Fall Creek, Ithaca, N.Y., *RMS June 27, 1945;* 3, *RMS 10987,* Letchworth State Park, N.Y.; 5–8, 11, Crystal Springs, Wisc., *RMS,* from Schuster, 1953.]

Sweden, and Finland, southward through Denmark, England, and Ireland, Germany (type), Belgium, Holland, Switzerland to France, Austria, northern Spain and Italy, the Dalmatian coast, and eastward to the USSR and the Ural Mts. In Europe regarded as a lowland species (Kaalaas, 1893), but reputedly in the Alps to 2200 m (Müller, 1951–58). Eastward extending into Asia Minor and Iran (Elbrus R. and at 2600 m in Prov. of Mazanderan), also in Siberia (Yenisey R. valley).[73]

In North America transcontinental in the Taiga, extending northward into the most northerly extension of the Tundra (82°26′ N. in Ellesmere I.), and with scattered though not infrequent stations southward in the cooler portions of the Deciduous Forest Region (chiefly in the Hemlock-Hardwoods Region). Westward occurring from British Columbia and Alberta southward to Montana and Washington. Eastward as follows:

ELLESMERE I. S. end of Hilgard Bay, 82°26′ N., 63°25′ W. (*RMS 35230, 35234a;* atypical material); Harbour and Goose Fiords, on S. coast (Bryhn, 1906).   N. GREENLAND. Heilprin Land, Brønlund Fjord, 82°10′ N., 31° W. (*Holmen 6595, 217, 231!*); Herlufsholm Strand, E. coast of Peary Land, 82°40′ N., ca. 21° W. (*Holmen 7449, p.p.?*);[74] Kap E. Rasmussen, E. coast of Peary Land, 82°30′ N., 20° W. (*Holmen 7178;* traces on old musk-ox feces on gravelly coastal plain). Also reported by Arnell (1960) from Kap Glacier (= *L. heterocolpa harpanthoides!*).   NE. GREENLAND. Maria Ø, 72°58′ N., 24°50′ W. (*Holmen 19-213;* among *L. rutheana, L. pellucida*).[75]   W. GREENLAND. Hollaenderbugt, Nûgssuaq Pen., slopes below Qârssua, 70°47′ N., 53°47′ W. (*RMS & KD 66-1512, 66-1520*); Magdlak, Alfred Wegeners Halvö, 71°06′ N., 51°40′ W. (*RMS & KD 66-1357a*); Umiarfik Fjord, Svartenhuk Pen. (*RMS & KD*).

QUEBEC. Pont Rouge; Pic Champlain; Saint Fabien; Côté de Rimouski (Lepage, 1944–45); Montmorency R. (*Macoun, fide* Evans, 1910); The Grotto, Mt. Ste Anne, Percé, Gaspé (*RMS 44010a*); Labelle and Montcalm Cos., near St. Jovite (Crum & Williams, 1960).   NOVA SCOTIA. Ingonish, Cape Breton I. (Brown, 1936a); Cape Breton Natl. Park, near N. end of Presque Isle, S. of Cap Rouge, ca. 250 ft (*RMS 432020*). ONTARIO. Nakitawisagi (L.) R., 54°20′ N. (Lepage, 1953); S. of Craigleth, Grey Co. (*Cain, 4254!*); Algoma and Dufferin Distrs. (Williams & Cain, 1959).   NEWFOUND-LAND. N. of Daniels Harbour, N. Pen. (*RMS 68-1468*); Squires Park, Humber R. (*RMS 68-1628*).

NEW HAMPSHIRE. Beaver Falls, Colebrook (*Evans*).   VERMONT. Hartford, Quechee Gulf (Lorenz, 1911).   CONNECTICUT. Salisbury, Litchfield Co. (Evans, 1910b; erroneously cited as *L. "muelleri"* by Evans & Nichols, 1908, and by Frye & Clark, 1937–47).   MASSACHUSETTS. W. side of Green R. gorge, Colrain (*RMS 68-106*).

[73] Perhaps identical at the species level is *Lophozia mayebarae* (Hatt.) Schust., comb. n. (*Cephalozia mayebarae* Hatt. Jour. Hattori Bot. Lab. no. 3:37, 1948 = *Hattoriella mayebarae* Inoue, ibid; no. 23:40, 1960); this plant also has a seta with ca. 8 epidermal, 4–6 inner, cell rows.

[74] *Holmen 7449* includes mostly *Lophozia heterocolpa.* A few stem fragments lack underleaves; they may represent *L. badensis* but are almost indeterminable.

[75] The remote-leaved plants referred to *Lophozia badensis* were small (900–980 μ wide) and clear green, in contrast to the dense-leaved and vigorous (3.2–3.5 mm wide) admixed *L. rutheana.* Although 2–3 juxtaposed stalked slime papillae were at times distinct, no lamellate underleaves occurred; trigones and cuticular papillae were both perceptible, and cells in the lobes averaged merely 24–25 μ, suggesting a small form of *L. collaris.* However, the size, the very broad and broadly attached leaves, and the aspect of the plants are those of *L. badensis,* and the obsolete underleaves also suggest this species.

NEW YORK. Sixmile Cr., Enfield Glen, Buttermilk Glen, Fall Cr., Coy Glen, Lick Brook, all in Tompkins Co. (Schuster, 1949a); Watkins Glen, Schuyler Co. (*Schuster, 1954*); Letchworth State Park, Wyoming Co. (*Schuster, 1949*); Paine Cr. near Aurora, Cayuga Co. (Schuster, 1949a); some of the New York reports very close to Pennsylvania, in which the species will certainly be found.   TENNESSEE. Alum Cave Bluffs, Mt. LeConte (Sharp, 1939); Roan Co. Westward extending as follows: OHIO. Cuyahoga, Greene, Ottawa Cos. (Miller, 1964).   WISCONSIN. Near Crystal Springs (*RMS 1948, plate*); Iron, Douglas, Ashland, and Bayfield Cos. (Conklin, 1929).   MICHIGAN. Tahaquemon Falls, Luce Co. (Nichols, 1935); Pictured Rocks, Munising Co. (*RMS*); Mackinac I.; Presque Isle, Marquette (*RMS*); Scoville Pt., Isle Royale; Gogebic and Luce Cos. (see Evans & Nichols, 1935).   MINNESOTA. Grand Marais, Cook Co. (*Conklin 2683, 2668*); S. of La Crescent, Houston Co. (Schuster, 1953); Lower Falls of Manitou R., L. Superior, Lake Co. (*RMS 18092*); Fond du Lac, St. Louis Co. (*Conklin 1750, 1784*); Stillwater, Washington Co. (*Holzinger, s.n.*; see Schuster (1953).   IOWA. Lacey-Keosauqua State Park, Van Buren Co. (*Conard, 1939*); Boone Co. (Conard, 1945). ARKANSAS. Washington Co., 12 mi E. of Springdale (*Anderson 12296*).[76]

Ecology

An obligate calciphyte, restricted, except in the far north, to ledges and cliffs (often in gorges or on lake shores where atmospheric humidity is high) and usually to rather strongly sheltered sites where there is little or no direct sunlight. Particularly common in areas with basalt or sedimentary rocks, such as soft sandstones or friable calcareous shales, where the species commonly occurs in damp, minute crevices, on decaying shale, or over seepage-moistened shale talus. Associated in such sites are usually *Gymnostomum* (either *aeruginosum* or *curvirostre*, sometimes both, or *calcareum*), most often also *Preissia* (rarely *Mannia rupestris* and *M. pilosa*), and very frequently *Aneura pinguis* and *Lophocolea minor*. Northward *Scapania gymnostomophila*, less often *S. cuspiduligera*, *Pellia fabbroniana*, *Lophozia gillmani*, *Solenostoma pumilum polaris*, *Mannia pilosa*, and *Tritomaria scitula* are also consociated. Also common on shaded north- or northwest-facing gypsum and dolomite ledges.

Particularly common in the postglacial gorges in the Finger Lakes region of central New York (Schuster, 1949a); there associated with a number of relicts of the immediate postglacial period, among them *Primula mistassinica*, *Pinguicula vulgaris*, and *Saxifraga aizoön*.[77] The latter, indeed, may undergo ecesis in the sods formed by *Gymnostomum* and *Lophozia badensis*. The occurrences of the species in the Hemlock-Hardwoods Forest, therefore, may represent in large part relict occurrences, as may also the stations in the Driftless Area of Wisconsin, Minnesota, and Iowa (or in its periphery).

[76] The Arkansas specimen may just possibly be closer to *L. turbinata*, not known outside of the oceanic and Mediterranean portions of Europe and North Africa. Better material is needed to resolve this problem. See Fig. 172:12–16.

[77] These last two angiosperms also associated with *Lophozia badensis* in west Greenland (e.g., at Umîarfik Fjord).

Lorenz (1911) also reports the species from gorges, occurring on damp rocks with *Rhabdoweisia denticulata*. Kaalaas (1893), however, found the species to occur on slightly damp and sandy soil or on thinly soil-covered cliffs, the substrate always containing some lime. In the North American Arctic the species often is similarly terricolous, rather than saxicolous.

The high arctic occurrences stand in sharp contrast to those in the subarctic and temperate regions. In the latter areas the species is found in damp rock crevices or on rock walls, almost always as a mod. *viridis* with $\pm$ spreading leaves, and often attaining a width of 500–1000 μ. By contrast, in the high Arctic it occurs on clayey soil along temporary rills and streams and on damp, insolated slopes, in calcareous moss-tundra (associated with very few angiosperms, among them *Saxifraga caespitosa*, *S. oppositifolia*, *Salix arctica*, and *Papaver radicatum*). The species is usually found on steep slopes below persistent snow banks, or along drainage channels on clay-shale slopes, associated with a variety of other Hepaticae, among them some which also occur with it far to the south (*Scapania gymnostomophila*, *Lophozia gillmani*, *Solenostoma pumilum polaris*) but also a number of arctic taxa, such as *Arnellia fennica*, *Cryptocolea imbricata*, *Lophozia pellucida*, *L. quadriloba*, and *L. heterocolpa* var. *harpanthoides*, *Cephaloziella arctica*, and *Blepharostoma trichophyllum brevirete*. Under such conditions the plants, although leptodermous, occur as a strongly fuscous purple or blackish-tinged, small-leaved phase (mod. *parvifolia-colorata-leptoderma*), which may be only 300–500 μ wide. Leaves of such plants are usually somewhat suberect and often up to 1.25 × as long as wide, on ♂ plants.

According to Macvicar (1926, p. 169), *L. badensis* is a spring-"fruiting" plant. However, I have seen it with capsules as late as August 2 (Wisconsin) and October 18 (Paine's Creek, Cayuga Co., N.Y.), suggesting a fall "fruiting" period. In the high Arctic of Ellesmere I. it occurs with archegonia and antheridia, but apparently never with capsules.

Differentiation

Often largely sterile and then difficult to identify, unless the rather rigid ecological restrictions are carefully noted. Sterile plants may give the impression at first of a small, gemmae-free phase of *Lophocolea minor*, from which they differ at once in the lack of underleaves and scattered rhizoids. Fertile plants, with their tubular, shortly rostrate perianths are at once placeable in *Leiocolea*.

Lophozia badensis is smaller and more delicate than our other species of *Leiocolea* and usually recognizable by its very delicacy, the translucent, slender stems, the usually remote leaves, and the tendency to grow as scattered, creeping plants among *Gymnostomum* and other calciphilous, tufa- or sod-forming mosses. The very large cells, for the small size of the plant, are distinctive. However, *L. collaris* occasionally occurs in habitually similar, equally minute phases, which are very difficult to separate (Schuster, 1953). Both Stephani and Conklin (1929) maintain that these two species are not sharply separable. This often appears to be true of arctic phenotypes, which may be almost indeterminable.

Supposedly, *L. badensis* differs from all forms of *L. collaris* (and from all other species of *Leiocolea* in North America) in lacking underleaves, except rarely in conjunction with the gynoecia, or at the bases of branches.[78] Furthermore, androecia in *L. badensis* are elongate and quite loosely spicate (Fig. 171:1–2), in *L. collaris* compactly spicate and formed of imbricate bracts (Fig. 169:5). Schiffner (Lotos, p. 31, 1905) first emphasized the distinction between the two species on the basis of the loosely spicate androecia of *L. badensis* vs. the compactly spicate ones of *L. collaris*. In *L. badensis* the androecium is intercalary with age, and even when young (and terminal) shows nonimbricate bracts. Finally, *L. badensis* has conspicuously larger leaf cells than does *L. collaris* and rarely, if ever, shows perceptible trigones (Fig. 171:11).[79]

Unlike the other species of *Leiocolea*, including *L. collaris*, *L. badensis* usually has a smooth cuticle; only rarely are the basal halves of the leaves perceptibly striolate-verruculose. In the blackish, high arctic phases the leaves may be exceedingly delicately striolate, and the stems more distinctly striolate. Even in such phases the fuscous cell walls remain very thin and the trigones are almost imperceptible (whereas in admixed *L. gillmani* the trigones are coarse and bulging). Apparently *L. badensis* is quite unable, no matter under what physiological conditions it exists, to form even a distinct mod. *mesoderma*, not to mention the mod. *pachyderma*. In this respect, its behavior differs markedly from that of other American species of *Leiocolea*.

Lack of gemmae (as well as the leptodermous cells with few but large oil-bodies) separates sterile plants from those of subgenera *Lophozia*, *Isopaches*, and *Massula*, as well as from the superficially similar (and often consociated) *Lophocolea minor*. In living plants the large oil-bodies, occurring usually only 2–5 per cell, distinguish the plant from other small species with bilobed leaves but no underleaves. Similar oil-bodies occur in other species of *Leiocolea* (which all possess underleaves), but in many cases our underleaf-bearing species of *Leiocolea* may superficially appear to be underleaf-free, because of the dense rhizoid mat, which may quite obscure them. Before material is relegated to *L.*

[78] Ventral merophytes in *Lophozia badensis* are normally only 2 cells broad. Even in small phases 300–400 μ wide, these narrow merophytes rather regularly produce stalked slime papillae, consisting of a stalk cell and terminating in a single slime papilla, or frequently of 2 stalk cells, free or connate only at base, each terminated by a slime cell. Therefore, in the strictest sense, underleaves are not totally lacking. However, in none of our other species are they reduced to mere stalked slime papillae. In some specimens from northern Ellesmere I., perhaps representing a reduced form of *L. collaris*, underleaves of intermediate size occur. Similarly, vestigial underleaves occur in compact, purplish green phases on clay, from Hollaenderbugt, west Greenland (*RMS & KD 66-1512*).

[79] Very occasionally small, brownish-tinged sun forms occur with a distinctly verruculose cuticle, and with the cells of the leaf middle and lobe bases only 25–32 μ. Such plants are almost impossible to separate from the fo. *pumila* of *Lophozia collaris*, except on the basis of the lack of underleaves. An example of this type of plant is *Cain 4254* (Grey Co., Ontario).

badensis, a number of shoot apices must be studied under the microscope. *Lophozia badensis*, often has relatively sparse rhizoids, under no conditions as abundant as in *L. gillmani* or *L. rutheana*.

Associated with the reduced size of the gametophyte, *L. badensis* has a delicate and small sporophyte. The seta is commonly formed as in *Anastrophyllum hellerianum*, i.e., of 8 rows of epidermal cells surrounding 4 rows of internal cells; this number is not sharply fixed, however, and setae occur with as many as 10–12 epidermal and 5–8 interior cell rows. The variability in this respect suggests that not too much emphasis can be placed on the structure of the setae in the Lophoziaceae; see p. 754. In other species of *Leiocolea*, the setae are of the "general type" of Douin, i.e. formed of indefinite and numerous cell rows (compare Fig. 163 : 2, 5, 13).

Variation

I have already indicated that in the Arctic small, often more compact, purplish-tinged, dark phases of *L. badensis* occur with broad and concave leaves and often with vestigial underleaves. Such plants must be separated with great care from *L. collaris*, from which they differ principally in the larger and more leptodermous cells, softer stem with thin-walled and almost pellucid cortical cells, and almost—but often not quite—smooth cuticle.

A possibly polyploid, larger-celled, and deviant form of *L. badensis*, similar in most respects to these clay-inhabiting arctic forms, occurs, also in the Arctic. The two extremes of the species may be separated as follows:

1. Stem with cortical cells ca. (20)22–25(35) μ wide; leaves with subapical cells and marginal cells (25)30–35(38) μ, median cells 30–36 × 32–40(45) μ; cells with 2–5(6–7) oil-bodies each; leaves with lobes rarely distinctly apiculate although usually sharp, margins never with teeth . . . . *L. badensis* var. *badensis* (see above)
1. Stem with cortical cells ca. (25)30–42(48) μ wide; leaves with subapical cells 30–36 × 32–45 μ, marginal cells (30)33–42 μ, median cells (35)38–45 × 38–48(52) μ; median cells with (3–5)6–10(12) oil-bodies each; leaves with lobes often conspicuously apiculate or mucronate, very broad (width exceeding 1.2 × as wide as long), on larger leaves often sporadically sinuate-dentate . . . . *L. badensis* var. *apiculata*, var. n.

LOPHOZIA BADENSIS var. *APICULATA* Schust.,

var. n.

[Fig. 172:1–7]

Varietas a var. badense distincta cellulis corticalibus maioribus 30–42 *u* lat., cellulis foliorum maioribus, 30–36 × 32–45 *u* in areis subapicalibus, gutteiis olei 6–12 omni in cellula, apicibus loborum foliorum saepe apiculatis mucronatisve, et foliis latis (saltem 1.2 plo latioribus quam longa).

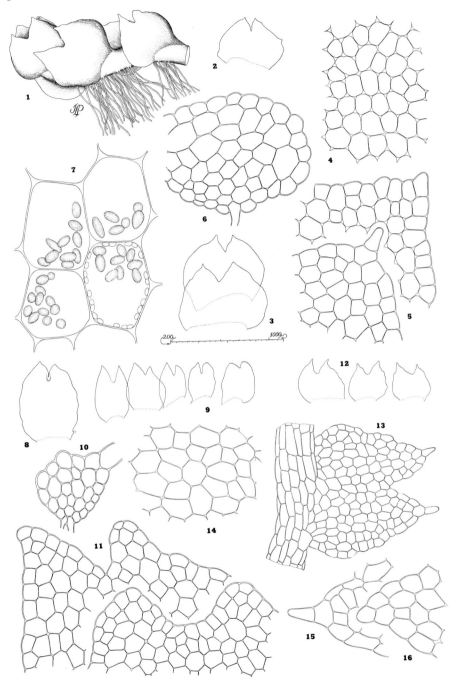

Green to olive-green, with distal parts of leaves often strongly violet or purplish; stem ca. 7 cells high, soft-textured, the cortical cells $(25)30–42(48)$ μ wide $\times$ $(50)70–110(125)$ μ long, the medullary cells, except ventrally, where smaller, mostly 36–48 μ in diam., leptodermous. Leaves antically $\pm$ assurgent, often almost connivent, very broadly ovate-orbicular, widest toward base, ca. 660 μ wide $\times$ 540 μ long to 860–980 μ wide $\times$ 610–650 μ long, narrowed toward the two lobes; sinus narrow, usually V-shaped, to 0.25 the leaf length; lobes erect, usually apiculate with a 1(2)-celled mucro formed often of an elongated cell; leaf margins convex-sided, sinuous to, occasionally, sinuous-dentate. Cells very large (at least on mature plants), tumid (toward leaf base 47–54 μ high, in cross section), the marginal $(30)33–42$ μ in the lobes, the sub-apical ca. 30–36 $\times$ 32–45 μ; median cells $(35)38–45 \times 38–48(52)$ μ; trigones small, concave-sided. Oil bodies $(3)5–7(9)$ per cell in lobes, $(3–5)6–10(12)$ *per cell in leaf middle*, usually ellipsoidal and 4–5 $\times$ 6–11 μ, a few to 6 $\times$ 12–13 μ, a few spherical and 5–6.5 μ, faintly granulate and smooth externally. Cuticle of leaves and stem very faintly striolate. Underleaves either of 1–2 stalked slime papillae, or occasionally with a few cells cut off basally and a minute lamella developed.

Type. West Greenland: Hollaenderbugt, northwest end of Nûgssuaq Peninsula, slopes below Qârssua, 70°47′ N., 53°47′ W. (*RMS & KD 66-1518, Aug. 10, 1966*). The type is mixed with *L. pellucida.*

This variant occurred on clay-shale-sandstone slopes, along small temporary rills that dry out rapidly after snow melt, associated with *Arnellia fennica, Lophozia pellucida, Distichium capillaceum,* and other Ca-tolerant taxa.

Lophozia badensis var. *apiculata* would tend to key to *L. turbinata* in Müller (1951–58, p. 689), because of the large cortical cells of the stem—those of *L. badensis* are stated to be only 20–25 μ in diam. Indeed, the existence of forms like var. *apiculata* throws doubt on the distinction of *L. turbinata.* Compare Fig. 172:1–5 with 172:12–16.

I have seen other arctic forms of *L. badensis* with, at least on some plants, larger than "normal" cells; however, in the present plants both cell size and oil-body number are unusually high.

FIG. 172. *Lophozia (Leiocolea) badensis* and *Lophozia (?) turbinata.* (1–7) var. *apiculata;* (8–11) typical var. *badensis;* (12–16) questionable *L. turbinata.* (1) Shoot apex; note dense rhizoids and lack of underleaves ($\times$27). (2, 3) Leaves ($\times$27). (4) Median cells ($\times$155). (5) Cells of leaf lobe apices ($\times$155). (6) Stem cross section ($\times$155). (7) Median cells with oil-bodies and, lower right, chloroplasts ($\times$475). (8) ♀ Bract ($\times$27). (9) Leaves ($\times$27). (10) Stem cross section ($\times$155). (11) Apices of two leaves ($\times$155). (12) Leaves ($\times$27). (13) Leaf, *in situ*, on stem ($\times$77). (14) Median leaf cells ($\times$155). (15–16) Leaf lobe apices ($\times$155). [1–7, *RMS & KD 66-1518,* Hollaenderbugt, Nugs-suaq Pen., W. Greenland; 8–11, *RMS 10987,* Letchworth State Park, Wyoming Co., N.Y.; 12–16, from plants from Arkansas, *Anderson;* 1–3, 8–9, 12, all drawn to top scale; 4–6, 10–11, 14–16, all drawn to bottom scale.]

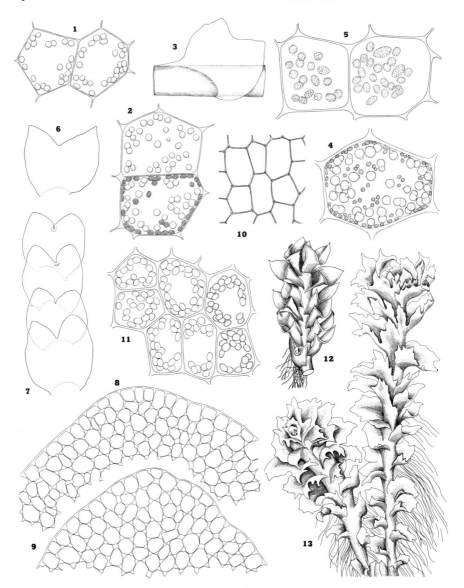

Fɪɢ. 173. *Lophozia* subg. *Massula:* internal variability. (1) Leptodermous cells with homogeneous oil-bodies, *Lophozia capitata* (×ca. 460). (2) Leptodermous cells with oil-bodies and (lower cell, stippled) chloroplasts, *L. marchica* (×492). (3–4) *L.grandiretis* subsp. *grandiretis:* 3, leaf, *in situ*, lateral aspect, to show the oblique insertion (×15); 4, cell with oil-bodies, and, stippled, chloroplasts (×465). (5) Cells showing the botryoidal oil-bodies, *L. grandiretis* subsp. *parviretis* (×660). (6–9) *L. obtusa:* 6, atypical leaf with bluntly pointed lobes; 7, typical leaves (×14.5); 8–9, leaf lobe apices, of a collen-chymatous extreme (×175). (10–13) *L. incisa:* 10, median cells (×ca. 200); 11, median cells of leptodermous phase with oil-bodies (×410; compare to pachydermous

Subgenus *MASSULA K. Müll. emend. Schust.*
[Fig. 173]

Lophozia subg. *Massula* K. Müll., Ber. Deutsch. Bot. Gesell. 57:341, 1939; Schuster, Amer.
    Midl. Nat. 45(1):50, 1951; K. Müller, Rabenh. Krypt.-Fl. ed. 3, 6:667, 1954; Kitagawa,
    Jour. Hattori Bot. Lab. no. 28:287, 1965.
Barbilophozia subg. *Obtusifolium* Buch, Mem. Soc. F. et Fl. Fennica 17:289, 1942 (new synonymy).
Leiocolea subg. *Obtusifolium* Buch, *ibid.* 17:289, 1942.
Obtusifolium S. Arn., Illus. Moss Fl. Fennosc. 1:133, 309, 1956 (new synonymy).
Lophozia subg. *Schistochilopsis* Kitagawa, Jour. Hattori Bot. Lab. no. 28:289, 1965 (new
    synonymy).

Moderate-sized, with creeping or slightly ascending stems, erect only when
caespitose and crowded. *Stem ± flattened, 10–12 cells high or more, with medulla
of large, thin-walled, hyaline cells; cells of ventral layers nearly as large as or* (rarely) *a
little larger than those of dorsal layers; cortical cells thin-walled or nearly so, scarcely
differentiated from medullary;* cortical layer occasionally with slight mycorrhizal
infection ventrally, but *ventral medullary layers usually nearly or completely free of
mycorrhizae.* Branching ± furcate, often frequent, *terminal; no intercalary
branches seen.* Rhizoids forming a dense ventral mat. Leaves ± spreading laterally, 0.3–0.5 *symmetrically to asymmetrically bilobed, but often in part 3–4-lobed, often
undulate,* occasionally with ± distinct marginal teeth or conspicuous cilia (more
often distinct on perigonal and perichaetial leaves); leaf *insertion oblique,* sometimes very much so and approaching horizontal (*L. obtusa*); *ventral leaf bases
never ciliate.* Cells usually *noncollenchymatous, ± pellucid, usually thin-walled*
(becoming somewhat uniformly thick-walled in *L. grandiretis*), *generally large*
(mostly 30–60 μ), *nonpapillose;* oil-bodies colorless, hyaline, strongly refractive,
nearly or quite homogeneous within and appearing smooth externally, *very
small* (2–4 μ) but up to 5–8 μ in *L. grandiretis* (where often double or few-segmented), *very numerous, mostly 25–60 per cell. Ventral stem sectors mostly* (2–3)*6–12
cells wide, usually lacking ability to develop underleaves.*[80] *Gemmae commonly produced,*
ovoid and smooth or angulate to stellate, their formation usually limited to the
youngest leaves (resulting in formation of apical gemmae masses).

Dioecious (exc. *L. elongata*). ♂ Bracts sometimes with an antical, incurved
lobe or tooth; *1–2, 2–3, or even 3–5-androus;* antheridial stalk *1-* or *2-seriate.*
Perianth unmodified, *plicate distally, not beaked.* Sporophyte with seta of many
cell rows; capsule wall 3–5-stratose, inner layers with tangential and radial

---

[80] The generally wide ventral stem sectors of *Massula* are a primitive characteristic which ally
it with the previous three subgenera (*Orthocaulis, Barbilophozia, Leiocolea*) and separate it from the
next two (*Isopaches, Lophozia*). All of the species possess ventral stem sectors that have "lost"
the ability to develop discrete underleaves (in spite of their great width), except *L. obtusa* and
L. elongata, in which normal sterile shoots have occasional ciliate and very well-developed
underleaves (much as in the preceding subgenera).

---

extreme in Fig. 178:7–8); 12, small sterile shoot showing asymmetrically bifid leaves
(×ca. 16); 13, vigorous plant, showing tooth formation (×ca. 12.5). [1, *RMS;* 2,
from material from northern Wisconsin, *RMS;* 3, *Marr 661a,* Great Whale R., Quebec;
4, Cook Co., Minn., *RMS;* 5, from type of *L. grandiretis* subsp. *parviretis;* 6–9, from plant
from Nanortalik, Greenland, *Vahl;* 10–12, after Schuster, 1949a, 1953; 13, after Müller,
1905–16.]

bands (with semiannular thickenings), outer layer of much higher cells, with radial (columnar) thickenings only (i.e., with "nodular" thickenings).

Lectotype. Lophozia marchica (Nees) Steph.

Massula includes a small complex of species, ranging from taxa which show a distinct approach to subg. *Lophozia* to very highly derivative species, such as the Himalayan *L. setosa* (Mitt.) Steph., with leaves and perianth margins armed with highly specialized, 1-celled, setose cilia. Except for *L. patagonica* Herz. & Grolle, of southern South America, the subgenus appears to be holarctic in range, with most species boreal to arctic, only *L. capitata* extending southward to Florida.

Massula agrees with subg. *Lophozia* in the (apparently) exclusively terminal-lateral branching of normal plants,[81] the general lack of under-leaves, and the ability of almost all species to develop gemmae in some abundance. The nondifferentiated medulla suggests subg. *Isopaches*, but *Massula* has a less firm stem, often with more elongated cortical cells, and usually has leptodermous leaf cells. Nevertheless, with antipodal taxa, such as *L. autoica* Schust. (see p. 478), a seeming transition between *Massula* and *Isopaches* occurs.

Massula differs from the related subgenera *Isopaches* and *Lophozia* chiefly in the numerous oil-bodies that usually appear to be nearly or quite homogeneous (hence lacking discernible oil globules or spherules)[82] and in the entirely homogeneous nature of the large-celled medulla of the dorsiventrally flattened stem.[83]

In addition to rather remote affinities to *Isopaches* and *Lophozia s. str., Massula* bears closer affinities to the small subg. *Protolophozia* Schust. (see p. 255), so far known only from Tierra del Fuego and some Pacific islands. The two subgenera agree in axial anatomy and in the strongly succubous, often polymorphous

---

[81] Branching in *Massula* is often repeatedly and closely furcate, as in, e.g., *L. grandiretis, L. opacifolia,* and *L. hyperarctica* (see Fig. 177:9); as a consequence, small compact colonies are often formed. In subg. *Lophozia* branching is usually more remote and almost never pseudo-dichotomous; more diffuse patches usually result. Only *L. excisa*, which (particularly in var. *elegans*) may show free furcate branching, demonstrates a clear transition.

[82] Müller (1939) stressed the small size of the oil-bodies as a subgeneric character. As is evident in *L. grandiretis*, oil-bodies may be up to 5–6 μ (when simple) or 6–8 μ (when double or compound). Hence I believe that the large number of oil-bodies is more characteristic than their form or size. Even this feature fails to sharply delimit *Massula*, since in *L. grandiretis* many phases have conspicuously botryoidal oil-bodies and some species of subg. *Lophozia* may have as many as 25 oil-bodies per cell.

[83] The stem structure appears fundamental to me, on a subgeneric basis. Therefore, I would exclude *L. excisa* and *L. perssonii* from *Massula* and place them in subg. *Lophozia*, into which they fit better because of the dorsiventrally differentiated and conspicuously mycorrhizal medulla. These two species also have relatively nonhomogeneous oil-bodies, rather distinctly granular-botryoidal in appearance, and rather fewer per cell than in *Massula* proper. On the basis of the oil-bodies, they belong, I think, fully as much in *Lophozia* as in *Massula*; hence I believe that the stem structure must be the deciding factor.

leaves, with a marked tendency in some species (e.g., the regional *L. capitata*, *L. marchica*) to develop deeply 2–3–4-lobed, undulate, irregular leaves. They agree further in the large, pellucid cells of the leaves, usually developing only small trigones, and in the wide ventral merophytes. In *Massula*, however, the ventral merophytes are generally unable to develop underleaves, or develop vestigial ones only, whereas *Protolophzoia* has large, bifid to lanceolate to trilobed underleaves, often of diverse form. These underleaves may be from 0.35 to 0.6 the lateral leaves in size. In some respects *Protolophozia* thus connects *Massula* with the smaller-celled *Barbilophozia-Orthocaulis* complex, constituting another of the numerous links forged by exotic taxa between the subgenera here accepted.

Classification

In the following key *Massula* is divided into four sections.[84] Some workers have questioned the inclusion of *L. obtusa* in *Massula* (e.g., Buch, 1942; S. Arnell [1956] went so far as to place it into a new "genus," *Obtusifolium*). However, *L. obtusa* agrees with *Massula s. str.* in stem structure, as in the numerous oil-bodies and the cell structure in general. The tendency to develop marked trigones is somewhat at variance with the other species (but *L. incisa* may develop even larger trigones in the var. *inermis* K. Müll.). Two characters that supposedly isolate *L. obtusa* from other species of *Massula* are the often papillose cuticle and the 5-stratose capsule wall. In the other species of *Massula* the cuticle is smooth and the capsule wall is usually not more than 4-stratose. *Lophozia capitata*, with, in part, a 5-stratose capsule wall, forms an exception.

All species placed in *Massula* agree in having a distinct aspect: they are, in general, more fleshy and more flaccid plants than the species of subg. *Lophozia*;[85] they have wavier, broader leaves (broader than long in all but forms of *L. incisa*), with generally thin-walled cells. They also have gemmae (when produced in abundance) at the shoot tips, in large masses. This latter characteristic is important: ability to produce gemmae

---

[84] All but the Elongatae were first diagnosed in a previous key (Schuster, 1951a, pp. 54–55), although *L. grandiretis* was there placed in a section of its own, the Grandiretae, which is here regarded as unnecessary and merged with the Incisae. Müller (1954, *in* 1951–58, p. 669) also accepted these groups, although sectio Marchicae is called the "*Capitata*-Gruppe." The characterization of these groups in Müller is partly incorrect. For example, sectio Marchicae ("*Capitata*-Gruppe") supposedly differs from the Obtusae in the 4-stratose capsule wall. Actually, the capsule wall in sectio Marchicae is commonly at least locally 5-stratose (Fig. 187:3), while Müller (*loc. cit.*, fig. 205f) illustrates it as only 3-stratose (supposedly characteristic of the Incisae, according to the diagnoses given in Müller). Müller also states that the Marchicae ("*Capitata*-Gruppe") have a 2-seriate antheridial stalk, as contrasted to the Obtusae, where it is supposedly 1-seriate. However, *L. capitata* has the antheridial stalk commonly wholly or largely 1-seriate (Fig. 187:11, 12).

[85] But compare the very fleshy *L. excisa* var. *elegans*, in which the stem becomes flattened, broadened, and, in cross section, distinctly broadly elliptical; see Fig. 194:1, 7.

seems strictly limited to embryonic leaves and hence is localized near the shoot apex, frequently resulting in large, mealy, terminal masses of gemmae, while older leaves, as they approach maturity, seem to lose the ability to produce gemmae. In subg. *Lophozia* this is not so; gemmae production, though initiated before maturity of the leaves, continues at least to some degree until the leaves have attained their full size—hence gemmae occur in obviously individual masses at the lobe apices of leaves of the younger portions of the stems.

Perhaps one of the most valuable criteria distinguishing *Massula* from related subgenera are the well-developed ventral stem sectors in *Massula* (much as in the *Orthocaulis-Barbilophozia-Leiocolea* complex), generally at least 5–12, often to 16, cells broad. In *Isopaches* and *Lophozia* ventral stem sectors are narrow and generally only 2–3 cells (more rarely 4 cells) wide.[86] In the narrow ventral stem sectors (as in stem structure) *L. excisa* agrees with subg. *Lophozia*; hence I would place this species in *Lophozia* rather than in *Massula*, where Müller (1939) placed it. In one species of *Massula* (*L. obtusa*) the wide ventral stem sectors still retain ability to develop underleaves on sterile shoots.[87] In *L. elongata*, although the ventral merophytes are reduced to a width of 2–4 cells on sterile axes, small underleaves are also usually developed.

Key to Sections and Species of Massula[88]

1. Gemmae usually present; dioecious; ventral merophytes normally over 5 cells broad; usually no underleaves . . . . 2.
   2. Gemmae angulate to stellate, greenish (walls never pigmented, even when plants pigmented); leaves often with scattered marginal teeth, or lobes ending in cusps or, alternatively, rounded; dorsal cortical cells of stem usually 2–4× as long as wide. Arctic-alpine to subarctic-alpine . . 3.

---

[86] Sectio Heteromorphae of subg. *Lophozia* connects *Massula* to *Lophozia* s.str. *Lophozia heteromorpha*, the type species, has the polymorphous 2–3-lobed leaves and soft-textured leaves and axis, with elongated cortical cells, of *Massula*; the ventral merophytes are also usually 4–6 or more cells broad and often produce underleaves—suggestive of some species of *Massula*, rather than of subg. *Lophozia*. However, the stem cross section is typical of *Lophozia* (Fig. 191). Without a stem section, the species can be confused with *Massula* and might be sought in the Incisae, because of the angulate greenish gemmae. However, the smaller cells, distinct trigones, distinct underleaves, and stem anatomy preclude the species from being closely allied to the Incisae.

[87] These are quite frequently present but are hidden among the dense rhizoid mat on older shoots. On younger parts of the plants they are often obvious and strikingly similar to those of *L. barbata*. Underleaves occur with about the same frequency in both species, are about equal in size, and are equally difficult to demonstrate.

[88] Modified from Schuster (1951a). An additional section is Schistochilopsis (Kitagawa) Schust., stat. n. (basionym: *Lophozia* subg. *Schistochilopsis* Kitagawa, Jour. Hattori Bot. Lab. no. 28:289, 1965). This differs from other sections of *Massula* in the development of a strongly asymmetrically bilobed leaf, with the smaller dorsal lobe folded over the ventral, and with a wide wing developed connecting the two "lobes." I have, ultimately, reduced *Schistochilopsis* to a sectio because, except for the peculiarly winged leaf keel reminiscent of *Schistochila*, all other criteria show a close approach to those here given to define the sectio Incisae. Indeed, the group appears most closely allied to this section.

3. Leaf lobes almost always 2, rounded or obtuse, separated by a narrow, sharply gibbous sinus; cuticle usually finely papillose; sterile shoots with distinct underleaves; leaves almost horizontally inserted, widely patent; cells small, 23–30 μ medially; gemmae rare; cortical stem cells subequal in width to basal leaf cells; stem relatively slender, little flattened. [sect. Obtusae] . . . . *L. obtusa* (p. 427)

3. Leaf lobes acute or apiculate, exceptionally obtuse, separated by a flat, obtuse to rectangular sinus; many leaves 2–3(4–5)-lobed; cuticle smooth; underleaves lacking; leaves subvertically to obliquely inserted and oriented; gemmae invariably present, usually abundant; stems usually stout, fleshy, often strongly flattened, sometimes nearly fasciated . . . . [sect. Incisae] 4.

4. Cells 40–50 × 50–75 μ in leaf middle, 40–50 μ in lobes and on margins; postical face of stem (and sometimes bases of leaves) usually purplish; leaves much broader than long, 2–3-lobed, lobes usually entire-margined, sometimes obtuse; dorsal cortical cells ca. 0.5× width of basal leaf cells. (Oil-bodies homogeneous to segmented, 5–8 μ long.) . . . . *L. grandiretis* (p. 456)

4. Cells 30–35 × 30–40 μ or smaller in leaf middle; postical faces of stem and leaves never purplish; dorsal cortical cells ca. 0.8–1 × width of basal leaf cells . . . . 5.

5. Leaves narrower, 1.0–1.5× as wide as long, deeply (2)3–5-lobed and with lobes usually ending in spinous teeth, lobe margins ± similarly armed; leaves 1–2-stratose to the base, soft-textured, their length greatly exceeding width of stem; cells mostly with 20–35 minute homogeneous oil-bodies only 2–3 μ long; stem fleshy, to 600 μ wide, 1.2–1.5× as wide as high. (Plants opaque, pale bluish green, never brownish in sun.) Chiefly subarctic-subalpine . . . . *L. incisa* (p. 441)

5. Leaves broad and short, 1.5–2.5× as wide as long, their lobes obtuse or ending in 1–2-celled teeth, but margins not otherwise spinous-dentate; leaves often 2–4(5)-stratose at base and quite rigid, their length usually not or hardly exceeding stem width; stem stout, very fleshy, (550)600–1200 μ wide, often flattened. Arctic to arctic-alpine . . . . 6.

6. Leaves (2)3–5-lobed, somewhat fluted and undulate-crispate, their lobes often ending in 1–2-celled sharp cusps, the sinuses sometimes reflexed; plants opaque, bluish green (as in *L. incisa*), stems without fuscous pigmentation, leaves never pigmented; cells with oil-bodies (25)30–60 or more per cell, minute, 2–2.5 × 2–3 μ, homogeneous . *L. opacifolia* (p. 448)

6. Leaves usually 2–3(4)-lobed, lobes obtuse, never apiculate or cuspidate, not fluted, without gibbous sinuses; plants a clear pellucid green, stems blackish-pigmented ventrally (at least with age), leaf lobes often yellowish brown in sun forms; cells with 6–21 ovoid to spherical, fine-segmented, large oil-bodies (3.5–7.5 μ to 4–7 × 5–8 μ) . . . . . *L. hyperarctica* (p. 437)

2. Gemmae ovoid, lacking protuberances; leaves usually lacking marginal teeth, lobes acute to obtuse; cortical cells 4–8× as long as wide, merely ± 0.5 as wide as basal leaf cells (or gemmae purplish) . . . . . .
. . . . [sect. Marchicae]   7.
    7. Gemmae greenish, rarely weakly reddish-tinged, scarcely forming globular masses at stem apices; cortical stem cells 4–8× as long as wide, only half as wide as basal leaf cells; cells of leaf middle large, 40–50 μ; leaves variable, flaccid and undulate, 2–5-lobed, the lobes usually acute. Subtropical to subarctic (rarely in the Arctic) . . 8.
        8. Plants green,[89] except for the deeply purplish-pigmented ventral side of the stem and rhizoids; perianths green throughout, mouth denticulate with short, 1-celled teeth; plants flexuous and flaccid, the leaves strongly undulate; gemmae ca. 16–20 μ; only in bogs.
. . . . *L. (M.) marchica* (p. 464)
        8. Plants and perianths largely reddish-tinged, but postical side of stem not purplish, green; perianth mouth denticulate with teeth largely 2–3 cells long or longer; plants scarcely flexuous, stems more rigid, leaves not flaccid, somewhat undulate; gemmae about 25 μ; usually on moist, sandy or gravelly acid soil . . . .
. . . . *L. (M.) capitata* (p. 469)
    7. Gemmae purplish, often forming distal terminal masses; cortical cells 2–4× as long as wide, no or little narrower than basal leaf cells; cells of leaf middle 20–35 μ; leaves uniformly 2-lobed, lacking marginal teeth. Arctic-alpine . . . . See *Scapania* subg. *Jensenia*

1. Gemmae constantly lacking; paroecious; ventral merophytes narrow (2–4 cells broad) but with small underleaves distinct nearly throughout. (Plants lack brown or vinaceous pigments; stem greenish beneath; leaves 2-lobed, but often with a tooth of antical or postical base.) . . . .
. . . . [sect. Elongatae]   *L. (M.) elongata* (p. 431)

Supplementary, Artificial Key to Species of Massula

1. Cells of leaf middle small: 25–35 × 30–40(45) μ or smaller . . . . . 2.
1. Cells of leaf middle large: 40 × 50 to 50(60) × 60–80 μ . . . . . . 8.
    2. Plants with stem purplish to purplish black beneath (at least with age); gemmae abundant, stellate-polygonal, 26–35 × 34–48 μ . . . .
. . . . *L. grandiretis* var. *parviretis*
    2. Plants with stem green or blackish beneath . . . . 3.
    3. Leaves (2)3–5-lobed, lobes acute or cuspidate to obtuse, sinuses not regularly gibbous; stem elliptical in cross section, soft and fleshy; gemmae ubiquitous, colorless, angulate . . . . 4.
    3. Leaves almost constantly bilobed; lobes edentate, acute, blunt, rounded or obtuse; stem not conspicuously fleshy, ± circular in cross section; gemmae smooth or (if angulate) rare (or absent) . . . . 6.

---

[89] The wholly green color of the leaves has been stressed in the literature as separating *L. marchica* from *L. capitata*. This character must be used with caution, since in some material of *marchica* the purplish pigmentation extends to the basal halves of the leaves (p. 469).

4. Plants clear green, often with brownish-pigmented lobes; stem blackish-pigmented beneath with age; leaf lobes often obtuse, never ending in cusps, entire-margined or sinuous; oil-bodies 6–20 per cell, fine-segmented. (Calciphyte; on clay). . . . . *L. hyperarctica*

4. Plants opaque bluish green, never with secondary pigments of leaves or postical stem surface; leaf lobes often cuspidate at apex, the teeth formed of enlarged cells; oil-bodies 25–60 per cell, minute, homogeneous . . . 5.

5. Leaves little wider than long, acute lobes usually sharply spinose-dentate; leaf base 1–2-stratose. Subarctic-subalpine. (Ca intolerant) . . *L. incisa*

5. Leaves 1.5–2.5 × as wide as long, lobes usually entire, obscurely cuspidate at apex; leaf bases often 3–4(5)-stratose. Arctic-alpine. (Ca tolerant) . . . . *L. opacifolia*

6. Gemmae rare, colorless, angulate; leaves almost horizontally inserted, the obtuse to rounded lobes separated by a narrow, gibbous sinus . . *L. obtusa*

6. Gemmae absent or smooth and ovoid (and colorless to purplish); leaves obliquely to subvertically inserted, never with gibbous sinuses . . . . 7.

7. Gemmae absent; leaf lobes sharp, leaf bases often with a sharp accessory tooth; paroecious; green . . . . *L. elongata*

7. Gemmae common, often purplish; leaf lobes blunt to rounded, leaf bases edentate; dioecious; often pigmented . . . . See *Scapania* subg. *Jensenia*

8. Stem pale green below; leaves often largely 3–4-lobed, with lobes often reddish to reddish violet; gemmae 1-celled, smooth. Nonarctic . . . . *L. capitata*

8. Stem violet to reddish black ventrally; leaves usually green, except sometimes basally, usually 2(3)-lobed. Arctic-alpine to subarctic . . . . 9.

9. Gemmae smooth, 16–20 μ; oil-bodies minute, 2–4 μ; cells ca. 40 × 50 μ in leaf middle . . . . *L. marchica*

9. Gemmae angular, 25–35 μ or larger; oil-bodies to 5 × 7–8 μ, often of 2–several segments; cells 40–50(60) × 60–80 μ in leaf middle. *L. grandiretis*

Sectio *OBTUSAE* Schust.

Barbilophozia subg. *Obtusifolium* Buch, Mem. Soc. F. et Fl. Fennica 17:289, 1942.
Leiocolea subg. *Obtusifolium* Buch, *ibid.* 17:289, 1942.
Lophozia subg. *Massula* sect. Obtusae Schust., Amer. Midl. Nat. 45(1):54, 1951.
Obtusifolium S. Arn., Illus. Moss Fl. Fennosc. 1:133, 309, 1956.

Folia subhorizontalia, omnia bilobata, lobis plerumque obtusis ad rotundatos, sinu plerumque reflexo, cellulae parvae, cuticula asperata; amphigastria saepius distincta.

Plants with *very oblique to subhorizontal*, normally bilobed *leaves*; leaf lobes usually *rounded to blunt*, the margins normally entire, *lobes noncuspidate*. *Underleaves usually distinct*, although often obscured by the rhizoids. *Cells relatively small, able to develop weak verrucae or striolae;* oil-bodies small, numerous. Gemmae rare, angulate, without pigmented walls. Capsule wall 5-stratose. Dioecious. Antheridia 2–3; paraphyses present.

Type. *Lophozia obtusa* (Lindb.) Evans, the only species of the section.

Lophozia obtusa remains a critical species in subg. *Massula*. It was placed in *Leiocolea* by Buch (1933a); in this he was followed by Frye and Clark

(1944, *in* 1937–47), who, however, saw no material. It had been previously placed by Warnstorf (1903) in the species of *Lophozia* now segregated as *Leiocolea*, and by Müller (1910, *in* 1905–16) in the *Orthocaulis* section of subg. *Barbilophozia s. lat.* None of these positions was natural, as Müller (1939) indicated in placing the species in a novel position, within subg. *Massula.* Meylan (1939) also questioned Buch's disposition of the species, suggesting it was closely related to *L. barbata* (and should perhaps be placed in a separate subgenus with the latter). Buch (1942), reiterating his belief that on the basis of stem structure the species clearly belonged in *Leiocolea* and not in *Lophozia* (incl. *Massula* and *Lophozia*), conceded it might best be separated subgenerically from *Leiocolea* (as "subg. *Obtusifolium* Buch").[90]

This, I think, is unnecessary. Material from Michigan and Germany (Feldberg, Baden, *leg.* K. Müller)[91] agrees quite closely in stem structure, leaf form, and leaf insertion with *Massula.* The species is possibly related most closely to *L. grandiretis*, agreeing in the wide leaves with obtuse or blunt lobes. The reflexed or gibbous sinuses are reminiscent of *L. marchica*, although the small cells separate *L. obtusa* from other *Massula* species. *Lophozia obtusa* does not occupy a more isolated position in *Massula* than such species as *L. grandiretis* and *L. incisa*; it is, admittedly, a synthetic species. It is the only member of the *Massula-Isopaches-Lophozia* complex which retains the ability to develop underleaves and virtually the only species of *Massula* which retains the ability to develop a papillose cuticle; it also retains relatively small leaf cells (and is closer to *Orthocaulis* than to *Massula* in this regard). Like the other *Massula* species, it has wide ventral stem sectors and wider-than-long leaves that regularly possess the ability, on robust shoots, to develop 3–4 lobes or marginal teeth. In all of these characteristics *L. obtusa* approaches species of *Orthocaulis.* The leaf shape of *L. obtusa*, with the somewhat reflexed sinuses, closely approaches that of *L. (Orthocaulis) kunzeana*: a species which both Buch (1933a) and I consider as archaic and reminiscent (at least) of the archetype of the genus *Lophozia.*

Lophozia obtusa, therefore, stands below other species of *Massula* and suggests a derivation of this subgenus from *Orthocaulis*-like ancestors (independent of the *Lophozia-Tritomaria-Saccobasis* developmental lines). It may therefore be found advisable eventually to separate *Massula* (including the allied subg. *Protolophozia*) generically from *Lophozia* (especially if the more radical scheme on p. 231 is accepted.[92]

When Buch (1942) reiterated his belief that *L. obtusa* should go near or in

---

[90] Arnell (1956, p. 133) states that the species is "difficult to place" and that "the simplest solution to these difficulties is to let it form a genus of its own." This, in my opinion, is no solution. If we were to place each isolated species into a monotypic genus, the concept of genus would be so severely diluted that its utility would disappear.

[91] And, recently, abundant collections of fresh material, which I collected near Col de Faucille, French-Swiss border, in the Jura.

[92] On a phylogenetic basis a good argument could be made in favor of accepting a genus *Lophozia* (including *Lophozia, Tritomaria, Saccobasis*) as discrete from a genus *Massula.* The existence, however, of the annectant *Isopaches* element (with stem anatomy nearer that of *Massula* and oil-bodies as in *Lophozia*) makes such a treatment difficult.

Leiocolea (on the basis, in part, of stem anatomy) he overlooked the fact that all species of *Massula* have the stem structure characteristic of *L. obtusa*. The often thick-walled cortical cells of *Leiocolea*, collenchymatous leaf cells, large oil-bodies, and inability to develop 3-lobed or dentate leaves, as well as the obviously specialized perianth, indicate that *Leiocolea* represents a distinct developmental line whose sole relationship to *Massula* seems to lie through *Orthocaulis*.

LOPHOZIA (MASSULA) OBTUSA (Lindb.) Evans
[Figs. 173:6–9; 174]

Jungermannia obtusa Lindb., Musci Scand., p. 7, 1879.
Lophozia obtusa Evans, Proc. Wash. Acad. Sci. 2:303, 1900.
Lophozia (subg. *Barbilophozia*) *obtusa* K. Müll., Rabenh. Krypt.–Fl. 6(1):648, fig. 305, 1910; Macvicar, Studs. Hdb. Brit. Hep. ed. 2:205, figs. 1–3, 1926.
Leiocolea obtusa Buch, Mem. Soc. F. et Fl. Fennica 8(1932):288, 1933; Frye & Clark, Univ. Wash. Publ. Biol. 6(3)(1944):382, figs. 1–9, 1945.
Lophozia (subg. *Massula*) *obtusa* K. Müll., Ber. Deutsch. Bot. Gesell. 57(8):341, 1939; Schuster, Amer. Midl. Nat. 45(1):54, pl. 9, figs. 1–8, 1951; Schuster, *ibid.* 49(2):341, pl. 10, figs. 1–8, 1953; Müller, Rabenh. Krypt.–Fl. ed. 3, 6:669, figs. 205d, 206, 1954.
Barbilophozia (subg. *Obtusifolium*) *obtusa* Buch, Mem. Soc. F. et Fl. Fennica 17(1940–41):289, 1942.
Obtusifolium obtusum S. Arn., Illus. Moss Fl. Fennosc. 1:133, 1956.

Usually scattered or in loose patches, green to yellowish green, medium-sized, ca. 1.5–2.5 mm wide × 2–5(6) cm long. Shoots prostrate or (with crowding) ascending, flexuous, simple or nearly so. Stems rather slender but fleshy, slightly dorsiventrally flattened (and hardly elliptical in cross section), ca. 300–330 μ wide, 12–14 or more cells high, often reddish purple or vinaceous ventrally; 1–2 cortical layers of cells *not or slightly thick-walled, subequal in width to basal leaf cells, short-rectangulate and only 2–3(3.5)× as long as wide* [ca. (16)19–26(30) μ wide × 42–76 μ long], *distinctly smaller in diam. than medullary cells;* medulla uniform, of very leptodermous cells, ca. 32–38 μ in diam., *without mycorrhizal infection.* Ventral merophytes 10–12 or more cells broad, *frequently developing rather large underleaves on sterile shoots,* bearing numerous rhizoids that are *often vinaceous or reddish at base.* Lateral leaves *very obliquely inserted* (antical ⅔ of insertion at mere 20–30° angle with stem), scarcely decurrent antically, *distant or contiguous* to weakly imbricate, widely, laterally spreading, *nearly plane* or barely concave, *quadrate-rotundate* and about as wide as long (ca. 970–1030 μ long × 1100–1250 μ wide) to occasionally slightly longer than wide or distinctly broader than long (1275 μ wide × 935 μ long), occasionally to 3000 μ wide × ca. 2000–2400 μ long, *bifid* (very exceptionally isolated leaves trifid) for 0.25–0.4 their length, the *sinus acute or subacute but sharply reflexed at base, thus somewhat gibbous; lobes broadly ovate or ovate-triangular, all or mostly rounded to broadly obtuse at apex* (isolated leaves, or sometimes all leaves of isolated shoots, acute or sub-acute at apex; occasional leaves of this type with 1–2 low accessory marginal teeth); *margins usually entire.* Cells *relatively small,* 23–30 μ on leaf margins, leaf apices, and leaf middle, at leaf middle occasionally 30–35 μ, almost isodiametric at and above leaf middle, near base 20–30(36) μ wide × 30–34(38) μ long, polygonal, *thin-walled, with minute to medium-large trigones;* cuticle smooth or very faintly striolate-verruculose; *oil-bodies minute,* spherical to ovoid, ca. 2–3 μ,

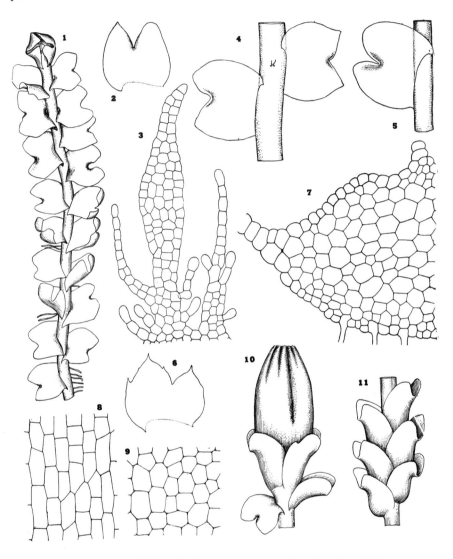

Fig. 174. *Lophozia* (*Massula*) *obtusa*. (1) Sterile shoot (×13). (2) Narrow leaf (×16.4). (3) Large underleaf (×148). (4) Postical shoot sector (×19). (5) Lateral sector of shoot (×21). (6) Leaf with acute lobes and accessory teeth, sporadic and more typical of *Massula s. lat.* (×16.4). (7) Stem cross section (×140). (8) Dorsal cortical stem cells (×155). (9) Basal leaf cells (×155). (10–11) Perianth- and androecium-bearing shoot sectors (×16–18). [1, after K. Müller, 1905–16; 2, 5, from Copper Harbor, Mich., *Steere;* 3–4, 6–9, from Feldberg, Baden, Germany, leg. Müller; 10–11, after Meylan, 1924.]

numerous, 15–50 per cell (Müller, 1948), no larger than chloroplasts. *Underleaves occasionally rather large*, locally rudimentary, on sterile stems sometimes to 450 μ long and formed of 2–3 laciniae (up to 3–5 cells wide), and bearing 3–4 or more stalked slime papillae or cilia. *Gemmae rarely produced, pale green, angulate,* 1(–2)-celled, ca. 18–20 μ.

Dioecious. ♂ Plants more slender; androecia becoming intercalary, of 6–12 (rarely to 15–20) pairs of bracts, spicate; bracts somewhat smaller than leaves, asymmetrically bilobed, *strongly concave and with lobes erect* or even incurved; 2–3-androus. Antheridia with *stalk uniseriate*. *Gynoecia rarely present*; bracts subtransverse, erect or suberect at base, the lobes sometimes reflexed, as large (or scarcely) as leaves, irregularly 2–4-lobed, the lobes acute to obtuse, often slightly dentate, the sinuses reflexed. Bracteole large, often bifid, variable. Perianth slenderly cylindrical, ellipsoid to subclavate, plicate only in distal ⅓, narrowed to the slightly dentate or serrate mouth. Sporophyte rare, *wall 5-stratose*; epidermal layer of large cells (20 μ thick) with nodular thickenings; inner layers each ca. 10 μ thick, with semiannular bands. Spores 11–14 μ; elaters 7–9 μ in diam. (according to Müller, 1951–58).

Type. Vicinity of Stockholm, Sweden.

Distribution

Possibly imperfectly holarctic, but to date found but once in Asia. A northern or nordic-alpine species, largely arctic-alpine in distribution, of sporadic occurrence (especially at its southern periphery in the northern portions of the Spruce-Fir Forest Region). Weakly "calciphilous" and therefore restricted in distribution. In North America very rare, except in the West.

In Europe sporadically distributed, from Scandinavia (Sweden, Norway, Finland), Russia, southward to England (Wales to Scottish Highlands), Germany (Black Forest; Eifel Mts.; Westphalia; Harz Mts.; the Thuringer Forest, etc.), Czechoslovakia, the Tatra Mts., Hungary, Bulgaria, Bucovina, Belgium, south to France (Cévennes, Cantal, Auvergne, the Vosges, common in the Jura!), south to the Pyrenees (in Spain), Switzerland (chiefly Jura!), northern Italy (Piedmont, Trentino). Also in the Faroes and Iceland. Most recently reported from Japan (Toyama Pref., Honshu; Kitagawa, 1967a).

In North America relatively widespread in the West, especially in the Rocky Mts., ranging from Alaska southward to British Columbia, Washington, Oregon, and California in the Sierras and Cascades?, and to Idaho and Wyoming, in the Rocky Mts., with a hiatus in its range from there eastward to the Great Lakes region, which will surely be filled by further collecting.

ELLESMERE I. Lastraea Valley, ca. 78°45′ N. (northernmost station; rare; Bryhn, 1906, p. 32). S. GREENLAND. Nanortalik, ca. 60°07′ N. 45° W. (*Vahl 1107!*).[93]

[93] This specimen was labeled "*J. floerkei*" by Jensen, but published as "*J. lycopodioides*" by Lange & Jensen, 1887. The plants are poorly developed, mostly androecial, and weathered, but show the gibbous sinuses and mostly round or blunt lobes (see Fig. 173:6–9).

QUEBEC. Cairn I., Richmond Gulf, E. coast of Hudson Bay (Wynne & Steere, 1943).
NORTHWEST TERRITORIES. Southampton I. (Jennings, 1936, p. 14).   NEWFOUNDLAND.
Kitty's Brook, 49°08′ N. and 56°57′ W. (Buch & Tuomikoski, 1955); Dicks Brook, E.
Arm, Bonne Bay (*RMS 68-1440*); River of Ponds Prov. Pk., N. Pen. (*RMS 68-1481*).
MAINE. Round Mountain L., Franklin Co. (*Lorenz*; see Evans, 1912d, p. 212).   MICH-
IGAN. Ft. Wilkins, Copper Harbor, Keweenaw Co. (*Steere, 1935*; duplicate verified!).
    The report of the species from Minnesota in Buch & Tuomikoski (1955) is erroneous.

Ecology

    "Mostly singly between other mosses, for example, *Barbilophozia* and
Hylocomium species, but occasionally also in pure mats on soil, grassy sites,
over fir needles, on siliceous and calcareous cliffs, in woods in the mountains. Almost always totally sterile. Sporophytes extremely rare, gemmae
rare" (Müller, 1954, *in* 1951–58). In the mountains of central Europe
scattered between 1800 and 2000 m, up to 2350 m. In Newfoundland
reported from along a spring, in woods, with *Riccardia multifida*, *Calypogeia
muelleriana*, and *Bryum ventricosum*. In Greenland mixed with *Cephalozia
leucantha*. Conspicuously lime-tolerant, and common in the limestone
mountains in central Europe (e.g., the Jura).

Differentiation

    The most obvious differential features of *L. obtusa* are the almost
horizontally oriented, uniformly bilobed, rather lax leaves, with rounded
or obtuse lobes, the generally narrow sinus conspicuously if narrowly
reflexed at base. In the rounded to obtuse leaf lobes *L. obtusa* superficially
recalls *Gymnocolea* and *Cladopodiella*, but in both of these the leaves are
generally much more obliquely inserted and narrow in shape, and in
neither is the leaf sinus reflexed; both are also strongly oxylophytic.
Often, but not uniformly, *L. obtusa* has the stem and rhizoid bases characteristically reddish or reddish purple postically.

    Lophozia obtusa is a sharply isolated species, without close affinities. The
varied positions assigned to it (subgenera *Barbilophozia*, *Leiocolea*, and *Massula*;
"genus" *Obtusifolium*) emphasize its anomalous character. Although the stem,
with its scarcely thick-walled cortical cells and lack of dorsiventral differentiation of the medulla, is similar to that of both *Leiocolea* and *Massula*, the other
features of the species are principally those of *Massula*. Among these are the
(*a*) form of the perianth, which is plicate above and not rostrate at the apex;
(*b*) leptodermous cells, with ability to develop only small trigones; (*c*) numerous
minute, spherical to ovoid oil-bodies; (*d*) relatively broad leaves, which on
occasion may bear a few marginal teeth and in isolated cases a third lobe.
    In contrast, several features isolate *L. obtusa* from the other species of *Massula*:

(*a*) unusually small leaf cells and their ability to develop a slightly roughened cuticle; (*b*) relatively short cortical cells of the stem, only 2–3.5× as long as wide, which are as broad as the basal leaf cells; (*c*) nearly horizontal insertion and orientation of the leaves; (*d*) generally rounded leaf lobes and reflexed, gibbous sinus; (*e*) frequent ability to develop relatively distinct underleaves. In the last of these characters, as well as in the ability of cells to develop papillae, *L. obtusa* clearly approaches *Leiocolea*.

Sectio *ELONGATAE* Schust., sect. n.

Plantae paroeciae; gemmae nullae; folia obliqua, profunde bilobata, in marginibus anticis saepe unidentata; cellulae parvae (mediae minores quam 30 μ); monandrae; pedicellus antheridialis 1-seriatus.

Plants with oblique, deeply bilobed leaves, whose antical margin often bears an accessory basal tooth; lobes sharply triangular but noncuspidate; sinus angular, narrowly rounded or obtuse at base, not gibbous. *Cells medium-sized* (under 30 μ in leaf middle) with small but distinct trigones; oil-bodies?; cuticle smooth. Small *underleaves usually present*, although ventral merophytes only 2–4 cells broad. *Gemmae absent. Paroecious*; *monandrous*, antheridial stalk 1-seriate.

Type. *Lophozia elongata* (Lindb.) Steph., the only species.

Sectio Elongatae is established for a single species whose disposition has aroused divergent opinions: Stephani considered it hardly distinct from *Lophozia ventricosa*; Evans (*in* Buch, Evans, and Verdoorn, 1938) placed it in *Orthocaulis* and was followed by Frye and Clark (1937–47) and Arnell (1956); Müller (1905–16) emphasized that it was closely allied to *L. excisa* and (1951–58) placed it into synonymy with this species! All of these workers overlooked the fact that axial anatomy prohibits placing the plant in subg. *Lophozia* (i.e., with *L. ventricosa* or *L. excisa*), while the weakly differentiated cortex of long cells, smooth cuticle, leaf form and very oblique insertion, narrow ventral merophytes, pellucid appearance and soft texture, and other characters separate it from *Orthocaulis*. I would suggest (unless the oil-bodies demonstrate a lack of logic in this disposition) that *L. elongata* be put into *Massula*, to which stem anatomy, leaf form and texture, leaf cells, and general facies suggest an affinity. In this subgenus it admittedly has no clear allies and therefore must be segregated into a separate section.

LOPHOZIA (MASSULA) ELONGATA (Lindb.) Steph.

[Figs. 175, 176]

Jungermannia elongata Lindb., Meddel. Soc. F. et Fl. Fennica 9:162, 1883.
Lophozia elongata Steph., Spec. Hep. 2:141, 1902; K. Müller, Rabenh. Krypt.–Fl. 6(1):692,
    1910; 6(2):767, figs. 204, 1916.
Orthocaulis elongatus Evans, *in* Buch, Evans, & Verdoorn, Ann. Bryol. 10 (1937): 4, 1938; Arnell,
    Illus. Moss Fl. Fennosc. 1:140, fig. 53, 1956.

Small, light to yellowish green, occasionally faintly brownish-tinged locally, 1–2 cm high × 1000–1250 μ wide (sterile shoots; to 1400–1550 μ wide at apices of gynoecial axes), sparingly branched, usually caespitose, occasionally creeping, often with small- and remote-leaved shoots. Stem somewhat elliptical in cross section, *light green* (no pigmentation of cortex or of medulla), ca. 200–240(260) μ in diam., from (9)10 to 12 cells high, rather *soft-textured and fleshy*; cortical cells rectangulate, *narrow*, little to *hardly thick-walled*, to (16)18–22(24) μ in diam. × (50)56–85 μ long (length usually *2.5–4:1 dorsally*), smooth, the *ventral cells with age often brownish* (because of fungal infection, which extends to rhizoids); medulla colorless, of hyaline cells with distinct trigones (in cross section), *no dorsiventral differentiation*, without fungal infection, the cells variable in size, to 20–24 μ high, mostly little larger than cortical and hardly differentiated from

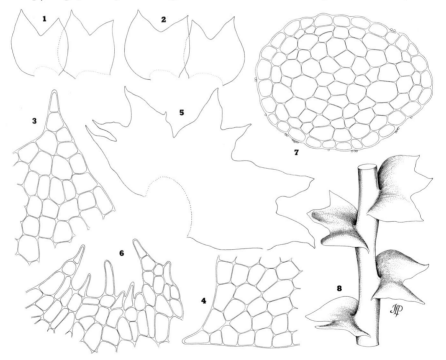

FIG. 175. *Lophozia (Massula) elongata*. (1–2) Leaves (×30). (3–4) Leaf lobe apices (×225). (5) Bracts and bracteole (×34). (6) Perianth mouth cells (×225). (8) Shoot sector, antical view (×30). [All from type, G.]

them. Rhizoids abundant, often brownish with age (because of fungal infection, *not pigment*). Leaves strongly succubous, obliquely oriented, rather patent, *often remote* on sterile stems, contiguous to weakly imbricate above on fertile axes, soft-textured, rather hyaline, usually asymmetrically rounded-quadrate to broadly ovate-quadrate, *usually slightly wider than long* (540 μ wide × 530 μ long to 640–670 μ wide × 620–630 μ long), occasionally a little longer than wide (430 μ wide × 550 μ long), rather *deeply bilobed* (0.3–0.35, rarely to 0.45); antical lobe smaller, the antical margin often nearly straight; postical lobe larger, the postical margin arched, the postical base ± ampliate; sinus obtusely to subacutely angulate, rounded but not gibbous at base; lobes triangular, sharp, blunt sometimes on upper leaves; occasional leaves with ventral lobe divided, the leaf then asymmetrically trifid as in *Orthocaulis*; *leaves frequently with a basal or suprabasal tooth of antical and/or* (more rarely) *postical margin*. Cells of lobe margins ca. (19)21–25 μ, of lobe bases and leaf middle ca. 21–25 × 23–28 μ; cuticle smooth; walls thin or feebly thickened, trigones *small and concave-sided, rather discrete*. Ventral merophytes usually only 2–4 cells broad, but rather regularly *with small* (often appressed and then difficult to see) *underleaves*, which are lanceolate, or lobed, with 1–2 marginal teeth or slime papillae. *Gemmae absent*.

Paroecious. ♂ Bracts like large leaves, usually with a sharp antical tooth or lacinium, the antical base a little concave (never strongly so), (always?) *monandrous; antheridial stalk 1-seriate*, ca. 85 × 18–20 μ; body ellipsoidal, ca. 100 × 125 μ. Gynoecial bracts rather irregular, wavy or crisped, variable, free from bracteole or united with it to 0.3–0.4 their length. Bracts 2(3)-lobed, narrow, *ovate to oblong-ovate*, ca. 650–730 μ wide × 750–950 μ long to 870–900 μ wide × 1100–1250 μ long, divided 0.3–0.45 their length, the lobes sharply acute, lanceolate to acutely triangular, *edentate*; margins edentate or sinuous, occasionally with stalked slime papillae or a lacinium below. Bracteole large, variable, free and lingulate to lanceolate with 1–2 laciniae above (then ca. 450 × 1050 μ) to bifid and oblong (then ca. 450 × 750 μ). Perianth obovoid to ovoid-cylindrical, plicate distally, the *contracted mouth plurilobulate*, each small lobe *denticulate*, the larger (usually terminal) teeth of the lobes *formed of 1–2 cells* (2-celled teeth 10–14 × 36–48 μ; 1-celled teeth *up to 10–12 × 36–42 μ*). Spores 10–14 μ, finely papillose; elaters 6 μ in diam. (*teste* Arnell, 1956).

Type. Norway: Hedemarken, Osterdalen, Tronfield (*Lindberg*); *teste* Müller.

Distribution

A rare and sporadically distributed northern species, reported only from Norway and Sweden and from the Tyrol (formerly Austria, now Yugoslavia; Schiffner's Hep. Eur. Exsic. no. 1400); also once, erroneously, from Greenland, as follows: Fleming Inlet (Kruuse); *teste* Jensen (1906a).[94]

[94] Since the species has been reported from North America—even if erroneously so (the specimen is *L. quadriloba* of the "*heterophylla*" extreme)—I give an illustrated account based on the type specimen (G).

Ecology

A rare species of marshes and bogs, associated at times with *Lophozia ventricosa*, *L. binsteadii*, and *Cephalozia media* (under evidently acid conditions), but the single Norwegian specimen I have seen (Lille Elvedalen, *Lindberg*, G) is evidently from a weakly calcareous site, judging from the associates (*Lophozia grandiretis*, *L. heterocolpa*, *Blepharostoma trichophyllum*).

Differentiation

Lophozia elongata has been—and remains—an enigmatical species. Müller (1916, p. 767, and 1910, p. 693, *in* 1905–16) emphasized the similarity to *L. excisa*. It shares with the latter the soft texture, large and rather pellucid leaf cells, and paroecious inflorescence, and has much the facies of the swamp form of *L. excisa*. My immediate reaction, when I first saw the plant, was that a gemma-free hygrophytic phase of *L. excisa* was at hand. It can readily be mistaken for such phases of *L. excisa*—*L. excisa* var. *cylindrica* (Dum.) Müll. Indeed, Müller (1951–58, p. 666) later considered it a synonym of this variety, claiming that Schiffner agreed with this viewpoint. Stem anatomy and perianth mouth armature, however, prohibit such a viewpoint. Müller also criticized as "incomprehensible" the disposition of *L. elongata* as a species of *Orthocaulis*, as, e.g., in Buch, Evans and Verdoorn (1938) and Frye and Clark (1937–47). It has also been placed in *Orthocaulis* by Arnell (1956), who, however, states that it is "closely related to *L. excisa*." Arnell (*loc. cit.*, p. 122) keys the plant in the Excisae, a section he places in *Lophozia* (p. 114) and diagnoses as paroecious, even though (p. 122) he assigns to it two species, *L. capitata* and *L. marchica*, which are dioecious! The fact that *L. elongata* is juxtaposed in his key to these two species (which belong to *Massula*) and to *L. excisa* (which belongs to subg. *Lophozia*) seems significant.

I agree with Müller that, in spite of the frequent presence of small underleaves and lack of medullary differentiation, *L. elongata* cannot go into *Orthocaulis*. Like *L. obtusa*, it fails to "fit" well into any of the subgenera of *Lophozia*. However, the medullary anatomy, the pellucid and rather large cells with small trigones, the occasional underleaves, the tendency for accessory lobes and/or teeth of the leaves, the thin-walled, elongated cortical cells, and the polymorphic, mostly edentate, perichaetial bracts can all be matched, individually or collectively, in *Massula*. Until the oil-bodies are known, even this disposition remains provisional.

Obvious similarities to sectio Excisae of subg. *Lophozia* cannot be denied. The deeply bilobed leaves suggest particularly *L. latifolia* Schust., and "normal" leaves of *L. elongata* look much like the leaves of slender ♂ stems of *L. latifolia*. Yet the ♀ bracts of *L. elongata* are usually longer than wide, are much more deeply lobed, and have narrower lobes. Of more importance, both stem

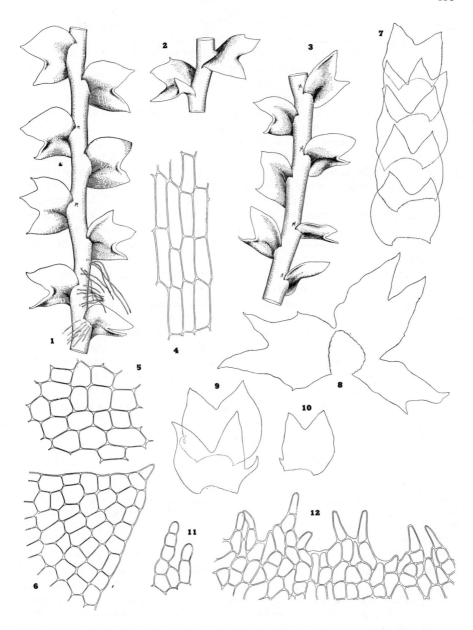

FIG. 176. *Lophozia (Massula) elongata.* (1) Shoot sector, postical aspect, rhizoids drawn in near base only (×26.5). (2) Shoot sector, antical aspect (×26.5). (3) Shoot sector, ventral aspect (×26.5). (4) Cortical cells of stem (×235). (5) Median leaf cells (×235). (6) Cells of leaf lobe apex (×235). (7) Leaves, sterile shoot sectors (×32). (8) Bracts and bracteole (×24). (9–10) Leaves with basal teeth (×32). (11) Underleaf (×315). (12) Perianth mouth cells (×235). [All drawn from type, leg. Lindberg, G.]

anatomy and perianth-mouth armature prohibit any close affinity to *L. latifolia*. In *L. latifolia*, as in *L. excisa*, the stem has a ventral, smaller-celled medullary band, mycorrhizal with age, and in both species the perianth mouth is, basically, fringed with elongated cells, united except for the apices (Fig. 197:1). By contrast, in *L. elongata* there is no small-celled mycorrhizal band in the medulla (although the ventral and lateral cortical cells may be infested with fungal hyphae, as is also the case in various *Massula* spp.), and the lobulate perianth mouth is denticulate to short-ciliate—the longest teeth two cells, rarely three cells, long; below these teeth, the perianth cells are hardly elongated.

Thus, in spite of the superficial similarity to the *L. excisa-latifolia* complex, *L. elongata* must be adjudged a good species. It seems to be constantly fertile, and every gynoecial shoot seen had antheridial bracts below the gynoecia. The ♂ bracts were, at times, difficult to find; the antical base of these bracts is often somewhat transverse, feebly ventricose, and armed with an accessory tooth; not far from the dorsal base a single antheridium (I saw only monandrous bracts) occurs. The antheridial stalks seen were 1-seriate. No purely androecial branches were seen.

Although the facies, growth habit, and paroecious inflorescence, among other features, suggest *L. excisa*, the axial anatomy effectively appears to prohibit our considering this species a member of subg. *Lophozia* sect. Excisae. In any event, the rather leptodermous cells, the soft-textured stem with a feebly thick-walled, 1-stratose cortex and no ventral mycorrhizal medullary band, the occasional occurrence of small underleaves, and the rectangulate, narrow cortical cells suggest *Massula*. The dorsal cortical cells may average only ca. (13)14–18 × (48)50–72 μ, averaging about 3–4× as long as wide.

Distinctive for the species is also the strong tendency for the leaves, even of sterile shoots, to develop a basal tooth of the antical (rarely also postical) margin.

Sectio *INCISAE* Schust.

Lophozia subg. *Massula* sect. Incisae and sect. Grandiretae Schust., Amer. Midl. Nat. 45(1):54, 55, 1951.
Incisa-Gruppe K. Müll., Rabenh. Krypt.-Fl. ed. 3, 6:669, 1954.

Plantae dioeciae, gemmae viridulae plerumque praesentes, angulatae ex una vel duabus cellulis constantes; folia subverticalia asymmetrice 2–3(4–5)-lobata, in cuspidibus ciliisve saepe terminata, saepe dentata ciliatave; plantae succulentes crassae, molles, plerumque caeruleovirides opacae aut virides pellucidae; amphigastria nulla.

Plants with oblique to subvertical 2–5-lobed, often rather asymmetric, leaves (dorsal lobe or lobes then smaller); lobes usually sharply pointed, often apiculate or mucronate, or ending in cusps, or variously armed with teeth or cilia (absent or vestigial in *L. opacifolia*); usually clear or bluish green, often opaque, sometimes (*L. grandiretis*) with purplish pigmentation. No underleaves. Gemmae always present, greenish, angulate, 1–2-celled.

Type. *Lophozia incisa* (Schrad.) Dumort.

Sectio Incisae includes two subsections, as follows:

a. Leaves with teeth (when present) little or not modified, formed of cells less than 2–4× as long as wide . . . . . . subsect. Incisae (type: *L. incisa*)

b. Leaves *Temnoma*-like, 3–4–5-lobed, lobes ending in and armed with numerous
1-celled, slender, long cilia (cells 6–10 × as long as wide) . . . .
. . . . subsect. Setosae (type: *L. setosa*)

Subsectio Incisae includes, besides the local species, *L. incisa*, *L. opacifolia*, *L. hyperarctica*, and *L. grandiretis*, also a Formosan species, *L. nakanishii* Inoue (Bull. Natl. Sci. Mus. Tokyo 9[1]:37, fig. 1:10–18, 1966). This differs from our species in the asymmetrically bilobed leaves with "teeth" which are blunt to subacute and formed of many leptodermous, hardly elongated cells—thus the bilobed leaves are sharply, coarsely plurilobulate, rather than armed with sharp teeth per se.

Sectio Incisae stands rather isolated in the genus *Lophozia* s. *lat.*, in its proclivity to develop sharply toothed or ciliate leaves; this is carried to an extreme in *L. setosa* (Mitt.) Steph. of Sikkim, in which strikingly aciculate, arched, 1-celled, thick-walled setose cilia are developed.

LOPHOZIA (MASSULA) HYPERARCTICA Schust.

[Fig. 177]

Lophozia (Massula) hyperarctica Schust., Canad. Jour. Bot. 39:967, fig. 1, 1961.

Scattered, or in small patches, erect or ascending (when growing amidst caespitose mosses) or prostrate (when on bare soil), *pure green or lobes of upper leaves golden brown*, rather pellucid, fleshy and tender. Shoots 1.5–1.8(2.2) mm wide, 5–12 mm long, simple or furcate, the subtending leaf hardly modified. *Stem flattened*, ± *fasciated, elliptical in cross section, stout, soft, and fleshy*, whitish green, the postical side *blackish*-pigmented (but *never purplish*) with age, 450–550 (rarely 650–800) μ wide × 350–450 μ high, of thin-walled cells throughout, 14–16 cells high; dorsal cortical cells narrowly rectangulate, ca. 22–28 to 28–32(36) μ wide, *2.5–4.5(5.5)* × *as long as wide* (basal leaf cells 1.5–2 × as wide). Rhizoids frequent, colorless, ca. *18–20 μ in diam.* Leaves obliquely inserted, obliquely spreading, *small for size of stem* (length hardly exceeding stem width), irregular, shallowly 2–3-lobed (for 0.15–0.3 their length) but sometimes shallowly 3–4-lobed (particularly near apices of gemmiparous shoots), *strongly transverse*, 750 μ wide × 500 μ long to ca. 1150–1400 μ wide × 600–620 μ long (*from a minimum of 1.5 up to 1.9–2.3* × *as wide as long*), rectangulate, *firm and often rigid*, somewhat fleshy at least at base (basal half 2-stratose, *above juncture with stem often 3–4-stratose*); margins entire, or with 1–2 low teeth, with prolonged gemmae formation sometimes bearing scattered teeth; *teeth and lobe apices never sharp and cuspidate;* sinus obtuse to crescentic; lobes rectangular with obtuse or subacute tips to obtuse or rounded at apex. Cells of lobe margins and of leaf tips 25–32(35) to (25)30–32 μ, almost isodiametric; *median cells* ca. *35–38(42–44 μ)*, almost isodiametric; basal cells 35–38(44) μ × 40–45(50) μ; cells thin-walled or uniformly slightly thick-walled, virtually *without trigones*, strongly convex; cuticle smooth; oil-bodies from *6 to 18 (rarely 20 to 21) per cell, finely but distinctly papillose-segmented*, grayish, spherical to ovoid, *from 4 × 5–6 to 5 × 6–8 μ* or (*3.5*)*5–7.5 μ*, averaging much larger than chloroplasts. Underleaves lacking.

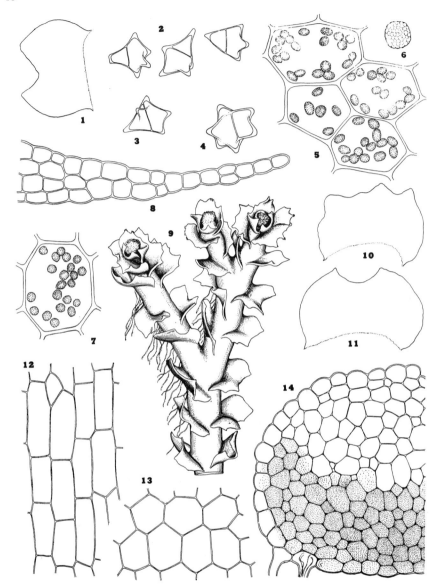

FIG. 177. *Lophozia hyperarctica.* (1) Leaf (×34). (2–4) Gemmae (×400). (5) Cells from base of leaf lobe, showing oil-bodies (×548). (6) An individual oil-body (×1400). (7) Cell from leaf middle (×570). (8) Longitudinal section of an unusually thick leaf, from base to sinus (from xeromorphic, brownish, compact phase of the species, with rigid leaves (×150). (9) Lax-leaved, green, gemmiparous, terminally branched shoot (×12.3). (10–11) Leaves (×24). (12) Dorsal cortical stem cells (×220). (13) Basal leaf cells (×220). (14) Cross section of one-half of small, nonfasciated stem, the fuscous-purple medullary cells stippled (×128). [All from type and cotype material, from near Alert, Ellesmere I.; 1–3, 5–7, *RMS 35061*; 4, 9–13, *RMS 35077*; 8, *RMS 35072a*; 14, *RMS 35142*. From Schuster, 1961.]

Gemmae *pale green*, at apices of uppermost leaves, 2-celled when mature, *polygonal to stellate, from 25 × 30 to 28–30 × 33–35 μ, up to 30 × 48 μ.* Sex organs unknown.

Type. Moist clay-shale polygon soil at Alert Weather Station, 82°30′ N., northeast Ellesmere I., Northwest Territories (*RMS 35142, 35061, 35063a, 35077, 35072a.* Material from these five collections, from the same locality, is extremely sparse and can be considered the collective type).

Known, to date, only from type material; evidently a high arctic species, apparently restricted to extremely calcareous sites.

Ecology

Occurring around the edges of soil polygons, on highly calcareous clay-shale sites, associated with a series of mosses and the following Hepaticae: *Scapania gymnostomophila, S. cuspiduligera, Solenostoma polaris, Cephaloziella arctica, Lophozia collaris, L. gillmani, L. pellucida, Athalamia hyalina, Sauteria alpina, Aneura pinguis,* and *Preissia quadrata.* Associated angiosperms were *Alopecurus alpinus, Eriophorum scheuchzeri* and *E. angustifolium* var. *triste, Juncus biglumis,* etc. The plant occurs in low, poorly drained areas and is likely to be submerged during the spring snow melt (in mid-June). It is found either as isolated individuals at the bases of *Alopecurus* culms or straggling among mosses; it seems to occur always as isolated stems or small tufts, over silty clay, often in admixture with coarser mosses. It then grows erect with the small leaves remote on the fleshy, flattened stems. Occasionally it is a pioneer on damp polygon soil, in full sun, then very short, only 3–8 mm long and compact, with imbricate leaves which are, at least toward their apices, golden-brown pigmented.

Differentiation and Variation

In the field giving the impression of a small *L. grandiretis*, having the same clear green color (but totally lacking all purplish pigmentation of the underside of the stem). The fleshy stems of the compact plants are about twice as wide as high (440 μ high × 800 μ wide); the thin-walled cortical cells ca. 28–32(36) μ in diam. dorsally and laterally; ventral cortical cells somewhat smaller, 25–30 μ in diam., with age blackish- or brownish-walled (and in mature stems the ventral ⅖ of the medulla also brownish-walled). However, ventral medullary cells are scarcely smaller than dorsal medullary cells (25–36 vs. 35–40 μ, respectively). The bilobed leaves on upper portions of the plants have brown-pigmented lobe apices; such pigmentation is unknown in *L. incisa.* Cells of the lobes and leaf middle, ranging from (25)30–32 μ up to 35 × 35–42 μ, show slightly thickened walls and faintly indicated trigones. The bilobed leaves are small compared to the fleshy stem, ca. 750 μ wide × 550 μ long and entire-margined, much like those of *L. grandiretis*; they are 2-, rarely 3-, stratose at the base. Cells have 15–21 finely papillose-granular oil-bodies, ranging from 3.5–5(7.5) μ and spherical to 4–5 × 5–8 μ and ovoid, quite unlike either *L. incisa* or *grandiretis*. The greenish, polygonal-stellate gemmae range from a low dimension of 25–28 × 30–35 μ to 33 × 30 μ, up to 30 × 48 μ.

Occasionally, when growing sheltered among caespitose mosses, plants have largely 1-stratose leaves, except at the base, where 2-stratose. By contrast, compact sun phases, with pigmented, brownish leaves, have more rigid leaves, which are 2-stratose to above the middle, and 3–4-stratose in the basal 0.2–0.3 of the leaf; such a leaf is figured in cross section in Fig. 177:8.

Although the distinctness of the species is difficult to establish with such sparse material, the plants are very different from *L. incisa*. Discrepancies in leaf form, in stem thickness, in oil-bodies, and in color (pure green in *L. hyperarctica*, in sun phases clear brownish on the leaf lobes; pale, bluish green in *L. incisa*, with sun phases identical in color) strongly support the treatment of *L. hyperarctica* as a well-marked species (Schuster, 1961c). Its occurrence over calcareous clay is in sharp contrast to the habitats of *L. incisa*, which is a decided oxylophyte. *Lophozia hyperarctica* not only agrees more closely with *L. grandiretis* in coloration, in the stout and flattened stems, and in the often distinctly segmented oil-bodies of superior size, but also approaches it in ecology.[95]

There thus appears to be a closer affinity to *L. grandiretis* (in which most phases show 2-several segmented oil-bodies). Although leaf cells in *L. hyperarctica* are scarcely larger than those of *L. opacifolia* and *L. incisa*, they differ greatly in content. In the latter two species, numerous, minute, homogeneous oil-bodies occur, giving the cells an opaque, milky appearance (and the whole plant, as a consequence, a peculiar, opaque, grayish to bluish green color). When living plants are available, this distinction is easily perceptible. Associated with the minute, numerous oil-bodies is a tendency for *L. opacifolia* and *L. incisa* to blacken in drying. In *L. hyperarctica* (and *L. grandiretis*) no such tendency exists.

Lophozia hyperarctica, although similar to small phases of *L. grandiretis*, differs in (*a*) smaller leaf cells, (*b*) fewer and different oil-bodies, (*c*) lack of vinaceous pigmentation, and (*d*) ability to develop golden brown pigmentation of at least the lobes of the uppermost leaves. The vinaceous pigmentation, often suffusing all of the ventral half of the stem (cortical and medullary cells alike)

[95] Among the distinctive features of *L. hyperarctica* are its oil-bodies. These are finely granular-papillose or segmented, formed of a large number of barely protuberant spherules, as in subg. *Lophozia*. Moreover, their size closely agrees with those of subg. *Lophozia*. I have studied oil-bodies in five different collections of this species and found no perceptible variation.

in *L. grandiretis*, is replaced in *L. hyperarctica* by black or fuscous pigmentation. The ventral half of the stem in *L. hyperarctica* may seem, at first glance, to be mycorrhizal. However, the dark discoloration is principally (or entirely) due to a cell wall pigment. The ventral 5–6 tiers of pigmented medullary cells appear to form a storage system. When these cells are macerated, it is evident that they are densely filled with opaque, whitish starch grains. The ventral cortical cells and the rhizoids bear a few fungal hyphae; it is possible that limited mycorrhizal infection also characterizes the lower medullary cells, but no evidence of cellular distintegration associated with mycorrhizae has been found.

LOPHOZIA (MASSULA) INCISA (*Schrad.*) *Dumort.*

[Figs. 173:10–13; 178]

Jungermannia incisa Schrad., Syst. Samml. Krypt. Gewächse 2:5, 1797.
Lophozia incisa Dumort., Rec. d'Obs., p. 17, 1835; Pearson, Hep. Brit. Isles 2:pl. 144, 1902; K. Müller, Rabenh. Krypt.-Fl. 6(1):708, fig. 323, 1910; Macvicar, Studs. Hdb. Brit. Hep. ed. 1, figs. 1–4, 1912 (ed. 2:192, figs. 1–4, 1926).
Jungermannia viridissima Nees, Naturg. Eur. Leberm. 2:134, 1836.
Jungermannia supina Tayl., Lond. Jour. Bot. 5:273, 1846 (new synonymy).
Cephalozia supina Steph., Spec. Hep. 3:321, 1908.
Lophozia (subg. *Massula*) *incisa* K. Müll., Ber. Deutsch. Bot. Gesell. 57:341, 1939; Schuster, Amer. Midl. Nat. 42:572, pl. 8:11–13, 1949; *ibid.* 45(1):55, pl. 11, figs. 6–10, 1951; *ibid.* 49(2):344, pl. 12, figs. 6–10, 1953; K. Müller, Rabenh. Krypt.-Fl. ed. 3, 6(1):679, fig. 211, 1954.
Lophozia cornuta var. *spinosa* Kamimura, Acta Phytotax. Geobot. 14:112, 1952.

Usually in compact, often extensive, and pure patches, loosely prostrate or with crowding suberect. Shoots small, (1.0)1.3–1.6(2.2) mm wide × 4–10 mm long, prostrate to strongly ascending, simple or rarely furcate, *fleshy, a milky, opaque, bluish green (in drying usually becoming blackish), easily bruised. Stems fleshy, stout, soft, elliptical in section,* 400–500 μ in diam., 9–10 to 12–15 cells high. Rhizoids abundant, colorless, 15 μ in diam. Leaves moderately obliquely inserted, distant to crowded, *the upper with dorsal half virtually transversely inserted* (especially in dense-leaved phases), *usually crowded, often forming a crispate head,* erect-spreading to spreading, *fleshy in texture, oblong, obrhombic to obtrapezoidal, polymorphic,* variously 2–3-lobed below; bifid leaves often with a large tooth of antical and/or postical margin; up to 3–5-lobed on the upper leaves (especially below gynoecia) and *quite asymmetric,* with postical lobe the largest; bilobed leaves usually about as long as wide, the 3–5-lobed ones 1.2–1.5× as wide as long (leaves, as a whole, ca. 700–950(1600) μ long × 500–1300(1700) μ wide); *the bilobed leaves often loosely complicate, variously crispate and plicate to undulate,* especially upper leaves; lobes various, unequal, on bilobed leaves the antical ± smaller, acute or apiculate to obtuse, *often ending in a 1-celled spinous tooth and with supplementary, similar spinous marginal teeth;* sinuses descending 0.3–0.5 the leaf length, acute to V- or U-shaped, their bases often narrowly reflexed; leaves unistratose distally, the *basal* ⅕ *often 2-, locally 3-, stratose.* Cells polygonal, thin-walled, with *trigones absent or vestigial* (except in var. *inermis*), the marginal 25–38 μ, the median 30–40 μ; *cells opaque,* somewhat whitish by reflected light, with numerous chloroplasts and *17–35(50) oil-bodies per cell; oil-bodies minute,*

Fig. 178. *Lophozia (Massula) incisa.* (1) Medium-sized ♂ shoot (×25). (2–4) ♀ Bract with bracteole (top) and two subinvolucral leaves (×21). (5) Perianth-bearing shoot tip with two innovations (×21). (6) Apex of vigorous shoot, ♀ plant (×26). (7–8) Leaf lobe apices, fig. 8 from lobe in fig. 9 (×162). (9) Series of leaves from sterile plants (×21). (10) Perianth mouth, juvenile (×122). (11) Gemmae (×ca. 225). (12) Perianth cross section (×15.5). [1–10, *RMS 45333*, Mt. LeConte, Tenn.; 11, from Schuster, 1951; 12, from Pictured Rocks, Alger Co., Mich., *RMS*.]

spherical to ovoid, to 2 × 2–4 μ, appearing essentially *homogeneous*; cuticle smooth. Underleaves absent, except near gynoecia. *Gemmae* almost constantly present, *pale green, 1–2-celled, tetrahedral to polyhedral,* (16)18–20 × 18–25 μ, produced in relatively moderate numbers from apices of upper leaves, not rendering these markedly erose.

Dioecious. Androecia terminal at first, forming a compact head or spike, eventually intercalary; bracts saccate at base, usually 3-lobed and with an antical, basal, often incurved tooth, with more or less dentate margins. Antheridia 1–2, large, with stalk short, 2-seriate. Gynoecia with bracts larger than leaves, otherwise similar to upper leaves, broadly obtrapezoidal-reniform, 1700–2400 μ wide × 1250–1600 μ long usually, irregularly 3–5-lobed for ca. 0.5 their length, the *lobes at least somewhat spinose-dentate* (like leaves, but usually more so), *crispate and undulate*; bracteole variable, ovate-pointed to lanceolate and free usually, up to 1100 μ wide × 1500 μ long, often much smaller and sometimes lacking. Perianth cylindrical-obovoid to obpyriform, fully half emergent at maturity, 1100–1500 μ in diam., deeply 5–6-plicate in distal ⅓, 1-stratose above, 2–3-stratose below, narrowed to mouth; mouth lobulate, the lobes denticulate to ciliate-dentate *with teeth usually 1–3(4) cells long; longer cells thick-walled,* ca. 18–22 × 50–75 μ. Capsule short, ovoid, carmine red, the wall 3–4-stratose; epidermal cells large, ca. 25 μ thick with nodular thickenings; inner layers each 10 μ thick, with semiannular bands. Elaters (7)8–10 μ in diam.; spores (10)12–15 μ, minutely papillose, brown.

Type. European.

Distribution

A widespread species of holarctic distribution, extending over the entire cool to boreal region of the Northern Hemisphere. Found from the southern portions of the Tundra, throughout the Taiga, including southern outliers of it, and in much of the northern edges of the Deciduous Forest Region. Occurring from sea level to above 6600 feet in eastern North America, and to 7500 feet in the European Alps (these last reports perhaps transferable to *L. opacifolia*).

Found throughout almost all of the northern two-thirds of Europe, from Spitsbergen (79° N; Arnell and Mårtensson, 1959),[96] northern Sweden, Norway, and Finland southward through England and Ireland, the Low Countries, Denmark, Germany, and France, Czechoslovakia, Poland and Russia to Switzerland, to the Pyrenees, the north Italian Alps, Istria, south-eastward to Bulgaria, Bucovina, and the Caucasus. Also occurring southward to the Azores (Allorge, 1948). Eastward into Asia (Siberia: Yenisey between 56 and 70° N., Lena R.; Novaya Zemlya) and to Japan (Kitagawa, 1965; Hattori, 1952), China, the Himalaya, Korea, Taiwan (Formosa), and Sakhalin (Kitagawa, 1965).

---

[96] Arnell and Mårtensson make a point of emphasizing that their collection is true *L. incisa*, not *L. opacifolia*.

In North America clearly transcontinental, ranging in the West from Alaska and the Yukon to British Columbia, Washington, Alberta, Montana, Wyoming, Oregon, Idaho, and southward in the mountains to California, Colorado, Nevada, and New Mexico (Shields, 1954; Clark, 1957). Recurring, at high altitudes, in Mexico. In eastern North America as follows:

GREENLAND. Apparently widespread in S. Greenland (K. Damsholt, personal communication), but the separation from *L. opacifolia* has not been consistently attempted, so the distribution is unsure. Plants approaching *L. incisa* occur as far north as Siorapaluk, but are better placed as *L. opacifolia*; see p. 455. Reported from E. Greenland as far north as Scoresby Sund (Jensen, 1897; Macoun, 1902) as a "forma *robusta J. grandireti* Lindb. aemulans"; also reported from Kap Greg, on *Sphagnum girgensohnii* (Harmsen, 1933). LABRADOR. Forteau (*Waghorne, 1894!*).

QUEBEC. Anticosti I. (*Victorin, 1902*); Gaspé Coast (Macoun, 1902); Mt. Lac des Cygnes, Charlevoix Co. (Kucyniak, 1947); Beauceville; Tadoussac; Bic; Ste-Irène, Matapédia; R. Ste-Anne-des-Monts (all Lepage, 1944–45); N. of Malbaie R., 10–12 mi NW. of Bridgeville, Gaspé (*RMS 43506*); Terrebonne and Montcalm Cos., near St. Jovite (Crum & Williams, 1960). NEWFOUNDLAND. Placentia Bay (*Waghorne 40!, 41!*); Chance Cove (*Waghorne 161!*); Spread Eagle (*Waghorne 227!*); Hopeall (*Waghorne 211!*); Avalon, S. coast, W. Nfld., N. Pen., C. and NE. Nfld. districts (Buch & Tuomikoski, 1955). ONTARIO. Algonquin Park; Ottawa; Belleville; Nipigon R., L. Superior; Algoma; Carleton; Cochrane; L. Timagami (Macoun, 1902; Cain & Fulford, 1948); Moosonee, James Bay; Cold Cr., Nobleton, York Co. (*Cain 4261!*); Sturgeon R. near Beardmore, Thunder Bay (*Cain 1887!*); Richard L., Dill Twp., Sudbury Distr. (*Cain 2086!*); Friday's, L. Timagami (*Cain 2450!*). MIQUELON I. (Lepage, 1944–45). NOVA SCOTIA. Cape Breton I. (*Nichols 36*); Truro; Margaree and Louisburg, Cape Breton I. (Macoun, 1902); Port Mouton (Brown, 1936a). NEW BRUNSWICK. Grand Manan (Lorenz, 1923).

MAINE. Mt. Katahdin (*RMS*); Flint I. (Frye & Clark, 1937–47); Mt. Desert I. (*Lorenz*). NEW HAMPSHIRE. Mt. Washington (*RMS*); Zealand Falls, White Mts. (*RMS*); Alstead (*Greenwood 274*); Franconia Mts. (*Lorenz, 1908*). VERMONT. Jamaica (*Lorenz, 1901*); Jericho; Willoughby. MASSACHUSETTS. Worcester (*Greenwood, 1908*); Bear R., Conway, ca. 0.5 mi above jct. with Deerfield R. (*RMS*); Hawley Bog, Hawley (*RMS*). RHODE ISLAND. Fide Evans, 1903. CONNECTICUT. Winchester, Litchfield Co.; Stafford, Tolland Co.; Hamden and Woodbridge, New Haven Co. (Evans & Nichols, 1908).

NEW YORK. Tompkins, Genesee, Ontario, Chenango, Cattaraugus Cos. (Schuster, 1949a); Slide, Wittenberg, and Cornell Mts., ca. 3600–4000 ft, Catskill Mts. (*RMS, s.n.*); Little Moose L., Herkimer Co. (Haynes, 1906); southern West Ft. Ann, L. George (*Burnham*). PENNSYLVANIA. Ridgway, Elk Co. (*Lanfear, 1933*); Cameron, Forest, McKean, Potter Cos. (Lanfear 1933a; in thesis). WEST VIRGINIA. Bald Knob, Pocahontas Co. (*Gray 71140*); Preston Co. (Ammons, 1940); Cranberry Glades, 9–10 mi W. of Mill Pt., Pocahontas Co. (*RMS 61217b*). VIRGINIA. White Top Mt. and Mt. Rogers, Grayson and Smyth Cos. (*RMS 38088, 38100, 38035, 38043a*, etc.); see Schuster & Patterson, 1957.

NORTH CAROLINA. Clingmans Dome, Swain Co. (*RMS 28107*); Andrews Bald, Swain Co. (*RMS 36613*); Mt. Mitchell, Yancey Co. (*RMS 24810, 24826a, 24812, 23262*); Grandfather Mt. at 5700 ft, Caldwell Co. (*RMS 30188a*); Avery, Jackson, Watauga, and Caldwell Cos. (Blomquist, 1936); Roan Mt., Mitchell Co. (*RMS 40302, 40377a, 40310, 36937b*). TENNESSEE. Near Myrtle Pt., Mt. LeConte, Sevier Co.

(*RMS 45322c, 45322e*); Roan High Bluff, 6100–6200 ft, Carter Co. (*RMS 61282c*); Clingmans Dome, Sevier Co. (*Sharp 516!*); summit of Mt. LeConte, Sevier Co. (*RMS 45333, 45326*).

Westward extending as follows. OHIO. Hocking Co. (*M.S. Taylor, 1921*).   ILLINOIS. Clark Co. (Arzeni, 1947).   KENTUCKY: Natural Bridge (*Taylor 60*); Powell Co. (Fulford 1934, 1936).   WISCONSIN. Vilas, Oneida, Adams, Bayfield, Douglas, Iron, Ashland, Barron, Grant Cos. (Conklin, 1929); Superior (*Conklin 1909*); Sand I., Apostle Isls., Bayfield Co. (*RMS*); Superior (*Conklin 775*); Rocky Arbor Roadside Park, NW. of Wisconsin Dells, S. edge of Juneau Co., in the Driftless Area (*RMS*). MICHIGAN. Copper Harbor, Keweenaw Co. (*RMS*); Amygdaloid I., Isle Royale (*RMS*); Cheboygan (*Ammons 277*); Pictured Rocks, Alger Co. (*RMS*); L. Lily, near Ft. Wilkins, Keweenaw Co. (*RMS 39162*); Sugar I., Chippewa Co. (Steere, 1934). MINNESOTA. Cook, Clearwater, Itasca, Lake, St. Louis, Winona Cos. (Schuster, 1953; 59 collections).

Lophozia incisa becomes rare or infrequent in the southern edge of the Tundra, and above tree level in the mountains, at least in eastern North America. Greenland reports need confirmation and may apply to *L. opacifolia* (which, in turn, may be only an arctic-alpine ecotype or ecospecies). As emphasized under the latter, the two "species" appear to converge in Labrador and at similar latitudes. Although *L. incisa* is frequent in the Hemlock-Hardwood Forest in New England and New York, it becomes essentially limited to the Spruce-Fraser Fir Zone of the highest peaks in the Southern Appalachians, almost entirely to elevations above 4800 feet. It is rare in the Driftless Area of Wisconsin (associated with *Lycopodium selago* var. *patens*).

Ecology

Almost exclusively a mesophytic (rarely mesohygrophytic) species limited to strongly acid sites; absent or rare in regions of basic rock outcrops, and there limited to ecologically advanced sites, where it occurs over peat, decaying wood, and similarly acid, organic substrates. Because of the restriction to distinctly acidic substrates, usually limited to three types of sites. (1) On decaying, moist logs, generally in shaded and humid areas, associated with several *Cephalozia* species (*catenulata*, *media*, *lacinulata*), *Lophozia porphyroleuca*, *L. ascendens*, *L. longidens*, and *L. ventricosa silvicola*, *Scapania umbrosa*, *S. apiculata*, *S. glaucocephala*, *Calypogeia suecica* (occasionally also *C. neesiana*), *Blepharostoma*, *Harpanthus scutatus*, and *Riccardia palmata* and *R. latifrons*. The species here occurs under a pH ranging from 5.2 to 4.6 or lower, usually on "soft" logs in a fairly advanced state of decay (as a consequence rarely found with such pioneers as *Ptilidium pulcherrimum* and *Jamesoniella*). (2) On peat-covered, sphagnous banks or humus-covered moist rocks (under a pH ranging from 3.9 to 4.8); associated here are a large variety of species. Under the most acid, sphagnous, moist conditions *Calypogeia sphagnicola*, *Odontoschisma denudatum*, *Mylia anomala*, *Lophozia ventricosa silvicola*, *Cephalozia media*, *pleniceps*, *leucantha, bicuspidata, Lepidozia reptans,* and *Blepharostoma* commonly occur consociated, especially on cold cliffs and ledges in the Taiga. In such sites, where there is usually considerable seepage and the species may be exposed to relatively high light intensities, the xeromorphic, collenchymatous phase ("var. *inermis*") may occur. Although often over living *Sphagnum* in such sites, forming characteristic, often nearly round, blue-green patches, *L. incisa* never occurs on *Sphagnum* in bogs [here it appears restricted to decaying logs in bog margins]. (3) Less frequently, over damp to moist rocks, often on sandstone or shale, as a pioneer (and only when the rocks are essentially Ca-free). Associated then may be *Tritomaria exsectiformis* and *Scapania nemorosa* or *Lophozia groenlandica, Gymnocolea, Cephalozia bicuspidata.* Also on exposed sandstone ledges in the Driftless Area of Wisconsin (with *Bazzania trilobata, Tetraphis, Lycopodium selago, Lophozia ventricosa silvicola*).

Although *L. incisa* is a decidedly oxylophytic plant, it may rarely occur over basic rocks (basalts; diorites) *after* a humus or peat layer is laid down—then with humicoles that tolerate basic conditions, such as *Lophozia heterocolpa* (rarely *L. grandiretis*; see Schuster, 1953) and *Blepharostoma trichophyllum.* Reimers (1940a) also reports the species on humus over gypsum, also with *L. heterocolpa* and *Blepharostoma.*

Differentiation

A characteristic species, without close relatives except for *L. opacifolia,* offering virtually no systematic problems (except as regards the relationship to, and separation from, *L. opacifolia*).

This species is one of the easiest of the Lophoziaceae to learn to recognize in the field: the plants are small (0.5–1 cm long) and form dense patches whose pale, opaque, bluish-green color is very characteristic.[97] The crowded, crispate, wavy leaves, the lobes of which are acute to acuminate . . . , and often supplied with supplementary teeth or lobes will suffice to separate it from all our other species . . . . The restricted occurrence . . . will also aid in its identification. Under the microscope the densely chlorophyllose, very large, and thin-walled cells are diagnostic; these, as in other species of *Massula,* contain numerous gleaming oil-bodies (Schuster, 1953).

In most keys this species is considered to have the leaf lobes spinose-dentate. This holds only for robust plants and is not a good diagnostic character. "However, leaf lobes, in all cases, are quite unequal (with frequent development

---

[97] The bluish green color cannot be preserved in drying. In most cases the plants turn blackish if dried slowly, indicating a strong tendency to swift decay (much more marked than in the various species of *Lophozia* subg. *Lophozia* which may occur admixed); even when rapidly air-dried, all portions which were even slightly pressed turn brown or black, indicating that the species "bruises" very readily. In fact, the poor state of dried, herbarium material is usually indicative of the species and affords an obvious mode of recognition.

of supplementary lobes), and end in acute to acuminate, sharp teeth,"
(Schuster, *loc. cit.*). The terminal cells of these teeth, whether occurring at the
apices or margins of the lobes, are usually very characteristically enlarged and
thick-walled (Fig. 178:8), a feature not recurring in the other species of *Massula*,
except in phases transitional to *L. opacifolia*. Gemmiparous plants usually show
no trace of the thick-walled spinose teeth, and neither do weak phases of
nongemmiparous plants.

At one time, *L. grandiretis* was considered identical with *L. incisa*. However,
the latter differs in (*a*) smaller cells; (*b*) smaller, never segmented oil-bodies;
(*c*) stem never purplish beneath; (*d*) leaves often with spinous teeth; and (*e*)
peculiar color. When the two occur together (as is very exceptionally the case,
considering that their ecological "requirements" are almost mutually exclusive;
see Schuster, 1953) they differ at a glance, in color (*L. grandiretis* a purer green,
without the opaquely pale bluish green cast), size (*L. grandiretis* much more
robust), and pigmentation (*L. grandiretis* with stem purplish beneath), as well
as in many microscopic features.

Variation

In addition to extraordinary variability in size and dimensions of the
leaves, depending on nutritional conditions, *L. incisa* also shows correlated
variation in number of leaf lobes and degree of development of the
spinous teeth. Smallest phases have leaves almost exclusively asymmetrically 2-lobed, as long as or barely longer than broad, usually loosely
complicate-canaliculate, with isolated leaves bearing a third lobe or large
tooth (Fig. 178:9); spinous teeth are absent or sporadic in occurrence.
In "normal," robust phases a high incidence of uppermost leaves are 3-4,
or even 5-, lobed, and leaf apices and margins bear enlarged, sharp,
spinous teeth. Extremes of these two types at first suggest wholly different
species but are connected by all types of transitional phases.

Lophozia incisa shows conspicuous variation in orientation and form of the
leaves. On some plants almost all leaves below gynoecial regions are bilobed,
with isolated and sporadic development of a triangular or spinous tooth of the
antical and/or postical margin, the teeth occasionally enlarged and showing
transition to accessory lobes. Such bifid leaves, whose antical lobe is always
somewhat smaller than the postical, are distinctly, if loosely, complicate-
bilobed, with almost transversely oriented dorsal lobes. Small plants with this
type of leaf insertion and orientation may superficially approach *Scapania* in
facies. The perianth mouth is equally variable, even in plants from the same
population: the mouth may be clearly armed with 1-, rarely 2-, celled teeth
formed of acute, tapering, triangular cells 25–35 μ wide at the base × 60–70 μ
long. According to Müller (1951–58, p. 679, fig. 211f), these 1-celled teeth may
be acuminate and up to 130 μ long; I have seen none that long. In other cases
the perianth mouth is shortly fimbriate with teeth 3-4 cells long (length
150–200 μ) and 2 cells wide at the base, the distal 2–3 cells occurring in a

uniseriate row; then the distal cells of the fimbriae are slender and tapering, varying from 60 μ long × 18–24 μ wide up to 85 μ long × 15 μ wide. Terminal cells are normally strongly thick-walled, at least distally, hence somewhat spinous. Equally variable is the form of the ♀ bracts, which may be subsymmetrically 4-lobed or strongly asymmetrically 5-lobed, with the antical lobe small and the postical lobe the largest. Associated with this are wide discrepancies in the size of the bracteole, which may be absent in isolated cases, small and lanceolate in others, or large (to ⅓ the bracts in area) and ovate, with sparingly dentate margins. All—or most—of this variability occurs within the population and hence is systematically inutile.

Perhaps of more importance, but still possibly of environmental origin, are small phases with large trigones, described by Müller as follows.

LOPHOZIA (MASSULA) INCISA var. INERMIS
K. Müll.

[Fig. 173:12]

Lophozia (Dilophozia) incisa var. *inermis* K. Müll., Rabenh. Krypt.-Fl. 6(1):710, 1910.

Similar to small phases of *L. incisa* described above, but cells with prominent, bulging trigones; leaves unequally bifid, usually without accessory teeth, often nearly transversely oriented.

Type. Bulgaria. (A fragment of the type studied, through the courtesy of the late Dr. Müller.)

Distribution

Perhaps sporadically distributed throughout the range of the species, but the distribution virtually unknown:

MICHIGAN. Moist, peaty wood, Amygdaloid I., Isle Royale, Keweenaw Co. (*RMS 13789a*).   MINNESOTA. Belle Rose I., Susie Isls., near Grand Portage, Cook Co. (*RMS 194307*; over peat on basic rocks, with *Lophozia grandiretis, Scapania carinthiaca, Blepharostoma, Lepidozia, Odontoschisma macounii*).   MAINE. Davis Pond, NW. Basin, Mt. Katahdin (*RMS;* see under *Nardia insecta*).

LOPHOZIA (MASSULA) OPACIFOLIA Culmann
[Figs. 179–182]

Lophozia opacifolia Culmann, Rev. Bryol. 47:21, 1920; Meylan, Beitr. Krypt.-Fl. Schweiz 6(1):174, 1924; K. Müller, Rabenh. Krypt.-Fl. ed. 3, 6(1):681, fig. 212, 1954; Jones, Trans. Brit. Bryol. Soc. 3(2):180, 1957; Schuster, *in* Schuster et al., Natl. Mus. Canada Bull. 164:30, 1959; Schuster, Canad. Jour. Bot. 39:972, figs. 2–3, 1961.

Similar to L. incisa, but *often more rigid, more robust*, forming *opaque, bluish green* patches, or creeping among other Hepaticae. Plants 1.5–2 mm wide, to 1–1.5 cm long, occasionally furcately branched, *very fleshy, opaque; stems elliptical in cross section, green* (the postical surface brownish with age, because of mycorrhizal invasion of ventral cortical cells), brittle, ca. 200–230 μ high × 400–450 μ

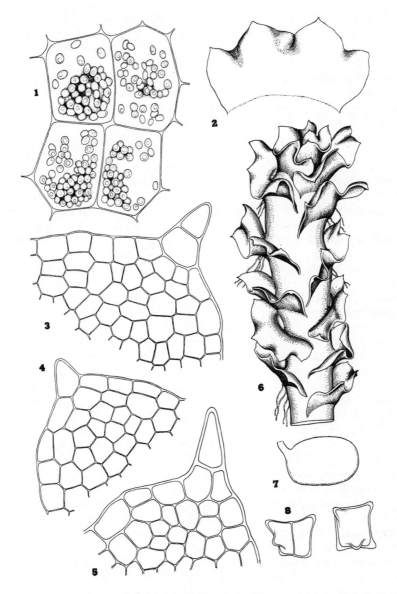

FIG. 179. *Lophozia opacifolia*. (1) Median cells (×735). (2) Leaf (×31). (3–5) Leaf lobe apices, showing weakly cuspidate lobes (×260). (6) Robust, leafy shoot, with strongly elliptical, almost fasciated stem (×19). (7) Stem in cross section (×23.5). (8) Two gemmae (×480). [All from *RMS 35104a*, Mt. Pullen, Ellesmere I.]

wide; dorsal cortical cells $(15)18–23(26)$ μ × $(65)110–150$ μ, thin-walled; medullary cells $25–35(45)$ μ in diam., pellucid, thin-walled. Leaves barely contiguous to ± imbricate, *polymorphous*, very obliquely inserted, 2- or partly 3- to 2–4-lobed for 0.25–0.35, *broad* (2-lobed leaves ca. 850 μ wide × 720 μ long to 940 μ wide × 650 μ long; 3-lobed, ca. 1050 μ wide × 500 μ long to 1000 μ wide × 600 μ long; 4-lobed, ca. 1000 μ wide × 540 μ long), *width averaging 1.2–2.0(2.5) × the length;* lobes broadly triangular to acute or sharply angulate, often incurved, *obtuse or pointed, entire* (rarely with 1–2 teeth), *rarely weakly mucronate* (apical cells then 26–28 × 38–46 μ long); *lower portions of leaves 2–3-, occasionally 4–5-, stratose.* Cells leptodermous, delicate, with minute or no trigones, 23–27 μ at lobe apices and margins, 23–28(30) × 25–35(45) μ medially; occasional apical teeth formed of 1 thick-walled cell, to 25 × 65 μ; oil-bodies 1.8–2.7 × 2.7–3.6 μ to 2–3 μ, homogeneous and pellucid, 25 or more, usually 30–45, per cell, occasionally even more numerous. Gemmae pale green, 1–2-celled, angulate, 20–25 μ or 15–17 × 15–20 to 18 × 26 μ.

Dioecious. ♂ Bracts weakly ventricose at base, the undivided base suberect, only lobes patulous; 2–3 or more antheridia per bract; stalk 1-2-seriate. ♀ Bracts similar to vegetative leaves, mostly somewhat asymmetrically 3–4-lobed for 0.25–0.4 their length, ± fluted, often with sinuses narrowly reflexed or gibbous, 1385 μ wide × 875 μ long (when bilobed) to 1450–1550 μ wide × 890–1000 μ long (when 3–4-lobed); margins edentate or with a few rather thin-walled, differentiated, large-celled, sharp, unicellular teeth, less often ± spinulose-dentate; lobe apices at times with well-differentiated large-celled teeth. Perianth mouth slightly lobulate, subentire to remotely denticulate, *with 1-celled teeth often only 20–60 μ long,* occasionally (Jones, 1957b) 60–80 μ long or more; perianth mouth cells occasionally even more differentiated [Greenland plants], the terminal cells then free for 0.5–1.0 their length, thick-walled, spinous, from 18–21 × 85–92 μ to 23 × 100–140 μ (4–6 × as long as wide). Seta 300–330 μ (8–10 cells) in diam., with 35–36 rows of epidermal cells (each ca. 24–28 μ in diam.) and gradually somewhat larger interior cells (the central appreciably larger, 42–50 μ in diam.). Capsules *nearly or quite spherical,* 700 μ in diam. or more; wall 3–4-stratose, 45–59 μ thick (inner 2 layers each 10–12 μ thick; outer 23–29 μ thick). Epidermal cells ± quadrate to short-rectangulate, ca. 25–28 to 23–25 × 30–38 μ, with moderately strong nodular thickenings, 2–3(4) per longer (often longitudinal) wall, 0–1(2) per shorter (usually transverse) wall; nodular thickenings usually extended as short spurs on the outer (free) tangential walls. Inner layers of elongated, less regular cells, where rectangulate 12–15 μ broad usually, with numerous rather narrow, complete semiannular bands. Elaters 7.2–8.5(9.0) μ in diam. × 100–135 μ long, loosely bispiral. Spores large, 14.5–18 up to 18–21 μ in diam.

Type. Susten, Alps, Canton Bern, Switzerland.

Distribution

A restricted known range in the Alps of central Europe (Culmann, 1920, 1926; Meylan, 1924), from Savoy and the Tyrol (Darmstädter Hut) to Switzerland (several stations; type) and possibly in Austria (Steiermark), according to Müller (1951–58). Arnell (1954) reports it as common in the Torneträsk area of Sweden and states it is "probably common in the fjeld area" of Norway (Arnell, 1956, p. 126). Jones (1957b) reports it as locally common in Scotland.

In North America known from:

NW. GREENLAND. Kânâk, Red Cliffs Pen., Inglefield Bay (*RMS 45603a*); Siorapaluk, Robertson Bugt (*RMS 45720*). W. GREENLAND. Flakkerhuk, Mudderbugten, E. end of Disko I. (*RMS & KD 66-173*); below Lyngmarksfjeld, near Godhavn, Disko I. (*RMS & KD 66-271, 66-176*); basalt cliffs near shore, below Skarvefjeld, E. of Godhavn, 69°16′ N., 53°28′ W. (*RMS & KD 66-314*); Lyngmarksbugt, Godhavn, ca. 25–125 m (*RMS & KD 66-567*); Sondrestrom Fjord, near shore of L. Ferguson, near head of Fjord, 66°5′ N. (*RMS & KD 66-076a*). S. GREENLAND. Agduitsoq Fjord, Sletten, 60°33′ N., 45°30′ W. (*C. A. Jørgensen, Aug. 9, 1957*; trace among *Sphagnum*, with *Cephalozia leucantha, Mylia anomala*).[98] ELLESMERE I. N. slope of Mt. Pullen, 5 mi S. of Alert, ca. 82°25′ N., NE. Ellesmere I. (*RMS 35104*). LABRADOR. "L'anse au Clair" (*Waghorne 3, Sept. 7, 1894*!; a very few plants, admixed with *Anastrophyllum minutum, Mylia taylori, Cephalozia leucantha, Blepharostoma trichophyllum*. These plants disagree in some respects with the type; for the discrepancies, see under "Differentiation").

Ecology

Supposedly restricted to montane areas, from altitudes of 1700 m upward, "on earth with a prolonged period of snow-cover" but in Scotland found at elevations above 950 m, also in areas with prolonged snow cover.

In Ellesmere I. the species occurs within 525 miles of the North Pole, in moss-tundra on calcareous clay-shale slopes fed by permanent snow banks, in areas where snow persists until after mid-June; here associated with *Lophozia quadriloba, L. heterocolpa, Anthelia juratzkana, Blepharostoma trichophyllum brevirete, Odontoschisma macounii, Tritomaria heterophylla*, etc. The plant occurs on ground where permafrost is never more than 4–8 in. below the surface and where air temperatures during the snow-free season (50–60 days or less) range usually between 28 and 38°F.

Unlike the closely allied *L. incisa*, the species evidently is tolerant of calcareous conditions. In both Ellesmere I. and west Greenland the species is typically a humicole over basic rocks (calcareous shales and slates or basalt), often associated with *Odontoschisma macounii, Blepharostoma trichophyllum brevirete*, etc.

---

[98] The Jørgensen specimen is badly preserved; hence a clear separation of the material from the temperate-subarctic *L. incisa* is hardly feasible. In so far as the characteristically blackened (in drying) and soft-textured, fleshy plants allow any judgment, they represent a poorly developed, compact form of *L. opacifolia*. The leaves are much broader than long; stems are flattened and rather elliptical in section. Also, the leaves are all bilobed, the lobe tips with weak to moderately distinct 1(2)-celled apiculi, which are often lacking on some lobes. Leaf bases are merely 2-stratose, but the plants are so feebly developed that, on the basis of overall vigor, thicker leaves could hardly be expected. Nevertheless, it may be evident, when mature material is collected in the same area, that a disposition of these plants under *L. incisa* would have been wiser.

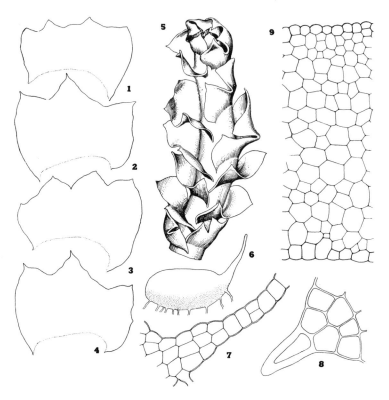

Fig. 180. *Lophozia opacifolia*. (1–4) Leaves, showing variation in form (×24.5). (5) Sterile shoot (×18). (6) Stem cross section, the stippled area more or less mycorrhizal (×41). (7) Base of leaf seen in cross section in fig. 6 (×133). (8) Mucronate apex of leaf lobe (×260). (9) Section through stem middle (×143). [1–4, 9, *RMS 35104a*, Mt. Pullen, Ellesmere I.; 5–8, *Waghorne no. 3, 1894*, from Labrador, transitional to normal *L. incisa*.]

Differentiation

Culmann (1920) was hesitant about the status of *L. opacifolia*, terming it a "subspecies aut varietas nova Lophoziae incisae." It was described as differing from "typical" xylicolous *L. incisa* in having (*a*) leaves with the base, up to the middle, 2–5-stratose, thus opaque and fleshy; (*b*) perianth dentate, not ciliate, at mouth; (*c*) larger spores, 18–19 μ in diam. *Lophozia opacifolia* has been variously considered to be either a distinct species or a variety of *L. incisa*. The latter disposition would be warranted, if it can be shown that the spore size is not constant, since *L. incisa* occasionally has leaves somewhat fleshy at the base. This is admitted by Culmann (1926, p. 40), who states that in lignicolous "typical" *L. incisa* he has seen plants with most leaves unistratose to the base, but with

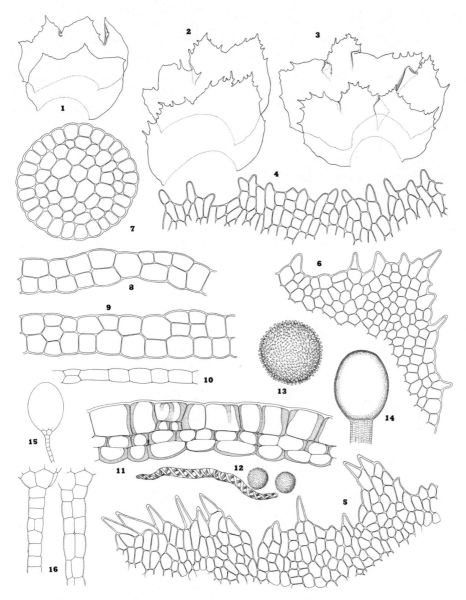

FIG. 181. *Lophozia* (*Massula*) *opacifolia*. (1) Above, ♂ bract, below, leaf (×15). (2) Subgynoecial leaves (×19). (3) ♀ Bracts (×19). (4–5) Perianth mouth sectors, showing extremes in tooth differentiation (×100). (6) ♀ Bract lobe (×100). (7) Seta cross section (×98). (8–9) Cross sections of perianth wall, respectively one-fourth and one-half from apex (×190). (10) Calyptra wall (×190). (11) Capsule wall cross section (×280). (12) Elaters and spores (×280). (13) Spore (×1060). (14) Capsule and seta apex (×16.5). (15) Antheridium (×70). (16) Antheridial stalks (×175). [1, *RMS & KD 66-567;* 2–16, *RMS & KD 66-271,* all from near Godhavn, W. Greenland.]

"une feuille présentant quatre assizes des cellules vers la base." The claims for recognition of *L. opacifolia* are thus not very strong.

Lophozia opacifolia can be confused only with *L. incisa*, of which it appears to be an arctic-alpine and high arctic derivative. The plant has the facies of an opaque, nonspinose form of *L. incisa* but differs in several significant features, in addition to the leaves that "tend" to be 3–5-stratose toward the base and much broader than long, frequently attaining a width of twice the length: (*a*) the triangular and often broad leaf lobes are edentate, and acute or subacute to weakly mucronate at the apex, the apical cell frequently thick-walled and differentiated from the other cells, but never long and spinous; (*b*) the perianth mouth is subentire or denticulate, with scattered 1-celled teeth 20–60, occasionally 60–80, μ

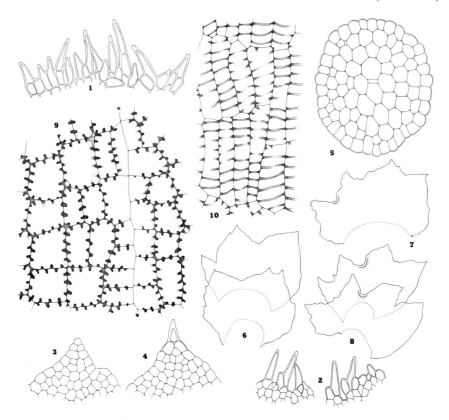

FIG. 182. *Lophozia (Massula) opacifolia*. (1–2) Sectors from perianth mouth, juvenile perianths with cells below apex immature (×100). (3–4) Lobe apices of leaves(×100). (5) Seta cross section (×110). (6) Leaves (×18). (7–8) Subfloral ♀ leaves (×18). (9) Epidermal cells of capsule wall, at asterisk margin between two valves (×265). (10) Inner cells of capsule wall (×265). [1–8, *RMS 45720*, Siorapaluk, NW. Greenland; 9–10, *RMS & KD 66-271*, Godhavn, W. Greenland.]

long or even larger, but not with the narrow setose cilia typical of *L. incisa;* (*c*) spores are larger, ranging from 14.5–18 *μ* to 18–19 up to 20 *μ* (vs. 12–15 *μ* in *L. incisa*). Admittedly, transitional forms between the two species occur.

As previously indicated (Schuster et al., 1959), the cells of *L. opacifolia* agree with those of *L. incisa* in the oil-bodies. The Ellesmere plant had leaves which were often 3–4-lobed and 2–2.5× as wide as long, and edentate leaf lobes, although these were often mucronate. However, the leaves of the Ellesmere plant were only 2-, locally 3-, layered at the base, rather than 3–5-stratose basally. I do not think that this feature is constant; in any event, the abundant Greenland collections fail to show 3–5-stratose leaf bases, even though they have larger spores than *L. incisa*. In fact, most of the characters of *L. opacifolia* are subject to a good deal of variation, and it might prove wiser to regard it merely as an arctic subspecies of *L. incisa*.

Plants from northwest Greenland also have the 1-celled cilia of the mouth extraordinarily differentiated and formed of cells prominently thick-walled distally, as much as 23 × 100–140 *μ* long; such cilia agree with those of *L. incisa* (see, e.g., Müller, 1951–58, fig. 211f).[99] Jones (1957b) shows that in the Scottish plants of *L. opacifolia* the distal perianth cells are 60–80 *μ* long, while Müller states that they are only 20–60 *μ* long in the Swiss material. Differences between the two "species" here also appear to be quantitative and subject to intergradation.

The northwest Greenland plants also show spherical capsules, while the sporangium in *L. incisa* is described as "ovoid" (Frye and Clark, 1937–47). The spores of *L. incisa* are supposedly 12–15 *μ*, those of *L. opacifolia* 18–19 or 18–20, *μ* in diam. However, the northwest Greenland plants have them 14.5–18 *μ* in diam.! These collections thus combine, in various ways, the characters of both *L. incisa* and *L. opacifolia*.[100] They also do not show the high incidence of extremely broad, 3–4-lobed leaves of the Ellesmere collection of *L. opacifolia*.

---

[99] Both of the cited Disko I. plants that bear perianths are much closer to the usual concept of *L. opacifolia* in their perianth mouths (compare Fig. 181): the scattered, remote, 1-celled teeth are not often sharply tapered, are rarely free for most of their length, and range mostly from 25–28 × 50–60 *μ*, a minority ranging to 24 × 82 *μ* (and these may be free only at their tips). As Figs. 181:4 and 5 show, there may be much variation from perianth to perianth, even in the same populations.

[100] The sporophyte of the Godhavn plant (*66-271*) shows some basic differences. (1) The seta, which is ca. 325 *μ* in diam., has only 26–27 rows of epidermal cells; it is 7–8 cells in diam. (2) The capsules range from almost spherical and 1335 *μ* wide × 1400 *μ* long to short-ellipsoidal and 925–1170 *μ* in diam. × 1125–1350 *μ* long. (3) The capsule wall is 3–4-stratose and 50–59 *μ* thick. (4) Epidermal cells are mostly short-oblong and 25–28(30) × 30–38(42) *μ*, some to 40–43 × 40–43 *μ* and quadrate or polygonal; they are 26–29 *μ* high and mostly at least as high as the 2(3) inner strata. (5) Inner cells are rather irregular, where rectangulate mostly 12–16(24) × 60–96 *μ*, with numerous semiannular bands. (6) Spores average much larger, 18–21 *μ* in diam. (7) Elaters are strongly contorted, 8–9.5 *μ* in diam., with the 2 spirals each 2.3–2.6 *μ* wide.

The sporophyte differences are mostly minor, except for the spore size, which in the case of the Godhavn plant is typical of *L. opacifolia* (as given in Müller, 1951–58, p. 681), while the Siorapaluk plants have spores only slightly larger (14.5–18 *μ*) than given by Müller for *L. incisa* ("Sporen . . . 12–15 *μ*").

LOPHOZIA (MASSULA) GRANDIRETIS (Lindb. ex Kaal.)
Schiffn.

[Figs. 173:3–5; 183–184]

Jungermannia grandiretis Lindb., Meddel. Soc. F. et Fl. Fennica 9:158, 1883 (*nomen nudum*);
  Kaalaas, Nyt Mag. Naturvid. 33:322, 1893.
Lophozia grandiretis Schiffn., Lotos 51(7):232, 1903 [reprint, p. 20]; Schiffner, Österr. Bot.
  Zeitschr. 57:5, 1907; K. Müller, Rabenh. Krypt.-Fl. 6(1):705, fig. 322, 1910; Evans,
  Rhodora 16:65, 1914; Meylan, Beitr. Krypt.-Fl. Schweiz 6(1):174, fig. 111, 1924.
Lophozia (subg. *Massula*) *grandiretis* Schust., Natl. Mus. Canada Bull. 122:9, (1950) 1951;
  Schuster, Amer. Midl. Nat. 45(1):54, pl. 10, pl. 11, figs. 1–5, 1951; Schuster, *ibid.* 49(2):
  341, pl. 11, pl. 12, figs. 1–5, 1953; K. Müller, Rabenh. Krypt.-Fl. ed. 3, 6(1):676, figs.
  209–210, 1954.

Scattered or forming compact patches, often roughly circular in shape, among other bryophytes, *brittle, turgid, robust* (shoots 1.2–2.4 mm wide). Stems usually 1–2, occasionally 2–5, cm long, *stout and fleshy*, ca. 450–850 μ in width, 380–460 μ and 12–14 cells high, somewhat flattened, sparingly terminally furcate or nearly simple, *usually vinaceous to purplish brown to purplish black ventrally* (in shade forms rarely green throughout; in sun forms the purplish pigmentation involving almost all of stem and leaf bases); cortical cells thin-walled, strongly elongated, (20)25–28(36) μ wide × 80–160(175) μ long to 30–40 × 160–260 μ, not or scarcely differentiated from medullary cells (in cross section), averaging 3.5–5 × as long as wide; medullary cells (26)32–48(56) μ in diam., free of mycorrhiza. Rhizoids colorless or purplish at base, 18–23 μ in diam. Leaves delicate, strongly obliquely inserted, varying from somewhat wavy and loosely imbricate, then somewhat flaccid and undulate (bog phases), to rather dense, slightly decurrent at antical base, commonly moderately to strongly concave or even cupped, 2- or occasionally 3-lobed for ($\frac{1}{5}$)$\frac{1}{4}$–$\frac{2}{5}$ their length, rectangulate and clearly *broader than long (averaging 1.2–1.5× as wide as long)*, to ca. 1850–1900 μ wide × 1300–1400 μ long, *pure green to* (at least basally) *often purplish*, sometimes purplish throughout; leaf margins entire to sinuous or, occasionally, bearing a few irregular small teeth; sinus obtuse to lunate to rectangulate, rarely acute at base; *lobes broadly triangular*, sometimes apiculate or subacute, sometimes obtuse or blunt. Cells *very large, thin-walled* or weakly but uniformly thick-walled, almost *without trigones*, those of leaf margins and leaf lobes usually from 36–42 × 38–48 μ to (40)45–54 μ, of the leaf middle (35)40–50 × 50–75 μ to 45–65 × 50–78 μ, at base from 40 × 60 to 48–54 × 72–84(100) μ, *to twice width of cortical cells*, moderately chlorophyllose; oil-bodies varying from homogeneous and spherical and 4–5 μ to (commonly) twinned and 4–5 × 6–9 μ, occasionally coarsely botryoidal and of sharply discrete spherules and 5–6 × 8–11 μ, usually (12–30)36–50 *per cell*, averaging larger than chloroplasts. *Gemmae polygonal, with sharply protruding angles, often stellate, pale green* (even when plant otherwise vinaceous or purple), 1–2-celled, (20)22–27 × 24–32(36) μ to 27 × 38 μ to 34–39 × 34–45(58) μ to 45–47 × 50–57 μ, usually produced *in conspicuous masses* at apices of shoots.[101] Chromosome no: *n = 14*(?).

---

[101] Gemma size appears to undergo much variation, and perhaps only the larger figures (representing optimal or mature size) have much significance.

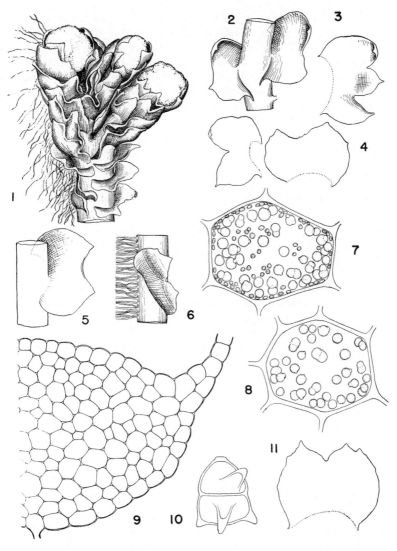

Fig. 183. *Lophozia (Massula) grandiretis.* (1) Compact, highly gemmiparous, sterile shoot of mod. *colorata* (×12). (2) Shoot sector, antical view (×19). (3) Trifid leaf (×19). (4) Two leaves (×ca. 19). (5) Leaf, *in situ* (×19). (6) Leaf, *in situ*, lateral aspect to show insertion and dense rhizoids (×19). (7) Cell with oil-bodies and chloroplasts (×515). (8) Cell with oil-bodies (×515). (9) Stem cross section (×130). (10) Gemma (×600). (11) Leaf (×15). [1–6, 9, from Big Susie I., Minn., *RMS;* 7–8, 10, 11, Grand Portage, Minn., *RMS.*]

Dioecious. Androecia similar to vegetatives portions of shoots, intercalary, the bracts similar to leaves, somewhat more concave at base than leaves; 3–5 antheridia, their stalks 2-seriate. Perichaetial bracts large, 1.5–2 × as wide as long, 2–3-lobed, the margins entire or subentire; bracteole variable in size, irregularly 2–3-lobed, united with one or both bracts at base. Perianth short, pyriform, short-emergent, contracted to subentire mouth; mouth armed with teeth formed of thin-walled, partially projecting cells. Spores 15–18 μ; elaters bispiral, 10–12 μ thick.

Type. Finland.

Distribution

A local to rare species of arctic-alpine distribution, occurring rarely southward into the Coniferous Forest Region. Until recently known from only two regional stations, but the distribution considerably amplified by Schuster (1951, 1953). Undoubtedly of much wider distribution in North America than the following reports indicate.

In Europe of sporadic distribution, especially in Scandinavia (Finland, Norway, Sweden), northward to 79°N. in Spitsbergen (Arnell and Mårtensson, 1959); in Denmark; in the Alpine regions (up to over 2000 m) of central Europe (Switzerland, Austria, and the Allgäuer Alps in Germany), but lately also reported at relatively low elevations (100 m) from Baden, Germany (Müller, 1951–58).

NW. GREENLAND. Kangerdluarssuk, Bowdoin Bugt (*RMS 45777a*); Kanderdlugs-suak, S. side Inglefield Bay (*RMS 45831a, 45903*); Kânâk, Red Cliffs Pen., Inglefield Bay (*RMS 45610*); Nucluet, Booth Sd. (*RMS, s.n.*). W. GREENLAND. Flakkerhuk, E. tip of Disko I., near Mudderbugten (*RMS & KD* 66-120); Fortune Bay, Disko I. (*RMS & KD 66-142*); below Lyngmarksfjeld, near Godhavn, Disko I. (*RMS & KD 66-176, 66-172, 66-161b, 66-163, 66-162b*); Magdlak, Alfred Wegeners Halvö, 71°06′ N., 51°40′ W. (*66-1352, 66-1359, 66-1364b*); mouth of Simiutap Kûa Valley, near pond, Umîarfik Fjord, 71°59′ N., 54°35–40′ W. (*RMS & KD 66-917, 66-919, 66-916b, 66-914c*). QUEBEC. S. side of mouth of Great Whale R., Hudson Bay (*Marr 661a; fide* Schuster, 1951). MANITOBA. Tundra along Hudson Bay, Churchill (*RMS 35001, 35006*). VERMONT. Willoughby (*A. Lorenz & A. W. Evans; fide* Evans, 1914, p. 65). MINNE-SOTA. Lucille I., Belle Rose I., Big Susie I., all in Susie Isls., NE. Cook Co. (*RMS 13657c, 12231, 12232a, 14898*); Grand Portage, Cook Co. (*RMS 11901, 11977*); Grand Marais, Cook Co. (*Conklin 2652, 3062*).

Also reported, to our west, from Banff, Alberta (*Brinkman 606; fide* Evans, 1914, p. 65). Müller (1951–58, p. 679) reports it (?incorrectly) from Alaska.

Ecology

I have pointed out (Schuster, 1953) that this rare plant is a "sub-calciphyte on . . . somewhat peaty soil over . . . calcareous ledges" and also occurs occasionally in "marly bogs." Over peaty moist soil lying over calcareous rocks, usually near water, commonly with *Odontoschisma macounii, Scapania degenii, Cephalozia pleniceps,* and *Blepharostoma tricho-phyllum* (both in Minnesota and Manitoba), with *Geocalyx graveolans, Mylia anomala* (Minnesota and Quebec stations), and less commonly with *Riccardia palmata, Lophozia incisa, Scapania carinthiaca,* and *S. mucronata* (Minnesota). The occurrence on moist peaty soil over basic rocks appears to be very characteristic; the similarity in habitat requirements to those of *Odontoschisma macounii* and *Scapania degenii* is particularly marked, and the association with these two rare species in the field is noteworthy; pH conditions range from 6.5 to 5.5. Associated vascular plants are commonly *Primula mistassinica* and *Pinguicula vulgaris,* occasionally *Polygonum viviparum.*

In bogs, underlain by marl, the species may occur with *Lophozia rutheana* and *Aneura pinguis* (Minnesota; marly tundra at Churchill, Manitoba), *Riccardia multifida, Moerckia hibernica,* and *Chiloscyphus pallescens.* Along pool margins in the Tundra sometimes in less distinctly calcareous sites, then with *Cephalozia leucantha, Mylia anomala, Lophozia wenzelii,* etc. (Quebec).

In Greenland the species is often very common but then occurs mostly as scattered, small patches or isolated individuals on peaty soil, mostly over basaltic rocks, usually adjoining flowing water. Associated may be most of the species cited above (*Scapania degenii, Odontoschisma macounii*) but also *Lophozia heterocolpa, L. gillmani, Tritomaria polymorpha, Nardia geoscyphus,* and *Solenostoma subellipticum.* Occurrences in irrigated, peaty meadows, with *Aulacomnium, Lophozia excisa, Cephalozia pleniceps, L. alpestris polaris,* under medium acid conditions are also occasional. Indeed, in Greenland the ecological tolerances of the species appear to expand perceptibly.

Differentiation

Lophozia grandiretis is a distinctive species. It has a facies similar to *L. incisa* (and the very similar *L. opacifolia*) and possesses angulate gemmae like these species. It further agrees in the thick and fleshy, somewhat flattened stem and the often brittle texture. However, *L. grandiretis* is quite distinct in both its ecology and morphological characteristics.[102] It is a species confined largely to somewhat subbasic, if peaty, sites, rather than being strongly oxylophytic (as are both *L. incisa* and *L. opacifolia*). With accumulation of peat over basic ledges, the pH is lowered to the point

---

[102] Stephani (1902, *in* 1898–1924, p. 16), who at times showed an extraordinarily narrow concept of species, included *L. grandiretis* as a synonym of *L. incisa.* Both Schiffner (1907a) and Müller (1905–16) express the opinion that the two are distinct. Indeed, Warnstorf (Hedwigia 53:209, 1913) states that *L. grandiretis* is more closely allied to *L. marchica* than to *L. incisa.* The form of the gemmae (angular in both *L. grandiretis* and *L. incisa,* smooth and 1-celled in *L. marchica*) suggests that such an implied relationship is incorrect, and that a closer affinity to *L. incisa* must be assumed. Since I (Schuster, 1953) have found *L. grandiretis* and *L. incisa* growing together and remaining obviously distinct, even to the naked eye, two fully different species must be at hand. In west Greenland, also, *L. opacifolia* and *L. grandiretis* may be found growing juxtaposed and remaining distinct to the naked eye.

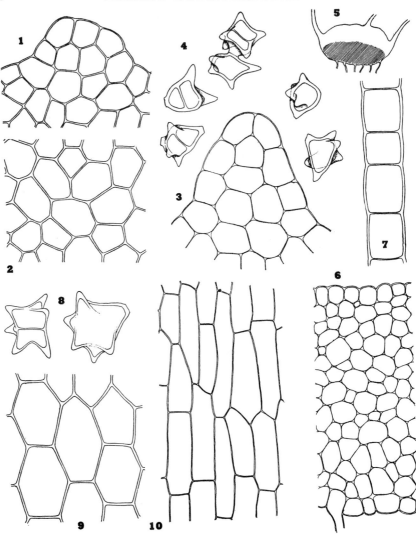

FIG. 184. *Lophozia* (*Massula*) *grandiretis*. (1–2) Cells of lobe apex and leaf middle, respectively, of phase with somewhat thickened walls (×175). (3) Cells, lobe apex, leptodermous phase (×190). (4) Gemmae (×400) (5) Stem cross section, the purplish pigmented zone cross-hatched (×25). (6) Vertical section through stem middle (×150). (7) Leaf section (×ca. 250). (8) Gemmae (×560). (9) Basal leaf cells (×235). (10) Dorsal cortical cells of stem (×235). [1–7, from Big Susie I., Minn., *RMS;* 8–10, Great Whale R., Quebec, *Marr 661a.*]

where *L. grandiretis* may still persist, but *L. incisa* has already undergone ecesis (Schuster, 1953). The two species then occur together and are readily separable with the naked eye, *L. grandiretis* being purer green (in shade), rather than opaque bluish green (as is *L. incisa*), and considerably more robust than *L. incisa*. *Lophozia grandiretis* is also readily distinct in the fleshier stem, usually purplish beneath (except in extreme bog forms from shaded sites), and in the tendency toward mass production of gemmae, which form veritable yellow-green or greenish "heads" at the apices of the shoots (Fig. 183:1). Under the microscope the much larger cells and larger, often segmented oil-bodies readily separate *L. grandiretis* from *L. incisa* (and *L. opacifolia*). Lack of dentition of the very broad leaves is an added distinctive feature of *L. grandiretis* but may also characterize arctic and alpine phases of *L. incisa* and is equally diagnostic of *L. opacifolia*.

In the large cells, *L. grandiretis* approaches *L. marchica* and *L. capitata*. The stem anatomy and the strongly elongated, thin-walled cortical cells further enhance this similarity. As in these species, cortical (and medullary) cells of the stem are inferior in diam. to basal leaf cells, averaging about half the diam. (compare Fig. 184:9 and 10). However, *L. grandiretis* differs from both of these species in the (*a*) polygonal to stellate gemmae; (*b*) somewhat larger cell size; (*c*) larger oil-bodies, often distinctly 2-several segmented. The restriction to somewhat basic sites further separates *L. grandiretis* from *L. marchica* and *L. capitata*, both of which are oxylophytes.

On peat-covered ledges *L. grandiretis* often occurs

... as a dense-leaved, pigmented modification that produces gemmae in extraordinary abundance.... The plants form small, almost rosette-like patches, of small extent, of which the stems are deep reddish black and the older leaves have a similar but less intense reddish pigmentation. Contrasted to these are the shoot-apices, which are yellowish green and bear large, very conspicuous masses of large (25–35 μ) stellate yellow-green gemmae.... On moister, less exposed peaty ledges ... the species is green throughout, except for the stems, which are purplish beneath. When the species grows in bogs it often forms a more lax-leaved modification, in which the stems are largely or entirely green .... (Schuster, 1953).

Such lax phases are often 3–5 cm long, bear remote, almost horizontally oriented leaves that are flaccid and extremely delicate. Plants of this type have been described as "variety *proteida*" by H. W. Arnell (1925) and are illustrated by Müller (1954, *in* 1951–58, fig. 210). In my opinion they represent a mere environmental modification (mod. *viridis-laxifolia*), and not a systematic entity.

Variation

Lophozia grandiretis shows considerable variation not only in pigmentation, as noted above, but in cell size. In typical var. *grandiretis* (to which the preceding diagnosis applies) median cells range from 40–50 × 50–75 μ up to 45–65 × 50–78 μ; basal cells show analogous variation and may reach the enormous size of 48–54 × 90–100 μ. By contrast, in the same region in which we may find typical *L. grandiretis*, a much smaller (perhaps haploid) phase exists, separable by the following key:

1. Median cells (36)40–65 × (45)50–78 μ; marginal cells of lobe, unless tangentially elongated, 36–48 up to 48–54 μ; larger gemmae to 45–47 × 50–57 μ . . . . var. *grandiretis*
1. Median cells 25–32(35) × 30–36(40–48) μ; marginal cells of lobe (where subisodiametric) 25–34 μ; gemmae up to 32–39 × 39–48 μ . var. *parviretis*

LOPHOZIA GRANDIRETIS var. *PARVIRETIS* Schust., var. n.

Plants of this variety may represent the haploid race of what is presumed to be a polyploid (chromosome no. = 14?) "typical" *L. grandiretis* and may be described as follows:[103] fleshy, stems thick, brittle, with age purplish black to (by reflected light) black; younger portions of shoots and majority of leaves (except toward base) bright green; uppermost leaves sometimes rather weakly purplish-tinged. Dorsal cortical cells 26–38 to 38–50 μ wide × 60–100 μ long. Leaves rather dense and small for size of stem, typically bilobed, often somewhat loosely folded (as in *L. incisa* and *L. opacifolia*) and with a slightly angulate-folded sinus, much wider than long (ca. 990 μ wide × 750 μ long to 1250 μ wide × 835 μ long, up to 1400–1500 μ wide × 1000–1100 μ long), bilobed by an obtuse-angulate sinus for usually less than 0.25 their length; lobes broad but shortly apiculate; margins entire or with mere obsolete denticulation by projecting cells. Marginal cells, where isodiametric or nearly so, 25–32(34) μ, median cells 25–32(35–40) × 30–36(40–48) μ; cells with small to rather distinct but concave trigones, with walls tending to be only slightly thickened, oil-bodies mostly 18–25 per cell in median cells, distinctly botryoidal, from 3.2–3.6 × 4 to 4 × 4.5–5 μ up to 4.5 × 7.2 μ. Gemmae strongly angulate-stellate, often with longly produced and almost spinose-extended angles, ca. 26 × 34 to 32 × 39–42 μ up to 39 × 48 μ. [Fig. 173:5.]

Type. Northwest Greenland: Kangerdlugssuak, on moist, peaty soil, irrigated by presumably somewhat calcareous seepage, slope above the shore just east of settlement, Inglefield Bay (*RMS 45831a, p. p.*).

Plants of *RMS 45831a* occurred in some quantity, admixed with *Odontoschisma macounii, Cryptocolea imbricata, Fissidens, Mnium, Lophozia heterocolpa harpanthoides, Tritomaria quinquedentata,* and *Aneura pinguis.* They occurred scattered over a large peaty *Dryas integrifolia-Cassiope tetragona*-grass-sedge-moss mat.

---

[103] Varietas var. grandireti similis, cellulis, autem, multo minoribus (mediis plerumque 25–32 × 30–36 μ, marginalibus plerumque 25–34 μ), gemmae maiores ad 32–39 × 39–48 μ.

Comparison of measurements of this collection and of "typical" *L. grandiretis* shows that differences exist in cell size, oil-body number, and gemma size. Only gemma size intergrades with that of the "normal" phase of the species. The type plants grew together with the normal, large-sized form of the species. In the latter, marginal cells, when not tangentially elongated, averaged 36–48 μ up to 48–54 μ; median cells ranged from 45–52 × 45–60 μ up to 50–54 × 65–75 μ. Dorsal cortical cells of the large-celled, typical form averaged little larger than those of var. *parviretis*, ranging from 38 to 50 × 80 to 125 μ. Study of admingled plants of the two extremes showed that var. *parviretis* had a strong tendency for rhizoids to be purplish at base and sometimes for much of their length, and the uppermost leaves were often purplish (even when the stem had not yet developed this coloration). By contrast, var. *grandiretis* showed largely colorless or bleached rhizoids, rarely purplish at the base, and usually had pure green upper leaves, exceptionally with traces of reddish purple pigmentation.

In cell size and in color of stem and rhizoid bases, *L. grandiretis* var. *parviretis* approaches *L. marchica*, of more temperate climates. The report by Bryhn (1906) of the latter species from Ellesmere I. may very well be based on plants of the present variety. Gemmae allow a ready distinction: *L. marchica* has smooth, spherical to ellipsoidal gemmae; all forms of *L. grandiretis* have prominently angulate-stellate to almost spinous-stellate gemmae.[104]

Sectio *MARCHICAE* Schust.

Lophozia subg. *Massula* sect. Marchicae Schust., Amer. Midl. Nat. 45(1):55, 1951.
Capitata-Gruppe K. Müll., Rabenh. Krypt.-Fl. ed. 3, 6:669, 1954.

Plantae dioiciae; viridulae ad rubellas, gemmae unicellulares, sphaericae ad ellipsoideas semper praesentes; folia obliqua, laxissima, undulata crispatave, 2-3-4-lobata, nunquam ciliata cuspidatave, aliquot dentibus (gemmis evolventibus plerumque consociatis) raro praedita; plantae molles sed graciles, saepe rubellae vinaceaeve.

Creeping in growth, exceedingly *lax, often rather flaccid*, with strongly undulate-crispate leaves, subpellucid, but often with purplish or reddish pigmentation (of leaves and/or ventral side of stem). Cortical cells *exceedingly elongated, narrow*, usually 4–8:1, 0.5–0.8 as wide as basal leaf cells. Leaves 2–3–4-lobed, broader than long, *wavy, delicate*, lacking differentiated teeth. No underleaves.

---

[104] There is a slight possibility that *Lophozia kiaerii* Joerg. (1934, p. 156), the type of which appears to be lost (Müller, 1951–58, p. 674; Arnell & Mårtensson, 1959), may be identical with *L. grandiretis* var. *parviretis*. Müller places the plant into the synonymy of *L. capitata*, which has smooth gemmae. Arnell and Mårtensson (1959, p. 114) maintain the species and state that it differs from *L. capitata* in the "angular, sometimes 2-celled gemmae." They place under *L. kiaerii* a plant from Spitsbergen that becomes "purple when old," with marginal cells 30 × 30 to 30 × 40 μ, median cells ca. 40 × 40 μ. Color and gemma form suggest *L. grandiretis s. lat.*, and the cell size more nearly approaches the type of var. *parviretis* than that of var. *grandiretis*. Unless Joergensen's type can be found, the status of *L. kiaerii* will remain speculative, especially since, as Müller notes, it was described as paroecious or also auto- or dioecious.

Gemmae greenish to reddish, *spherical to ovoid, smooth, 1-celled*.

Type. *Lophozia marchica* (Nees) Steph.

The group is quite isolated from other groups within *Massula*, by both the very narrow, elongated cortical cells (Fig. 185:3) and the smooth, 1-celled gemmae. Indeed, the gemmae are unique within the entire genus *Lophozia*. Both our taxa are extreme oxylophytes and are usually helophytic, although *L. capitata* often occurs on moist, sandy, acid ground.

LOPHOZIA (MASSULA) MARCHICA (Nees) Steph.[105]

[Figs. 173:1, 185]

Jungermannia socia var. *obtusa* Nees, Naturg. Eur. Leberm. 2:72, 1836.
Jungermannia marchica Nees, *ibid.* 2:77, 1836; Limpricht, Flora 65:45, 1882; Evans, Bull. Torrey Bot. Club 23:13, pls. 254–255, 1896.
Jungermannia polita Aust., Proc. Acad. Nat. Sci. Phila. 21(1869):220, 1870 (not of Nees, Naturg. Eur. Leberm. 2:145, 1836).
Jungermannia laxa Lindb., Acta Soc. Sci. Fennica 10:529, 1875; Lindberg, Musci Scand., p. 7, 1879 (footnote).
Lophozia marchica Steph., Spec. Hep. 2:148, 1902 (in part); Evans, Rhodora 12:199, 1910; K. Müller, Rabenh. Krypt.-Fl. 6(1):702, figs. 320–321, 1910; Haynes, The Bryologist 9:99, 1906 (in part only).
Lophozia (subg. *Massula*) *marchica* K. Müll., Ber. Deutsch. Bot. Gesell. 57:355, 1939; Schuster, Amer. Midl. Nat. 42(3): 571, 1949; Schuster, *ibid.* 45(1):55, pl. 12, 1951; Schuster, *ibid.* 49(2):347, pl. 13, 1953; K. Müller, Rabenh. Krypt.-Fl. ed. 3, 6(1): 675, fig. 208, 1954.

Delicate, creeping, usually scattered, over Sphagnum, pellucid pale to pure green, except for the undersurface of the stems (and sometimes leaf bases), *which are deep violet*. Shoots simple to slightly branched, *delicate and flaccid*, to 1.85–2 mm wide, up to 2.5 cm long. *Stems strongly flexuous*, 320–450(500) μ wide × 300–360(420) μ high, 11–14 cells high, soft-textured, green above and *deep violet below; cortical cells thin-walled* or imperceptibly thick-walled, narrowly rectangulate (ca. 14–20

---

[105] The citation of Stephani as the author of this combination is in some respects misleading, since the *Lophozia marchica* of Stephani includes also *L. capitata*. Equally misleading is Frye & Clark's (1945, *in* 1937–47) inclusion of *Jungermannia novae-caesareae* Evans (= *Lophozia novae-caesareae* Steph.) as a synonym of *L. marchica*. Evans (1893, p. 308) specifically mentions that the plants are "pale green, varying to dark green or yellowish . . . bearing scattered, whitish radicles." This surely cannot refer to *L. marchica*, characterized by the intense purplish color of the stem underside and of the rhizoids. *Lophozia novae-caesareae* is a mod. *viridis* of *L. capitata*. The citation of the range of *L. marchica* is equally confused in Frye & Clark. Unfortunately, I have been unable to verify the accuracy of all the dubious, older reports (and specimens) they cite. An additional element of confusion is introduced *vis-à-vis* the "*Jungermannia polita*" of Austin. Although Frye and Clark (loc. cit., p. 355) correctly list this as a synonym of *L. marchica*, they cite Austin's report of "*J. polita*" from New Jersey under *Saccobasis polita*, thus giving the latter a misleading range. It is necessary to again call attention to these errors since in European literature (i.a., in Müller, 1951–58, p. 718) they are carried over! Evans (1896a) long ago noted that the "*J. polita*" from New Jersey was *Lophozia marchica* s. str., emphasizing that his own *J. novae-caesareae* Evans differed from *J. marchica* "in its pale and delicate stems . . . and in its perianth, which is keeled to below the middle and more deeply denticulate at the mouth." He admits that its "nearest European ally" is *L. mildeana* (*L. capitata*). Indeed, Evans (1910b, p. 199) places his *J. novae-caesareae* as a simple synonym of *L. mildeana* (= *L. capitata*).

Grolle (Trans. Brit. Bryol. Soc. 5(3): 543, 1968) has just shown that *L. laxa* (Lindb.) Grolle must supplant *L. marchica* (Nees) Steph.

μ wide, occasionally to 22–26 μ wide × 76–160 μ long; *averaging 4–8× as long as wide and ca. half as wide as basal leaf cells*), in cross section nearly iso-diametric and moderately smaller than medullary cells; medullary cells leptodermous, to ca. 30–36 × 35–38 μ in diam. Rhizoids numerous, *purplish at least at base.* Leaves distant to contiguous, *polymorphous, delicate, flaccid and undulate,* dorsally hardly to slightly decurrent, very obliquely inserted, widely and somewhat horizontally spreading, *varying from 2- to 4-lobed,* the lowermost (and those of weak shoots) commonly largely bilobed, then from 1.0–1.3 × as wide as long, the upper leaves and those of robust shoots commonly 3–4-lobed (antical lobe often a mere tooth) and then 1.5–1.8× as wide as long, ca. 850 μ long × 950–1000 μ wide to 800–1050 μ long × 1200–1500 μ wide, subquad-rate to obtrapezoidal to transversely rectangular-obtrapezoidal in general outline; margins usually entire, except for lobes, but sometimes with a few accessory, small teeth; lobes irregular, triangular to ovate, usually pointed, sometimes obtuse, the antical (and sometimes postical) a mere tooth; sinuses usually descending *0.4–0.65 the leaf length,* acute to obtuse, their bases *usually recurved and gibbous.* Cells *extremely leptodermous, large,* of lobes and margins ca. 34–38 μ, *of leaf middle* 40 × 50–60 μ, near leaf base ca. 34–40 μ wide × 40–60(76) μ long, sharply polygonal, *without trigones;* oil-bodies minute, ca. 2–4 μ, spherical or subspherical, 30–60 per cell, appearing almost or quite homo-geneous; cuticle smooth. Underleaves lacking, except contiguous to gynoecia. Gemmae frequent, at apices of upper leaves (which are hardly malformed but may bear scattered marginal teeth), usually in relatively moderate numbers, *not forming compact apical "heads"; gemmae spherical to ovoid, smooth, 1-celled, green,* thin-walled, ca. (13)15–17 μ.

Dioecious, somewhat heteromorphic, ♂ plants usually more slender. Andro-ecia of 4–6 pairs of 2-lobed bracts, the antical lobe erect and often incurved; 1–2-androus; without paraphyses; androecium innovating at apex, the shoot continuing its growth (the innovations often very slender). Perichaetial bracts larger than leaves, 1000–1250 μ long × 2000–2500 μ wide, erect, deeply 3–5-lobed, somewhat crispate and undulate, the lobes sometimes spreading and usually narrow and lanceolate, the sinuses usually descending to the middle or deeper; bract margins subentire to irregularly dentate. Perianth obovoid-clavate to elongate-ovoid at maturity, to 1–1.5 mm wide × 3 mm long, more than half emergent, somewhat narrowed and plicate in distal ⅓; mouth, at least in part, *with teeth short and blunt, 1–2-celled.* Capsule ovoid, its wall 3–4-stratose, as in *L. capitata.* Elaters bispiral, ca. 8 μ in diam.; spores 13–16(18) μ, finely roughened, brown.

Type. Near Landsberg, Brandenburg, Germany (*von Flotow, 1822*).

Distribution

Lophozia marchica is a rare to infrequent species, found throughout most of the Taiga (spruce-fir forests), occurring evidently as a rarity in the Tundra, and very rarely in spruce- and tamarack-bordered peat bogs lying in the northern periphery of the Deciduous Forest Region (Schuster, 1949a). The species appears confined largely to western Europe (Opland,

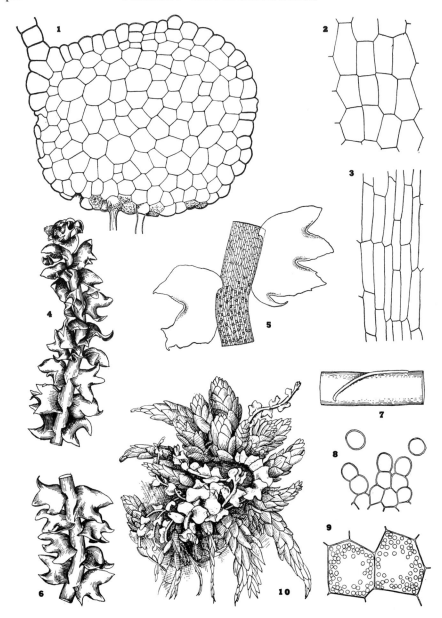

Fig. 185. *Lophozia* (*Massula*) *marchica*. (1) Stem cross section (×150). (2) Basal leaf cells (×150). (3) Dorsal cortical cells, the two complete cells at right 240 μ long (×150). (4) Gemmiparous shoot, antical aspect (×11.5). (5) Postical aspect of shoot sector, rhizoid initials drawn in below only (×28). (6) Shoot sector, antical aspect (×11.5). (7) Stem sector, lateral view, to show leaf insertion (×28). (8) Gemmiparous leaf lobe and gemmae (×ca. 365). (9) Two cells with oil-bodies (×340). (10) Plants, *in situ*, on head of *Sphagnum* plant (×ca. 4). [1–3, 8 mi. SW. of Gheen, Minn., *RMS*; 4–9, *RMS*, *1945*, Junius Bog, N.Y.; 10, from K. Müller, 1905–16.]

Norway; scattered in southern Sweden, but also at 63° N.; Finland), southward to the north German plains (Hamburg, Brandenburg), the Baltic Sea region, Pomerania, Denmark, etc., to Bavaria, Switzerland, the Tyrol (Arlberg), and to the eastern half of North America:

ELLESMERE I. Beitstad Fiord in Hayes Sd. region, and Goose Fiord, on S. coast (Bryhn, 1906); these reports need verification, the species being unknown elsewhere from well within the Tundra; see p. 463.   ONTARIO. L. Timagami, in bog with *Cladopodiella fluitans* (Cain & Fulford, 1958).   QUEBEC. Old Factory (Lepage, 1953). NEWFOUNDLAND. Upland bog, N. of Robinson's, W. Nfld. (*RMS 68-1269*). NOVA SCOTIA. S. of Milford, Annapolis Co. (*RMS 68-500*); W. of Charlotte, Halifax Co. (*RMS 68-574*); Summit French Mt., Cape Breton, Inverness Co. (*RMS 68-1049*).

MAINE. Mt. Desert I. (*Greenwood 113*); near Schoodic L. (*Evans*); near Mt. Katahdin (*Lorenz*); Beach Mt., Mt. Desert I. (*Rand; fide* Evans, 1910b).   NEW HAMPSHIRE. Waterville (*Lorenz*).   VERMONT. Franklin (*Lorenz, 1908*); also cited from Jericho (Evans, 1905; this is *L. capitata*, which was not then distinguished).   MASSACHUSETTS. Boston (*Clarke, 1907*) [Also reported from Woods Hole, but this represents *L. capitata*]. CONNECTICUT. Bethany (*Lorenz 996*) [Also reported by Evans (1902d) from East Haven, but this represents *L. capitata*].   NEW YORK. Junius Bog, near Geneva, Seneca Co. (*Schuster, 1945;* see Schuster, 1949a).   NEW JERSEY. Closter (Austin, listed in Austin [1870] as *Jungermannia polita*; I have [1954] verified that this plant is *L. marchica*). [Also reported by Frye & Clark (1937-47, p. 356) from Fairmount (*A. Dantum, 1908*), Atsion and Pleasant Mills, but these are *L. capitata*, judging from Evans' description.]

MICHIGAN. Isle Royale, in *Sphagnum* bog; Phoenix, Keweenaw Co., and Whitefish Bay, Chippewa Co. (Evans & Nichols, 1935); Reese's Bog, Cheboygan Co.   MINNESOTA. Pigeon R., Cook Co. (*Conklin 2650*); Black Bay near Island View, Rainy L., Koochiching Co. (*RMS 14617*); bog 4 mi SE. of Ericsburg, Koochiching Co. (*RMS 14629*); in bogs 2 mi N. of Orr and 8 mi SW. of Gheen, St. Louis Co. (*RMS 13485, 6848a, 14643, 13451, 5121a*, etc.).

Also reported from Delaware and West Virginia (Evans and Nichols, 1908). It is highly unlikely that either of these reports represents this northern species, which is almost strictly confined to areas which underwent Pleistocene glaciation (with consequent formation of poorly drained systems in which peat bogs formed). *Lophozia capitata* is surely represented (see Evans, 1910b, who discusses the range of both *L. marchica* and *L. capitata* [*L. mildeana*]).

Ecology

Confined exclusively to peat bogs. The report of the species from "wet sandy soil" (Frye and Clark, 1937-47, p. 356) is due to confusion with *L. capitata*, which was at one time considered a synonym of *L. marchica*.

The species (according to Schuster, 1953) is to be found in many of the better developed, more open *Sphagnum* bogs of the region. It occurs here largely along the deer trails or holes in the bogs, on the vertical sides or the wet bases of which it may form considerable patches. Often it occurs only as small isolated mats and may be overlooked. Associated species, outside of *Sphagnum* and the Ericaceous shrubs (*Ledum, Chamaedaphne, Andromeda*, etc.), are *Mylia anomala, Cladopodiella fluitans, Calypogeia sphagnicola, Cephaloziella elachista,* and such *Cephalozia* species as *connivens* and *pleniceps*. More rarely (and usually in the wettest places) *Scapania irrigua* and *paludicola* may occur with it.

The species occurs only in acid bogs: it has a very low tolerance for lime and also is an early-invading species, when conditions are still quite wet. By the time the bog level is raised, the bog forest becomes dense, and such mesophytes as *Lophocolea heterophylla* invade, the *Lophozia* has begun to disappear. Associated with the usual peat bog angiosperms, such as *Drosera rotundifolia, Sarracenia purpurea, Vaccinium oxycoccos*, as well as, occasionally, *Utricularia cornuta* and *Lycopodium inundatum* (Schuster, 1949a).

Unlike some of the preceding, helophytic associates (such as the *Mylia, Calypogeia, Cephalozia connivens* and *pleniceps*), *L. marchica* appears exclusively confined to bogs, never growing over wet, *Sphagnum*-covered crests or ledges or cliffs or on peaty banks. Unlike the closely allied *L. capitata*, *L. marchica* never appears to grow in sandy-moory sites or over mineral substrates of any kind. The plant is usually more hygrophytic and more distinctly helophytic; it most often occurs in the wettest parts of bogs, often covered by water during part of the growing season.

Differentiation

Lophozia marchica, although extremely variable in leaf shape, is a stenotypic species, causing few taxonomic problems. The sharp ecological restrictions usually serve to identify it, only the closely allied *L. capitata* occasionally occurring under identical conditions.

Lophozia marchica has a distinctive facies, the pure grass-green but pellucid color (except for the purplish postical face of the stem and bases of the rhizoids), together with the crisped and undulate, highly variable, delicate leaves, being diagnostic. The broad, lax and distant, wavy, irregularly 2- (seldom 3-4-) lobed leaves, with reflexed and gibbous sinuses, the contorted, creeping mode of growth, and the delicate and fleshy stems (with delicate and strongly elongated cortical cells), recall only *L. capitata*, and to a lesser extent lax bog forms of *L. grandiretis*. The purplish postical face of the stem is shared with *L. grandiretis*, but the latter differs at once in the (*a*) much larger leaf cells, often 60–80 μ long in the leaf middle, with considerably larger, frequently segmented, oil-bodies; (*b*) broader and larger cortical cells of the stem; (*c*) angulate to stellate and much larger gemmae. Furthermore, *L. grandiretis* is confined to peaty sites which are underlain by marl or basic rocks, while *L. marchica* is a strictly obligate oxylophyte.

Separation of the similar and closely allied *L. capitata* is more difficult. The latter almost never has the ventral side of the stem vinaceous or purplish (in shade forms the whole plant is green; in sun forms the perianth and distal portions of the leaves, sometimes the whole plant, become reddish). In *L. marchica*, by contrast, even shade forms show the vinaceous pigmentation of the postical stem surface, which may extend to the basal halves of the leaves, leaving only distal portions greenish. *Lophozia marchica* also is described as having the perianth mouth shortly denticulate (according to Müller, 1951–58, the teeth are short and 1-celled; this probably represents their minimal development); *L. capitata* has a ciliate perianth mouth, the apices of the teeth often being uniseriate for a length of 5–9 cells. Gemmae of *L. marchica* are also smaller, usually 13–17 μ, those of *L. capitata* being 24 μ to 20 × 27 or even up to 27 × 32 μ.

The preceding distinctions suggest that two sharply discrete species are at hand. However, study of a series of specimens from northern peat bogs reveals intergradations, particularly in sandy areas, in depressions filled with *Chamae-daphne calyculata* and with some *Sphagnum* (as, i.a., in *RMS 19898*, from 5 miles east of Tahaquemon Falls, near Paradise, Mich.). Such plants are partly wholly pellucid green, partly have reddish-tinged leaves and bracts. In the latter case (the plants evidently representing sun forms) there is a marked tendency for stems to be ± purplish ventrally, and some leaf bases show similar pigmentation (much as in "normal" *L. marchica*). However, the gemmae are large (spherical and 24 μ to oval and 20–22 × 27–36 μ or 27 × 32 μ); the oil-bodies are fewer and larger (20–40 per cell in lobes and leaf middle; 3.5–5.4 μ and up to 5 × 6–7 μ); the perianth mouth is ciliate-laciniate, with the teeth ending in uniseriate cilia to 5–9 cells long. Thus, in spite of the occurrence of purple pigmentation on the stem underside (a "*marchica* character") other characters approximate those of *L. capitata*, to which it seems wise to refer such plants. This evident intergradation, of course, has been responsible for the unwillingness of some workers (Stephani, 1902, *in* 1898–1924; Evans, 1902d, p. 211) to maintain two species, although others [and occasionally the same worker] held the opposite view (Müller, 1910, *in* 1905–16; Warnstorf, 1903, p. 200; Evans, 1910b, p. 199). If there are two species, their separation will have to depend on a consideration of a constellation of characters.

LOPHOZIA (MASSULA) CAPITATA (Hook.) Boulay
[Figs. 186, 187]

Jungermannia capitata Hook., Brit. Jungerm., pl. 80, 1816.
Jungermannia mildeana G., Verh. Zool.-Bot. Gesell. Wien 17:626, pl. 16, 1867.
Jungermannia novae-caesareae Evans, Bull. Torrey Bot. Club 20:308, pl. 163, 1893.
Jungermannia mildei Kaal., Nyt Mag. Naturvid. 33:333, 1893.
Lophozia mildeana Schiffn., Lotos 51(7):266 [reprint, p. 54], 1903; K. Müller, Rabenh. Krypt.-Fl. 6(1):699, fig. 319, 1910; Frye & Clark, Univ. Wash. Publ. Biol. 6(3):359, figs. 1–9, 1945.
Lophozia capitata Boulay, Musci de Fr. 2:112, 1904 [as to name only; taxonomically = *L. excisa*]; K. Müller, Trans. Brit. Bryol. Soc. 1(4):356, 1950; Jones, *ibid.* 1(4):353, fig. 1, 1950.
?*Lophozia kiaerii* Joerg., Bergens Mus. Skrifter No. 16:156, 1934.
Lophozia marchica Frye & Clark, Univ. Wash. Publ. Biol. 6(3):355, 1945 [not of Stephani; in part at least, as to synonymy, citations, and figures].
Lophozia (Mass.) mildeana Schust., Amer. Midl. Nat. 42(3):571, 705, 1949; Schuster, *ibid.* 45(1):55, 1951.
Lophozia (Mass.) capitata Schust., Amer. Midl. Nat. 49(2):346, Fig. 76, Pl. 10, figs. 9:12, 1953; K. Müller, Rabenh. Krypt.-Fl. ed. 3, 6(1):672, fig. 207, 1954.

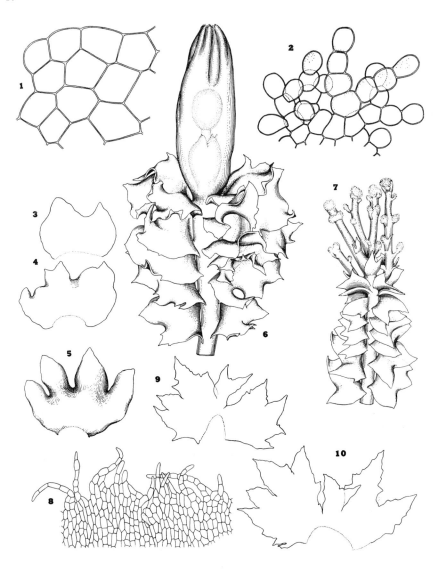

FIG. 186. *Lophozia* (*Massula*) *capitata*. (1) Cells of leaf lobe apex (×235).
(2) Gemmiparous leaf tip (×330). (3–5) Leaves (×17). (6) Mature shoot
of mod. *viridis* with capsule ready to emerge (×10.5). (7) Shoot with
unusually free development of small-leaved gemmiparous innovations
(×15). (8) Cells of somewhat immature perianth mouth, the cilia often
deciduous with maturity (×61). (9–10) Bracts and bracteoles (×11.5).
[1, bog in Itasca Co., Minn., *RMS;* 2, *RMS 14121*, bog near Jaynes,
Minn.; 3–5, bog near Tahaquemon Falls, Mich., *RMS, 1949;* 6–10,
Orange Co., N.C., *RMS.*]

In distinct patches or mats, often crowded and caespitose, or as isolated suberect stems among caespitose mosses, or creeping over other mosses, whitish
green or pale green (in shade) but *often reddish-tinged or reddish violet, never* (or
exceptionally) *with underside of stem violet or purple.* Shoots 1.4–2.5 mm wide, on
♀ plants to 4.5 mm wide above, but ♂ plants usually more slender and only
1–1.5 mm wide, to 1–3 cm long; sparingly branched (but often with subfloral
innovations and with *1–3 or more slender, remote-leaved gemmiparous innovations*
from apices of normal shoots, these capitate and *densely gemmiparous* at apex).
Stems 160–275 μ in diam., soft in texture, scarcely dorsiventrally compressed;
cortical cells scarcely thick-walled, isodiametric in cross section, scarcely smaller
(in cross section) than medullary cells, which are extremely thin-walled and
delicate; *no mycorrhizal medullary infection;* cortical cells narrow and elongate,
averaging (2.5–4)4–8× as long as wide, ca. 0.6–0.8 the width of basal leaf cells.
Rhizoids abundant, *colorless.* Leaves hardly decurrent, the lower horizontally
spreading, the upper often antically somewhat secund, *extremely polymorphic,*
varying from largely 2-lobed (on weak and ♂ shoots) to largely 3–4-lobed
(upper portions of robust and of ♀ shoots), the 2-lobed as long as broad and
subquadrate, the 3–4-lobed leaves at least 1.4–1.7× as wide as long, obtrapezoidal to rectangulate, from 900–1100 μ long × 1150–1360 μ wide to 1020–
1200 μ long × 1630–1800 μ wide, *thin, delicate, and flaccid, strongly undulate or
somewhat crispate;* lobes ovate to triangular, often ± unequal, usually bluntly
acute at apex, entire or subentire; *sinuses descending 0.35–0.6* the leaf length,
narrowly rounded and *reflexed or gibbous below.* Cells ca. (30)32–38(40) μ on
margins and in leaf apices, *32–36(40) × 38–45(50) μ medially,* 40–50 × 54–
63 μ at leaf base, polygonal, *thin-walled* (or slightly and equally thickened),
essentially *without trigones,* quite pellucid; oil-bodies usually from 16–28 to
35–60 (basal cells) per cell, small, ca. 2.4–4 to 3.6–5.4 μ and spherical, a few
ovoid and to 5 × 6–7 μ, appearing homogeneous or formed of a few barely
perceptible segments, which (if at all perceptible) do not protrude through the
common membrane. Underleaves absent. Gemmae at apices of upper leaves,
thin-walled, spherical to ovoid, smooth, 1-celled, *relatively large,* ca. 18–20(24) ×
22–33 to 20–22 × 36 or 27 × 32 μ, *usually greenish* but in sun forms sometimes
slightly reddish-tinged, each bearing numerous minute, spherical oil-bodies.
Gemmae commonly (but not always) in *large (often globular) masses,* sometimes
from reduced leaves of slender, etiolated branches or innovations, sometimes
in smaller numbers from margins of hardly modified leaves.

Dioecious. ♂ Plants slender; androecia becoming intercalary, often violet-
tinged, of 3–5 pairs of bracts usually; bracts ventricose at base, more nearly
transversely inserted than leaves, bilobed, somewhat similar to leaves in size,
usually approximated, 1–2-androus. Antheridia ovoid, 230 × 200 μ wide;
stalk 1–2-seriate, 125 μ long. ♀ Bracts (and adjacent vegetative leaves) larger
than normal leaves, commonly irregularly (2)3–5-lobed, not or hardly broader
than long, from 1600 × 1600 μ to 2100–2200 μ long × 1500–2100 μ wide,
strongly crispate and fluted, commonly widely spreading from perianth; lobes
acute to subacute, often sparingly and irregularly, obscurely dentate. Bracteole
variable, much smaller than bracts, usually bilobed-divaricate, 600–900 μ
wide × 1000–1100 μ long, margins similar to those of bracts. Perianth large,

cylindrical to ovoid-clavate, from 1.8 to 2 mm wide × 3.6 to 4.5 mm long, ca. 0.7–0.85 emergent, often *reddish-tinged* (but mouth sometimes decolorate), upper half distinctly plicate, contracted to mouth. Mouth lobulate and *ciliate to dentate*, the teeth 2–3, *partly 5–10, cells long*, sinuate to tortuous, deciduous with age, formed of delicate, thin-walled cells.

Capsule ovoid, wall of (3)4–5 cell layers. Epidermal layer of short rectangulate to subquadrate large cells (ca. 20 to 25–27 μ thick × 14–16 × 35–38 μ to 19–23 × 26–30 μ, some ca. 30 × 30 μ, with (1)2–3(4) strong nodular thickenings per longitudinal wall and (0)1–2 thickenings per transverse wall; bearing usually (8)10–16 oil-bodies per cell, each 3–4 μ and indistinctly segmented. Inner cells, in 3–4 layers, each ca. 10 μ thick; cells 13–16 μ wide × 65–75 μ long to 16–20 × 60 μ, often irregular, with commonly 6–8 very slender, complete to often incomplete semiannular bands. Spores reddish to reddish violet, 12.5–15 μ (14–16 μ; *teste* Müller), finely but sharply verruculose; elaters 7–8.5 × 90–110 μ, 2-spiral; spiral 2.5–3 μ wide.

Type. England: Cadnam Bog (*Hooker*); type of *L. mildeana* from near Breslau, Silesia, Germany (*J. Milde, 1866*).

Distribution

Lophozia capitata is sporadically, superficially rather peculiarly, distributed. It occurs scattered through much of the more temperate portions of western Europe [from the north German plains and the Baltic Sea region, south to the Thüringer Forest, Saxony, etc.; westward to England (very local) and Ireland, to the southwest coast of Norway and southern Sweden; also in Finland], and in the warmer to temperate, slightly into the boreal, portions of eastern North America. The species "enjoys" the distinction of occurring farther southward than other members of its family in eastern North America: in the "Oligocene Island" area of central Florida, into the Sabalian Region.

The distribution appears to coincide largely with sandy regions or dunes, formerly the beaches of old lakes, and with the coastal plain of eastern North America, where old beaches are of widespread distribution. In the interior (Michigan, Minnesota) largely in peaty meadows or bogs underlain by sand, in old sand dunes, or around old beaches (from lakes of immediate postglacial age). To a larger extent than any other species of the family, *L. capitata* is restricted to arenaceous sites, chiefly between sea level and 1200 feet.

In North America evidently absent from the western half of the continent and lacking also from eastern Asia; in the eastern half of North America ranging as follows:

ONTARIO. Manitoulin I. (*RMS & Cain, Sept. 1, 1949; Cain 4087*).   NEWFOUNDLAND. Ditch, edge of upland moor, N. of Robinson's, W. Nfld. (*RMS 68-1251*).   NOVA SCOTIA. Fishing Cove L. track, Cabot Trail, French Mt., Cape Breton (*RMS 68-1098*). MAINE. Biddeford Pool (*Lorenz*; Evans, 1910b).   NEW HAMPSHIRE. Franconia Mts., with *Microlepidozia setacea* (*Haynes*; Evans, 1910b); Waterville (*Lorenz*; Evans, 1910b). VERMONT. Jericho (Evans, 1910b). RHODE ISLAND. Westerly (*Lorenz, 1924*). MASSACHUSETTS. Woods Hole (Evans, 1910b; earlier listed as *L. marchica*); disused gravel bank, Breckenridge Rd., and ski slope on N. side of Holyoke Range, Hadley (*RMS 43901a, 43908, 43903a, 43908a*); pond margin near summit of Mt. Greylock (*RMS, s.n*); Guilder Pond, Mt. Everett, Berkshire Co. (*RMS 68-164, p.p.*). CONNECTICUT. East Haven, Huntington, Orange, and Milford (*Evans; Lorenz;* Evans, 1910b).

NEW YORK. Plymouth Bog, Chenango Co. (Schuster, 1949a); forest of Dean L., near Ft. Montgomery, Rockland Co. (*RMS 24234a*); McLean Bogs, Tompkins Co. (*A. L. Andrews*); South Hill Marsh, Ithaca, Tompkins Co. (Schuster, 1949a); Sixmile Creek (*A. L. Andrews*); Fishers I., Long Island Sd. (*Evans*). PENNSYLVANIA. Delaware Co. (*Knout, 1914*). OHIO. Ashtabula and Hocking Cos. (Miller, 1964). NEW JERSEY. Atsion and Pleasant Mills (Evans, 1893; as *J. novae-caesareae*); Great I. near Elizabethport (*Haynes, 1906*). DELAWARE. *Teste* Evans (1910b, p. 199). WEST VIRGINIA. Near Easton (*Andrews, 1905*; det. Evans).

NORTH CAROLINA. Near Trenton, Trenton Co. (*Schuster, 1954*); ca. 5 mi W. of Durham in Orange Co. (*RMS 32012; plate*); 1 mi NE. of Folkstone, Onslow Co. (*RMS 34181*); Bennet Memorial, near Durham, Durham Co. (*Channell, Apr. 1954*). SOUTH CAROLINA. NW. of Sumter, Sumter Co. (*RMS 18709*). FLORIDA. 15 mi SW. of Sanford, Seminole Co. (*Rapp; fide* Evans, 1923f).

Westward occurring as follows. MICHIGAN. Near Paradise, 5 mi E. of Tahaquemon Falls, Luce Co. (*RMS 19898*); near Deer Park (*Schuster, 1949*); Little L. Sixteen, Cheboygan Co.; E. of Shingleton, Alger Co. MINNESOTA. Peaty meadow at Ham L., Anoka Co. (*RMS 17462*); along Rte. 38 near Jaynes, Itasca Co. (*RMS 14121; northwesternmost report!*); see Schuster, 1953. IOWA. Moore Station, Poweshiek Co. (*Conard, Apr. 15, 1946!*).

Ecology

As stressed above, *L. capitata* appears largely (but not exclusively!) confined to old dune regions or at least to areas where wave action, and in part glacial action, result in mineral soils with poor water-retaining capacity. Combined with this is restriction to areas where the plants are continuously damp or wet. As a consequence, *L. capitata* is found chiefly in two types of sites. (1) In depressions of poorly drained sand barrens, especially in glaciated areas (chiefly northern Michigan and Minnesota in our area), where the water table is near the surface. Such areas are usually characterized by peat formation, the occurrence of Ericaceae, chiefly *Chamaedaphne* (but rarely or never *Ledum*), of cranberries, sedges, *Lycopodium inundatum*, *Drosera*, *Xyris* (usually *X. torta*), and *Eriocaulon* (usually *E. septangulare*), and orchids (*Pogonia ophioglossoides*). The less depressed portions of these areas of "coastal plain" affinity are usually jack-pine barrens, while the wetter portions develop peat bogs. (2) In damp ditches or on moist banks on the sandy outer coastal plain (chiefly from North Carolina to northern Florida), associated with *Scapania nemorosa*, various mosses (often *Aulacomnium palustre*, *Polytrichum*), *Cephaloziella*, and such coastal plain genera as *Xyris*, *Eriocaulon*, various Ericads,

Pogonia ophioglossoides, Drosera spp., *Lycopodium carolinianum, L. adpressum,* etc., locally also with Venus flytrap (*Dionaea*). A related occurrence is on acid, exposed loamy soil of north-facing slopes (Hadley, Mass.), associated with *Cephaloziella rubella, Polytrichum, Ditrichum pusillum, Lophozia bicrenata,* and locally the tracheophytes *Lycopodium inundatum* and *Drosera rotundifolia.*

Less commonly the species occurs on loamy or clayey wet soils around the borders of lakes or swamps, in insolated sites; then sometimes with *Solenostoma crenulatum, Scapania nemorosa* (or *irrigua*), *Cephaloziella* (often *rubella*), *Nardia insecta, Calypogeia neesiana* and *fissa,* etc. Even less frequently found in mature peat bogs, together with members of the *Mylia-Cladopodiella* Associule (see Schuster, 1957, p. 285), among them *Mylia anomala, Cladopodiella fluitans, Cephaloziella elachista,* and *Calypogeia sphagnicola.* Such occurrences are identical with the "normal" occurrences of the similar *L. marchica.*

In the Piedmont of the Southeast part of the early stages in succession of old fields. Bliss and Linn (1955) show that fallow fields undergo succession from an initial *Sphaerocarpus-Bryum-Physcomitrium-Weissia* stage, followed by dominance of *Bryum-Physcomitrium* the following year, *Bryum-Pleuridium-Physcomitrium* the second year, and *Ditrichum-Physcomitrium* during years 3-5, when *Andropogon virginicus* (broom sedge) becomes dominant. During years 9-15 a broom sedge-pine stage (*Pinus echinata* and *taeda* chiefly) supersedes, with the ground-level community formed by a *Lophozia capitata-Eurhynchium serrulatum-Ditrichum pallidum* community.[106] The *L. capitata* predominates under young pines, especially in the transition zone "from exposed ground in the open aspect to heavy needle cover under the pines," in my experience chiefly where partially shaded by young pines or *Andropogon,* on clayey-loamy soil, often associated with *Cladoniae,* traces of *Cephaloziella hyalina, Fossombronia* sp. The *Lophozia* is a sporadic component of such old field succession but may persist until open mature oak-hickory forest is attained, occurring under oaks, on sterile acid soil, as scattered slender plants among *Dicranella heteromalla,* under rather xeric conditions.

In the Southeast the species is spring fruiting (as early as Apr. 9 in the Piedmont of North Carolina).

Differentiation

Lophozia capitata bears all the marks of a typical species of *Massula:* mostly broader than long leaves, with the lobe number varying from 2 to 4, depending largely on robustness; leptodermous and large cells without obvious trigones; numerous, minute oil-bodies; a stem with almost no cortical differentiation, and cortical cells up to 4-8 × as long as wide,

---

[106] The species similarly occurs as a near-pioneer on the loamy-gravelly soil on the steep sides of disused gravel pits (as at Hadley, Mass.); here it forms a community with *Ditrichum pusillum, Polytrichum, Pogonatum pensilvanicum,* with associated tracheophytes, chiefly *Hypericum gentianoides* and *Comptonia peregrina.*

regularly rectangulate; a soft and delicate texture, with often undulate leaves; dioecious inflorescence; free production of gemmae.

The species is subject to confusion with both "*L. marchica* and *L. grandiretis*, because of the large, thin-walled and pellucid leaf cells (averaging over 40 μ in leaf middle), the very numerous, minute, [nearly] homogeneous oil-bodies, ... as well as the irregular, somewhat wavy leaves that are 2–3- or even irregularly 4-lobed. It is not likely to be mistaken for other species" (Schuster, 1953). The smaller leaf cells, the smaller, smooth gemmae, the lack of vinaceous pigmentation of the stem, and the considerably less fleshy stem (which, furthermore, is scarcely dorsiventrally flattened) all separate it from *L. grandiretis*. The normal lack of vinaceous pigmentation of the undersurface of the stem and rhizoids separates *L. capitata* from *L. marchica;* but see p. 469. *Lophozia capitata* agrees with *L. grandiretis* in the frequent tendency for leaves and distal portions of the perianths to become reddish-tinged—the pigmentation in these two species extends, so to speak, from the apices of these appendages down. In this respect, *L. capitata* (and *L. grandiretis*) disagree with *L. marchica*, in which the vinaceous—not red—pigmentation occurs on the undersurface of the stem, spreading (in extreme cases) from there onto the basal halves of the leaves. Except for these differences in type and localization of pigmentation, *L. capitata* is closely similar to *L. marchica*. The two were for many years confused with each other. Transitional phases, discussed under *L. marchica*, occur.

Although not closely allied to *L. excisa*, *L. capitata* was once considered to be a synonym of this species. After the name dropped into synonymy under *L. excisa*, *L. mildeana* ("*Jungermannia mildeana*") was sharply differentiated from both *L. excisa* and *L. marchica*. This name was universally used for the present taxon, in spite of the fact that Schiffner many years ago studied the type of *L. capitata* and correctly pointed out its identity with *L. mildeana*. Only in 1950 did Müller make the needed combination in conjunction with the appropriate synonymy. As I have pointed out (Schuster, 1953, p. 346), many features separate *L. capitata* from *L. excisa*, among them (*a*) nearly spherical, smooth gemmae, usually pale green in color; (*b*) larger leaf cells, 35–50 μ in the leaf middle; (*c*) polymorphic leaves, often 3–4-lobed and deeply divided, and much wider than long; (*d*) stems with cortical cells extremely narrow and elongate, and with the medulla not becoming mycorrhizal.

The oil-bodies of *L. capitata* appear to be variable in size and number. In some cases (*RMS 34181*, Folkstone, N. C.) there may be only 18–32 minute, spherical oil-bodies per cell, each 2.3–3.4 μ. In others (i.a., *RMS 17467*, Tahaquemon Falls, Mich.) there are 20–40 oil-bodies per cell in the leaf lobes and middle, gradually increasing to 35–60 per cell near the leaf base; oil-bodies range from 3.5–5.4 μ to ovoid and (perhaps exceptionally) up to 5 × 6–7 μ. On the other hand, some collections (i.a., *RMS & Cain*, Manitoulin

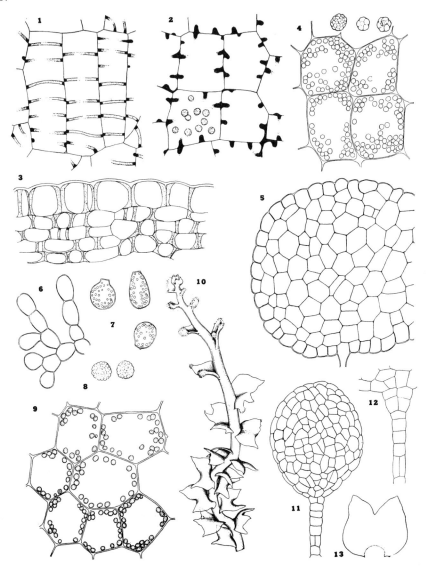

Fig. 187. *Lophozia* (*Massula*) *capitata*. (1) Cells of innermost layer of capsule wall (×560). (2) Epidermal cells, capsule wall, the persistent oil-bodies drawn in in one cell (×560). (3) Cross section, capsule wall (×365). (4) Cells above leaf base, with oil-bodies (×260), and above, several more enlarged oil-bodies (×925). (5) Stem cross section (×225). (6) Gemma fascicle (×ca. 275). (7) Gemmae (×ca. 290). (8) Oil-bodies, from cells in fig. 9 (×ca. 1250). (9) Cells with oil-bodies (×ca. 250). (10) Suberect gemmiparous shoot of mod. *viridis* (×12). (11) Antheridium (×135). (12) Antheridial stalk (×200). (13) Leaf (×12). [1–3, 10–12, from Orange Co., N.C., *RMS, Apr. 1954;* 4, 6, five miles E. of Tahaquemon Falls, Luce Co., Mich., *RMS;* 5, 13, Poweshiek Co., Iowa, *Conard;* 7–9, Manitoulin I., Ontario, *RMS, Sept. 1, 1949.*]

I., Ont.) show only 15–22, occasionally as few as 11 or up to 24–26, oil-bodies per median cell; the oil-bodies average 3.5–4.2 μ, a few to 2.5–3.4 μ. They are never strictly homogeneous, but are formed of relatively few spherules, which do not or hardly protrude through the common bounding membrane (Fig. 187:4, 8). Similar oil-bodies also occur in the epidermal cells of the capsule wall (drawn in for one cell in Fig. 187:2).

Depending on degree of maturity, *L. capitata* is very variable as to leaf-lobe number. Although robust, fertile forms have a high percentage of 3–4-lobed, wide, undulate leaves, the weaker, strongly gemmiparous forms found under difficult conditions may have almost uniformly bilobed leaves. Gemmae often occur in large, swollen-appearing masses; they frequently turn reddish to reddish purple with maturity, even in plants from rather sheltered sites that lack other pigmentation. Young gemmae, however, are at all times pale greenish in color. The spherical to ovoid form of the gemmae is distinctive.

Müller (1951–58, p. 674) considers *L. kiaerii* Joerg. a synonym of *L. capitata*, stating that it agrees with the latter in all important features, except in the sexuality, which is considered to be paroecious or also autoecious and dioecious in *L. kiaerii*. The plant, known only from Opland, Norway, deserves further careful study; see also p. 463.

Subgenus *ISOPACHES* (*Buch*) *Schust.*

Isopaches Buch, Mem. Soc. F. et Fl. Fennica 8:287, 1933 (as genus); Evans, The Bryologist 38:63, 1935; K. Müller, Rabenh. Krypt.-Fl. ed. 3, 6:683, 1954.
Lophozia sect. Compactae Joerg., Norges Leverm., Bergens Mus. Skrifter no. 16:146, 1934.
Lophozia subg. *Isopaches* Schust., Amer. Midl. Nat. 45(1):56, 1951; Schuster, *ibid.* 49(2):348, 1953; Jones, Trans. Brit. Bryol. Soc. 3(3):360, 1958; Kitagawa, Jour Hattori Bot. Lab. no. 28:271, 1965.

Small or medium-sized, 5–12 mm long, closely *prostrate*, the apex strongly ascending. Stem *thick, somewhat fleshy, merely 8–10 cells high,* with medulla of large, *somewhat thick-walled, uniform cells not* (or only tardily) *attacked by mycor-rhizae,*[107] the ventral medullary cells as large as dorsal; cortical cells little different from medullary cells, *thick-walled.* Rhizoids long, colorless, forming a moderately to very dense ventral mat. Leaves somewhat antically assurgent, ± erect or obliquely patent, ± imbricate, never horizontally spreading, ± asym-metrically to *nearly symmetrically bilobed* (exceptional leaves 3–4-lobed near gynoecia), ± quadrate to orbicular, concave, the margins entire (except on gemmiparous shoots and subfloral leaves); leaf insertion oblique, but not strongly so, the leaves subvertical and appearing somewhat subtransverse; *leaf bases never ciliate. Cells strongly and equally pachydermous,* lumina often appearing *guttulate,* almost noncollenchymatous, with rounded angles, *small,* about 25–32 μ (or less) in leaf middle; oil-bodies distinctly grayish because of the numerous, distinct, included oil-globules, little refractive, appearing granulate externally, moderate to rather large in size (mostly 4–5 μ to 6 × 9–10 μ), *mostly (2)3–6 to 6–12 per cell* (much as in subg. *Lophozia*); *absent from marginal* (and lobe apex)

---

[107] Ventral cortical cells (especially rhizoid-initials) are early attacked by mycorrhizal fungi, and the cortical infection often extends to the medullary layer of cells immediately above.

cells on mature leaves. Ventral stem sectors obsolete, *never producing underleaves.*
Gemmae ± strongly angulate-stellate, rust-red to red-brown, *common; gemmi-*
parous leaves becoming strongly dentate.

Paroecious or autoecious or dioecious. *Often* (or usually) *monandrous; an-*
theridial stalk 1(2)-seriate. Perianth unmodified, *deeply plicate,* gradually narrowed
to the *ciliate mouth, not beaked.* Sporophyte with seta of many rows of cells (often
reduced to 16 external rows in *L. bicrenata*).[108] Capsule wall *2-stratose* (rarely
3-stratose in *L. bicrenata*) ; *inner and outer layers both with incomplete tangential* and
complete radial *thickenings* (i.e., both with nodular to incompletely semiannular
thickenings).

Type. Lophozia bicrenata, by present designation.

Isopaches (after the exclusion of "*I.*" *hellerianus*) stands in some ways
between *Massula* and *Lophozia.* As in the former, the medulla of the stem
is unspecialized and the leaf cells are noncollenchymatous; as in the
latter, the oil-bodies are relatively few in number, generally moderately
large, not homogeneous or highly refractive in appearance. *Isopaches*
cannot be considered as truly annectant between these two subgenera,
however, since it is sidewise specialized in the reduced capsule wall
(normally bistratose), as well as in the development of spurlike tangential
extensions of the columnar thickenings of the radial walls of the outer
layer. The three included holarctic species are also in general more xeromorphic than the other species of the complex, as is obvious from the
guttulate cell net and the most often closely imbricate leaves.

Isopaches has been held to include only the widespread, imperfectly holarctic
L. bicrenata and the local European *L. decolorans* (an alpine species, differing
from *L. bicrenata* in its dioecious inflorescences). However, the New Zealand *L.*
pumicicola Berggr. (N.Z. Hep. 1:21, fig. 15, 1898) also must surely be referred
to *Isopaches.*[109] Like *L. bicrenata,* it is paroecious and often monandrous; the
antheridial stalks are short and 1-seriate. More important, it has a similar
axial anatomy, although with age the medullary cells may be nearly all infected
by fungal hyphae and become brownish. This antipodal species differs from
L. bicrenata in having lobed perichaetial bracts which may bear 1–2 additional
lateral teeth but are otherwise edentate, and in lacking gemmae; it also has
more ovate leaves, clearly longer than broad, with more closely juxtaposed lobes
separated by a sharper sinus.

An allied species, also from New Zealand, is *L. autoica* Schuster (1969b).[110] It
differs from other taxa of *Isopaches* as follows: constantly autoecious; stem

---

[108] The seta is supposed to be formed of 16 epidermal rows of cells, *fide* Buch (1933a, 1942);
there may, however, be up to 21 outer cell rows. *Isopaches* thus does not differ from *Lophozia*
in this regard.

[109] *L. innominata* Hodgs. (Trans. Roy. Soc. N. Z. 76:69, 1946) is a synonym. Hodgson incorrectly states that *L. pumicicola* is "dioicous," although the type (*Lund*) is paroecious; her plant is
described as paroecious.

[110] *Type.* Leith Valley, Morrisons Creek, Dunedin, New Zealand (*RMS 48593a*); see Schuster,
(1969b).

9–11 cells high, but medulla agreeing in lacking any ventral mycorrhizal band; leaves ovate, soft-textured, 0.35–0.45 bilobed, ± squarrose, the lobes often strongly so; cells thin-walled with very small trigones, the median cells 24–32 × 36–48 μ, each with 20–45 small (2.8–4.5 × 3.6–4.8 μ, usually) oil-bodies; gemmae greenish, polygonal. Monandrous; antheridial stalk 1-seriate. Bracteole free from bracts; bracts 2–3–4-lobed, lobes usually entire, occasionally paucidentate below. Perianth crenulate-denticulate with distal cells free for 0.3–1.0 their length.

This species seems to form a transition point between *Isopaches* and *Massula* (the larger, leptodermous cells and numerous smaller oil-bodies suggest *Massula*) and also seems allied to subg. *Protolophozia* (p. 255).

In the Northern Hemisphere, only three species of *Isopaches* occur. One of these, *L. decolorans*, is restricted to the European Alps and Norway; it differs from our two species in the densely imbricate leaves, giving the plant a julaceous aspect, and in the entire-margined to crenulate perianth mouth.

Key to Species of Isopaches

1. Paroecious; leaves orbicular, mostly lunately 0.25–0.35 bilobed, lobes broader than long, usually nearly symmetrical; cells each with (3–5)6–12(15) oil-bodies; leaf lobes mostly subacute or feebly apiculate, usually ending in solitary cells. Perianth mouth lobulate-ciliate, the longer teeth 3–4-celled, slender. Nonarctic; oxylophytic . . *L. bicrenata* (p. 479)
1. Dioecious; leaves subquadrate usually, angularly 0.3–0.5 bilobed, lobes typically as long or longer than broad, the dorsal usually considerably narrower and sharper; cells each with (2)3–6(7–9) oil-bodies; leaf lobes mostly apiculate to sharply acute, ending in (1)2–3 superposed cells. Perianth mouth feebly lobulate, sharply spinose-dentate to shortly ciliate-dentate, teeth 1–2(3) cells long. Arctic; Ca tolerant . *L. alboviridis* (p. 487)

LOPHOZIA (ISOPACHES) BICRENATA (Schmid.)
Dumort.

[Figs. 188, 189]

Jungermannia bicrenata Schmid., Icones Pl. ed. 2, 3:247 or 250, pl. 64, fig. 2, 1797 [not seen].
Jungermannia commutata Hüben., Hep. Germ., p. 192, 1834.
Lophozia bicrenata Dumort., Rec. d'Obs., p. 17, 1835; Evans, Rhodora 4:209, 1902; K. Müller, Rabenh. Krypt.-Fl. 6(1):687, fig. 315, 1910; Macvicar, Studs. Hdb. Brit. Hep. ed. 2:188, figs. 1–5, 1926.
Jungermannia excisa Underw. *in* Gray, Manual of Botany ed. 6:720, 1889 (not of Dickson, 1793), and of other American authors before ca. 1900.
Isopaches bicrenatus Buch, Mem. Soc. F. et Fl. Fennica 8(1932):288, 1933; Frye & Clark, Univ. Wash. Publ. Biol. 6(3):373, 1945; K. Müller, Rabenh. Krypt.-Fl. ed. 3, 6(1):684, fig. 214, 1954.
Lophozia (subg. *Isopaches*) *bicrenata* Schust., Amer. Midl. Nat. 42(3):572, pl. 8, figs. 14–17, 1949; Schuster, *ibid.* 45(1):57, pl. 13, 1951 (as *L. bicrenatus*); Schuster, *ibid.* 49(2):348, pl. 14, 1953.

Creeping, closely attached to substrate, forming thin, often compact sods, *usually gregarious*, more rarely occurring isolated among other bryophytes, *small, compact, stout, and fleshy*, green to reddish brown, odorous. *Shoots 500–1000 μ wide* × 5–10 mm long, prostrate except for the ascending tips, usually unbranched or bearing only 1–2 branches, occasionally with gynoecial innovations. Stems ca. 175–265 μ in diam., *thick and rather fleshy, only 8–9(10) cells high*; cortical cells, in cross section, similar to medullary, rounded, somewhat thick-walled, with rounded, thickened angles, the dorsal ca. 22–30 μ in diam.; medullary cells slightly less thick-walled or thin-walled except at angles, 22–32 μ, similar to cortical cells in diam.; ventral 2–3 cell layers, including the cortical, mycorrhizal with age, somewhat smaller in diam. *Rhizoids· numerous*, long, forming a dense mat to near shoot tips. *Leaves moderately to closely imbricate, shoots consequently often stoutly vermiform*, the leaves somewhat antically secund, obliquely erect to erect-spreading, *somewhat to strongly concave, rounded-quadrate to broadly rotundate*, averaging 1.0–1.2 × as wide as long, almost invariably bilobed (isolated leaves rarely trilobed); sinus descending 0.25–0.3 the leaf length, *crescentic or obtuse* to rectangular; *lobes ± incurved*, broadly triangular, acute to subacute, entire, or rarely with 1–2 isolated teeth. *Cells thick-walled, usually strongly so, more so at angles, the cell lumen often becoming guttulate*, rounded; marginal cells ca. 19–24 μ, the subapical 19–24 × 25–32 μ, median cells 22–27 × 30–34 μ, near base ca. 28–30 μ; cuticle smooth or faintly verrucose; *oil-bodies absent in isolated or many cells of the leaf margins* and lobes, in other marginal and lobe cells small and inconspicuous, 2–6 per cell; *median cells with (3–5)6–12(15) spherical to ovoid oil-bodies*, finely granular and grayish in appearance (in cells with 3–6 oil-bodies, these often large, to 5 × 9–10 μ, but usually smaller; in cells with 10–15 oil-bodies, ca. 4–7.4 to 6 × 9 μ). *Gemmae constantly in large numbers, reddish yellow to reddish brown*, polygonal to stellate, 20–25 μ, somewhat thick-walled and with thickened angles or protuberances, 1–2-celled, at least some cells *with 2–4 large oil-bodies* (and often several smaller ones) which persist after drying. Gemmiparous shoots not strongly ascending, their *leaves becoming erose-lacerate with prolonged gemma formation*, but hardly reduced in size.

Paroecious (but sometimes with supernumerary androecia); *producing perianths (and capsules) usually throughout the growing season;* almost always abundantly fertile, even when freely producing gemmae. Androecia below gynoecia, but sometimes isolated androecia on branches from below the gynoecia (then with small, bilobed, simple ♂ bracts); ♂ bracts of normal androecia imbricate, larger than leaves, somewhat concave, subequally bilobed with ventral lobe somewhat larger, *the margins remotely and obsoletely to distinctly but finely dentate*; 1–2-androus; antheridial stalk 1(2)-seriate. Gynoecia with bracts somewhat similar to ♂ bracts, larger, bilobed or with 1–2 supernumerary lobes, the *lobes usually dentate*, rarely only remotely so; bracteole lanceolate to obcuneate or oblong, sometimes shallowly 2–3-lobed at summit and with a few scattered teeth, united with one (or both) bracts at base. Perianth ovoid to cylindrical-ovoid, ca. 0.6–0.75 emergent, distal half distinctly plicate, cells below mouth mostly 16–19 × 18–27 μ, hyaline, thick-walled; mouth narrowed, *often decolorate, spinose-ciliate with acuminate teeth in large part 3–4 cells long* × 1–3 cells wide at

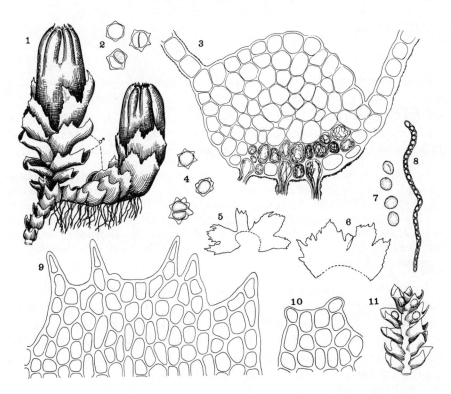

FIG. 188. *Lophozia (Isopaches) bicrenata*. (1) Fertile plant, the ♂ bracts labeled (×ca. 15).
(2) Gemmae (×ca. 260). (3) Stem cross section (×ca. 190). (4) Gemmae (×ca. 260).
(5–6) ♀ Bracts and bracteole (×ca. 12). (7) Spores (×ca. 275). (8) Elater (×135).
(9) Perianth mouth cells (×150). (10) Cells of gemmiparous leaf lobe (×ca. 150).
(11) Androecial innovation from paroecious main stem (×25). [1, 4–8, 10, 11, from
Taughannock Falls, N.Y., *RMS;* 2, Delaware Water Gap, Pa., *RMS; 1943;* 9, Great
Palisade, Minn., *RMS.*]

base, the largest cilia often with 1–2 basal 1-celled teeth; terminal cells of
cilia usually sharply tapering, 35–48 μ long, 3–5× as long as wide. Capsule
ovoid, reddish brown, its *wall 2-, locally 3-, stratose.* Epidermal cells subquadrate
or short-rectangular, ca. 30–35 (if subdivided, 18) μ wide × 27–36 μ long,
each wall with (1)2–3(4) prominent radial bands *which are distinctly extended
across part of the free tangential walls as short spurs.* Inner cell layer of rectangulate
cells, 16(18)20 × 32–50 μ long, their longitudinal radial walls bearing usually
4–6 narrow and delicate radial bands, extending usually for 0.2–0.5 across the
tangential face, but sometimes subcomplete or complete, *forming incomplete or
subcomplete, slender, semiannular bands.* Seta ca. 230 μ in diam., of ca. 16–21
epidermal cell rows, surrounding an innermost group of ca. 4 cell rows and a
peripheral group of ca. 8–12 cell rows, the seta thus ca. 6–7 cells in diam.
Elaters bispiral, ca. 8 μ in diam.; spores 12–15 μ, reddish brown, granulate-
verruculose.

Type. European.

Distribution

An extraordinarily widespread species, ranging across almost the entire holarctic region, from the edge of the Tundra zone, throughout all but the southern periphery of the Deciduous Forest Region. In eastern North America, at least, extending southward into the Lower Austral Coastal Plain. Occasionally in alpine, almost never in arctic, Tundra.

In Europe[111] distributed throughout most of Scandinavia (Norway, Sweden, Finland, Denmark), the Shetland Isls., England and Ireland, throughout almost all of central and western Europe (Germany, Austria, Hungary, Poland, France, the Low Countries), north into Italy, northwest Spain, southern Portugal, and the Balkans (Dalmatia), east to Estonia and Lithuania. Extending eastward into Asia (Siberia: on the Yenisey R. from 60–70° N.; Ob R.). Not known from eastern Siberia or Japan.

In North America possibly transcontinental,[112] in the western half extending from Alaska (Griggs, 1935) to (doubtfully) British Columbia (Brinkman, 1934). Apparently more frequent eastward, where often a weed in eroded and cleared ground, being one of the few members of its family to remain abundant after disturbance by man, and in many cases depending upon such disturbance for maintenance.

LABRADOR. Goose Bay, edge of RCAF Base (*RMS, s.n.*). NEWFOUNDLAND. Near St. John's, Avalon Distr. (Buch & Tuomikoski, 1955). QUEBEC. Near mouth of Great Whale R., E. coast of Hudson Bay (Schuster, 1951); Rivière à Martre; Salmon R., Spider R., and Tadoussac, all in S. Quebec (Lepage, 1944–45); Ft. George, lake N. of Wawicho L., and Roggan R., all ca. 53–54° N. (Lepage, 1953); L. Tesekau, Rupert R., 76° W. ONTARIO. L. Nipissing; North Bay, Algoma (Macoun, 1902); Thunder Bay Distr. (Cain & Fulford, 1948); Dill and Blezard Twps., Sudbury Distr. (*Cain 3052!, 2129!*); Paradise Bay, L. Timagami (*Cain 3005*). NOVA SCOTIA. Halifax (*M. S. Brown*); below Mary Ann Falls, S. of Neils Harbour, Cape Breton (*RMS 42660, 42669, 66316a*); between Campbell and Craigmore, Inverness Co., Cape Breton (*RMS 43979a*); E. of Barrington, Shelburne Co., on Rte. 3 (*RMS 43066a*).
MAINE. Mt. Katahdin (*Evans, 1908*); Mt. Desert I. NEW HAMPSHIRE. Jackson; Crawford Bridle Path (Evans, 1902d); Mt. Washington, ca. 4500 ft (*RMS 23055c*, with *Marsupella ustulata*). VERMONT. Andover (Evans, 1902d); Mt. Mansfield, 4000 ft (*RMS 43813*). MASSACHUSETTS. Cape Cod, near Plymouth (*RMS*); Woods Hole (Evans, 1902d); Leicester; Holden; Nahant (*Seymour 7553!*); Breckenridge Rd., Hadley (*RMS 43903, 43905*); Mt. Toby, Sunderland (*RMS, s. n.*). RHODE ISLAND. North Kingston (Evans, 1906b). CONNECTICUT. W. Goshen (*Underwood, 1889*); Orange, Beacon Falls, Meriden, Seymour, and Hamden, New Haven Co.; Goshen,

---

[111] Reports from Spitsbergen are not confirmed in Müller (1951–58) or in Arnell & Mårtensson (1959).
[112] *Lophozia bicrenata* was reported from north Greenland by Hesselbo (1915); the specimen represents *L. latifolia* (confirmed by the author and S. Arnell).

Litchfield Co.; Bolton, Vernon, Tolland Co.; Huntington, Fairfield Co. (Evans & Nichols, 1908).

NEW YORK. Wildwood State Park, Long I. (*RMS*); Cold Spring Harbor, Long I.; Mt. Moriah, Cattaraugus Co. (*Boehner; det.* Schuster); Tompkins, Seneca, Ontario, Cattaraugus Cos. (Schuster, 1949a); Letchworth Park, Genesee R., Wyoming Co. (*RMS 16549*). PENNSYLVANIA. Near Zelienople, Butler Co.; Philadelphia (*B.B.K.!*). KENTUCKY. Boone, Lewis, Powell Cos. (Fulford, 1934). WEST VIRGINIA. Monongalia Co. (Ammons, 1940). VIRGINIA. Mountain L., Giles Co. (*RMS*); Chesterfield, King and Queen, and Spottsylvania Cos. (Patterson, 1950). OHIO. Hocking Co. (Miller, 1964).

NORTH CAROLINA. Bluff Mt., 4500 ft, near Jefferson, Ashe Co. (*RMS 29537*); Nigger Mt., Ashe Co., 4200 ft (*RMS 29606*); Cascades, near Hanging Rock State Park, Stokes Co. (*Blomquist & RMS*); E. of Reidsville, Rockingham Co. (*Blomquist 10978*); near Durham, Durham Co. (*Blomquist 10216*); State Park, 10 mi NW. of Raleigh, Wake Co. (*RMS 36291*); Falls of Chattooga R., Jackson Co. (*Conard C-2!*); Wayah Bald, ca. 0.5 mi W. of summit, 4900–5100 ft, 8 mi W. of Franklin, Macon Co. (*RMS 39291*); Lake View, Moore Co. (*Blomquist 10220*); near Asheboro, Randolph Co. (*Anderson 7316*); New Hope Cr., Orange Co. (*RMS 28480*). SOUTH CAROLINA: Whitewater R., near Jocassee, Oconee Co. (*RMS, s. n.*). TENNESSEE. Roan Mt., Carter Co. GEORGIA. Brasstown Bald, 4750 ft, Towns Co. (*RMS 34322*); Rabun Bald, ca. 4000 ft, Rabun Co. (*RMS 40687*). MISSISSIPPI. Oxford, campus of Univ. of Mississippi, Lafayette Co. (*RMS*); Tombigbee State Park, near Tupelo, Lee Co. (*RMS M-290*).

Westward extending as follows. WISCONSIN. Amnicon Falls, Douglas Co. (Conklin, 1929). MICHIGAN. Keweenaw [near Central], Marquette, Luce [near Newberry] Cos. (Evans & Nichols, 1935; Steere, etc.). MINNESOTA. Thompson, Carlton Co.; Grand Portage and Pigeon R., Cook Co; Rainy L. near Black Bay, Koochiching Co.; Great Palisade on L. Superior, Lake Co.; Oneota Ravine, Duluth, St. Louis Co. (Schuster, 1953). KANSAS. W. of Linn, Washington Co.; Woodson County State Park, Woodson Co. (McGregor, 1955).

The species is of rare and sporadic distribution in arctic and alpine regions, but frequent in the Coniferous Forest. It is perhaps most common and most nearly "weedy" in the Oak-Hickory and Oak-Pine-Hickory Forests of temperate portions of eastern North America, from Connecticut and New York southward to North Carolina, extending southward even into the Coastal Plain (in North Carolina and Mississippi).

Ecology

The only species of its family with truly "weedy" propensities, becoming very common on acid, leached soils, as on loamy roadside banks, the compacted soil of paths, disused gravel pits, eroded, abandoned farm land, burned and cleared forest land, and even N-free volcanic ash (Griggs, 1935). The plant is often bound to substrates low in organic matter and usually occurs in sites poor in Ca and with a low pH. In eastern North America, it may occur almost wherever broken, leached, partially insolated sloping ground occurs, where the soil is relatively compact, either loamy or clayey, and acid. Frequently associated are *Melampyrum*, *Desmodium* spp., various Ericaceae, particularly *Vaccinium* and *Gaylussacia*, species of *Cladonia*, *Dicranum*, etc.[113] Common on soil adjacent to rock outcrops, where the species may actually occur on shale talus and detritus below cliffs (Taughannock Gorge, N.Y.; with *Cephalozia bicuspidata* and *Gymnocolea inflata*), but equally common on clayey exposed soils in openings in forests. The restriction, in the East, to various phases of the Oak-Hickory Forest is marked, and the species can be classified as almost consistently a xerophyte.

In leached, sterile sites *Cephaloziella* (usually *rubella*; more rarely *byssacea*) is commonly consociated, an easily recognized pioneer *Lophozia bicrenata-Cephaloziella* "associule" resulting (Schuster, 1949a, 1957, p. 286); also commonly associated are *Diphyscium foliosum*, *Polytrichum*, and *Ditrichum* spp. Less commonly found on sterile, sandy soils or depressions in sand plains, with *Cephaloziella* and *Polytrichum*, or around swamps and bogs; in either case, the pH ranges from a maximum of 6.0–5.0 downward. When occurring on soil over cliffs, *Gymnocolea* may be associated in sites where much intermittent seepage occurs, while *Lophozia ventricosa silvicola*, *Tritomaria exsectiformis*, and other oxylophytic taxa are associated on relatively dry cliff faces.[114] When on poorly drained sandy to loamy soil, as around swamps, *Cephaloziella rubella*, *Solenostoma crenulatum*, *Lophozia excisa*, and rarely *L. capitata* may also be consociated. Similar occurrences from around the elevated borders of lakes are known.

Very little is known about the habitat of the species in the periphery of the Arctic. It appears to occur here in wet tundra, as on the "margin of pool in granite near mouth, Great Whale River" among *Anastrophyllum minutum*, *Chandonanthus setiformis*, *Calypogeia*, *Ptilidium ciliare*, and *Cephaloziella? arctica* (Schuster, 1951). In alpine situations, found in compact sods with *Marsupella ustulata* on exposed mineral soil (i.a., on Mt. Washington and Mt. Katahdin), often in only small quantity. There at elevations up to 4500–5000 feet (but in the European Alps recorded as high as 1950 and 2300 m). In northern Quebec occurring similarly on "sandy beaten track" with *Jungermannia pusilla*, *Cephalozia ambigua*, *Nardia geoscyphus*, and *Dicranella subulata* (Lepage, 1953).

Throughout most of its range the species is most common over loamy to gravelly broken soils, associated with some or all of the following species: *Ditrichum pallidum* and *pusillum*, *Scapania nemorosa*, *Solenostoma crenulatum*, *Polytrichum* spp., *Diplophyllum apiculatum*. Southward (i.a., Whitewater R., S. C.;

---

[113] Such sites are particularly common in earlier stages of forest succession. In glaciated areas (as in central New York) often on exposed, leached ravine crests, occurring either in the open *Pinus rigida-P. resinosa* stage, or the oak-chestnut stage immediately following it, associated with an ericaceous understory and shrub layer. Similarly, further southward very common on eroding banks and slopes in the Oak-Hickory Forest.

[114] The association with *Gymnocolea*, *Lophozia ventricosa silvicola*, and various *Cephaloziella* species is, northward, a striking one. It may occur in Deciduous Forest Regions (Schuster, 1949a; Central New York), northward as far as the alpine Tundra zone (Mt. Mansfield, Vt.).

Rabun Bald, Ga.; Wayah Bald, N. C.) *Diplophyllum andrewsii* may replace *D. apiculatum* or occur with it. Northward, on relatively moist sites, on loamy or gravelly north-facing slopes, *Lophozia capitata* may also be consociated, at times, with *Drosera rotundifolia* and *Lycopodium inundatum*. However, occurrences on sites as moist as this are relatively infrequent, the species possessing not only marked toleration of strongly intermittent moisture conditions but also a high level of direct insolation; therefore it is often a primary invader. Associated with this, it usually is more or less brownish- to reddish-tinged and has rather concave, erect, and imbricate leaves. Restriction to essentially inorganic substrates and acidic sites is very marked; in this respect the species differs strongly from the more tolerant *L. excisa*.

An additional marked feature of the species is its ability to tolerate disturbance. Along a disused wood road, through a mixed oak-hickory-birch woods (Hadley, Mass.), I have observed the species almost weekly for several years; here it thrives in spite of being almost constantly walked over. Similarly, in a recently disused gravel pit it appeared within four years after grading!

Differentiation

Once recognized, *L. bicrenata* causes no further difficulty. Both ecological characteristics and facies are distinctive. The small size, compactness (due to the imbricate, concave, somewhat erect and antically secund leaves), closely creeping growth pattern, and almost constant presence of masses of reddish stellate angular gemmae at apices of sterile shoots all lend an unmistakable facies to the plant.

The very thick-walled cells, especially under very xeric conditions [Fig. 189:3-4] ..., the occurrence on acid, sterile habitats, and the paroecious inflorescences (with the bracts and the leaves below the female bracts concave and always distinctly dentate) ... are highly characteristic. Perianths are always developed in abundance, and have a distinctly ciliate-dentate mouth [Fig. 189:6] .... (Schuster, 1953.)

The ciliate-dentate perianth mouth, the free development of the orange-red or rusty-red gemmae, and the thick-walled, guttulate cells separate the species from small compact phases of *L. excisa* (with which there has been much past confusion). Unlike *L. excisa*, fresh plants of *L. bicrenata* are aromatic. Commonly *L. bicrenata* shows development of strongly erose-lacerate leaf margins, as a consequence of prolonged gemmae formation. This is never the case in *L. excisa*, in which gemmae are produced sparingly. Usually *L. bicrenata* has some or many marginal cells lacking oil-bodies (or with them reduced to minute spherules); such cells appear to be nearly or quite dead (Fig. 189:4). In *L. excisa*, marginal cells are closely similar to median cells and have oil-bodies similar to them (Fig. 193:5). Finally, *L. bicrenata* has a stem which is only 8-9 cells high and lacks a "mycorrhizal band" of small medullary cells; by contrast,

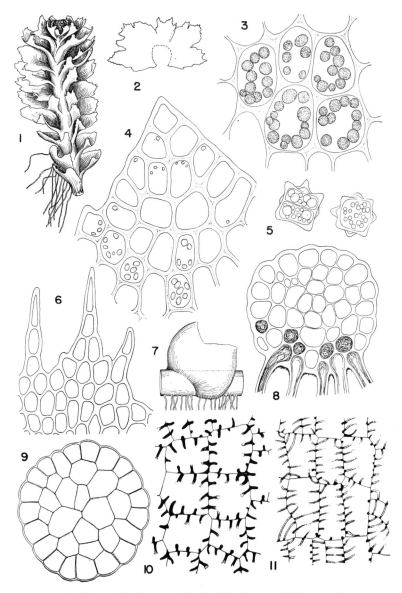

FIG. 189. *Lophozia (Isopaches) bicrenata*. (1) Sterile shoot, with leaves characteristically erose-dentate with prolonged copious gemma formation (×22). (2) Bracts and bracteole (×15). (3) Cells of leaf middle, with oil-bodies (×500). (4) Same, leaf lobe, showing many oil-body-free cells (×300). (5) Gemmae, one with the characteristic large oil-bodies (×440). (6) Perianth mouth cells (×270). (7) Leaf, *in situ* (×28). (8) Stem cross section, mature area, with mycorrhizal infection (×ca. 200). (9) Seta cross section (×ca. 200). (10) Epidermal cells, capsule wall (×355). (11) Inner cells, capsule wall (×355). [1, Delaware Water Gap., Pa., *RMS;* 2, 8–11, Taughannock, N.Y., *RMS;* 3–6, Great Palisade, Minn., *RMS.*]

the stem of *L. excisa* is 15–22 or more cells high and has distinct dorsiventral differentiation of the medulla, the smaller ventral medullary cells becoming mycorrhizal with maturity.

Lophozia bicrenata occasionally produces somewhat spicate accessory androecia (Fig. 188:11). Such androecial branches are rarely formed in abundance, although Schiffner (1913) describes a highly gemmiparous, almost purely masculine phase, in which there is almost total suppression of gynoecia. He also describes a similar phase of *L. excisa*, which, however, should be studied in connection with the closely allied *L. latifolia*.

The antheridial stalk of *L. bicrenata* is usually 1-seriate (Grolle, 1966d) and only exceptionally 2-seriate (although Müller, 1951–58, cites it as 2-seriate).

A distinctive feature of *L. bicrenata* is lack of periodicity in sporophyte production. Although Macvicar (1926) states that it "fruits" from May to July, capsules are produced from May until late October; I have seen them as late as Oct. 25 as far north as Massachusetts. Indeed, sporophyte production throughout the growing season appears to be the rule. Production of sex organs and sporophytes appears to have no effect on gemmae formation, since the latter also occurs throughout the growing season.

LOPHOZIA (ISOPACHES) ALBOVIRIDIS Schust., *sp. n.*

[Fig. 190]

Plantae dioeciae; folia plerumque subquadrata, angulariter 0.3–0.5 bilobata; lobis tam longis quam lati aut longioribus; omnis cellula (2)3–6(7–9) guttas olei habens; lobi foliorum plerumque acuti, in (1)2–3 cellulis superpositis terminati; os perianthii dentes solum 1–2(3) cellulis longos praebens. Plantae arcticae; calcium tolerantes.

Very small (shoots ca. 700–1000 μ wide in vegetative areas, 1300–1500 μ wide at gynoecia × 3–7 mm long), *whitish green*, upper parts of leaves often ± pale, clear brown, rarely with dull reddish pigmentation (at perianth mouth), prostrate with ascending tips, attached by numerous very long, pale rhizoids. Stems as in *L. bicrenata*, 8–10 cells high, brittle and terete, firm-celled, the ventral 2(3) cell layers eventually mycorrhizal but cells firm, not disorganized with age. Leaves loosely to moderately imbricate, typically rather pellucid light green, subvertical to moderately obliquely inserted, rather concave on sterile shoots, occasionally very loosely subcomplicate (plants then with a *Scapaniella*-like facies), mostly obliquely subquadrate, 510 μ wide × 530 μ long (longer ventral half) to 580 μ wide × 540–575 μ long, divided by an *angular* (rarely lunate-angular) *sinus* with rounded base into two *triangular, acute to apiculate lobes, usually asymmetrically* (often strikingly) so, the dorsal lobe shorter and narrower, often longer than wide at base; both lobes ending in *sharp tips formed usually by 2–3 superposed cells*. Cells, particularly of lobes and marginal and submarginal sectors, *thick-walled, smooth*, with rounded angles; cells rather smaller than in *L. bicrenata*, the marginal (tangentially measured) *only 15–18(19–20)* μ; sub-apical variable, from 13–17 × 19–24 μ to 17–22 × 22–27 μ; median becoming considerably larger, (19)20–24(25) × 20–28 μ; median cells and cells at lobe bases usually with only (2)3–6(7–9) oil-bodies [rarely a few cells with 10–15(18)

smaller ones, to 4–4.5 μ only] which are quite large, however, and usually
spherical to very short-ovoid (ca. 5–6 × 6–7 μ to 6.5 × 7–8 μ up to 7–8 ×
10 μ), almost fill the cell lumen, and are much larger than the minute chloroplasts; oil-bodies smooth, grayish, their content finely granular. Gemmae
(1)2-celled, red-brown, sharply stellate or cubical-stellate, 16–20 × 19–23 to
24 × 25–26 μ.

Dioecious. Only young gynoecia known. Subfloral leaves 2–3-lobed, with a few
sharp teeth. Bracts 2–3(4)-lobed, sharply and rather coarsely paucidentate
with sharp, triangular, variable teeth that often end in 2–4 superposed cells,
the teeth formed of rigid, thick-walled cells; bracteole ± ovate to lanceolate,
paucidentate, fused to one bract. Perianth mouth hardly to shallowly lobulate
and spinose-dentate to very shortly spinose-ciliate, the teeth to 2(3) cells long, the
terminal cell 33–38 × 12–14.5 μ, acuminate.

Type. West Greenland: Magdlak, Alfred Wegeners Halvö, 71°06′ N.,
51°40′ W., Aug. 8, 1966 (RMS & KD 66-1375, p.p.).

The type occurred with Scapania and Anastrophyllum minutum.

Again known only from northwest Greenland: Kangerdlugssuak,
Inglefield Bredning, ca. 77°23′ N., 67° W. (RMS 45833c; trace only,
with Cephalozia bicuspidata ambigua, Gymnomitrion corallioides, Prasanthus
suecicus, Anastrophyllum minutum grandis, Scapania praetervisa s. lat.).

Differentiation

This small species is clearly allied only to L. bicrenata and L. decolorans.
I have carefully dissected eight young gynoecia, at the stage where open
and yet unopened archegonia are present, and saw no trace of antheridia
in any subfloral leaves, nor could separate androecial branches be demonstrated; all subfloral leaves were nonventricose. Furthermore, all
gynoecia seen showed no trace of fertilization, suggesting that a dioecious
plant is definitely at hand. This is indicative of L. decolorans, but the
present species differs from this (and approaches L. bicrenata) in the toothed
perianth mouth and much less dense leaves. It is segregated from the
predominantly nonarctic L. bicrenata chiefly on the basis of the criteria in
the key (p. 479).

The plants, found in small quantity only, are if anything a little smaller
than L. bicrenata and mostly lack all trace of secondary pigments (except
at the mouth of young perianths), even though the associated Scapania and
Anastrophyllum showed pigmentation; they have the stellate gemmae
produced in some quantity. The tendency for the leaves to become very
loosely conduplicate on lower sectors of stems, with the sharp and narrow
dorsal half loosely folded over the larger ventral portion, is striking. Even
upper leaves and bracts tend to show this marked asymmetry.

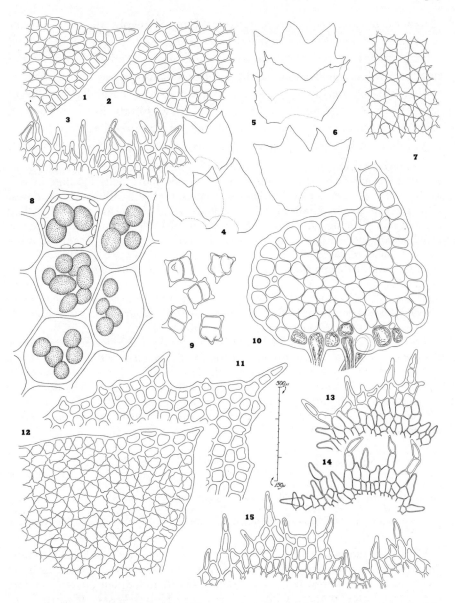

Fig. 190. *Lophozia (Isopaches) alboviridis.* (1–2) Apices of lobes of uppermost leaf in fig. 4 (×170). (3) Mouth of young perianth (×170). (4) Leaves (×38). (5) ♀ Bracts (×38). (6) Subfloral leaf from ♀ shoot (×38). (7) Median cells of collenchymatous extreme (×170). (8) Median cells with oil-bodies and, upper left, chloroplasts (×855). (9) Gemmae (×310). (10) Stem cross section (×250). (11) Upper part of lobe of ♀ bract (×170). (12) Lobe of leaf of collenchymatous extreme (×170). (13–15) Parts of perianth apices, from young perianths (×170). [1–3, 7, 11–15, all drawn to left scale; 4–6, drawn to right scale; all from *RMS & KD 66-1375*, type.]

Leaves, by and large, are considerably more deeply and more asymmetrically bilobed than in normal *L. bicrenata*, with sharper lobes. Leaf cells, particularly the marginal and submarginal, tend to be somewhat smaller, and the leaf shows a diagnostic contrast between the thicker-walled, smaller peripheral cells and a median and basal field of larger cells. With age, all of the smaller cells, i.e., all within the lobes and several rows of marginal cells below the lobes, lose their oil-bodies; on younger leaves these cells have very small oil-bodies (mostly under 3–4 μ in diam.).

The several young perianths studied showed a striking degree of agreement: the mouth is hardly lobulate—and certainly never so distinctly lobulate-incised as shown for *L. bicrenata* by Müller (1951–58, fig. 214e), with the teeth almost all 1–2-celled, a very small minority 3-celled; some perianths show no 3-celled teeth at all. Although *L. bicrenata* exhibits considerable variation in the degree to which the perianth mouth is armed, I have seen none so weakly spinose-ciliate as in *L. alboviridis*.

In spite of the distinctions that emerge from the key and diagnoses, the status of *L. alboviridis* will need further confirmation, after study of more copious material. If, as is conceivable, it proves to be a mere arctic subspecies of *L. bicrenata*, that species would then acquire an extraordinary level of plasticity—far beyond its already phenomenal malleability.

The small fragment from Inglefield Bay is sterile; it is referred here, rather than to *L. bicrenata*, chiefly because of the smaller median cells, only 19–22 × 21–25 μ usually. It agrees perfectly with the type of *L. alboviridis* in oil-body form and size; however, some of the larger median and basal cells may have 10–18 smaller oil-bodies, rather than fewer, larger ones. Both collections have the cuboid-stellate to stellate gemmae with the sharp projections often decolorate.

Subgenus *LOPHOZIA* Dumort. emend. *K. Müll.*

Lophozia Dumort., Rec. d'Obs., p. 17, 1835 (*p.p.*, as genus).
Subg. *Dilophozia* K. Müll., Rabenh. Krypt.-Fl. 6(1):622, 659, 1910, in part (with exclusion of subg. *Massula*); K. Müll., Ber. Deutsch. Bot. Gesell. 57:341, 1939 (in largest part, but after inclusion of *L. excisa*); Schuster, Amer. Midl. Nat. 45(1):57, 1951; Schuster, *ibid*. 49(2):350, 1953.

Medium-large to small, *usually 0.9–2.5 mm wide* × 8–60 mm long, generally prostrate with ascending tips (suberect in sectio Longidentatae and in other species when much crowded). Branching sporadic to frequent, *lateral-terminal; no flagella.*[115] Stem of mature shoots *10–16 or even 18–22 cells high, relatively slender and slightly or hardly flattened,* with *cortical cells usually at least slightly thick-walled;* dorsal cortical cells 0.5–0.9 as wide as basal leaf cells, varying from 2 to 4× as long as wide usually,[116] their diam. essentially identical with that of adjacent dorsal medullary cells; *medulla distinctly dorsiventrally differentiated,* ventral half

---

[115] Exceptionally lateral-intercalary branches occur, as in *L. heteromorpha* (p. 510); I have also seen one instance of basiscopic lateral-terminal branching of the *Radula* type, in *L. excisa* (p. 525).

[116] See Buch, 1933a; in *L. pellucida*, however, to 3–5× as long as wide.

of *small, short, narrow cells* (less in diam. than ventral cortical cells) *soon mycor-rhizal and in part destroyed, becoming brownish,* dorsal half (or more) of large, pellucid cells that remain colorless. Leaves quite *obliquely inserted,* less so in the Longidentatae, obliquely spreading to suberect, *almost uniformly bilobed, little or no wider than long* (usually 0.75–1.25× as wide as long), *relatively firm, entire-margined, not or only loosely imbricate;* postical base never ciliate. *Cells with small to distinctly bulging trigones* but intervening walls remaining thin; *cuticle smooth; cells usually small and firm, relatively opaque,* ca. 20–35 μ in leaf middle (except sectio Excisae); *oil-bodies distinctly granular or papillose-segmented, moderate in size* (4–6 μ to 4 × 9 μ), *mostly 8–24 per cell. Ventral merophytes narrow,* usually 2 (rarely 2–4) cells broad, *unable to produce underleaves* except adjacent to gynoecia or rarely with vestigial ones or stalked slime papillae. *Gemmae produced in all species,* angular, tetrahedral to polyhedral to stellate, 1–2-celled, usually abun-dantly produced from upper leaf apices.

Dioecious (only *L. excisa* paroecious). Antheridial stalk 1- or 2-seriate; paraphyses lacking. Perianth unmodified, *gradually narrowed and plicate* near the crenulate to ciliate mouth; *mouth never beaked.* Sporophyte with *capsule wall 3–4-stratose;* epidermal cells little thicker than those of inner layers; *inner layer with tangential bands largely complete.*

Type. Lophozia ventricosa (Dicks.) Dumort.

Plants of this subgenus are likely to be mistaken for those of other subgenera with bilobed leaves (*Isopaches, Massula, Leiocolea*). *Lophozia* differs from these subgenera in the specialized, dorsiventrally differenti-ated medulla, whose ventral portion forms a small-celled mycorrhizal tissue. In this respect subg. *Lophozia* agrees with *Tritomaria,* with which it may possibly have shared a common origin. It further differs from *Isopaches* and *Massula* in its ability to develop distinct leaf collenchyma. It agrees with both of these subgenera in the essentially smooth cuticle and in the lack of underleaves; in both of these respects, however, it differs from subg. *Leiocolea.* The numerous oil-bodies (mostly 8–24 per cell), smooth cuticle, plicate and erostrate perianths, and abundant development of gemmae further serve to separate subg. *Lophozia* from *Leiocolea.*

The relatively moderate size, more opaque cells, comparatively slender and rigid stem, regularly bilobed leaves whose width never exceeds 1.2–1.4× the length, collenchymatous cells (only slightly so in the Excisae!), and relatively few oil-bodies give the plants an aspect different from that of *Massula,* with which subg. *Lophozia* has been united until recently. The two are admittedly close, with *L. (Massula) hyperarctica* having finely segmented, relatively few oil-bodies, while *L. (Lophozia) pellucida* has the large, pellucid leaf cells and relatively elongated and narrow cortical cells of *Massula.* Nevertheless, separation of the two subgenera is warranted, each taxon possessing a distinctive constellation of characters.[117]

Lophozia excisa seems to form a transition between the subgenera *Lophozia* and *Massula*. Some arctic phenotypes of this species (in particular var. *elegans*; see p. 522) are freely and repeatedly furcately branched; they have chlorophyllose, rather turgid *L. incisa*-like leaves, and—in particular—a thick, considerably flattened stem, elliptical in section. Such extremes of *L. excisa* may easily be mistaken for subg. *Massula*, since they have the facies of sectio Incisae; yet, if the stem is sectioned the early and extensive fungal infection of much or most of the medulla will separate such plants from *Massula*.

Subgenus *Lophozia* (*Dilophozia* of Müller, 1905–16, in largest part) has been varyingly circumscribed. Until exclusion of taxa now placed in *Massula* it was large and unwieldy. Müller (1939), in establishing *Massula*, transferred the *L. excisa* complex to it (i.e., sectio Excisae of this work); Schuster (1949a, 1951a) transferred this back to *Lophozia s. str.*, a procedure followed by Müller (1954, *in* 1951–58). However the subgenus is circumscribed, it is a most difficult and complex group. Experimental work (such as that of Buch, 1933) is often the only final criterion in establishing species or at least genotypes. Species characters are few, and the characteristics which must be relied upon are often untrustworthy, unless utilized with caution and good sense.

Characters that initially appear of primary importance, such as leaf dimension and form (used extensively in Buch, 1936; Schuster, 1951a, 1953), are clearly subject to allometric variation and may show an additional, superposed pattern of differences caused by environmental factors (light intensity, etc.). Of primary taxonomic significance is the genetically determined growth habit (whether autonomously strongly ascending or prostrate), as shown by Buch (1933). However, in actual use, this character is often difficult to apply, since normally prostrate species, with crowding and competition for light, become adaptively ascending or erect in growth. Perhaps equally important is cell size (allowing, i.a., the ready recognition of *L. alpestris* by the unusually small leaf cells). Yet putatively polyploid races (or perhaps microspecies) are being discovered (Arnell, 1950c, etc.) in which the cell size is much larger. Thus, separation of sterile plants of the high arctic race of *L. alpestris* (subsp. *polaris*) with large cells from sterile plants of the normally large-celled *L. excisa* and

---

[117] Müller (1951–58, p. 642) stated that "vielleicht wird *Massula* später, wenn die unnötige Gattungszersplitterung nicht unterbunden wird, als Gattung aufgefasst werden, da sie sich im Gametophyten und Sporophyten von *Dilophozia* (*Lophozia*) unterscheidet." Such a generic distinction (aside from the existence of *L. excisa*) appears impossible for two reasons: (a) existence of the subg. *Protolophozia*—a group which intimately connects *Massula* with, apparently, *Barbilophozia* and *Lophozia*; (b) existence of *Lophozia heteromorpha*, a Greenland species discussed on p. 507. This latter species has the mycorrhizal medulla and collenchymatous leaf cells of subg. *Lophozia* but the polymorphous 2-3-4-lobed leaves of *Massula*, the soft and elongated cortical cells of species of that subgenus, and the distinct underleaves of some species (e.g., *L. obtusa*) of *Massula*.

L. latifolia is at times almost hopelessly difficult. Perhaps the most acute limiting factor in identifying material of this subgenus is the extremely marked "somatic adaptation," i.e., environmental "response." Most widespread species thus exist in varying manifestations, showing differences in color, degree of development of collenchyma, leaf shape and orientation, and general aspect.

As Buch has shown, and as is evident from field experience with the group, phenocopy production is remarkably frequent. This, together with the very few characteristics which actually exist for the discrimination of species, results in a deplorable situation for the average worker. Discrimination of species is, and will remain, a matter of careful judgment, involving an appreciation of the individual variation patterns of the species and the habitat in which a particular individual developed, a study of the living cell, and particularly the development of a high level of "taxonomic intuition"—the almost subconscious result of prolonged struggle and acquaintance with a group.[118]

In order to become familiar with the species of this complex, it is almost necessary to study living plants (to establish oil-body size, form, and number) and (at least at first) to concentrate on the determination of only fertile, copious materials. The reality of species in this complex subgenus becomes readily evident when four taxa (*L. ascendens, L. "silvicola," L. porphyroleuca,* and *L. longidens*) are found intimately admixed, *yet remain clearly distinct* (see p. 549). Yet the demonstration of such reality remains in many cases a matter of the utmost difficulty.[119]

In subg. *Lophozia* gemmae are almost constantly produced, except in *L. porphyroleuca.* Their color is usually independent of the amount of insolation, although the intensity of secondary pigments is affected by light intensity. In species with vinaceous or red-brown gemmae, these secondary pigments are produced in both shade and sun forms; in species with greenish gemmae, the whole plant may become pigmented, while the gemmae remain devoid of secondary, or wall, pigments (see Buch, 1933).[120] In this respect *Lophozia* differs to some extent from *Massula,* in which the gemmae are usually greenish but sometimes (i.a., in *L. capitata*)

---

[118] The fact that two experienced and acute workers on the European hepatics, H. Buch and K. Müller, could not agree as to the status of "*Lophozia silvicola*" suggests that even "taxonomic intuition" does not always overcome the almost overwhelming difficulty of this group.

[119] The taxonomic difficulties increase considerably as one goes northward. In the Arctic completely new series of phenotypes come to hand (Schuster, 1961c, p. 965), and new problems must be dealt with. In essence, concepts based on the limited suite of phenotypes produced in temperate zones become largely inapplicable in the Arctic. For this reason a supplementary key to phenotypes found in the North American Arctic is given (p. 502), based partly on Schuster (*in* Schuster et al., 1959). The status of many arctic phenotypes remains subjective in the absence of transplant experiments and cultivation, side by side, of the various taxa.

[120] Minor exceptions occur with arctic phenotypes of *L. wenzelii* (p. 599).

may become slightly pigmented when the leaves become reddish-pigmented.

In this respect *L. pellucida*, which is here placed in subg. *Lophozia*, behaves like *Massula*. In plants without deep orange-brown pigmentation, the gemmae are pale orange. In this instance, pigmentation pattern and response to insolation are nearly identical in normal leaf cells and in gemmae. Such identity of response is rare in species of subg. *Lophozia;* therefore, gemma color is of fundamental importance as a species character.

As is indicated in the key on p. 495, subg. *Lophozia* is divided into what I consider three sharply definable species groups or sections,[121] which are of interest in that they appear, superficially at least, to show diverse relationships. Sectio Longidentatae appears to connect *Lophozia* with *Tritomaria*. It has the nearly subvertical leaves dorsally often nearly transversely inserted and a leaf shape somewhat reminiscent of *Tritomaria* (especially *T. scitula*). Sectio Excisae at first glance appears to connect *Lophozia* with *Isopaches*, sharing with the last similarly dentate perigonal and perichaetial bracts, a similar leaf shape and leaf insertion, and the inability to develop very distinct trigones. The thick, fleshy stems of the included species, the small size, and the similarity in facies seem suggestive but may be due to convergence. Sectio Ventricosae, of the three sections, appears to occupy the most isolated position. Through *L. alpestris* it may be connected to sectio Excisae, but it has no apparent relatives in any other genus or subgenus.

The three main sections may be briefly reviewed as follows:

SECTIO LONGIDENTATAE. Including only *L. longidens* and *ascendens*; differs from other sections in the narrow and subvertically oriented leaves with drawn-out, narrow lobes; the relatively subtransverse line of insertion of the upper portions of the leaves; the very obviously ascending to suberect gemmiparous shoots; the smaller stems, nearly free of mycorrhizal infection and only 9–12 cells high at maturity, with little dorsiventral differentiation of the medulla; the longly ciliate-dentate perianth-mouth lobes; and the more or less obviously dentate perichaetial bracts.

SECTIO EXCISAE. Including *L. excisa* and *L. latifolia* (and perhaps the rare *L. perssonii*). Characterized by the following constellation of features: (*a*) from generally (6)8–20 oil-bodies per median cell, formed of fine spherules, ranging from 4–6 μ up to 4–5 $\times$ 9 μ; (*b*) slightly or hardly collenchymatous cells, even under xeric conditions, with persistently thin walls; (*c*) relatively large cell size, ranging from 20–25 μ or more on the lobe margins to 30–40 μ in the leaf middle; (*d*) a strongly mycorrhizal ventral, smaller-celled medullary "band" often encompassing most of the medulla; (*e*) little or hardly thick-walled cortex of the stem; (*f*) leaves generally 1.0–1.25 (rarely to 1.5 $\times$) as wide as long, soft in texture; (*g*) reddish, purple or vinaceous gemmae; (*h*) ♀ bracts often with sparse accessory teeth; (*i*) ♂ bracts usually with an antical tooth near the base.

[121] Two of these (Ventricosae and Excisae) are also accepted by Müller (1954, *in* 1951–58, p. 647). He, however, would merge the Longidentatae with the Ventricosae. More recent study has convinced me that the continued recognition of the Longidentatae is well warranted.

A fourth and still poorly understood section is the Heteromorphae (p. 506). The single species, *L. heteromorpha*, may prove to be an atypical taxon of the Ventricosae.

SECTIO LOPHOZIA (= VENTRICOSAE). The largest and most difficult section, including taxa with a decumbent or prostrate growth pattern (unless crowded), much as in the Excisae. It agrees with the Excisae in the generally relatively broad leaves (only rarely as little as 0.75–0.95 as wide as long), with the lobes never drawn out, and in the size and number of oil-bodies per cell. The species have relatively obliquely inserted leaves (at an angle of about 30–45°) with rather broad, often obtuse or rectangular lobes; a decumbent to slightly ascending growth (even of gemmiparous shoots); high stems (mostly 16–22 cells high), with strong dorsiventral differentiation of the medulla; a shortly ciliate-dentate or merely denticulate-crenulate perianth mouth, with teeth merely 1–3, rarely 4–5, cells long; usually entire margins of the ♂ and ♀ bracts, with the antical base of the ♂ bracts normally lacking a tooth. Plants of this section are habitually distinct from those of the Excisae in the smaller and more collenchymatous cells (18–22 μ in the lobes; ca. 20–25 × 25–30 μ in the leaf middle, except in polyploid races) and less fleshy, more rigid texture.

Key to Sections and Species[122] *of Subgenus Lophozia*

1. Plants erect or suberect in growth with gemmae formation; leaves sub-vertical, merely 0.6–0.75(0.85–0.9) as wide as long, rectangular, with narrow lobes; stems with dorsiventral differentiation of medulla slight, only 9–12 cells high at maturity, medulla usually lacking mycorrhizal infestation; perianth at mouth laciniate, the cilia up to 6–8 cells long . . . .
. . . . [sect. Longidentatae] 2.
  2. Gemmae orange- to reddish brown at maturity (even in diffuse light); oil-bodies obviously composed of numerous, somewhat protuberant oil-globules . . . . *L. longidens* (p. 531)
  2. Gemmae yellowish green to yellowish (even in intense light); oil-bodies composed of difficultly perceptible oil-globules, appearing nearly smooth
. . . . *L. ascendens* (p. 541)
1. Plants creeping in growth (except shoot tips, and where strongly crowded), the gemmiparous shoot apices merely ascending; leaves ± oblique, 0.75–1.35(1.45) × as wide as long, ovate-rectangular to quadrate or orbicu-lar, with broad, obtuse or triangular lobes; stems 16–22 cells high usually, with strong dorsiventral differentiation of medulla; perianth mouth crenulate or with teeth 1–4 cells long (exc. *L. perssonii*) . . . . 3.
  3. Paroecious or dioecious; ♂ bracts with sharp tooth at antical base; gemmae deep purplish, often relatively infrequently produced; perianth mouth merely *crenulate* by the *rounded tips* of projecting *thin-walled* cells, *never sharply dentate; cells leptodermous,* developing ± *small, concave-sided* trigones even under xeric conditions; leaves brittle and fleshy to *soft-textured,* on robust stems 1.0–1.3(1.45) × as wide as long; cells of leaf lobes mostly 23–28(30) μ . . . . [sect. Excisae] 4.

[122] I have taken the unparalleled step of giving three keys to the chief taxa (species; varieties), since the group is almost hopelessly complex and hardly lends itself to organization in key form. Gemmiparous plants are often best organized in the second key (p. 499); arctic phenotypes, which pose special problems (p. 493), are best organized in the last key (p. 502), in which are treated certain exclusively arctic, perhaps polyploid, large-celled extremes which have been omitted in the first two keys.

4. Dioecious; ± soft to flaccid; cells of lobes 24–28(30–32) μ; spores
14.5–18 μ; ♂ bracts in many pairs, small, forming slender spikes,
without marginal teeth except for a single basal tooth; always
monandrous; gemmae rare, mostly obtusely polygonal. · Rare;
chiefly arctic . . . . *L. latifolia* (p. 525)

4. Paroecious; often fleshy and brittle; cells of lobes 20–25 μ on margins;
spores 12–17 μ; ♂ bracts similar to ♀ bracts, large, in 1–2 pairs, more
or less sharply dentate, very broad, usually 2–3-androus; gemmae
frequent to common, often sharply stellate. Very widespread. . . .
. . . . *L. excisa* (p. 511)

3. Dioecious; ♂ bracts unarmed at base, merely bilobed; gemmae in most
species yellowish green;[123] both ♂ and ♀ bracts with lobes usually entire;
perianth mouth generally with at least scattered sharp, thick-walled
teeth 1–2(3–4) cells long; plant mostly 1–4 cm long, often brownish- or
reddish-pigmented . . . . [sect. Lophozia]   5.

5. Gemmae greenish to nearly colorless (even in dark pigmented sun
forms);[124] leaves on well-developed sterile shoots from subequally long
and wide to longer than wide (or leaves strongly concave) . . .   6.

6. Leaves normally all 2-lobed; underleaves consistently lacking;
without purplish pigments of distal parts of leaves. . . . . .   7.

7. Leaves ± strongly concave, saucer-shaped or cupped (on sterile
shoots), 1.05–1.25× as wide as long, essentially orbicular to
orbicular-quadrate, widest at or usually slightly above middle,
mostly with strongly arched sides, with ± incurved, obtuse
lobes, divided 0.12–0.2 by a *very wide* crescentic or obtuse,
shallow sinus; often brownish (red pigments, exc. on venter of
stem, normally absent); cells 22–25 μ in leaf lobes, 20–25 ×
30 μ in leaf middle or larger.[125] Arctic-alpine to subarctic.   8.

8. Oil-bodies very variable in number, size, and shape, (2–3)4–
9(10–12) per cell, in large part ovoid and up to 4–5 × 7–9 μ,
formed of spherules of uniform size, appearing papillose;
leaves generally strongly saucer-shaped, mostly subvertical
and often suberect (toward shoot apex). *L. wenzelii* (p. 595)

8. Oil-bodies 15–25 per cell, spherical or subspherical, usually
4–5 μ and similar in size, biconcentric, with a large to small,
distinct, highly refractive central globule, nearly smooth
externally; leaves weakly to moderately saucer-shaped, quite
obliquely inserted and antically secund (rather than suberect)
. . . . *L. groenlandica* (p. 587)

7. Leaves flat or scarcely concave (but becoming somewhat
canaliculate) with distinctly spreading lobes,[126] 0.75–1.05(1.10) ×
as wide as long, oblong-ovate to rotund-ovate, rarely rotundate,

---

[123] Reddish brown to brown in *L. alpestris* (with cells 18–20 μ in lobes, and small gemmae,
only 16–20 × 16–24 μ) and orange to yellowish brown in sun forms of *L. pellucida* (with a
characteristic pale brownish or golden pigmentation); reddish to red-brown also in the arctic
L. rubrigemma and *L. alpestris polaris*, plants with very large cells.
[124] Very rarely slightly fuscous-tinged (but never reddish) in arctic phases of *L. ventricosa s. lat.*
[125] Larger in *L. wenzelii* var. *lapponica*.
[126] Except, rarely, in the extreme mod. *parvifolia-densifolia-colorata* of *L. ventricosa* var. *confusa*.

widest at or below middle, divided usually (0.2)0.25–0.3 by an obtuse to rectangular (rarely crescentic) sinus; sun forms often locally reddish or carmine-tinged; cells of leaf lobes 16–22 or 25 μ . . . . 9.

9. Perianth mouth with teeth in large part 2–4(5) cells long; leaves narrow, 0.75–0.95 as wide as long, even on plants from insolated sites; cells (even in mesophytic extremes) normally strongly collenchymatous, with very large, bulging trigones; usually with 5–10 oil-bodies per cell. (Only on decaying logs and humus; gemmae usually absent or very few; leaves usually red at postical base) . *L. porphyroleuca* (p. 550)

9. Perianth mouth with teeth only 1–2(3–4) cells long; cells usually moderately collenchymatous, trigones varying from little to not bulging; oil-bodies usually 10–22 per cell; gemmae always abundant . *L. ventricosa s. lat.* (p. 558) 10.

   10. Leaves 0.75–0.95(1.0) as wide as long, even in densely insolated sites, pale green in shade, bright red in sun; oil-bodies chiefly circular or nearly, (12)15–22(24) per cell, each consisting of a large, inner, highly refractive sphere surrounded by a mantle of obscure, smaller spherules; usually small . . . .
   . . . . *L. ventricosa* var. *silvicola* (p. 570)

   10. Leaves 0.95–1.05× as wide as long (except juvenile plants from densely shaded areas), ± deeper green (sometimes reddish-tinged); oil-bodies more or less ovoid, (5–9)10–16 per cell, larger, each consisting of a greater number of smaller spherules (without a larger sphere within, in all but exceptional cases) . . . . . 11.

      11. Perianth mouth not or shallowly lobulate, with teeth normally 1(2)-celled, never of strongly elongated cells; leaves divided for 0.2–0.3 their length, lobes never rectangulate; cells (isolated smaller than normal cells omitted) with 9–16(20) oil-bodies. 12.

         12. Plants small, 1–2 mm wide × 1–2.5 cm, rarely reddish-tinged; perianth less than 1 × 3 mm in size; mouth closely dentate; leaves smaller, softer, sinuses never reflexed; gemmae abundant
         . . . . . . *L. ventricosa* var. *ventricosa* (p. 565)

         12. Plants larger, 2.5–4 mm wide × 2.5–6 cm long; perianths from 1 × 2.5–5 mm long, longly exserted, carmine-pigmented; mouth not lobed, often with remote 1-celled teeth; leaves ± plicate, with ± gibbose sinuses, stiff; gemmae few, often sporadic. Rare, montane and arctic
         . . . . . . *L. ventricosa* var. *longiflora* (p. 579)

      11. Perianth mouth divided into many narrow, small lobes, mostly ending in 2–3(4)-celled elongated teeth or short cilia, the lobes often with additional

1(2)-celled teeth; leaves variable, divided 0.2–0.3
or sometimes 0.3–0.4 their length, the lobes then
angulate; cells, even the larger, mostly with 5–10
(in isolated exceptional cells 11–13) oil-bodies; most
often with little or no reddish pigments, even on
perianths. Arctic. *L. ventricosa* var. *confusa* (p. 581)

6. Leaves 2–3(4)-lobed, 1.1–1.3× as wide as long, wavy but not
cupped, the bilobed ones often with a sharp tooth toward the base;
occasional lamellate underleaves; plants soft, lax, with a mixture
of brownish and purplish pigments of distal parts of leaves (never of
base) . . . . [See sectio Heteromorphae] *L. heteromorpha* (p. 507)

5. Gemmae becoming reddish, orange-brownish, or reddish brown (at
least in sun forms); leaves brownish to golden-brownish, except in
shade forms, orbicular to transverse-oblong, 0.95–1.5× as wide as
long on mature shoots; cells either very small (18–20 μ) in leaf lobes
and on margins or very large (23–50 μ); oil-bodies uniformly formed
of small spherules, without a central highly refractive oil-globule . 13.

13. Gemmae *reddish brown or reddish* (*even in shade forms*), normally much
deeper or more intense in color than leaf cells; cells of leaf apices
and margins very small, 18–20(21) μ (in arctic phenotypes to
30–36 μ), in leaf middle ca. 20 × 25 μ (in arctic phenotypes to
34–38 × 35–42 μ or even larger); leaves typically subrotund
and 0.95–1.25× as wide as long; gemmae 16–20 × 16–24 μ to
23–26 × 23–27 μ, or smaller . . . . 14.

14. Perianth mouth ciliate, with teeth mostly 2–5(10) cells long
and to 180 μ long; gemmae 12–19 × 17–21 μ, with 1–2
large, yellowish, persistent oil-bodies in each cell; ♀ bracts
with lobes often ± coarsely toothed. (Small, to 1 mm wide ×
5 mm long; strictly calciphilous) . . . *L. perssonii* (p. 635)

14. Perianth mouth denticulate, teeth of 1–3 cells only; gemmae
without large, persistent, yellowish oil-bodies; ♀ bracts with
lobes edentate (except in the large-celled *L. rubrigemma*).
(Larger plants; avoiding Ca-rich areas, except sometimes *L.
alpestris* subsp. *polaris*) . . . . 15.

15. Cells of leaf margins always under 30 μ; median cells
25–30 × 25–35 μ or less; perianth mouth largely with
teeth of 2–3 superimposed cells, the terminal cells not
greatly enlarged (under 20–24 × 52–62 μ); often with
vinaceous pigmentation, at least of leaf bases; ♀ bracts
with lobes edentate, never crispate. *L. alpestris*[127] (p. 607)

15. Cells of leaf margins (26)30–36 μ or larger; median cells
(32)34–43 × 35–45 μ, each with 16–25(30) oil-bodies;
perianth mouth armed with 1–2-(rarely a few 3-) celled

---

[127] In the high Arctic largely replaced by the large-celled subsp. *polaris*. This approaches
L. pellucida in the broader leaves and deeper sinus, as well as the larger cells and gemmae. It
differs at once in the intensely reddish, rather than golden or golden-brown, gemmae; see
p. 618.

teeth, distal cells very thick-walled and large, 20–28 ×
50–72 to 25–30 × 75 μ; without vinaceous pigmentation,
except sometimes of postical leaf bases and apices of
perianths; ♀ bracts with 2–4 lobes more or less dentate,
somewhat crispate. Arctic only.   *L. rubrigemma* (p. 621)

13. Gemmae identical in color with leaves, varying from greenish
(rare shade forms) to golden brownish; cells very large, on leaf
margins and in apices 35–40 μ (only 23–32 μ in var. *minor*), in leaf
middle 35–42 μ wide (25–30 μ in var. *minor*); leaves typically
broader than long and 1.1–1.5× as wide as long or more; gemmae from 25–28 × 28–30 to 32–35 × 35–40 μ. Arctic . . . . .
. . . . *L. pellucida* (p. 625)

Artificial Key to Species and Varieties of Subgenus Lophozia: Gemmiparous Living Material[128]

1. Gemmae reddish brown, reddish yellow, red, or purplish in color (usually
even in shade forms)   .   .   .   .   2.
   2. Plants autonomously erect or suberect;   leaves subvertical, ovate-rec-
   tangulate, usually 0.65–0.85 (rarely to 0.95) as wide as long, often ±
   parallel-sided, with straight, hornlike lobes . . . . *L. longidens* (p. 531)
   2. Plants prostrate or creeping except where crowded, only shoot apices
   suberect; leaves ± horizontal, broad, 0.95–1.5(1.7) × as wide as long,
   with broad, obtuse to rectangular or rounded lobes . . . . 3.
      3. Marginal cells of nongemmiparous lobes averaging less than 30 μ;
      median cells averaging less than 34 μ wide;   leaves with acute or
      subacute lobes 4
         4. Cells larger, usually with minute to small trigones: apical and
         marginal cells (22)23–30 μ, median cells 24–30 μ wide; gemmae
         largely over 23–26 × 24–28 μ . . . . 5.
            5. Paroecious;   almost constantly fertile;   plant soft-textured,
            fleshy, ± brittle; widespread and abundant, polymorphous;
            ♀ bracts with sharp teeth . . . . *L. excisa* (p. 511)
            5. Dioecious, often largely sterile; plants usually firmer;   rare,
            largely or entirely arctic   . . . . 6.[129]
               6. Gemmae (usually much) deeper in color than the leaf cells,
               red to purplish; rarely on strongly calcareous sites . . 7.
                  7. Gemmae always present, reddish to carmine; perianth
                  mouth with scattered 1–2-celled teeth; ♂ bracts without

---

[128] With the highly technical nature of most taxa in subg. *Lophozia*, sterile (or sometimes even fertile) dead material is often essentially indeterminable. This key is intended to meet the need for some means of determining (*a*) sterile material;   (*b*) the highly derivative, often polyploid arctic taxa. Existing keys such as that of Müller (1954, *in* 1951–58) quite fail in this respect. The student should be warned against attempting to use the "Comparison" in Frye & Clark (*loc. cit.*, pp. 342–343), since some characters given are only temporal (position of androecia) or environmental (color of rhizoid bases, etc.), while others (gemmae color) are confused. See in this connection the discussion of this "Comparison" in Schuster (1951a, footnote, pp. 58–59).

[129] Plants probably belonging to *L. perssonii* have been found a single time in east Greenland; they would key here. For their separation see the following key to arctic phenotypes.

antical tooth; leaves 0.2–0.25 bilobed; plants green or
slightly brownish; almost always sterile . . . .
. . . . *L. alpestris* subsp. *polaris* (p. 614)
7. Gemmae rare, vinaceous or dark purplish; perianth
mouth crenulate with thin-walled, projecting, rounded
cells; ♂ bracts with antical base armed with a sharp
tooth; leaves 0.3–0.5 bilobed (exc. weak and ♂ plants);
plants becoming somewhat reddish brown in sun; usually
fertile (and with slender, long androecia) . . . .
. . . . *L. latifolia* (p. 625)
6. Gemmae nearly identical in color with leaf cells, pale
yellowish brown to golden. Habitually sterile; obligate
calciphyte . . . . *L. pellucida* var. *minor* (p. 634)
4. Cells small and opaque, with large to bulging trigones: apical and
marginal cells 18–20 μ, median cells averaging ca. 20 × 25 μ;
gemmae 16–20 × 16–24 μ, reddish brown; dioecious. Widespread
in subarctic–subalpine to arctic-alpine . . *L. alpestris* (p. 607)
3. Marginal cells of mature leaves 30–36 to 35–40 μ; median cells from
(32)34–42(45) to (35)40–45(55) μ; gemmae large, 23–26 × 23–27 to
(25)28–32 × 30–36 μ. Plants soft-textured. High arctic (70–82° N.)[130]
. . . . 8.
8. Plants calcicolous; a clear orange- to chestnut-brown (in sun
forms), with gemmae similar in color to lobes of upper leaves or
± orange-yellow; no trace of vinaceous pigmentation; marginal
cells of lobes 35–40 μ, median cells 35–42(45) × 40–50(55) μ;
large forms with occasionally blunt or even rounded leaf lobes,
often with blunt to crescentic sinuses . . . .
. . . . See *L. pellucida* [couplet 18]
8. Plants of acid substrates; with distal portions of leaves yellow- to
chestnut-brown, sometimes reddish-tinged, postical leaf bases and
perianths often vinaceous; marginal cells of lobes usually (26)30–36
μ, median cells (32)34–43 × 35–45 μ; leaf lobes consistently acute
and sharp-pointed, sinuses usually acute. . *L. rubrigemma* (p.621)
1. Gemmae with walls colorless, thus appearing greenish; dioecious; male
bracts without a tooth on antical margin . . . . 9.
9. Mature leaves oblong to oblong-ovate, with antical and postical margins
little arched, 0.6–0.95(1.0) as wide as long, divided by an angular sinus
for ¼–⅓ their length . . . . 10.
10. Plants suberect in growth, small, usually 3–6 mm high; vegetative
leaves not with a reddish or carmine base, often only 0.6–0.8 as wide
as long, with hornlike lobes; perianth mouth ciliate-laciniate.
. . . . *L. ascendens* (p. 541)

---

[130] The large-celled phenotypes of subg. *Lophozia* in which pigmented gemmae are developed
are not well understood. In addition to the two species keying here, *L. longidens* subsp. *arctica*
may be sought here; this differs in the narrow leaves with long-drawn-out lobes. Also. *L.
alpestris* subsp. *polaris* may be sought here; this has marginal leaf cells of the lobes 25–30 μ in
tangential diam.

10. Plants prostrate (except when crowded), 1–3 cm long usually; vegetative leaves (at least in exposed phases) with bright red to carmine postical bases, usually 0.8–0.95 as wide as long, with obtuse to rectangular lobes . . . . 11.

    11. Oil-bodies (4–7)8–12 per cell, formed of distinct, small globules throughout, papillose in appearance, irregular in shape, often ovoid; postical leaf base becoming carmine even in phases growing in deep shade; gemmae not produced except in plants of wet, shaded sites; perianth mouth with some teeth 3–5 cells long . . . . *L. porphyroleuca* (p. 550)

    11. Oil-bodies (12)15–24 per cell, biconcentric, smooth or nearly so externally, usually ± spherical; postical leaf base green in shade, brighter red in sun forms; gemmae always abundant; perianth mouth with 1(2)-celled teeth . . . . *L. ventricosa* var. *silvicola* (p. 570)

9. Mature leaves rounded-quadrate to rotundate, to transverse-oblong, their margins usually ± rounded, (0.95)1.0–1.2 × (rarely to 1.4–1.7 × in arctic phases) as broad as long; prostrate (except when crowded); perianth mouth with teeth normally 1–2- (rarely to 3–4) celled . . . 12.

  12. Gemmae 20–24 × 25–30(32) μ, or less, uniformly yellow-green or greenish; leaves with cells of apex to 20–23 μ usually, on margins; median cells to 23–26 × 24–32 μ; leaves always with acute to bluntly angular lobes . . . . 13.

    13. Leaves quite polymorphous, 2–3(4)-lobed, the lobes irregular in size and shape on many leaves; ventral merophytes often 4–6 cells wide or more, then producing (on sterile shoots) linear to lanceolate underleaves. Soft-textured arctic plants with aspect of Excisae . . . . *L. heteromorpha* (p. 507)

    13. Leaves rather uniformly 2-lobed (exceptional 3-lobed leaves on some shoots); the lobes rather uniform, except antical usually slightly smaller; ventral merophytes usually 1–2 cells wide and lacking underleaves (sessile or slightly stalked slime papillae aside) . . . . 14.

      14. Oil-bodies variable in size and form, largely ovoid, formed of numerous protuberant spherules, papillose-segmented in appearance, without a central refringent globule . . 15.

        15. Leaves (except in plants from wet sites) concave, saucer-shaped, their lobes strongly incurved (cannot be flattened without tearing); leaves subtransversely oriented, bilobed by a shallow, crescentic sinus. Arctic-alpine . . . . *L. wenzelii* (p. 595)

        15. Leaves from nearly flat to somewhat canaliculate (in forms from dry sites), their lobes not or hardly incurved; leaves quite obliquely inserted and oriented, usually laterally patent to antically slightly secund, bilobed ¼–⅓ by a rectangulate (rarely crescentic) sinus. *L. ventricosa* 16.

          16. Plants moderate in size, 1–2 mm wide; perianths 2–3 mm long . . . . 17.

        17. Perianth mouth with cilia formed of to 2–4 cells;
        median cells with $(3–5)6–12(13–15)$ oil-bodies.
        Arctic . . . . *L. ventricosa* var. *confusa* (p. 581)
        17. Perianth mouth with teeth of 1–2(3) cells;
        median cells with $(10)11–18$ oil-bodies each.
        Widespread. . . . .
        . . . . . *L. ventricosa* var. *ventricosa* (p. 565)
    16. Plants robust, 2.5–4 mm wide; perianths 4–5 mm
    long. Rare and arctic-alpine . . . .
    . . . . *L. ventricosa* var. *longiflora* (p. 579)
  14. Oil-bodies biconcentric and subspherical (exceptional ones
    ovoid and to $5 \times 9\,\mu$), with glistening central oil globule;
    leaves strongly concave and bilobed by a shallow, crescentic
    sinus (often cannot be flattened without tearing), quite
    obliquely inserted and strongly antically secund . . .
    . . . . *L. groenlandica* (p. 587)
12. Gemmae $(25)28–32 \times 30–36\,\mu$ or larger, usually slightly yellow-
brownish (green in shade forms only!); cells of leaf apices $23–30\,\mu$ or
larger, of leaf middle $25–30 \times 30–40\,\mu$ or larger; leaf lobes often
blunt or rounded . . . . *L. pellucida* 18.
  18. Cells of leaf apices $35–42\,\mu$, of leaf middle $(32)35–42(45) \times$
    $40–50(55)\,\mu$; leaf lobes usually blunt or rounded, often with a
    gibbous sinus; plants extremely pellucid . . . .
    . . . . *L. pellucida* (typical) (p. 625)
  18. Cells of leaf apices $23–27 (30–35)\,\mu$, of leaf middle $25–30 \times$
    $30–38\,\mu$; leaf lobes usually subacute to blunt, the sinus rarely
    gibbous; less pellucid . . . . *L. pellucida* var. *minor* (p. 634)

Supplementary Key to Arctic Taxa of Subgenus Lophozia

1. Gemmae yellow-brown or orange to reddish to red-brown with maturity
(typically constrasted to the often $\pm$ greenish leaves); dioecious or paroecious . . . . 2.
  2. Plants (unless closely crowded) prostrate to procumbent in growth; leaf
  lobes triangular to broadly triangular, as wide as or wider than long,
  rectangulate (exceptionally $\pm$ acutely triangular) to obtuse or occasionally
  rounded; leaves orbicular or orbicular-quadrate to broadly oblong,
  nearly as wide as or wider than long; usually with local reddish or brownish pigmentation (when grown exposed) . . . . 3.
    3. Almost always fertile; gemmae usually rare or absent; perianth
    mouth evenly crenulate with cells not projecting for more than 0.5
    their length; cells $\pm$ leptodermous, with minute to small (rarely
    conspicuous) trigones; oil-bodies almost all equal in size, small, mostly
    18–30 per cell; cells relatively large ($23–30\,\mu$ on margins and in
    lobes); never with vinaceous pigmentation of postical leaf bases but
    distal parts of leaves (or their margins) $\pm$ dull purplish . . . . 4.
    4. Paroecious; almost always with mature perianths present; ♂

bracts, or most of them, 2–3-androus; without narrow, slender antheridial spikes; usually succulent, brittle, very fleshy, the larger (green or green and carmine) phases crispate; gemmae (except on fertile plants) sometimes frequent . . . . *L. excisa*

4. Dioecious; ♂ bracts normally all 1-androus; with narrow, slender antheridial spikes; usually less stout, not succulent or brittle, with aspect of *L. alpestris* fo. *sphagnorum*; leaves soft and lax, divided 0.3–0.5 by an angular or sharp, often gibbose sinus; gemmae on fertile phenotypes rare or absent . . . . *L. latifolia*

3. Usually sterile (or with unfertilized gynoecia with rudimentary perianths); gemmae always abundant; perianth mouth setulose or denticulate to ciliate with teeth largely formed of at least 1 or 2 elongated, often ± thick-walled cells; plants normally not very fleshy . . . . 5.

5. Cells of leaf margins large, pellucid, on mature plants $(25)28–36 \mu$ or larger, in leaf middle $(32)34–38 \times 35–42 \mu$ or larger; without vinaceous pigmentation, except sometimes of postical leaf bases and perianth mouth (the leaves ± golden to brownish) . . . 6.

6. Cells of leaf margins $(25)28–36 \mu$ or sometimes larger; median cells each with 16–25(30) oil-bodies; leaf lobes acutely triangular; leaves no or hardly wider than long; gemmae $23–27 \times 23–31$ to $32–38 \times 38–45 \mu$, bright to deep red (even when plant is a mod. *viridis*). Perianth mouth mostly armed with teeth that are 1–3-celled, the terminal cells very thick-walled, enlarged, $20–28 \times 50–72$ to $23–30 \times 75–88 \mu$. ♀ Bracts 2–4-lobed, the lobes usually crispate, with ± conspicuous teeth. Acid substrates, usually between rocks . . . . *L. rubrigemma*

6. Cells of leaf margins $(30)35–40 \times (33)35–50 \mu$; median cells each with (7)9–18 oil-bodies; leaf lobes broadly triangular, often blunt or rounded; leaves much wider than long on mature shoots; gemmae $25–26 \times 28–35$ up to $34–35 \times 36–43 \mu$, orange to pale orange-yellow (in sun forms; their color identical to that of leaf cells; in mod. *viridis* hardly pigmented). Perianth mouth crenulate to shortly ciliate, apical cells thin-walled, ca. $22 \times 50–70 \mu$, not tapered, not unusually enlarged. ♀ Bracts 2–3-lobed, lobes unarmed. Only over extremely calcareous substrates, usually on clay-shale . . . . *L. pellucida*

5. Cells of leaf margins (except near gemmiparous areas) always under 30μ; median cells $(24)25–30 \times (24)25–40 \mu$ or less in size; perianth mouth armed, in large part, with teeth formed of 2, 3, or even more superposed cells, the terminal cell not exceeding $20 \times 52 \mu$ ($24 \times 62 \mu$ in extreme cases) . . . . 7.

7. Perianth mouth with 1–2–3-celled teeth; oil-bodies variable in number and size, (3)4–14(16) per cell. Usually with vinaceous pigmentation, at least of leaf bases. ♀ Bracts with lobes edentate, rarely crispate; usually sterile. Usually on noncalcareous substrates . . . . *L. alpestris s. lat.* 8.

8. With ± concave leaves and a mostly shallow, lunate sinus; not amidst *Sphagnum*; oil-bodies variable in size, many irregular to ellipsoidal, 4–6 × 6–8(9) μ, (3)4–15 per cell.   9.
  9. Marginal cells of leaf tips (14)16–20 μ; gemmae (15)16–18(20–24) μ. Oil-bodies (3)4–10(12) per cell, nearly occluding lumen. Ca-intolerant . . . .
  . . . . *L. alpestris* subsp. *alpestris*
  9. Marginal cells of leaf apices 23–28(30) μ; gemmae 22–28 × (17)20–26 μ or larger. Oil-bodies 8–17 per cell, mostly 10–13. Ca-tolerant . . *L. alpestris* subsp. *polaris*
8. With flat, ± remote leaves, often divided by an angular sinus for 0.2–0.4 their length; often amidst *Sphagnum*; oil-bodies small, of few coarse segments, 3–3.6 × 4.5 to 4.5 × 6 μ, occurring (3–6)7–14(16) per cell usually; always with highly vinaceous leaf bases, often almost blackish; with *crispate* ♀ *bracts irregularly lobed*. (Marginal cells ca. 22–25(27) μ, the median 24–27 × 30–35 μ.) . . . *L. alpestris* fo. *sphagnorum*
  7. Perianth mouth ciliate with cilia 2–5(10) cells long (terminal cell never enlarged); oil-bodies variable in size, some to 4 × 9 μ, 3–8 per cell. Bright green, except for the contrastingly red gemmae, the postical leaf bases not vinaceous. ♀ bracts irregularly paucidentate, but not crisped. On calcareous substrates.
  . . . . *L. perssonii*
2. Plants erect to strongly ascending, with ± subtransverse leaves; leaf lobes long drawn out, hornlike to narrowly triangular, acute (omitting from consideration the very tip, which is usually blunt through gemmae formation) but sinus ± rounded at base; leaves oblong to oblong-ovate, 0.35–0.45(0.5) bilobed, narrower than wide; pale green to green, rarely ± brown. Leaves suberect at base, often with ± squarrose lobes . . .
  . . . . *L. longidens*   10.
  10. Cells relatively small: marginal 20–26 μ, median 24–27(29) × 25–30 μ; with 6–12 polymorphic oil-bodies per cell, many ovoid to ellipsoidal and up to 4.5–5 × 8.5–10 μ, appearing nearly to occlude lumen at times; gemmae 20–24 × 22–30(33) μ . . . .
  . . . . *L. longidens* subsp. *longidens*
  10. Cells large: marginal mostly 27–30 μ, median 30–35 × 34–42 μ; with (9)13–24 smaller (to 3.5–5 × 4.5–8.5 μ), usually spherical, rarely ellipsoidal, oil-bodies; gemmae 25–34 to 31 × 34 to 36 × 42 μ
  . . . . *L. longidens* subsp. *arctica*
1. Gemmae green or whitish to yellowish green, *even when leaf lobes deeply brownish-tinged, purplish to fuscous;*[131] bases of leaves exceptionally with reddish pigmentation, occasionally with bluish purple color . . . . 11.
  11. Leaves uniform: almost all bilobed; usually without vinaceous pigmentation of lobes (in arctic phenotypes); relatively firm plants; no underleaves . . . . 12.

[131] See the exceptions noted under *L. ventricosa* (which rarely has infuscated gemmae) and *L. wenzelii* (sun forms of which may have salmon-tinged gemmae).

12. Leaves typically nearly flat to weakly concave or weakly canaliculate: the lobes not or rarely feebly incurved;[132] sinus descending
(0.15)0.2–0.35(0.45) the leaf length, often angular . . . . .
. . . . *L. ventricosa* 13.
13. Cells with (3–5)6–12(13–15) or 11–18 polymorphous, elliptical
to fusiform to ovoid to spherical oil-bodies, each formed of
minute spherules; leaves usually 0.9–1.1 × as wide as long,
ovate-subquadrate to rotundate, with arched, convex sides, the
sinus often subrectangulate to obtusangular . . . . . . 14.
    14. Median cells each with (3–5)6–12(13–15), average ca.
8–10, oil-bodies, often to 6–7 × 7–9 μ, appearing to nearly
fill the cell lumen; leaf lobes normally with a warm brownish to yellow-brown to purplish brown (rarely reddish at
shoot tips) coloration. Perianth mouth with cilia, formed
mostly of 2–4 superposed single cells, a few small laciniae or
shallow lobes also developed which are 2–4 cells wide at
base, ending in a cilium each . . . . 15.
        15. Leaves no or little wider than long, bilobed less than
0.3 their length . . . . . . *L. ventricosa* var. *confusa*
        15. Leaves (mature shoots) broadly quadrate, ca. 1.1–1.2 ×
as wide as long, with angular sinus descending 0.35–
0.4(0.45) the leaf length. Sterile. . . . .
. . . . *L. ventricosa* var. *confusa* fo. *discoensis*
    14. Median cells each with (10)11–20 (average 12–15) oil-
bodies or more, usually somewhat smaller and hardly
filling lumen; pigmentation, if any, variable. Perianth
mouth denticulate with teeth 1 or 2, rarely 3, cells long,
without laciniae, not or imperceptibly lobed . . . . 16.
        16. Cells small: marginal 20–25 μ, median ca. 20–25 ×
23–28 μ; at least sporadically with bright red to
carmine pigmentation; oil-bodies mostly 9–16 (to
17–21) per cell . . . .
. . . *L. ventricosa* var. *ventricosa* and *L. v.* var. *longiflora*
        16. Cells large: marginal 30–38 μ; median ca. 32–37 ×
35–50 μ; stem black-purple ventrally, leaf bases dull
purplish, leaf lobes ± light brown, without reddish
pigments at all; oil-bodies 14–20, occasionally 21–25,
per cell . . . . *L. ventricosa* var. *rigida*
13. Cells with mostly 15–24 small, spherical (exceptionally ellipsoidal), biconcentric oil-bodies; often with carmine pigmentation of leaf lobes and/or ventral leaf bases; perianth mouth with
short 1(2)-celled teeth . . . . *L. ventricosa* var. *silvicola*
12. Leaves typically strongly concave, mostly with disc saucer-shaped
or cupped, the blunt lobes incurved; sinus descending 0.1–0.2 the

---

[132] Rarely, in very compact, often ± reddish purple to purplish brown arctic-montane sun
forms of very exposed sites (e.g., with *Gymnomitrion concinnatum*) the ± antically connivent
leaves may have ± incurved lobe apices; the leaf sinuses remain somewhat angular.

leaf length, on mature leaves very broadly and shallowly crescentic; never with carmine or reddish pigmentation (ventral side of stem sometimes excepted); perianth mouth with 1–2-celled teeth . 17.

17. Cells with up to (18)20–25(30), less often in xeromorphic forms only 12–15, spherical, rarely some ellipsoidal, minute oil-bodies which are clearly biconcentric; leaves usually (1.0)1.1–1.25× as wide as long, broadly quadrate, usually with weakly arched sides; leaves quite succubous, obliquely oriented, except in hygromorphic forms (which have numerous very small oil-bodies, up to 25–30–32 per cell), strongly concave, antically secund, saucerlike to almost hemispherical, with blunt lobes incurved; pigmentation, if any, warm brown to yellow-brown on distal portion of leaves . . . . *L. groenlandica*

17. Cells with variable, mostly ovoid to ellipsoidal oil-bodies, the larger to 5–7 × 8–10 μ, which (living plants!) are never biconcentric, formed of minute, identical spherules; leaves broadly orbicular, usually with rather strongly arched sides, weakly succubous, vertically oriented (so that concave face of leaf is turned towards the shoot apex), not or almost imperceptibly antically secund; pigmentation of leaves brownish to fuscous. Gemmae in sun forms salmon-pink, yellow-green in shade forms . . . . *L. wenzelii* 18.

18. Smaller plants, ca. 1.2–1.6 mm wide; cells of lobes and lobe margins ca. 20–25 μ, of leaf middle 22–27 × 24–30 μ; median cells with (2–3)4–9(10–12) oil-bodies; gemmae ca. 20–24 × 21–25 μ . . . . *L. wenzelii* var. *wenzelii*

18. Robust plants, ca. 2.2–2.5 mm wide; cells of lobes and lobe margins ca. 30–38 μ, of leaf middle 30–38 × 30–46 μ; median cells with (9)12–20(23) oil-bodies; gemmae ca. 26–28 × 30–32 μ . . . . *L. wenzelii* var. *lapponica*

11. Leaves polymorphic, some simply bilobed, others bilobed with a prominent tooth of antical margin, others irregularly 3–4-lobed; vinaceous pigmentation of leaf lobes (but not of postical leaf bases); large, lax, soft plants; often with linear to lanceolate underleaves . . *L. heteromorpha*

Sectio *HETEROMORPHAE* Schust., sect. n.

Plantae ?dioeciae, repentes; folia subhorizontalia polymorpha, 2–3(4) lobos et saepe dentem basalem habentia; amphigastria lamellata sparsaque; gemmae viridulae, planta localiter, autem, ± purpurascens aut brunneolo-purpuream.

Soft, lax, with aspect of sectio Excisae and of subg. *Massula* but ventral ⅓ of medulla smaller-celled, becoming mycorrhizal and brownish. Leaves remote to subimbricate, very variable, predominantly 2-lobed for 0.3–0.5 their length, but some 3(4)-lobed, or 2-lobed but with a lobelike lacinium, wavy, often with sinuses slightly gibbous, soft-textured. Ventral merophytes variable in width, on many shoots to 4–5 or more cells wide and developing lanceolate to subulate underleaves up to 550 μ long × 10 or more cells wide. Cortical cells little differentiated, not thick-walled, usually 2.5–5× as long as wide dorsally and laterally, narrower than basal leaf cells (on an average). Gemmae polygonal to quadrate, remaining unpigmented (even when leaf lobes are strongly purplish- to brownish-pigmented).

Including only *L. heteromorpha* Schust. & Damsholt of Greenland.

This section appears to represent a contact point between *Lophozia s. str.* and *Massula*. It is of interest that only the ventral ⅓ of the medulla is smaller-celled and mycorrhizal, while in *Lophozia s. str.* usually the ventral ½–⅔ becomes small-celled and mycorrhizal. The ability, sporadically expressed, to develop conspicuous underleaves suggests that a primitive taxon is at hand. The polymorphous leaves and occasional underleaves suggest *L. (Massula) elongata*, an equally rare and poorly known plant, but the latter has smaller cells, a nonmycorrhizal medulla and also lacks the purplish pigmentation of *L. heteromorpha*.

LOPHOZIA (LOPHOZIA) HETEROMORPHA
Schust. & Damsholt, sp. n.

[Fig. 191]

Plantae molles, laxe foliaceae, partim brunneolae ad purpurascentes, basibus, autem, foliorum, nunquam pigmentiferis; folia laxa, polymorpha, dentibus accessoriis saepe praedita; sinus variabiles, partim descendentes per 0.4–0.7 longitudinem folii; amphigastria filiformia ad lanceolata localiter evoluta.

Prostrate to (with crowding amidst mosses) erect in growth, medium-large, *soft-textured and laxly leafy*, with leaves to 1200–1750(2100) μ wide × 1–2(3?) cm long, a *pale*, pellucid green but uppermost parts of plants with distal parts of leaf lobes, particularly in and peripheral to leaf margins, *having brownish to purplish pigmentation* (the 2 pigments varyingly combined, or occurring separately) with age, stem sometimes ventrally with a little purplish pigmentation, but this never extended to postical leaf bases, which are always *pellucid and colorless;* rhizoids copious, colorless, except occasionally at or near juncture with axis, where purplish. Branching free, polymorphous, terminal and narrowly furcate and lateral-intercalary; the intercalary branches very frequent, often lax and slender. Stem elliptical in cross section, *soft-textured*, 250–350(365) μ wide, the dorsal and lateral cortical cells normally narrowly oblong, ca. (14)16–25(26) μ wide × (50)60–105 μ long (ca. 2.5–5:1), feebly thick-walled (in cross section similar to dorsal medullary cells). Ventral cortical cells much shorter, ca. (13)15–24(25) μ wide × 30–54 μ long, often ± purplish; medulla with distinct dorsiventral differentiation, the *ventral ⅓ or less* of smaller cells (ca. 13–17 μ in diam.) which in part or all become *yellow or brown with age* and filled with fungal hyphae, the hyaline dorsal medullary cells ca. 24–26 μ in diam. *Leaves lax*, at most contiguous, *soft, often wavy and undulate, extremely polymorphous* even on 1 stem, very obliquely attached to stem and moderately to conspicuously decurrent on it, *unistratose to base*, pellucid, *varying from 2- to 3(4)-lobed* but predominantly 2-lobed, the mature 2-lobed leaves to ca. 940–1100 μ wide × 830–860 μ long, exceptionally 730–800 μ wide × 780–880(920) μ long [width-length ratio usually ca. (0.9)*1.1–1.28:1*]; 3–4-lobed leaves to ca. 1150–1400 μ wide × 800–820 μ long (width-length ratio ca. *1.45–1.75:1*); lobes usually acute, entire-margined or occasionally with 1-several smaller but *never spinose* accessory teeth, occasional leaves very strongly asymmetric and then the third (or fourth) lobe laciniiform; sinuses acute to crescentic, *very variable*, descending 0.2–0.5(0.6–0.7) the leaf length, the deeper ones often somewhat gibbous. Ventral merophytes variable in width, *usually 4–5 (rarely more) cells wide*, sporadically (but not consistently) *producing distinct underleaves*. Underleaves *filiform to lanceolate* usually, generally 50–100 μ wide × 200–300 μ long but often larger and with lamina to 9–10 cells broad (and then to 180 × 520 μ), entire-margined but often ending in a slime papilla, in some cases on one or both sides at base with a sessile or stalked slime papilla, *never ciliate*. Cells rather pellucid, the cuticle smooth and thin-walled (in lobes, particularly of gemmiparous leaves, sometimes feebly thick-walled) but with rather *distinct to moderately large, never bulging trigones;* marginal cells of lobes ca. 18–23 μ, often subquadrate; cells in lobe middle ca. 17–24 × 18–25(27) μ; median cells ca. (18)23–25(26) × 23–30(35) μ; basal cells ca. 24–30 × 25–36(40) μ (their width ca. 1–1.5× that of cortical cells). Occasional to frequent shoot tips with *colorless* (pale green in life) *gemmae*, their formation restricted to lobe tips of uppermost leaves (thus not long continued); gemmae polygonal to quadrate, thin-walled, angles slightly produced, hardly thickened, mostly 2-celled at maturity, ca. 14–17 × 14–18 μ to 15–19 × 20–24 μ.

Presumably dioecious (2 decayed gynoecial apices seen, without subtending androecia discernible). Otherwise unknown.

Type. Hackluyt I. in Smith Sound, northwest Greenland, 77°24′ N., 72°31′ W. (*G. Seidenfaden 85*). Determined and reported by Harmsen and Seidenfaden (1932, p. 16) as "*Lophozia barbata.*"[133]

Distribution

Known only from the type.

Ecology

Evidently a high arctic taxon, which occurred in a "tuft of *Polytrichum alpinum* and *Webera nutans*."

Differentiation

A species which combines to varying degrees characters of subgenera *Lophozia* and *Massula*. It agrees with *Lophozia* in the stem, which has a mycorrhizal, smaller-celled ventral medullary band, even though mycorrhizal infestation is usually tardily developed, and in the distinctly collenchymatous leaf cells. It agrees with *Massula* in the polymorphous

---

[133] A second specimen of "*Lophozia barbata*" cited by Harmsen and Seidenfaden, from Bjørling I., Cary Isls., is *Lophozia (Barbilophozia) hatcheri!*

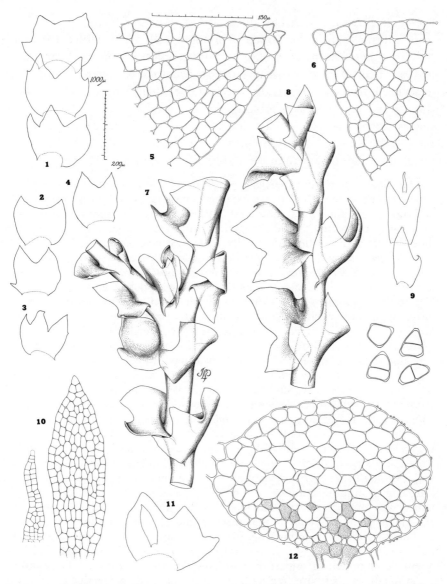

FIG. 191. *Lophozia* (*Lophozia*) *heteromorpha*. (1–4) Leaves, showing the diagnostic variability (×17.5). (5–6) Two leaf lobes, in fig. 5 from a gemmiparous leaf (×185); (7–8) Shoot sectors, in fig. 7 with a *Frullania*-type branch, but bifid supporting leaf (×23). (9) Underleaves, showing variation in size and form (×23). (10) Two underleaves, the one at right also shown in fig. 11 (×95). (11) Leaf and adjoining underleaf (×23). (12) Stem; mycorrhizal, brownish cells stippled (×185). [All from type.; 1–4, drawn to scale at left; 5–6, 12, drawn to top scale; 10, to scale left of fig. 5.]

leaves—a feature which has given rise to the species name—and in the wide ventral merophytes, often 5 or more cells wide, which frequently develop lanceolate to lanceolate-subulate, unlobed and edentate under-leaves. The soft texture (leaves are so soft that it is hardly possible to dissect one from the stem), the wavy and pellucid leaves, and the rather thin-walled and narrowly oblong dorsal cortical cells of the leaves all suggest subg. *Massula*. I have left the species in subg. *Lophozia* because the combination of a mycorrhizal medulla and collenchymatous cells seems more significant than the habitual similarities to *Massula*.

The plants often have purplish pigments of distal sectors and margins of leaves, but never at the postical leaf bases; other plants show brownish pigments of distal parts of the leaves; and on still others the two pigment types are mixed to varying proportions. This, together with the texture and the apparent dioecism, as well as the rather pellucid leaves with weakly collenchymatous cells, suggest *Lophozia latifolia*. Sterile plants may initially be sought under this species, especially since the shoots often have almost entirely bilobed leaves that are broader than long and frequently have a gibbous sinus and sharp lobes. Yet on tips of plants with brownish to purplish leaf lobes the gemmae are uniformly colorless. Furthermore, two other criteria prohibit a closer degree of affinity to *L. latifolia*: the frequently 3- or even 4-lobed leaves, with many leaves quite asymmetric and developing occasional accessory teeth, and the sporadic presence of lanceolate underleaves of sterile shoots.

Underleaves when present vary greatly in size—another reason for the name *heteromorpha*. Often mere stalked slime papillae are present, but in other cases there are lanceolate underleaves to 200–300 μ long $\times$ 50–100 up to 120–180 μ wide and formed of a lamina 5–10 cells wide or more; such underleaves are inserted on ventral merophytes at least 4–5 cells wide and, although entire, bear a terminal slime papilla and may bear basal sessile or stalked slime papillae. Cilia such as occur on the underleaves of the subgenera *Leiocolea*, *Barbilophozia*, and *Orthocaulis* are never developed. Development of the underleaves often hinges on gemma formation; on gemmiparous shoots they are often conspic-uously developed, yet at times lacking or vestigial for inexplicable reasons. Yet completely gemma-free shoots may develop underleaves which may be equally as well developed as on gemmiparous plants. Sporadically, large bifid underleaves may occur, which are obcuneate and to 420 μ wide $\times$ 770 μ long. The frequent development of underleaves isolates this species within subg. *Lophozia*.

Although gemmae are frequent, they are confined to the tips of otherwise normal uppermost leaves. Branching is abundant; frequent furcate terminal branches are present, but more often—in fact often abnormally abundantly—small, slender lateral-intercalary branches occur. Underleaf production is in no way associated with branching. The frequent purplish red pigmentation, soft texture, polymorphous leaves, and pellucid appearance suggest *L. capitata*. From this *L. heteromorpha* differs at once in the angulate gemmae, the stem cross section, and the collenchymatous leaf cells.

Sectio *EXCISAE* Schust.

Lophozia subg. *Dilophozia* sect. Excisae Schust., Amer. Midl. Nat. 45(1):59, 1951.
Excisa-Gruppe K. Müll., Rabenh. Krypt.-Fl. ed. 3, 6:647, 1954.

Plantae paroeciae aut dioeciae, repentes, bractea mascula dente basali praedita; perianthium crenulatum, cellulas digitiformes, per maiorem partem longitudinis raro discretas habens; plantae molles ad succulentes, saepe rubentes; gemmae rubrae ad purpurascentes.

Prostrate or creeping, often rather fleshy, soft or brittle in texture, pure green to green with reddish or purplish to brownish pigments; stem, toward maturity, with *almost the entire medulla brown and mycorrhizal; cortical cells thin-walled*, similar to medullary in diam. Upper leaves and bracts often crispate; ♂ bracts normally *with a tooth at antical base*. Perianth mouth fringed with *fingerlike, thin-walled cells*, usually united for most of their length but, sporadically, some free for all their length. Antheridial stalk varying from 1- to 2-celled. Gemmae 1–2- to 2–4-celled, angulate, *red to purplish*. Paro- or dioecious.

Type. *Lophozia excisa* (Dicks.) Dumort.

Including besides the type only *L. latifolia* Schust. (p. 525).

The Excisae combine, to some extent, characteristics of subgenera *Lophozia* and *Massula*. The mycorrhizal stem is like that of the former, but the leptodermous cortical cells, identical in diam. to the medullary cells, recall those of *Massula*; the tendency for the entire medulla, rather than only the ventral portions, to become mycorrhizal is unique. The thin-walled cells suggest *Massula*; so does the marked tendency to produce *both* -1 and 2-seriate antheridial stalks and the tendency toward production of flattened, elliptical stems and leaves that are bistratose basally, like those found in *L. excisa* var. *elegans*. However, in the one critical feature separating *Massula* and *Lophozia*—the mycorrhizal medulla—the Excisae are clearly nearer *Lophozia s. str.*

LOPHOZIA (LOPHOZIA) EXCISA (Dicks.) Dumort.[134]
[Figs. 192–194]

Jungermannia excisa Dicks., Pl. Crypt. Brit. Fasc. 3:11, 1793.
Jungermannia excisa var. *crispata* Hook., Brit. Jungerm., pl. 9, fig. 11, 1816.
Jungermannia intermedia Lindenb., Nova Acta Acad. Caes. Leop.-Carol. Nat. Cur. Suppl. 14:83, 1829.
Jungermannia cylindracea Dumort., Syll. Jungerm., p. 54, 1831.
Lophozia cylindracea Dumort., Rec. d'Obs., p. 17, 1835.
Lophozia excisa Dumort., Rec. d'Obs., p.17, 1835; K. Müller, Rabenh. Krypt.-Fl. 6(1):693, figs. 317, 318, 1910; Haynes, Bryologist 9:99, pl. 9, figs. 10–13, 1906; Macvicar, Studs. Hdb. Brit. Hep. ed. 2:189, figs. 1–6, 1926; Schuster, Amer. Midl. Nat. 45(1):59, pl. 14, 1951; Schuster, *ibid.* 49(2):354, pl. 15, 1953; K. Müller, Rabenh. Krypt.-Fl. ed. 3,6:664, fig. 204, 1954.
Jungermannia arenaria Nees, Naturg. Eur. Leberm. 2:132, 1836.

[134] *L. excisa* var. *grandiretis* (Arnell, 1954) is a synonym of *L. pellucida* (p. 630).

Jungermannia socia Nees, *ibid.* 2:72, 1836.

Jungermannia limprichtii Lindb., Musci Scand., p. 7, 1879.

Jungermannia excisa var. *crispa* [*sic* !] Underw., *in* Gray, Manual of Botany ed. 6:720, 1889.

Jungermannia propagulifera G., *in* Neumayer, Deutsch. Polar Exped. 2:451, pl. 1:6–12, 1890; Evans, Bull. Torrey Bot. Club 25:418, pl. 346:8–18, 1898 (new synonymy).

Jungermannia alpestris var. *latior* Jens., Öfversigt Kgl. Vetensk.-Akad. Förhandl. 1900(6):800, 1900 (not of Nees, which, *fide* H. W. Arnell, is *L. wenzelii*).

Lophozia propagulifera Steph., Spec. Hep. 2:139, 1901.

Jungermannia alpestris var. *major* Jens., Meddel. om Grønland 30:306, 1906.

Lophozia jurensis Meyl. ex. K. Müll., Rabenh. Krypt.-Fl. ed. 2, 6(2):767, fig. 205, 1916 (*teste* K. Müller, 1954, p. 666); Meylan, Beitr. Krypt.-Fl. Schweiz 6(1):170, fig. 106, 1924.

Very variable in size and appearance, in small, compact turfs or patches, then small (to 5–8 mm high) or creeping among robust caespitose mosses [then to 1.2–1.6(2.5) mm wide × 2–3 cm high], *fleshy, rather soft-textured* and sometimes almost flaccid, prostrate to suberect, *pure green (shade forms) or tinged, especially above, with brick-red, dull carmine, or brownish red,* simple or with sparing branching; usually with 1–several subfloral innovations. Stems to 250–300(425) μ in diam., fleshy, 15–18 cells high, the *cortical cells not or scarcely thick-walled,* barely smaller or subequal in size to dorsal medullary cells; medulla with 6–9 ventral cell layers smaller-celled, with age strongly mycorrhizal and brownish, the larger and equally leptodermous dorsal medullary cells colorless; stems green, but with mycorrhizal infection the ventral side becoming brownish. Rhizoids numerous to shoot apex, colorless, long. Leaves distinctly succubous, erect-spreading to spreading, distant to slightly imbricate, *almost uniformly bilobed,* somewhat asymmetrically *ovate-quadrate* and from 0.95 to 1.1 × as long as wide (smaller phases) *to broadly ovate-quadrate* and from *1.1 to 1.25(1.45)* × as wide as long, ca. 900 μ wide × 650–850 μ long to 1150 μ wide × 800–850 μ long, becoming semiamplexicaul on upper portions of shoots, entire-margined (except below or near ♂ bracts, then often with an antical tooth), *soft-textured and rather delicate,* often somewhat undulate; *sinus shallow and crescentic* and *descending only* ⅕–¼ *the leaf length,* but on larger leaves sometimes rectangulate and descending to ⅓ the leaf length; lobes broadly triangular, obtusely apiculate to subacute, the dorsal usually slightly smaller; antical and postical margins both strongly convex, the postical margin somewhat more strongly so, particularly above the somewhat dilated postical base. Cells of leaf margins in *lobes 23–27(28)* μ, median cells (27)28–30(32) × 30–35(36–40) μ, at the base 28–32 × 35–40 μ, polygonal to quadrate, *thin-walled and with minute to very small trigones* (even when growing in xeric sites); cuticle smooth; oil-bodies varying, in small plants from dry sites occasionally only 11–18 per cell and up to 5 × 9 μ, but usually spherical or subspherical, *4–5(5.5)* μ, and a few ovoid and to 5 × 7.5 μ, occurring *(16)17–24(28) per cell,* somewhat larger than chloroplasts, faintly papillose (the spherules little or not protuberant). *Gemmae usually rather infrequent and produced in small numbers, vinaceous to purplish or brownish purple at maturity,* pyramidal to polyhedral-polyangulate, 1–2-celled, *thin-walled,* ca. 23 × 26 μ to 25–27 × 28–32 μ, a few to 26–30 × 32–40 μ, each cell with 9–20 small spherical oil-bodies; *gemmae-producing leaves (and shoots) hardly modified,* the leaf margins not becoming erose-lacerate.

Paroecious. Androecia below gynoecia, usually of 2–4(5) pairs of bracts; bracts somewhat larger than leaves, from 850 μ wide × 720 μ long to 950–1150

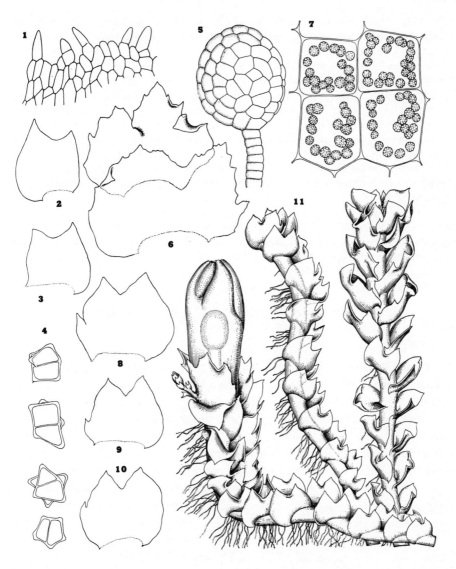

Fig. 192. *Lophozia (Lophozia) excisa.* (1) Cells of perianth mouth, with rare and exceptional development of 1–2-celled leptodermous teeth (×133). (2–3) Medium-sized leaves (×25). (4) Gemmae (×350). (5) Antheridium (×215). (6) Two ♀ bracts (×25). (7) Median cells with oil-bodies (×535). (8–10) ♂ Bracts (×25). (11) Three shoots, two with mature androecia (and, in each case, archegonia in the shoot apex), the other with mature perianth (×15). [1–3, 7–11, from *RMS 35003*, Churchill, Manitoba; 4–6, *RMS 35118*, Mt. Pullen, Alert, Ellesmere I.]

μ wide $\times$ 720–760 μ long to 1500 μ wide $\times$ 950 μ long, broadly ovate-orbicular, with an antical tooth somewhat above base, and sometimes with scattered smaller teeth distally, otherwise similar to leaves, but more erect, with antical base somewhat concave; *1–2(3–4)-androus*. Antheridia ovoid, the stalk 1-seriate, sometimes locally (rarely entirely) 2-seriate, the body ca. 115–125 μ wide $\times$ 135 μ long. ♀ Bracts variable, larger than leaves, to 1400–1700 μ wide $\times$ 800–1000 μ long, somewhat sheathing perianth at base, *often wavy or crispate on margins, frequently fluted*, irregularly 3–5-lobed, slightly to much broader than long, the margins of lobes varying from *nearly edentate* (robust swamp forms) *to finely dentate* (xeric, small phases); bracteole much smaller than bracts, oblong to bilobed, to lanceolate, with free margins often slightly dentate, often united with one or both bracts for up to half its length. Perianths cylindrical to cylin-drical-clavate, ca. 1.25 mm wide $\times$ 2.5 mm long, 0.5–0.75 emergent, strongly plicate in distal 0.3–0.5, somewhat to moderately contracted to mouth; mouth somewhat lobed, *the shallow lobes crenulate with* rather elongate (length 1.5–3:1), *thin-walled, finger-shaped cells, whose distal ends only are free.* Capsule ovoid, reddish brown, its wall 3–4-stratose. Epidermal layer with nodular thickenings; inner layers with incomplete to subcomplete semiannular bands. Elaters ca. 7.5–9 μ thick, bispiral; spores verrucose, brown, *(12–14)15–17 μ*.

Type locality. Holt and Edgefield Heaths, England (*Dickson*).

Distribution

Basically bipolar in range. In Antarctica and to its north found in Tierra del Fuego, southernmost Chile (Ultima Esperanza; Magellanes), south Argentina (Lapataia), and also on South Georgia and on Antarctica itself (Canal de la Belgica sur les corniches de la falaise, *233f;* G). In the last few years also found in New Zealand (Schuster, 1969b). In the Northern Hemisphere a very wide-ranging boreal and arctic species, evidently holarctic in distribution; generally rather scarce and scattered. The "species has a wide distribution in the subarctic-subalpine and northernmost of the Deciduous . . . Forest regions . . . ranging north . . . into the Arctic-Alpine Zone" (Schuster 1953). Recent experience has shown that the species is quite frequent in much of the Arctic, extending northward at least to 82°26′ N. in Ellesmere I. (*RMS*).

In Europe widespread, throughout most of Scandinavia (Norway, Sweden, Finland), the Faroes, southward to Denmark, Belgium, France, England, and Ireland, Germany, Switzerland, northwest Spain, northern Italy, Hungary, Dalmatia, Bucovina to the Caucasus Mts., northward to Lithuania and Estonia, and northwest USSR. Extending eastward into Siberia and probably the alpine region of Japan (Kitagawa, 1965).

In North America clearly transcontinental, from the Aleutian Isls. to con-tinental Alaska to British Columbia, Alberta, Washington, Idaho, California, and Wyoming in the West. Eastwards as follows:

E. GREENLAND. Fleming Inlet (*Kruuse;* det. Jensen, 1906a, p. 306, as *L. elongata*); Scoresby Sund, Röhss Fjord, Turner Sd., and Ingmikertorajik (*Jensen;* all as *L. excisa* var. *cylindrica*); E. coast at 76° N. (Simmons; as fo. *limprichtii*); Scoresby Sund at Danmarks Ø (Jensen, 1897; as fo. *limprichtii*), Kap Brewster, 70°10′ N., 22°10′ W. (*Hartz, 1900*).[135] N. GREENLAND. Centrum Ø, 82°44′ N. (*Th. Wulff, July 2, 1917!*); Kap Glacier, S. coast of Independence Fjord, dry stony field with *Hierochloë alpina* (*Holmen, 7960*).[136] NW. GREENLAND. Kânâk, Red Cliffs Pen., Inglefield Bay (*RMS 45604, 45639c, 45565c, 45611, 45603d, 45607, 46154,* etc.); Kangerdlugssuak, Inglefield Bay (*RMS 45859*). W. GREENLAND. SE. of Jacobshavn, Disko Bay, 69°12′ N., 51°05′ W. (*RMS & KD 66-224, 66-225*); Eqe, near Eqip Sermia (Glacier), ca. 1 mi from Inland Ice, 350–450 m, 69°36′ N., 50°10′ W. (*RMS & KD 66-293c,* compact extreme, ±julaceous; *66-336b,* mod. *parvifolia*). Disko I.: Godhavn, mouth of Blaesedalen, ved Lyngmarksfjeld, in and near small ravine running into Rødeelv, 69°15′ N., 53°30′ W. (*RMS & KD 66-281c*); Fortune Bay (*RMS & KD 66-201a*); Godhavn, below Lyngmarksfjeld (*RMS & KD 66-166a*); Flakkerhuk, near Mudderbugten, 69°39′ N., 51°55′ W. (*66-173b*). Sondrestrom Fjord, sunny S.-facing slopes, near air strip, 66°5′ N. (*RMS & KD 66-046, 66-052*). ELLESMERE I. Mt. Pullen, ca. 5 mi S. of Alert, NE. Ellesmere I., 82°25′ N. (*RMS 35118*); The Dean, 5–6 mi SE. of Alert (*RMS 35158, p.p.*); Harbour and Goose Fiords, S. coast (Bryhn, 1906).

LABRADOR. *Fide* Macoun (1902). NEWFOUNDLAND. Rose Blanche, S. coast; Benoit Brook, W. Nfld. (Buch & Tuomikoski, 1955). QUEBEC. Tadoussac, Saguenay Co. (*Evans 8!,* as *L. longidens,* det. Evans; also reported from Tadoussac by Lepage, 1944–45); lake N. of Wawicho L., Ungava, 53°51′ N., 77°57′ W. (Lepage, 1953). NOVA SCOTIA. Halifax (Brown, 1936a). ONTARIO. Ottawa "on rotten wood" and Belleville "on rotten logs" (Macoun, 1902; both reports very doubtful!); L. Penage, Sudbury Distr. (Cain & Fulford, 1948); Algonquin Park; Halton; Hastings, Cliff Bay West, Manitoulin I. (*Cain 4992!*); Wanapitei R., Sudbury Distr. (*Cain 2113!*); Muskoka; Nipissing; Peterborough (Williams & Cain, 1959). MANITOBA. RCAF Base at Churchill, Hudson Bay (*RMS 35003*).

MAINE. Lubec (Evans, 1912c). NEW HAMPSHIRE. Thorn Mt., Jackson (Evans, 1906b, with discussion). RHODE ISLAND. Hopkinton (Evans, 1913). CONNECTICUT. Madison (*Lorenz, 1925*); North Haven (Evans, 1906b). NEW YORK. Peaked Mt., Warren Co. (*Burnham*); South Hill Marsh, Ithaca, Tompkins Co., and near Woodhull, Steuben Co. (Schuster, 1949a). [Also reported from Syracuse (Underwood & Cook, Hep. Amer. no. 112; Frye & Clark, 1937–47, p. 348), but this contains *L. porphyroleuca* and *L. incisa* (Schuster, 1949a, p. 570).] NEW JERSEY. Passaic, Delaware R., on "shaded banks" (var. "*crispa*" Hook.; evidently referring to var. *crispata*; Austin, *in* Britton, 1889). [Also reported from "sterile ground in open woods" but referring certainly to *L. bicrenata,* which at that time was not distinguished in North America.]

MICHIGAN. Devils Washtub, near Copper Harbor, Keweenaw Co. (*Schuster, 1949*); Grand Ledge; Phoenix, Keweenaw Co.; Burt L., Cheboygan Co. (*Steere, Nichols*). WISCONSIN. Cornucopia (*Conklin 1853*); Siskiwit R., Bayfield Co. (Conklin, 1929).

---

[135] Lectotype (present designation) of *Jungermannia alpestris* var. *major* Jens. (1906a); this is a mod. *laxifolia-latifolia* of *L. excisa,* essentially identical to var. *cylindrica,* approaching *L. latifolia* in facies and color. On two plants with gynoecia at the mature-archegonial stage, I was able to find (1)2–3 antheridia per bract in 2 pairs of subfloral leaves. The second specimen cited by Jensen for his var. *major,* from Hurry Inlet, is *L. latifolia* (see p. 528).

[136] The specimen from Centrum Ø is fertile, has mature capsules, and is clearly paroecious; Hesselbo (1923a) published it as *L. alpestris,* while Arnell (1960) calls it "poor material, not possible to determine." The crenulate perianth mouth, dull carmine pigmentation, and reddish gemmae leave no doubt as to its disposition. *Holmen 7960* is also definitely paroecious; it was published by Arnell (1960) as *L. alpestris.*

MINNESOTA. Cook Co.: Mt. Rose, Grand Portage (*RMS*, in part); Temperance R. (*RMS 19712*); Hungry Jack L. and Grand Marais (*Conklin*). Minneapolis, Hennepin Co. (*RMS 17235*); Rainy L. near Black Bay, Koochiching Co. (*RMS 18115*); Oak I., Lake of the Woods Co. (*RMS 17260*); Oneota, Fairmount Park, and Chester Cr., Duluth, St. Louis Co. (*Conklin 887, 2123, 1860, etc.*); see Schuster, 1953.

Southward rarely occurring as a disjunct in the Oak-Chestnut Forest Zone of the Southern Appalachians, at 3000–4000 ft and higher: NORTH CAROLINA. Nigger Mt., near Jefferson, Ashe Co., 4200 ft (*RMS 29610a*).   TENNESSEE. Sevier Co. (Sharp, 1939).   WEST VIRGINIA: Kates Mt. (*E. G. Britton, Oct. 13, 1914!*; typical, gemmiparous traces among *Cephaloziella hampeana*).

Material that appears to be either wholly identical or at best represents an antipodal race was described by Gottsche (1890) as *Jungermannia propagulifera* from South Georgia. Evans (1898) admits its close affinity to *L. excisa* but maintains it as distinct because of the supposedly more deeply bilobed leaves, divided to about ⅓ by a subacute sinus, the more leptodermous cells, and slightly different ♀ bracts. Although the somewhat more deeply bilobed leaves (much like those of the closely allied *L. latifolia*!) may warrant a separate variety, the supposedly differential features derived from the bracts and cells fail to hold. Material of *L. excisa*, virtually identical with *L. propagulifera* in these respects, has been repeatedly studied. Both Evans (1898) and Herzog (1954, p. 32) report *L. propagulifera* from West Patagonia (Chile); I collected it in both Patagonia and Tierra del Fuego [42 km north of Puerto Natales, Distr. Magellanes, Chile (Feb. 10, 1961) and Lapataia, Beagle Channel, west of Ushuaia, Tierra del Fuego, Argentina (Feb. 22, 1961)]. It is abundant in the Antarctic Peninsula (Livingston, Hook, Deception, Anvers, Argentine Isls.; *RMS*). *L. excisa*, therefore, acquires a bipolar distribution, closely analogous to that of *Lophozia hatcheri*.

Ecology

Lophozia excisa is a very complex, variable species which occurs under a wide variety of circumstances, but usually associated with mineral substrates: either acid sandy or loamy soils, over rock outcrops, or directly on rock surfaces, occasionally on loamy banks away from rock outcrops. The species has a wide toleration for variation in pH, occurring occasionally on subcalcareous shales or sandstones, with a pH of about 7.0, as well as on basic intrusive rocks, and on more acid intrusive rocks, where the pH may be as low as 4.8 (Schuster, 1953).

Very often a pioneer or invading soon after ecesis by pioneer species. However, at least in the Arctic, a frequent component of the damp tundra heath, occurring scattered among such mosses as *Dicranum*, as a minor member of well-established communities, together with *Lophozia quadriloba* (another "weak calciphile").

On calcareous rocks, when a pioneer, it may be associated with *Lophocolea minor*, *Scapania mucronata* (and, more rarely, *S. microphylla* and *Reboulia*, as in Minnesota and New York). In contrast, found in the

Keweenaw Peninsula on basic rocks with the arctic *Lophozia quadriloba* and with *L. barbata* and *Tritomaria quinquedentata;* all these species show toleration for calcareous conditions. Under such conditions also often found with various lichens (*Umbilicaria, Peltigera,* etc.). However, when scattered among *Dicranum* and *Polytrichum* in wet tundra turf (over basic rocks), frequently associated with *Cephaloziella arctica* and *Anastrophyllum minutum* (Churchill, Manitoba; northeast Ellesmere I.).

When growing over exposed loamy, acid soil, the species may occur with *Cephaloziella rubella, Lophozia bicrenata,* and *Solenostoma crenulatum* (i.e., members of the *L. bicrenata-Cephaloziella* community of leached, acid, sterile soil); on such soil, under somewhat moister conditions, *Scapania irrigua* and *Lophozia capitata,* as well as various mosses (chiefly *Polytrichum commune*), may be associated.

At the southern extreme of its range, over soil in crevices of acidic rocks, together with the equally disjunct *Scapania mucronata*—forming a matrix in which the fern *Polypodium virginianum* undergoes ecesis.[137]

In the Arctic *L. excisa* becomes very widespread under a wide range of conditions, mostly over peat or peaty soil, at times amidst mosses (such as *Aulacomnium turgidum*) over peaty, circumneutral soil, associated with *Scapania brevicaulis*; at times actually over *Sphagnum*—a site where, more often, it is replaced by the allied *L. latifolia.*[138] On peaty soil in sedge-grass meadows *Cephalozia pleniceps, Riccardia palmata* (or *sinuata*), *Lophozia kunzeana,* etc., may be associated. Also, over peaty soil between rocks, near the Greenland ice cap and in sites irrigated by melt water, *L. excisa* may occur with *Gymnomitrion concinnatum, G. apiculatum, G. corallioides, Anastrophyllum minutum grandis, Lophozia groenlandica, L. alpestris s. lat., Marsupella sprucei,* etc.

Perhaps the most extreme conditions under which the species is found is on cold, exposed bluffs, covered by sandy silt from deposits from the retreating ice cap (as at Eqe, Greenland); here, in full sun, a pioneer associated with *Gymnomitrion concinnatum, G. corallioides, G. apiculatum, Prasanthus suecicus, Tritomaria*

---

[137] Reported by Frye & Clark (1937–47, p. 347) from "rotten wood"; this is never the case *unless* the logs have first been charred by fires. The report is perhaps based on Underwood & Cook's Hep. Amer. Exsic. no. 112, from Syracuse, N.Y., which is listed by Frye & Clark under their "Examinations"; this material represents *Lophozia porphyroleuca* (Schuster, 1949a).

[138] The occasional occurrences over *Sphagnum* polsters, in the Arctic, are usually as succulent, fleshy, often rather lax phases, sometimes bright green, sometimes partly tinged with dull red. Typical is a specimen from Kânâk, northwest Greenland (*RMS 45603d*), which, even though clearly paroecious, was laxly leafy and approached *L. marchica* in facies. Such plants may also approach *L. latifolia* but differ in, i.a., the paroecious inflorescences. In *RMS 45603d* the ♂ bracts were 1–2–3-androus and either occurred immediately below the ♀ ones, or several pairs of sterile leaves intervened. The last type of plant seems to correspond perfectly to *L. jurensis* Meyl., which Müller (1951–58) regarded as a variety of *L. excisa.*

heterophylla, and *Scapania perssonii*. A more exposed site could hardly be imagined, and the plant is found here as a pigmented, reddish or purplish julaceous extreme with rigid stem and small, antically secund, strongly concave leaves, giving it the aspect of a small *Lophozia bicrenata* or of *L. decolorans*.

Differentiation

Lophozia excisa is extremely polymorphous yet fundamentally strongly isolated from all other species of the subg. *Lophozia*, except *L. latifolia*. It differs from the latter, as well as other regional species of *Lophozia* subg. *Lophozia*, in the paroecious inflorescences. The species is compared with *L. ventricosa* (*L. "silvicola,"* in part) by both Müller (1954, *in* 1951–58) and Macvicar (1926), the latter stating that it is "only likely to be mistaken for *L. ventricosa*. . . ." It differs from this species and the allied *L. groenlandica*, however, in the softer texture, the more delicate and leptodermous cells, the usually sparing, pigmented gemmae, and typically the "crispate heads" formed by the bracts and upper leaves at the apices of the fertile shoots (before elongation of perianths). This feature, however, is marked only in more compact phases, while bog and swamp phases (var. *cylindracea*) show it to a lesser extent; note, e.g., the scarcely crispate apices of the two fertile shoots illustrated (Fig. 192:11, the two right-hand shoots, both with mature archegonia and antheridia).

The relatively large and leptodermous leaf cells, the generally pure green color (reddish-tinged on distal portions of leaves and perianths only when exposed), broad, orbicular leaves with a generally shallow and crescentic sinus, and often sparing production of vinaceous or dull purple gemmae are diagnostic. *Lophozia excisa* is paroecious and almost without exception abundantly fertile; it usually has more or less dentate perichaetial (and sometimes also perigonal) bracts. In these respects it resembles *L. bicrenata* but differs from our other species of subg. *Lophozia*, all of which are dioecious.

Lophozia excisa is so variable that it has given rise to a much confused synonymy. At one time (in North America until the early years of this century) it was confused with *L. bicrenata*. Plants forming compact patches on dry soil, producing gemmae freely [fo. *limprichtii* (Lindb.) Mass.], are easily confused with *L. bicrenata*, which they resemble in size, fleshy stems, pigmented gemmae, leaf shape, and paroecious inflorescences. However, a large number of differences exist between the two taxa, all phases of *L. excisa* differing from *L. bicrenata* in the (a) crenulate to feebly denticulate perianth mouth; (b) somewhat larger, leptodermous cells, bearing at most small trigones; (c) absence of any tendency for the leaves to become erose-lacerate with gemmae formation; (d) much less conspicuous gemmae masses; (e) more thin-walled, purple—rather than rusty red—gemmae, whose cells bear many minute oil-bodies; (f) stem anatomy.

Confusion with other species of *Lophozia* is unlikely, except with the rare *L. latifolia*. This latter species gives the impression of being almost identical with laxer, more robust phases of *L. excisa* ("var. *cylindracea*"), but differs from all phases of *L. excisa* in being dioecious, with the androecia intercalary on otherwise vegetative shoots. ♀ Plants of *L. latifolia*, furthermore, generally have broader leaves, averaging over 1.25 × as wide as long, with the sinus descending at least 0.3–0.45 the leaf length. In the high Arctic, *L. excisa* often occurs in the same area as the large-celled subsp. *polaris* of *L. alpestris*. The two, when sterile, may be difficult to separate. Their separation is discussed under *L. alpestris* subsp. *polaris*.

Variation

Lophozia excisa is a species with a tremendously wide range, both geographic and ecological, correlated with a high level of polymorphism, much of which is certainly nongenetic. Somatic adaptability of this species appears to be markedly high, yet it is usually scarce in most of its range. Variation occurs, in general, from a small, compact phase, a pioneer on damp loamy or clayey soil, in which the ♂ and ♀ bracts are relatively strongly and closely dentate (fo. *limprichtii*, grading into typical *L. excisa*), to a larger, lax, soft, and pure green phase, often 2–3 cm high, in which the bracts are undulate-crispate, lobed but otherwise entire or subentire ("var. *cylindracea*" and "var. *crispata*"). Small, compact phases of exposed soil are often reddish-pigmented above; laxer phases, which frequently grow scattered amidst caespitose mosses, are pellucid and pure green, except that sometimes the very margins of the leaves may be reddish brown or carmine red.

In the high Arctic, phases of *L. excisa* commonly occur which, though relatively large, are compact and very fleshy (leaves often 2-stratose in their basal portions). These plants, of strongly insolated sites, have concave leaves with brownish purple pigmentation of the distal halves which is so intense that the plants, when in patches, look blackish on examination by reflected light. Cell walls, in such phases of extremely sunny sites, are no thicker than usual. The gemmae, however, are deep brownish purple and slightly thick-walled.

It is almost always possible to find gemmae in this species, but they are rarely abundant, except in some arctic phases. Material from northeastern Ellesmere I., representing a large phase with broad, undulate leaves and crispate bracts, bore purplish gemmae in great quantity. In such cases the plant is of very striking aspect, the bright purplish red gemma masses standing in sharp contrast to the soft green color of the rest of the plant.

Leaves and bracts in this species, as in other Lophoziaceae, undergo marked heterogonic variation. As a consequence, small, compact forms of the species may have leaves (and ♂ and ♀ bracts) which are only slightly, if at all, wider

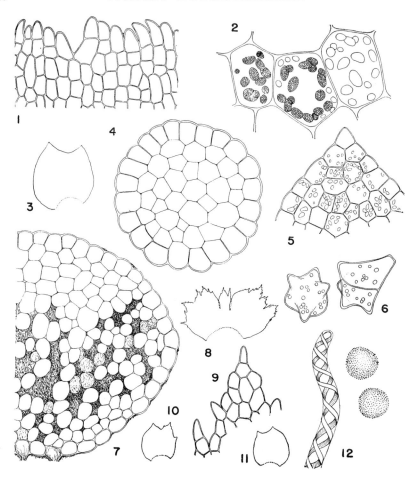

FIG. 193. *Lophozia* (*Lophozia*) *excisa*. (1) Cells of perianth mouth (×200). (2) Median cells with oil-bodies, in middle cell also chloroplasts (×500). (3) Leaf (×24). (4) Seta cross section (×ca. 200). (5) Cells of leaf lobe, note oil-bodies in essentially all cells (×210). (6) Gemmae; note uniformly small oil-bodies (×500). (7) Cross section of mature axis (×200). (8) Bracts and bracteole (×12). (9) Apex of bract lobe (×ca. 200). (10–11) Leaves (×12). (12) End of elater and spores (×480). [1–3, 5–6, 10, 12, Grand Portage, Minn., *RMS;* 4, 7, South Hill, Ithaca, N.Y., *RMS;* 8–9, 11, Woodhull, N.Y., *RMS.*]

than long. Robust, tender phases, by contrast, may show very strongly transverse leaves, in some cases as much as 1.2–1.45× as wide as long; ♂ and ♀ bracts may be equally, or even more strongly, transversely developed.

   Antheridial stalks are variable; in one of the collections from Ellesmere I. (*RMS 35118*) all of dozens of antheridia examined had uniseriate stalks; in the other collection from this area (*RMS 35158*) most antheridia had uniseriate stalks, a few had isolated cells longitudinally divided, and one or two had almost wholly biseriate stalks, suggesting that Müller (1948) placed too much emphasis on this feature in the classification of the leafy Hepaticae.

Two other supposed variants of *L. excisa* have been a source of taxonomic confusion. One, *L. elongata* (Lindb.) Steph., has been placed in *Orthocaulis* (Evans, *in* Buch, Evans, and Verdoorn, 1938; Frye and Clark, 1937–47). Müller (1954, *in* 1951–58, pp. 666–667) claims that this plant is identical with the larger, lax phase of *L. excisa*, bearing clavate perianths ["var. *cylindracea* (Dum.) K. Müll."]. However, *L. elongata* is a good species, with the axial anatomy of subg. *Massula* (see p. 436).

The other plant, *L. jurensis* Meyl. ex K. Müll., has been reduced to a variety of *L. excisa* (Müller, *loc. cit.*, p. 666). This is a bog plant, occurring among *Sphagnum*, differing habitually from *L. excisa* in the lax facies, approaching that of *L. marchica*. It appears to produce isolated androecial shoots and is perhaps partly autoecious, although normally paroecious. In the occasional intercalary androecia it approaches *L. latifolia*, to which the more deeply divided leaves (0.25–0.33 the leaf length) also suggest an affinity. The virtually uniform dioecious inflorescences of *L. latifolia* and the occurrence of the latter as normal, soil- and rock-inhabiting forms with strong pigmentation show that *L. latifolia*, unlike the var. *jurensis*, cannot be regarded as a mere phase of *L. excisa*.

I have placed, after much hesitation, *J. propagulifera* G. (*L. propagulifera* Steph.) in the synonymy of *L. excisa*. I have not seen Gottsche's type, but his diagnosis, so far as it goes, is quite specific. Evans (1898) states that the species has a paroecious inflorescence; this I can confirm from personally collected samples (Lapataia, Tierra del Fuego, Argentina), and from the plant from Antarctica (Canal de la Belgica; Exped. ant. belg., 1 février 1898; G). The latter plant has exactly the stem anatomy of the northern *L. excisa*, with eventually the lower ⅔ of the medulla brown, the cells largely destroyed through fungal activity. Stem cross sections from the antarctic plant quite match Fig. 193:7 in all respects. I was furthermore impressed with the Tierra del Fuegan material and material I collected in Chile (north of Puerto Natales toward Ultima Esperanza) by the vivid, grass-green color of phases from sheltered sites, by the fleshiness of the plants, and by the similarity in facies and ecology to northern *L. excisa* as I knew it. The statement in Stephani (1901, *in* 1898–1924, p. 139) that *L. propagulifera* is dioecious is in error, as material in his own herbarium shows. Evans (1898) recognized the similarity of *L. propagulifera* to *L. excisa* but believed *L. excisa* had "less deeply bifid" leaves, "so that the sinus is obtuse or lunate, the leaf cells have thicker walls, and there are slight differences in the bracts." However, material of *L. propagulifera* from the Antarctic, cited above, has shallowly lobed leaves, often to an even greater degree than northern *L. excisa*. Also, the polymorphism in the bracts of *L. excisa s. lat.* is notoriously great and, I think, embraces that illustrated for *L. propagulifera* by Evans. I cannot see any differences in leaf cells; they are thin with very small trigones in both "species." Furthermore, in the antarctic plant, the 2–3-androus perigonal bracts were identical to those of northern *L. excisa*. Antheridial stalks within a single androecium also showed variation from completely biseriate to largely uniseriate, although in the limited material studied I found no wholly uniseriate stalks. The variability of *L. propagulifera* is thus identical to that of *L. excisa*.

In addition to the already bewildering variability portrayed above, all or most of which appears to be either purely environmentally induced or minor (and taxonomically negligible), at least one extreme of *L. excisa* diverges so strikingly that it should be separately treated; it is separable as follows:

1. Gemmae 1–2-celled, usually ± purplish, dull; stem terete to feebly dorsiventrally flattened; leaves usually quite oblique; plants with very slight branching . . . . var. *excisa* (see above)
1. Gemmae 2–3(4)-celled, bright red, in large masses; stem strongly flattened, 2.0–2.4× as wide as high; leaves dense, pectinate-distichous, dorsal half nearly vertical; with free, furcate, terminal branching . . . .
. . . . var. *elegans*, var. n.

LOPHOZIA EXCISA var. *ELEGANS* Schust., var. n.

[Fig. 194]

Plantae succulentae, ad friabiles variantes, libere terminaliter furcate ramosae; Massulae foliis subverticalibus lobos dorsales minores habentibus similes; omnis cellula solum 5–12 guttae olei saepe habens; gemmae scarlatinae ad carmesinas, usque ad 32–35 × 34–38 μ, e 2–3(4) cellulis constantes.

Fleshy and soft-textured, varying to brittle, prostrate, closely attached by numerous colorless, long rhizoids, medium-sized (shoots with leaves ca. 2–2.45 mm wide × 8–15 mm long), rather freely, furcately branched, with terminal branching, *clear and rather pellucid green* (only the older mycorrhizal parts of the stems light brownish), *secondary pigments lacking* except for the *intensely red gemmae.* Stem *flatly elliptical in cross section* (ca. 250–275 μ high × 510–665 μ wide), *very fleshy, flattened,* ca. 16–18 cells high; *cortical cells thin-walled* like medullary, free walls excepted, short- or medium-oblong, ca. 16–24(26) μ wide × (40)56–72 μ long, their diam. subequal to that of medullary cells; medulla not clearly dorsiventrally differentiated, a few dorsal strata of medullary cells ± larger, but, *with age, the medulla as a whole brownish and mycorrhizal*; rhizoids 14–17 μ in diam., colorless, with age brownish. Leaves rather *dense, often with a 2-stratose basal field, soft-textured and fragile,* pectinate-distichous, *subvertical (dorsal half nearly vertical), with an L. incisa-like aspect,* the dorsal half with insertion only moderately oblique, arcuately inserted, ± decurrent, the leaves rather loosely and very obtusely folded or slightly plicate, asymmetrically subquadrate to broadly oblong-ovate, very *variable,* from longer than wide (940 μ wide × 1100 μ long to apex of ventral lobe) to, usually, *much wider than long* (1025–1325 μ × 940–965 μ), *often unequally bilobed,* the sinus descending 0.2–0.4 the leaf length, bluntly angular to sublunate; *lobes* ± *unequal,* the dorsal "half" of leaf often both *much shorter and smaller than the ventral,* both bluntly to, rarely, subacutely pointed, entire or (usually associated with gemmae formation) with 1–2 small, 1-celled teeth; *leaf lobes not ending in cusps,* and teeth, if present, undifferentiated; leaf margins below lobes entire, the *dorsal usually diagnostically narrowly recurved,* much as in *Marsupella emarginata.* No underleaves; ventral merophytes at least 3–4 cells wide. Cells *medium-sized,* the marginal of lobes ca. 25–30 μ, tangentially measured; subapical cells ca. 23–26 × 24–27 μ; median cells irregularly 4–6-sided, ca. 24–29 × 25–36(40) to 25–33 × 25–38 μ; cuticle smooth; *trigones minute and walls thin.* Oil-bodies of subapical and median cells rather few, *mostly (5)7–15(18–20) per cell*, variable, rather large, 4.2–5.5 μ, and spherical to 4.2–5 × 4.8–5.5 μ up to 5.5–6 × 6–8 (rarely to 7 × 9) μ and ellipsoidal, *opaque, granular-botryoidal.* Gemmae abundant, *sharply tetrahedral to polyhedral to cuboid*, 25–28 × 25–28 to 32–35 × 34–38 μ usually, *intense red, 2–3(4)-celled.* Otherwise unknown.

Type. Slopes above Kangerdlugssuakavsak, head of Kangerdluarssuk Fjord, west Greenland, 71°21′ N., 51°40′ W. (*RMS & KD 66-1287, Aug. 6-7, 1966*).

The type occurred over humus-covered, clayey, calcareous soil, amidst rock fragments, over an intermittently rather dry but well-vegetated rocky slope. Associated Hepaticae were *Tritomaria scitula, T. heterophylla, T. quinquedentata, Solenostoma subellipticum, S. sphaerocarpum, Anthelia juratzkana*, and traces of *Cryptocolea imbricata.*

Differentiation

Although known only from a few individuals, this distinctive plant cannot easily be confused with any other. Superficially like *Massula* in the fleshy, flattened stem and thick leaves, it differs from all *Massula* species with angulate gemmae in having these bright and intense red—and diagnostically contrasted to the light and clear green leaves and stem; the gemmae, also, are unique in being to a large extent 3-celled at maturity (Fig. 194:2–3).

It approaches *L. incisa* in the very soft and fleshy stems and in the leaf form, almost all the leaves being rather asymmetrically bilobed, as in weaker forms of *L. incisa;* it further agrees with this species in the cells and in the rather dense and subvertical orientation of the leaves, as well as in their rather sharp lobes. Unlike *L. incisa*, accessory dentition and cusps of the leaf lobe tips are lacking. Also, unlike *L. incisa*, the cells bear few (often only 5–12) oil-bodies per cell, and these are relatively large and quite obviously granular-botryoidal, rather than glistening and smooth. Stems are quite fleshy and soft-textured, elliptical in section; as they mature, the entire medulla becomes very conspicuously mycorrhizal, the stem turning light brownish.

The combination of red gemmae, leptodermous cells, and clear green color, plus the very copiously mycorrhizal stem, strongly suggest the protean *L. excisa.* The facies of the plant, however, is that of *Massula;* the very flattened and fleshy axis, averaging about twice as wide as high, recurs commonly in *Massula*, not in subg. *Lophozia;* the subvertical, dense, polymorphic leaves, with the dorsal and usually smaller half vertical

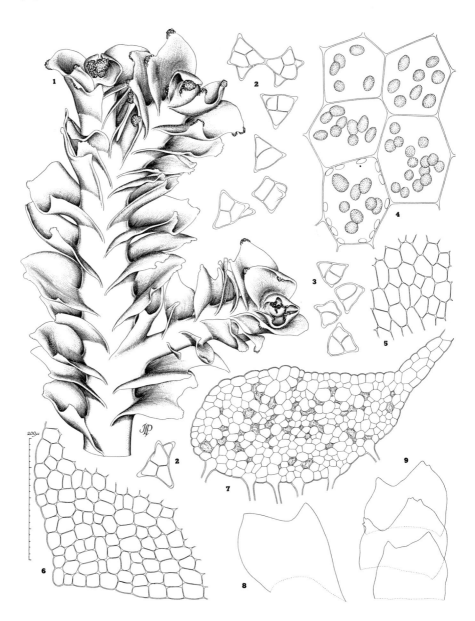

FIG. 194. *Lophozia* (*Lophozia*) *excisa* var. *elegans*. Gemmiparous shoot with two ♀ *Radula*-type branches (×21.5). (2–3) Gemmae (×ca. 345). (4) Median cells (×695). (5) Median cells (×165). (6) Cells of lobe apex (×165). (7) Stem cross section (×130). (8–9) Leaves (×27). [All from type, *RMS & KD 66-1287*, Kangerdluarssuk Fjord, W. Greenland; 5–6, drawn to scale at left.]

or almost so—much as in *Tritomaria*—recur again in *L. incisa*, but hardly in *L. excisa*. Finally, the high incidence of 3-celled gemmae is a unique feature, not encountered in any other *Lophozia* known to me. However, a highly derivative extreme of *L. excisa* is probably at hand, and varietal treatment under this species seems judicious. Until more copious material is available, a definitive answer must remain in abeyance.

The few plants studied were of great interest in that, on two of them, weak, widely patent lateral branches were present which originated far above the axils of normal, unmodified leaves. These branches originated juxtaposed to the abaxial base of a leaf above and were clearly *Radula* type; in addition, *Frullania*-type branches, at an acute angle to the parent axis, with a lanceolate supporting leaf, were seen.

LOPHOZIA (LOPHOZIA) LATIFOLIA Schust.

[Figs. 195–197]

Jungermannia alpestris var. *major* Jens., Meddel. om Grønland 30:306, 1906 (*p.p.*; Hurry Inlet plants).
Lophozia (Dilophozia) latifolia Schust., Amer. Midl. Nat. 45(1):60, 1951; Schuster, The Bryologist 56:258, pls. 1–2, 1953; Schuster, Amer. Midl. Nat. 49(2):369, fig. 21:9–15, 1953.
Lophozia latifolia Arn., Svensk Bot. Tidskr. 48:796, 1954; Arnell, Illus. Moss Fl. Fennosc. 1:121, 1956.
Lophozia excisa var. *palustris* Arn. (in sched.; *fide* Arnell, 1956, p. 121).

Growing as isolated, creeping to ascending stems among other bryophytes, very seldom in small patches, pale green to (in exposed situations) a ± *brown or purplish brown*; rarely a little purplish pigmentation of underside of stem and leaf bases. Mature sterile shoots to 2.0–2.4 mm wide, flexuous, generally pale green with the facies of *Lophozia marchica* or *excisa*, the *leaves distant to approximate*, obliquely inserted and *generally horizontally spreading*. Stems to ca. 300–325 μ in diam.; cortical cells only slightly thick-walled, the dorsal 19–24 to 23–26 × 34–44 μ in diam.; medulla with dorsal band of large, hyaline, thin-walled, noncollenchymatous cells, 19–24(32) μ in diam., lacking mycorrhizal infestation; ventral band (occupying lower ⅓–⅔ of medulla) 6–8 cells high, of much smaller cells, ca. 9–12 μ in diam., largely disintegrated at maturity by mycorrhizal activity, becoming brownish and opaque. Rhizoids numerous, long, scattered, colorless. Leaves of mature sterile and ♀ shoots *1.1–1.5× as wide as long*, transversely rectangular to rectangular-ovate, ca. 750–760 μ long × 1000–1155 μ wide (to 900 μ long × 1050–1200 μ wide), horizontally spreading, bilobed by a rectangular to subobtuse sinus for about ⅓–⅖ their length, the *sinus often somewhat reflexed*; lobes acute to subacute, terminated by a single cell. ♂ Plants often slenderer, with more nearly orbicular, more shallowly bilobed leaves. *Cells 23–25 to 24–28(32) μ in leaf tips and along margins*, 25–30 × 28–36, occasionally to 40 × 40 μ, medially; cells colorless to brown, never reddish, *thin-walled with small concave trigones* (rarely slightly bulging in mod. *colorata*); oil-bodies of median cells small, spherical and 3.5–6 μ, or ovoid and 3.5–5.5 × 4–7 μ, of many uniform, minute, rather indistinct spherules which scarcely

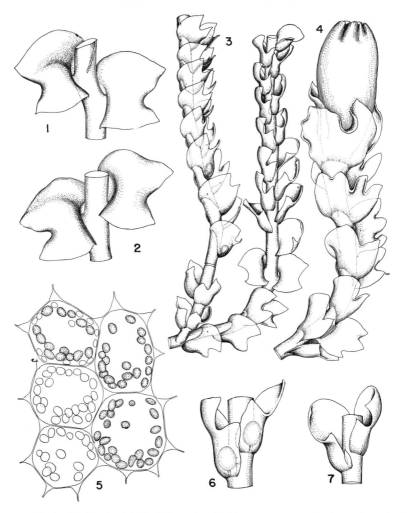

FIG. 195. *Lophozia (Lophozia) latifolia.* (1–2) Shoot sectors, ♀ plant (×20). (3) Androecial plant (×16). (4) Gynoecial plant (×13.3). (5) Median cells with oil-bodies and, upper left, chloroplasts (×550). (6–7 Sectors of androecia, in fig. 6 with the monandrous nature of the bracts evident (×25). [All from *RMS 19468*, Grand Marais, Minn., type; all from Schuster, 1953a.]

protrude through the common membrane; median cells with 9–15 or 15–24 oil-bodies. *Gemmae relatively rare*, limited to juvenile leaves of shoot apex, forming in small, often inconspicuous purplish clusters; gemmae irregularly pyriform or tetrahedral to few-angulate, 24–26 × 22–32 μ long (a few to 36–38 μ), somewhat thick-walled, ± *deep purplish.*

Dioecious; usually fertile. ♂ and ♀ Plants often in the same patch. ♂ Plants with androecia eventually intercalary, *slenderly spicate.* ♂ *Bracts much smaller than vegetative leaves, gibbous at base*, contiguous to subimbricate, forming a *distinct, if not compact, slender androecium; monandrous.* ♀ Plants somewhat more robust, to 2.5–3 mm wide; subinvolucral leaves usually more *strongly undulate* than vegetative leaves, varying on individual plants from spreading and somewhat adaxially convex to suberect and somewhat concave. ♀ Bracts wide, ± undulate, or crispate, 1.2–2× as wide as long, slightly united at base with each other and bracteole, ⅓–½ bilobed into *broad, obtuse lobes*, the sinuses often acute, usually ± *gibbous*; lobes obtusely, irregularly, coarsely undulate-crenate, *devoid of sharp teeth;* bracteole ½–⅔ as large as bracts, 2–3-lobed. Perianth ± brownish or rarely purplish, elongate-clavate to cylindrical, plicate, contracted distally into a scarcely lobed mouth, the *apex crenulate with thin-walled cells 2–3× as long as wide* (ca. 12–16 × 32–50 μ). Capsule wall 3-stratose; outer layer of rectangular cells ca. 25 μ wide × 45 μ long (occasionally divided by a transverse wall and then 20–30 μ long); radial longitudinal walls of undivided cells with 2–3 reddish brown vertical bands, slightly extending onto a tangential wall; radial transverse walls mostly with 0–1 vertical band. Inner layer of less regular, often shallowly sigmoid cells, mostly 18–21 μ wide × 62–80 μ long, with numerous vertical bands of the longitudinal walls (visible externally as nodular thickenings), which most often extend onto the exposed tangential wall as incomplete tangential bands; in many cases with tangential bands complete. *Spores 14.5–18 μ in diam.*, reddish brown, finely verruculose; elaters ca. 110 μ long × 7–7.5 (rarely 8) μ in diam., bispiral, the spiral reddish brown, 2 μ wide. Seta 300 μ in diam., with 26–27 epidermal rows of cells, mostly 8–9 cells in diam.

Type. The Point, Grand Marais, Cook Co., Minn., growing in peaty depressions (*RMS 19468*).

Distribution

Lophozia latifolia is very predominantly an arctic species, although the type came from the "Tundra strip" along Lake Superior (see Schuster, 1957, p. 259), occurring northward as far as there is exposed land.

The species is supposedly rather frequent in northern to northernmost Scandinavia (Norway, Sweden; Arnell, 1956, p. 121) and in Spitsbergen (Arnell and Mårtensson, 1959). However, Grolle (1967) has shown that all European reports, except for one from Spitsbergen, are based on errors in identification. It occurs in North America in Alaska, on the west Pacific and Arctic coasts (Persson, 1962, p. 8), westward to Bering Strait (Persson, 1963, p. 7).

ELLESMERE I. Steep, weathered slope of The Dean, ca. 82°26′ N., 62°05′ W. (*RMS 35405*); steep scree slope, ca. 2000 ft, 6–7 mi WNW. of Mt. Olga (*RMS 35579b, 35559, 35576*). N. GREENLAND. "Sidste Naes," Independence Bay, Peary Land, 81°52′ N. (*P. Freuchen, June 18, 1912*!; reported by Hesselbo, 1915, as "*L. bicrenata*").[139] NW.

---

[139] The plant reported by Arnell (1960) from Herlufsholm Strand, Peary Land (*Holmen 7263*), as *L. latifolia* represents *L. (Leiocolea) heterocolpa* var. *harpanthoides*, as shown by the distinct underleaves and occasional brownish ovoid gemmae on erect shoots. A second plant, from Cap Salor, 82°54′ N. (*Th. Wulff*), reported by Arnell (1960) as *L. latifolia*, is sterile and seems almost impossible to determine; it is probably *L. alpestris* subsp. *polaris*.

GREENLAND. Siorapaluk, Robertson Bugt, 77°48′ N., 71° W. (*RMS 45706*); Thule (Dundas), Wolstenholme Fjord (*P. Freuchen, July 31, 1919!*; mixed with *Plagiochila arctica, Aulacomnium turgidum*).   W. GREENLAND. Søndre Strømfjord (*Böcher, Aug. 29, 1946!*; published as *L. alpestris* var. *major* in Böcher, 1954); shore of L. Ferguson, head of Søndre Strømfjord, *Salix*-moss tundra (*RMS & KD 66-055, 66-050a, 66-048, 66-082,* etc.); Tupilak I., W. of Egedesminde, 68°42′ N. (*RMS & KD 66-107a*).   E. GREEN-LAND. Hurry Inlet, 70°51′ N. (Dusén); reported by Jensen (1906a) as *L. alpestris* var. *major*.[140]   LABRADOR. Great Whale R., 55°17′ N., 77°47′ W. (Persson & Holmen, 1961, p. 181).   MINNESOTA. The Point, Grand Marais, Cook Co. (*RMS 19468,* type; *13268b, 13277,* isotopes).

Ecology

Lophozia latifolia is predominantly a plant of peaty, acid sites. I have collected it in northwest Greenland on moist, peaty, steep, vegetated slopes, manured by leachings from dovekie roosts, associated with *Cephalo-ziella elegans* s. lat., *Tritomaria exsectiformis arctica, Polytrichum,* and other mosses. It occurs in west Greenland in wet, sphagnous areas, often over and amidst *Sphagnum,* associated at times with *Lophozia* (*O.*) *kunzeana* and *binsteadii.* The occurrence over *Sphagnum,* as straggling isolated stems, often conspicuously brownish or purplish brown (at least on the androecia), is diagnostic.

However, the species is not rare in basaltic or weakly basic sedimentary regions, *after* humus or peat deposition. At the type locality, in a basaltic area, in peat-filled, insolated rock crevices where there is runoff, associated with *Polytrichum, Lophozia* (*O.*) *kunzeana,* and *Sphagnum.* In the Arctic restricted to "noncalcareous or very weaky calcareous areas" and associated with *Anastrophyllum minutum grandis, Lophozia excisa, Tritomaria quinquedentata,* and various mosses. The ecology of the species is discussed more fully in Schuster (1953d) and Schuster et al. (1959).

Differentiation

A critical and difficult species, characterized by the following ensemble of features: (*a*) dioecious inflorescences, but with androecia and gynoecia (and perianths) usually freely produced; (*b*) androecia slenderly, often laxly spicate; (*c*) perianth with mouth crenulate with fingerlike, thin-walled cells free only distally; (*d*) gemmae production sporadic— especially when, as is usually the case, sex organs are freely produced; (*e*) ♂ bracts monandrous; (*f*) cells rather large, leptodermous, with small (rarely feebly bulging yet still small) trigones, the wall pigments

---

[140] The Dusén collection, of which I have seen two slides at Copenhagen, is mostly sterile (hence the Cap Brewster plant was selected as the lectotype of var. *major*) but bears a few gynoecia.

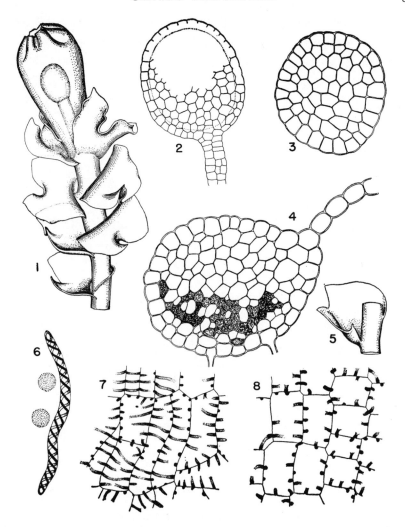

FIG. 196. *Lophozia* (*Lophozia*) *latifolia*. (1) Apex of gynoecial plant (×13.3). (2) Antheridium (×133). (3) Seta cross section (×106). (4) Stem cross section (×180). (5) ♂ Bract, somewhat spread out to show basal tooth (×25). (6) Elater and spores (×330). (7) Inner cells of capsule wall (×343). (8) Epidermal cells of capsule wall (×343). [All from type, *RMS 19468*, Grand Marais, Minn.; all from Schuster, 1953a.]

usually brownish, rarely with a little dull purplish pigmentation of ventral face of plant and/or perianths.

In criteria *c*, *d*, and *f*, *L. latifolia* very closely approaches *L. excisa*, and, sporadically, plants which closely correspond to *L. latifolia* (e.g., *RMS & KD 66–050a*) may produce, in addition to numerous unisexual individuals,

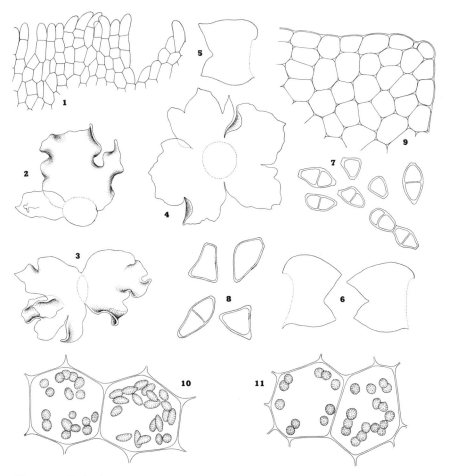

Fig. 197. *Lophozia (Lophozia) latifolia.* (1) Perianth mouth (×ca. 140). (2) Bract and bracteole; the bracteole with a surface lamella (×21.5). (3) Ring of bracts and bracteole (×18.5). (4) Bracts and bracteole, flattened (×16). (5–6) Leaves (×16). (7) Gemmae (×260). (8) Gemmae (×465). (9) Leaf lobe apex (×210). (10) Median cells with oil-bodies (×ca. 680). (11) Median cells with oil-bodies (×ca. 650). [1–9, from type; 10, *RMS 45869*, Inglefield Bay, NW. Greenland, a phenotype with one or two paroecious individuals; 11, *RMS 45706*, Siorapaluk, NW. Greenland.]

an isolated individual in which the slenderly spicate androecium, with monandrous bracts, is followed by several pairs of vegetative leaves and then a gynoecium.

The rare formation (and the purple color) of gemmae eliminates almost all other species of *Lophozia s. str.*; the free fertility, the slender androecia, and the crenulate perianth mouth collectively eliminate all other *Lophozia* species except *L. excisa.* Unlike the latter, in which paroecious inflorescences are the norm (and the large, leaflike ♂ bracts are usually 2–3-androus), *L. latifolia* often occurs with ♂ and ♀ shoots in separate small patches. Also, *L. latifolia* never occurs in the crispate, rigid, and rather fleshy phase so common in *L. excisa*. The usual presence of brown pigmentation, at least of the upper parts of leaves (and lack of purplish pigmentation, except occasionally on the venter of the plant), serves additionally to separate *L. latifolia* from *L. excisa*. *Lophozia latifolia* also sometimes has somewhat larger leaf cells and slightly larger spores than *L. excisa*; it often has less distinctly polygonal gemmae; the ♂ bracts, aside from the apical lobes and basal tooth, are always edentate. Nevertheless, there are clear and close affinities to the lax bog forms of *L. excisa*, "var. *cylindrica*" and "var. *jurensis*," both of which may develop broader than usual leaves that are more than normally bilobed.

Confusion of *L. latifolia* with peat-inhabiting forms of *L. alpestris* subsp. *polaris*, especially the fo. *sphagnorum*, is very readily possible. However, all arctic forms of *L. alpestris* are almost constantly sterile and never produce the diagnostically slender androecia of *L. latifolia*; they usually bear abundant bright red gemmae. When gynoecia are present, the very considerable difference in the perianth mouth will immediately separate all forms of *L. alpestris* from *L. latifolia*.

Sectio *LONGIDENTATAE* Schust.

Lophozia subg. *Dilophozia* sect. *Longidentatae* Schust., Amer. Midl. Nat. 45(1):58, 1951.

Plantae dioeciae, erectae suberectaeve, folia subverticalia oblonga elongataque lobos angustos acutosque habentia, os perianthii laciniis ciliisque usque ad 6–8 cellulas longis armatum.

Type. *Lophozia longidens* (Lindb.) Macoun.

Including, besides the type, *L. ascendens* (Warnst.) Schust.

This section is briefly diagnosed in the discussion (p. 494). It includes the only species of subg. *Lophozia* in which the shoots are strict and stand away from the substrate (independent of crowding), and the leaves are subvertical and narrow, with characteristically elongated leaf lobes. The perianth mouth is more conspicuously ciliate to ciliate-laciniate than in any other of our taxa of subg. *Lophozia*.

LOPHOZIA (*LOPHOZIA*) *LONGIDENS* (*Lindb.*) *Macoun*
[Figs. 202:5; 198, 199]

Jungermannia porphyroleuca var. *attenuata* Nees, Naturg. Eur. Leberm. 2:80, 1836.
Jungermannia longidens Lindb., Musci Scand., p. 7, 1879.
Lophozia longidens Macoun, Cat. Canad. Pl. 7:18, 1902; Evans, Rhodora, 9:59, 1907 (in part; in part *L. ascendens*); K. Müller, Rabenh. Krypt.-Fl. 6(1):661, fig. 308, 1910; Macvicar, Studs. Hdb. Brit. Hep. ed. 2:177, 1926; Schuster, Amer. Midl. Nat. 42(3):568, pl. 8, figs. 7–9, 1949; Schuster, *ibid.* 45(1):59, pl. 16, figs. 1–3, 1951; Schuster, *ibid.* 49(2):356, pls. 17, 19, figs. 1–3, 19, figs. 1–8, 1953; Schuster, The Bryologist 55:177, 1952; K. Müller, Rabenh. Krypt.-Fl. ed. 3, 6(1):662, fig. 203, 1954.

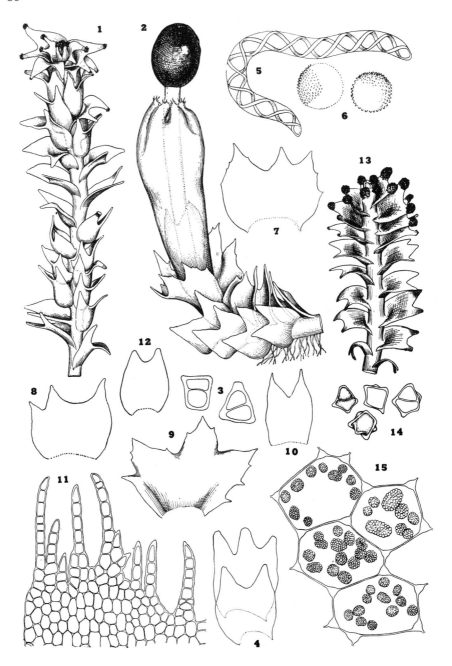

FIG. 198. *Lophozia* (*Lophozia*) *longidens*. (1) Shoot with intercalary androecia and gemmae (×23). (2) Perianth-bearing shoot apex, lateral view (×23). (3) Gemmae (×380). (4) Two leaves of mod. *parvifolia* (×39). (5–6) Elater and spores (×885).

In erect to strongly ascending patches or tufts, usually pure to dark green, rather slender (shoots 1.2–1.8 mm wide × 8–15, occasionally 20–30, mm high), the shoots subsimple, arising from a creeping, more or less branched base; stems 9–12, occasionally 12–16, cells high, 200–275 μ thick, erect portions usually with sparse rhizoids; dorsiventral differentiation distinct (slight on weak stems), ventral, smaller medullary cells heavily mycorrhizal with age, at least on prostrate shoot sectors; stems concolorous, with age often somewhat brownish beneath, never reddish. Leaves approximate to somewhat imbricate, spreading (upper erect-spreading, often with somewhat spreading lobes; lowermost often squarrose, with reflexed lobes), their insertion weakly oblique, the leaf planes almost transversely oriented on the upper leaves; leaves ovate-quadrate to ovate-rectangular, on weak and normal shoots clearly longer than wide (averaging 0.65–0.75, occasionally to 0.85, as wide as long; on robust shoots occasionally to 0.95–1.0 × as wide as long), occasionally to 850–890 μ wide × 930–1000 μ long on robust plants, the antical and postical margins nearly equally, slightly arched, scarcely decurrent antically; apex ¼–⅓, occasionally to ⅖, 2-lobed (rarely 3-lobed on individual leaves), the sinus deeply lunate or obtuse usually, the lobes narrowly triangular, acute, straight or barely divergent, hornlike (on gemmiparous leaves the very tips normally obtuse). Cells with small, concave-sided to rather small but bulging trigones and thin walls usually quite lacking all trace of secondary pigmentation; cells at bases of lobes 20–23 × 23–28(32) μ, becoming 23–27(30) × 28–35 μ medially and (24)25–30 × (26)32–40 μ at base; oil-bodies 4–10 (occasionally to 14) per cell, spherical (and 4.5–6 μ) to irregularly ovoid (and then occasionally up to 4 × 9 to 4.5–6.5 × 10 μ), formed of numerous quite discrete, peripherally bulging spherules, appearing papillose, without a central homogeneous sphere; usually rather strongly chlorophyllose, the cells somewhat opaque. Gemmae in reddish yellow to reddish brown, conspicuous masses, polygonal to quadrate to trapezoidal, large, varying from 14–18 × 20–28 μ up to 20–22(24) × 24–27(35) μ, isolated ones to 16 × 32 μ, 1–2-celled.

Dioecious; rarely with perianths. ♂ Plants more slender than ♀, often more freely branched, denser-leaved, the leaves erect-spreading; bracts in 5–6, up to 6–8, pairs, at first terminal, eventually intercalary; bracts imbricate, saccate at base; 1–2-androus. ♀ Bracts larger than leaves, usually 2–4(5)-lobed for ⅓–⅔ their length, the lobes unequal in size, from triangular to hornlike and sharply acute, more or less coarsely dentate to sparsely serrate on the margins, exceptionally subentire, broader than long; bracteole lingulate to bilobed or dentate, relatively small, often united with 1 bract. Perianth obclavate to obpyriform, longly exserted, plicate near the contracted mouth, the mouth lobulate and rather closely ciliate or ciliate-dentate, the longer cilia 3–6, occasionally to 8–9, cells long, usually numerous and somewhat crowded. Spores 10–13 μ, verruculose, relatively coarsely so (compared to L. ascendens).

(7–9) ♀ Bracts (×21.5). (10) ♀ Bracteole (×21.5). (11) Perianth mouth (×188). (12) Leaf (×ca. 24). (13) Gemmiparous shoot (×ca. 20). (14) Gemmae (×300). (15) Median cells with oil-bodies (×650). [1–4, RMS 17329, Amygdaloid I., Isle Royale, Mich; 5–11, Belle Rose I., Minn., RMS; 12–14, Blue Mt., Adirondack Mts., N.Y., RMS; 15, Susie Isls., Minn., RMS.]

Type. Finland (*Lindberg*).

Distribution

Essentially a species of the "subarctic-subalpine Coniferous Zone (Canadian-Hudsonian of Merriam). It is extremely limited in occurrence in the region to the south of that zone" (Schuster, 1949a, p. 568).[141] Supposedly "hardly reaching the Arctic" (Buch and Tuomikoski, 1955), but several recent collections of typical *L. longidens* (subsp. *longidens*) and the large-celled subsp. *arctica* extend the species well into the Arctic. Apparently imperfectly circumboreal, and with "continental tendencies," occurring in Europe (from northern Italy, the Tyrol, Austria, the Carpathian Mts., in Bucovina, northern and eastern France, Switzerland, the montane portions of Germany north to Pomerania; northward to Denmark, Scotland, Sweden, Finland; northward to 70°). Eastward to Siberia (58–61°) and northward supposedly to Spitsbergen[142]; not known from Ireland, Iceland, or the Faroes. Also in Siberia, Shensi Prov., China, and in Asia Minor (Trapezunt). Not known so far from Japan (Kitagawa, 1965), but recently reported from the Himalaya.

In North America transcontinental in the Spruce-Fir Biome, occurring westward in British Columbia and Alberta, southward to Washington and Oregon; in the Rocky Mts. to Idaho, Montana, and Colorado. In our area as follows, largely at median and moderate elevations (1800–4000 feet; but in Europe reported from as high as 3350 m).

NW. GREENLAND. Kekertat, Harward Øer, near the Heilprin Gletscher, E. end of Inglefield Bay, 77°31′ N., 66°40′ W. (*RMS 46005;* northernmost report!). w. GREENLAND. Lyngmarksbugt, Godhavn, Disko I., 69°15′ N., 53°30′ W., ca. 25–125 m (*RMS & KD 66-564*).

QUEBEC. E. end of Gaspé Pen., N. of Malbaie R., 10–12 mi NW. of Bridgeville (*RMS 43503, 43516*); arkose hillside near The Narrows, Cairn I., Richmond Gulf (*Marr 631!*); near mouth of Great Whale R., Hudson Bay (*Marr 653!* see Schuster, 1951); Tadoussac, Saguenay Co. (*Evans 80!*; a second specimen, same locality, *Evans 8*, is largely or entirely *L. excisa*); Bic, Rimouski Co. (*Evans 147!*). NEWFOUNDLAND. Gorge of Back R., Biscay Bay, Avalon Distr.; Big Bay, Rencontre West, S. coast; Great Harbour Deep, N. Pen.; Hampden and Sandy L., C. and NE. Nfld. (Buch & Tuomikoski, 1955). NOVA SCOTIA. Mary Ann Falls, S. of Neils Harbour, Cape Breton Natl. Park (*RMS 42665b, 42671, 42683, etc.*) and 3–3.5 mi S. of Mary Ann Falls (*RMS 66326, 66329,* etc.); valley of Barrasois R., Cape Breton I. (*Nichols 302!*); Yarmouth (Macoun, Can. Hep. no. 34!); Smoky Mts., Cape Smoky (Brown, 1936a). MAINE. Mt. Katahdin: at foot of Hamline Ridge, ca. 3000 ft (*RMS 32907*, mixed with *L. ascendens!*); Saddle Trail (*RMS 17014a*). Streaked Mt., Hebron (*Allen 5; fide* Evans,

[141] Steere (1937) characterizes it as "arctic-alpine"; both Schuster (1949a, 1953) and Müller (1951–58) emphasize the virtual restriction to subarctic-subalpine regions.

[142] It is relevant that the most recent thorough account of Spitsbergen Hepaticae (Arnell & Mårtensson, 1959) does not cite the species from there.

1907b); Mt. Desert I. VERMONT. Mt. Mansfield, Smuggler's Notch (*Lorenz, 1917!*), Willoughby (*Lorenz, 1915!*); Stratton (*Howe, July 17, 1899!*); Pico Peak, near Gifford Woods (*RMS, s.n.*). NEW HAMPSHIRE. Ellis R. (*Underwood & Cook, 1889!*); Crystal Cascade, Mt. Washington (*Underwood & Cook, 1899!*); Waterville (*Lorenz 396!*); trail to Lonesome L., Franconia Mts. (*Evans et al., 1908!*); also Mt. Osceola, Mt. Tecumseh, Mt. Carrigain, in White Mts. (*fide* Lorenz, 1910a); krumholz zone at treeline, Davis Trail, S. of L. of the Clouds, Mt. Washington (*RMS 23100, 23115a*).[143] MASSACHUSETTS. Near summit of Mt. Greylock (*RMS, s.n.*). CONNECTICUT. Pistapaug L., Durham, 340 ft (*Lorenz, 1914!*).

NEW YORK. Blue Mt., near summit, and Raquette L., Adirondack Mts. (*Schuster, 1945*); Gowanda, Cattaraugus Co. (Boehner, 1943, p. 12; this report needs verification); Little Moose L., Herkimer Co. (*Haynes, 1910*); between Cornell and Wittenberg Mts. (*RMS 17586a*) and Slide Mt., Catskill Mts. (*Britton, 1901!*); Lyon Mt., Adirondack Mts. (*RMS A-224, A-227*); crest of N.-facing cliff, Peasleeville, E. side of Adirondack Mts. (*RMS A-218*). MICHIGAN. Ft. Wilkins (Steere, 1937); Copper Harbor and L. Lily, Keweenaw Co. (*RMS 39155*); Amygdaloid I., Isle Royale (*RMS 17329a, 17329, 17429a*, with *L. ascendens*); Phoenix, Keweenaw Co.; Marquette, Gogebic, Chippewa Cos. (*Steere*). WISCONSIN. Sand I., Apostle Isls., Bayfield Co. (*RMS, s.n.*).[144] MINNESOTA. The Point, Grand Marais (*RMS 13260*); Belle Rose I., small island between Susie and Lucille Isls., Big Susie I., Lucille I., all in Susie Isls., Grand Portage, Cook Co. (*RMS 12136, 12145, 11984, 12133, 12142, 12100, 13675, 13661*); Hat Pt., Grand Portage (*RMS 13208*); Pigeon R., Ontario border, N. of Grand Portage (*RMS 13426, 13413*); Pigeon Pt. (*RMS 17364, 17369*); near Lutsen, Cook Co. (*RMS 15094, 15088*); Grand Portage I. (*Holzinger*); Great Palisade, L. Superior, Lake Co. (*RMS 12013*); Safety I., Lake Co. (*Holzinger*); Oak I., L. of the Woods (*RMS 17196a, 17262*); Nopeming, St. Louis Co. (*Conklin 1382*).

Although a common corticole and saxicole of moist and humid regions of the northern Spruce-Fir Forest, it is totally lacking from the Appalachian extension of this from Virginia to North Carolina and Tennessee, exactly as is its close ally, *L. ascendens*.

Ecology

The species inhabits three types of niches: acidic (never basic!) rocks, the lower trunks of trees, and decaying logs and stumps. According to Evans (1907b, p. 60), "it seems to attain its best development on rocks, but also occurs on logs," while Lorenz (1910a) finds it most common on living trees in moist areas, especially on north and east sides of trees, accompanied by *Dicranum, Plagiothecium, Radula*, and etiolated *Jamesoniella*, usually a foot or two above the ground rather than at the very base; it frequently occurs on yellow birch (*Betula lutea*), and above the birch zone often on scrub balsam. I have similarly observed it on birch bark (both

---

[143] New Hampshire specimens (*Farlow*) reported as *L. longidens* by Evans (1907b, p. 59) have green gemmae; they belong to *L. ascendens*.

[144] Frye and Clark (*loc. cit.*) give Conklin (1914a) as an authority for the occurrence of this species in Wisconsin. The Conklin paper deals with Isle Royale hepatics only. Conklin (1929) in his comprehensive account of the Hepaticae of Wisconsin does not cite it from the state. I have seen it only from Sand I.

yellow and paper birch), associated at times with the preceding species, at times with *Ptilidium pulcherrimum* (Mt. Katahdin).

On shaded, humid to damp cliffs and ledges found with several other Lophoziae, especially *L. barbata*, *L. ventricosa* var. *silvicola*, *Tritomaria quinque-dentata*, and various mosses (often *Dicranum*). It always occurs on acidic rocks (arkose, granite, gneiss), with the measured pH from 5.6 to 4.5 (Schuster, 1953). In oceanic regions, such as Newfoundland, it may occur with *Frullania tamarisci* (Buch and Tuomikoski, 1955).

Perhaps equally frequent is the occurrence on damp to moist decaying logs, where associated (Schuster, 1953, p. 356; Schuster, 1952, pp. 181–182) chiefly with *Lophozia ascendens*, *L. ventricosa* var. *silvicola*, *L. porphyroleuca*, *Jamesoniella autumnalis*, *Jungermannia lanceolata*, *Nowellia curvifolia*, *Tritomaria exsectiformis*, *Calypogeia suecica*, e.g., various members of the *Nowellia-Jamesoniella* Associule; see especially Table 1 (pp. 212–213) and p. 277 in Schuster, 1957. Its repeated and frequent occurrence on decaying logs, admixed with *L. ascendens*, is of considerable systematic significance.

With virtually no exceptions, occurring in dense shade and in sites with relatively high humidity. In our area, invasion of the butts of trees occurs only in the mountains in sites with a persistently low saturation deficit. Usually a pioneer or near pioneer in nature, disappearing when a dense moss mat is formed (either over rocks or logs); occurring only very exceptionally on soil.

In the Arctic, where *L. longidens* is very rare and always sterile (but gemmiparous), it is apparently strictly limited to very sheltered but rather dry—if humid—sites. I have collected it twice, both times in sheltered clefts in bedrock, once with *Dicranum* and *Rhacomitrium*, the other time with *Polytrichum*, *L. hatcheri*, *L. excisa*, *Cladonia*, and *Stereocaulon*, both times on a thin soil layer over rocks. The plants are then usually rather small (leaves 680–725 μ long × 650–660 μ wide to 770–815 μ long × 575–595 μ wide), but have typically elongated leaves (1.2–1.35× as long as wide). In such sites snow lies long, but during the growing season the plants hardly become moistened by rain. In the far north, thus, the species is an "ecological specialist" and seems to occur in only a limited series of biotypes.

Differentiation

Lophozia longidens fruits rarely, but constantly and abundantly produces gemmae. Gemmae occur on erect or stiffly ascending shoots bearing ± crowded leaves which are much more nearly transversely oriented than the lower leaves and are erect or slightly spreading basally, with the lobes erect, varying to strongly and characteristically squarrose. The orange-red to reddish brown gemmae masses are characteristic, occurring in conspicuous clusters at lobe apices of uppermost leaves; in exceedingly shaded sites, or when juvenile, they may be only slightly reddish (the gemmae masses then appearing orange in color), but mature gemmae are never totally devoid of secondary pigmentation.

Meylan (1924) presumed the species was a "subspecies" of *L. ventricosa*, but both Buch (1933) and Müller (1938) have presented cogent arguments against this viewpoint and I have indicated (Schuster, 1951a) that *L. longidens* (and the closely affiliated *L. ascendens*) belong in a section of their own.

Lophozia longidens is a distinctive species, showing a close relationship only to *L. ascendens*, sharing the following diagnostic features: (*a*) erect or strongly ascending growth; (*b*) narrow, subrectangulate leaves with sharp, narrowly triangular to hornlike lobes, separated by a relatively deep sinus; (*c*) mostly 4–10 (in occasional cells 12–14) oil-bodies per cell, without a central highly refractive sphere; (*d*) a tendency for the upper leaves to be suberect, but often with somewhat spreading lobes, while the distal portions of lowermost leaves are strongly spreading to squarrose; (*e*) relatively slightly oblique insertion of the leaves, with the leaf surfaces often subtransverse; (*f*) perianth mouth armed with cilia or laciniae to 6–9 cells long.

The relatively slightly oblique leaves, their hornlike lobes, and the ascending growth give the plant a distinctive facies, shared only by *L. ascendens*; on this basis the two species can be separated in the field from other *Lophozia* species. *Lophozia longidens* and *L. ascendens* typically bear narrower leaves than the other species of *Lophozia*, normally merely 0.6–0.75(0.8) as wide as long (Schuster, 1951a, p. 58), although robust plants may have them considerably broader, rarely approaching as wide as long.[145]

Lophozia longidens differs from *L. ascendens* in only a few vegetative features (Schuster, 1952, pp. 177–179), chiefly the reddish brown color of the mature gemmae. Secondary pigmentation of the gemmae takes place only with maturation, the younger ones lacking pigments.

Consequently, if only dry, much handled herbarium material, which has lost mature gemmae, is studied, confusion may arise. However, in the field gemma color is an excellent character, differentiating the two species without difficulty. Material of the two taxa, growing intermingled, has been collected many times (chiefly on Mt. Katahdin, Me., where the two are almost invariably consociated), and in each case gemmae differences were found to hold. Admittedly, juvenile plants of *L. longidens* may show gemmae with only a faint brownish

---

[145] This was noted by Jones [Trans. Brit. Bryol. Soc. 2(2): 342, 1953] for "robust stems," while "the leaves of the more slender stems are narrower and more typical." Since leaf dimensions in this species, as in other members of the family, are allometrically controlled, such unusually broad leaves on markedly robust individuals are to be expected.

tinge; such plants alway occur with mature, more typical individuals. There is also a distinct difference in gemma size; gemmae of *L. longidens* average 20–22 × 24–27 *μ*, with a few narrower and up to 16–17 × 32 *μ*; by contrast gemmae of intermingled plants of *L. ascendens* (in *RMS 32907*) average 11–15 × 15–19 *μ*, a few to 19 × 20 *μ*.

Correlated with differences in gemmae are slight discrepancies in the oil-bodies. Both species show the same number of oil-bodies per cell (4–10; in isolated cells 10–12 or even 14), in both varying from spherical (in cells with over 8) to ovoid or ellipsoid (chiefly in cells with fewer oil-bodies). However, two slight differences are apparent. (1) In *L. longidens* oil-bodies appear rather distinctly fine-botryoidal, the component spherules protruding distinctly through the peripheral bounding membrane, giving the surface a perceptibly papillose appearance; in *L. ascendens* oil-bodies appear almost smooth, with barely perceptible component spherules, not distinctly protruding, lying in what appears to be a matrix of nearly identical refractive index (and thus almost invisible), the surface not distinctly papillose. (2) In *L. longidens* oil-bodies average larger, and more of the cell lumen (usually ca. ⅔) is obscured by them, with spherical bodies from 4.5 to 5.5(6) *μ*, ovoid or ellipsoid ones to 4 × 9 or even 6.5 × 10 *μ* in isolated cases; in *L. ascendens* spherical oil-bodies average 4–5 *μ*, occasional ones only 3.6 *μ*, elongated ones from 3.6 × 7 *μ* to 4 × 8 *μ*. These differences are slight but appear to be constant; they are best appreciated when the two species grow intermingled.[146]

Other differences have been given (Schuster, 1953, p. 352): (*a*) more robust size of *L. longidens*, which attains a height of 8–25 mm vs. mostly 3–10 mm in *L. ascendens;* and (*b*) a somewhat deeper green color in *L. longidens*, *L. ascendens* being pale yellow-green. The dark green color of *L. longidens* has also been cited by Macvicar (1926, p. 178), who emphasizes the "very chlorophyllose" leaves. *Lophozia ascendens* never becomes as deep green as typical *L. longidens*. However, small phases of the latter, often only 4–7 mm high, intermingled with *L. ascendens* on shaded decaying wood (as in *RMS 32907*), are scarcely larger than *L. ascendens* and relatively light green, scarcely different from *L. ascendens* in color. Slender, juvenile phases of *L. longidens* may, therefore, be difficult to separate from relatively mature *L. ascendens*.

When fertile plants are available (only infrequently, as both species reproduce almost exclusively by gemmae), *L. longidens* can be separated from *L. ascendens* readily on the basis of (*a*) broader, more deeply 2–5-lobed ♀ bracts, whose acute lobes are usually irregularly dentate; (*b*) the ciliate, rather than lobulate-laciniate perianth mouth.

Variation

The preceding account is based exclusively on normal (subsp.) *longidens*. In the high Arctic it is in part replaced by the following subspecies (separable by the key on p. 504).

---

[146] As in *RMS 32907* (Mt. Katahdin, Me.). The preceding measurements and descriptions are based on plants growing intermingled with *L. "silvicola."* Figure 202 clearly illustrates these differences.

LOPHOZIA LONGIDENS subsp. *ARCTICA* Schust.,
subsp. n.

[Fig. 199:1–8]

Plantae relative dilute virides, e basi prostrata erectae; cellulae maiores et membranas tenuiores habentes quam in subsp. *longidente;* in marginibus loborum plerumque 26–30 μ, media in parte folii plerumque 30–35 × 34–42 μ; gemmae magnae, 27–34 ad 31–36 × 34–42 μ.

Relatively light and translucent green, leaves *sometimes brownish, erect in growth* from a procumbent base, to 1.5–1.65 mm wide × 1 cm tall. Stem with dorsal cortical cells slightly thick-walled, (26)30–33 × 45–65 to 26–30 × 65–92 μ. Leaves appear almost subtransversely oriented from an obliquely inserted base, suberect to almost erect basally and somewhat concave, with lobes reflexed or squarrose, the plant thus with a typical *L. longidens* facies. Leaves polymorphic, typically oblong-ovate and bilobed, but not infrequently with an antical and/or postical tooth which may be so elaborated that the leaf appears 3- or 4-fid; 2-fid leaves ca. 570 μ wide × 690 μ long to 680 μ wide × 750–775 μ long to 650 μ wide × 840 μ long (length usually from 1.1 to 1.3 × the width); lobes hornlike or narrowly triangular; sinus 0.3–0.5 leaf length. Cells *large and rather thin-walled*, those of lobe margins below tips (25)26–30(33) μ, tangentially measured; cells of lobe bases and median cells variable but usually (27)30–35 μ *wide* × (32)34–42 μ *long*, polygonal to quadrate, thin-walled (walls usually colorless, rarely yellow-brown), with small trigones; oil-bodies numerous, (9)13–24 *per cell*, spherical to rarely ovoid, 3.3–4.5 to 3.5–3.8 × 5.5–6.4 μ, with the typical rather evidently botryoidal form found in the species, *s. str;* cuticle smooth. *Gemmae large*, red-brown to reddish, usually almost spherical or quadrate-spherical with relatively low but many tuberculate extensions, hardly stellate, (25–26)27–34 μ to 31 × 34 μ, a few to 36 × 42 μ.

Type. Kânâk, Red Cliff Peninsula, NW. Greenland, in deep, vertical, dry clefts formed in bedrock by ice action, plateau between coastal hills and ice cap, 1500–1600 feet, 4 miles northwest of village (*RMS 45649*).

Cotype. Dundas, North Star Bay, in *Cassiope*-moss tundra, gentle slope above village (*RMS 46267, p.p.*).

Known only from acid sites in northwest Greenland, 75°–77°30′ N.

The type was associated with *Lophozia hatcheri, Anastrophyllum minutum, Gymnomitrion corallioides*, and various mosses and lichens. The cotype from Dundas grew associated with *Lophozia hatcheri, L. barbata, Plagiochila asplenioides, Anastrophyllum minutum grandis*.

The larger leaf cells, with many more oil-bodies, together with the greater size of the gemmae, suggest a polyploid derivative of *L. longidens*. Since there is nearly total allopatry, it seems appropriate to consider the two as distinct.

Subspecies *arctica* shows all of the usual features which allow us to recognize *L. longidens s. lat.* at a glance: reddish gemmae formed at the apices of unusually long, acutely triangular, almost hornlike, drawn-out lobes;

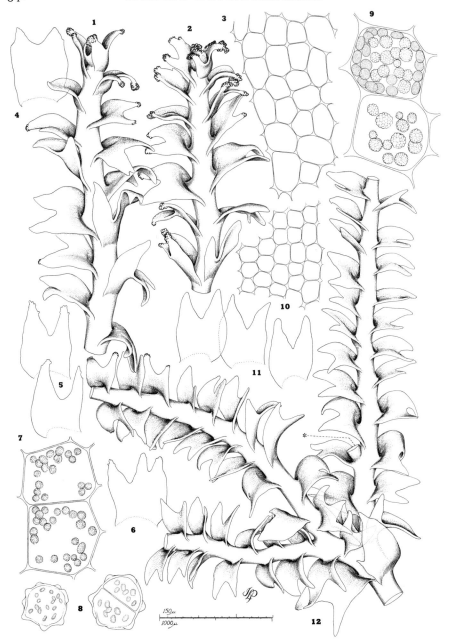

FIG. 199. *Lophozia* (*Lophozia*) *longidens*. (1–8, subsp. *arctica*; 9–12, subsp. *longidens*).
(1–2) Gemmiparous shoot apices (×24). (3) Median cells (×212). (4) Leaf from older,
mature stem (×24). (5) Normal leaves from gemmiparous stems (×31). (6) Atypical
trifid leaf (×24). (7) Median cells with oil-bodies (×545). (8) Gemmae (×530).
(9) Median cells with oil-bodies and (top cell) chloroplasts (×805). (10) Median cells

a deep sinus, descending at times to 0.5 the leaf length; leaves in part almost transversely oriented, often nearly erect at the base, with lobes gradually reflexed or squarrose distally (so that, in edge view, a leaf may look semicircularly arcuate).

Lophozia longidens subsp. *arctica* closely approaches in cell size, oil–body number, degree of pellucidity, and gemma size and color the plants referred to *L. rubrigemma* (p. 621). Occasional individuals may prove difficult to separate.

LOPHOZIA (LOPHOZIA) ASCENDENS (Warnst.) Schust.[147]
[Figs. 200, 201, 202:1–2]

Lophozia longidens Schiffn., Österr. Bot. Zeitschr. 56:26, 1906 (in part, not of Lindberg & Macoun); Evans, Rhodora 9:59–60, 1907 (in part, at least, not of Lindberg & Macoun); Frye & Clark, Univ. Wash. Publ. Biol. 6:342, 1945 (as per description, in part).
Lophozia porphyroleuca K. Müller, Rabenh. Krypt.-Fl. 6(1):669, fig. 310b, 1910; Evans, Rhodora 19:263, 1917 (*p.p.*, not of Nees).
Sphenolobus ascendens Warnst., Hedwigia 57:63, fig. 2, 1915.
Lophozia gracillima Buch, *in* Bryophyta Nova, Ann. Bryol. 6:129, figs. 1:6–13, 2:4–6, 1933; Schuster, Amer. Midl. Nat. 45:59, 61, pl. 15, 1951.
Lophozia ascendens Schust., The Bryologist 55:180, 1952; K. Müller, Rabenh. Krypt.-Fl. ed. 3, 6(1):652, fig. 198, 1954; Kitagawa, Jour. Hattori Bot. Lab. no. 28:279, fig. 10, 1965.

Small (*smaller than other species of Lophozia*), usually merely *3–6, rarely 8–15, mm high, the shoot* (*600*)*800–1250 μ wide* (somewhat wider below perianth) or often less, forming small, diffuse patches or growing as isolated plants, soon *erect or suberect* from an originally prostrate base; shoots characteristically *pale yellowish green* throughout, *in direct sunlight the older portions of the plants reddish to purplish brown* (*androecia becoming pigmented in diffuse light*). Stem rather rigid and stout, compared to total shoot width, from 245 μ high × 265 μ wide up to a maximum of 300 μ high × 400 μ wide, only slightly flattened, *at maturity only 7–11 cells high*; cortical cells largest on dorsal stem surface, (22)25–28(30) × 42–56 μ, ventrally only 15–17 × 37–50 μ; medulla dorsally formed of large and elongate cells 87–150 μ long × 17–30 μ in diam., a few to 35–38 μ in diam.; *ventral medullary band obscure*, medullary cells gradually somewhat smaller and narrowed postically, *in ca. 4 layers*, the cells 16–18 × 18–20 μ in diam., often attacked by mycorrhiza (at least in prostrate portions of shoots); *stems not becoming reddish or purplish postically*. Leaves little oblique in insertion or orientation, the line of insertion slightly acroscopically arched, *almost transverse dorsally, upper leaves suberect* (often with more spreading lobes), *the lower often with squarrose lobes*, weakly canaliculate, *narrowly rectangular to ovate-rectangular*, usually from 560–630 μ wide × 750–835 μ long up to 759–770 μ wide × 800–860 μ long, on weak shoots narrower and smaller (width on slender shoots usually only

---

[147] The citation *L. ascendens* (Warnst.) Schust. ex K. Müll., *in* Kitagawa (1963b) is incorrect.

(×212). (11) Leaves from gemmiparous shoot (×31). (12) Part of gemmiparous plant, with two *Frullania*-type terminal branches; at asterisk base of a terminal branch with associated unmodified leaf (×26). [1–8, *RMS 45649*, Kânâk, Greenland, type of subsp. *arctica;* 9–12, *RMS & KD 66-594*, Godhavn, Disko I., Greenland, subsp. *longidens;* 3, 10 to upper scale; 5, 11, to lower scale.]

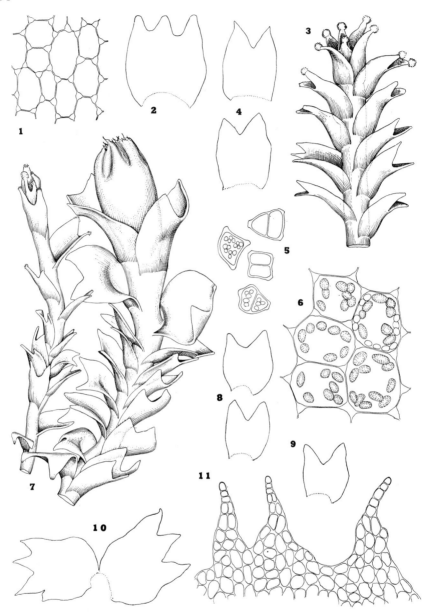

Fig. 200. *Lophozia* (*Lophozia*) *ascendens*. (1) Cells near leaf base (×285). (2) Large, trilobate leaf (×31). (3) Antical aspect of gemmiparous shoot (×24). (4) Leaves (×22). (5) Gemmae (×500). (6) Median cells with oil-bodies and, upper right, chloroplasts (×460). (7) Perianth-bearing and associated slender, gemmiparous ♂ plant (×25). (8–9) Leaves (×18). (10) ♀ Bracts (×18). (11) Apex of perianth mouth (×205). [1–3, Pigeon Point, Minn., *RMS*; 4–6, *RMS 24436*, Cornell Mt., Catskill Mts., N.Y.; 7–11, *RMS 32901 and 32902b*, Mt. Katahdin, Me.]

0.55–0.65 the length, on robust extremes up to 0.9 the length, especially associated with formation of sex organs; on upper leaves of slender shoots sometimes only 0.38–0.45 as wide as long); *sinus descending from 0.3 to 0.45, occasionally to 0.55, the leaf length,* the 2 (rarely 3) *lobes narrow, and often hornlike.* Cells of leaf middle averaging 20(24)–27 μ × 27(30)–35 μ, producing (even under humid and shaded conditions) *distinct, usually bulging trigones; oil-bodies mostly 6–10 per cell,* with extremes of 4 and 12–14, colorless and filling ca. ½ the lumen or less, spherical to ovoid, varying mostly from 4–5 μ to 5 × 8–9 μ, *formed of numerous barely protuberant, minute globules, appearing barely granular or papillose, never with a central homogeneous sphere.* Gemmae abundant, in spherical masses at apices of upper leaf lobes, *yellowish green,* their walls unable to develop secondary pigments even in direct sunlight, angular to somewhat stellate because of protuberant angles, 1–2-celled, *small,* from 14–17(19) μ × 15–19(23) μ to 10–13.5 × 16–22 μ, to a maximum of 16 × 18–25 μ.

Dioecious; rarely with perianths but often with androecia. Androecia becoming intercalary, bracts usually in only 2–4 pairs, approximate to weakly imbricate, as large as or slightly larger than juxtaposed leaves, erect and strongly saccate at base, *usually reddish* (even when rest of plant pure green). ♀ Bracts larger than leaves, *longer than wide, subrectangular,* ca. 925 μ wide × 1270 μ long to 1050 μ wide × 1300 μ long, *divided for (0.2)0.25–0.33 into 2–3(4) triangular, entire lobes;* bracteole usually absent. Perianth terete and cylindrical, to 1500 μ long × 700 μ wide, plicate and contracted to mouth; *mouth divided into (8)12–15 narrowly triangular small lobes or laciniae,* 3–5(7) cells wide at base and usually 6–10 cells long, the distal 2–5 cells uniseriate, at the base often with 1–2 short teeth. Spores finely punctate-papillose, 9.5–10.5 μ (*fide* Buch, 1933).

Type. Waterville, N.H. (*Lorenz!*); the type of *L. gracillima* from "Prov. Savonia australis, par. Lappee," Finland (*H. Buch!*).

Distribution

Restricted to the Spruce-Fir Biome of Europe and eastern North America; with recent reports from Asia (Takinosawa, Sakhalin; Kitagawa, 1963b, p. 147); also in subalpine forests in Siberia and Japan (Honshu; Kitagawa, 1965). Perhaps with further study proving transboreal in North America and perhaps circumboreal. In North America absent (as is the related *L. porphyroleuca*) from the Red Spruce-Fraser Fir extension of the southern Appalachian Mts.

In Europe found from the Pyrenees (Bagnière-de-Luchon) and Roumania (Topoluga Valley) northward to Lorraine (Dagsburg), Germany (Baden!), Switzerland, the Tyrol, Austria, the Carpathian and Tatra Mts. (Czechoslovakia; Poland), northward to Finland (type of *L. gracillima!*), Sweden (Mårtensson, 1955–56, p. 33) and northwestern USSR.

In North America occurring in the eastern Spruce-Fir Climax Forest, as follows:

NOVA SCOTIA. Near Indian Brook, Cape Breton I. (*Nichols, Aug. 10. 1909;* NY sub *Scapania subalpina*!) ;[148] decaying log at Warren L., N. of Ingonish, Cape Breton (*RMS 66301, 66303, 66304, 66306b*); E. side of Cape Breton Natl. Park, 3–3.5 mi S. of waterfall of Mary Ann Brook, (*RMS 66331, 66310, 66346*).   NEWFOUNDLAND. South Branch and Steady Brook in W. Nfld.; Port au Choix, Eddie's Cove West, near Doctor Hill, St. Barbe Bay, and Englee, all on N. Pen.; Hampden in C. Nfld. (Buch & Tuomikoski, 1955).   QUEBEC. Decaying logs, N. of R. Malbaie, ca. 10–12 mi NW. of Bridgeville, Gaspé (*RMS 43503, 43518, 43521, p.p.,* among *L. longidens, L. porphyroleuca, L. "silvicola"*; *43514, 43516, c. per.,* etc.).   ONTARIO. Gull L. Portage, L. Timagami (*Caine 2448;* trace!); Algoma; L. Opeongo, Algonquin Park; Durham (Williams & Cain, 1959).
    MAINE. Mt. Katahdin: slope of Northwest Basin, 3500 ft (*RMS 17040, 17015, 17038*); Saddle Trail, ca. 3500 ft (*RMS 17014a*); below Chimney Pond (*RMS 32901, c. per.; 32902b*); trail to North Basin of Hamlin Peak (*RMS 32905, c. per.*).   NEW HAMPSHIRE. Mt. Moosilauke, Coos Co. (*RMS 17980a*); Waterville (*Lorenz,* type of *L. ascendens*!; reported by Evans, 1917e, as *L. porphyroleuca*) ;[149] near Zealand Falls and on Mt. Lafayette (*RMS*).   VERMONT. Stratton (*M. A. Howe, July 18, 1899*!; among *Riccardia palmata, Nowellia*); Pico Peak (*RMS;* abundant); Mt. Mansfield, on Long Trail below Taft Lodge, SE. of summit, 3200–3300 ft (*RMS 43823*).
    NEW YORK. Whitney Preserve, N. of Long L., Adirondack Mts. (*RMS 48*); Arnold L., Mt. Marcy, Adirondack Mts. (*L. R. Wilson 6*!); slope of Cornell Mt., on trail to Slide Mt., Catskill Mts., ca. 3500 ft (*RMS 24436; southernmost known station for the species*!).   MICHIGAN. Deer L., Barraga Co. (*RMS 16101, 16104*); Copper Harbor, Keweenaw Co. (*RMS 18209a*); Amygdaloid I., Isle Royale (*RMS 17429a, 17329 17329a, 17304, 17396a, 17398a,* etc.); Ft. Wilkins, N. of Copper Harbor (*RMS 39304, 39310,* etc.).   WISCONSIN. Sand I., Apostle Isls., Bayfield Co. (*RMS 17523a, 17552*). MINNESOTA. Pigeon Pt., near Grand Portage, Cook Co. (*RMS 8001a, 5009c*).

Ecology

Almost exclusively restricted to moist, often rather wet, decaying logs or stumps (most frequently in quite an advanced state of decay), with a measured pH of 4.6–5.2 (Schuster, 1953). Occasionally on peaty soil along paths or on humus adjacent to decaying logs, but such occurrences are exceptional. Unlike the closely related *L. longidens,* the species never appears to undergo ecesis on shaded damp rock faces, nor on the bark of the lower trunks of living trees; as a consequence, it is usually much rarer.

On decaying logs almost always in areas with high humidity, such as moist spruce-fir forests; often found adjacent to either running water (e.g., mountain brooks) or to lakes, perhaps because of the high humidity requirement (stressed by Schuster, 1952). Although growing occasionally

---

[148] The *Lophozia* is sparse, a mod. *colorata-parvifolia,* evidently from a wet but sunny log; shoots are only 500–650 μ wide (stems 150–165 μ) × 3–6 mm high, almost flagelliform, very slender, with all but the uppermost leaves appearing almost blackish (because of an intense purplish pigmentation!), but the gemmae remaining colorless and typical. Associated Scapaniae are *S. irrigua* and *S. nemorosa* (no *S. subalpina*!); also *Lophozia porphyroleuca.*

[149] Müller (1954, *in* 1951–58, p. 654) cites the type, mistakenly, as from New Haven, Conn.

in nearly full sun, on wet decayed logs, occurrences at $\frac{1}{12}$–$\frac{1}{6}$ of full light are more frequent; the bulk of collections comes from sites with a light intensity below $\frac{1}{50}$ of full sunlight.

Associated species are commonly other *Lophozia* spp. (chiefly *L. longidens, L. porphyroleuca, L. ventricosa silvicola*, less often *L. incisa* and *L. attenuata*), *Tritomaria exsectiformis, Anastrophyllum hellerianum, Blepharostoma trichophyllum, Jamesoniella autumnalis, Scapania umbrosa* (less often *S. nemorosa, S. apiculata, S. glaucocephala*), *Harpanthus scutatus, Cephalozia media, Lophocolea heterophylla, Riccardia latifrons, Nowellia curvifolia, Jungermannia lanceolata, Calypogeia suecica, Geocalyx graveolans*. The species is clearly a member of the xylicolous *Nowellia-Jamesoniella* Associule. As emphasized (Schuster, 1952; 1957, p. 277), it is usually a member of the more mesic (and ecologically more advanced, less pioneer) *Cephalozia-Riccardia* facies. Under extremely exposed conditions rarely a member of the more xeric, pioneer *Nowellia-Jamesoniella-Ptilidium* facies, and then found with *Ptilidium pulcherrimum*. The species usually occurs in admixture with a large number of associated species, often with a frequency of less than 5 %, often with "a negligible coverage." It is exceptional to find large patches of any size. Apparently only under very favorable conditions, when such patches are formed, do perianths occur.

Differentiation

A stenotypic species, of distinctive aspect, recognizable in the field *after* its diagnostic features are once appreciated. Among distinctive features are chiefly the very small size (rarely over 6 mm high); the strongly ascending to erect growth; almost transversely inserted suberect upper leaves, with at most the lobes spreading; abundant, yellowish green gemmae masses of the upper leaves; very narrow, rather deeply bilobed leaves with narrow, often hornlike lobes; the pale and yellowish green color, except for older portions of insolated plants. Until "revived" as a species (Schuster, 1952), it was variously confused in collections in North America with *L. longidens, L. porphyroleuca*, and *L. "silvicola."* Characters differentiating it from these species are detailed in Schuster (1952) and only briefly recapitulated here.

Lophozia ascendens differs from the allied *L. longidens* in the color and size of the gemmae (p. 538). These are smaller and remain yellowish green even in direct sunlight, while those of *L. longidens* become slightly reddish brown in even exceedingly diffuse light and conspicuously red-brown in direct light.[150] Other differences between the two species lie in (*a*) the

---

[150] The validity of this criterion has been demonstrated many times in nature since the two taxa very often occur together on decaying logs. On the slopes of Mt. Katahdin, Me., I have found the two species consociated at least a dozen times, yet remaining quite distinct and recognizable with the hand lens on the basis of gemma color.

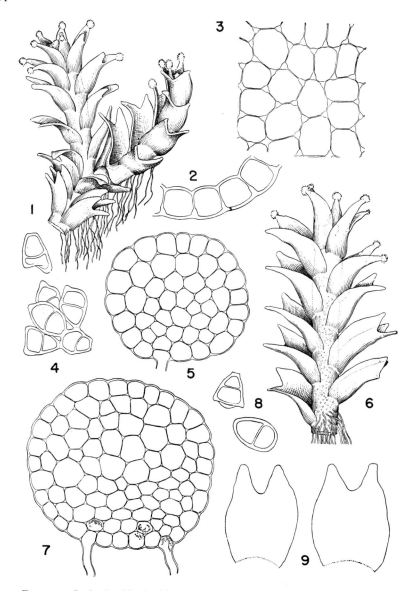

FIG. 201. *Lophozia* (*Lophozia*) *ascendens*. (1) Gemmiparous shoot with terminal branch (×20). (2) Leaf section (×250). (3) Median cells (×310). (4) Gemmae (×520). (5) Stem section, small stem (×175). (6) Gemmiparous shoot apex, ventral aspect (×26). (7) Large stem, cross section (×190). (8) Gemmae (×520). (9) Leaves (×35). [1–4, 6–9, Pigeon Point, Minn., *RMS;* 5, Schwarzwald, Baden, Germany, *K. Müller.*]

smoother, finely granular-appearing oil-bodies of *L. ascendens*:[51] in *L. longidens* oil-bodies are rather coarsely papillose; (*b*) the smaller gemma size in *L. ascendens* averaging 14–17(19) × 15–19(22) μ; (*c*) a tendency for ♂ bracts and often older leaves to become carmine to deep red in *L. ascendens*, but the apical leaves and stem and gemmae are always a diagnostic pale yellowish green; in *L. longidens* the entire plant, *except* the gemmae, is a pure green, rarely as pale as that of *L. ascendens*.

Two diametrically opposed tendencies are evident in regard to pigmentation: in *L. longidens* leaves remain green, but gemmae become brownish at maturity (no matter how exposed the plant may be), although lower portions of the plants, and particularly the ♂ bracts, become reddish, even in somewhat diffuse light. The perianth mouth of *L. ascendens* bears laciniae, or narrow lobes, usually 3–6 cells wide at base; in *L. longidens* the perianth mouth usually bears elongate, narrow cilia. The oblong to quadrate, shallowly 2–4-lobed or dentate ♀ bracts whose margins are otherwise entire are also distinctive, those of *L. longidens* being obtrapezoidal to transversely rectangular, with the longer lobes more or less spinous-dentate.

A distinction between *L. ascendens* and *L. porphyroleuca* is at times more difficult to maintain. However, when the two occur intermingled,[52] habitual differences between them allow a distinction in the field on the basis of the: (*a*) strongly acroscopically arched and suberect shoots of *L. ascendens* vs. largely prostrate shoots of *L. porphyroleuca*, with merely the very shoot apex suberect;[53] (*b*) much less robust size of *L. ascendens*; (*c*) absence of any tendency for the postical face of the stem, and of adjacent leaf bases, to acquire reddish pigmentation; such a tendency is marked in *L. porphyroleuca*, even in shade forms; (*d*) copious development of gemmae under all conditions; in *L. porphyroleuca* only extreme shade forms of wet decayed logs develop gemmae (such gemmiparous shoots are also prostrate); (*e*) laciniate-lobulate perianth mouth, that of *L. porphyroleuca* being merely short-ciliate; (*f*) erect-spreading upper leaves, usually at a 40–55° angle with the stem, and little oblique in insertion, appearing almost transversely oriented, but with the long, hornlike lobes often distinctly spreading (on older leaves often even squarrose, as in *L. longidens*); in *L. porphyroleuca*, as in *L. ventricosa* var. *silvicola*, the leaves are

[51] Kitagawa (1965, p. 280) claims that the oil-bodies of *Lophozia ascendens* are homogeneous; as Fig. 200:6 shows, they are smooth externally but faintly botryoidal internally.

[52] This is the case in numerous collections made on Mt. Katahdin. In one collection (*RMS 32903*) all four related species, *L. ascendens*, *L. longidens*, *L. ventricosa* var. *silvicola*, and *L. porphyroleuca*, were associated; these four taxa again recur, intimately admixed, in material from the Gaspé (*RMS 43503*, etc.).

[53] With excessive crowding all species of subg. *Lophozia* may become suberect in growth. This, however, is a reaction to an *extrinsic* condition, while in *L. ascendens* the strongly ascending growth pattern is *intrinsic*, being exhibited even when the plants are not crowded.

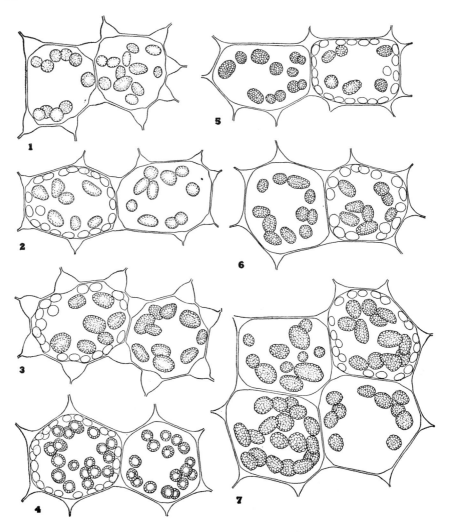

FIG. 202. *Lophozia* subg. *Lophozia:* median leaf cells with oil-bodies and (in isolated cells) chloroplasts (all ×ca. 800). (1–2) *L. ascendens.* (3) *L. porphyroleuca.* (4) *L. ventricosa* var. *silvicola.* (5) *L. longidens.* (6) *L. ventricosa* var. *ventricosa.* (7) *L. wenzelii.* [1, *RMS 32902b,* 2–5, *RMS 32905,* growing intermingled on a single log; 6, *RMS 33031,* all from Mt. Katahdin, Me.; 7, from Nissontjårro, Torne Lappmark, Sweden, leg. S. Arnell, 1954.]

obliquely inserted and oriented and spread quite widely, usually at over a 65–75° angle with the stem apex, and the lobes are short, broadly triangular, and never strongly spreading or squarrose. The discrepancies in leaf insertion and orientation and in depth of the sinus of the leaves (and consequent form of the lobes) give a habitually different facies to the two species. In size, form, and number of oil-bodies the two agree very closely.

Lophozia ascendens and *L. ventricosa* var. *silvicola* differ in much the same fashion as does the former from *L. porphyroleuca*. In fact, *L. ventricosa* var. *silvicola* and *porphyroleuca* are sometimes not separable in the field. Particularly in the form of the leaves, and their insertion and orientation, the latter two taxa are nearly identical (and differ considerably from *L. ascendens*). However, *L. ventricosa* var. *silvicola* agrees with *L. ascendens* in the ability to produce gemmae abundantly under a wide diversity of environmental conditions (whereas *L. porphyroleuca* produces them sparingly, and only under exceedingly shaded and humid conditions). *Lophozia ventricosa* var. *silvicola* further differs from *L. ascendens* in the slightly denticulate perianth mouth and the oil-bodies, which occur up to 15–25 (often only 12–16) per cell and are almost invariably spherical and distinctly biconcentric. Similar, biconcentric oil-bodies do not occur in *L. ascendens*, nor are the oil-bodies as numerous.

TABLE 5

Lophozia ventricosa var. silvicola-porphyroleuca-ascendens Complex*

Criterion	L. ventricosa var. silvicola	L. porphyroleuca	L. ascendens
Perianth-mouth teeth	1–2-celled	(1–2)3–4-celled	Laciniate-lobulate, 5–9 cells long × 3–6 cells wide at base
Leaf width: length	0.75–0.95:1	0.75–0.9:1	0.6–0.85:1
Growth pattern	Prostrate to decumbent	Closely prostrate	Suberect
Gemmae	Sparse 16–23 × 16–26 (35) μ	Absent	Abundant 14–17(19) × 15–19 (22) μ
Shoot width	1.35–1.8 mm	1.05–1.75 mm	0.8–1.25 mm
length	1–2 cm	1–2 cm	2–6 mm
Oil-bodies	11–16(20) per cell Biconcentric	(4–5)6–10 per cell Of small spherules throughout, somewhat protuberant	(4–5)6–10(12) per cell Of very delicate spherules, appearing nearly homogeneous
Leaf lobes	Broad-triangular	Broad-triangular	Narrowly triangular
Trigones	Scarcely bulging	Strongly bulging	Scarcely bulging
Stem and postical leaf bases	Carmine basally, in part	Carmine basally, in part	Green throughout

* All from *RMS 32901* and *32902*, from two decaying logs, lying side by side, under identical conditions, from Mt. Katahdin, Me.

Sectio *LOPHOZIA*

Lophozia subg. *Dilophozia* sect. *Ventricosae* Schust., Amer. Midl. Nat. 45(1):60, 1951.

Type. *Lophozia ventricosa* (Dicks.) Dumort.

The "Ventricosae" are a large and incredibly difficult complex in which species concepts have not hardened. It is possible that, associated with sporadic gametophytic mutation and rather frequent sexual reproduction, the mechanisms exist in this group for rapid adaptive evolution, and the various "small species"—to use an old-fashioned term once current in hepaticology—and minor taxa may be relatively recent in origin.[154]

LOPHOZIA (LOPHOZIA) PORPHYROLEUCA (Nees) Schiffn.

[Figs. 202:3; 203; 205:1–4]

Jungermannia porphyroleuca Nees, Naturg. Eur. Leberm. 2:78, 1836.
Jungermannia ventricosa var. *porphyroleuca* Hartm., Skand. Fl. ed. 10, 2:138, 1871; Husnot, Hep. Gall. p. 36, 1875; Limpricht, Krypt.-Fl. Schles., p. 280, 1876.
Jungermannia guttulata Lindb. & Arn., Kgl. Sv. Vetensk.-Akad. Handl. 23(5):51, 1889.
Lophozia guttulata Evans, Proc. Wash. Acad. 2:302, 1900; K. Müller, Rabenh. Krypt.-Fl. 6(1):668, fig. 310, 1910 (in part),[155] Macvicar, Studs. Hdb. Brit. Hep. ed. 2:181, figs. 1–5, 1926.
Jungermannia porphyroleuca var. *guttulata* Warnst., Krypt. Fl. Prov. Brandenburg 1:182, 1903.
Lophozia porphyroleuca Schiffn., Krit. Bemerk. Eur. Leberm. *in* Lotos 51(7):61, 1903; Macvicar, Studs. Hdb. Brit. Hep. ed. 2:180, figs. 1–4, 1926; Schuster, Amer. Midl. Nat. 42(3):569, pl. 8, figs. 1,2, 1949; Schuster, *ibid.* 49(2):363, pl. 19, figs. 9–15, 1953; K. Müller, Rabenh. Krypt.-Fl. ed. 3:650, fig. 197, 1954.
Lophozia ventricosa var. *porphyroleuca* K. Müll., Rabenh. Krypt.-Fl. 6(1):666, 1910.[156]
Lophozia fauriana Steph., Spec. Hep. 6:112, 1917.
Lophozia porphyroleuca var. *guttulata* Meyl., Beitr. Krypt.-Fl. Schweiz 6:177, 1924; K. Müller, Rabenh. Krypt.-Fl. ed. 3, 6:651, 1954.[157]

Usually green or (in sun) pale green, often locally to largely reddish-tinged, rather small, almost always on decaying wood, (0.8)*1.0–1.5*(1.75) *mm wide*, 1–2 cm long, ♂ plants often smaller, *closely prostrate* with obscurely ascending apices, stems creeping, *reddish or carmine* to (with age) red-brown beneath; ventral 0.3 of medulla small-celled, mycorrhizal. Rhizoids abundant, colorless, but in sun forms *often carmine at base*. Leaves *often carmine at base*, moderately imbricate, rather subtransversely oriented, somewhat concave but not loosely conduplicate-canaliculate (*lobes often obscurely incurved*, never spreading or squarrose), *oblong or rectangular*, mostly ca. 650–750 μ wide × 825 μ long (*width-length ratio ca. 0.75–0.9:1*), but the subinvolucral leaves occasionally subquadrate-subrotundate and ca. 950 × 950 μ,[158] bilobed for ca. 0.25–0.3 their length by means of a generally rectangulate sinus, the *lobes triangular to broadly triangular*, acute. Underleaves lacking. Cells ca. (19)21–25 × 24–30 μ medially, very strongly collenchymatous *even under shaded, moist conditions, the trigones always strongly bulging* and (in lobes) often subconfluent, the lumen sometimes nearly stellate; oil-bodies distinctly granular-botryoidal, *papillose-appearing*,

---

[154] For example, I regard the Japanese *L. fauriana* Steph., kept as distinct from *L. porphyroleuca* by Kitagawa (1965), as at best a minor variant of this (initially European) species. Similarly, *L. silvicoloides* Kitagawa seems at most a variety parallel with *L. ventricosa* var. *silvicola*. I think that the Ventricosae will prove to include a limited suite of highly polymorphic species; my treatments of *L. ventricosa* and *L. alpestris* reflect this belief.

[155] Too late to make the needed changes it has been shown by Grolle (1968) that *J. porphyroleuca* was never effectively published by Nees; the correct name must be *L. guttulata*.

[156] Müller accredits the combination to Hartman (1871), who, however, formed it under the genus *Jungermannia*.

[157] Müller accredits the combination to Warnstorf (1903), who, however, formed it under the genus *Jungermannia*.

[158] Müller (1951–58, p. 650) states "oft doppelt so lang wie breit." I have seen no leaves less than 0.65 as wide as long.

without central spherule, variable in size and shape, ranging from subspherical and
4–5 × 4.5–6 μ to ovoid or ellipsoidal and 4–4.5 × 5–7 μ to 5(6–7) × 9–10 μ,
only (3–4)5–9, in isolated cells to 10–12 per cell; cuticle smooth. *Gemmae
normally lacking,* rarely produced in small quantity from uppermost leaf lobes,
yellow-green and as in *L. ventricosa.*

Dioecious, but *abundantly fertile and very often with capsules,* the capsules produced into late September. Androecia becoming intercalary; bracts saccate or
gibbous basally, usually in 4–5 pairs, often pigmented; usually diandrous.
♀ Bracts narrow, oblong to ovate, 2–3(4)-lobed for (0.15)0.2–0.3 the length,
ca. 920 μ wide × 1050 μ long or longer; bracteole small and ± lanceolate,
united with one of the bracts, or sometimes apparently obsolete or absent.
Perianth (except in shade forms) often reddish-pigmented except at base and
mouth, elongate ovate, at the mouth shallowly lobed into 8–12 small lobes,
which, in addition to 1–2-celled teeth end in and/or bear cilia *formed of 3–4(5)
superimposed cells.* Capsule ovoid, red-brown. Spores 8–10 μ; elaters 6–7 μ in
diam.

Type. Central Europe.

Distribution

Holarctic in range, although somewhat imperfectly so (rare or lacking
in most of the more continental areas; with a weakly suboceanic range);
absent in the Tundra.[159]

In Europe, throughout most of Scandinavia, southward to the Alps
(Austria, Switzerland, but not known south of the Alps) and to the
Pyrenees; also in Great Britain and Ireland. Widespread in northern and
northwestern USSR, eastward to northern Asia (Yenisey R. and Ob
R.) also on Sakhalin (Kitagawa, 1963b), in Korea (Hong, 1966) and
Japan (Hattori, 1958, p. 40), but Hattori confused several plants under the
name *L. porphyroleuca,* as developed below; only a portion of his reports
under that name (and, earlier, under the synonym *L. fauriana*) belong
here.[160] In North America widespread on the west coast (Alaska, British

---

[159] Frye and Clark (1937–47, p. 363) and Müller (1951–58, p. 651) report the species from
Greenland and Ellesmere I., as well as from Spitsbergen. Since *Lophozia porphyroleuca* is so
nearly uniformly restricted to xylicolous sites, these reports, based on Harmsen (1933) and
Bryhn (1906), respectively, cannot be accepted. Similarly, the report from Spitsbergen (based
on Watson, 1922) is surely erroneous; Arnell and Mårtensson (1959) do not report the species
from this locality.

[160] The type of *L. fauriana* Steph. (*Faurie 1131,* Miyokosan, Japan) is a mod. *angustifolia-
pachyderma,* lacks gemmae, has sharp leaf lobes, and has a perianth mouth with cilia to 4–5 cells
long. Trigones are nodose, as is typical of *L. porphyroleuca.* There is no question but that *L.
fauriana* is a synonym of *L. porphyroleuca,* as regards the type. *Lophozia fauriana* of Hattori (1953,
pl. 2, fig. 16) is clearly *L. "silvicola,"* however. Kitagawa (1965) attempts to maintain *L.
fauriana* as a species separate from *L. porphyroleuca* on the basis of supposedly longer perianth
mouth cilia ("usually 5–7 cells in length"), but his fig. 8:2 shows them 2–4 cells long. I am
convinced that no real distinction occurs.

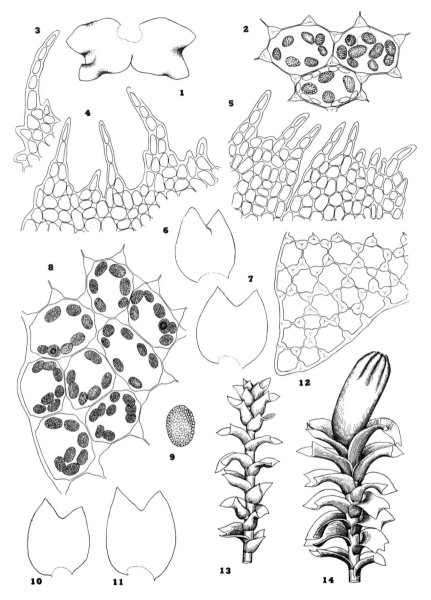

FIG. 203. *Lophozia* (*Lophozia*) *porphyroleuca*. (1) ♀ Bracts; no bracteole seen (×10.5). (2) Median cells with oil-bodies (×603). (3–4) Perianth mouth cilia (×218). (5) Sector of perianth mouth, least ciliate extreme (×250). (6–7) Leaves, in fig. 7 a large, subinvolucral one (×20). (8) Cells of lobe base near leaf margin (×620). (9) Oil-body (×1275). (10–11) Leaves (×27). (12) Leaf lobe cells (×325). (13) ♂ shoot (×21). (14) ♀ shoot (×ca. 24). [1, 5–7, *RMS 24441*, Cornell Mt., Catskill Mts., N.Y.; 2–4, *RMS 24453*, same loc.; 8–11, Manitoulin I., Ontario, *RMS, Sept. 1, 1949*; 12–14, Schwarzwald, Baden, Germany, *K. Müller*.]

Columbia, Washington, Oregon, California, eastward to Montana, Idaho, Utah, Wyoming, and Colorado) and equally common eastward, although rare or infrequent in the central portions of the continent. The species is usually an excellent indicator of the Spruce-Fir Biome, occurring southward barely into local bogs and swamps in the Northern Hardwoods Forest (Schuster, 1949a); typical forms appear to drop out in the northern edge of the Taiga, and the species fails to penetrate into the Southern Appalachian extension of the Spruce-Fir Biome.

NEWFOUNDLAND. Biscay Bay, Avalon Distr.; St. Alban's, Pushthrough, Big Bay at Rencontre East, on S. coast; South Branch, Piccadily; St. George's and Steady and Benoit Brooks, W. Nfld.; Cow Head, and St. John's I., St. Barbe, N. to Cook Harbour, to Williamsport, N. Pen.; Hampden, Sandy L., Humber, and Kitty's Brook, C. Nfld. (Buch & Tuomikoski); St. Pierre (Le Gallo, 1951).   QUEBEC. N. of R. Malbaie, ca. 10–12 mi NW. of Bridgeville, E. end of Gaspé (*RMS 43521, 43524b, 43503 p.p.*, among *L. longidens, L. ascendens, L. ventricosa silvicola*); Gulf of St. Lawrence, Gaspé (Macoun, Can. Hep. no. 59!); Spider R.; Lac Chicdos, ca. 20 mi S. of Rimouski; R. Ste-Anne-des-Monts (Lepage, 1944–45); Mt. Lac des Cygnes, Charlevoix Co. (Kucyniak, 1947).   ONTARIO. Cain & Fulford (1948), but judging from associated species cited, such as *Lophozia bicrenata*, and ecology given, all or most of these reports are clearly erroneous.[161] Belleville; N.E. coast of L. Superior; L. Nipigon and on Humboldt I. ("on dry rocks"—therefore doubtful), etc. (Macoun, 1902; these reports in large part questionable); 5 mi E. of West Bay, Manitoulin I., in bog (*Cain & RMS 4098!*). NOVA SCOTIA. Sandy Cove (Lowe, 1909); Barrasois, Victoria Co.; Musquodoboit (Brown, 1936a); Cape Breton Ntl. Park, 3 mi S. of Mary Ann Brook waterfall, S. of Neils Harbour (*RMS 66346b, 66323, 66310*); Warren L., N. of Ingonish, Cape Breton Natl. Park (*RMS 66301a, 66304, 66306c*).

MAINE. Hamlin Ridge, North Basin Trail, 3000 ft, Mt. Katahdin (*RMS 32913, 32913a*); Mt. Bigelow; Mt. Desert I. (*Lorenz*).   NEW HAMPSHIRE. White Mts., abundant in Spruce-Fir Zone of Mt. Washington, Mt. Clinton, Mt. Adams, South Twin Mt., and elsewhere (*RMS*); Franconia Mts. (Lorenz, 1908c); Guileford; also reported from Waterville (Evans, 1917e = *L. ascendens;* see Schuster, 1952).   VERMONT. Pico Peak (*RMS*); Long Trail near Taft Lodge, 3200–3300 ft, Mt. Mansfield (*RMS 43823, 43923*); Willoughby (*Lorenz, 1904*); Haselton Trail, Mt. Mansfield, 3000–3200 ft (*RMS 43842a*).   MASSACHUSETTS. Mt. Greylock, Adams (*Andrews*); Ram Pasture, Nantucket; Hawley (cranberry) Bog, Franklin Co. (*RMS, s.n.*).   CONNECTICUT. Stafford, with *Riccardia latifrons, Calypogeia suecica!*   NEW YORK. L. Placid (*Britton!*); Ringwood and Freeville Swamps, Tompkins Co.; Bergen Swamp, Genesee Co.; Plymouth Pond, Chenango Co. (all Schuster, 1949a); near Syracuse, Onondaga Co. (Underwood & Cook, Hep. Amer. no. 112, as "*Jungermannia excisa*"!); Jamesville, Onondaga Co. (Underwood & Cook, as "*L. guttulata*"!); near top of Cornell Mt. and between Cornell and Slide Mts., 3200–3600 ft, Catskill Mts. (*RMS 24441, 24453, 24436, 24474b, 7640b*; southernmost reports from eastern North America!).

[161] As a control, I examined three collections determined by Fulford, which served as a basis for reports of *Lophozia porphyroleuca* in Cain & Fulford (1948): (a) on *Sphagnum* on rock cliff, Sturgeon R. near Beardmore, Thunder Bay Distr., no. 2045 = *L. ventricosa silvicola*, a mod. *viridis-gemmipara* with largely oblong leaves; (b) exposed moist rock cliff, Gull L. Portage, L. Timagami, no. 3063 = surely not *L. porphyroleuca*, probably *L. ventricosa silvicola*, although not certainly identifiable; (c) on peat in black spruce bog, 10 miles east of Pagwa R., Cochrane Distr., no. 1990 = *L. (Leiocolea) heterocolpa*; a few plants have the characteristic gemmae and all plants distinct underleaves.

Westward in our area reported as follows. MICHIGAN. Near Montreal R. and Phoenix; W. of Strongs and near Whitefish Bay (Evans & Nichols, 1935); Rapid R., 6–7 mi NNE. of Kalkaska, Kalkaska Co. (*RMS 38940*); Reese's Bog, Cheboygan Co.; Porcupine Mts., Ontonagan Co. (Nichols & Steere, 1937); Isle Royale, Keweenaw Co.; Gogebic and Chippewa Cos. (Sugar I.) (Steere, 1934, etc.). WISCONSIN. L. Nebagamon, Douglas Co. (as *L. guttulata*, Conklin, 1929); Vilas, Oneido, Ashland Cos. (Conklin, 1929). MINNESOTA. Cook, Lake, and St. Louis Cos. (Conklin, 1929; Schuster, 1953; citations only in part correct!).

Many, if not most, reports from eastern North America of *L. porphyroleuca* are based on confusion with *L. ventricosa* var. *silvicola* and (at times) *L. ascendens* (Schuster, 1952). This includes even some of my own reports from Minnesota (*RMS 11910, 11742*, which represent *L. ventricosa silvicola*, at least in largest part; *RMS 17438a*, which was reported as "intermediate with *L. silvicola*" is actually the latter). Plants reported from Long I., Susie Isls., Cook Co. (*RMS 7171a*) are critical; these, as well as others assigned to *L. porphyroleuca* on the basis of vegetative features (marked reddish pigmentation of the leaf bases and sometimes even lobes; bulging trigones; narrow, rectangulate leaves), possess gemmae. In plants assigned now to *L. ventricosa silvicola* the perianth mouth has 1–2-celled teeth (i.a., in *RMS 11910*), but in *RMS 7171a* the presence of gemmae in the mod. *colorata-pachyderma-angustifolia* is linked with perianth-mouth teeth in part 3 or even 4 cells long, as is normal for *L. porphyroleuca*. Such intermediate plants can only be identified by means of the oil-bodies, although the linkage of gemmae with the mod. *colorata-pachyderma* simply does not appear to occur in "normal" *L. porphyroleuca*, suggesting that a phase of *L. ventricosa silvicola* is at hand with unusually long cilia of the perianth mouth. I therefore would now regard the Minnesota reports as suspect and am dubious about reports from Wisconsin and Michigan (excepting the citation from Kalkaska Co., which is treated separately below).

Lophozia porphyroleuca appears to have a range basically similar to that of *Scapania umbrosa*: frequent in the more nearly oceanic portions of the northeast, southward in the Appalachian system to the Catskill Mts., N.Y., and no further; common in the Far West; recurring rarely, in disjunct fashion, in the Great Lakes area. The many citations from the Rocky Mts. need re-examination. Presumably many, if not most, will prove to represent *L. ventricosa silvicola* (or partly *L. ventricosa*) in the mod. *angustifolia-pachyderma*.

The latter type of modification appears to be the basis for all or most reports of *L. porphyroleuca* from the Arctic. The species has been reported from Ellesmere I. (Skräling I. and Eskimopolis, Hayes Sound region, *in* Polunin, 1947, after Bryhn, 1906) and from Thule, northwest Greenland; Röhss Fjord; Igmikertorajik (without perianths, as var. *porphyroleuca* of *L. ventricosa*; see Jensen, 1906a); Scoresby Sund (Hekla Havn, Gaaseland, Runde Fjeld, Cape Stewart); from a high moor at Tasiusak, "among *Cynodontium strumiferum*," and probably from other places (Jensen, 1906a; Harmsen, 1933; etc.). I do not believe that any of these reports can be accepted and would restrict *L. porphyroleuca* to plants of the Spruce-Fir Biome, admitting that veritable phenocopies of *L. porphyroleuca*, assigned to *L. ventricosa*, may occur in the Arctic. All or most of these plants belong to *L. ventricosa confusa* (p. 581). Mårtensson (1955, p. 33) also states that the species does not appear to go above the forested belt in northern Sweden and that its "occurrence [is] concentrated in the conifer forest areas."

Ecology

Almost restricted to damp or moist, decaying logs and stumps, with exceptional occurrences over peat in bogs. There are numerous literature citations from other habitats, ranging from "peaty soil" (Frye and Clark, 1937–47, p. 363) to "moist rock cliffs" (Cain and Fulford, 1948, p. 180). However, *if* one examines material from such habitats, the plants almost always prove to be a mod. *pachyderma* of either *L. ventricosa* or, more often, *L. ventricosa silvicola*. Buch (1933) has shown that *L. porphyroleuca* may be almost impossible to separate from *L. ventricosa* when only sterile, dead material is studied. In any case, none of the specimens seen labeled *L. porphyroleuca*, from rocks, soil-covered rocks, or peat over ledges, can be referred safely to *L. porphyroleuca*. I remain firmly convinced that, in general, the almost constant "occurrence locally on . . . decaying logs, as well as the large bulging trigones, and the ciliate perianth mouth . . . , easily differentiate this species" (Schuster, 1949a).[162]

On decaying logs the species is abundant chiefly in the Spruce-Fir Biome, although found scattered in the Northern Hardwoods Forest, in peat bogs and swamps (Schuster, 1949a). It is particularly abundant on moist, shaded, decaying spruce and fir logs, often with several other Lophoziae, chiefly *Lophozia ascendens*, *L. ventricosa silvicola*, and *L. longidens* (as in *RMS 32913*, Mt. Katahdin, Me., and *43503*, Gaspé, Quebec), less often with *L. incisa*, *L. attenuata*, but also with *Calypogeia suecica*, *Geocalyx graveolans*, *Scapania apiculata*, *S. glaucocephala*, various *Cephalozia* spp., *Blepharostoma*, and occasionally *Lepidozia reptans*; the last two species usually invade somewhat later, at the same time as *Tetraphis pellucida*. Southward often confined to logs at bog edges, often in insolated sites, and then on moist or even nearly wet logs with *Nowellia curvifolia*, *Cephalozia connivens*, *Ricardia latifrons*, *Scapania irrigua*, and, rarely, *S. paludicola*.

---

[162] There are rare, sporadic exceptions, so far as I know confined to montane areas (where the species is ubiquitous on damp, decaying logs in the Spruce-Fir Zone). Here damp rocks may exceptionally be invaded—but always with a "reservoir" population on nearby logs. An example is the occurrence on logs on Mt. Mansfield, Vt. (*RMS 43802, 43802a*), admixed with *Lophozia ventricosa silvicola, L. attenuata, Calypogeia suecica*, and *Cephalozia media*. Within 2 feet, on a very thin soil layer over an exposed boulder, a few individuals of *L. porphyroleuca* recurred.

Perfectly "normal" *L. porphyroleuca* may also occur in *Sphagnum* bogs, where it is usually confined to decaying peaty wood, but may spread to nearby peat, in full sun; associated then may be *Cephaloziella rubella, Polytrichum*, and *Dicranum* (i.a., in *Cain & RMS 4098*).

All, or virtually all, of the many reports (e.g., Hattori, 1958, p. 40) of collections from "humus-covered banks," "humus-covered rocks or in crevices" are due to confusion, probably most often with *L. ventricosa silvicola*.

The behavior pattern of this species is one of its most distinctive characters. On decaying logs it usually occurs as the "*guttulata*" modification, a plant with very coarse trigones, quite lacking gemmae. As Buch has shown, gemmae—and usually very few of them—develop only under exceptional, moist conditions. Inversely, plants are almost always provided with sex organs and produce capsules exceedingly freely. In all cases, the trigones are very large and bulging, even in humid and shaded sites, and the mod. *leptoderma* or *mesoderma* is hardly to be encountered. Stems also tend to develop reddish pigmentation ventrally, even in plants from very shaded sites, and the dark color of the stems is usually evident even in the field, under the hand lens. By contrast, *L. ventricosa silvicola*, which is often found associated on decaying logs and freely occurs intimately intermingled, "behaves" very differently: it is usually provided with small or moderate trigones, under conditions prevailing on shaded logs; yellow-green gemmae are abundantly produced; inversely, sex organs are usually absent or infrequent, and capsules are exceptionally produced. The stems, under the stated conditions, also remain greenish. Finally, there is an obvious difference in size: the *L. porphyroleuca* is always much smaller in size than *L. ventricosa silvicola*. (On living plants there are also notable differences, discussed below, in the oil-bodies.)

It is thus obvious that the very specialized and (for a *Lophozia*) restricted ecology of the species, together with its behavior pattern, is of the utmost importance in properly recognizing it.[163]

As is emphasized (Schuster, 1953), "all material that can be clearly determined as this species has been found on acid, organic substrata." In addition to the usual occurrence on "decaying, moist to very moist, often quite sunny and exposed logs" the species is occasionally found over peat, "associated with *Jamesoniella*, various *Riccardia* species (*latifrons, palmata*), occasionally *Odontoschisma denudatum* and various *Cephalozia* species (*media, connivens, catenulata*) . . . . The pH range is from about 4.8 to 3.5." Plants with the aspect of *L. porphyroleuca* found over insolated, peat-capped, moist ledges, usually on living *Sphagnum*, must be separated with great care from *L. ventricosa silvicola*. Indeed, most such plants, when cytologically examined, prove to be a mod. *colorata-pachyderma* of *L. ventricosa silvicola*, which is almost inseparable from *L. porphyroleuca*, except on the basis of the perianth mouth and the oil-bodies (see below). Critical in such cases is the presence or absence of gemmae: in *L. ventricosa silvicola* a few, at least, are always present, whereas in bona fide *L. porphyroleuca* the mod. *colorata-pachyderma* is unable to develop them.

The species is abundant in spruce-fir forests in the moister, more nearly oceanic portions of North America, particularly in the montane area from New England to the Gaspé; it seems to virtually disappear in the otherwise similar forests of the middle portion of the continent.

Differentiation

Lophozia porphyroleuca has been almost consistently misunderstood and misidentified, even in recent times. Ordinary modifications (mod.

---

[163] I have repeatedly been able to separate the species from its congeners, with the naked eye or the hand lens, in the field. Subsequent cytological study has demonstrated clearly that correct field identification of the species is possible, *after* one is well acquainted with its ecology.

angustifolia-pachyderma-viridis vel colorata) may closely resemble the phases of
L. ventricosa silvicola found over insolated, peaty sites, as in bogs or over
exposed, peat-covered ledges. When such a distressingly similar modification is at hand, it is separated from *silvicola* principally by means of the
oil-bodies, the absence of gemmae, and the more ciliate perianth mouth.
The almost constant lack of gemmae is a notable distinguishing feature of
this species, much of the gemma-bearing material referred to it belonging
to other species. Some of the problems attending the correct circumscription of *L. porphyroleuca* have already been touched upon above.

In spite of the aura of confusion, *L. porphyroleuca* is a characteristic
species, well circumscribed by the following ensemble of characters:
(*a*) restriction to organic, and usually xylicolous, sites: (*b*) smaller size
than *L. ventricosa*, but superior to *L. ascendens*; (*c*) prostrate to procumbent
growth; (*d*) inability to develop gemmae freely under any conditions and
total inability to develop them under almost all conditions; (*e*) greenish
gemmae color, in the rare instances when gemmae are present; (*f*) leaves
narrow, never as stiffly subpectinate as is often the case in dense-leaved
forms of *L. ventricosa silvicola*, often somewhat concave rather than loosely
canaliculate; (*g*) stem, and often rhizoid bases tending to become carmine, even under shaded conditions;[164] (*h*) trigones coarsely bulging and
often almost confluent; (*i*) perianth-mouth cilia elongate, in part 3–4(5)
cells long; (*j*) abundant fertility and general presence of capsules even into
October (in New England)—a much greater fertility than in the *L.
ascendens-longidens* complex, in which sporophytes and perianths are rare
or infrequent, and in the *L. ventricosa* complex; the linkage of this with
the almost constant lack of gemmae is diagnostic; (*k*) cells with uniformly
granular-botryoidal oil-bodies of varying shape—many ovoid to ellipsoid
—and size, usually occurring only 3–10 per cell.[165]

Except for pigmentation, *L. porphyroleuca* appears to be less variable than
most other species of the subgenus; this is perhaps associated with the restricted

---

[164] It must be emphasized that on shaded, decaying logs associated *Lophozia silvicola* is green
throughout, the stem at most brownish with age, owing to development of the mycorrhizal band;
in *L. porphyroleuca* a reddish to carmine pigmentation *of the cell walls* develops. However, in
strongly insolated sites, *L. silvicola* also can develop reddish coloration of stem and rhizoids, but a
much higher light intensity is necessary to effect this response.

[165] Hattori (1958, p. 40) cites the Japanese *Lophozia fauriana* Steph. (Spec. Hep. 6:112, 1917)
as a synonym of *L. porphyroleuca* and states that "*L. ventricosa* seems to be conspecific with *L.
porphyroleuca*, too." There has been confusion of the Japanese plants, as evident from illustrations of the cells and oil-bodies in Hattori's papers (1951b, 1953). The first paper illustrates
(pl. 7) a cell with 10 ovoid oil-bodies lacking a central spherule; this figure is conceivably
based on "normal" plants of *L. porphyroleuca*. In the 1953 paper the figures (pl. 1:16–20, pl.
2:16–17) show cells of a different plant, with smaller, usually spherical oil-bodies, ranging
from 21–22 to 27 per cell; the oil-bodies, where drawn in detail, are biconcentric. The plants
figured in Hattori (1953) appear to belong to *L. ventricosa silvicola* rather than to *L. porphyroleuca*;
hence the ecological observations cited in Hattori (1958, p. 40) cannot all be accepted.

sites to which it is usually confined. Owing to almost identical ecological requirements, the species often occurs with *L. ascendens* (which see for the separation of the two).

Much previous difficulty, fully documented in the literature, regarding the
discrimination between *L. ventricosa* and *L. porphyroleuca*, has been resolved by
the segregation of *L. ventricosa silvicola* from *L. ventricosa*. The larger size and
broader leaf shape should suffice generally to separate typical *L. ventricosa* from
the smaller *L. porphyroleuca*. Furthermore, *L. ventricosa* has usually 10–15 oil-
bodies per cell, although certain arctic phases (var. *confusa*; see p. 581) have
fewer (see, i.a., Schuster et al., 1959), while most phases of *L. porphyroleuca* have
only 5–9(10) oil-bodies per cell. On the basis of oil-body number there appears
to be obvious intergradation between *L. ventricosa* and *L. porphyroleuca*. For
example, material from Michigan (Kalkaska Co., *RMS 38940*), with the narrow,
sharp, 2–3-celled perianth-mouth teeth of *L. porphyroleuca* and the lack of
gemmae characteristic of that species, as well as oblong, parallel-sided leaves of
sterile stems, has 8–15 oil-bodies per cell. These plants are a mod. *mesoderma-
viridis*. Evidently in both species oil-body number decreases with greater
exposure; hence some intergradation in this feature, bridging the distinctions
emphasized in the keys, is to be expected.

Although *L. ventricosa* and *L. porphyroleuca* are ordinarily separable on the
basis of leaf shape, *L. ventricosa silvicola* and *L. porphyroleuca* share a nearly identical leaf shape, even though *L. ventricosa silvicola* often has a stiffer, more subpectinate orientation of the leaves, with the lobes almost always spreading.
The occasional difficulties in separating *L. ventricosa silvicola* and *L. porphyroleuca*
are dealt with above; they are also discussed under *L. ventricosa silvicola*. In
general, the lack of gemmae or their rarity, the smaller overall size, and in particular the fewer (usually 5–10, but often locally 3–5), often larger and ellipsoidal, nonbiconcentric oil-bodies separate *L. porphyroleuca* from *L. ventricosa
silvicola*. In critical cases the oil-bodies are the surest method of distinguishing
the two, although, with practice, separation under the low power of the wide-
field binocular is feasible.

Authors occasionally attempt to separate a smaller, more xeromorphic,
gemma-free phase (mod. *angustifolia-pachyderma-colorata-vel viridis-egemmipara*) as
a separate species (*L. guttulata*) or variety (*L. porphyroleuca* var. *guttulata*). I
think that this plant is a mere environmental modification and does not deserve
taxonomic recognition.

LOPHOZIA (LOPHOZIA) VENTRICOSA (*Dicks.*) Dumort.[166]

[Figs. 202: 4, 6; 204–210:1–6]

Jungermannia ventricosa Dicks., Fasc. Pl. Crypt. Brit. 2:14, 1790.[167]
Jungermannia globulifera Roth, Fl. Germ. 3:379, 1803.

---

[166] Fuller citations are given under the several varieties and forms recognized. More complete
diagnoses will also be found under *each* of the varieties recognized.

[167] The Dickson type—if still extant—should be ignored, and conventional concepts of what is
L. ventricosa, sensu Buch (1933), must be followed, since definite identification of dead material
probably is hardly possible.

Lophozia ventricosa Dumort., Rec. d'Obs., p. 17, 1835; Macvicar, Studs. Hdb. Brit. Hep. ed. 2:179, 1926; Schuster, Amer. Midl. Nat. 49(2):361, pl. 18, 1953; K. Müller, Rabenh. Krypt.-Fl. ed. 3, 6:647, figs. 195a, 196, 1954.

Jungermannia longiflora Nees, Naturg. Eur. Leberm. 2:95, 1836.

Lophozia longiflora Schiffn., Lotos 1903(7):45 [of reprint], 1903; K. Müller, Rabenh. Krypt.-Fl. 6(1):671, 1910; *ibid.* ed. 3, 6:654, fig. 199, 1954; Macvicar, Studs. Hdb. Brit. Hep. ed. 2:182, 1926.

Lophozia confertifolia Schiffn., Österr. Bot. Zeitschr. 55:47, 1905(?); et auct., *p. maj. p.*

Lophozia silvicola Buch, Ann. Bryol. 6:125, 1933; Buch, Mem. Soc. F. et Fl. Fennica 17(1940–41):290, 1942; Schuster, Amer. Midl. Nat. 49(2):358, pl. 17:4–9, 1953.

Lophozia silvicoloides Kitagawa, Jour. Hattori Bot. Lab. no. 28:276, fig. 9, 1965 (provisional synonymy).

Exceedingly variable in size and pigmentation: from 0.8–2 to 2.5–3(4) mm wide usually × 1–5 cm long, *prostrate or creeping* (rarely suberect with crowding), green or yellow-green with, often, either brownish or reddish pigments present (sometimes varied combinations of both). Stem ca. 16–22 cells high, with well-developed, eventually brown, ventral mycorrhizal medullary band. Leaves from contiguous and typically *oblique to subhorizontal* on creeping or prostrate phenotypes, to rather dense, almost subvertical and caniculate (in *"conferti-folia"* phenotypes), varying from longer than broad (0.75–0.95 as broad as long) to broader than long (1.0–1.25 × as broad as long), in part on the basis of allometric growth but in part with genetic differences, oblong-ovate to broadly orbicular-ovate, *margins usually weakly arched* (typically more so than in the oblong-leaved *L. porphyroleuca*, less so than in the orbicular-leaved *L. wenzelii*), somewhat concave toward base, the *lobes widely spreading* (laxer forms) to loosely folded over each other (dense, *"confertifolia"* phenotypes), but *never conspicuously incurved;* sinus variable but usually *obtusely angulate to rectangulate* with base rounded, descending 0.15–0.25(0.35–0.4) the leaf length; lobes bluntly acute to acute usually, often subequal, the dorsal a little smaller. No underleaves. Rhizoids long, colorless, rarely reddish at base. Cells medium-sized: marginal in lobes ca. 20–25 μ, the median ca. 20–25 × 23–28 μ, up to 25 × 30–38 μ (except in the large-celled, ?polyploid var. *rigida*), with distinct but *concave-sided to moderately bulging trigones*, never approaching in coarseness the coarsest ones produced by *L. porphyroleuca;* oil-bodies variable in number, form, and size (see individual varieties). *Gemmae always present*, usually abundant, even in extremes of insolated sites, *greenish, mostly* ca. *18–20 × 22–30 μ* (much larger in var. *rigida*).

Dioecious. ♂ Bracts usually loosely imbricate or contiguous, in 4–7 pairs, ventricose at base; antheridia 1–2 per bract, with 1-seriate stalk. ♀ Bracts variable, noncrispate, 1 usually 2-lobed, the other usually 3(4)-lobed, the *lobe margins entire;* bracteole 1–2-lobed, fused with 1 bract for much or most of its length. Perianth long-emergent, ovoid-cylindrical, the pluriplicate mouth variably (from faintly to distinctly) lobulate, the obscure to perceptible lobes denticulate with 1–2-celled teeth (and, in some varieties, in part with 3–4–5-celled teeth). Capsule wall 3–4-stratose. Spores mostly (10)11–15(16) μ, finely verruculose. Elaters 7–9 μ in diam.

Type. England; a type, which has not been designated, presumably exists in the Dickson Herbarium (BM).

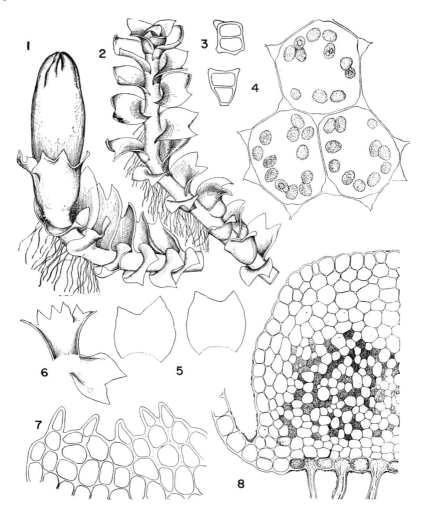

FIG. 204. *Lophozia* (*Lophozia*) *ventricosa* var. *ventricosa*. (1–2) Perianth-bearing and androecial plants (×14). (3) Gemme (×380). (4) Cells with oil-bodies, leaf middle (×720). (5) Leaves (×12.5). (6) Bracts and bracteole, fused to one bract (×9). (7) Perianth mouth cells (×260). (8) Stem cross section, with free development of mycorrhizae (×150). [1–3, 5–8, Pigeon Point, Minn., *RMS*; 4, small island between Lucille I. and Susie I., Susie Isls., Minn., *RMS*.]

Distribution

Circumboreal and circumpolar. Our most difficult, most variable, and taxonomically intractable species; extending northward to the northern edge of land (at least to 82°25′ N.; Schuster et al., 1959), southward in North America to Tennessee and North Carolina in the East, to California

in the West. The distribution is detailed under each of the several taxonomic entities recognized.

Ecology

Concurrently with the wide range of this *typus polymorphus*, the ecological
distribution is correspondingly broad and difficult to circumscribe. It is
easier to state where the species, *s. lat.*, fails to occur: it is never truly
corticolous; it is never aquatic, although it may occur in peat bogs—
especially on the sides of deer tracks—where submersion does not occur,
and also on rocks along streams where submersion is rare and sporadic;
it is rarely found on purely mineral soils. Basically a saxicole—usually
after preparation by other taxa, rarely a pioneer—and humicole, and
frequently a xylicole (but usually not a pioneer on decaying xylem);
mostly a mesophyte, occasionally a mesoxerophyte, or rarely a true
hygrophyte. Able to tolerate both constantly very diffuse light (then often
in dry but humid sites, as on rock walls) and considerable direct, intense
sunlight (on very moist sites, such as along trails in bogs, on peaty soils
in the Tundra, or on exposed peaty, seepage-moist ledges).

More detail on the associates and ecology of the several varieties is given, in so
far as this is distinctive.

Differentiation

Lophozia ventricosa, as broadly delimited, basically includes the Northern
Hemisphere *Lophozia* phenotypes, with green gemmae, that (*a*) lack the
vertical, strongly concave leaves with incurved lobes of *L. wenzelii*;
(*b*) are larger and less erect in growth than *L. ascendens*; (*c*) lack the purple
color and underleaves of *L. heteromorpha* and have almost constantly
bilobed leaves; (*d*) lack the very broad and shallowly bilobed, antically
assurgent leaves of *L. groenlandica;* (*e*) lack the ability to develop as coarse
trigones as *L. porphyroleuca*, but possess under essentially all ecological
conditions the ability to develop gemmae. A considerably more extensive
suite of differential features exists and is discussed under the species
enumerated above. However, the point I wish to make here is that, to
some extent, *L. ventricosa* consists of what is left of the green-gemma
Lophoziae, *after* the more definitely delimited and, usually, much less
polymorphic allied taxa have been removed. This is hardly a satisfactory
definition of the species, but the subjoined treatment of variation will, I
think, resolve the question of why this situation exists.

Variation

Lophozia ventricosa, broadly delimited (to include *L. longiflora* and *L. silvicola* as varieties), is widespread and immensely variable. It grows in subarctic sites intermingled with *L. porphyroleuca*, *L. longidens*, and *L. ascendens*, yet remains quite distinct (see p. 549). Similarly, in the Arctic, *L. ventricosa* var. *confertifolia* may grow admixed with *L. wenzelii* var. *wenzelii*, and *L. ventricosa* var. *confusa* may grow admixed with *L. wenzelii* var. *lapponica*. In all these cases the various taxa are distinct in the manner in which the leaves are held and/or their color, and often in other, less palpable features. However, I have never found two plants admixed that I would regard as varieties of *L. ventricosa*. (The one possible exception is a seeming transition of *L. v.* var. *ventricosa* and *L. v.* var. *silvicola*, in the Southern Appalachians; here, I think, is a case of actual intergradation of populations rather than admixture.)

As a consequence of over 20 years of field observations and careful microscopic comparison of numerous gatherings, usually studied alive, I have concluded that anything except a broad treatment of *L. ventricosa* is impractical and untenable. This is particularly true when one works in the Arctic, where *L. ventricosa* is extraordinarily variable, and the usual "lines" drawn between *L. ventricosa s. str.*, "*L. silvicola*," and "*L. longiflora*" become nearly meaningless. Rather, there are different gene combinations in the arctic plants, and their study demands that we either adopt a broad and variable species concept, with the more recognizable entities distinguished as varieties, or create even more species. The latter proposal is absurd. Admittedly, culture work may eventually show that most of the entities here recognized remain distinct when grown together, as Buch (1933) demonstrated for *L. ventricosa s. str.* and *L. "silvicola."* However, this merely proves that we are dealing with distinct genotypes, not the broad, polymorphous series of populations that genetics has taught us constitutes a normal biological species.

After over 10 years of attempting to maintain *L. silvicola* as a species (see Schuster, 1953), I reluctantly concede defeat. *Lophozia ventricosa s. lat.* appears to consist of a series of phenotypes, corresponding apparently only in part to distinct genotypes, of which the "*silvicola*" extreme is but a single one. I believe it will be necessary to maintain *L. ventricosa* in a broader sense to include not only the *silvicola* extremes but also arctic extremes with only 9–12 or fewer oil-bodies that are never biconcentric, as well as the "normal" form with broad, rotund-quadrate leaves, typically with an obtuse to crescentic sinus, and with 10–15 uniformly granular-botryoidal oil-bodies.

Key to Varieties[168] *of L. ventricosa*

1. Leaves generally broader (on mature sterile axes), usually 0.95–1.25× as wide as long; oil-bodies (3–5)6–20 per cell usually, variable, many ellipsoidal to ovoid and to 5–7 × 6–9 μ (rarely larger), *never* biconcentric, lacking a central homogeneous "eye"; rarely with extensive carmine pigmentation of leaves (and especially of leaf bases) . . . . 2.
  2. Cells smaller: marginal 20–25 μ, median mostly 20–25 × 23–28 to 25–30 × 30–38 μ; gemmae mostly 18–20 × 20–25 μ; at least sometimes with reddish pigmentation . . . . 3.
    3. Perianth mouth armed with 1(2–3)-celled teeth, hardly lobed or with few, broad lobes; cells with mostly 9–16 oil-bodies each, occasional cells with to 17–21 oil-bodies . . . . 4.
      4. Smaller, usually 0.8–2.2 mm wide; perianths mostly 0.75 × 2–3 mm at maturity; leaves almost all 2-lobed, lobes usually sharp; oil-bodies to 5–6 × 6–8 μ usually, almost never larger. Common in temperate areas, rare (or lacking?) in Arctic . . . var. *ventricosa*
      4. Robust, 2.5–3(4) mm wide, usually dense-leaved, with subvertical leaves, occasionally with gibbous sinuses, often with some lobes blunt, frequently some 3-lobed leaves; many oil-bodies to 6 × 8–10 μ, a few 7.5–8 × 9–12 μ. Arctic and alpine; often reddish or at least locally reddish to pinkish, with usually carmine perianths bleached at mouth . . . . var. *longiflora*
    3. Perianth mouth divided into many small, narrow lobes, each usually ending in a 2–3(4)-celled tooth or short cilium, with interspersed 1-celled teeth frequent; cells with (3–5)6–12(13–15) oil-bodies, to 7 × 10 μ, often nearly appearing to occlude lumen; usually with warm brown pigments of leaf apices, occasionally reddish ventrally and on stem. Arctic . . . . var. *confusa*   *a.*
      a. Leaves not or hardly wider than long (except for sporadic 3–4-lobed ones), bilobed to 0.3 or less . . . . fo. *confusa*
      a. Leaves (mature shoots) broadly quadrate, 1.1–1.2× as wide as long, divided 0.35–0.45(0.5) into broad triangular lobes . . . . .
        . . . . fo. *discoensis*
  2. Cells large: marginal 30–38 μ in distal half of leaf, median 32–37 × 35–50 μ (each with 14–20, occasionally 21–25, oil-bodies, the larger to 5–7 × 9–10 μ); gemmae mostly 26–28 × 28–38 to 29–33 × 35–38 μ. (Stem black-purple ventrally, leaf bases dull purplish; leaf lobes ± light brown; no reddish pigments at all. Leaves usually rather pectinate, with mostly sharp lobes and an angular sinus. Plants rigid, vigorous, often 2–2.4 mm wide; leaves 1.05–1.25× as wide as long). Arctic . . . . .
    . . . . var. *rigida*
1. Leaves usually oblong to oblong-ovate, to sides mostly little arched, usually (0.75)0.8–0.95(1.05)× as wide as long, the sinus nearly always angular, mostly descending 0.25–0.35 the leaf length; smaller (to 1.5 mm wide),

---

[168] See also the key on p. 505.

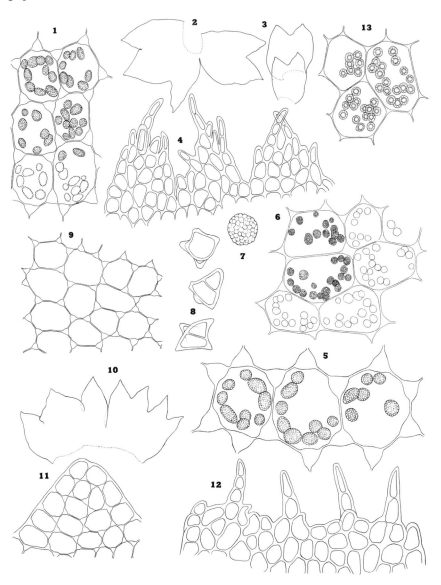

FIG. 205. *Lophozia ventricosa-porphyroleuca* complex. (1–4, *L. porphyroleuca;* 6–12, *L. ventricosa ventricosa*). (1) Median cells (×580). (2) Bracts, both bifid, and lanceolate bracteole (×19). (3) Leaves from sterile shoot (×19). (4) Perianth mouth (×220). (5) Cells with oil-bodies, of an intermediate phase between *L. ventricosa* and *L. porphyroleuca*, with leaves as wide as long, concave, of the "*confertifolia*" phenotype, over *Sphagnum;* lobes ± brown, stems below, rhizoid and leaf bases vinaceous (×770). (6) Cells with oil-bodies (×515). (7) Individual oil-body (×1700). (8) Gemmae (×560). (9) Median cells (×320). (10) Bracts and, between them, ovate-lanceolate bracteole (×14). (11) Leaf lobe apex cells (×260). (12) Cells of perianth mouth, extreme with

often light green plants, frequently with brilliant red pigmentation; oil-bodies 15–24 per cell, all or almost all spherical, small, under 5–6×5–7μ, biconcentric, with a homogeneous "eye"; cell and gemma size and perianth mouth as in var. *ventricosa*. Widespread . . . . var. *silvicola*.

LOPHOZIA VENTRICOSA var. *VENTRICOSA* (Dicks.)
Dumort.

[Figs. 202:6; 204; 205:6–12; 206: 1–3]

Jungermannia ventricosa Dicks., Fasc. Pl. Crypt. Brit. 2:14, 1790.
Lophozia ventricosa Dumort., Rec. d'Obs., p. 17, 1835; Macvicar, Studs. Hdb. Brit. Hep. ed. 2:179, 1926; Schuster, Amer. Midl. Nat. 49(2):361, pl. 18, 1953; K. Müller, Rabenh. Krypt.-Fl. ed. 3, 6:647, figs. 195a, 196, 1954.

Green to yellow-green, occasionally partly brownish or locally reddish, *rarely carmine, medium-sized:* usually 0.8–2.2(2.5) mm wide $\times$ 1–2.5 cm long. Stems usually brownish to purplish beneath (except in extreme shade forms); rhizoids usually colorless to base. Leaves somewhat *oblique to subhorizontal, patent*, mostly 850–975 μ long $\times$ 1030–1140 μ wide, somewhat concave at base but *otherwise usually nearly flat*, but frequently very obtusely complicate and canaliculate (but the *lobes never incurved*), *broadly ovate-quadrate to quadrate-rotundate, 0.95–1.15(1.25)* $\times$ *as wide as long*, with sides only moderately arched, bilobed *for 0.15–0.25 their length; sinus usually broad, lunate to obtusely angulate; lobes acute to bluntly acute*, broad and usually low. Cells *ca. 20–25(29) μ on lobe margins* and in apices, 20–25 $\times$ 23–28 up to 25–30 $\times$ 30–38 μ medially, with distinct but *never coarsely nodose trigones*. Oil-bodies *mostly (9)10–16* (exceptionally 17–23) *per cell, variable, many or most ovoid to ellipsoidal, to 5–7 $\times$ 6–9 μ* (exceptional ones larger), formed *throughout of distinct minute spherules*. Gemmae pale to yellow-green even in sun, mostly 18–20 $\times$ 20–25 μ, *always abundant*.

Androecia often reddish or purplish red. Perianths *mostly ca. 0.75 $\times$ 2–3 mm long at maturity, very shallowly lobulate at mouth*, the subtruncate lobes *with 1- to 1–2(3)-celled teeth*. Spores 10–14 μ; elaters ca. 7–9 $\times$ 85–125 μ.

Type. England (Dickson Herbarium; BM).

Distribution

Variety *ventricosa* and var. *silvicola* are the dominant forms of nonarctic and relatively lowland to subalpine situations. I have studied large suites of living material from the high Arctic (north Ellesmere I.; northwest Greenland) and mid-Arctic (west Greenland), and found that almost none of the material could be safely referred to var. *ventricosa* as circumscribed as a species by Buch (1933); I have followed this narrow circumscription

---

cilia most copiously developed ($\times 235$). (13) *L. ventricosa silvicola*, cells with oil-bodies ($\times 580$). [1, 13, from *RMS 24436a* and *24444*, respectively, Cornell Mt., Catskill Mts., N.Y.; 2–4, *RMS 32913a*, Mt. Katahdin, Me.; 5, *RMS 33020*, Mt. Katahdin, sun form over peat; 6–7, Grand Portage Bog, Cook Co., Minn., *RMS;* 8–12, *RMS 5747*, Pigeon Point, Minn., also shown in Fig. 204: 1–3, 5–8.]

here, as has Arnell (1956; as the species *L. ventricosa*). The exact distribution of var. *ventricosa* is unresolved since it has not been sharply separated from other variants here recognized: in essence, for many years all Lophoziae with angulate yellow-green gemmae and a non-*L. wenzelii* aspect were referred here.

As narrowly delimited, var. *ventricosa*, although given as "circumpolar?" by Buch and Tuomikoski (1955), appears absent from eastern Asia (it is not reported from Japan by Kitagawa, 1965), is rare or lacking in the Arctic (e.g., the extensive list of Arnell and Mårtensson, 1959, does not cite it from Spitsbergen), but extends much farther southward than other varieties, except for var. *silvicola;* both of these, in our area, extend in the Appalachians to North Carolina and Tennessee. It appears to be widespread in Europe [but a reasonably accurate portrayal of its range is difficult, since in standard works, such as Müller (1951–58), "*L. ventricosa*" is treated as a species, including, i.a., var. *silvicola*]. Arnell (1956, p. 119), whose "*L. ventricosa*" is narrowly delimited to exclude *L. silvicola* and *L. longiflora*, gives the range as "common, but rarely above the upper limits of the forest" in Sweden, and cites it from Norway, Finland, Russia, Denmark, and Siberia.

The following include only *reasonably* reliable literature citations:

NW. GREENLAND. Kânâk, Red Cliffs Pen., Inglefield Bay (*RMS 45476, 45722b;* among *Dicranum* in *Cassiope tetragona-Dryas chamissonis* heath); Dundas, near Thule, Wolstenholme Fjord (*RMS 46265a;* in a rich moss-*Cassiope* tundra on damp, exposed soil).[169]   NEWFOUNDLAND. Avalon, S. coast, W. Nfld., C. and NE. Nfld., and N. Pen. (Buch & Tuomikoski, 1955, p. 13).[170]   QUEBEC. Cairn I., Richmond Gulf; mouth of Great Whale R.; Ft. Chimo Air Base, Ungava Bay; 80 mi up Leaf R., Ungava Bay (Schuster, 1951); also reported from George R. near 57°21′ N. (Kucyniak,

---

[169] These collections have ca. 11–18-oil-bodies per median cell; all tend to develop purplish-fuscous pigmentation of distal portions of the leaves and a fuscous to purplish-fuscous ventral face to the stem. The Dundas plant deviates from the two from Kânâk, which retain greenish gemmae to maturity, in one notable respect: the initially greenish gemmae tend to develop with age the same dull brown to purplish brown pigmentation as the leaf lobes. A genetically distinct form is surely at hand, no other phenotypes of *L. ventricosa* having an identical gemma response. However, in leaf dimensions (leaves ca. 0.9–1.1 × as wide as long) and oil-body size and form, these plants fit closest to var. *ventricosa*.

[170] Buch and Tuomikoski, who report the species from some 18 localities in the above districts admit that these are reports "in the collective sense, i.e., including specimens that may belong to *L. silvicola* but have been difficult to identify with certainty from herbarium specimens." Apparently, without having oil-bodies available, Buch could not surely separate his own "*L. silvicola*" from *L. ventricosa s. str.* (= var. *ventricosa* of the present treatment), even though he used (Buch, 1936) leaf dimension and form as primary criteria for separating these taxa.

From north of Newfoundland there are many old, and some newer, reports of *L. ventricosa:* from Labrador (Macoun, 1902); Ellesmere I. (Greely, 1888; Bryhn, 1906); Melville Pen. (Polunin, 1947); Alaska (Evans, 1900); Yukon (Evans, 1903); Baffin I. (Polunin, 1947); Vansittart I. (Polunin, 1947); and from many localities in Greenland (see Harmsen, 1933; Jensen, 1906a; Lange & Jensen, 1887; Arnell, 1960, does not record the species from northernmost Greenland, although Hesselbo, 1923a, had reported it from Lemming Fjord, Somerdalen, and Cape May, from 82°27′ to 82°53′ N.).

All or most of these reports are very questionable; some indeed, may refer to *L. groenlandica* or *L. wenzelii*, whereas the bulk probably refer to the arctic *L. ventricosa* var. *confusa* and/or var. *longiflora*.

1949), Mt. Shefford (Fabius, 1950), Tadoussac and Saguenay (Evans, 1916), and many other localities in S. Quebec (Lepage, 1944–45); all these reports somewhat doubtful. ONTARIO. Wanapitei R., Hart Twp., and Bear L., all Sudbury Distr. (*Cain 2113, 1872, 2087*); Sturgeon R. near Beardmore, Thunder Bay Distr. (*Cain 1873*); Bruce Mines, Algoma Distr. (*Cain 3443*); Loon L. Portage, L. Timagami (*Cain 3010*); old reports by Macoun (1902) and several by Cain & Fulford (1948) need revision; also cited from Moose R., James Bay. ANTICOSTI I. (Macoun, 1902). NEW BRUNSWICK. Woodstock (Macoun, 1902). NOVA SCOTIA. Cape Breton Natl. Park, Mary Ann Falls, S. of Neils Harbour (*RMS 42656b, 42672a, 42673, 42692a*); 3–3.5 mi S. of Mary Ann Brook waterfall, N. of Ingonish (*RMS 66315, 66331, 66335*); also reported from Baddeck, Margaree, Louisburg, Halifax, Dartmouth, Cole Harbour (Macoun, 1902; Brown, 1936a).

MAINE. Mt. Katahdin, ca. 5000 ft, side of Saddle Slide (*RMS 32997*); Mt. Desert I. (Lorenz, 1924). NEW HAMPSHIRE. Mt. Washington, widespread (*RMS, s.n.*); Crystal Cascade, White Mts. (*Underwood & Cook*, Amer. Hep. no. 90); Franconia Mts. (Lorenz, 1908c). MASSACHUSETTS. Evans (1913). RHODE ISLAND. Evans (1923g). CONNECTICUT. Salisbury, Litchfield Co. (*Evans*). NEW YORK. Catskill Mts.: Wittenberg Mt., Cornell Mt, and between Cornell and Slide Mts. (*RMS 17586, 17572, 17578b, 17619a, 24241;* only the last studied cytologically).

NORTH CAROLINA. Grandfather Mt., Avery Co., ca. 0.2 mi from summit parking lot (*RMS 44508, 44605, 44609, 44507*); near summit of Roan Mt., near Roan High Bluff, Mitchell Co. (*RMS 40301*); near summit of Mt. Mitchell, Yancey Co., 6600 ft (*RMS 23160, 23188*). TENNESSEE. Exposed cliffs, Roan Mt., 6100–6200 ft, Carter Co. (*RMS 61284*).

Westward occurring as follows: MICHIGAN. Reported variously from Pictured Rocks, Alger Co.; Luce and Chippewa Cos.; Keweenaw, Marquette, Ontonagon Cos.; Isle Royale, Keweenaw Co.; Sugar Loaf Mt., Marquette; Huron Mts. (Evans & Nichols, 1935; Steere, 1937; Nichols, 1935; Nichols & Steere, 1936, 1937, etc.). WISCONSIN. Sauk, Ashland, Bayfield Cos. (Conklin, 1929); Sand I., Apostle Isls., Bayfield Co. (*RMS 17529*). MINNESOTA. Carlton, Cook, Koochiching, Lake, Lake of the Woods, and St. Louis Cos. (Schuster, 1953; numerous locs.).

Also reported from IOWA (Frye & Clark, 1937–47; based on a Shimek collection which Conard, 1945a, p. 109, has shown to represent *Lophocolea heterophylla*).

Of the above reports, *all* literature reports and those based on dead material must be regarded with a residuum of doubt; they may belong in part to var. *silvicola*. With isolated and noted exceptions, my collections cited by number have been cytologically examined and have the var. *ventricosa*-type oil-bodies.

Ecology

Largely unresolved, because of widespread confusion with other varieties. Arnell (1956, p. 119) states that the plant occurs mostly in shade and ± moist places, on rock ledges and outcrops, on the ground in forests, especially on foot paths, decaying logs, but not among *Sphagnum* in bogs. This well summarizes the ecology and agrees with Schuster (1953, p. 361), who states the plant "occurs usually associated with moist, shaded rocks, in crevices on the sides of which it may form extensive mats. . . . ." It is also reported (Schuster *loc. cit.*) in the Great Lakes region as from cool ledges, with *Scapania nemorosa, Lophozia kunzeana,*

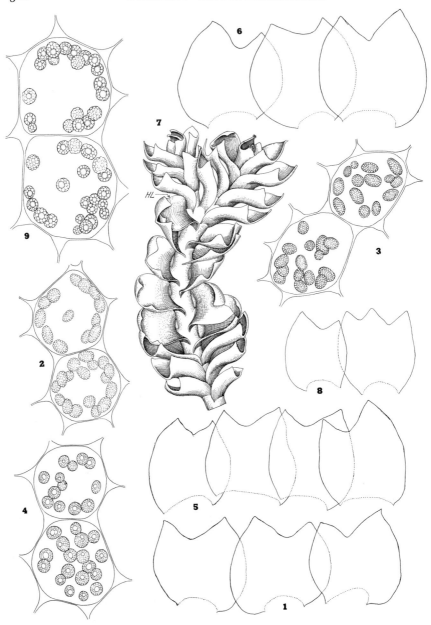

FIG. 206. *Lophozia* (*Lophozia*) *ventricosa s. lat.*: (1–3) var. *ventricosa*; (4–9) var. *silvicola*. (1) Leaves (×22). (2) Cells with oil-bodies from median cells of leaf in fig. 1 (×765). (3) Median cells with oil-bodies (×735). (4) Median cells with oil-bodies, from leaf in fig. 5 (×765). (5) Four leaves (×22); (6) Leaves, from orbicular-leaved extreme of var. *silvicola* (×22). (7) Gemmiparous shoot of mod. *densifolia-colorata*, the "*confertifolia*" phenotype (×ca. 15). (8) Two leaves, the right one atypical and trilobed (×15.5).

Blepharostoma, Lepidozia reptans, etc., where it occurs chiefly at a "pH below 6.0, but appears to be unable to compete when the growth of the moss-mat and subsequent decay result in a pH of below 3.8." Reported as "abundant on thin soil over rocks and as a chasmophyte" at a pH of between 4.0 and 6.0, while on "sunny, wet *Sphagum* hummocks capping the crests of ledges," under conditions of greater acidity, it is replaced by a "red-pigmented modification with large trigones" that often approaches "*L. confertifolia*" in the somewhat concave leaves. Some of these red-pigmented modifications are better referred to var. *silvicola*. The plant also occurs "on decaying wood, though it is here often replaced by *L. porphyroleuca*" (and, especially in the New England mountains, often also by *L. ascendens, L. ventricosa silvicola,* and/or *L. longidens*). Generally speaking "one of the least critical factors governing the occurrence . . . appears to be the amount of humic acids in the substrate . . . the plant often being an initial or subinitial invading species, wherever mesic conditions prevail; under such conditions it can persist until a dense peat layer is produced" (Schuster, *loc. cit.*). Admittedly, this taxon rarely occurs "over peat in bogs, far from any rock outcrops."

In the Appalachian region variously occurring associated with rocks, as, e.g., among *Dicranum* on dry cliff faces and over rather dry ledges, with *Bazzania trilobata, B. denudata, Mylia taylori, Lepidozia reptans, Anastrophyllum michauxii,* or over *Sphagnum* and/or with *Lophozia attenuata, Anastrophyllum saxicolus, and S. michauxii* on seepage-moistened vertical rock walls and ledges, or on shaded ledges with *Scapania nemorosa*.

Differentiation

When living plants are available, usually readily separable from related taxa. The oil-body characters first emphasized by Buch (1933) are mostly rather constant. Living plants have the oil-body characteristics of *L. wenzelii* and *L. porphyroleuca;* e.g., each cell has polymorphic oil-bodies, formed of a large number of weakly protuberant globules without trace of a central spherule in fresh material. The oil-bodies are not biconcentric, thus differing from those of *L. ventricosa* var. *silvicola*. They also agree in being usually—but not always—fewer in number than in var. *silvicola* (where they average 15–22 per cell). In var. *ventricosa* there are usually

---

(9) Median cells with oil-bodies, arctic phenotype of var. *silvicola* (×755). [1–2, 4–5, all drawn from intermingled plants from Roan Mt., N.C., *RMS 38656;* 3, from plant found among *Gymnocolea* in sunny rock pool, Porcupine I., Susie Isls., Minn., *RMS, Aug. 16, 1948;* 6, *RMS 38657,* Roan Mt., N.C.; plants with var. *ventricosa*-type leaf dimensions, but var. *silvicola* oil-bodies; 7–8, *RMS 13657e,* from Lucille I., Susie Isls., Minn., of a mod. *densifolia-colorata;* in fig. 7, HL = half leaf.]

(9)10–16, or occasionally more, oil-bodies per cell; in this respect it approaches *L. wenzelii* but differs from the otherwise closely allied *L. porphyroleuca* (which has from 5–9 in most cells).

Careful study of oil-bodies of living plants, therefore, facilitates the separation of var. *ventricosa* from immediate relatives, except for the closely allied *L. wenzelii*.

Buch (1933) claimed that vars. *ventricosa* and *silvicola* differed in color. *Lophozia ventricosa s. str.* is supposedly deeper green, with less pellucid cells, while *silvicola* is pale green (except for the carmine-pigmented portions, often in fact whitish green), with more pellucid cells. This difference, according to Buch, gives the plants, when compared side by side,[171] a perceptibly different facies.

LOPHOZIA VENTRICOSA var. SILVICOLA (Buch) *Jones*

[Figs. 202:4; 206:4–9; 207]

Lophozia silvicola Buch, Ann. Bryol. 6:125, figs. 1:1–5, 2:1–3, 1933; Banwell, Trans. Brit. Bryol. Soc. 1(3):194, fig. 1, 1949; Schuster, Amer. Midl. Nat. 45:60, pl. 16, figs. 4–9 1951; Schuster, *ibid.* 49:358, pl. 17, figs. 4–9, 1953.
Lophozia ventricosa auct. (in part); K. Müller, Rabenh. Krypt.-Fl. ed. 3, 6(1):647, fig. 196c,e-f, 1954 (in part, at least).
Lophozia fauriana Hatt., Jour. Hattori Bot. Lab. no. 7:46, 1952; Hattori, *ibid.* no. 10:66, pl. I:16–20, pl. II:16–17, 1953 (*p.p.*; not of Stephani, 1917, p. 112).
Lophozia alpestris fo. *granditexta* Hatt., Bot. Mag. Tokyo 58:40, fig. 17, 1944.
Lophozia ventricosa var. *silvicola* Jones, Trans. Brit. Bryol. Soc. 3(3):359, 1958.
Lophozia silvicoloides Kitagawa, Jour. Hattori Bot. Lab. no. 28:276, fig. 9, 1965 (new, provisional synonymy).[172]

Closely prostrate, except when crowded, only the apices ascending, *pale green in shade, in sun postical stem surface and adjacent portions of the leaves becoming carmine-red;* shoots subsimple to sparingly branched, terminally so, *mostly 1350–1800* (2100–2500) μ wide × 1–2(3–5) cm long. Stems elliptical in cross section, from a minimum of 200–220 μ wide to 250–300 μ in diam.; stem section mostly 14–16 cells high; dorsal cortical and ventral cortical cells nearly similar, scarcely elongate, somewhat thick-walled, 17–26 μ wide × 20–30 μ long, the

---

[171] There may also be a slight difference in pigmentation pattern. Under somewhat xeric conditions, *L. ventricosa* var. *ventricosa* may develop dull purplish pigmentation of the leaf lobes, which may extend virtually to the leaf base. This pigment is much less brilliant than the carmine pigmentation of the sun forms of *L. ventricosa* var. *silvicola*. However, in var. *ventricosa* there may also be traces of the carmine pigmentation of the leaf bases, although apparently never as spectacularly developed as in extremes of var. *silvicola* and in *L. porphyroleuca*.

[172] *Lophozia silvicoloides* Kitagawa is the plant with biconcentric oil-bodies which Hattori (1951b, 1953) illustrated as *L. fauriana*. Possibly this plant, which differs from var. *silvicola* chiefly in the ciliate-laciniate perianth mouth with teeth 3–6 cells long, should be regarded as a discrete variety, parallel with var. *silvicola*. In the more strongly armed perianth mouth (but not in the narrow leaves and biconcentric oil-bodies) this plant approaches *L. ventricosa* var. *confusa* (p. 581).

postical becoming brownish; dorsal 3–4 layers of hyaline, delicate medullary cells 17–25 μ wide × 55–95 μ long, the ventral half or more of medulla of smaller, less elongate cells (10–20 μ wide × 30–50 μ long). Leaves *weakly or scarcely canaliculate, rectangulate to broadly ovate-rectangular* to (rarely) suborbicular, from a minimum of 765 μ wide × 725–765 μ long to 870 μ wide × 920 μ long, to a maximum of 1000–1160 μ wide × 1230–1400 μ long (*width 0.75–0.95 the length, more rarely to 1.05× the length*), widest at or barely above middle; *lobes subacute to rectangulate*, occasionally obtuse, 2 (sometimes 3), the postical usually distinctly larger; *sinus descending usually 0.25–0.3(0.35)*, rectangulate to slightly obtuse (rarely crescentic). Cells collenchymatous, in sun forms trigones often bulging, 22–29 μ in leaf lobes, toward middle larger and 25–27 × 30–35 μ, toward base 25–27 × 28–45 μ; oil-bodies (*12*)*15–24*(*28*) (averaging 17–20) *per cell*, obscuring ½–⅔ the cell lumen, *spherical*, occasional ones short-ovoid or oval, *4–5(6) μ in diam.*, a few to 5–6 × 6.5–7.2 μ and short-ovoid, *formed of a centric or eccentric, highly refractive sphere, surrounded by a mantle of less refringent, minute, scarcely protuberant, barely perceptible spherules.* Gemmae *abundant* (in both shade and sun forms; except bog forms), *pale greenish*, normally 2-celled, pyriform to rhombic to quadrate to stellate, with protuberant angles, *rather large*, (17–22 × 25–35 μ to 16–23 × 18–26 μ).

♂ Bracts ± imbricate, often ± ventricose, usually reddish (even in otherwise green shade forms), bilobed like leaves, in 3–7 pairs usually, the lobes spreading to erect and imbricate. ♀ Bracts somewhat larger than leaves (shoot at bracts to 3.0–4.5 mm wide), 3–4-lobed to ¼–⅓ at apex, sometimes merely bilobed, otherwise entire; bracteole small and lanceolate, or wanting. Perianth cylindrical to obpyriform, plicate near the contracted mouth, to ca. 2 mm long, the *mouth armed with sharp, 1–2-celled teeth*, whose walls are ± thickened. Spores weakly papillose, 11–13 μ.

Type. Finland (Prov. Nylandia, ad opp. Lovisa, *H. Buch, Aug. 1915*).

Distribution

Abundant and widespread, southward largely montane or submontane (but descending to sea level northward); abundant in the Spruce-Fir Biome but with restricted and relict stations in the Northern Hardwoods Forest (Beech-Maple-Hemlock-Yellow Birch Climax); apparently widespread (but much less common) at least in the edge of the alpine Tundra; rare or lacking in the high Arctic but extending north to at least 70° in west Greenland.

The range not exactly known, because of confusion until recently with the apparently less common *L. ventricosa* var. *ventricosa*; in Europe found from the central portion (Germany and Switzerland) northward to England (Banwell, 1949), Sweden, Norway, and Finland (Buch, 1933b, p. 128); presumably reports from southern Europe (Spain, Sardinia, Bulgaria

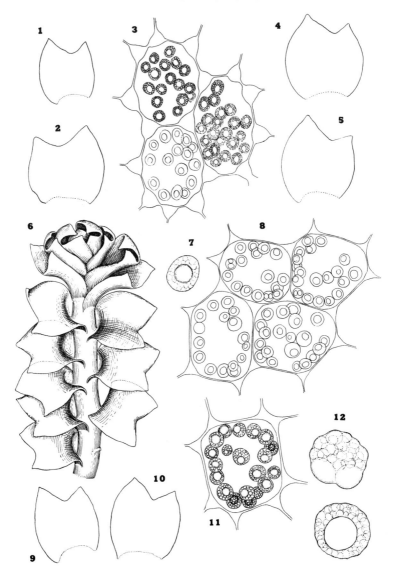

Fig. 207. *Lophozia* (*Lophozia*) *ventricosa* var. *silvicola*. (1) Normal leaf
(×16.5). (2) Leaf of broadest, atypical extreme (×25). (3) Median cells
with oil-bodies (×665). (4–5) Leaves (×25). (6) Shoot sector (×15.8).
(7) Oil-body (×ca. 2350). (8) Cells with oil-bodies (×715). (9–10) Leaves
(×23.5). (11) Cell with oil-bodies (×845). (12) Two oil-bodies, the
upper disintegrating (×ca. 2850). [1, 3, *RMS 18534b*, Susie Isls., Minn.;
2, 4–12, Great Palisade, Lake Superior, Minn., *RMS*.]

northward) attributed to *L. ventricosa* refer, to at least a large degree, to
L. ventricosa var. *silvicola*.

The North American range confused with that of typical *L. ventricosa*. First
reported by Nichols and Steere (1936) from Michigan, by Schuster (1951a,
p. 105; 1953, p. 361) from Minnesota, and Persson (1962, p. 8) from Alaska.
With recent investigation proving widespread in the Coniferous Biome of
eastern North America. The following stations do not adequately reflect the
abundance of the plant.

W. GREENLAND. Fortune Bay, Disko I., 69°16′ N., 53°50′ W. (*RMS & KD 66-208a*);
Ritenbenk, Agpat, 69°46′ N., 51°20′ W. (*RMS & KD 66-257a*); Tupilak I., W. of
Egedesminde, 68°42′ N. (*RMS & KD 66-014b, 66-006*); E. end of Agpat Ø, neck of
Umiasugssup Ilua, 69°52′ N., 51°38′ W. (*RMS & KD 66-1422a*).   LABRADOR.
Ungava Pen. (Dix, 1956).   NEWFOUNDLAND. Widespread; Avalon, S. coast, W.
Nfld., N. Pen., and C. and NE. Nfld. distrs. (Buch & Tuomikoski, 1955).   NOVA
SCOTIA. Mary Ann Falls, Cape Breton Natl. Park, E. side, S. of Neils Harbour (*RMS
42687c, 42681, 42662b*); 3–3.5 mi S. of waterfall, Mary Ann Brook, N. of Ingonish
(*RMS 66314, 66331*).   QUEBEC. N. of Malbaie R., 10–12 mi NW. of Bridgeville,
Gaspé Pen. (*RMS 43513b, 43514, 43503a, 43056a, 43603, 43524, 43522b*).   ONTARIO.
Dreamers Rock, Manitoulin I. (*Cain 4086*). Algonquin Park; Kenora; Manitoulin I.;
Muskoka; Thunder Bay (Williams & Cain, 1959).

MAINE. Mt. Katahdin: abundant, at foot of Hamlin Ridge; Chimney Pond; Saddle
Trail below and above treeline; Cleft Rock pool, etc. (*RMS 32901b, 32905, 32944b*).
NEW HAMPSHIRE. Mt. Washington, below treeline (*RMS*); Mt. Moosilauke (*RMS*).
VERMONT. Mt. Mansfield: Haselton Trail, 3000–3200 ft (*RMS 43802, 43844b, 43802a,
43801a, 43844*); Long Trail, near summit, 4100–4300 ft (*RMS 43814*), and near treeline at 3600–4100 ft (*RMS 43870*).   NEW YORK. Catskill Mts.: Spruce-Fir Forest on
slope of Cornell Mt; between Cornell and Slide Mts., 3600–3700 ft; near summit of
Cornell Mt., 3500 ft; near summit of Slide Mt., 4100–4200 ft; near summit of Wittenberg Mt., 3400–3500 ft (*RMS 17632, 17582, 17580c, 17586, 24470a, 17615, 24307, 24308,
24318, 24714, 24427, 24471a, 24444, 24424*, etc.). Indian Falls, Mt. Marcy (*Wilson 20!*);
cliff wall near Peasleeville, E. edge of Adirondack Mts. (*RMS A-216, A-211, A-213*).
Tompkins Co.: Fall Creek ravine, Ithaca; Enfield Glen; Lick Brook, S. of Ithaca
(*RMS*). Briggs Gully, Richmond Twp., Ontario Co. (*RMS*); E. of Woodhull and SE. of
Haskinville, Steuben Co. (*RMS*). (Most of the latter reports listed as *L. ventricosa* by
Schuster, 1949a.)   MASSACHUSETTS. Summit, Mt. Everett (*RMS 68-136, 137*).

VIRGINIA. Peaks of Otter, Bedford Co., at summit (*RMS 28597;* see Schuster &
Patterson, 1957).   NORTH CAROLINA. NW. side of Roan Mt., Mitchell Co., on cliff
wall, 6200 ft (*RMS 36935, 40302, 36918, 37973, 40337a, 40337, 40310, 40319, 40302a,
36937a*); Mt. Mitchell, Yancey Co., 6630 ft (*RMS 23682, 23160a, 23159, 23151a,
24810a, 34541, 34601, 24602, 24808*, etc.); Grandfather Mt., Avery Co., ca. 5300 ft,
near end of toll road (*RMS 44507, 44609*); "Flat Rock" on Blue Ridge Pkwy., near
Linville (*RMS 30169*).   TENNESSEE. Ledges, SW. summit of Roan Mt., 6200 ft
(*RMS 36946, 36938, 38660, 36933*); Myrtle Pt., Mt. LeConte, Sevier Co. (*RMS*).

MICHIGAN. Near Negaunee (Nichols & Steere, 1936); Pictured Rocks, near Munising
(*RMS 39291*).   WISCONSIN. Near Dells of Wisconsin R. (*RMS 39205*).   MINNESOTA.
Cook Co.: Grand Marais (*RMS 13265*); Pigeon Pt., near Grand Portage (*RMS 17266,
17350, 17438*); Long I., Big Susie I., Lucille I., Porcupine I., all on Susie Isls. (*RMS
7169b, 7163a, 7167a, 7160a, 7173a, 7174a, 11731, 9967, 13378b, 18534b, 19402*); Rainy
L., E. of Island View, Koochiching Co. (*RMS 18111, 18133, 18136*); Great Palisade, L.
Superior, Lake Co. (*RMS 12012, 11785a, 11781*).

Ecology

Lophozia ventricosa var. *silvicola* is widespread under noncalcareous conditions, often associated with rock outcrops, either on sandstones and shales or on igneous or metamorphic rocks; mostly on damp, shaded rocks, forming a pure green mod. *viridis*, often with various other Lophoziae (*Anastrophyllum minutum*, *A. michauxii*), *Scapania nemorosa*, *Blepharostoma*, *Lepidozia reptans*, rarely *Lophozia alpestris*, etc. On exposed acidic sandstone bluffs (as at Pictured Rocks) with *Gymnocolea inflata*, *Tritomaria exsectiformis*, *Lepidozia*, *Blepharostoma*, *Cephalozia bicuspidata*, i.e., a member of the *Gymnocolea-Cephalozia bicuspidata* Associule. Also often over exposed peat-covered, moist, basaltic ledges, associated with *Sphagnum*, *Odontoschisma denudatum*, *Calypogeia sphagnicola*, *Mylia anomala*, *Cephalozia* spp., etc., i.e., as a member of the *Mylia-Odontoschisma* phase of the helophytic *Mylia-Cladopodiella* Associule. Under such conditions often in partial sun as a rather pellucid phase with coarse trigones and strongly reddish-tinged leaf bases. Such forms closely mimic *L. porphyroleuca*, except for the presence of gemmae and the characteristic biconcentric oil-bodies.

This variety is frequent, under somewhat similar conditions, on several of the higher peaks of the Southern Appalachians; chiefly the mod. *viridis* occurs southward. Northward it is found to, and occasionally above, timber line in New England. It is absent southward from decaying logs, although rather frequent on them in the New England mountains, where associated with mesic phases of the *Nowellia-Jamesoniella* Associule, chiefly with *Lophozia porphyroleuca*, *L. ascendens*, *L. longidens*, *L. attenuata*, *Blepharostoma*, *Anastrophyllum hellerianum*, *Calypogeia suecica*, *Cephalozia media*, and other Cephaloziae.

Although phases of decaying logs are usually light yellow-green throughout (even when associated *L. porphyroleuca* shows reddish leaf bases), *L. ventricosa* var. *silvicola* often occurs over peat on ledges, as sun forms, frequently on *Sphagnum*, often deeply carmine-pigmented (and then often with a tertiary brown pigment of the distal parts of the leaves, which may mask much of the carmine). Shade forms may occur in bogs, straggling between *Sphagnum*, and are light green except for the ventral side of the stem and the ventral leaf bases, which, even with diffuse light, acquire a certain amount of reddish pigment.

Buch and Tuomikoski (1955) report the taxon in Newfoundland occurring on "moist to wet humus-covered siliceous rocks, peaty soil, also on decayed wood, etc., with *Cephalozia media*, *Dicranum fuscescens*, *Jamesoniella autumnalis*, *Lepidozia reptans*, *Orthocaulis attenuatus*, *Cephalozia leucantha*, *Lophocolea heterophylla*, *Mylia taylori*, *Ptilidium pulcherrimum*, *Anastrophyllum michauxii*, *Bazzania trilobata*, *Diplophyllum albicans*, etc."

In the Catskill Mts. abundant on dry to damp rock walls and peaty soil over rocks, and on moist ledges, associated with *Blepharostoma, Bazzania trilobata, B. tricrenata, Cephalozia media, C. pleniceps, Anastrophyllum michauxii, A. minutum, Lepidozia reptans, Dicranum, Ptilidium pulcherrimum, P. ciliare, Lophozia attenuata, Mylia taylori*—thus with almost the same complex of species as in Newfoundland.

Much less frequent in the Arctic, but there occurring at times as a long, sterile, sparingly gemmiparous phase among *Sphagnum* around boggy pools and tarns (as on Agpat Ø, west Greenland, *RMS & KD 66–1422a*). Such plants occur with *Mylia anomala, Calypogeia sphagnicola, Cephalozia pleniceps*, etc. The plants are then freely furcately branched and show gross variation in leaf form, with a high incidence of leaves rectangulately bifid for 0.3–0.35 and rather many trifid leaves.

Differentiation

Lophozia ventricosa var. *silvicola* is a critical taxon, about whose value there has been a diversity of opinion; Banwell (1949), Schuster (1953), and Arnell (1956) accepted it, admitting its exceedingly close relationship to *L. ventricosa*, but Müller (1939) considered it at first to be a form of the polymorphic *L. ventricosa*, later (1954, *in* 1951–58) claiming that "*L. silvicola*" represents the "old" *L. ventricosa*, while *L. ventricosa sensu* Buch was partly (or entirely) *L. porphyroleuca*. The latter appears improbable to me.

Lophozia ventricosa silvicola differs from other taxa with greenish gemmae largely in the oil-bodies, formed of a highly refractive central (or occasionally eccentric) sphere, surrounded by a layer or "mantle" of small, less refractive spherules. Similar oil-bodies are absent in our other species, except in *L. groenlandica*. The oil-bodies are numerous, averaging at least 15 (with a range of 12, more commonly 15–20, occasionally to 25); they average only 10–15 per cell in typical *L. ventricosa*, 6–9 per cell in *L. porphyroleuca*, 6–12, rarely 14, per cell in *L. ascendens* and *L. wenzelii s. str.* Additional differential features are (*a*) decumbent growth, strongly ascending only with crowding and exceedingly moist and shaded conditions; (*b*) relatively narrow, rectangulate leaves. Buch (1936, p. 60) separates *L.* "*silvicola*" from *L. ventricosa s. str.* on the basis of the narrow leaves (0.75–0.95 × as wide as long in "*L. silvicola*," vs. 0.95–1.20 × as wide as long in *L. ventricosa*). *L.* "*silvicola*" may have leaves 0.95–1.05 × as wide as long in sun forms (Schuster, 1953); even broader leaves are figured for this taxon by Banwell (1949, fig. la). Furthermore, extreme shade forms of *L. ventricosa s. str.* have narrow leaves, varying from 0.9–1.05 × as wide as long (Schuster, 1953, p. 360). This distinction, therefore, is of slight practical value; see also p. 579. However, Buch's culture work (see Buch, 1933, p. 9, figs. 6 and 7) shows that, *under identical conditions*, the *silvicola* and *ventricosa* phenotypes differ in leaf shape, the latter tending to have broader leaves. The amplitude of somatic variation in this respect is unfortunately so high that the slight genetic differences are often masked.

On the basis of the yellow-green gemmae and rather narrowly rectangulate leaf form (leaf width usually 0.75–0.95 the length) with a usually angular sinus, var. *silvicola* is likely to be mistaken for *L. ascendens* and *L. porphyroleuca*. The three taxa may occur intimately intermingled (on decayed logs, Mt. Katahdin, Me., *RMS 32905*). The latter collection is critical, making possible a sharp definition of these three taxa. From this collection and two others made immediately nearby (*RMS 32901, 32902*) are derived Table 5 (p. 549).

As these collections demonstrate, var. *silvicola* differs from *L. porphyroleuca* (growing on a moist shaded decayed log) as follows: (1) Gemmae are developed abundantly; in *L. porphyroleuca* only sparingly, on many plants not at all. (2) Leaves are laterally patent, nearly flat except at base, not distinctly canaliculate; in *L. porphyroleuca* leaves are less distant, more erect-spreading, slightly dorsally secund, and have the dorsal lobe loosely folded over the ventral, the leaves being canaliculate; the lobes, instead of being strongly spreading, are often faintly incurved. (3) The stem in such shade forms is postically not or slightly reddish, with vestigial reddish pigmentation not extended to leaf and rhizoid bases; intermingled shade plants of *L. porphyroleuca* have rhizoid bases, postical face of stem, and postical leaf bases carmine-red. (4) Cells have small, quite concave-sided trigones; in *L. porphyroleuca* even the shade extremes associated with the *silvicola* plants show prominent, bulging trigones. (5) Oil-bodies are small (ca. 4–5 μ and spherical, a few to 5 $\times$ 6.5 μ and short ovoid), numerous (averaging 15–20, occasionally to 25, per cell), and distinctly biconcentric; in *L. porphyroleuca* oil-bodies are larger, usually ovoid or irregularly ellipsoid, ca. 4.5 $\times$ 6 to 5–6 $\times$ 6–9 μ, rarely to 6 $\times$ 10 μ, few per cell (averaging 4–8, with extremes of 1–3 in some cells, up to 10 in others), not biconcentric.

Intermingled var. *silvicola* differs from *L. ascendens* chiefly in (*a*) laterally patent leaves, averaging over 0.8 $\times$ as wide as long, with broader lobes and a shallower sinus; (*b*) more oblique leaf orientation; (*c*) prostrate growth; (*d*) presence of slight postical pigmentation of the stem; (*e*) form of the oil-bodies (see above), those of the associated *L. ascendens* being uniformly formed of minute spherules, and occurring mostly 6–12, rarely to 14, per cell; (*f*) larger size, the width of the plants (with leaves) averaging well over 1 mm. These differences are discussed in detail under *L. ascendens*, where the more significant differences in fertile plants are also cited (see p. 549).

Critical study of plants growing intermingled demonstrates that (*a*) a higher saturation deficit is necessary in var. *silvicola* than in *L. porphyroleuca* to produce the mod. *pachyderma*; (*b*) with diffuse light, *L. ascendens* fails to produce pigmentation; var. *silvicola* produces at most a very slight pigmentation of the postical stem surface, not involving the rhizoid bases and not or barely extending onto the postical leaf bases; by contrast, in *L. porphyroleuca* even low light intensity suffices to stimulate production of a rather intense, carmine-red pigmentation of the postical stem surface, which extends to the leaf bases and rhizoid bases; (c) var. *silvicola* and *L. ascendens* retain the ability to produce gemmae abundantly under humid and shaded as well as under dry and insolated conditions; in *L. porphyroleuca* gemma production is restricted to modifications of very moist, deeply shaded sites, while with increased light intensity (and saturation deficit) the species soon loses this ability; (d) *L. ascendens* is strongly ascending in growth; var. *silvicola* and *porphyroleuca* are prostrate, except for the weakly curved shoot apices; (e) var. *silvicola* produces many small, spherical, bicon-centric oil-bodies, *L. ascendens* and *L. porphyroleuca* smaller numbers of variable, ovoid to ellipsoidal oil-bodies, uniformly formed of fine globules. Oil-body number and form clearly remain diagnostic when the plants occur intermingled (Fig. 202).

The different responses of these three taxa demonstrate that three dis-tinct species are involved and that var. *silvicola* differs from the other two in a series of morphological, cytological, and physiological features. Such a clear distinction of var. *silvicola* from var. *ventricosa* cannot always be main-tained.

Although the preceding discussion indicates that *L. ventricosa* var. *silvicola* is clearly distinct from *L. ascendens*, extreme forms of these two taxa cannot always be separated when dead, perianth-free plants are at hand. As pointed out (Schuster, 1953), phases of "*L. silvicola*" with narrow, dense leaves occur on rather dry rock walls. These are formed particularly under crowded conditions, when growth is conditioned by competition with erect mosses. The most ex-treme of these *L. ascendens*-like phases of var. *silvicola* seen (*RMS 39205*, Rocky Arbor Park, near Wisconsin Dells, Wisc.) grew on shaded, rather dry sandstone ledges. In xeromorphic phases (mod. *viridis-densifolia-angustifolia*) the shoots were strongly ascending to suberect, and the dense leaves very narrow, with *L. ascendens*-like narrow, almost hornlike lobes. Such plants had many leaves varying from 580 μ wide × 850 μ long to 600 μ wide × 750 μ long (width-length ratio ± 0.68–0.80:1). The biconcentric oil-bodies served to definitely place these plants as *silvicola*. Equally indicative was the tendency for the lower stem faces to become reddish brown.

The critical nature of *L. ventricosa* var. *silvicola* is well illustrated by material collected on Roan Mt., N.C. and Tenn. Most collections are clearly var. *silvicola*. However, a single collection (*RMS 38656*) apparently consists of two kinds of plants, *both* light green, and almost or quite inseparable except for one feature: some plants had rather flat or loosely complicate leaves, averaging as long or slightly longer than broad; others had somewhat concave, more trans-versely oriented leaves which were clearly broader than long. Narrow-leaved plants gave the impression of being a less xeric phase, yet had somewhat larger trigones than the denser-leaved plants in which the trigones were not at all bulging. Correlated with the broad leaves (width:length ca. 1100:980 to 1140:1000 to 1240:1000 μ) was (a) a tendency toward formation of ovoid oil-bodies, although some were spherical, ranging from 8 to 19 (average 15) per cell; (b) a definite lack, in most cases, of a central spherule, almost all the oil-bodies appearing to be composed of uniformly sized spherules. The plants

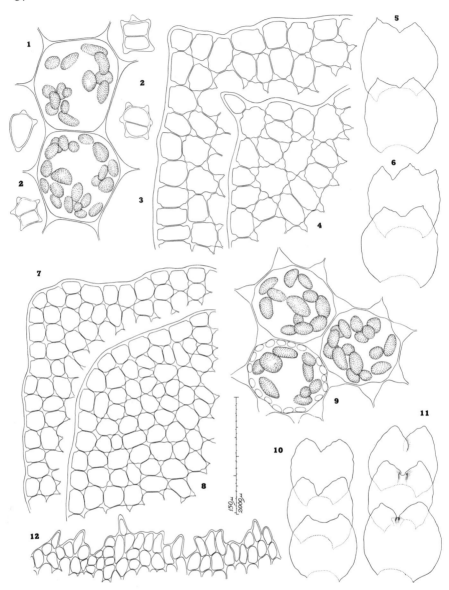

FIG. 208. *Lophozia ventricosa* s. *lat.* (1–6, var. *rigida*; 7–12, var. *longiflora*). (1) Median cells with oil-bodies (×555). (2) Gemmae (×310). (3–4) Apices of leaf lobes (×208). (5–6) Leaves (×16). (7–8) Apices of leaf lobes (×208). (9) Median cells with oil-bodies and, lower left, chloroplasts (×710). (10–11) Leaves (×16). (12) Perianth mouth sector (×208). [1–6, *RMS & KD 66-565*, Godhavn, Disko I., Greenland, type of var. *rigida*; 7–12, all from *RMS & KD 66-551*, same loc.; 3–4, 7–8, 12, all drawn to left scale; 5–6, 10–11, drawn to right scale.]

with broad leaves, fewer oil-bodies which were often ovoid and usually lacking a central spherule, and a tendency for hollow leaves, clearly conform with *L. ventricosa sensu* Buch.

Mingled with these plants were narrow-leaved ones (width:length 1000:950 to 1050:1100–1150 to 1100:1050 μ, i.e., ca. 0.95–1.05× as wide as long). Narrow leaf size was in turn correlated with (*a*) more spherical oil-bodies, only exceptional ones being ovoid, ranging from 13 to 28 (average 18.7) per cell and (*b*) a well-defined tendency for each oil-body to possess a small but distinct central "eye." On the basis of leaf size, oil-body number, and oil-body shape and composition, these plants correspond with *L. ventricosa* var. *silvicola*.

In this same area, however, plants can be found with intermediate characters. For example, in *RMS 38657*, collected within 100 feet of the last collection cited, we find plants (admixed with *L. incisa*, again with *Dicranum*), which may bear somewhat concave leaves that are clearly broader than long (Fig. 206:6), the width-length ratio ranging from 1280:1100 to 1300:1200 to 1310:1240 μ. Using the leaf-width criteria established for separating "*L. silvicola*" and *L. ventricosa* by Buch (1936), we would look for these plants under *L. ventricosa*. However, they have the oil-bodies of var. *silvicola*, with the central "eye" quite distinct.

Such obvious transitions between vars. *silvicola* and *ventricosa* have been seen, to date, only in the Southern Appalachians.

LOPHOZIA VENTRICOSA var. *LONGIFLORA* (Nees) Macoun

[Figs. 208 : 7–12; 212:10–11]

Jungermannia longiflora Nees, Naturg. Eur. Leberm. 2:95, 1836.
Lophozia ventricosa var. *longiflora* Macoun, Cat. Canad. Pls. 7:17, 1902.[173]
Lophozia longiflora Schiffn., Lotos 1903(7):45, 1903; Evans, Rhodora 10:189, 1910; Lorenz, The Bryologist 13:39–40, pl. 4, 1910; Macvicar, Studs. Hdb. Brit. Hep. ed. 2:182, 1926; Müller, Rabenh. Krypt.-Fl. ed. 3, 6:654, figs. 195:3, 199, 1954.
Jungermannia ventricosa var. *longiflora* Macvicar, Studs. Hdb. Brit. Hep. ed. 2:182, 1926 (in synonymy accredited to Macoun).

Vigorous, often over *Sphagnum* or peat, yellowish green, when insolated with yellow-brown, warm pigmentation of distal parts of leaves, $\pm$ mixed with reddish or pink (thus often salmon-colored), at least at the ventral *leaf bases normally carmine red or vinaceous; stem ventrally deep red* to reddish black; rhizoids often reddish at base. Plants usually 2.5–3(4) mm wide × 2.5–6 cm long. Leaves mostly rather dense, usually subvertical and loosely folded to very bluntly conduplicate, at times canaliculate, the lobes widely spreading, almost never incurved, the leaves occasionally stiff and rather plicate, the *sinuses occasionally gibbous* (particularly near gynoecia); leaves variable, 1025–1060 μ long to 1065 μ wide × 925–950 μ long (width usually 1.05–1.15× the length), 2(3)-lobed for 0.2–0.3(0.35) their length, the *lobes often blunt or rounded*, the sinuses variable, deeply lunate to widely obtuse-angular to bluntly angulate.

---

[173] Cited, in error, as "*Jungermannia ventricosa* var. *longiflora* Macoun" *in* Müller (1905–16, p. 671) and Macvicar (1926, p. 182). On that basis, I regarded the combination under *Lophozia* as new (Schuster, 1953, p. 361).

Cells as in *L. ventricosa* var. *ventricosa*, ± strongly collenchymatous, marginal in distal 0.5 of leaf ± quadrate, 22–26 μ; median cells ca. 24–30(32) × (25)28–36 μ; the median and subapical with *9–16(17–21)* *(6–7 in smaller cells) oil-bodies each*, often nearly occluding the cell lumen, *variable but all without central "eye,"* from a minimum of 4.6–5.5 × 4.8–6–7 μ up to *5.5–6 × 8–10 μ, isolated ones to 7.5–8 × 9–12 μ*, of uniform, small, weakly protuberant spherules. Gemmae pale green, apparently never with wall pigments, *rare*.

♂ Bracts in up to 5 pairs, saccate and gibbous at base, with spreading and more sharply tapered lobes than normal leaves, often reddish. ♀ Bracts larger than leaves, erect or erect-spreading, 2–3–4-lobed to 0.25–0.35 their length, the ovate-triangular *lobes blunt to subacute at the tips*, entire-margined, the sinuses often gibbous. *Perianths cylindrical, red below, long-emergent, large*, from 920–1175 μ in diam. × 2.5–2.8 mm long (Disko I. phenotype) *up to 1–2 × 3–4(5) mm long; mouth ± decolorate, hardly lobulate, with remote to locally subcontiguous, small, almost uniformly 1-celled teeth* that are not conspicuously elongated and are *often free for only part of their length*. Spores 14–16 μ; elaters 7–9 μ in diam.

Type. Germany: Riesengebirge (*Nees*).

Distribution

Poorly known; apparently largely alpine, but at least locally occurring in the Arctic.

Known from central Europe (chiefly Switzerland, Austria, south Germany, but also France: Alsace; as "var. *uliginosa*"), northward recurring in Norway (Joergensen, 1934; Arnell, 1956).

For North America there are many reports, but perhaps most of these are based simply on the mod. *colorata* of either var. *ventricosa* or var. *silvicola*. The following are the more reliable reports:

W. GREENLAND. Lyngmarksbugt, Godhavn, Disko I., 69°15′ N., 53°30′ W. (*RMS & KD 66-551, 66-557*).   QUEBEC. "On old logs, coast of Gaspé" (Macoun, 1902; ambiguous report).   MAINE. In tufts of *Sphagnum*, Schoodic L. (Evans, 1908c); Mt. Desert I., Backers I. (Lorenz, 1924); Mt. Katahdin (Lorenz, 1917).   NEW HAMPSHIRE. Eagle L., Mt. Lafayette; Mt. Osceola, 3800 ft; Carrigains Pond, 3100 ft, all in White Mts. (Lorenz, 1910a).

Also reported as follows: WISCONSIN. Black R., Douglas Co. (Conklin, 1929). MINNESOTA. Cook Co.: Pigeon Pt. (*RMS 5747*), Beaver Dam and Hungry Jack Trails (*Conklin 2356*) and Grand Marais (*Conklin 2665*); see Schuster, 1953.

Ecology

Typically a low-alpine to subalpine plant of *Sphagnum* and boggy ground, although Arnell (1956) also cites it from "shady rocks and cliffs." Lorenz (1917, p. 45) reported it from the peaty banks of a mountain pond, associated with *Lophozia kunzeana*, *Gymnocolea inflata*, *Mylia anomala*, *Cladopodiella francisci*, and *Cephalozia bicuspidata*.

Differentiation

"Clearly allied to *L. ventricosa*, from which it appears to be not always separable," according to Macvicar (1926). Evans (1908c) emphasized its similarity to *L. porphyroleuca*, because of the tendency to form carmine pigments of the ventral face of the stem and ventral leaf bases; if differs from this species in the much larger size, the wider leaves, the large but still considerably smaller trigones, and the unlobed perianth mouth with minute, usually 1-celled, teeth.[174] Müller (1905–16, p. 673) regarded it as a "erheblich constante und darum also kleine Art anzuerkennende Form der *L. ventricosa*" which, in association with the moist, sunny sites that it inhabits, has developed an unusually vigorous size. He considered most typical, *vis-à-vis L. ventricosa s. str.* (var. *ventricosa*), the larger size, the somewhat canaliculate-complicate leaves, the deep red perianths with their bleached apices, the freely reddish-pigmented leaves, the 3–5-lobed perichaetial bracts, the relatively rare gemmae, and the rather larger spores.

In my opinion, "*L. longiflora*" is best regarded as a variety or perhaps eco-subspecies of *L. ventricosa*. The oil-body number (Müller, 1939, 1951–58, reports 15–20 oil-bodies per cell; his figures closely approximate those in my material from Greenland) is that of *L. ventricosa s. str.*[175] An identical, if less intense, pigmentation pattern also characterizes var. *ventricosa*—leaving, in essence, chiefly vigor as a distinguishing criterion. However, on my material—as well as on European plants, which both Müller (1951–58) and Macvicar (1926) describe as having blunt or obtuse lobes—the leaf lobes are usually blunt or even rounded (Fig. 208:10–11). Perhaps equally diagnostic, the sinuses tend to be gibbous.

LOPHOZIA VENTRICOSA var. CONFUSA Schust., var. n.

[Figs. 209; 210: 1–6]

Varietas var. *ventricosae* similis, differens autem, ut (*a*) omnis cellula 5–12 guttis 6–7 × 8–10 μ olei praedita; (*b*) folia 0.25–0.4(0.45) bilobata, sinu angulato; et (*c*) os perianthii lobulatum, dentibus longioribus, ad 3–4 cellulis long, praeditum.

Similar in vigor to var. *ventricosa*; green to warm brown on leaf lobes and distal portions of the leaves; red pigments *usually absent* or (sun forms) ordinarily mixed with brownish, or confined to ♂ bracts; leaf bases colorless, rarely bluish

---

[174] Lorenz (1910a) also compares var. *longiflora* to *L. porphyroleuca* but admits that "in the field it looks like a robust *L. ventricosa*, much tinged with carmine, and with fat perianths." Perhaps Macoun (1902) was right in considering the plant as "scarcely a variety," although there appears to be some ecological segregation *vis-à-vis* var. *ventricosa*—var. *longiflora* inhabiting, as Lorenz noted, chiefly "high and cold localities, in the sun . . . ."

[175] Perhaps the strongest argument in favor of the *reality* of a var. *longiflora* is the fact that it can occur in areas (e.g., west Greenland) where as least two other varieties (*rigida*, *confusa*) of *L. ventricosa* are found and is easily separable from them.

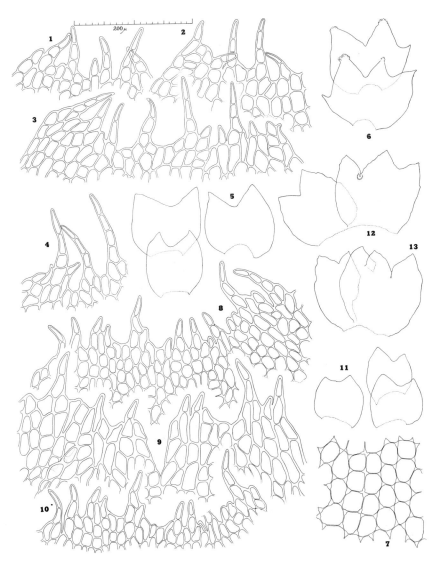

FIG. 209. *Lophozia ventricosa* var. *confusa*. (1–4) Sectors of perianth mouth (×164). (5) Leaves (×14). (6) Atypical leaves, gemmiparous shoot tip (×14). (7) Median cells (×7). (8–10) Sectors of perianth mouth, in fig. 9 of a ? polyploid individual with larger than normal cells (×164). (11) Leaves (×11.8). (12) Bracts and bracteole (×11.8). (13) One bract and, at left, bracteole (×11.8). [1–7, *RMS & KD 66-219*, Jacobshavn, W. Greenland; 8–13, *RMS & KD 66-311a*, Disko I., W. Greenland: 1–4, 8–10, drawn to top scale; 11–13, drawn to one scale.]

purple, rarely vinaceous. Leaves flat, with base concave, usually rather patent, distal portions often rather squarrose, the lobes never incurved (except in the mod. *parvifolia-densifolia-colorata*); leaves basically broadly rotund-quadrate to transversely oblong-quadrate, with rounded sides, from 660–670 μ wide $\times$ 640 μ long to 725 μ wide $\times$ 625 μ long, occasionally 835 μ wide $\times$ 620 μ long, usually 0.25–0.4(0.45) 2-lobed, exceptional ones 3(4)-lobed; *lobes usually angular*, the sinuses from angulate-crescentic to *strongly angulate*. Marginal cells ca. 20–25 μ, median (18)20–25(26) $\times$ 23–28 μ, with trigones usually flat- to concave-sided; oil-bodies in leaf middle ranging from (2–4)5–10(11) to (6)8–12(13) per cell, diverse in shape and size: from 3–3.5 to 4–4.5 μ and spherical to ovoid and 3.6–4.5 $\times$ 6.5 to 5–6.5 $\times$ 7–8 μ, *or occasionally larger* (to 6–7 $\times$ 8–10 μ); oil-bodies uniformly of small spherules (living plants; in dying material 1–2 central spheres develop). Greenish 1–2-celled *gemmae abundantly developed* in both shade and sun phases.

♀ Bracts 2–3(4)-lobed for 0.25–0.35 their length, lobes usually acute. Perianth mouth *shallowly plurilobulate, with the longer teeth 2–3–4-celled,* $\pm$ elongated, thus short-ciliate, much as in *L. porphyroleuca*; teeth often situated at the apices of shallow lobes, formed of $\pm$ elongated cells (to 9–12 $\times$ 30–42 μ).

Type. Jacobshavn, west Greenland (*RMS & KD 66–220, July 16, 1966*). The type has a few mature extruded and some enclosed, green to black capsules.

Distribution

Variety *confusa* is still so poorly known that its distribution cannot be adequately portrayed. Apparently, this is the common arctic replacement of "ordinary" var. *ventricosa*, although some perianth-free collections, which show the oil-body characters of var. *ventricosa*, are here identified provisionally with the latter.

N. ELLESMERE I. Mt. Pullen, S. of Alert, 82°25′ N. (*RMS*; reported by Schuster et al., 1959, as *L. ventricosa*). NW. GREENLAND. Kangerdlugssuak, dry peaty soil between rocks, Inglefield Bay (*RMS 45897, 45959, c. caps; 45893a, 45879a*; also several other collections probably here, e.g., *RMS 45642*); Kânâk, Red Cliffs Pen., Inglefield Bay (*RMS 45642*); Siorapaluk, Robertson Bugt (*RMS 45674b*). W. GREENLAND. Mouth of Blaesedalen, near Godhavn, Disko I. (*RMS & KD 66-311a*); Eqe, near Eqip Sermia (Glacier), 69°46′ N., 50°10′ W. (*RMS & KD 66-328*); SE. of Jacobshavn, Disko Bay, 69°12′ N., 51°05′ W. (*RMS 66-222, 66-219*); Ritenbenk, Agpat, 69°46′ N., 51°20′ W. (*RMS & KD 66-263*); Qalagtoq, Upernivik Ø, Inukavsait Fjord, 71°14′ N., 52°35′ W. (*RMS & KD 66-1102b*); Lyngmarksbugt, Godhavn, ca. 25–125 m (*RMS & KD 66-558*).

Differentiation

This extreme variant has the longer teeth of the perianth mouth of *L. porphyroleuca* and the variable and relatively few oil-bodies of that species. However, the abundant development of gemmae, the broad and frequently 2–3-lobed leaves with often deep sinuses, the rounded flanks of the

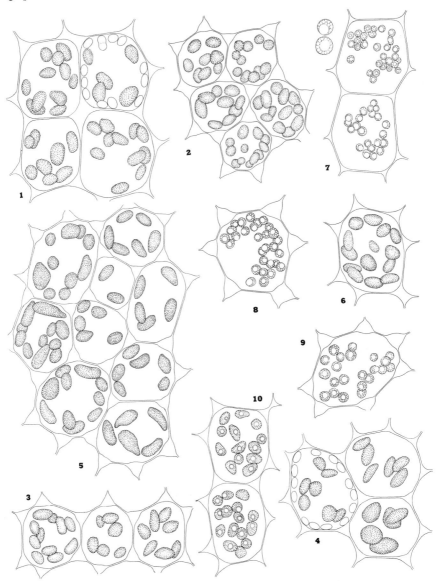

FIG. 210. *Lophozia ventricosa s. lat.* (1–6) and *Lophozia groenlandica* (7–10) (median cells of arctic varieties). (1–4) var. *confusa:* (1) Type, *RMS & KD 66-220*, Jacobshavn, W. Greenland (×800). (2) *RMS 45959*, Kangerlugssuak, Greenland (×688). (3) *RMS 35401*, The Dean, Alert, Ellesmere I. (×660). (4) *RMS & KD 66-328*, Eqe, W. Greenland (×750). (5) var. *confusa* fo. *discoensis:* type, *RMS & KD 66-020* (×620). (6) var. *confusa:* *RMS 45722b*, growing with plants of *L. groenlandica* in fig. 8 (×810). (7) *RMS 46160a*, Kânâk, Inglefield Bay, Greenland, a fo. with faintly brownish gemmae (×680). (8) *RMS 45722b*, Kânâk, Greenland, growing with *L. ventricosa* var. *confusa* in fig. 6 (×810). (9) *RMS 45632a*, Kânâk, Greenland (×940). (10) *RMS 35393*, Alert, Ellesmere I., drawn from plants growing near those in fig. 3 (×640). [figs. 1 and 4 with single cells with chloroplasts; others with only oil-bodies drawn in.]

leaves, and other characters effectively exclude the plant from the steno-typic and highly characteristic *L. porphyroleuca*.

In the type plants no red pigments occur; in the very exposed-growing paratype material (Blaesedalen) the ventral side of the stem and leaf bases may be ± reddish; in the extreme form, on exposed sandy ridges near the Inland Ice at Eqe (*RMS & KD 66–328*), shoot tips may be ± reddish purple, with dull purplish older leaves. Also, in *RMS 45863a* (Kangerdlugssuak), the ventral surface of the stem, many postical leaf bases, and the perianth mouth develop a reddish coloration and thus show an approach to *L. porphyroleuca*. However, *usually* the leaves tend to have warm brown to yellow-brown pigmentation and the postical leaf bases are colorless (occasionally, as in *RMS 45897*, bluish-purple-tinged, and then the stem may become purplish brown to almost black beneath)—hence a *"porphyroleuca"* pigmentation pattern is generally absent.

Lophozia ventricosa var. *confusa* includes a variable series of arctic and high arctic phenotypes which, collectively, differ from var. *ventricosa* in two features: the perianth mouth tends to be more strongly dentate, and the oil-bodies are larger, on an average, and fewer per cell. Obviously dead sterile herbarium material can hardly be placed. The variety also "tends" to have more deeply bilobed leaves (sinuses descending 0.25–0.4 the leaf length; in fo. *discoensis* mostly 0.35–0.45 the leaf length).

Locally, in west Greenland, var. *confusa* occurs as a phenotype with unus-ually broad leaves divided by a deep, angular sinus; this is perhaps best segregated as a distinct form:

LOPHOZIA VENTRICOSA var. CONFUSA fo. DISCOENSIS Schust., fo. n.

[Fig. 210: 5]

Forma var. *confusae* similis, folia, autem, profunde bilobata (plerumque 0.35–0.45), lobis satis longioribus, in 2 singulis cellulis saepe terminantibus; guttae olei solummodo 3–10 omni in cellula, nonnullae ad 5.5–9 × 12–14 μ.

Differing chiefly in the larger and usually broader leaves, broadly quadrate-ovate, widest toward base, ca. 960 μ wide × 780 μ long, up to 1025 μ wide × 840–880 μ long on mature shoots; leaves green, but upper leaf lobes often purplish brown, the stem remaining greenish below until maturity. Leaves *deeply bilobed* [usually 0.35–0.4(0.45) the length] by an angulate sinus, the lobes relatively long and triangular, *usually ending in 2 superposed cells*. Cells along lobe margins and in lobes mostly 20–23 μ, in leaf middle mostly 20–24 × 25–30 μ, with moderately coarse, usually weakly bulging, pale trigones; oil-bodies excessively variable, but all uniformly formed of fine spherules throughout (as in var. *ventricosa*), *occurring only* (2)3–10(12) *per cell*, spherical and 4.5–6 μ to regularly or irregularly ovoid and 4.8 × 7 μ to 5.5–6 × 12–14 μ, occasional ones 7 × 8 or 9 × 12 μ, relatively large, often largely obscuring the cell lumen.

Type. Tupilak I., west of Edegesminde, west Greenland, 68°43′ N., 52°55′ W. (*RMS & KD 66–020;* among *Blepharostoma trichophyllum brevirete* and *Lophozia groenlandica*).

The leaves, usually widest toward base, often much broader than long, and divided by an angular sinus into two sharp-pointed, triangular lobes, give this plant a very different facies from "normal" *L. ventricosa*. The excessively variable oil-bodies average larger than in var. *ventricosa* and in many cases approach those of *Leicolea* in size.

The distinctness of this relatively vigorous plant remains to be fully demonstrated. However, there are smaller admixed plants, with quadrate-orbicular leaves with a shallow, lunate sinus, (descending to 0.15) also with yellow-green gemmae, having numerous (16–25 per cell or more) minute (under 5.5 μ long), and biconcentric oil-bodies; these plants correspond closely to *L. groenlandica*. Obviously two distinct entities with yellow-green gemmae are involved.

LOPHOZIA VENTRICOSA var. *RIGIDA* Schust., var. *n.*

[Fig. 208: 1-6]

Varietas var. *ventricosae* similis, differens, autem, quod (*a*) planta vigens, fragilis, rigida, (*b*) cellulae permagnae, marginales 36–38 μ, mediae 32–37 × 35–50 μ, (*c*) guttae olei numerosae saepissime 14–20 omni in cellula et (*d*) gemmae magnae, ad 29–33 × 35–38 μ.

Vigorous, brittle and rigid, prostrate to creeping, scarcely branched (branches terminal), shoot width ca. 2–2.4 mm × 1–3 cm long, opaque, pale to rather whitish green but in sun forms with weak brownish pigmentation of distal parts of leaves; stem ventrally dull purplish black, the *dull purplish* pigmentation extended onto postical leaf bases; *no trace of red pigments*. Leaves stiffly sub-vertical, weakly antically assurgent on laxer shoots, then ± oblique, rather dense, the plants *"confertifolia"*-like, the leaves loosely or hardly folded, stiffly pectinate, widely spreading, the *lobes patent and never incurved*; leaves inter-mingled 2–3-lobed, all clearly broader than long, the 2-lobed leaves broadly ovate-orbicular, somewhat asymmetrical (ventral half with margin more ampliate, especially toward base; lobe broader), ca. 1330–1340 μ broad × 1120 μ long to 1440 μ broad × 1170 μ long (ca. *1.05–1.25× as wide as long*),[176] the 3-lobed leaves very broadly orbicular-subquadrate, ca. 1430–1450 μ wide × 1100 μ long; sinuses ± *obtuse-angular to angularly rounded or broadly V-shaped*, descending ca. 0.2–0.25 the leaf length, often rather narrow (on 2-lobed leaves 600–700 μ wide, width not more than 0.5 leaf width); margins both convex normally, the ventral more so; *lobes angular, often ± apiculate*. Cells *very large*; *marginal cells* of distal ½ of leaf *36–38 μ*; *median cells 32–37 × 35–50 μ*; trigones moderate in size, not strongly bulging; oil-bodies variable in size and form, spherical and 4.2–6 μ to more often irregularly ellipsoidal and 4.5–5.5 × 6.5–10 to 7 × 9 μ, uniformly formed of small spherules (lacking central "eye"), *mostly (10)14–20(21–25) per median cell*. Gemmae common, greenish, polygonal to polygonal-substellate, *ca. 26–28 × 28–38 μ to 29–33 × 35–38 μ*. Dioecious. Androecia locally dull purplish-tinged.

Type. Lyngmarksbugt, Godhavn, Disko, I., west Greenland (*RMS & KD 66–563*).

[176] Near androecia often becoming narrowly oblong-ovate and ca. 950 μ wide × 1300 μ long.

Plants of this variety are an analogous (presumably polyploid) derivative of *L. ventricosa*, as *L. wenzelii lapponica* is of *L. wenzelii*. They are strikingly similar to *L. wenzelii* in pigmentation: green, with at most limited brownish pigmentation of the upper parts of leaves; younger parts of stem and ventral leaf bases a very dull purplish, the older stem black beneath. However, the stiff, almost pectinate manner in which the rather widely patent leaves are held is very different; the plant has all the facies of *L. ventricosa s. lat.*, of which it appears to be simply a large-celled extreme.

Besides the large cells of the leaves, the much bigger gemmae are also distinctive. Oil-bodies are, as is common with polyploid varieties, much more numerous, few cells having less than 12, occasional cells having more than 20, oil-bodies. The plants are ♂ only; ♂ bracts, which are normally ventricose at the base, have the basal pocket often dull purplish.

Superficially this variant may be similar to *L. ventricosa* var. *longiflora*, but it lacks the bright red pigmentation of the ventral leaf bases and stem apices of that taxon and, of course, has larger cells and gemmae.

LOPHOZIA (LOPHOZIA) GROENLANDICA
(Nees in G., L., & N.) Macoun
[Figs. 210: 7–10; 211: 4–10; 216: 1–4]

Jungermannia groenlandica Nees *in* G., L., & N., Syn. Hep., p. 114, 1844; Jensen, Öfversigt Kgl. Vetensk.-Akad. Förhandl. no. 6:800, figs. 1-5, 1900.
Sphenolobus groenlandicus Steph., Spec. Hep. 2:164, 1902; K. Müller, Rabenh. Krypt.-Fl. 6(1):617, 1910, and 6(2):762, 1916.
Lophozia groenlandica Macoun, Cat. Canad. Pls. 7:19, 1902.
Lophozia murmanica Kaal., *in* Bryhn, Rept. 2nd Norwegian Arctic Exped. in the "Fram" 1898–1902, 2(11):34, 1906 (*teste* Müller, 1905–16).
Jungermannia murmanica Arn., Ark. Bot. 15(5):34, 1917.
Lophozia wenzelii, in part, of authors, including Schuster, Amer. Midl. Nat. 49(2):367, pl. 20, figs. 4-10, 1953, and K. Müller, Rabenh. Krypt.-Fl. ed. 3, 6(1):655, 1954.

Usually in spongy to compact patches, often densely caespitose and erect or suberect in growth, pure green but often with *leaf lobes somewhat yellowish brown, with ventral side of stem brownish purple or reddish brown;* shoots 1.3–1.8(2) mm wide × 1–3(4) cm long, sparingly, terminally, furcately branched, rather flexuous. Stems ca. (340)425–480 μ in diam., the dorsal and lateral cortical cells ca. 23–27(29) μ wide × 42–60(80) μ long, hardly thick-walled; medulla very strongly dorsiventrally differentiated, the 4–6 dorsal layers colorless and pellucid, large-celled (cells ca. 25–33 μ in diam.), the 10–15 ventral tiers very small-celled (cells ca. 10–15, rarely to 20, μ in diam.); ventral medullary cells somewhat dull rose to reddish-tinged and with age strongly mycorrhizal. Rhizoids often rose-tinged near base. Leaves *somewhat to strongly antically secund,* slightly to quite distinctly imbricate, *quite obliquely inserted (and oriented),* slightly decurrent antically, *rotund-quadrate,* averaging (on mature shoots) (930–1050) 1150–1200 μ wide × (920–980)1000–1100 μ long, occasionally to 1300–1350 μ

FIG. 211. *Lophozia* (*Lophozia*) *wenzelii* (1–2), *L. ventricosa* (3), and *L. groenlandica* (4–10). (1) Two mature leaves, of mod. *laxifolia-viridis* (×15). (2) Two leaves, mod. *densifolia-viridis* (×15). (3) Leaf (×16). (4–5) Leaves of mature shoot; note tendency to tear in flattening (4, ×21.5; 5, ×18). (6) Stem cross section (×155). (7) Gemmae (×800). (8) Median leaf cells with oil-bodies (×600). (9) Juvenile leaf from small shoot, a mod. *angustifolia* (×19). (10) Mature plant with gemmae (×21). [1, from Black Forest, Germany, swampy area between Herzogenhorn and Spiesshorn, *Oct. 4, 1903*, *K. Müller*, as *L. wenzelii*; 2, Feldberg, Baden, Germany, 1460 m, *K. Müller, Sept. 5, 1937*, as "*L. wenzelii* fo. *confertifolia*"; 3, Zastlerwand, Feldberg, 1350 m, *K. Müller, Sept. 27, 1903*, labeled by Müller as identical with what Buch calls *L. ventricosa* and identified by Buch as such, but considered by Müller to be *L. wenzelii*; 4–10, *RMS 19409*, Porcupine I., Cook Co., Minn., identified by Müller with *L. wenzelii*, by Buch as probably *L. groenlandica*.]

wide × 1200–1225 μ long or 1550 μ wide × 1300 μ long [*averaging from 1.05 to 1.15(1.25)× as wide as long;* on small-leaved, juvenile shoots sometimes less broad], *widest commonly above middle, quite strongly concave* (although scarcely hemispherical, *never canaliculate*), *with incurved lobes, very shallowly bilobed; sinus extending for 0.05–0.15* (rarely to 0.25) *the leaf length, usually crescentic* and very broad, the *lobes broadly obtuse;* leaf margins usually moderately to strongly arched, antically often less so. Cells distinctly collenchymatous, those of leaf tips and margins with their walls commonly *yellowish to yellowish brown,* 20–24 × 21–27 μ; median cells 23–26 μ wide × 24–32(33) μ long; cuticle smooth; oil-bodies (in all but exceptional individuals) *biconcentric* (with a centric or excentric, interior, pale bluish-appearing, homogeneous oil-spherule, surrounded by a granular peripheral mantle of minute spherules), *usually (8–10)12–15(17–18), occasionally 20–30, per median and submedian cell,* often largely spherical and 4.5–6 μ, but frequently in large part ovoid to irregularly ellipsoid and to 4.5–5 × 7–10 μ. *Gemmae pale greenish,* in relatively small quantities, from apices of uppermost leaves, 1–2-celled and polygonal, or quadrate to stellate, 18–20 × 21–25 μ up to (occasional max.) 23–24 × 28–32 μ, their walls slightly, their angles prominently, thickened.

Dioecious. ♀ Bracts usually dimorphic: one 2-lobed (ca. 1500–1550 μ wide × 1300 μ long), the other 3–4-lobed (ca. 1400 μ wide × 1200 μ long), erect and somewhat sheathing perianth at base, but with spreading lobes; bracteole usually distinct as a lanceolate lobe, up to 500 μ wide × 1000–1200 μ long, united for up to 0.5–0.6 its length with one of the bracts; margins of bracts and bracteole entire. Perianth at mouth *with remote 1–2-celled teeth,* 35–55 μ long; end cell of teeth narrowly triangular, ca. 25–35 μ long × 13–15 μ wide.

Type. Greenland. The type of *L. murmanica* from Litsa, Murmania, Lapland.

Distribution

This poorly known and long-forgotten species appears to be essentially arctic-alpine in distribution,[177] with a disjunct extension into the Great Lakes region. It occurs northward virtually to the northernmost periphery of the Northern Hemisphere (82°26′ N.). The range is much confused in the literature with that of *L. wenzelli* and *L. ventricosa*: the following reports are the only certain ones known to me:

NW. GREENLAND.[178] Kânâk, Red Cliffs, Pen., Inglefield Bay (*RMS 45632a, 45635e, 45722b, 46160a*); Kekertat, Harward Øer, near Heilprin Gletscher, E. end of Inglefield Bay (*RMS 46020, 46021*); Siorapaluk, Robertson Bugt (*RMS 45668; M. P. Porsild, Aug. 15, 1943!,* probably). W. GREENLAND. Narsak Fjord, W. coast (*Vahl, Breutel,* type); Tupilak I., W. of Egedesminde, 68°43′ N., 52°55′ W. (*RMS & KD 66-017a,*

---

[177] Aside from the American stations, reported again only from Spitsbergen by Arnell & Mårtensson (1959, p. 113, fig. 1), who state that, next to *L. latifolia, L. groenlandica* is the commonest species of the genus. Persson (1962) also reports it from the arctic coast of Alaska.

[178] Two collections reported by Arnell (1960) from Peary Land have been examined; one represents *Lophozia pellucida,* the other *L. collaris. Lophozia groenlandica* thus is not known, to date, from northernmost Greenland.

with *Tritomaria exsectiformis arctica*; oil-bodies biconcentric and leaves mostly very broad with shallowly crescentic sinuses. *RMS & KD 66-023*, with *Anastrophyllum minutum*, a ± xeromorphic form). s. GREENLAND. (Böck). E. GREENLAND. Hurry Inlet (Jensen, 1900, p. 800, ♂ plant). ELLESMERE I. King Oscar's Land, Gaasefjord; Cape Rutherford, 78°50′ N. (Bryhn, 1906; as *L. murmanica*); The Dean, 5 mi S. of Alert, 82°26′ N., 62°05′ W., NE. Ellesmere (*RMS 35393;* northernmost station for species!). BAFFIN I. Frobisher Bay, 63°45′ N., 68°34′ W. (Persson & Holmen, 1961). SOUTHAMPTON I. Coral Harbour, 64°09′ N., 83°05′ W. (Persson & Holmen, 1961). LABRADOR. Great Whale R. (Persson & Holmen, 1961).

MICHIGAN. Pictured Rocks, Alger Co. (*RMS 39294, p. p.*, among *Lophozia incisa, Gymnocolea inflata, Cephalozia bicuspidata*, on sunny, damp sandstone ledges). MINNESOTA. Porcupine I., Susie Isls., L. Superior shore near Grand Portage, Cook Co. (*RMS 19409*); reported and illustrated in Schuster (1953) as *Lophozia wenzelii*.

Also reported from Spitsbergen (*teste* Bryhn), from Lapland (near Litsa, type of *L. murmanica* Kaal.), and from alpine situations in central Europe (Feldberg, Baden, *Müller*; Monte Coglians, Friaul, 2100 m, *Kern*); see Schuster (1961c). The last two reports need verification.

Ecology

Associated with ledges and cliffs, and with exposed rocks, on either north or northwest-facing cliffs, in direct sun to permanent shade (Lake Superior), or on exposed peaty soil over weathered mountain slopes (Ellesmere I.). Associated in the former case with *Dicranum*, and in the latter with *Rhacomitrium lanuginosum, Tritomaria quinquedentata, Anastrophyllum minutum, Lophozia alpestris* subsp. *polaris, Cephaloziella arctica.* In Greenland reported as occurring with *Anthelia*; in Michigan on cold lake bluffs, over sandstone, with members of the oxylophyte *Gymnocolea-Cephalozia bicuspidata* Associule. In the far north on steep slopes, over non-calcareous rocks, in areas where snow lies late into the spring (to June 15–30), and where the moss-tundra is kept damp by melting snow banks during the remainder of the season.

In northwest Greenland, where I have repeatedly collected the species, it occurs varyingly as follows: (*a*) on the soil of an exposed, steep, partially consolidated slope adjoining a glacier, with *Gymnomitrion apiculatum, G. corallioides, G. concinnatum, Lophozia alpestris s. lat., Scapania perssonii, Prasanthus suecicus*, and *Polytrichum;* (*b*) in a site close to (*a*) but where subject to irrigation from the melt water of the glacier, then in swelling tufts with *Lophozia alpestris polaris, Scapania spitsbergensis, Cephalozia bicuspidata, Anthelia juratzkana;* (*c*) on a wet, peaty, steep slope manured by leachings from Dovekie roosts, forming a mod. *laxifolia-viridis-lepto-derma-latifolia*, with *Sphagnum, Aulacomnium, and Polytrichum;* (*d*) on peaty soil on the margin of a small rill, with *Cassiope tetragona*, associated with *Prasanthus, Blepharostoma, Anastrophyllum minutum grandis, Lophozia excisa,* and *L. ventricosa.*

Differentiation

The pale greenish gemmae and broad, often soft leaves, normally bilobed (rarely sporadically trilobed) by a shallow and crescentic sinus, suggest the commoner *L. wenzelii* and phases of *L. v. ventricosa.*[179] When living material is at hand, confusion with these two is hardly possible, because of the numerous, small, biconcentric oil-bodies, very similar to those of *L. ventricosa* var. *silvicola.* However, the much broader leaves of mature plants, quite antically secund and concave, and with incurved lobes, and the usually yellowish brown pigmentation of the leaf lobes give the plants a different facies from var. *silvicola.* There is seemingly no intergradation between these two taxa. Furthermore, their pigmentation pattern appears to differ considerably, var. *silvicola* tending to have (in sun forms) a bright carmine pigmentation of the leaf bases.

The situation with respect to *L. wenzelii* and typical *L. ventricosa* is very different, and it is questionable whether *L. groenlandica* can always be distinguished except on the basis of living plants.[180] *Lophozia groenlandica* closely approaches *L. wenzelii* in the rotund-quadrate leaves with shallow sinus, and in the tendency for the leaves to be strongly concave. There are, however, distinct differences in facies: *L. groenlandica* has considerably more obliquely inserted leaves, which are usually more antically secund (in *L. wenzelii* they are subvertical and concave toward the shoot apex). Also, there is a considerable difference in the oil-bodies: those of *L. groenlandica* are biconcentric and similar to those of *L. ventricosa* var. *silvicola,* whereas *L. wenzelii* agrees with typical *L. ventricosa* and its var. *confusa* in having oil-bodies formed throughout of minute spherules.

Oil-bodies are more polymorphous in *L. groenlandica* than in *L. ventricosa* var. *silvicola.* In some cases (i.a., in the Lake Superior plants) almost all the oil-bodies in *L. groenlandica* are spherical, and the larger, included oil-spherule is usually nearly centric (Fig. 211:8). In Ellesmere I. plants, however, many, if not most, oil-bodies are ovoid or irregularly ellipsoidal in shape, although spherical ones are frequent, and the included oil-globule is often eccentric. The oil-bodies average somewhat less numerous than in var. *silvicola* (where they are usually 15–22, occasionally 25, per median cell). In most *L. groenlandica*

---

[179] The leaf dimensions among arctic plants do not allow a certain separation between *L. groenlandica* and these two species. Thus, in admixed plants of *L. groenlandica* and *L. ventricosa* (*RMS 45722b*), *L. groenlandica* has leaves mostly 0.95–1.0× as wide as long, and *L. ventricosa* has them 1.0–1.25× as wide as long. In *RMS 45668,* a lax, broad form of *L. groenlandica,* the leaves were 1.05–1.3× as wide as long.

[180] When first "exposed" to this difficult species, on the basis of Lake Superior material, I confused it with *L. wenzelii.* When, 4 years later, it was seen again a second time, in Ellesmere I., the specific identity of the Minnesota and Ellesmere I. plants was at once suspected; the similarity in facies was quite marked. Probably, after having once been correctly recognized, the species can be identified again even on the basis of dead material.

plants there are 12–15 oil-bodies per median cell, although not uncommonly only 8–10; very exceptionally there may be as many as 17 or 18. In oil-body number, therefore, *L. groenlandica* closely agrees with *L. ventricosa* and *L. wenzelii*.

I have studied *L. groenlandica* intermittently for over 15 years. At first it appeared impossible to keep this species apart from *L. wenzelii*, and material of it was initially reported (Schuster, 1953) as *L. wenzelii*. The species is also generally considered in the literature (i.a., Müller, 1954, *in* 1951–58; Frye and Clark, 1937–47) as a simple synonym of *L. wenzelii*.[181]

With study of living plants of this taxon (1955, in northern Ellesmere I.; 1961, 1966, in Greenland) I found repeated correlation between a broad, *L. wenzelii*-like leaf with a very shallow sinus, pale green gemmae, and *silvicola*-like oil-bodies, *exactly as in the Minnesota material*. Therefore, it hardly appears possible to retain this taxon as a synonym of *L. wenzelii* (Schuster, 1961c).[182]

The application of the name *L. groenlandica* to this taxon is perhaps not above criticism. Müller (1910, *in* 1905–16, p. 617) discussed this species briefly and came to the conclusion that one "can be in doubt . . . whether it is to be placed in *Lophozia* or *Sphenolobus*" but later (1916, p. 762) concluded that "*Jg. groenlandica*" belongs in the "Formenkreis" of *L. wenzelii*. Furthermore, *Lophozia murmanica* Kaal. is considered to be "very close" to *L. groenlandica* and is placed as a synonym of this species, which in turn is regarded as identical with *L. wenzelii*. This conclusion has been generally accepted since (Frye and Clark, 1937–47; Schuster, 1953; Müller, 1951–58), and *L. groenlandica* is not listed in the American literature of the last 40 years. The name *L. groenlandica* is revived for the present taxon for two reasons: (*a*) Nees's *Jungermannia groenlandica* (as regards description) is inseparable from the present plants; (*b*) the application of Nees's name for the species effectively removes the necessity for proposal of a new name.[183] The identity of the type of *J. groenlandica* with the present plants will never be susceptible to unquestionable and final proof, since the oil-body characteristics of the former cannot be determined. An additional element of uncertainty is introduced by the fact that Nees's type shows a high incidence

---

[181] The Minnesota plants were determined by the late Dr. Müller as *L. wenzelii*. H. Buch did not consider the Minnesota material to be *L. wenzelii*. He thought it a large-celled phase of *L. alpestris* (*L. alpestris* "*major*" C. Jensen), but admitted that if it differed from *L. alpestris* not only in cell size, but also in its *silvicola*-like oil-bodies, it "könnte dann vielleicht als Art behandelt werden. Dann muss sie aber allem Anschein nach *L. groenlandica* (Nees) benannt werden." Since the "var. *major*" is partly *L. latifolia*, partly *L. excisa* (see p. 515) no affinity to *L. groenlandica* seems possible.

The separation of *L. groenlandica* and *L. wenzelii* is often admittedly almost impossible, *if* only dead plants are available for study. For example, plants from Minnesota, provisionally assigned to *L. groenlandica* (Fig. 216:1–4) are identical in aspect to plants from the type region of *L. wenzelii* (compare Fig. 215:1–6).

[182] It should be noted that *L. alpestris* ssp. *polaris* is abundant in northern Ellesmere I. Indeed, a few plants were found admixed with *L. groenlandica*. This variety is always readily separable by (*a*) reddish gemmae; (*b*) much smaller trigones; (*c*) flatter leaves with a usually deeper, angular sinus; (*d*) nonbiconcentric oil-bodies. Buch's attempt to equate *L. alpestris* ssp. *polaris* with *L. groenlandica* is impossible to follow.

[183] Also, the usage is in accord with that recommended to me by Buch (see footnote 181). and, as far as I can judge, with that in Arnell & Mårtensson (1959), who also restrict the name to plants with broad leaves, pale gemmae, and (at least in part) biconcentric oil-bodies.

of trilobed leaves, while the material assigned to this species by Jensen (1900) and Bryhn (1906), as well as that assigned to it by the writer, has almost uniformly bilobed leaves, only sporadic leaves being trilobed.

The distinction of *L. groenlandica* from *L. ventricosa* is at times difficult, unless living plants are available. This is clearly illustrated by the Ellesmere I. plants, which occurred in a small area within 2 feet of a patch of *L. ventricosa* (var. *confusa*). The plants, in both cases, had concave, antically secund leaves, with somewhat incurved lobes. The xeromorphic phase of *L. ventricosa*, indeed, had much the facies of a small *L. groenlandica*. However, the following distinctions occurred. (1) *Lophozia groenlandica* had the ventral side of the stem and sometimes the adjacent portion of the postical bases of the leaves, as well as the rhizoids, rose to reddish-tinged; in *L. ventricosa* such pigments were lacking, the stem, with age, becoming brownish from mycorrhizally induced breakdown of its ventral portion. (2) *Lophozia groenlandica* had an average of 12–15, occasionally 17–18, biconcentric oil-bodies per cell; *L. ventricosa* had only (5–7)9–15, rarely 16, oil-bodies per cell, all formed of spherules of uniform size. (3) *Lophozia groenlandica* had broad leaves, occasionally 1.15–1.2× as wide as long, with a notably shallower and broader sinus and thus more obtuse lobes; in *L. ventricosa* leaves averaged only 0.95–1.1× as wide as long, and many had a more distinct, if obtuse, sinus and subacute lobes. Although these distinctions, except for (2), are slight, the two taxa are separable by the difference in color (indicating different physiological responses to identical insolation!) and were recognized as distinct in the field. In both cases the leaf cells, at least of the distal half of the leaf, tend to develop a light golden or yellowish pigmentation, although in *L. ventricosa* a slight reddish tinge on the leaf lobes is locally present.[184]

Plants from the Lake Superior shore of Michigan are critical in some respects. When typically developed they bear all the features characteristic of *L. groenlandica*, including the biconcentric oil-bodies; very broad, rotund-quadrate leaves with a wide, shallowly crescentic sinus; and a concave leaf form, with the blunt and broad lobes distinctly incurved. Such plants have leaves varying from 1.09 to 1.33× as broad as long (typical measurements: 1150 × 950 μ; 1200 × 900 μ; 1250 × 1050 μ; 1480 × 1150 μ). However, a single case has been seen of a slender *silvicola*-like shoot bearing spreading, almost squarrose

---

[184] I have also collected *L. groenlandica* growing mixed with *L. ventricosa* in northwest Greenland (*RMS 45722b*). The cytological distinctions remain obvious, the *L. groenlandica* having (17)20–30(32) small, subspherical biconcentric oil-bodies per cell (Fig. 210:8), the *L. ventricosa* (8)11–18(20) larger, often ovoid, uniformly fine-botryoidal oil-bodies (Fig. 210:6).

Oil-body number in *L. groenlandica* appears to vary considerably, depending in part on environmental conditions.

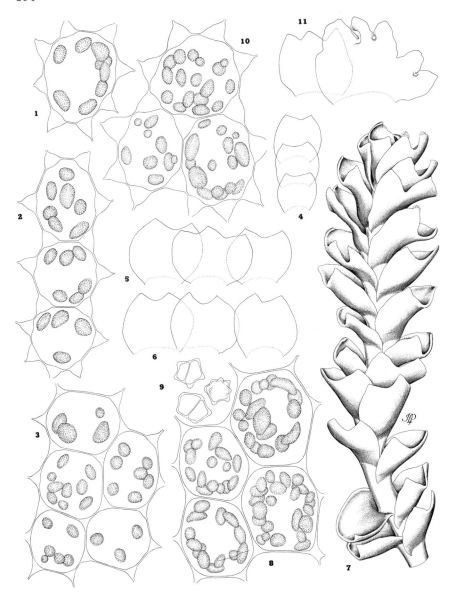

FIG. 212. *Lophozia (Lophozia) wenzelii* (1–9) and *L. ventricosa* var. *longiflora* (10–11).
(1–5) var. *wenzelii:* 1, cell with oil-bodies ($\times 655$); 2, cells with oil-bodies ($\times 655$);
3, cells with oil-bodies ($\times 700$); 4, leaves, from an xeromorphic small mod. *densifolia-colorata* ($\times 14$); 5, leaves from normal, rather fleshy mod. *viridis* ($\times 14$). (6–9) var.
lapponica: 6, leaves, dense-leaved form ($\times 13$); 7, medium-sized shoot, moderately lax
phase ($\times 23$); 8, median cells with oil-bodies ($\times 585$); 9, gemmae ($\times 310$). (10)
Median cells with oil-bodies ($\times 655$). (11) ♀ Bracts and, at right, bracteole ($\times 16$).
[1–2, 10–11, from intermixed plants of *RMS & KD 66-551*, Godhavn, Disko I.,

leaves, giving rise to a branch with concave, broad *L. groenlandica*-like leaves. The weak parent axis had leaves only 830–860 × 880–890 µ to 900 × 820 µ to 1025 × 1000 µ long (width 0.93–1.1× the length), although the sinus was shallow and crescentic; the strong branch had leaves from 960 µ × 780 µ long to 1100 µ × 800 µ long (1.26–1.37× as wide as long). Evidently, therefore, under presumably "difficult" conditions *silvicola*-like extremes may be produced that possibly cannot be separated from *L. ventricosa* var. *silvicola* except by culture experiments.

LOPHOZIA (LOPHOZIA) WENZELII (Nees) Steph.

[Figs. 202:7; 211:1–2; 212:19; 213–216]

Jungermannia wenzelii Nees, Naturg. Eur. Leberm. 2:58, 1836.
Lophozia wenzelii Steph., Spec. Hep. 2:135, 1902; K. Müller, Rabenh. Krypt.-Fl. 6(1):675, fig. 312, 1910 (ed. 3, 6:655, fig. 200, in part, 1954); Macvicar, Studs. Hdb. Brit. Hep. ed. 2:187, 1926.
Lophozia confertifolia Schiffn., Österr. Bot. Zeitschr. 55:47, 1905 (*p.p.*; at least as regards subsequent interpretations).

Usually hygrophytic, *over boggy or swampy ground*, often on moist peaty but exposed sites, prostrate or, with crowding, suberect. Shoots green, leaves green to (in sun) having distal parts (at least) usually rather intensely reddish brown or brownish (not yellow-brown as in *L. groenlandica*); stems reddish to purplish brown beneath, green above, at maturity often ± reddish throughout. Shoots mostly 1.2–1.6 mm broad with leaves, 1–4(6) cm long, the stems ca. 360–400 µ broad, with numerous colorless rhizoids that may be reddish or reddish brown at base; branching sparing, furcate, terminal. Leaves obliquely inserted, *mostly rather dense* (in bog forms much laxer), *subvertically oriented and diagnostically cupped or saucer-shaped* (the concavity turned more *toward stem apex* than in *L. groenlandica*), *broadly orbicular*, with *both margins rather strongly arched*, widest near, slightly above, or below the middle, when flattened ca. 1105–1210 µ wide × 840–940 µ long to 1230–1250 µ wide × 820–840 µ long (*width:length ca. 1.2–1.48:1*, less only on weak shoots); the incurved and usually equal lobes very *broadly triangular and mostly blunt to obtuse*, the sinus broad (width ca. 0.5–0.66 leaf width; ca. 0.7–0.9 leaf length), *shallowly crescentic*, descending only to 0.15–0.2(0.25); dorsal lobe almost transverse, except in very lax bog forms, where more oblique. No underleaves. Cells with thin walls and moderate, usually concave-sided trigones (mod. *pachyderma* never produced); marginal cells ca. (18)20–23(24) µ; *subapical cells of lobes ca. 22–25 µ*; median cells 22–25(27) × 24–30 µ; cuticle smooth; *oil-bodies (2–3)4–9(10–12)* per cell, *very variable, several usually quite large and ellipsoidal, to 5–7 × 9–10 µ*, mostly smaller and 4.5–5 × 5.5–8 µ, some spherical and 4–5.5 µ; oil-bodies uniformly of fine spherules, *never with a central "eye."* Gemmae pale green (*in sun forms often ± pinkish or salmon colored*), mostly 2-celled, some 1-celled at maturity, weakly polygonal to quadrate, 20–24 × 21–25 µ, a few to 17–22 × 26–32 µ.

Greenland; see also in Fig. 208:7–12; 3–4, *RMS & KD 66-001a*, Tupilak I., W. Greenland; 5, *RMS & KD 66-223*, Jacobshavn, W. Greenland; 6–9, *RMS & KD 66-253b* and *66-254*, Ritenbenk, W. Greenland.]

Dioecious. Androecia loosely spicate, the lax to contiguous bracts numerous, in to 15 pairs, ± strongly ventricose, often smaller than leaves, with deeper sinuses. Gynoecia with bracts ± ovate, one usually 2-lobed to 0.25–0.3, the other appearing 3–4-lobed (but 1–2 lobes apparently represent the bracteole, fused to the bract), lobes ± triangular, bluntly rounded to *blunt to acute*, entire-margined. Perianth obpyriform, deeply plicate at least above, mouth shallowly lobulate with rather close to *remote, normally 1-celled teeth*, varying to irregularly toothed with 1–2(3)-celled teeth arising from low lobes.

Type. Germany: Riesengebirge, Koppenplan, "am Quellbache der Aupa, 1400 m" (*v. Flotow, 1824*); material from the general type locality was issued in Schiffner's Hep. Eur. Exsic. no. 171 and is portrayed in Fig. 215:1–9.

Distribution

The range is poorly known because of confusion with (*a*) *L. groenlandica*, a plant that is surely distinct at the species level (see p. 591) and (*b*) *L. confertifolia*, a concept (rather than a species) that appears to involve the mod. *densifolia* of both *L. ventricosa* and *L. wenzelii*. Thus the range given in Müller (1905–16) is more reliable than his later one (Müller, 1951–58), since in the first he did not include these elements in *L. wenzelii*.

In Europe widespread from Spitsbergen (Arnell and Mårtensson, 1959) and Scandinavia (Sweden, Norway, Finland, also in the Faroes and in Denmark), to Russia southward to Great Britain, central Europe (Germany, Switzerland, Austria; the Vosges Mts. of France), and northern Italy; also in Siberia. In Asia in Sikkim, Sakhalin, Formosa and the alpine zone in Japan.[185] In western North America reported from Alaska (base of Mt. Eielson, along Thoroughfare R., *Weber & Viereck 10160*) and Yukon.

True *L. wenzelii* appears to occur chiefly near the summits of a few New England mountains, on or near the alpine summits, and in non-calcareous parts of the Arctic.

W. GREENLAND. Ritenbenk, Agpat, 69°47′ N, 51°20′ W. (*RMS & KD 66-253b, 66-254a*); SE. of Jacobshavn, Disko Bay, 69°12′ N., 51°05′ W. (*RMS & KD 66-223*); Fortune Bay, Disko I., 69°16′ N., 53°50′ W. (*RMS & KD 66-145a, 66-133, 66-130*); Lyngmarksbugt, Godhavn, ca. 25–125 m, (*RMS & KD 66-557b*; among *L. ventricosa* var. *longiflora*); Sondrestrom Fjord, near shore of L. Ferguson, in *Salix*-moss tundra, 66°5′ N. (*RMS & KD 66-051*, questionable); Tupilak I., near Egedesminde, 68°42′ N. (*RMS & KD 66-001a*). Also reported from Tasiusak (*Berggren*), S. Kangerdluarsuk, Sukkertoppen, and Kobbefjord (*Warming & Holm*), *fide* Lange & Jensen (1887, p. 97).

[185] First reported from Japan by Kitagawa, 1965, p. 284; however, as he remarks, the Japanese plants differ perceptibly in the smaller cells (15–18 μ in the leaf tips; 16–18 × 18–22 μ in leaf middle). If Kitagawa's measurements are accurate, the Japanese plant should perhaps be segregated. Kitagawa (1963a) has also reduced *L. formosana* Horik., from Taiwan, to a synonym of *L. wenzelii*.

E. GREENLAND. Hekla-Havn, Scoresby Sund, Runde-Fjeld, and Cap Stewart (Jensen, 1897); Rypefjeld, 77° N. (Hesselbo, 1948a); all these reports need verification.   NW. TERRITORIES.   QUEBEC. Great Whale R.; Leaf Bay near the Narrows, and 80 mi. up Leaf R., Ungava Bay (Schuster, 1951).   NEWFOUNDLAND. Biscay Bay, Avalon Distr.; Pushthrough and Rencontre West, S. Coast (Buch & Tuomikoski, 1955).   ONTARIO. L. Timagami; 8 mi. S. of Sudbury (Williams & Cain, 1959).

MAINE. Mt. Katahdin, at Caribou Springs (*RMS 15954*).   NEW HAMPSHIRE. Mt. Washington, plateau between summit and L. of the Clouds, 5100–5500 ft (*RMS 66-1749*).

Ecology

Typically a species of boggy sites, and usually of boggy depressions or peaty margins of tarns or small pools, either above treeline or north of the Taiga. In boggy areas associated usually with *Sphagnum* and *Scapania hyperborea* (more rarely *S. paludicola* or *S. irrigua*), *Gymnocolea inflata*, *Cephalozia bicuspidata*, *Lophozia* (*O.*) *kunzeana*, *Pleuroclada albescens*; in nonarctic areas (e.g., the White Mts. of New England) bog forms of *Calypogeia muelleriana*, *Lophozia kunzeana*, and *L. alpestris* may be consociated.

The species varies considerably in aspect, with alpine forms often quite strikingly reddish beneath (including rhizoid- and ventral leaf-bases), although otherwise green in shade forms; in sun forms the entire plant may be reddish to reddish brown, although lacking the carmine pigmentation so often typical of forms of *L. ventricosa*. Frequently the plant is quite handsome, because of the contrast between the red to deep red stem and green leaves (whose distal portions and ventral bases may also become conspicuously reddish, as opposed to the pure green of the rest of the leaves). Arctic forms, on the other hand, often lack the reddish pigments of the leaves, brownish pigmentation forming on distal leaf sectors.

Differentiation

Lophozia wenzelii has been much confused by American students.[186] Part of the confusion has arisen from attempting to follow Müller in placing *L. groenlandica* here—a "trap" into which I myself fell (Schuster, 1953)—or from trying to follow treatments (Schuster, 1951a; Müller, 1951–58) based on Buch (1933), who placed the controversial *L. confertifolia* Schiffn. into the synonymy of *L. wenzelii*. It now seems clear that under this name a heterogeneous series of plants belonging in most cases to a mod. *colorata-densifolia* has been placed; these plants belong in

---

[186] The confusion does not seem limited to this side of the Atlantic. Jones (1958a, p. 371) admits that in England "the status of *L. wenzelii* is . . . very obscure." He also points out that "the belief has grown among Scandinavian hepaticologists . . . that *L. wenzelii* and *L. confertifolia* are conspecific, representing hygrophytic and xerophytic states, respectively. But some alpine plants which have been named *L. confertifolia* are very close to *L. ventricosa*, while *L. wenzelii* itself could well be treated as a variety of *L. ventricosa* . . . ."

Fig. 213. *Lophozia* (*Lophozia*) *wenzelii*. (1–6) Sectors from perianth mouth, showing variations in ciliation and dentition; drawn to left scale (×180). (7–8) Leaves (×19.5). (9–10) Bracts and, at right in fig. 9, bracteole (×19.5). (11) Gemmae (×345). (12) Median cells with oil-bodies and, lower right, chloroplasts (×740). (13) Gemmiparous shoot apex of medium size; drawn to scale at right (×27). (14) Two leaves, one so concave it tore in flattening (×15). [1–10, *RMS & KD 66-145a*, Fortune Bay, Disko I., Greenland; 11–12, *RMS & KD 66-130*, same loc.; 13–14, *RMS & KD 66-223*, Jacobshavn Glacier, Greenland.]

part as dense-leaved forms of *L. ventricosa* (and its var. *silvicola*), but in part perhaps to true *L. wenzelii*. Probably figs. 200a, b, d in Müller (1951–58), cited as *L. wenzelii* (these figures were first published in Müller, 1905–16, as fig. 314, under *L. confertifolia*), cannot belong here; they should be carefully compared with "*L. longiflora*" in Müller (1905–16, fig. 311a–b)—a plant I would place as a variety of *L. ventricosa* (see Schuster, 1951a).

If we eliminate from *L. wenzelii* the dense-leaved plants with non-incurved leaf lobes ("*confertifolia*" phenotype) and with rather deeper and angular sinuses, what remains is a rather distinctive plant, almost entirely of arctic and arctic-alpine distribution, at least in North America. Such plants ordinarily have rather stiff leaves, diagnostically subvertical or almost vertical—the dorsal "lobe" is usually quite vertically oriented, often rather stiffly pectinate and with a ladderlike, regular arrangement, strongly concave to saucerlike, with usually strongly incurved and very broad lobes. The shallow sinuses are very wide and never angulate on mature plants.

On weak individuals—often to be found with mature ones in the same patch—or on lower, small leaves of stems, these distinctions are not clear. Here, as elsewhere in *Lophozia* (and in other critical genera like *Scapania*) persistent attempts to found our taxonomy on the mod. *parvifolia-angustifolia*, developed under difficult conditions, can result only in chaos. Also, fertile plants, especially when growing closely crowded, are hardly typical. The gynoecial shoots may then develop subcomplicate bracts, as well as subinvolucral leaves, almost scapanioid in form, with deeper sinuses. Such leaves and bracts may be quite a bit longer than wide and often have blunt or even rounded lobes. Associated sterile shoots may be similarly deviant. Usually one can find associated with such fertile plants ordinary and often almost terete sterile shoots with normal, broadly orbicular, concave, shallowly lobed leaves. The development in exposed forms of such terete, almost wormlike shoots, rigid and often rather straight, frequently with the rigid leaves quite dense and contiguous, gives these sterile plants a very distinct aspect, quite unlike any other *Lophozia*. Furthermore, such dense-leaved plants hardly resemble "*L. confertifolia*," leading to the suspicion that this taxon has nothing to do with *L. wenzelii s. str.*

In arctic sun forms, which develop at times rather conspicuous brownish pigmentation of the leaves, the gemmae may be salmon-colored or yellowish pink but never develop the bright red or vinaceous color of *L. alpestris* or *L. excisa* and their allies. Such arctic phenotypes may be abundantly gemmiparous, but most alpine phases only sparingly develop gemmae.

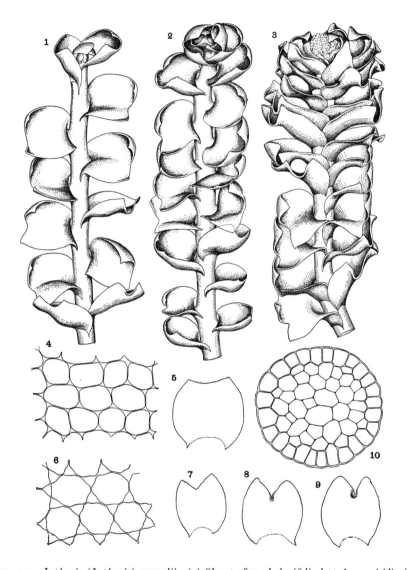

FIG. 214. *Lophozia* (*Lophozia*) *wenzelii*. (1) Shoot of mod. *laxifolia-leptoderma-viridis*, from periodically submerged niche on wet rocks (×16). (2) Shoot of mod. *subcolorata-meso-derma*, from vertical sides of wet *Sphagnum* polsters 6–8 in. above water level (×16). (3) Shoot of the "*L. confertifolia*" phase, mod. *colorata-pachyderma-densifolia*, strongly gemmi-parous (×18). (4) Cells, leaf middle, from plant in fig. 2 (×255). (5) Leaf from fig. 2, flattened (×16). (6) Cells from plant in fig. 3 (×310). (7–9) Leaves from plant in fig. 3 (×ca. 14). (10) Seta cross section (×ca. 140). [All from *RMS 15958* and *15958a*, Caribou Springs, Mt. Katahdin Tableland, Me.]

The perianth mouth of this species is much more variable than prior accounts and figures suggest; Figs. 215:11–13 and 15 (drawn from intermingled plants belonging to a single evident phenotype) show the extent of the variability seen.

Because of the greenish gemmae and broad leaves, *L. wenzelii* can be confused only with *L. ventricosa* var. *ventricosa* and *L. groenlandica*. The cells, which bear mostly 4–9 oil-bodies each—isolated median cells may have only 3 oil-bodies—are usually quite helpful in separating typical forms of *L. wenzelii*, since the other two species have more numerous oil-bodies (only the arctic form of *L. ventricosa*, *L. ventricosa confusa*, may have as few oil-bodies as *L. wenzelii*, but this never has cupped leaves). The concave leaves, often strongly saucerlike, of *L. wenzelii* usually will separate it from any form of *L. ventricosa*; also, the latter has ordinarily a narrower, deeper sinus, often—but not invariably—bluntly angular.[187] In *L. wenzelii* the leaf width-sinus width-leaf length ratio typically ranges about as follows: 1:0.52–0.66:0.72–0.88. In other words, the sinus width varies from slightly more than ½ to ⅔ the leaf width and from ca. 0.7 to 0.9 the leaf length. Only in *L. groenlandica* may the sinus width be over 0.7 the leaf length.

Confusion of *L. wenzelii* is, thus, most readily possible with *L. groenlandica*. When living plants are available, the two are easily separable: the former, with few and relatively large, very variable oil-bodies, each formed of numerous spherules, is at once different from *L. groenlandica* with its numerous minute, usually biconcentric, mostly spherical to short-ovoid oil-bodies, all, as a rule, very similar in size. With dead plants, the orientation of the leaves offers the only certain points of distinction: in *L. wenzelii* the subvertical leaf position is usually marked—the concave face of the leaf is turned toward the stem, when the shoot is observed from its antical face; in *L. groenlandica* the concave leaves are more antically secund and more oblique, so that, when plants are lying flat on the slide with the antical face to the observer, one looks down into the leaf cavities. With very lax phenotypes these distinctions are much less surely applicable, but *L. wenzelii* usually has more arched leaf margins, and the leaves are hardly every subquadrate; *L. groenlandica* commonly has "squarer"-appearing leaves. Compare Fig. 216:1 and 12. Even this distinction may "fail"—*if* the plants attributed to *L. wenzelii* by Schiffner (Fig. 215:1) are correctly assigned.

---

[187] Weak plants and etiolated shoots of *L. wenzelii* quite lack the cupped and broad leaf form normal to that species. With allometric growth leaves of such plants may be ovate and clearly longer than wide to ovate-subquadrate; they then are distressingly similar to those of *L. ventricosa s. lat.*

In addition to the confusingly complex relationships of *L. wenzelii* to *L. ventricosa* and *L. groenlandica*, *L. wenzelii* has been said "not to appear to be always separable from *L. alpestris* . . ." (Macvicar, 1926, p. 188). Jensen (1901, p. 132) also emphasized that *L. wenzelii* appeared to be [translated]:

only an extreme form of *J. alpestris*, which occurs in polymorphous forms, especially in alpine districts. The principal differences between them are only:

J. alpestris	*J. wenzelii*
Greatest breadth of leaves is below or in the middle	Greatest breadth of leaves is above the middle
Gemmae brownish	Gemmae pale green or brownish

Jensen further believed that *both* species differed from *L. ventricosa* in their smaller cells (marginal cells of lobes only 16–22 μ, and their walls yellowish, in the *L. alpestris-wenzelii* complex; marginal cells 19–32 μ, with walls not yellowish, in *L. ventricosa*). However, it seems that the small-celled, soft-textured plants with brownish gemmae which he referred to *L. wenzelii* actually represent hygromorphic forms of *L. alpestris*. Gemma color (as Buch, 1933, showed) *seems* to afford a quite unambiguous mode of separation. In all phenotypes of *L. alpestris*, even in shade forms, red-brown to brownish gemmae occur (or, in subsp. *polaris*, deep red gemmae); in almost all phenotypes of *L. wenzelii* the gemmae remain greenish, even though extreme sun forms in the Arctic produce tardily salmon-colored to pinkish gemmae.[188]

Occasionally *L. wenzelii* and robust forms of *L. ventricosa* (var. *longiflora*) may key out together if the *L. ventricosa* exists as a dense-leaved phase with almost subvertical leaves; when growing exposed, it may occasionally have weakly incurved leaf lobes.

The distinction of these two taxa clearly emerges, however, when they grow intimately intertwined (as in *RMS & KD 66–557b*, Godhavn, Greenland). Then *L. wenzelii* stands out because of the dense and ballooned-out leaves, usually all perfectly saucer-shaped—the plants almost subjulaceous; leaf bases are a deep, dull purple, and the stem (where these pigments are dense) is blackish purple; oil-bodies occur mostly (3)4–9(11) per median cell; gemmae are frequent; leaves have a rather dull brownish pigmentation, unrelieved by pinkish or reddish coloring. Contrastingly, *L. ventricosa* var. *longiflora* has the variable leaves chiefly plicate-canaliculate, mostly with spreading or erect-spreading lobes; leaves are never saucer-shaped, usually less dense; leaf bases, of at least the younger leaves, are bright carmine; traces of a similar pigmentation suffuse much of the distal parts of some upper leaves, although most of the secondary pigmentation is a warm, light brown, unlike the deeper, duller brown

---

[188] In alpine situations in New England (e.g., in *RMS 66–1749*, Mt. Washington, N.H.), *L. wenzelii* and *L. alpestris* subsp. *alpestris* may occur admixed. The latter is then a smaller plant with much less concave leaves and with rust-colored gemmae even on plants that otherwise largely lack cell wall pigments, as contrasted to the more vigorous *L. wenzelii*, with more cupped leaves and darker pigments of at least the ventral side of the axis, but with greenish gemmae.

In my opinion, the closeness of *L. wenzelii* and *L. alpestris* (as well as their seeming proclivities at "intergradation") is much exaggerated.

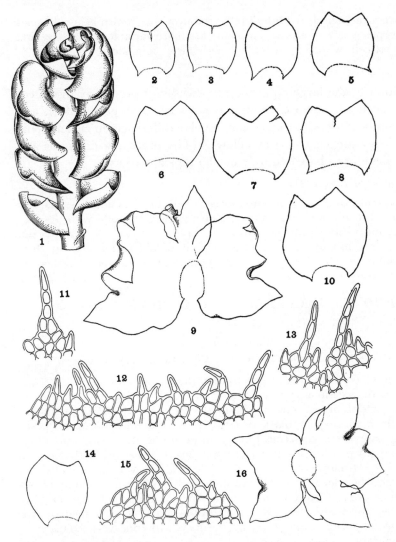

FIG. 215. *Lophozia (Lophozia) wenzelii*. (1) Shoot apex (×16). (2–8) Leaves (×16). (9) Bracts and bracteole (×15). (10) Leaf (×15). (11–13) Perianth mouth sectors (×180). (14) Leaf (×15). (15) Perianth mouth sector (×180). (16) Bracts and bracteole (×15). [1–9, from plant from type locality of *L. wenzelii*, issued by Schiffner in Hep. Eur. Exsic. no. 171; 10–16, *RMS 15954*, Mt. Katahdin, Me.]

of *L. wenzelii;* the stem is reddish distally, red-brown to brown below, but not blackish purple as in *L. wenzelii;* oil-bodies are more variable in both size and number, most cells with (8)9–16(17–19) oil-bodies; gemmae are rare.

Variation

In addition to the largely phenotypic variability previously discussed, *L. wenzelii* occurs in two extremes as regards cell size, gemma size, and oil-body number; these extremes may involve differences in chromosome number. They are separable as follows. (The preceding diagnosis and discussion have been based entirely on the typical variety.)

Key to Varieties of L. wenzelii

1. Smaller plants, usually 1.2–1.6 mm broad; cells of lobes and lobe margins 20–25 μ, of leaf middle 22–27 × 24–30 μ; median cells with (2–3)4–9(10–12) oil-bodies; gemmae mostly 20–24 × 21–25 μ . . var. *wenzelii*
1. Robust plants, usually 2.2–2.5 mm wide; cells of lobe margins ca. 30–38 μ, of leaf middle 30–38 × 30–46 μ; median cells with (9)12–20(23) oil-bodies; gemmae mostly 26–28 × 28–32 μ . . . . var. *lapponica*

LOPHOZIA WENZELII var. *LAPPONICA* Buch & S. Arn.

[Fig. 212:6–9]

Lophozia wenzelii var. *lapponica* Buch & S. Arn., Svensk Bot. Tidskr. 44:81, 1950.

Rigid, fleshy, generally more vigorous than the species proper, often 2.2–2.5 mm wide, with occasional furcate branching. Shoots typically with leaves brownish to reddish brown, rarely reddish, at least on distal portions, the stems purplish brown to purplish black (rhizoid bases vinaceous to dull purplish; ventral leaf bases often vinaceous to dull purplish-tinged). Leaves ca. 1065–1175(1425–1500) μ wide × 820–935(1065–1180) μ long (width:length ca. 1.1–1.4:1), ± strongly concave and subvertical or vertical, divided for 0.12–0.2(0.25) their length by a broad and shallowly lunate or rarely obtusangular sinus; lobes usually bluntly triangular. Marginal cells ca. 30–35 μ, varying to 36–38, occasionally 38–42, μ in distal parts of leaf; median cells averaging 30–38 μ wide × 30–46 μ long, normally with concave-sided trigones; oil-bodies variable as in the species proper, from spherical and 4–6.5 μ to ellipsoidal or ovoid and 5–6 × 8 μ to 3–6.5 × 5–7 μ, to 3 × 8 to 5–8.5 × 11 μ, finely granular-botryoidal, without central "eye," occurring (9)12–20(23) per cell. Gemmae greenish, 1–2-celled, polygonal to quadrate to weakly stellate, mostly 26 μ to 27–29 × 28–30 to 26–28 × 30–32 μ.

Type. Lule Lappmark, Sweden.

Distribution

Known from Finland and Sweden (Buch and Arnell, *in* Arnell, 1950c; Arnell, 1956) and reported as mostly "on rather sunny places on tops of high mountains . . . ."

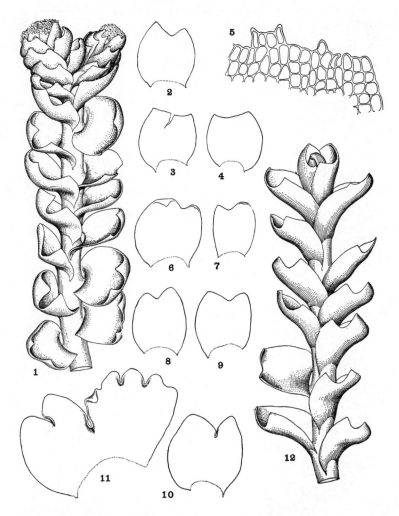

FIG. 216. *Lophozia groenlandica* (?) (1–4) and *Lophozia wenzelii* (5–12). (1) Shoot, of brown sun form, with gemmae and furcate terminal branching (×18.5). (2–4) Leaves (×16). (5) Perianth mouth sector (×ca. 180). (6–9) Leaves, in part from an unusually narrow-leaved ♀ plant (×16). (10) Subinvolucral leaf (×16). (11) Bracts and the connate bracteole (×16). (12) Sterile, weakly gemmiparous shoot (×17). [1–4, *RMS 13657d*, Susie Isls., Minn.; 5–12 from plants from Finland, leg. et det. H. Buch, of a green phase with stems purplish beneath, perianths below mouth often carmine.]

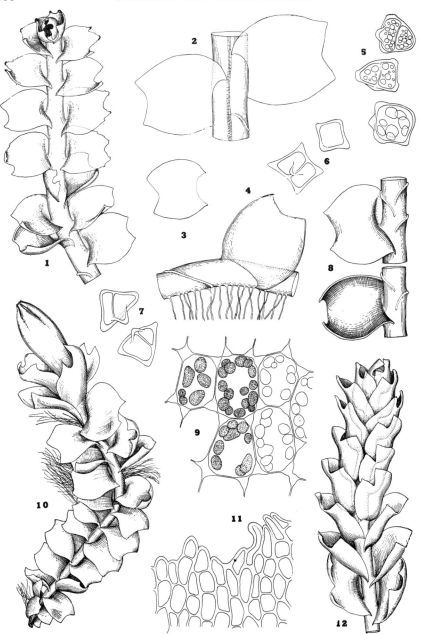

FIG. 217. *Lophozia alpestris* subsp. *alpestris*. (1) Small gemmiparous shoot of the "*gelida*"
phase (×20). (2) Shoot sector, postical view, showing ventral merophytes (×32).
(3) Leaf of mod. *parvifolia* (×15). (4) Leaf, mod. *parvifolia*, lateral aspect (×32). (5)
Gemmae (×610). (6–7) Gemmae (×625). (8) Shoot sectors, with leaf insertions on
stem, the upper leaf unusually explanate, the lower "normal" (×26). (9) Cells with

In our region, reported only from:

W. GREENLAND. Jacobshavn, Disko Bay, 69°12′ N., 51°05′ W. (*RMS & KD 66-246*), growing in intermittently submerged depressions in rocky arctic tundra, fringed by boulders and rocks, associated with *Gymnocolea inflata, Lophozia ventricosa*, and *Scapania hyperborea*. Ritenbenk, 69°46′ N., 51°20′ W. (*RMS & KD 66-262*). QUEBEC. Knob Lake (Lepage, 1960).

The description has been prepared from the Greenland plants. The very large cells, the strongly concave and broad leaves, and the large gemmae will separate this plant from other Lophoziae (subg. *Lophozia*) with green gemmae. Arnell states that the plant is frequently "strongly red" colored, and Müller (1951–58) that it agrees with *L. alpestris* in the "braunen bis rötlichen Zellwänden." Such coloration is less developed in the Greenland plants (although shoot tips may be rather reddish), but the purplish to vinaceous color of rhizoid bases, ventral stem base, and postical leaf bases is well marked. The Ritenbenk plant (*RMS & KD 66-262*) is particularly vigorous and is superficially identical with "*L. confertifolia*" in the dense, vertically oriented, and less than normally cupped leaves—the leaf apices rarely being conspicuously incurved.

LOPHOZIA (LOPHOZIA) ALPESTRIS (Schleich.)
Evans
[Figs. 217, 218]

Jungermannia alpestris Schleich., *in* Weber, Hist. Musc. Hep. Prodr., p. 80, 1815.
Jungermannia sudetica Nees ex Hüben., Hep. Germ., p. 142, 1834.
Jungermannia goeppertiana Hüben., *ibid.*, p. 254, 1834.
Jungermannia curvula and *J. sicca* Nees, Naturg. Eur. Leberm. 2:117 & 118, respectively, 1836; and *J. tumidula* Nees, *ibid.*, 2:233, 1836.
Jungermannia gelida Tayl., London Jour. Bot. 4:277, 1845; Pearson, Hep. Brit. Isl, 1:334, 2:pl. 143, 1902.
Cephalozia alpestris Cogn., Bull. Soc. Roy. Bot. Belgique 10:282, 1871; Hep. Belg., p. 35, 1872.
Jungermannia alpestris var. *gelida* Cooke, Hdb. Brit. Hep., p. 186, 1894.
Lophozia alpestris Evans, *in* Kennedy & Collins, Rhodora 3:181, 1901; Lorenz, The Bryologist 13:69, pl. 8, 1910; Macvicar, Studs. Hdb. Brit. Hep. ed. 2:185, figs. 1–6, 1926.
Lophozia gelida Steph., Spec. Hep. 2:136, 1902.
Lophozia alpestris var. *gelida* Macvicar, Ann. Scot. Nat. Hist., p. 49, 1904.
Lophozia (subg. Dilophozia) alpestris K. Müll., Rabenh. Krypt.-Fl. 6(1):679, fig. 313, 1910; Schuster, Amer. Midl. Nat. 45(1):60, pl. 18, figs. 1–8, 1951; *ibid.* 49(2):365, pl. 21, figs. 1–8, 1953; K. Müller, Rabenh. Krypt.-Fl. ed. 3, 6(1):658, fig. 201, 1954.

Extremely polymorphous, in thin patches or mats or small compact tufts, but often only scattered, varying from *dull and deep green* or bronzed to a *warm brown to reddish brown, leaf bases often purplish*. Shoots sparingly terminally branched, to 1500–2000 μ wide × (0.5)1–4 cm long. Stems rather rigid, 220–320 μ in diam.,

---

oil-bodies (×508). (10) Perianth-bearing shoot (×ca. 10). (11) Perianth mouth cells (×315). (12) Androecial shoot tip (×22). [1–4, 6–7, *RMS 5800*, Pigeon Point, Minn., fo. "*gelida*"; 5, 8, 9, Letchworth State Park, N.Y., *RMS*; 10, from Müller, 1905–16; 11–12, *RMS 17568*, Sand I., Wisc.]

prostrate with ascending tips (or ascending to suberect when crowded), flexuous, the ventral side brownish with age, *often with a purplish or vinaceous secondary pigmentation;* rhizoids numerous, colorless or reddish at base. Leaves weakly to distinctly imbricate, barely antically decurrent, *quite obliquely inserted,* from antically secund to widely and often nearly horizontally spreading except at the shoot apices, *broadly ovate-quadrate to rotund-quadrate,* mostly ca. 675–700 μ long × 680–720 μ wide up to 725–820 μ long × 750–950 μ wide, *(0.85)0.95–1.15(1.3)× as wide as long* except on the most slender shoots, usually distinctly and *often strongly concave to canaliculate-concave,* often *shallowly saucerlike, shallowly bilobed at apex, the sinus usually descending only* ⅒–⅕ *the leaf length,* less commonly ± angular and descending to ¼, typically *shallowly crescentic to obtuse; antical and postical leaf margins both strongly arched;* lobes very short and broad, obtuse to weakly acute, *usually ± incurved.* Cells *rather opaque,* of the margins and middle of the leaf lobes *only (15)18–21 μ, of leaf middle ca. 20 × 25 μ,* with thin walls and *distinct to moderately bulging trigones;* oil-bodies formed of nearly uniform, small spherules, appearing weakly papillose (under oil immersion), without a central "eye," *very variable in shape and size,* subglobose to ovoid or ellipsoid, (4)6–9 (rarely 10–16) per median cell, ca. 4–6.5 and spherical to 4–5 × 6–9 μ and ovoid, a few to 5–6.5 × 10–11 μ; chloroplasts 2.0–3.5 μ. Underleaves lacking. *Asexual reproduction always freely present, the gemmiparous leaves often with somewhat irregularly lacerate lobes ("var. gelida"); gemmae rust-red to reddish brown* (even in forms from extremely shaded sites), 1–2-celled, quadrate to polygonal, *small,* (15)16–20 × 16–24 μ, *averaging 18–20 × 18–20 μ,* the cells rather occluded *with large to small, polymorphous oil-bodies.*

Dioecious. Androecia 700–800 μ wide, intercalary with age, usually carmine-tinged, elongate, *spicate, of 5–7(10) or more pairs of closely approximate to imbricate bracts, quite compact;* bracts broadly ovate, erect-spreading, strongly saccate at base, with incurved lobes, otherwise like leaves; 2–3 antheridia, their stalks 1–2-seriate. Gynoecia with bracts larger than leaves, erect or erect-spreading, but with apices often spreading, shallowly 2–3-lobed at apex, broadly rotundate to obtrapezoidal, *their margins entire,* lobes subacute to obtuse; bracteole large, usually bilobed and united for some distance with one (rarely both) bract. Perianth cylindrical-ovoid, 0.5–0.75 emergent at maturity, smooth below but obtusely plicate in the gradually narrowed distal ¼; *perianth mouth crenulate-denticulate, the teeth at least in part 1–2 cells long,* elsewhere projecting only for part of the length of the terminal cell and merely crenulate. Capsule with wall firm, 3-stratose; outer layer with nodular, inner with semiannular thickenings. Elaters ca. 7 μ in diam.; spores 12–14 μ, reddish to purplish brown, delicately papillate.

Type. Switzerland (Schleicher's Pl. Crypt. Exsic. Helv. II, no. 59).

Distribution

An extremely widely distributed holarctic and circumpolar species, restricted to arctic-alpine and subarctic-subalpine regions. Absent from most major southward extensions of the Spruce-Fir Forests (such as the

Fraser Fir-Spruce Forests of the Southern Appalachians), but òccurring very rarely, as a late Pleistocene relict, in the northern portions of the Deciduous Forest, the Hemlock-Hardwoods Forest (Schuster, 1949a). The species is almost ubiquitous in the southern and central portions of the Tundra, and in alpine Tundra, as well as in the northern portions of the Spruce-Fir Forests (Taiga).

In Europe widespread, from northernmost Europe (Spitsbergen; throughout northern Scandinavia) to northern England and Scotland, Denmark, and the Faroes, alpine and subalpine portions of central Europe (southern and western Germany; adjacent Czechoslovakia and Hungary, portions of montane Poland; Switzerland; eastern and montane France; Austria), southward to central and northwest Spain (Casares-Gil, 1919), to Portugal and Madeira; also in the French Pyrenees, the Auvergne, the north Italian Alps, the Carpathians (in Bucovina), the Transylvanian Alps, and Asia Minor. Common in the central European Alps (but rare in the lowlands of northern Germany) and in the Tatra Mts.

Occurring eastward into Asia (Lena and Yenisey R. valleys; Novaya Zemlya), in the Himalayas, and in Japan (Hattori, 1958, p. 40). Also on Sakhalin I. (Kitagawa, 1963b).

Also in Iceland.

In North America presumably transcontinental at high latitudes, in the West from Alaska to British Columbia, southward in the Sierras and coastal ranges to Washington and California, and in the Rocky Mts. to Alberta, Montana, Wyoming, and Colorado. Eastward as follows:

GREENLAND. Reported from: Godthaab, and Thule (76°30′ N.) in W. Greenland; Scoresby Sund, Röhss Fiord, Fleming Inlet, Liverpool Kyst, Cape Stewart, Turner Sd., Cape Dalton, Nualik, Lille Ø, Cape Hildebrandt, Akiliarisek, Kangerdlugsuarsik, Odesund, Kingorsuak (fo. *amphigastriata*), Ikerasarsuak, Misotuk Smalsund, Ingmikertorajik, Joern Ø, Angmagsalik Fjord, Tasiusak, Hurry Inlet (Jensen, 1906a, etc.); Kangerdlugssuak Fjord (Harmsen, 1933); all in E. Greenland. Dove Bugt (76°40′ N), Cape Salar (82°54′ N.), and Centrum Ø (82°44′ N.), in NE. Greenland (Hesselbo, 1923a); see p. 515; most of these reports from the far north probably refer to subsp. *polaris*. NW. GREENLAND. Kânâk, Red Cliffs Pen., Ingelfield Bay (*RMS 45729*, *46160b*, *46161a*); Bjørling Ø, Cary Isls., Smith Sd., 76°43′ N., 72°22′ W. (*Seidenfaden*; published questionably as *L. kunzeana* by Harmsen & Seidenfaden, 1932).[189]

S. GREENLAND. Narssarq (*Vahl 37, 1829*! det. by Jensen as *J. floerkei* var., but typical of *L. alpestris* in the rust-red, small gemmae, ca. 16–18 μ in diam., and the small, ca. 17–18 μ, subapical cells of the leaf lobes). ELLESMERE I. Near Ft. Juliana, Skräling I., Cape Viele, Cape Rutherford, Bedford Pim I., and Fram Harbor, all in Hayes Sd. region (Bryhn, 1906); Fram, Harbour, and Goose Fiords, S. coast (Bryhn, 1906);

[189] The Seidenfaden plant approaches fo. *sphagnorum* in the black to purplish black stems and vinaceous postical leaf bases, the brown to purplish brown leaves, and the abundant reddish purple gemmae. Marginal cells of lobes are, however, only 14–18 μ, and gemmae only 14.5–18 × 18–20(24) μ. The plants are a lax extreme, mod. *laxifolia-colorata-latifolia*, with leaves unusually broad, to 1080–1420 μ broad × 750–960 μ long (1.4–1.5× as wide as long).

Reindeer Cove, W. coast (Bryhn, 1906); damp humus, basic weathered slopes of Mt. Pullen, Alert (*RMS 35302a*).[190]   DEVON I. REGION. North Kent I. (Bryhn, 1906). BAFFIN I. Arctic Bay, N. Baffin (Polunin, 1947); Cumberland Sd. (Steere, 1939).

QUEBEC. Cairn I., Richmond Gulf, and near mouth of Great Whale R., both on E. coast of Hudson Bay (Schuster, 1951); Wakeham Bay (Wynne & Steere, 1943); Rupert R. above L. Némiskau, Wakeham Bay, Great Whale R., L. Mistassini, Ste-Thérèse, Iberville, Roberval, Tadoussac, Bic, and Mt. Albert (Lepage, 1945).   LABRA-DOR. Hopedale (*Perrott, 1925*).   NOVA SCOTIA. Cape Breton I. (*Nichols, 1909*); Barrasois, Purcell's Cove, and Halibut Cove (Brown, 1936a); Mary Ann Falls, S. of Neils Harbour, Cape Breton Natl. Park (*RMS 42656a, 42668, 42673, etc.*).   NEW-FOUNDLAND. Green's Harbour (*Waghorne 191, p.p.*!, among *Diplophyllum albicans, Sc. nemorosa*); St. John's, Avalon Distr.; Rencontre West, on S. coast; Cow Head and Great Harbor Deep, N. Pen.; Gaff Topsail, Central Nfld. (Buch & Tuomikoski, 1955). ONTARIO. S. Twin I., James Bay; E. coast of L. Superior (Macoun, 1902); N. of Capreol, Sudbury Distr. (*Cain 2965*!); Otter Head, Thunder Bay Distr.; Cochrane Distr. (Cain & Fulford, 1948); Long Pt., and Gull L. Portage, L. Timagami (*Cain 3028, 3019*!).

MAINE. Mt. Katahdin: ubiquitous on Mt. Pomola, the Knife Edge, Saddle Slide, Baxter Peak, Chimney Pond Basin, North Basin, Northwest Basin, Garfield Ridge, etc. (*RMS 33007, 32966a, 15973a, 15988, 15984a, 15988, 15985. 15969a, 15970, 17302a, 15817, 17025, etc.; 15969a c. caps.*!); Mt. Desert I. (*Lorenz*).   NEW HAMSPHIRE. Ubiquitous on Mt. Washington, Mt. Monroe, Mt. Clinton, and elsewhere in the White Mts., and at Zealand Notch (*Schuster, 1944, 1953, 1966*); Waterville (*Lorenz, 93*); "White Mts." (*Oakes*, Austin, Hep. Bor.-Amer. Exsic. no. 39, 1873); Mt. Lafayette (Lorenz, 1908c); Waterville and West Branch near Osceola Camp (Lorenz, 1910, p. 70).   VERMONT. Mt. Horrid, near Rochester (*Dutton 157*); Rochester; Smuggler's Notch, Mt. Mansfield (*RMS 66-1851, p.p.*): Haselton Trail, 3000 ft, and Long Trail, Mt. Mansfield, 4000–4100 ft, near summit (*RMS 43812, 43843a*).   MASSACHUSETTS. Guilder Pond, Mt. Everett, 2065 ft, Berkshire Co., very rare (*RMS 68-158*); Sheffield. CONNECTICUT. Salisbury.   NEW YORK. Letchworth State Park, in gorge of Genesee R., Wyoming Co., in W. New York (*Schuster, 1949*; a relict occurrence in the Deciduous Biome!); Peaked Mt., Warren Co. (*Burnham*); Little Moose L., Herkimer Co. (*Haynes, 1904*); Peasleeville, E. edge of Adirondack Mts. (*RMS A-256, A-258*).

Westward recurring as follows: MICHIGAN. Amygdaloid I., Isle Royale; Copper Harbor, Keweenaw Co.; Pictured Rocks, Alger Co. (*RMS, s.n.*); Sugar Loaf Mt., Marquette; Gogebic, Phoenix, and Keweenaw Cos. (*Steere; Nichols*).   WISCONSIN. Sand I., Apostle Isls., Bayfield Co. (*RMS 17529, 17568, c. per., andr.*); Goodrich, Marathon Co.; Oak and Wilson I., Ashland Co.; Squaw Pt. and Orienta Falls, Bayfield Co. (Conklin, 1929).   MINNESOTA. Cook Co.: The Point, Grand Marais; Pigeon R. W. of "Sextus City"; Pigeon Pt. near the Tip (*RMS 13268b, 13426, 17441, 5800, etc.; Conklin 2670*); Porcupine I. and Hat Pt., near Grand Portage (*RMS 14500, Conklin 3128*). Two Island R., Lake Co. (*Conklin 2623*); Oneota, Chester Creek, Nopeming, all near Duluth, St. Louis Co. (*Conklin 790, 880, 1380, etc.*).

---

[190] In three months of intensive collecting in northern Ellesmere I. I found typical *L. alpestris* only rarely. The putative polyploid race or variety "var. *major* (C. Jensen) K. Müll." (sensu auct.; = subsp. *polaris* Schust.) was abundant and almost ubiquitous in many places. It is probable that at least most of the Bryhn reports should be transferred to this. Similarly, in both northwest and west Greenland, in four months of field work, very few collections safely attributable to subsp. *alpestris* could be found; those from Kânâk seem typical (marginal cells above leaf middle 15–18 μ; gemmae [15]16–18[20] μ).

Ecology

The species proper is usually more xerophytic than most other species of subg. *Lophozia*. It is almost uniformly confined to rocks or to soil over rocky slopes or ledges, occurring rarely on purely organic substrates (such as peat or decaying wood). In lowland areas, as in the Great Lakes region, most commonly on rather dry, but shaded and sheltered, north- and northwest-facing cliff walls and ledges, where it is either a pioneer or comes in soon after ecesis of the first pioneer species; more rarely it occurs on sheltered moist rocks. It is associated here with *Diplophyllum taxifolium, Anastrophyllum minutum,* and occasionally *A. michauxii,* more rarely *Lepidozia reptans, L. barbata,* and *Tritomaria quinquedentata.* The pH range is about 4.5–5.4 (Schuster, 1953). On moist sandstone also with *Cephalozia bicuspidata, Scapania curta, S. nemorosa,* and *Lophozia ventricosa.* The lowland occurrence in the gorge of the Genesee R., in the Hemlock-Hardwoods Forest of western New York, is equally restricted to damp vertical noncalcareous shale rock walls, associated with *Scapania nemorosa, S. curta, Cephalozia bicuspidata lammersiana, Gymnocolea inflata,* and *Solenostoma crenuliformis* (Schuster, 1949a).

Much more ubiquitous in the alpine tundra (where Lorenz, 1917, aptly characterizes it as occurring in "fifty-seven varieties");[191] in alpine tundra in New England found under xeric conditions, as a small, strongly gemmiparous phase ("var. *gelida*") with somewhat denticulate leaves, associated with abnormally strong development of gemmae. Here associated with *Gymnomitrion concinnatum,* less often *G. corallioides, Chandonanthus setiformis, Diplophyllum taxifolium, Rhacomitrium, Andreaea,* and other mosses, on thin exposed soil over boulders—or even scattered amidst the foregoing, directly over exposed rocks. On mineral soil, often among boulders, in equally dry and exposed sites, similar small phases of *L. alpestris* may be found with *Marsupella ustulata, Cephalozia ambigua, Cephaloziella byssacea, Lophozia bicrenata,* associated with such distinctive alpine plants as *Diapensia lapponica* and *Arenaria groenlandica.* At the bases of boulders, on damper soil, *Lophozia attenuata,* rarely *L. atlantica,* and occasionally *L. ventricosa silvicola* (in the "*confertifolia*" phase) are associated.

In alpine, as in arctic, tundra, *L. alpestris* also becomes common in wetter sites, and then often occurs as a larger, more luxuriant form, with fewer gemmae (but occasionally with androecia and perianths). Such phases occur over moist rock walls, often along mountain brooks, associated sometimes with *Gymnocolea*

[191] Bernet (1888) also emphasizes that it is polymorphous, "as the great number of synonyms attests."

inflata, Marsupella sphacelata, Scapania subalpina and *undulata, Tritomaria quin-quedentata, Lophozia ventricosa, L. incisa, Solenostoma sphaerocarpum,* and *Cephalozia bicuspidata* (i.a., in the Northwest Basin of Mt. Katahdin). Under such con-ditions, *L. alpestris* and associated species form thick mats and eventually lie over a thin soil-peat substratum, rather than directly on rock. Buch and Tuomikoski (1955) also report *L. alpestris* as mostly "on wet siliceous rock cliffs," with many of the preceding species, as well as with *Anastrophyllum michauxii, Mylia taylori, Lophozia wenzelii, L. attenuata, Nardia scalaris, Scapania scandica.* Less often, *L. alpestris* undergoes ecesis on peat-covered stones or in pronouncedly peaty sites, as along the sunny banks of alpine streams, associated with *Nardia geoscyphus, N. insecta, Pellia epiphylla, Odontoschisma elongatum,* etc. (Northwest Basin, Mt. Katahdin).

Closely similar to such occurrences are those around the margins of pools in granitic rocks, and in crevices of wet granite (with *Gymnocolea inflata, Odonto-schisma elongatum, Cephalozia ambigua, Anthelia, Scapania paludicola;* coast of Hudson Bay; Schuster, 1951).

L. alpestris s. str. (subsp. *alpestris*) is almost restricted to acidic rocks (most often granitic rocks, occasionally arkose, relatively rarely noncalcareous sand-stones or shales) but, according to Müller (1910, p. 682), is also occasionally found on calcareous rocks. No such stations are known to me; Kaalaas (1893, p. 335) also states that it is "always on a siliceous substratum." The species is often found in deeply shaded sites (then dark green except for the rust-red gemmae), but also in strongly insolated sites (then rusty brown, with somewhat deeper and more reddish gemmae; occasionally becoming carmine-tinged, especially the androecia and leaf bases). Always gemmiparous, rarely with perianths, very rarely with capsules.

Differentiation

Normal phases of *L. alpestris* are characterized by a whole ensemble of features that readily distinguish it from allied species. In fact, it has no immediate relatives in the genus (allied species, such as *L. ventricosa, L. wenzelii,* and *L. groenlandica* normally having pale, greenish gemmae). The nearly rotundate and concave leaves, generally with a shallow and broad, usually lunate sinus; the reddish or rusty brown gemmae, uniformly present; the usual presence of conspicuous brownish pigmentation of the leaves,[192] often associated with vinaceous pigmentation of the leaf bases and postical face of the stem; and the unusually small leaf cells serve to give it a distinctive facies. Gemmae-free plants are rarely encountered; they can be recognized by the pronounced tendency

[192] Extreme shade forms of *L. alpestris*, often found with *Diplophyllum taxifolium*, may be rather dull, chlorophyllose green throughout, except for the pale brown gemmae and tardily brownish ventral side of the stem. The brown gemmae, alone, usually separate such phenotypes in boreal areas.

toward brownish secondary pigmentation (only extreme shade plants being a characteristic deep green or olive-green) and by the small cells.

There are many reports of forms transitional to *L. ventricosa* and *L. wenzelii* (see, i.a., Müller, 1910, *in* 1905–16, p. 681; Macvicar, 1926, p. 186). Admittedly, *L. alpestris* is extremely polymorphous. However, actual transition to the preceding species is improbable. Even the most extreme shade phases of *L. alpestris* which I have seen, with deep green leaves and scarcely any indication of brownish pigment, bear the typical rusty red-brown gemmae. In both *L. ventricosa* and *L. wenzelii*, extreme sun forms, with strongly pigmented leaves, retain greenish gemmae (i.e., have colorless walls).[193]

Differentiation of the putative polyploid, subsp. *polaris*, with large cells is more difficult. This is discussed on p. 618.

Variation

Like other species of the sectio Ventricosae, *L. alpestris* undergoes an extraordinary amount of variation; much is clearly environmentally induced. For instance, on dry, somewhat sunny rock walls, and on dry soil and boulders in alpine tundra, plants are very small, hardly over 600–800 μ wide and 4–10 mm long. Such plants are bronzed even in shaded sites and become blackish brown in sun. They are often reversionary in developing underleaves. These plants, habitually sterile, were described as *J. gelida* [= var. *gelida* (Tayl.) Macvicar, but hardly more than an xeromorphic phase], *J. curvula*, *J. sicca*, and *J. tumidula*. Such phases are often freely gemmiparous, may have slightly dentate leaves, and often have somewhat narrower leaves (because of allometric growth) and a deeper sinus, sometimes descending to $\frac{1}{4}$–$\frac{1}{3}$ the leaf length. In such cases, the sinus may become almost rectangular. Such mod. *parvifolia-colorata-gemmipara* of *L. alpestris* "go over" in shaded but damp sites to a mod. *parvifolia-angustifolia-viridis* in which the brownish gemmae may be very few. *Both* of these small forms often have the leaves hardly saucer-shaped, often clearly longer than broad, and even on the same stem show wide variation from leaves with shallow, crescentic sinuses—the "normal" condition—to leaves with deeper, angular sinuses.

[193] The "var. *littoralis*" (Arn.) Schiffn. (Lotos 59:17, 1911) is supposedly a hygrophytic, soft, yellowish green phase of *L. alpestris*. It is stated (i.a., by Macvicar, 1926) to have colorless gemmae or, rarely, some pale reddish. I have seen no material referable to *L. alpestris* in which the gemmae color is not well developed. This is true even of the small, otherwise almost uniformly dull green, extreme shade forms often encountered on damp, shaded cliffs in our boreal forests.

However, in arctic extremes of *L. wenzelii*, from insolated sites on peaty ground, salmon-piak pigmentation of the gemmae may tardily occur. *All* such plants have the larger cells and gemmne of normal *L. wenzelii*.

On wet sites *L. alpestris* supposedly "goes over" into *L. wenzelii*. Although the plants become larger, green and more pellucid, and softer, gemmae color separates them effectively from *L. wenzelii*, as does the smaller cell size. The var. *littoralis*, supposedly with colorless gemmae ("or with a few pale reddish") should be re-examined; it has not been found in North America.

On wet cliffs in the mountains, *L. alpestris* occasionally produces lax-leaved, highly pigmented phases which may be difficult to identify. They may have the stem, ventrally at least, intensely violet-purple to vinaceous, and this pigmentation extends undiminished to the ventral leaf bases. On the distal portions of the leaves this pigment is diluted or absent, being in part replaced (or complemented by) a brownish red pigmentation. (This brownish or cinnamon-brown pigmentation is the dominant one formed in *L. alpestris* in shaded sites; the deep vinaceous pigments seem a response to intense illumination.) On such lax-leaved phases the leaves, although typically very broad and wider than long, may show an angular notch at the apex and are never saucer-shaped, but are often flat or somewhat undulate.[194]

Superposed on the great environmentally induced variation is geographically correlated genetic variation, possibly a consequence of polyploidy. In the far north (68–83° N.) *L. alpestris* is often largely replaced by the following subspecies.

LOPHOZIA ALPESTRIS subsp. *POLARIS* Schust., subsp. n.
[Fig. 218:10–23]

Lophozia alpestris var. *major* (Jens.) K. Müll., of authors, incl. Schuster et al., 1959, p. 36 (*nec J. alpestris* var. *major* Jens., 1906a).[195]

Plantae vigentes plerumque virides, textura ± molles; cellulae magnae chlorophyllosaeque, membranas tenues atque trigonas parvas habentes; cellulae loborum marginales 22–28 μ, mediae 23–33 × 25–35 μ; omnes guttae olei parvae, usque ad 4.5–5.5 × 6–8.5 μ, plerumque 10–13 in omni cellula; gemmae ad 24–28 × 24–32 μ.

[194] Large phases of this mod. *colorata-latifolia-laxifolia* have been confused with *L. longiflora* by some workers. For example, a specimen labeled as *longiflora* from Huntington Ravine, N.H., Aug. 7, 1917 (*A. W. Evans*) is, in my opinion, a phase of *L. alpestris*. It has reddish brown gemmae, averaging only 16–21 μ. The smaller leaves have apical cells 18–24 μ in the lobes, 16–19 μ on the lobe margins; larger leaves have subapical cells (18)20–24 μ, on the margins 16–20 μ. Admittedly the cells are somewhat larger than in the mod. *parvifolia* of *L. alpestris*; nevertheless, the pigmentation pattern of such plants is clearly that of *L. alpestris*—and so are the gemmae.

[195] The lectotype of *J. alpestris* var. *major*, from Kap Brewster (*N. Hartz, 1900*), the lectotype by present designation, is paroecious and clearly *L. excisa* of the lax phase often called var. *cylindrica*. [This is the only specimen so named, and so cited by Jensen (1906a), in the Copenhagen herbarium.] I could find (1)2–3 antheridia in subfloral leaves of two plants. I have seen ♀ plants of the second specimen, from Hurry Inlet, cited by Jensen (1906a) in his original diagnosis; they belong to *L. latifolia*.

By common convention, the name *L. alpestris* var. *major* has been adopted since the time of Müller (1910, *in* 1905–16, p. 681), for the large-celled greenish plant that represents the mod. *leptoderma-laxifolia-latifolia-viridis* of *L. alpestris* subsp. *polaris*.

Rather vigorous, often ± soft-textured (green phases) to rather rigid (small, dense- and small-leaved, brown-pigmented phenotypes), the shoots to 1.4–2.2(2.35) mm wide × 10–30 mm long. Stems 320–390(400) μ in diam., the thin- to weakly thick-walled dorsal cortical cells ca. (16)18–24 × 40–65(80) μ long, oblong to oblong-hexagonal. Leaves on more robust plants widely laterally patent to somewhat antically assurgent, inserted on a very oblique line, never approaching the subvertical insertion of *L. wenzelii*, broadly orbicular, typically with strongly convex antical and even more convex postical margins, and with a *shallow*, crescentic, or widely obtusangular sinus descending 0.15–0.2 (rarely 0.25) the leaf length, occasional leaves with an obtusangular sinus going to 0.3 the length; leaf outline broadly orbicular-quadrate, to (860)1440–1575 μ wide × (715)990–1275 μ long (width:length ca. 1.24–1.45:1). Cells *large, chlorophyllose*, lax, thin-walled, with small, concave-sided trigones; cell walls colorless or ± red-brown on distal parts of leaves, the postical leaf bases sometimes vinaceous; marginal and subapical cells from 22–26 to 25–28(30) μ, median cells from 23–33 × 25–35 μ. Oil-bodies per median cell mostly 8–17 (usually 10–13), variable, *relatively small for cell size*, 3.5–4.5 × 4–4.5 μ and nearly spherical to 4.5–5.5 × 6–8.5 and ellipsoidal, of small to large spherules, little or distinctly protruding; no central "eye" developed. Gemmae always present in abundance, *bright red*, the gemmae masses forming a sharp contrast to the usually green leaves; gemmae cuboidal to polygonal-stellate, occasionally tetrahedral, (1)2-celled, from (19)20–24 × 20–26 μ to 24–28 × 24–32 μ.

Dioecious. *Habitually sterile.* ♂ plant very rare. A few young gynoecia only seen, with immature bracts and no mature perianth.

Type. Blaesedalen near Lyngmarksfjeld, Godhavn, Disko I. (*RMS & KD 66–281, July 1966*).

Distribution

A strictly arctic plant, whose taxonomic relationships to *L. alpestris* subsp. *alpestris* remain problematical in the absence of fertile material. A very common, often abundant taxon in calcareous and circumneutral sites, becoming rare or absent in strongly acid environments, although occasionally found on *Sphagnum*. The rather wide pH tolerance is characteristic.

So far surely known only from the American Arctic.

N. GREENLAND. I.P. Koch Fjord, 82°48′ N. (*Th. Wulff, June 19, 1917!;* poorly developed and a little doubtful, with reddish gemmae; reported as *L. muelleri* by Hesselbo, 1923); Cap Salor, 82°45′ N. (*Th. Wulff, June 29, 1917;* very poor and rather doubtful); Mascaret Inlet, Sverdrup Isl., 82°55′ N., 44°30′ W. (*Ellitsgaard 7978!,* typical). NW. GREENLAND. Kânâk, Inglefield Bay (*RMS 45636, 43639h, 45650a, 45639g*); ca. 4 mi NW. of Kânâk (*RMS 45634a*); Kangerdluarssuk, Inglefield Bay (*RMS 45869*). W. GREENLAND. Blaesedalen, near Lyngmarksfjeld, near Godhavn, Disko I., 69°15′ N., 53°30′ W. (*RMS & KD 66-281*); Hollaenderbugt, Nugssuaq Pen., 70°47′ N., 53° W. (*RMS & KD 66-1500*), Lyngmarksbugt, Godhavn (*RMS & KD 66-342, 66-569*);

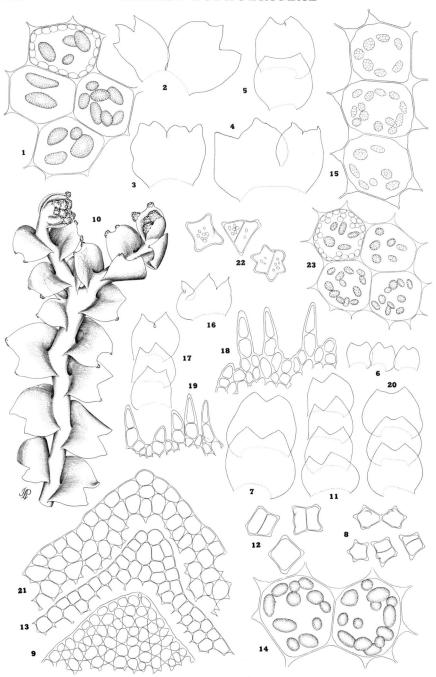

FIG. 218. *Lophozia (Lophozia) alpestris.* (1–9) subsp. *alpestris* and (10–23) subsp. *polaris.*
(1) Median cells with oil-bodies and, upper cell, chloroplasts; mod. *leptoderma* (×920).
(2) ♀ Bract and, at left, bracteole (×11). (3) Bract (×11). (4) ♀ Bract, and at right,
bracteole (×11). (5) Leaves (×11). (6) Leaves of a mod. *colorata-parvifolia* (×11).

Fortune Bay, Disko I., 69°16′ N., 53°50′ W. (*RMS 66-203a*); Flakkerhuk, near Mudderbugten, E. end of Disko I., low, wet, sandy-peaty flat plain, 69°39′ N., 51°55′ W. (*RMS & KD 66-173b*); Tupilak I., W. of Egedesminde, 68°42′ N. (*RMS & KD 66-013*, xeric form); Sondrestrom Fjord, near shore of L. Ferguson, in *Salix*-moss tundra, 66°5′ N. (*RMS & KD 66-087*); Qalagtoq, Upernivik Ø, Inukavsait Fjord, 71°14′ N., 52°35′ W. (*RMS & KD 66-1101*); Eqe, near Eqip Sermia (Glacier), ca. 1 mi from Inland Ice, 350-450 m, 69°46′ N., 50°10′ W. (*RMS & KD 66-336, 66-323, 66-322c*). ELLESMERE I.   Mt. Pullen and The Dean, 5–6 mi S. of Alert, 82°25′ N., 62°05–15′ W. (*RMS 35116, 35123, 35111a, 35409b, 35180, 35186*, etc.); scree slope, 9 mi W. of Mt. Olga, E. edge of U.S. Range, 82°24′ N., 65°20–30′ W. (*RMS 35555a*).

Ecology

A tolerant and widespread taxon in the Arctic, rare or lacking only under the relatively xeric and acid conditions found in the *Rhacomitrium lanuginosum-Lophozia hatcheri-Chandonanthus setiformis* bryophyte community of the fellmark, where snow accumulation is often slight.

The species may occasionally occur on *Sphagnum* but is more usually found on circumneutral soils, often between crevices in basic rocks (such as basalt, as is the case with the type; then associated with *Scapania cuspiduligera, Lophozia excisa, Tritomaria quinquedentata,* and other taxa tolerant of basic or circumneutral soils). Also often on more acid soils, between rocks, where kept wet (and sometimes submersed) by melt from adjoining ice caps or firn ice; then often varying, depending on exposure, from green to partly purplish red. Associated under such conditions may be *Lophozia groenlandica, Gymnomitrion concinnatum, Marsupella revoluta, Scapania spitsbergensis,* etc.

On Ellesmere I. Schuster et al. (1959) report a wide variety of associates, often in damp clay soil depressions in scree or amidst boulders, that is suggestive of strongly to weakly basic conditions, e.g., *Tritomaria heterophylla, T. quinquedentata, Scapania praetervisa polaris, Arnellia fennica, Cephaloziella arctica, Lophozia quadriloba, L. heterocolpa, Anthelia juratzkana,* occasionally *Cryptocolea imbricata,* as well as a long list of cited mosses. The species occasionally occurs in west

---

(7) Same (×31). (8) Gemmae (×370). (9) Cells of lobe (×175). (10) Shoot with gemmae and a *Frullania*-type branch (×14). (11) Leaves (×11). (12) Gemmae (×370). (13) Cells of lobe apex (×175). (14) Cells with oil-bodies, a mod. with fine-botryoidal oil-bodies and rather large cells, 25–28 μ in lobe apices and margins (×690). (15) Cells, somewhat below average size, with oil-bodies (×825). (16) ♀ Bract (×11). (17) Leaves (×11). (18–19) Cells from mouth of juvenile perianths (×180). (20) Leaves (×11). (21) Cells of apex of lobe (×175). (22) Gemmae (×420). (23) Cells with oil-bodies and, upper left, chloroplasts (×545). [1, *RMS 43803*, Mt. Mansfield, Vt.; 2–4, Sand I., Wisc., *RMS;* 5, Huntington Ravine, Mt. Washington, Evans, 1917, as *L. confertifolia,* but with red-brown gemmae; 6–9, *RMS 45724b*, Kânâk, NW. Greenland; 10–13, *RMS & KD 66-281*, Godhavn, Greenland; 14, doubtful phase, *RMS 45636*, Kânâk, NW. Greenland; 15–19, *RMS 45610c,* fo. *sphagnorum,* Kânâk, NW. Greenland; 20–21, *RMS 45634a*, Kânâk, NW. Greenland; 22–23, *RMS 35116*, Mt. Pullen, Alert, Ellesmere I.]

Greenland (as at Hollaenderbugt, Nugssuaq Pen.) on fast-drying highly calcareous clay-shale slopes, associated with such very distinct calciphytes as *Lophozia pellucida* and *L. badensis, Tritomaria scitula,* and *Scapania gymnostomophila,* as well as *Arnellia fennica.*

Small and hardly typical forms may occur on weakly or medium acid soils (with *Gymnomitrion apiculatum, G. corallioides, G. concinnatum, Lophozia groenlandica, L. excisa, Anastrophyllum minutum,* and *Marsupella sprucei; RMS 45639h, 45650a*), while luxuriant, sterile forms may be found in swelling tufts over soil between rocks kept wet by melt water from the adjoining ice cap. Such forms (e.g., *RMS 45636,* Kânâk) may be partly purplish-tinged, with limited gemmae production and with some vinaceous postical leaf bases; associated are *Lophozia groenlandica, Scapania spitsbergensis, Marsupella revoluta,* etc.

In the far north this taxon is almost never directly on rocks and is often strongly hygrophytic. Particularly common among various *Brya* at the permanently wet heads of small mountain valleys filled with firn ice and snow. The plant here may be freed from snow only for 10–12 weeks each year.

Differentiation

Lophozia alpestris subsp. *polaris* differs from typical *L. alpestris* in a number of features, all associated with the larger cells (presumably related in turn to polyploidy): (*a*) cells of leaf tips ca. 23–24 μ, of leaf middle ca. 23–25 $\times$ 24–30 μ or larger, vs. 18–20 μ and 20 $\times$ 25 μ, respectively, for "typical" *L. alpestris;* (*b*) gemmae 20–25 $\times$ 24–30 μ vs. 16–20 $\times$ 16–24 μ in "typical" *alpestris.* In addition to these differences, readily ascertainable with the microscope, there is a very distinct discrepancy in facies. *Lophozia alpestris* subsp. *polaris* "looks" like large, lax forms of *L. excisa* or *L. latifolia,* rather than like *L. alpestris s. str.* It is a more robust plant, with the pellucid, flat or scarcely concave and lax leaves of *L. latifolia.* Associated with the more robust size, the leaves are generally at least 1.1–1.2 $\times$ as wide as long, often divided by a rectangulate sinus extending for ¼–⅓ their length, or even farther. In this, also, subsp. *polaris* has the facies of *L. latifolia.* Although often found in relatively dry moss patches, *L. alpestris polaris* seems unable to produce sharply bulging trigones, and even when occurring in strongly insolated sites (with often 24 hours of sunlight daily during the arctic summer!) it fails to produce intense pigmentation, only the distal halves of the leaves becoming light brown, with occasional vinaceous pigmentation of the ventral leaf bases. In both of these respects the plant again differs from *L. alpestris s. str.*

and approaches the *L. latifolia-excisa* complex. As a consequence, the practical problem of identification involves separation of this taxon from the latter complex, rather than from typical *L. alpestris*.[196] Three features stand out. (1) In both typical *L. alpestris* and subsp. *polaris* the perianth mouth is very unevenly terminated, varying locally from irregularly crenulate to short-denticulate with scattered 1–2-celled teeth: in *L. latifolia* and *excisa* the mouth of the perianth is evenly crenulate. (2) The cells generally have fewer than 13 oil-bodies, in many cases as few as 3–5; in the *L. latifolia-excisa* complex there are usually 15–24(28) oil-bodies per cell, in only isolated cells as few as 11–17. (3) The gemmae are carmine red, rarely with a faint trace of a vinaceous tinge; in the *L. excisa-latifolia* complex gemmae are mostly distinctly purplish or vinaceous.

Sterile plants of *L. excisa* are also likely to be confused with *L. alpestris* subsp. *polaris*. However, *L. excisa* (at least the robust, pure green Arctic phase of the species) is usually softer, more fleshy, and tender, and the gemmae are duller and more purplish. It also often has larger median cells (averaging 28–30 or even 32 × 30–35 μ, occasionally even 40 μ long) and generally more numerous (17–24, sometimes 28) oil-bodies per cell. It is, furthermore, difficult to find totally sterile *L. excisa*.

Having had considerable experience with both typical *L. alpestris* and subsp. *polaris*, I am inclined to consider the latter worthy of specific rank. If differences in capsule-wall anatomy and spore size are found to occur between the two, elevation of the subspecies to a distinct species will prove justified.

Although *L. alpestris polaris* is widespread under basic or circumneutral conditions, an extreme of this plant occurs (mostly in basaltic regions where peat has lowered the pH) which seems so different that it is difficult to reconcile with "normal" subsp. *polaris;* this is described as follows:

LOPHOZIA ALPESTRIS subsp. POLARIS fo. SPHAGNORUM Schust., fo. n.

[Fig. 218:15–19]

Caules folia remota habentes, infra nigri aut picei; bases foliorum saepe intense subpurpureae ad caeruleopurpureas; gemmae clare rubrae, ad 18–23 × 18–24 μ.

Habitually sterile and straggling over *Sphagnum;* stems flexuous, usually remote-leaved, commonly blackish to piceous beneath. Leaf bases often purplish to bluish purple, but distal portions brownish-pigmented or with reddish and brownish pigments mixed, then deeply colored, to (occasionally) pale green; leaves *L. latifolia*-like, nearly flat, broad, 925–1150 μ wide × 860–1000 μ long, bifid for 0.35 their length, the sinus usually angular (exc. on slender, ♂ shoots,

---

[196] It is possible that sterile, dead plants cannot always be certainly separated.

where often shallower). Marginal cells of leaf lobes averaging 22–25(27) μ, median cells 24–26 × 30–35 μ; walls thin, trigones distinct, at times faintly bulging; (3–5)7–14(18–20) *rather coarsely botryoidal* oil-bodies that average only 3 × 4.5 to 3.6–4.5 × 4.5–6 μ. Gemmae bright red, frequent, rather small: 14–18 × 16–20 up to 18–23 × 18–24 μ. ♀ Bracts crispate as in *L. latifolia;* perianth mouth (in *RMS 45610c*) dentate with sharply pointed teeth, in large part formed of 1–2 elongated, tapered, thick-walled cells, in part merely denticulate with partly projecting cells; the sharp terminal cells ca. 18–20 × 42–52 up to 24 × 62 μ in extreme cases.

Type. Kânâk, Red Cliffs Peninsula, northwest Greenland (*RMS 45610c*).

Distribution

Known only, to date, from: NW. GREENLAND. Kânâk, Red Cliffs Pen., Inglefield Bay (*RMS 45610c, 45604*); Kekertat, E. end of Inglefield Bay, near Heilprin Gletscher (*RMS 46003b;* probable).

Ecology

Among bright red *Sphagnum* polsters, on peaty ground irrigated from snow fields or glaciers above, near sea level; associated with *Vaccinium uliginosum microphyllum, Cassiope tetragona, Pedicularis lanata, Cephaloziella uncinata, Tritomaria quinquedentata, Lophozia grandiretis, Riccardia palmata, Cephalozia pleniceps,* and *Anastrophyllum minutum.*

The fo. *sphagnorum* may, possibly, be a mere environmental phase of subsp. *polaris,* but it is a difficult plant to evaluate. The rather coarsely botryoidal and small oil-bodies seem distinctive; the very intense pigmentation—sometimes almost reddish black to purplish black, distal parts of leaves excepted—is very striking.

The plant is likely to be confused with forms of *L. excisa* which may occasionally occur in similar sites (and, indeed, this species is admixed in *RMS 45604;* lax, helophytic extremes of it differ from the *L. alpestris sphagnorum* in the paroecious inflorescences and very sparing gemmae development). Confusion is also possible with *L. latifolia,* a plant with, usually, a very similar ecology. *Lophozia latifolia* may in extreme cases have vinaceous leaf bases and, locally, a more or less piceous stem, with brownish pigmentation, often admixed with some reddish, on the distal portions of leaves. Such extremes are almost inseparable when sterile.[197]

[197] However, in such an extreme of *L. latifolia* (from Sondrestrom Fjord, west Greenland, *RMS 66–048*), the perianths were distinct and had ordinary, crenulate apices similar to those of *L. excisa;* also, numerous ♂ plants with slender, long androecia were present, and gemmae were quite lacking.
   In spite of its *L. latifolia*-like appearance, this *Sphagnum*-inhabiting extreme of *L. alpestris* is quite distinct: particularly the very infrequent fertility, linked with the free development of gemmae, is diagnostic. In *L. latifolia,* in contrast, free fertility is linked with rare gemma formation.

LOPHOZIA (LOPHOZIA) RUBRIGEMMA Schust., *sp. n.*

[Figs. 219, 222:1–7]

Species calciphoba; plantae prostratae, partes foliorum superiores flavo-brunneae ad cas-
taneas; folia 1.2–1.3 plo latiora quam longa, 0.25–0.35(0.4) bilobata, omni lobo acuto; cellulae
pellucidae, permagnae; marginales plerumque 30–36 μ, mediae plerumque 40–43 × 40–45 μ;
guttae olei saepissime 16–25 omni in cellula; ♀ bracteae valde crispatae, parce dentatae, os
perianthii dentibus armatum 1(2–4)-cellularibus, cellula terminali magna, ad 23–30 × 75–88 μ.

Of xeric sites, *noncalcicolous, small to medium-sized, prostrate* (or, with competition, suberect), the *upper portions and especially margins and lobes* of at least
upper leaves a characteristic yellow-brown to occasionally chestnut-brown,
never purplish or copper-red (but the usually decolorate leaf bases may show
some, usually weak, vinaceous pigmentation at the postical base). Stem with
the usual dorsiventral differentiation of the medulla. Leaves uniformly bilobed,
± pellucid, variable, oblique, never squarrose, *widest near base*, from ca.
835–860 μ long × 1015–1075 μ wide, up to 1000 μ long × 1225 μ wide (width
1.2–1.3× the length), on weak shoots narrower and as long as broad; sinus
descending 0.25–0.35(0.4) the leaf length, obtuse to usually *bluntly acute; lobes
triangular,* usually acute to bluntly subacute, *never rounded.* Underleaves lacking.
Cells very large, rather pellucid: marginal cells of lobes typically (26)30–36(39) μ;
median cells typically (28–30)32–38 × (32)35–42 up to 40–43 × 40–45 μ,
thin-walled but with distinct trigones which are not coarse (except sometimes in
pigmented cells). Oil-bodies (*14*)*16–25*(*30*) *per median cell, relatively small,*
spherical and 3–3.5(4–4.5) μ to ellipsoidal and 3–3.5 × 5–6.5 up to 4.5 × 7(9)
μ, formed of uniform, fine, only weakly protuberant spherules.

Gemmae abundant on tips of upper leaves, *deeper colored than leaf cells, reddish
chestnut to rust-red to bright red,* polygonal to quadrate, ca. *23–28*(*30*) × (*23*)*25–
32*(*34*) μ, occasionally to 33–36 × 38–40 to 38 × 38–45 μ.

Dioecious. Gynoecia with *strongly crispate bracts,* from 1300 μ wide × 990 μ
long to 1325–1600 μ wide × 1120–1150 μ long, irregularly (2)3–4-lobed for
0.25–0.35 their length, the triangular lobes and margins below *sparingly,
coarsely armed with bluntly acute to quite sharp teeth* that range from 1–3-celled to
almost lobelike, but never spinous, their terminal cells not prominently elongated; the 1-celled teeth usually hardly elongated, quite blunt; apices of main
lobes ending in 1–2(3) superposed, single cells, the end cell often acute, from
24 × 28 to 28–36 × 45 μ. Bracteole oblong to obovate, ca. 615 μ wide ×
925 μ long to 770 μ wide × 875 μ long, fused for ca. 0.5–0.65 its length with
1 bract, 2-lobed 0.3–0.45 its length, the narrow sinus reflexed, margins sparingly
dentate with blunt teeth. Leaves immediately below bracts transitional in
nature and size, ca. 1475–1540 μ wide × 1075–1120 μ long or larger, somewhat
asymmetrically 2-lobed for 0.35–0.4 their length with triangular, acute lobes
bearing several obtuse teeth (similar to those of bract margins). Only juvenile
perianths known: these with mouth *often vinaceous,* armed with rather closely
juxtaposed 1(2–4)-celled teeth, whose *terminal cell is usually slenderly tear-drop
shaped* (widest just above the somewhat narrowed base, pointed distally, quite
attenuated above), and *very large* (from a minimum of 15–20 × 50–55 μ to
20–25 × 60–72 μ to 23–30 × 75–88 μ). ♂ Plant unknown.

FIG. 219. *Lophozia (Lophozia) rubrigemma.* (1) Leaves (×21). (2) Gemmiparous plant (×18). (3) Cell with oil-bodies (×500). (4–6) Teeth from young perianth apices (×210). (7) Perianth mouth (×142). (8) Perianth mouth (×105). (9) Bract and bracteole, above; bracts and bracteole, below (×26). (10) Gemmae (×440). (11) Cells with oil-bodies (×460). [1–3, 7, *RMS 45817c*, Bowdoin Bugt, Red Cliffs Pen., NW. Greenland; 4–6, *RMS 45901*, NW. Greenland; 8–9, *RMS 45961c*, NW. Greenland; 10–11, *RMS 45787a*, Kangerdluarssuk, NW. Greenland.]

Type. Kangerdlugssuak, Inglefield Bay, northwest Greenland (*RMS 45961c*); the type plants were segregated from a mat of *Rhacomitrium, Chandonanthus setiformis, Ptilidium ciliare,* and *Lophozia hatcheri.*

Distribution

To date known only from Greenland:

NW. GREENLAND. Inglefield Bay region. Paratypes are as follows: Kangerdluarssuk, Bowdoin Bugt (*RMS 45787a; 45817c; 45793a,* trace); Kangerdlugssuak (*RMS 45901;* apparently an xeromorphic extreme, growing exposed, amidst *Cephalozia pleniceps* and *Anthelia juratzkana,* in a depression between rocks); Kekertat, near Heilprin Gletscher, E. end of Inglefield Bay (*RMS 46079*).   W. GREENLAND.   SE. of Jacobshavn, Disko Bay, 69°12′ N., 51°05′ W. (*RMS & KD 66-227*); Ritenbenk, Agpat, 69°46′ N., 51°20′ W. (*RMS & KD 66-270*).

Ecology

Growing usually in relatively arid sites, generally on peaty soil between boulders or in mats between clefts in rocks, scattered amidst strongly xerophilous byrophytes such as *Rhacomitrium canescens, Dicranum, Chandonanthus setiformis, Lophozia hatcheri, Ptilidium ciliare,* as well as *Tritomaria quinquedentata* var. *grandiretis,* less often *Lophozia ventricosa, Blepharostoma trichophyllum brevirete.* In spite of the occurrence in xeric sites, the plants are usually rather lax and lax-leaved, with the cells quite leptodermous and with small trigones only (except sometimes in and near the leaf margins, where trigones may become weakly nodular).

Differentiation

A difficult and problematical taxon—the plants were initially found only in small quantity and in a sterile condition—understandable only in the light of a clear comprehension of its ecology. It is almost wholly restricted to sites in which little snow accumulates and which, during the short arctic summer, may undergo prolonged exposure to drought and wind. The rather constant association with *Polytrichum, Lophozia hatcheri,* xeromorphic forms of *L. kunzeana,* and *Rhacomitrium* is symptomatic. Although not limited to fell fields, *L. rubrigemma* is typical of that restricted and difficult habitat. The plants often grow erect amidst *Dicranum* and *Cephaloziella byssacea s. lat.,* rarely forming pure patches. In sheltered sites, as between rock clefts, the plants may be pure green, the deep rust-red gemmae excepted. With limited insolation the upper leaves show yellow-brown to chestnut pigmentation of their margins, and with more intense illumination this pigmentation spreads to the distal portions of the

leaves. Some forms tend to show vinaceous pigmentation of the postical leaf bases, but this may be quite lacking (as in *RMS & KD 66–
227*).

Because of the sparing production of reddish gemmae, the plants were
placed in juxtaposition first to *L. alpestris* and then to *L. latifolia*. With
these two species *L. rubrigemma* shares a dioecious inflorescence and
approaches the latter and *L. alpestris polaris* in the large leaf cells and
rather small trigones. However, with the discovery of gynoecia in several
collections, it was possible to definitely eliminate *L. latifolia* from consideration, in spite of the frequent similarity in facies, for the following
reasons: (*a*) the perianth of *L. rubrigemma* bears a peculiarly armed mouth,
with 1–2(3–4)-celled teeth, whose extremely large, differentiated terminal cells are lanceolate in outline and conspicuously thick-walled,
tending to become vinaceous-pigmented; (*b*) the bracts and (to a lesser
extent) the subinvolucral leaves are crispate, and bear on the margins
several coarse but rather blunt teeth, in addition to the primary 2–4
lobes. Criterion (*b*), in particular, serves effectively to separate the plant
from all forms of *L. alpestris*. Unlike all forms of the latter, the leaves of
L. rubrigemma lack reddish pigmentation (postical leaf bases sometimes
excepted) but tend to develop a characteristic brown to chestnut pigmentation, particularly of the marginal and submarginal cells. Also,
even though the various forms of *L. alpestris* vary widely in oil-body
number, the average ranges from ca. 4 to 15 oil-bodies per cell (with usually
diagnostic, marked differences in size and shape within each cell);
by contrast, oil-bodies in *L. rubrigemma* occur 16–25(30) per cell, are more
nearly uniform, and are usually ovoid to ellipsoidal in shape.

Lophozia rubrigemma often has much larger leaf cells than those in any form of
L. alpestris or in *L. latifolia*, although there is sufficient environmentally induced
variation so that occasionally the smaller-celled phases have cells not much
larger than those of *L. alpestris* subsp. *polaris*. Because of the large leaf cells the
species may be confused with the calciphilous *L. pellucida*, described from
calcareous sites in Ellesmere I. (Schuster et al., 1959). The latter, now known
from Sweden, Alaska, and Greenland as well, has even larger cells, gemmae
whose color corresponds to that of the leaf cells, and leaves whose lobes are
often blunt and whose sinuses are usually shallower, often lunate. *Lophozia
rubrigemma*, in addition to the much more intensely pigmented gemmae,
which have given the plant its name, also has more numerous oil-bodies per
cell [usually 16–25(30), as contrasted to only (7)9–18 oil-bodies per cell in *L.
pellucida*]. The ecological differences between the two taxa are major: *L.
rubrigemma* is diagnostically a species of acid rocks; *L. pellucida* is confined
strictly to strongly calcareous, sedimentary regions. No immediate affinity
between the two is believed to exist.

LOPHOZIA (*LOPHOZIA*) *PELLUCIDA* Schust.

[Figs. 220-221]

Lophozia excisa var. *grandiretis* Arn., Svensk Bot. Tidskr. 48:796, 1954.
Lophozia pellucida Schust., *in* Schuster et al., Natl. Mus. Canada, Bull. 164:34, 1959 [*nomen nudum*]; Schuster, Canad. Jour. Bot. 39:978, 1961; Arnell & Persson, Svensk Bot. Tidskr. 55:376, 1961.

Decumbent to prostrate, with crowding ascending to suberect, slender to rather compact and fleshy, scattered or in loose patches, *pellucid, light green or* (with maturity) *yellow-brown or clear chestnut brown, the postical leaf bases rarely lightly vinaceous.* Shoots slender and small, 1750–2000(2200) μ wide × 1–2.5 cm long, sparingly and very diffusely branched. Stems ± fleshy, slender to stout, almost terete to very slightly flattened, 245–325(425–500) μ in diam., *10–12(13–14) cells high,* flexuous and sinuous, *rather soft* and somewhat translucent, pale green to (with age) brownish; *ventral cortical cells linear-rectangulate,* (55)90–175 μ long × 16–20(23) μ wide (length:width ca. 4–8:1); dorsal cortical cells *elongate,* (48)50–110(120–160) μ *long* × (16–18)21–28(32) μ wide, rectangulate, ca. 2.5–6× as long as wide, *scarcely thick-walled, pellucid;* dorsal medullary cells pellucid, 23–26 μ high, in 5–6 tiers; ventral medullary cells notably smaller, usually 13–19 μ in diam., in 3–5 tiers, partly destroyed by mycorrhizae, distinctly smaller than ventral cortical cells. Rhizoids scattered, not abundant, colorless to pale brownish, their bases rarely slightly vinaceous red, *17–23 μ in diam. Leaves distant to contiguous, very obliquely inserted, and quite evidently decurrent antically,* broadly orbicular-ovate to quadrate-rotundate, from ca. 850 μ wide × 750 μ long to 1050–1150 μ wide × 800–950(1100) μ long, to 1200–1220 μ wide × 700–850 μ long (*width 1.4–1.7× length*) up to 1300–1350(1650) μ wide × 950–1000(1350) μ long (*width ca. 1.1–1.35× length*), *nearly flat* or slightly concave to saucer-shaped, but not canaliculate, *usually 2-lobed* (a very few 3-lobed), entire-margined or faintly crenulate, caused locally by projecting cells; sinus descending ⅛–⅙ in terrestrial, to ⅕–⅓ in hygric phases, *obtuse to rounded,* sometimes crescentic, often somewhat reflexed (hygric forms); lobes broadly triangular, *often one or both obtuse or blunt to distinctly rounded* (hygric forms) or mostly or all acute to subacute (mesic forms). *Cells pellucid, with walls becoming distinctly clear yellowish to chestnut brown with age, extremely thin, the trigones small to moderate but never bulging;* cells very large, *the marginal ones* (30)35–40 μ, the subapical 35–40 × 35–50 μ, the *median (33)35–42(45)* × (36)40–50(55)μ; cuticle smooth; oil-bodies usually (7)9–18 per cell, *very small for cell size,* spherical and 4.5–5.5 μ to ovoid or ellipsoid and 4.5 × 7–9 μ or 5–5.5 × 6.5–7.5 μ, a few to 5.5 × 8 or 5 × 9 μ, finely papillose-segmented in appearance, nearly smooth; chloroplasts few and smaller than oil-bodies. *Underleaves frequent, minute,* of 1–2 stalked slime papillae. Gemmae in *very pale green to orange-yellow to very pale brownish (never red or red-brown!)* masses from the uppermost leaves, 1–2-celled, *polygonal to stellate,* rather thin-walled, and with only weakly thickened protrusions, *large,* from 25–26 × 28 μ to 25 × 32–35 μ to (30)34–35 × 36–40(43) μ.

Dioecious. ♀ Bracts connate with bracteole, 2–3(4)-lobed and plurilobulate, polymorphic, ± fluted and crisped; bracteole small, usually distinct. Perianth longly exceeding bracts, ca. 2 mm long, cylindrical-ovoid to cylindrical-clavate,

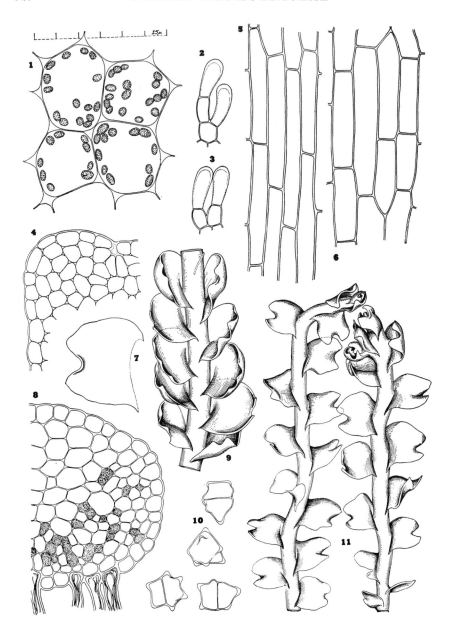

Fig. 220. *Lophozia (Lophozia) pellucida.* (1) Median cells (×400). (2–3) Underleaves from sterile shoot (×288). (4) Apex of blunt-lobed leaf (×109). (5) Ventral cortical cells (×250). (6) Dorsal cortical cells (×250). (7) Leaf, typical (×23). (8) Stem cross section (×163). (9) Sector of robust shoot with saucer-shaped, antically connivent, imbricate leaves, with *L. alpestris*-like facies (×15.5). (10) Gemmae (×340). (11) Two typical lax-leaved sterile shoots (×12.5). [All drawn from type specimen, *RMS 35141*, Parr Inlet, Dumbbell Bay, Ellesmere I. Fig. 1 to scale = 90 μ.]

inflated, smooth, only apical portion faintly plicate, *mouth crenulate-ciliolate, with terminal cells thin-walled*, ca. *22 × 50–70 μ*; cells of perianth middle ca. 24–26 × 26–50 μ to 40 × 40 μ.

Type. South end of Parr Inlet, Dumbbell Bay, on snow-fed, calcareous solifluction slope, near Alert (82°30′ N.), northeast Ellesmere I. (*RMS 35141*).

Distribution

In addition to the type locality and a series of sites from the same general region in northern Ellesmere I. (all confined to damp calcareous slopes, usually below "hanging," persistent snow banks), known from Greenland, Alaska, and Sweden (Schuster et al., 1959; Arnell and Persson, 1961).

N. GREENLAND. Brønlund Fjord, Heilprin Land, in moss-cushions along rivulet, 82°10′ N., 31° W. (*Holmen 238f;* reported by Arnell, 1960, as *L. groenlandica*). W. GREENLAND. Head of Tasiussaq Fjord, 71°04′ N., 51°17′ W. (*Holmen 13,523, p.p.!.* among *Arnellia fennica, Blepharostoma*); Fjord S. of Marmoralik, 71°05′ N., 51°12′ W; (*RMS & KD, s.n.*); Ingia, N. tip of Ubekendt I., 71°18′ N., 53°35′ W. (*RMS & KD, s.n.*); Hollaenderbugt, Nugssuaq Pen., 70°47′ N., 53°47′ W. (*RMS & KD 66-1512, 66-1518;* in phenomenal quantities!); Port Rd., Sondrestrom Fjord (*RMS & KD, s.n.*). NE. GREENLAND. Maria Ø, 72°58′ N., 24°50′ W. (*Holmen 19,213;* trace among *Lophozia rutheana, L. badensis*); Labben Hills, Mesters Vig (*H. Raup 718!;* trace with *Scapania cuspiduligera, Lophozia quadriloba, Blepharostoma*). ELLESMERE I. Clay-shale slope below snow bank, Mt. Pullen, 800–900 ft, 82°25′ N., 62°20′ W. (*RMS 35786, 35312*); clay-shale calcareous flat, outwash plain N. of Mt. Pullen, 82°26′ N., 62°15′ W. (*RMS 35382a*); steep clay-shale slope, S. end of Hilgard Bay, 82°26′ N., 63°25′ W. (*RMS 35931, 35934a*); calcareous clay slope below hanging snow bank, 2.5 mi W. of Cape Belknap, adjoining mouth of Colan Bay, 82°32′ N. (*RMS 35205a, 35213, 35205, 35207, 35028*); calcareous clay-shale slope, valley between Mt. Olga ("Peak 1") and Mt. Erica ("Peak 2"), ca. 82°22′ N., 64°15′ W., ca. 5–6 mi SW of head of Hilgard Bay (*RMS 35549, 35520*); damp clay-shale slope adjacent to polar ice cap, 1.5–2 mi W. of Cape Belknap, 82°32′ N., 62°20′ W. (*RMS 35634b, 35626,* and *35638a, p.p.,* among *L. pellucida minor*); head of Parr Inlet, S. end of Dumbbell Bay, 82°30′ N., 62°20′ W. (*RMS 35141c, p.p.,* with *Arnellia fennica*); on calcareous solifluction slope, below persistent snow bank, NW. side of The Dean, 82°25′ N., 62°10′ W. (*RMS 35167*). SWEDEN. Jukkasjarvi socken, Abisko Natl. Park, Karsavaggejokk, Torne Lappmark (*Persson, July 29, 1944!*).

The Swedish plants are not quite typical; they show the pale brownish to golden brown pigmentation of the upper portions of the leaves; have generally quite large cells (although leaves on some plants may possess cells averaging less than 35–38 μ wide); are nearly leptodermous; have broad and transversely suborbicular-ovate leaves clearly wider than long, with shallow sinuses that may be slightly reflexed; thin-walled, elongated cortical cells; and almost colorless gemmae that become light brownish with age. The gemmae show a great variation in size and may be as small as 25–27 × 29–32 μ (perhaps somewhat immature), but the largest are up to 30–35 × 35–46 μ; they tend to be more sharply angulate than in the type.

At least in the North American Arctic *L. pellucida* is chiefly a species of recently emerged, highly calcareous lowlands, giving the impression of having evolved not long ago by polyploidy. In the summer of 1960 I repeatedly searched for it in northwest Greenland (77–78° N.) in a noncalcareous region, without being able to find any trace.

Ecology

Basically a high arctic and arctic species of difficult, exposed, highly calcareous sites. In western Greenland often on steep clay-shale soli-fluction slopes where few other taxa survive (such as *Salix arctica s. lat.* and/or *S. glauca, Dryas integrifolia, Polygonum viviparum, Saxifraga oppositifolia*). In such plants it may be associated with *Arnellia fennica, Tritomaria scitula,* and *Lophozia alpestris polaris* (the last at once distinct because of the deep red gemmae). Under "ideal" conditions for this species (very sterile and exposed, calcareous clay slopes that dry out readily in summer) *L. pellucida* may occur in enormous quantities, as on the north side of the Nugssuaq Peninsula, western Greenland, in widespread, diffuse, orange-colored patches that are distinct at some distance. The plants here form a frieze, conspicuous because of their unique color, along all the multifarious small drainage rills that have been eroded in the clay.

In Ellesmere I., equally sharply restricted to highly calcareous sites, found (in the type region) exclusively on moss-covered, clay-shale slopes or (more commonly) on seepage-moistened clay-shale slopes situated below hanging, persistent snow banks or firn-ice-filled valleys. The plants occur often among the conspicuous, reddish brown or cinnamon-colored moss *Holmgrenia chrysea,* sometimes with *Equisetum variegatum* (at its northernmost known limit at 82°32′ N.) and with other Hepaticae: *Arnellia fennica, Blepharostoma trichophyllum brevirete, Cephaloziella artica, Aneura pinguis, Lophozia gillmani, L. collaris,* and *L. badensis, Athalamia hyalina* (rarely *Sauteria alpina*), *Preissia quadrata.* The preceding taxa occur admixed in very damp, often hygric sites. On steeper, better drained slopes, some of the preceding species may be associated, as well as *Lophozia alpestris polaris, L. quadriloba, Scapania gymnostomophila, Solenostoma pumilum polaris, Cryptocolea imbricata,* rarely *L. heterocolpa harpanthoides.* Associated angiosperms are usually *Papaver radicatum, Salix arctica, Saxifraga caespitosa* and *oppositifolia* more rarely *S. flagellaris.* Associated mosses are, i.a., *Drepanocladus revolvens, Hypnum bambergeri, Distichium capillaceum, Cinclidium arcticum, Didymodon recurvirostris.*

Differentiation

Distinguished principally by the combination of large and leptodermous cells, light and pellucid green color (where the light to warm yellow- or orange-brown pigmentation does not prevail), and the large, stellate or polygonal gemmae that are greenish on greenish shoot tips but turn a bright orange or yellowish orange with maturity, occasionally with faint orange brown pigmentation. Gemmae that are slightly similar in color may occasionally be formed in *L. wenzelii* (see p. 599), but that species is otherwise very different.

The color, the peculiar pellucidity, and the massses of usually orange-colored gemmae give this species a distinctive aspect, making it readily recognizable in the field. This, along with the restriction to highly calcareous sites, often accompanied by *Equisetum arvense* and *E. variegatum*, and the almost constant sterility, easily identify the plant. Although the type came from a well-irrigated clayey solifluction slope, occurrences in crevices on clay slopes that dry out are much more frequent, here often with *Arnellia fennica*, *Distichium capillaceum* and *Tritomaria* (*scitula* or *heterophylla*).

Because of the colored mature gemmae and large cell size, this species is likely to be mistaken by the uninitiated for *L. alpestris* subsp. *polaris*, but the latter plant produces deep red gemmae even in extreme greenish shade forms (and, where mixed with *L. pellucida*, is at once different in the red gemma color) and never has the clear transparent coloring of *L. pellucida*. The most welcome proof of the specific distinction of these two taxa comes from plants of *RMS & KD 66–1503b*, Hollaenderbugt, Nugssuaq Peninsula, where they occur together, the *L. pellucida* in its var. *minor*, the compact form most likely to be confused with *L. alpestris polaris*.[198] The following comparison was made from two plants growing within 2 mm of each other:

Criteria	L. pellucida minor	L. alpestris polaris
Gemmae		
Color, shape	Light orange; sharply stellate	Deep red; cuboid-stellate
Size	30–32 × 30–32 to 30–34 × 45 μ	20–23 × 20–24 to 24–27 × 26 μ
Number	In immense masses	Few; in small clumps
Leaves		
Dimensions	1230 × 800 μ long (1.35–1.45:1)	920–1050 × 740–820 μ (1.2–1.3:1)
Sinus	Lunate > obtusangular	Obtusangular
Marginal cells	30–36 μ	23–27 μ
Median cells	(25)28–35 × 30–38 μ	25–32 × 30–34 μ
Cortical cells	(21)22–30(32) × (48)50–95(125) μ	19–24 × 35–72 μ

Perhaps the most striking distinction between forms such as the above occurring under difficult, exposed conditions is the differences in degree of gemmae productivity. In *L. pellucida* gemmae occur in immense masses, and

[198] I have also collected *L. pellucida* and *L. alpestris polaris* admixed in Ellesmere I. (*RMS 35167*); exactly the same range of differences persists as outlined for the Greenland plants.

the patches stand out to the naked eye, even at a distance of 5–6 feet, because of the orange-colored gemma masses; in *L. alpestris polaris* the modest amount of gemma production that characterizes phases of sheltered, damp sites persists in such forms of exposed environments. Under the microscope, the drastic differences in gemmae form and size equally preclude confusion of the two taxa: gemmae of *L. pellucida*, besides being much larger, have more sharply drawn out angles and are rarely ever purely cuboidal; in *L. alpestris polaris* the gemmae, besides being smaller, are less angular, often cuboidal, with the angles less acutely produced.

Lophozia pellucida was first recognized by Arnell (1954) as a variety of *L. excisa*. The dioecious inflorescence, gemma color, and highly distinctive ecology all preclude there being a distinct affinity to *L. excisa*. Indeed, when I first discussed *L. pellucida* (in Schuster et al., 1959) no allusion was made to Arnell's *L. excisa* var. *grandiretis* since the almost constant sterility and demonstrably dioecious inflorescence, plus all the many other differential criteria, seemed to preclude any real relationship between these taxa. However, *L. pellucida* may possibly be sought under subg. *Massula*. It has the strongly oblique, pellucid, broad leaves characteristic of this subgenus; it also has leptodermous to slightly thickened, strongly elongated, narrow cortical cells, much as in *Massula*. The ventral cortical cells, 4–8 ×, rarely to 10 ×, as long as wide, are without parallel in subg. *Lophozia;* they occur in lax as well as in rather dense-leaved phases. Hygromorphic extremes of the species also may have flexuous, soft, straggling, sinuous stems as in *L. (Massula) marchica*. Yet the axial anatomy is that of subg. *Lophozia*, as are the oil-bodies. With age, in sun, plants acquire a clear brownish pigmentation, occasionally supplemented by a slight vinaceous tinge of the postical leaf bases, unlike most other species of subg. *Lophozia;* furthermore, the rhizoids are broad in diam. and unusually sparse. The leaves are also distinctive in being usually nearly plane, but with the rounded sinus often narrowly reflexed.

From the foregoing it is clear that *L. pellucida* occupies an isolated position in subg. *Lophozia*, showing some characteristics of *Massula*. If, on the basis of the large thin-walled leaf cells, it is sought in *Massula*, it may be looked for under *L. obtusa* (because of the often obtuse or rounded leaf lobes and somewhat reflexed sinus). It differs at once from this species, however, in form and number of oil-bodies, even larger leaf cells, narrower and more elongated cortical cells of the stem, and stem anatomy.

On the basis of the leaves with two blunt or rounded lobes, weak shoots of *L. pellucida* are subject to possible confusion with *Gymnocolea*. The leaf cells, however, are much larger than in the latter and possess thin walls and more distinct trigones. This resemblance to *Lophozia (Massula) obtusa* and *Gymnocolea inflata,* based on the leaf shape, is enhanced by the fact that minute and inconspicuous underleaves may occur in *L. pellucida*—another anomalous feature for a species of subg. *Lophozia.* These are inserted on normally wide ventral merophytes (usually 2–3, rarely 4, cells broad) and consist either of a single clavate slime papilla on a 1-celled stalk, or of 2 such papillae, each on a stalk cell. In the latter case, the stalk cells are united for a short distance at base. Occasionally a single stalk cell may bear 2 slime papillae inserted on it. Similar vestigial underleaves commonly occur in *Gymnocolea,* while considerably larger and more elaborated ones are frequent in *L.* (*Massula*) *obtusa.*

Among species of subg. *Lophozia,* a distant relationship with *L. alpestris* and *wenzelii* is possible. Although the majority of less robust shoots of *L. pellucida* have obtuse and blunt to rounded leaf lobes, separated by a rounded sinus, large, mature shoots bear at least some leaves in which the sinus is shallow and crescentic, descending at most for ⅙ the leaf length. Leaves of this type are also somewhat antically secund and slightly to distinctly concave, with rather incurved leaf lobes. Such *L. alpestris*-like leaves occur sporadically and sometimes rather frequently; they may be very broad (1200–1220 μ wide × 700–850 μ long, or even 1500 × 1050 to 1550 × 900 μ), thus ca. 1.4–1.7 × as wide as long, particularly on robust shoots with stout stems. Associated with the ability to develop such leaves is the usual development of a concave, weakly saucer-shaped leaf form, and the general presence of warm, chestnut-brown coloration, recalling that of *L. alpestris.* It is hardly possible, however, that *L. pellucida* can be derived from *L. alpestris.*

Gemmae of *L. pellucida* are abundant in xeromorphic forms but may be sparingly produced in lax-leaved hygromorphic forms. They are often largely greenish, even on lightly brown-pigmented plants. In deeply chestnut-brown plants, the gemmae may be slightly brownish-pigmented rather than orange, but are never the bright reddish-brown or carmine characteristic of gemmae of *L. alpestris polaris.* This is the only species of subg. *Lophozia* where the pattern and nature of pigment production of the gemmae nearly correspond with those characterizing normal leaf cells although a slightly higher light intensity may be necessary to initiate pigment production in the gemmae than is needed in the leaf cells. At least, it is common to find distinctly pigmented plants in which the gemmae have nearly colorless walls.

Lophozia pellucida is definitely dioecious. The few ♀ plants seen had produced archegonia, evidently not fertilized; neither bracts nor bracteole had fully matured, nor had any trace of a perianth developed. Such unfertilized gynoecia were always subtended by 1–2 innovations. More robust stems occasionally undergo almost perfectly dichotomous forking, the branching being terminal. In such cases, it is difficult to distinguish the branch from the continuation of

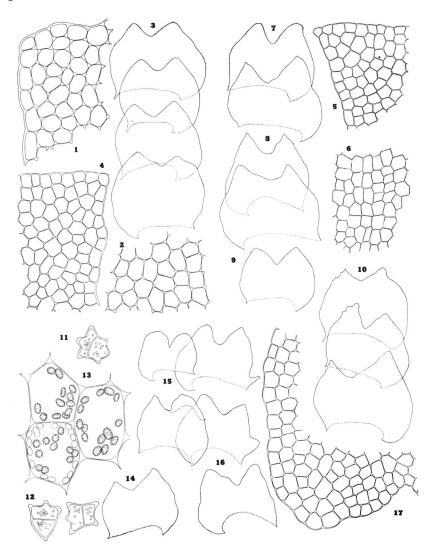

FIG. 221. *Lophozia* (*Lophozia*) *pellucida* (1–3, 15–17) and *Lophozia pellucida minor* (4–14). (1) Cells of leaf apex (×115). (2) Median cells (×115). (3) Four leaves (×21.5). (4) Cells of lobe apex (×115). (5) Lobe apex (×115). (6) Median cells (×115). (7–9) Leaves (×21.5). (10) Three leaves, the upper two from gemmiparous area (×21.5). (11–12) Gemmae (×285). (13) Cells with oil-bodies (×475). (14) Leaf of yellow-brown form (×15). (15–16) Leaves (×21.5). (17) Cells from ventral lobe of leaf in fig. 16, left (×115). [1–14, all from Ellesmere I.: 1–4, 6, 10, *RMS 35626*, Alert; 5, 7–9, 11–14, *RMS 35237*, Hilgard Bay; 15–17, *Holmen 13523, p.p.*, Tasiussaq Fjord, Greenland.]

the main axis, since the leaf subtending the branch is neither narrowed nor otherwise modified.

Although the extraordinarily large size of the cells is diagnostic of "typical" *L. pellucida*, deviations from the norm are frequent and may be partially the result of differences in environmental conditions. Plants from wet sites are generally lax-leaved and have cells 35–40, occasionally 35–45, μ in the lobes. Occasional forms from steep, well-drained slopes (i.a., *RMS 35931, 35934a*) may be equally robust, although relatively fleshy, with denser leaves; in such plants the apical and subapical cells may be 35–40, occasionally even 45 μ, or (in other cases) average 35–42 μ.

On clay-shale slopes that are merely damp, *L. pellucida* may occur as a smaller, more compact phase, with apical cells 33–45 μ, admixed with plants which are similar but have smaller cells (23–28, occasionally 30, μ in leaf apices, in i.a., *RMS 35634c, 35638a, 35638*). A few plants from similar sites, however, are quite transitional in cell size; e.g., in plants of *RMS 35820* cells may be 28–36 μ in diam. in the leaf apices, thus approaching "typical" *L. pellucida*, or even (*RMS 35823*) up to 30–38 μ in the leaf tips. Except for the cell size of these latter plants (locally admixed with "normal" large-celled *L. pellucida*), it would appear that two distinct species were at hand. In fact, the smaller-celled plants bear a close similarity (in green, nearly leptodermous phases) to *L. ventricosa* and were at first tentatively referred to this species. A separation between these smaller-celled plants, referred to *L. pellucida* var. *minor*, and typical *L. pellucida* is not always easy to effect. This is especially true when gemmiparous leaf lobes, in which the cells are often considerably enlarged, are studied.

Both phenotypes produce a warm brownish pigmentation of the leaf lobes, at times extending almost to the leaf bases (but becoming less intense basally); both have gemmae that are greenish when immature and may remain yellowish green in shade extremes, but almost always develop a warm, light brown pigmentation (identical with that of the leaf lobes). *The identity in reaction of the cells of the leaf lobes and of the gemmae, as regards pigment formation, is one of the notable features of this complex*, isolating it from others in subg. *Lophozia*. Both plants also agree in the form, size, and number of oil-bodies and in gemma size. Leaf shape and cell size thus remain the only effective means of distinguishing the two taxa. These distinctions are clear from *RMS 35626* (Cape Belknap, Ellesmere I.), where two plants, growing within 1–2 mm of each other, show the following features (compare also Fig. 221:1–3, 4, 6, 10):

	L. pellucida (typical)	*L. pellucida* var. *minor*
Marginal cells	(33)35–40(45) μ	23–28(30) μ
Subapical cells	32–38 × 38–46 μ	23–28(30) × 23–30(34) μ
Median cells	(30)32–36 × 35–48 μ	23–30 × 23–34(38) μ
Basal cells	35–40 × 40–48 μ	28–32 × 32–38 μ

Both plants grew under identical relatively exposed and xeric conditions (much more so than those characterizing the type locality of *L. pellucida*). The distinctions between the two taxa may be associated with polyploidy and therefore appear to be similar in nature to those distinguishing *Plagiochila asplenioides* and *arctica* (considered specific) and *Tritomaria quinquèdentata* and *T. quinquedentata* var. *grandiretis* (considered merely varietal).[199]

The status of the difficult, smaller-celled phase remains to be determined on the basis of culture studies and chromosome counts.

LOPHOZIA PELLUCIDA var. *MINOR* Schust.

[Fig. 221:4–14]

Lophozia pellucida var. *minor* Schust., Canad. Jour. Bot. 39:984, 1961.

Like *L. ventricosa*, but commonly paler, often suffused by a clear, yellowish or chestnut-brown coloration; without reddish pigmentation of stem or postical leaf bases; *leaf lobes (and sometimes entire distal halves of leaves) clear yellowish brown in sun forms.* Stem rather soft-textured; dorsal and lateral cortical cells (18)20–24 × 65–90(100–120) μ, ca. 3–5× as long as wide, rectangulate, thin-walled. Leaves broadly ovate to transversely oblong, from 930–1050 μ wide × 650–970 μ long to 1100–1175(1400) μ wide × 750–875(1050) μ long (ca. *1.08–1.48 × as wide as long*), nearly flat to weakly concave, ¼–⅓(⅖)-bilobed; sinus obtuse to rectangulate, occasionally rounded. Cells with smaller trigones (not distinctly bulging); the marginal 23–27 μ (locally 30–35 μ long); the subapical 23–28(30) × 27–32(35) μ; the median (23)25–30 × (27)30–38 μ. Oil-bodies (5–8)9–15 per median cell, spherical and 4–5 μ in diam. to ovoid or ellipsoid and 4 × 4.5 to 5 × 6.5–8 μ. *Gemmae of the same color as leaf lobes* (or sometimes barely more intensely pigmented), their walls colorless (then gemmae masses pale green) to yellowish brown (then masses appearing yellowish brown), 2-celled, ca. 28 × 30 to 32 × 30–36 μ when mature, rarely to 30–34 × 45 μ.

Type. Damp, weakly basic steep slope, south end of Hilgard Bay, 82°26′ N., 63°25′ W., northeast Ellesmere I. (*RMS 35237*).

Distribution

Known only from northern Ellesmere I. and western Greenland. In addition to the type, the following collections belong here:

W. GREENLAND. Hollaenderbugt, Nugssuaq Pen. (*RMS & KD 66-1503b*).
ELLESMERE I. S. end of Hilgard Bay, ca. 82°26′ N., 63°25′ W., NE. coast (*RMS 35235a, p.p., 35251*); low slope, edge of polar ice cap, 1.5–2 mi W. of Cape Belknap, 82°32′ N., 62°20′ W. (*RMS 35638, 35632, 35634c, 35638a, 35823, 35820, 35824*).

In the high Arctic of North America confined to cold, often snow-fed slopes, associated with tolerant species or calciphytes (*Lophozia gillmani, L. collaris, L. badensis, L. quadriloba, Cryptocolea imbricata, Aneura pinguis, Cephaloziella arctica, Arnellia fennica, Blepharostoma trichophyllum brevirete* and *Solenostoma pumilum polaris.*

---

[199] It seems relevant to point out that in the *L. pellucida* complex the two phases have essentially the same number of oil-bodies per cell. In the two analogous cases cited in *Plagiochila* and *Tritomaria*, the putative polyploids average about twice as many oil-bodies per cell as occur in the haploid stem form.

Differentiation

A baffling taxon. Superficially at times suggesting a derivative of *L. ventricosa*, with which it may agree in form, size, and number of oil-bodies, in leaf shape, and in depth and form of sinus. However, *L. pellucida minor* differs from *L. ventricosa* in that the gemmae closely parallel the leaf cells as regards pigmentation of their walls: the gemmae are colorless when the leaf cells have colorless walls, and are brownish-pigmented when the cell walls of the leaf lobes bear a similar pigment.[200]

Pigmentation pattern, form, size, and number of oil-bodies, and gemmae size agree closely with those of *L. pellucida* var. *pellucida*, a common plant in the same area, as do the dimensions (and proportions) of the cortical cells. However, typical *L. pellucida* has more polymorphous leaves, some markedly broader, on robust phases bearing a much more shallow, crescentic sinus (and on weaker phases commonly with rounded leaf lobes). In spite of the marked differences in cell size, it remains an open question whether *L. pellucida* and var. *minor* are fully distinct. A situation parallel to that of the *Cephalozia ambigua-bicuspidata-lammersiana* complex may be at hand.

LOPHOZIA (LOPHOZIA) PERSSONII Buch & S. Arn.
[Fig. 222:8–13]

Lophozia perssonii Buch & S. Arn., *in* Buch, Bot. Notiser 1944:382, figs. 1–2, 1944; K. Müller, Rabenh. Krypt.-Fl. ed. 3, 6:660, fig. 202, 1954.

Small, to *ca.* 1 mm wide × 5 mm long, green. Stem to 500 μ in diam., to 16 cells or more high, the medulla dorsiventrally differentiated into a ventral region (eventually embracing fully ½ to ⅔ of the medulla) which becomes brownish and mycorrhizal, and an upper, dorsal, hyaline region; dorsal cortical cells leptodermous, hardly smaller than medullary, short-oblong (length:width ca. 2–4:1). Rhizoids colorless. Leaves subvertical in orientation, inserted at a ca. 45° angle, rounded-quadrate to broadly ovate, 400–700 μ broad and approximately as long, concave, lunately to angularly bilobed for 0.25–0.3 their length; lobes sharply pointed, the tips ending in 1 cell or 2(3) cells superposed, or occasionally blunt. Cells (20)21–26(30) μ on margins of lobes, becoming moderately larger toward the leaf middle [there 25–30 × 30–40(50) μ], thin-walled but *with distinct, triangular trigones*. Oil-bodies present in all leaf cells, spherical to ellipsoidal, 4–8 to 4 × 9 μ, *3–8(10) per cell*. Gemmae *in reddish yellow to red-brown masses*, reddish, tetrahedral to cuboid, with weakly protuberant angles, 12–19 × 17–21 μ, 2-celled at maturity, each cell *with 1–2 large (to 10 μ) and often also several small yellowish, persistent oil-bodies*.

---

[200] Although it might, at first, appear desirable to dispose of var. *minor* as a larger-celled, arctic derivative of *L. ventricosa*, this designation appears impossible to maintain. Both *L. pellucida* (typical) and *L. ventricosa* occur on The Dean in northeast Ellesmere I. The former is a plant of low, calcareous, wet slopes; the latter of high, noncalcareous, peat-covered scree slopes. The strongly calciphytic restriction of *L. pellucida* vs. the distinctly oxylophytic restriction of *L. ventricosa* clearly suggests that the calciphytic var. *minor* belongs with *L. pellucida* rather than with *L. ventricosa*.

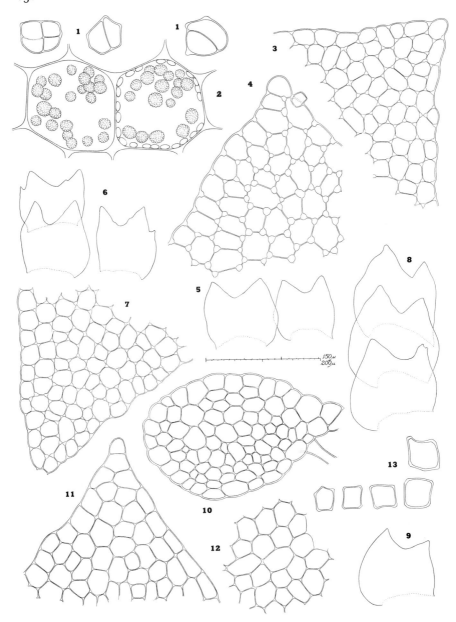

FIG. 222. *Lophozia* (*Lophozia*) *rubrigemma* (1–7) and *Lophozia* (*Lophozia*) *perssonii* (8–13). (1) Gemmae (×285). (2) Median cells with oil-bodies and, at right, chloroplasts (×590). (3) Cells of lobe apex (×153). (4) Cells of lobe apex (×200). (5) Leaves (×18). (6) Leaves (×18). (7) Leaf lobe apex (×153). (8–9) Leaves (×35). (10) Stem cross section (×190). (11) Leaf lobe apex (×200). (12) Median cells (×200). (13) Gemmae (×190). [1–3, *RMS & KD 66-270a*, Ritenbenk, W. Greenland; 4–5,

Dioecious. ♂ Bracts 1–2-androus. ♀ Bracts loosely sheathing perianth, bilobed, to twice as large as leaves, the lobes irregularly ± paucidentate; bracteole united with one of the bracts. Perianths half-emergent at maturity, slenderly ovoid, narrowed toward apex, ca. 5–7-plicate above, the narrowed *mouth ciliate;* cilia uniseriate to 180 μ long, formed of *2–5* (rarely to 10) *superposed cells.* Epidermal cells of capsule wall with nodular thickenings; inner layer with semiannular bands. Spores finely papillose, 14–18 μ.

Type. Sweden, Prov. Dalecarlia, Sörvik *(Buch & Arnell, Aug. 1940).*

Distribution

Previously known from Fennoscandia, where found, according to Mårtensson, as a rarity in the subalpine belt; it occurs in northern Sweden in the Torneträsk region, besides in the type locality, and in Finland (Regio Kuusamoënsis). Also reported from Spitsbergen (Arnell and Mårtensson, 1959) and Britain (Paton & Birks, 1968). A single, unfortunately somewhat questionable, occurrence in Greenland.

E. GREENLAND. Gåseland, Faxe Sø, Scoresby Sund, 70°15′ N., 29° W. *(Holmen 18595, p.p.).* Mixed with *Blepharostoma trichophyllum brevirete, Tritomaria scitula* and ?*Sauteria alpina.*

Ecology

Buch (1944) stressed the restriction to bare, calcareous soil, and Mårtensson (1955-56) also gives it from "loose," "bare," calcareous soil, associated with *Distichium capillaceum, Gymnostomum recurvirostrum,* and *Leiocolea* spp. Arnell (1956, p. 120) states that it occurs on "calcareous soil or soil-covered lime rocks."

Differentiation

This small calcicolous species is likely to be mistaken only for *L. excisa* (which is at least somewhat Ca-tolerant), *L. bicrenata,* and, less probably, *L. alpestris;* the last two taxa usually "avoid" Ca-rich sites.

The dioecious inflorescence, smaller size, ciliate perianth mouth, and distinct trigones, as well as the fewer (3–8 per cell) oil-bodies and the much smaller gemmae containing long-persistent, large oil-bodies, separate *L. perssonii* from *L. excisa.* Although the ciliate perianth mouth and dentate bracts suggest *L. bicrenata* (and the closely allied *L. alboviridis*), *both* of these species differ in the thick-walled leaf cells with poorly demarcated trigones, as well as in the smaller stems, at most 8–9 cells high (and with a

RMS 45961c, NW. Greenland; 6–7, *RMS & KD 66-227,* Jacobshavn, W. Greenland; 8–13, from *Holmen 18595,* E. Greenland, questionable *L. perssonii;* 3, 7, drawn to lower scale; 4, 11–12, drawn to upper scale.]

collenchymatous, persistent medulla with limited mycorrhizal infection). *Lophozia alpestris* has a perianth mouth with only 1–2-celled teeth, ♀ bracts without dentition of the lobes, and usually shows brownish to vinaceous pigmentation.

Buch (1944, p. 387), in separating sterile plants, placed major emphasis on the gemmae, which bear characteristically large, long-persistent oil-bodies; these are also shown by Paton & Birks (1968, fig. 1d).

The sparing and sterile Greenland plants, in so far as they allow a determination at all, strongly suggest *L. perssonii*. The plants are a lax shade form, growing individually admixed with *Blepharostoma*. Shoots are to 850–1150 μ wide with leaves, stems 220–250 μ broad and very fleshy and soft, light green throughout, except brownish with eventual mycorrhizal activity. Cortical cells dorsally are leptodermous, irregularly to regularly oblong, ca. 24–32(35) × 50–100 μ (length:width 2–4:1). The ventral ⅓ to ½ of the medulla is strongly mycorrhizal with age; there are, from the start, smaller cells with very early fungal infection. With age, much of the medulla is ventrally destroyed, as in *L. excisa*. Leaves are contiguous, oblique, scarcely concave to almost flat, broadly quadrate-rotund, ranging from 570 μ wide × 470–500 μ long to 600–700 μ wide × 520–580 μ long; they are 0.2–0.25 bilobed by a crescentic to less often bluntly rectangulate sinus; lobes are broad, ± acute to blunt. Gemmae are frequent on uppermost leaves, reddish with maturity, and (22)23–28 × 24–28(32) μ in diam., with strongly produced, thickened angles; walls are otherwise thin. No oil-bodies were retained in the dry herbarium plant (6 years old). Older, swollen, brownish, opaque gemmae occurred on older leaves, of much larger size; these apparently had started germinating. Leaves are very delicate and pure green, with leptodermous cells. Marginal cells are (22)23–26 to 25–30 μ, varying from leaf to leaf; median cells range from 24–26 × 24–28 μ to 23–25 × 28–34 μ up to 25–30 × (25)28–45 μ.

Cell size *seems* to eliminate *L. excisa* and *L. latifolia*. Also, the plants were much smaller, with flatter leaves, spreading more horizontally. It thus appears, purely through a process of elimination—and because of the occurrence under distinctly calcareous conditions—that *L. perssonii* is at hand. The more angular gemmae and their rather larger size (compare diagnosis, taken from type) represent a considerable deviation. Yet, weighing all criteria and considering the calcareous habitat (and calciphyte associates), only *L. perssonii* seems possible.

TRITOMARIA Schiffn.

Sphenolobus, of Stephani (1902) and other authors (in part).
Lophozia, of authors before 1908 (in part).
Tritomaria Schiffn., Berichte Naturw. Ver. Innsbruck 31:12, 1908; Frye & Clark, Univ. Wash. Publ. Biol. 6(3):416, [1944] 1945; Schuster, Amer. Midl. Nat. 45(1):62, 1951; Schuster, *ibid.* 49(2):376, 1953; K. Müller, Rabenh. Krypt.-Fl. ed. 3, 6:732, 1954.
Sphenolobus subg. *Tritomaria* K. Müll., *ibid.* 6(1):606, 1910; Macvicar, Studs. Hdb. Brit. Hep. ed. 2:214, 1926.
Jungermannia sect. *Exsectae* Jens., Danmarks Mosser, Bryofyter 1:105, 1915 (in part).
Jungermannia subg. *Tritomaria* Arn., Ark. Bot. 19(10):15, 1925 (in part).
Saccobasis Buch, Mem. F. et Fl. Fennica 8:291, 1933; Schuster, Amer. Midl. Nat. 45(1):65, pl. 25, 1951; K. Müller, Rabenh. Krypt.-Fl. ed. 3, 6:715, 1954 (new synonymy).

Prostrate with ascending apices or suberect to erect in growth, sparingly branched (branches of the *Frullania* or *Radula* type or, at least in *T. quinquedentata*, lateral-intercalary); no stolons or flagella; freely rhizoidous, except sometimes toward shoot apices. Stem 11–24 cells high, with a firm, sometimes strongly thick-walled cortex in 1–2(3) strata; cortical cells distinctly elongate, usually narrowly oblong to linear (length:width ca. 2.5–8:1); *medulla dorsiventrally differentiated*, the ventral 0.25–0.45 of smaller cells, eventually brown and mycorrhizal. Distichous-leaved: *ventral merophytes obsolete*, at most 2 cell rows broad, *unable to develop underleaves.*[201] Leaves vertically or subvertically oriented, the *dorsal half of insertion* (decurrence along dorsal midline, if any, neglected) *transverse or inclined toward stem apex; leaves typically (2)3(4)–lobed*, often asymmetrically so (with the third, antical "lobe" shorter than the 2 subequal ventral "lobes"); dentition of leaves absent (exc. near gynoecia of *T. heterophylla*) and leaf bases never ciliate. Cells firm, usually distinctly collenchymatous and with bulging trigones (in *T. exsecta* with walls becoming nearly equally thick); median and basal cells away from marginal sectors often conspicuously elongated (length:width ca. 2–3.5:1); cuticle usually quite papillose to striolate; oil-bodies several per cell, medium-sized, botryoidal. Asexual reproduction by gemmae rare to ubiquitous: *gemmae yellowish brown or golden to reddish brown or red*, elliptical to polygonal, 2-celled at maturity.

Dioecious. Androecia intercalary, the bracts concave or saccate at base, otherwise leaflike; bracts usually 2–4-androus; antheridial stalk 1- or 2-seriate. ♀ Bracts nearly leaflike usually, often rather broader, sometimes (*T. heterophylla*) ciliate-dentate, commonly lobed like leaves; bracteole absent or (at least in *T. scitula*) narrow, oblong, fused for most of its length with one bract. Perianth plicate, the mouth considerably (subg. *Tritomaria*) or hardly (*Saccobasis*) contracted; mouth entire or ciliate-dentate. Capsule wall 3–5-layered.

Lectotype. Tritomaria exsecta (Schmid.) Schiffn.

Tritomaria, as broadly delimited here, includes ca. seven species only, all of which are holarctic and predominantly arctic and/or arctic-alpine; only *T. exsecta* and *T. exsectiformis* occur into the Deciduous Forest. The only exception is *T. camerunensis* (Arnell, 1958), found at 3400 m in equatorial west Africa; this may represent a disjunct phase of *T. exsectiformis*. The species are almost always over substrates rich in humic acids (peaty soil, although rarely *Sphagnum*; decaying logs and stumps;

---

[201]Müller (1951–58, p. 736) states for *T. quinquedentata* "Unterblätter fehlen oder nur am Stengelende, klein." I have never seen underleaves in any *Tritomaria*, away from gynoecia, and even there they may be lacking.

rarely the bark of trees), occasionally over rocks or mineral soil; however, the calciphilous *T. scitula* and *T. heterophylla* may occur on thin, largely mineral soil over rocks.

Following Joergensen (1934), *Tritomaria* is defined here to include *Saccobasis* as a mere subgenus; some of the reasons are given in the discussion after subg. *Tritomaria*, some after subg. *Saccobasis*. In essence, when the subequally trilobed-leaved species of *Tritomaria*, such as *T. scitula* and especially *T. heterophylla*, are studied, a close approach to *Saccobasis* in leaf form is evident. Since weak plants of *Saccobasis* lack, or only show a suggestion of, the peculiar leaf insertion emphasized by Buch (1933a) as the chief generic feature, this sole distinction hardly suffices to warrant a separate genus. Each of the two subgenera is fully described so that the worker with a narrow generic concept is provided with adequate diagnoses. It is noteworthy that both subgenera agree in sporadic development of *Radula*-type branches.

The genus *Tritomaria* bears affinities chiefly to *Lophozia* subg. *Lophozia*, and it is not inconceivable that it evolved from *Lophozia*-like antecedents. More likely the two groups merely share a common ancestral type that had already (*a*) lost underleaves; (*b*) developed dorsiventral differentiation of the medulla, with the ventral mycorrhizal band distinct. This ventral mycorrhizal band may be very conspicuously developed (as in *T. scitula*, Fig. 227:1) or may be only 3–4 cell strata high (as in *T. quinquedentata*, Fig. 232:6). Its presence adequately separates the genus except from *Lophozia* subg. *Lophozia*.

Tritomaria as here restricted is admittedly very closely related to *Lophozia* subg. *Lophozia* but differs from it in that the trilobed character of the leaves is fixed (in *Lophozia* only incidental and usually confined to robust shoots), and the leaves become vertically oriented and more nearly transversely inserted. In many other respects *Tritomaria* closely agrees with *Lophozia* subg. *Lophozia*, although the cuticle of the leaves and stem is almost always distinctly verruculose to striolate in *Tritomaria*, whereas it is quite smooth in subg. *Lophozia*. Also, the tendency for the cortical cells of the stem to become narrow and prominently elongated does not commonly recur in *Lophozia* subg. *Lophozia* (although it may be marked in *Lophozia* subg. *Massula*).

Buch (1933a) was the first to place *Jungermannia* or "*Lophozia*" *quinquedentata* into *Tritomaria*, with which it agrees in basic leaf form, absence of underleaves, stem anatomy, and the transverse nature of the dorsal end of the leaf insertion. The species was at one time placed in subg. *Barbilophozia* (Müller, 1905–16), and Meylan (1939) maintained it might well have been left there. I agreed with Buch in placing it within *Tritomaria* (Schuster, 1951a), although emphasizing the distinctive and isolated position occupied by this taxon by placing it into a separate subgenus. This isolated position has also been pointed out by Müller (1948, 1954, *in* 1951–58). However, I now feel that sectional segregation is more correct; Kitagawa (1966) would accept *Trilophozia* in the subgeneric rank.

Key to Subgenera and Sections of Tritomaria

a. Leaves dorsally transversely inserted, insertion at end not decurrent; leaves with sharp lobes; perianth strongly dentate at mouth; stem with cortical cells (length:width) 2–6:1, never extremely thick-walled. Cells of leaves (exc. in *T. quinquedentata grandiretis*) small to medium-sized, the median 8–22 × 15–35 μ . . . . subg. *Tritomaria*   *b.*

    b. Gemmae always abundant, red-brown; small to medium-sized plants, never with ventral lobe strongly ampliate, with leaves usually 0.75–1.25× as wide as long; antheridial stalk 1-seriate or largely so; capsule wall 3–4-stratose; stem ca. 14–16 cells high . . . . sect. *Tritomaria*

    b. Gemmae rare, yellow-brown to yellowish; vigorous prostrate plants, usually 3–5 mm wide, usually with ventral lobe strongly ampliate (and leaves then very asymmetrical), leaves usually 1.2–1.4× as wide as long; antheridial stalk 2–seriate; capsule wall 5-stratose; stem ca. 20–22 cells high . . . . sect. *Trilophozia*

a. Leaves inserted on a broadly V-shaped line, the ± decurrent apices of the line of insertion excepted, the insertion of dorsal half arched toward stem apex; leaf lobes very blunt to rounded, 2–3–4, very shallow, equal; perianth truncate and entire at mouth; stem with cortical cells ca. (3)4–6(8):1, often very thick-walled. Cells large, median ca. 25–30 × 40–50 μ . . . . subg. *Saccobasis*

Artificial Key to Species of Subgenera Tritomaria and Saccobasis[202]

1. Leaves strongly asymmetrically (2)3-lobed, the ventral margin usually 1.4–2× as long as antical . . . . 2.

    2. Rust-red gemmae always abundantly developed; leaves typically as long as or longer than wide, with ventral margin usually weakly ampliate. Small plants, 1.2–2 mm wide . . . . 3.

        3. Gemmae smooth, elliptical, ca. 8–12 × 11–18 μ; cells of leaves lacking coarse trigones, usually ± thick-walled . . *T. exsecta* (p. 647)

        3. Gemmae angular, ca. (10–13)14–18 × (11)15–26 μ; cells of leaves with coarse trigones, intervening walls thin . . . . *T. exsectiformis* (p. 653)   *a.*

            a. Leaves asymmetrically ovate-lanceolate to ovate, longer than broad; cuticle coarsely papillose; gemmae mostly 14–18 × 17–26 μ. Nonarctic . . . . subsp. *exsectiformis*

            a. Leaves on mature shoots ovate-orbicular to broadly orbicular, mostly wider than long; cuticle feebly verrucose; gemmae mostly 10–16 × 11–24 μ. Arctic . . . . subsp. *arctica*

    2. Gemmae usually lacking (rarely present, then yellow-brown); leaves typically ± fluted, wider than long, the ventral "lobe" usually greatly ampliate. Usually large, (1.5)2–3(4) mm wide . *T. quinquedentata s. lat.*

---

[202] See also separate keys to taxa under the two subgenera (pp. 644, 696).

1. Leaves equally or subequally 3-lobed (isolated leaves 2- or 4-lobed at times); ventral leaf margin usually 1.0–1.3× length of antical . . . . 4.
    4. Leaf lobes (or most of them) sharp; leaf insertion dorsally transverse; basal leaf cells not strongly elongated . . . . 5.
        5. Gemmae absent (or rare; then yellow-brown); opaque, green to yellow-brown plants, lacking vinaceous pigments usually; leaves uniformly 3-lobed . . . . *T. quinquedentata* fo. *gracilis* (p. 691)
        5. Gemmae abundant, light to deep reddish; pure green to whitish green plants, sometimes vinaceous locally or clear brownish or golden; leaves (2)3(4)-lobed . . . . 6.
            6. Leaves usually 0.8–1.15× as wide as long, lobes never sharply cuspidate; trigones small to slightly bulging, the leaves usually green and opaque to subhyaline and light brown, almost never with vinaceous pigments of leaf bases; gemmae deep reddish even in shade forms. ♀ Bracts trifid, lobes entire; perianth mouth with cilia mostly 4–5 cells long . . . . *T. scitula* (p. 664)
            6. Leaves mostly 1.25–1.45× as wide as long, lobes often or usually mucronate or cuspidate; trigones always coarse, often nodose; subhyaline to whitish green, at least leaf bases often (but not always) strongly vinaceous; gemmae light reddish to purplish, occasionally only faintly colored. ♀ Bracts 2–4-fid, spinose-dentate to ciliate; perianth mouth with teeth 0.5–1(2)-celled . . . . . .
        . . . . *T. heterophylla* (p. 671)
    4. Leaf lobes blunt to rounded; leaf insertion dorsally arched toward stem apex; cells of leaf bases strongly elongated . . . .
    . . . . subg. *Saccobasis* 7.
        7. Leaves very variable, (2)3(4)-lobed, normally much wider than long (width:length ca. 1.1–1.4:1); gemmae frequent, yellow-brown or golden, smoothly ellipsoidal; ♀ bracts broad, subreniform, much wider than long . . . . *T. polita* subsp. *polymorpha* (p. 700)
        7. Leaves as long as or longer than wide, all 3-lobed; gemmae very rare, angulate, brown to purplish; ♀ bracts oblong to obcuneate, longer than broad . . . . *T. polita* subsp. *polita* (p. 696)

Subgenus *TRITOMARIA* Schiffn.

Tritomaria (as a genus) of Buch (1933a), Müller (1951–58), Schuster (1951a), etc.[203]
Tritomaria subg. *Tritomaria* Schust., Amer. Midl. Nat. 45(1):65, 1951.

Gametophyte growing rather decumbent (*T. scitula* and *quinquedentata*) or sometimes strongly ascending (*T. exsecta, exsectiformis*) but always with *shoot apices strongly autonomously arched away from substrate.* Sparingly, irregularly branched, the branches predominantly lateral-terminal, sometimes lateral-intercalary. Stem mostly 12–16, occasionally to 20–24, cell layers high, with the *cortical cells distinctly elongate* (2.5–6× as long as wide), usually *only moderately thick-walled*, slightly to distinctly narrower than basal leaf cells; medulla *with distinct dorsiventral differentiation* into a dorsal region of large, elongate, hyaline

---

[203] The synonymy under the genus *Tritomaria* applies, except for *Saccobasis*.

cells, and a ventral region of smaller, shorter, compact cells in large part attacked and destroyed by mycorrhizal fungi, becoming brownish (mature stems, in ventral aspect, therefore always deeply pigmented, even when dorsal side is green). Rhizoids numerous, *in a dense ventral mat.* Leaves *generally trilobed:* evolved from an originally bilobed leaf in which the ventral lobe is considerably *larger than the dorsal,* with ventral leaf margin longer, more strongly arched, as in subg. *Lophozia; leaves more or less asymmetric,* the ventral margin usually longer than the dorsal (little longer in *T. scitula*); leaf insertion basically oblique on ventral ⅔, becoming virtually or quite *transverse on dorsal* ⅓, the leaves usually subtransversely oriented, ± canaliculate (i.e., leaf plane at right angles to axis); with the *dorsally transverse and nondecurrent insertion* the leaves become ± dorsally secund (as in *Anastrophyllum*), *especially in drying.* Leaf cells always ± distinctly collenchymatous, able to develop large, bulging, sometimes confluent trigones, moderate in size (width 8–22 μ), but *often becoming notably elongate* (unlike in *Lophozia*) and 36–48(56) μ long and rectangular; *cuticle able to develop distinct papillae* (in the region of cell elongation delicately rugose-striolate); oil-bodies as in subg. *Lophozia,* moderate in size (4–8 μ), of numerous distinct globules, appearing coarsely granulose or papillose, moderate in number (5–12 per cell, but up to 20 or more per elongate cell). Gemmae abundantly and constantly produced in all species, rare only in *T. quinquedentata.*

Perianth large, distally plicate, with a distinct dorsal fold, *dentate to ciliate at mouth,* ± contracted; bracts 3–4(5)-lobed, lobes entire, usually asymmetric and like leaves; *bracteole absent* or (*T. scitula*) small, oblong, fused with one bract. Androecia with bracts saccate, each usually with 2 antheridia; antheridia with stalk 1-seriate (*T. exsecta*) or largely 1-seriate with occasional cells divided by vertical walls (*T. scitula*) or uniformly 2-seriate (*T. quinquedentata*). Capsule-wall 3-stratose (*T. exsecta, exsectiformis*) to 5-stratose (*T. quinquedentata*), the inner walls with semiannular bands of thickening, the outer wall with them uniformly nodular (i.e., confined to radial walls); seta of numerous rows of cells of unspecialized type.

The subg. *Tritomaria* is basically holarctic in range; it appears to include only the five species here treated, plus *T. camerunensis* S. Arn. (Arnell, 1958), from central Africa, at 3400 m. The group is small and sharply defined, appearing to bear a distinct relationship only with *Tritomaria* subg. *Saccobasis* and *Lophozia* subg. *Lophozia.* At one time (i.a., in Müller, 1905–16, following Stephani, 1902) the species of *Tritomaria* were placed in the form-genus "*Sphenolobus,*" agreeing with this in the essentially transversely inserted, usually canaliculate leaves and the strongly narrowed, underleaf-free ventral merophytes. However, the stem anatomy of subg. *Tritomaria* is very different, agreeing in the dorsiventrally differentiated medulla with that of *Lophozia* subgenus *Lophozia* and *Saccobasis.* Indeed, subg. *Saccobasis* agrees very closely with subg. *Tritomaria,* the preceding diagnosis also holding largely for *Saccobasis.* The chief distinguishing features of *Saccobasis* cited by Meylan (1924) and Buch (1933a) are the leaf insertion and relative length of the cortical stem cells.[204] The first is a valid differential feature and serves to separate *Saccobasis* not only from subg. *Tritomaria* but from other Lophoziaceae as well. As shown under *Lophozia* subg. *Massula*, length of cortical stem cells (and their width relative to the basal leaf cells) is likely to vary very widely among otherwise closely related species. I am therefore not willing to concede this to be a generic character. The other supposed difference cited by Buch (1933a) to separate *Saccobasis* is the subequally trilobed leaves, which, however, occur again in *T. scitula*, and from a *scitula*-like common ancestral form it appears possible to derive subg. *Tritomaria* as well as *Saccobasis*. Such an ancestral form closely approaches *Lophozia* subg. *Lophozia* (as previously stressed) in the following characters: (*a*) tendency for development of trilobed leaves on strong shoots, but bifid leaves on weaker shoots; (*b*) ventral half of leaf obliquely inserted, dorsal less obliquely inserted; (*c*) stem with a ventral mycorrhizal tissue of small cells in the medulla; (*d*) perianth unmodified. The ability of *T. scitula* to still develop a gynoecial bracteole (Fig. 228:4), fused with one bract (exactly as in *Lophozia* subg. *Lophozia*), is probably significant, in view of the general lack of a bracteole in other species of *Tritomaria*.

Key to Species and Subspecies of Subgenus Tritomaria[205]

1. Mature leaves 0.75–1.0(1.15) × as wide as long;[206] always with reddish brown gemmae abundantly developed; cells of leaf middle from 8–10 × 10–16 μ up to 17–22 μ wide × 20–30(36) μ . . . . 2.
2. Mature leaves strongly asymmetrically (2)3-lobed, the antical lobe a mere sharp tooth situated ca. midway between base and apex of leaf; postical leaf margin strongly arched and much longer than the nearly straight or weakly convex antical margin. Oxylophytic . . . . . . 3.
   3. Cells of leaf middle 8–12 μ wide × 10–18 μ long, generally with small trigones (and often somewhat thick-walled, the lumens becoming guttulate); cells of middle of postical margin 8–12 × 10–16 μ; gemmae elliptical, smooth, 8–12 × 11–18 μ; perianth mouth with

---

[204] The cited differences in the form of the cortical cells of the stem almost disappear when a series of individuals of the various species of *Tritomaria* is studied. As was pointed out (in Schuster, 1951a, p. 24), *Tritomaria* has dorsal cortical cells which are more elongate than is customary in *Lophozia* subg. *Lophozia*. They average between 2.5 and 3.5 or even 4.0 × as long as wide; furthermore, their width is distinctly less than that of the basal leaf cells. Study of a further series of specimens reveals that the cortical cells may be even more elongate, in *T. heterophylla* ranging from 19–22 × (60)75–100 μ long (ca. 3–5:1) and in *T. quinquedentata* ranging from a minimum of 15–17 × 35–45 μ (ca. 2–3:1) up to as high as 14–17 × (38)45–72(80) μ (ca. 3.5–5.5:1). It is thus evident that the elongated cortical cells in *Saccobasis* "are but an extreme development of a condition that also characterizes *Tritomaria*" (Schuster, 1951a).

[205] Varieties and forms of *T. quinquedentata* are keyed on p. 687.

[206] The arctic *T. exsectiformis* subsp. *arctica* has mostly broadly orbicular, asymmetrically (2)3-lobed leaves on mature plants; it has conspicuous reddish gemma masses.

rather numerous, crowded cilia, mostly 1(2) cells wide at base . . .
. . . . *T. exsecta* (p. 647)

   3. Cells of leaf middle 17–22 μ wide × 20–36 μ long, usually with
conspicuous trigones, the intervening walls thin; cells of middle of
postical margin 17–20 μ; gemmae angulate, (10–13)14–18 ×
(11)15–26 μ; perianth mouth with more distant cilia, often 2 to
several cells broad at base . . . . . *T. exsectiformis* (p. 653)   4.

      4. Leaves basically ovate-lanceolate to narrowly ovate in outline;
cuticle coarsely papillose; gemmae mostly 14–18 × 17–26 μ.
Temperate and boreal . . . . *T. exsectiformis* subsp. *exsectiformis*

      4. Leaves on mature shoots broadly orbicular to ovate-orbicular,
wider than long; cuticle weakly verrucose; gemmae mostly
10–16 × 11–20 μ. Arctic . *T. exsectiformis* subsp. *arctica*, subsp. n.

  2. Mature leaves subequally (2)3-lobed; antical lobe virtually as well
developed as the others; postical leaf margin hardly more distinctly
arched than antical, subequal to it in length or little longer; cells and
gemmae as in *T. exsectiformis*. Calciphyte . . . . . *T. scitula* (p. 664)

1. Mature leaves 1.15–1.45× as wide as long; gemmae angulate; cells of
leaf middle 21–25(25–32) μ wide × 24–30(30–42) μ long, little elongated;
trigones usually large, bulging; sinuses between lobes obtuse to rectangulate,
subequal, shallow. . . . . 5.

   5. Leaves subequally trilobed, the antical lobe (and margin) almost as long
as the postical; stem postically, and the postical leaf bases, mostly vinaceous pigmented (only the distal portions of the leaves sometimes purplish
to yellowish brown); leaves below gynoecia, and bracts, sharply dentate
to spinose-dentate; cells usually with extremely large, often subconfluent,
trigones, thus pellucid; subapical cells 21–25 × 24–30 μ, virtually equal
in size to median cells; gemmae always freely developed, reddish brown,
the larger 27–36 μ long; small, 0.8–1.5 mm, with facies of *T. scitula*,
often in part with bilobed leaves . . . . *T. heterophylla* (p. 671)

   5. Leaves usually quite asymmetrically trilobed, the antical lobe (and
margin) markedly shorter than the postical; stem (and sometimes leaves)
postically clear brown pigmented, the leaves distally often yellowish
brown in color, rarely vinaceous; leaves below gynoecia, and bracts,
with lobes entire-margined or merely sinuate-dentate; gemmae very
rare, yellowish to pale brown; robust, 2–3.6 mm wide usually, never with
simply bilobed leaves . . . . *T. quinquedentata* (p. 678)

Sectio *TRITOMARIA*

Plants small to medium-sized, mostly 1–1.5(2) mm wide, autonomously
suberect from a prostrate base, leaves vertically or subvertically oriented,
usually loosely conduplicate or canaliculate. Branching (mostly?) terminal.
Rust-red to red-brown gemmae always abundantly present. Antheridial stalk
1- (or, rarely, locally 2-) seriate; capsule wall 3–4-stratose.

Type. *Tritomaria exsecta* (Schmid.) Schiffn. Including all the species of subg. *Tritomaria*, *T.
quinquedentata* excepted. The type species, at least, often has lateral-intercalary branches.

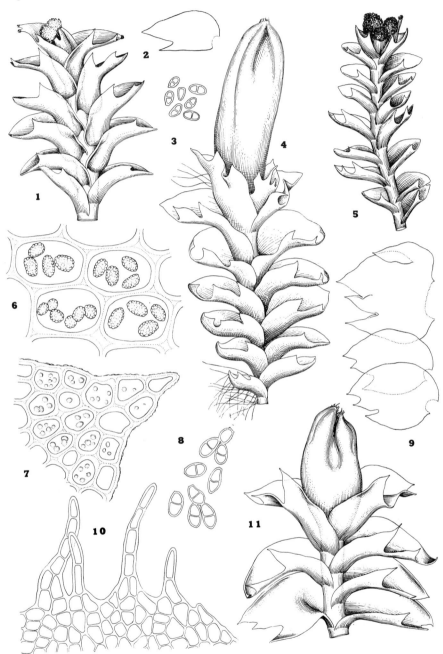

Fig. 223. *Tritomaria (Tritomaria) exsecta*. (1) Apex of gemmiparous ♂ shoot (×22). (2) Leaf (×ca. 20). (3) Gemmae (×220). (4) Perianth-bearing shoot (×24). (5) Sterile, gemmiparous shoot (×ca. 14). (6) Median cells with oil-bodies (×1000).

TRITOMARIA (TRITOMARIA) EXSECTA (*Schmid.*)
Schiffn.
[Fig. 223]

Jungermannia exsecta Schmid., Icones Pl. et Anal. Part., ed. 2:241, 1797 (not of Hooker, Brit. Jungerm., pl. 19, 1816!).

Jungermannia donniana Hüben. [*sic*]!, Hep. Germ., p. 116, 1834 (except synonymy; not *J. doniana* of Hooker, Brit. Jungerm., pl. 39, 1816).

Lophozia exsecta Dumort., Rec. d'Obs., p. 17, 1835.

Scapania exsecta Aust., Hep. Bor.-Amer. Exsic. no. 21, 1873.

Diplophyllum exsectum Thériot & Monguillon, Bull. Soc. Agr. Sci., Le Mans (1899): 199, 1899; Warnst., Krypt.-Fl. Mark Brandenburg 1:160, 1902.

Sphenolobus exsectus Steph., Spec. Hep. 2:170, 1902.

Sphenolobus (subg. *Tritomaria*) *exsectus* K. Müll., Rabenh. Krypt.-Fl. 6(1):606, fig. 294, 1910; Macvicar, Studs. Hdb. Brit. Hep. ed. 1:209, figs. 1–4, 1912 (ed. 2:216, figs. 1–4, 1926).

Tritomaria exsecta Schiffn., Ber. Naturw. Ver. Innsbruck 31:12, 1908; Loeske, Hedwigia 49:13, 1909; K. Müller, Rabenh. Krypt.-Fl. ed. 3, 6(1):740, fig. 242, 1954.

Tritomaria (subg. *Eutritomaria*) *exsecta* Schust., Amer. Midl. Nat. 45(1):65, 1951; Schuster, *ibid.* 49(2):380, 1953.

In dense patches, or occasionally scattered among mosses, pale green or slightly brownish-tinged; shoots 1.2–1.8 mm wide (on ♀ plants to 2.35 mm wide) × 5–18 (rarely to 25) mm long, *erect or ascending from a prostrate base*, at least the shoot apex strongly acroscopically arched, usually nearly unbranched, but often with subfloral innovations. Stem 220–260 μ in diam., the cortical cells 12–15 μ wide, medullary 16–20 μ; rhizoids ca. 10 μ in diam. Leaves approximate to distinctly imbricate, somewhat antically secund and suberect when dry, their distal portions widely spreading when moist, *asymmetrically ovate to broadly oval*, to ca. 950–1000 μ long × 760–810 μ wide (averaging 1.2–1.3 × as long as wide), antical half transversely inserted, the leaves therefore *conspicuously canaliculate-complanate, obliquely (2)3-lobed at apex* (antical lobe a mere sharp tooth, situated between base and apex of leaf; the much larger apical part of leaf usually shallowly bilobed, lobes subcontiguous but the upper commonly slightly shorter than the lower, lowermost lobe thus forming apparent leaf apex); sinuses shallow, crescentic or obtuse; all lobes acute or short-acuminate; antical and postical margins both arched, the antical often only slightly so, *postical margin much longer than antical, strongly dilated and arched. Cells minute, generally with small trigones, the walls becoming conspicuously thickened;* cells of leaf apices 12–16 μ;[207] *marginal cells 8–12 × 10–16 μ, the median cells 8–12(15) μ wide × 10–18(24) μ long*, rounded rectangulate; cuticle distinctly, sometimes strongly, verruculose, near leaf bases striolate; oil-bodies 3–6(8) per cell in leaf apices, to 15–18 per cell in basal cells, in the enlarged and elongated cells below the gemmae-initials up to 24 per cell, filling the entire

---

[207] The distal cells of gemmiparous leaf lobes suddenly 2–3 times as long as "normal" cells.

---

(7) Cells of lobe apex, oil-bodies drawn in except where lacking (×460). (8) Gemmae (×380). (9) ♀ Bract (and united bracteole?), at top and two leaves below (×22). (10) Perianth mouth sector (×300). (11) Perianth-bearing shoot tip (×22). [1, *RMS 19071*, Balsam Gap, N.C.; 2–3, 5, Olean Rock City, N.Y., *RMS;* 4, after K. Müller, 1905–16; 6–7, Dry Falls, N.C., *RMS*, a mod. *pachyderma;* 8–11, *RMS 29399*, North Carolina.]

lumen, spherical to ovoid and 3–5 μ to 3–3.5 × 4.5–5 μ, finely segmented and appearing somewhat papillose. *Gemmae always present* at apices of all or almost all sterile plants, in *rust-red masses* at tips of uppermost leaves, small, *usually 8–12 × 11–18 μ*, 2-celled when mature, smoothly *elliptical in outline*, in isolated cases to 10 × 17 up to 12–13 × 20–22 μ.

Dioecious; ♂ and ♀ plants almost always in separate patches. Androecia becoming intercalary, with bracts numerous, usually in 4–8(10) pairs, similar to leaves, but *strongly ventricose* at base and more markedly canaliculate usually, their antical margins usually sharply erect, at least near base; 1–2-androus; antheridial body ca. 150–180 × 200–240 μ; antheridial stalk 1-seriate, ca. 50–60 × 15 μ. ♀ Bracts not larger than upper leaves, sometimes slightly smaller, often loosely sheathing perianth base, except for lobes which may spread obliquely, rounded-quadrate to broadly, obliquely ovate, to 1225 μ wide × 1100 μ long, 3–5-lobed at summit (often obliquely so), lobes acute to short-acuminate, often with a few obscure supplementary teeth; sinuses descending to 0.25–0.3 the leaf length. Perianth cylindrical to shortly clavate, 4–6-plicate in distal ⅓, contracted to mouth; mouth highly variable, unequally lobulate and spinose-dentate to spinose-ciliate, the *cilia sometimes crowded,* usually formed of 2–4 (but *frequently of 5–6*) superposed cells, each ca. 9–12 × 20–30 μ. Capsule wall 3-stratose, the epidermal cells larger, with nodular (radial) thickenings, the inner layers of smaller cells (the 2 inner layers together somewhat thinner than epidermal cells), with annular or locally semiannular thickenings. Elaters ca. 8 μ in diam., bispiral; spores 9–12 μ, finely papillate. Chromosome no. *n* = 9.

Type. Bavaria, Germany.

Distribution

A holarctic species confined largely to the Taiga (Spruce-Fir Pan-climax), with scattered and isolated stations in the northern portions of the Deciduous Forest; apparently lacking from the Tundra or Tundra-Taiga ecotone, and absent or very rare above treeline in alpine situations.

Widespread throughout much of Europe, from Scandinavia (southern parts of Norway, Sweden, Finland, where rather rare), southward to England (Wales to West Inverness) and Ireland, rare in the north German plain, becoming widespread in the montane portions of central Europe (southern Germany, Austria, Hungary, Switzerland, eastern France, northern Italy), extending southward to the French Pyrenees (Allorge, 1955). Eastward extending into the Baltic Provinces of Russia, to northern Russia, and the Caucasus. Also occurring eastward into Asia (Siberia;[208] Himalaya; China; Formosa; also in the alpine region of Japan and Korea, Hattori, 1952; Hong, 1966). Reported from Madeira.

---

[208] Arnell (1906, p. 150) states that *T. exsecta* occurs in Siberia north to Lebjedevo, 65°5′ N., in the Yenisey R. valley.

Recurring as a disjunct on Mt. Kinabalu, Borneo (Kitagawa, 1967).

In North America probably transcontinental in the Spruce-Fir Forest Region, from southern Alaska to British Columbia, Alberta, and Washington, southward to Colorado. Reported as a disjunct from Mexico (Mirador; Gottsche, 1863). In eastern North America widespread, with its distribution largely overlapping that of *T. exsectiformis*, but showing a considerably expanded extension southward into the Southern Appalachians, and becoming rare and sporadically distributed in much of the northern edge of the range of *T. exsectiformis*. Not attaining as high a latitude as *T. exsectiformis* (rare, e.g., in Newfoundland).

NEWFOUNDLAND. Squires Memorial Pk., Humber R. (*RMS 68-1652a, 68-1660, c. per.*). NOVA SCOTIA. Halfway Brook, Cape Breton I.; Halifax; Folleigh Mt.; MacNab's I.; L. Charlotte, Guys Co. (Brown, 1936a, etc.); Dartmouth Barrens; Ship Harbour L.; Barrasois, Cape Breton I. (*Nichols*). QUEBEC. La Tuque; Carignan; Montmorency R.; Gaspé Coast (Lepage, 1944-45); near St. Jovite, Terrebonne Co. (Crum & Williams, 1960). ONTARIO. Hemlock L., Ottawa and Algonquin Park (*fide* Macoun, 1902; but both of these collections, as in the Yale Herbarium, are *T. exsectiformis, fide* Cain & Fulford, 1948); Algonquin Park; Muskoka; L. Timagami (Williams & Cain, 1959); Gomphidius Bay, L. Timagami (*Cain 3055!*). ANTICOSTI I. Southwest Pt. (Macoun, 1902). NEW BRUNSWICK. Southern Head, Grand Manan.

MAINE. Mt. Katahdin, near Roaring Brook campground (*RMS 32957*); Mt. Desert I. (*Lorenz*); Greenville. NEW HAMPSHIRE. Franconia Mts. (Lorenz, 1908c); Shelburne; Glen Ellis. VERMONT. Willoughby; E. side of Willoughby L., ca. 1500 ft (*RMS 46207, 46213*); Haselton Trail, 3000 ft, Mt. Mansfield (*RMS 43841*). MASSACHUSETTS. Bear R. flume, Conway (*RMS*); Chesterfield Gorge, Hampshire Co. (*RMS 41302a*); Rattlesnake Gutter, Leverett, Franklin Co. (*RMS 43918*). CONNECTICUT. Branford and Naugatuck, New Haven Co.; New Milford, Litchfield Co. (Evans & Nichols, 1908). NEW YORK. Lyon Mt., Adirondack Mts., 2300-2800 ft (*RMS A-255*); Little Moose L., Herkimer Co. (*Haynes!*); Avalanche Gap, Mt. Marcy, Adirondack Mts. (*Wilson 22!;* gemmae to 21 μ long); Olean Rock City, Cattaraugus Co. (Schuster, 1949a); Enfield Glen, Tompkins Co. (Schuster, 1949a; trace only); "Catskill Mts." (Peck, 1866; quite possibly *T. exsectiformis!*); near summit of Cornell Mt., Catskill Mts. (*RMS 24413, p.p.*).

Southward, more or less as a disjunct, as follows: WEST VIRGINIA. Grant, Pocahontas, Preston, Tucker Cos. (Ammons, 1940.) VIRGINIA. White Top Mt., Smyth Co., 5678 ft (Evans, 1893a); above The Cascades, Little Stony Cr., near Mountain L., Giles Co. (*RMS 40213*). KENTUCKY. Powell Co. (*Fulford*). NORTH CAROLINA. Macon Co.: above Glenn Falls, 3500-3600 ft, SW. of Highlands (*RMS 40604a*); Dry Falls, Cullasaja R. (*RMS 25245, 25252*); Satulah Falls, near Highlands (*Anderson 8328!*). Haywood Co.: summit of Richland Balsam, E. of Rich Mt., 6300-6400 ft (*RMS 39668b*); Bubbling Spring, N. of Beech Gap, near Blue Ridge Pkwy., 5100-5200 ft (*RMS 39371*). Yancey Co.: Mt. Mitchell (*RMS 23260, 23146, 23142, p.p., 23687a, 23266, 24602a, 23263, 34824a,* etc.); Mt. Craig, Black Mts. (*RMS 24479a, 24774*); Balsam Cone Mt., Black Mts., 6500 ft (*Anderson 10955!*). Swain Co.: Clingmans Dome (*RMS 28107, 28108, 27117c,* etc.); Andrews Bald, ca. 5800 ft, Smoky Mt. Natl. Park (*RMS 36600a, 36613a*). Ashe Co.: Bluff Mt., 4600-5000 ft, near W. Jefferson (*RMS 30077, 30088, 30074, 29542*); Nigger Mt., near Jefferson, ca. 4200 ft (*RMS 29624, 29620*). Devils Courthouse, Blue Ridge Pkwy., 5600 ft, Transylvania Co. (*RMS 39984b*);

above Linville Caverns, Rte. 221, McDowell Co., 2900 ft (*RMS 29041*); Grandfather Mt., Caldwell Co. (*RMS 30188a*); Linville Gorge, below Falls, Burke Co., and 1.5–2.5 mi below Falls (*RMS 28986, 28887a, 28959a, 29399, c. per.*); near summit, Roan Mt., 6100 ft, Mitchell Co. (*RMS 36967a, 40337a*); narrows of Chattooga R., Jackson Co., 2500 ft (*RMS 39403c*); Balsam Gap, 5260 ft, and Craggy Gardens, Craggy Mts., ca. 5800 ft, Blue Ridge Pkwy. (*RMS 19071, 19148, 19070a, 19061, 24500d, 24767a*). TENNESSEE. Sevier Co.: summit, Alum Cave Trail and near Myrtle Pt., Mt. LeConte (*RMS 24190a, 45326, 45322c, 45332, 45333*); near summit of Clingmans Dome, 6600 ft (*RMS 34717a*). Carter and Sevier Cos. (Sharp, 1939); Roan Mt., 6100 ft, Carter Co. (*RMS 37971a, c. per.*). SOUTH CAROLINA. Ca. 1 mi above confluence of Whitewater R. and Thompson R., near Jocassee, Oconee Co., 1200 ft (*Anderson 12586!, p.p.*). GEORGIA. SW. of Ellicott Rock, on Chattooga R., barely within Rabun Co., 1900–2000 ft (*RMS 39870b, 39839e; 39837d, p.p.*, among *Plagiochila sharpii* and *P. caduciloba*, etc.); ca. 0.2 mi below High Falls of Big Cr., 4–5 mi SSE. of Highlands, N.C., in Rabun Co. (*RMS 40706a;* southernmost reports!).

Westward occurring as follows: MICHIGAN. Pictured Rocks, near Munising; Munising Falls, Alger Co.; Gogebic, Ontonagon, Keweenaw Cos. (Evans & Nichols, 1935). WISCONSIN. Lafayette, Sauk, Adams, Bayfield, Ashland, Douglas, and Grant Cos. (Conklin, 1929); Sand I., Apostle Isls., Bayfield Co. (*RMS 17569*). MINNESOTA. Big Susie I., Susie Isls., Cook Co. (*RMS 14901*, etc.); Ravine near Lutsen, The Point at Grand Marais, Lutsen, Pigeon R., Hungry Jack L., all in Cook Co. (*RMS 15086, 13780, 6514*, etc.; *Conklin 1189, 2271*, etc.; *Holzinger, s.n.*); Manitou R., Great Palisade on L. Superior, and Two Island R., all in Lake Co. (*RMS 13231, 12010, 12013a; Conklin 2618*); Oneota, French R., and Fairmount Park, near Duluth, St. Louis Co. (*Conklin 774, 614, 1609*, etc.); Lamoille, Winona Co. (*Holzinger*). IOWA. Girard, Clayton Co. (*Conard 1927!*); Allamakee Co. (Conard, 1945); Pictured Rock Canyon (*Shimek*). Also cited from Ohio: Hocking Co. (Miller, 1964).

Tritomaria exsecta is an infrequent species in the Spruce-Fir Forests of northeastern North America. In my experience, out of 10 collections made of the *T. exsectiformis* complex, in the north, 8–9 prove to be *T. exsectiformis*. By contrast, *T. exsectiformis* becomes very rare south of New York and Pennsylvania, *T. exsecta* almost wholly replacing it (and becoming very abundant in the Spruce-Fir Zone and Hemlock-Hardwoods Forest, i.e., at 4000 feet and higher, in the Southern Appalachians).

Ecology

In the North and at high elevations in the Southern Appalachians on soil over rocks, or in soil-filled crevices in rocks, less frequently over bare rocks (but even then rarely as a pioneer), often over peat associated with rock exposures, and frequently also on decaying logs. Limited exclusively to acid sites; either a mesophyte or somewhat xerophytic, never occurring where subject to submersion on cliffs. Usually in total or partial shade. In the North the ecology appears to be virtually identical with that of the commoner *T. exsectiformis*, and it occurs with the same group of associates. Consequently, it is impossible to tell the species from *T. exsectiformis* by its "behavior." On logs a typical member of the *Nowellia-Jamesoniella* Associule.

By contrast, in the Southern Appalachians the species acquires a much wider ecological range. It occurs there quite rarely on decaying logs (a frequent habitat northward), but often over damp ledges or cliff faces, associated with *Diplophyllum apiculatum* (rarely, near the mountain summits, also *D. taxifolium*), occasionally in dry sites with *Andreaea rupestris*, while in damper sites *Anastrophyllum minutum*, *Blepharostoma trichophyllum*, *Lepidozia reptans*, *Bazzania tricrenata* and *B. denudata* are associated. At lower elevations (ca. 4000 feet) *Plagiochila caduciloba*, *P. austini*, and *P. sullivantii* may also be consociated. Perhaps even more frequent, and certainly more conspicuous, are the occurrences on bark. *Tritomaria exsecta* becomes a characteristic and abundant secondary species on the bark of *Abies fraseri* and sometimes of *Betula lutea* at elevations of over 5000 feet in Tennessee and North Carolina (and on a few high peaks in Virginia). It follows usually the *Bazzania nudicaulis-Herberta* pioneer community (consisting of these two taxa, as well as *Leptoscyphus cuneifolius*, *Plagiochila tridenticulata*, occasionally *Cephaloziella pearsoni* and *Frullania asagrayana*), undergoing ecesis with *Blepharostoma trichophyllum*, *Lepidozia reptans*, and *Anastrophyllum michauxii*. Most frequently this secondary community is limited to the basal 1–2 feet of the trunks but under favorable conditions may extend to a height of 5–8 feet or more (where it may be intimately admixed with the *Bazzania-Herberta* community). Corticolous occurrences here depend, in large part, on the high incidence of fog and mist, high constant atmospheric humidity, and low temperatures (hence low evaporation rate). The species usually is found only on mature trees and in rather deep shade.

Similar communities, consisting of *Tritomaria, Bazzania nudicaulis, Herberta Anastrophyllum michauxii, A. minutum, Lophozia ventricosa silvicola*, and *Diplophyllum taxifolium*, also occur on humus-covered rocks or often on virtually bare, shaded rocks, at the summits of the highest Southern Appalachian peaks, at 6000–6680 feet. Occasional accessory species, with the above, are *Blepharostoma trichophyllum, Bazzania tricrenata, Scapania nemorosa*, and *Lepidozia reptans*.

One of the southernmost stations known to me is on ledges around Dry Falls, northwest of Highlands, N.C., in the spray and seepage-moist area around the falls, with *Plagiochila austini* and *caduciloba* (species of tropical affinity) and such other northern types as *Blepharostoma trichophyllum, Pellia neesiana, Harpanthus scutatus*, and the lycopod, *Lycopodium selago*, at an unusually low elevation. Plants from this site are atypical in that, though occurring under shaded and very moist conditions, they show excessively thickened cell walls, rendering the cell lumen guttulate (Fig. 223:7). Similar extreme modification of the cell net in other collections is rare. Oil-bodies of such plants are also smaller than those of *T. exsectiformis*, averaging (in the leaf middle) only 3–3.5 μ (when subspherical) to 3×4.5–5 μ, rarely $3.5 \times 5\,\mu$; they occur, except in gemmiparous leaves, mostly 3–6 per cell and (because of the small cell size) appear to be larger than they actually are.

The most interesting collection, phytogeographically, is that from the Whitewater R., S.C., in hemlock forest at only 1200 feet. The plant occurred here with two mosses of tropical affinity, *Schlotheimia lancifolia* and *S. rugifolia*, and with *Metzgeria crassipilis*, *Harpalejeunea ovata*, and traces of *Scapania nemorosa*, on the bark of hemlock. Equally interesting is the collection from the Chattooga R. in Rabun Co., Ga., at 1900–2000 feet, from a shaded boulder, associated with *Plagiochila sharpii*, *P. caduciloba*, *Harpalejeunea ovata*, *Bazzania denudata*, and *Acrobolbus ciliatus*!

Differentiation

The moderate size, the canaliculate, transversely oriented leaves strongly spreading from an oblique or suberect base, and their asymmetric form (with mature, nongemmiparous leaves obliquely tridentate or trilobed), and the constantly abundant, ferruginous gemmae at the shoot apices serve to give this species a distinctive aspect. It is superficially indistinguishable from *T. exsectiformis* but hardly subject to confusion with any other. *Tritomaria exsecta* and *T. exsectiformis* share a rather similar size, usually with the shoots 1–2 cm high at most, often smaller, and possess in common a tendency to form extensive pure, compact patches or mats. This tendency toward mat formation is undoubtedly a consequence of very abundant gemmae formation, resulting in development of extensive colonies of plants, all of about equal age and vigor.

The plants are usually green, often rather pale green (because of the thick cell walls), with the rust-red or ferruginous gemmae forming obvious, contrasting masses at the shoot tips. In sunny sites the plants are often somewhat brownish-tinged, at least above. Since *T. exsectiformis* has an identical coloration pattern, it is quite impossible to separate these two species in the field.

Under the microscope *T. exsecta* differs from *T. exsectiformis* in a whole series of features: (*a*) leaf cells are much smaller, the median only about 8–12 μ wide × 10–24 μ long; (*b*) cell walls tend to become rather prominently thickened, the trigones not or hardly bulging, the lumen therefore rounded (in *T. exsectiformis*, by contrast, the walls are relatively thin and trigones tend to project or bulge into the cell lumen); (*c*) gemmae are small, averaging usually at most 16–18 μ long × 8–12 μ wide, often somewhat smaller, and are ellipsoidal and smooth (rather than angular and ca. 18–24 μ long, as in *T. exsectiformis*). The difference in gemma shape is the most readily observed character and is evidently totally constant and reliable.

Size differences in the gemmae, cited as separating *T. exsecta* and *T. exsecti-formis*, are not always as well marked as is generally supposed. For example, material from Manitou Falls, Minn. (*RMS 13231*), has large gemmae, mostly 10 × 17 to 12 × 20 μ, a few to 13 × 22 μ! Leaf cells are also somewhat larger, averaging 13–14 μ wide along the postical leaf margin, at the midpoint between apex and base of leaf, and 13 μ wide × 18–24 μ long in the leaf middle. Also, cells have trigones locally quite prominent and sometimes bulging, although this feature is variable. Such plants deviate considerably from what Müller (1951–58, p. 734) diagnosed as typical *T. exsecta*: with the gemmae 8 × 18 μ and the median cells 8 × 12 μ! Material of this type may have been the basis for the contention of Meylan (Bull. Herb. Boissier 6:496, 1909) and other authors that the two taxa intergrade. Similar plants occur in Newfoundland (*RMS 68-1652a*). However, the gemma shape appears to me to be wholly stable, even if deviations in size of cells and of gemmae occur.

Müller (1905–16) also speaks of differences in the ♀ bracts and perianths between these two species. When larger series of specimens are studied, these differences tend to disappear. For example, *T. exsecta* is supposed to have "very long, narrow, cilia-like teeth" at the perianth mouth, while *T. exsectiformis* has them only "2–3 cells long." A comparison of Fig. 223:10 and Fig. 224:10 shows that the two species may be identical in the perianth-mouth dentition. Müller also describes the perianth of *T. exsectiformis* as "broadly ovate, not so long as in *S. exsectus.*" Again, a comparison of the figures cited shows no distinction between the two species on this basis. The ♀ bracts are so very variable in both taxa that study of a series soon results in the conviction that no constant differences occur in these organs. In both species variation from 3 to 5 lobes per bract occurs, and the bracts range from almost symmetrical to strongly asymmetrical. In both species, lobes are often acuminate to cuspidate at the tips (a feature which Müller would restrict to *T. exsecta*). There are also no differences between the two species in diam. of spores and elaters. Both are also clearly dioecious, although Müller (1951–58, p. 141) states that "one should expect that *T. exsectiformis* is not dioecious, as previously stated, but monoecious."

Apparently *T. exsecta* is found with perianths (and even capsules) more frequently in Europe than in North America. At least, among the hundreds of field examinations of *T. exsecta* in the Southern Appalachians (where the confusingly similar *T. exsectiformis* does not occur), I have found perianths only twice, although ♂ plants are frequent. By contrast, *T. exsectiformis* rather frequently has perianths and androecia (the two sexes usually in separate patches), while capsules are rather rare, although not nearly as infrequent as Müller (1905–16, 1951–58) suggests.

TRITOMARIA (TRITOMARIA) EXSECTIFORMIS (Breidl.) Schiffn.

[Figs. 224–226]

Jungermannia exsecta of Hooker, Brit. Jungerm., pl. 19, 1816, and (in part) of numerous other authors before 1893 (not of Schmidel, Icones Pl. et Anal. Part. ed. 2:241, 1797).
Jungermannia exsectaeformis Breidl., Mitt. Naturw. Ver. Steiermark 30:321, 1894.

Diplophyllum exsectiforme Warnst., Krypt.-Fl. Mark Brandenburg 1:161, 1902.
Sphenolobus exsectaeformis Steph., Spec. Hep. 2:170, 1902.
Sphenolobus (subg. *Tritomaria*) *exsectiformis* K. Müll., Rabenh. Krypt.-Fl. 6(1):609, fig. 295, 1910;
    Macvicar, Studs. Hdb. Brit. Hep. ed. 1:211, figs. 1–4, 1912 (ed. 2:217, figs. 1–4, 1926).
Lophozia exsectiformis Boulay, Musc. France 2:92, 1904.
Tritomaria exsectiformis Schiffn., Ber. Naturw. Ver. Innsbruck 31:12, 1908; Loeske, Hedwigia
    49:13, 1909–1910; Buch, Mem. Soc. F. et Fl. Fennica 8(1932):285, fig. 1:35, 1933; K.
    Müller, Rabenh. Krypt.-Fl. ed. 3, 6(1):738, fig. 241, 1954.
Tritomaria (subg. *Eutritomaria*) *exsectiformis* Schust., Amer. Midl. Nat. 45(1):65, pl. 23, 1951;
    Schuster, *ibid.* 49(2):378, pl. 26, 1953.

Superficially identical to T. exsecta, in patches or growing scattered, green to
brownish (♂ bracts often conspicuously vinaceous), forming patches 5–25
(rarely to 30) mm high. Shoots to (1.2)1.6–2.0 mm wide, simple or with 1–2
branches from the prostrate base, *ascending or* (with crowding) *almost erect*. Stems
ca. (200)240–275 μ in diam., green above, brown beneath (from mycorrhizal
infection); cortical cells ca. 20 μ in diam., medullary cells ca. 30 μ dorsally.
Rhizoids ca. 20 μ in diam. Leaves contiguous to moderately imbricate,
transversely oriented, strongly concave-canaliculate to loosely canaliculate-complicate, erect-
spreading to spreading (55–85° with stem) from a suberect and somewhat
sheathing base, asymmetrically and obliquely ovate or ovate-lanceolate,
obliquely tridentate at summit of mature leaves, antical margin nearly straight, *no
more than ⅔ the length of the strongly arched and dilated postical margin;* antical lobe
small, toothlike, situated slightly above middle of leaf, the larger postical "half"
of leaf usually bilobed or bidentate (often blunt and entire on gemmiparous
leaves), the apical teeth small and approximated, the *leaf thus unequally trilobate*
(but on gemmiparous leaves the 2 postical lobes or teeth not or hardly developed;
leaf then strongly asymmetrically bilobed). Cells *much larger than in T. exsecta*,
usually with *thin walls but prominent, bulging trigones*, the marginal *17–20 μ* (middle
of postical margin), *the median 17–22 μ wide × 20–36(40) μ long;* cuticle
usually strongly verruculose; oil-bodies of median cells *mostly (6–9)12–15 per
cell*, in the lobes *mostly (5)7–13 per cell*, spherical to ovoid, 3.0–4.5 μ to 3.6 ×
4.2–5.4 μ, formed of distinct, somewhat protruding segments, thus appearing
papillose. *Gemmae abundantly developed in rust-red masses* from uppermost lobes of
sterile and ♂ shoots, 2-celled usually, *irregularly polygonal to pyriform, (13)14–
18(20) μ wide × (16)17–26 μ long.*

Dioecious; *usually sterile.*[209] Androecia as in *T. exsecta.* ♂ Bracts often vinaceous, broadly oblong-ovate, strongly saccate at base, 2–3-lobed like leaves, ca.
(750)800–1050 μ wide × (800)1000–1100 μ long, the antical lobe or tooth
prominent, situated often near apex of leaf (the bracts thus subequally 2–3-
lobed); antical margin narrowly reflexed or erect, at least near base. Antheridia 1–2 per bract, their stalks uniseriate. Gynoecia much as in *T. exsecta*, the
bracts subequal to leaves or slightly smaller than upper leaves, erect-spreading
to suberect, sheathing perianth at base, ovate or subovate, often broadly so,
ca. 1190 μ long × 935 μ wide to 1050–1400 μ long × 1020–1300 μ wide,
distally obliquely and often asymmetrically 3–5-lobed for ca. 0.2–0.25 their

[209] Müller (1951–58, p. 141) supposed that *T. exsectiformis* is paroecious, regarding it evidently
as a polyploid derivative of *T. exsecta*. This is incorrect; I have seen both perianth-bearing
and androecial plants and reported (Schuster, 1951a, 1953) capsules for the first time.

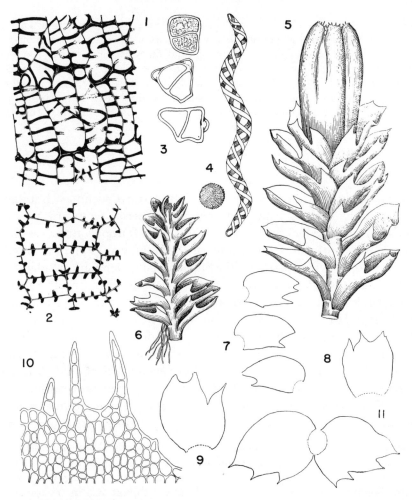

FIG. 224. *Tritomaria* (*Tritomaria*) *exsectiformis*. (1) Cells of innermost layer of capsule wall (×290). (2) Epidermal cells, capsule wall (×290). (3) Gemmae (×600). (4) Spore and elater (×550). (5) Perianth-bearing shoot tip (×ca. 18). (6) Sterile shoot tip with gemmae (×ca. 16). (7) Leaves (×ca. 18). (8–9) ♂ Bracts (×18). (10) Perianth mouth sector (×175). (11) Perichaetial bracts (×18). [1–5, 8–11, from Big Susie I., Minn., *RMS;* 6, Delaware Water Gap, Pa., *RMS;* 7, Olean Rock City, N.Y., *RMS & Rader*; from Schuster, 1953.]

length, the lobes sharp, often merely toothlike. Perianth cylindrical to cylindrical-clavate, strongly 4–5-plicate above and contracted to mouth, the mouth hardly half the maximal width of perianth; perianth with a median longitudinal sulcus and marked lateral sulci, separated by strong dorsolateral folds; mouth lobulate and provided with *rather remote cilia and teeth*, from 2–3 cells wide at base × 2–5(6) cells long, the teeth sometimes reduced and rather blunt, at other times sharp and ciliiform, the cells below cilia, at mouth,

15–20 μ wide × 18–28 μ long. Capsule wall 3(4)-stratose; epidermal cells short-rectangulate to polygonal, ca. 1.5–2× as long as wide, with nodular thickenings extended slightly onto tangential faces as short spurs (1–3 thickenings per longitudinal wall; 0–2 per transverse wall). Inner cell layer of narrowly rectangulate, often somewhat sigmoid cells, ca. 22–26 μ wide × 32–38 μ long, bearing semiannular bands, locally weak or incomplete at middle of tangential face. Elaters ca. 7–9.5 μ, closely bispiral, the spiral 2.2–3 μ wide; spores distinctly papillate, ca. 10.5–12.6 μ.

Type. European.

Distribution

Imperfectly holarctic, largely in the Spruce-Fir Panclimax, but known to date only from Europe eastward into Siberia, westward to North America, where evidently transcontinental. In Europe found over almost the entire continent from southern Scandinavia (Finland, Sweden, Norway north to Finnmarken), the Shetland Isls., Denmark, England, and Ireland, southward to central Europe (Germany, Switzerland, Austria), southern France, and the French Pyrenees (Allorge, 1955); eastward to Poland, Hungary, Moravia, the Tatra Mts. of Czechoslovakia and the Carpathians, from there to the Caspian region (Astrakhan) and into the Caucasus Mts. (see Müller, 1951–58), eastward across Siberia.[210] In the Alps up to 2300 m, but generally not above 2000 m. Reported (Arnell and Mårtensson, 1959) from Spitsbergen, but may represent subsp. *arctica*. Typical *T. exsectiformis* absent from Arctic; not known from Greenland, Iceland, or Faroes. In Greenland replaced by subsp. *arctica* (see p. 661).

It seems likely that *T. camerunensis* (Arnell, 1958), from equatorial west Africa, at 3400 m. on Mt. Cameroon, is only a disjunct phase of *T. exsectiformis*.

In North America apparently sporadically distributed in the West, known from Bonanza and Hunker Creeks, Central Yukon R., in Yukon-Alaska (Persson, 1946) to British Columbia and Alberta to Montana, Idaho, and Colorado (i.e., restricted virtually to the Rocky Mts.), but widespread in the cooler portions of eastern North America and regionally an abundant species. Apparently lacking in the arctic Tundra (although rarely in alpine Tundra, i.a., Mt. Washington and Mt. Katahdin in New England), and quite infrequent south of the Taiga, although reported from the Hemlock-Hardwoods Forest Region of New York (Schuster, 1949a), and southward also in the Oak-Hickory Region (Pennsylvania; North Carolina), where very rare and of disjunct distribution, occurring at 2000–4000 feet.

NEWFOUNDLAND. Pushthrough, Ramea, and Grand Bruit, on S. coast; St. John I. and Eddie's Cove West, on N. Pen.; Hampden, C. Nfld. (Buch & Tuomikoski, 1955); Squires Memorial Park, on Humber R., C. Nfld. (*RMS 68-1655b, p.p.*). ONTARIO.

[210] Arnell (1906, p. 150) states that *T. exsectiformis* occurs northward in Russia to Archangelsk and in Siberia to Nikulina (60°20′ N.) and Fatjanova (64°5′ N.), in the Yenisey R. valley.

Gorge Cr., Thunder Bay Distr. (*Cain 2013!*); High Rock I. and Long Pt., L. Timagami (*Cain 2489!, 3021!*); Durham; Halton; Manitoulin I.; Petersborough; Nipissing; Algonquin Park; Carlton; Hastings; Sudbury (Cain & Fulford, 1948; Williams & Cain, 1959); E. of Janetville, Pigeon R. Flats (*Williams 935!*); W. of Millbrook, Cavan Twp. (*Williams 875!*). QUEBEC. Tadoussac; Bic; Baie-des-Sables, côté de Matane; R. Petite-Cascapédia, côté de Bonaventure (Lepage 1944–45); between Ft. George and Goose Bay, 53°54′ N. (Lepage, 1953); Terrebonne, Montcalm Co., near St. Jovite (Crum & Williams, 1960). NOVA SCOTIA. Ship Harbour L., Dartmouth Barrens, Guys Co. (Brown, 1936a); Mary Ann Falls, S. of Neils Harbour, Cape Breton (*RMS 42692, 42681*).

MAINE. Chimney Pond, Mt. Katahdin (*RMS 32901c, p.p.*); Mt. Desert I. (*Lorenz*). NEW HAMPSHIRE. Echo L., Franconia Mts. (Lorenz, 1908c); Randolph; Laconia; Mt. Washington (*RMS, s.n.*). VERMONT. Jericho. MASSACHUSETTS. NE. side of Mt. Toby, Sunderland (*RMS 46246*). CONNECTICUT. Lantern Hill, North Stonington. NEW YORK. Jamesville, Onondaga Co. (Underwood & Cook, Hep. Amer. no. 77, as *Jungermannia exsecta*); Olean Rock City, Cattaraugus Co.; Lick Brook, S. of Ithaca, Tompkins Co.; Briggs Gully, Honeoye L., Ontario Co. (all Schuster, 1949a); near Haskinville, Steuben Co. (*RMS 13999*); Little Moose L., Herkimer Co. (*Haynes*); Slide and Cornell Mts., Catskill Mts. (*RMS 18401, 18400, 17579*); Peasleeville, Adirondack Mts. (*RMS A-256, A-258*). PENNSYLVANIA. Delaware Water Gap (*Schuster, 1943*). OHIO. Geauga Co. (Miller, 1964).

NORTH CAROLINA. Nigger Mt., near Jefferson, ca. 4200 ft, Ashe Co. (*RMS 29616*; southernmost report of the species!).[211]

Westward occurring as follows: MICHIGAN. Amygdaloid I., Isle Royale (*RMS 17329*); shore of L. Lilly, near Copper Harbor, Keweenaw Co. (*RMS 39162b*, etc.); Houghton and Keweenaw Cos. (Steere, 1937); Pictured Rocks, Alger Co. (*RMS, s.n.*); Tobin Harbor and Ryan I., Isle Royale; Douglas L., Cheboygan Co.; Ontonagan, Marquette, Gogebic, Luce, and Chippewa Cos. (Evans & Nichols, 1935, etc.); Rapid R., 6–7 mi NNE. of Kalkaska, Kalkaska Co. (*RMS 38968*). WISCONSIN. Sand I., Apostle Isls., Bayfield Co. (*RMS, s.n.*); Vilas, Bayfield, Ashland, Douglas Cos. (Conklin, 1929); Rocky Arbor Roadside Park, NW. of Wisconsin Dells, Juneau Co. (*RMS 39215*). MINNESOTA. Carlton, Carlton Co. (*Conklin 601a*); Curtain Falls, Taylors Falls, Chisago Co. (*RMS 14600*, etc.). Cook Co.: The Point, Grand Marais; Cascade R.; Hat Pt. at Grand Portage; L. Superior, 1 mi S. of Temperance R.; Pigeon R., W. of Sextus City; Lutsen; Hungry Jack L.; Moss L.; Duncan L.; Arrowhead (Brule) R., etc. (*RMS 13793, 17370, 11412, 13209, 13240, 13406*, etc.; *Conklin 3114, 3047, 2568, 2305, 2306*, etc.); Lucille, Sailboat, Porcupine, Belle Rose, Big Susie, Little Susie, and Long Isls., all in Susie Isls., NE. of Grand Portage, Cook Co. (*RMS 12101, 13650, 11717, 11759, 13600, 12000, 12114, 12243, 13509, 7235a, 11003*, etc.); Rainy L. near Black Bay, Koochiching Co. (*RMS 18136, 18123*); Encampment R., Knife R., Two Island R., all in Lake Co. (*RMS 13454; Conklin 619, 2605*); Oak I., Lake of the Woods Co. (*RMS 17267*, etc.); Oneota, Nopeming, and French R., near Duluth, St. Louis Co. (*Conklin 784, 636, 1344*, etc.). (Of these, *RMS 11003, 12101, 12114, 13240, 17267, 17370* bear perianths and, in part, androecia; *13454* and *11717* bear capsules). IOWA. Winneshiek and Clayton Cos., in NE. Iowa (Conard, 1945).

The species is much more frequent, indeed in many places very abundant, in the Great Lakes region, as compared with *T. exsecta*. Although usually found in the Taiga, the southern reports (Iowa, Pennsylvania, western and central

---

[211] Reports in Blomquist (1936), based on Andrews (1921), of this species, from North Carolina have been found to represent *Anastrophyllum minutum*. Dr. Andrews (*in litt.*) confirmed my disposition of this material.

New York, North Carolina) are all from deciduous forests (Hemlock-Hard-woods, Oak-Hickory, and Basswood-Maple Climaxes). Nevertheless, the species is fundamentally of more northerly overall distribution than *T. exsecta*. There is a striking ecological reversal in the Southern Appalachians, where *T. exsecta* occurs mostly in the Spruce-Fir Forest.

Ecology

Found under a rather wide variety of conditions:

on decaying logs, on thin peaty soil over cliffs (especially on their crests and in their crevices), and occasionally on soil among mosses on the vertical sides of cliffs . . . more rarely on sandstone cliffs and boulders. On decaying logs associated most often with *Jamesoniella autumnalis, Nowellia, (Blepharostoma), Lophozia ascendens, (L. incisa), L. longidens, Scapania apiculata*, occasionally with *Lophozia porphyroleuca*, etc. The species is [usually] not a pioneer on logs, but may persist until they have virtually disintegrated to humus; the pH varies from 3.9 to 5.2 under such conditions. Though very often associated with cliffs and ledges, the species usually occurs over at least a thin layer of organic matter . . . (Schuster, 1953).

It is infrequently found on weakly calcareous rocks, such as sandstones, and appears to lack toleration for a pH above 6.5, occurring over basic rocks only after deposition of a humus or peat layer, then very common, associated with *Lophozia incisa, L. ventricosa, Blepharostoma trichophyllum, Lepidozia reptans*, occasionally *Scapania mucronata*, etc. Under such conditions the species often forms "pure mats, the erect, caespitose plants growing closely together" (Schuster, 1953). Essentially a humicolous and chasmophytic species, less frequently saxicolous and corticolous, with a rather high toleration for direct sunlight and a moderate toleration for intermittent moisture conditions and a high saturation deficit.

Montane near the southern limit of its range (Pennsylvania, North Carolina), associated with *Lophozia barbata*, and (in the latter state) also with *Scapania mucronata, Bazzania trilobata*, and *Lophozia excisa*, forming a turf in exposed, soil-filled, rock crevices in which *Polypodium virginianum* undergoes ecesis. The plant here occurs as a distinct xerophyte.

Differentiation

Inseparable from *T. exsecta*, except under the microscope, but scarcely to be confused with other species. On gemmiparous shoots the two ventral lobes are frequently united or not developed (a condition which also often characterizes many leaves on weak shoots), and then the trilobate leaf shape is not realized. This should lead to no serious problems, since a high percentage of leaves, even under such conditions, show the strongly asymmetrically trilobed form.

Although *T. exsectiformis* is sometimes stated to be more robust than *T. exsecta*, this is hardly frequent enough to warrant any generalization. The species has the transversely inserted and oriented, canaliculate, elongate-appearing, asymmetrical leaves characteristic of *T. exsecta* and agrees in facies with this species, as well as in the characteristic suberect or ascending mode of growth. Apparently it occurs more frequently with perianths and androecia than *T. exsecta*, at least in North America; very rarely with capsules. Both species reproduce almost exclusively by the ever-present, characteristic rust-red gemmae.

Under the microscope *T. exsectiformis* differs from *T. exsecta* chiefly in the (*a*) considerably larger cell size; (*b*) larger, angulate and polygonal gemmae;[212] (*c*) tendency for the cells to develop strongly bulging trigones, at times confluent, rather than nearly equally thickened walls. Also, *T. exsectiformis* appears to have, in many cases, a less laciniate-fimbriate perianth mouth, with more remote cilia. However, the few collections with perianths hardly justify generalization, and no sharp distinction between the two species exists in this respect. However, *T. exsectiformis* often bears perianths whose laciniate-lobulate mouth is hardly produced as cilia (Fig. 225:13). Such perianths, with virtual suppression of the cilia, may occur with plants having "normally" ciliate perianth orifices.

Oil-bodies of this species, associated with the larger cells, appear to average greater in number per cell than in *T. exsecta*. This distinction is so subject to variation as to make it of dubious practical value in separating the two taxa. Extreme variability also characterizes the ♂ bracts. On robust, dense-leaved plants these are broadly, somewhat asymmetrically ovate to ovate-triangular, averaging nearly or quite as broad as long; they are then widest barely above the base, strongly ventricose at the base, with a distinctly, if narrowly, reflexed antical margin. However, simple oblong-ovate to oblong ♂ bracts also occur which are clearly longer than broad and subequally trilobed near the apex (compare Figs. 224:8–9 and 225:4, 8).

Variation

Normal *T. exsectiformis*, as previously treated (subsp. *exsectiformis*), is characterized basically by its asymmetrically ovate to ovate-lanceolate leaves that average conspicuously longer than broad (Fig. 224:5–7); this subspecies, except perhaps in the West, hardly penetrates north of the Taiga. It is replaced, under strictly arctic conditions between latitudes 69° and 78° N., by a broad-leaved subspecies (for separation of the two taxa see the key, p. 641).

---

[212] Gemma shape appears more constant than gemma size in separating *T. exsectiformis*. In many collections gemmae average only from 13 × 16 to 14 × 17 up to a maximum of 13 × 22–23 μ (i.a., in *RMS 13406*). They thus are hardly larger than in *T. exsecta*, yet show the obviously polygonal to tetrahedral to pyriform outline of normal *T. exsectiformis* gemmae.

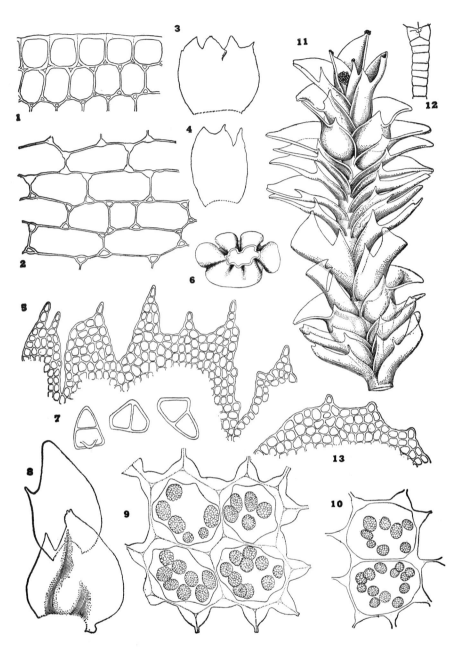

Fig. 225. *Tritomaria (Tritomaria) exsectiformis*. (1) Cells along middle of ventral leaf margin (×424). (2) Median leaf cells (×424). (3) ♀ Bract (×20). (4) ♂ Bract (×20). (5, 13) Sectors of perianth mouth (×150). (6) Cross section of perianth just below apex (×28). (7) Gemmae (×650). (8) Leaf, above, and ♂ bract, below (×32). (9) Cells above leaf middle (×780). (10) Cells above leaf middle (×935). (11) ♂ Plant with two intercalary androecia and gemmae (×24). (12) Antheridial stalk (×265). [1–6, 13, *RMS 11717*, Big Susie I., Minn.; 7–8, 11–12, *RMS 13406*, Minnesota; 9, Belle Rose I., Minn., from a mod. *pachyderma*, *RMS;* 10, *RMS 13999*, Steuben Co., N.Y.]

TRITOMARIA EXSECTIFORMIS subsp. *ARCTICA*
Schust., subsp. n.
[Fig. 226]

Folia late, asymmetrice ovato-orbicularia, latiora quam longa, margine ventrali saepe semi-circulatim arcuato; trigonae magnitudine parvae moderataeve; cuticula leniter verrucosa; gemmae parvae, plerumque 10–14 × 11.5–15 μ.

Erect, only occasionally furcate, usually scattered, rather fleshy, pure green or ± brownish-tinged, with contrasting, carmine-red masses of gemmae. Shoots 8–12 mm high, 850–1350(1650) μ wide with leaves. Stems 210–275(285–330) μ in diam., somewhat fleshy, flexuous to rigid, green to, with age, whitish; cortical cells occasionally weakly brownish, rarely weakly and locally reddish brown beneath; dorsal cortical cells rather regularly rectangulate, slightly thick-walled to almost leptodermous, 19–22 × (25)40–80 μ (length-width ratio 2–4:1), varying to 18–23 × 48–80 μ to 20–26 × 45–65 μ. Leaves transverse (antical half sometimes barely decurrent and not strictly transversely inserted), merely concave to concave-canaliculate, suberect to loosely sheathing basally, portions above base obliquely (rarely widely) spreading to suberect and even somewhat loosely convolute about the stem, ± antically secund, variable, on weak stems asymmetrically ovate and then often somewhat longer than wide (620–705 μ wide × 705–775 μ long); leaves on mature stems *broadly and asymmetrically ovate-orbicular and clearly wider than long* (725–815 μ wide × 660–770 μ long, to 835 μ wide × 680 μ long, to 880 μ wide × 835 μ long, to 1100 μ wide × 1000 μ long), asymmetrically trifid (the 2 lower or distal lobes often united, or separated by a slight notch only, the leaf then strongly asymmetrically bilobed). Dorsal lobe usually a mere spinous tooth varying to a distinct spinous lobe, situated ca. 0.6–0.7 from base to apex of leaf; *ventral margin strongly, often almost semicircularly, arcuate,* antical margin nearly straight to moderately (or, on broad leaves, strongly) arcuate. Cells variable in size: on narrower-than-long leaves, median cells (12–13)14–16 × (18)20–25 μ, basal cells 15.5–18 × (20)22–26(30) μ, cells of middle of postical margin (14)15–17(18) μ; on unusually broad leaves, median cells more nearly isodiametric, (16)17–18.5(23) × (18)20–26 μ, basal cells (18)20–24 × 22–26 μ, cells of postical margin 13–15 to 14–16 μ, sometimes 16–21 μ. Trigones *small or moderate in size* to rather large, hardly to moderately bulging; intervening walls thin; *cuticle weakly verrucose.* Median cells usually with (6–7)8–14(16) coarsely papillose-botryoidal oil-bodies, with distinctly projecting spherules, spherical (and 2.5–4 μ), ovoid or ellipsoidal (and 2.4 × 4.5 to 3.8 × 6.5 μ). Gemmae bright reddish or claret-red with usually a tinge of reddish brown, mostly 2-celled, rounded to angularly tetrahedral to polygonal, small; *from 9.5 × 11 μ to 10–11 × 11.8–13.5 μ to 13.5–14 × 13.5–15 μ, a few to 13.5 × 18.5 μ and 15–16 × 20 μ.* Otherwise unknown.

Type. Siorapaluk, northwest Greenland (*RMS 45682*).

Distribution

NW. GREENLAND. Siorapaluk, on moist, peaty, enriched (from leachings of dovekie roosts above) slope above village, 150–250 ft (*RMS 45679, 45674, 45682, 45705*).

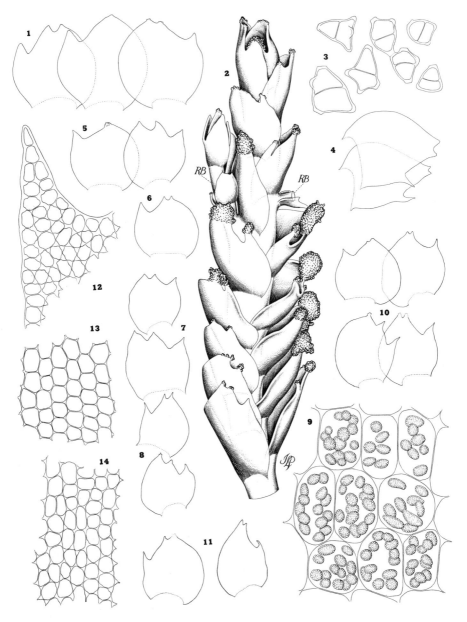

Fig. 226. *Tritomaria (Tritomaria) exsectiformis* subsp. *arctica*. (1) Leaves (×27). (2) Strongly gemmiparous shoot, with two *Radula*-type infra-axillary branches, at RB (×28). (3) Gemmae (×800). (4) Smaller, relatively narrow leaves (×28). (5–6) Leaves (×18.5). (7–8) Leaves (×18.5). (9) Median cells with oil-bodies (×850). (10–11) Leaves (×18.5). (12) Antical lobe of leaf (×208). (13) Lower median cells (×208). (14) Upper median cells (×208). [1, *RMS 45679*, Siorapaluk, NW. Greenland; 2, 5–6, *RMS 45682*, same loc.; 3–4, 9, *RMS 45674*, same loc.; 7–8, 12–14, *RMS & KD 66-226*, Jacobshavn, W. Greenland.]

W. GREENLAND. Tupilak I., near Egedesminde, 68°42′ N. (*RMS & KD 66-017*); SE. of Jacobshavn, Disko Bay, 69°12′ N., 51°05′ W. (*RMS & KD 66-226, p.p.*, with *Scapania microphylla*).

Ecology

Tritomaria exsectiformis arctica occurs as isolated plants, less often in small, pure patches, at the type locality amidst *Lophozia opacifolia*, *L. groenlandica*, *Cephaloziella byssacea*, *Anastrophyllum minutum grandis*, *A. sphenoloboides*, *Tritomaria quinquedentata*. At Jacobshavn the plant was on wet peat around a small tarn, together with *Cephalozia bicuspidata* and *Scapania microphylla*.

The plants stand out conspicuously from associated species because of the bright red gemmae masses and usually occur, not in patches, but as scattered individuals; hence herbarium specimens are sparing.

Differentiation

Tritomaria exsectiformis arctica belongs to the *T. exsecta-exsectiformis* complex and in some ways is midway between these species. Leaf cells are typically rather small, almost as in *T. exsecta*, and gemma size (omitting rare individual exceptions) is almost exactly as in that species [8–12 × 11–18 μ in *T. exsecta*; 9.5–13.5(15–16) × 11–18.5(20) μ in *T. exsectiformis arctica*]; in these two respects it differs clearly from subsp. *exsectiformis* although its gemma size overlaps that of the latter. However, gemmae are polygonal or at least angulate, as in subsp. *exsectiformis*; in this, however, gemmae are larger, mostly ranging from 14–18 × 17–24 μ. *Tritomaria exsectiformis arctica* differs from both *T. exsecta* and *T. exsectiformis* subsp. *exsectiformis* in that the leaves of normal and mature plants clearly average wider than long, ranging from 1.05 to 1.25 × as wide as long, with a broadly orbicular outline.

Cell dimensions in *T. exsectiformis arctica* show considerable variation, depending on leaf width. On narrow leaves, which are comparable to those of *T. exsecta* in shape, cell width ranges as in that species, i.e., (12–13)14–16 μ; on the unusually broad leaves, cells average 17–18.5 μ wide. Evidently leaf width, in cell rows, remains approximately fixed, and cell width is controlled by the leaf dimensions. At times the cells are considerably larger on individual leaves and/or individual plants. In *RMS 45705*, e.g., some leaves (which average longer than wide, ca. 670 μ wide × 790 μ long) bear cells that are considerably larger and possess strong, almost bulging trigones. Cells of the middle of the postical margin here range from 17 to 20 μ and median cells from 18–19 μ wide × 24–28 μ long up to 20–24 or even 24–26 μ wide × 24–26 μ long. In cell size such plants clearly overlap and almost match well-developed *T. exsectiformis* subsp. *exsectiformis*, yet they have the characteristically small gemmae of subsp. *arctica*. In the type collection many leaves show narrow strips of tissue, usually longitudinally arranged, which are bistratose. Often such bistratose strips are only 2–3 cells wide. Similar locally bistratose leaves also occur in associated plants of *Lophozia groenlandica*. The stimulus producing such polystratification is uninvestigated.

Gemma form and size are particularly variable in *T. exsectiformis arctica*. In the abundant type collection, although the gemmae are usually no more than 13 × 15 μ, isolated gemmae range up to 18 × 26 μ, typical of subsp. *exsectiformis*. Yet in this collection plants can be found with gemmae averaging no larger than in *T. exsecta*, and with the angles hardly developed, the gemmae ovoid or ellipsoid with only slight angular protrusions. Such plants might readily be mistaken for *T. exsecta*, yet they occur mingled with others with sharply angular gemmae. In contrast, in *RMS 45674*, no gemmae ranged above 13–14 × 14–15(18) μ and all were sharply angular. Fortunately, gemma form and size, which so sharply separate the temperate zone *T. exsecta* and *T. exsectiformis*, are not "needed" to differentiate *T. exsectiformis arctica*, which stands well defined on the basis of both the broadly orbicular leaves of mature, normal plants and the more nearly leptodermous cells with numerous and rather small oil-bodies.

Pigmentation patterns in this plant agree perfectly with those of the *T. exsecta-exsectiformis* complex. In the relatively rarely developed mod. *colorata* a small sector of the leaf bases is rather deeply brownish-pigmented (often vinaceous in the Jacobshavn and Tupilak I. specimens), much as in "normal" *T. quinquedentata*, and upper portions of the leaf are slightly brownish (in transmitted light cell walls and trigones are yellowish).

The Tupilak I. specimen is typical: leaves are mostly strongly cupped, rather than canaliculate, with both antical and postical margins markedly convex; trilobed leaves are much broader than long; marginal cells are 15–18 μ; median cells are 17–18 × 22–26 μ; the rust-red gemmae are angulate and up to 16–17 × 21 to 21 × 24–26 μ. Oil-bodies are (7)8–14 per cell.

TRITOMARIA (TRITOMARIA) SCITULA (Tayl.) Joerg.
[Figs. 227; 228: 1–6]

Jungermannia scitula Tayl., London Jour. Bot. 5:274, 1846.
Sphenolobus scitulus Steph., Spec. Hep. 2:168, 1902; Evans, The Bryologist 15:56, pl. 2, figs. 1–5, 1912; K. Müller, Rabenh. Krypt.-Fl. 6(2):761, 1916.
Lophozia exsecta var. *scituta* (sic!) Pears., *in* Macoun, Cat. Canad. Pls. 7:23, 1902.
Diplophylleia exsectiformis var. *aequiloba* Culmann, Rev. Bryol. 32:73, figs. 1–8, 1905.
Sphenolobus politus var. *acuta* Kaal. ex Müll., Rabenh. Krypt.-Fl. 6(1):615, 1910.
Sphenolobus exsectiformis var. *aequiloba* K. Müll., *ibid.* 6(1):611, fig. 296, 1910.
Tritomaria scitula Joerg., Bergens Mus. Aarbok, Naturv. [1919–20] 7:3, 1921; Buch, Mem. Soc. F. et Fl. Fennica 8(1932):28, fig. I:32–34, 1933; K. Müller, Rabenh. Krypt.-Fl. ed. 3, 6(1):734, fig. 239, 1954.
Tritomaria (subg. *Eutritomaria*) *scitula* Schust., Amer. Midl. Nat. 45(1):65, pl. 22, 1951; Schuster, *ibid.* 49(2):387, pl. 25, 1953.

Usually scattered or in small patches, clear green and chlorophyllose to occasionally clear brownish (sun forms; these may be ± vinaceous at postical leaf bases);

medium-sized [shoots 1.5–1.85(2.0) mm wide × 5–15 mm long usually], *relatively prostrate*, but with apices ascending, simple or almost so, except for isolated *terminal* branches from bases of older shoots. Stems ca. 300–360 μ wide, ca. 14–15 cells high; cortical cells somewhat (but never prominently) thick-walled, shortly rectangular (ca. 16–20 μ wide × 36–68 μ long; averaging 2.5–3.5 × as long as wide and *only slightly narrower than basal leaf cells*); medulla dorsiventrally differentiated, the ventral 5–7 cell layers of small, eventually brownish and mycorrhizal cells, the dorsal 5–7 layers of larger, pellucid cells, all of the medullary cells thin-walled and weakly thickened at the angles. Rhizoids very abundant on prostrate shoot sectors. Leaves *subtransversely oriented*, somewhat pectinate and stiffly antically secund, often ± plicate, the *dorsal half transversely inserted*, not decurrent, slightly to conspicuously imbricate, erect-spreading to widely spreading (50–85°), mostly *ovate-oblong to short rectangulate*, ca. 715–850(1000) μ long × 630–680(900) μ wide, on robust forms occasionally quadrate-orbicular or quadrate-obdeltoid and 1175–1275 μ long × 1250–1320(1430) μ wide [width 0.9–1.15(1.2) × length], *strongly canaliculate, subequally 3-lobed at the broad summit*, occasional leaves 2- or 4-lobed, the antical leaf margin not or hardly shorter than postical, usually almost equally arched (occasional 2-lobed leaves with ventral lobe larger); *lobes acute to subacute*, their tips occasionally obtuse; sinuses descending 0.2–0.3 the leaf length, U- to V-shaped with base rounded. Cells *moderately collenchymatous*, in leaf middle ca. *22–27 × 24–36 μ*, subquadrate to shortly rectangulate, on leaf-margins ca. 17–25 μ, at base 20–25 μ wide × 36–48(56) μ long (averaging 1.8–2.5 × as long as wide and hardly wider than cortical stem cells); cuticle ± strongly verruculose to striolate;[213] trigones large, *usually weakly bulging*, never subconfluent, intervening walls usually thin; oil-bodies usually (3)5–8(10) per cell, very variable in size, ca. (4)5–7 × 5–9 μ, spherical to ovoid, finely papillose-segmented. *Asexual reproduction constantly present*, by means of *sharply polygonal* or tetrahedral 1–2-celled gemmae, *ca. 18–25(26) μ wide × 20–28(32) μ long, reddish brown to rust-red*, with somewhat thickened walls and angles; gemmiparous leaves unmodified, usually trilobed at summit, but the lobes often obtuse.

Dioecious; *rarely fertile*. Androecia eventually intercalary; bracts diandrous, like leaves but ventricose at base; antheridia ovoid, ca. 152–160 μ, the *stalk* 50 μ long, *uniseriate* (but with isolated cells vertically septate). ♀ Bracts slightly larger than leaves below them, ca. 1230–1300 μ long × 1200–1520 μ wide, broadly oblong to *orbicular-subquadrate, subequally 3- (to 4-) lobed at apex* (the antical lobe distinctly shorter, the postical longest), the lobes acute to shortly acuminate, never cuspidate, *otherwise entire*; bracteole lacking, or a small (300 × 475 μ) oblong, ± bidentate structure, united broadly with one bract. Perianth cylindrical to narrowly ovoid, to 3 mm long × 1 mm in diam., 4–5 plicate toward the only moderately contracted mouth; mouth irregularly, shallowly lobulate and with *short teeth or cilia*, the majority *(1)2–3-celled, the longest 4–5-celled*, from a broad base. Sporophyte (*fide* Müller) very rare, the

---

[213] The cuticle is described as smooth by Frye & Clark (1937–47, p. 421) and by Müller (1951–58, p. 734). I have found that the median and basal cells, at least, are strongly verruculose to striolate, the leaf apices less so, much as in other species of the genus.

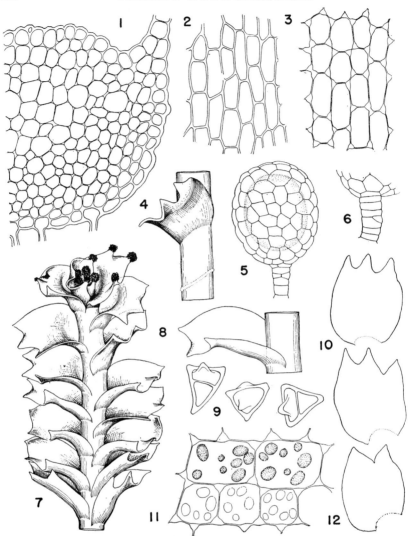

Fig. 227. *Tritomaria (Tritomaria) scitula.* (1) Stem cross section, fungal hyphae omitted (×170). (2) Dorsal cortical cells of stem, mod. *densifolia* (×200). (3) Basal leaf cells (×200). (4) Lateral aspect of shoot sector, rhizoids omitted (×28). (5) Antheridium (×145). (6) Antheridial stalk (×220). (7) Gemmiparous plant (×21). (8) Leaf, *in situ* (×28). (9) Gemmae (×600). (10, 12) Leaves from a single shoot, showing variation from bilobed to trilobed (×28). (11) Median leaf cells with oil-bodies (×450). [All from small island between Lucille I. and Susie I., Susie Isls., Minn., *RMS.*]

capsule yellow-brown, oval. Epidermal layer thick (18–20 μ) with nodular thickenings; usually 3 inner layers, each ca. 8 μ thick, with nodular thickenings, extended onto the tangential walls as fingerlike extensions; capsule wall to 45–50 μ thick. Spores brown, 14–16 μ. Elaters short, 90–110 μ long $\times$ 9 μ in diam., with 2 brown spirals each 3 μ broad.

Type. Western North America (*Drummond*); a portion of the type (Mitten Herbarium, NY) studied.

Nomenclature

Müller (1910, *in* 1905–16, p. 615) refers to a "var. *acuta* Kaalaas (in litt. ohne Namen beschrieben in Beitr. z. Lebermoosfl. Norw. S. 19, 1898)," which he regards as the gemmae-bearing "form" of *Saccobasis politus*.[214] Some years later, Müller (1954, p. 734, *in* 1951–58) cites "*Jungermannia polita* var. *acuta* Kaalaas bei K. Müller, Leberm. 1(1910):615," in synonymy of *T. scitula*. In Kaalaas (1898) there is no reference to a var. *acuta*; neither the name *acuta* nor the combination *Jungermannia polita* var. *acuta* is established in Müller (1910, *in* 1905–16), since "*acuta*" is there referred to under "*Sphenolobus politus.*"

To make this situation even more chaotic, Frye and Clark (1945, p. 420, *in* 1937–47) refer, in synonymy of *T. scitula*, to a "*Sphenolobus politus* var. *acuta* Schiffn., Österr. Bot. Zeitschr. 58:1908, according to Joergensen . . . ." No such reference is cited in the Schiffner paper!

Distribution

In general, rather rare and sporadically distributed, largely of "low" arctic-alpine distribution, with the range much dissected by the obligate restriction to basic or subbasic sites. The scattered sites away from the Tundra (such as those peripheral to Lake Superior) are probably relict in nature.

In Europe found both in the far north, in Scandinavia (Norway, Sweden, Finland), and in the Alps of the central portion (Switzerland, northern Italy, southern Germany, Austria), also in Bucovina, the Carpathians (Borsa), Bulgaria, and the USSR (Waigatsch I.; central Urals), and known from Iceland.

In North America transcontinental, common only in west Greenland, the western range poorly known, the species known only from a few reports from Alberta and British Columbia (Evans, 1912a), and from "W. North America" (*Drummond*; the type), Alaska, central Yukon (Hunker Creek, *fide* Persson, 1946; Bering Strait, Persson, 1962). Also known from Jan Mayen I., east of Greenland, and (in our area) from:

---

[214] There is, however, an allusion in Kaalaas (Beiträge zur Lebermoosflora Norwegens, *in* Vidensk. Skrift. I. Math.-naturw. Kl. 1898, no. 3:17, 1898) to a gemmae-bearing form of *Jungermannia polita* Nees—without any new name being formed. This must be the reference Müller had in mind, although he cites the wrong page.

NW. GREENLAND. Kangerdlugssuak, S. side of Inglefield Bay near Academy Bugt, 77°23' N., 67° W (*RMS 45898*). W. GREENLAND. Ritenbenk, Agpat, 69°46' N., 51°20' W. (*RMS & KD 66-264*); basalt cliffs, near shore, below Skarvefjeld, E. of Godhavn, Disko I., 69°16' N., 53°28' W. (*RMS & KD 66-314*); Godhavn, below Lyngmarksfjeld (*RMS & KD 66-150, 66-166*); Kangerdlugssuakavsak, head of Kangerdluarssuk Fjord, 71°21' N., 51°40' W. (*RMS & KD 66-1286, 66-1287b*); mouth of Simiutap Kûa Valley, near pond, Umîarfik Fjord, 71°59' N., 54°35-40' W. (*RMS & KD 66-917, 66-916c*); E. side of Hollaenderbugt, Nugssuaq Pen., slopes below Qârssua, 70°47' N., 53°47' W. (*RMS 66-1506a, 66-1507, 66-1517*); head of Marmoralik Fjord, N. of Akuliarusikavsak, 71°05' N., 51°12' W. (*RMS & KD 66-1206*); Magdlak, Alfred Wegeners Halvö, 71°06' N., 51°40' W. (*RMS & KD 66-1357g, 66-1364a*). E. GREENLAND. Gaaseland, Faxe Sø. 70°15' N., 29° W. (*Holmen 19,462; 18,595*, a single plant among *Sauteria* and ?*Lophozia perssonii*); Clavering Ø, Eskimonaes, 74°06' N., 21°20' W. (*Gelting, Sept. 21, 1932!*).[215]

NORTHWEST TERRITORIES. Tukarak I., Belcher Isls., Hudson Bay (Schuster, 1951). ONTARIO. Reports of this species from Ontario in Macoun (1902) are surely incorrect, since they are based on material from "dead wood" and "old stumps and logs," except for a citation from Thunder Bay, L. Superior, which may be correct; Bruce Pen., N. shore of Emmett L. (*H. Williams 1890!*); according to Williams (*in litt.*) abundant on Bruce Pen. along the shores, on Manitoulin I., and along L. Superior in Ontario. QUEBEC. Island near mouth of Seal R., Cape Jones, and near mouth of Fishing Lake Cr., Richmond Gulf, both stations on E. coast of Hudson Bay (Schuster, 1951; Lepage, 1944–45, states that previous reports [Macoun, 1902] of the species from Quebec are incorrect); Cascapédia R., mouth of R. Charles-Vallée, Gaspé Pen. (*Crum 10956*); Parc de la Gaspésie, Riv. au Diable, and R. Ste-Anne near hotel (*H. Williams, in litt.*). NEW BRUNSWICK. In herb. Austin (*fide* Macoun, 1902, as *Lophozia exsecta* var. "*scituta*," but these specimens questionable, *fide* Evans, 1912a).

MICHIGAN. Near Marquette, on L. Superior (*Schuster, 1949*); "Devils Washtub," shore of L. Superior, SW. of Copper Harbor, Keweenaw Co. (*RMS 17513*); Amygdaloid I., Isle Royale (*RMS 17434, 17426, 17417, 17330, 17407a*); Bolton Sink, Alpena Co. (*A. J. Sharp, June 1955!*). WISCONSIN. Sand I., Apostle Isls., Bayfield Co. (*RMS 17541*). MINNESOTA. Cook Co.: smallest island between Lucille and Susie Isls., Big Susie I., Porcupine I., Little Susie I., all NE of Grand Portage, in L. Superior (*RMS 11967, 11982, 12002, 11731, 12250b, 14869, 14870, 7249a, etc.*); The Point, Grand Marais (*RMS 7607, 13798, 6558, etc.*); Grand Marais (*Conklin 2653*); Temperance R. ravine (*RMS 19711a*). Upper Falls, Gooseberry R., Lake Co. (*RMS 13246, 13245, etc.*); see Schuster (1953).

Almost all stations around Lake Superior (where the species is locally common) are in the narrow, microclimatic "Tundra Zone" (Schuster, 1957) within a few feet of the cold waters of Lake Superior.

Ecology

An obligate calciphyte, at least in eastern North America; often found on thin soil over basaltic rocks (such as diorite and diabase), often in soil-filled rock crevices, and occasionally on rock itself, under pH conditions

---

[215] I am slightly uncertain as to the determination of the Clavering Ø plant. A whitish green phase, with few, reddish gemmae (16–20 × 22–24 μ to 24 × 26 μ), is at hand; it occasionally shows feeble vinaceous pigmentation of the ventral leaf bases, much as in *T. heterophylla*. The leaves, as in typical *T. scitula*, average at least 1.1–1.2 × as long as wide.

ranging from 5.8 to 7.0 (and possibly higher). Associated species in the Lake Superior and Hudson Bay areas are almost all calciphiles (*Preissia quadrata, Scapania gymnostomophila, S. cuspiduligera, Odontoschisma macounii, Lophozia gillmani* and *heterocolpa*) or tolerate weakly calcareous conditions (*Plagiochila asplenioides, Blepharostoma trichophyllum, Cephalozia pleniceps, Lophozia hatcheri, Tritomaria quinquedentata, Solenostoma pumilum, Scapania mucronata*). Unlike *Scapania degenii*, a species with much the same distribution in eastern North America, *T. scitula* occurs in areas never subject to inundation, as a mesophyte or meso-xerophyte, and (below the Arctic) appears restricted largely to areas with diffuse light, "preferring" north-west-facing cliffs or north-facing sides of ravines.

In the northern portions of its range a wider variety of species is sometimes found consociated (in addition to the preceding ones), including *Anthelia juratzkana*, and *Scapania curta* (Schuster, 1951).

Tritomaria scitula rarely forms the extensive cushions that are made by the allied *T. exsectiformis* and infrequently occurs crowded, in the fashion so common to the latter species. Associated with lack of cushion formation, is the prostrate growth, except for ascending apices of the shoots. ♂ Plants are rare in our area, while perianth-bearing ones have been seen only once (*RMS & KD 66–1505a*, Nugssuaq Peninsula, Greenland).

Differentiation

A distinctive species.

The nearly uniformly trilobed leaves (even on gemmiparous shoots), with gemmae nearly always present in reddish brown masses at the leaf apices . . . , remind one of *T. exsecta* and *exsectiformis*. However, the almost symmetrical trilobed leaves . . . are quite unlike those found in the latter two species, or in *T. quinquedentata*. The species looks superficially more like a species of *Lophozia*, for instance, *L. longidens* (Schuster, 1953).

On the basis of the angulate gemmae and relatively large cells *T. scitula* appears allied to *T. exsectiformis* (of which it was formerly considered to be a mere variety). Indeed, only the difference in leaf shape serves to adequately separate the two species, vegetatively. However, *T. scitula* has a somewhat less longly toothed perianth mouth than *T. exsectiformis*. In spite of the few well-marked differential features, *T. scitula* is certainly a well-founded species, as is apparent from its very different ecology.

Lax, green, gemmae-free plants are of rare occurrence; they may be extremely similar, at times, to slender plants of *T. quinquedentata* (Buch, 1933a; Schuster, 1951, p. 13). This is especially the case when, as occurs frequently, the leaves are slightly asymmetric (Fig. 227:10, lower leaf). Furthermore, *T. quinquedentata* has a similarly prostrate manner of growth. However, the broader, even more opaque leaves of *T. quinquedentata* are usually sufficient to

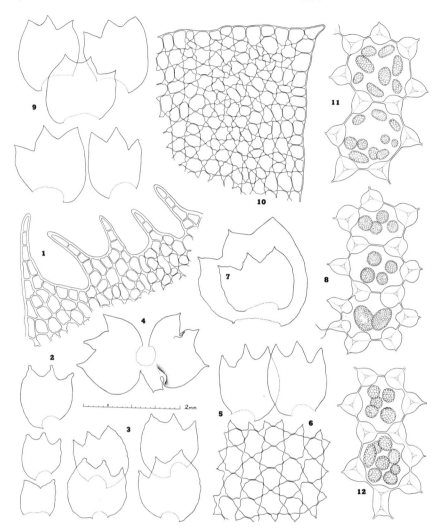

FIG. 228. *Tritomaria*. (1–6) *T. scitula:* 1, perianth mouth cells (×156); 2, 3, leaves from gemmiparous plant (×13.1); 4, two bracts and bracteole (×13.1); 5, two leaves (×13.1); 6, median cells (×200). (7–8) *T. heterophylla:* 7, leaf of *T. heterophylla* drawn within leaf of *T. quinquedentata*, from admixed plants (×16); 8, median cells of *T. heterophylla* (×685). (9–10) *T. quinquedentata* fo. *gracilis;* 9, at top, three leaves of type of fo. *gracilis;* at bottom, two leaves of type of var. *tenera* (×26.5); 10, cells of leaf lobe (×200). (11–12) *T. quinquedentata* var. *turgida:* 11, two median cells with oil-bodies (×790), from phenotype with coarsely verrucose cuticle and violet postical leaf base; 12, two cells from lobe middle (×910), from phenotype with ± elongate cells of postical leaf base. [1–4, *RMS & KD 66-1505a*, W. Greenland; 5–6, from type of *T. scitula;* 7, *RMS 35602a*, Ellesmere I.; 8, *RMS 45891*, NW. Greenland; 9, top 3 leaves from type of *T. quinquedentata gracilis*, Cap Brewster, E. Greenland; bottom 2 leaves from type of *T. quinquedentata tenera*, Ruopok, Greenland; 10, from type of *T. quinquedentata tenera;* 11, *RMS 45705d*, NW. Greenland; 12, *RMS 45790*, NW. Greenland.]

effect a separation, although this is not always the case with lax, green, slender phases (mod. *parvifolia-laxifolia-viridis*) of *T. quinquedentata*, in which the leaves may be longer than wide.

The nearest relative of *T. scitula* is surely the exclusively arctic *T. heterophylla*. *Tritomaria scitula* can be separated from the latter at once by the (*a*) generally more opaque appearance, caused largely by the much less nodose trigones, with the postical leaf bases almost never vinaceous;[216] (*b*) leaves quadrate to slightly to distinctly longer than wide, their lobes never cuspidate; (*c*) ♀ bracts and leaves below them as long or nearly as long as wide, the lobe margins quite entire; (*d*) perianth mouth much more distinctly ciliate-dentate, some of the small laciniae formed being terminated by 4 or even 5 single, superposed cells. In the rarity of perianths in *T. scitula*, this distinction, although seemingly definite, is of little practical significance—however fundamental it may be. However, leaf shape alone in most cases will allow a separation of the two taxa.

Although I have had occasional doubts about the degree of separation existing between *T. scitula* and *T. heterophylla*, the collection of *RMS & KD 66-1287b* (west Greenland: Kangerdluarssuk Fjord) provides an unambiguous answer. Here both species occur admixed in phases from a rather exposed, dry site. The *T. scitula* is normal in size, rather unusually light green with clear brownish pigments and rare traces of vinaceous on the ventral side of the stem; the *T. heterophylla* is a dwarfed, xeromorphic mod. *parvifolia-viridis* with whitish green leaves and greenish, in part light reddish, gemmae. The relevant dimensions of the two taxa follow:

	T. scitula	*T. heterophylla*
Trilobed leaves, width × length	700 × 940; 900 × 900 μ 950 × 965; 965 × 1145 μ	740 × 650; 760 × 635 μ 880 × 675 μ
Quadrilobed leaves, width × length	860 × 965 μ	900 × 655 μ
Leaf lobes	Blunt to subacute	± Cuspidate
Trigones	Weakly bulging	Coarsely nodose
Gemmae	22–24 × 26–29 μ Deep brown to red-brown	21–24 × 24–26(30) μ Pale green to faint red-orange

TRITOMARIA (TRITOMARIA) HETEROPHYLLA Schust.

[Figs. 228:7–8; 229–230; 231:1–7]

Tritomaria heterophylla Schust., Canad. Jour. Bot. 36:272, 1958.

[216] Occasionally one finds (e.g., Hollaenderbugt, Nugssuaq Peninsula, western Greenland) large masses of *T. scitula* on exposed, calcareous clay; such plants may have the ventral leaf bases rather distinctly vinaceous, but then the entire plant acquires a warm brown coloration—never the vinaceous pigmentation that suffuses almost the entire plant in most sun forms of *T. heterophylla*.

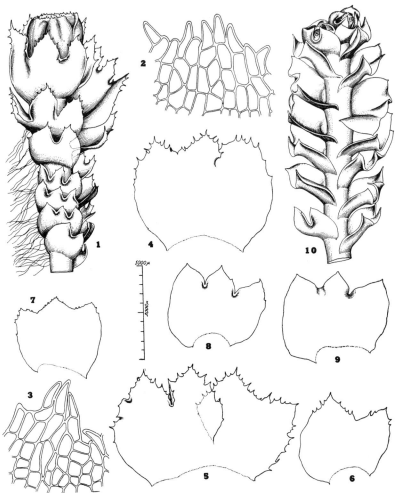

FIG. 229. *Tritomaria heterophylla.* (1) Shoot with perianth, unfertilized plant (×ca. 18). (2–3) Portions of perianth mouth (×235). (4) ♀ Bract (×25). (5) ♀ Bract (×25). (6) Leaf below gynoecium (×25). (7) Gemmiparous leaf (×25). (8–9) Sterile leaves (×25). (10) Sterile shoot (×ca. 18). [All from *RMS 35103*; 4–9 drawn to scale.]

Similar to T. scitula, but *lacking the clear green* color of that species, forming *prostrate*, often ± *brownish or purplish brown patches*, only apices of shoots ascending, unless strongly crowded, when suberect. Shoots simple or rarely terminally furcate (supporting leaf at base of branch bilobed, lacking postical lobe), (700)1000–1350(1650) mm wide × 8–16 mm long, at least distal portions of leaves *pellucid and pale green* (shade forms) or brownish; leaf margins sometimes decolorate distally; *postical leaf bases slightly to prominently purplish-pigmented*, the *distal portions of leaves often also with a rosy hue* (exc. mod. *viridis*), the underside of the *stem ± purplish black*. Stems rather rigid, 275–300(380) μ in diam.; dorsal cortical cells markedly elongated, ca. 19–22 μ wide × (60)75–100 μ long, their longitudinal walls strongly thickened (*length-width ratio ca. 3–5:1*). Leaves weakly to moderately imbricate, stiffly and obliquely laterally spreading, often loosely complicate (especially when bilobed), virtually transverse, from 575–620 μ wide × 480–490 μ long, the *larger or uppermost somewhat fluted, transversely oblong*, ranging from 820–950 μ wide × 700–720 μ long (*1.17–1.32× as wide as long*) to 1050–1200 μ wide × 750–820(900) μ long (*1.40–1.46× as wide as long*), *subequally 3-lobed* for 0.2–0.3 their length (*occasional leaves subequally 2- or 4-lobed*) ; lobes ovate-triangular, acute to subobtuse, often *distinctly apiculate or cuspidate, on upper leaves of ♀ plants often spinose-dentate,* frequently with slightly reflexed margins, on upper leaves of gemmiparous shoots occasionally with *scattered, sharp to aciculate teeth*; sinuses obtuse to somewhat acute, *narrowly but sharply reflexed and gibbous* at base. Cells *strongly collenchymatous,* on margins of lobes 20–25(27) μ tangentially measured; subapical cells and those of lobe middle 21–25 μ wide × 24–30 μ long, *hardly smaller than median cells,* which are (22)23–27 × 25–30(35) μ; basal cells 23–27 × 30–38 μ; cuticle weakly verruculose to striolate; oil-bodies spherical to ellipsoidal, 2–7(8) per cell in leaf lobes, (2)3–10 per cell in median cells, from 5 to 6 μ up to 6 × 9, occasionally 7 × 13 μ, finely papillose-segmented; chloroplasts ca. 3–3.5 μ. Gemmae 1–2-celled, *reddish brown to purplish,* 18–21 × 28 μ to 23 × 24–27 μ *up to a maximum of 24–32 × 35 μ.*

Dioecious. ♂ Plants with bracts hardly different from vegetative leaves, lacking paraphyses. Antheridia orange-yellow, body ca. 150–190 μ, the uniseriate stalk 20–22 μ in diam. ♀ Plants with leaves gradually larger above, *their margins often with scattered spinose teeth.* ♀ Bracts exceedingly polymorphous, varying from obtrapezoidal to transversely oblong, larger than leaves, from 1350 μ wide × 1000 μ long to 1700 μ wide × 1300 μ long (*width 1.3–1.35× length*) up to 1900 μ wide × 1150 μ long (ca. 1.6–1.7× as wide as long), erect or suberect (at least when perianth is immature), *strongly plicate and undulate-crispate,* (2)3–4(5)-lobed, to ¼ to ⅓ their length, the sinus narrow at base, *strongly reflexed to gibbous*; lobes ovate-triangular, their *margins and margins of disk* ± *sharply spinose-dentate to ciliate-dentate for some distance below the lobes,* ending in spinose cusps; teeth slender, mostly 2–3-celled (terminal cell 15–18 × 45–64 μ), absent or few on weak plants. Perianth ovoid, plicate, contracted to mouth, which is usually decolorate, *crenulate-dentate to crenulate-setulose,* bearing tapering, narrow teeth (formed of cells ca. 12–13 × 30–48 μ long which are free for their distal 0.2–0.5, or frequently wholly free), occasionally with scattered 2-celled teeth (of which the terminal cell is slender and acuminate).

Type. Foot of Mt. Pullen, on northeast slope, ca. 5 miles south of Alert, northeast Ellesmere I., ca. 82°30′ N. (*RMS 35103*).

Distribution

A typically high arctic species, so far known only from the North American Arctic, rather generally restricted to basic areas (in regions with acidic rocks usually local and confined to sites near basalt dikes or with some basic drift).

N. GREENLAND. Heilprin Land, Brønlund Fjord, Peary Land, 82°10′ N., 31° W. (*Holmen 281e*, trace among *Scapania praetervisa, Lophozia heterocolpa*; *Holmen 8839*, among *Anthelia juratzkana*); Low Pt., 83°06′ N. (*Th. Wulff, June 14, 1917;* reported by Hesselbo, 1923a, as *T. quinquedentata*, but by Arnell, 1960, p. 4, as *Lophozia latifolia*!); Kap Glacier, S. coast of Independence Fjord, 81°48′ N., 31°45′ W. (*Holmen 6328, p.p.*, with *T. quinquedentata*, etc.); B. G. Schley Fjord, 83°00′ N., 24°30′ W. (*Troelsen 8762, p.p.!*). NW. GREENLAND. Kânâk, Inglefield Bay (*RMS 45551*); Kekertat, near Heilprin Gletscher, E. end Inglefield Bay (*RMS 45891*). W. GREENLAND. Eqe, near Eqip Sermia (Glacier), peaty soil along a small rill, ca. 200–250 m.s.m., 69°46′ N., 50°10′ W. (*RMS & KD 66-251, 66-249c, 66-289b*); Kangerdlugssuakavsak, head of Kangerdluarssuk Fjord, 71°21′ N., 51°40′ W. (*RMS & KD 66-1281, 66-1287b, 66-1280b, 66-1283*); Uvkusigssat, 71°03′ N., 51°53′ W. (*RMS & KD 66-1393*, mod. *parvifolia-viridis*). E. GREENLAND. Runde Fjeld, W. of Scoresby Sund, ca. 70°30′ N., 28°40′ W. (*Hartz*; reported by Jensen, 1897, p. 379, as *J. gracilis = Lophozia attenuata*). ELLESMERE I. On NW. slope of The Dean, about 1 mi E. of Mt. Pullen (*RMS 35102, 35120, 35156a, p.p.*); E. side of Mt. Pullen (*RMS 35315a, 35307c, p.p., 35133b, p.p.*); exposed, soil-covered, damp, snow-fed scree slope, 2000–2400 ft, E. edge of U.S. Range, 9–10 mi due W. of Mt. Olga, ca. 82°24′ N., 65°20–30′ W., NE. Ellesmere I. (*RMS 35615a, p.p.*). SOUTHAMPTON I. Coral Harbour, 64°09′ N., 83°05′ W. (Persson & Holmen, 1961).

The latitudinal range of the species, thus far, is from about 64°09′ N. to 83°06′ N., making this one of the most typical high arctic taxa of Hepaticae known.

Ecology

In addition to the collections made in Ellesmere I. (see above), I have collected the species in Inglefield Bay, where it occurred in peaty meadows and on peaty soil between rock clefts, sometimes in *Eriophorum* meadows, close to rivulets, associated with *Riccardia palmata, Aneura pinguis, Solenostoma pumilum polaris, Anthelia juratzkana,* and *Cephalozia pleniceps*—all taxa normally tolerating subbasic conditions, except perhaps for the *Riccardia*.

In western Greenland, at Eqe, a small quantity of plants was found on peaty soil along a rivulet, where not subject to inundation, associated with *Prasanthus suecicus, Cryptocolea imbricata, Gymnomitrion concinnatum, Cephalozia bicuspidata s. lat., Anthelia juratzkana, Tritomaria quinquedentata, Anastrophyllum minutum, Odontoschisma macounii, Cephaloziella arctica,* and *Blepharostoma*—all taxa which, in the Arctic at least, tolerate subbasic conditions.

At Kangerdluarssuk Fjord, the species occurs as a small, very pale form (mod. *parvifolia-viridis*), in part with some vinaceous pigmentation (mod. *colorata*), on calcareous but rather dry soil, with such hardy associates as *Saxifraga oppositifolia*; associated hepatics are *Tritomaria scitula, T. quinquedentata, Scapania gymnostomophila, S. praetervisa, Solenostoma subellipticum, S. sphaerocarpum, Cryptocolea, Blasia,* and the moss *Distichium capillaceum*.

In the type region on weakly calcareous, moss-covered rubble and weathered rock, moistened by runoff from persisting snow banks, in full sun, associated with *Lophozia heterocolpa*, *L. quadriloba*, *L. alpestris* subsp. *polaris*, *Blepharostoma trichophyllum brevirete*, *Anthelia juratzkana*, *Cephaloziella arctica*, *Tritomaria quinquedentata*, *Scapania praetervisa*, *Arnellia fennica*, and *Plagiochila arctica*, as well as *Cryptocolea imbricata*, *Gymnomitrion concinnatum*, and *Odontoschisma macounii*. In its toleration for basic conditions evidently similar to *T. scitula* but, unlike this greener, more opaque species, usually confined to well-lit sites.

The species appears to be most common in locally moist situations, in full sun (during the short arctic summer), contiguous to snow banks which "feed" into the moss-herb tundra lying below. Although evidently associated with calcareous rock outcrops, the species usually undergoes ecesis only *after* a rather distinct humus layer is deposited. In this respect, it is very similar to *Odontoschisma macounii*.

Differentiation

Often recognizable with a hand lens on the basis of the somewhat crispate, trilobed, broad leaves, the rather pellucid appearance (much more marked than in *T. quinquedentata*, which is a relatively opaque plant), and the generally vinaceous pigmentation of the leaf bases, which may extend $\pm$ to the rest of the leaves; these may be somewhat rosy in color, although the purplish pigmentation may be largely masked by a brownish secondary pigment. The mod. *viridis* lacks secondary pigments and is distinct in the very pellucid aspect, due in part to the very coarse trigones. ♀ Plants are most distinctive; they bear somewhat larger leaves distally, imbricate, strongly crisped and undulate, whose lobes are spinose-dentate to spinose-ciliate, except on weak plants (on which dentition is sparing). No other *Tritomaria* has such highly differentiated bracts. The species also differs from others of subg. *Tritomaria*, except *T. quinquedentata*, in the broad, transversely rectangulate leaves, ranging from about 1.15 to 1.45× as broad as long (measured to apex of ventral lobe, which is slightly the longest in most cases).

Type material was always admixed with *T. quinquedentata*. The two taxa differ in a large number of features, e.g.: (*a*) plants only 1–1.35, rarely to 1.5–1.65, mm wide in *T. heterophylla*, (1.5)2–3.5 mm wide in *T. quinquedentata*; (*b*) aspect pellucid in *T. heterophylla*, opaque in *T. quinquedentata*; (*c*) trigones and subapical cells larger in *T. heterophylla* than in *T. quinquedentata*; (*d*) ventral stem surface and postical leaf bases typically vinaceous in *T. heterophylla*; in *T. quinquedentata* stems brownish postically with age, ventral leaf bases in extreme cases dark chestnut-brown-pigmented, at most with a trace of reddish color; (*e*) leaves

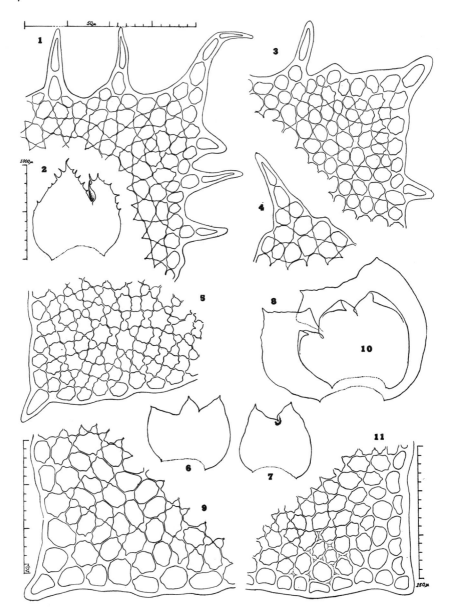

FIG. 230. *Tritomaria heterophylla* (1–7), *T. quinquedentata* var. *grandiretis* (8–9), and *T. quinquedentata* (10–11). (1) Lobe of ♀ bract (×235). (2) Entire ♀ bract (×25). (3–4) Apices of vegetative leaves of ♀ plant (×235). (5) Lobe of leaf of sterile plant (×235). (6–7) Two leaves from sterile plant (×25). (8) Leaf (×25). (9) Cells of apex of middle leaf lobe (×235). (10) Leaf (×25). (11) Apex of middle lobe of leaf (×235). [1–7, *RMS 35103;* 8–11, *RMS 35593;* 1, 3–5, 9, 11 drawn to scales juxtaposed to figs. 1, 9, 11; 2, 6–8, 10, to scale to left of fig. 2.]

approaching a symmetrical form in *T. heterophylla*, asymmetric in *T. quinquedentata*; (*f*) gemmae always present in *T. heterophylla*, normally lacking in *T. quinquedentata*. In addition to these differences, *T. heterophylla* differed from xeromorphic phases of *T. quinquedentata* present at the type station in the spreading and loosely canaliculate leaves (vs. concave and often suberect leaves in *T. quinquedentata*), giving the two taxa distinct facies. Most of these distinctions are clear from Fig. 231, drawn from intermingled plants.

In several of the above features (particularly *a*, *e*, and *f*) *T. scitula* agrees with *T. heterophylla*. Both taxa grow closely adnate to the substrate unless crowded, both often produce isolated 2- or 4-lobed leaves, and both bear gemmae in abundance. However, *T. heterophylla* lacks the characteristic clear green color of *T. scitula*, being a more pellucid, less chlorophyllose plant; also the gemmae of *T. heterophylla* are often more sharply stellate and up to 35 μ in longer diam.; the leaves of gemmiparous plants occasionally or often bear sharply spinose, scattered teeth; the cuspidate-lobed leaves are always markedly broader than long, the inverse situation prevailing in most forms of *T. scitula*. In addition to these discrepancies, there is the marked difference in facies, caused by the tendency toward extensive production of purple pigmentation of the postical leaf bases in *T. heterophylla*, which does not recur to any extent in other species of subg. *Tritomaria*. In this, *T. heterophylla* approaches subg. *Saccobasis*, which it further resembles in the subequally trilobed leaves. However, the leaf insertion in *T. heterophylla* is of the usual *Tritomaria* type and therefore much less complex than in *Saccobasis*.

Perhaps *T. heterophylla* is a high arctic derivative of *T. scitula*. However, where (as in west Greenland) both taxa occur, they are readily separated. When material with inflorescences is available, *T. heterophylla* is sharply separable from *T. scitula* (as well as all other species of the genus) by the ♀ bracts and subinvolucral leaves. Both bracts and leaves subtending them have the lobes cuspidate or spinose-dentate at the apex, and most often bear scattered, sharp, spinose to spinose-ciliate teeth. Strikingly enough, in spite of the strongly armed bracts, the perianth mouth is distinct from that of all other species of subg. *Tritomaria* in being very short-dentate, with the teeth at most 1–2-celled, while the other Tritomariae, with edentate ♀ bracts, have strongly ciliate perianth apices.

Very small forms of *T. heterophylla* may occur under difficult and rather xeric conditions and then lack secondary pigments; they are rather dense- and small-leaved and are a very pale green because of the development of coarse, knotlike trigones, which in extreme cases may approach the cell lumen in area. The mod. *viridis* of this species, hence, is exceedingly different from the much greener, more opaque, and very common—indeed, almost "normal"—mod.

viridis of *T. scitula*. Such small-leaved green forms may, at times, have rather poorly pigmented gemmae, ranging from very pale orange to pale rusty red.[217]

The mod. *parvifolia-colorata* of *T. polita polymorpha* may approach *T. heterophylla* in size and form. However, this plant usually has blunt leaf lobes (at least in large part) and, on well-developed plants, a very different leaf insertion.

Sectio *TRILOPHOZIA* (Schust.) Schust., stat. n.

Tritomaria subg. *Trilophozia* Schust., Amer. Midl. Nat. 45(1):65, 1951; Kitagawa, Jour. Hattori Bot. Lab. no. 29:117,1969

Sectio sect. *Tritomariae* similis, plantae, autem, plerumque prostratae, robustae (ad 4 mm lat.); gemmae rarae, flavo-brunneae; pedicellus antheridii 2-seriatus; membrana capsulae 5-stratosa.

Plants variable, usually vigorous, mostly 2–3(4) mm wide, unless crowded; prostrate or procumbent with only apices ascending; leaves oriented mostly obliquely, less often subvertically, usually explanate, less often strongly concave or canaliculate. Branching (at least in part) lateral-intercalary. Gemmae extremely rare, yellowish brown, in very small quantity. Antheridial stalk 2-seriate; capsule wall 5-stratose.

Type. *Tritomaria quinquedentata* (Huds.) Buch, the only species.

As Müller (1948) points out, *T. quinquedentata* differs in several rather fundamental features from the other taxa here assigned to subg. *Tritomaria:* it has a biseriate antheridial stalk and a 5-stratose capsule wall. In the latter it differs also from *Barbilophozia*. Since, in addition, the nearly horizontal manner of growth, as well as the less transversely inserted dorsal half of the leaf, gives the species a very different facies, I believe that it should be removed from the *T. scitula-exsecta* complex as a separate section.[218]

TRITOMARIA (*TRITOMARIA*) *QUINQUEDENTATA* (*Huds.*) *Buch*
[Figs. 228:9–12; 230:8–11; 231:8–13; 232]

Jungermannia quinquedentata Huds., Fl. Angl. ed. 1:433, 1762; *ibid.*, ed. 2:511, 1778.
Jungermannia barbata var. *quinquedentata* Nees, Naturg. Eur. Leberm. 2:196, 1836.
Jungermannia lyoni Tayl., Trans. Proc. Bot. Soc. Edinb. 1:116, pl. 7, 1844.
Lophozia quinquedentata Cogn., Bull. Soc. Bot. Belgique 10:279, 1872.
Jungermannia trilobata Steph., Hedwigia 34:50, 1895.
Sphenolobus trilobatus Steph., Spec. Hep. 2:167, 1902.
Lophozia lyoni Steph., *ibid.* 2:149, 1902; Evans, Rhodora 4:210, 1902.
Barbilophozia quinquedentata Loeske, Verh. Bot. Ver. Brandenburg 49:37, 1907.

[217] For the separation of such forms and *T. scitula*, see p. 671.

[218] It should be noted, however, that the antheridial stalk is rather variable. In *T. quinquedentata* the stalk is apparently uniformly biseriate; in *T. exsecta*, uniseriate. In *T. scitula*, which serves, in some manner, as a common ancestral form, the stalk may be entirely uniseriate or isolated cells may be longitudinally septate (i.e., biseriate). Such variation, recurring in *T. exsectiformis*, indicates the lack of true phylogenetic importance of this character, at least in Lophoziaceae.

Lophozia (subg. *Barbilophozia*) *quinquedentata* K. Müll., Rabenh. Krypt.-Fl. 6(1):624, fig. 298,
   1910; Meylan, Beitr. Krypt.-Fl. Schweiz 6(1):160, fig. 96, 1924; Macvicar, Studs. Hdb.
   Brit. Hep. ed. 2:193, figs. 1–4, 1926.
Lophozia verrucosa Steph., Spec. Hep. 6:114, 1917 (probably).
Tritomaria quinquedentata Buch, Mem. Soc. F. et Fl. Fennica 8(1932):290, 1933; K. Müller,
   Rabenh. Krypt.-Fl. ed. 3, 6(1):736, fig. 240, 1954.
Lophozia asymmetrica Horik., Jour. Sci. Hiroshima Univ., Ser. B., Div. 2, 2:153, fig. 16, 1934.
Tritomaria (subg. *Trilophozia*) *quinquedentata* Schust., Amer. Midl. Nat. 45(1):65, pl. 24, 1951;
   Schuster, *ibid.* 49(2):382, pl. 27, 1953.

Prostrate, often in extensive patches or mats, pure green to yellowish brown or
pale brown; *vinaceous pigments lacking* (rarely a little at postical leaf bases);
shoots creeping, ascending only at the very apex (unless crowded), robust,
2–3(3.6–4) mm wide × 1.5–5(6) cm long, remotely branching. Stems subterete, green to somewhat brownish, deep brown or purplish brown ventrally
(because of mycorrhizal infection), 350–450 μ in diam., ca. 18–22 cells high;
cortical cells in 2–3 layers, thick-walled and of smaller diam. [ca. (10)14–17(20) μ]
than medullary cells, subisodiametric in cross section, from 35–45 to (38)45–
72(80) μ long (length:width ca. [2]3–6:1); medullary cells in upper 0.6–0.8
of stem rather large and pellucid, thin-walled and with small trigones, ca.
22–36 μ in diam., the *lower 5–8 layers of medullary* cells much smaller in diam.,
ca. 18–24 μ (nearly similar to cortical cells), thin-walled and with small trigones, becoming *mycorrhizal with age and brownish*. Rhizoids long and dense,
forming a conspicuous ventral mat. Leaves weakly to distinctly imbricate, the
antical end transversely inserted, the postical ⅔ inserted ca. 36–45° obliquely (from
horizontal), not decurrent, *usually horizontally and widely spreading*, but (at least
when crowded) *sometimes with antical lobe loosely folded over the larger postical*
"*half*," the leaf then loosely complicate-canaliculate, *strongly asymmetric, orbicular-ovate* to subreniform, *averaging wider than long* (*ca.* 1.2–1.6 mm wide × 1–1.4
mm long), rarely as wide as long (1020 × 1020 μ), often *somewhat fluted*,
strongly unequally trilobate (the ventral lobe longest, to ca. 1300–1400 μ; median
lobe ca. 1100–1200 μ; antical lobe ca. 1000–1050 μ in length × 1500–1600 μ
in total leaf width), the *lobes broadly ovate-triangular*, spreading (or the *postical
often incurved*), *apiculate to cuspidate* from an obtuse base; sinuses descending
0.1–0.25 the leaf length, *broadly obtuse or occasionally crescentic*, often somewhat
gibbous, *those between antical and median lobes narrower or subequal to those between
median and postical lobes*. Cells strongly collenchymatous, thin-walled and with
distinct to bulging trigones, of leaf margins and leaf apices ca. (*16*)*18–23* μ,
subisodiametric; median cells ca. (18)20–24 × 21–28(30) μ, at leaf base ca.
23–28 × 28–38 μ; *cuticle weakly verruculose* (exc. subsp. *papillifera*); *oil-bodies
usually 2–7(9–10) per cell*, spherical (and 4.5–6 μ) to ovoid or ellipsoid (and
3.5 × 5.5 to 4.5 × 8 to 7 × 8–10 μ), finely granular-papillate; chloroplasts
ca. 3–3.5 μ. *Asexual reproduction rare*, via small groups of *yellowish to yellowish
brown, 1–2-celled gemmae* at apices of the uppermost leaves; gemmae tetrahedral
to polyhedral, 15–20 μ.
   Dioecious, but often with sex organs. ♂ Plants usually in separate patches;
androecia loosely spicate, bracts in 4–12 (rarely 16–20) pairs, more transversely
oriented than leaves, somewhat similar to them in shape and size but strongly
ventricose at base, antical (and sometimes median) lobes erect or suberect,

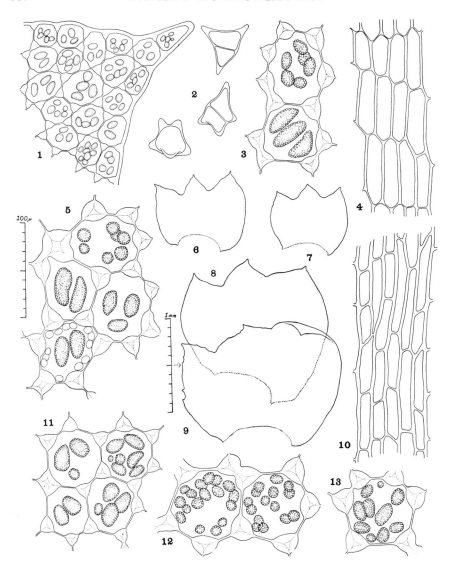

FIG. 231. *Tritomaria heterophylla* (1–7), *T. quinquedentata* (8, 10–11), and *T. quinque-dentata* var. *grandiretis* (9, 12–13). (1) Cells of lobe, with oil-bodies drawn in (×345). (2) Gemmae (×450). (3) Median cells with oil-bodies (×645). (4) Dorsal cortical cells of stem (×250). (5) Cells of subapical part of ventral lobe of leaf (×645). (6–7) Two sterile leaves (×25). (8) Leaf (×25). (9) Leaf (×25). (10) Dorsal cortical stem cells (×300). (11) Subapical cells of lobe (×645). (12–13) Subapical cells with oil-bodies drawn in (×645). [1–7, *RMS 35102*; 8, 10–11, *RMS 35102a*, admixed with *35102*; 9, 12–13, *RMS 35593*; 3, 5, 11–13, drawn to scale to left of fig. 5; 6–9 drawn to scale to left of fig. 9.]

sharp; antheridia 2–3, their stalk 2-seriate; a few linear to lanceolate paraphyses usually present. Gynoecia with bracts broader and slightly longer than leaves, erect with spreading lobes, 3–5 lobed to ¼ to ⅓ their length, plicate or crispate, *lobes rarely with obscure accessory teeth;* bracteole small or obsolete to rather large, when distinct at all fused for much of its length with bracts. Perianth 0.65–0.75 emergent, cylindrical-obovoid, 2–3-stratose at least in basal fourth, pluriplicate above middle, rounded and narrowed to mouth; mouth lobulate, lobes with polymorphous teeth and cilia up to 4–5 or more cells long; cells near mouth ca. 20–25 μ. Capsule ovoid, its wall (4)5-stratose; outer layer of larger cells with nodular thickenings; their thickness equal to that of inner layers; the inner cell layer with semiannular bands (partly incomplete or obsolete in middle of tangential wall). Elaters 6–7 μ in diam.; spores 12–15 μ, finely verruculose, yellowish brown.[219]

Type. England.[220]

Distribution

A widespread, essentially circumpolar species found throughout most of the Tundra and Taiga regions of the Northern Hemisphere, but absent in southward extensions of the Taiga and in the Southern Appalachians and Cordilleras.

Found throughout northern Europe (northern Sweden, Norway, Finland, Russia) and the Spitsbergen Isls., southward into England and Ireland, Denmark, the Faroes, Germany, and to Spain in the Pyrenees and their periphery and to Switzerland and northern Italy. Also to Austria, Bucovina, and Bulgaria, and east to the USSR. Eastward extending into Asia: north and east Siberia, *teste* Arnell; Novaya Zemlya; Japan; Formosa (as *Lophozia asymmetrica* Horik.); Manchuria; China, in Yunnan Prov. (*Lophozia verrucosa* Steph.); and northwestern India and the Himalayas (*Jungermannia trilobata* Steph.).

Also on Jan Mayen I. and Iceland (Hesselbo).

In North America transcontinental, from Alaska and the Yukon southward to British Columbia and Alberta, and eastward to the District of Keewatin and Manitoba. Eastward as follows:

E. GREENLAND.   Cape Dalton; Fjord N. of Cape Dalton; Cape Daussy; Cap Ravn; also between 74°30′ N. and 66°58′ N.; Danmarks Ø, Hekla Havn, Gaaseland, Gaasefjord, Kobberpynt, Röde-Ö's Vestskraent, Runde-Fjeld, and Cape Stewart, all in Scoresby Sund; Hurry Inlet; Cape Parry; Kaiser Frans Josef Fjord; Röhss Fjord; Clavering Ø (Hesselbo, 1948); Cape Mary; Lille Pendulum Ø; Dove Bugt, 76°40′ N. (Jensen, 1910); Rypefjeld and Gefions Havn, 77° N. (Hesselbo, 1948a).   NW. GREENLAND.   Kânâk, Red Cliffs Pen., Inglefield Bay (*RMS 46155e, 45732, 45503, 45488,* etc.);

---

[219] The preceding diagnosis is so worded that italicized characters separate the taxon from both *T. heterophylla* and the following varieties of *T. quinquedentata.*

[220] For the synonymy of *Lophozia asymmetrica, L. verrucosa, Jungermannia trilobata,* and *Sphenolobus trilobatus,* see Kitagawa (1963a) and Schuster (1965).

Kangerdlugssuak, Inglefield Bay (*RMS 45870, 45843, 45899, 45963,* etc.); Kangerd-luarssuk, Inglefield Bay (*RMS 45793c, 45820*); Kekertat, Harward Øer, E. end Ingle-field Bay (*RMS 46048, 46009, 46003c*); Siorapaluk, Robertson Bugt (*RMS 45753, 45755*); Dundas, Wolstenholme Sd. (*RMS 46272, 46266, 46252,* etc.); Uvdle, 76°35′ N.; Hackluyt I., 77°24′ N.; Pandora Harbour, 78°14′ N. (Harmsen & Seidenfaden, 1932); Thule (*Freuchen;* Hesselbo, 1923).   w. greenland. Ritenbenk, Agpat, 69°46′ N., 51°20′ W. (*RMS & KD 66-264, 66-273a*); Eqe, near Eqip Sermia (Glacier), 69°46′ N., 50°10′ W. (*RMS & KD 66-249, 66-333*); Fortune Bay, Disko I., 69°16′ N., 53°50′ W. (*RMS & KD 66-139, 66-141, 66-131*); Godhavn and vic., Disko I., 69°21′ N., 52°55′ W. (*RMS & KD 66-146, 66-271*); Tupilak I., W. of Egedesminde, 68°42′ N. (*RMS & KD 66-111, 66-117a, 66-122,* etc.); Kangerdlugssuakavsak, head of Kangerdluarssuk Fjord, 71°21′ N., 51°40′ W. (*RMS & KD 66-1287b*); mouth of Simiutap Kûa Valley, Umiarfik Fjord, 71°59′ N., 54°35-40′ W. (*RMS & KD 66-914g*).   n. greenland. Heilprin Land, Brønlund Fjord, 82°10′ N., 31° W. (*Holmen 6749, p.p.*); Kap Glacier and Saxifragadal, S. coast Independence Fjord; Herlufsholm Strand, and G. B. Schley Fjord, E. coast Peary Land; Jørgen Brønlund Fjord, Heilprin Land (Arnell, 1960); also Low Pt., 83°06′ N., Cape Benet, 83°02′ N., and Lemming Fjord, 82°53′ N. (Hesselbo, 1923a); Sverdrup I., Mascaret Inlet, 82°55′ N., 44°30′ W. (*K. Ellitsgaard;* among *Lophozia alpestris* subsp. *polaris*). Several of the reports of this species from N. Greenland (in Arnell, 1960) represent *Tritomaria heterophylla* or *T. scitula* (under which the appropriate notations appear).   s. and sw. greenland. General (see Lange & Jensen, 1887).

ellesmere i. Mt. Pullen, S. of Alert, NE. Ellesmere, 82°25′ N.; The Dean, SE. of Alert, 82°25′ N., 62°05-15′ W.; E. edge of U.S. Range, 9–10 mi W. of Mt. Olga, 82°24′ N., 65°10-30′ W. (partly var. *grandiretis*); "Grant Land," 82°27′ N. (as *Lophozia gracilis,* in NY; *L. J. Wolf,* of Peary Polar Exped., *July 1906!;* see Schuster et al., 1959); "very common" in the Hayes Sd. region, S. coast, and SW. coast, with perianths, at Goose Fiord (Bryhn, 1906).   devon i. *Fide* Steere.   distr. of keewatin. *Fide* Steere.   baffin i. Pond Inlet, on N. Baffin (Hesselbo, 1935) and Exeter Sd., 66°27′ N., in C. Baffin I. (Harmsen & Seidenfaden, 1932); Pangnirtung (*Edwards 6!*). melville pen. Igloolik, Cape Elizabeth, Itividleriaq and Vansittart I. (Hesselbo, 1935).

labrador. Macoun (1902); Attikamagen L., Iron Arm, NW. end (*Harper 3558!*); SE. Slimy L. (Dix, 1956).   quebec. Oka and Rigaud; Cap à l'Aigle; Sacré-Coeur, côté de Rimouski; Rimouski; Ste-Anne-de-la-Pocatière, etc. (Lepage); S. of Cairn I., Richmond Gulf; Manitounuck Sd.; near Fishing Lake Cr., Richmond Gulf; Great Whale R.; all on E. coast of Hudson Bay; Cape Aigle (*Macoun 3157*); Mt. Albert, Gaspé (*O. D. Allen, 1881;* as *J. incisa?!*); Ste-Anne, Kamouraska Co. (*Lepage 224!*). newfoundland. Benoit Brook, W. Nfld.; Cow Head, Port au Choix, St. John I., St. Anthony, all in N. Pen.; Hampden and Norris Arm, C. and NE. Nfld. (Buch & Tuomikoski, 1955).   nova scotia. Margaree, Cape Breton I.; Aspy Bay, etc. (Brown, 1936a).   new brunswick. Woodstock (Macoun, 1902).   ontario. Sudbury Jct.; Otter Head, L. Superior; L. Nipigon; Dawson Rte., W. of L. Superior (Macoun, 1902); L. Penage, Sudbury Distr. (*Cain 2124*); Coppermine Pt., W. of Pancake Bay, Algoma Distr. (*Cain 5308*); Sturgeon R. near Beardmore, Thunder Bay Distr. (*Cain 2003*); 14 mi E. of Nipigon (*Cain 4790*); Long Pt., L. Timagami (*Cain 4274*).

maine. Summit of Cathedral Trail, 4500 ft; Pomola Peak, 3800–4500 ft; Northwest Basin, 2800 ft; Hamlin Ridge, 4500 ft; North Basin, ca. 2800 ft, all Mt. Katahdin (*RMS 32980a, 15803, 17010, 32906,* etc.); Mt. Kineo, Moosehead L.; Round Mountain L. and vic., Franklin Co. (*Lorenz*).   new hampshire. Jackson.   vermont. Mt. Mansfield, 0.2–0.3 mi S. of summit, ca. 4000 ft (*RMS 43806e, c. gemmae*); Smugglers

Notch, NE. of Mt. Mansfield (*RMS*).   CONNECTICUT.   Meriden, New Haven Co. (*Evans*; southernmost station at 41°30′ N., at edge of Upper Austral!).   NEW YORK. Catskill Mts., 2–3 mi E. of Haines Falls, Greene Co. (*RMS 44506*); "Huckleberry Hill" near Peasleeville, 1800–2000 ft (*RMS A-257, A-258, A-260*). Essex Co.: Algonquin Peak, 5000 ft, North Elba; Cascades, between Upper and Lower Cascades Ls., Keene; Indian Head, Lower Ausable L., Keene; Indian Pass, North Elba; summit of Mt. Marcy; Rainbow Falls, Keene; Mt. Whiteface; Wilmington Notch; etc. (N. G. Miller, 1966).

Westward occurring as follows: MANITOBA. *Fide* Macoun (1902).   MICHIGAN. Amygdaloid I., Isle Royale (*RMS*); "Devils Washtub," near Copper Harbor, Keweenaw Co. (*RMS 17512, c. gemmae*); Isle Royale (C. E. Allen et al., 1901); Negaunee, Marquette Co. (Nichols, 1935); Sugar Loaf Mt., Marquette; Mountain Stream, Huron Mts.; Negaunee, Marquette Co.; Tobin Harbor, Rock Harbor, Keweenaw Co. (Steere, 1937; Nichols, 1935); Ontonagon Co. (Steere, 1937, etc.).   WISCONSIN. Sand I., Apostle Isls., Bayfield Co. (*RMS*); St. Croix Falls, Polk Co. (*RMS*); Douglas, Ashland, Iron Cos. (Conklin, 1929).   MINNESOTA. Carlton, Carlton Co. (*Conklin 656*); Gunflint L. (*RMS 13403*); Sailboat, Big Susie, Little Susie, Porcupine, Long, Belle Rose Isls., Susie Isls., NE. of Grand Portage, Cook Co. (*RMS 11956, 11992, 14911, 14739*, etc.); The Point, Grand Marais (*RMS 13781, 6511*, etc.); Temperance R., Pigeon R. near Sextus City, Pigeon Pt., ravine near Lutsen, Grand Portage I., Duncan L., Hungry Jack L., Little Caribou R., Stair Portage, Mt. Josephine and Hat Pt., near Grand Portage, Lutsen, Cascade R., Little Caribou R., etc., all in Cook Co. (*RMS 14710, 13426, 10012, 5889, 15086*, etc.; *Holzinger 8*; *Conklin 1191, 1477*, etc.); Rainy L. at Black Bay, Koochiching Co. (*RMS 18123*, etc.); Gooseberry R., Pipestone Rapids on Basswood L., Knife R., and Two Island R., all in Lake Co. (*RMS 13245; Conklin 876, 2503; Holzinger, s.n.*); Oak I., Lake of the Woods Co. (*RMS 17274*); Oneota, Chester Cr., Fairmount Park, Condon Park, all near Duluth, St. Louis Co. (*Conklin 82, 930*); Spirit L. and Nopeming, St. Louis Co. (*Conklin 1636*, etc.).

Ecology

Widespread almost wherever there are basic rock outcrops. Infrequent in the southern edge of the Spruce-Fir Zone, with rare outliers to the south of this, but extends to the northern edge of the Tundra (to north of 83° N.). The distribution probably somewhat affected by a "weak affinity" for basic rock (perhaps most frequent on basic intrusive rocks, such as diabases and diorites, and other Ca- and Mg-rich basalts), with the consequence that the species is rare in most of the New England mountains, except where local outcrops of basic rocks occur.[221] Although often found on densely shaded cliffs, the species also occurs in open sunny sites.

[221] Regarded by Reimers (1940, p. 207) as a "Silikatmoos"; this is not correct, and even Reimers reports the species from humus lying over gypsum. Loeske states that it is distributed only in granitic areas in the Harz Mts. Such a restricted distribution pattern is certainly not generally the case in North America. Steere (1937) finds that it is common in areas of basic rocks, reporting it from "moist trap rock ledges" in Michigan, with such calciphiles as *Scapania cuspiduligera* and *Lophozia gillmani*. It recurs with exactly these two taxa in west Greenland, and as far south as Smuggler's Notch, Vt., there in addition with *Saxifraga aizoön*.

Found in boreal and subboreal localities under a great variety of conditions, as long as there is some integral connection between its substrate and rock exposure . . . essentially epipetric, but . . . only occasionally a pioneer species on rocks: it usually occurs on the crests of cliffs, over thin soil; from these it spreads into the crevices: reacting like a typical chasmophyte. From the crevices and cliff-crests it has limited ability to spread over exposed vertical rock faces, forming eventually dense mats whose size is limited by ice-action and their own weight when wet (Schuster, 1953).

On rock outcrops, particularly in the Great Lakes area, usually associated with various species of the *Lophozia-Scapania* community complex, among them *Anastrophyllum minutum* and *Diplophyllum taxifolium* (under conditions at the xeric extreme for the *Tritomaria*), as well as *Lepidozia reptans*, *Anastrophyllum saxicolus*, *A. michauxii*, *Lophozia ventricosa*, *L. kunzeana*, *L. attenuata*, *L. barbata*, *L. hatcheri*, *Scapania undulata*, *S. nemorosa*, *S. subalpina*, and sometimes *S. mucronata*. The "requirements" and tolerances of *T. quinquedentata* are extremely similar to those of *Lophozia barbata*, and these two robust species often occur together. The occurrences noted above are at the more acid extreme of the species range, with a pH from 4.8 to 5.5. At the more nearly basic end of its range, on basaltic rock walls, *T. quinquedentata* occurs with *Lophozia hatcheri*, *L. quadriloba*, and *Scapania mucronata*, as well as occasionally *T. scitula* (Keweenaw Co., Mich.; frequently in west Greenland).

In tundra and around rock-pools (in the microclimatic "Tundra Zone" around much of Lake Superior), under subcalcareous conditions, the species may occur with *Scapania degenii*, *Odontoschisma macounii*, and even *Ptilidium ciliare*, together with *Selaginella selaginoides*, *Primula mistassinica*, *Euphrasia* sp., and *Scirpus caespitosus callosus*, at a pH of 5.5–6.8. Under similarly subcalcareous conditions, on thin soil over basic rocks, it may occur with *Tritomaria scitula*, *Scapania mucronata*, *Blepharostoma trichophyllum*, etc. Although the species has been reported from old logs, I have never seen it as a pronounced humicole. In the high Arctic on mossy, snow-fed slopes, in full sun on weakly calcareous talus, together with *Lophozia alpestris* subsp. *polaris*, *L. heterocolpa*. *L. hyperborea*, *L. quadriloba*, *Anastrophyllum minutum*, *Scapania praetervisa polaris*, *Tritomaria heterophylla*, *Cephaloziella arctica*, *Solenostoma pumilum polaris*, *Odontoschisma macounii*, and *Cryptocolea imbricata*, often as a compact sun form, with coarsely bulging trigones and yellowish cell walls. The leaves are then dense, imbricate, strongly concave, and somewhat antically secund, suberect to erect-spreading, giving the plant a very different facies from the lax-leaved phases with nearly horizontally patent leaves which are commoner in less arctic stations. *Tritomaria quinquedentata* very rarely produces gemmae (as noted by Arnell, 1906, p. 150). Arnell also notes that mature capsules may often be found in late summer, which corresponds with my experience; I have seen sporophytes only from mid-August to late September.

Differentiation

Usually easily recognized and hardly to be confused with any other species of *Tritomaria*, especially when mature and typically developed. Normally much more robust than our other species, averaging fully twice the width (for the small phases, see p. 691). *Tritomaria quinquedentata* differs from the other species of the genus not only in size but also in the (*a*) very rare occurrence of gemmae, and their pale yellow-brown color; (*b*) frequent development of sex organs and capsules; (*c*) often explanate leaves, which have a diagnostic form. The leaves vary from fully as broad as long (the length measured to the apex of the ventral lobe, which is by far the largest) to, more commonly, conspicuously broader than long.

Tritomaria quinquedentata is a strongly isolated species, without close relatives. The very broad leaves (wider than long on all but extreme, lax, small-leaved modifications) and their tendency to spread almost horizontally, nearly in one plane, together with the almost uniform absence of gemmae, give it an aspect quite different from our other species of *Tritomaria*. It usually has a more prostrate growth pattern (although *T. scitula* may be scarcely more ascending).

Frequently *T. quinquedentata* forms large, almost pure mats, which often bear perianths or androecia but almost never gemmae, differing thus from other taxa of the genus, in which there is abundant gemmae production but infrequent or rare development of sex organs. The yellowish or yellowish brown gemmae in *T. quinquedentata* are produced in such small numbers that they are often overlooked. Our other species produce gemmae in large, rust-red masses, forming a conspicuous identifying feature.

Because of the robust size and generally somewhat explanate, scarcely canaliculate leaves, *T. quinquedentata* is likely to be confused with *Lophozia barbata*. Indeed, it was once placed into *Barbilophozia* with *L. barbata* and even regarded as a mere variety of *L. barbata*. However, the constantly 3-lobed, asymmetric leaves, usually cuspidate at the lobe apices, distinguish *T. quinquedentata* from *L. barbata* (in which the nearly symmetrical leaves are commonly 4-lobed). Furthermore, *T. quinquedentata* has the antical end of the leaf essentially transversely inserted; in *L. barbata* the antical end (like the rest of the insertion) is strongly oblique.

In the high Arctic, small, narrow-leaved forms (fo. *gracilis*) occur that are at times difficult to separate from *T. scitula* and from the broad-leaved *T. heterophylla*. The differentiation between these plants is treated under *T. heterophylla* (p. 675) and *T. scitula* (p. 669); see also below.

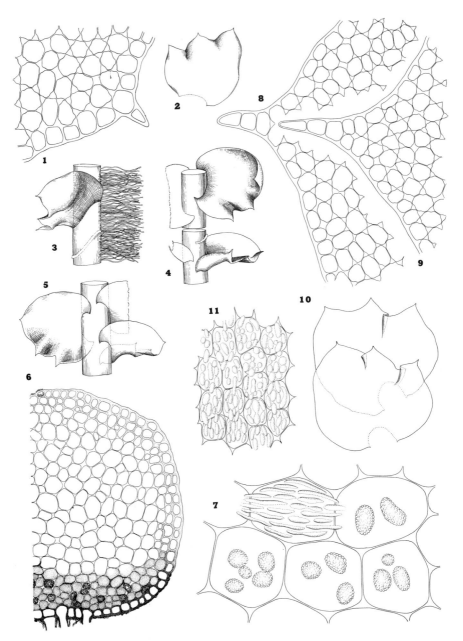

FIG. 232. *Tritomaria quinquedentata* subsp. *quinquedentata* (1–7), subsp. *papillifera* (8–11). (1) Cells of leaf lobe apex (×278). (2) Leaf (×20). (3) Shoot sector, lateral aspect (×20). (4) Shoot sectors, dorsal aspect, above with explanate leaves, below with them more subvertical (×20). (5) Explanate shoot sector, postical aspect (×20). (6) Stem cross section (×135). (7) Median cells from xeromorphic form with maximal development of cuticular papillae; near fo. *gracilis* (×1050). (8–9) Ventral and dorsal leaf

Variation

Although in the Arctic a veritable *typus polymorphus*, *T. quinquedentata* is generally a relatively stenotypic species southward, and there has a constantly recognizable facies. However, weak forms occasionally are found which bear narrowly obtrapezoidal or oblong leaves closely similar to those of *T. scitula*.

Southward usually with the plants somewhat prostrate, mat-forming rather than caespitose, and often with patent, fluted leaves, only the ventral lobe apex generally being distinctly incurved. In close-leaved extremes, the leaves are more transversely oriented but do not become strongly concave. By contrast, in the far north the dominant phases of the species are strongly concave-leaved, the transversely oriented and frequently suberect, imbricate leaves being often almost hemispherical. Associated with this, they lack the distinctly fluted appearance so often marked in the more patent-leaved southern phases. These differences hardly warrant systematic recognition, being probably primarily environmental in origin.

By contrast, the variants in the following key are presumably genetic in origin.

Key to Intraspecific Variants of T. quinquedentata

1. Cells with small, often very inconspicuous papillae . . . . 2.
   2. Leaves broader than long, usually much so (width usually 1.15–1.25× length), markedly asymmetric, the ventral lobe strongly ampliate, much longer than dorsal . . . . 3.
      3. Cells smaller: in margins 15–21 μ usually; trigones (even in xeromorphic forms) usually rather small, little or not bulging; usually quite opaque. Oil-bodies 3–7(8–10) per cell . . . var. *quinquedentata*
      3. Cells larger: marginal cells 18–25 μ or usually larger; trigones often very coarse; plants usually subpellucid . . . . 4.
         4. Leaf margins edentate . . . . 5.
            5. Marginal cells of lobes (16)18–22 μ; median cells 22–26 × 25–30(36) μ, with (3–6)7–11(14) oil-bodies each . . . . var. *turgida*
            5. Marginal cells of lobes 25–30 μ; median cells 26–30 × 28–42 μ, with 9–17(18–19) oil-bodies each . . . . . . var. *grandiretis*
         4. Leaf margins (especially of dorsal lobe) toothed, the teeth variable, 1–6 cells long; marginal cells averaging 22–28 μ; trigones rather small . . . . [var. *dentata*][222]

---

[222] Not found, to date, in our region; see Arnell (1956, p. 148); known only from Sweden.

---

apices, respectively (×270). (10) Two leaves (×17.5). (11) Median cells, with cuticular papillae drawn in (×330). [1–6, from Gunflint Lake, Minn., *RMS*; 7, *RMS & KD 66-1467*, Agpat, W. Greenland; 8–11, from type of subsp. *papillifera*, from "Baltistan" (= type of *Jungermannia trilobata* Steph., nec L.).]

2. Leaves oblong to quadrate or rounded-quadrate, as long or nearly as long as broad (width 0.85–1.15× length), subsymmetrical: ventral lobe not strongly ampliate . . . . fo. *gracilis*

1. Cells with coarse, close, cuticular papillae; cells small, the marginal 13–15 μ, the median (12)13–16 × 13–17 μ. (Asia only) . . . subsp. *papillifera*[223]

TRITOMARIA QUINQUEDENTATA var. *GRANDIRETIS* Buch & Arn.

[Fig. 230:8–9; 231:9, 12–13]

Tritomaria quinquedentata var. *grandiretis* Buch & Arn., *in* Arnell, Svensk Bot. Tidskr. 44(1):84, 1950; Schuster, Canad. Jour. Bot. 36:284, figs. 2:8–9, 3:12–13, 1958.

Similar to typical *T. quinquedentata*, but somewhat *paler and more pellucid* in appearance. *Cells much larger*: marginal cells of lobe apices ca. 23–28(30–35) μ; subapical and lobar cells (21)23–28 × 24–35 μ; median cells (23)25–32 × 30–42 μ; basal cells 23–32(35) × 35–48(56) μ. Oil-bodies more numerous, 9–19 per median cell.

Type. The first locality mentioned by Arnell, from Novaya Zemlya, 71°20′ N., may be so considered.

Distribution

Known from northern Siberia and Scandinavia (Sweden: Torne Lappmark at 68°07′ N.) and as follows:

NW. GREENLAND. Kangerdluarssuk, E. side of Red Cliffs Pen., Inglefield Bay (*RMS 45817d*). W. GREENLAND. Tupilak I., W. of Egedesminde, 68°42′ N. (*RMS & KD 66-914*); Lyngmarksbugt near Godhavn, Disko I., 69°15′ N., 53°30′ W. (*RMS & KD 66-343a*). ELLESMERE I. Moist, snow-fed scree slope, on humus, ca. 2000–2400 ft, E. edge of U.S. Range, ca. 9–10 mi due W. of Mt. Olga, ca. 82°24′ N., 65°20–30′ W. (*RMS 35572a, 35593*).

Occurring admixed with typical *T. quinquedentata*, *Anastrophyllum minutum* var. *grandis*, *Lophozia quadriloba*, *Scapania praetervisa polaris* (in Ellesmere I.) and with *Lophozia ventricosa*, *L. hatcheri*, *L. rubrigemma*, *Chandonanthus setiformis*, and *Ptilidium ciliare* (in northwest Greenland).

The var. *grandiretis* has been considered to be a putative polyploid derivative of normal *T. quinquedentata*. The two Ellesmere I. collections, in which the

---

[223] See Schuster, Rev. Bryol. et Lichén. 34(1–2):275, 1966. The subspecies is known, with certainty, only from northwest India, the Himalayas, and Formosa. Synonyms are *Jungermannia trilobata* Steph. (nec Linnaeus), *Sphenolobus trilobatus* Steph., *Lophozia verrucosa* Steph., and *L. asymmetrica* Horik.; see Fig. 232:8–11.

Both Kitagawa (1963a) and Schuster (1965) have dealt independently with the coarsely tuberculate Asiatic extreme of *T. quinquedentata* (the paper by Schuster, although submitted to Rev. Bryol. et Lichén. *in February 1964* was not published until 1966!). If the plant is segregated as a mere variety, it must bear the name *T. quinquedentata* var. *asymmetrica* (Horik.) Kitagawa (1963a); if, as seems to me more appropriate, as a subspecies, it should properly be called *T. quinquedentata* subsp. *papillifera* Schust.

plants are admixed with typical *T. quinquedentata*, also suggest such an explanation, the intermingled plants of the two taxa showing the following differences in cell size:

	var. *quinquedentata*	var. *grandiretis*
Marginal cells of lobes	17–21(23) μ	23–28(30–35) μ
Subapical and median cells of lobes	16–21(23) × 18–33 μ	(21)23–28(30) × 24–35 μ
Median cells	21–24 × 25–35 μ	(23)25- 32 × 30–42 μ
Basal cells	20–25 × 35–45 μ	23–32(35) × 35–48(56) μ
Oil–bodies, number per subapical and median cell	3–7(8–10)	9–17(18–19)

Of these differences, the discrepancy in oil–body number is at once striking and systematically important (Schuster, 1958).

TRITOMARIA QUINQUEDENTATA var. TURGIDA
(*Lindb.*) *Weimark*
[Fig: 228:11–12]

Jungermannia quinquedentata var. *turgida* Lindb., Kgl. Sv. Vetensk.-Akad. Handl. 23(5):59, 1889.
Tritomaria quinquedentata var. *turgida* Weimark, Lunds Bot. Förening 2:9, 1937; Weimark, Svensk Bot. Tidskr. 31(3):375, 1937; Schuster, Canad. Jour. Bot. 36:285, 1958.

Turgid arctic form of *T. quinquedentata*, often with facies of *Tritomaria* (*Saccobasis*) *polita*, normally growing in fens, bogs, or wet polygon ground (with such species as *Calliergon sarmentosum*, *Drepanocladus revolvens*, *D. uncinatus*, *Oncophorus wahlenbergii*). Large, to 4 mm broad; yellow green to usually brownish-tinged; leaf lobes commonly very short and broad, obtuse or virtually so, the two antical lobes hardly cuspidate, the postical normally hardly or not at all cuspidate. Cells often slightly larger than in typical *T. quinquedentata*, the marginal cells (16)18–22 μ, with coarse, bulging trigones (in this respect often closely approaching var. *grandiretis*, with which it may prove necessary to unite var. *turgida*).

Type. Yenisey R., Siberia.

Distribution

A high subarctic and arctic phase of *T. quinquedentata*. Persson (1947) states that this plant is "rather common" in northern Scandinavia "but seems to be absent from Central Europe." Müller subsequently reports it from one station in the Tyrolese Alps (Spiegelkogel, 3400 m). Mårtensson (1955) doubts the validity of the taxon. In North America reported as follows:

E. GREENLAND. Hurry Inlet (*Dusén, Aug. 4, 1899*!; see also Jensen, 1906a); Cape Stewart, at Scoresby Sund, and Dove Bugt, 76°40′ N. (*fide* Jensen, 1910). NW. GREENLAND. Kangerdluarssuk, S. side of Inglefield Bay (*RMS 45790*); Dundas, Wolstenholme Fjord (*RMS 46265*); Siorapaluk, Robertson Bugt (*RMS 45705d*). ELLESMERE I. Supposedly numerous localities in the southern half (Bryhn, 1906); see, however, the appended discussion.

Widespread in Alaska, from the eastern Pacific to the central Pacific Coast, the Alaska Range, the Aleutian Isls., and the Bering Sea district (St. Lawrence I.), according to Persson (1947).

Persson states that he tried to determine "whether this variety might merit a higher systematic rank . . . the statements about a bigger size of the cells, which, combined with the northern distribution, suggested it possibly to be a substituting species with a higher number of chromosomes." He was unable to come to any definite conclusion, "the characters, especially the size of the cells, being too varying. Yet this form is of interest . . . because of its likeness to *Saccobasis*." In fact, Clark and Frye (1942) erroneously reported *Saccobasis* from Thumb Bay, Alaska, the material proving to be *T. quinquedentata* var. *turgida* (Persson, 1947).

Persson states (*loc. cit.*) that "this interesting variety is reported to be one of the most common hepatics brought home by the 'Fram' expedition, for Arctic North America (Bryhn, 1906) but I have not seen it again reported for North America." The dubious status of this variant led to a study of the Ellesmere I. collections reported as var. *turgida* by Bryhn, with the following results (Schuster, 1958): Framshavn plants (*Simmons 753, 754, 739, 740*) = *Lophozia hatcheri*; plants from Bedford Pim I., off the east coast of Ellesmere I., 78°40' N. (*Simmons 1357*), labeled simply "*Lophozia lyoni*" (= *Tritomaria quinquedentata*) on the small (inner) packet, but "*Lophozia quinquedentata turgida*" on the outer packet = a yellow-brown, thick-stemmed, small-leaved modification of typical *Tritomaria quinquedentata* (mod. *parvifolia-pachycaulis-colorata*). I cannot consider this plant varietally distinct from *T. quinquedentata*. Associated with their smaller size, the leaves are more nearly symmetrically trilobate, somewhat narrower than usual. The lobes are typically all cuspidate with a mucro formed of 1–2 hyaline, enlarged cells. Occasionally the postical lobe is blunt; this usually is associated with development of androecia, ♂ bracts commonly possessing blunt or rounded postical lobes. In shape and size of leaves these plants approximate *Tritomaria heterophylla;* they differ from this in the smaller cell size (16–18 μ wide in the leaf middle and at bases of leaf lobes, only rarely and locally to 18–20 μ wide in the median cells, 15–18 μ on the lobe margins), the absence of vinaceous pigmentation, and the lack of gemmae. Associated evidently with the habitat, the plants have antically secund, saucerlike, concave leaves, which are suberect, thus giving the plants a facies much like that of *Lophozia binsteadii*. Another collection, same data as the preceding (*Simmons 1325*), labeled "*Lophozia lyoni turgida*," consists of several stems of *Tritomaria quinquedentata*, a phase with rather narrow, concave, suberect leaves. The leaf lobes are relatively narrow and in general all sharply cuspidate; these plants cannot be assigned to var. *turgida* unless that taxon is delimited very differently from Persson's 1947 treatment. A final collection, from Bedford Pim I. (*Simmons 1320*), supposedly *Lophozia lyoni turgida* = *Lophozia hatcheri*. Thus Bryhn had no clear concept of this variant.

Plants approaching var. *turgida* occur as far south as the Hudson Bay region. I have seen sparse, somewhat brownish yellow plants from North-west Territory (Farmer I., 58°25′ N., 80°48′ W., *A.R.A. Taylor 200, Aug. 24, 1944, p.m.p.*), in which the ventral lobes are blunt to obtuse, often actually rounded, but the two antical lobes are slightly cuspidate. In these plants the marginal cells of the lobes average 20–23 μ, thus being somewhat intermediate in size between those of "normal" *T. quinquedentata* and those of var. *grandiretis*; the median leaf cells are also very strongly collenchymatous.

TRITOMARIA QUINQUEDENTATA fo. GRACILIS
(Jens.) Schust., comb. n.
[Fig. 228:9–10]

Jungermannia quinquedentata fo. *gracilis* Jens., Öfversigt Kgl. Vetensk.-Akad. Förhandl. 1900, no. 6:798, 1900.
Jungermannia quinquedentata var. *tenera* Arn. & Jens., Naturw. Untersuch. des Sarekgebirges in Schwedisch-Lappland 3:111, 1907 (lectotype, present designation: Ruopsok, July 19, 1902).
Lophozia quinquedentata var. *tenera* Meyl., Beitr. Krypt.-Fl. Schweiz 6(1):160, 1924.
Lophozia tenera Meyl., ibid. 7(2):164, 1933.

Small, ¼ to ½ as large as normal, yellowish brown, opaque, a mod. *densifolia-angustifolia-parvifolia-subaequiloba*: the *leaves concave, with lobes ascending*, little plicate, sharp but rarely sharply apiculate, entire-margined, often as long or almost as long as wide (leaf width from ca. 0.85 to 1.15 × the length, usually) because of the reduction in size of the ventral lobe (which has become much narrower; *ventral margin little ampliate*, rather weakly convex in most cases); *sinuses usually sharp*, ± *rectangulate. Leaf lobes subequal* in size or the ventral only 0.1–0.2 longer than dorsal, but sinus between dorsal and middle lobe remaining usually deeper and sharper than sinus between middle and ventral lobes. Gemmae (constantly?) lacking.

Type. Scoresby Sund, eastern Greenland, 14 km west of Cape Stewart at mouth of Hurry Inlet, 70°25′ N., ca. 23° W. (*Hartz, Aug. 3, 1891!*); lectotype.

Distribution

E. GREENLAND. Kangerdlugssuak, middle of Fjord, ca. 68° N. (*Böcher 904!, p.p.*, with *Cephalozia pleniceps, Nardia geoscyphus, Blepharostoma trichophyllum brevirete*); Scoresby Sund, Charcots Land,[224] 71°54′ N., 29° W., 700 m (*Holmen 19000!*); Hekla Havn,

[224] The plants from Charcots Land (*Holmen 19000*) are a mod. *parvifolia-densifolia*; they are unusual in having almost symmetrically trilobed leaves (the ventral lobe has the base unusually weakly dilated and is only a little longer than the dorsal). Yellow-brown color is normal; there is no vinaceous pigmentation; leaves are quite opaque, as is normal for *T. quinquedentata*. Leaf measurements clearly show plants with relatively narrow and small leaves:

Width	Length of Ventral Lobe	Length of Dorsal Lobe
800 μ	730 μ	670 μ
830	770	720
980	1050	820

Scoresby Sund, 70°30′ N., 26° W. (*Hartz, Apr. 1892*, det. Jensen as *J. quinquedentata* fo. *gracilis*; also *Hartz, Feb. 1892*, det. Jensen as fo. *gracilis*); Danmarks Ø, Scoresby Sund, 70°30′ N., ca. 26° W. (*Hartz, June 1892*, det. Jensen as fo. *gracilis*; this is a larger plant, grading into typical *T. quinquedentata*). W. GREENLAND. Godhavn, below Lyngmarksfjeld, Disko I., 69°15′ N., 53°30′ W. (*RMS & KD 55-195*); Fortune Bay, Disko I. (*RMS & KD 66-126, 66-127*); Tupilak I., W. of Egedesminde, 68°42′ N. (*RMS & KD 66-037a*); E. end of Agpat Ø, E. of Umiasugssup ilua, 69°52′ N., 51° 38′ W. (*RMS & KD 66-1497*); Eqe, near Eqip Sermia (Glacier), 69°46′ N., 50°10′ W. (*RMS & KD 66-325*, transitional).

The fo. *gracilis* is first alluded to by Jensen (1897, p. 378); he refers a plant from Hekla Havn, Scoresby Sund, to "fo. *gracilis*," but without a diagnosis. In 1900 a proper diagnosis was published in which Jensen refers to fo. *gracilis* two east Greenland specimens, one from Cape Stewart and the other from Kaiser Franz Joseph Fjord. I designate the Cape Stewart plant as lectotype. In 1907 Arnell and Jensen considered this plant to merit varietal status and described it as var. *tenera*, citing fo. *gracilis* as a synonym. Meylan (1933, p. 164) finally considered this plant as an independent species, or "une sous-espèce, si l'on veut," proposing the name *Lophozia tenera* (Jensen) Meylan. Meylan regarded the plant as having the same relationship to "*L. quinquedentata*" as *L. hatcheri* has to *L. lycopodioides*. I disagree with this taxonomic conclusion; I can regard "*Lophozia tenera*" only as an xeromorphic form or ecotype which is widespread in the high Arctic, often in the continental portions. Meylan (*loc. cit.*) also regarded the "var. *aquatica*" of Pearson as a synonym of "*Lophozia tenera*."

Müller (1951–58, p. 734) complicated this situation further by incorrectly placing "var. *tenera*" or "*Lophozia tenera*" as a synonym of *Tritomaria scitula*, to which it certainly does not belong. It admittedly approaches this species in the occasionally unusually narrow leaves, but

---

The collection of *Böcher 904* is one of several phenotypes seen with unusually narrow leaves whose ventral lobes are not expanded (the ventral lobe not being greatly dilated; in "var. *turgida*," the opposite extreme is attained and the ventral lobe is very broadly ampliate). The plants are yellowish to yellow-brown, with deep brown ventral stem surface, rarely when young with a little reddish pigmentation discernible. Associated with the dense leaves and small size (mod. *densifolia-parvifolia-angustifolia*), the leaves are rather similar to those of *T. scitula*, in which sporadically 2-lobed leaves are found. Leaf measurements are as follows:

	Width	Length of Ventral Lobe	Length of Dorsal Lobe
Trilobed leaves	780 μ	690 μ	560 μ
	600	630	550
	560	600	530
Bilobed leaf	530	600	530

Thus it is evident that there is allometric growth, the widest leaves being broader than long and more asymmetric, and the narrower leaves longer than wide and nearly symmetric.

differs in the more convex ventral leaf margin, the more opaque, usually yellow-brown color, the more fluted leaves (those of *T. scitula* are canaliculate but never fluted), and, in particular, the constant lack of gemmae. In extreme cases fo. *gracilis* may form veritable phenocopies of *T. scitula*, but such phenocopies seemingly are always gemma-free. Also, in fo. *gracilis* the leaf sinuses are largely sharply angular, often rectangulate or even acute; in *T. scitula* the sinuses are rounded or blunt at the base. The often angular to sharp sinuses also serve to distinguish fo. *gracilis* from typical *T. quinquedentata*. In plants that show a transition to the main species, which has shallow and rounded sinuses, transitions in sinus form occur. Such transitions even are found on some forms of the type of fo. *gracilis*.

Plants from Hekla Havn (*Hartz, Apr. 1892*) are exceedingly similar to those of *Böcher 904*, representing also an extreme mod. *densifolia-angustifolia-parvifolia-subaequiloba*, with concave and only feebly fluted leaves. They grew mixed with *Gymnomitrion concinnatum*, an extreme xerophyte, which would suggest that these small, subequally trilobed phases of *T. quinquedentata* are mere environmental phases or at best ecotypes of dry sites. These phases are "evolved" by the reduction of the ventral lobe, which becomes shorter as well as narrower—the ventral margin not being strongly ampliate. However, it is quite clear from studying a large series of forms that fo. *gracilis* is not a purely xeromorphic form, since in boggy ground large, soft and lax, robust, green forms occur (*RMS & KD 66-037a*, Tupilak I., W. of Egedesminde) in which the ventral lobe is also hardly ampliate, and the leaves are subsymmetrical. Hence a genetically determined fo. *gracilis* exists.

Occasional extremes (e.g., *RMS & KD 66-195*) may have vinaceous leaf bases and exceedingly coarse trigones. Such forms may also have cells with only 2–5 oil-bodies per median cell, with extremes of 1 and 6–7 oil-bodies to be found per cell.

Subgenus *SACCOBASIS* (Buch) Schust., comb. n.

Tritomaria auct. (*p.p.*, incl. Joergensen, 1934).
Saccobasis Buch, Mem. Soc. F. et Fl. Fennica 8:291, 1933; Schuster, Amer. Midl. Nat. 45(1):65, 1951; K. Müller, Rabenh. Krypt.-Fl. ed. 3, 6:715, 1956; Arnell, Illus. Moss Fl. Fennosc. 1:144, 1956.

Shoots large, up to 3–5 cm tall, *erect or ascending* in growth (more rarely decumbent), *not* or inconspicuously autonomously dorsally *curved at apices*. Stem in cross section about (11)12–16 cells high, the cortical 1–3 strata more thick-walled than the interior; dorsal cortical cells extremely elongate, more or less *thick-walled*, *4–8× as long as wide*, somewhat narrower but longer than the basal leaf cells, as long as underlying medullary cells; ventral cortical cells short, in 1–2 layers, associated with development of rhizoid initials; medulla with ventral region of several layers of smaller cells that may become mycorrhizal, occupying ca. the basal fourth of the medulla; dorsal portion of medulla of larger, more hyaline cells, of 2–4× the length of ventral medullary cells; ventral stem sectors obsolete, merely 1–2 cell rows wide, not able to develop underleaves. Rhizoids long and merely moderately numerous (unlike *Tritomaria*). Leaves scarcely succubous, broadly quadrate or rectangular when spread out and as long as wide to wider than long, erect-spreading, at least on basal half distinctly concave-canaliculate (the apical portions often strongly spreading or reflexed outward and spread out in a plane); leaf nearly *equally*, *shallowly* (2)3(4)-*lobed* at the truncate apex, the *lobes broad and obtuse*, the 2 ventral lobes apparently originating as in *Tritomaria*. Leaf insertion essentially *forming an inverted catenary curve*: both lateral halves of the line of insertion *curving toward stem apex*, forming a deep curve open to the stem apex (resulting in the pocketlike leaf base); dorsal half of insertion usually reaching further toward the stem apex than ventral half, at the extreme dorsal end of the insertion *slightly decurrent*; ventral end of insertion suddenly, strongly curved toward stem base (in a hooklike manner; the *base of the ventral lobe therefore strongly decurrent*). Leaf cells *strongly collenchymatous*, isodiametric near apices, but near base *becoming extremely elongate*, somewhat wider than dorsal cortical cells. Gemmae rare, occurring on younger leaves of shoot apices, smooth or angulate, 1–2-celled.

Dioecious. ♀ Bracts in two pairs, somewhat larger than leaves, sometimes more deeply lobed. Perianth as in *Tritomaria*, but *wide and entire at mouth*. ♂ Bracts like vegetative leaves, differing only in the somewhat saccate or ventricose bases; *antheridia 2–4 per axil*, the stalk 1-seriate. Capsule wall 3–4-stratose, thickenings as in subg. *Tritomaria*.

Type. Tritomaria polita (Nees) Schiffn. = *Saccobasis polita* (Nees) Buch.

The peculiar type of leaf insertion (which does not recur in any other genus of Lophoziaceae) warrants the separation of *Jungermannia* (or *Sphenolobus*) *polita* as a discrete subgenus. *Saccobasis* is much more strongly specialized than subg. *Tritomaria* in the following ways: leaf insertion has become extremely specialized, with the line of insertion curving toward the shoot apex both dorsally and ventrally, but suddenly curving basad again dorsally and ventrally at the extreme ends of the line of insertion (with consequent abrupt decurrence of both dorsal and ventral lobes); the ventral leaf base is especially strongly decurrent. The apical curvature of the dorsal and ventral portions of the line of insertion (exc. the extreme ends) results in an axillary "pocket," such as is not found in other Lophozioid taxa. The leaf itself has, secondarily, become equally trilobed (probably from a *T. scitula*-like type of leaf), and the lobes have become relatively short, broad, and obtuse. The stem structure is of the *Lophozia-Tritomaria* type, i.e., there is dorsiventral differentiation of the medulla, into a ventral band of smaller, shorter, cells,[225] and a dorsal region of larger

---

[225] Buch (1933a) stressed the fact that in *Lophozia* (subg. *Lophozia*) and *Tritomaria* the ventral medullary band is attacked by mycorrhizae, becoming brownish and discolored. Apparently, however, he did not observe similar mycorrhizal infection in *Saccobasis*. In Norwegian material

more elongate cells. As in *Tritomaria*, the dorsal cortical cells are elongate, but they are more elongate than in that subgenus (attaining an average length equal to that of the underlying medullary cells), and they may become extremely pachydermous.[226] Another feature characterizing *Saccobasis* is the lack of any autonomous dorsal curvature of the stem apices, as stressed by Buch.

I have previously suggested (Schuster, 1951a, p. 66) that a person with a conservative generic concept might very well place *Saccobasis* as a mere subgenus of *Tritomaria*, stating (*loc. cit.*, p. 24) that "there seems to be no reason for considering *Saccobasis* as other than a highly sidewise specialized . . ." taxon which is postulated to have been "derived from a common ancestral form with *T. scitula* . . . ." Since then I have described the collenchymatous-leaved, purplish, symmetrically lobed *T. heterophylla* (Schuster, 1958), a species which approaches *Saccobasis* even more closely. A careful comparison of the two taxa, *T. heterophylla* and *T. polita* subsp. *polymorpha*, has convinced me that a generic distinction between the two can hardly be upheld. I have already noted that the attempt (as in Frye and Clark, *loc. cit.*) to separate *Saccobasis* on the basis of the supposed bast-fiberlike, elongated, thick-walled cortical cells is untenable. Weak forms of *T. polita polymorpha* have cortical cells almost leptodermous, ranging from 15–17 × 60–95 μ (1:4–6) to 16.5–24 × 40–75 μ (1:2–4.5). If this is compared to Fig. 231:4 (*T. heterophylla*) and 231:10 (*T. quinquedentata*), it will be seen that no real distinctions can be drawn. Other supposed generic distinctions of *Saccobasis* [the strict growth of the shoot tip, without curvature, and the peculiar leaf insertion (Fig. 233:1–5)] also "fail" with weaker, arctic phases. The compact mod. *parvifolia-densifolia* may grow prostrate, with the shoot tip distinctly arched away from the substrate, and the leaf insertion of the mod. *parvifolia* is much simplified, the insertion nearly transverse in the dorsal half. Such mod. *parvifolia-densifolia* phenotypes are, indeed, extremely hard to separate from *T. heterophylla*, since they may have similarly broad leaves (to 1230 μ wide × only 900 μ long).

There remains, as a generic distinction, only the leaf insertion. However, even in *Lophozia* and other *Tritomaria* species, the ♂ bracts may have the dorsal end of the insertion arched upward, as, e.g., in *Lophozia gillmani* and *L. excisa*.

---

of *T.* (*S.*) *polita* the medulla is mycorrhizal ventrally, as in *Lophozia* and *Tritomaria*. The dorsiventral differentiation, however, is poorly indicated (see Fig. 233:8). On comparing stem sections of *L.* (*Orthocaulis*) *kunzeana* and *L.* (*Barbilophozia*) *barbata* (Figs. 148:5 and 162:8) it is obvious that the dorsiventral specialization is hardly more distinctly indicated in *Saccobasis* than in these two taxa.

[226] Buch (1933a) depicted the cortical cells with walls nearly as wide as the lumen, but did not stress the pachydermous nature of the cells as a generic character, obviously because his own experience showed that the thickness of the cell wall is largely a reflection of environmental conditions. Frye and Clark (1937–47, p. 337) separate *Saccobasis* from other genera of Lophoziaceae because of the "cortical cells of the stem with walls about as thick as the diameter of the cell cavity." In Fig. 233:9 I show the dorsal cortical stem cells of the genus: here the walls range from ⅓ to ⅙ the thickness of the lumen. Obviously the thickness of the cell walls varies greatly with differences in environmental conditions and is thus taxonomically nearly useless.

Hence this distinction, also, hardly seems to me to warrant a separate genus.

Branching also does not allow us to draw any distinctions between *Tritomaria* subg. *Tritomaria* and subg. *Saccobasis*. In the first, branching is predominantly terminal, furcate, of the *Frullania* type, accompanied (at least in *T. quinquedentata*) by occasional lateral-intercalary branches (see, e.g., figs. 110 and 238g, in Müller, 1951–58). This is exactly as in *T. (Saccobasis) polita polymorpha.*[227]

Key to Subspecies of Subgenus Saccobasis

1. Leaves trilobed, averaging ca. 0.9–1.25× as long as wide; median cells with 2–7 oil-bodies each; plants with a greasy, glistening appearance. ♀ Bracts ovate to spatulate and at least as long as wide, divided more deeply than leaves into 3 lingulate lobes. Gemmae red-brown, polygonal, ca. 28 μ . . . . *T. (S.) polita* subsp. *polita* (p. 696)
1. Leaves polymorphous: mostly 3-lobed, some 4-, some 2-, lobed, all averaging 0.6–0.8× as long as wide; median cells with mostly (5–6)8–12 oil-bodies each; plants ± dull-textured. ♀ Bracts strongly transverse, subreniform, twice as wide as long, obsoletely 3–4-lobed, with lobes 3–5× as wide as long. Gemmae smooth, ovoid or ellipsoidal, yellow-brown, 20–22 × 22–31 μ . . . . *T. (S.) polita* subsp. *polymorpha* (p. 700)

TRITOMARIA (SACCOBASIS) POLITA (*Nees*)
Schiffn.

[Figs. 233, 234]

Jungermannia polita Nees, Naturg. Eur. Leberm. 2:145, 1836.
Diplophyllum politum Dumort., Hep. Eur., p. 50, 1874.
Jungermannia medelpadica Arn., Rev. Bryol. 18:12, 1891.
Sphenolobus politus Steph., Spec. Hep. 2:169, 1902; K. Müller, Rabenh. Krypt.-Fl. 6(1):613, fig. 297, 1910; Macvicar, Studs. Hdb. Brit. Hep. ed. 2:215, figs. 1–3, 1926.
Sphenolobus medelpadicus Steph., Spec. Hep. 2:169, 1902.
Lophozia polita Boulay, Musc. France 2:102, 1904.
Tritomaria polita Schiffn., Ber. Naturw. Ver. Innsbruck 31:12, 1908.
Saccobasis polita Buch, Mem. Soc. F. et Fl. Fennica 8(1932):292, fig. II:1–10, 1933; Schuster, Amer. Midl. Nat. 45(1):68, pl. 25, 1951; Schuster, Natl. Mus. Canada Bull. 122:11, 1951; K. Müller, Rabenh. Krypt.-Fl. ed. 3, 6:716, figs. 231–232, 1954.

In loose patches or scattered among other bryophytes, rather robust, 2–3 mm wide × 1.5–5 cm long, greenish or brownish green to, *usually, reddish brown or purplish black, with a somewhat greasy luster.* Stems ascending to erect from a decumbent or prostrate base, postically brown to *purplish black*, rigid, the elongate cortical cells often bastlike, simple or rarely branched; rhizoids long, frequent below, sparse or none above, violet-red at base usually. *Leaves symmetrical, lax* and *soft, shining, virtually transversely oriented*, attached by a complex and sinuous line (see subgeneric diagnosis), *at least slightly decurrent both antically and postically*, distant to weakly imbricate, the *base concave and pocketlike but the equally trilobed and truncate summit widely spreading*, subquadrate to short-rectangular, ca. 1720 μ wide × 1880 μ long, *trilobed 0.1–0.25 their length at their broadly*

---

[227] Müller (*loc. cit.*, fig. 231d) illustrates a branch that *seems* to be a basiscopic and *Radula*-type for *T. polita*. I have seen only one such branch in *T. polita* subsp. *polymorpha* (p. 703). Exactly similar *Radula*-type branches may occur in subg. *Tritomaria* (see, e.g., Fig. 226:2).

truncate apices, the lobes and sinuses similar, *obtuse to rectangulate*. Cells of leaf middle *large, ca. 25–32 × 45–50 μ*; *basal cells distinctly rectangulate*, to ca. 60–100 μ long (ca. *2–4× as long as wide*), averaging slightly wider than cortical cells of stem; cell walls *thin, but trigones large*, more or less *bulging*, often sub-confluent; oil-bodies 2–7 per cell, spherical to ellipsoidal, 5–10 μ, formed of spherules; cuticle weakly verruculose or verrucose-striolate, at least at base. *Gemmae reddish brown to purplish*, at apices of upper leaf lobes (leaf lobes then frequently acute); gemmae *tetrahedral to polyhedral, often stellate*, ca. 28 μ, very rarely produced [I have never observed them].

Dioecious. Androecia at first terminal, bracts similar to leaves but more concave at base, erect-spreading, contiguous to imbricate; 2–4-androus. Gynoecia with bracts erect and somewhat sheathing, their apices spreading or erect-spreading, *oblong-ovate to oblong-spatulate*, nearly *similar to leaves* but 3-(5-) lobed at summit, *more deeply lobed*, the lobes undulate. Perianth 0.5–0.65 emergent, cylindrical to the somewhat contracted, entire or subentire mouth. Elaters ca. 8 μ in diam.; spores ca. 15 μ, reddish brown.

Type. Radstädter Tauern, near Salzburg, Austria *(Funk)*.

Distribution

Essentially an arctic-alpine species, rather rare and perhaps restricted in distribution because of a weak "affinity" for calcareous substrates (at least in eastern North America). In Europe found both in the far north, ranging from Spitsbergen (Arnell and Mårtensson, 1959) and Norway, Sweden, and Finland southward to Scotland and Belgium (the Ardennes), recurring in the Alpine chain, from Switzerland (Meylan, 1924) to Austria (type), and in the northern Italian Alps; in Savoy; the Tyrol; Carinthia; eastward to the USSR.

In North America apparently transcontinental at high latitudes, the distribution uncertain because of confusion with subsp. *polymorpha*; supposedly ranging in the West from Alaska and the Yukon (Persson, 1947; Frye and Clark, 1937–47; Persson and Weber, 1958) to British Columbia and Alberta. In the east a very rare species south of the main body of the Tundra; absent from the various alpine localities in New England, as well as from the Great Lakes region.[228] Also reported from Ontario (Cain and Fulford, 1948). The duplicate collection (and, on inquiry, the original) sent by Dr. Cain, determined by M. Fulford, contains only *Marsupella sphacelata* var. *erythrorhiza*! Frye and Clark also issue a number, purporting to be this species, from the West; this material (at least in the

---

[228] Reported (Frye & Clark, 1937–47, p. 425) from New Jersey, based on Austin (1870); this mistake has already been perpetuated in Buch & Tuomikoski (1955). As has been pointed out (Schuster, 1951, p. 68), this is an error for *Lophozia marchica*, the "*Jungermannia polita*" Austin nec Nees being based on nomenclatural confusion. The Austin specimen, under that name (NY), is *L. marchica*.

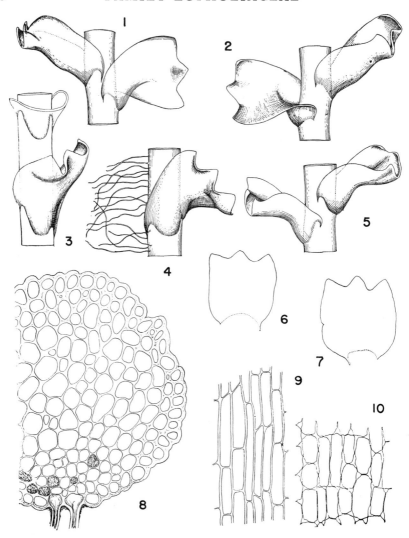

Fig. 233. *Tritomaria* (*Saccobasis*) *polita* subsp. *polita*. (1) Shoot sector, postical aspect (×18). (2) Same, antical aspect (×18). (3–4) Lateral aspects of short shoot sectors, showing arcuate insertion and, in fig. 4, rhizoids (×21). (5) Shoot sector, postical aspect (×16). (6–7) Leaves (×ca. 21). (8) Stem cross section (×160). (9) Dorsal cortical cells of stem (×ca. 140). (10) Basal leaf cells, showing the diagnostically elongate form (×ca. 140). [All drawn from Austrian plants collected by Loitlesberger.]

University of Minnesota herbarium) contains only *Lophozia floerkei* (Schuster, 1951).

GREENLAND. Nordostbugt, in bogs (Hartz; *fide* Jensen, 1906a). ELLESMERE I. Hayes Sd. area, on E. coast (Bryhn, 1906, p. 40). MELVILLE PEN. Steere (1941). NORTHWEST TERRITORIES. Tukarak I., Belcher Isls., Hudson Bay (Wynne & Steere, 1943; Schuster 1951). QUEBEC. Ft. Chimo Air Base, Ungava Bay (Schuster, 1951); mainland S. of Cairn I., Richmond Gulf, E. coast of Hudson Bay (Schuster, 1951); Gaspé Pen. (Lepage, 1944–45); between N. coast of Gaspé Pen. and Murdochsville (*RMS*); N. slopes, Mt. Albert, Gaspé (*H. Williams 10685!*). NEWFOUNDLAND. St. Barbe Bay and Cook Harbor, N. Pen. (Buch & Tuomikoski, 1955).

Ecology

Although reported from "weathered granite," the species appears to occur much more frequently on sedimentary basic rocks or diabase and other basaltic rocks. Evidently mostly either directly on damp or wet rocks or on thin humus or peaty muck over such rocks; in the Gaspé occasionally in cedar swamps. "Judging from the accompanying species S. [*Tritomaria*] *polita* is one of a large group of pioneer or near-pioneer species that undergoes ecesis on basic rocks or thin soil over basic rocks" (Schuster, 1951). Accompanying species in eastern North America are such "weak calciphytes" as *Lophozia heterocolpa* and *Cephaloziella arctica* var. *alpina*, as well as pronounced calciphytes like *Preissia quadrata* and *Lophozia bantriensis*; also species, such as *Blepharostoma trichophyllum* and *Cephalozia pleniceps*, which may occur over thin soil shielding basic rocks. Buch and Tuomikoski (1955) report *T. polita* from a deep moist cleft in limestone, associated with *Blepharostoma*, *Campylium stellatum*, *Dicranum* sp., *Lophozia heterocolpa*, *Pellia* sp., *Preissia*, *Odontoschisma macounii*, and *Timmia norvegica*; also with *Plagiochila asplenioides*.

Variation and Differentiation

Although with transversely oriented and subequally trilobate leaves, like *Tritomaria scitula*, *T. polita* can hardly be confused with this species because of the more elongated cortical cells; the greasy lustre; usually purplish or purplish brown color; straight stems, without the shoot apices acroscopically arched; strongly elongated basal leaf cells; and, particularly the peculiar insertion of the leaf. The last feature is quite without parallel except for its presence in subsp. *polymorpha*. However, gemmiparous phases of *T. polita* may have acute leaf lobes (" = var. *acuta* Kaal.") and, according to Müller (1905-16, p. 615), may have an aspect closely approaching that of *T. scitula*. Normal phases of *T. polita*, however, differ from all other species of *Tritomaria* in the extremely elongated, narrow cortical cells of the stem.

Buch (1933a) illustrated the cortical cells of this species as extremely thick-walled, the walls amost equal to the lumen in diameter. Frye and Clark (1937–47, p. 337), evidently generalizing from Buch's figure, separate *Saccobasis* from all genera of Lophoziaceae on the basis of the supposedly bast- or fiberlike nature of these cortical cells. This is evidently justified only for material from xeric sites. Austrian plants (collected by Loitlesberger) here depicted show cortical cell walls that are only moderately thickened, averaging ⅓ to ⅙ the diam. of the lumen. Similarly Schuster (1951a) reports material from Quebec and Northwest Territories with the lumen 3–8× the diam. of the cell walls. Evidently, the amount of wall material laid down is subject to a great deal of variation under differing environmental conditions and is accordingly meaningless as a diagnostic character.

TRITOMARIA (SACCOBASIS) POLITA subsp. *POLYMORPHA* Schust., subsp. n.

[Fig. 234:1–11]

Subspecies a subsp. *polita* differens quod (*a*) textura hebes, (*b*) folia variabiliter 2-3-4-lobata, typice multo latiora quam longa; (*c*) gemmae ellipsoideae, leves; (*d*) ♀ bractae latissimae subreniformes, ad 2 plo latiores quam longae, solummodo obsolete 3-4-lobatae.

Procumbent to prostrate in growth, except where crowded, yellow-green to blackish, reddish brown ventrally (leaf bases often brown to red-brown ventrally), moderate in size; small, scorched phenotypes only 1–1.6 mm wide, larger yellow-green to olive-green phenotypes 2–2.4(2.5) mm wide × 8–20 mm long; plants sparingly branched, the branches predominantly terminal, furcate, *Frullania* type (dorsal half-leaf usually oblong, shallowly bilobed), occasionally intercalary from lower half of leaf axil; *surface texture dull.* Stem to 325–400 μ in diam., terete, ca. 11–12 cells high, dorsally and laterally with 1(2) strata of smaller, thicker-walled cells with yellowish to yellow-brown walls, ventrally with the cortex less defined but with several layers of smaller (13–17 μ) medullary cells that, in part, become fuscous, disorganized, mycorrhizal; upper medullary cells collenchymatous, large, 24–32 μ in diam.; dorsal and lateral cortical cells variable, in small phenotypes 15–17 × 60–95 μ (1:4–6) ranging to 16.5–24 × 40–75(1:2–4.5), in robust plants (14)15–21 × (48)60–80 μ (ca. 1:3.5–5), the longitudinal walls *somewhat but not strongly incrassate,* wall layer perhaps 0.15–0.2 diam. of lumen; rhizoids ± purplish or red-brown at base. Leaves vertical or almost so, *stiffly* obliquely to widely patent, *usually rather dense* and subimbricate, imbricate above, ± *plicate and undulate* but not or hardly canaliculate, the base little or moderately concave, inserted on an obtusely, broadly V-shaped line, when flattened *transversely oblong,* on mature shoots 1475–1530 μ wide × 1125–1230 μ long to 1640–1680(1850) μ wide × 1065–1130(1230) μ broad (ca. *1.25–1.55× as wide as long*), variously *4-, 3-,* or *2-lobed* on the same, mature shoots, *for 0.15–0.25(0.3) their length, the lobes and sinuses both usually blunt to rounded,* occasionally ± angular; antical leaf sector with insertion obliquely running up toward stem apex, at the very end at most feebly decurrent; lobes often incurved, sinuses frequently somewhat gibbous, resulting in crispate leaf apices. Cells rather pellucid, very strongly collenchymatous, with coarsely bulging trigones usually, large: marginal upper cells

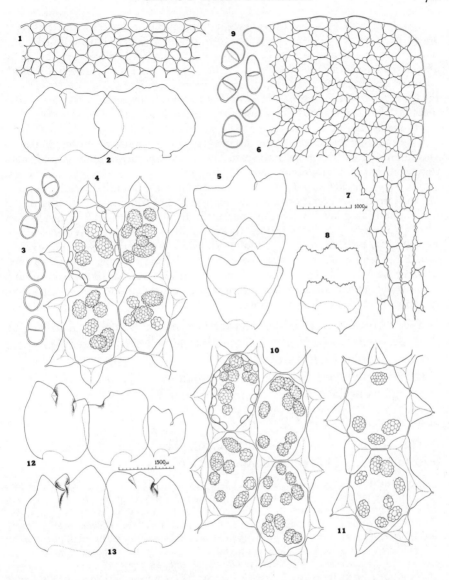

FIG. 234. *Tritomaria (Saccobasis) polita* subsp. *polymorpha* (1–11) and *Tritomaria (Saccobasis) polita* subsp. *polita* (12–13). (1) Cells of perianth mouth (×210). (2) Two ♀ bracts (×13). (3) Gemmae (×240). (4) Cells from upper one-third of leaf (×550). (5) Leaves (×14.8). (6) Leaf lobe apex (×150). (7) Basal leaf cells (×150). (8) Two gemmiparous leaves (×14.8). (9) Gemmae (×310). (10) Median cells (×565). (11) Cells from above leaf middle (×585). (12) Three leaves (×10). (13) Two ♀ bracts (×10). [1–2, *RMS & KD 66-153*, Godhavn, W. Greenland; 3–4, *RMS & KD 66-568*, Godhavn; 5–9, *RMS & KD 66-1384*, Magdlak, W. Greenland; 10, *RMS & KD 66-144*, Fortune Bay, Disko I., 11, *RMS & KD 66-196*; 12–13, *Williams 10685*, Quebec; 12, 13 drawn to lower scale; 5, 8, drawn to upper scale.]

subquadrate and 25–29 μ to somewhat tangentially elongated and 27–31 ×
20–25 μ; subapical cells 25–30 × 27–32 μ, polyhedral; median cells becoming
oblong-hexagonal and (23)25–32 × 35–50 μ; basal cells elongated, 27–30 ×
40–60(70) μ; *cuticle smooth; oil-bodies mostly 8–12 per median cell, coarsely botryoidal,*
variable, spherical and 5.5–6.5(7.5) μ to more often ellipsoidal and 5 × 7–9.5
to 6.5 × 9–10.5 μ. Gemmae usually lacking, rarely traces of a few on uppermost leaf lobes, in *yellow-brown* fascicles, identical to leaf lobes in color, *ellip-*
soidal, smooth, 20–22 × 26–31 μ.

Dioecious. ♂ Bracts becoming intercalary, in 3–4 pairs or more, similar to
vegetative leaves but more imbricate and more ventricose at base and often
somewhat more pigmented; usually 2-androus. Antheridia with body ca.
180 × 160 μ, jacket of polyhedral cells less than 20 μ in diam.; stalk 1-seriate,
ca. 26 μ in diam. Perianth-bearing shoots with 1–3 innovations (in absence of
fertilization; apparently normal perianths form even then). ♀ Bracts *very*
broad, transverse oblong-elliptical to subreniform and 2000–2100 μ wide ×
1100–1300 μ long, obsoletely 3–4-lobed *for only 0.1–0.15*, the vestigial lobes very
broad and sinuous, the sinuses lunate to flatly lunate; bracts strongly plicate.
Perianth stoutly oblong-clavate, ca. 920–980 μ in diam. × 1800–2100 μ long,
the distal 0.35–0.6 bluntly but strongly plicate, the moderately contracted,
often decolorate mouth quite entire.

Type. Fortune Bay, Disko I., west Greenland (*RMS & KD 66–144,*
June 27, 1966). *Paratype.* Lyngmarksfjeld toward Blaesedalen, near
Godhavn, Disko I. (*RMS & KD 66–153, c. per.*).

Known only from Greenland, but probably to be found elsewhere in
the North American Arctic; doubtfully south to Quebec.

Distribution

This taxon appears widespread on the west coast of Greenland, at
least to 70° N.; I have not found it north of 75°.

E. GREENLAND. Cap Ravn (*T. Böcher 817, 1932*), det. Harmsen as *Lophozia barbata*
and published as such. S. GREENLAND. Dyrnaes, 60°57′ N., 46°02′ W. (*K. Damsholt,*
June 29, 1963). W. GREENLAND. Ubekendt Ejland, Igdlorssuit, 71°14′ N., 53°35′ W.
(*Holmen & Steere, July 2, 1962*); Marrait, 70°30′ N., 54°12′ W. (*Holmen, Sept. 3, 1956*)
and Ikorfat, Nugssuaq Pen., 70°45′ N., 53°07′ W. (*Holmen 62-1574*, in small part among
L. quadriloba); Tupilak I., Egedesminde, 68°42′ N., 52°55′ W. (*Holmen, June 30, 1956,*
p.p., among *L. quadriloba; RMS & KD 66-026, 66-028, 66-113*); Kobbefjord (*Warming*
& Holm, 1884; det. Jensen as *J. lycopodioides* var. *floerkei*, and published as such in Lange
& Jensen, 1887; mixed with *Lophozia collaris*); Fortune Bay, Disko I., 69°16′ N.,
53°50′ W. (*RMS & KD 66-142, 66-144*); Lyngmarksbugt and below Lyngmarksfjeld,
Godhavn, Disko I. (*RMS & KD 66-568, 66-173a, 66-196*); Magdlak, Alfred Wegeners
Halvö (*RMS & KD 66-1354, 66-1364a*). QUEBEC. N. slopes, Mt. Albert, Gaspé (*H.*
Williams 10650; doubtful).

Several of the above reports, based on sparing and sterile plants, are not surely
placeable and may represent subsp. *polita*. I have arbitrarily placed all west
Greenland collections in subsp. *polymorpha*, since those with gemmae have
smooth "*polymorpha*-type" gemmae, and virtually all collections tend to develop
broad leaves.

Ecology

A hygrophytic taxon, of rills and brook edges, seemingly restricted to peaty but not acid areas where there is some basic seepage (e.g., in the Disko Bay area, where common, often in the gneiss and granite zone but below sandstone outcrops or thick basalt cliffs, from which basic seepage occurs). Associated is a series of tolerant species (*Blepharostoma trichophyllum brevirete*, *Odontoschisma elongatum*, *Cephalozia pleniceps*, *Tritomaria quinquedentata*) or species widespread on organic substrates in areas with calcareous seepage (*Lophozia gillmani*, *L. heterocolpa*, *Scapania degenii*, and *Paludella squarrosa*).

Differentiation

The most reliable diagnostic feature of subsp. *polymorpha* is the variable leaves, which have given the plant its name; these are mostly 3–4-lobed on the same stem but are in part 2-lobed; on weak axes a predominance of 2-lobed leaves may occur. Whether the leaves are 2- or 4-lobed, their width is normally greater than their length. On mature axes the leaves commonly vary from 1530 μ wide × 1230 μ long up to 1640–1680 μ wide × 1065–1130 μ broad, and the length is then only 0.62–0.78 the width. Median cells seem to have more oil-bodies than in subsp. *polita*, most cells having (5)8–12 oil-bodies, rather than 2–6(7). Finally, plants lack the "eigenartigen Fettglanz" which Müller states to be characteristic of typical *T. polita*. Perhaps most important of all are the striking differences in gemmae, those of subsp. *polymorpha* being smooth, not polygonal.

This taxon is occasionally difficult to identify when juvenile or weak stems are examined (e.g., the specimen from Cap Ravn), on which many or almost all leaves may be 2-lobed. However, the peculiar insertion is usually discernible even on weak stems, although the decurrence may be then quite obscure. The elongated basal leaf cells and the dark-colored stems with strongly elongate cortical cells are distinctive even on such weak plants. In all cases, equally and shallowly 3-lobate leaves can be found, which usually have the diagnostically blunt lobe tips.

Branching in this taxon appears to be nearly always *Frullania*-type and terminal; occasional lateral-intercalary branches also occur. One *Radula*-type branch was seen in the Tupilak I. specimen: here the branch issued smoothly, without a collar, from below the insertion of a normal leaf and some distance above the next lower, quite unmodified, leaf on the same side of the axis.

Because of the symmetrically 2–3-lobed leaves on weak plants, which are subquadrate or broader than long, and the coarse trigones, weak phases may

be confused with *T. heterophylla*. The latter species, however, always has gemmae (rare in *T. polita polymorpha*), is a lighter color, is never scorched, and
has sharp and often cuspidate leaf lobes (in *T. polita polymorpha* normally
blunt or even rounded, even on weak forms). However, on very weak plants the
"*Saccobasis*"-type leaf insertion may hardly be evident.

Variation

In the Arctic small, scorched phases occur, only 1000–1600 μ wide.
These are prostrate in growth and have feebly ascending apices (in contradistinction to Buch's statements) and have blackish brown leaves,
at times with no trace of vinaceous pigmentation (when this occurs, it is
found only near the postical leaf bases, often masked by the deep brownish
pigments). Rhizoids may be quite fuscous, although bleached at the
tips. Such small plants may not develop the generally diagnostic leaf
insertion and may have cortical cells that, dorsally, are only ca. 15–17 ×
60–95 μ, with the walls only feebly thickened (less than 0.1 the thickness
of the lumen). Dwarf plants, which have mostly shallowly 2-lobed, less
often 3-lobed, leaves, may be only 1000 μ wide; these then have thin-walled
cortical cells ca. 20 × 65 μ in diam.

ANASTROPHYLLUM (Spr.) Steph.

[Figs. 235–249]

Jungermannia subg. (or sect.) *Sphenolobus* Lindb., Not. Sällsk. F. et Fl. Fennica Förhandl. 13:369,
　　1874 (*p.p.*).
Jungermannia subg. *Anastrophyllum* Spr., Jour. Bot. 14:234, 1876.
Anastrophyllum Steph., Hedwigia 32:140, 1893 (in part); Schuster, Amer. Midl. Nat. 45(1):68,
　　1951.
Sphenolobus Berggr., N.Z. Hep. 1:22, 1898 (name only; see p. 752); Stephani, Spec. Hep. 2:156,
　　1902 (*p. min. p.*; see p. 751); K. Müller, Rabenh. Krypt.-Fl. 6(1):587, 1910 (*p.p.*;
　　subg. *Eusphenolobus* only).
Eremonotus Lindb. & Kaal. ex Pears., Hep. Brit. Isles 1:200, 1902.
Crossocalyx Meyl., Bull. Soc. Vaud. Sci. Nat. 60:266, 1939.

Gametophyte *erect or ascending* (especially if and when gemmae are
formed), *usually rather rigid* (as contrasted to *Lophozia s. lat.*), minute to
robust, mostly 0.5–4.5 mm wide but sometimes smaller or larger, usually
with brownish to rose-red or vinaceous secondary pigments developed even
in shade forms. Stems 6–16 or more cells high, with a firm to rigid
cortex, the medullary cells ± collenchymatous, *without distinct dorsiventral
differentiation* (but, in some taxa, eventually mycorrhizal and brown);
branching variable, *Frullania*-type lateral-terminal branches frequent in
some taxa, absent in others; lateral- or ventral-intercalary branches
usually present (but, apparently, rarely both in one species); *stems
never or very exceptionally becoming flagelliform; without antical-intercalary*

branching. Purely distichous-leaved, the *underleaves reduced to slime papillae* which are rarely stalked; ventral merophytes *usually 1–2 cells broad*. *Leaves firm, uniformly bifid* (rarely, sporadically, a ventral lobe again divided; Fig. 239:1), usually for 0.2–0.55 their length, *vertically oriented, mostly concave to canaliculate* (but one or both lobes sometimes incurved, the leaves then ± cupped), the ventral half of the insertion oblique, the *dorsal half (at least in part) virtually transverse* (and, as a consequence of the differences in insertion between ventral and dorsal halves, usually drawn into a position lying loosely over the ventral half); line of insertion dorsally extended to stem midline and either stopping there or ± decurrent on the stem (the innermost sectors of the line of insertion then again becoming oblique). Cells *firm*, either ± thick-walled or collenchymatous to varying (often considerable) degrees; *cuticle smooth or, usually, weakly (rarely coarsely) papillose;* uppermost cells always subisodiametric, *usually small* (mostly under 20 μ in diam.), medially becoming slightly to distinctly elongated, at base sometimes prominently elongated; oil-bodies 3–4 or more per cell, small to moderate in size, botryoidal. *Gemmae absent in most taxa* (exc. subgroups *Acantholobus, Isolobus, Crossocalyx*; there reddish or vinaceous, their formation ± inhibiting further growth of leaves).

Dioecious, rarely auto- or paroecious. Androecia loosely to distinctly spicate; bracts ± ventricose at base, the antical base often with a ± inflexed tooth; 1–2-androus usually; no paraphyses; antheridial stalk 1–2-seriate. ♀ Bracts ± larger than leaves, varying from 2- to 4-lobed, lobes entire to varyingly toothed. Perianth large, longly emergent from bracts, distally plicate, never beaked, *often with a distinct antical furrow* that may extend to near base, terete or slightly dorsiventrally flattened; mouth lobulate and usually varyingly ciliate to ciliate-dentate. Sporophyte with seta of numerous cell rows or (reduced subgenera) with 8(9) epidermal + 4(5) inner cell rows. Capsule wall 2(3) or 3–4- to 5-stratose, at least 22 μ thick; epidermal cells with strong columnar (nodular) thickenings, inner with semiannular bands (occasionally feebly distinct in their median sectors). Spores 10–12 μ or larger, ca. twice the diam. of the bispiral elaters (which are 6–7.5 μ or greater in diam.).

Type. Anastrophyllum donianum (Hook.) Steph.

Anastrophyllum is a rather large genus, with ca. 25–30 species (see p. 710; including several species recently described by Herzog and by Horikawa). It is largely distributed in tropical or montane rain forests under sub-tropical to cool-antipodal conditions, with isolated species extending northward to Japan and to humid, oceanic portions of western Europe and western North America [*A. assimile* (Mitt.) Steph., *A. donianum* (Hook.) Steph., *A. joergensenii* Schiffn.; the last not known from the Western Hemisphere]. The species are almost all limited to "old" regions, and those found outside the tropics (*A. assimile*, *A. joergensenii*, *A. donianum*) occur either in or peripheral to unglaciated regions. Associated with the tropical-oceanic range of this old genus is a sharp restriction to acidic, usually igneous, rocks.

Several regional species of continental distribution, *A. hellerianum*, *A. minutum*, *A. saxicolus*, and *A. michauxii*, represent exceptions to the preceding generalizations. Correlated with this, they represent isolated elements in the genus.

Anastrophyllum is here broadly delimited, exactly in the sense adopted by Schuster (1951a, pp. 69–75; 1953, pp. 370–376) and for the reasons given in these papers. This "broad" definition of the genus has been criticized by some (e.g., Müller, 1951–58, pp. 616, 719),[229] and as a consequence I adopted (Schuster, 1961b, 1965) a somewhat narrower, alternative generic concept (compare columns 1 and 2 in Schuster, 1951a, pp. 30–31), recognizing the "*Eurylobus-Euanastrophyllum* complex" (Schuster, 1951a, p. 71) as the genus *Anastrophyllum s. str.*, and the "*Sphenolobus-Crossocalyx-Eremonotus* complex" (*loc. cit.*, p. 72) as the genus *Eremonotus*. Jones (1958a), who did not adopt my broad delimitation of *Anastrophyllum*, retained a narrower generic concept essentially because he felt that a broad spectrum of intergradation had to occur before the two groups could be merged. In my 1951 paper, dealing solely with the North American taxa, I did not cite in full the bases for a broad delimitation of *Anastrophyllum*, and no subsequent worker has attempted to evaluate them except, in part, Kitagawa (1962b, p. 251), who admitted that "it is reasonable to combine *Sphenolobus* with *Anastrophyllum*, for an almost complete series of intermediates is recognizable between these two genera."[230] Ironically, Kitagawa (1965, p. 253) then failed to draw the obvious conclusion since he attempted to retain both "*Sphenolobus*" and *Anastrophyllum*! The interested student, however, should compare the diagnoses of "*Sphenolobus*" and "*Anastrophyllum*" in Kitagawa (1965). I have looked in vain for even a single difference that can reasonably be interpreted as generic in nature.

As emphasized elsewhere (pp. 710–716), it is exactly the existence of a

---

[229] The statement in Müller (p. 719) that I (Schuster, 1949a) had united *Anastrophyllum* and *Sphenolobus* with *Lophozia* is incorrect.

[230] Kitagawa (1963a, p. 171) also states that "*Sphenolobus acuminatus* offers an example of intermediates standing between *Anastrophyllum* and *Sphenolobus*. The strongly secund leaves widest near the base suggest ... *Anastrophyllum*," but "the dorsal base of the leaf is neither dilated nor decurrent," and on that account the species could "remain" in *Sphenolobus*.

wide spectrum of intermediate taxa, evident as soon as the nonholarctic flora is studied, which ultimately defeats any attempt to separate two (or more) genera. It we are to split *Anastrophyllum*, as I have here broadly delimited it, it must be divided into a rather large number of genera, not merely into two or three.

Detailed reinvestigation of the whole problem has convinced me that even the "usual" criteria (cited, e.g., in Müller, 1951–58) used to justify narrow generic concepts fail when actual material is studied, *even with respect to the well-known holarctic taxa.*

Thus, I had adopted the "usual" concept of a complex, including *Crossocalyx*, "*Sphenolobus*" (*minutus*), and *Eremonotus* (*myriocarpus*), as plants with a seta of the 8 + 4 row type and a capsule with a bistratose wall. However, even "*Crosso-calyx*" *hellerianus* may have a seta with 10 epidermal + 4–5 inner cell rows (Fig. 235:8–9) and a capsule wall that is in part tristratose (Fig. 235:11). Obviously these size-related distinctions are subject to more variation than the prior literature allows for—and thus their significance must be regarded as secondary.

Similarly, Evans (1900) speaks of *A. reichardtii* as hardly typical of its genus, and Evans and Nichols (1935) consider *A. michauxii* as closely allied to *A. reichardtii*. In turn, *A. michauxii* produces phenotypes which can hardly be separated, as regards facies, from "*Sphenolobus*" *minutus*. The occurrence of such phases is, unquestionably, a vigorous argument in favor of a conservative generic concept. In these phases, which are commonly only 700–850(900) μ wide when mature and fertile, leaves are usually little longer than wide (averaging ca. 530 μ wide × 600 μ long to 580–620 μ wide × 625–650 μ long), although they are widest usually below the middle and thus ovate rather than oblong. The leaves, furthermore, are canaliculate, with the dorsal lobe lying loosely over the ventral, rather than being elevated above the stem or squarrose; the ventral lobe has the apex spreading rather than incurved. Plants of this type have a size and facies essentially identical with those of "*Sphenolobus*" *minutus* and in extreme cases are so similar that only careful microscopic examination will serve to separate them from this species (p. 730). Furthermore, such plants may be fertile (as, i.a., in *RMS 45070*, Waterrock Knob, N.C.); the oblong-cylindric perianths then are almost exactly as in "*S.*" *minutus*, with the distal 0.5–0.65 strongly plicate and weakly narrowed toward base, rather than obovate to clavate; there is always a subfloral innovation which is usually soon again fertile and may again innovate, resulting in the characteristic pseudolateral orientation of the perianths that is so common in "*S.*" *minutus*. ♂ Plants, which may have androecia only 750–825 μ wide, commonly show ± imbricate bracts more strongly saccate than in "normal" *A. michauxii*, with the lobes not or infrequently spreading to squarrose but often suberect, thus closely approaching those of "*S.*" *minutus* in form.[231]

---

[231] Separation of such phases from "*S.*" *minutus* hinges on these crucial features: (*a*) leaves on the larger stems show distinct antical decurrence; (*b*) cells possess distinct and—at least in lobes—bulging trigones; (*c*) cells have from 2 to 7 oil-bodies per cell, with many possessing 4–6; basal cells commonly have from 10 to 15 oil-bodies per cell; (*d*) ♀ bracts spread from the perianth, rather than sheath it, and are uniformly bilobed.

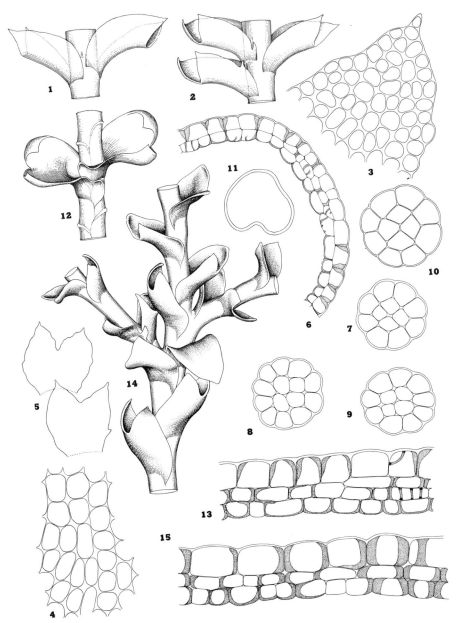

FIG. 235. *Anastrophyllum* (generic criteria). (1–11) *A. (Crossocalyx) hellerianum:* 1, robust shoot tip, ventral aspect (×50); 2, same dorsal aspect, from below gynoecium (×50); 3, cells of leaf lobe (×255); 4, median and submedian cells (×255); 5, two leaves (×55); 6, section of capsule, in cross section (×275); 7–9, cross section of setae, showing variation in epidermal cell row number (7, ×175; 8–9, ×140); 10, cross section of seta (×245); 11, perianth cross section, 1-stratose, one-fourth from apex

If the generic limits adopted in Buch (1933a) and Müller (1951–58), which were based solely on study of the *restricted and impoverished* holarctic flora as represented in Europe, are accepted, the placement of numerous tropical and antipodal taxa becomes impossible, *unless a whole suite of additional genera is founded for their reception.* Where would one place, e.g., (*a*) the minute "*Sphenolobus*" *intricatus* (Lindenb. & G.) Steph. of Mexico (Schuster, 1965, p. 269), which has a leaf form, insertion, and orientation as in *Eremonotus* [in which tentatively placed (Schuster, 1965)] and yet has ventral-intercalary branching and collenchymatous cells, as in *Anastrophyllum, s. str.*;[232] (*b*) "*Sphenolobus*" *pearcei* Steph. [*Anastrophyllum pearcei* (Steph.) Schust., 1965, p. 271], which diverges in the relatively obliquely inserted leaves, armed cuticle, small trigones, and (apparently) entirely *Frullania*-type branching; (*c*) *Anastrophyllum novazelandiae* Schust. of New Zealand and *A. crenulatum* Schust. of Tierra del Fuego— minute species, with a "*Sphenolobus*"-like transverse, nondecurrent leaf insertion and yet collenchymatous and in part elongated cells, as in *Anastrophyllum*, and with solely *Frullania*-type branching (see Schuster, 1965, p. 282, and fig. 4)? These examples are in addition to that cited by Kitagawa (1963a; *A. acuminatum*), and others I mention later (see, i.a., p. 714, *A. mayebarae* and p. 746, *A. sphenoloboides*). How many more transitions must one find to demolish "accepted" —and largely mythological—distinctions between *Anastrophyllum* and "*Sphenolobus*"?

Anastrophyllum, as thus broadly delimited, is of world-wide range and— much like *Lophozia* and *Plagiochila*—only imperfectly susceptible to division into well-defined subgenera and species groups. The delimitation of the genus, as broadly circumscribed as here attempted, has been made easier because a whole suite of taxa which Stephani placed into "*Sphenolobus*" and "*Anastrophyllum*" has been removed to, i.a., *Acrobolbus* Mitt., *Andrewsianthus* Schust., *Cephalolobus* Schust., *Gymnomitrion* Nees, *Lophozia* subg. *Hypolophozia* Schust., *Lophozia* subg. *Orthocaulis* (Buch) Schust., *Gymnocolea* Dumort., *Tritomaria* Schiffn., and other genera (see Schuster, 1961b, 1965). After this major "cleansing" operation, the remaining taxa are perhaps divisible, although imperfectly so, into two poorly

---

[232] With renewed adherence to a broad generic concept in *Anastrophyllum* I would now regard this plant as *Anastrophyllum intricatum* (Lindenb. & G.) Schust., comb. n. (basionym: *Jungermannia intricata* Lindenb. & G. *in* G., L., & N., Syn. Hep., p. 679, 1847).

(×50). (12–13) *A.* (*Isolobus*) *michauxii*: 12, shoot sector, antical aspect, with merophyte borders indicated (×28); 13, cross section of capsule wall (×570). (14) *A.* (*Anastrophyllum*) *assimile*: shoot sector with two terminal, *Frullania*-type branches (×40). (15) *A.* (*Anastrophyllum*) *piligerum*: capsule wall in cross section (×570). [1–2, from plants from Cascade R., Minn., *RMS;* 3–5, all from type of *A. hellerianum;* 6–9, 11, all from type of "*Jung. verruculosa* var. *compacta*" of Lindberg, from Smaland: 10, Lake Lilly, Keweenaw Pen., Mich., *RMS;* 12, Susie I., Minn., *RMS;* 13, *RMS 61230*, Devil's Courthouse, N.C.; 14, Inglefield Bay, NW. Greenland, *RMS;* 15, from Monte Verde Cuba, *Wright, March 3.*]

bounded units equivalent to *"Sphenolobus"* and *"Anastrophyllum s. str."*

The bases for a division into two such units have, to some extent, varied. Stephani (Spec. Hep. 2:167, 1902), e.g., referred taxa with lateral branching to *Sphenolobus* and taxa with ventral branching to *Anastrophyllum*, although his stated reason (*ibid.*, pp. 109–110) for a generic division is rather different. Any such dichotomy on the basis of branching is quite impossible to justify. Buch (1933a) emphasized that "der Hauptunterschied . . . liegt im Blatte . . . Bei *Anastrophyllum* ist das Blatt an der Basis am breitesten und greift beiderseits etwas über den Stamm. Ausserdem ist die Insertion bogenförmig . . . und . . . läuft wenigstens auf der Stammdorsalseite der Segmentsseitengrenze entlang herab." By constrast, *"Sphenolobus"* was limited to taxa with the leaf insertion "auf der Stammdorsalseite eine kurze Strecke quer." I added criteria derived from the cells (Schuster, 1951a); in *Anastrophyllum* the cells are thin-walled but have conspicuous trigones (and, typically, the basal cells become very much elongated); in *"Sphenolobus" s. lat.* the cells are small and equally thick-walled, and the basal cells are hardly elongated. Study of a wide range of materials soon shows little or no correlation among these three criteria—or between any others that I have investigated. Reluctantly, hence, I return to my earlier (1951a) conservative classification of this complex.

For the benefit of the more "radical" worker, however, I give separate diagnoses and discussions for the *"Eurylobus-Anastrophyllum"* and *"Crossocalyx-Eremonotus"* complexes (pp. 713, 750) and define each unit fully.

There are perhaps no more than 25–30 species in *Anastrophyllum s. lat.* Stephani (1898–1924) placed 31 species in *Anastrophyllum*, but a number are synonyms (e.g., *A. reichardtii*; see p. 717) and others belong in *Anastrepta* (see Grolle, 1961c, pp. 83–84).[233] Of the 29 species Stephani (*loc. cit.*) placed in *"Sphenolobus"* only two, *"Sph. minutus"* and *"Sph. hellerianus,"* can conceivably remain in this group, however defined.

Regional taxa are divisible into subgenera as follows:

Key to Subgenera, Sections, and Species of Anastrophyllum s. lat.

1. Leaves typically widest toward base, *in situ* usually arching across the stem, ovate to oblong-ovate, usually longer than broad, the dorsal half inserted on an arched line whose inner edges *usually* are obliquely or longitudinally

---

[233] *"Anastrophyllum"* recurvifolium of Stephani, Hedwigia 32:140, 1893 (basionym: *Jungermannia recurvifolia* Nees, Enumeratio Pl. Crypt. Javae 1:32, 1830) is an *Andrewsianthus, A. recurvifolius* (Nees) Schust., comb. n.; and *"Anastrophyllum" sundaicum* Schiffn., Denkschr. Math. Nat. Cl. K. Akad. Wiss. (Wien) 67:202, 1898, is also an *Andrewsianthus, A. sundaicum* (Schiffn.) Schust., comb. n. *"Anastrophyllum" yakushimense* Horik. (1934, p. 149) has been made the type of the genus *Hattoria* Schust. (Schuster, 1961b, p. 70). The genus *Hattoria*, with unlobed leaves, is perhaps more nearly allied to *Jamesoniella* according to Inoue (1966). Inoue (1966f) has segregated an additional *"Anastrophyllum," A. speciosum* Horik., also with unlobed leaves, into the genus *Scaphophyllum*, placed in the Jungermanniaceae. Perhaps remotely allied to the last two genera is *A. schizopleurum* (Spr.) Steph., with interlocking dorsal merophytes; this is the type of *Gottschelia* Grolle [(Jour. Hattori Bot. Lab. (1968)].

decurrent along the stem midline; capsule wall 3–4(5)-stratose; leaves with cells collenchymatous, the trigones sometimes confluent; basal cells *usually* somewhat, often strikingly, *elongated*; perianth not antically longitudinally sulcate, terete. Gemmae absent (exc. in *A. michauxii*; there gemmae formation not strongly inhibiting leaf size) . . . .
. . . . *Eurylobus-Anastrophyllum* complex (p. 713)   2.

2. Plants with facies of "*Sphenolobus*," relatively small (to 1 mm wide); leaves dorsally transversely inserted and not or feebly decurrent, not dilated at base, not or barely extending across stem, hardly longer than wide, divided to 0.35–0.45 or 0.45–0.7; ♀ bracts bifid like leaves; cells small (median 12–15 or 16–20 μ wide); stem only 7–9 cells high, with cortical cells 2–4× as long as wide; mostly with *Frullania*-type branching; gemmae unknown . . . .
. . . . . . subg. *Schizophyllum* Schust. (*A. sphenoloboides* Schust., p. 741)

2. Plants wholly unlike "*Sphenolobus*" in facies, robust (1–6 mm wide when mature); leaves strongly arcuately inserted, the dorsal base dilated, extending fully across (sometimes beyond) stem, decurrent, divided 0.1–0.5; stem 12–16 cells or more high, with usually strongly elongated cortical cells . . . . 3.

   3. Leaves much broader than long, dilated near base, cupped and nearly hemispherical; leaf cells unable to produce strongly bulging trigones; both dorsal and ventral leaf lobes ± incurved; antical leaf base only weakly decurrent. ♀ Bracts erect, divided usually into 3–5 dentate lobes. Gemmae absent. [♂ Bracts with paraphyses; antheridial stalk 2-seriate.] . . . .
   . . . subg. *Eurylobus* Schust. (*A. saxicolus* [Schrad.] Schust., p. 734)

   3. Leaves as long to much longer than broad, widest just above base, never cupped and hemispherical but often somewhat complicate to canaliculate; leaf cells able to produce coarse, often strongly confluent trigones; dorsal lobes squarrose to erect, but never incurved, strongly decurrent at base (on mature shoots). ♀ Bracts various, in our species simply bilobed . . . . subg. *Anastrophyllum*   4.

      4. Basal leaf cells hardly elongated (length:width ca. 1.5–2:1 at most); gemmae usually present, purplish or reddish; ♀ bracts simply bilobed, like leaves, usually spreading from perianth base and perianth ± clavate (in ours); leaves 0.3–0.45 bilobed. [♂ Bracts with ± distinct paraphyses; antheridial stalk 1-seriate.] . . . .
      . . . . . sect. *Isolobus*, sect. n. (*A. michauxii* [Web.] Buch, p. 724)

      4. Basal leaf cells strongly elongated (in part 2–4:1 or more); gemmae never present; ♀ bracts often variously modified and differing from leaves in form; perianth never clavate . . . . 5.

         5. Cells with coarse (and often confluent) but never nodose trigones; *Frullania*-type branching often frequent, but lateral-intercalary branches common; leaves 0.25–0.3 bilobed, the conspicuous lobes separated by a wide sinus. Antheridial stalk 1-seriate.
         . . sect. *Assimiles*, sect. n. (*A. assimile* [Mitt.] Steph. p. 717)

         5. Cells with nodular, sharply defined trigones (and occasionally intermediate thickenings) separated mostly by narrow, thin

walls; typically without *Frullania*-type branching; leaves (holarctic taxa) less than 0.15 bilobed, the small lobes separated by a narrow sinus or small notch. Antheridial stalk 2-seriate (type at least) . . [sect. *Anastrophyllum*; no regional taxa]

1. Leaves typically oblong-ovate to rotund-quadrate, not conspicuously ampliate at base, *in situ* with antical base usually not or hardly extended across stem, not decurrent on stem midline; capsule wall 2(3)- or 3-stratose; leaves with cells firm, ± thick-walled, with weak and concave-sided trigones at best, the basal cells never elongated; antheridial stalk often or usually 1-seriate; perianth often somewhat dorsiventrally compressed, with antical face longitudinally sulcate. Gemmae (in our taxa) present, prolonged formation resulting in flagelliform, reduced-leaved axes . . . . *Crossocalyx-Eremonotus* complex   6.

   6. Seta with many cell rows; stem 9–10(11) cells high in cross section; gemmae at maturity 2(3–4)-celled, formed on little to moderately reduced, erect stems. Plants 0.5–1.5 mm wide with leaves; very rarely with stolons; with *both* terminal and intercalary lateral branching . . . . subg. *Acantholobus* (p. 755) (*A. minutum* [Cr.] Schust. p. 757)

   6. Seta with 8(9–10) outer, 4 inner, cell rows; stem only 5–7 cells high; gemmae 1–2-celled or lacking. Plants minute, ca. 0.08–0.2 mm wide with leaves . . . . 7.

      7. With erect, filiform, gemmiparous shoots (developing quadrate, vinaceous 1–2-celled gemmae) bearing reduced, ± appressed leaves; flagella or stolons lacking; gynoecium with a large bracteole. Branches lateral, more rarely also ventral-intercalary . . . . subg. *Crossocalyx* (Meyl.) Schust. (p. 769)   8.

         8. Oxylophyte: on decaying wood or bark; branching ventral- and lateral-intercalary, sparing; gemmae vinaceous, ± cubical, only 10–11 μ usually; ♂ bracts with sharp-pointed lobes often faintly toothed . . . . *A. (C.) hellerianum* (Nees) Schust. (p. 772)

         8. Calciphyte: on basic rocks only; with terminal, *Frullania*-type furcate branching; gemmae purplish to red, tetra- to polyhedral, mostly 14–16 × 14–19 μ; ♂ bracts with lobes entire-margined, blunt or rounded at tips . . . *A. (C.) tenue* H. Williams (p. 779)

      7. Without gemmae and without erect, flagelliform shoots with reduced leaves; axillary flagella or stolons present; gynoecium without a bracteole or with it vestigial. Branches (leafy and stoloniform) all lateral-intercalary. (On rocks and soil) . . . . [subg. *Eremonotus*][234]

---

[234] This monotypic subgenus represents an end-point in reduction of *Anastrophyllum s. lat.* It is isolated from other subgenera by the markedly dorsiventrally compressed and antically sulcate perianths (a condition hinted at in subg. *Acantholobus* and sometimes rather distinct in subg. *Crossocalyx*); by the abundant development of microphyllous flagelliform branches, some of which are geotropic; by the lack of gemmae. The cells are minute, usually very thick-walled, without becoming guttulate. The cuticle, at least of bracts and subfloral leaves, bears distinct if inconspicuous, low, often rather close papillae, reminiscent of *Crossocalyx*. Branches are usually abundantly developed, and gynoecia (which develop perianths freely in absence of fertilization) may bear 1–2 subfloral innovations; *all branches and innovations seen were lateral-intercalary.*

The compressed perianth and seta cross section are suggestive of *Crossocalyx*, a subgenus that is very different in aspect, in part because of abundant gemmae formation. Nonetheless, the

EURYLOBUS-ANASTROPHYLLUM Complex[235]

Gametophyte mostly *relatively robust* [shoots usually (1) 1.5–5 mm wide × 2–12 cm long], simple or sparingly branched (but often with gynoecial innovations). Stems ± oval in cross section, rigid and stiff (*10*) *12–16 cells high* or more; cortical cells thick-walled in 1–2(3) layers, relatively elongate, (2)3–4× as long as broad, able to develop very thick walls, subequal in width to basal leaf cells, subequal in diam. to medullary cells (or at most moderately smaller). Medulla nearly or quite homogeneous, 8–12 cells or more high, the cells rounded in cross section (walls nearly thin, except at the angles), *without dorsiventral differentiation* (at most ventral 1–3 layers of somewhat smaller cells, not forming a mycorrhizal band). *Rhizoids relatively sparse*, often nearly none on suberect portions of shoots. *Leaves uniformly bilobed*, the ventral lobe usually slightly to distinctly larger (rarely isolated leaves with ventral lobes secondarily divided, then antical lobe largest; see Fig. 239:1), antically secund (*especially in drying*), *transversely or subtransversely oriented*, the postical half obliquely inserted, the *antical half inserted nearly transversely but the line of insertion usually distinctly, if briefly, decurrent along the inner edges of the lateral merophytes* (i.e., along the dorsal midline of stem; Fig. 235:12, 14), *the dilated antical base usually extending nearly or fully across stem;* leaves more or less complicate or complicate-canaliculate, in a few cases merely strongly concave, *the two halves never explanate in the same plane*, the dorsal half usually loosely lying over the ventral. Cells subisodiametric in lobes, typically becoming oblong and elongated in leaf middle and below, *strongly collenchymatous, the trigones often confluent*, frequently yellowish- or reddish-tinged. Oil-bodies (where known) rather large, *mostly 4–9 μ long, only 2–5(7–9) per cell*, finely granular or papillose externally. Sometimes with asexual reproduction by pigmented, 1–2-celled gemmae (*A. michauxii*), but *usually without asexual reproduction.*

Dioecious, often sterile, rarely monoecious. ♂ Bracts saccate at base, but often with spreading lobes, 2–4-androus; antheridial stalk 1–2-seriate. ♀ Bracts variable (in the various subgenera), from 2-lobed with entire lobes, to 3–5(6–7)-fid, with dentate or spinose-dentate margins. Perianths exserted or long-emergent, ovoid to cylindrical to obovoid-clavate, *not at all dorsiventrally compressed*, inflated and terete, plicate in upper fourth at least, typically *with a distinct, deep, antical, longitudinal sulcus, never with an antical fold;* mouth lobulate, usually ciliate. Sporophyte (where known) with *seta of numerous cell rows;* capsule ovoid, the wall 3–4(5)-stratose, firm. Outer layer of capsule wall with nodular thickenings; inner layers with annular to semiannular bands.

---

strict, microphyllous gemmiparous shoots of *Crossocalyx* are comparable to the remote-leaved microphyllous shoots and stolons of *Eremonotus*.

A distinct, if remote, affinity also occurs to *Cephalolobus* Schust. of the cold Antipodes. This similarity is enhanced by the equally small size, the lobulate perianth mouth, the copious development of lateral-intercalary flagella, the purely lateral-intercalary branches and innovations, and the lack of gemmae. However, in *Cephalolobus*, cells are larger and thin-walled with large, often bulging trigones, and bear coarse papillae. Furthermore, the perianth is subterete and not dorsiventrally compressed, and lacks the antical deep, broad sulcus characteristic of *Eremonotus*.

[235] Approximately equivalent to the genus *Anastrophyllum* (*sensu* Müller, 1951–58), with deletion of certain taxa and addition of "*Sphenolobus*" *saxicolus*. For students with a "narrow" generic concept, this constitutes *Anastrophyllum.*

A key to the elements contained is given on pp. 710–712 in which the chief types represented are separated as subgenera and sections. Possibly further study will demonstrate that the subgenera proposed should be given only the status of sections. *Eurylobus* was accepted by Müller (1951–58) as a subgenus, but of *Sphenolobus*. Wherever this monotypic taxon is placed, it remains a disturbingly strange element. The stem anatomy, collenchymatous cells, and distinct decurrence of the dorsal half of the leaf all suggest *Anastrophyllum*, rather than "*Sphenolobus*." The solution to this problem, of course, is to combine the genera *Anastrophyllum* and "*Sphenolobus*" (or *Eremonotus*) into a broadly circumscribed genus *Anastrophyllum*, a procedure which I follow here and adopted earlier (Schuster, 1949a, 1951a, 1953). I admitted (Schuster, 1951a) that a division into two generic groups, *Anastrophyllum s. str.* and *Eremonotus*, could be attempted. Buch (*in litt.*) considered *Anastrophyllum s. lat.* (as in Schuster, 1951a) too broadly delimited, and Müller (1951–58) expressed a similar opinion. In reworking both regional and exotic taxa, however, I have again been struck by the fact that *any* distinction between *Anastrophyllum s. str.* and *Eremonotus* ("*Sphenolobus*") is difficult to maintain. The continued disagreement as to the correct disposition of *A. saxicolus* is a case in point; the disposition of the Japanese *A. mayebarae* Hatt. is equally difficult; see p. 740.

Anastrophyllum mayebarae [see Hattori, Jour. Jap. Bot. 28(5):141–143, fig. 64, 1953; similar to *A. bidens* (Nees) Steph., from Java to the Philippines] illustrates the problem of generic lines in the *Anastrophyllum-Eremonotus* complex and illuminates the fact that, when authors who deal with only the few European taxa adopt a narrow generic concept, it is based merely on European species, and no others. *Anastrophyllum mayebarae* keys to the "genus" *Sphenolobus* (as delimited in Müller, 1951–58, key, p. 617) because of the folded, virtually transversely inserted and uniformly bilobed leaves and the small size. Yet it disagrees with the diagnosis of that genus in Müller (*loc. cit.*, p. 719) in the strongly collenchymatous cells, characteristic of *Anastrophyllum* (*sensu* Müller and Buch). Buch (1933a, p. 289), admitting that *Sphenolobus s. str.* "ist . . . näher verwandt" with *Anastrophyllum*, claims this last differs in the (*a*) dilated leaves, widest at base, reaching across the stem both antically and postically; (*b*) arched leaf insertion, the arching not being limited to the acroscopic end of the insertion "sondern läuft wenigstens auf der Stamm-dorsalseite der Segmentseitengrenze entlang herab." *Anastrophyllum mayebarae* (and Hattori's illustrations show this) clearly has a transverse, nondecurrent dorsal leaf insertion, limited to the acroscopic end of the segment, as in "*Sphenolobus*" *minutus*, which both Müller and Buch regard as "typical" of the taxon "*Sphenolobus*"! The species also does not show the "generic innovations" Buch (*loc. cit.*, p. 297) uses to separate *Sphenolobus* and *Anastrophyllum*, e.g., "development in *Sphenolobus* of a mycorrhizal cortex" and the "dilation and decurrence of the leaf base in *Anastrophyllum*" (loosely translated).

Müller (1951–58) stated that my broad treatment of *Anastrophyllum* to include *Sphenolobus* was due to the fact that I had "nur die nordamerikanischen Arten *Anastrophyllum michauxii* und *A. reichardtii* im Auge." Actually, my treatment was based on the fact that I could not clearly place a considerable suite of tropical-oceanic and antipodal species, such as *A. bidens*, which is closely similar to *A. mayebarae*. Müller's acceptance of *Eremonotus*, *Crossocalyx*, and *Sphenolobus* as distinct genera, separate from *Anastrophyllum*, could well be attributed to his having only the European species "im Auge." This is clear from his statement that the

hauptsächlich aus tropisch-ozeanischen Arten bestehende Gattung *Anastrophyllum* unterscheidet sich von *Sphenolobus* durch die stattlichere Grösse, die Art der Blattanheftung, wobei das Blatt über den Stengel übergreift und der Vorderlappen am Stamm kurz herabläuft, durch die Blattform (länger als breit, am Grunde am breitesten) das stark verdickte Zellnetz und durch die langestreckten Stengelzellen (bei *Sphenolobus* kurz-rechteckig).[236]

Although a prolonged controversy over so subjective a matter as generic limits in the Lophoziaceae (see, i.a., the Introduction in Schuster, 1951a, where the infinite variety of generic concepts in the group is summarized) is pointless, it must be emphasized in any treatment of even the well-known European and American species that there is little agreement as to generic placement of several species. Both Müller (1948, 1951–58) and Andrews (1948) expressed reservations about Buch's

---

[236] It is illuminating to use one of the "tropical-oceanic" elements that Müller speaks of when he discusses *Anastrophyllum*, the *A. mayebarae-bidens* complex, and to apply Müller's criteria (which are basically those of Buch). To follow Müller's sequence, *A. mayebarae* has (*a*) a small size, ca. 1 mm wide; leaves that (*b*) do not reach across and beyond the stem at base, (*c*) are not decurrent dorsally, (*d*) are hardly longer than wide, being basically similar in outline to those of "*Sphenolobus*" *minutus*, and (*e*) are widest distally when flattened out; (*f*) cells that are very prominently collenchymatous, with bulging confluent trigones; (*g*) cortical cells that are only 2–4× as long as wide, but strongly thick-walled.

Of these supposed generic differences, *a-e* and *g* place this species in "*Sphenolobus*," using Müller's criteria. Yet *f* (cell type) clearly places the species in *Anastrophyllum*. More importantly, *A. mayebarae* has some other features that suggest *Anastrophyllum* rather than "*Sphenolobus*," such as (*a*) lack of gemmae; (*b*) oval cross section of the stem, a feature that Müller (*loc. cit.*, p. 616) insists separates *Anastrophyllum* from "*Sphenolobus*"; (*c*) strongly squarrose dorsal lobes, reminiscent of *A. michauxii*; (*d*) a distinct "*Anastrophyllum*" facies that is hard to describe.

Thus, if *A. mayebarae* is accepted as an *Anastrophyllum*, one is reduced to using, as the single major generic character of *Anastrophyllum* (vs. "*Sphenolobus*"), the strongly collenchymatous cells of the former, vs. the thick-walled, not or hardly collenchymatous cells of "*Sphenolobus*"! This hardly suffices for separating genera. A consideration of the tropical-oceanic species to which Müller alludes, therefore, does not result in a strong conviction that sharp generic lines exist.

(1933a) transfer of *"Sphenolobus" michauxii* to *Anastrophyllum.* Müller considered my (Schuster, 1951a, 1953) transfer of *"Sphenolobus" saxicolus* even more dubious, although Andrews (1948) clearly considered *A. michauxii* and *"S." saxicolus* congeneric.

Subgenus *ANASTROPHYLLUM* Spr.

Jungermannia subg. *Anastrophyllum* Spr., Journ. Bot. 14:234, 1876.
Anastrophyllum Steph., Hedwigia 32:140, 1893 (in part).

Typically erect and caespitose, mostly tropical and antipodal in range (a few in not or imperfectly glaciated regions in the north), *usually vigorous,* brown to reddish to blackish or piceous, often rigid, with the *shoot apices strongly upcurved or almost coiled, the stiff leaves antically secund* (strikingly so when dry), the plants thus quite convex in ventral aspect. Usually almost unbranched or with *sparing intercalary branching* (from ventral portions of leaf axils or from ventral face of axis) but at least occasionally with *Frullania*-type terminal branches.[237] Stem with a (1)2–3-stratose cortex of thick-walled cells, narrower than the medullary cells. Leaves slightly to much longer than broad, usually ovate, strongly canaliculate to canaliculate-conduplicate, at the *ampliate base extended across the stem* (often on both sides) and often largely obscuring stem in superficial view, ± strongly *decurrent antically;* leaves ca. 0.1–0.35 bilobed, the lobes usually spreading, never lying in the same plane. Cells firm, walls thin to somewhat thickened, *trigones usually coarse and often confluent;* cells within lobes isodiametric, toward leaf middle and especially *toward base progressively more elongated,* often greatly so.

Type. Anastrophyllum donianum (Hook.) Spr.

Including three species of holarctic range, the regional *A. assimile* (*A. reichardtii*) and *A. donianum* (Europe; rare in western North America) and *A. joergensenii* (west Europe), plus a wide variety of handsome, vigorous taxa, mostly in montane tropics and the Antipodes; as outliers are placed *A. michauxii* and its allies, constituting sectio *Isolobus.* This section differs from sectio *Anastrophyllum* (type: *A. donianum*), sectio *Assimiles* (type: *A. assimile*), and several purely tropical and antipodal sections in the ability to develop gemmae—this is strikingly absent in all other taxa of subg. *Anastrophyllum.*

Subgenus *Anastrophyllum* consists, basically, of a series of erect and caespitose, rather rigid, mostly vigorous taxa, in which *Frullania*-type branching is rare or lacking, although present, e.g., in *A. michauxii* and *A. assimile* (species which should on this basis and others be placed in sections of their own, apart from the typical taxa of sectio *Anastrophyllum*); shoot tips are usually strikingly arcuate and leaves are typically strongly

---

[237] In *A. assimile* and *A. michauxii* at least (see pp. 723, 724).

antically assurgent or secund. As a consequence, the plants usually have a distinctive facies. Under the microscope the conspicuous collenchymatous cells, often with immense, nodular trigones that may be confluent or separated only by narrow, thin walls, are diagnostic; even more diagnostic is the very marked tendency (absent in sectio *Isolobus* and several other small sections) for the basal leaf cells to become narrowly oblong and strongly drawn out.

A key to the holarctic sections and species is provided on p. 711.

Only two taxa (*A. assimile, A. michauxii*) occur regionally. Both have, apparently consistently, uniseriate antheridial stalks, but Müller (1951–58) describes the stalk of *A. donianum*, the type, as biseriate.

Sectio *ASSIMILES* Schust., sect. n.

Plantae brunneae; cellulae basales ± elongatae (2–2.5:1); gemmae o; pedicelli antheridii 1-seriati; bracteae (1)2-andrae, paraphysibus parcis praeditae.

Plants (in the mass) fuscous, relatively moderate in size (mostly 1–1.5 mm wide), glossy; cortical cells 2–3× as long as wide; branching lateral-intercalary and lateral-terminal; leaves slightly longer than wide, conspicuously bifid; cells with non-nodose but large, often confluent trigones; basal cells very distinctly elongated. Gemmae lacking. Dioecious; ♂ bracts (1)2-androus; antheridial stalk 1-seriate.

Type. Anastrophyllum assimile (Mitt.) Steph.; I know of no other taxa belonging in this section. As Evans (1900) emphasized, this species is hardly typical of the genus (here subgenus) *Anastrophyllum*.

ANASTROPHYLLUM (*A.*) *ASSIMILE* (*Mitt.*) Steph.
[Figs. 235:14; 236]

Jungermannia assimile Mitt., Jour. Proc. Linn. Soc. Bot. 5:93, 1861.
Jungermannia reichardtii G., in Juratzka, Verh. Zool.-Bot. Gesell. Wien 20:168, pl. 3B, 1870; Hedwigia 9:34, 1870.
Jungermannia nardioides Lindb., Musci. Scand., p. 8, 1879 (*fide* Kaalaas, 1898).
Jungermannia minuta var. *robusta* Massal. & Carest., Nuovo Giorn. Bot. Ital. 12:333, 1880.
Anastrophyllum reichardtii Steph., Hedwigia 32:140, 1893; Evans, Proc. Wash. Acad. Sci. 2:299 (353), pl. 16, figs. 4–17, 1900; K. Müller, Rabenh. Krypt.-Fl. 6(1):583, fig. 289, 1910; ed. 3, 6:746, fig. 245, 1954; Arnell, Illus. Moss Fl. Fennosc. 1:87, fig. 41, 1956.
Anastrophyllum assimile Steph., Hedwigia 32:140, 1893.
Anastrophyllum nardioides Kaal., Skrift. Vid. Selsk. Christiana Math.-Nat. Klasse 1898(9):18, 1898.
Anastrophyllum japonicum var. *otianum* Hatt., Jour. Hattori Bot. Lab. no. 9:16, 1953.

In extensive mats, or scattered, varying from *red-brown or deep brown or brownish green to almost black*, rarely yellowish brown, *quite glossy, medium-sized*: shoots (0.75)1.0–1.5 mm wide × 2–5 cm long. Stem rigid, blackish to brownish black, usually ascending, simple or with few (and *usually postico-lateral or lateral-intercalary*) branches; occasionally with *Frullania*-type terminal branches;

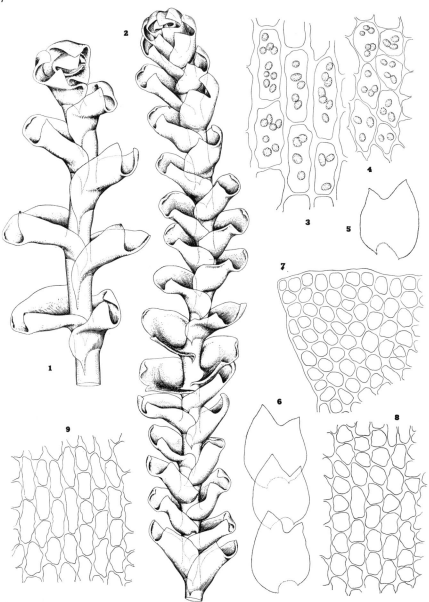

FIG. 236. *Anastrophyllum (Anastrophyllum) assimile.* (1) Lax-leaved shoot apex, showing maximal antical leaf decurrence (×36). (2) ♂ Shoot (×25). (3) Basal leaf cells with oil-bodies (×560). (4) Median cells with oil-bodies (×560). (5–6) Leaves (×22). (7) Leaf lobe apex (×270). (8) Median cells (×270). (9) Basal leaf cells (×270). [All from plants from NW. Greenland: 1–2, Kânâk, *RMS 46155, RMS 45534a;* 3–4, *RMS 45356;* 5–9, *RMS 46155.*]

always with 1–2 innovations from below unfertilized gynoecia; stem 150–260 μ in diam., usually characteristically and *strongly acroscopically curved*, especially in drying; cortical cells regularly oblong, thick-walled, the dorsal 12–13(14) × 25–40(45) μ, averaging 2–3 × as long as wide. Rhizoids frequent below, usually but not always few or absent above, colorless to (with age) brownish. Leaves contiguous to weakly imbricate, semiamplexicaul, the very base somewhat sheathing the stem, inserted by an arcuate line, ventral half obliquely inserted, dorsal half ± transversely so, except the *insertion at dorsal inner edges oblique, distinctly decurrent*; leaves antically somewhat to rather strongly *secund*; leaves spreading to obliquely or erect-spreading, ovate to asymmetrically broadly oblong-ovate to subrotund-quadrate when flattened, ca. 700–810 μ wide × 700–835 μ long to 750 μ wide × 625–675 μ long, occasionally 800 μ wide × 900 μ long, loosely *conduplicate-canaliculate, averaging about as wide* (at widest point, in basal ⅓) *as long*, the ± dilated postical and antical bases arching largely or quite across the stem; antical margin weakly arcuate, postical more strongly so (especially in the ± dilated basal portion); *sinus descending 0.25–0.3 the leaf length*, slightly acute to moderately obtuse, angulate or somewhat rounded at base; lobes usually quite *asymmetric* (the ventral larger) and triangular, the antical normally acute, the postical acute or blunt, *both lobes typically weakly incurved*. Cells, except in peripheral portions of leaf, tending to be *in regular rows*, small, the marginal ca. 13–15 μ, the median (12)13–16(17) × (15)16–22(25) μ, the *basal rather elongated*, from 14–15 × 24–30 μ up to 16–18 × 32–36(40–45) μ, averaging 2–2.5× as long as wide; *cells with coarse, often confluent, irregular trigones*, no or few intermediate thickenings, the *lumen of the cell usually stellate* or rounded-stellate and irregular (except, sometimes, in peripheral and basal cells); cuticle smooth; oil-bodies 2–3 to 4–6 per cell, small, finely botryoidal, or "spherical, [appearing] homogeneous, 2–4 μ" (Arnell, 1956). Gemmae uniformly *absent*.

Dioecious. Androecia compact, becoming intercalary; bracts imbricate, in (5)6–10 pairs, similar to leaves in shape but somewhat smaller, ca. 700 μ wide × 750 μ long, strongly concave, the antical base arching across stem and often with a small basal tooth; (1)2-androus; antheridial stalk uniseriate to in part 2-seriate; sometimes with small or minute paraphyses, which are polymorphous and range from subulate to oblong, with the apices acuminate to blunt. Perichaetial bracts similar to, but larger than, vegetative leaves, *sheathing the basal fourth of perianth*, the lobes tending to be more acute, usually broader than long (ca. 1050 μ wide × 950 μ long), entire or sinuate to obsoletely angular-dentate; bracteole absent. Perianth slenderly ovoid to cylindrical-ovoid, *widest at or below middle*, gradually narrowed in distal half and *with a deep antical sulcus, deeply 4–5-plicate*, the decolorate *mouth lobulate-lacerate* with ciliate divisions, the longer cilia in part formed of 4–7 superposed cells, variously twisted or curved; some of the subapical cells of the perianth mouth projecting and *forming obtuse to subacute papillae 1 or 2 cells long;* archegonia 10–12. Sporophyte unknown.

Type. Sikkim: Lachen, 3300–3600 m (*Hooker 1321*, 1849; NY); type of *A. reichardtii*, Steiermark, Austria (*Dr. H. W. Reichardt*).

Distribution

A rather rare and very locally distributed species, which previously was found (except for Queen Charlotte Isls. and Alaska) chiefly in Europe. There restricted in range to southern Norway (59–61°51′ N.; Kaalaas, 1898, and Joergensen, 1934), the Alps of central Europe, where usually a species of high altitudes (1500–2400 m), in Austria, Switzerland, and northern Italy (see map, fig. 246, in Müller, 1951–58; p. 748). Müller (*loc. cit.*, p. 749) hypothesizes that the species is restricted in the Alps to sites which in "aller Wahrscheinlichkeit nach während der letzten Eiszeit nicht vom Eis bedeckt waren"; he considers it a relict that survived the last glacial period in the Alps.

In addition to the restricted European range, reported from Nepal and Sikkim, in the Himalaya (Grolle, 1964f, p. 176), and also from Korea (Grolle, 1964f) and central Japan (Hattori and Inoue, 1960, p. 90).[238] Disjunct on Mt. Kinabalu, Borneo (Kitagawa, 1967).

When we add to these disjunct occurrences the very disjunct stations in Alaska and on the west coast of Greenland, it becomes evident that *A. assimile* must be regarded as one of the old, "critical" Tertiary relict species, and thus Müller's hypothesis that it occurs only in areas that escaped Pleistocene glaciation is substantially correct. The stations, mostly within 1–2 miles of local, *extant* ice caps—thus under currently Pleistocene conditions—in west Greenland are seemingly impossible to explain. However, all four stations now known share one crucial feature: they are on isolated peninsulas (Kânâk; Kangerdlugssuak) or islands (Upernivik Ø; Umanak) that just possibly escaped total glaciation, at least nunataks and headlands remaining exposed.

The species has been previously reported from North America only from coastal portions of Alaska [Columbia Fiord, Orca, and Port Wells (Evans, 1900), Virgin Bay on the central Pacific Coast, and from south of Povohok on the Bering Sea (Persson, 1946)] and British Columbia (Queen Charlotte Isls.; Grolle, 1964f). It occurs far north of previous reports in northwest Greenland, as follows:

---

[238] There remains a problem as regards *A. japonicum* Steph., Spec. Hep. 6:106, 1917 (nec. *A. japonicum* Steph., *ibid.* 2:160, 1902; see Schuster, 1965, p. 271, with respect to this "new height in taxonomic confusion"). I checked the type of the 1917 *A. japonicum*, a plant from Zizogatake (*Faurie 1330*), which clearly occurred on decaying wood, with *Nowellia* and a *Scapania*. In my opinion, a plant identical to *A. michauxii*, which is often xylicolous, was at hand (Schuster, *loc. cit.*, p. 271), not of *A. assimile*—a species that in my experience is under no conditions xylicolous. Hence the citation of "*A. japonicum* Steph." (1917) as a synonym of *A. assimile*, by Hattori & Inoue (1960) and Grolle (1964f, p. 176), is in error. See also footnote, p. 732.

NW. GREENLAND. Kânâk and Kangerdlugssuak, Inglefield Bay 77°23′–77°32′ N. and 67–68°35′ W. (*RMS 45552, 46155, 45956,* etc.).   W. GREENLAND. Qualagtoq, Upernivik Ø, Inukavsait Fjord (*RMS & KD 66-1100*); Umanak (*RMS & KD 66-1480, 66-1479*).

The plant occurs, in the Inglefield Bay area, chiefly in sites at some distance above sea level (500–1400 feet) and within, in general, 0.5–1.5 miles of the existing ice cap, which were most certainly covered within the last several thousand years by the presently obviously retreating ice cap. This does not eliminate the possibility that the species survived the Pleistocene in relict sites on coastal nunataks or exposed headlands.

Ecology

In Europe, forming "pure carpets or [growing] among other mosses on siliceous, shady and wet rocks and stones, seldom on slate" (Arnell, 1956). Müller (1954, p. 749) notes occurrences on siliceous cliffs or the rock detritus formed by their decomposition, in moist sites, where the species may form almost black mats.

In west and northwest Greenland the species is locally abundant, although extremely narrowly restricted in occurrence. It is found commonly with *Marsupella arctica, M. revoluta, Scapania crassiretis, Gymnocolea inflata, Odontoschima macounii,* occasionally *Tritomaria quinquedentata;* in drier sites occasionally with *Gymnomitrion concinnatum.* The species occurs almost without exception between acidic rocks, emergent or submerged in drainage channels or rills constantly irrigated by snow or ice melt. The surrounding areas may be exceptionally barren and arid, with, locally, isolated patches of *Cassiope tetragona* and *Dryas chamissonis,* rarely *Vaccinium uliginosum microphyllum.*

Anastrophyllum assimile seems less tolerant of running water than are *Marsupella arctica* and *M. revoluta,* which typically occur with it in Greenland, and hence is often absent from sites where it may be covered by water, except during the period of snow melt.

Variation and Differentiation

Anastrophyllum assimile is easily recognized, showing little or no similarity to other regional taxa. It differs from all our other species in its glossy texture when dry and in its more or less fuscous coloration. Müller mentions a similarity to *Anastrophyllum minutum,* but the latter differs at once from *A. assimile* in the (*a*) nonelongated basal leaf cells; (*b*) equally thickened cell walls; (*c*) very different color, which never verges on the fuscous or blackish; (*d*) nondecurrent antical leaf bases. The two species often occur mixed in the Greenland collections reported above but are distinguishable at a glance by color alone—the *A. assimile* being almost black, the *A. minutum* a warm yellow-brown to chestnut color. *Anastro-phyllum assimile* differs at once from the other arctic species of the genus found in North America in the fuscous coloration, the plants ranging from blackish green, when grown in shade, to brownish black. As will be pointed out below, this color distinction appears valid only for the some-what deviant Greenland populations. Other differential characters of value are the (*a*) quite distinctly decurrent antical leaf base, whose inser-tion recalls that of *A. michauxii* but is much more obviously decurrent than in *A. sphenoloboides*; (*b*) relatively patent leaves, suggesting *A. michauxii*, but somewhat different in shape from the leaves in that species; (*c*) absence of gemmae; (*d*) perianth widest below the middle, rather narrowed in the distal portions, quite distinctly lacerate-lobulate to lacerate-laciniate, with dentate to ciliate divisions, and with the external surface tending to develop, distally 1–2-celled superficial tubercles or teeth; (*e*) strongly glistening texture, when dry.

Evans (1900, pp. 300–301) carefully discusses the discrepancies in previous descriptions of the perichaetial bracts and perianths. He points out that the diagnosis of Limpricht (*in* Cohn, 1876, p. 279), as well as descriptions by Stephani and by Massalongo and Kaalaas, diverge in important details from his own description. His study of European materials has suggested that the earlier diagnoses were erroneous; the description of the bracts and perianth given above is based on Evans (1900). The Greenland material is uniformly ♂ or sterile.

Greenland plants are strikingly uniform: all represent a relatively slight phase, the shoots usually only 750–850(900) μ wide, with the stems ca. 150–180 μ in diam. The plants deviate in color from Alaskan and Norwegian specimens studied, all of which are more distinctly brownish in color, with less distinct infuscation, if any is detectable. Arnell (1956) describes the plants as red-brown to almost black, seldom yellowish brown, and Evans speaks of them as, i.a., blackish purple. No trace of reddish color is ever present in the Greenland plants. These plants also have leaves which are narrower and less ampliate at the base than in the European and Alaskan populations, although weak plants of the latter show a decided transition in form. In the Greenland plants typical leaves average ca. 800–835 μ long × 715–750 μ wide, on weaker shoots 515–530 μ wide × 510–550 μ long. The leaves also spread more widely and, in hygric phases, may be almost squarrose; they nearly always have spreading and, in some cases, even subsquarrose lobes, unlike the Alaskan and European populations. Furthermore, the cells may have less coarse trigones, with the lumina never becoming stellate. These differences, together with the peculiar geographical restriction to a strongly glaciated region, suggest long isolation from the other populations. Possibly segregation as a separate subspecies is warranted, but such recognition should be delayed until fertile material can be found.

Kaalaas (1898, pp. 18–19) has shown that *A. "reichardtii"* is a variable species, ranging in one site from plants with "grüner oder trüb-grüner Färbung mit reichlicher Entwicklung von Wurzelhaaren an dem Stengel und mit ziemlich weichen Blättern, deren Zellwände nur wenig oder kaum sternförmig verdickt sind" to plants which are of "schwarz-brauner Farbe mit rigiden Blättern, ausserordentlichen verdickten Zellwänden und sternförmigen Zelllumen." Both forms may grow admixed. The first type represents *Jungermannia nardioides* of Lindberg, the latter typical *A. assimile.* Equally variable is the degree to which collenchyma is developed. In the Greenland plants the cells tend to be slightly smaller than in the Alaskan specimens studied (lower vs. upper ranges given for cell sizes in the diagnosis) and to develop somewhat thick walls, with coarse, often confluent trigones, having no intermediate thickenings. In the Alaskan material, cell walls are thin, with trigones very coarse, irregularly knotlike, and, in the basal half of the leaf, very irregular and occasionally with intermediate thickenings—the walls becoming extremely undulate-sinuous.

In addition to the Greenland material alluded to above, I have studied a series of European plants and two from Alaska: Port Wells, June 26, 1899 (*Trelease & Saunders "A"*) and Columbia Fiord, Prince William Sound (*Coville & Kearney 1389*). The latter specimen has abundant perianths and also androecia. All ♂ bracts seen were diandrous. Antheridial stalks, although predominantly uniseriate, were often biseriate in the distal fifth, and sometimes 1 or 2 basal cells were longitudinally divided. The stalk was quite long and slender (120–130 μ long × 20–21 μ in diam.) as contrasted to the broadly ellipsoidal body (140 × 165 μ long). Although the northwest Greenland plants show only sporadic branching, the plants from Alaska are relatively freely branched; in all cases the branches are uniformly intercalary and postico-lateral in origin—from barely within the ventral angle of the leaves, or just below this point.[239]

Sectio *ISOLOBUS* Schust., sect. n.

Plantae brunneae subbrunneaeve, cellulae semper sine trigonis nodosis, cellulae basales numquam conspicue elongatae; gemmae vulgares, in surculis erectis quorum folia saepe reducta, sitae; pedicelli antheridii 1-seriati; ♂ bracteae 2–3-andrae, paraphysibus parcis praeditae.

Brown or brownish, moderate in size (mostly 1.2–2.0 mm wide), dull or almost so; branches ventral-intercalary and lateral-terminal; cells with trigones moderately coarse, occasionally somewhat bulging but never nodose, often ± confluent; *basal leaf cells little or hardly elongated. Gemmae common,* on erect, often reduced-leaved shoots, purplish. Dioecious; ♂ bracts 2–3-androus; antheridial stalks 1-seriate; paraphyses occasional.

---

[239] Branching in the north Greenland plants is almost always lateral-intercalary; in the west Greenland plants, however, occasional *Frullania*-type lateral branches are present—sometimes two, almost opposed, from the two sides of a single axis (Fig. 236).

Type. Anastrophyllum michauxii (Web.) Buch; closely allied and perhaps merely separate subspecies are the Mexican *A. gottscheanum* Schust. (see Schuster, 1965, p. 270) and the Japanese *A. japonicum* Steph. (1897; nec 1917; see Schuster, 1965, p. 271).

In any event, an isolated species complex is at hand, which differs from all other sections of *Anastrophyllum* known to me in the (*a*) ability to develop gemmae and (*b*) usually squarrose dorsal lobe, contrasted to a typically incurved ventral lobe. Distinctive is the dilated leaf base; *in situ*, on all but the most juvenile forms, the dorsal leaf base extends *across or even beyond the stem* (rather than merely to the stem midline, as in normal forms of *A. minutum*). Antheridia occur usually in pairs in the bracts; the stalk seems to be uniformly 1-seriate. Amidst the antheridia are linear to lanceolate paraphyses, sometimes bearing isolated branches.

Branching is usually sparing, of two types, lateral-terminal and postical-intercalary. Lateral branching is more common. Occasional phases of *A. michauxii* (as from Roan Mt., Tenn.) may show rather frequent lateral-terminal branching, often with a branch issuing from each side of the stem at closely juxtaposed points. In such cases, which occur most often in prostrate-growing individuals, postical branching is absent or rare.

The capsule wall is 3-, locally 4-, stratose; epidermal cells and the internal strata bear nodular (radial) thickenings with feebly to distinctly dilated tangential spurs; the innermost stratum bears semiannular bands.

The position of *A. michauxii* in *Anastrophyllum* is isolated, hence the segregation into a distinct section. The species agrees with *A. (Eurylobus) saxicolus* in several respects: cortical cells of stem little elongated, ca. 2–4 times as long as wide in most cases; leaves divided virtually to the middle and averaging, on fully mature, robust plants, most often as broad as to slightly broader than long; stem in cross section nearly terete. *Anastrophyllum michauxii* further differs from *Anastrophyllum s. str.* (sectio *Anastrophyllum*) in the presence of gemmae. The reasons listed by Müller (1951–58, p. 616) for excluding *A. saxicolus* from *Anastrophyllum* very largely apply also to *A. michauxii*. Since the leaf shape of *A. michauxii* almost wholly matches that of *A. assimile* (compare Figs. 237:5–6 and 238:2), although commonly somewhat more deeply bilobed, no basis exists on this ground to exclude it from *Anastrophyllum*.[240] As a consequence, it is extremely difficult, in my mind, to put *A. michauxii* into *Anastrophyllum* but to omit *A. saxicolus*, as Müller (1951–58) and Buch (1933a) have done.

ANASTROPHYLLUM (A.) MICHAUXII (Web.) Buch
[Figs. 235:12–13; 237–238]

Jungermannia michauxii Web., Hist. Musc. Hep. Prodr., p. 76, 1815.
Jungermannia densa Nees, Naturg. Eur. Leberm. 2:143, 1836.

[240] Evans and Nichols (1935) also comment on the fact that *A. michauxii* "is related to" *A. assimile* (= *reichardtii*).

Jungermannia minuta var. *procera* Nees, *ibid.* 2:444, 1836.
Jungermannia anacampta Tayl., London Journ. Bot. 5:273, 1846.
Jungermannia fertilis Lindb., Acta Soc. Sci. Fennica 10:261, 1872.
Sphenolobus michauxii Steph., Spec. Hep. 2:164, 1902; K. Müller, Rabenh. Krypt.-Fl. 6(1):600, fig. 292, 1910.
Diplophyllum michauxii Warnst. *in* Loeske, Moosfl. Harzes, p. 64, 1903.
Lophozia michauxii Boulay, Musc. France 2:105, 1904.
Anastrophyllum japonicum Steph., Spec. Hep. 6:106, 1917 [nec Steph., Bull. Herb. Boissier 5:85, 1897].
Anastrophyllum tamurae Steph., *ibid.* 6:108, 1917.
Anastrophyllum michauxii Buch, Mem. Soc. F. et Fl. Fennica 8(1932):289, 1933; Schuster, Amer. Midl. Nat. 42(3):577, pl. 7, figs. 12–16, 1949; K. Müller, Rabenh. Krypt.-Fl. ed. 3, 6(1): 749, fig. 247, 1954.
Anastrophyllum (subg. *Euanastrophyllum*) *michauxii* Schust., Amer. Midl. Nat. 45(1):72, pl. 21, figs. 10–20, 1951; Schuster, *ibid.* 49(2):375, pl. 24, figs. 10:20, 1953.

Equally robust and somewhat similar to *A. saxicolus*, usually forming extensive mats or patches, *olive-green to brown, rather robust,* (1.0)1.2–1.4–2.0(2.5) mm wide × 1.5–5 cm long. Shoots usually *ascending to suberect,* simple or rarely branched, usually with subfloral innovations (often several from beneath unfertilized gynoecia); branches usually lateral-terminal, *Frullania* type, and (less frequently) ventral-intercalary. Stems 180–250(300) μ in diam., rather rigid, straight, yellow-brown, 10–14 cells high, nearly terete but slightly flattened; cortical cells in 1–2 strata, ca. 12–16 × 23–55 μ, thick-walled (averaging 3–4× as long as broad); medullary cells 17–24 μ in diam., collenchymatous, not mycorrhizal, not dorsiventrally differentiated. Rhizoids frequent, especially below. Leaves contiguous to weakly imbricate, *subequally bilobed, strongly spreading from a somewhat sheathing base, broadly ovate to ovate-quadrate,* averaging (0.85)0.95–1.15× as wide as long, usually *widest shortly above the dilated, semiamplexicaul base,* ca. 725–900(980) μ wide × 750–825(925) μ long (or larger), the antical half virtually transversely inserted, *extending across, and sometimes slightly beyond stem, distinctly decurrent,* the postical half obliquely inserted but not decurrent; sinus descending 0.3–0.45 the leaf length, acute to rectangulate; *lobes acute, the postical often somewhat incurved, the antical usually squarrose, not lying over postical. Cells rather small,* their walls often yellowish, those of the margin and lobe apices ca. (10)11–13(15–16) μ, those of the leaf middle subisodiametric, ca. 13–16(18) × 16–21(25) μ, *with large to coarse, bulging, often subconfluent trigones;* basal cells only sporadically distinctly elongated, averaging 15–17 × 24–30 μ; cuticle delicately verruculose; oil-bodies mostly (2)3–6(7–8) per cell, subspherical to ovoid or ellipsoid, 4 × 5 μ to 5–6 × 9 μ, finely granular-papillose externally, of minute spherules; chloroplasts ca. 3.5–4 μ. Gemmae at margins of upper leaves, light red (deeply shaded extremes) to *dull purplish (in masses often deep purplish),* 1–2-celled, ovoid and smooth in part but *largely 3–4-angled* in profile, thick-walled, ca. (12)13–18 × 20–27 μ; gemmiparous shoots often with leaves ± reduced, but *gemmiparous leaves always ± patent, contiguous to weakly imbricate, never erect or erect-appressed,* not or hardly more strongly ascending than normal shoots.

Dioecious, but *often fertile.* Androecia compactly spicate, of 4–6 pairs of bracts, becoming intercalary; bracts similar in size to vegetative leaves, hemispherically concave at base, with spreading lobes, divided ca. 0.3–0.5 and often with an antical, basal, incurved long tooth (which may bear 1–2 small cilia);

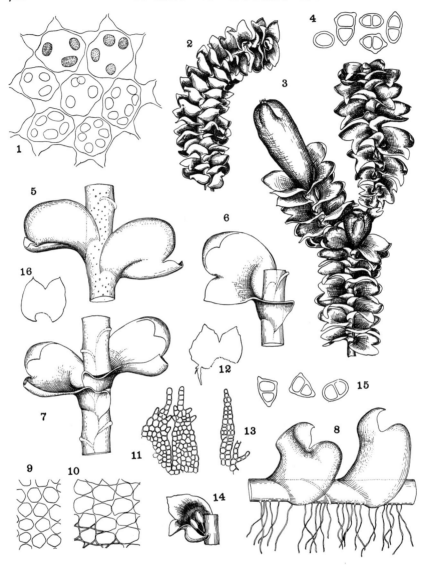

FIG. 237. *Anastrophyllum* (*Anastrophyllum*) *michauxii*. (1) Cells of leaf middle, with oil-bodies (×610). (2) ♂ Shoot (×ca. 10). (3) Plant with old gynoecial axis with two innovations, one with perianth (×ca. 10). (4) Gemmae (×ca. 300). (5) Shoot sector, ventral aspect; rhizoid initials drawn in (×27). (6) Same, dorsal aspect (×27). (7) Shoot sector, dorsal aspect, leaf insertion and merophyte boundaries indicated (×24). (8) Shoot sector, lateral aspect (×28). (9) Cells within leaf lobe (×ca. 300). (10) Submedian cells (×ca. 300). (11) Lobes of perianth mouth (×ca. 72). (12) ♂ Bract, flattened, to show basal appendage (×ca. 18). (13) Basal tooth of ♂ bract (×ca. 50). (14) ♂ Bract, slightly flattened to show antheridia (×18). (15) Gemmae (×ca. 300). (16) Leaf (×ca. 18). [1, Cascade River ravine, Minn., *RMS*; 2–3, 11–16, Raquette Lake, Adirondack Mts., N.Y., *RMS*; 4, 9, 10, Olean Rock City, N.Y. *RMS*; 5–8, Big Susie I., Minn., *RMS*; in part from Schuster, 1951, 1953.]

antheridia 2–3, among distinct to rudimentary paraphyses, their stalks uni-seriate. Gynoecia often with subfloral innovations (when sterile); *bracts little larger than leaves below them, similar to them* in shape (but rarely with a third lobe), *widely spreading to squarrose and not at all sheathing perianth at base;* bracteole absent or vestigial. *Perianth large, almost wholly exserted, clavate to obpyriform-clavate,* to 0.8–1.0 mm wide × 2.4–3 mm long, smooth except in distal ¼ where weakly 5–6-plicate and contracted strongly to mouth, *with a short antical median sulcus* (more marked, however, than other sulci); mouth divided into narrow, dentate to laciniate lobes which end in cilia. Seta with ca. 30 epidermal cell rows and numerous interior rows. Capsule ovoid, the wall 3–4-stratose; epidermal cells and internal cells both with radial (nodular) thickenings that are ± feebly extended as tangential spurs; innermost stratum with semiannular bands. Elaters bispiral, ca. 8 μ in diam.; spores 10–13 μ, finely papillate, brownish.

Type. In Canada and the Carolinas (*A. Michaux*).

Distribution

A continental, nearly circumpolar species, ranging from near sea level (in the Taiga) up to over 6600 feet (Mt. Mitchell, N.C.), occurring apparently across the Coniferous Panclimax from Europe to Asia and across North America. In Europe found scattered, and rarely, in the central portion [Bohemia; Upper Bavaria; Voralberg; Steiermark; Saxony; Harz Mts.; Silesia (Oder Region); the Riesengebirge; Czechoslovakia; the Carpathians], Hungary; Bucovina; Roumania; northern Italy; southernmost Sweden, Norway, and Finland; absent in England and Ireland, the Faroes, Iceland, and northernmost Europe, thus generally "avoiding" oceanic areas. Also occurring eastward in Asia (alpine regions of Japan, *fide* Hattori, 1952; Hokkaido, Honshu, Shikoku [Kitagawa, 1966]), and questionably in Shensi Prov., China (Nicholson, 1930a).

In North America, transcontinental, occurring in the West from Alaska and Yukon southward through British Columbia and Alberta to Idaho, Washington, and Wyoming; absent from Greenland. In eastern North America as follows;

LABRADOR. Seal L. (Macoun, 1902).   NEWFOUNDLAND. Holyrood, Avalon Distr.; Hare Bay, Rencontre West, Big Bay, Port aux Basques, all on S. coast; South Branch, Piccadilly, Spruce Brook, in W. Nfld.; *inter* Cow Head and Stanford R., and Great Harbor Deep, N. Pen.; Gaff Topsail, C. Nfld. (Buch & Tuomikoski, 1955).   NOVA SCOTIA. North West Arm (*Macoun 98, 1313*); Barrasois, Cape Breton.   QUEBEC. Labelle, Montcalm Cos., near St. Jovite (Crum & Williams, 1960); Seal L., N. Quebec (Buch & Tuomikoski, 1955); Lead L. (*Low 1637*); Ste-Anne R., Gaspé (*Macoun 33*); L. Némiskau; Mont Albert, Gaspé (*Lepage 3923a!*).   ONTARIO. Tea L., Algonquin Park (*Williams & Cain 5552!*); Belleville (*Macoun 43*); Cache L., Algonquin Natl. Park (Macoun, 1902); L. Nipigon (*Macoun 63*); North Hastings; Tilley L., Algoma Distr. (*Williams & Cain 5672b!*); Little Fluor I., L. Superior; Dreamer's Rock, Manitoulin Distr. (*Cain 4083!*): Algoma, Algonquin Park, Manitoulin I., and Thunder Bay

(Williams & Cain, 1959). MAINE. Mt. Desert I. (*Lorenz*); NE. slope of Baxter Peak
at 5000 ft, Northwest Basin at Davis Pond, Cathedral Trail, near Chimney Pond, and
Northwest Basin, all in Baxter State Park (*RMS 15976, 33026, 32996, 32901c, 15945,
17016,* etc.). NEW HAMPSHIRE. Mt. Washington (*RMS H-2143*); Franconia Mts.
(*Lorenz, 1908*); Mt. Lafayette, 3000 ft (*G. B. Kaiser, 1910*). VERMONT. *Fide* Evans
(1913): Pico Peak (*RMS*); near summit, 4250 ft, Mt. Mansfield, and Haselton Trail at
2800–3200 ft (*RMS 43806c, 43861a, 43869, 43844,* etc.). MASSACHUSETTS. Mt.
Greylock (*Andrews*); Windsor Jambs (*RMS*); Tannery Falls, Franklin Co. (*RMS*).
CONNECTICUT. Litchfield, Salisbury Co. (Evans & Nichols, 1908). NEW YORK.
Martiny Rocks, Cattaraugus Co. (*Boehner, 1948!*); Little Moose L., Herkimer Co.
(*Haynes*); Olean Rock City, Cattaraugus Co. (Schuster, 1949a); Blue Mt., Adirondack
Mts. (*RMS*); N. of Raquette L., Adirondack Mts. (*RMS*); Slide Mt., Cornell Mt.,
Wittenberg Mt., Catskill Mts. near Phoenicia (*RMS 24412, 17579, 44704b,* etc.).

Westward occurring as follows: MICHIGAN: Amygdaloid I., Isle Royale (*RMS*);
Copper Harbor, Keweenaw Pen. (*RMS*); Pictured Rocks, Alger Co.; Sugar Loaf
Mt., Marquette; Cliff R. and Mountain Stream, Huron Mts.; Ontonagon, Ontonagon
Co. and Whitefish Bay, Chippewa Co. (Evans & Nichols, 1935); Reese's Bog,
Cheboygan Co.; Phoenix, Keweenaw Co., etc. (Steere, 1937, etc.; Nichols & Steere,
1937). WISCONSIN. Sand I., Apostle Isls., Bayfield Co. (*RMS*); State Line, Vilas
Co.; White R., Bayfield Co.; Solon Springs, Douglas Co.; Mellen, Ashland Co.
(Conklin, 1929). MINNESOTA. Big Susie, Little Susie, Lucille, Sailboat, and Porcupine
Isls., Susie Isls., near Grand Portage, Cook Co. (*RMS 14913, 13511, 7190a, 13667,
10045, 13601, 9962,* etc.); Cascade R. ravine (*RMS 11415*); Stair Portage (*MacM.,
L. & B.*); Gunflint Trail, Little Devils Track Trail, Lutsen, Hungry Jack L., all in
Cook Co. (*Holzinger; Conklin 1123, 2250,* etc.); Pigeon Pt., Cook Co. (*RMS 13033*);
Encampment R., L. Superior (*RMS 13438,* etc.); Great Palisade, L. Superior, and
Two Island R., Lake Co. (*RMS 11254; Conklin 2621*); Briery, Pike Lake Rd., St.
Louis Co. (*Conklin 1444*).

Southward occurring as a disjunct on the higher peaks in the Appalachian Mountains,
rarely ranging along river valleys, to 2500 ft. WEST VIRGINIA. Boone, Monongalia,
Pocahontas, Preston Cos. (Ammons, 1940); Cheat View. VIRGINIA. Mountain
L., Giles Co. (*Sharp*; Patterson, 1949); White Top Mt., 5678 ft (Evans, 1893a; *RMS*);
Mt. Rogers, Grayson Co., ca. 5700 ft (*RMS & Patterson 38001, 38009a*); The Cascades,
Salt Pond Mt., Giles Co. (*RMS 40383*).

NORTH CAROLINA. Grandfather Mt., Caldwell & Avery Cos. (*RMS 30188, 44609b*).
Yancey Co.: Mt. Mitchell (*RMS 23140, 23146a, 23268b, 23162, 23262a, 34201a,* etc.);
Mt. Craig, Black Mts. (*RMS 24479a*). Swain Co: Clingmans Dome (*RMS 28117d,*
etc.); Andrews Bald (*RMS 36612f*). Flat Rock, near Linville, on Blue Ridge Pkwy.,
Avery Co. (*RMS 30168, 30167a, 30165*); Linville Gorge, Burke Co. (*RMS 28887,
28891b, 28905a*); Roan Mt., Mitchell Co. (*RMS 36935, 40316; Anderson 4150*);
Devils Courthouse, near Beech Gap, 5900 ft (*RMS 61230, 61232*); Balsam Gap, 25 mi
N. of Asheville, 5260 ft (*RMS 19064, 19076b*); Craggy Gardens, Craggy Mts., 5800 ft
(*RMS 24500c*); Waterrock Knob, ca. 6100 ft (*RMS 45070*), all on Blue Ridge Pkwy.;
primeval forest, Highlands, Macon Co. (*Blomquist 11179!*); Dry Sluice Gap, 5300 ft,
near Appalachian Trail (*Schofield 9825!*); Narrows of Chattooga R., W. of Bull Pen Mt.,
2500 ft, Jackson and Macon Cos. (*RMS 39400a, 39421, 39668a*); Richland Balsam, E.
of Rich Mt., 6400 ft, Haywood Co. (*RMS 39700a*). TENNESSEE. Sevier Co.:
Clingmans Dome (*RMS 34713, 34712a; Sharp 5328!*); Boulevard Trail, 6000 ft, Mt.
LeConte (*RMS 45242a*); summit, 6400 ft, and near Myrtle Pt., Mt. LeConte (*RMS
45322, 45328a, 45333c, 45332b, 45334d*); Mt. Kephart, 6000–6100 ft (*RMS 45375a*);
Pinnacle, near Greenbrier (*Sharp 34591,* as "*Marsupella sphacelata*"). Roan Mt., Carter
Co. (*RMS 61282b, 38660b*).

Anastrophyllum michauxii appears to be an excellent indicator species of the Spruce-Fir Forest Region.[241] Although rarely found where spruce-fir forest is absent (i.a., Cattaraugus Co., N.Y.; Schuster, 1949a), it may occur in such areas *if* the growing season is less than 125 days. The species is equally rare in alpine situations, above treeline, where it occurs very sporadically, often in poorly developed phases (i.a., on Mt. Katahdin, Me., at up to 5200 feet, some 1200–1500 feet above tree line), while it is totally lacking in the Tundra and even in the Tundra-Taiga ecotone.

Ecology

In all but the disjunct Southern Appalachian portion of its range, *A. michauxii* is largely confined to vertical faces of exposed to sheltered cliffs and ledges, where it may form extensive, rather thick red-brown to greenish brown mats. It is here a pioneer species, found most commonly with *Lophozia alpestris*, *Diplophyllum taxifolium*, and occasionally *Anastrophyllum saxicolus*, *A. minutum*, etc. Eastward, *Bazzania tricrenata* and *B. denudata* are also often consociated. The species is most often somewhat of an xerophyte, occurring with *Anastrophyllum minutum* on relatively shaded cliff faces and ledges which are not quite as exposed as those invaded by *Frullania asagrayana*-Lichen communities. Occasionally also on ledges and cliff faces where likely to be thoroughly soaked and perhaps submersed by run-off water during and after rains; under such conditions *Scapania nemorosa*, *Tritomaria quinquedentata* (occasionally *T. exsecta* or *exsectiformis*), *Lophozia attenuata*, *L. ventricosa*, and other species of the variable *Lophozia-Scapania* epipetric community may be associated. The species appears to lack tolerance for calcareous substrates and does not have a high tolerance for direct sunlight. Occasionally, on shaded siliceous rocks, also with *Mylia taylori*, *Lophozia alpestris*, *L. ventricosa silvicola*, *L. longidens*, *Cephalozia bicuspidata*, *Gymnocolea inflata* (Buch and Tuomikoski, 1955). Both northward and in the Southern Appalachians, *A. michauxii* also occurs on shaded, decaying logs, here associated with *Odontoschisma denudatum*, *Anastrophyllum hellerianum*, *Nowellia curvifolia*, *Riccardia palmata*, *Cephalozia catenulata* and *C. media* (rarely *C. lacinulata*), *Lophozia incisa*, *Jamesoniella autumnalis*, *Calypogeia suecica*, and various mosses, such as *Dicranum fuscescens* and *Tetraphis pellucida*. Occurrences on decaying logs are more frequent in the Southern Appalachians, where *A. michauxii* is a typical species on such sites (Sharp, 1939).[242]

[241] Striking exceptions are the plants from "The Narrows" and to their south, at ca. 2500 feet, on the Chattooga R., N. C., within 3–4 miles of the South Carolina and Georgia borders, occurring in typical "cove forest."

[242] Populations on decaying logs are usually freely fertile—capsules are generally produced in abundance, and seemingly during the entire summer season—the species not being "spring fruiting." In the spruce-fir forests of the Southern Appalachians I have found it "in fruit" from early June until mid-July and into early August.

In the Southern Appalachians, the species is found on bark of Fraser fir (*Abies fraseri*) perhaps even more frequently than on decaying logs. *Anastrophyllum michauxii* may occur anywhere from 1 to 10 feet high on bark, on the fog-enshrouded mountain summits (5000–6400 feet), associated with the *Bazzania nudicaulis-Herberta* community. Although often associated with members of this community (i.e., *B. nudicaulis, H. adunca tenuis, Plagiochila tridenticulata, Cephaloziella pearsoni, Leptoscyphus cuneifolius, Frullania asagrayana*, etc.) it usually undergoes ecesis on bark at about the same time as *Blepharostoma trichophyllum, Tritomaria exsecta*, and *Lepidozia reptans*, i.e., under more mature and mesic conditions. Bark populations are usually sterile but freely gemmiparous.

Somewhat similar to the occurrences on Fraser fir at high elevations are those at the southern edge of the Blue Ridge, at lower elevations (2500 feet) on the bark of hemlock, at heights of 3–8 feet. The plant occurs here as a mod. *parvifolia*, producing reddish gemmae in abundance and occasional perianths. Associated under such conditions may be other species often found on rocks, such as *Bazzania trilobata, Microlepidozia sylvatica, Cephalozia media*, and *Blepharostoma*, as well as typical corticoles such as *Radula obconica* and *Lejeunea ulicina*.

Variation and Differentiation

Occurring as two distinct environmental phases. Under optimal conditions forming thick, pure stands (particularly on moist logs and partly isolated damp rock walls); it is then abundantly fertile, robust, and nearly or quite free of gemmae. Such pure stands usually abundantly bear perianths and capsules, while androecia are often intermingled. Such robust phenotypes can be confused with *Lophozia kunzeana* and *Anastrophyllum saxicolus*, from both of which *A. michauxii* differs in the sharper leaf lobes, of which the antical is normally diagnostically squarrose (Fig. 237:3, 6). The long, clavate, long-exserted perianths with spreading to squarrose bracts are also distinctive (Fig. 237:3).

Under xeric conditions, either on bark or on dry, shaded rocks, the species often occurs as a sterile, small, strongly gemmiparous modification (mod. *parvifolia-lepto-vel mesoderma-viridis-vel-colorata*). This can easily be confused with *Anastrophyllum minutum* and in extreme instances perhaps with *A. hellerianum*, since in such phases, which may be only 0.25–0.8 mm wide, the decurrence of the antical leaf bases is hardly perceptible. Such atypical, xeromorphic, gemmiparous forms are readily recognizable since they possess (*a*) cells with distinct trigones, usually slightly bulging even in gemmiparous, small-leaved phases; (*b*) spreading, but frequently somewhat concave, gemmiparous leaves, whose often somewhat erose margins bear (*c*) dull rose-red to purplish to purplish brown gemmae. In *A. michauxii*, malnourished, small-leaved phases often appear to be exceptionally freely gemmiparous. In such cases gemma formation inhibits formation of leaves of mature size or form. However, these leaves invariably still spread from the stem, whereas in *A. minutum* leaves with prolonged gemma formation may become erect or erect-appressed.[243]

Since the equally robust *Lophozia kunzeana* and *Anastrophyllum saxicolus* are similarly suberect in growth, normally bear bilobed, subtransverse leaves, and may be similarly pigmented, they may be confused. However, *L. kunzeana* differs in the presence of underleaves, the nonsquarrose dorsal lobes, the more obtuse leaf lobes, and the greenish brown to cinnamon-brown gemmae. *Anastrophyllum saxicolus* is quite different in the cupped leaves with incurved, blunt dorsal lobes, and in the broader leaves, as well as in the lack of gemmae. In *A. michauxii* the leaves are more antically secund when dry, unlike those of *A. saxicolus* or *L. kunzeana*.

The leaf shape of *A. michauxii* is variable. In the Southern Appalachians, at least, the leaf is as broad or slightly broader than long (to 925 μ long × 980 μ wide in average, mature plants). By contrast, material from Quebec (*Lepage 3923a*) shows distinctly more elongate leaves, 900–1000 μ long × 850 μ broad. Appalachian plants have marginal cells of the lobes ca. 11.5–13 μ; the Quebec plants have them 12.5–16 μ. Other differences are lacking, but Quebec plants are denser-leaved than is common in Appalachian phases. Perhaps equally variable is the gemma color. In extreme shade phases (as on Fraser fir in dense fog forests of the Southern Appalachians) gemmae are light red; in "normal" phases, from slightly to distinctly insolated rock walls gemmae are "darker, more purplish" (Schuster, 1949a) and acquire a much deeper hue than in the sometimes superficially similar *Anastrophyllum minutum*, in which they are usually "bright red, occasionally almost scarlet."

Japanese material (Mt. Hayachine, 1400 m, northern Honshu, on bark of fallen tree; with *Nowellia curvifolia* and *Scapania ampliata*; *leg.* D. Shimizu) is relatively small, the larger shoots 950–1100 μ wide, strongly brownish-tinged but with the gemmae a bright red. Except for the gemma color, the plants approach a typical mod. *parvifolia* of the species. Cells of the lobe margins average 13 μ, median lobe cells are ca. 15 μ wide; trigones are coarse and often subconfluent; the leaves have spreading, although rarely squarrose, dorsal lobes—as is typical of many smaller modifications—and the dorsal lobes are obliquely inserted, showing only slight decurrence, again typical of mod. *parvifolia*. Even the untypical gemma color can be matched in American plants, as, i.a., on specimens from Mt. Rogers, Va., on bark of *Abies fraseri*.

Similar phases of *A. michauxii* are frequent in gorges of the Blue Ridge Escarpment, e.g., at 2500 feet in the Chattooga R. gorge, N.C. (*RMS 39418*). Such plants, confined almost exclusively to the acid bark of *Tsuga canadensis* at heights of 3–10 feet, are associated with *Bazzania trilobata*, *Blepharostoma*, *Microlepidozia sylvatica*, *Cephalozia media*, and such more typical corticoles as *Lejeunea ulicina* and *Radula obconica*.

---

[243] Exceptionally, on trunks of *Abies fraseri* in the Southern Appalachians, extremely reduced, slender phases occur, only 250–500 μ wide, in which the leaves are almost appressed. Such plants superficially approach *Anastrophyllum hellerianum* but differ in the red (rather than purplish) gemmae, of superior size.

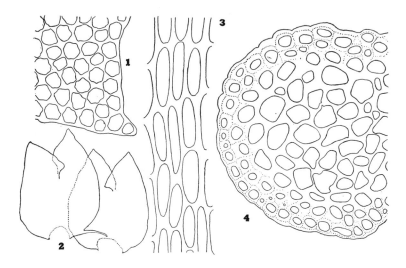

Fig. 238. *Anastrophyllum (Anastrophyllum) michauxii.* (1) Cells of leaf lobe (×300). (2) Leaves of sterile shoot (×25). (3) Dorsal cortical cells of stem (×300). (4) Stem cross section (×300). [All from *Lepage 3923a*, Mt. Albert, Quebec.]

Relationships

In spite of the aforementioned tendency for weak phases to occur as phenocopies of *A. minutum*, it is closely allied only to *A. japonicum* (*sensu* Hattori, Hep. Jap. no. 301, 1956); according to Hattori, *A. japonicum* and *A. tamurae* of Stephani (Spec. Hep. 6:106 and 108, 1917) are identical. The latter is described as having the perianth "anguste clavata (4.5 mm longa, medio 1.5 mm lata)...inflata...," much as in *A. michauxii*. Stephani's description of the leaf ("ad ⅕ biloba" for *A. japonicum*; "ad ⅙ inciso-biloba" for *A. tamurae*), however, would eliminate them from near affinity to *A. michauxii*, *if* Stephani's descriptions were correct. However, in specimens of *A. japonicum* issued by Hattori (Hep. Jap. no. 301, 1956) leaves are bilobed 0.3–0.4 their length and agree exactly in shape with those of many forms of *A. michauxii*. Specimens determined by Stephani (*Faurie 1791, 1302*) as *A. japonicum* Steph., "sp. n.," also have leaves 0.3–0.35 bilobed. I am inclined to regard the *A. japonicum-tamurae* complex as a synonym of *A. michauxii* or at best as forming a weakly defined subspecies.[244]

---

[244] I have studied two Faurie collections from Japan, labeled "*Anastrophyllum japonicum* Steph. sp. n." and determined by Stephani. *Faurie 1791* (Dist. Nagano, Norikura, 2500 m. Aug. 1905), a xeromorphic phase with small, conduplicate leaves, has the facies of *Anastrophyllum minutum*, except for the dorsal lobe insertion; it probably represents an xeromorphic phase of *A. michauxii*. *Faurie 1302* (Zizogatake, 2500 m, prope Kofu, July 1903) bears juvenile perianths and gemmae; it closely matches *A. michauxii*, except for the perhaps slightly smaller average cell size (marginal

"*Anastrophyllum japonicum*" of Stephani (1917, p. 106) is a later homonym of the "*Anastrophyllum japonicus*" of Stephani (Bull. Herb. Boissier 5:85, 1897). The 1897 "*Anastrophyllum japonicus*" has clear priority; hence if *A. japonicum* Steph. (1917) is a good species distinct from the old *A. michauxii*, it cannot bear that name, but will have to be known as *A. tamurae* Steph. Until types of all these taxa are compared, further nomenclatural changes can only add to the chaos introduced by Stephani.

Thus *A. michauxii* appears to be replaced, at least in part, in Japan by an allied taxon, *A. japonicus* Steph. (Bull. Herb. Boissier 5:85, 1897; see Schuster, 1965, p. 271, for the involved synonymy), fertile plants of which appear to differ from *A. michauxii* "in two ways: the perianth is widest in the median half and is never clavate . . .; the perichaetial bracts, which are 2–3-lobed, loosely sheath the perianth at base, rather than being strongly spreading." I have stated (*loc. cit.*) that "this taxon will prove at best a subspecies of *A. michauxii s. str.*"

I have described a similar plant (Schuster, 1965, p. 270), *A. gottscheanum* Schust., from Pico de Orizaba, Mexico, which differs from *A. michauxii* in the smaller cells (7–11 μ in lobe margins; median cells 8–12 × 9–14 μ), apparent lack of gemmae, seemingly uniform *Frullania*-type branching, smaller size (shoots 550–650 μ wide), and leaves with *both* lobes typically incurved (hence mature shoots bear rather strongly cupped leaves).

Subgenus *EURYLOBUS* Schust.

Anastrophyllum subg. *Eurylobus* Schust., Amer. Midl. Nat. 45(1):71, 1951.
Sphenolobus subg. *Eurylobus* K. Müll., Rabenh. Krypt.-Fl. ed. 3, 6:725, 1954.

Subgenus a subg. *Anastrophylla* distinctum ut (*a*) pigmenta valde subrubra subbrunneave nulla, (*b*) folia latissima, multo latiora quam longa, valde cyathiformia hemisphericaque, (*c*) cellulae solidae, trigonis, autem, male evolutis, et (*d*) cellulae basales numquam elongatae.

Robust, suberect to erect and caespitose or loosely procumbent; boreal to arctic, green to brownish, *lacking reddish pigments*, rather rigid, with shoot apices feebly upcurved, the leaves not conspicuously antically secund (plants little convex in postical aspect). Stem ca. 13–16 cells high. Branching infrequent, often *all ventral-intercalary but sometimes with Frullania-type terminal branches* (see p. 739). *Leaves much broader than long*, when flattened broadest beyond middle, *strongly concave or cupped*, bilobed to 0.5 their length, the broadly ovate lobes erect or almost so, *tips blunt and often incurved;* ampliate antical leaf base *extended across and often a little beyond stem*, obliquely inserted on an arched line and *somewhat decurrent*. Cells firm, walls thin to slightly thickened, *with distinct (but hardly bulging) trigones;* basal cells not elongated. *Gemmae absent.* ♀ Bracts (2)3–5-lobed, lobes ± toothed. Antheridia 2–3 per bract, with paraphyses; *antheridial stalk 2-seriate.*

---

cells 10–13 μ; median 13–15 μ wide). Larger leaves show the squarrose to erect dorsal lobes diagnostic of mature *A. michauxii*. I fail to see how *A. japonicum* can be regarded as more than a slightly smaller-celled race and do not feel certain that cell size will allow a constant separation.

Type. Anastrophyllum saxicolus (Schrad.) Schust., the only species.

The subg. *Eurylobus* has been assigned to both "*Sphenolobus*" (Müller, 1951–58; Buch, 1933a) and *Anastrophyllum* (Schuster, 1951a, 1953). This is simply another of the numerous cases (see pp. 709, 714) in which transition between these two groups occurs. As I pointed out (Schuster, 1951a, p. 71), *Eurylobus* and *Anastrophyllum* share a whole ensemble of characters, among them the diagnostic decurrence of the ampliate antical leaf bases (Fig. 239:1).

An extraordinary variant, with many shoots triseriate and isophyllous, is discussed on p. 739.

ANASTROPHYLLUM (EURYLOBUS) SAXICOLUS
(Schrad.) Schust.
[Figs. 239–240]

Jungermannia resupinata L., Fl. Suecica, p. 400, 1755 (in part); also of Wahlenberg, Fl. Lapp., 1812, and Nees, Naturg. Eur. Leberm. 1:243, 1833.
Jungermannia saxicola Schrad., Syst. Samml. Krypt. Gewächse, no. 97, 1796.
Diplophyllum saxicolum Dumort., Rec. d'Obs., p. 16. 1835.
Lophozia saxicola Schiffn., *in* Engler & Prantl, Nat. Pflanzenfam. 1(3):85, 1893.
Sphenolobus saxicolus Steph., Spec. Hep. 2:160, 1902; K. Müller, Rabenh. Krypt.-Fl. 6(1):603, fig. 293, 1910; Macvicar, Studs. Hdb. Brit. Hep. ed. 2:207, 1926.
Anastrophyllum (Eurylobus) saxicolus Schust., Amer. Midl. Nat. 45(1):71, pl. 19, 1951; Schuster, *ibid.* 49(2):371, pl. 22, 1953.
Sphenolobus (Eurylobus) saxicolus K. Müller, Rabenh. Krypt.-Fl. ed. 3, 6(1):725, fig. 235, 1954.

In extensive olive-green to brown or yellowish brown mats or patches, the *individual plants suberect or strongly ascending*, when as isolated stems sometimes creeping; shoots *robust, 1.2–2.5(3) mm wide* × 2–5 cm high, *simple or very sparingly branched* (almost invariably so from beneath ♀ inflorescences). Stems rigid, rather stout, mostly 240–380 μ thick; cortical cells thick-walled in 1–2 layers, somewhat smaller than the weakly collenchymatous medullary cells, gradually grading into these in size (of cross section). Leaves weakly to closely imbricate, *very regularly, steplike in insertion*, the plants with a consequent pectinate appearance, *convex and closely shingled postically*, somewhat antically secund (the shoots thus convex in ventral aspect), *virtually transversely inserted*, but postical half somewhat oblique, *antical half shortly decurrent, deeply unequally bilobed* ($\frac{1}{3}$–$\frac{3}{5}$), *concavely complicate*, often hemispherically concave, *very broad*, varying from 850 μ wide (widest point near apices of lobes) × 500 μ long to apex of ventral lobe, up to 1020–1200 μ wide × 765–820 μ long, occasionally larger (*width of mature leaves 1.3–1.7×* the length!); *dorsal lobe erect to erect-spreading*, its apex at an angle of 0–15° with stem, broadly ovate to suborbicular, varying from 0.9 to 1.1 as wide as long, obtuse to subacute, the *apex commonly slightly incurved;* ventral lobe spreading, larger than antical, *strongly concave and with apex strongly incurved*, broadly ovate and from 0.9 to 1.2(1.4) × as wide as long; *keel weakly to semicircularly arched;* sinus often *gibbous and recurved.* Cells roundish-quadrate, with concave to weakly convex-sided trigones, and ± thickened walls; marginal

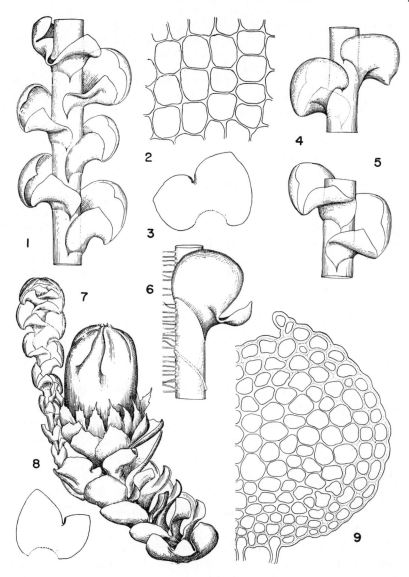

FIG. 239. *Anastrophyllum* (*Eurylobus*) *saxicolus*. (1) Shoot sector, antical aspect, uppermost leaf trilobed and *Orthocaulis*-like (×23). (2) Median leaf cells of ventral lobe (×340). (3) Leaf (×26). (4–5) Postical and antical views of shoot sectors (×25). (6) Lateral view of shoot sector and of leaf insertion (×25). (7) Shoot with rather immature perianth, unfertilized, and subfloral innovation (×18). (8) Small leaf (×26). (9) Stem cross section (×240). [All from plants from Pigeon Pt., Minn., *RMS*; from Schuster, 1953.]

cells ca. 17–20 μ; median cells subisodiametric, (15)18–22 × 18–24 μ; each cell with 2–4, *occasionally* 5, *oil-bodies*, each 4–6 μ and spherical to 5 × 7–10 μ and ovoid, formed of numerous, minute, rather discrete spherules; cuticle weakly verruculose. *Underleaves lacking. Without gemmae.*

Dioecious, *usually sterile.* ♂ Bracts somewhat saccate at antical base, otherwise similar to leaves and equal to them in size, the androecium barely distinct, intercalary with age; antheridia 1–3, accompanied by a few paraphyses. ♀ Bracts erect, concave, sheathing base of perianth, ¼–⅓ divided into 2–5 narrowly triangular, acuminately pointed lobes, whose sides are remotely dentate or spinous-dentate (usually one bract 2-lobed, the other 4–5-lobed); subinvolucral leaves 2–3-lobed to the middle or deeper, the ovate lobes often remotely denticulate; bracteole distinct, united with one bract. Perianth becoming cylindrical-oblong to cylindrical-ovate at maturity, very large, longly exserted beyond bracts, to 3–4 mm long × 1.5 mm wide, narrowed and deeply plicate toward mouth, with a deep antical groove; mouth lobulate-ciliate, the numerous laciniae to 9–10 cells long, terminating in uniseriate cilia 4–6 cells long. Capsule with wall 3-stratose, the epidermal cell layer larger (16 μ thick); inner layers with semiannular bands, the innermost 10 μ thick; outer layer with nodular thickenings only. Seta ca. 9 cells in diam., the epidermal cells in ca. 25 rows; epidermal cells ca. half the size of inner (cross section). Spores 12–14 μ, weakly verrucose; elaters 8 μ in diam.

Type. Meissner, in Hesse, Germany (*Schrader*).

Distribution

A rather rare and sporadically distributed circumpolar species, in general of arctic-alpine distribution. In Europe rare in the southern edge of the Alps (Switzerland; north Italian Alps), northward through Central Europe (Austria; Salzburg; the Carpathian Mts.; the Tatra Mts.; Czechoslovakia; the Riesengebirge; Thuringia; Rhone region; the Harz), becoming more frequent northward (Scotland; Shetland Isls.; Norway; Sweden; Finland); eastward to Siberia (Lena and Yenisey R. valleys) and the alpine region of Japan (see Hattori, 1952 and Kitagawa, 1966).

In North America of sporadic distribution in the West from Alaska (Persson and Weber, 1958) and the Yukon (Klondike, *R. S. Williams, 1898!*; "The Dome," Yukon, *Macoun 45!*), southward to ?British Columbia [Frye and Clark (1937–47, p. 379) report no *A. saxicolus* in the Brinkman collections from British Columbia, suggesting this extension in range may be an error]. In our area as follows:

E. GREENLAND. Danmarks Ø, Scoresby Sund (Jensen, 1897); also without data (*Vahl*). W. GREENLAND. Tupilak I., near Egedesminde (*Holmen*); Tunugdliarfik Fjord (Lange & Jensen, 1887).

BAFFIN I. Pangnirtung (*Wynne-Edwards 66!*). QUEBEC. Tadoussac; Mt. la Table (Lepage, 1944–45); Mt. Albert, summit (*Macoun, 1882;* issued in Carrington & Pearson, Hep. Brit. Exsic.!). ONTARIO. Moose Factory, James Bay (*Greville, 1843!*;

ex herb. Taylor, NY); Ouimet Canyon, N. of Dorion, Thunder Bay (Williams & Cain, 1959); Sutton Ridge, Hudson Bay 54°25′ N. (*Sjörs*).

MAINE. Near summit of Cathedral Trail, Mt. Katahdin, 5200 ft (*RMS 32966*). VERMONT. Mt. Mansfield, near summit, 4000 ft (*RMS 43806*). NEW YORK. Wilmington Notch, N. and S. sides of W. branch of Ausable R., 6.5 mi SW. of Wilmington, Essex Co.,1500–2000 ft (N. G. Miller, 1966). MINNESOTA. Pigeon Pt., ca. 1.5 mi from tip, N.-facing cliffs (*RMS 7000a, 10012, 8007, 10013*; see Schuster, 1953).

Southward recurring as a disjunct at high elevations in the Southern Appalachians: NORTH CAROLINA. Cliff faces, summit of Roan Mt. ca. 6000–6200 ft, Mitchell Co. (*RMS 38657, 36918, 36932c, 40316*). TENNESSEE. Cliff faces, Roan Mt., Carter Co., ca. 6200 ft (*RMS 36921, 61280, 61283a, 61283, 36933*); Charlies Bunion, 5300 ft, along Appalachian Trail NE. of Mt. Kephart, Sevier Co. (*Schofield 9774*!; southernmost report).

Ecology

The name of the species clearly emphasizes its epilithic occurrence. Either *A. saxicolus* is a pioneer on sheer rock faces, usually in shaded and humid to damp sites, occasionally under exposed conditions, or it undergoes ecesis after an initial group of consociated species has come in, including in the Spruce-Fir Zone such pioneers as *Tritomaria quinquedentata, Anastrophyllum minutum, Lophozia ventricosa, Diplophyllum taxifolium* (Schuster, 1953). In alpine Tundra common associates are *Chandonanthus setiformis* (Mt. Katahdin, Mt. Mansfield, Mt. Albert, and also Baffin I.), *Lophozia alpestris, Anastrophyllum minutum* var. *grandis* and occasionally *Scapania nemorosa*. Apparently only on acidic granitic and metamorphic rocks (with the measured pH 4.6–5.2, *fide* Schuster, 1953); on such rocks the species usually occupies shaded sites.

Anastrophyllum saxicolus also occurs in more exposed areas such as the crests of cliffs. Here it may occur on thin soil with various Cladoniae and *Polytrichum* (Schuster, 1953), or in boggy areas at the summits of the mountains with *Dicranum* and other mosses (Mt. Albert, Quebec). In the alpine Tundra (as on Mt. Katahdin) the species is often restricted to the lower edges of the vertical faces of large, dry boulders, where it occurs as an extreme xerophyte.

Differentiation

A most distinctive species, with an unmistakable facies, due to the (*a*) very regularly, pectinately inserted leaves; (*b*) very broad leaves, with the ventral lobes incurved and appearing rounded *in situ*, the dorsal never squarrose, the leaves strongly concave, somewhat cupped or hemispherical (Fig. 239:1, 4–6). The species never produces gemmae and is very rarely fertile.[245]

---

[245] The scattered but very wide distribution of *A. saxicolus*, therefore, cannot easily be explained. Where the plant occurs it often forms very extensive clones, purely as a consequence of vegetative spreading. For instance, on Pigeon Point, Minn., it forms extensive mats on a single cliff face, exerting a high degree of dominance in many cases. By a conservative estimate, several hundred square feet are here carpeted by almost pure mats of the species.

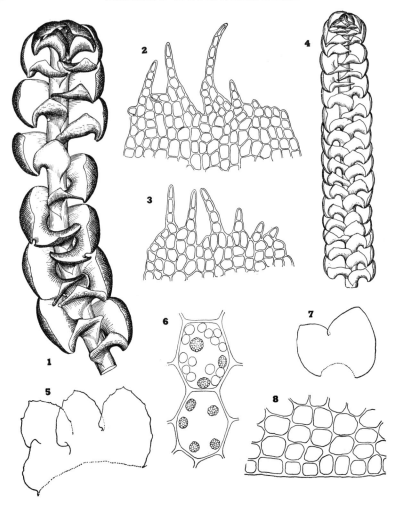

FIG. 240. *Anastrophyllum (Eurylobus) saxicolus.* (1) Shoot sector, antical view (×18.5).
(2–3) Sectors of perianth mouth (×150). (4) Sterile shoot (×ca. 12). (5) Subinvolucral bract (×23). (6) Median cells with oil-bodies and, upper cell, chloroplasts
(×680). (7) Leaf (×18.5). (8) Marginal leaf cells (×340). [1–5, 7–8, Pigeon Point,
Minn., *RMS*; 6, *Wynne-Edwards 66*, Baffin I.]

This stenotypic, old species is sharply isolated from other Lophoziae.
This isolation was emphasized (Schuster, 1951a), and the species was
relegated to a monotypic subgenus (*Eurylobus*) within *Anastrophyllum*;
Müller (1951–58), while accepting the subgeneric disposition, preferred
to leave the group within "*Sphenolobus*" (to which it has only a tenuous
relationship).

Anastrophyllum saxicolus differs from *A. michauxii* and *A.* ("*Sphenolobus*") *minutum*, the somewhat remotely related regional taxa, in facies and in the inability to produce gemmae, as well as in the much broader leaves (whose width greatly exceeds the length of the larger, ventral lobe). The greater size, the obtuse, ovate leaf lobes, the more evidently collenchymatous cells, and the arcuately inserted, somewhat decurrent dorsal lobes separate *A. saxicolus* from *A. minutum*. The squarrose, sharply pointed dorsal lobes of *A. michauxii* give it a totally different facies from *A. saxicolus*, readily appreciated even with the naked eye.

The species is more likely to be confused on superficial examination with the denser-leaved, complicate-leaved phases of *Lophozia* (*O.*) *kunzeana*, which may have a surprisingly similar facies. The total lack of underleaves will separate *A. saxicolus* from the *Lophozia*. A superficial similarity also exists to *Scapania*, but the concave, cupped leaves of *A. saxicolus* are not at all keeled.

Variation

A stenotypic species, varying only in growth habit from creeping and closely prostrate when a pioneer on bare rock walls, to erect and caespitose when crowded and forming dense patches. In the Tennessee plant (*RMS 61280*) there occur interspersed amidst normal stems individuals showing a most striking modification, in which the ventral merophytes broaden out locally on the stems, with postical half-leaves developed or, more frequently, bilobed underleaves which may be identical to the lateral leaves in form. Hundreds of such stems have been seen; well-developed ones are perfectly triquetrous and triangular in outline when examined apically. Such reversions occur chiefly on ♀ plants, although perfectly sterile stems may also show them!

Often *A. saxicolus* is caespitose and then infrequently branched. In *RMS 61280*, of about 150 branches seen, all were postical in origin and intercalary. Gynoecia, if unfertilized, often develop one, rarely two, innovations but occasionally are innovation-free. By contrast, plants of *RMS 61281*, from the same area, showed occasional terminal branching (*Frullania*-type branches; supporting stem leaf ovate, unlobed) and what appeared to be lateral-intercalary fasciculate branching from (apparently) injured shoot apices. In one case, two *Frullania*-type branches, one from each side of the stem, occurred only two leaf pairs apart.[246]

Subgenus *SCHIZOPHYLLUM* Schust., subg. *n.*

Plantae parvae, pigmenta subrubra aut subpurpurea habentes; ramificatio omnino aut praevalenter terminalis, type *Frullaniae*; caulis tantummodo 6–9 cellulis alt.; folia dorsaliter vix aut non decurrentia, numquam valde secunda, basis anticalis non dilatata (trans caulem non extensa).

[246] The observations on branching in this species are of utmost interest, suggesting that environmental conditions may affect the type of branching which occurs. In one case, with crowding (*RMS 61280*), only postical intercalary branches occurred; with less crowded conditions (*RMS 61281*) some—but not all—of the branches were terminal! The conclusion to be drawn here, with respect to the significance of branching types, is too obvious to need spelling out!

Mostly erect or suberect, generally boreal or arctic and subantarctic, *usually small* (mostly under 1 mm wide), red-brown or purplish (or, at least locally, with some reddish pigments), often flexuous, rarely rigid, *shoot tips not or slightly arched*, the leaves *never strikingly antically secund*; *branching wholly or predominantly of the Frullania type*. Stems slender, *6–9 cells high only;* cortex 1-stratose. Leaves $\pm$ longer than broad, oblong-ovate (unless with lobes divaricate), 0.35–0.45 or more deeply bilobed, lobes widely separated; leaves loosely concave-condu-plicate, *antical base never dilated, not extended across stem, transversely inserted or almost imperceptibly decurrent.* Cells with walls $\pm$ thin, with strongly developed, mostly bulging (sometimes confluent) trigones; basal cells, at least in a small group, conspicuously elongated. No asexual reproduction.

Type. Anastrophyllum sphenoloboides Schust.

The subg. *Schizophyllum* includes, among taxa I have studied, three species complexes or sections, united by small size, linked with an almost nondecurrent antical leaf base that is not ampliate (hence fails to extend across and beyond the stem), and free *Frullania*-type branching. These are features of the *Eremonotus-Crossocalyx* complex, from which *Schizophyllum* differs in the consistently 3-layered capsule wall (in investigated taxa)[247] and the strongly collenchymatous cells, more or less elongated at base.

The single regional species is only obscurely related to the other two sections, as the following key shows:

Key to Sections of Subgenus Schizophyllum

1. Leaves ovate to oblong-ovate, 0.35–0.45 bilobed, the lobes erect . . .   2.
   2. Leaf margins crenulate (septae between marginal cells dilated). Sub-antarctic . . . . sect. *Crenulatae* Schust., sect. n.[248]
   2. Leaf margins noncrenulate, the septae between marginal cells not pro-truding. Arctic . . . .
     sect. *Schizophyllum* Schust., sect. n. (*A. sphenoloboides* Schust., type, p. 741)
1. Leaves widest distally, divided 0.6–0.7, the lobes divaricate; leaf margins not crenulate. Dioecious. Japan to Indomalaya . . . .
   . . . . sect. *Bidentes* Schust., sect. n.[249]

[247] Note, however, that *Acantholobus* (*A. minutum*), which is closely allied to this complex, may have a 3-stratose capsule wall.

[248] With *A. crenulatum* Schust., sp. n. (type: Cerro Garibaldi, Tierra del Fuego, Argentina, *RMS 58304a.* Sterilis; folia contigua, patula, transverse inserta, 0.25–0.35 bifida, lobis sub-aequalibus, acutis, crenulatis. Cellulae ca. 8–12 × 10–15 μ, margine noduloso-incrassatis) and *A. novazelandiae* Schust. (New Zealand; Schuster, 1965, p. 282, fig. 4). *Anastrophyllum pusillum* Steph. (Bolivia; Stephani, Spec. Hep. 6:108, 1917) also belongs here; it has *Frullania*-type branches and crenulate leaves.

Sectio Crenulatae: Sectio sectionibus *Schizophylli* aliis differens quod (*a*) margo folii valde crenulatus et (*b*) distributio solum antipodalis; sectioni *Schizophyllo* forma folii ovata similis.

[249] With *A. mayebarae* Hatt. (1953b, pp. 141–143, fig. 64) of Japan and *A. bidens* (Nees) Steph. of Java; see p. 714. Kitagawa (1966, p. 126) considers *A. mayebarae* to be a synonym of *A. bidens.* He also, erroneously, places here as a synonym *Anastrophyllum cephalozielloides* Schiffn., Denkschr. Akad. Wissen. Wien 67:201, 1898; this plant is an *Andrewsianthus, Andrewsianthus cephalozielloides* Schust., comb. n.

Sectio a sectionibus *Schizophylli* aliis differens quod (*a*) folia latissima distalliter, lobis lanceo-latis divergentibusque; et (*b*) sinus per 0.6–0.7 longitudinem folii descendentes.

ANASTROPHYLLUM (SCHIZOPHYLLUM)
SPHENOLOBOIDES Schust., *sp. n.*

[Figs. 241, 242, 243: 1–10]

Plantae paro- aut heteroeciae; parvae, rubro-purpureae aut subrubrae; rami typi *Frullaniae*; folia ovata, longiora quam lata, 0.35–0.45 bilobata, antice vix decurrentia; gemmae nullae; pedicellus antheridii 2-seriatus; membrana capsulae 3-stratosa.

Growing erect, 1–2 cm tall, unbranched or having an occasional terminal branch of the *Frullania* type; 1–2(3–4) innovations from beneath the perianths. Shoots green, the upper portions of at least the upper leaves *purplish brown to reddish purple or ± reddish pigmented*, the leaf bases and stem usually remaining greenish; shoots (500–750)790–900(950) μ wide. Stems *abundantly provided with rhizoids* below, even to near the shoot apex; stems (120–140)175–225 μ in diam., nearly straight and rather rigid; cortical cells thick-walled, basically *narrowly oblong*, from (8)9–11(12) × (24)28–40(45) μ to 10–13(14–17) × (30)–40–75(80) μ. Leaves *concave, sheathing basally and nearly erect*, the distal portions suberect to loosely spreading or very loosely convolute, the lobes often ± incurved; leaves regularly distichous, slightly distant to contiguous (subimbricate only when suberect), rather broadly ovate *to ovate-oblong*, 410–475 μ × 430–510 μ long to ca. 575 μ wide × 575–600 μ long, occasionally 580–600 μ wide × 550–610 μ long (length measured from insertion), thus varying from ca. 0.95 to 1.05× as long as wide, widest at the rounded and rather conspicuously narrowed base, *bilobed for 0.35–0.45 their length by a sharp, V-shaped sinus;* lobes ovate-triangular, essentially equal, *considerably longer than wide*, the ventral acute or blunt to narrowly rounded at the tip, the dorsal ± blunt to acute, occasionally sharply acute; insertion arcuate, dorsally nearly transverse, but at the *inner margins normally slightly but distinctly decurrent.* Upper leaf cells with the walls usually reddish-purple pigmented, often (but not always) becoming strongly collenchymatous, *with coarse, often confluent trigones; marginal cells ca. 13 to 15–19 μ in lobes;* subapical cells of lobes ca. (12)13–16 × (13)14–17(18–20) μ; median cells usually with somewhat thickened walls, but with trigones hardly produced to conspicuous, the cells from 14–19(22) × 18–24(26) μ, ± *rectangulate;* basal cells becoming largely *distinctly elongated, 14–16 × (20)25–30 μ to 17–18 × 28–35 μ, averaging twice as long as wide, locally to 2.5× as long as broad;* cuticle verruculose-striolate in basal half of leaves. Oil-bodies of cells in lobes small, 2.7–3 × 3.8–4 μ to 3.6–4 × 4–5.5 μ, 2–5(6) per cell; median cells also with (2)3–5(6) oil-bodies each, the oil-bodies of spherules, subbotryoidal and 2.8–3 × 4.5–5 to 3.8–4.2(5–6) × 6.5–7.5(9–11) μ, some spherical and 3.6–4 μ. Underleaves lacking. *Gemmae apparently consistently absent.*

Heteroecious (paro- and autoecious). Androecia 650–700 μ wide, becoming intercalary, of 3–6 to 7–12 (16–25) pairs of bracts; bracts similar to leaves but broader than long, somewhat more imbricate, conspicuously concave at base, with the antical base transversely inserted and armed on the margin just above with an erect or inflexed, small, sharp tooth or lobe; lobes erect, 2 (rarely 3), broader than on vegetative leaves, usually acute to rectangulate; antheridia 1–2(3); stalk 2-seriate; with 1–3 slender, usually ciliiform paraphyses. Gynoecia terminal, usually producing 1–2(3–4) innovations if fertilization does not take place; bracts *suberect and loosely sheathing the perianth* at base, the acute lobes

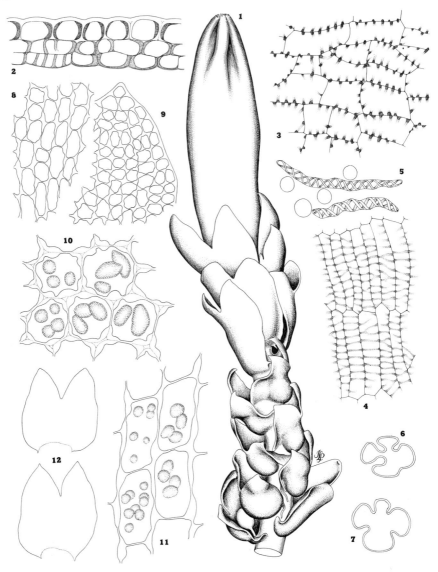

FIG. 241. *Anastrophyllum* (*Schizophyllum*) *sphenoloboides*. (1) Paroecious shoot, with several sterile leaves between ♀ and ♂ bracts (×41). (2) Capsule wall cross section (×400). (3) Epidermal cells, capsule wall (×310). (4) Inner cells of capsule wall (×310). (5) Spores and elaters (×310). (6–7) Two perianth cross sections, one-fifth from apex, dorsal face at top (×30). (8) Basal leaf cells (×200). (9) Cells of lobe apex (×200). (10) Subapical cells (×705). (11) Basal cells with oil-bodies (×580). (12) Two leaves (×35). [All from northwest Greenland: 1–9, 11, paroecious phenotype, *RMS 45984*; 10, *RMS 46003*, paroecious phenotype; 12, *RMS 45676*.]

erect or slightly incurved, similar to leaves, but broader than long, broadly, transversely ovate and usually bilobed for 0.4 their length, ranging from 730–770 μ wide × 640–725 μ long to 925–1060 μ wide × 770–775(815) μ long (rarely trilobed and to 1100 μ wide × 875 μ long); bracts when bilobed divided unequally, the dorsal lobe considerably narrower though not shorter than the ventral lobe; lobes acute to acutish; rarely with a ventral tooth of 1 lobe; margins of bracts otherwise entire, similar to those of leaves. *Perianth longly emergent*, at least ¾ exserted, *slenderly cylindrical to cylindrical-subclavate*, (650)700–800(880) μ in diam. × 2.3–2.9 mm long, terete and smooth except in the contracted distal ⅕–¼, where becoming *deeply 3-plicate (2 plicae lateral; 1 ventral; with a deep antical sulcus)*, rarely with weak development of supplementary plicae near the mouth; mouth contracted, shallowly lobulate, the margins denticulate to dentate with close to contiguous short, acute teeth 1–3(4–5) cells long and 1–2 cells wide at base; terminal cells narrowly triangular, acute, ca. 13–14 × 20–34 to 16 × 25–30 μ, thick-walled. Seta strongly elongated, 240 μ in diam., with 22–24 epidermal cell rows, 6–7 cells in diam. Capsule reddish brown, ovoid; *wall 3-stratose*, to 28–35 μ thick, the epidermal cells 11–12 to 13–17 μ thick, with firm outer tangential walls; inner 2 layers each 6.5–7 to 7.5–10 μ thick, thin-walled. Epidermal cells oblong to oblong-subquadrate, 18–22(25–26) × 22–34(36–45) μ, relatively regular in shape, with conspicuous nodular thickenings extended for 0.2–0.3 the cell width as gradually obscured tangential spurs or bands, whose thin and ill-defined outer ends are often dilated and confluent; longitudinal walls with 2–4(5) tangential spurs, cross walls with 0–1(2) spurs or bands. Innermost cell layer of narrower, elongated, rather irregular cells, ca. 12–17(–18–20) × 50–75 μ, bearing numerous, often oblique or curved, complete semiannular bands. Spores (only a few seen; perhaps postmature) 12–14.5 μ; elaters hardly tortuous, with scarcely attenuated ends, 6.5–8.8(9–10) μ in diam. × 110–160 μ long, with two reddish brown spirals.

Type. NW. Greenland: Kangerdlugssuak, Inglefield Bay (*RMS 45984*).

Distribution

Known only from the type station and two others in the Inglefield Bay region and one from Alaska:

NW. GREENLAND. Kangerdlugssuak; damp, peaty soil, between rocks, growing with *Lophozia*(?)*groenlandica, Anastrophyllum minutum grandis, Dicranum*, etc., between rocks; steep, N.-facing rocky slope (*RMS 45984, Aug. 9, 1960*). Kekertat, Harward Ø, near Heilprin Gletscher; peaty soil around shallow lake just behind the village, ca. 100 ft (*RMS 46003, Aug. 10, 1960;* paratype). Siorapaluk, Robertson Bugt (*RMS 45676*).
ALASKA. W. of Cyril Cove, Amchitka I., Aleutian Isls. (*Shacklette 7258!*).

Ecology

Anastrophyllum sphenoloboides appears restricted to peaty soil, usually in richly vegetated ground. The Siorapaluk specimen grew on peaty, steep slopes below dovekie (little auk) roosts, where fertilized by leachings from guano; admixed were *Anastrophyllum minutum* var. *grandis, Lophozia*

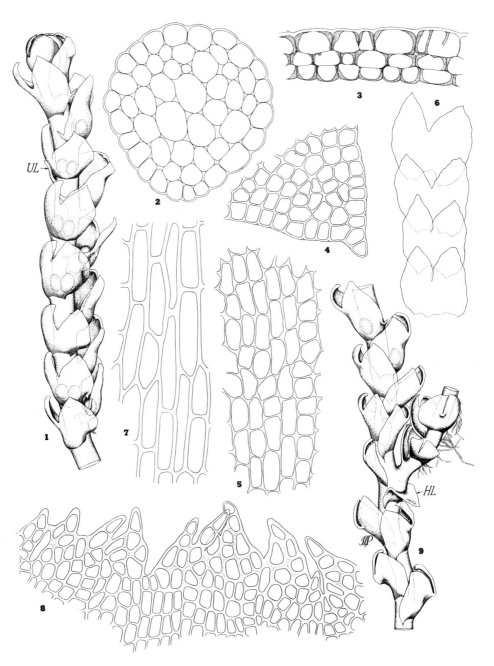

Fig. 242. *Anastrophyllum (Schizophyllum) sphenoloboides*. (1) Androecial shoot, with mostly diandrous bracts and, at UL an underleaf (×37). (2) Seta cross section (×196). (3) Capsule wall cross section (×530). (4) Leaf lobe cells (×340). (5) Cells of median part of leaf base (×340). (6) Leaves (×36). (7) Dorsal cortical stem cells (×475). (8) Perianth mouth (×276). (9) ♂ Shoot with *Frullania*-type branch, at HL the half stem leaf (×31). [All from *RMS 45676*, Siorapaluk, NW. Greenland.]

opacifolia, L. groenlandica, and (adjacent) *L. ventricosa* and *Tritomaria exsectiformis arctica*; a quantity of a sterile *Cephaloziella* (*C. elegans?*) and of mosses was also admixed.

Differentiation

The most distinctive constellation of characters which circumscribes this species is as follows: the longly cylindrical perianths, with loosely sheathing bracts that hardly invest more than the basal ¼ of the perianth; the somewhat sheathing leaf bases, with the leaf often suberect rather than widely spreading, with incurved lobes; the *A. minutum*-like facies, but with reddish to purplish brown, rather than yellow-brown to castaneous, coloration; the dorsally hardly but still perceptibly decurrent leaves; the lack of gemmae; the ability to develop distinct collenchyma; and the more or less elongated cells of the leaf bases, rather longer than in *A. minutum*, but less elongated than in *A. assimile.*

In the field, *A. sphenoloboides* was confused with *A. minutum* var. *grandis.* The two taxa are strikingly similar in many respects, both growing erect, with pectinately distichous bilobed leaves that are somewhat concave to canaliculate-concave. The plants show the same size variation; both tend to have the triangular leaf lobes blunt to acute, and the leaf lobes slightly incurved. Indeed, were it not for striking differences in color, the plants would probably have been referred to *A. minutum.* However, *A. sphenoloboides* has at least the distal portions of the leaves, on the upper parts of shoots, reddish-purple to brownish-purple-pigmented; in *A. minutum* var. *grandis* the leaves (at least the distal portions) are a warm, clear yellow-brown to mustard yellow, in extreme cases a warm yellow-brown to chestnut brown. Since the two taxa occur intimately admixed in the type collection, the color differences of adjacently growing plants are immediately perceptible.

On dissection a series of other characters stand out. In *A. sphenoloboides* the leaves are less spreading laterally, the bases being more nearly sheathing, the distal portions of the leaves, except for the lobe apices, almost suberect to loosely spreading, with the lobes again incurved. The leaves are also differently shaped, being ovate, with the width and length usually essentially equal; the lobes are erect and clearly longer than wide at the base; the sinus is sharp, often almost closed at the base, and descending more deeply (0.35–0.45 the leaf length); the dorsal bases of the leaves are arcuately inserted and slightly but distinctly decurrent, rather than transversely inserted. There are also excellent cellular differences, the basal portion of the leaf in *A. sphenoloboides* having a high incidence of

elongate, oblong cells that are 2–2.5× as long as broad (rather than virtually isodiametric as in *A. minutum*). In the more highly colored lobes (of
leaves that have grown relatively exposed) the cells are very strongly
collenchymatous, with the intervening walls also thickened (essentially
equally thick-walled in *A. minutum*, with concave-sided trigones at best).
Anastrophyllum sphenoloboides abundantly develops colorless rhizoids; in
A. minutum rhizoids of the upper stem are rare.

Even in neighboring plants, the degree of development of collenchyma in
A. sphenoloboides varies tremendously, evidently to some extent with age and
exposure. On some of the upper, most deeply pigmented leaves the cells may
show only weak and nearly evenly thickened walls; by contrast, on subfloral
leaves and ♂ bracts the lobe cells may bear extremely coarse, salient, bulging,
and almost confluent trigones. Leaves with such prominent trigones are readily
found and, in themselves, allow an easy separation of this species from *A.
minutum*. There appear to be only trivial differences between *A. sphenoloboides*
and *A. minutum* in the number, form, and size of the oil-bodies. In both species
the cells within the upper portions of the leaf lobes bear usually 2–4(5) rather
small oil-bodies; in *A. sphenoloboides* I have seen as many as 6 in isolated cells,
but never in *A. minutum*. The median cells in both species bear mostly from
(3)4 to 6 oil-bodies, although in *A. minutum* some cells have as many as 7.
These variations are insignificant and may prove meaningless after study of
more collections. There is no significant difference in form and size of the oil-
bodies in median cells (mostly 3–3.5 μ to 2.7–3.6 × 4.5–7.5 μ in *A. sphenoloboides*; 3–4 μ to 2.5–3.5 × 4–6.5 μ in *A. minutum*).

The dorsal and lateral cortical cells of the stem in *A. sphenoloboides* are basically
narrowly oblong and range from (8)9–11(12) μ wide × (24)28–40(45) μ long
to 10–15(17) × (30)40–75 μ, depending on the stem studied. They are more
strongly elongated than in *A. minutum*, and in this feature the species much more
closely approaches "typical" *Anastrophyllum*.

Relationships

Anastrophyllum sphenoloboides is of utmost interest because it represents
another segment in the complex of species that connect "*Sphenolobus*"
with *Anastrophyllum*. Study of the present species and of such others as
the Japanese *A. mayebarae* Hattori confirms the fact that the so-called
generic distinctions in this complex have been too narrowly drawn;
see p. 709. Buch (1933a) considered as basic, in this complex, the leaf
insertion and leaf type. In *Anastrophyllum* the leaf is basically ovate and
usually longer than wide; it is arcuately inserted, with the antical end of
the insertion consequently distinctly decurrent. Also, the basal leaf cells
tend to become strongly elongated, as do those of the cortex of the stem,
and the leaf cells tend to develop coarse and in some cases confluent
trigones. By contrast, in "*Sphenolobus*" the leaf varies from ovate to

broadly ovate-quadrate and is often wider than long, although not necessarily so; it is transversely inserted in its dorsal end and thus not decurrent, and the cells are hardly elongated anywhere in the leaf, are basically quadrate to short oblong, and do not form conspicuous trigones but may develop nearly equally thickened walls. The cortical cells of the stem remain relatively abbreviated and short-oblong. By these criteria *A. sphenoloboides* belongs to *Anastrophyllum*, in Buch's sense, rather than to his "*Sphenolobus*," for the following reasons. (1) Its leaves are ovate and basically as long as wide, or even slightly longer than broad, with the maximal width basal, the leaf rounded basally. (2) Leaf insertion is arcuate, although so little dorsally that the material must be carefully examined from cleared specimens for this to be ascertained; the degree of dorsal decurrence is usually so slight as to be of more theoretical than practical significance. (3) Dorsal cortical cells are strongly elongated and average 2.5–5, locally 4–6×, as long as wide. (4) Leaves, at least with age, tend to develop strong trigones, which become confluent because of the fact that the intervening walls may become thickened as well.

Other criteria significant in disposing of *A. sphenoloboides* are to be derived from the capsule-wall anatomy. It shows 3 cell layers, the outer of somewhat larger cells (average thickness 13–15 μ), inner layers thinner (ca. 7.8–8.5 μ thick). This is the same basic condition found in *Anastrophyllum saxicolus*. The seta of *A. sphenoloboides* is also relatively massive, with ca. 20–24 epidermal cell rows, and is 6–7 cells (ca. 240 μ) in diam.

Anastrophyllum sphenoloboides shows remote affinities to *A. michauxii*, from which it differs in the much more slender, hardly clavate perianth, nearly sheathing perichaetial bracts, incurved dorsal leaf lobes, less decurrent dorsal leaf bases, much smaller size, lack of gemmae, much less spreading distal portions of the leaves, coloration, and *A. minutum*-like facies. The color, the leaf shape, particularly depth of sinus, and the much less elongated basal leaf cells, together with their less collenchymatous form, suffice to separate the species from *A. assimile*.

Herzog (1934a, p. 85) describes an *A. minutirete* from Columbia, South America, which approaches *A. sphenoloboides* in the general appearance, leaf size (700 μ long × 600 μ wide), shape and form of sinuses, and small cell size. This is known only from ♂ plants and is so briefly described that its relationship to *A. sphenoloboides* remains ambiguous.

Among the species of *Anastrophyllum* which I have been able to study, *A. sphenoloboides* shows a remote affinity to *A. leucostomum*. It agrees with this species in developing terminal, lateral branches; in the reddish coloration that is produced; in the basic leaf shape (although this is somewhat narrower in *A. sphenoloboides*) and cell form; in the soft-textured stem with a relatively weakly differentiated cortex and oblong, angulate cortical cells; and in the basic form of the perianth. However, *A. leucostomum* has the typical leaf insertion of subg. *Anastrophyllum*—the antical half of the leaf being basically weakly succubous,

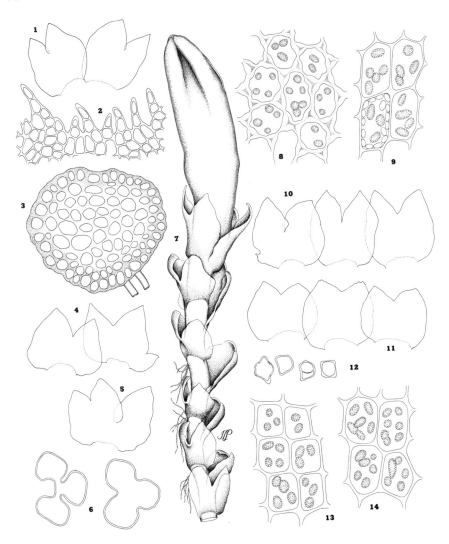

Fig. 243. *Anastrophyllum* (*Schizophyllum*) *sphenoloboides* (1–10) and *A. minutum grandis* (11–14). (1) Two ♀ bracts (×17). (2) Perianth mouth cells (×195). (3) Stem cross section, pigmented cortex stippled (×195). (4) Two ♀ bracts, the one at right with ? vestigial bracteole (×19). (5) Trilobed ♀ bract (×19). (6) Cross sections of perianth, at left one-fourth, at right one-third from apex; dorsal groove in each case at left (×30). (7) Shoot with perianth, lacking androecium at base (×44). (8) Subapical cells (×650). (9) Basal cells (×570). (10) Mature leaves (×32). (11) Three leaves, the median with old gemmae stalks (×32). (12) Gemmae (×275). (13) Cells from lobe middle (×700). (14) Cells from near leaf base (×560). [1–3, *RMS 45984;* 4–10, from *RMS 45676*, Siorapaluk, Greenland; 11–14, from *A. minutum grandis*, mixed with material drawn in figs. 4–10, *RMS 45676*, Siorapaluk, Greenland.]

with the antical base suddenly and strongly decurrent along the dorsal midline of the stem.

Variation

Androecia on paroecious plants of this species may lie directly below the perichaetial bracts, but more often are separated from them by a variable zone, frequently involving 2–6 pairs of sterile, nearly normal vegetative leaves; subfloral innovations, which are usually single, may also be androecial. The androecial spike is relatively conspicuous, consisting usually of 5–12 to 10–16 pairs of somewhat imbricate bracts; these are very conspicuously gibbous in the central portion of the base, which forms a sort of pocket for the reception of the 1–2, less often 3, antheridia. Paraphyses are relatively rarely present and are usually, if developed at all, in the form of cilia. ♂ Bracts normally possess a single basal cilium which is curved and often slightly hooked apically; this is occasionally so reflexed that it points to near the base of the stem. Antheridial stalks, which are 14–16 μ in diam. × 40–55 μ long, appear to be uniformly biseriate in the Siorapaluk plants (uniseriate in the type and paratype); the body averages 110–125(130–135) μ long × 85–95(105) μ in diam.

Initially I believed that the type and paratype plants were not identical with those from Siorapaluk. In the latter it was difficult to demonstrate an auto-ecious inflorescence (and few unquestionably paroecious inflorescences were found). Pseudodioecism, caused by decay of older shoot sectors connecting ♂ and ♀ branches, was common. But in the type plants, paroecious inflores-cences were freely developed. This fact, together with the differences in antheridial stalks, initially suggested two distinct taxa were at hand. However, agreement in other details is sufficiently close that this seems improbable. Also, the type plants themselves showed very much variation in placement of sex organs; I have studied several scores of plants from the point of view of the distribution of sex organs. Not a single, unquestionably dioecious individual could be found, as is evident from the fact that almost every plant with an unfertilized perianth with subfloral innovation developed an innovation that was at least at first androecial, rarely purely gynoecial. In many cases a perianth-bearing shoot has what appear to be only vegetative leaves below the perichaetial bracts. Occasional truly autoecious individuals appear to occur, but often an androecium is followed by numerous pairs of sterile leaves and then by a gynoecium, with the time interval so great between the production of the two that the androecial shoot sector had, in the meantime, nearly decayed. There are all sorts of transitions between this putative condition and one in which ♂ bracts are found just below the ♀ bracts. In a few cases as many as 16–25 pairs of ♂ bracts occurred below the gynoecium; such extremes are rare. I have also seen two individuals, clearly paroecious, that developed subfloral innovations which were, in one case, gynoecial only, in the other, paroecious. Evidently the plants exhibit a tendency toward a heteroecious condition such as characterizes *Radula obconica*.

CROSSOCALYX-EREMONOTUS COMPLEX

Jungermannia sect. *Sphenolobus* Lindb., Not. Sällsk. F. et Fl. Fennica Förhandl. 13:369, 1874
    (*type: J. verruculosa* Lindb. = *J. helleriana* Nees).
Sphenolobus Berggr., N.Z. Hep. 1:22, 1898 (name only).
Sphenolobus Steph., Spec. Hep. 2:156, 1902 (*p. min. p.*).
Eremonotus Lindb. & Kaal. ex Pears., Hep. Brit. Isles 1:200, 1902 (*type: Jungermannia myriocarpa*
    Carr.).
Crossocalyx Meyl., Bull. Soc. Vaud. Sci. Nat. 60:266, 1939 (*type: Jungermannia helleriana* Nees).[250]
Anastrophyllum Schust., Amer. Midl. Nat. 45(1):68, 1951 (*p.p.*; subgenera *Sphenolobus, Eremonotus,*
    Crossocalyx only).

Minute or small, usually 0.25–1.0(1.5) mm wide, green to deep brown (*never reddish*), strongly negatively geotropic, erect or suberect (especially with formation of gemmiparous shoots), simple to sparingly branched; branches at least in part lateral-intercalary;[251] *stolons or flagella usually lacking.*[252] Stem elliptical in cross section, *normally 5–10(11) cells high;* cortical cells in 1(2) strata somewhat thicker-walled than the medullary cells, but *not forming a sharply defined cortex,* the firm medullary cells without a mycorrhizal ventral band; mycorrhizae in 1–3 cortical cell strata, if present. Rhizoids few to rare, scattered. Lateral leaves usually *pectinately oriented,* distant to contiguous or weakly imbricate, sometimes becoming hemispherical, ± *complicate-canaliculate,* U-shaped in cross section, almost *consistently bilobed,* inserted along an oblique line ventrally, *transversely inserted dorsally, not decurrent;* leaves broadly quadrate to oblong-ovate, *usually widest near middle, without ampliate bases* (the *antical base usually extending to midline of axis, but not across stem*) ; leaf margins entire, except sometimes near gynoecia, lobes broad and blunt to narrow and sharp. *Underleaves lacking* or reduced to stalked slime papillae, up to 2–4 cells long (*ventral merophytes 1–2 cells broad*). *Leaf cells small or minute, firm,* chlorophyllose, the cuticle nearly smooth to weakly papillose with low papillae; *cells subquadrate above, short-oblong* (length 1–2 × width) *basally;* cell walls becoming nearly or quite *evenly thick-walled, trigones* (*and intermediate thickenings*) *lacking or obscure;* oil-bodies (2)3–4(5) to 4–7 per cell, finely granular-botryoidal, medium-sized for the cell size. Asexual reproduction lacking or by *carmine-red to vinaceous* 1–2(3–4)-celled angular *gemmae;* shoots, with prolonged gemmae formation, *becoming attenuate, filiform.*

Dioecious. Gynoecia and androecia usually on long, leading stems (androecia becoming intercalary; gynoecia sometimes with a subfloral innovation,

---

[250] Other synonymy is indicated under the various subgenera accepted.

[251] The branches arise (see, e.g., Müller, 1954, *in* 1951–58, pp. 721, 726) laterally from near the base of the ventral lobe. In *A. myriocarpum,* as Wollny (1909) has indicated, they are also lateral and intercalary. In *A. minutum* they are *both* lateral-intercalary and lateral-terminal (in some cases all the branches appear to be lateral-terminal); in *A.* (*Crossocalyx*) *hellerianum* I have seen, to date, *only* lateral-intercalary branching and occasional postical-intercalary branches. In *A. minutum* the intercalary branches may arise from near the middle of the leaf axil, or may in many cases be ventrally displaced so that they arise from the ventral angle of the leaf insertion. Such branches are still lateral in origin, even though they issue from the ventral face of the stem.

[252] Microphyllous geotropic stoloniform axes regularly occur in all phases of *A. myriocarpum.* Initially this would seem to sharply separate the *A. myriocarpum* element from the other related elements (subgenera *Crossocalyx, Acantholobus*). However, some collections of *A.* (*Acantholobus*) *minutum* (e.g., *Seidenfaden 45,* from Bjørling Ø, Cary Isls., Smith Sound, Greenland) have occasional rhizoidous geotropic, ventral flagella or stolons.

even with perianth development). Androecia with bracts 1–2-androus; antheridial *stalk 1-seriate*, occasionally (subg. *Eremonotus*) in part 2-seriate. Gynoecia usually developing subfloral innovations in absence of sporophyte development. Bracts usually 2–3-lobed (often with 1-several accessory teeth; sometimes sharply spinose-dentate); bracteole usually oblong or lanceolate, *small but distinct* (rarely more than 0.5 × size of bract)[253] commonly fused at base with one bract. Perianth cylindrical-ovoid, contracted, and 3–5-plicate above (*with a distinct and often deep, long antical furrow*), usually slightly to distinctly *dorsiventrally flattened*, the narrowed *mouth lobulate*, the small, short *lobes* ± *dentate or ciliate*. Capsule ovoid, the *wall 2-stratose* or (*A. minutum*) varying to (2)3-stratose. Epidermal cells with strong nodular (radial) thickenings; inner cells with semiannular bands. Seta with many cell rows (subg. *Acantholobus*) varying to with 8 epidermal + 4 interior cell rows.

A key to the contained elements is given on p. 712.

Since, with a very narrow approach, one may consider the *Crossocalyx-Eremonotus* complex to represent an autonomous genus, for which Grolle (1963a) proposed we use *Sphenolobus* (Lindb.) Berggr., I have given it a rather full diagnosis.

Nomenclature

The status and typification of "*Sphenolobus*," if adopted as a genus, pose major problems. For a variety of reasons, some of which are enumerated below, it has proved desirable to suppress this name entirely, as a *nomen ambiguum* (Bonner and Schuster, 1964).

The nomenclature of *Eremonotus* and *Sphenolobus* is greatly complicated, in part by procedures sanctioned in the International Rules. The following discussion supplements those in Schuster (1961b), Bonner and Schuster (1964), and Grolle (1963a, 1964).

With regard to the typification of *Sphenolobus*, the following points should be mentioned. (1) The first mention of *Sphenolobus* is in Lindberg, Not. Sällsk. F. et Fl. Fennica Förhandl. 13:369, 1874.[254] Here Lindberg *refers to* a section *Sphenolobus* and *describes* a "*Jungermannia (Sphenolobus) verruculosa* Lindb." The later description of *Jungermannia* sect. *Sphenolobus* by Lindberg (Musci. Scand., p. 9, 1879) is not relevant. (2) Müller, on the basis of study of Lindberg's type demonstrated [Rabenh. Krypt.-Fl. 6(1):590, 1910] that *J. verruculosa* is a synonym of *J. helleriana* Nees, placed by Stephani (Spec. Hep. 2:158, 1902) into *Sphenolobus* when he elevated the latter to generic rank; this species was placed by Buch (1933a) into the heterogeneous "genus" *Isopaches*, and later divorced from this by Meylan (1939, p. 266) and placed into the new monotypic genus *Crossocalyx*. (3) Schuster (1961b, p. 59) pointed out that Stephani (1902, *in*

---

[253] The form of the ♀ bracts and the relative size of the bracteole need further investigation.

[254] Lindberg describes "*Jungermannia (Sphenolobus) verruculosa*" in detail; this species is not only treated first, but the name appears in boldface *with citation of author*. Later—*and merely in a footnote*—Lindberg assigns several other taxa, *without diagnoses*, to *Sphenolobus*, *without citation of authorities* (except for "*rigida*" Lindb., *a synonym*). From the internal evidence it is, I think, clear that Lindberg based his group on *J. verruculosa*, which must be regarded as the type, and *only peripherally* added the other taxa later.

1898–1924) did not propose a generitype of *Sphenolobus*, and, even if he had done so, the *J. verruculosa* element would have had to serve as the type of *Sphenolobus*—in whatever rank that group is accepted. (Int. Code Bot. Nom., Art. 7, Utrecht, 1961). Since *J. verruculosa* (= *Sphenolobus hellerianus*) must be considered the type of *Sphenolobus*, in whatever rank we wish to accept that group, *Crossocalyx* is technically a synonym of *Sphenolobus*. Grolle (1963a, p. 18) is thus wrong in considering *J. minuta* Schreb. ex Cranz [= *Sphenolobus minutus* (Schreb. ex Cranz) Steph.] as the type of *Sphenolobus*. Berggren (1898, p. 22) admittedly stated that "*S. (Jungermannia) minuta*" was "allied to" *Sphenolobus perigonialis*; this cannot be taken as selection of *S. minutus* as the generitype of *Sphenolobus*. (4) On the preceding bases, the *S. minutus* element, whether placed in a separate subgenus or within *Eremonotus*, must bear the subgeneric name *Acantholobus* Schust. (1961b), for reasons detailed in that paper.

With regard to the use of "*Sphenolobus*" rather than "*Eremonotus*" as the overall name for the complex of subgenera *Acantholobus-Crossocalyx-Eremonotus* these points are relevant. (1) The generic name *Sphenolobus* had always been attributed to Stephani, 1902 (Spec. Hep. 2:158); see, i.a., Müller (1910, p. 587, *in* 1905–16), Macvicar (1912, 1926), Frye and Clark (1937–47), Buch, Evans, and Verdoorn (1938), Evans (1939), Schuster (1951a, 1953, 1961b). (2) Berggren (1898) cites, without mention of basionym or other direct reference, or any description, "*Sphenolobus* Lindb." in a manner that suggests he considers it to be a distinct genus. Article 32, Note 1, of the current version of the International Code (p. 33) sanctions this practice, stating that "an indirect reference is a clear indication, by the citation of the author's name, . . . that a previously and effectively published description applies to the taxon to which the name is given." The reference in the Code to the case of *Hemisphace* strengthens this conclusion. (3) On that basis, *Sphenolobus* (Lindb.) Berggr. (1898) predates the generic name *Eremonotus* Lindb. & Kaal. ex Pears. (1902). (4) Thus the *Crossocalyx-Eremonotus* complex, *if recognized as a genus* distinct from *Anastrophyllum*, would in theory have to bear the epithet *Sphenolobus*, rather than *Eremonotus*, which I (Schuster, 1961b) employed. My treatment (*loc. cit.*, 1961) of this complex under *Eremonotus* is based on the specific and unambiguous statement in Müller (1954, p. 616, *in* 1951–58) that "wenn man *Sphenolobus*, *Crossocalyx*, und *Eremonotus* zu einer Gattung zusammenzieht, ist der älteste Name dafür *Eremonotus*." (5) Although I may eventually have to reluctantly adopt this usage of *Sphenolobus*, I would do so with grave reservations as to its wisdom. As demonstrated earlier (Schuster, 1961b, 1965), "*Sphenolobus*" *sensu* Stephani is a mélange of at least 15 genera! In his interpretation, Stephani was obviously influenced by Lindberg (1879, pp. 7–8), who placed in his "*Jungermannia* sect. c, *Sphenolobus*" taxa now referable to 5 genera (*Douinia*, *Saccobasis*, *Orthocaulis*, *Anastrophyllum*, and *Scapania*), as well as several actually referable to *Sphenolobus s. lat.* (6) As a consequence, there has been no *generic* concept of *Sphenolobus*, or if there has been one (such as that of Buch, 1933a; Meylan, 1939; or Müller 1951–58) it has been nomenclaturally indefensible. Thus, apart from the nomenclatural "rights" of *Sphenolobus* for continued existence, it might reasonably be argued that the name—even as early as in Lindberg (1879)—was such an unnatural admixture that its *use* in modern taxonomy seems completely senseless.

I have stated (Schuster, 1965, p. 241) that the group *Sphenolobus* "is clearly an artifact" and have shown (1961b, 1965) that under *Sphenolobus sensu* Stephani are found taxa belonging in the following groups: *Cephalolobus* Schust., *Acrobolbus* Mitt., *Crossocalyx* Meyl., *Orthocaulis* Buch [= *Lophozia* subg. *Orthocaulis* (Buch) Schust.], *Cephaloziopsis* (Spr.) Schiffn. or *Cephaloziella* subg. *Cephaloziopsis* (Spr.) Douin, *Anastrophyllum* (Spr.) Steph., *Lophozia* subg. *Lophozia* Dumort., *Saccobasis* Buch, *Tritomaria* Schiffn., *Roivainenia* Perss., *Gymnocolea* Dumort., *Lophozia* subg. *Hypolophozia* Schust. [= *Sphenolobus subinflatus* (Spr.) Steph.], *Acantholobus* Schust. If we add the additional discordant taxa placed in "*Sphenolobus*" by Lindberg (1879), we have, as further elements, *Douinia* Buch and *Anastrophyllum* subg. *Eurylobus* Schust. This leaves out of consideration wholly irrelevant placements into *Sphenolobus*, such as "*Sphenolobus*" *crispifolius* Steph. (1924, *in* 1898–1924 p. 433), based on a *Plagiochila*; "*Sphenolobus*" *laceratus* Steph. (1901, p. 173, *in* 1898–1924) and "*Sphenolobus*" *incompletus* (G.) Steph., based on species of *Gymnomitrion*; "*Sphenolobus*" *perigonialis* (Hook. f. & Tayl.) Berggr., as to diagnosis or material (not name), based on *Andrewsianthus confusus* (Schust.) Schust.; see Schuster, 1965, p. 281.

Under such conditions, suppression of *Sphenolobus* as a name *in any rank* seems the only rational course and appears in clear accord with both Articles 69 and 70 of the International Code.

Taxonomy

The *Crossocalyx-Eremonotus* complex is here restricted to a limited group of minute or small species, usually with regularly pectinately inserted, transversely oriented, bilobed, complicate leaves. Leaf insertion and orientation give the plants, in general, a distinctive facies. Linked with this is the following ensemble of characters which I regard as crucial in separating the complex:[255] (*a*) leaf cells noncollenchymatous or with small trigones, becoming almost evenly thick-walled; (*b*) basal leaf cells not markedly elongated, at best short-oblong and twice as long as broad; (*c*) no reddish or purplish coloration, but a brown secondary pigment commonly developed; (*d*) cuticle almost smooth or with low papillae; (*e*) antheridial stalk normally 1-seriate; (*f*) capsule wall 2-stratose, at least in part, although often partly or largely 3-stratose; (*g*) axis reduced, at most 9–11 cells high; (*h*) leaf with insertion dorsally transverse, the antical leaf base not ampliate (thus extended to stem midline and *not across stem*); (*i*) underleaves lacking or a mere 1–2-seriate, minute filament, never present throughout; (*j*) postical-intercalary branching either absent or at least exceptional, the intercalary branching seemingly always lateral, but at times from ventral end of leaf axil. The only exception I currently know to this rule (I have no illusions that others will not be

---

[255] These criteria were earlier used to define this group (Schuster, 1961b, pp. 72–73, under the name "*Sphenolobus-Crossocalyx-Eremonotus* complex").

discovered) occurs in the actual type of *J. helleriana* itself, in which I have seen two ventral-intercalary branches.

As thus defined, the *Crossocalyx-Eremonotus* complex corresponds exactly to *Eremonotus* as circumscribed by Schuster (1961b) and includes taxa placed in three genera by Müller (1954, *in* 1951–58): *Eremonotus s. str.*, *Sphenolobus*, and *Crossocalyx*. I am of the opinion that it is nearly pointless to regard the three monotypic or stenotypic groups mentioned above as separate genera.[256]

The *Crossocalyx-Eremonotus* complex is also allied to the antipodal stenotypic genus *Cephalolobus* Schust. (Schuster, 1963, 1965). This genus differs in its well-developed trigones, the intervening cell walls remaining thin; in the retention of small to distinct underleaves; in consistently lacking gemmae; in the development of coarse—and usually extremely salient—cuticular papillae; and in the consistently lateral-axillary branches, terminal branching of any type being either absent or exceptionally rare. Also, *Cephalolobus* is characterized by the development of long, slender, conspicuous flagella, representing both microphyllous extensions of the main stems and axillary branches, in which it closely approaches *Andrewsianthus* Schust.

Evolution within the *Crossocalyx-Eremonotus* complex appears to have involved chiefly reduction and simplification. Thus the most primitive element, *A. minutum* (subg. *Acantholobus*) approaches subg. *Anastrophyllum*; it still retains terminal *Frullania*-type branches and appears never to develop true ventral-intercalary branches; it retains an unspecialized seta and a largely or entirely 3-stratose capsule wall.[257] The next step in evolution involves subg. *Crossocalyx*, in which *Frullania*-type branching may be lost (*A. hellerianus*) and branches are predominantly lateral-intercalary (but a few ventral-intercalary ones occur), while the seta is reduced to 8 (or 9–10) epidermal + 4 internal cell rows, and the capsule

---

[256] Müller (1947) admitted that the three taxa here recognized as typifying three discrete subgenera could have been placed within a single genus; he again states (1954, *in* 1951–58) that such a consolidation is "möglich" but then cites a series of criteria which he holds sufficient for generic segregation. These criteria, taken individually, (*loc. cit., p.* 617) are basically all *species criteria*, not of generic significance. Only the difference in the form of the seta (used as the primary criterion in the key on p. 712 to subgenera) is possibly of greater significance. This, in turn, shows correlation with the vigor of the plant. Even the seta character is not absolute; e.g., in *A. (Crossocalyx) hellerianum*, which usually agrees with subg. *Eremonotus* in having a seta with 8 + 4 cell rows (Fig. 235:7, 10), the type of "*J. verruculosa* var. *compacta*"—a synonym of *A. hellerianum*—may have setae with 9 + 4 or 10 + 4 cell rows.

Grolle (1963a, p. 18) also agrees with the broad generic delimitation I adopted (Schuster, 1961b), although he attempts to maintain the name *Sphenolobus* for this grouping.

[257] Müller (1951–58, p. 722) states that the wall is 2-stratose. At least in vigorous plants I find it to be regularly 3-stratose (as is typical of most taxa in *Anastrophyllum s. str.*). Indeed, *if* a cleavage between the *Crossocalyx-Eremonotus* complex and *Anastrophyllum s. str.* were to be attempted, *A. minutum* could logically be transferred to the latter group. Ironically, it is my detailed investigation of this very species, also carefully studied by Buch (1933a), which has convinced me that any attempt at a generic cleavage is futile. *Both* branching and capsule-wall anatomy link *A. minutum* to *Anastrophyllum s. str.*, as does seta anatomy, whereas cell type, axial anatomy, and leaf form and insertion are more suggestive of the *Crossocalyx-Eremonotus* complex.

wall becomes 2-stratose, only locally 3-stratose. The ultimate step in evolution involves subg. *Eremonotus* (*A. myriocarpum*), with branches constantly lateral-intercalary, some of them becoming stoloniform; the seta is apparently uniformly of 8 + 4 cell rows and the capsule wall uniformly 2-stratose. Within this sequence the *Acantholobus* element has almost cylindrical perianths, the *Crossocalyx* element has them moderately compressed, while they are strongly dorsiventrally compressed in *Eremonotus s. str.*, accompanied by, within this sequence, progressive development of a more and more salient antical longitudinal median sulcus. Gemmae are retained in the first two groups but are lost in the last one.

Branching

In two species, *A. minutum* and *A. hellerianum*, lateral-intercalary branches occur, and in *A. minutum* lateral-terminal, *Frullania*-type branching is found as well. Neither type of branching is frequent. In *A. minutum* the intercalary branches seen mostly issue from the ventral ¼ or ⅕ of the leaf axil; in *A. hellerianum* they are seemingly more nearly median (see Müller, 1947).

In the lateral-intercalary branching, the *A.* (*Acantholobus*) *minutum* element closely approaches the small antipodal genus *Cephalolobus* Schust. (Schuster, 1963, 1965). In this genus, however, terminal branching is—a single exception aside—hardly known (although with study of larger suites of specimens than I have seen, such branching may yet prove more frequent). By contrast, typical species of *Anastrophyllum* studied show only postical-intercalary branching, supplemented (in taxa such as *A. assimile*, *A. michauxii*, and *A. saxicolus*) by lateral-terminal branching of the *Frullania* type. In a few primitive taxa of *Anastrophyllum*, such as *A. novazelandiae* and *A. sphenoloboides*, only such terminal, *Frullania*-type branching is known.

Leaf Form and Insertion

Characteristic of all taxa is a leaf which is not dilated basally and hence, *in situ*, does not extend across the stem; exceptions occur only in extremes of *A. minutum* (var. *grandis*). The leaves are almost constantly bilobed, but *A. minutum* shows rare and exceptional leaves with the dorsal lobe bifid, hence with isolated asymmetrical trifid leaves—a condition also occurring in *Anastrophyllum* (e.g., *A. saxicolus*).

Subgenus *ACANTHOLOBUS* (Schust.) Schust., comb. n.

Sphenolobus (genus) Steph., Spec. Hep. 2:156, 1902 (*p. min. p.*); K. Müller, Rabenh. Krypt.-Fl. 6(1):587, 1910 (*p. min. p.*); Macvicar, Studs. Hdb. Brit. Hep. ed. 2:207, 1926 (*p. min. p.*); K. Müller, Rabenh. Krypt.-Fl. ed. 3, 6(1):718, 1954 (subg. *Eusphenolobus* only!).
Anastrophyllum subg. *Sphenolobus* Schust., Amer. Midl. Nat. 45(1):74, 1951.
Eremonotus subg. *Acantholobus* Schust., Rev. Bryol. et Lichén. 39(1–2):73, 1961.

Usually rigid, brownish to yellow-brown (green in shade forms), moderately small, usually 0.5–1.2(1.5) mm wide × 10–25 mm high, ± rarely branched;

branches lateral, axillary-intercalary and terminal, *Frullania* type; no stoloniform branches usually present.[258] Stem ca. 9–11 cells high, elliptical in section, with ventral and lateral 2–3 peripheral strata at maturity usually mycorrhizal; cortical cells short-oblong. Leaves small, pectinately distichous, nearly contiguous to imbricate, conduplicate-concave to almost hemispherical, obliquely to widely spreading and ± antically secund, *dorsal half transverse, nondecurrent*; ca. 0.35 2-lobed (rarely a minority 3-lobed), the lobes usually subequal, or dorsal lobes smaller. Cells ± small (medially 15–22 × 17–24 μ), with cuticle smooth or almost so, the walls ± equally thick, oil-bodies 3–7 per cell. Gemmae 1–2(3–4)-celled, reddish orange to scarlet, polygonal, usually formed from reduced, ± appressed leaves of erect, often filiform stems.

Androecia with bracts ventricose; antheridia 1 or 1–2 per bract; antheridial stalk 1-seriate. Gynoecia, if not maturing capsules, always innovating; bracts 2–3-lobed, the lobes entire or weakly paucidentate; bracteole usually lanceolate or oblong, connate with one bract. Perianth subcylindrical, somewhat dorsiventrally compressed, plicate (and with a deep antical furrow), the contracted mouth lobulate-ciliate. *Seta formed of numerous cell rows.* Capsule wall 26–30 μ thick, 2- *to 3-stratose*; outer layer with very strong nodular (columnar) thickenings of all walls, inner with semiannular bands. Spores 11–15 μ; elaters 6.5–8 × 140–175 μ.

Type. Jungermannia minuta Schreb., *in* Cranz, Fortsetz. Hist. Grönland, p. 285, 1770 (Schuster, 1961b, p. 73).

This subgenus corresponds exactly to *Sphenolobus sensu* Buch (1933a) and Müller (1951–58), after the exclusion of *"Sphenolobus" saxicolus* (Schuster, 1961b).

Acantholobus is the least reduced element in the *Crossocalyx-Eremonotus* complex, as is evident from both the gametophytic axis (9–11 cells in diam.) and that of the sporophyte (seta unreduced, of the "general type" of Douin). As the most primitive element in this complex, the group shows convergence to *Anastrophyllum* subg. *Schizophyllum* (see p. 745) and to the antipodal *Andrewsianthus-Cephalolobus* complex. These groups, however, have distinctly collenchymatous leaf cells, with conspicuous, often bulging trigones. In *Anastrophyllum* subg. *Schizophyllum*, even in the most primitive taxa, the basal leaf cells are diagnostically elongated, in contrast to the lack of such differentiation in the *Crossocalyx-Eremonotus* complex. In *Cephalolobus*, the axis regularly retains the ability to develop underleaves.

Only one Northern Hemisphere species appears referable to *Acantholobus*, but a second species (*Lophozia cavifolia*) approaches so very closely to *A. minutum*, the type of *Acantholobus*, that the two are likely to be confused; to avoid this error, their characters are contrasted below:

1. Cells smaller: ca. 12–14 to 16–18 μ on margins of lobes, 15–20(22) × 17–21(24) μ in leaf middle, equally thick-walled; underleaves consistently

---

[258] I have seen rhizoidous ventral geotropic flagella or stolons only once, in a small form of *A. minutum* from Greenland: Bjørling Ø, Cary Isls., Smith Sound, 76°43′ N., 72°22′ W. (*Seidenfaden 45*!).

lacking; leaves 0.9–1.4× as wide as long, 0.25–0.4 bilobed, greatly variable in form. Oil-bodies normally 3–4(6–7) per cell. Ubiquitous . . . . . .
. . . . *Anastrophyllum (A.) minutum* (p. 757)
1. Cells large: ca. 20–25 μ on lobe margins, 20–27 × 21–30 μ in leaf middle, with conspicuous but ill-defined trigones; sporadically with minute under-leaves; leaves hemispherical, 1.3–1.4 to 1.4–1.65× as wide as long, only 0.25 bilobed to 3–4-lobed. Oil-bodies 5–8(9–10) per cell. Rare, arctic . . . .
. . . . See *Lophozia (O.) cavifolia* (p. 330)

ANASTROPHYLLUM (ACANTHOLOBUS) MINUTUM

(Schreb. ex Cranz) Schust.

[Figs. 243:11–14; 244–246]

Jungermannia minuta Schreb., *in* Cranz, Fortsetz. Hist. Grönland, p. 285, 1770.[259]
Jungermannia minuta Dicks., Fasc. Sec. Pl. Crypt. Brit. 2:13, 1790.
Jungermannia bicornis Web. & Mohr, Bot. Taschenb., p. 423, 1807.
Jungermannia weberi Mart., Fl. Crypt. Erlang. p. 157, 1817.
Jungermannia gypsophila Wallr. *in* Bluff & Fingerhuth, Comp. Fl. Germ. Sect. 2(3):63, 1831.
Sarcoscyphus gypsophilus Nees, Naturg. Eur. Leberm. 1:140, 1833.
Jungermannia treviranii Hüben., Hep. Germ. p. 240, 1834.
Diplophyllum minutum Dumort., Rec. d'Obs., p. 16, 1835.
Jungermannia saccatula Lindb., Meddel. Soc. F. et Fl. Fennica 9:162, 1883.
Cephaloziopsis saccatula Schiffn., *in* Engler & Prantl. Nat. Pflanzenfam. 1(3):85, 1893.
Lophozia minuta Schiffn., *ibid.*, 1(3):85, 1893.
Diplophyllum gypsophilum Loeske, Moosfl. Harzes, p. 65, 1903.
Sphenolobus gypsophilum Loeske, Verh. Bot. Ver. Brandenburg 46:167, 1904.
Sphenolobus minutus Steph., Spec. Hep. 2:157, 1902; K. Müller, Rabenh. Krypt.-Fl. 6(1):594, fig. 291, 1910.
Anastrophyllum (Sphenolobus) minutum Schust., Amer. Midl. Nat. 42(3):576, 1949; Schuster, *ibid.*, 45(1):74, pl. 20, 1951; Schuster, *ibid.*, 49(2):373, pl. 23, 1953.
Eremonotus (Sphenolobus) minutus Schust., *in* Schuster, Steere, & Thomson, Natl. Mus. Canada Bull. 164:40, 1959.
Eremonotus (Acantholobus) minutus Schust., Rev. Bryol. et Lichén. 30(1–2):73, 1961.

Erect or ascending in growth from a prostrate or procumbent base, in patches or mats or scattered amidst erect-growing mosses, rather *small*, shoots 0.7–1.2, rarely to 1.5, mm wide × 1–3 cm long, subsimple or very sparsely branched (but often with subfloral innovations), green (shade phases) or usually brownish, *often golden-brown*. Stems 125–175 μ wide, rather fragile and brittle when dry, filiform; *stem 9–11 cells high*, in cross section with cortical cells nearly subequal to medullary cells, the cortical cells rather short, 2–3× as long as wide, aver-aging subequal in width to basal leaf cells, moderately, evenly thick-walled; cortical cells ca. 12–16(18) μ in diam., the medullary to 16–18 × 20–26 μ; outer 1–3 strata of cells mycorrhizal with age. Rhizoids very sparse, usually only on basal portions of stems. Leaves distant to imbricate, variable in size and shape, *typically ± as long as to longer than wide* (1.0–1.33× as long as wide), *regularly pectinately inserted and oriented*, spreading and canaliculate to somewhat erect-spreading and nearly hemispherically concave (var. *grandis*), uniformly bilobed

---

[259] Grolle (1961c, pp. 81–82) has shown that the usual accreditation of the name to Crantz is incorrect.

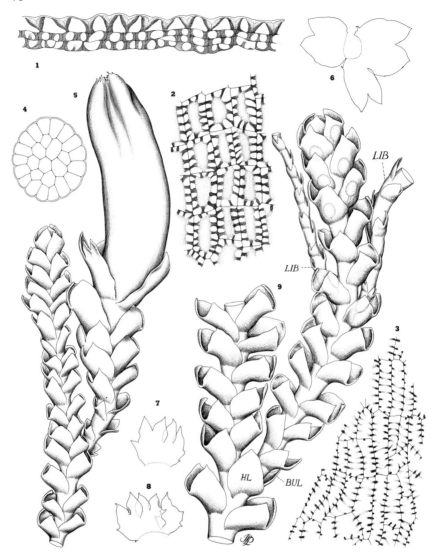

FIG. 244. *Anastrophyllum* (*Acantholobus*) *minutum*. (1) Capsule wall cross section (×310). (2) Epidermal cells in surface view (×300). (3) Internal cells (×300). (4) Seta cross section (×160). (5) Sterile shoot with lateral-ventral perianth-bearing innovation (×30.5). (6) ♀ Bracts and a bracteole (×17). (7) ♀ Bract and (8) ♀ bract and bracteole (×17). (9) Sterile axis with a ♂ branch, of the *Frullania* type which bears two lateral-intercalary branches arising from the ventral angles of lateral leaves; at HL, dorsal half-leaf associated with branch; BUL, first branch appendage, an underleaf (×30.5). [All from *RMS 45505*, Kânâk, NW. Greenland.]

¼–⅖ their length, short-rectangular to subquadrate (more rarely wider than long and rounded-quadrate; in var. *grandis*), the dorsal half transversely inserted and not decurrent; lobes triangular to broadly triangular, acute to apiculate, rarely blunt, spreading to incurved, the *ventral usually distinctly larger*; sinus acute to obtuse. Cells *subquadrate*, ± regularly concentrically arranged in the lobes, 12–17(18) μ on margins and in lobes, in leaf middle 15–20 × 18–24 μ, near base to 16–19 × 20–25 μ; walls becoming yellowish in sun, *subequally thickened, often strongly so* (particularly on margins), *without evident trigones*, except in basal half of leaf, where weakly collenchymatous and subequally thickened; cuticle faintly verruculose-striolate, at least in basal half of leaf; *oil-bodies 2–5(6) per cell*, papillose-segmented, subovoid and 2.5–3 × 4.5–6 μ to 5 × 6–7.5 μ, a few to 6 × 9 μ. Gemmae at apices of ± attenuated shoots, *often with reduced, erect to erect-appressed leaves, in bright red to scarlet masses*, irregularly angular to tetrahedral or polyhedral, (1)2-, occasionally 2–4-, celled, 15–30 μ long.

Dioecious, usually sterile. ♂ Bracts in 4–10, occasionally to 12, pairs, intercalary with age; bracts like leaves, but somewhat more concave; 1(2)-androus; antheridia ca. 130–140 × 150–170 μ, stalk 1-seriate. ♀ Bracts somewhat larger than leaves, polymorphic, one of each pair often 3–4-lobed, the other often 2–3-lobed, erect-sheathing, the *lobes entire* and acute to apiculate; bracteole seemingly absent to large and united at base to one bract. Perianth ⅔–¾ emergent, subcylindrical but *somewhat dorsiventrally compressed*, little narrowed to mouth, to 1 mm wide × 2 mm long, strongly plicate distally, with a deep antical sulcus and 3–4 additional, less marked sulci; the slightly but suddenly contracted mouth shallowly lobed, the lobes with teeth 2–3 cells long. Seta of many cell rows; capsule wall 2- or 3-stratose. Spores 11–12 to 12–15 μ, brown, verruculose; elaters bispiral, to 8 μ in diam.

Type. Greenland, in herb. Dillenius (Oxford); see Grolle, 1961c.

Distribution

A very common, highly polymorphic, widely distributed, holarctic species, found throughout much of the Tundra and Taiga of the Northern Hemisphere, with isolated, disjunct stations in the upper edge of the Deciduous Forest Region (in eastern North America, in the Hemlock-Hardwoods Region).

Found throughout northern Europe (Scandinavia; Shetland Isls.; northern Russia), southward in Scotland and England, Ireland, Germany, as far south as northern Italy and Yugoslavia; also in Bulgaria, Spain (the Pyrenees), eastward to northern Siberia (64–70° N.), Novaya Zemlya (70° N.), and eastern Siberia (64° N.).

Northward to Spitsbergen (Arnell and Mårtensson, 1959), Jan Mayen I. (Hesselbo, 1923b), Iceland. Also the Azores (Allorge and Allorge, 1948) and Japan,[260] and reported from South Africa (Arnell, 1963, p. 356) and Kerguelen I. (G., L., & N., Syn. Hep., 1844–47.)

[260] Hattori (1952); in the alpine region. *Sphenolobus acuminatus* Horik. is placed in synonymy.

Kitagawa (1966, pp. 130–131) points out that the Japanese plants, and those from the Himalaya, diverge from European-North American phenotypes in the more often divergent and abruptly acute leaf lobes; the more strongly thick-walled marginal leaf cells; and the more irregularly lobed and spinose-dentate ♀ bracts; possibly a different subspecies should be established for these plants. The disjunct New Guinea plants are even more different (p. 765). Kitagawa regards "*Sphenolobus*" *acuminatus* as a Taiwanese endemic; in 1963a he placed it in *Anastrophyllum*, in 1966 in *Sphenolobus*!

In North America transcontinental, found in the West from Alaska (Clark and Frye, 1942; Sherrard, 1957) and the Aleutian Isls. eastward to the Yukon (Evans, 1903), southward to Keewatin, Alberta, British Columbia, Idaho, and Washington. Recurring as a disjunct at high elevations in Mexico (Mirador; Gottsche, 1863). In our area as follows:

ELLESMERE I. See under var. *grandis*; reported as very common by Bryhn (1906, with "fruiting" plants from Cape Viele [Hayes Sd. region] and Goose Fjord [S. coast]). DEVON I. Philpots I. (Polunin, 1947). N. GREENLAND. Jewell Fjord, Nansen Land, 83°06′ N., 42°30′ W.; Kap Glacier, S. coast of Independence Fjord, 81°48′ N., 31°35′ W.; Saxifragadal, 81°51′ N., 31°15′ W.; Herlufsholm Strand, E. coast of Peary Land, 82°40′ N.; 21°W.; Brønlund Fjord, Heilprin Land, 82°10′ N., 31° W. (Arnell, 1960; "much can be referred to var. *grandis*."). NW. GREENLAND. Foulke Fjord, 78° N. (Harmsen & Seidenfaden, 1932); Uvdle in Wolstenholme Fjord; Bjørling I. in Cary Isls.; Hackluyt I., Smith Sd., Pandora Harbour, Inglefield Land (Harmsen & Seiden-faden, 1932); Kânâk, Red Cliffs Pen. (*RMS 45506a, 45487b, 45536a*); Kekertat, Harward Øer, E. end of Inglefield Bay (*RMS 46046a, 46034b, 46004c*); Siorapaluk, Robertson Bugt, Inglefield Bay (*RMS 45667a*); N. side of Booth Sd., N. of Nucluet, Steensby Land (*RMS 46204a*); Dundas, near Thule, Wolstenholme Fjord (*RMS 46267b*). W. GREENLAND. SE. of Jacobshavn, Disko Bay, 69°12′ N., 51°05′ W. (*RMS & KD 66-245, 66-243, 66-235*); Lyngmarksbugt, Godhavn, ca. 25–125 m, 69°15′ N, 53°30′ W. (*RMS & KD 66-557*); Fortune Bay, Disko I. (*RMS & KD 66-209a, 66-123*); Tupilak I., W. of Egedesminde (*RMS & KD 66-108, 66-023a, 66-013b, 66-033b*); Eqe, near Eqip Sermia (Glacier), 69°46′ N., 50°10′ W. (*RMS & KD 66-289*); head of Sondrestrom Fjord, near shore of L. Ferguson, in *Salix*-moss tundra, 66°5′ N. (*RMS & KD 66-101*); mouth of Simiutap Kûa Valley, near pond, Umîarfik Fjord, 71°59′ N., 54°35–40′ W. (*RMS & KD 66-919*); E. end of Agpat Ø, neck E. of Umi-asugssup Ilua 69°52′ N., 51°38′ W. (*RMS & KD 66-146b*); Umanak, 70°47′ N., 53°47′ W. (*RMS & KD 66-1481*). E. GREENLAND. Clavering Ø (Hesselbo, 1948); Danmarks Ø, Hekla Havn and Gaaseland, all in Scoresby Sund; Dove Bugt, 76°49′ N.; Röhss Fjord; Cape Parry; Hurry Inlet; Ingmikertorajik; Tasiusak; Lille Pendulum Ø; Kaiser Franz Josef Fjord (Jensen, 1900, etc.).

BAFFIN I. Steere (1939); Pond Inlet in N. Baffin and Exeter Sd., C. Baffin (Polunin, 1947; Harmsen & Seidenfaden, 1932). MELVILLE PEN. No loc. (*Edwards*, ex Hooker, 1825a, p. 420; see Polunin, 1947). HUDSON BAY. Nottingham I.; Southampton I. (Polunin, 1947). MANITOBA. Near RCAF Base, Churchill, Hudson Bay (*RMS 35002, 35018*, near var. *grandis*). NORTHWEST TERRITORY. Tukarak I., Belcher Isls. (Wynne & Steere, 1943). LABRADOR. Macoun (1902); L'Anse an Clair (*Waghorne 3!*; det. and publ. by Underwood, 1892a, as "*Diplophyllum dicksoni*"). NEWFOUNDLAND. Chance Cove (*Waghorne 156, 166, 165!*; reported by Underwood, 1892a, as "*Diplophyllum dicksoni*"; in part var. *major*); Rencontre West, S. coast; Benoit Brook, W. Nfld.; Griguet, N. Pen.; Norris Arm, C. and NE. Nfld. (Buch &

Tuomikoski, 1955); New Harbor (among *Diplophyllum albicans;* Underwood & Cook, Hep. Amer. no. 127!); Humber R., Squires Mem. Pk. (*RMS 68-1644*); Steady Brook (*RMS 68-1393*).

QUEBEC. Wakeham Bay; Oka; La Tuque; Ste-Anne-des-Monts R.; Table-top Mt.; St. Colomban; Tadoussac; Bic (Evans, 1916; Lepage, 1944–45); L. Némiskau, Rupert R., 76°50′ W.; L. Mistassini (Lepage, 1944–45); E. of George R. near 55°09′ N. and ca. 7 mi W. of Lake Indian House, ca. 56°20′ N., 64°54′ W. (Kucyniak, 1949); Great Whale R.; Richmond Gulf; Port Harrison; Cape Smith (Wynne & Steere, 1943; Schuster, 1951); Leaf Bay, Ungava Bay, and ca. 80 mi up Leaf R. (Schuster, 1951). NOVA SCOTIA. Ship Harbour L., Dartmouth Barrens, Cape Smoky Mts., Purcell's and Sandy Cove (Brown, 1936a); Smoky Mt., Cape Breton (Macoun, 1902); Mary Ann Falls, S. of Neils Harbour, Cape Breton I. (*RMS 42693, 42688a, 42687*, etc.; *RMS 42686* = fo. *cuspidata*). ONTARIO. Câche L., Algonquin Park; Otter Head, L. Superior; Sudbury Jct.; L. Nipigon (Macoun, 1902); L. Timagami (Cain & Fulford, 1948); Muskoka, Cape Henrietta Maria, Kenora, Thunder Bay, etc. (Williams & Cain, 1959); N. of Agawa Bay, L. Superior, Algoma Distr. (*Williams & Cain 5359!*).

MAINE. Mt. Katahdin, near summit of Cathedral Trail, 5000 ft (*RMS 32966a*; near var. *grandis*); Mt. Desert I. (Lorenz, 1924). NEW HAMPSHIRE. Franconia Mts. (Lorenz, 1908c); Mt. Washington (*RMS, s.n.*); Mt. Clinton (*RMS, s.n.*) VERMONT. Long Trail, 4000–4100 ft, Mt. Mansfield (*RMS 43809a, 43864a, 43821*). MASSACHUSETTS. Tannery Falls, Franklin Co. (*RMS, s.n.*); Mt. Everett, town of Mt. Washington (*Evans*).

NEW YORK. Summit, Wittenberg Mt., Cornell Mt. at 3600 ft and between Cornell and Wittenberg Mts., Catskill Mts. (*RMS 17589, 24715, 24699, 24413a*, etc.); Blue Mt., Adirondack Mts. (*RMS, s.n.*); Coy Glen, Ithaca, Tompkins Co. (*Wiegand, 1893!*, NY); Little Moose L., Herkimer Co. (*Haynes*); Olean Rock City, Martiny Rocks, W. branch of Four Mile Rd., Allegany, all in Cattaraugus Co. (Schuster, 1949a). WEST VIRGINIA. Monongalia Co. (Ammons, 1940); "Cooper's Rock" (Sheldon, 1910a). VIRGINIA. Walker Mt.; near summit, White Top Mt., Smyth Co., 5400 ft (*RMS 38099*); Mt. Rogers, ca. 5200 ft, Grayson Co. (*RMS 38021*; the green fo. with cuspidate lobes = fo. *cuspidata*). NORTH CAROLINA. Avery and Jackson Cos.; Jones Knob (Andrews, 1921); Dry Sluice Gap on Appalachian Trail, Swain Co., 5300 ft (*Schofield 9825*); 40–50 ft below summit of Mt. Mitchell, Yancey Co. (*RMS 34543, 23180c, 23182, 23167, 24827*); near summit of Roan Mt., Mitchell Co. (*RMS 40331*); Grandfather Mt., ca. 5700–5800 ft, Caldwell Co. (*RMS 30185, 30181*); Craggy Gardens, Craggy Mts., Blue Ridge Pkwy., Buncombe Co. (*RMS 24765b*); Balsam Gap, 5220–5260 ft, Blue Ridge Pkwy. (*RMS 19060*). TENNESSEE. Trail to Myrtle Pt., Mt. LeConte, and SE. slopes of Mt. LeConte below Myrtle Pt., Sevier Co. (*RMS 45325a, 45243, 45240a*, etc.); Roan Mt., Carter Co., high cliffs, 6100–6200 ft (*RMS 61279*). MICHIGAN. Isle Royale (Conklin, 1914a); Mountain Stream, Huron Mts. (Nichols, 1938); N. of Phoenix, Keweenaw Co. (Evans & Nichols, 1935). MINNESOTA. Pigeon R. and Pigeon Pt., Cook Co.; Little Susie I., Big Susie I., Belle Rose I., Lucille I., Porcupine I., all in Susie Isls., Cook Co.; Lutsen, Grand Maris, Cascade R., Hungry Jack L., Cook Co.; Great Palisade, Lake Co.; Nopeming, St. Louis Co. (Schuster, 1953).

Ecology

In the far north common and ubiquitous, occurring everywhere in the Tundra, usually on peaty soil, except under very xeric conditions. Frequent even in shallow tundra vegetation over basic rocks (then sometimes with such lime-tolerant species as *Scapania degenii, Plagiochila asplenioides,*

Odontoschisma macounii, *Cephaloziella arctica*, and the more ubiquitous *Blepharostoma trichophyllum*), but usually more common in acidic sites, often among *Dicranum* and other relatively xerophytic mosses, where it occurs as straggly, isolated, erect, golden-brown filaments. In the Arctic the species is almost always found as the concave and broad-leaved, dense-leaved, golden brown, gemmae-free phase (var. *grandis*).

Further southward, in the Spruce-Fir Zone, and in alpine situations, often over damp, shaded rocks (then as a green phase, freely gemmiparous, with distant, unequally bilobed, canaliculate-complicate leaves distinctly longer than broad, the "fo. *cuspidata*"). In such sites a member of the variable *Lophozia-Scapania* community, usually on shaded and protected cliffs. "It normally occurs as a distinct xerophyte . . . associated here most often . . . with *Anastrophyllum michauxii*, *Lophozia alpestris* (occasionally *ventricosa*), *Diplophyllum taxifolium*, and occasionally *Anastrophyllum saxicolus*, *Lepidozia reptans*, and *Tritomaria quinquedentata*" (Schuster, 1953). Although in the high Tundra, the var. *grandis* is often helophytic, the narrow-leaved, $\pm$ typical phase (and the "fo. *cuspidata*") usually "lacks tolerance for any direct sunlight; it may form large mats on shaded, northwest-facing cliffs, or in narrow cuts, where there is little or no seepage, and where atmospheric moisture is high" (Schuster, 1953). Such a behavior pattern also characterizes the species at the southern limits of its range, in the Appalachian Mts. (where it is local or rare). Two other factors have been cited as limiting the occurrence of *A. minutum*: lack of toleration for much moisture, the species being rare or absent in wet, peaty places and in the boggy portions of the Tundra, and lack of toleration for calcareous substrates (Schuster, 1953; Buch and Tuomikoski, 1955). However, in the far north the species becomes more "tolerant" and may occur over peaty soil overlying calcareous rocks; it has also been reported from gypsum, in Europe.[261]

The species at its southern limits in eastern North America, at high elevations in the Appalachians, is restricted to damp or at least humid, shaded rock walls, often to deeply recessed chasms. The common associates are, under such conditions, *Herberta adunca tenuis*, *Bazzania denudata*, *Lepidozia reptans*, occasionally *Plagiochila sullivantii* or *P. tridenticulata*.

In the Tundra and northern edge of the Taiga often a pioneer on barely damp rocks, with lichens, mosses (often *Dicranum*), and the hepatics *Chandonanthus setiformis*, *Lophozia longidens*, *L. ventricosa*, occasionally *L. wenzelii*, and *Ptilidium ciliare*, and sometimes with *Mylia taylori* and *Anastrophyllum saxicolus*. When in moister sites, among mosses, in damp tundra (or, as is rarely the case,

---

[261] See, e.g., Reimers (1940, 1940a); the ecotype over gypsum has been described as a distinct species, *Sphenolobus gypsophilus* (Wallr.) Loeske.

in wet tundra, as around pool margins), associated species include *Cephaloziella arctica, Cephalozia bicuspidata ambigua, Lophozia wenzelii, L. atlantica,* and *Ptilidium ciliare,* as well as the ubiquitous *Blepharostoma trichophyllum.*

Differentiation

Generally easily recognized because of the small or moderate size, the very regular, pectinate-distichous orientation of the transverse leaves, the constantly bilobed leaves, and the small, subquadrate, thick-walled cells. "This, combined with the rigid, flexuous form of the stems, is quite diagnostic. The leaves are quite unlike those of any related form except robust plants of the much smaller . . . *Anastrophyllum hellerianum.*" In "normal" phases, with canaliculate-concave leaves, this similarity to *A. hellerianum* is quite evident (compare Figs. 244 and 247), but in the phases from the Tundra, with cupped, suberect leaves, no such affinity is evident. In the "normal" Taiga phenotypes gemmae are freely produced. Such "gemmae-bearing plants (which are found nearly universally in patches occurring under relatively xeric conditions) are very diagnostic: the erect or ascending shoots become attenuate, with the leaves somewhat reduced in size and erect or erect-appressed" (Schuster, 1953), (Fig. 244:6). In this regard *A. minutum* approaches, in some measure, the related *A. hellerianum,* which is a relatively minute plant, however. The strong "tendency" for xeric shade forms to produce scarlet, prominent gemmae masses stands in strong contrast to the behavior of the arctic var. *grandis,* in which the xeric phases are more robust and almost invariably free of gemmae.

The larger tundra phases (var. *grandis*) bear a close similarity to the rare and local *Lophozia (Orthocaulis) cavifolia* (p. 768). However, the smaller cells (13–15, rarely 15–18 μ, wide in lobes and leaf middle) with equally, often markedly thickened walls, the transverse dorsal insertion of the leaves, and the stem anatomy separate *A. minutum* from *L. cavifolius.* This similarity is so marked that Müller (1951–58) placed the latter next to *A. minutum* in subg. *Sphenolobus,* emphasizing again the fact that the generic limits in this family remain subjectively conceived.

Variation

Associated with its extremely wide distribution, *A. minutum* also exhibits marked polymorphism. The polymorphism is geographically correlated in part, suggesting that this ancient and isolated species has evolved into a series of ecotypes (or perhaps even ecospecies). In the far north the common phase is a brown, rather robust, often fertile, dense-leaved plant bearing cupped, broader-than-long leaves, almost invariably devoid of gemmae. This has been considered a distinct species ("*Jungermannia*

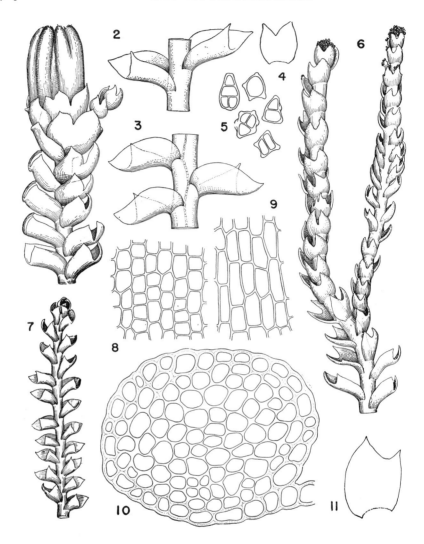

FIG. 245. *Anastrophyllum* (*Acantholobus*) *minutum*. (1) Perianth-bearing shoot apex with innovation (×20). (2–3) Dorsal and ventral shoot sectors, respectively (×40 and ×33). (4) Leaf (×ca. 25). (5) Gemmae (×280). (6) Gemmiparous shoot of extreme shade form (×45). (7) Sterile shoot (×18). (8) Basal leaf cells (×240). (9) Dorsal cortical cells of stem (×240). (10) Stem cross section (×360). (11) Leaf of shade form with narrow leaves and sharp lobes (×45). [1, Great Palisade, Lake Co., Minn., *RMS*; 2–3, 8–11, small island between Lucille and Susie Isls., Minn., *RMS*; 4, 7, Blue Mt. Lake, Adirondack Mts., N.Y., *RMS*; 5–6, Olean Rock City, N.Y., *RMS*; from Schuster, 1953.]

rigida" or "*Sphenolobus saccatulus*") but is more probably a mere variety (var. *grandis*), described separately below.

In the Taiga and in the isolated stations representing Pleistocene relict occurrences in the northern edge of the Deciduous Forest (as, e.g., in western New York), the species occurs principally as a canaliculate-leaved phase with usually more distant leaves. The leaves, furthermore, are frequently longer than broad (sometimes only 0.8–0.9 as wide as long) and often unequally bilobed, with spreading and acute lobes. This form, under moist and shaded conditions, develops sharply apiculate leaf lobes ("var. *cuspidata* Kaal.," but is probably merely an environmental modification).[262] This phase of the species appears almost unable to tolerate direct sunlight and never has hemispherically cupped leaves or incurved leaf lobes. It seems to be constantly sterile but almost invariably is very freely gemmiparous (gemmiparous extremes constitute "var. *denticulata*" of Anzi, 1881) and, particularly with gemma formation, is much less robust than var. *grandis*, often only 0.5–0.75 mm wide. Under very shaded and damp conditions, however, even this variant may be nearly or quite gemma-free. It then is often provided with apiculate lobes, the apiculus being commonly prominent and formed of two superposed cells, of which the terminal may be elongate (and up to $35–39 \times 13–15 \mu$).

These two extremes, and the shade form, "var. *cuspidata*," are separable as follows:

1. Leaves distant to contiguous, from 0.8–0.95 to 1.0–1.15 as wide as long, complicate-canaliculate, easily flattened out (and without tearing), ± folded in sinus, the lobes spreading, acute to short-acuminate, triangular; sinus descending 0.25–0.35 the leaf length, the lobes often quite unequal; gemmae freely produced; marginal cells of lobes 12–14(15) μ usually . . . *a*.
   a. Leaf lobes merely acute [Fig. 246:8–9] . . . . "typical" *A. minutum*.
   a. Leaf lobes cuspidate to short-acuminate [Fig. 245:2–4, 11] . . . . . .
   . . . . fo. *cuspidata* (Kaal.), comb. n.
1. Leaves ± imbricate, often rather closely so, typically (1.2)1.3–1.4 as wide as long and hemispherically cupped, impossible to flatten out (without tearing), not folded in sinus, the lobes both erect and incurved, blunt, broadly ovate-triangular [Fig. 246:1–3]; sinus descending 0.35–0.5 the leaf length, the lobes often subequal; gemmae normally absent; marginal cells of lobes 16–18(20) μ . . . . var. *grandis* (G. ex Lindb.), comb. n.

In addition to these two holarctic extremes, a related plant occurs at 13,500–14,000 ft in New Guinea, subsp. *novoguineanensis* Schust. (Schuster, 1969b); this differs from the holarctic subspecies in the orange-colored gemmae,

---

[262] *Jungermannia minuta* var. *cuspidata* Kaal., Nyt Mag. Naturvid. 33:376, 1893. Kaalaas quotes, as a possible synonym, *J. minuta* var. *protracta b*. Nees, Gottsche, & Lindenberg, Syn. Hep. p. 121, 1844.

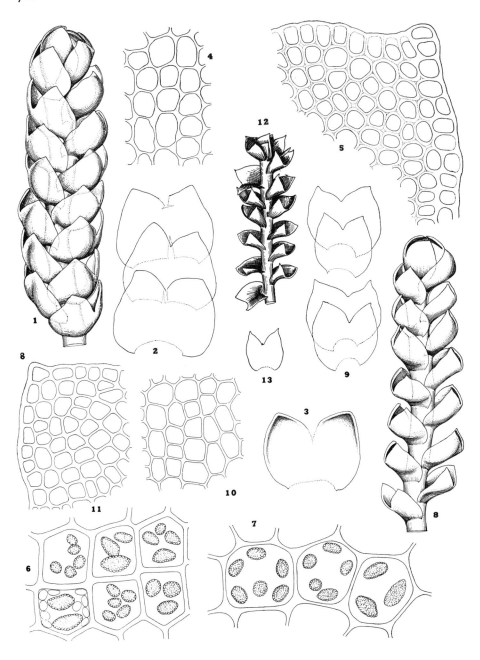

FIG. 246. *Anastrophyllum* (*Acantholobus*) *minutum* (1–7, var. *grandis;* compare also with *Lophozia cavifolia* in Fig. 156; 8–11, var. *minutum;* 12–13, fo. *cuspidata*). (1) Vigorous sterile shoot (×31). (2) Flattened leaves, all tearing in flattening (×36). (3) Leaf, only weakly flattened (×39). (4) Median cells (×282). (5) Leaf lobe cells (×282).

essentially identical in color to the leaves. The plant reported by Kitagawa (1967) from Mt. Kinabalu, Borneo, may belong here.

ANASTROPHYLLUM MINUTUM var. *GRANDIS*
(*G. ex Lindb.*), *comb. n.*

[Fig. 246:1–5, 7]

Cephalozia rigida var. *grandis* G. ex Lindb., Bot. Notiser, p. 165, 1872.
Jungermannia rigida var. *grandis* Lindb., Musci Scand., p. 8, 1879.
Jungermannia saccatula Lindb., Meddel. Soc. F. et Fl. Fennica 9:162, 1883.
Jungermannia minuta var. *grandis* Lindb., *in* Lindberg & Arnell, Kgl. Svensk. Vetensk.-Akad.
    Handl. 23:60, 1889; *fide* Buch (1951).
Sphenolobus saccatulus K. Müll., Rabenh. Krypt.-Fl. 6(1):599, 1910.
Sphenolobus minutus var. *grandis* Frye & Clark, Univ. Wash. Publ. Biol. 6(3):377, [1944] 1945.
Sphenolobus minutus fo. *maior* Schiffn., Krit. Bemerk., *in* Lotos 1905:57 (of reprint), 1905 (?prob-
    ably; *fide* Müller, 1951–58).

Robust (0.8)1.0–1.2(1.5) mm wide, golden brown to blackish brown, *compact, almost amentiform*, because of the densely imbricate leaves. *Leaves hemispherically concave*, cupped, suberect, appearing inflated, ca. 600–700 μ wide × 510–570 μ long or larger, averaging *1.1–1.2× as wide as long, varying to 1.3–1.4× as wide as long*, tearing when flattened; lobes *broadly* ovate-tri-angular, blunt or subobtuse, more or less *incurved*; sinus descending to 0.35–0.5 the leaf length, usually less, the leaves not folded in the sinus. Cells larger than in var. *minutum* and often more thick-walled (lumina sometimes nearly guttulate); marginal cells 16–18(19–20) μ; cells of free lobes 18–20 × 18–23 μ; median cells (18–19)20–23(24) × (20)21–25(26) μ, each with (2)3–6(7) oil-bodies, ranging from 4.5–5.5 μ to 3–4.5 × 5–7 μ, up to 6–6.5 × 9–11 μ (if only 2 per cell). *Gemmae usually lacking.* Often fertile. Chromosome no. = 8 (Vaarama), requiring verification.

Type. Buch (1951a, p. 75) cites two "Originale" for var. *grandis*: Lapponiae Pitensis, Sweden (*Lindberg*) and Muonio Lapponiae, Finland (*Norrlin*); I suggest that the first be regarded as the lectotype.

Distribution

Essentially arctic; absent or virtually so from alpine sites. In Europe reported from alpine regions of Sweden and Norway. Similar plants (fo. *major*) are reported from Switzerland and Italy (Müller, 1951–58).

In North America probably transcontinental, but the distribution west of our region is not clarified.

---

(6) Median cells with oil-bodies and, lower left, chloroplasts (×878). (7) Submedian cells with oil-bodies (×735). (8) Shoot apex (×31). (9) Leaves (×36). (10) Median cells (×282). (11) Leaf apex cells (×282). (12) Sterile shoot (×20). (13) Leaf (×25). [1–2, 4–5, 7, *RMS 46267a* and *RMS 46267*, Dundas, NW. Greenland; 3, 6, Churchill, Manitoba, *RMS*; 8–11, Dundas, NW. Greenland. *RMS*; 12–13, New York, *RMS*.]

N. GREENLAND. According to Arnell (1960), most of the collections (cited on p. 760) can be referred to var. *grandis*. NW. GREENLAND. Kânâk, Red Cliffs Pen. (*RMS 45639, 45491a, 45525a, 45619, 46155d*); Kangerdluarssuk (*RMS 45793b, 45811, 45806, 45819a*); Iglorssuit (*RMS 45777*); and Kangerdlugssuak (*RMS 45984, 45868a, 45986, 45962, etc.*), all in Inglefield Bay; Kekertat, E. end of Inglefield Bay (*RMS 46017, 46003*); Dundas, Wolstenholme Sd. (*RMS 46267a, 46264, 46265*); N. side of Booth Sd., N. of Nucluet, Steensby Land (*RMS 46203c*). W. GREENLAND. Eqe, near Eqip Sermia (Glacier), ca. 1 mi from Inland Ice, 350–450 m, 69°46' N., 50°10' W. (*RMS & KD 66-293b*); Fortune Bay, Disko I., 69°16' N., 53°50' W. (*RMS & KD 66-140, 66-139, 66-136*); Tupilak I., W. of Egedesminde, 68°42' N. (*RMS & KD 66-032c, trace*); Sondrestrom Fjord, near shore of L. Ferguson, near head of Fjord, in *Salix*-moss tundra, 66°5' N. (*RMS & KD 66-075*); Qalagtoq, Upernivik Ø, Inukavsait Fjord, 71°14' N., 52°35' W. (*RMS & KD 66-1100b*); Umanak (*RMS & KD 66-1561*).

Ecology

In the Arctic, var. *grandis* is associated with a wide variety of other bryophytes, under several different ecological conditions. It may occur in strongly insolated run-off crevices irrigated by melt water from local ice caps, there associated with *Anastrophyllum assimile, Marsupella revoluta, M. arctica, Tritomaria quinquedentata, Scapania perssonii*, etc. When on peaty soil at margins of such rills, it may occur with *Gymnomitria* (*concinnatum, corallioides, apiculatum*), *Scapania perssonii, Lophozia alpestris polaris, L. excisa, L. groenlandica, Marsupella sprucei*, etc. Or it may occur in damp moss-*Cassiope tetragona* tundra, associated with *Plagiochila asplenioides subarctica, L. hatcheri, L. barbata*, etc.

The plant appears, in general, to occur in more mesic and more nearly optimal sites than arctic phases of "normal" *A. minutum*, and I suspect that (aside from south Greenland), *all arctic forms of A. minutum are closer to var. grandis* than to the gemmiparous boreal shade forms with sharp leaf lobes ("var. *cuspidata*").

Differentiation

There are no sharp boundaries that distinguish var. *grandis*, and, to some extent, any attempt to assign the arctic forms to var. *minutum* or var. *grandis* must involve an arbitrary element. As Fig. 243 shows, var. *grandis* is a more robust, densely leaved plant; the leaves are cupped, rather than loosely conduplicate as in var. *minutum*. Although well-developed var. *grandis*, first compared with the plants of var. *minutum* found under subarctic conditions, seems drastically distinct, the differences lose their significance when large suites of this ubiquitous arctic species are studied.

Table 6, based on plants collected at Dundas, Greenland, illustrates distinctions between typical forms and contrasts these with *Lophozia* (*O.*) *cavifolia*, a plant that may be easily mistaken for var. *grandis* (see p. 330).

TABLE 6

Typical Anastrophyllum minutum and Lophozia (O.) cavifolia

Criterion	A. minutum minutum (RMS 46267b)	A. minutum grandis (RMS 46267a)	L. (O.) cavifolia (RMS 46264a)
Leaf width:length	1.0–1.15:1	(1.2)1.3–1.4:1	(1.2)1.3–1.4:1 when 2-lobed, to 1.4–1.65:1 if 3(4)-lobed
Leaf form	Concave-conduplicate ± folded in sinus 2-lobed	Hemispherical Not folded in sinus 2-lobed*	Same as var. *grandis*, but sporadical leaves 3(4)-lobed
Marginal cells of lobes	12–14(15) μ	15–18(19–20) μ	(19)20–25(26) μ
Cells within free lobes	15–17 × 15–17 μ	18–20 × 18–23 μ	18–23 × 18–23 μ
Cells of leaf middle	15–18 × 17–20(22) μ	20–22(24) × 21–25 μ	20–27 × 21–29 μ
Oil-bodies of median cells	(2)3–6 per cell	(2)3–6(7) per cell	5–8(9–10) per cell
Collenchyma	Lacking; walls equally thick	Feebly distinct; angles rounded	Distinct; often rather conspicuous
Stem	8–10 cells high	9–10 cells high	13–14 cells high
Gemmae	Absent	Absent	Frequent; 20–24 × 21–27 to 27 × 33 μ

* If hundreds of plants of var. *grandis* are examined, one can often find an isolated individual that bears a single trifid leaf.

Occasional forms of *A. minutum grandis* may have the cells still somewhat larger, e.g., the specimen from Peary Land, referred to *L. cavifolia* by Arnell (1960); here cells of the leaf-lobe margins are 15.5–18.5, locally to 18–21(23), μ; cells within the free lobes are 18–20 × (17)18–20 μ; median cells are (18)20–24(25) × 20–27 μ; leaves, as in var. *grandis*, are always 2-lobed, 1.2–1.3× as wide as long. Such plants *and all others* of *A. minutum* (including var. *grandis*) differ from *L. cavifolia* in showing a gradual but marked increase in cell size from the margins of the leaf lobes toward the leaf base.

Subgenus *CROSSOCALYX* (*Meyl.*) *Schust.*

Jungermannia sect. *Sphenolobus* Lindb., Not. Sällsk. F. et Fl. Fennica Förhandl. 13:369, 1874 (*type: J. verruculosa* Lindb. = *J. helleriana* Nees).
Sphenolobus Steph., Spec. Hep. 2:156, 1902 (*p. min. p.*).
Crossocalyx Meyl., Bull. Soc. Vaud. Sci. Nat. 60:266, 1939 (*type: J. helleriana* Nees); K. Müller, Rabenh. Krypt.-Fl. ed. 3, 6(1):726, 1954.
Anastrophyllum subg. *Crossocalyx* Schust., Amer. Midl. Nat. 45(1):74, 1951.
Eremonotus subg. *Sphenolobus* Schust., Rev. Bryol. et Lichén. 30(1–2):63, 1961.
Sphenolobus subg. *Crossocalyx* Grolle, Rev. Bryol. et Lichén. 32:163, 1963; Kitagawa, Jour. Hattori Bot. Lab. no. 29:134, 1966.

Gametophyte minute, up to 0.5–0.85 mm wide, stiffly erect or suberect from a short, creeping rhizoidous basal portion; branching sporadic, *intercalary,* ventral-intercalary and/or from axil of leaf[263] (in *A. tenue*, terminal, *Frullania* type). Stem elliptical in section, only 6–7(8) cells high, without a distinctly differentiated cortex; *no stolons*. Mature, nongemmiparous vegetative sectors with relatively large, spreading, often loosely complicate-canaliculate, bilobed leaves; with gemmae formation the leaves becoming strongly reduced, shallowly bilobed, flat or weakly concave, erect-appressed, the *gemmiparous shoots thus slender and filiform*. Cells small, with maturity often somewhat thick-walled but the corners (angles) somewhat to obviously rounded off; oil-bodies 2–4 per cell, rarely more, rather small. Underleaves absent, but gynoecial plants sometimes with them vestigial, formed of 1 row of cells (Fig. 247). Gemmiparous shoots erect, strict, rigid, filiform, with characteristically reduced leaves and bright *vinaceous or purplish gemmae masses at the shoot tips*; gemmae usually 1-celled and cubical (on essentially unreduced shoots with limited gemma formation, occasional gemmae 2-celled) or 2-celled and tetra- to polyhedral (*A. tenue*).

♀ Bracts usually 2(3–4)-lobed like leaves, *margins weakly to strongly spinose-dentate*, not or vestigially so on weak plants; leaves below bracts often with basal spinous teeth. Perianth moderately dorsiventrally compressed, at least above, plicate and with a long, well-developed antical sulcus, unistratose throughout; mouth with 2–4-celled cilia. Capsule broadly ellipsoidal, the *wall 2- to 2(3)-stratose*; inner layer 8–9 μ thick, with semiannular bands; epidermal cells with strong nodular (radial) thickenings, ± tangentially extended. Seta, in cross section, *with 8(9–10) epidermal, 4 inner, rows of cells*. Androecia with bracts ventricose, *monandrous*; antheridia with *stalk 1-seriate*, body largely of nearly isodiametric, subquadrate to hexagonal cells.

Type. Jungermannia helleriana Nees.

Since *J. helleriana* Nees is the type of *Crossocalyx* Meyl. (1939), this last group is clearly a synonym of *Sphenolobus* (Müller, 1951–58, p. 719; Schuster, 1961b; Bonner and Schuster, 1964). Thus "*Sphenolobus*" should be used instead of "*Crossocalyx*," and the species placed in *Sphenolobus* by, i.a., Buch (1933a) and Schuster (1951a) have been relocated in a new taxon, for which the subgeneric name *Acantholobus* Schust. (1961b) is used. Bonner and Schuster (1964) have given reasons why "*Sphenolobus*" should be suppressed as a *nomen confusum*; hence the later but unambiguous name *Crossocalyx* is used here; see also p. 751.

Crossocalyx, so delimited, is restricted to chiefly subarctic and boreal sectors of the Northern Hemisphere. The group is rather remotely allied to subg. *Acantholobus*, from which it differs as follows: (*a*) axis of gameto-phytes more reduced, only 6–7(8) cells high; (*b*) gemma formation inducing even more extreme malformation and reduction of the leaves,

---

[263] Müller (1954, *in* 1951–58, p. 726) states that the branches arise "from the base of the postical leaf margin." However, Müller (1947, p. 41, fig. 2a) illustrates a branch arising from the *axil* of the leaf, and seemingly from the median portion of the axil. Thus branching in this taxon agrees with that of, i.a., the allied genus *Cephalolobus*. It is significant that the good illustration of "*Sphenolobus*" *filiformis* (= *Eremonotus myriocarpus*) by Wollny (1909, pl. 16) shows a ♀ plant with three axillary, lateral-intercalary branches. Branching in all of these taxa, therefore, is essentially identical.

the gemmiparous shoots extremely slender and highly modified; (c) sporophytic axis reduced, the cells in only two layers, the outer of 8(9–10), the inner of 4, rows. The reduced nature of the type species has led to past misinterpretations, the species having been placed in *Diplophyllum* (because of the weakly conduplicate leaves), in *Cephalozia* (because of the seta cross section), *Prionolobus*, *Isopaches* (by Buch, 1933a, 1942, on the basis of the ascending gemmiparous shoots and thick-walled cells), and *Crossocalyx*. Meylan (1939), who placed the type species in *Crossocalyx*, indicated a phylogenetic position in the Cephaloziaceae because of the *Cephalozia*-like seta. Meylan (*loc. cit.*) and Müller (1947, p. 42) also admitted a close affinity of *Crossocalyx* to *Eremonotus*.

The affinity of *A. (Crossocalyx) hellerianum* to *A. (Acantholobus) minutum* is clear from leaf form and insertion (compare Figs. 245 and 247); from the similar tendency for gemmiparous shoots to become slender and reduced-leaved, with leaves erect to erect-appressed (compare Figs. 245:6 and 247:2); from the 2–3-stratose capsule wall of both taxa; from the similarly incrassate, small cells and similar oil-body form, size, and (approximately) number (compare Figs. 246 and 247); from the rather similar axial anatomy, with a tendency in both groups for the ventral and lateroventral cells of the cortical 2(3) strata to become brownish and fungus-infested; from the identical form of the perianth, with a similar, diagnostically deep and long dorsal, longitudinal sulcus (compare Figs. 245:1 and 247:14; see discussion in Schuster, 1951a, p. 27). In turn, seta anatomy of *A. (C.) hellerianum* and perianth form agree exactly with those of *A. (Eremonotus) myriocarpus*, and both share an identical leaf form and branching modes, to the point that both taxa lack terminal branching. In some measure, *A. hellerianum* thus occupies a midground between *Acantholobus* and *Eremonotus*, making a generic distinction almost futile to maintain.[264]

In addition to the type species, *A. hellerianum*, I tentatively assign here the poorly known *A. tenue*, known to date only from sparing plants. The two taxa are separable as follows:

Key to Species of Crossocalyx

1. Oxylophyte: on decaying wood or bark; branching ventral- and lateral-intercalary, sparing; gemmae vinaceous, ± cubical, only 10–11 μ usually; ♂ bracts with sharp-pointed lobes often faintly toothed . . . .
. . . . *A. (C.) hellerianum* (p. 772)
1. Calciphyte: on basic rocks only; with terminal, *Frullania*-type furcate branching; gemmae purplish to red, tetra- to polyhedral, mostly 14–16 × 14–19 μ; ♂ bracts with lobes entire-margined, blunt or rounded at tips. .
. . . . *A.(C.) tenue* (p. 779)

---

[264] The fact that Wollny (1909, p. 345) described "*Eremonotus*" *myriocarpus* as "new" under the name "*Sphenolobus*" *filiformis* is also suggestive.

ANASTROPHYLLUM (CROSSOCALYX) HELLERIANUM
(Nees) Schust.

[Figs. 235:1–11; 247]

Jungermannia helleriana Nees ex Lindenb., Nova Acta Acad. Caes. Leop.-Carol. Nat. Cur. 14, Suppl.:64, 1829.
Diplophyllum hellerianum Dumort., Rec. d'Obs., p. 16, 1835.
Jungermannia verruculosa Lindb., Not. Sällsk. F. et Fl. Fennica Förhandl. 13:369, 1874.
Diplophylleia helleriana Trev., Mem. R. Ist. Lomb., Ser. 3, 4:420, 1877 [reprint p. 38].
Jungermannia verruculosa var. *helleri* Lindb., Musci Scand., p. 8, 1879.
Cephalozia helleri Lindb., Meddel. Soc. F. et Fl. Fennica 14:65, 1887.
Prionolobus hellerianus Schiffn., *in* Engler & Prantl, Nat. Pflanzenfam. 1(3):98, 1895.
Sphenolobus hellerianus Steph., Spec. Hep. 2:158, 1902; K. Müller, Rabenh. Krypt.-Fl. 6(1):590, fig. 290, 1910; Macvicar, Studs. Hdb. Brit. Hep. ed. 2:211, figs. 1–7, 1926.
Lophozia helleriana Boulay, Musc. France 2:94, 1904.
Isopaches hellerianus Buch, Mem. Soc. F. et Fl. Fennica 8(1932):288, 1933; Frye & Clark, Univ. Wash. Publ. Biol. 6(3):371, figs. 1–8, (1944) 1945.
Crossocalyx hellerianus Meyl., Bull. Soc. Vaud. Sci. Nat. 60:266, 1939; K. Müller, Rabenh. Krypt.-Fl. ed. 3, 6(1):727, fig. 236, 1954.
Anastrophyllum hellerianum Schust., Amer. Midl. Nat. 42:575, pl. 7, figs. 4–7, 1949.
Anastrophyllum (subg. *Crossocalyx*) *hellerianum* Schust., *ibid.* 45(1):74, pl. 21, figs. 1–9, 1951; Schuster, *ibid.* 49(2):373, fig. 13:6–7, pl. 24, figs. 1–9. 1953.

Minute, forming dull greenish (in direct sunlight brownish) patches. Shoots (when nongemmiparous) *to a maximum of 500–850 μ wide* × 2–5(8) mm long, but usually the *majority of shoots slender and strongly gemmiparous, only 100–200 μ wide;* shoots *erect or suberect* from a prostrate, creeping base, simple or with sparse furcate branches, often with gynoecial innovations. Stems slender, rather rigid and flexuous, to 110 μ in diam., filiform, *only 6–7(8) cells high;* cortical cells ca. 15–20(24) μ wide × 24–50 μ long, in cross section subisodiametric, moderately thick-walled; medullary cells in 4–5 layers usually, somewhat round-lumened (the walls and angles somewhat thickened), their diam. not or little greater than that of cortical cells; mycorrhizal infection of ventral 2–3 strata taking place with age. Leaves (nongemmiparous and fertile shoots), distant to contiguous, widely spreading, *uniformly 2-lobed for 0.3–0.45 their length, subquadrate* to oblong-ovate, averaging as long as or barely longer than broad, ca. 260–275(360) to 280–300(350) μ, *strongly canaliculate and subcomplicate,* the generally slightly smaller antical half of leaf loosely folded over the postical half; lobes acute, sometimes barely cuspidate or apiculate, the postical often incurved; leaf margins entire, but upper leaves of fertile shoots often with a small subbasal antical tooth, more rarely with a ventral tooth; sinus rectangular to narrowly V-shaped. Leaves of gemmiparous plants not, or less evidently, complicate, becoming progressively smaller distally, the *gemmiparous shoots usually becoming highly reduced-leaved,* the leaves oblong, erect-appressed, imbricate or in extreme cases distant, often virtually obsolete rudiments. Cells ca. (12)13–16(17) μ on leaf margins and in lobes, ca. 14–18 × 16–25 μ in leaf middle, 17–19(20–25) × (17)18–25(35) μ at base, *their walls feebly to moderately thickened, the angles indistinctly or moderately so; cuticle verruculose,* often strongly so; oil-bodies subspherical to ovoid, 2–4, usually 3, per cell, finely granular-papillose in appearance, rather small, ca. 3–5 × 4–6(8) μ. Underleaves absent, but upper portions of fertile shoots often with stalked slime papillae, the stalk

1- or rarely 2-celled (Fig. 247:12). *Gemmae always developed in great abundance, at apices of stiff, erect, slender, often highly attenuated and sometimes crowded, reduced-leaved shoots* (up to 150–210 μ wide); *gemmae nearly cubic, vinaceous, 1-celled, ca. 10–12 μ.* (Rarely a few gemmae at apices of nearly normal-leaved shoots; these then rarely in part 2-celled and slightly larger; such shoots evidently only beginning to form gemmae and yet hardly modified.)

Dioecious. Androecia terminal, at least initially, compactly spicate; bracts imbricate, *in 4–6 pairs,* erect-spreading to suberect, subequally bilobed, conspicuously ventricose, the lobes often erect or incurved, *acute to apiculate;* margins of bracts otherwise entire, except the *dorsal lobe sometimes obscurely sinuous-dentate;* 1-androus. ♀ Bracts conspicuously larger than leaves, 2(3)-lobed for 0.3–0.4 their length, loosely complicate and suberect, their margins irregularly (sometimes freely and strongly, sometimes only sparsely) dentate to spinose-dentate; bracteole usually small, dentate to spinose-dentate, united at base with 1 or both bracts. Perianth 0.55–0.65 emergent, cylindrical-ovoid, narrowed to less than half its width at mouth, with a deep *antical sulcus* extending for at least half its length, and weaker, shallow sulci laterad of this, the apex of perianth usually obscurely 5-plicate near mouth; perianth mouth lobulate, the lobes irregularly spinose-dentate to spinose-ciliate, the cilia 2–8-celled. Seta delicate, ca. 85–145 μ in diam., of 8(9–10) epidermal and 4 inner cell rows. Capsule ovoid, ca. 300 μ long, reddish brown, its wall 2-stratose (and 21 μ thick) to locally 3-stratose (and then 25–28.5 μ thick); epidermal cells 12–13 μ high, innermost cells 8–10 μ high; internal stratum (if present) 7.5–9 μ high; epidermal and internal cells with columnar, vertical (nodular) thickenings; innermost stratum with semiannular bands. Elaters ca. 7 μ in diam., 90–135 μ long, with 2 closely wound spirals each 2.5 μ wide; spores 10–12 μ in diam., granulate-vermiculate.

Type. Near Amorbach, Bavaria, Germany (*Dr. Heller*).

Distribution

Essentially circumboreal, largely continental. An index species of the Spruce-Fir Panclimax, with sporadic, isolated stations in the northern portions of the deciduous forests (in eastern North America in the Hemlock-Hardwoods Forest, more rarely in the Mixed Mesophytic or Cove Forest); totally lacking from the Tundra and apparently from the Tundra-Taiga ecotone.

Found sporadically throughout much of northern and central but not northernmost Europe, from Scandinavia (Norway, Sweden, Finland) southward through England and Scotland (rare), but absent from Ireland; recurring in the Low Countries, much of the cooler or more elevated portions of Germany, France, Switzerland, to Austria, Czechoslovakia (Tatra Mts.), the Carpathians, eastward into Poland and Russia, and into Asia (Siberia, at the Yenisey R., and Lena R.). Reported from

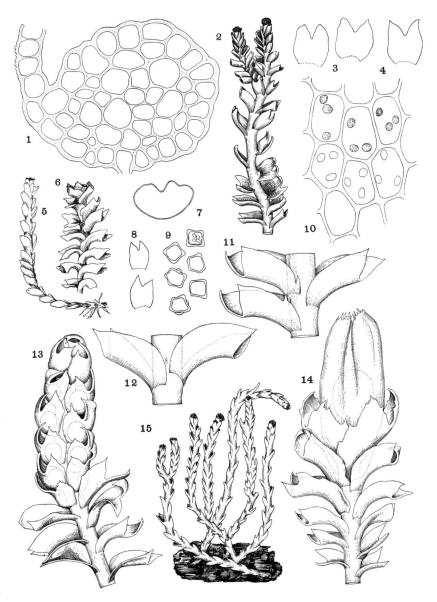

FIG. 247. *Anastrophyllum* (*Crossocalyx*) *hellerianum*. (1) Stem cross section (×275). (2) Gemmiparous shoot, with slight modification of shoot apices associated only with early stages of gemmae formation (×22). (3–4) Leaves (×27.5). (5–6) Strongly gemmiparous and feebly gemmiparous shoots, showing differences in form (×ca. 22). (7) Perianth cross section, antical face up (×33). (8) Leaves from near tip of gemmiparous shoot (×27.5). (9) Gemmae (×ca. 450). (10) Median cells with oil-bodies (×460). (11) Dorsal view of shoot sector, upper part of gynoecial shoot (×52). (12) Same, postical aspect, with vestigial underleaf (×52). (13) Androecial shoot apex (×ca. 44). (14)

Japan (*Abies veitchii-A. mariesii* forest at 1900 m, Ontake Mt., Middle Honshu; see Hattori, 1958, p. 40; see also Kitagawa, 1966).

In North America appearing to be rare in the West, known only from Washington and British Columbia (but certainly more widespread, and surely extending eastward into Manitoba, thus probably transcontinental in the Taiga); reported by Clark and Frye (1942) from Thumb Bay, Alaska. Eastward often common, ranging as follows:

W. NEWFOUNDLAND. South Branch, Codroy R. (Buch & Tuomikoski, 1955). NOVA SCOTIA. Barrasois, Cape Breton (*Nichols*; Brown, 1936a); Warren L., N. of Ingonish, Cape Breton (*RMS 66302a*); E. side of Cape Breton Natl. Park, 3–3.5 mi S. of Mary Ann Brook falls, S. of Neils Harbour (*RMS 66310b, 66310*). QUEBEC. Trail to Table Top Mt.; Bic (Lepage, 1944–45); N. of Malbaie R., 10–12 mi NW. of Bridgeville, E. end of Gaspé Pen. (*RMS 43521, 43520, 43525*); Mt. Shefford (Fabius, 1950). ONTARIO. Muskoka Distr., Sinclair Twp. (*Williams*); Cliff L. and Cedar L., 28 mi N. of Vermilion Bay, Kenora Distr. (*Cain 5569, 5597, 5613*); near MacGregors Cove, Algoma Distr. (*Williams 5221*); mouth of Agawa R., Algoma Distr. (*Cain & Williams 5650!*); Belleville (*Macoun 182*); L. Superior (*Macoun 217*); Quetico Prov. Park (*Drexler 2146*); Moosonee, S. end of James Bay (Buch & Tuomikoski, 1955); Ottawa and mouth of Nipigon R., L. Superior (Macoun, 1902); Carlton, Hastings, L. Timagami, and Thunder Bay (Cain & Fulford, 1948).

MAINE. Northwest Basin, Mt. Katahdin (*RMS 17016, 17024*, etc.); Schoodic L., Piscataquis Co.; Mt. Desert I. (*Lorenz*); Round Mountain L. and vic., Franklin Co. (*Lorenz*). NEW HAMPSHIRE. Mt. Washington, near Pinkham Notch (*RMS, s.n.*); Franconia Mts. (*Lorenz*, 1908c). VERMONT. Willoughby; Pico Peak (*RMS, s.n.*); Haselton Trail, Mt. Mansfield, 2800–3200 ft (*RMS 43844a, 43842*). MASSACHUSETTS. Sheffield; Tannery Falls, Franklin Co. (*RMS, s.n.*); Hawley Cranberry Swamp (bog), Franklin Co. (*RMS, s.n.*). RHODE ISLAND. *Fide* Evans (1923g). CONNECTICUT. Salisbury.

NEW YORK. Slope of Cornell Mt., in spruce forest, ca. 3600–3900 ft (*RMS 17577, 24412a*); spruce forest between Slide and Cornell Mts., Catskill Mts. (*RMS 17641a*); Bald Mt., near Heart L., N. of Mt. Marcy, Adirondack Mts. (*RMS & Matthysse 119, 125*); Coy Glen, Ithaca, Tompkins Co. (*RMS & Steere*); Sixhundred Ravine, near Slaterville, Tompkins Co. (Schuster, 1949a); Lick Brook, near Enfield, Tompkins Co. (Schuster, 1949a); Briggs Gully, S. end of Honeoye L., Richmond, Ontario Co. (Schuster, 1949a); Little Moose L., Herkimer Co. (*Haynes*). WEST VIRGINIA. Pocahontas Co. (Ammons, 1940); Cranberry Glades (Sheldon, 1910a). VIRGINIA. The Cascades, Little Stony Cr. near Mountain L., Giles Co. (*RMS 40262, 40270, 40262a, 43083*).

NORTH CAROLINA. Linville Gorge, 1.5–2 mi below falls, McDowell Co. (*RMS 34743, 28894a*); Neddie Cr. above confluence with Tuckasegee R., Rte. 281, Jackson Co. (*RMS 29415*); Chattooga R. along W. bank, at bridge, and 0.5–1.0 mi below Narrows, in Macon Co. (*RMS 39451, 39461, 39434a*); Chattooga R., 0.2 mi above Ellicott Rock, E. side, Jackson Co. (*RMS, s.n.*). SOUTH CAROLINA. Lower end of Thompson R. gorge, 3.5–4 mi N. of Jocassee, ca. 1–1.5 mi S. of North Carolina boundary (*RMS 45199b*). GEORGIA. Chattooga R., W. bank, S. of Ellicott Rock, NE. corner

---

Perianth-bearing shoot apex (×ca. 44). (15) Typical flagelliform gemmiparous shoots, *in situ*, on wood (×12). [1, 7, 9, 10–14, Cascade River, Minn., *RMS*; 2–3, 8, Briggs Gully, Honeoye Lake, N.Y., *RMS*; 4–6, from N. of Mt. MacIntyre, N.Y., *RMS*; 15, *RMS 24112*, Mt. LeConte, Tenn.]

of Rabun Co., 1900–2000 ft (*RMS & Bryan 39833*); High Falls of Big Cr., 4 mi SSE. of Highlands, N.C., 0.2 mi S. of North Carolina border in Rabun Co., 1900–2000 ft (*RMS 39948*); southernmost reports!   TENNESSEE.   Mt. LeConte, above Alum Cave parking lot, Sevier Co. (*RMS, s.n.*); Mt. LeConte (*A. L. Andrews*); Mt. LeConte trail, Sevier Co. (Clebsch 1954, as *Lophozia helleriana*); below Arch Rock, Smoky Mt. Natl. Park, Sevier Co., 4000 ft (*Sharp 4148!*); along W. branch of Little Pigeon R., ca. 4000 ft, below the Chimneys, Smoky Mt. Natl. Park, Sevier Co. (*RMS 36542*).   OHIO. Champaign, Hocking Cos. (Miller, 1964).

Westward extending as follows: MICHIGAN.   Mott I. and Amygdaloid I., Isle Royale (*RMS 17329*, etc.); Copper Harbor, Keweenaw Co. (*RMS, s.n.;* Steere, 1937); Deer L., W. of Munising, Barraga Co. (*RMS, s.n.*); Reese's Bog, Cheboygan Co.; Gogebic, Ontonagon, Keweenaw, Luce, Chippewa Cos. (Evans & Nichols, 1935, etc.); L. Lily near Ft. Wilkins, NW. of Copper Harbor, Keweenaw Co. (*RMS 39150*; *c. caps.*). WISCONSIN.   Sand I., Apostle Isls., Bayfield Co. (*RMS, s.n.*); Vilas, Oneida, Marathon, Bayfield, Douglas and Ashland Cos. (Conklin, 1929).   MINNESOTA.   Little Susie and Big Susie Isls., near Grand Portage, Cook Co. (*RMS 7235, 12103a*); ravine near Lutsen, Cascade R. ravine, along L. Superior N. of Cascade R., Pigeon Pt., Hat Pt. near Grand Portage, Grand Marais, on Portage from N. of S. lakes, Old Iron Trail, Gunflint Trail, Lutsen and Temperance R., all in Cook Co. (*RMS 15087, 11408, 12268, 8001d, 10011, 13207a, 11408, 5880b, 5100b*, etc.; *Conklin 3048, Cheney 27, Holzinger, s.n.*); Carlton, Carlton Co. (*Conklin 600,p.p.*); W. of Togo, Itasca Co. (*RMS 14014*, etc.); Encampment R. near L. Superior, Gooseberry R. ravine, and U.S. Pen. on Basswood L., all in Lake Co. (*RMS 13458, 13249b*, etc.; *Holzinger, s.n.*); Angle Inlet, L. of the Woods, Lake of the Woods Co., 3–4 mi from Manitoba border (*RMS 13600*, etc.); French R., and Briery, Pike Lake Rd., St. Louis Co. (*Conklin 601, 1053, 1231a*).

The species is found mostly within the Taiga with isolated stations, as in New York State (see Schuster, 1949a), in the Hemlock-Hardwoods Forest, south of the main body of Spruce-Fir Forest. Surprisingly, in the Southern Appalachians the species is lacking at high elevations in the Fraser Fir-Spruce Forest, being confined either to Hemlock-Hardwoods Forest, or Mixed Mesophytic Cove Forest, in both cases at elevations of only 1900–4200 feet. As emphasized below, the behavior of the species changes considerably near the southern end of its range. *Anastrophyllum hellerianum* extends southward, along the Chattooga R., barely into Georgia on the west bank and to within 0.2 mile of South Carolina on the east bank. Along Big Creek it extends again about 0.2 mile into Georgia, south of Highlands, N.C. The critical elevation at which the species appears to drop out is between 1900 and 2100 feet.

Ecology

Northward almost without exception restricted to damp or rather dry decaying logs, usually in shaded (rarely in partially insolated) sites associated with a wide variety of taxa of the *Nowellia-Jamesoniella* Associule, chiefly *Nowellia curvifolia*, *Blepharostoma*, *Tritomaria exsectiformis* (rarely *exsecta*), *Lophozia porphyroleuca*, *L. ventricosa silvicola*, *L. ascendens*, *Lepidozia reptans*, more sporadically *Scapania apiculata*, *S. glaucocephala*, and *S. umbrosa*. Usually *A. hellerianum* appears unable to

invade recently decorticated logs, and is never a pioneer species, but invades after preparation by mosses and such pioneer Hepaticae as *Lophocolea heterophylla*, *Ptilidium ciliare*, *Jamesoniella*. On such "ripe" logs it is often very abundant as an accessory species, often associated with these preceding species . . . . The species apparently requires a substrate rather rich in humic acids, but is unable to persist after the log has fallen into humus. It occurs under a measured pH of 4.6–5.2 (Schuster, 1953).

Exceptionally the species occurs over humus lying over exposed rocks, near tree level in the mountains (trail to Northwest Basin, Mt. Katahdin, *RMS 17024*).

In deciduous forests, where the species is rare, its ecology undergoes profound changes. Here it may occur on bark of *Tsuga canadensis*, up to 5–8 feet or more high, associated with reduced forms of *Bazzania trilobata*, *Metzgeria furcata*, etc. (Schuster, 1949a). In the Southern Appalachians the species is largely restricted to such habitats, both on *Tsuga* (then with corticolous extremes of *Bazzania trilobata*, *B. denudata*, *Jamesoniella autumnalis*, *Cephalozia media*, and even *Nowellia curvifolia*, as well as with *Metzgeria furcata*, *Frullania asagrayana*, *Lejeunea ulicina*, and also *Drepanolejeunea appalachiana* and *Harpalejeunea ovata*, species of tropical affinity!) and on *Betula lutea* (with *Metzgeria*, *Radula obconica* and *tenax*) or the branches of *Rhododendron maximum* (with *Lejeunea ruthii* and *ulicina*), vary rarely on old trees of *Pinus strobus*. These corticolous phases may be almost unrecognizable, consisting of scattered erect, stiff but slender shoots with almost obsolete leaves, bearing vinaceous red to purplish gemmae masses at the apices. Fortunately, the polygonal to quadrate, small, vinaceous gemmae are absolutely distinctive. Vegetative shoots of such phases may be only 0.8–2 mm high × 120–160 μ in diam. (stems 45–55 μ), with erect to erect-appressed tiny leaves 100 μ wide. Leaves of such phases are bifid to the middle, with lobes only 3–4 cells broad. Corticolous phases are largely confined to run-off crevices of bark of *Tsuga*. Here they usually occur as microscopic, persistently juvenile forms, creeping except for the erect or ascending gemmiparous apices, found either by accident or as a consequence of careful systematic search. The species is evidently widespread on old mature trees of *Tsuga* in ravines, since a large percentage of such trees that were systematically examined bear traces of it. Usually only such minute amounts are found—often a few scattered isolated stems—that the plant will be overlooked.

On decaying moist logs *A. hellerianum* often produces both androecia and perianths and not infrequently mature capsules (early June to August); on rocks and bark only gemmae are produced.

Differentiation

The ecology, coupled with the constant occurrence of erect, slender, attenuate gemmiparous shoots bearing small, cubic, wine-colored gemmae at their tips, is diagnostic. Very frequently, particularly on shaded, moist, decayed logs, gemmiparous shoots are produced in immense numbers, becoming densely crowded and forming brushlike mats, the individual shoots with the appearance of stiff, purplish-tipped bristles. More commonly the species occurs in very small quantity, the isolated purplish-tipped gemmiparous shoots standing up stiffly from the substrate, readily evident among the more prostrate species which usually accompany it. "Though the species is very minute and often overlooked, it manages to compete with the larger species (such as *Jamesoniella*) largely because of the peculiar growth" (Schuster, 1953).

Only four taxa can possibly be confused with *A. hellerianum*: *Scapania apiculata*, *S. glaucocephala*, *S. carinthiaca*, and *Lophozia longidens*. All of these are found on decaying logs and agree in producing erect, slender, gemmiparous shoots tipped with pigmented gemmae. However, *S. apiculata* and *Lophozia longidens* have, respectively, reddish brown and orange-red gemmae. Furthermore, all three species of *Scapania* have ovoid or ellipsoid, smooth gemmae. Finally, all of these species usually have much more robust gemmiparous shoots, with less obviously reduced leaves (except for *S. apiculata*, in which the cells bear bulging trigones, however).

In the Southeast both *Anastrophyllum hellerianum* and *A. michauxii* may occur on the bark of *Tsuga*, frequently to heights of 8–16 feet above ground level. Both taxa are then often highly reduced, the latter to the point where the gemmiparous shoots may be confused with those of *A. hellerianum*. However, *A. michauxii* has very different gemmae, and the gemmiparous shoots are never wormlike as in *A. hellerianum*. Indeed, *A. hellerianum* under such conditions looks like a juvenile *Cephaloziella* with ascending to erect shoot tips bearing gemmae. The totally etiolated form of such shoots renders them virtually unrecognizable, except on the basis of the subquadrate, 1-celled, vinaceous gemmae.

Relationships and Variation

Except for the still unclarified relationships to *A. tenue* (see p. 782), *A. (Crossocalyx) hellerianum* is perhaps most closely allied to *A. (E.) myriocarpum;* however, it differs from the latter in a number of respects. (1) Although some branches are lateral-intercalary and axillary (see, e.g., Müller, 1947, fig. 2a), others are postical-intercalary, issuing from the ventral side of the axis.[265] (2) Leaf cells are in actuality thin or slightly thick-walled but produce tolerably distinct trigones with concave or rarely even feebly convex sides. (3) Erect gemmiparous shoots are abundantly

---

[265] Two such branches were found in the sparing type of *J. helleriana* ("Spessart: leg. Heller!"), a fragment in herb. G., ex herb. Müller Arg. Müller (1947, p. 42) has argued that the lateral branching of this taxon separates it from the Cephaloziaceae and depicts a lateral-intercalary branch; however, unquestionable postical-intercalary branches occur in addition.

produced. (4) Geotropic flagella are lacking. (5) The perianth is weakly or sometimes hardly compressed (with sporophyte maturation it always becomes cylindrical). These differences are in contrast to marked likenesses, as, e.g., in the (a) lack—at least normally—of terminal branching; (b) similar 2-stratose capsule-wall cross section (but see below); (c) seta with usually 4 inner, 8 outer, cell rows; (d) similar leaf insertion and often orientation.

I have checked the few type plants (at G), and they correspond well to extant concepts of this species. The larger leaves, which are only feebly canaliculate-complicate, range to 350 μ wide × 330 μ long (thus slightly wider than long) to 320 μ wide × 360 μ long (and then slightly longer than broad); they are about 0.35–0.4 bilobed, with essentially equal lobes ca. 9–10 cells wide at the base; the lobes tend to be submucronate with 1–2 sharp terminal cells. The cells vary considerably from the lobe apices toward the base, becoming progressively larger: in the lobe apices they are ca. 12–15(17) × 16–23 μ, in the leaf base ca. 20–25 × 24–29(35)μ. Unlike in *A. minutum*, the cells tend to be quite irregular in size and shape.

The type of "Jung. *verruculosa* var. *compacta* Lindb." (Smaland, *K.A. Th. Seth*) (G) has also been checked. It represents a freely fertile phenotype, with numerous capsules—but no trace of gemmae! Leaves are all longer than broad and distinctly verruculose. The cells are much larger than in the type: in the lobe apices ca. 21–24 × 22–25 μ, with thin walls and very distinct or even slightly bulging, well-defined trigones; toward the base the cells become larger and less collenchymatous, ca. 24–28(30) × 35–48 μ. In these plants, no lateral-intercalary branches were seen. The leaves were not at all complicate but merely canaliculate and longer than wide. The seta, which consists of 2 rings of cells in all cases seen, is formed either of 4 inner and 8 outer cell rows (then 11 μ in diam.) or of 4 inner and 9–10 outer cell rows (and is then 145 μ in diam.). The capsule wall is variable; it is usually described as 2-stratose (Müller, 1951–58; Schuster, 1951). In actuality it varies from 2- to 3-stratose on the same capsule. When 2-stratose, the epidermal cells are 12–13 μ high, the inner 8–10 μ high (total ca. 21 μ). When 3-stratose, a thinner inner layer ca. 7–9 μ high is intercalated, and the wall diam. increases to 26–29 μ. Epidermal cells bear strong nodular thickenings; inner cells bear semiannular bands. Spores are verruculose-papillose, 10–12 μ in diam. Elaters are ca. 6–7.5 × 90–135 μ, feebly tortuous, rather tapered at the ends, closely 2-spiral; spirals ca. 2.5 μ wide.

ANASTROPHYLLUM (CROSSOCALYX) TENUE H. Williams

[Fig. 248]

Anastrophyllum tenue Williams, The Bryologist 71(1):34, 1968.

Plantae *A. helleriano* similes (minutae; erecta aut ascendentes; gemmae vinaceae, ad apices surculorum qui folia parva habent sitae); plantae differentes, autem, quod (a) rami terminales typi *Frullaniae*, (b) gemmae saepissime 14–16 × 14–19 u, et (c) ♂ bracteae lobos in margine integros, in cacuminibus obtusos rotundatosve habentes.

Minute, clear green, erect to strongly ascending in growth (particularly with gemmae formation), the sterile shoots with leaves only 430–550 μ wide × 4–6(8) mm high. Branching frequent to common, *always acutely furcate and terminal, Frullania type.* Stems 120–155 μ in width, soft-textured, (6)7–8(9) *cells high,* of a layer of rather leptodermous or feebly thick-walled, quadrate to oblong cortical cells (15–19 μ wide × 24–48 μ long in surface view), similar in diam. to the almost leptodermous medullary cells (cortical, 15–19 μ in diam.; medullary, 12–17 μ in diam.); no dorsiventral medullary differentiation, but ventral cortical and 2–3 layers of medullary cells with age mycorrhizal, brown. Rhizoids frequent, colorless, 8.5–12 μ in diam. Leaves broad-based, remote to subcontiguous, vertical to subvertical (then somewhat antically inclined), rather concave (strongly so near androecia, there ± cupped at times) but hardly canaliculate and never conduplicate, broadly oblong-ovate to ovate-orbicular, 300–310 μ wide and long to ca. 340–360(420) μ wide × 300–325(335) μ long, divided for (0.2)0.25–0.3 by a bluntly angular to rectangulate sinus; lobes triangular, acute to (especially near androecia) *often blunt at tips,* entire-margined; dorsal half of leaf inserted to stem midline, *not dilated at base, not extended beyond stem midline,* somewhat oblique to subtransversely inserted, *not decurrent antically.* No underleaves. Cells *smooth* (or very faintly striolate at base), *leptodermous* to moderately thick-walled or with *trigones* ± *small* to large but concave-sided; marginal cells 14–18 μ in lobes; median cells (12)14–16 μ wide × (14)15–20 μ long; basal cells 15–18 μ wide × 18–28 μ long; oil-bodies 5–7 or more per cell, small, faintly botryoidal (ill-preserved in type plants). Asexual reproduction always present, by red-brown or rust-red gemmae (forming *purplish, deeply pigmented small clusters* at tips of uppermost, usually erect, reduced leaves); gemmiparous shoots, with prolonged gemmae formation, filiform, reduced-leaved, only 140–175 μ in diam., the minute, scalelike, quadrate, bifid leaves of few cells, erect to erect-appressed. *Gemmae 2-celled at maturity,* 12–16 × 14–19 μ to 18–19 × 19–20 μ, tetrahedral to polyhedral.

Dioecious. Androecia conspicuous, 520–585 μ broad, of usually 2–3(4) distant to contiguous, strongly inflated bracts; bracts subrotundate or broadly rotund, ca. 440 μ wide × 370–400 μ long, broad-based, ca. 0.25–0.3 bilobed; the *lobes blunt or rounded,* incurved, often faintly brownish-pigmented at the tips; antical base of bract with a sharp, linear to lanceolate inflexed tooth; monandrous; no paraphyses; antheridium with globose or subglobose body ca. 165 × 175 μ, wall of subquadrate to polygonal cells ca. 20–28 μ in diam.; stalk 20–22 μ in diam., rather short, 1-seriate. Gynoecial plant rare; ♀ bracts bilobed, with entire margins. Perianth (from unfertilized gynoecia only) pyriform, with a dorsal furrow, contracted to apex but not beaked.

Type. Bruce Peninsula, Ontario: north side of dolomite block, shaded faces, St. Edmund Twp. (*H. Williams 1848!*).

I also found a trace of this plant with *Lophozia heterocolpa,* near Tobermory, at the north end of the Bruce Peninsula, in 1949. Williams (*in litt.*) also reports it from Little Cove, ca. 4 mi. E. of Tobermory, to ca. half way towards Cabot Head; also at Driftwood Cove, all on Bruce Peninsula. Also at Eugenia Falls, Grey Co. (*Williams 1226*).

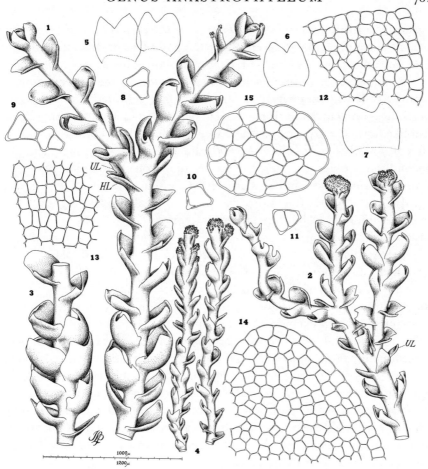

FIG. 248. *Anastrophyllum (Crossocalyx) tenue.* (1) ♂ Shoot with terminal branch; at HL the stem half leaf, and at UL the first branch appendage, an underleaf (×38). (2) Gemmiparous plant with two terminal branches; at UL first branch appendage visible (×31.5). (3) Androecial sector of shoot (×38). (4) Two slender, erect gemmiparous shoots (×38); (5–7) Leaves (×38); (8–11) Gemmae (×420). (12) Leaf lobe apex (×220). (13) Median cells (×220). (14) Ventral lobe apex (×220). (15) Stem cross section (×285). [All drawn from type specimen; figs. 1, 3–7, drawn to top scale; 2, drawn to bottom scale.]

Ecology

A decided calcicole, insofar as present reports go. Associated with the type are *Scapania gymnostomophila* and *Tritomaria scitula.*

Differentiation

Distinctive for this minute plant is the green color, with the shoot tips (particularly of the flagelliform shoots) adorned with deep purplish masses of gemmae. In the intermingled *Tritomaria scitula* and *Scapania gymnosto-mophila* the gemmae masses, seen under the hand lens or dissecting micro-scope, are much lighter orange-red or rust-red. The species *superficially* is a close match for *A. hellerianum* but differs from it in (*a*) the larger gemmae, usually 2-celled at maturity; (*b*) the abundant terminal, *Frullania*-type branching—on one plant I have seen 9 such branches, resulting in a shrublike fastigiate aspect; (*c*) the fewer ♂ bracts, their blunt lobe apices and consistently entire margins; (*d*) the consistently noncaniculate, nonconduplicate leaves; (*e*) the more numerous oil-bodies per cell. Admittedly somewhat similar sterile plants often occur in *A. hellerianum* (e.g., Fig. 247:2), but the androecial plants of that species (Fig. 247:13) are quite different. The very different ecology (*A. hellerianum* is rather consistently corticolous or xylicolous) alone precludes the two plants from being identical. Indeed, until gynoecial plants and plants with sporophytes are found, the subgeneric placement of *A. tenue* remains uncertain. The lack of trifid leaves and the lack of medullary dorsiventral differentiation exclude *Tritomaria;* the vertical or subvertical leaves and strictly erect growth exclude *Lophozia.*

This questionable plant is placed in *Crossocalyx* because (*a*) it produces purplish gemmae on apices of small-leaved, slender flagelliform shoots; (*b*) the bilobed leaves are often somewhat oblique and closely approach those of *A. hellerianum* in form; (*c*) the stem, which shows mycorrhizal activity in both lower parts of cortex and medulla, is only 6–8 cells high. No dorsiventral differentiation of the medulla occurs, as regards cell diameter, and the cortical cells, in cross section, are hardly differentiated from the medullary.

The plants have occasional diffuse androecia with 1–2(3) pairs of subcontiguous or contiguous ♂ bracts that are ventricose almost through-out, with erect and usually blunt lobes; the few mature bracts seen were monandrous. ♂ Bracts and leaves adjacent to them tend to develop a sharp, often lanceolate, frequently incurved tooth of the antical base.

The gemmae, although deep purplish *en masse*, are actually rust-red by transmitted light; they are 2-celled, quadrate to polygonal, mostly 14–16 × 14–19 μ. They occur, exactly as in *A. hellerianum*, in terminal masses at the tips of flagelliform, reduced-leaved erect shoots. Gemmip-arous shoots often develop some underleaves.

The basis for proposing this as a new species is circumstantial and not quite satisfactory. The plant has been pronounced to be a form of *Tritomaria scitula,* but this can hardly be the case; the stem anatomy (lack of a small-celled ventral medullary strand; stem only 6–8 cells high); the gemma size and the purple color of the gemmae *en masse;* the small-leaved shoots; the very small size of the plants; the almost constantly bilobed leaves and ♂ bracts—all these preclude *T. scitula.*

Similarly, the occurrence (over dolomite) and the 2-celled, relatively large gemmae preclude *A. hellerianum.* The plants admittedly look like a very small and slender form of *A. michauxii,* such as sometimes occurs on bark of trees, but the nearly leptodermous, small cells rule out this species.

GYMNOCOLEA (Dumort.) Dumort.

[Figs. 249–251]

Jungermannia sect. *Gymnocolea* Dumort., Syll. Jungerm. Eur., p. 52, 1831 (in part).
Gymnocolea Dumort., Rec. d'Obs. p. 17, 1835; K. Müller, Rabenh. Krypt.-Fl. 6(1):738, 1910;
    Schuster, Amer. Midl. Nat. 45(1):76, pl. 26, 1951.
Lophozia Howe, Mem. Torrey Bot. Club 7:103, 1899 (in part).

In deep green to brown to *scorched-appearing blackish patches,* medium-sized [0.75–1.5(2.6) mm wide × 8–30 mm long]. Stems loosely decumbent to ascending, suberect with crowding, the apices always ascending, (6)7–8(10) *cells high,* the cortical cells ± thick-walled, little differentiated from medullary, subequal to them in diam., ca. 21–25 μ in diam. × (25)35–48 μ long [length (1.0)1.5–2.0× width]; medullary cells ± thin-walled, without mycorrhizal infection, without dorsiventral differentiation except that the lower *2–3 tiers are somewhat narrower in diam.* than the upper tiers, ± mycorrhizal; branching very diverse, either terminal (and then from the lower halves of lateral segments, the branch replacing ventral half of a leaf; supporting leaf ovate to lanceolate, un-lobed or subtending leaf normal, except for a narrower base, bilobed) or intercalary, the intercalary branches lateral or postical. *Rhizoids usually very sparse,* scattered. Leaves remote to subimbricate, *obliquely inserted (at an angle of ca. 40–45°* from transverse, except at the postical, somewhat arcuate end of the line of insertion), not or hardly decurrent, suborbicular to shortly ovate-oblong, *uniformly bilobed,* nearly plane to ± concave, *never complicate; lobes obtuse to rounded, rarely subacute.* Cells relatively small, ca. 25–30 μ in leaf middle, *not or weakly collenchymatous, the walls becoming equally or nearly equally thickened;* cuticle smooth or obscurely verruculose; *oil-bodies small to large, spherical to ovoid,* 2.4–4 × 4 to 3–4 × 5 to 5–6 × 6–7 μ, ca. (2–3)4–8(9–10) or occasionally only 1–2 per cell, faintly papillose-segmented, almost smooth externally. *Ventral merophytes narrow,* usually 2–4 cells broad, *not or rarely with distinct underleaves* (except near bases of branches), the occasional underleaves usually formed of 1–3 basally connate stalked slime papillae. Asexual reproduction by gemmae absent or very rare,[266] the gemmae tetrahedral-polyhedral, 1–2-celled.

Dioecious (our taxa) or autoecious. Androecia eventually intercalary, loosely spicate, often longly so, of up to 5–8(10–12) pairs of bracts; bracts scarcely smaller than leaves, somewhat to strongly ventricose, otherwise similar to leaves; *1-androus;* antheridial stalk *1-seriate;* paraphyses absent. Gynoecia with *bracts almost exactly like leaves, bilobed, no larger than upper leaves (sometimes barely smaller);* bracteole absent or a lingulate vestige, larger (and ovate to bilobed) only in *Gymnocoleopsis.* Perianths almost wholly exserted, of two types: (*a*) *unfertilized subglobose or shortly ovoid,* attached by a constricted base, from which they are *freely caducous;* (*b*) fertilized, becoming ellipsoidal to clavate-obpyriform. *Perianths* (in either case) *smooth and essentially eplicate, rounded and contracted to mouth, not beaked;* mouth dentate and lobulate. Seta of numerous cell rows, epidermal a little larger (subg. *Gymnocolea*), or of 8 epidermal and 4(5) internal cell rows (subg. *Gymnocoleopsis*). Sporophyte with *capsule wall bistratose* or (*Gymnocoleopis*) 2–3-stratose; capsule oblong-ovoid. Epidermal cells large, subquadrate to short-rectangular, *with nodular thickenings that are somewhat tangentially extended as spurs.* Inner cell layer of narrowly rectangulate cells with nodular (radial) thickenings *extended across the tangential faces as short spurs* (less frequently partly or wholly extending across tangential faces as weakly defined bands, then becoming semiannular). Elaters to 120–200 μ long × 6–9 μ in diam., with 2 reddish brown spirals. Spores 12–18 μ, finely granular-papillate, brown.

Type. Jungermannia inflata Huds. = *Gymnocolea inflata* (Huds.) Durmort.

Gymnocolea is a small genus, represented throughout the cooler and more boreal to arctic portions of the Northern Hemisphere; three antipodal taxa, two Andean and the other from Kerguelen I., are known. Although three species are recognized from the Northern Hemisphere, one [*G. marginata* (Steph.) Hatt.] is known only from Japan;[267] all three may prove to be mere extremes of a single, highly polymorphous species. The genus was not recognized for many years after Dumortier first proposed it, largely because the extremely superficial methods of this author led to a certain reluctance, during the last century, to accept many of the groups he proposed. However, Müller (1910, *in* 1905–16) revived the genus, and it has been universally accepted since.

---

[266] Most authors, including Frye & Clark (1937–47, p. 367), state that the gemmae are unknown. However, Müller (1951–58) says that 3–5-angled, 1–2-celled dark brown gemmae occur in *G. acutiloba.*

[267] For this, see Hattori (1952b, p. 318, and 1953a, p. 41, fig. 52:a-f) and Schuster (1965, p. 274). Apparently identical is the Japanese *Gymnocolea montana* (Horik.) Hatt.

Differentiation

Gymnocolea differs from all other genera of our region, except the wholly unrelated *Chonecolea*, in the unique method of asexual reproduction. Unfertilized perianths, attached by a weak and constricted line to the apices of the stems, freely break off and from them arise rejuvenations, usually from near the broken edge. Schiffner (1904c) gives full details as to this mode of asexual reproduction. The common species, *G. inflata*, the only one in which this mode of reproduction has been studied, usually occurs around wet or damp rocks, or near waterfalls, where plants are subject to inundation, at least sporadically or during the spring. The caducous perianths, each trapping a bubble of air within, are carried away by flowing water, resulting in a wide dispersal of the species. Capsules are only rarely produced, and dissemination of the species is evidently largely a consequence of production of the caducous balloon-like, smooth perianths.

Müller (1905–16, p. 740; 1951–58, p. 705) has suggested an affinity of *Gymnocolea* to *Lophozia* subg. *Leiocolea*, but I doubt that such a relationship exists, because of the following (Schuster 1951a, p. 76): stem merely 6–8 cells high; noncollenchymatous leaf cells; smaller and more numerous oil-bodies; ventral stem sectors only 1–3 cells broad usually, able to develop at most only rudimentary underleaves; 2-stratose capsule wall, the inner layer with thickenings more or less reduced; few and scattered rhizoids; leaves more obliquely, less horizontally, inserted.

These features were stated not only to isolate the genus from *Leiocolea* but also to indicate a possible, distant affinity to *Lophozia* subg. *Isopaches*. The reduced stems, with no dorsiventral modification of medulla; leaf cells with a tendency to become equally thick-walled; oil-bodies of similar size and constitution; equally narrowed ventral merophytes and a common reduction of the capsule wall to a bistratose condition, with a common reduction of the tangential bands of the inner cell layer, all suggest at least a superficial similarity. Whether this similarity is truly indicative of any close relationship is not clear. The reduced stems and capsule walls may merely reflect parallel reduction and simplification. In any case, *Gymnocolea* differs from *Isopaches* in a series of characters, among them (*a*) obtuse to rounded leaf lobes, rarely acute or subacute; (*b*) eplicate, peculiar perianths; (*c*) edentate, small perichaetial bracts; (*d*) much sparser rhizoids; (*e*) monandrous ♂ bracts.

Unlike *Isopaches*, *Gymnocolea* has developed several anomalous features, making it impossible to retain it readily as a subgenus of *Lophozia*. Although a small group, it appears to be wholly isolated.

Gymnocolea is divisible into two subgeneric complexes (Schuster 1967c), as shown in the following key; only subg. *Gymnocolea* occurs in the Northern Hemisphere.

Key to Subgenera of Gymnocolea

a. Terminal branching of the *Frullania* type common; postical-intercalary branching frequent; perianth 1-stratose, usually caducous in absence of fertilization; dioecious; seta with numerous cell rows; capsule wall 2-stratose, its cells little elongated; gynoecium with bracteole absent or vestigial . . . . subg. *Gymnocolea* Dumort.

a. Terminal branching of the *Frullania* type lacking; postical-intercalary branching lacking; all branches lateral-intercalary; perianth 2–3-stratose to middle or higher, persistent; autoecious; seta with 8 epidermal, 4(5) internal, cell rows; capsule wall 2–3-stratose, of strongly elongated cells; gynoecium with a large, ovate to bilobed bracteole united with both bracts . . . . subg. *Gymnocoleopsis* Schust.[268]

Key to Species of Gymnocolea

1. Perianths common, often sterile and caducous; ♀ bracts equal in size to leaves or somewhat smaller; leaves suborbicular to broadly ovate, with rounded or obtuse lobes (lobes never terminated by 2 superposed cells); terminal branching varying in the same clone, sometimes associated with bilobed leaves, at other times with a lanceolate, unlobed leaf; leaves with sides evenly arched, edentate. Never with gemmae. . . *G. inflata* (p. 786)

1. Perianths rare, noncaducous; ♀ bracts larger than leaves; leaves subquadrate to ovate-rectangular, with subacute to acute lobes (often terminated by 2 superposed cells); branches almost invariably arising from below an unlobed leaf; leaves (at least on robust shoots) often with margins 1–2-angulate or dentate. Occasionally gemmiparous, rare. *G. acutiloba* (p. 795)

GYMNOCOLEA INFLATA (*Huds.*) Dumort.

[Figs. 249–250; 251:14–18]

Jungermannia inflata Huds., Fl. Angl. ed. 2:511, 1778.
Gymnocolea inflata Dumort., Rec. d'Obs., p. 17, 1835; K. Müller, Rabenh. Krypt.-Fl. 6(1):741, figs. 332–333, 1910 (ed. 3, 6(1):706, figs. 224–225, 1954); Macvicar, Studs. Hdb. Brit. Hep. ed. 2:161, figs. 1–3, 1926; Schuster, Amer. Midl. Nat. 42(3):577, 705, pl. 9, figs. 11–14, 1949; Schuster, *ibid.* 45(1):77, pl. 26, 1951; Schuster, *ibid.* 49(2):384, pls. 28, 29, figs. 1–3, 1953.
Lophozia inflata Howe, Mem. Torrey Bot. Club 7:110, 1899.

Very variable, in patches or growing scattered (as over *Sphagnum*), often forming interwoven mats, *deep green to brownish or purplish black*, often scorched in appearance, with a slightly oily luster. Shoots mostly 750–900(1100–1500) μ wide, occasionally 1200–2650 μ wide, 6–25 mm long, loosely prostrate or ascending when crowded, often intricately interwoven; branching free to sparing, either terminal (the lateral branches then subtended by either an ovate to lanceolate, unlobed leaf or by a normal bilobed leaf), but *with age*

---

[268] *Type. Gymnocolea multiflora* (Steph.) Schust. Also belonging near here a Southern Hemisphere plant from Kerguelen I. (Royal Sound, *Eaton*), *Jungermannia cylindriformis* Mitt. (Jour. Proc. Linn. Soc. 15:196, 1877), placed by Stephani (Spec. Hep. 2:143, 1901) in *Lophozia*, as *L. cylindriformis* = *Gymnocolea cylindriformis* (Mitt.) Schust.

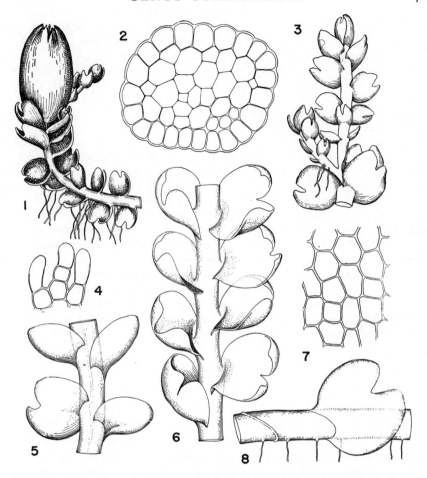

Fig. 249. *Gymnocolea inflata.* (1) Shoot apex with sterile perianth and innovation (×16). (2) Small stem, cross section (×200). (3) Apex of shoot, with two subfloral innovations, postical view (×18). (4) Underleaf (×ca. 275). (5-6) Postical and antical shoot sectors, respectively (×30). (7) Median cells (×240). (8) Lateral shoot sector, showing leaf insertion and rhizoids (×45). [1-2, 5-6, 8, Taughannock Ravine, New York, *RMS*; 3-4, 7, Gunflint Lake, Minn., *RMS.*]

often intercalary from the underside of robust, older stems; gynoecia when sterile often with subfloral innovations. Stems slender, 130–200(225) μ wide; cortical cells ca. (17)21–25 × (25)35–48 μ; stems (6)7–8(9) cells high. Leaves obliquely inserted and oriented, varying from horizontally spreading and virtually flat to somewhat suberect and strongly concave, *distant or rarely barely imbricate*, varying from subrotund and barely wider than long to shortly oblong-ovate and (0.95)1.1–1.3× longer than wide, ca. 350–700 (800–1000) μ wide × 400–900(975) μ long, entire-margined, *uniformly bilobed for ¼–⅖(½) their length*, the sinus V-shaped to narrowly obtuse; lobes broadly triangular to ovate, *obtuse to narrowly rounded* (rarely subacute), sub-equal.[269] Cells ca. 21–24(25–27) μ on margins and in lobes, 24–27(28) × 25–28(30–35) μ in leaf middle, polygonal, *walls slightly to distinctly thickened*, usually brownish, with trigones absent or minute; *cuticle smooth* or obscurely verruculose; oil-bodies finely papillose-segmented, spherical to ovoid, 2.5 × 4 to 3–4 × 5 to 5–6 × 6–7 μ, mostly (2–3)4–8(9–10) per cell. *Underleaves usually lacking*, occasionally rudimentary ones present, formed of 1–3 short, stalked slime papillae. *Gemmae absent.*

Dioecious, but *usually fertile.* Androecia slenderly and loosely spicate, becoming intercalary, of 4–6 or up to 8 pairs of slightly distant to contiguous bracts; bracts concave, bilobed, monandrous, almost transversely inserted. ♀ Bracts similar to vegetative leaves, but often more deeply bilobed, concave, *often slightly smaller than the larger vegetative leaves;* ca. 600–700 μ wide × 700–800 μ long; bracteole absent or small and lingulate. *Perianths at first globose*, with fertilization maturing and becoming oblong-obovoid to obpyriform, smooth, to 850–1200 μ in diam. × 2–2.7 mm long, contracted to the mouth and some-times slightly plicate apically; *mouth copiously lobulate*, the lobuli in the form of acute teeth 2–4 cells wide at base × 5–8 cells long. Sterile perianths remaining *subglobose or shortly obovoid, caducous.* Capsule with epidermal cells with nodular thickenings. Inner cell layer of rectangulate cells, bearing radial (vertical) bands evident at nodular thickenings, these variously extended partly across the tangential faces as spurs or semicomplete, although weak, bands. Elaters 6–9 μ in diam.; spores 10–18 μ in diam.

Type. Great Britain.

Distribution

A widely dispersed species, occurring throughout much of the arctic-alpine region (but evidently rare or absent in the northern edge of the Tundra) and the Taiga, and in scattered isolated sites in the northern portions of the Deciduous Forest Region; in eastern North America also with isolated disjunct sites in the Oak-Hickory or Oak-Pine-Hickory Region, in the outer Coastal Plain, as far south as North Carolina.

Widely dispersed throughout northern Europe, ranging from northern portions of Scandinavia (Norway, Sweden, Finland) to Denmark, southward through England and Ireland, to France, Germany, Austria, Switzerland (throughout the Alps), to northern Italy, Spain (*fide* Casares-Gil, 1919). Also on the Azores (Allorge & Allorge, 1948).

The species appears to be widespread in Japan, from which it has been variously reported as *Cephalozia montana* (Horikawa, 1932), *Gymnocolea*

---

[269] Deviations in leaf size and shape are frequent. In *RMS 43079* (Nova Scotia), e.g., many if not most leaves are somewhat unequally bilobed, much as in *Cladopodiella fluitans.* Furthermore, on some essentially mature plants the leaves are quite narrow (ca. 500 μ wide × 860 μ long; thus 1.5–1.6× as long as wide), again as in *C. fluitans.* Such plants occur with others on which the leaves are 0.9–1.05× as wide as long.

montana, or *G. inflata* subsp. *montana* (Hattori, 1958, p. 41). I have been unable to find consistent differences between the Japanese and American plants.[270]

In North America transcontinental in distribution, equally frequent in both East and West. In the West ranging from Alaska and the Yukon southward through British Columbia, Alberta, Washington, and Oregon to California and Wyoming and the Black Hills region of South Dakota; southward confined to the mountains, at median and high elevations. In the East ranging as follows:

NW. GREENLAND. Kangerdlugssuak, S. side of Inglefield Bay, 77°23' N., 67° W. (*RMS 45956a, 45985*); Kekertat, Harward Øer near Heilprin Gletscher, 77°31' N., 66°40' W. (*RMS 46050*); Kangerdluarssuk, W. side of Bowdoin Bugt, 77°33' N., 68°35' W. (*RMS 45791*); Kânâk, Red Cliffs Pen., Inglefield Bay (*RMS 45726a, 45728*). W. GREENLAND. SE. of Jacobshavn, Disko Bay, 69°12' N., 51°05' W. (*RMS & KD 66-246a*); Lyngmarksbugt and below Lyngmarksfjeld, Godhavn, Disko I., 69°15' N., 53°30' W. (*RMS & KD 66-197, 66-192, 66-102c, 66-344, 66-555*); Tupilak I., E. of Egedesminde, 68°42' N. (*RMS & KD 66-024a, 66-011, 66-009a, 66-002a, 66-016,* etc.); E. end of Agpat Ø, neck E. of Umiasugssûp ilua, 69°52' N., 51°38' W. (*RMS & KD 66-1468*); Ritenbenk (*Berggren*); Maneetsok (*Warming & Holm*); Holstensborg, Ameralik, and Baals Revier (*Vahl, fide* Lange & Jensen, 1887). E. GREENLAND. Hurry Inlet, Ryders Dal, 70°51' N., and Smalsund, 65°59' N. (Jensen, 1906a); Hekla Havn, Scoresby Sund (Jensen, 1897). BAFFIN I. Strathcona Sd., in N. Baffin, and Kingua Fiord, in C. Baffin (Steere, 1939; Polunin, 1947, p. 501, footnote). LABRA-DOR. Blanc Sablon (*Waghorne 23, 1891!*, det. and cited by Underwood, 1892a, as "*Marsupella adusta*"; a form with ± acute or subacute lobes, *trans. ad G. acutiloba*).

QUEBEC. Cairn I., Richmond Gulf, and mouth of Great Whale R., both on E. coast of Hudson Bay (see Schuster, 1951); mountain E. of George R., 55°09' N. (Kucyniak, 1949); Tadoussac, Saguenay Co. (*Evans 43*); Marten R., Ungava (*Lepage 4357a!*); Rigaud; Tadoussac; St.-Ulric de Matane; Mt. Albert and Mt. La Table; Rupert R., above L. Némiskau; Oatmeal Portage; R. à Martre (Lepage, 1944–45); Mt. Lac des Cygnes, Charlevoix Co. (Kucyniak, 1947). NEWFOUNDLAND. Avalon Distr., S. Coast, W. Nfld., N. Pen., C. and NE. Nfld. (Buch & Tuomikoski, 1955). MIQUELON I. (Buch & Tuomikoski, 1955). NOVA SCOTIA. Dartmouth (*Brown 211*); edge of spruce-tamarack bog, between Campbell and Craigmore, Inverness Co., Cape Breton (*RMS 43070, 43084, 43079, 43073,* etc.); Mary Ann Falls, S. of Neils Harbour (*RMS 42657a*) and 3 mi S. of Mary Ann Falls, N. of Ingonish, Cape Breton (*RMS 66349*); Smoky Mt. near Aspy Bay; Louisburg, Cape Breton; Sable I. (Macoun, 1902).

ONTARIO. Manitoulin I., 5 mi E. of West Bay (*Cain 4066!*); Algoma; L. Timagami; High Park; Toronto; York (Williams & Cain, 1959); mouth of Agawa R. and point N. of Agawa Bay, Algoma Distr. (*Williams & Cain 5661, 5310!*). MANITOBA. Seal R., W. of Great I., 59°03' N., 96°47' W. (*Ritchie, 1850!*).

MAINE. Davis Pond and vic., northwest Basin; Chimney Pond, above Basin Ponds; Hamlin Peak and base of Hamlin Ridge, Northwest Basin, all on Mt. Katahdin (*RMS*

---

[270] *Gymnocolea marginata* (Steph.) Hatt., however, may be distinct in the slightly larger cells; the median cells in the type (G) are 25–34 × 25–36 μ. Cell size is so variable, however, that this may prove an inadequate basis for retaining the species; see Schuster (1965, p. 274). Kitagawa (1966, p. 113) considers *G. marginata* a valid species and would place *G. montana* under it as a simple synonym.

32910, 17032a, 15983, 15978, 32935a, 17004, etc.); White Cap Mt., Rumford; Grafton, Oxford Co. (*Adams, 1937*); Mt. Desert I. (*Lorenz*). NEW HAMPSHIRE. Hermit L., Mt. Washington (Underwood & Cook, Hep. Amer. no. 75); Lake of the Clouds, Mt. Washington, Coos Co. (*RMS, s.n.*); Mt. Pleasant (Lorenz, 1904); Mt. Monadnock (*RMS 19451, c. caps.*); Eagle L., Mt. Lafayette (*RMS, s.n.*) and the Flume, Franconia Mts. (Lorenz, 1908c). VERMONT. Mt. Mansfield, along Long Trail, S. of summit, 4000 ft (*RMS 43805a, 43813, 43869a*). MASSACHUSETTS. Hadley, on N. slope of Holyoke Range (*RMS, s.n.*); Sandwich; Still River; Harvard. CONNECTICUT. Bolton, Tolland Co.; Branford, Naugatuck, New Haven Co. (Evans & Nichols, 1908).

NEW YORK. Fall Cr., Ithaca and Taughannock Falls, Tompkins Co.; Letchworth Park, Genesee R., Wyoming Co. (all Schuster, 1949a); Ft. Edward, near L. George (*Burnham*). PENNSYLVANIA. Gulf Mills (*Kaiser, 1910*). NEW JERSEY. Long Branch (Haynes, 1906b, p. 75, in bog with *Pallavicinia*); "low sandy barrens" (Austin, Hep. Bor.-Amer. Exsic. no. 34, 1873); Batsto, Burlington Co., and Atsion, Atlantic Co. (Britton 1889).

Southward occurring as a disjunct in NORTH CAROLINA. Cascades, near Hanging Rock State Park, Stokes Co. (*RMS 28260*); Flat Rock, Blue Ridge Pkwy., near Linville (*RMS 30166, 30171*); Devils Courthouse, on Blue Ridge Pkwy., Haywood Co., 5600 ft, southernmost montane locality in North America (*RMS 39310, 61235, 45310*). Also, totally out of the "normal" range in the Coastal Plain, as follows: swamp near Little R., S. of Vass, Moore Co. (*Blomquist 7235!*); edge of Angola Bay, on Hwy. 53, Pender Co. (*Blomquist 10103!*); White L., Bladen Co. (*Blomquist 7114!*). TENNESSEE. Near Alum Cave, Mt. LeConte, Sevier Co. (Sharp, 1939; *Sharp 5562!*; Schuster, 1951); SE. slopes below Myrtle Pt., Mt. LeConte, 5900–6000 ft (*RMS 45375c*); at Charlies Bunion, Appalachian Trail, Smoky Mt. Natl. Park (*RMS, s.n.*); Bledsoe Co. (Clebsch, 1954).

Westward occurring as follows: MICHIGAN. Sandstone bluffs, just E. of Miners Castle, Pictured Rocks, Alger Co. (*RMS, s.n.*). MINNESOTA. Cook Co.: Little Susie (Oley) I., Porcupine I., Belle Rose I., Big Susie I., all in Susie Isls., near Grand Portage (*RMS 14863, 14660, 12246, 13606, 5179b, 13027,* etc.); Gunflint L. on Gunflint Trail, Cook Co. (*RMS 13408*). ARKANSAS. Camp Bard, S. of Mena, Ouachita Mts., Polk Co. (*Anderson 11421!*).

The isolated stations from the warm Coastal Plain of the Southeast (North Carolina) are totally anomalous. Careful study of the material reveals no structural discrepancies.

Ecology

Except northward, where usually helophytic, most frequently epipetric or associated with rock outcrops, virtually without exception on non-calcareous rocks. Usually a pioneer on wet to moist, often rather strongly insolated acidic rocks, either on cliff crests or seepage-moist talus, or over peaty soil around rock pools lying in partial or full sun, forming blackish or brownish black mats. Associated in both types of sites very commonly with other oxylophytes such as *Cephalozia bicuspidata, Cephaloziella byssacea, Lophozia alpestris,* and *L. kunzeana,* often forming a distinct *Gymnocolea-Cephalozia bicuspidata* or *Gymnocolea-Cephaloziella byssacea* community (Schuster, 1957, p. 263) characterized by (*a*) a high level of exposure; (*b*) intermittently high moisture, due to either seepage or spray; (*c*) strongly acidic substrate. The measured pH in such sites ranges from 3.7 to

4.8, rarely to 5.2. Occasionally (south to North Carolina) on intermittently moist cliff faces, with *Marsupella sphacelata media, Andreaea rupestris, Scapania nemorosa*. More frequent in or near the spray zone, or in the zone of periodic submersion, in cascading streams, or around water falls, in the New England mountains. Here associated with *Cephalozia bicuspidata, Marsupella sphacelata* (particularly the equally scorched appearing var. *media*), *Scapania undulata*, etc. Allied to such occurrences are those at the margins of rock pools or in the spillways of streams running over granite, usually in full sun. The plant occurs here as a black extreme, together with *Odontoschisma elongatum, Lophozia alpestris, L. wenzelii, Cephalozia bicuspidata*, occasionally also *Mylia taylori, Ptilidium ciliare, Chandonanthus setiformis* (see Schuster, 1951).

Further southward (Mt. LeConte, Tenn.) also over seepage-moistened, insolated rocks, with *Scapania nemorosa, Lophozia attenuata, Sphagnum, Harpanthus scutatus*, or (Flat Rock, N.C.) in shallow, intermittently wet depressions of granitic exposures, associated with *Selaginella tortipila, Lophozia attenuata, Anastrophyllum michauxii*.[271]

The occurrences in the North Carolina Coastal Plain are quite without parallel: there found on intermittently moist banks (as of the Tomahawk R.) and on decaying logs in swamps.

In addition to the preceding occurrences, mostly characterized by the proximity of exposed rock, *G. inflata* is found in peat bogs, particularly in those developing in depressions in sand dunes or sand plains, often being common in such moors. Here it may occur under conditions closely paralleling those typical for the superficially similar *Cladopodiella fluitans*. Associated then are various members of the *Mylia-Cladopodiella* Associule, chief among them *Lophozia capitata* (or *marchica*), *Cephalozia connivens* or *loitlesbergeri, Mylia anomala, Calypogeia sphagnicola, Cephaloziella elachista*, etc. On peaty ground, on the loamy-peaty margins of bogs, the species is also frequently associated, northeastward, with *Cladopodiella francisci* and *Nardia scalaris*, less often with *Scapania irrigua*. Similarly, the species occurs commonly in the mountains along the peaty to sandy or gravelly margins of ponds or lakes (e.g., those lying in the glacial cirques of Mt. Katahdin, or at Eagle Lake, Mt. Lafayette, N.H.); associated here are most often *Pellia epiphylla, Nardia insecta* and *geoscyphus, Sphagnum, Scapania irrigua, S. paludicola, Odontoschisma elongatum*.

A review of the preceding diverse occurrences reveals three obvious characteristics in common which distinguish these various sites: (*a*)

[271] It should be emphasized that the European authors uniformly report *G. inflata* to be largely confined to "wet moors and heathy places" or to "moorigen Boden, auf nacktem Torf, sandiger Erde und in Moorlöchern" (Macvicar, 1926; Müller, 1954). Occurrences on moist rocks are "seldom" or "sehr selten." By contrast, in North America the plant is characterized by very frequent occurrences on mineral substrates and on rocks.

high light intensity, usually much direct sunlight; (*b*) a (at least intermittent) high level of moisture, and often periodic submersion; (*c*) very acid substrate. Often *G. inflata* occurs in areas subject to alternatingly very dry and wet conditions, as on exposed cliff faces.

Differentiation

Gymnocolea inflata is not likely to be mistaken for other species, except for the intimately allied *G. acutiloba* (discussed under the latter), and the often very similar *Cladopodiella fluitans*. It is usually distinct because of the tendency toward strong fuscous pigmentation, the obliquely inserted leaves with two obtuse or rounded lobes, and the noncollenchymatous leaf cells, bearing finely granular or papillose oil-bodies. These features are shared with *Cladopodiella fluitans*. Since helophytic occurrences of *Gymnocolea* may be virtually identical with those of *Cladopodiella*, confusion of these taxa is possible. The likelihood of confusion is accentuated because in *Gymnocolea* older plants may, at least under certain conditions, produce numerous postical-intercalary branches, strongly suggesting *Cladopodiella* with its postical branching.

Sterile plants of *Gymnocolea* differ from those of *Cladopodiella fluitans*, however, in the smaller, more subisodiametric cells (those of *Cladopodiella* $\pm$ elongate and ca. 35–42 μ in the leaf middle); the mostly subequally bilobed leaves with a relatively open sinus (in *Cladopodiella* the leaves often with antical lobe considerably smaller, the sinus often narrow or almost closed); the much more sporadic and smaller underleaves; the pronounced differences in stem anatomy (compare Fig. 250:14–15, Fig. 249:2, 4), in particular the less elongated cortical cells of *Gymnocolea;* the occurrence of at least a large number of terminal-lateral branches; and the usually thick-walled leaf cells (those of *Cladopodiella* usually leptodermous). When fertile, of course, the diagnostic inflated perianths of *Gymnocolea*, situated terminally, at once separate it from *Cladopodiella* (in which the trigonous perianths are on short postical branches).

Small, scorched forms of *Marsupella sphacelata media* (*M. sullivantii*) often bear, on juvenile shoots, distinctly oblique leaves that are closely similar in form to those of *Gymnocolea*. Such plants may be confused with *Gymnocolea*. Mature, plants, with the transverse, loosely complicate leaves, are usually associated and serve to readily separate the *Marsupella;* the distinct trigones and consistent presence of only 2–3 oil-bodies per cell, as well as the different stem structure also serve to distinguish the *Marsupella* from all forms of *Gymnocolea*.

Variation

Concurrently with its wide geographical distribution and almost equally extensive ecological range, *G. inflata* is a highly polymorphic species. Schiffner discusses its polymorphism in exhaustive detail, issuing a long series of supporting specimens in his Hep. Eur. Exsic.

Although *G. inflata* typically has blunt to rounded leaf lobes, phases with acute to subacute leaf lobes are frequent, and I am not certain as to how to keep these distinct from *G. acutiloba*. Probably the best solution is to follow Arnell (1956, p. 94) and consider the latter a mere variety of *G. inflata*.

Possibly also deserving of varietal distinction are large arctic plants lacking sterile perianths (e.g., *RMS & KD 66–105a*, Tupilak I., west Greenland), which are quite different from the smaller, scorched-appearing "normal" arctic forms of the species; they are perhaps twice as large (shoots to 2.6 mm wide with leaves or more) and intensely brown rather than fuscous, with horizontally patent leaves that are often slightly adaxially convex, or flat, rather than concave. Sterile perianths are, apparently, never produced, but fertile ones are common. These disagree with the usual diagnoses (as, e.g., in Müller, 1951–58) in being strongly 5–6-plicate in the distal ⅓ or ¼, in being terete below rather than obovoid, and in the decolorate, hyaline, lobulate-laciniate mouth with longer than normal cilia and teeth. Many leaves of such plants may possess fewer oil-bodies, often 1–5 per cell, with some cells having 1–2 very large oil-bodies.[272]

In addition to purely environmentally induced variation, there is a putatively genetic variant ("var. *heterostipa*"), as well as the closely allied "species" *G. acutiloba*. The latter, here reluctantly retained, differs essentially in the acute-lobed leaves, often bearing 1-several obscure accessory marginal teeth. It is very possible that this is only an extreme occurring under relatively xeric conditions directly over rocks.

Typical *G. inflata* and the var. *heterostipa* are separable as follows:

1. Branches few, arising as innovations beneath the perianths and associated with bilobed lateral leaves (rarely with supporting leaves unlobed); postical-intercalary branches absent or rare . . . . Typical *G. inflata*
1. Branches of mature and old stems numerous, arising in large part as postical-intercalary innovations, partly laterally, the subtending or branch leaf almost always unlobed . . . . *G. inflata* var. *heterostipa*

GYMNOCOLEA INFLATA var. HETEROSTIPA
(Carr. & Spr.) K. Müll.
[Fig. 250:10–12]

Cephalozia heterostipa Carr. & Spr., *in* Spruce, On *Cephalozia*, p. 55, 1882.
Jungermannia inflata var. heterostipa Lindb. & Arn., Kgl. Sv. Vetensk.-Akad. Handl. 23(5):47, 1889.
Gymnocolea inflata var. heterostipa K. Müll., Rabenh. Krypt.-Fl. 6(1):743, 1910; Schuster, Amer. Midl. Nat. 49(2):386, pl. 29, figs. 1–3, 1953.

[272] Such robust plants, reminiscent of *Leiocolea* rather than *Gymnocolea*, also diverge in occurring, at least at times, under conditions where there is basic seepage. *RMS & KD 66–105a*, e.g., occurred on thin, seepage-wet soil along a rill, together with such Ca-tolerant or demanding taxa as *Aneura pinguis* and *Tritomaria* (*Saccobasis*) *polita polymorpha* and the moss *Drepanocladus*.

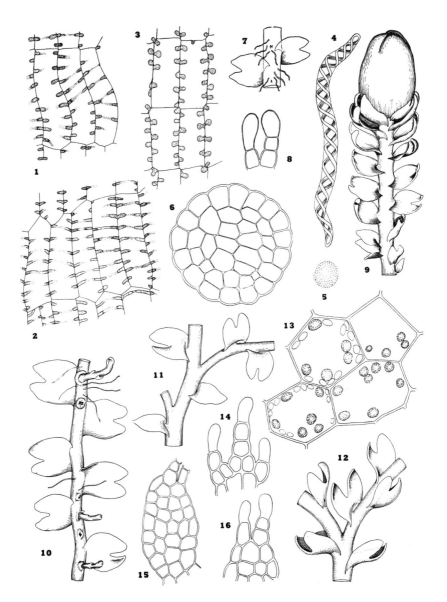

Fig. 250. *Gymnocolea inflata.* (1–2) Cells, inner layer of capsule wall (×435). (3) Epidermal cells, capsule wall (×365). (4–5) Spore and elater (×520). (6) Seta cross section (×132). (7) Shoot sector, ventral aspect, with vestigial underleaves (×ca. 18). (8) Underleaf (×ca. 275). (9) Shoot with potentially caducous perianth (×ca. 16). (10) Old stem, ventral view, with ventral-intercalary innovations (×20). (11) Shoot sector, antical view, with two terminal branches and unlobed half leaves (×23). (12) Shoot sector, with terminal branch and bilobed "half leaf" (×23). (13) Median cells with oil-bodies and, at left, chloroplasts (×535). (14–16) Underleaves, that of fig. 15 of maximal size (×265). [1–5, Mt. Monadnock, N.H., *RMS*; 6, *RMS 15987*, Davis Pond, Mt. Katahdin, Me.; 7–9, Taughannock, Ithaca, N.Y., *RMS*; 10–16, *RMS 13606*, Porcupine I., Susie Isls., Minn. (of the dubious var. *heterostipa*).]

Differing from the typical phases of the species in the free branching, in large part ventral and intercalary in origin (the branches clearly smaller, surrounded at base by a distinct ring, indicative of their endogenous origin). Lateral branches freely developed, usually replacing the ventral half of the supporting leaf, which is thus ovate to ovate-lanceolate and unlobed; occasional lateral branches with subtending leaf bilobed. Rhizoids usually rather frequent. Underleaves often more distinct.

Type. Glyders, North Wales, England (*Holmes, 1876*).

Distribution

Of sporadic occurrence, possibly throughout the range of the species. Known from a few localities in Europe (England; also Norway, *fide* Arnell, 1956) and from Washington in the western United States. In eastern North America twice reported:

E. GREENLAND. Danmarks Ø, Scoresby Sund (Jensen, 1898, p. 381). MINNESOTA. Porcupine I., Susie Isls., near Grand Portage, Cook Co. (*RMS 13606*).

Although usually considered to occur in *Sphagnum* bogs, the Minnesota plants were from shaded, wet cold rocks in a run-off cut through northwest-facing cliffs, closely juxtaposed to Lake Superior; associated were *Scapania nemorosa* and *undulata* and *Cephalozia bicuspidata*.

The var. *heterostipa* is, in my opinion, of dubious validity. The production of numerous postical-intercalary branches may be environmentally induced. More typical phases of *G. inflata* also often possess lateral branches arising from "above" unlobed lateral leaves.

GYMNOCOLEA ACUTILOBA (*Schiffn.*) K. Müll.[273]
[Fig. 251:1–13]

Jungermannia acutiloba Kaal., Nyt Mag. Naturvid. 40:250, 1902. [Not *J. acutiloba* Hook. f. & Tayl., Lond. Jour. Bot. 4:90, 1845.]
Lophozia acutiloba Schiffn., Hedwigia 48:187, figs. 1–6, 1909.
Pleuroclada acutiloba Steph., Bot. Centralbl. 110:317, 1909.
Gymnocolea acutiloba K. Müll., Rabenh. Krypt.-Fl. 6(1):745, fig. 334, 1910; Macvicar, Studs. Hdb. Brit. Hep. ed. 2:163, cum fig., 1926; K. Müller, Rabenh. Krypt.-Fl. ed. 3, 6(1):709, fig. 226, 1954.
Gymnocolea inflata var. *acutiloba* Arn., Illus. Moss Fl. Fennosc. 1:94, 1956.

Prostrate, in fuscous brown to *black or blackish green* (in shade rarely deep green) intricate patches, *on rocks*. Shoots 1–1.5 cm long × *650–1500 μ wide*. Stems 100–180 μ in diam., slender, sparsely branched; branches terminal, from lower half of acroscopic end of segment (thus *with an ovate-lanceolate, unlobed leaf associated with branch*), very rarely from axil of bilobed leaf, only occasionally intercalary and postical. Leaves *distant* to weakly imbricate, *subvertical*, little oblique in insertion, erect-spreading (in small xeromorphic forms) to widely spreading (in larger, laxer forms), *nearly flat*, but the lobe apices sometimes incurved, *subquadrate to ovate-quadrate*, from 320 μ wide × 370 μ long

---

[273] *Jungermannia acutiloba* Kaal. being a later homonym, this taxon, if recognized as a species, would have to be regarded as established by Schiffner (1909c).

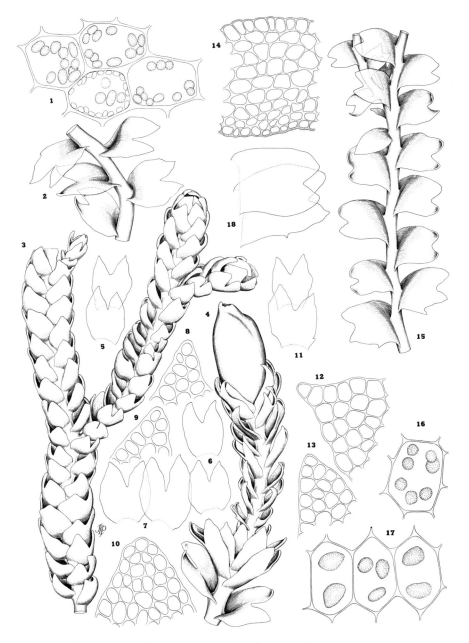

FIG. 251. *Gymnocolea acutiloba* (1–13) and *G. inflata* (14–18). (1) Median cells with oil-bodies and, lower cell, chloroplasts (×555). (2) Shoot sector (×24). (3) ♂ Plant with two *Frullania*-type branches and (upper right) lateral intercalary branch from axil of ♂ bract, with androecial apex destroyed; lower supporting stem leaf bifid, upper unlobed (×24). (4) Gynoecial shoot, with old gynoecium (which has shed perianth) and new

(brown xeromorphic forms) up to 500–560 μ wide × 660–700 μ long (green shade forms), narrowed basally, ⅓, occasionally to ½, bilobed, exceptional leaves trilobed, the *sinus narrow, obtuse to acute; lobes somewhat divergent*, narrowly ovate to triangular, *acute or subacute* (terminated by 1, *often 2, superposed cells*), rarely the very tip obtuse; lateral margins at or below middle *often with a small tooth or angulation*, particularly on antical (but often also on postical) margin, more occasionally some leaves 2–3-dentate. Cells with walls usually more or less brownish, somewhat to strongly *equally thickened;* trigones minute; cuticle smooth to weakly verruculose; apical and marginal cells subquadrate, 21–24 μ; median cells almost isodiametric, 21–24 × 22–28(32) μ; oil-bodies in median cells varying from (3–4)5 *to 10, occasionally 12, per cell,* spherical and (4)5–6(7) μ to ovoid and 4–5 × 5–8(9) μ, *appearing almost homogeneous* (the spherules exceedingly inconspicuous, nonprotruding); chloroplasts ca. 3.5 μ. Underleaves absent or minute, small and subulate or formed of 2 cilia ending in slime papillae. Asexual reproduction *absent* (caducous perianths not developed), or rarely present, *via* gemmae; gemmae on the lobe apices, *dark brown,* 3–4–5-angulate, 2-celled, ca. 14–18 μ.

Dioecious, *almost invariably sterile.* ♀ Bracts *larger than leaves*, otherwise identical (but more often 3-lobed, usually with rounded lobes). Perianth rare, inflated, broadly clavate, eplicate, the lobulate mouth with short teeth.

Type. Hardanger, Norway (*J. Havaas, 1898*); the type illustrated by Schiffner (1909c, figs. 1–6, p. 189).

Distribution

A rare, poorly understood taxon, possibly merely a variety of the polymorphous, more ubiquitous *G. inflata.* Known in Europe, from the Alps in Germany, Austria (Vorarlberg; Salzburg; South Tyrol; Ortlaer region), and again in Norway (type) and middle and northern Sweden; also in Wales, England, and Asia Minor. Previously reported a single time from eastern North America:

W. GREENLAND. Godhavn, Disko I. (Persson, 1942a). MAINE. Summit of Cathedral Trail, at shaded base of large boulders, Mt. Katahdin, ca. 5000–5200 ft (*RMS 32994a*, 1954).

Plants from Labrador and those from Charlies Bunion, The Jumpoff on Mt. Kephart, and from Mt. LeConte, Tenn.,[274] closely approach *G.*

---

[274] Tennessee: Sevier Co.: Charlies Bunion, 5300–5375 ft, Appalachian Trail NE. of Mt. Kephart (*RMS 45308, 45307c*); Mt. Kephart, 5950–6100 ft, at The Jumpoff (*RMS 45384a*).

---

perianth-bearing innovation (×24). (5–7) Leaves (×24). (8–10) Leaf lobe apices (×210). (11) Leaves (×24). (12–13) Leaf lobe apices (×210). (14) Part of stem cross section (×180). (15) Lax sterile shoot with *Frullania*-type branch and ovate, unlobed supporting leaf (×11). (16–17) Leaf cells with oil-bodies (×715). (18) Two leaves (×22). [2–7, 11, 18, all drawn to scale of fig. 2; 8–10, 12–13, all to scale of fig. 12. 1–2, 11–13, *RMS 32994a*, Mt. Katahdin, Me.; 3–10, *RMS 45304*, Charlies Bunion, Tenn.; 14–18, *RMS 45791*, NW. Greenland.]

acutiloba in most respects and should probably be referred to it. The Tennessee plants are associated with the "copper hepatic," *Cephaloziella massalongoi*, but analysis of the substrate revealed mere traces of Cu.

The species has recently been found to be widespread in Cu-containing sites on Latouche I., Alaska (Shacklette, 1961), where it forms a "*Gymnocolea acutiloba* community" characterized by its dark brown to almost black color, and the "almost total absence of any plant except the . . . *Gymnocolea* . . . [which formed a] liverwort peat . . ." for a depth up to 8–10 in.

The very limited distribution perhaps reflects a rather general restriction to Cu-containing substrates, where, Müller (1951–58, pp. 280–281) states, it may be associated with *Cephaloziella massalongoi* and *C. phyllacantha*. Schatz (1955) mentions the same association and speculates that "copper ores are the principal or sole substrate" for these species. In this conjunction, see, however, the discussion under *Cephaloziella massalongoi*.

Ecology

Restricted, apparently, largely to igneous rocks (on which it either is a pioneer or comes in soon after the pioneer species; this is clear from the single local collections where the plants occur directly over rock faces admixed with *Gymnomitrion concinnatum* and reduced forms of *Lophozia alpestris* and *Chandonanthus setiformis*). Also on moist slate and other non-calcareous metamorphic rocks, or over the detritus from copper mines. Unlike the closely related *G. inflata*, the species never occurs over peat.

Differentiation

Although exceedingly closely allied to *G. inflata* (and not considered specifically distinct from this by Arnell), *G. acutiloba* differs in the acute or subacute leaf lobes, and supposedly in the virtual restriction of branches to the lower half of the acroscopic ends of the merophytes (with the branch subtended, consequently by a leaf of which only the dorsal half has developed),[275] and in the occasional development of teeth of the leaves. *Gymnocolea acutiloba* also has, on an average, somewhat smaller leaf cells. Even smaller celled is *G. andina* of Peru (Buchloh, 1961).

The plant has been previously described only from the mod. *colorata-parvifolia-pachyderma*, in which the shoots rarely attain a width of 1 mm. Such plants are admittedly often difficult to distinguish from similar modifications of *G. inflata*. However, the material from Maine from a shaded site represents largely a mod. *viridis-megafolia-lepto- vel mesoderma*. It is more robust and has larger, laxer leaves (the maximal sizes given in the description are based on this; the minimal on European plants of the "var. *heterostipoides*"). These plants show that *G. acutiloba* cannot be dismissed as merely a small-leaved modification

---

[275] I do not consider the supposed distinction in form of the supporting leaf useful in separating *G. acutiloba* from *G. inflata*.

resulting from a "difficult" environment. They display uniformly lateral branching, the branches always "subtended" by an ovate-lanceolate, entire leaf. The leaves on many plants show a high incidence of lobes ending in 2 superposed cells, a feature which also characterizes the type material, judging from Müller (1951–58, fig. 226c).

A feature of *G. acutiloba* which has not been sufficiently emphasized is the tendency for the leaves to develop small, usually obtuse marginal teeth. This is noted by Müller (1905–16, p. 746) and Macvicar (1926, p. 164) for the antical margin. However, both margins may be dentate, and this dentition is particularly well marked in robust plants that I collected in Maine (Fig. 251). Similar dentition does not appear to occur in "typical" *G. inflata*.

MESOPTYCHIA (*Lindb. & Arn.*) *Evans*
[Figs. 252–253]

Jungermannia (sect. *Mesoptychia*) Lindb. & Arn., Kgl. Sv. Vetensk.-Akad. Handl. 23(5):39, 1889.
Lophozia Steph., Spec. Hep. 2:132, 1901 (in small part, only as to *M. sahlbergii*).
Mesoptychia Evans, Ottawa Nat. 17:15, 1903; K. Müller, Rabenh. Krypt.-Fl. 6(1):716, fig. 362, 1910; Frye & Clark, Univ. Wash. Publ. Biol. 6(2):305, 1944; Schuster, Amer. Midl. Nat. 45(1):78, pls. 27–28, 1951.

Robust, large, 3.5–4.5 mm wide or more, decumbent to suberect in growth, sometimes caespitose. Stem 15–16 or more cells high in cross section; ventral medullary region composed of somewhat smaller cells, in 6–9 layers, merely 18–24(27) μ in diam.; dorsal medullary cells ca. 24–32(38) μ; ventral portion of medulla not mycorrhizal, gradually grading into the dorsal region; cortical cells little thick-walled, elongate, 3–5 × as long as wide, 18–20 μ wide on an average dorsally. Rhizoids rather numerous, scattered. *Ventral merophytes broad,* occupying ⅕–¼ of the stem perimeter, developing large, *bifid, pluriciliate underleaves.* Lateral leaves uniformly, *shallowly bilobed,* much broader than long, the width 1.2–1.3 × or more the length of the ventral lobe; sinus obtusely angulate to rounded, *descending for about* ¹⁄₁₀–⅛ *the leaf length,* dividing the leaf into a large ventral and a much smaller dorsal lobe; ventral lobe broad, the ventral half of the leaf as a whole ovate, with the apex rounded to obtuse, flat to slightly concave adaxially; dorsal "lobe" represented by a mere small, acute tooth on what appears to be otherwise an entire, obliquely ovate-triangular leaf, terminated generally by 1–2 or more cells forming an apiculum; dorsal half of leaf strongly adaxially convex, forming an oblique convex dorsal fold, the antical leaf margin thus moderately to strongly deflexed. Leaf insertion *extremely oblique,* the lower portions somewhat arched, the upper half or more nearly horizontal, more or less obliquely decurrent on the lateral merophytes. Leaf cells thin-walled, distinctly to strongly collenchymatous, the marginal cells averaging 25 μ

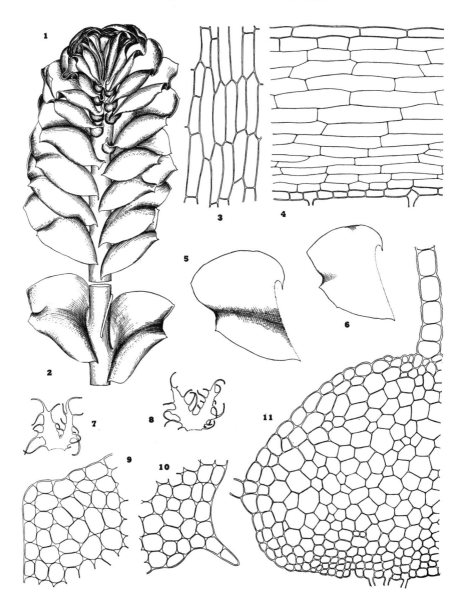

Fig. 252. *Mesoptychia sahlbergii.* (1) Androecial shoot apex (×9.5). (2) Shoot sector, dorsal view (×9.5). (3) Dorsal cortical stem cells (×165). (4) Longisection of stem (×132). (5–6) Leaves (×11). (7–8) Underleaves (×16.5). (9) Ventral lobe apex (×170). (10) Dorsal lobe apex (×170). (11) Stem cross section, intercepting two leaf bases (×155). [All from plants from Lena R., Siberia (ex Stockholm).]

in the lobes; median cells of lobes 23–32 μ (averaging 28 μ); cuticle rather weakly verruculose; oil-bodies undescribed. *Gemmae unknown.*

Dioecious. ♀ Bracts like leaves, except for somewhat larger size and more obvious apiculate lobes; bracteole similar to the underleaves but larger and less deeply lobed and less ciliate.[276] Perianth well-developed, occurring at the summit of a *well-developed rhizoidous perigynium, at right angles to the axis of the gametophyte*, its height (before maturity) equal to that of the fleshy perigynium;[277] perianth mouth shortly beaked and suddenly constricted, the beak formed of short cilia 3–4 cells long. ♂ Plants like sterile plants, except for the androecial regions. Androecia formed of 4 or more pairs of bracts, at first terminal but becoming intercalary by subsequent terminal growth. Bracts similar to leaves but the antical half of the leaf base strongly ventricose, and the antical base produced into a distinct, often dentate or ciliate, small lobe; antheridia therefore limited to the axils of the dorsal half of the bracts, which are modified, while the rest of the bracts are quite similar to leaves. Antheridia generally 2, rarely 3, per bract, ivory white;[278] antheridial *stalk biseriate*.

Type. Jungermannia sahlbergii Lindb. & Arn. [= *Mesoptychia sahlbergii* (Lindb. & Arn.) Evans]. The genus is monotypic.

The genus *Mesoptychia* represents an extremely isolated element in the Lophoziaceae whose affinities are correspondingly obscure. The presence of a well-developed perigynium occurring at right angles to the stem is the most important generic character (although not even mentioned in the diagnosis in Frye and Clark, *loc. cit.*, p. 305): such a perigynium does not recur again in the family. *Mesoptychia* has, at times, been placed near

[276] Frye and Clark (1937–47, p. 305) describe the bracteole as "similar to the leaves but larger" and again as "similar to the bracts, larger"—apparently in error. The bracteole is usually larger than the underleaves below it on the shoot, but never becomes larger than the lateral leaves or the associated bracts. The inferior size of the bracteole appears to be a constant feature of all anisophyllous Jungermanniales.

[277] As far as I am aware, just the immature perianth is known. The only illustration in the recent literature of the ♀ plant is pl. 28, fig. 6, in Schuster, 1951a, based on Evans (1903). Fertile ♀ plants were not present in material that I studied. The immature perianth is described as ± laterally compressed, slightly so ventrally and there with an obtuse, rounded fold, more distinctly so dorsally and there with a sharper fold or blunt carina.

I wonder whether the lateral compression of the perianth may not be due to drying of the plants under pressure. If the species has a laterally compressed perianth normally, with dorsal and ventral keels, the genus should perhaps be removed from the Lophoziaceae. Since the bulk of the other characters, however, clearly ally the genus to the Lophoziaceae rather than the Plagiochilaceae, I tentatively retain it in its present position. Inoue (1966) would place *Mesoptychia* in the Marsupidiaceae Schust. (Schuster, 1963, 1967b = Adelanthaceae subf. Marsupidioideae); such a disposition is, I think, untenable.

[278] Frye and Clark (*loc. cit.*, p. 305) state "antheridia up to 16." I have never seen more than 3 antheridia per bract, and to my knowledge none of the Jungermanniales ever has as many as 16 antheridia per bract.

Acrobolbus, since both genera possess ventrally bulging perigynia, or marsupia. However, *Acrobolbus* belongs in a separate family, the Acrobolbaceae (for which see Vol. III). *Mesoptychia* differs from *Acrobolbus* (and almost all other North American genera of Jungermanniales, except the unrelated *Arnellia*) in retaining a well-developed perianth at the summit of the pendent, rhizoidous perigynium. The genus differs further from *Acrobolbus* in the very well-developed underleaves produced by the broad ventral merophytes of the stem; in the more robust stems, over 12 cell layers high; in the extremely shallowly bilobed leaves; and in the development of a "free" calyptra. Study of the sporophyte and of the living cell will undoubtedly disclose other important differentiating characters, but these structures were not available to me for comparative study. *Mesoptychia sahlbergii*, the only known species, differs from *Acrobolbus ciliatus* (the only North American species of *Acrobolbus*) also in the inability to develop rhizoids from the marginal leaf cells, as well as in the much greater robustness.

Frye and Clark placed *Mesoptychia* in their "tribe" Nardioideae, together with such totally unrelated elements as *Mylia* and *Pedinophyllum*. The genus, however, has many distinct Lophozioid features, as recognized by Lindberg and Arnell (1889), Evans (1903), and Müller (1939–40); the last author formally put *Mesoptychia* in the family Lophoziaceae. The more obvious Lophozioid features are the uniformly, though shallowly, bilobed leaves, as well as the broad ventral stem sectors that develop large, bifid, pluriciliate underleaves on sterile as well as fertile shoots; in addition, the irregularly scattered rhizoid initials, the large size of the ♂ bracts (which are subequal in size to normal vegetative leaves), and the eventual intercalary position of the androecium (e.g., lack of specialized form and occurrence of the androecia) indicate that the genus is related to the Lophoziaceae. Leaf cells are also obviously collenchymatous, generally forming strongly bulging trigones, a feature typical of (though not peculiar to) the Lophoziaceae. Finally, the facies of sterile plants is obviously "Lophozioid," and quite unlike that of the Plagiochilaceae or Jungermanniaceae (e.g., the "Nardioideae" of Frye and Clark). All the Jungermannioid or "Nardioid" genera agree in having minute or no underleaves (never bilobed and ciliate), narrow ventral stem sectors, and entire, occasionally retuse leaves. Frye and Clark (*loc. cit.*, p. 267) state that their Nardioid group includes genera with "essentially unlobed alternate leaves." *Mesoptychia* would scarcely agree with this diagnosis. Frye and Clark further appear to have quite misunderstood the genus, since they separate it (in their key, p. 268) from *Pedinophyllum* on the basis of the supposedly "transverse to little succubous" leaves of *Mesoptychia* vs. the "quite succubous" leaves of *Pedinophyllum*. The leaves in *Mesoptychia* are extremely obliquely inserted, except for the arched lower part of the line of insertion (Schuster, 1951a, pls. 27:1–2; 28:2). The sterile shoot, both in leaf form and in the very oblique line of insertion, is strongly reminiscent of *Lophozia* subgenera *Leiocolea* and

Barbilophozia. The resemblance in leaf form between *L.* (*Leiocolea*) *rutheana* and *Mesoptychia* is particularly striking.[279]

It seems probable, therefore, that *Mesoptychia* can be derived from an unspecialized *Orthocaulis*-like type with large underleaves, which had developed the following *Leiocolea*-like features: very oblique, subhorizontally inserted, uniformly and shallowly bilobed leaves. The essential derivation from an unspecialized type of this kind would involve the evolution of the perigynium, at right angles to the axis of the stem. This perigynium development has, as is well known, occurred repeatedly in such families as Arnelliaceae, Gymnomitriaceae, and Jungermanniaceae—entirely homoplastically. It is probable that the development of the perigynium in *Mesoptychia* has proceeded quite independently.

The perigynium of *Mesoptychia* bears many similarities to that of *Arnellia.* In both genera the perigynia are at right angles to the axis, becoming subterranean and densely rhizoidous, and in both they are surmounted by a distinct perianth. Also, they are similar in that there not only is development of a perigynium by elaboration of the stem tissues lying well below the gynoecium, but the receptacular tissue underlying the archegonia also proliferates, resulting in a calyptral perigynium as well. This is clearly evident from the fact that the archegonia "appear to be carried up onto the calyptra" during development of the sporophyte. Actually what happens is that the calyptra proper (i.e., the old archegonial wall of the fertilized archegonium) grows to only a limited degree, the lower portions of the "calyptra" being formed by proliferation of the ring of tissue on which the unfertilized archegonia, lying peripheral to the fertilized archegonium, are situated. The "calyptra" in these cases, as also the perigynium below it, clearly consists largely of tissue derived from the shoot apex.

MESOPTYCHIA SAHLBERGII
(*Lindb. & Arn.*) *Evans*
[Figs. 252–253]

Jungermannia (sect. *Mesoptychia*) *sahlbergii* Lindb. & Arn., Kgl. Sv. Vetensk.-Akad. Handl. 23(5):40, 1889.
Lophozia sahlbergii Steph., Spec. Hep. 2:132, 1901.
Mesoptychia sahlbergii Evans, Ottawa Nat. 17:15, pl. 1, 1903; K. Müller, Rabenh. Krypt.-Fl. 6(1):716, fig. 362, IV, 1910; Schuster, Amer. Midl. Nat. 45(1):81, pls. 27–28, 1951.

Very robust, 3.5–4.5(5) mm wide, brownish green to brownish purple; *even in otherwise green phases the postical faces of the stem, the rhizoids, and the underleaves generally violet or purple.* Stems 4–6 cm long, loosely decumbent. *Leaves much wider than long, the postical half almost perfectly ovate and abruptly to obtusely pointed,* separated by a *very shallow sinus* from the shorter antical half of the leaf; antical lobe commonly more *sharply apiculate or mucronate*, often with a sudden, sharp, 2-celled tooth, from an obtuse base. Cells usually with somewhat bulging trigones; the marginal ca. 25 μ (or up to 33 μ); the subapical 23–32 μ; the

---

[279] The immature perianth of *Mesoptychia* is described as beaked; if this is correct, the similarity to *Leiocolea* may prove more than coincidental.

median and basal up to 40 μ, nearly isodiametric; cuticle verruculose. *Underleaves very large, bifid for 0.7–0.8 their length, the lobes and bases strongly ciliate;* cilia commonly 8–10 cells long and uniseriate.

Reproductive characters as under generic diagnosis.

Lectotype. Mjelnitsa, Yenisey, Siberia, ca. 65°30′ N. (*J. Sahlberg, 1876*).

Distribution

Probably holarctic but usually only at very high latitudes (56–77° N.). So far unknown from the arctic extremes of Europe but known from both the Yenisey and Lena R. valleys in Siberia (Arnell, 1913; Persson, 1949). Material from the latter locality has been studied (Figs. 252–253).

In North America, locally abundant in the arctic, northern portions of Alaska (see map in Persson, 1949, p. 511; Persson, 1962, p. 9; Persson and Gjaerevoll, 1957); abundant and fertile materials from a number of other Alaskan localities, collected by W. C. Steere, have also been studied; found in the Yukon as well. In the eastern North American Arctic evidently very rare, unknown from the Greenland coast, and known only as follows.

ELLESMERE I. Goose Fiord, on S. coast, and Reindeer Cove, on W. coast, ca. 76°40′ N. (Simmons; ex Bryhn, 1906, p. 40). I have not seen these collections; they should be compared with *Lophozia rutheana*.

The distribution of *Mesoptychia* is somewhat similar to that of *Arnellia*, but the latter is a much more ubiquitous taxon. In both cases the range appears to radiate out from the unglaciated or imperfectly glaciated portions of the high Arctic. The unglaciated northern portions of Alaska and Yukon, the probably only locally glaciated coast of Ellesmere I., and the unglaciated northern portions of Siberia appear to be the only regions where *Mesoptychia* occurs.

Ecology

Occurring in loose mats over damp, calcareous soil and also over limestone. Associated species (in Alaska) are *Arnellia fennica, Campylium stellatum, Drepanocladus revolvens, Hylocomium splendens, Lophozia quadriloba,* and (in marshes) *Calliergon cordifolium* and *Cirriphyllum cirrosum.*

Persson (1949, p. 510) notes that the species is "confined to more or less calcareous substratum . . . (with) . . . *Arnellia fennica,* species of *Distichium, Encalypta,* and *Timmia,* etc." The species was diligently searched for in northern Ellesmere I. (1955) and Greenland (1960, 1966) but not found.

Differentiation

Mesoptychia sahlbergii is one of the largest and most conspicuous of the arctic Hepaticae. It can scarcely be confused with any other species,

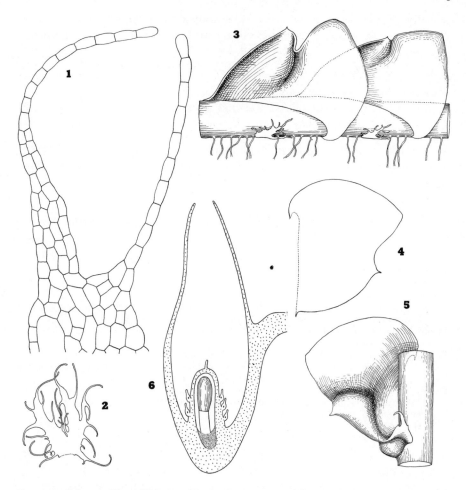

FIG. 253. *Mesoptychia sahlbergii*. (1) Underleaf lobe apex (×000). (2) Underleaf (×000). (3) Shoot sector, lateral aspect (×000). (4) Leaf (×000). (5) ♂ Bract (×000). (6) Longisection through young gynoecium (×ca. 000). [1–5, from plants from Lena R., Siberia (ex Stockholm); 6, after Evans, 1903.]

even when sterile. However, large phases of *Lophozia rutheana* are similar. These, however, lack the vinaceous pigmentation of the underleaves, have plurilaciniate and irregularly ciliate underleaves (in *Mesoptychia* they are regularly bifid, with the margins regularly ciliate), and have more distinctly and much less asymmetrically bilobed lateral leaves. Leaf cells in *L. rutheana* are also consistently larger in size.

M. sahlbergii has a relatively massive stem, 340–380 μ high or more (Fig. 252:4); cortical cells average 18–25 μ wide × 90–125(150–160) μ long, dorsally (1:4–7) and are slightly thick-walled, with often yellowish walls. Medullary cells exhibit only feeble dorsiventral differentiation; dorsal medullary cells average 24–33(38) μ in diam., while the 6–9 ventral strata of medullary cells are merely 18–24(27) μ in diam., but hardly form a distinct ventral band and are free of mycorrhizal infection. Several ventral medullary cell rows tend to be slightly thick-walled, with yellowish walls.

The distinctive leaves of *M. sahlbergii* average 2.5–2.65 mm wide × 2.0–2.15 mm long (to apex of the ventral, longer lobe); length to apex of the dorsal lobe is ca. 1.5–1.85 mm. The very large, bifid, ciliate underleaves average 800–900 μ long and are, most often, diagnostically purplish pigmented.

Family JUNGERMANNIACEAE *Dumort.*

Tribe Jungermanniaceae Dumort., Comm. Bot., p. 112, 1829 and Syll. Jungerm., pp. 24, 35, 1835.
Subfamily Nardiae Joerg., Bergens Mus. Skrifter 16:90, 1934.
Family Epigonianthaceae Buch, Evans & Verdoorn, Ann. Bryol. 10(1937):4, 1938 (in part, after exclusion of Lophoziaceae, Arnelliaceae, and miscellaneous genera).
Family Jungermanniaceae Evans, Bot. Rev. 5:91, 1939 (in part, after exclusion of Lophoziaceae and Arnelliaceae).
Family Nardiaceae K. Müll., Rabenh. Krypt.-Fl. 6 (Ergänzungs Bd.):184, 1939–40.
"Tribe" Nardioideae Frye & Clark, Univ. Wash. Publ. Biol. 6(2):267, (1943) 1944 (in part, with elimination of some genera of Plagiochilaceae and Lophoziaceae).
Family Jungermanniaceae auct., emend. K. Müll., *ibid.* ed. 3, 6(1):197, 1951; Schuster, Amer. Midl. Nat. 45(1):9, 1951 (notes on delimitation); Schuster, *ibid.* 49(2):386, 1953; Amakawa, Jour. Hattori Bot. Lab. no. 21:268, 1959.

Small to medium-sized (usually 0.5–4 mm wide × 8–50 mm long), usually rather chlorophyllose and opaque, varying from green (in shade) to blackish green, brownish, reddish brown, purplish, or carmine (in partial sun). Stems erect to creeping, typically simply organized, but the cortex sometimes modified as a hyalodermis, the cortical cells mostly weakly differentiated, rectangulate, 1.5–4.5(6)× as long as wide, their diam. ± approaching that of adjacent medullary cells; medulla not strongly dorsiventrally modified, at most ventral 2–4 strata of medullary cells progressively somewhat smaller in diam. *Branching either terminal* (of *Frullania* type, the subtending leaf narrower than normal; at times appearing furcate, with the subtending leaf unmodified) *or intercalary* (from lower half of leaf axil, or lateroventral or ventral, and then sometimes stoloniform). *Rhizoids scattered over the postical face of stem* (rarely more frequent near postical leaf base, but never in distinct fascicles from

underleaf bases; occasionally from leaf surfaces), colorless to vinaceous.[280]
Shoots distichous or strongly anisophyllous. Lateral leaves alternate, unlobed
(in some cases retuse; in *Nardia* sometimes regularly but shallowly
bilobed), *always succubously oblique,* usually strongly so (insertion at
35–65° from transverse), not or short- to long-decurrent antically,
usually not or short-decurrent postically, the *line* of insertion *not or weakly*
(rarely strongly) *acroscopically arched, extended to stem midline dorsally. Leaf
margins entire. Underleaves small and lanceolate, not or rarely emarginate* (bilobed only in the extraterritorial *Notoscyphus), never divided into cilia or
ciliate marginally, or absent;* if present, normally not connate with lateral
leaves. *Cells collenchymatous,* usually *thin-walled, but trigones distinct to
bulging;* cuticle often striolate or verruculose; oil-bodies varying from
botryoidal to finely granulose (formed of many distinct to very many minute,
hardly protuberant oil-globules) to homogeneous (the component minute
spherules in a matrix of similar refractive index, hence scarcely resolvable);
if homogeneous, often transversely barred with age, *varying usually from
2 to 25* (more rarely 1) per cell, normally in all leaf cells (except in
Jungermannia tetragona), conspicuously larger than chloroplasts. *Usually
without any means of specialized asexual reproduction;* rarely [*J. lanceolata,
Mylia, Solenostoma rubripunctum* (Hattori), c.n.] with 1 or 2-celled gemmae
from the margins or surfaces of the leaves or (*S. caespiticium*) with 1-celled
gemmae produced endogenously within swollen shoot tips.

Dioecious or paroecious, rarely sporadically autoecious. ♂ Bracts
forming a conspicuous androecium, *situated on leading branches, becoming
intercalary;* bracts ventricose at base, otherwise hardly differing from
leaves in size or shape (except in *Jamesoniella,* in which ♂ bracts may have
an antical, basal tooth); *with usually 1–3 antheridia per bract,* usually
without paraphyses. Antheridia shortly ovoid to subglobular, *short-
stalked; stalk 2(4)-seriate, except 1-seriate in Mylia; jacket cells numerous,*
not in tiers. *Gynoecia at apices of main shoots, anisophyllous,* with generally
1(2–3) pairs of bracts and a small or no bracteole; archegonia numerous.
Perianth normally *present,*[281] various but *not trigonous,* ranging from long and
longly emergent (then normally without a trace of subtending perigynium) to abbreviated and nearly or quite included within bracts (then

[280] Rhizoids often (as in *Mylia,* in *Solenostoma* subg. *Plectocolea,* and in *Jamesoniella*) arise chiefly
from near the postical leaf bases, at least early in their development; in various *Solenostoma*
subg. *Plectocolea* spp. (chiefly in several oriental taxa, less so in *S. sphaerocarpum* and its allies)
scattered rhizoid initials are found on the leaf lamina; in *Nardia* and *Notoscyphus* most rhizoids
may arise from underleaf bases, but scattered rhizoids are always produced as well.

[281] *Cryptocoleopsis* Amak., a monotypic Japanese group, lacks a perianth; *Notoscyphus,* probably
in Jungermanniaceae, is also perianth-free.

normally subtended by a well-developed tubular "stem" perigynium, lying in axis of stem or at an angle to it); perianth smooth and terete to variously plicate, usually narrowed to mouth, *never strongly laterally compressed and wide at mouth* (exc. in *Mylia*)*; perigynium, if present, not becoming strongly subterranean, not or hardly rhizoidous.*

Sporophyte with *seta of numerous cell rows* (a minimum of 10–16 epidermal rows, with at least 7–12 rows within, and usually a central strand of 4 rows; thus a minimum of 5–6 cells in diam.); epidermal cells rectangulate, not tiered.[282] *Capsule ellipsoidal to almost spherical,* 4-valved to base, the valves straight; *capsule wall usually 2(3–5)-stratose (6–7-stratose in Cryptochila).* Epidermal layer with nodular thickenings; inner layer normally with semiannular bands, complete or subcomplete across the interior tangential walls. Spores brown to red-brown, usually ± finely papillate, rarely irregularly reticulate, small (10–20 μ), approximately twice the diam. of the elaters; elaters free, ± tapering on each end, normally 2-spiral, less often 1- or 3–4-spiral.

Type. Jungermannia [Ruppius] L.

The family Jungermanniaceae is troublesome, as to both its own delimitation and the delimitation of genera within it. I approximate here, in large part, recent treatments, e.g. of Müller (1940, 1951–58), Schuster (1951a, 1953), and Amakawa (1960a).[283] In these publications the family Jungermanniaceae is, collectively, essentially circumscribed to include plants with the following characters: (*a*) leaves alternate and usually unlobed, succubously inserted, insertion extending to stem midline; (*b*) collenchymatous cells, provided with oil-bodies formed of fine globules, thus appearing granulose or homogeneous, rarely distinctly segmented; (*c*) a generally simple, unspecialized stem structure; (*d*) underleaves reduced or absent; (*e*) perianth present, with or without a subtending perigynium, usually not compressed, basically terete or 4–5-plicate; (*f*) branching, both terminal and intercalary, normally lateral; (*g*) gynoecia and androecia on unmodified shoots; (*h*) sporophyte with an unreduced seta, but the capsule wall usually reduced to a 2–4-stratose condition; (*i*) spore-elater diam. ratio approximating 2:1. The inclusion of *Mylia*, with a laterally compressed, truncate perianth, somewhat distends these family limits, as does that of *Gottschelia*, with dorsally interlocking merophytes; both genera develop gemmae.

In most of these features (*b, d, e, f, g, i*) the Jungermanniaceae closely approach the Lophoziaceae and, indeed, might well be considered as

---

[282] See Amakawa (1960a, fig. 4) for illustrations of a wide range of seta types.
[283] The treatment by Amakawa is particularly useful for students of holarctic Jungermanniaceae.

"merely a specialized offshoot or subfamily of the Lophoziaceae" (Schuster 1951a). From this they can be derived by (*a*) development of entire or merely emarginate leaves; (*b*) in general a higher level of suppression of the underleaves; (*c*) loss, in general, of the ability to develop fasciculate gemmae from leaf tips or leaf margins; (*d*) reduction of the capsule wall to a 2-stratose condition. However, *Jamesoniella* has a 4–5-layered capsule wall, and in the allied antipodal genus *Cryptochila* Schust. (Schuster, 1963) it is 6–7-stratose.

The Jungermanniaceae are difficultly derivable directly from [existing] Lophoziaceae. They differ from them in almost uniformly lacking gemmae . . . . The Lophoziaceae almost uniformly are able to produce polygonal, characteristic gemmae (smooth only in a few species of *Massula*; absent only in *Chandonanthus*, *Mesoptychia* . . .).

On the other hand,

There are several good reasons why a family separation between the Lophoziaceae and Jungermanniaceae is difficult to maintain. The lobed vs. entire nature of the leaves does not represent, in any way, an approach to an absolute criterion. In the Lophoziaceae there have been described . . . species of *Anastrophyllum* with entire leaves; in other species of *Anastrophyllum* the sinus becomes very shallow, and the leaves are merely emarginate . . . there can be no doubt that they belong in the Lophoziaceae. Therefore, with *Anastrophyllum*, *Anastrepta*, and *Mesoptychia*, the Lophoziaceae approach or even attain the entire-leaved condition characteristic of the Jungermanniaceae. In addition, these advanced types (except for *Anastrepta*) approach the Jungermanniaceae in their inability to develop gemmae. Furthermore, less specialized Jungermanniaceae . . . such as *Nardia* . . . and *Jamesoniella autumnalis* show a well-defined tendency to develop emarginate leaves (Schuster, 1951a).

It is perhaps also more than mere coincidence that these two genera retain such primitive features as underleaves, and (in *Jamesoniella*) a 4–5-stratose capsule wall. Kitagawa (1965), partly for these reasons, has recently transferred *Jamesoniella* to the Lophoziaceae.

Since the nature of antheridia and antheridial stalks also appears to show no significant differences between the two groups, and since no significant differences occur between the two groups in oil-body characteristics, the separation between the Lophoziaceae and Jungermanniaceae must be regarded as either one of convenience, or merely as a tentative one. It should be noted that some recent workers, counter to the rules of priority, use the term Nardiaceae for the Jungermanniaceae (Müller, 1939–40, etc.).

Perhaps the best argument in favor of a retention of two families is that in both the tendencies toward obsolescence of leaf sinuses and underleaves, the reduction in thickness (in layers) of the capsule wall, and the development of a perigynium—all specialized features—appear to represent parallel tendencies.

A linear arrangement from existing types of Lophoziaceae to the Jungermanniaceae is thus difficult to perceive. Furthermore, the very few Jungermanniaceae able to develop gemmae either are derivative types (*Mylia*, *Jungermannia lanceolata*) or possess gemmae of a different type and ontogeny from those in the Lophoziaceae (*Solenostoma rubripunctum, S. caespiticium*).

Since the initial typesetting of this volume, Inoue (1966, pp. 178–79) proposed removing the *Jamesoniella* (Spr.) Schiffn.-*Hattoria* Schust.-*Cryptochila* Schust. complex from the Jungermanniaceae and placing it into the Lophoziaceae, as a subf. Jamesonielloideae, together with genera (*Syzigiella* Spr., *Cuspidatula* Steph., *Anastrepta* (Lindb.) Schiffn.) which, I think, are not allied to *Jamesoniella*. Hodgson (1962) also had proposed a "*Jamesoniellaceae*" [nomen nudum!] for *Jamesoniella* and *Cuspidatula*. I believe the antipodal *Cuspidatula* is allied to *Anastrophyllum*—as is obvious from the occasionally bifid leaves— and has been derived from *Anastrophyllum*-like ancestral types by reduction or loss of one leaf lobe. The free formation of gemmae in *Anastrepta* places this genus nearer the Lophoziaceae, away from the *Jamesoniella-Hattoria-Cryptochila* complex. Finally *Syzigiella* is an isolated unit, whose affinity to the Lophoziaceae-Jungermanniaceae complex is obscure. There are, I think, two alternative classifications of all these units possible, as follows:

Conservative Classification	More Radical Classification
Family Jungermanniaceae	
Subfam. Lophozioideae	Family Lophoziaceae s. str.
	Family Jungermanniaceae
Subfam. Jamesonielloideae	Subfam. Jamesonielloideae
[*Jamesoniella, Hattoria,*	[*genera as at left*]
Cryptochila; Scaphophyllum?]	
Subfam. Gottschelioideae	Subfam. Gottschelioideae
[*Gottschelia*]	[*Gottschelia*]
Subfam. Jungermannioideae	Subfam. Jungermannioideae
[*Jungermannia, Nardia, Soleno-*	[*genera as at left*]
stoma, Cryptocolea, Diplocolea,	
Cryptocoleopsis]	
Subfam. Mylioideae [*Mylia*]	Subfam. Mylioideae [*Mylia*]
Family Plagiochilaceae	Family Plagiochilaceae
Subfam. Syzigielloideae	Subfam. Syzigielloideae
[*Syzigiella*]	[*Syzigiella*]
Subfam. Plagiochiloideae	Subfam. Plagiochiloideae

Since I shall shortly discuss the problems involved in classifying these units, any further discussion here is premature. However, I strongly feel that the obvious similarities of *Syzigiella* to the Plagiochilaceae remove this genus from the Jungermanniaceae. Among such features, found in some or all Plagiochilaceae, are: (*a*) rhizoids in fascicles from bases of leaves and/or underleaves; (*b*) opposite leaves; (*c*) obviously Plagiochiloid facies, with Plagiochiloid leaves and cnemis of several *Syzigiella* spp. [e.g., *S. pectiniformis* Spr., *S. tonduzana* Steph., *S. subintegerrima* (Nees) Spr. = *S. plagiochiloides* Herz. nec Spr. = *S. herzogiana* Schust., 1959–60, nomen novum; see discussion in Schuster, 1959– 60]; (*d*) occasionally laterally compressed young perianths of, e.g., *S. setulosa*

Steph. [*S. purpurea* (Steph.) Schust.; see Inoue, 1966, fig. 18F; see also Schuster, 1959–60]; (*e*) uniform lack of asexual reproduction by gemmae. Admittedly the toothed bracts and bracteole—the latter fused with the bracts— that occur in *Syzigiella* suggest certain Lophoziaceae. However, all of the preceding criteria cited distinguish *Syzigiella* from the Lophoziaceae, while separating it from most or all Jungermanniaceae. *Syzigiella* is possibly rather remotely allied to *Plagiochilidium* Herz., and perhaps more closely to *Plagiochilion* Hatt. It shares with the latter the ventral-intercalary branching with rare and sporadic *Frullania*-type branches; opposed leaves; lack of asexual reproduction; tendency towards toothed leaves of some taxa; rhizoids in well-defined fascicles at leaf bases. Unlike Inoue, who claims the perianth of *Syzigiella* is typically Lophozioid, I detect some similarity to that of the Plagiochilaceae, at least on young perianths, which, before the sporophyte matures may be laterally compressed.

The relevance of this appraisal, as regards our flora, hinges largely on the fact that the genus *Syzigiella* suggests that the Plagiochilaceae may have evolved from *Jamesoniella*-like taxa. Inoue has already suggested that *Plagiochilidium* Herz. has some affinities to the Plagiochilaceae, others to *Jamesoniella*. Hence, if any demonstrable connection between Jungermanniaceae and Plagiochilaceae exists, it is likely that the contact points once resided in this general complex.

Whatever the exact affinity of *Syzigiella*, the genus should go into a separate subfamily:

Subf. Syzigielloideae Schust., subf. n. Plantae a Jungermanniaceis differentes quod folia opposita atque rhizoidea fasciculata; a Plagiochilaceis differentes quod (*a*) perianthia typice non aut paulo lateraliter compressa; (*b*) cortex non differentiatus; et (*c*) bracteae bracteoleae que diverse lobatae dentataeque, foliis eo dissimiles. Typus: *Syzigiella* Spr.

In addition to a marked affinity to the Lophoziaceae, the circumscription of the Jungermanniaceae in various works (Buch, 1936; Buch, Evans and Verdoorn, 1938; Evans, 1939; Frye and Clark, 1937–47) implies a close relationship to the Arnelliaceae and Plagiochilaceae, since these families (or genera here attributed to these families) are placed in the Jungermanniaceae (or "Epigonianthaceae" or "Nardioideae") in these publications.

The separation from the Arnelliaceae affords few difficulties, since the latter family is distinct from the Jungermanniaceae in (*a*) opposite leaves; (*b*) production of a deep, tubular, rhizoidous, subterranean perigynium; (*c*) a 1:1 spore-elater diam. ratio; (*d*) nature of branching; (*e*) homogeneous, glistening oil-bodies (at least in *Arnellia*); (*f*) a mostly slenderly ovoid to cylindrical capsule. These and other differences are discussed in detail under the Arnelliaceae.

On the other hand, separation of the Plagiochilaceae from the Jungermanniaceae is almost as arbitrary as separation of the latter from the Lophoziaceae. The Plagiochilaceae are here separated from most Jungermanniaceae on the basis of the (*a*) 3–7-stratose capsule wall; (*b*) laterally compressed, wide-apertured perianth; (*c*) almost universal retention of underleaves provided with slime papillae; (*d*) tendency, in some groups, for restricted disposition of the rhizoids; (*e*) tendency toward asexual reproductive modes involving fragmentation of the leaves, or dropping of them *in toto*. On this basis, such genera as *Pedinophyllum* belong to the Plagiochilaceae, rather than Jungermanniaceae, while *Mylia* does not fit well into either family. With the Bornean monotypic genus *Plagiochilidium* Herz., a distinction between the two families becomes difficult to maintain, the genus showing some features of *Jamesoniella*, others of *Syzigiella* and *Plagiochila*. Similarly the monotypic *Cryptocolea* shows similarities to both the Jungermannian genus *Nardia* and to *Solenostoma* (subg. *Plectocolea*), with which it agrees in lack of underleaves, distinct perigynium, etc., and to the Plagiochilaceae, which it approaches in the segmented oil-bodies and the wide-apertured perianth. The anomalous position of these genera and consequent difficulty in sharply differentiating between Jungermanniaceae and Plagiochilaceae render such a distinction less valuable. The obscure boundaries between the two families have been recently re-emphasized (Schuster, 1959–60), indicating that in stem anatomy such genera as *Mylia* and *Pedinophyllum* are closer to the Jungermanniaceae than to the Plagiochilaceae. *Mylia*, for reasons given on p. 813, is here placed in a special subfamily of its own, in the Jungermanniaceae. Careful study of the sporophytes, spore germination patterns, and possibly cytology of these taxa may resolve present difficulties.

By the criteria previously established, the Jungermanniaceae become a rather small family, including the regional genera *Jungermannia*, *Jamesoniella*, *Nardia*, *Cryptocolea*, and *Solenostoma* (including, as a subgenus, *Plectocolea*, often considered a discrete genus) and—in a separate subfamily—*Mylia*. The monotypic *Gyrothyra* Howe, often placed in the "Epigonianthaceae" or Jungermanniaceae (Buch, Evans and Verdoorn, 1938; Müller, 1951–58) but incorrectly assigned to the Ptilidiaceae ("Ptilidioideae") by Frye and Clark (1937–47), is considered (Schuster, 1951a, 1955c) to represent a unigeneric family, the Gyrothyraceae. *Schiffneria* Steph., assigned to the "Epigonianthaceae" in Buch, Evans and Verdoorn (1938), is doubtfully assigned to the Cephaloziaceae, *s. lat.* Therefore, in addition to the regional genera listed above, only the exotic genera *Notoscyphus* Mitt., *Cryptochila* Schust., *Scaphophyllum* Inoue, *Cryptocoleopsis* Amak., *Diplocolea* Amak., and possibly *Stephaniella* Jack are assigned to the Jungermanniaceae.[284] As emphasized elsewhere, the anomalous, monotypic *Cryptocolea* Schust. appears to belong at the summit of the Jungermanniaceae, rather than in the Plagiochilaceae.

The Jungermanniaceae, so constituted, are largely mesophytic plants of bare soil, humus, decaying logs, or bare or soil-covered rocks. Few are truly helophytic, virtually none are arborescent (in our area *Jamesoniella autumnalis* is rarely found on bark), and almost none are truly aquatic

---

[284] For recent treatments of *Notoscyphus*, see Amakawa (1960a, p. 271); for *Cryptochila* see Schuster (1963); for *Cryptocoleopsis* and *Diplocolea* see Amakawa (1960a, p. 274; 1962, 1963); for *Scaphophyllum*, see Inoue (1966f).

(although *Solenostoma pumilum* and *cordifolium* may occur in sites subject to inundation).

Key to Regional Subfamilies of Jungermanniaceae

1. Perianths ± narrowed to the mouth (mostly conspicuously so), not sharply laterally compressed and strongly bilabiate; asexual reproduction usually absent, rarely by endogenous gemmae, or (very exceptionally) by gemmae from apices of leaves lacking enlarged cells; capsule wall 2–3(4)-stratose, in *Jamesoniella* often 4–5-stratose; antheridia with stalk 2(4)-seriate, the jacket of numerous small cells. Capsule with epidermal cells undergoing a "one-phase" development, normally essentially all longitudinal (and most long, transverse) walls with isolated nodular thickenings . . . .
. . . . subf. Jungermannioideae
1. Perianths long-emergent, Plagiochiloid, strongly laterally compressed and keeled dorsally and ventrally, truncate at the wide, vertical, and closed mouth; asexual reproduction by gemmae from apices of leaves with ± enlarged cells; capsule wall 3–4-stratose; antheridia with stalk 1-seriate, jacket layer of few and large cells. Capsule with epidermal cells undergoing a "two-phase" development, the primary walls (alternating longitudinal; all or almost all transverse) lacking pigmented secondary thickenings, the secondary walls (mostly alternating longitudinal walls) with few, strong vertical thickenings that are tangentially extended and may coalesce (on the tangential walls) with neighboring bands . . . . subf. Mylioideae

Subfamily JUNGERMANNIOIDEAE

Plants very diverse, typically with a terete or feebly dorsiventrally compressed perianth (absent in *Cryptocoleopsis* and *Notoscyphus*, in which fleshy perigynia are developed) that is *contracted to the narrowed apex*. Underleaves absent or lanceolate, rarely (*Notoscyphus*) bifid. *Asexual reproduction usually absent*, rarely by endogenous 1-celled gemmae or by gemmae from the ± *reduced leaves of erect shoots*. Antheridia with *stalk 2- or 4-seriate*, the ovoid jacket formed of *numerous small cells*. Capsule wall 2- or 2–4-stratose (6–7-stratose in *Cryptochila*), the epidermal cells with ± numerous noncoalescent, well-defined vertical (columnar) thickenings *on all or almost all longitudinal and most longer transverse walls* (thus without evident "two-phase" development).

The number of genera and their distinction constitute a technical and as yet unresolved problem. *Cryptocolea* is sharply isolated (except, perhaps, from the more recently described Oriental genera *Cryptocoleopsis* Amak.

and *Diplocolea* Amak., 1962), and so is the *Jamesoniella* (Spr.) Schiffn.-*Cryptochila* Schust. complex—the latter so strongly so that Kitagawa (1965) has transferred *Jamesoniella* to the Lophoziaceae! *Nardia* also appears to be relatively sharply definable, as is the exotic genus *Notoscyphus*. However, from this point on, agreement ceases.

The remaining taxa are classified by almost all workers into 2 or 3 genera; *Jungermannia* L., *Plectocolea* (Mitt.) Mitt., and *Solenostoma* Mitt. are recognized by, i.a., Evans (1939), Buch, Evans, and Verdoorn (1938), Frye and Clark (1937–47), and Jones (1958a), while Zerov (1964) recognizes not only these 3 genera but additionally elevates the group Luridae or *Luridaplozia* to generic standing as "*Haplozia* Dum." For reasons detailed on p. 835. I would recognize only *Jungermannia* L. (as narrowly delimited by. e.g., Joergensen, 1934; Jones, 1958a; Müller, 1951–58; and Schuster, 1953) and *Solenostoma*; into the latter group the *Plectocolea* and *Luridaplozia* elements are placed as subgenera. Amakawa (1960a) and Grolle (1966e) go even further, putting all these elements into the single genus *Jungermannia*, a proposal I cannot follow (see pp. 899–901).

Key to Genera of Subfamily Jungermannioideae

1. Leaves (and bracts) at shoot apices not conspicuously valvate. Perianth externally evident, longly emergent or included within bracts, but bracts always more or less spreading, at least distally; perianth usually narrowed to mouth, never markedly laterally compressed; ♀ bracts not margined with slime papillae; cells with oil-bodies finely granular to homogeneous . . 2.
  2. Perianth mouth crenulate-denticulate or ciliate; ♀ bracts laciniate or at least with basal teeth or cilia; bracteole well developed, ciliate at least at base; perigynium lacking; branching (at least in large part) ventral-intercalary; with vestigial underleaves, at least below gynoecia; leaves quadrate-rotundate to quadrate-rectangular, often retuse; oil-bodies 7–20 per cell. Capsule wall 4–5-stratose . . . . *Jamesoniella* (p. 815)
  2. Perianth mouth (or its shallow lobes) crenulate with projecting cells (the teeth never more than 1-celled); ♀ bracts entire; bracteole entire or absent; perigynium present or absent; branching normally lateral, terminal, and/or intercalary; leaves various. Capsule wall usually 2-stratose . . . . 3.
    3. Lanceolate to subulate underleaves present; leaves orbicular to quadrate, often emarginate or bilobed (sometimes only in gynoecia; sometimes wholly unlobed); perianth plicate, subtended by a high, fleshy perigynium (sometimes at an angle to the stem), the perianth mouth not contracted into a beak . . . . *Nardia* (p. 844)
    3. Underleaves lacking; leaves various but never bilobed (even in gynoecia); perigynium, if present, continuous with stem . . . . . 4.
      4. Perianth distinctly plicate, at least distally, gradually narrowed to

mouth (rarely lobulate and not at all contracted) or rounded into it, never with a recessed beak, never truncate at apex; leaves various, cordate to ovate to reniform, to elliptical, to rotundate, but not oblong and with parallel sides; never with gemmae in fascicles from the margins of reduced leaves . . . . *Solenostoma*[285] (p. 897)

4. Perianth smoothly cylindrical, terete to the suddenly truncated apex with a small beak set in a distinct depression; leaves oblong, tending to be parallel sided, rounded-truncate to truncate-retuse at apex; occasionally with elliptical, 1–2-celled gemmae from margins of reduced leaves . . . . *Jungermannia* (p. 834)

1. Leaves (and bracts) at shoot tips diagnostically appressed-valvate. Perianth wide at the lobulate mouth, hidden between the marginally closely connivent, ± appressed, large, mussel-shaped bracts, which form a turnip-shaped or bilabiate, somewhat laterally compressed terminal "head"; a short to tall shoot perigynium present; ♀ bracts with margins bearing slime papillae; cells with oil-bodies tending to break into coarse segments, of the "grape-cluster" type; no underleaves. [Leaves typically imbricate, usually antically contiguoconnivent to form a distinct, antical keel] . . *Cryptocolea* (p. 886)

JAMESONIELLA (Spr.) Schiffn.

[Figs. 254–256]

Jungermannia, in part, of authors before 1900.
Aplozia Dumort., Hep. Eur., p. 55, 1874 (in smaller part).
Jungermannia subg. *Jamesoniella* Spr., Journ. Bot. 14:230, 1876.
Jamesoniella Schiffn., *in* Engler & Prantl, Nat. Pflanzenfam. 1(3):82, 1893.

Robust, often forming extensive turfs or less often patches or mats, green to reddish brown, yellowish, or purple. Stems normally *smooth*, *nonparaphyllose*, erect or strongly ascending, rarely prostrate, with strongly autonomous, acroscopic curvature of shoot apex, usually rigid, branching sparing, *intercalary and postical*, less often lateral (terminal or intercalary), and often innovating from beneath gynoecia. Rhizoids scattered on ventral face of stem, often few. Leaves alternate, succubously obliquely inserted, ovate to rotundate or rotund-quadrate, *entire to retuse*, *nondentate*, at least the apical ones antically connivent; leaf margins flat or incurved, *never revolute*. Cells with several to numerous finely granulose to almost smooth oil-bodies, collenchymatous, with small and concave-sided to distinctly bulging trigones; marginal cells not sharply differentiated as a border. *Underleaves vestigial or very small*, except sometimes near the gynoecia. Asexual reproduction lacking.

Dioecious. Androecia and gynoecia at first terminal on leading shoots, androecia becoming intercalary through vegetative proliferation. ♂

---

[285] Including *Plectocolea* as a subgenus.

Bracts ventricose at base, the antical base dilated and forming an *in-folded*, 1–2-toothed *flap or lobule*. *Antheridia solitary* (in pairs only in *J. undulifolia*), their stalks 2- or 4-seriate. ♀ Bracts ± larger than leaves, *never with revolute margins*, at least the bases *more or less dentate, ciliate or laciniate; bracteole well developed, similarly armed as bracts*, rarely edentate. Perianth at least 0.5 emergent, *not normally subtended by a perigynium*, oblong- to fusiform-ovoid to obclavate, *smoothly terete and tubular below, only distally distinctly 4–5-plicate and never twisted, gradually narrowed to the mouth; mouth never beaked, usually* (in our species) *distinctly ciliate or dentate* to crenulate-denticulate. Sporangium with massive seta of numerous cell rows; *capsule ovoid, wall 4–5-stratose.* Epidermal layer with strong nodular thickenings; inner cell layers with semiannular to annular bands.

Type. Jamesoniella colorata (Lehm.) Spr. (Amakawa, 1960a, p. 271). *J. carringtonii* (Balf.) Schiffn. is the first species mentioned by Spruce (1876, p. 230) when he proposed the subg. *Jamesoniella;* it could perhaps, logically be considered the type species, but such a typification would only cause taxonomic chaos, since Grolle (1964i, pp. 653–656) has shown that *Jamesoniella carringtonii* belongs to *Plagiochila*!

Jamesoniella consists of a series of handsome, robust, erect-growing, caespitose species with antically secund, frequently erect-appressed, often longly decurrent leaves, with often strongly collenchymatous leaf cells, with entirely or almost entirely postical-intercalary branching, and with a different facies from that of *Jungermannia* and *Solenostoma*.

These species of *Jamesoniella*, typified by the type, *J. colorata* (Lehm.) Spr., and by *J. grandiflora* Lindenb. & G., bear little relationship to the ubiquitous, common *J. autumnalis* of cooler portions of the Northern Hemisphere. Typical species of the genus are predominantly antipodal and are confined to old land masses; they are part of the pre-Tertiary flora nearly lacking in the highly glaciated cold and temperate portions of the Northern Hemisphere. It is perhaps doubtful if these two types of plants should be considered congeneric. Evans (1915a, p. 82), e.g., stated:

The generic position of *Jungermannia autumnalis* DC. is far from established. Stephani transferred it to *Jamesoniella*, apparently on account of its toothed or lacerate perichaetial bracts, bracteole and perianth. It occupies an anomalous position in this genus, however, on account of its general habitat, its branches of the *Frullania*-type, its cell-structure, and its distinct underleaves. In typical members of *Jamesoniella* the plants tend to be erect and the leaves to be erect and appressed; the branches seem to be invariably ventral and intercalary; the leaf cells have large and nodulose trigones; and the underleaves of the stem are very rudimentary.... Possibly the wisest course would be to make *J. autumnalis* the type of a new genus, as K. Müller has already suggested.

It is believed that a conservative course would be a separation of *Jamesoniella* into two discrete subgenera, as follows:

a. Plants usually robust, erect and tufted in growth, often with antically connivent, erect to erect-appressed, long-decurrent leaves; cells often strongly collenchymatous, with coarsely bulging trigones; branching intercalary-postical; underleaves obsolete . . . . subg. *Jamesoniella*

a. Plants moderate in size, green to reddish brown, closely prostrate in growth with merely shoot apices ascending; leaves laterally more or less patent, widely spreading with antical bases short decurrent; cells slightly collenchymatous, with small to barely bulging trigones; branching in large part lateral, terminal, of *Frullania* type;[286] underleaves minute but usually distinct on shoot apices, with age disappearing . . . . . subg. *Crossogyna*, subg. n.

The subg. *Jamesoniella* does not occur in nontropical portions of the Northern Hemisphere; it is primarily antipodal, with a few species extended into montane portions of the tropics of the Northern Hemisphere.

The nearest relative of *Jamesoniella* is *Cryptochila* Schust. (Schuster, 1963, pp. 217, 284) of New Zealand, including only *C. pseudocclusa* (Hodgs.) Schust. Although this has the toothed bracts and bracteole of *Jamesoniella*, it differs in the *Plagiochila*-like facies, induced by the tightly revolute antical margin of the long-decurrent leaves and perichaetial bracts. The genus differs from *Jamesoniella* subg. *Jamesoniella* also in the regular occurrence of *both Frullania*-type lateral and ventral-intercalary branches, and from both subgenera of *Jamesoniella* in the paraphyllose stem and the long and slender perianths that are deeply pluriplicate for most of their length, and spirally twisted as in the Lophozioid genus *Roivainenia*. Sterile plants of *Cryptochila* are likely to be searched for in *Plagiochila*. *Cryptochila* has the thickest known capsule walls of any genus of Jungermanniaceae; they are 6–7-stratose, and are thus obviously thicker than in *Jamesoniella*! Perhaps also remotely allied are *Scaphophyllum* Inoue (1966f) and *Gottschelia* Grolle (1968).

Subgenus *CROSSOGYNA* Schust., subg. n.

Sectio a subg. *Jamesoniella* differens ut (*a*) rami saepissime laterali-terminales et (partim) lateriventrales intercalaresque; (*b*) plantae repentes et prostratae, (*c*) folia patentia, anticaliter brevidecurrentia, et (*d*) plantae virides ad dilute brunneas aut rubro-brunneas.

Plants green to brown to reddish brown, creeping and prostrate, with many rhizoids all along the stems. Branches *terminal and lateral, and* (*in part*) *intercalary and lateroventral.* Leaves laterally patent, usually widely so, with short-decurrent antical bases. Underleaves distinct but minute, ephemeral, usually hidden amidst rhizoids.

Type. *Jamesoniella autumnalis* (DeCand.) Steph.

Grolle (1964i, p. 662) would place *J. autumnalis* into sectio *Jamesoniella* together with the generitype, *J. colorata*, yet he states on p. 653 that "alle

[286] See Evans (1912, p. 17).

sicheren *Jamesoniella*-Arten . . . ventral-interkalare Verzweigung haben." I would rather agree with Spruce (who excluded *J. autumnalis* from his subg. *Jamesoniella*) and Evans and Müller (who queried the propriety of retaining this species in *Jamesoniella*) and isolate the *J. autumnalis* element from typical *Jamesoniella*.

I have repeatedly studied branching in *Crossogyna*. ♂ Plants show a preponderance of terminal, *Frullania*-type branches, although occasional intercalary branches occur. ♀ Plants show mostly intercalary branching, although occasional terminal branches occur. All or almost all of the intercalary branches, if carefully examined, appear to originate lateroventrally, juxtaposed to the ventral angle of a lateral leaf. I have not seen any branches which can be safely regarded as actually originating from ventral merophytes or from the axils of the vestigial underleaves—not even near the gynoecia. These intercalary branches are, therefore, not homologous in origin with, e.g., the truly ventral-intercalary branches of *Bazzania*.

Key to Sections and Species of Jamesoniella

1. Perianth mouth ciliate, the cilia usually 5–10 to 11–13 cells long; perianth 2-stratose below. Antheridia solitary, the stalk 2-seriate. Cells of leaf middle 25–30 μ. Leaves of sterile and mature shoot sectors ± laterally patent, never undulate; ♀ bracts usually strongly ciliate or ciliate-lacinate at base, not undulate. Leaves finely striolate toward base; usually red-brown (only shade forms green). Ubiquitous; rarely in peat bogs . . . . [sect. *Crossogyna*] *J. autumnalis* (p. 818)
1. Perianth mouth crenulate-denticulate to denticulate with 1-, rarely 2-, celled teeth; perianth 1-stratose throughout. Antheridia in pairs, stalk 4-seriate. Leaves of vigorous sterile shoots and ♀ bracts undulate, commonly somewhat antically secund, and rather appressed dorsally; ♀ bracts broad, at the base often subentire or merely with 1–2(3) small to vestigial cilia. Leaves with smooth cuticle. Rare, northern, helophytic, usually green (rarely somewhat brownish) . . . . [sect. *Biantheridion* Grolle] *J. undulifolia* (p. 830)

Our species have been the subject of a prolonged controversy between K. Müller (see especially 1905–16, pp. 753–760) and Schiffner (see Krit. Bemerk. to no. 415 of Hep. Exsic. Eur.). Schiffner correctly insisted that there are two species in the *J. autumnalis* complex, and Müller (*loc. cit.*) reluctantly followed Schiffner in admitting a second species, which Schiffner designated *J. schraderi* but which Müller called *J. undulifolia*.

JAMESONIELLA (CROSSOGYNA) AUTUMNALIS
(DeCand.) Steph.
[Figs. 254–255]

Jungermannia autumnalis DeCand., Fl. Française 6:202, 1815.
Jungermannia schraderi, of Ekart, Syn. Jungerm. Germ., p. 39, pl. 11, fig. 97, 1831; of Austin, Hep. Bor.-Amer. Exsic. no. 27; of Underwood & Cook, Hep. Amer. Exsic. no. 18 (not of Martius, Fl. Crypt. Erlang., p. 180, pl. 6, fig. 55, 1817).
Jungermannia subapicalis Nees, Naturg. Eur. Leberm. 1:310, 1833.

Aplozia schraderi and *subapicalis* Dumort., Hep. Eur., p. 56, 1874.
Jungermannia laevifolia Lindb., *in* Stephani, Bot. Centralbl. 50:71, 1892.
Aplozia autumnalis Heeg, Verh. Zool.-Bot. Gesell. Wien 43:80, 1893.
Jungermannia laevifolia Lindb., MS, *in* Stephani, Bull. Herb. Boissier 5:79, 1897.
Jungermannia rauiana Steph., Spec. Hep. 2:73, 1901.
Jamesoniella autumnalis Steph., *ibid.* 2:92, 1901; K. Müller, Rabenh. Krypt.-Fl. 6(1):576, fig.
    287, 1909; Macvicar, Studs. Hdb. Brit. Hep. ed. 2:154, figs. 1–6, 1926; Schuster, Amer.
    Midl. Nat. 42(3):578, pl. 10, figs. 1–4, 1949; Schuster, *ibid.* 49(2):391, pl. 29, figs. 4–12,
    1953; Amakawa, Jour. Hattori Bot. Lab. 21:271, 1959.
Jamesoniella subapicalis Schiffn., Lotos 59:68, 1911.
Jungermannia moriokensis Steph., Spec. Hep. 6:88, 1917 (after Hattori, 1952, p. 48).
Jungermannia variablis Steph., *ibid.* 6:95, 1917.
Jamesoniella nipponica Hatt., Jour. Jap. Bot. 19:350, fig. 26, 1943 (after Hattori, 1951, p. 76).
Jamesoniella autumnalis var. *nipponica* Hatt., Jour. Hattori Bot. Lab. no. 5:76, 1951 (after Hattori,
    1952, p. 48).

Prostrate, gregarious or forming patches, green and somewhat opaque (*rather strongly chlorophyllose*), in sun *becoming reddish brown to brownish carmine, quite opaque*. Shoots with ascending apices, ca. (1350)1700–2400 μ wide × 1–3(4) cm long, to 3.5 mm wide at apices of ♀ plants, simple or sparingly branched; branches lateral-intercalary, juxtaposed to lower half of leaf axil, and terminal, of *Frullania* type, subtended by a narrow, atypical leaf. Stems ca. 225–300 μ in diam., rather stout and fleshy; cortical cells *relatively thin-walled*, not forming a sharply differentiated cortex, rather regularly *short-rectangulate* dorsally, ca. 24–28(32) μ broad × 45–55(60) μ long (*averaging 1.6–2.5× as long as broad*); medulla in ca. 8–9 strata, undifferentiated dorsiventrally, not mycorrhizal, the cells similar in diam. as dorsal cortical [averaging 24–28(30) μ], *somewhat thick-walled*, hyaline. Rhizoids *numerous* up to near stem apex, *colorless*. Leaves quite obliquely inserted, the line of insertion wide, scarcely arched, rather weakly decurrent antically, weakly to distinctly imbricate, but not crowded, those of sterile shoots *nearly horizontally patent*, widely spreading, those of upper portions of ♀ shoots suberect and somewhat antically secund, *rotund-quadrate to very shortly oblong-oval*, averaging to 950–1050 μ long × 950–1160 μ broad, occasionally 1020–1100 μ long × 1100–1120 μ broad, *ca. 0.9–1.1× as long as wide* (those below the ♀ bracts becoming somewhat larger, more distinctly oblong-oval), the *apices usually rounded-truncate to retuse, occasionally emarginate* (and then with shallow, rounded lobes). Cells of leaf margins *hardly differentiated, subquadrate*, similar in size to submarginal cells, ca. 20–26 μ; median cells ca. 25–35 × 32–38(42) μ; basal cells to (30)35–38 × 40–50 μ; cell walls thin, *trigones small to barely bulging*, the cell lumen usually rounded-polygonal; cuticle nearly smooth, weakly verruculose, finely striolate at least near leaf bases; *oil-bodies 7–15(18–20) per cell*, finely granulose in appearance, 3 × 5 to 4 × 7 μ, up to 6 × 9–10 μ, ovoid to ellipsoidal, a few spherical and only 4–5 μ, colorless; chloroplasts averaging 4 μ. *Underleaves usually obsolete* except below gynoecia, occasionally evident at shoot apices (hardly discernible elsewhere and often not apparent at all), subulate, of a stalked slime papilla or minutely lanceolate, *rarely and only locally larger. Asexual reproduction lacking.*

Normally *dioecious*. Androecia *compactly spicate*, of 4–6, rarely more, pairs of bracts, at first terminal, eventually intercalary; bracts strongly ventricose, hardly as large as leaves, closely imbricate (the young androecia budlike, often *orange-red in color even when rest of plant green*), with usually 1–2 teeth of the

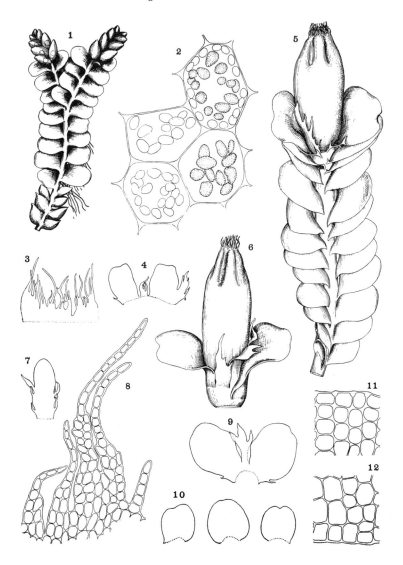

Fig. 254. *Jamesoniella autumnalis* (1) ♂ Plant with terminal branch (×ca. 7). (2) Median cells (×585). (3) Perianth mouth (×ca. 15). (4) Bracts and bracteole (×ca. 9). (5) Perianth-bearing shoot (×9). (6) Perianth-bearing shoot apex, ventral view, to show bracts and bracteole (×9). (7) ♀ Bracteole (×9). (8) Perianth mouth sector (×116). (9) Bracts and bracteole (×9). (10) Leaves (×ca. 7). (11–12) Cells of leaf margins (×ca. 225). [1, 11–12, Six-mile Creek, Ithaca, N.Y., *RMS*; 2, *RMS 31797c*, Pender Co., N.C.; 3–9, Angle Inlet, Lake of the Woods, Minn., *RMS*; 10, Togo, Minn., *RMS*.]

ampliate, inflexed antical base; *1-androus*, antheridial *stalk 2-seriate*. Gynoecia variable; bracts usually sheathing at base, spreading to reflexed distally, oblong to oblong-oval, larger than leaves, ca. 1.15–1.25× as long as wide, commonly 975–1125 μ broad × 1125–1200 μ long, *rounded to truncate or retuse at apex, at least their antical bases with 1–2 or more cilia or teeth*, rarely absent on material from southeastern sites (sometimes innermost bract smaller in size and ciliate-laciniate distally as well; Fig. 255:6, 12); *bracteole always large*, commonly 0.2–0.4 the size of bracts, variable, from lingulate with ciliate sides or trifid or laciniate at apex, or divided into several acuminate lobes, free or united for a short distance with one (rarely both) bracts. Perianth without a perigynium at base, slenderly cylindrical to fusiform-cylindrical, 2-stratose in basal 0.3–0.5, to 3.5 mm long, rather gradually narrowed in distal fourth to the rather narrow but somewhat open mouth, strongly 4–5-plicate in upper fourth; *mouth irregularly and shallowly multifid and laciniate*, the laciniae commonly 2–6 cells wide at base, running out into uniseriate *cilia 5–13 cells long*. Capsule ovoid, *the wall 4–5-stratose*, firm. Epidermal cells with strong nodular thickenings; inner 3 layers with semiannular bands. Spores 11–15 μ, granulate; elaters ca. 8 μ in diam., bispiral.

Type. France.[287]

Distribution

Generally common or even abundant; found throughout the Spruce-Fir Panclimax and much of the Deciduous Forest Regions of the Northern Hemisphere. Occurring throughout most nonarctic portions of Europe, from southern and central Scandinavia (Norway, Denmark, Finland, Sweden), southward to England and Ireland, the montane and north-western parts of France, Germany, the Baltic Republics (USSR), Austria, Switzerland, northern Italy, eastward to the Carpathians, Russia, Poland. Eastward to Siberia, Korea, Japan, Formosa, China and (incorrectly?) reported from India, Sumatra, Java, the Philippines, New Guinea (see, i.a., Allorge, 1955). Widespread in Japan (Hattori, 1952; Amakawa, 1960a), whence it was described as the synonymous *J. nipponica* Hatt. Not known from the Faroes and Ireland, giving the impression of being a continental species. Recently reported from the Sikkim-Himalaya (Hattori, 1966, p. 510).

In North America transcontinental in the Taiga, rare and evidently sporadic in the southern edges of the Tundra, ranging southward through the northern half of the Deciduous Forest Region, largely dropping out

---

[287] I have studied the type of *J. autumnalis* in the DeCandolle collection (G). It corresponds to the normal, rotundate-leaved phase of *J. autumnalis*. The perianth mouth bears numerous cilia, the longer to 11–13 cells long. The type, no. 13, is labeled simply "M. Mougeot 16 janv. 1815." DeCandolle [in Fl. Française, vol. 5(=6):202, 1815] cites only the Mougeot specimen from "les forêts de sapins des Vosges, près Bruyères." (The citation of the type as from "St. Bernhard" by Müller, 1951–58, is incorrect.)

in the Southeast in the upper portions of the Piedmont; then recurring as a disjunct in the outer portions of the Coastal Plain (particularly in Virginia and the Carolinas). Recurring in Mexico.[288] In the West found from Alaska (central Pacific Coast; Aleutian Isls.; see Persson, 1946), to British Columbia and Alberta, southward to Washington, Oregon, Idaho, Montana, and Wyoming, with a lacuna in the known distribution from there eastward to Minnesota, which will certainly be filled with further collecting in the intervening Taiga of Canada.[289] So common in eastern North America that only representative records and all peripheral citations are given:

NEWFOUNDLAND. Not uncommon, to 60°38′ N. in Avalon Distr. (Holyrood), on S. coast (St. Albans W. to Grand Bruit), on N. Pen. (Cow Head and Old Port au Choix, St. Barbe), and NE. coast (Millertown Jct. and Gander) (Buch & Tuomikoski, 1955). QUEBEC. Rupert R., 76° W., L. Mistassini (*Lepage*); S. Quebec only, N. to southernmost James Bay area (Buch & Tuomikoski, 1955); Granby (*Fabius 1566!*), R. à Martre; Chelsea; Rigaud; Oka; Iberville; Waterloo; Mt. Orford; La Tuque; Lac Wayagamack; Montmorency R.; Tadoussac; Bic; St. Cléophas; Matapédia; Mt. Albert; Mt. Shefford (Lepage, 1944–45; Fabius, 1950). ONTARIO. L. of Two Rivers, Algonquin Park (*Cain 1110!*); Ottawa; Belleville; Sudbury Jct., Algoma; Owen Sd. (Macoun, 1902); Little Macaulay L., Algonquin Park (*Cain 1046!*); L. Timagami (*Cain!*) NOVA SCOTIA. Pictou; Truro; Margaree, Cape Breton I. (Macoun, 1902); Halifax (*Brown 1918!*); E. side of Cape Breton Natl. Park, 3–3.5 mi S. of Mary Ann Brook waterfall (*RMS 66325a, 66335b, 66326*). NEW BRUNSWICK. Tobique R.; Woodstock (Macoun, 1902); Grand Manan, Southern Head. PRINCE EDWARD I. Near Brakley Pt. (Macoun, 1902).

MAINE. Mt. Katahdin, at and below 2800 ft (*RMS*); Mt. Desert I. (*Lorenz*). NEW HAMPSHIRE. Mt. Washington, and general elsewhere at lower elevations in the White Mts., Coos Co. (*RMS*); Waterville (*Lorenz*); Franconia Mts. (Lorenz, 1908c). VERMONT. Goshen (*Dutton 1918!*); E. of Bristol, Addison Co. (*Correll 7823!*). MASSACHUSETTS. Worcester; Oxford; Shrewsbury; Leicester; Hadley, Hampshire Co. (*RMS, s.n.*); Bear R., Conway, Franklin Co. (*RMS, s.n.*) RHODE ISLAND. *Fide* Evans (1906b). CONNECTICUT. Litchfield, Hartford, Tolland, New Haven, and Middlesex Cos. (Evans & Nichols, 1908).

NEW YORK. Jamesville, Onondaga Co. (*Underwood, 1888*); Risby L., Herkimer Co. (*Haynes, 1906*); Cold Spring Harbor (Andrews, 1931); L. George, at Anaquassacock Hills, SW. of Shushan, E. of Ft. Ann, etc. (*Burnham*); Tompkins, Chemung, Cattaraugus, Cayuga, Genesee, Onondaga, Allegany, Madison, Cortland, Tioga, and Chenango Cos. (Schuster, 1949a); Slide, Cornell, and Wittenberg Mts., Catskill Mts. (*RMS*). NEW JERSEY. PENNSYLVANIA. Cameron, Crawford, Elk, Fayette, Fulton, Lawrence, Potter, Warren Cos. (*Lanfear*); Lindell (*Kaiser, 1911!*).

MARYLAND. Near Washington, D.C. (Plitt, 1908). VIRGINIA. Cape Henry State Park, Princess Anne Co. (*RMS & Patterson*); L. Drummond, Dismal Swamp, Norfolk Co. (*RMS & Patterson 34527, 34534*); Mountain L., Giles Co. (*Blomquist!*); Greene,

---

[288] Mirador (*Liebmann no. 351*, det. Gottsche as *J. subapicalis* Nees).
The Mexican material seems safely referred here, although it represents the oval-leaved phenotype [var. *nipponica* (Hatt.) Hatt.]; I have seen a specimen at G.

[289] Reports from Greenland (Macoun, 1902, etc.) are all based upon the next species, as study of pertinent Vahl specimens has shown.

Norfolk, Rappahannock, Smyth Cos. (Patterson, 1949). WEST VIRGINIA.. Fayette, Marion, Mercer, Monongalia, Tucker, Preston, Randolph, Upshur, Webster Cos. (Ammons, 1940); Deer Cr., Pocahontas Co. (*Gray!*); Barbour Co. (*Gray!*). KENTUCKY. Laurel, Harlan, Carter, Letcher, Lewis, Powell Cos. (*Fulford*). TENNESSEE. Near Clarksville, Montgomery Co. (*Clebsch*); Blount, Johnson, Knox, Sevier, Washington Cos. (Sharp, 1939).

NORTH CAROLINA. Coastal Plain: swamp 1.5 mi E. of Whiteville, Columbus Co. (*RMS 29214a, 29483*, etc.); N. bank of Cape Fear R., ca. 10 mi N. of Wilmington, Pender Co. (*RMS 31797c*); Chowan R., 1 mi N. of Winton, Gates Co. (*RMS 34580a*); near Scranton, Hyde Co. (*RMS 28760b*); 5–8 mi S. of Manns Harbor, Rte. 264, Dare Co. (*RMS 28388a*); E. of Washington, Beaufort Co. (*Blomquist 8126!*); E. of Jamesville, Martin Co. (*Blomquist 10716!*); Waccamaw R., Hwy. 130, Columbus Co. (*Blomquist 10142!*); Smiths I., Brunswick Co. (*Anderson 6797!*); general in the Piedmont and mountain counties: Moore, Franklin, Durham, Forsyth, Ashe, Macon, Watauga, Caldwell, Henderson, Swain, Burke, McDowell, Yancey Cos. In the Appalachians to 5800–6600 ft or higher (as on Andrews Bald, Smoky Mts. Natl. Park, Swain Co. [*RMS 36600*] and Mt. Mitchell, Yancey Co., at 6600 ft [*RMS 24824a*]). SOUTH CAROLINA. Whitewater R. Gorge, below Lower Falls, Oconee Co. (*RMS, s.n.*); Middendorf, Chesterfield Co. (*Anderson 5604!*); Atomic Energy Comm. Reserv., Aitken Co. (*Batson!*). GEORGIA. Brasstown Bald, Towns-White Co. (*RMS 34345*). FLORIDA. Damp soil along Rock Cr., Torreya State Park, Liberty Co. (*RMS 44082*); southernmost American station! MISSISSIPPI. Tishomingo State Park, Tishomingo Co. (*RMS, s.n.*).

WESTWARD occurring to: OHIO. Lawrence to Clarke to Cuyahoga Cos. (Miller, 1964). INDIANA. Turkey Run State Park (*Drexler 1129*); Wayne, Montgomery, Putnam Cos. (*Underwood*, in part). ILLINOIS. Urbana (*Drexler 1267*). MICHIGAN. Amygdaloid I., Isle Royale (*RMS, s.n.*); Copper Harbor, Keweenaw Co. (*RMS, s.n.*); Sugar I., Chippewa Co. (*Steere*); Ann Arbor; Isle Royale; Keweenaw, Luce, Marquette, Leelanau, Cheboygan Cos. WISCONSIN. Sand I., Apostle Isls., Bayfield Co. (*RMS, s.n.*); L. Minocqua, Oneida Co. (*Culberson!*); near Merrill, Lincoln Co. (*Culberson!*); Lafayette, Vilas, Oneida, Marathon, Bayfield, Douglas, Iron, Ashland, Sauk, Sawyer, Superior, Grant, Polk, Rock Cos. (Conklin, 1929).

MINNESOTA. Carlton, Chisago, Clearwater, Cook, Goodhue, Itasca, Lake, Lake of the Woods, St. Louis, and Winona Cos. (Schuster, 1953). IOWA. Clayton, Mitchell, Winneshiek, Allamakee, Fayette, Muscatine, Wapello, Webster, Hardin, Boone, Linn, Poweshiek, Marion Cos. (Conard, 1945). MISSOURI. *Fide* Evans and Nichols (1908). KANSAS. Douglas, Leavenworth, and Woodson Cos. (McGregor, 1955).

The only peculiarity revealed by the preceding, broad distribution pattern is the occurrence as an epiphyte in broad-leaved evergreen swamps of the southeastern Coastal Plain. The latter occurrence, however strange at first glance, is correlated with that of other species of cool (but not arctic) regions, such as *Bazzania trilobata* and *Frullania asagrayana*. These species, e.g., are found with *Jamesoniella* in a *Nyssa aquatica-Taxodium-Planera* swamp in coastal North Carolina, near Whiteville. Yet they occur here with such tropical and subtropical epiphytes as *Epidendrum conopseum* and *Lopholejeunea muelleriana*!

Ecology

A common, indeed often ubiquitous, species, with a correspondingly wide ecological amplitude, occurring under a great variety of conditions. Northward it may be found on moist, somewhat calcareous, shaded sandstones, associated with . . . *Plagiochila, Jungermannia lanceolata, Ptilidium pulcherrimum* . . . [or even] on moist rocks at the edges of periodically wet depressions, associated with *Scapania degenii* and *irrigua* . . . . The species, on such sites, occurs on essentially mineral substrates, with little or no humic materials, and may be a pioneer species; it occurs here at a pH of from 5.5 to 7.0. Such occurrences are, however, more or less infrequent. More commonly the species occurs on decaying logs (and the resulting humus), associated with such xylicolous forms as *Blepharostoma, Lepidozia, Scapania apiculata* and *glaucocephala, Cephalozia media, Lophozia ascendens,* . . . *Lophozia porphyroleuca, Odontoschisma denudatum, Riccardia palmata, R. latifrons, Jungermannia lanceolata.* The measured pH range, under such conditions, varies from 4.5 to 5.2, but may possibly go considerably lower (Schuster, 1953).

Other associated species on logs are commonly *Nowellia curvifolia, Ptilidium pulcherrimum, Lophozia incisa, Anastrophyllum michauxii, Tritomaria exsecta, Harpanthus scutatus, Cephalozia catenulata,* and such mosses as *Dicranum fuscescens.*

Jamesoniella autumnalis also occurs frequently over acidic rocks, such as noncalcareous sandstones, quartz conglomerate, and granitic rocks. It also occasionally is found over moist, humus-rich soils on shaded banks (then sometimes with *Diphyscium sessile* and occasionally *Buxbaumia aphylla*); as at Sixmile Creek, N.Y. Also sometimes over *Sphagnum compactum* on wet, peaty banks in the southeast Piedmont and Coastal Plain.

Infrequent are occurrences on bark of living trees, where usually confined to exposed roots and the flaring butts of trees, in swamps, northward. However, in the southeastern Coastal Plain, particularly in southeastern Virginia and eastern North Carolina, such occurrences are the normal and almost exclusive habitat of the species. The plants here occur in deep swamps, sometimes in *Chamaecyparis-Taxodium* swamps (Chowan R., N.C.), on bark of both of these trees, together with *Leucobryum, Bazzania trilobata,* and *Frullania asagrayana!* At other times they are found in *Persea-Taxodium-Magnolia* swamps or *Acer rubrum-Nyssa-Planera* swampland on bark of *Nyssa aquatica* (then with such tropical and subtropical taxa as *Rectolejeunea maxonii, Leucolejeunea unciloba* and *conchifolia, Radula caloosiensis, Cheilolejeunea rigidula,* and even *Lopholejeunea muelleriana*). Such corticolous occurrences are usually of slender, atypical, highly pigmented plants, which only occasionally produce gynoecia.

The species is similarly corticolous only in the Southern Appalachians, occurring on a variety of trees (*Tsuga canadensis* at lower elevations; *Betula lutea* at higher ones), at higher elevations (to 6000 feet) with *Tritomaria exsecta, Plagiochila tridenticulata, Lejeunea ulicina, Frullania asagrayana,* and *Metzgeria fruticulosa* (Andrews Bald, Smoky Mt. Natl. Park, N.C., *RMS 36600*). In such sites it represents a casual, and late, invader in the *Bazzania nudicaulis-Herberta* Associule.

Differentiation and Variation

In spite of the variation in the environments inhabited, *J. autumnalis* is stenotypic through most of its range. Plants in deeply shaded localities are sometimes a clear but opaque green throughout, but usually show at least traces of the dull carmine-red or brownish red pigmentation that is diagnostic of the plant in the field. In extreme cases (as in corticolous phases) plants may be purplish brown, remote-leaved and atypical in facies; such forms are of rare, sporadic occurrence. Sterile xylicolous phases may be mistaken for *Odontoschisma denudatum* or *prostratum*, but differ at once in (*a*) lacking flagella, (*b*) in having more but smaller oil-bodies per cell.

Gynoecial *J. autumnalis* can hardly be confused with other species. The entire leaves, essential lack of underleaves and scattered rhizoids may suggest species of *Solenostoma* or *Jungermannia*. However, the cilia or teeth of ♀ bracts, and the ciliate or laciniate bracteole, as well as the ciliate perianth mouth, at once distinguish *J. autumnalis* from species of these genera. However, sterile material (unless bearing the characteristic reddish or reddish brown pigmentation) is

at times difficult to keep separate from that of some species which may occur with it. Confusion in the field is very likely with *Jungermannia lanceolata*, with which it often occurs, on both rocks and decaying logs. The latter has, on well-developed plants, considerably more elongate leaves . . . ; under the microscope the larger, distinctly bulging trigones of the *Jungermannia* . . . serve to separate it from *Jamesoniella*, which has the trigones always with concave sides (Schuster, 1953).

Both *Jamesoniella* and the *Jungermannia* occasionally have retuse or emarginate leaves. These, however, occur with a predictable frequency in *Jamesoniella*, while they are rare and sporadic in *Jungermannia*. The frequency of such notched leaves is generally an excellent, if not absolutely reliable, recognition feature of the *Jamesoniella*.

♀ Bracts and bracteoles of this species are very variable within the same population. Often both bracts are subentire, except for isolated teeth of the antical (and less often also postical) leaf bases (Fig. 254:9). However, the innermost bract may be much smaller and laciniate throughout, except at base; see, e.g., Fig. 255:6, 12. Such a bract is also illustrated by Müller (1909, *in* 1905–16, fig. 287h).[290] Occasionally deeply lobed and laciniate bracts occur (Fig. 255:12), with one of the innermost bracts much reduced to 1–2 lanceolate and acuminate divisions, the other formed of 4–5 such divisions, each bearing several stalked slime papillae. These deeply lobed bracts are then

---

[290] This type of bract is misinterpreted by Frye & Clark (p. 273), who copy the figure of Müller (their fig. 7) but misinterpret it as a bracteole! Perhaps from this has arisen their erroneous statement (p. 272): "bracteole larger than bracts." And perhaps also from this misinterpretation has arisen their designation of the bracteole depicted by Müller (*loc. cit.*, fig. 287i), which they redraw (their fig. 8), as "an underleaf near the inflorescence." If the lingulate bracteole, trilobed or trilaciniate at the apex, depicted by Müller is compared with that in the ring of bracts here illustrated (Fig. 255:9), it is evident that the bracteole may be small and merely trifid at the summit!

associated with a small, similarly deeply lobed and laciniate bracteole. Occurrence of such derivative bracts is of no taxonomic significance, since they are found on plants in populations that produce other individuals with "normal" bracts (such as those in Fig. 255:5, taken from a plant found admixed with the preceding one illustrated).

In addition to the preceding variations, in part caused by minor genetic heterogeneity and also by reaction to different environments, three more extreme variants stand out, which have been considered to be distinct species. These may be separated by the subjoined key:

1. Plants with underleaves, except in and below gynoecia, vestigial and sporadic in occurrence, never large and broadly lamellate . . . . . . 2.
   2. Uniformly dioecious; perianths not produced in unusually large numbers; cells thin-walled; perianth mouth longly ciliate. . . . . . . 3.
      3. Leaves primarily rotund-quadrate, (0.95)1.0–1.1 × as broad as long on mature, ♀ shoots, rounded-truncate to truncate-emarginate at apex in most cases; ♀ bracts, except for the occasionally laciniate-lacerate inner ones, oblong to quadrate-oblong, from 1.0 to 1.25 × as long as broad . . . . var. *autumnalis*
      3. Leaves narrowly ovate to oval-elliptical on mature (♀) shoots as well as on sterile shoots, 1.19–1.36 × as long as broad (0.74–0.85 as broad as long), widest in basal third, narrowed to apex and rounded, never emarginate or truncate; undivided ♀ bracts narrowly ovate to oblong, from 1.35 to 1.9(rarely 2.0) × as long as broad, exclusive of basal cilia . . . . var. *nipponica*
   2. At least partly monoecious; perianths produced in large numbers; cells thick-walled; perianth mouth with teeth only 1–3-celled . . . . . .
   . . . . var. *myriocarpa*
1. Plants with underleaves variable, in part large and lanceolate to ovate; dioecious . . . . var. *heterostipa*

JAMESONIELLA AUTUMNALIS var. HETEROSTIPA (Evans) Frye & Clark

Jamesoniella heterostipa Evans, The Bryologist 18:81, pl. 1, 1915.
Jamesoniella autumnalis var. *heterostipa* Frye & Clark, Univ. Wash. Publ. Biol. 6(2):274, figs. 1–13, (1943) 1944.

In depressed, *deep green patches on wet rocks*, somewhat brownish-tinged. Stems to 600 μ in diam., sparingly branched, observed branches terminal and of *Frullania*-type. Leaves more or less horizontal, but variously revolute and vaguely plicate, to 1500–2000 μ long × 1400–1900 μ broad, rarely emarginate at apex. Cells thin-walled, ca. 19 μ on margins, 24 μ in leaf middle, 34 μ at base, isodiametric; cuticle smooth. *Underleaves polymorphous*, some minute and evanescent, *some persistent and much larger*, varying from filiform to ovate, free or adnate at base with a lateral leaf, rounded to acuminate at apex, entire margined, *up to 1300 μ long × 750 μ broad*.

Dioecious. Outer bracts to 2400 μ long × 2000 μ broad, the innermost bract lacerate to laciniate-lobed. Perianth (immature) with mouth irregularly lobed, the acuminate lobes ciliate to dentate on margins.

Type. Barrington Passage, Cape Sable I., Nova Scotia (*Macoun 52*).

Distribution

Apart from the type, known only from Quebec: Oka (Lepage, 1944–45).

Evans (1915a) described this incompletely known taxon on the basis of the polymorphous underleaves. He admits that cells of typical *J. autumnalis* agree with those of the present plant, "while the perichaetial bracts and the mouth of the perianth are essentially the same . . . ," and also that the divergent characters (somewhat more robust size; polymorphous and partly lamellate underleaves) "may be caused by . . . (the) unusual habitat and that it may represent an aberrant form of *J. autumnalis*," but adds "in the absence of proof to this effect the plant . . . ought surely to be recognized as distinct." I would suggest that the polymorphous underleaves are at least partly a response to environment, and that therefore the plant does not deserve recognition as a species. Temporarily it seems wisest to follow Frye and Clark in relegating it to the rank of a variety.

JAMESONIELLA AUTUMNALIS var. MYRIOCARPA
(*Brinkman*) *Frye & Clark*

Jamesoniella myriocarpa Brinkman, The Bryologist 36:57, 1933.
Jamesoniella autumnalis var. *myriocarpa* Frye & Clark, Univ. Wash. Publ. Biol. 6(2):275, (1943) 1944.

Olive-green, on decaying wood, *freely branched*, often with gynoecial innovations. Leaves averaging 800 μ wide and long, subquadrate to rotundate, occasionally retuse. Cells near leaf apex ca. 19 μ, somewhat thick-walled, with small trigones. Underleaves constant only near gynoecia, *rare elsewhere and then small.*

Monoecious or heteroecious. ♀ Bracts subquadrate, 1100–1200 μ high × 900–1100 μ broad, truncate to retuse at apex, rarely slightly bilobed, their margins entire, rarely with short teeth. *Perianths numerous*, often closely approximated, pluriplicate, the mouth variously lobed and toothed (the lobes ending in a single tooth, or with lateral *teeth 1–3 cells high*). ♂ *Bracts situated some distance below perianths*, shorter than leaves, strongly concave at base, truncate or occasionally emarginate at apex.

Type. Pictou, Nova Scotia, (*J. Macoun 967, July 24, 1883*).

The plant has been inadequately described and never illustrated. According to Brinkman, who regarded it as a discrete species, it differs from typical *J. autumnalis* in that the cilia or teeth of the perianth are

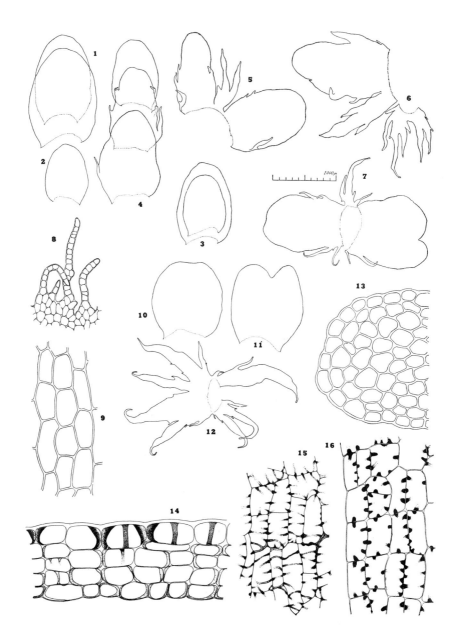

Fig. 255. *Jamesoniella autumnalis.* (1–3) Leaves (×17). (4) Gradation, top to bottom, from vegetative leaf to ♀ bract (×17). (5–7) Three cycles of ♀ bracts and bracteole, showing variation in one population (×17). (8) Cells of perianth mouth (×75). (9) Dorsal cortical cells (×180). (10–11) Leaves (×17). (12) ♀ Bracts and bracteole, showing maximally laciniate-divided form (×17). (13) Stem cross section (×180). (14) Cross section of capsule wall (×445). (15) Inner cells of capsule wall (×290). (16) Epidermal cells, capsule wall (×290). [1–8, *RMS 38060,* var. *nipponica,* White Top Mt., Va.; remainder all var. *autumnalis.* 9, 13, *RMS 32212,* N.C.; 10–12, *RMS 28886a,* Linville Gorge, N.C; 14–16, *RMS 11955a,* Susie Isls., Minn.; 1–7, 10–12, all to scale.]

only 3 cells high or less; in the multiplicate and lobulate perianth; and in the autoecious inflorescences. However, it seems more probable that an extreme variant of the polymorphous *J. autumnalis* is at hand. Frye and Clark (1944, *in* 1937–47, p. 275), who have seen no material, incorrectly consider the plant to be "unisexual, or sometimes bisexual," in spite of Brinkman's statement that it is monoecious or heteroecious. It is not clear from Brinkman's description how the numerous perianths are situated. Presumably they occur on abbreviated subfloral innovations which are soon again floriferous.

I have not studied the plant and, in the absence of any basis for an independent judgement, arbitrarily follow Frye and Clark.

JAMESONIELLA AUTUMNALIS var. *NIPPONICA*
(*Hatt.*) *Hatt.*
[Fig. 255: 1–8]

Jamesoniella nipponica Hatt., Jour. Jap. Bot. 19:350, fig. 26, 1943.
Jamesoniella autumnalis nipponica Hatt., Jour. Hattori Bot. Lab. no. 5:76, 1951.

In large patches on partly insolated, intermittently damp cliff walls, light green, with upper portions of leaves partly with a reddish brown tinge, the distal portions of *perianth carmine-red.* Stems freely branched, with subfloral innovations that are soon again floriferous. Leaves slightly to distinctly imbricated, somewhat concave and suberect or antically connivent, rarely widely spreading, *narrowly ovate to ovate-elliptical, widest usually in basal third,* the somewhat narrowed apex rounded, *never truncate or emarginate;* leaves varying from 680–700 μ wide × 825–830 μ long on lower parts of sterile shoots, on fertile shoots from 750 μ wide × 1000 μ long below, gradually larger above and becoming 875–900 μ wide × 1130–1220 μ long, to an extreme of 1100 μ wide × 1320 μ long (*averaging 0.74–0.85 as broad as long,* or 1.19–1.36× as long as broad). Underleaves virtually absent, present only in and just below gynoecia, and at bases of branches.

Dioecious. ♀ Bracts, and several pairs of leaves below them, imbricate and suberect or erect, subtransversely oriented, except for the spreading apices, *superficially suggesting a paroecious inflorescence.* Bracts sheathing perianth at base, only the apices erect-spreading to widely spreading, the outer varying from 600 μ wide × 1200 μ long (exclusive of cilia) to 875–980 μ wide × 1250–1300 μ long, up to 800 μ wide × 1525 μ long, with 1–2(3) teeth or cilia on one or both margins. Innermost bract often reduced in size and divided into 4–6 acuminate, irregular laciniae. Bracteole narrow, averaging less than half the width of bract, from 2–3-fid to lanceolate with 1–2 lateral teeth. Perianth mouth rubescent, cilia contorted, elongate (in part uniseriate for 9–16 cells).

Distribution

Described from Japan; recurring in Mexico (p. 822) and rare regionally:

Cliff face near summit of White Top Mt., Smyth Co., Va., ca. 5600 ft

(*RMS and P.M. Patterson 38060, June 9, 1956*). Known only from the above collection in our area.

This extreme variant deviates from normal *J. autumnalis* in the narrowly ovate-elliptical leaves, which appear never to develop the truncate-retuse to emarginate apices characteristic of all other phases of the species which have been seen. On this basis, difficultly recognizable when sterile, except by the reddish brown pigmentation, opaque appearance, and vestigial underleaves.

JAMESONIELLA UNDULIFOLIA (*Nees*) K. Müll.

[Fig. 256]

Jungermannia schraderi var. *undulifolia* Nees, Naturg. Eur. Leberm. 1:306, 1833.
Jungermannia autumnalis var. *schraderi* Pears., Hep. Brit. Isles, p. 305, 1902.
Jamesoniella schraderi Schiffn., Lotos (Krit. Bemerk. Ser. IX in Lotos) 59:70 (p. 14 of reprint),
  1911; Macvicar, Studs. Hdb. Brit. Hep. ed. 2:156, figs. 1–6, 1926.
Jamesoniella undulifolia K. Müll., Rabenh. Krypt.-Fl. 6(2):758, fig. 203, 1916; K. Müller, *ibid.* ed.
  3, 6:861, fig. 300, 1956.

Rather opaque green to somewhat brownish-tinged, locally reddish brown, growing erect and caespitose (with crowding), or loosely prostrate, *vigorous, often more robust than J. autumnalis,* the shoots 1–2(3–5) cm long × 1–1.5 mm wide, *appearing ± laterally compressed* because of the antically connivent leaves, simple or rarely branched (the branches at least frequently furcate and terminal). Stem greenish, flexuous, the cortical cells oblong to oblong-hexagonal, (13)14–18(20) × 38–52(64) μ, with thin, often somewhat brownish walls, 225–275 μ in diam. Rhizoids abundant, colorless. Leaves loosely imbricate, obliquely inserted, moderately to longly decurrent antically but not decurrent postically, inserted by a wide and oblique line, usually *antically secund and ± connivent,* broadly orbicular to transversely oblong-orbicular and almost without exception *considerably broader than long,* varying from 800–815 μ long × 970–1125 μ wide to 860 μ long × 1050 μ wide (ca. 1.15–1.4× as wide as long) up to 680 μ long × 1120 μ wide and to 640–750–900 μ long × 1150 μ wide (the *widest leaves thus 1.5–1.8× as broad as long*); leaves slightly to moderately concave, *soft-textured,* often the *distal portions of the margin somewhat incurved,* the leaves never convex, on robust plants and especially near the gynoecia *more or less undulate,* entire-margined and broadly rounded to *occasionally retuse or emarginate.* Underleaves relatively *distinct, but* small, usually *discernible* on sterile shoots, lanceolate, often ending in or armed with 1-several slime papillae. Cells moderately collenchymatous, the trigones rarely even slightly bulging, thin-walled or very slightly thick-walled, 19–24(25) μ on margins, the marginal not differentiated from interior cells; median cells 22–25(28) × 25–32(33) μ; basal cells ca. 25–32 × 26–34 μ; cuticle faintly verruculose. Oil-bodies 6–14, *mostly 8–10, per cell,* spherical and ca. 4–5.4(6.5) μ to ellipsoidal and 4–5 × 6.5–8 μ, a few to 6 × 9 μ, formed of distinct, somewhat protruding, moderate spherules. No asexual reproduction.

Dioecious. ♂ Plants with eventually intercalary androecia; bracts in several pairs, concave at base and with a ± incurved tooth of antical base; *2-androus; antheridial stalk 4-seriate.* ♀ Plants usually somewhat more vigorous, the *upper leaves ± undulate.* Bracts ± transversely oblong-quadrate to rounded-oblong, from 1010 μ wide × 775 μ long to 1100 μ wide × 890 μ long up to 1200–1350 μ wide × 850–900 μ long, similar in shape to the leaves, but *usually quite undulate,* often truncate-retuse to truncate-emarginate at the broad apex, the antical base varying from *commonly subentire* (with a minute basal tooth tipped by a slime papilla) *to bearing 1–2(3) small, often barely discernible, acute teeth,* which may be tipped with slime papillae when young (or when small). *Bracteole distinct,* free, variable in shape and size, usually no more than 0.2–0.4 the area of bract but sometimes larger, varying from lanceolate and entire-margined or with 1–2 basal teeth (and ca. 460 μ wide × 680 μ long) to oblong-ovate or even subquadrate and with crenulate-lobulate apex (and ca. 550 μ wide and long); lateral margins of bracteole often unarmed. Perianth inflated, *unistratose to base,* oblong-ovoid, obtusely pluriplicate in upper part, the mouth contracted and denticulate with 1–2-celled teeth (in European material) *or merely sinuate-crenulate,* with the terminal teeth free usually only at their rounded apices; terminal cells of latter type of perianth often nearly fingerlike, from 13–16 × 28–45(50) μ to 16–19 × 25–30 μ. Spores finely papillose, 14 μ in diam.; elaters 8 μ in diam., bispiral. Capsule wall as in *J. autumnalis.*

Type. Riesengebirge, near Warmbrunn and Herischdorf (*Nees;* in herb. Nees).

Distribution

A rare, sporadically distributed species, northern in range. Found occasionally in Scandinavia (Dalsland and Uppland, Sweden; Hordaland, Norway; Aland, Finland; Jylland and Bornholm, Denmark), Great Britain (Westmoreland, Argyll), and central Europe [Switzerland; Vosges; Odenwald; Saxony; Vogtland; the Riesengebirge; western Poland (formerly Ubedel); Eure-et-Loir, France].[291]

The species is excluded from the North American flora by both Frye and Clark (1937–47) and Grolle (1964i, p. 661). However, it occurs in west and northwest Greenland as follows:

W. GREENLAND. Sydostbugten ved Christianshaab, 69°35′ N. (*Vahl 111, Sept. 1835!*; reported by Lange & Jensen, 1887a, p. 98, as *Jungermannia schraderi,* cited as from "Moser, ofte mellem Sphagna").[292] NW. GREENLAND. Siorapaluk (Atikordlok), Robertson Bay, 77°48′ N., 71° W. (*RMS 45704, July 24, 1960*). Northernmost report of genus and species! The species is rare in Greenland; in 1966 I searched, without success, for it in west Greenland, from 69–72° N.

[291] According to Arnell (1956), also in Korea and Japan, but Grolle (1964i, p. 661) has shown that these reports derive from an earlier confusion between *J. autumnalis* and *J. undulifolia.*

[292] *Vahl 111* is typical, with crenulate perianth mouth; a small quantity was also detected in *Vahl 108,* mixed with *Lophozia binsteadii* and *Calypogeia sphagnicola,* from the same locality. Four other collections reported by Lange & Jensen as "*J. schraderi*" consist of *Mylia anomala* (Ameralik Fjord; Narssaq), *Solenostoma obovata* (Tunugdliarfik Fjord), and *Plagiochila* (Kobbefjord).

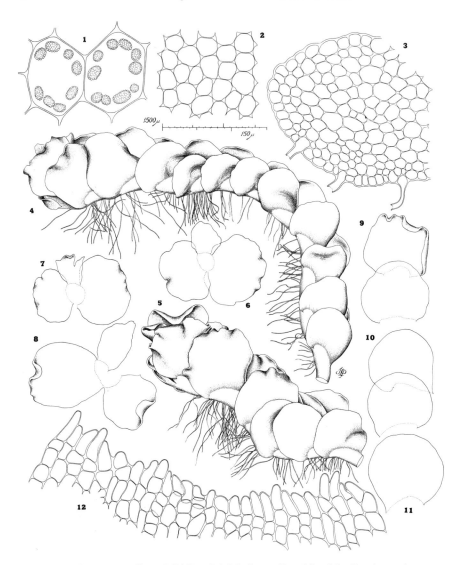

FIG. 256. *Jamesoniella undulifolia.* (1) Median cells with oil-bodies (×740).
(2) Median cells (×197). (3) Stem cross section (×180). (4–5) Apices of
gynoecial shoots, lateral aspect (×17.5). (6–8) ♀ Bract + bracteole cycles
(×17.5). (9) ♀ bracts (×17.5). (10–11) Leaves (×17.5). (12) Cells from
perianth mouth (×197). [All from *RMS 45756*, NW. Greenland; 2, 12,
drawn to lower scale; 4–11, drawn to upper scale.]

There are several old reports of *J. "schraderi"* by Sullivant and Underwood which are untrustworthy and refer surely to *J. autumnalis*, with which *J. "schraderi"* (= *J. undulifolia*) was confused for many years. Frye and Clark (*loc. cit.*, p. 274) for that reason refuse to accept this species as present in our area.

Ecology

Usually a helophytic species, growing commonly amidst *Sphagnum* in insolated sites.

At the single station in northwestern Greenland where I have found the plant, it was abundant, growing on a steep, peaty, wet slope, covered by a dense moss and moss-grass tundra, which was atypically richly developed because of continuous manuring by leaching from the guano deposited several hundred feet above by large numbers of dovekies. Associated here are *Sphagnum, Polytrichum* and the hepatics *Anastrophyllum minutum grandis, A. sphenoloboides, Lophozia groenlandica, Cephaloziella elegans, C.* aff. ad *C. byssacea,* and *Tritomaria exsectiformis* subsp. *artica.*

Differentiation

Usually larger and more vigorous than *J. autumnalis*, with a somewhat different aspect owing to the fact that the usually much broader leaves are antically secund, so that the plant looks laterally compressed. The leaves are perhaps somewhat softer, frequently more markedly decurrent antically, and often characteristically, if weakly, wavy or undulate on fertile plants. The less strongly pigmented, slightly brownish phases may have a striking similarity in size and aspect to *Odontoschisma sphagni,* less so to weakly pigmented *O. elongatum.* The lack of stolons, the less collen-chymatous cells, and the small and numerous oil-bodies at once separate *J. undulifolia* from any species of *Odontoschisma,* when sterile.

Like *J. autumnalis, J. undulifolia* is usually freely fertile. The undulate ♀ bracts and, in particular, the crenulate to short-denticulate perianth mouth readily separate it from any phase of *J. autumnalis.* The species appears to be sharply defined, in spite of statements in Müller (1905–1916, p. 759).

There has been confusion with regard to the correct name of this plant. Schiffner (1901–43) stated that *Jamesoniella schraderi* was the correct name for *J. undulifolia. Jamesoniella schraderi* is based on *Jungermannia schraderi* Martius (1817); of this no type seems to exist, but the diagnosis of the perianth mouth "ore breviter lacinulato" appears to indicate *Jamesoniella autumnalis,* described 2 years earlier by DeCandolle as *Jungermannia autumnalis.* Müller (*loc. cit.* and 1942, pp. 122–124) has shown that of 22 specimens in Nees's herbarium as

"*Jungermannia schraderi*" 18 belong to *Jamesoniella autumnalis* ("var. *communis*" or fo. "*subapicalis*"); thus Nees understood, under the name *J. schraderi*, essentially what is today universally known as *Jamesoniella autumnalis*. Therefore, Müller (1916) was justified in elevating Nees's var. *undulifolia* to species standing as *Jamesoniella undulifolia*.

JUNGERMANNIA [*Ruppius*] L.

[Fig. 257]

Jungermania L., Spec. Pl., p. 1131, 1753 (in part), and of post-Linnean authors.
Liochlaena Nees, *in* G., L., et N., Syn. Hep., p. 150, 1844; Meylan, Beitr. Krypt.-Fl. Schweiz 6(1):150, 1924.
Haplozia subg. *Liochlaena* K. Müll., Rabenh. Krypt.-Fl. 6(1):572, 1909.
Jungermannia Schust., Amer. Midl. Nat. 49(2):394, 1953.

Prostrate, with ascending shoot apices, green, with brownish (rarely reddish brown) secondary pigmentation; branching lateral, normally intercalary and from lower halves of leaf axils. Stems slender, with scattered colorless (or brownish) rhizoids postically, the unistratose cortex ill-defined, of short-rectangulate, only slightly thick-walled cells subequal in diam. to those of the medulla; medulla homogeneous. Leaves strongly succubously inserted and oriented, spreading, *averaging longer than broad, oblong-rectangulate to oblong-ovate*, rounded to *occasionally* retuse at apex, entire-margined. Cells chlorophyllose, collenchymatous, the walls thin, trigones distinct to slightly bulging; cuticle distinctly *verruculose;* oil-bodies moderate in size (5–10 μ long × 5–7 μ wide), finely granular in appearance, 7–11 per cell. *Underleaves quite lacking*, except on erect, gemmiparous shoots. *With asexual reproduction by means of 1–2-celled, ovoid to ellipsoid gemmae, arising in fascicles* from margins and abaxial face of reduced leaves; gemmiparous shoots erect or suberect, attenuate.

Paroecious (ours) or dioecious. Gynoecia with ♀ *bracts essentially like leaves*, entire and without basal cilia or teeth; *bracteole absent. Perianth not subtended by perigynium*, longly emergent, *cylindrical and smoothly terete, at apex suddenly constricted into a short, slender beak set in a shallow depression.* Sporophyte seta of numerous cell rows; capsule ovoid, *wall 2-stratose.* Epidermal cells larger than inner cells, with strong nodular thickenings. Inner layer of narrower, more elongate cells, bearing semiannular bands. Elaters 8–9(10) μ in diam., bispiral; spores 12–14 μ, obscurely verruculose or granulate.

Type. Jungermannia lanceolata L. emend. Schrad.

Jungermannia s. str., as here delimited (in conformity with usage of the vast majority of prior workers, e.g., Joergensen, 1934, Müller, 1951–58, Schuster, 1953, Jones, 1958a), appears to include only 2 species, the European-North American *J. lanceolata*, with 17 chromosomes and a monoecious inflorescence, and the Asiatic *J. cylindrica* (Steph.) Hatt. (see Grolle, 1965a, pp. 211–212), which has 9 chromosomes and is dioecious.[293]

The genus differs from all other genera of Jungermanniaceae in the unique, terete and smooth perianth, whose apex is suddenly, truncately contracted into a short beak, set in a slight depression. Fully as important as a "generic character" is the ability to produce gemmae from the leaf margins (and distal portions of the abaxial leaf faces), much as in the Lophoziaceae. Although *J. lanceolata* is considered by some workers (Evans, 1940; Buch, Evans, and Verdoorn, 1938; Frye and Clark, 1937–47) as congeneric with the species here classed in *Solenostoma* subg. *Solenostoma*, I agree with Müller (1940) and Meylan (1924) in believing *J. lanceolata* to be without close relatives. Meylan (*loc. cit.*) followed Nees in placing this isolated species in the genus *Liochlaena*. Howe (1899, p. 97), among others, discusses the synonymy of this species and also arrives at the conclusion that "in the spirit at least of the 'law of residues'" we must consider *J. lanceolata* (and its relatives, if any) as "the logical heir to the generic name *Jungermannia*."[294]

Although *Jungermannia* Linnaeus (1753), based on *Jungermannia* of Ruppius, originally embraced all so-called "scale mosses" or leafy Hepaticae, excepting *Porella*, the genus was greatly narrowed down during the first half of the nineteenth century. Except for Scandinavian authors who used *Jungermannia* for the group of species with lobed leaves now placed in *Lophozia* and allied genera, the generic name *Jungermannia* eventually came to be associated with the entire-leaved species today comprising the genera *Nardia*, *Jungermannia* s. str., and *Solenostoma* (including *Plectocolea*). This association has become so firmly established that adoption now of any other course would result in nomenclatural chaos.

Yet there are drawbacks to limiting the name *Jungermannia* to the *J. lanceolata* element. The most serious objection is that the Linnean *J. lanceolata* appears to be a composite species, none of whose components can be proved to be conspecific with *J. lanceolata sensu* Schrader.[295] According to the strictest

[293] Possibly allied to *Jungermannia* is *J. nuda* Lindenb. & G. of Mexico, *in* Gottsche, 1863, pl. 18. This species has a slightly rostrate and depressed perianth apex, exactly as in *Jungermannia*. Gottsche, however, states that there are lanceolate-subulate underleaves toward the apices of ♀ shoots and on innovations.

[294] In this connection the vexatious problem of whether to spell *Jungermannia* with one, or two *n*'s may be examined. I believe, with Howe (1899, p. 98), that one should "not interfere with the spelling adopted by Ruppius, Micheli, and Linnaeus," i.e., use *Jungermannia*, since Bescherelle (Jour. Bot. 7:191, 1893) states that he "has been informed by other botanists, among them Herr Stephani, that Jungermann wrote his name with a single *n* or with two *n*'s." Others, however, have taken the stand that Ludovicus Jungerman "never wrote his name Jungerma*nn* but always Jungerma*n*" [Wilbrand, Flora 9(2):518, 1826] and therefore designate the genus *Jungermania*. Among recent writers to adopt this spelling is Müller (1951–58).

[295] There is no Linnean type specimen; the several specimens of Dillenius at Oxford, which served *in part* for Linnaeus' concept of *J. lanceolata*, consist of various taxa and, in my opinion, can hardly serve to typify *J. lanceolata*.

interpretation of the International Rules, which *at present* make no provision for *nomina specifica conservanda*, *J. lanceolata* Schrad. (1797) could be considered as a later homonym of *J. lanceolata* L. (1753) and therefore would be illegitimate. Yet *J. lanceolata sensu* Schrad. (nec Linnaeus?) has been in continuous use, in exactly our present sense, since 1797. It is almost unique among older species of Hepaticae in lacking any synonymy, indicating a singularly even and uneventful history. It therefore seems scarcely defensible to apply the rules rigorously in such a case, since this would result in chaos where there is now order—a course of events which the rules are supposed to prevent. Therefore, I suggest that we informally consider *J. lanceolata* L. emend. Schrad. to be a *nomen conservandum*, whether or not our present rules sanction such a course. Any other procedure is clearly out of the question. *All* recent taxonomic treatments implicitly accept such a "conserved" *J. lanceolata*.

Acceptance of such a course is inherent in the nomenclatural usage of most recent workers (Buch, Evans, and Verdoorn, 1938; Frye and Clark, 1937–47; K. Müller, 1939; Schuster, 1949a, 1953, etc.). The only other logical alternative is that *Jungermannia* be given up altogether, a course advocated by Schiffner (1893–95) yet neither sanctioned by the International Rules nor consistent with the fact that the groups Jungermanniaceae, Jungermanniinae, Jungermanniales, and Jungermanniae are all predicated on the existence of a genus *Jungermannia*!

Amakawa (1960a) has returned to the conservative position of treating *Jungermannia* in a broad sense, including not only the single species here treated but also species that I would relegate to *Solenostoma*. This point of view is quite defensible. Yet, as I have pointed out (Schuster, 1953), *J. lanceolata* is sharply isolated, without close relatives in *Solenostoma*. Since no transitional taxa between *Solenostoma* and *Jungermannia* are known, I retain the now almost customary generic distinction between them.

Since the preceding paragraphs were written, Grolle (1966e) has adopted the conservative generic delimitation of Amakawa and taken the drastic step—one which no other taxonomist has had the "courage" to do—of tampering with the functional typification of *Jungermannia*, a typification which has not been challenged since Howe (1899) stated explicitly that we must consider *J. lanceolata* (and its relatives, if any) as "*the logical heir to the generic name Jungermannia.*" Howe, as well as all other workers since at least the time of Hooker (1816), have uniformly interpreted *J. lanceolata in the sense here understood*. I cannot bring myself to adopt the radical proposals of Grolle, which, in effect would (*a*) alter the concept of *J. lanceolata* so that it would correspond with *J. riparia*; (*b*) throw *J. riparia* into the synonymy of the emended *J. lanceolata* and, most important, (*c*) change the entire taxonomic concept of *Jungermannia*. The majority of workers who have adopted a narrow generic concept of *Jungermannia* (to include *J. lanceolata*) would have to use *Liochlaena* for this.[296]

[296] With a "new" delimitation of *Jungermannia*, the opportunity for creation of many *nomina nova* would arise—of which several have already been proposed, e.g., *Jungermannia* "*karl-muelleri*" Grolle and *J.* "*jenseniana*" Grolle—not to speak of numerous "new combinations," e.g., *Jungermannia levieri* Grolle, *J. subelliptica* Grolle, *J. lignicola* Grolle, *J. paroica* Grolle. It seems necessary to add that all of us active in the group have been aware of the "*Jungermannia* problem" but have

The implicit typification of *Jungermannia* by *J. lanceolata*, as that species has been understood in every work dealing with Hepaticae since the time of Hooker (1816), should, for many reasons, therefore remain inviolable. Jones (1958a, p. 371) neatly sums up the status quo: "As Schuster remarks (1953, p. 388), it is by 'common consent' rather than on good nomenclatural grounds that *Jungermannia lanceolata* L. is accepted as the type of the genus *Jungermannia*; it is in fact the only one of Linnaeus' species of *Jungermannia* which was not subsequently transferred to some other well-defined genus; Joergensen was the first to treat it as the type of the genus." And, it must be added, every worker, including Howe (1899), Joergensen (1934), Jones (1958a), Schuster (1953, 1958c), McGregor (1955), Amakawa (1960a), Frye and Clark (1937–47), and Müller (1905–16; 1951–58), to name only a few, has adopted *J. lanceolata* in that sense—and not in Grolle's sense (equivalent to *J. riparia*). Attempts, at this late date, to use ambiguous pre-Linnean specimens and/or illustrations to alter long-existing nomenclature should be avoided at all costs.

JUNGERMANNIA LANCEOLATA L. emend. Schrad.[297]

[Fig. 257]

Jungermannia lanceolata L., Spec. Pl., p. 1131, 1753 (in part?); Schrader, Syst. Samml. Krypt. Gewächse 2:4, 1797; Frye & Clark, Univ. Wash. Publ. Biol. 6(2):278, figs. 1–10, (1943) 1944; Schuster, Amer. Midl. Nat. 42(3):579, pl. 11, figs. 8–9, 1949; Schuster, *ibid.* 49(2):394, pl. 30, figs 1–6, 1953.
Liochlaena lanceolata Nees, *in* G., L., & N., Syn. Hep., p. 150, 1844.
Aplozia lanceolata Dumort., Hep. Eur., p. 58, 1874; Macvicar, Studs. Hdb. Brit. Hep. ed. 2:153, figs. 1–3, 1926.
Solenostoma lanceolata Steph., Spec. Hep. 2:60, 1901.
Haplozia (subg. *Liochlaena*) *lanceolata* K. Müll., Rabenh. Krypt.-Fl. 6(1):572, fig. 286, 1909.
Jungermannia leiantha Grolle, Taxon 15: 187, 1966.

In prostrate mats or patches, light, rather *translucent green to slightly brownish* (rarely slightly reddish-brown-tinged), medium-sized; shoots 2.0–2.5 mm wide × 2–3 cm long. Stems irregularly and sparingly branched, the branches usually lateral-intercalary, from lower half of the leaf axil; stems 190–250(300) μ in diam.; cortical cells only slightly thick-walled, short-rectangulate, 20–25 μ wide × 35–60 μ long (length-width ratio ca. 2–3:1); medulla of cells subequal in diam. to those of the cortex, hardly different from these in cross section; stem 9–12 cells high, only slightly dorsiventrally flattened. Rhizoids pale

---

[297] Frye and Clark (1937–47, p. 276) state that the "Linnean packet of *J. lanceolata*, the type species, is a mixture, and no plant in it agrees with his description." Schiffner, however, showed long before that the Linnean *J. lanceolata* was based, not on an actual specimen, but on illustrations in Dillenius and Micheli.

---

shown restraint in the face of *ex post facto* rules that at times invite—when they do not demand—creation of nomenclatural chaos.

I would urge adoption of the alternative solution suggested by Grolle—one he eventually rejects—of creating a neotype for *J. lanceolata* (*sensu* Micheli, 1729) that conforms to current usage. All taxonomic harm would thus be avoided.

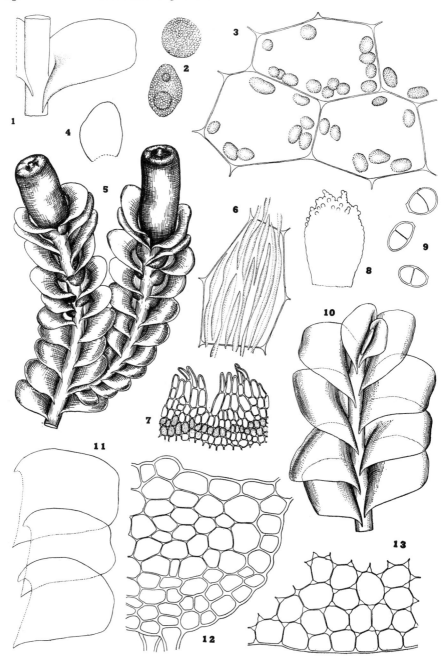

FIG. 257. *Jungermannia lanceolata*. (1) Leaf, *in situ*, of narrow-leaved swamp form (×ca. 15). (2) Two oil-bodies (×1550). (3) Median cells, leptodermous phase, with oil-bodies (×535). (4) ♀ Bract (×7.5). (5) Paroecious plant with perianths (×ca.

brownish with age, numerous, to stem apex. Leaves slightly to moderately imbricate, obliquely inserted, *nearly horizontally, widely spreading*, often with slightly deflexed apices, *oblong-rectangulate to oblong-oval* (if widest clearly at any one point, that point near or below leaf middle), ca. 920–1050 μ wide × 1100–1250 μ long (*1.1–1.25× as long as wide*), rounded to *rounded-truncate, frequently retuse at apex;* antical base somewhat decurrent. Cells quite chlorophyllose, collenchymatous, polygonal to rounded-polygonal, the trigones distinct (swamp forms) to *distinctly bulging* (normal forms); marginal and apical cells (24)27–30 μ; median cells ca. 28–32(35) × 30–38 μ; *cuticle distinctly verruculose*, near leaf bases becoming striolate; oil-bodies mostly *7–11 per cell*, slightly granular (because of the numerous slightly defined, contained oil-spherules), from ca. 5–7 μ and spherical to 5–6 × 10 μ (a few ellipsoid and 5.5 × 13 μ) and ovoid. Underleaves lacking. *Asexual reproduction* (rarely present!) *by means of gemmae from the apices of erect, slender, attenuated shoots* (2–5 mm high), which occasionally bear minute to large underleaves, and bear reduced oblong leaves (ca. 300–400 μ wide × 450–600 μ long), the upper scalelike and erect-appressed; gemmae greenish, more or less ellipsoidal, often slightly constricted medially, *2-celled*, not or slightly thick-walled, from *20 × 23–25 μ to 20–23 × 28–33 μ.*

Paroecious (rarely with some purely ♀ branches, or with purely ♂ branches). Perigonal bracts in 2–4 pairs, similar in size to leaves, but somewhat more broadly oval or ovate, *prominently ventricose at base only, the distal portions spreading*, antical base inflexed; antheridia 1–3. ♀ Bracts similar to leaves and ♂ bracts, somewhat smaller to somewhat larger than vegetative leaves, oval to somewhat ovate-oblong and widest below the middle, ca. 1200 × 1600 μ, erect and sheathing at base, the apices spreading to reflexed. Perianth conspicuous, 0.7–0.85 emergent, ca. 1 mm in diam. × 2.8–3.2 mm long, *longly tubular* (sometimes slightly increasing in diam. distally), *smooth, eplicate, precipitously contracted distally into a small beak set in a shallow, ringlike depression*, in profile virtually truncate at tip; cells in 1–2 apical rows elongated, elsewhere ca. 25–35 × 35–44 μ. Sporophyte as in generic diagnosis. Chromosome no. = 17.

Type Locality. Uncertain (see discussion under genus).

Distribution

A very widespread species, locally common but generally infrequent, of holarctic range, found throughout much or most of the Taiga, absent everywhere in the Tundra, extending southward into the southern portions of the Deciduous Forest Region, but rare and local in the Coastal Plain of the southeastern United States, as far south as North Carolina.

8.5). (6) Upper cell of fig. 3, showing surface striae (×535). (7) Perianth mouth cells (×ca. 52). (8) Gemmiparous leaf (×41). (9) Gemmae (×300). (10) Sterile shoot (×16). (11) Leaves (×21). (12) Stem cross section (×220). (13) Marginal and apical cells (×220). [1–3, 6, *RMS 12115*, Big Susie I., Minn.; 4–5, 7, Sixmile Creek, Ithaca, N.Y., *RMS*, after Schuster, 1953; 8–9, *Anderson 7330*, N.C.; 10–13, *RMS 36200a*, Crabtree Creek, Wake Co., N.C.]

In Europe widespread except in the far north, ranging from central Scandinavia (Norway, Sweden, Finland) southward through Denmark, England (very rare), the Low Countries, Germany, northwest France, French Basque, into Switzerland, Austria, Spain, Italy, the Baltic Provinces of Russia, northern Russia, the Caucasus.[298] Also on the Canary Isls. (Tenerife). Absent from Scotland, Ireland, and the Faroes, the distribution tending to be "continental."

In North America transcontinental in the Taiga, presumably, but with a hiatus in its range in the region west of Minnesota; westward known from southern Alaska to British Columbia, Washington, Oregon, south to California; in the Rocky Mt. region south to Montana, Idaho, and Utah. In our area as follows:

LABRADOR. Macoun (1902). NEWFOUNDLAND. Witlers Bay (*Waghorne 1951, p.p.*, among *Scapania nemorosa*); Hare Bay and Big Bay, Rencontre West, S. coast; Steady Brook, Benoit Brook, Glenbournie, in W. Nfld.; Cow Head, Stanford R., Port au Choix, St. John I., Eddie's Cove, St. Barbe Bay, Cook Harbor, St. Anthony, all on N. Pen.; Millertown Jct., Gambo, in C. and NE. Nfld. (Buch & Tuomikoski, 1955). MIQUELON ISLAND. (Delamare et al., 1888). QUEBEC. Mt. Lac des Cygnes, St. Urbain; La Tuque; Tadoussac; Bic; Pont-Rouge; Mt. Shefford (Lepage, 1944–45; Fabius, 1950; Kucyniak, 1947). ONTARIO. Belleville (*Macoun*); L. Timagami. NOVA SCOTIA. Margaree; Sandy Cove, Digby Co.; Blomidon (Brown, 1936a); E. side of Cape Breton Natl. Park, 3.5 mi S. of Mary Ann Brook waterfall (*RMS 66325*).

MAINE. Pleasant Ridge, Somerset Co. (*Chamberlain 3305*); Mt. Desert I.; ridge descending from Mt. Katahdin to NW. Basin (*Schuster, 1949*). NEW HAMPSHIRE. Franconia Mts. (*Evans, 1908*); Wildwood Path (*Lorenz, 1908*); Melburne (*Farlow!*). VERMONT. Newfane (*Grout, 1909!*); Willoughby (Lorenz); L. Dunmore; SW. slope of Willoughby Mt., Willoughby L. (*RMS 46213b*). MASSACHUSETTS. Worcester (Greenwood, 1910); Mt. Greylock (Andrews, 1902); Bear R., Conway (*RMS 66-1601*); Tannery Falls, Savoy State Forest (*RMS, s.n.*). RHODE ISLAND. *Fide* Evans (1923g). CONNECTICUT. Hamden, New Haven, Oxford, all in New Haven Co. (Evans & Nichols, 1908).

NEW YORK. Enfield Glen, Tompkins Co. (Schuster, 1949a); Rice Brook, Allegany State Park, Cattaraugus Co. (*Boehner 1941!*); Sixmile Cr., S. of Ithaca, Tompkins Co. (Schuster, 1949a); Warren and Washington Cos. (*Burnham, 1918!*); Kirkville, Onondaga Co. (Underwood & Cook, Hep. Amer. no. 114); Irondequoit Bay near Rochester, Monroe Co.; Briggs Gully, Ontario Co.; Plymouth Bog, Chenango Co; Van Etten, Chemung Co. (Schuster, 1949a); Little Moose L., Herkimer Co. (*Haynes*); Ft. Edward, L. George (*Burnham*). NEW JERSEY. Highlands, Monmouth Co. (*Haynes, 1905*). VIRGINIA. Great Falls (*Svihla 57*). WEST VIRGINIA. Tibbs Run; Monongalia, Preston Cos. (Ammons, 1940). KENTUCKY. Natural Bridge, Powell Co. (*Taylor 35*); Carter, Lewis, Whitley Cos. (*Fulford*). TENNESSEE. Blount Co. (Sharp, 1939); Overton Co. (Clebsch, 1954).

NORTH CAROLINA. Durham, in the Piedmont (*Blomquist, 1931!*); Hemlock Bluff, near Cary, Wake Co., in the Coastal Plain (*RMS 28709*, etc.); Crabtree Cr. State Park,

---

[298] Reports of the species, as *J. lanceolata* or as *J. l.* subsp. *stephanii* Amak. (1960a, p. 71), from Japan and India (the Himalayas) refer to the dioecious *J. cylindrica* (Grolle, 1965a, pp. 211–212.)

Wake Co. (*RMS 36200a*); Neddie Cr., E. fork of Tuckasegee R., Jackson Co. (*RMS 29446*); 5 mi E. of Troy, Montgomery Co. (*Anderson 4835!*); Shepherd Mt. near Asheboro, Randolph Co., with gemmae! (*Anderson 7330!*); Winston-Salem, Forsyth Co. (*Shallert!*); below Dry Falls, Cullasaja R., Macon Co. (*RMS 36089*); Lower Rock Bridge, above Bonas Defeat, E. fork of Tuckasegee R., Jackson Co. (*Anderson 10610!*); Roan Mt., Mitchell Co. (*Shallert!*).   GEORGIA. Dade Co. (Carroll, 1945).   ALABAMA.   Flat Cr., 0.5 mi S. of Lacon, Morgan Co. (*Correll 8272!*).

Westward to OHIO. Athens, Hocking, Jackson, Lake, Monroe Cos. (Miller, 1964). INDIANA. Putnam Co. (*Underwood, 1891*).   MICHIGAN. Copper Harbor, Keweenaw Co. (*RMS, s.n.*); Amygdaloid I., Isle Royale (*RMS 17329a*); Gogebic, Marquette, Ontonagon, Leelaneau Cos.; Sugar I., Chippewa Co. (*Steere, Nichols*; Darlington, 1938); Bart L. and Douglas L., Cheboygan Co. (*Nichols*).   ILLINOIS. Fide Hague (1937).   WISCONSIN. Sand I., Apostle Isls., Bayfield Co. (*RMS 17676, 17546, 17551*); L. Nebagamon! and Brule R., Douglas Co. (*Conklin, 1903, 1923*); near Bayfield, Bayfield, Co. (*RMS, s.n.*; *c. gemmae*); Lafayette, Lincoln, Douglas, Iron Cos. (Conklin, 1929).

MINNESOTA. Taylors Falls, Chisago Co. (*RMS 1323a*). Cook Co.: Porcupine and Big Susie Isls., Susie Isls. near Grand Portage (*RMS 13635, 12115*, etc.); Hat Pt. near Grand Portage and Grand Portage Bog (*RMS 11489*, etc.); Lutsen and Hungry Jack L. (*Conklin 2003, 2293*). Two Island R., Lake Co. (*Conklin 2597*); Oneota and French R., near Duluth, St. Louis Co. (*Conklin 772, 514*); Whitewater State Park, Winona Co. (*RMS 18025; c. gemmae!*).   IOWA. Clayton Co. (Conard, 1945).   KANSAS. NE. of Tonganoxie, Leavenworth Co. (McGregor, 1955).

As has been pointed out (Schuster, 1953), the range of *J. lanceolata* corresponds rather closely with that of the superficially somewhat similar *Jamesoniella autumnalis*. Both species drop out as the ecotone between Taiga and Tundra is attained; both become rare and sporadic in the Piedmont-Coastal Plain border of the southeastern states (although *Jamesoniella* occurs sporadically in the extreme Coastal Plain in Virginia and North Carolina).

Ecology

Not only is the range of *J. lanceolata* somewhat like that of *Jamesoniella autumnalis*, but the ecology is also quite similar. *Jungermannia lanceolata* may indeed occur with *Jamesoniella*, *Ptilidium pulcherrimum*, *Scapania mucronata*, etc., on damp sandstone rocks, under shaded, somewhat moist conditions, under a pH as high as 6.0–7.0; such occurrences are exceptional, the species usually being more oxylophytic and more common on damp soil, often over rocks, very frequently in areas with high atmospheric humidity (as around the shores of lakes, such as Lake Superior), or in ravines or gorges. The underlying soil may be purely mineral or may be high in humic materials. Under such conditions, *Plagiochila asplenioides*, *Trichocolea*, *Geocalyx*, *Blepharostoma*, *Lepidozia reptans*, *Cephalozia media*, and sometimes *Pellia epiphylla* are associated northward; in weakly calcareous sites sometimes *Lophocolea minor*. Southward (Wake Co., N.C.) the species occurs on sandy banks, soil-covered rocks, and bluffs above streams with

Plagiochila, Cephalozia media, Pellia epiphylla and *neesiana, Trichocolea tomentella,* and *Riccardia multifida.*

Jungermannia lanceolata also occurs on peaty ground in boggy areas (with *Mnium* and *Riccardia multifida*) but is absent from true peat bogs. It also is frequent over moist decaying logs, northward associated with *Nowellia, Geocalyx, Blepharostoma, Calypogeia integristipula, Lophozia porphyroleuca, L. in-cisa, Riccardia latifrons, R. palmata, Tritomaria exsectiformis, Cephalozia media, C. pleniceps, Lepidozia reptans,* etc.; the pH here is at its lowest extreme, ranging from 4.5–5.2 (Schuster, 1953). More rarely *Calypogeia suecica, Lophozia longidens,* and *L. ascendens* are associated on decaying logs.

In general, *J. lanceolata* "avoids" bare acidic rocks, sandy soils, and other loose substrates that are very intermittently moist. It is typically mesophytic and usually restricted to areas with indirect lighting, exceptionally occurring in diffusely illuminated, humid sites, as a pioneer on damp granitic rocks (as under deep recesses beneath Dry Falls, N.C.); under such conditions admixed with *Harpanthus scutatus,* growing adjacent to *Plagiochila caduciloba* and *sullivantii.*

Variation and Differentiation

In general, a stenotypic species, whose identification, when fertile, affords no problems whatsoever.[299]

The species is most common as a mod. *viridis,* rather rarely developing light brownish secondary pigmentation. Very exceptionally (Sand I., Wisc.), there occur over moist sandstone ledges phases which have the upper leaves and upper portions of the perianths suffused with a prominent reddish or carmine tertiary pigment, ± masking the brown. Such reddish pigmentation is uncommon (Schuster, 1953).

The plants have been stated to produce gemmae only very rarely. Indeed, the only detailed description and illustration of these I have seen are in Warnstorf (1903), those of Frye and Clark (1937–47) being derived from Warnstorf's. Gemmae appear to be produced rarely, if at all, through most of the North American range of the species (few of my collections from the East show gemmae), yet they are frequently found on plants from the Lake Superior district (Schuster, 1953), being reported from Michigan (Copper Harbor), Wisconsin (near Bayfield; Sand I., Apostle Isls.), and Minnesota (Taylors Falls; Whitewater State Park).[300] They also occur on Newfoundland plants (Squires Memorial Pk., *RMS 68-1659a*) from a dry boulder.

[299] American students appear to have difficulty in separating this unique species from *Mylia* spp. and from *Solenostoma obscurum.* I have seen *Mylia anomala* from Ontario labeled "*J. lanceolata;* det. Fulford," and plants of *J. lanceolata* labeled *Mylia taylori.* The much larger leaf cells and the conspicuous underleaves separate *Mylia* from *J. lanceolata.* The Swain Co., N.C., report of *J. lanceolata* (Blomquist, 1936) is based on a specimen of *Solenostoma obscurum* (det. "*J. lanceolata*" by Fulford). The intense claret-colored to violet-purple pigmentation of the rhizoids and postical leaf bases, and the wholly different perianths and inflorescences, at once separate *S. obscurum* from *J. lanceolata.*

[300] McGregor (1955, p. 66) also notes that his Kansas material seems "regularly to produce gemmae" and differs thus from "published descriptions." It is possible that in the Midwest there

The gemmiparous plants are distinctive, since they produce gemmae on erect or suberect shoots with vestigial leaves, much as in *Lophozia heterocolpa* or *Odontoschisma denudatum*. Such gemmiparous plants are usually rather infrequent, but in the material collected over damp humus at the side of a path in Whitewater State Park the sterile plants are largely gemmiparous (Schuster, 1953). Gemmiparous shoots are pale, often ivory-white or slightly brownish-tinged, slender in form with reduced, narrowly oblong, erect-appressed leaves. As in *L. heterocolpa*, the gemmiparous leaves are only slightly, if at all imbricate; as in that species, gemmae are ovoid, smooth, and 2-celled. *Jungermannia lanceolata* always appears to have pale, whitish gemmae, and their form is much more broadly ovoid.

Although underleaves are quite absent on normal shoots of *J. lanceolata*, the erect, gemmiparous shoots often show these very distinctly, their size in many cases approaching that of the lateral leaves. There appears to be some correlation between the occurrence of the species on intermittently damp substrates, such as well-drained slopes or soil, and gemmae production. At least, none of the plants I have seen from permanently damp sites produced gemmae. Breidler (1894) earlier described the gemmiparous phase as var. *prolifera*; such taxonomic recognition is not warranted.

When fertile, *J. lanceolata* is hardly mistakable for any other species. The truncately terminated, tubular perianths with a small beak set in a depression of the apical truncation are unique. Sterile plants are harder to identify, but when well developed are characterized by the (*a*) somewhat rectangular, parallel-sided leaves, whose apices are commonly truncate or even retuse; (*b*) rather numerous, granulose oil-bodies per cell; (*c*) verruculose to striolate cuticle; (*d*) usually somewhat bulging trigones of mature leaves; (*e*) lack of underleaves; and (*f*) colorless, scattered rhizoids. On robust plants, and those from wet, shaded sites, leaves may be prominently rectangular, with rounded-truncate to truncate-retuse apices (Fig. 257:1, 11). Such plants may be confusing because they may show only small trigones (Fig. 257:3). Sterile plants are perhaps most readily confused with *Jamesoniella* but normally have clearly more oblong and more elongated leaves, and a more distinctly verruculose or striolate cuticle. Also, *J. lanceolata* averages somewhat larger than *Jamesoniella*.

The only nearly allied species is *J. cylindrica* (Steph.) Hatt. of Japan and the Himalaya (see, e.g., Amakawa, 1960a, p. 71, figs. 1c, 2b, 41; as *J. lanceolata* subsp. *stephanii* Amak.); this differs in being dioecious, in having 9 as the haploid number of chromosomes, and, apparently, in the gemmae. In our plants they are 2-celled when mature and larger (20–23 × 23–33 μ) than in

---

has been selection for biotypes in which gemmae production is abundant; this may perhaps be related to the lower moisture levels of the sites to which the species is confined in this region. I have also seen gemmae on a plant from Randolph Co., N.C. (*Anderson 7330*).

the oriental plant, in which Amakawa describes them as 1-celled and only 14 × 18 μ.

NARDIA S. F. Gray

[Figs. 258–267]

Jungermannia, auct., *p.p.*

Nardius S. F. Gray, Nat. Arr. Brit. Pl. 1:694, 1821, *p.p.* (*N. scalaris*, type); Evans, Ann. Bryol.
     10:36–42, 1938 (as *Nardia*); Frye & Clark, Univ. Wash. Publ. Biol. 6(2):306, (1943)
     1944 (as *Nardia*).
Mesophylla Dumort., Comm. Bot., p. 112, 1822, *p.p.* (*M. compressa*, type).
Alicularia Corda, *in* Opiz, Beiträge 1:652, 1828 (1829); Nees, Naturg. Eur. Leberm. 2:448,
     1836, *p.p.*; G., L., & N., Syn. Hep., p. 12, 1844, *p.p.*; K. Müller, Rabenh. Krypt.-Fl.
     6(1):508, 1909; Macvicar, Studs. Hdb. Brit. Hep., ed. 2:123, 1926.
Solenostoma subg. *Alicularia* Mitt. *in* Godman, Nat. Hist. Azores, p. 319, 1870.
Nardia sect. *Mesophylla* Carr., Brit. Hep., p. 10, 1874.

Generally in mats or patches, medium-sized (shoots usually 1–2.5 mm wide) or rarely minute (*N. breidleri*), ± prostrate to ascending, the shoot tips ascending or suberect (exc. *N. compressa*); shoots sparingly branched, *branches largely intercalary from the ventral halves of the leaf axils*, rarely with terminal, furcate branching. Stems somewhat to *conspicuously fleshy, stout*; cortical cells leptodermous, *generally only weakly elongated* [ca. 16–28(35) μ wide × (27)32–38(45) μ long], not or hardly smaller in diam. than the medullary cells, which are equally leptodermous, but much elongated (to 125 μ long). Rhizoids frequent to numerous, *scattered.* Leaves *succubously inserted and oriented*, spreading or somewhat antically secund and suberect, *entire to bilobed*, becoming subtransverse near shoot apices, *orbicular to broader than long* and subquadrate to subreniform, *entire or shallowly bilobed, broadly attached* by a long and scarcely acroscopically arched line, and gradually going over into axial tissue; leaf margins entire. *Underleaves distinct, lanceolate to sublinear*, small, usually free from the lateral leaves. Cells *usually distinctly collenchymatous* (except *N. breidleri*), small to medium-sized (15–40 μ), with a *smooth* (rarely verruculose) *cuticle; oil-bodies solitary or from 2–3 to 2–5(6) per cell, very large*, either homogeneous (and then often faintly transversely barred with age) or finely granulate. *Without asexual reproduction.*

Dioecious or paroecious. Androecia becoming intercalary; bracts somewhat similar to and often hardly more concave than leaves, equal to them in size or (in paroecious species) commonly larger, often more distinctly emarginate than leaves, concave at least at base; antheridia usually (1)2–3 per bract, their stalk 2-seriate. Gynoecia terminal on leading shoots. ♀ Bracts similar to or larger than leaves, often shallowly bilobed, *erect and closely sheathing the immersed perianth. Perianth very short, included within bracts*, narrowed to mouth but not beaked, crenulate at apex, *situated at the tip of a prominent fleshy stem-perigynium* (which may be continuous with axis or else is elaborated more highly along ventral side of stem, forming a more or less rhizoidous, bulbous perigynium at an angle to the axis). Calyptra formed (except at the tip) from elaboration of the shoot apex (i.e., *with formation of a calyptral perigynium as well as shoot perigynium*), thus with the old archegonia situated well up on the "calyptra." Capsule ovoid-globose or (*N. geoscyphus*) ± globose, 4-valved to base, the wall largely or wholly 2-stratose. Epidermal cells large, with nodular thickenings, ca. $1.5\times$ as thick as inner cells; inner layer of smaller cells, bearing semiannular bands. Spores 9–22 μ in diam., 1.8–2 × the diam. of the elaters. Elaters (1)2 (rarely 3–4) spiral. Seta long, slender, of 16–18 or more epidermal cell rows similar in diam. to (or hardly larger than) the numerous interior cell rows.

Type. Amakawa (1960a, p. 276) considers *N. compressa* the type. This typification is untenable, since Evans (1938c, p. 36) has "definitely selected" *N. scalaris* as the type. Evans gives a carefully reasoned history of the genus.

The genus *Nardia* is a well-defined taxon, largely of holarctic distribution; it is particularly well developed in Japan (nine species; see Amakawa, 1960a). Evans (1938c) showed that the name *Nardia* (or *Nardius* of S. F. Gray) had to be resurrected for the group commonly called *Alicularia* during much of the preceding century; the latter generic name was in vogue partly because Gray's generic names, with their masculine endings, failed to gain general acceptance. Under the name *Alicularia* Stephani (1901) lists only seven species, although two adequately defined taxa (*N. geoscyphus* and *N. lescurii*) are erroneously given as synonymous with *N. insecta*; Grolle (1964a, p. 299) lists fourteen. The majority of species are holarctic in range, although *N. notoscyphoides* Schiffn. comes from Java and *Nardia lindmanii* (Steph.), comb. n., is from Brazil.[301] A single species, *N. arnelliana* Grolle (Grolle, 1964a), occurs in the high mountains of tropical Africa.

Of the seven North American species all but one (*N. cavana* Clark) occur in our area. *Nardia cavana* appears to be a west coast endemic.[302] *Nardia breidleri*, a small species of the west coast and Rocky Mountains, also occurring in Asia and Europe, has been found a single time in Greenland.

Nardia in some respects is among the most specialized of the Jungermanniaceae. In some species, it has the most complex perigynium development to be found in the family. The shoot tip is elaborated not only below the

---

[301] *Nardia lindmanii* (Steph.), comb. n. = *Alicularia lindmanii* Steph., Kgl. Svensk. Akad. Handl. 23:25, 1897.

[302] *Nardia cavana* Clark is regarded by Evans as merely a form of *N. scalaris*. I have seen only a single specimen of *N. cavana* and can find no clear basis for distinguishing it from *N. scalaris*.

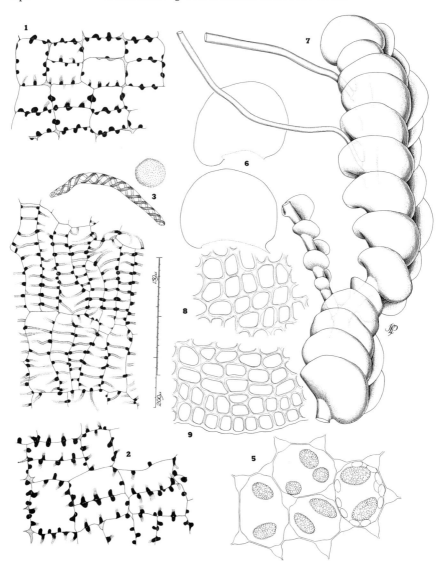

Fig. 258. *Nardia:* generic plate. (1–5) Sectio *Nardia*; (6–9) Sectio *Compressae.* (1–4) *Nardia insecta:* 1–2, epidermal cells of capsule wall (×280); 3, spore and elater (×280); 4, inner cell layer (×280). (5) *Nardia scalaris* subsp. *botryoidea:* median cells with oil-bodies and, at right, chloroplasts (×655). (6–9) *Nardia compressa:* 6, two leaves (×14); 7, shoot, lateral aspect, showing two leafless ventral-intercalary flagella and a ventral-intercalary leafy branch (×13); 8–9, median and apical cells, respectively; note thick walls with focus near leaf surface, thin walls with collenchyma in median focus (×165). [1–4, *RMS 43062*, Nova Scotia, all to scale at left; 5, *RMS & KD 66-027*, Tupilak I., W. Greenland; 6–9, from Narssaq, S. Greenland, *Vahl. Sept. 1829*; 8–9, drawn to scale at right.]

juncture of bracts and perianth (thus forming a tubular to bulbous perigynium at whose summit is situated the short perianth, closely sheathed by the much longer bracts), but also at the very apex of the shoot, in the archegonial region (thus resulting in a "calyptral perigynium"—a composite structure formed at the base of stem tissue, capped by the calyptra proper). In some species the stem perigynium is formed by the symmetrical elaboration of axial tissue beneath the perianth, the perigynium thus lying continuous with the axis, but sometimes the ventral face of the perigynium is more highly elaborated, as a somewhat swollen or bulbous "foot," often bearing rhizoids. To some extent the type of perigynium developed is environmentally controlled; *N. geoscyphus* and *N. insecta*, as well as *N. lescurii*, when crowded develop erect shoots with tubular perigynia that are not or scarcely more elaborated on the postical face of the stem. Inversely, when plants of these species grow prostrate, the ventral side of the perigynium is more highly elaborated, and a perigynium at a slight to distinct angle with the stem, swollen or bulbous ventrally, is formed. As a consequence, the perigynium form in this genus can be used as a systematic character only with some reservations. In *N. compressa* and *N. scalaris* (in both of which the leaves are uniformly rotundate to reniform) the perigynium always appears to lie in the axis of the stem; in our other species (in which the leaves have a tendency to be emarginate, at least toward the gynoecia) the form of the perigynium is much modified by environmental conditions.[303]

Although the reduced perianth and constantly well-developed perigynia represent derivative characteristics, *Nardia* is more primitive than other genera of Jungermanniaceae in the well-developed underleaves (and relatively broad ventral merophytes), as well as in the unspecialized axis without a defined cortex. This relatively primitive position is further enhanced by the strong tendency for some taxa to develop emarginate leaves (in *N. insecta* and *N. lescurii* this is carried to an extreme, all vegetative leaves being regularly bilobed). However, those species which seem primitive in having the most obviously lobed leaves (*N. insecta, N. lescurii, N. breidleri*) are among the most specialized in having the most highly distinctive perigynia.[304]

---

[303] Knapp (1930) and M. S. Taylor (1939) discuss the perigynium of *Nardia*. According to Knapp, development of the perigynium in *N. geoscyphus* and *N. breidleri* is closely similar to that of *Prasanthus* (i.e., follows the *Prasanthus* type of development), the longitudinal growth on the ventral side of the prostrate axis, after fertilization, being much more pronounced than growth on the dorsal side. Thus, the whole archegonial receptacle is displaced to the apparent dorsal side of the stem. Subsequent development induces vertical growth (and enlargement of the stem on the ventral side, into which the sporophyte "bores"), resulting in a perigynium lying at an angle to the axis (Fig. 264:12). As Taylor (1939, p. 96) has indicated, in erect-growing phases of *N. geoscyphus* ("fo. *erecta*" and "fo. *suberecta*") no ventral protuberance develops on sporophyte-bearing stems; therefore a perigynium of the *Isotachis* type forms, lying in the axis of the stem. This is also the case in similar, erect or suberect modifications of *Nardia insecta* (Fig. 264:10). In both cases, development of the "calyptral perigynium" is equally clearly expressed.

[304] Some of the taxa with consistently bilobed leaves are also specialized in their elaters. For example, *N. breidleri* is unique in the 3–4 spiral elaters, while the Japanese *N. unispiralis* has 1-spiral elaters.

Species of *Nardia* are generally associated with exposed rocks, although occurring oftener on soil over ledges or at the bases of rocks than as pioneers over bare rock surfaces. Less often they occur on sandy soils or sandy-peaty ground (as along ditches, or on margins of montane streams or shallow lakes). The species are never helophytic, in the strict sense, and rarely or never occur on purely organic substrates. Although six species occur locally, they are never a common component of the flora, most of them being encountered only occasionally. Species with bilobed leaves may be confused with *Harpanthus* (with which they agree in the scattered rhizoids and distinct underleaves), while the entire-leaved species may be mistaken for *Solenostoma*, from which, however, they are easily separable on the basis of the distinct underleaves.

Nardia, as represented locally, can be divided into three distinct sections, as in the subjoined key. The very small species with sharply but shallowly bilobed leaves and with specialized elaters (either 1- or 3–4-spiral) constituting the section *Breidlerion* include besides *N. breidleri* the Japanese *N. unispiralis* and the African *N. arnelliana*. They possess small leaf cells and have the underleaves vestigial in the gynoecia. Additional sections will be needed to include other exotic taxa. Some of these, such as the Japanese *N. subclavata* and *N. sieboldii*, have only a single oil-body per cell, and in the former many leaf cells may actually lack oil-bodies.

Key to Species and Sections of Nardia

1. Plants minute, to 500 μ wide $\times$ 2–4 mm high usually; leaves all 0.2–0.25 bilobed, lobes blunt or subacute, sinus angular; cells thin- or slightly and equally thick-walled, very small, 13–16(18) μ wide subapically and medially . . . . [sect. *Breidlerion* Grolle] *N. breidleri* (p. 883)
1. Plants vigorous, 820–1500 μ wide $\times$ 8 mm or more (usually over 15 mm) long; leaves usually entire (if bilobed, cells with $\pm$ coarse, nodose trigones); cells larger, 20–28 μ or larger subapically and medially . . . . 2.
   2. Leaves orbicular or suborbicular, not sharply erect-appressed; plants not appearing laterally compressed; leaves with marginal cells subequal to interior cells in size . . . . [sect. *Nardia*] 3.
      3. Leaves, at least adjacent to gynoecia and androecia, bilobed or emarginate; bracts bilobed or retuse; perigynia normally bulbous, at an angle to the stem; cells with oil-bodies formed of fine spherules, appearing granular-papillate . . . . 4.
         4. Paroecious; ♂ bracts usually $\pm$ strongly concave at base; cells usually with 2–3 oil-bodies per cell; leaves of sterile shoots entire or occasionally emarginate or deeply bilobed and with generally acute or subacute lobes. Not Southern Appalachian . . . . . 5.
           5. Leaves of vegetative shoot entire or barely retuse to emarginate, only those of reproductive regions emarginate (with obtuse or

rounded lobes); cells 20–28 μ medially, with smaller trigones
. . . . *N. geoscyphus* (p. 863)
5. Leaves uniformly, deeply bilobed, the lobes and sinuses both
acute or blunt, rounded in only isolated instances; cells 30–36 ×
32–38(45) μ medially, with sharper trigones . *N. insecta* (p. 871)
4. Dioecious; ♂ bracts spreading, hardly concave, the antheridia
externally readily obvious; cells in many cases with 4–5, rarely 6,
oil-bodies; leaves uniformly bilobed, the lobes blunt or broadly
rounded. Southern Appalachian . . . . . . *N. lescurii* (p. 877)
3. Leaves and bracts normally entire, broadly rounded (the perichaetial
bracts rarely retuse); perigynia never bulbous, always coextensive with
axis, tubular; cells with oil-bodies glistening, homogeneous (with age
coarsely segmented or faintly transversely barred) . *N. scalaris* (p. 854)
2. Leaves reniform, much broader than long, uniformly entire, sharply erect-
appressed; plants appearing laterally strongly compressed; leaves with
marginal cells notably smaller in size . . . .
. . . . [sect. *Compressae*] *N. compressa* (p. 850)

Key to Fertile Plants of Nardia

1. Dioecious, the ♂ plants usually in separate patches . . . . 2.
  2. Robust, shoots 1–3 mm wide; cells with distinct to coarse trigones . . 3.
    3. Leaves patent; plants prostrate in growth, rather small (1–2.4 mm
    wide) . . . . 4.
      4. Leaves uniformly entire; cells each with 2–3(4) glistening, homo-
      geneous oil-bodies which may become faintly transversely barred
      or segmented with age . . . . *N. scalaris*
      4. Leaves uniformly bilobed, the lobes rounded or blunt; cells each
      with (2)3–5(6) granular-segmented oil-bodies . . . . *N. lescurii*
    3. Leaves erect-appressed, appearing somewhat antically secund, reni-
    form, decurrent, entire; plants robust, suberect or ascending in
    growth, 2–3 mm wide (westward to Greenland only!) . *N. compressa*
  2. Minute, shoots under 0.5 mm wide; cells small (13–16 μ wide usually),
  without trigones; leaves bilobed, sinus and lobes blunt to angular . . .
  . . . . *N. breidleri*
1. Paroecious; cells each with 2–3(4–5) granular-segmented, opaque oil-
bodies; at least the uppermost leaves of fertile plants emarginate or bilobed
. . . . 5
  5. Leaves generally entire, sporadic ones retuse, except the ♀ bracts (and
  immediately subtending leaves, which are often emarginate); occasionally with isolated leaves emarginate, but these with obtuse to rounded
  lobes and usually sinuses; underleaves usually vestigial except near
  apices of fertile shoots; leaf cells small (marginal 18–24 μ) . . . . . .
  . . . . *N. geoscyphus*
  5. Leaves uniformly strongly emarginate to bilobed, even on sterile shoots;
  sinuses (and often lobes) acute to subacute; underleaves present throughout; leaf cells large (marginal 25–30 μ) . . . . *N. insecta*

Sectio COMPRESSAE Schust., sect. n.

Plantae vigentes, saepe erectae; caulis hyaloderma perspicuum, e cellulis angustis oblongisque compositum, habens, cellulis intra membranas crassas habentibus, in medullam cellulas maiores habentem gradatim abeuntibus; surculi lateraliter compressi; folia non lobata, longe decurrentia; stolones ventrali-intercalares frequentes.

Vigorous, often rigid, erect, to 5–15 cm tall; stem with a ± distinct hyaloderm of narrow, rectangulate cells (length:width *ca.* 3.5–5:1), the medullary cells immediately within hyaloderm prominently thick-walled and strongly elongated, gradually grading into the larger and more nearly leptodermous cells of the middle of the medulla. Rhizoids ± rare. Leaves erect-appressed, orbicular-reniform, entire, antically decurrent. Cells mostly 18–20 μ on margins, median cells with walls thin to feebly thickened and with small, never coarsely bulging trigones; oil-bodies homogeneous, glistening. Elaters 2-spiral.

Type. Nardia compressa (Hook.) S. F. Gray.

In my opinion, *N. compressa* is very distinct in facies and in stem anatomy from the other species of *Nardia* and must be placed into a section of its own.

The laterally compressed shoots are very distinctive; so, also, is the frequent presence of ventral-intercalary branching. The ventral branches are often leafless and flagelliform, or may be provided with gradually more distinct leaves, becoming eventually normally leafy. I have seen no terminal branching in *N. compressa*.

NARDIA COMPRESSA (*Hook.*) S. F. Gray
[Figs. 258:6–9; 259]

Jungermannia compressa Hook., Brit. Jungerm., pl. 58, 1816.
Nardius compressus S. F. Gray, Nat. Arr. Brit. Pl. 1:694, 1821.
Mesophylla compressa Dumort., Comm. Bot., p. 112, 1822.
Alicularia scalaris var. *compressa* Nees, Naturg. Eur. Leberm. 2:449, 1836.
Alicularia compressa G., L., & N., Syn. Hep., p. 12, 1844; K. Müller, Rabenh. Krypt.-Fl. 6(1):511,
    fig. 266, 1909; Macvicar, Studs. Hdb. Brit. Hep. ed. 2:124, figs. 1–4, 1926.
Alicularia pachyphylla De Not., Mem. Accad. Torino, Ser. 2, 18:487, 1859.
Nardia scalaris var. *compressa* Carr., Brit. Hep., p. 24, 1874.
Nardia compressa Carr., *ibid.* p. 29, 1875; Amakawa, Jour. Hattori Bot. Lab. no. 21:277, fig. 7:a–g,
    1960.

Robust, (1.5)2–3 mm wide, forming large spongy mats, often scorched in appearance, *reddish-brown to purplish* (in shade phases sometimes dull green); shoots erect or suberect, *2–15 cm high,* simple or subsimple. Stem *slender, flexuous,* ca. 220–360 μ in diam., cortical cells *narrowly rectangulate* (ca. 18–24 $\times$ (50)60–90(120) μ), *their diam. larger than that of medullary cells.* Medullary cells within hyalodermis *prominently thick-walled,* only 13–18 μ in diam., strongly elongate, the medullary cells gradually becoming slightly larger (15–20 μ) and more nearly thin-walled near the center of the axis. Rhizoids *sparse below, virtually or quite absent above.* Leaves *erect-appressed and imbricate,* the shoots appearing to be *laterally compressed, orbicular-reniform to reniform,* slightly convex, decurrent antically, *averaging 1.2–1.5× as wide as long*, ca. 0.9–1.0 mm long × 1.2 mm wide up to 1.8 mm long × 2–3 mm wide, *rounded and uniformly entire at apex.* Cells ca. 18–20 μ and subquadrate on margins, 25–32 × 30–40 μ in the leaf middle, $\pm$ thickened or thin-walled, with small to large trigones; oil-bodies homogeneous and shining, 1–2 per cell (*fide* Schiffner, 1901), 7–9(10) × 11–15 μ, often transversely barred (Amakawa, 1960a). *Ventral merophytes very narrow,* usually appearing only 2–4 cell rows broad. *Underleaves small, suberect,* usually difficult to demonstrate except near shoot apices, subulate or lingulate to lanceolate, rarely bidentate or bilobed, to 500 μ long × 300 μ wide.

Dioecious. Androecia with usually 3–4 pairs of bracts virtually identical in form and size to vegetative leaves, 2–3-androus. Perichaetial bracts like leaves, but broader, *nearly flat, erect-appressed.* Perigynium lying in axis of stem, the stem apex appearing somewhat clavately swollen distally; perianth very short, a somewhat conoidal extension of the perigynium, *only 0.3–0.5 the height of the perigynium,* hidden within the *broadly reniform* bracts, often purplish or vinaceous. Elaters 8–12 μ, bispiral; spores 10–13(16) μ, granular-papillate, reddish brown.

Type. Near Bantry, County Cork, Ireland (*Miss Hutchins*).

Distribution

An oceanic species, known from montane and western Europe (southwest France, the Alps of central Europe, from northern Italy, Switzerland, northern Spain, northward to Austria, Germany, Brittany, Belgium), northward to England, Ireland, Scotland, Norway, Sweden. Eastward to the Caucasus and known from Asia (Asia Minor; Kamchatka); reported from Yamagata, Japan (Amakawa, 1960a). Also in Iceland.

In North America reported originally from Atka I. in the Aleutian Isl., Alaska (Macoun, 1902), later from McDonald Lake and Yes Bay, eastern Pacific Coast, and Hinchinbrook I., on central Pacific Coast (Persson, 1946), British Columbia and Washington (Schofield, 1968).

In our area correctly reported a single time:

S. GREENLAND. Narssaq, ca. 60°57′ N., 46°03′ W. (*Vahl 223, Sept. 1829!*).

This is the only authentic collection of the species that I have seen from our region; the plants have dense, appressed reniform leaves, show some purplish pigmentation, and have the diagnostic, hyaline, often collapsed cortical cells normal to the species and the very short perianths surmounting erect perigynia. They are quite typical.

The two other reports of this species from Greenland (Lange and Jensen, 1887), from Kekertak (Kekertaq ved Ikerasarsuak, *Vahl 212!*) and from Nanortalik, both represent *N. scalaris.*

It is noteworthy that since the time of Vahl this species has not again been found in our region. Apparently it represents an oceanic-subarctic element in our flora, like *Gymnomitrion obtusum* and (perhaps) *Anthelia julacea,* that has not

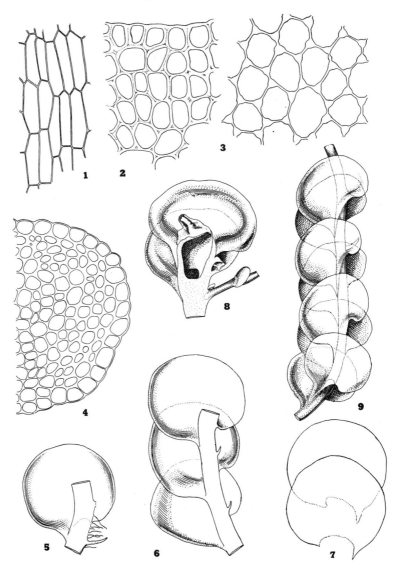

FIG. 259. *Nardia compressa*. (1) Dorsal cortical stem cells (×170). (2) Cells towards leaf apex (×290). (3) Median cells (×290). (4) Stem cross section (×210). (5–6) Shoot sectors, one row of leaves removed, showing reduced underleaves (×22.5). (7) Leaves (×22.5). (8) Apex of plant with perianth and perigynium (×ca. 25). (9) Lateral aspect of sterile shoot (×ca. 25). [1–7, from plants from Rutsdalen, Jämtland, Sweden, *T. G. Halle, 1924*; 8–9, from K. Müller, 1905–16, after Gottsche.]

penetrated further west in the Western Hemisphere than southern Greenland. I have searched for it in both west and northwest Greenland, without success.

Ecology

Frequently hygrophytic, and often submerged or on wet rocks, particularly along rocky streams and springs, associated with *Scapania* and *Marsupella*. Apparently constantly restricted to siliceous rocks. Also over peat in moors. Plants from peat bogs generally have large trigones and are reddish pigmented, the entire plant often being deep red or purple. Such a mod. *colorata-pachyderma* is considered as the typical phase of the species (Schiffner, 1901, p. 4). Such plants may be only 2–3 cm tall, occasionally higher. In smaller (particularly xeromorphic) phases, the leaves are not at all wavy when dry and are strongly laterally pressed against the stem and erect. In larger, laxer phases the leaves are undulate when dry and less strongly erect-appressed.

Plants from moist, shaded sites may be 3–10 or more cm tall, are often dull green, and may have erect-spreading leaves, occasionally nearly squarrose leaves. Cells of such forms show only minute trigones. Such extremes Schiffner (*loc. cit.*, p. 5) has unnecessarily separated as "var. *virescens*."

Differentiation

Although once considered to be a mere variety of *N. scalaris*, *N. compressa* is sharply isolated from other species of the genus. The robust size, purple color, and erect growth, with normally strongly appressed and erect leaves, resulting in sharply laterally compressed shoots, give the plants an unmistakable facies. Leaves are uniformly entire, as in *N. scalaris*, but tend to be broader and more reniform. In lateral view the shoots of the two species are very different. In *N. compressa* the broad leaves extend almost as far beyond the stem postically (for at least 1–2 × the width of the stem) as antically, i.e., the leaves are not distinctly antically secund. In *N. scalaris* (and all our other species of the genus), the leaves do not project beyond the stem postically and are to a ± extent antically secund; they are also spreading or erect-spreading, never erect-appressed.

Nardia compressa is habitually distinct because the "large, swollen, reddish-purple to dark purple tufts of the typical plant do not resemble those of other species of this genus" (Macvicar, 1926). Aquatic phases of *N. scalaris* occasionally resemble *N. compressa* but differ in stem anatomy (cortical cells slightly smaller in diam. than medullary cells), and in the fact that the leaves do not extend to any degree below the postical side of the stem and are less decurrent; *N. scalaris* is also never purplish in color.

The stem anatomy of *N. compressa* is unique among our species of the genus. The stem is firm and relatively rigid, quite slender and flexuous, rather than soft and fleshy. The cortical cells are rather hyaline, quite thin-walled, and relatively elongated, averaging usually 3–4(4–6)× as long as wide. By contrast, the medullary cells are clearly much narrower, the several exterior strata being much smaller in diam. They are, furthermore, quite strongly and equally thick-walled, thus appearing superficially to be even smaller than they actually are. As a consequence, the cortical cells are very sharply differentiated. In its stem anatomy, *N. compressa* approaches more closely such species of *Solenostoma* as *S. (Plectocolea) crenuliformis*.

Sectio *NARDIA*

Plants rather large, usually 12–30 or more mm long × 1–2.5 mm wide; leaves entire or bilobed, large (width usually 3–4× stem diam. or greater); cells large, the marginal mostly 20–30 μ, the median with thin walls and distinct, concave-sided to (usually) bulging trigones. Perigynium erect or ventrally bulbous and forming an incipient marsupium. Elaters 2-spiral.

Type. Nardia scalaris (Schrad.) S. F. Gray.

Including, besides *N. geoscyphus*, *N. geoscyphoides* Amak. of Japan and the regional *N. scalaris*, *N. insecta*, and *N. lescurii*. *Nardia japonica* Steph. and *N. hiroshii* Amak. (see Amakawa, 1960a, pp. 283–285, fig. 9) probably should be assigned here; both are known only from Japan, and both have uniformly bilobed leaves.

NARDIA SCALARIS (*Schrad.*) *S. F. Gray*
[Figs. 260–261]

Jungermannia scalaris Schrad., Syst. Samml. Krypt. Gewächse 2:4, 1797.
Nardia scalaris S. F. Gray, Nat. Arr. Brit. Pl. 1:664, 1821 (as *Nardius*); Lindberg, Acta Soc. Sci. Fennica 10:115, 1871; Evans, Rhodora 14:11, 1912; Frye & Clark, Univ. Wash. Publ. Biol. 6(2):309, figs. 1–9, (1943) 1944; K. Müller, Rabenh. Krypt.-Fl. ed. 3, 6:850, figs. 294, 295b-c, 1956.
Mesophylla scalaris Dumort., Comm. Bot., p. 112, 1822.
Alicularia scalaris Corda, *in* Opiz, Beiträge, p. 652, 1828; K. Müller, Rabenh. Krypt.-Fl. 6(1):514, fig. 267, 1909; Macvicar, Studs. Hdb. Brit. Hep. ed. 2:126, figs. 1–5, 1926.
Alicularia roteana De Not., Mem. Accad. Torino, Ser. 2, 18:484, 1859 (*teste* Massalongo, 1902).
Jungermannia wallrothiana Hüben., Hep. Germ., p. 85, 1834.
Mesophylla roteana Dumort., Hep. Eur., p. 130, 1874.
Nardia cavana Clark, The Bryologist 43:29, 1940; Frye & Clark, Univ. Wash. Publ. Biol. 6:311, figs. 1–13, (1943) 1944.
Nardia macrostipa Kaal., *in* Joergensen, Bergens Mus. Skrifter no. 16:95, 1934.
Nardia harae Amak., Jour. Jap. Bot. 32:38, fig. 4:A-L, 1957.

In flat patches or mats, pure light green to greenish brown to (occasionally) reddish brown. Rather large, shoots 1600–2400 μ wide × 10–25(45) mm long, prostrate except for ascending apices, sparingly branched. Stems 275–360(400) μ in diam., rather soft-textured, stout and rather fleshy; cortical cells from 1.0–2.0(3.0)× as long as wide (22–25 × 40–70 μ to 28–35 × 32–40 μ to 23–28 × 36–46 μ), little smaller in diam. than the almost uniform medullary cells, which are to 35–45 μ wide, leptodermous. Rhizoids copious, colorless to slightly brownish. Leaves contiguous to moderately imbricate, obliquely inserted and oriented, hardly decurrent antically, *erect-spreading to spreading, rotundate to reniform-rotundate*, 700–800 μ wide × 600–900 μ long to 1020–1120(1480) μ wide × 725–760(1200) μ long, usually *uniformly entire and rounded at the apex*, nearly plane to slightly or moderately concave, usually only the uppermost leaves of fertile shoots occasionally retuse (these larger, up to 1150–1200 μ wide × 800–850 μ long). Cells rather pellucid, thin-walled, with distinct to moderately bulging trigones; marginal cells (18)20–28(33) μ, somewhat smaller than median; median cells 24–32(35) μ × 30–35(36–38) μ; cuticle smooth; *oil-bodies 2–3(4) per cell, glistening and homogeneous*, with age often faintly to strongly transversely barred, ellipsoid to ovoid, ca. 6 × 8–11 μ to 7–7.5 × 12–17 μ; chloroplasts 2–3.2 μ. *Underleaves distinct throughout*, subulate to lanceolate, spreading, entire, acute to acuminate, occasionally barely connate at base with leaves on one side of stem.

Dioecious, the two sexes often in separate patches. Androecia becoming intercalary, the *bracts imbricate, concave*, in several pairs; 2–3-androus. ♀ Bracts larger than leaves (the subinvolucral leaves grading into them in size), strongly concave, hardly undulate, broadly ovate to subrotund to rotund-quadrate, ca. 1200–1250 μ wide × 950–1000 μ long, often retuse or emarginate at apex; bracts erect and closely sheathing the reduced perianth, often distally somewhat connivent (and then gynoecial region bilabiate as in *Cryptocolea*); bracteole subulate to lanceolate, usually slightly connate to bracts below, larger than underleaves. *Perianth short, remaining virtually or quite hidden and included within the erect and sheathing bracts;* broadly conical in shape, the mouth somewhat contracted, crenulate at apex. Perigynium *high and conspicuous*, fleshy, lying *in axis of stem*, never becoming bulbous (in subsp. *scalaris*), usually *2–3× as high as the abbreviated perianth*. Capsule deep brown, subglobose, the wall 2-stratose. Epidermal cells large and subquadrate or short-rectangulate, with strong nodular thickenings; inner cell layer with numerous semiannular bands. Spores (15)16–18(20) μ, finely papillate, yellow-brown. Elaters bispiral, 8–10(12) μ in diam. × 150–200 μ long, with 2 broad, deep brown spirals. Chromosome no. *n* = 9.

Type. Central Europe.

Distribution

An "old" species, with a somewhat relict and disjunct range; largely confined to the Spruce-Fir Forest Region and the northern portions of impinging deciduous forests, but also reported from the lower portions of the Arctic.

In Europe widely dispersed and often common (except in calcareous regions), ranging from northern Scandinavia (Norway, Sweden, Finland), southward to England, Ireland, the Faroes, Scotland, Belgium, the montane portions of central Europe (up to 2500 m!), eastward to the

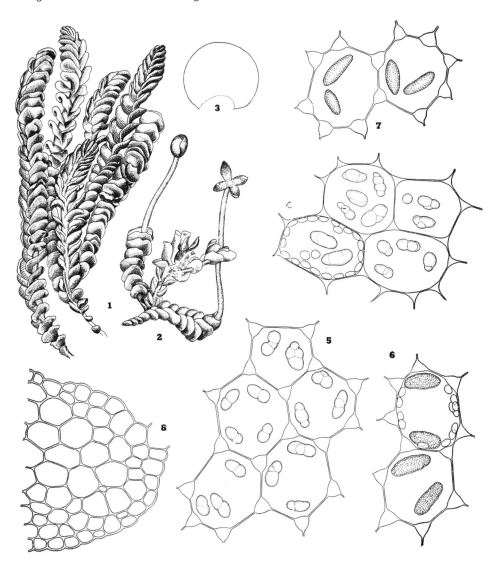

Fig. 260. *Nardia scalaris* (1–5, ssp. *scalaris;* 6–8, ssp. *botryoidea*). (1) Cluster of plants (×5.6). (2) Capsule-bearing plant (×5.6). (3) Leaf (×18). (4) Median leaf cells with oil-bodies and, lower left, chloroplasts (×535). (5) Median cells with oil-bodies (×535). (6) Cells with oil-bodies and, upper cell, chloroplasts (×535). (7) Cells with oil-bodies (×570). (8) Stem cross section (×200). [1–2, after K. Müller, 1905–16; 3–4, *RMS 39617*, Richland Balsam, N.C.; 5, *RMS 45370*, below Myrtle Point, Mt. LeConte, Tenn.; 6, *RMS 45371*, same loc., but not mixed with *RMS 45370*; 7–8, *RMS 43068-43069*, Nova Scotia.]

Transylvanian Alps and Bulgaria, southward to Italy (to Sicily and Corsica), Spain, Portugal, Madeira, Tenerife, and also to the Azores (Allorge, 1948). Also in Iceland. The species (as emphasized in Buch and Tuomikoski, 1955) has "oceanic tendencies" and is common in Iceland, Jan Mayen I., and western Europe. Also reported from Asia (central and northern Japan: Hokkaido, Honshu, Shikoku)[305] and at one time (erroneously) attributed to St. Paul I. in the Indian Ocean and to Java. These reports lack recent authentication.

In North America with an oceanic-Appalachian range, occurring in the Far West from Alaska (Attu I. eastward), the Yukon, and British Columbia, south to Washington, Oregon, and Wyoming. Eastward as follows:[306]

W. GREENLAND. Island at mouth of Sermilik Fjord, S. of Godthaab, 63°28′ N. (*Rosenvinge, May 26, 1888*; with *Nardia breidleri, Prasanthus suecicus*); Ameralik Fjord, S. of Godthaab, 64°03′ N. (*Vahl 179, Aug. 1830*!); Tupilak I., near Egedesminde, 68°42′ N. (*RMS & KD 66-027*). s. GREENLAND. Narssaq, ca. 60°57′ N., 46°03′ W. (*Vahl 42, Sept. 1829*!); Kekertaq ved Ikerasarsuak (*Vahl 212, Apr. 1829*!); Nanortalik, 60°07′ N. (*Vahl, Nov. 1828*!); all reported as *A. compressa* by Lange & Jensen, 1887. E. GREENLAND. Cap Ravn, in *Scirpetum;* Jarn I., 65°57′ N. (Jensen, 1906a); Scoresby Sund (Jensen).

MELVILLE PEN. No locality (Polunin, 1947; an old, very doubtful report). LABRADOR. Battle Harbor (*Waghorne!*). NEWFOUNDLAND. South Side Hill at St. John's, Aquaforte, Back R. at Biscay Bay, all in Avalon Distr.; Pass I., Hare Bay, Rencontre West, Ramea, Grandy Brook at Burgeo, all on S. Coast; Table Mt. of Cape Ray, South Branch, and Steady Brook, in W. Nfld.; Kitty's Brook in C. Nfld.; Lantern Cove (all *fide* Buch & Tuomikoski, 1955); Humber R., Squires Mem. Pk. (*RMS*). NOVA SCOTIA. Halifax (*Brown 263!*); Five Islands (*Brown!*); Arichat, Cape Breton (Evans, 1912c); E. side of Cape Breton Natl. Park, 3–3.5 mi S. of Mary Ann Brook Waterfall (*RMS 66316*); Cheticamp R. and Indian Brook, Cape Breton (*RMS*). NEW BRUNSWICK. Campobello I. (Evans, 1912c). MAINE. Eastport (*Evans 88!*).

Recurring at isolated stations at high elevations (3500–5900 ft) in the Southern Appalachians: NORTH CAROLINA. Craggy Gardens, above Blue Ridge Pkwy., Buncombe Co., ca. 5900 ft (*RMS 29653a, 29641, 29659, p.p.*, with *Diplophyllum obtusatum*); Richland Balsam, 6200–6400 ft, Haywood Co. (*RMS 39617, 39643*). TENNESSEE. Sevier Co.: Alum Cave, Mt. LeConte (Sharp, 1939; *RMS 24180, 24001*); Charlie's Bunion, on Appalachian Trail, 4 mi NE. of Newfound Gap (*RMS 45307a*); Roaring Fork, Mt. LeConte, 5500 ft (*Sharp 5614!*); summit, Mt. LeConte (*RMS 24010*).

The southernmost stations in eastern North America from the old "Arcto-tertiary" Forest Region, associated with *Rhododendron catawbiense, Aesculus octandra*, with rare extensions upward into the Spruce-Fir Region.[307]

---

[305] Amakawa (1960a, pp. 281–282) segregates the Japanese plant as subsp. *harae*, having earlier (1957, p. 38) considered it an autonomous species, *N. harae*. According to Amakawa, all fertile Japanese plants studied differ from subsp. *scalaris* in having the perigynium short-pendent.

[306] Arnell (1960) reported *N. scalaris* from Peary Land, north Greenland. All of the cited numbers (*Holmen 7449, 7709, 7905, 8714*) have been checked; all represent *Arnellia fennica*.

[307] Sharp (1939) indicates on his map (fig. 3) stations on the north shore of Lake Superior in Ontario, and a station in eastern Pennsylvania. These reports are very doubtful. Equally dubious is the report from Wyoming, all such "continental" stations needing verification.

Ecology

Generally on moist, igneous or metamorphic, noncalcareous rocks, often forming thin, pure, pale green mats in sheltered places (such as shaded dripping recesses) or on seepage-wet, sunny cliffs. Associated in the Southern Appalachians, in which it is very local, with *Scapania nemorosa, Calypogeia muelleriana, Diplophyllum apiculatum, D. obtusatum* (a vicariant of the oceanic *D. obtusifolium*), *Marsupella emarginata* (and *paroica*), *Gymnocolea inflata*, and *Sphagnum*; in the Spruce-Fir Region also with *Diplophyllum taxifolium* and (very rarely, in Tennessee) with *Marsupella funckii*.

The species is "common and often abundant on the south coast" of New-foundland, occurring "on wet acid rock and soil, especially by brooks and rivers, with *Cephalozia bicuspidata, Diplophyllum albicans, Pellia* (*epiphylla* and *neesiana*), *Marsupella emarginata, Scapania nemorosa, Calypogeia muelleriana, Gymnocolea inflata*, etc." (Buch and Tuomikoski, 1955). Evans (1912c) notes that in Nova Scotia and Maine the species is found on low cliffs by the sea, in some cases forming dense mats, "scarcely ten feet above the high water mark and only a few feet below the edge of the woods." In the Southern Appalachians usually in permanently damp recesses beneath cliffs, in dense shade.

In Europe, where the species is more ubiquitous, it occurs under less restricted conditions, frequently in large mats, and often forming sporophytes freely in the spring (March–June). American material, as noted by Evans, is usually sterile; I have seen only ♀ plants in the Southern Appalachians.

My Nova Scotia material is all from moist, loamy-clayey soil at the edge of a bog, associated with *Scapania irrigua* and *S. nemorosa, Cladopodiella francisci, Gymnocolea inflata, Drosera rotundifolia*, and *Lycopodium inundatum*.

Variation and Differentiation

In eastern North America, where the typical subspecies is local, showing only limited variation, but occasionally, with crowding, the plants may form caespitose patches, with the individual plants virtually erect. This is the "var. *procerior*" of Schiffner (1901, *in* 1901–43), which is hardly more than an environmental response. In our area, at least, the species usually occurs as a pure but rather pale green phase (mod. *viridis*), rarely showing extensive development of brownish pigmentation. Fertile plants often have the underside of the stem, below the gynoecial region, vinaceous to purplish carmine (underleaf bases and, rarely, rhizoid bases in this area may be similarly tinged), much as in *N. geoscyphus* and *N. insecta*. In the latter species the vinaceous pigmentation is often more extensive, involving all or most of the rhizoids and occasionally the leaf bases.

Nardia scalaris is easily recognized. It agrees with sterile material of *N. geoscyphus* (and *N. compressa*) in the orbicular, entire leaves and subulate or subulate-lanceolate underleaves. The generally prostrate growth and the shape and orientation of the leaves are totally unlike those of *N. compressa*, but agree with those of *N. geoscyphus*. *Nardia scalaris* differs from this latter species at once in the (*a*) dioecious inflorescences; (*b*) slightly to nonemarginate subinvolucral leaves; (*c*) very different, generally glistening, homogeneous oil-bodies; (*d*) inability (of our subspecies) to develop a bulbous perigynium; (*e*) larger size;. (*f*) underleaves distinct throughout; (*g*) pale rhizoids. The presence of distinct underleaves readily separates the species from superficially similar species in *Solenostoma*; the large, constantly distinct underleaves and the very large, glistening, occasionally segmented oil-bodies distinguish it from *Jamesoniella*.

On the basis of the rotundate, entire leaves and the glistening, homogeneous, weakly barred, or few-segmented oil-bodies, the species is subject to confusion with *Cryptocolea imbricata*. The latter, however, totally lacks underleaves and is very different in the form of the perichaetial bracts and in the even more markedly hidden and reduced perianth. Sterile shoots and gynoecial shoots of *Cryptocolea* also have diagnostically valvate apices.

Occasional extremes of *N. scalaris* show a very close if superficial resemblance to *Cryptocolea* in that the bracts are distally connivent, except for the slightly flaring margins (Fig. 261:5), resulting in a somewhat similar bilabiate gynoecial system. The leaves beneath the gynoecia in *Cryptocolea* are ovate to oval, however, and distinctly longer than broad (see Fig. 268:3, 6), and the perichaetial bracts and subtending leaves are never emarginate but bear marginal slime papillae, at least when young (Fig. 268:6). The gynoecial region in the *Cryptocolea* is, furthermore, even more closely imbricate, sheathing subinvolucral leaves (Fig. 268:1, 7–8). The vegetative plants also differ considerably in that the *Nardia* has somewhat spreading leaves, while those of *Cryptocolea* are concave, often almost spoonlike, and frequently suberect, with the antical margins appressed to form a feeble antical keel, at least in dense-leaved forms.

In Europe *N. scalaris* is much more polymorphous than in America, and aquatic to subaquatic phases occur that are considerably more robust. The more obvious are "var. *distans*" Carr. (Brit. Hep., p. 24, 1874), in which the plants form swelling tufts, 3–8 cm high, with distant leaves that are reduced in size and shrink in drying. This appears to be a mere modification, or at best an ecotype. Hardly more distinct is "var. *rivularis*" Lindb. (1875, p. 531), which may be equally tall, forming submerged tufts; it is the most robust extreme of the species, often to 2 mm wide. Such aquatic and subaquatic phases are to be expected in our area.

Nardia cavana (Clark, 1940) supposedly differs from *N. scalaris* in having the

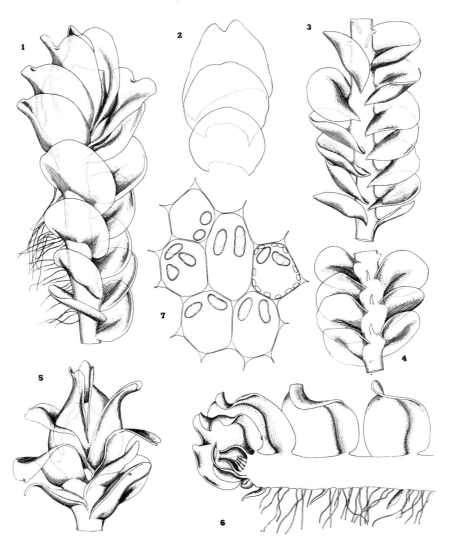

FIG. 261. *Nardia scalaris*. (1) Apex of suberect ♀ shoot with included perianth (×20).
(2) ♀ Bract, subinvolucral leaf, and leaf, from top to bottom (×20). (3) Antical aspect
of sterile shoot sector (×16). (4) Same, ventral aspect (×16). (5) Apex of suberect
shoot with mature, included perianth and *Cryptocolea*-like aspect (×20). (6) Apex of
procumbent shoot with postmature archegonia and young perianth (×18). (7) Median
cells with oil-bodies and, at right, chloroplasts (×432). [1–5, Craggy Gardens,
Blue Ridge Pkwy., N.C., *RMS;* 6, *RMS 24180*, Alum Cave, Mt. LeConte, Tenn.;
7, *RMS 24010*, summit of Mt. LeConte, Tenn., an emarginate-leaved phenotype.]

"trigones distinctly bulging; underleaves mostly united at base with a leaf; rhizoids colorless to straw-colored or reddish" (Frye and Clark, 1944, p. 307). However, as is clear from examination of typical *N. scalaris*, this species, too, often has distinctly bulging trigones—a feature subject to great environmental modification. The underleaves are often connate at their bases with the leaves of one side of the stem (this feature varying from plant to plant), and the fertile plants may at times have reddish rhizoid bases. No criterion, therefore, exists for recognizing this "species." The oil-bodies, as described for *N. cavana*, are identical in number and size with those of *N. scalaris*. The plant is smaller (to 1500 μ wide) than typical *N. scalaris*.

The plants from high elevations (5300–6400 feet) in the Southern Appalachians, and at least some of those from Nova Scotia, deviate from subsp. *scalaris*. A mod. *megafolia-viridis*, which occurs in the Southern Appalachians, departs widely from the ordinary concepts of the species in several ways. Such plants (as, i.a., *RMS 39617* from Richland Balsam, N.C.) may be as wide as 2.5 mm, with the leaves up to 1480 μ broad × 1200 μ long. Robust shoots possess a marked tendency to produce sharply emarginate leaves, bearing a shallow but broadly V-shaped sinus and broadly rounded lobes. Such larger phases also tend to have bigger leaf cells, with the marginal averaging 28–38 μ, tangentially measured, the median ranging from 28–32 × 30–40 μ to 30–36 × 32–38 μ. Most notable, in such plants, are the almost consistently coarsely segmented, rather than faintly barred, oil-bodies, which are 2–4(5) per cell and range from 6–7 × 11–12 μ up to 9 × 15 μ, rarely to 8 × 18–20 μ; they have mostly 2–5, more rarely 6–7, ovoid to spherical, glistening segments. Segmentation of the oil-bodies in these plants is not age-correlated, even the youngest leaves showing this feature.

Equally deviant are the Nova Scotian plants (*RMS 43068*, etc.) in which the laxer extremes, from wetter sites, show leaves that also are often retuse or emarginate on sterile stems. These plants all have finely granular to granular-botryoidal oil-bodies (Fig. 260:7), almost constantly 2, rarely 3, per cell. The oil-bodies do not have the distinctly glistening appearance normally present. Even more strongly botryoidal oil-bodies characterize the abundant ♂ plants found at Charlies Bunion in Tennessee. These plants have the brownish coloration, entire and rotund leaves, and dioecious inflorescences of typical *N. scalaris*, but their oil-bodies are almost exactly as in *N. geoscyphus*.

In so far as I have seen ♀ plants, the American material agrees with the European in possessing a perigynium continuous with the axes. Amakawa has segregated a Japanese plant, identical to normal *N. scalaris* in almost all respects except in the slightly bulbous perigynium, lying at a slight angle with the axis, as subsp. *harae*.

The three apparently valid extremes of *N. scalaris s. lat.* appear to represent at least good subspecies, separable by the following key:

1. Cells each with 2–4(5) glistening and homogeneous (with age faintly transversely barred) oil-bodies; leaves never retuse or emarginate; marginal cells mostly 20–28 μ, median 24–32 × 30–35 μ . . . . 2.

2. Perigynium erect, without a bulbous base. Europe, North America
. . . . subsp. *scalaris*
2. Perigynium at an angle with axis, short-pendent and gibbous below.
Japan . . . . subsp. *harae* (Amak.) Amak.[308]
1. Cells each with 2–3 to 4–6(7–8) opaque, dull-surfaced, granular-botryoidal
oil-bodies of the *N. geoscyphus* type; vigorous plants often with notched
leaves. Leaves with marginal cells to 28–38 μ, median to (28)30–36 ×
(30)32–40 μ. North America . . . . subsp. *botryoidea* Schust., subsp. n.

NARDIA SCALARIS subsp. *BOTRYOIDEA* Schust., subsp. n.

[Figs. 258:5; 260:6–8]

Planta dioecia; subspeciei *scalari* similis, sed cellulas relative maiores habens (cellulis mediis
28–32 × 30–40 μ), et omnis cellula 4–6(7–8) guttas olei opacas, perspicue botryoideas habens.

Similar to subsp. *scalaris* but often more vigorous (to 2.5 mm wide), green
to brown, occasionally somewhat scorched in appearance. Leaves up to 1480 μ
broad × 1200 μ long, on robust shoots often sharply emarginate with a
broadly V-shaped notch and broadly rounded lobes. Leaves on robust forms
with cells often *relatively large*; marginal ca. 28–38 μ, median to 28–32 × 30–40
up to 30–36 × 32–38 μ. Oil-bodies finely or coarsely *granular-botryoidal*, even
on young leaves opaque, often yellowish or brownish, from 6–7 × 11–12 to
8–9 × 15–20 μ, 2–3(4) per cell or smaller and 4–6(7–8) per cell. *Dioecious*.

Type. Smoky Mts., Sevier Co., Tennessee: Charlies Bunion, on
Appalachian Trail, 5300–5375 feet (*RMS 45305*; with a few mature
capsules, June 21, 1960). *Paratypes.* Tennessee: southeast slopes of
Mt. LeConte, below Myrtle Point, Sevier Co., 5900–6000 feet, on
sphagnous, seepage-wet, insolated ledges (*RMS 45375a, 45243b*); Mt.
Kephart, Smoky Mt. Natl. Park, 5950–6100 feet, at The Jumpoff, vertical
cliffs (*RMS 45383*).

Less surely here are plants from:

W. GREENLAND. Tupilak I. near Egedesminde (*RMS & KD 66–027*). NOVA
SCOTIA. Between Campbell and Craigmore, Inverness Co., Cape Breton I. (*RMS
43086, 43082b, 43069, 43068*).

All of these specimens, studied while fresh, had oil-bodies that ranged from
coarsely botryoidal when young to finely granular-botryoidal (all other
collections). The occurrence of subsp. *botryoidea* together with subsp. *scalaris*
(Fig. 260:5, 6) suggests the former may represent an independent species.

Ecology

The Tennessee plants occurred on thin soil over and amidst exposed
ledges, often where seepage-wet and with localized *Sphagnum* polsters;

---

[308] For subsp. *harae* see Amakawa (1960a, p. 280, figs. 1f, 8p-z).

associated were *Mylia taylori, Cephaloziella massalongoi, Gymnomitrion laceratum, Diplophyllum obtusatum* and *D. apiculatum* subsp. *taxifolioides, Scapania nemorosa, Marsupella funckii, Gymnocolea* cf. *inflata, Cephalozia bicuspidata.* The *Gymnocolea* and *Scapania irrigua* occurred with the *Nardia* in Nova Scotia, but the plants here grew on exposed, wet, peaty ground, in partial sun, at the edge of a spruce-tamarack bog.

Differentiation

If it were not for the easily ascertainable dioecious inflorescence (the plants usually occur in patches of a single sex), subsp. *botryoidea* could be confused with the paroecious *N. geoscyphus*, since it has the 2–3 large, opaque, segmented to botryoidal oil-bodies of the latter. When only dead plants are available for study, not surely separable from subsp. *scalaris.* The plant remains a taxonomic problem.

NARDIA GEOSCYPHUS (De Not.) Lindb. in Carr.

[Figs. 262; 263:6–11]

Jungermannia scalaris var. *minor* Nees, Naturg. Eur. Leberm. 1:281, 1833.
Alicularia scalaris var. *minor* Nees, *ibid.* 2:449, 1836.
Jungermannia haematosticta Nees, *ibid.* 2:453, 1836 (*nomen nudum*).
Alicularia geoscyphus De Not., Mem. Accad. Torino, Ser. 2, 18:486, fig. III (1–13), 1859; K. Müller, Rabenh. Krypt.-Fl. 6 (1):517, fig. 268, 1909; Macvicar, Studs. Hdb. Brit. Hep. ed. 2:128, figs. 1–4, 1926.
Sarcoscyphus anomalus G., *in* Gottsche & Rabenhorst, Hep. Eur. Exsic. no. 470, 1869 (*nomen nudum; teste* Pearson).
Jungermannia silvrettae G., *in ibid.* no. 470, 1869 (*nomen nudum*).
Nardia geoscyphus Lindb., *in* Carrington, Brit. Hep., p. 27, 1875; Evans, Rhodora 9:57, 1907 (in largest part, not *N. insecta*); Frye & Clark, Univ. Wash. Publ. Biol. 6(2):316, figs. 1–8, (1943) 1944.
Marsupella silvrettae Dumort., Hep. Eur., p. 128, 1874.
Nardia repanda Lindb., *in* Carrington, Brit. Hep., p. 27, 1875 (not. *J. scalaris* var. *repanda* Hüben., Hep. Germ., p. 81, 1934).
Alicularia minor Limpr., *in* Cohn, Krypt.-Fl. Schlesien 1:251, 1876.
Nardia hematosticta Lindb., Musci Scand., p. 8, 1879; Evans, Proc. Wash. Acad. Sci. 2:296, 1900.
Sarcoscyphus silvrettae Steph., Bot. Ver. Landshut 7:17, 1879.
Jungermannia dovrensis Limpr., Jahresb. Schles. Gesell. Vaterl. Kult. 61:10, 1884.
Nardia minor Arn., Leberm. Stud. Nordl. Norwegen, p. 39, 1892.
Mesophylla minor Bouvet, Bull. Soc. Étud. Sci. Angers, 1902:189, 1903.
Nardia silvrettae Pears., Hep. Brit. Isles 1:372, pl. 172, 1902.

   Small, polymorphous, in small flat patches, green (shade phases) but *usually brownish or reddish brown,* rarely somewhat purplish-tinged. Shoots 800–1300 μ wide $\times$ *5–10 mm* long, prostrate with ascending tips or, with crowding, suberect, simple or sparingly branched, the branches largely intercalary, *often carmine red to purplish beneath.* Stems 275–325 μ in diam., quite fleshy and soft. Rhizoids copious, colorless or *carmine to vinaceous.* Leaves obliquely inserted and oriented, not decurrent at antical base, varying from distant or contiguous below (on sterile shoots) to closely imbricate (especially on distal portions of fertile shoots), *orbicular to orbicular-reniform,* usually moderately *concave,* often somewhat

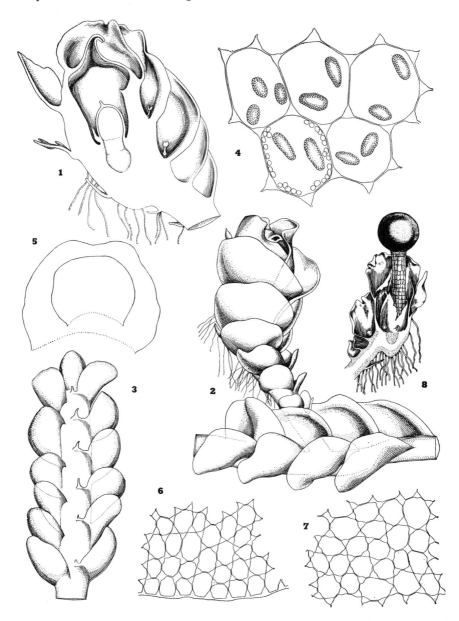

FIG. 262. *Nardia geoscyphus*. (1) Longisection of fertile shoot tip, bisecting, at upper left, the bracteole (×28). (2) Sterile stem sector with fertile branch with young gynoecium (×21). (3) Sterile shoot apex, postical aspect, rhizoids omitted (×21). (4) Median cells with oil-bodies, and, lower left, chloroplasts (×525). (5) ♂ Bract and, drawn within, sterile leaf (×28). (6) Marginal cells of leaf apex (×253). (7) Median leaf cells (×253). (8) Sporophyte-bearing shoot apex (×ca. 14). [1–3, 5–7, *RMS 15988*, NW. Basin, Mt. Katahdin, Me., also shown in Fig. 263, growing mixed with *N. insecta;* 4, *RMS 33007*, Mt. Katahdin, Me.; 8, after Schiffner, 1893–95.]

saucer-shaped and conspicuously concave, generally varying from 750–850 μ wide × 425–575 μ long up to 900–950 μ wide × 550–570 μ long (averaging 1.45–1.75× as wide as long), *entire* or broadly and shallowly retuse (*on sterile shoots and lowermost portions of fertile shoots*) *to shallowly emarginate* (*on uppermost vegetative leaves and on ♂ and ♀ bracts*); *lobes rounded or blunt on emarginate leaves*, the shallow and obtuse to rounded sinus descending at most ⅛–⅕ the leaf length, only exceptionally deeper and sharp (but *never descending ⅓–⅖ the leaf length*). Cells relatively *small*, thin-walled, with large to bulging trigones, those of the margins *averaging 18–24 μ*, the subapical averaging from 17–21 × 20–25(30) μ up to 22–26(28) × 24–32 μ, a little smaller than median cells, which range from 20–26 × 24–26(30) μ up to (22)25–28 × 28–33 μ, only isolated cells larger; cuticle smooth; *oil-bodies 2–3 per cell*, rather large, 6 × 10 to 7 × 15–20 μ, ellipsoidal to ovoid, distinctly fine-segmented and *appearing opaque and granular-papillose;* chloroplasts small, only 2–3.5 μ. Underleaves subulate to lanceolate, *often vestigial* or small and inconspicuous on lower portions of sterile shoots, becoming conspicuous and larger near apices of fertile shoots, frequently narrowly united with leaf bases on one side of stem.

Paroecious. ♂ Bracts in 2–3 pairs immediately below the gynoecia, larger than leaves, somewhat to *conspicuously concave, closely imbricate, erect or suberect* from a spreading base, subentire to *emarginate at apex*; usually 2-androus. Gynoecia in caespitose, crowded plants continuous with axis, but usually (in prostrate plants) at nearly or quite a right angle to axis. Bracts larger and broader than leaves, to 1100 μ wide × 625–750 μ long, rotundate to rotund-reniform, wider than long, often sinuous to somewhat crispate, *erect and sheathing perianth, only retuse or emarginate at apex*. Bracteole large, often lobulate. Perianth short, concealed within bracts, immersed, contracted to the crenulate-denticulate mouth. Perigynium high and conspicuous, massive and fleshy, in prostrate, "normal" plants *becoming gibbous and fleshy, shallowly bulbous*, densely rhizoidous, commonly 500–800 μ high (up to 2–4× as high as free perianth). Capsule broadly ovoid to subglobose, its wall 2-stratose. Epidermal cells with nodular thickenings. Inner layer with semiannular bands. Spores 14–16 μ, granular-papillate; elaters 2-spiral, 8–10 μ in diam. Chromosome no. $n = 18$.

Type. Europe.

Distribution

In eastern North America, at least, a species of strongly northern range, widespread and locally common in alpine situations and in the Arctic, southward becoming much rarer (and evidently largely replaced by the closely allied *N. insecta*). Because of confusion with this latter species and with the closely allied *N. lescurii*, the North American range of *N. geoscyphus* is still poorly defined. Some of the following reports (not followed by an!) are probably based on *N. insecta*.

Widespread, locally abundant, and polymorphic in Europe, where found from the Italian and French Alps northward to England and Scotland (not Ireland!), the Faroes, becoming very frequent in Scandinavia (Norway, Sweden, Finland), extending eastward to the Caucasus.

Also occurring on the Atlantic islands, to Madeira and to the Azores (Allorge and Allorge, 1948).

In North America evidently frequent to common in the Far West, from Alaska to British Columbia, Washington, Oregon, and California, and in the Rocky Mts. to Alberta, Montana, and Wyoming.

In the East as follows:

N. GREENLAND. Kap Glacier, S. coast of Independence Fjord, moist *Luzula confusa*-dominated heath, 81°48′ N., 31°45′ W. (*Holmen 6337, p.p.*!). E. GREENLAND. Scoresby Sund at Danmarks Ø (with *Dicranella secunda*) and Hekla Havn (Jensen, 1897); Turner Ø (Jensen, 1906a); Kangerdlugssuak, middle of Fjord, ca. 68° N. (*Böcher 904, p.p.*; fertile, with conspicuous, pendulous marsupia). NW. GREENLAND. Kânâk, Red Cliffs Pen., Inglefield Bay, 77°30′ N. (*RMS 45537a, 45477a*); Kekertat, near Heilprin Gletscher, E. end of Inglefield Bay (*RMS 46019*); Igdlorssuit, E. side of Red Cliffs Pen., Inglefield Bay (*RMS 45777c*). W. GREENLAND. Ameralik Fjord, S. of Godthaab, 64°03′ N. (*Vahl. 191* and *Vahl, s.n., Aug. 1830!*; reported as *N. scalaris* by Lange & Jensen, 1887; a third collection from this station is good *N. scalaris*); Godhavn, Disko I., at Lyngmarksfjeld, 69°15′ N., 53°30′ W. (*RMS & KD 66-154a, July 9, 1966; c. caps*); Eqe, near Eqip Sermia (glacier), peaty soil along a small rill, ca. 200–250 m.s.m., 69°46′ N., 50°10′ W. (*RMS & KD 66-250, 66-249f*); Fortune Bay, Disko I., 69°16′ N., 53°50′ W. (*RMS & KD 66-199a*); SE. of Jacobshavn, Disko Bay, 69°12′ N., 51°05′ W. (*RMS & KD 66-230*); Qalagtoq, Upernivik Ø, Inukavsait Fjord, 71°14′ N., 52°35′ W. (*RMS & KD 66-1101*); Kangerdlugssuakavsak, head of Kangerdluarssuk Fjord, 71°21′ N., 51°40′ W. (*RMS & KD 66-1285b*).

MELVILLE PEN. Questionable (Polunin, 1947). NOVA SCOTIA. Musquodoboit (Brown, 1936a, as "var. *suberecta*"); Yarmouth; Jeddore (Brown, 1936a). NEWFOUNDLAND. Table Mt. of Cape Ray, W. Nfld. (Buch & Tuomikoski, 1955). QUEBEC. Mainland S. of Cairn I., Richmond Gulf, Hudson Bay (Schuster, 1951; typical material!); Roggan R., Ungava, ca. 54°15′ N. and 78–79°44′ W.; below Anataiatch L.; Larch R., Ungava Bay area, 57°33–40′ N., 69°30′–70°07′ W.; Koksoak R., 57°42′ N., 69°27′ W. (Lepage, 1953).

MAINE. Mt. Katahdin: Chimney Pond, 2800 ft (*RMS*); west of summit of Baxter Peak, 5500 ft (*RMS 15815, 15816, 15996, p.p.*); above Davis Pond, 2850 ft, Northwest Basin (*RMS 15988, p.p.*. with *N. insecta*); Saddle Slide, 5000 ft (*RMS 33007, 33013, p.p.*). Mt. Desert I. (Lorenz, 1920). NEW HAMPSHIRE. Mt. Washington (*Evans, 1917; Schuster, 1951*); Lincoln (*Lorenz, 1912*); Crawford Bridle Path (Evans, 1902d); Franconia (Evans, 1907b); Davis Trail, 4500–5000 ft and Glen Boulder-Boot Spur Trail, Mt. Washington (*RMS 23055, 23112a*). MASSACHUSETTS. West Newbury (*Haynes 45, 1905*); this is the lowland form with emarginate sterile leaves, superficially approaching *N. insecta* but with cells smaller!; Worcester. CONNECTICUT. Bolton (Evans, 1914).

NEW JERSEY. Highlands, Monmouth Co. (*Haynes 1624;* see Haynes, 1913); Middletown, Monmouth Co. (*Haynes*); possibly both represent *N. insecta*. PENNSYLVANIA. Wilkes-Barre (*Lanfear 190d*); Ohiopyle, Fayette Co. OHIO. Red Rock Ravine, Hocking Co. (*M. S. Taylor, 1923*).

The species is also reported by Frye and Clark (1937–47) from Kentucky,[309] North Carolina, and Georgia, as well as from Virginia (Patterson, 1949, 1950).

---

[309] The Kentucky report is credited to Fulford (1936), who, however, reports only *N. lescurii* from Kentucky.

Apparently all of these reports from the Southern Appalachians refer to the endemic species, *N. lescurii*. I have checked the two reports from North Carolina: one from Blowing Rock (*Blomquist;* determined as *N. geoscyphus* by M. Fulford) is unquestionably the unrelated *Lophozia bicrenata!;* the other, from Cascades (*Schallert*), listed under *N. geoscyphus* in both Frye and Clark ("Examination") and in Blomquist (1936), is *N. lescurii*. Examination of specimens from Virginia, listed in Patterson (1949, 1950), showed them to be dioecious, with bilobed vegetative leaves; thus they represent *N. lescurii* as well. The Georgia report (accredited to Underwood, 1884) surely refers to *N. lescurii*.

I have not seen the material, attributed to *N. geoscyphus*, from New Jersey, Pennsylvania, and Ohio, but believe it may represent the allied *N. insecta*, which is apparently the common lowland representative in the East.

Ecology

Usually on thin soil over bare, granitic or metamorphic, acidic rocks; also commonly over peaty banks or peaty soil along rocky streams in the mountains, or at the rocky-peaty edges of mountain pools.

In the north, in and at the juncture with the Tundra, also occasionally on evidently noncalcareous sedimentary rocks and basalts, with *Lophozia alpestris*, *Pleuroclada albescens islandica*, *Mylia taylori*, and *Blepharostoma trichophyllum* (Schuster, 1951). Also, northward, on sandy paths and moist banks (with *Scapania subalpina*, *Solenostoma hyalinum*, *S. pusillum*, *Lophozia bicrenata*, *Cephalozia ambigua*, and *Dicranella subulata*), *teste* Lepage (1953).

In West Greenland often a common species, but mostly occurring as scattered individuals or gregariously in small patches. It may occur on damp, irrigated, peaty soil, in grass-*Eriophorum scheuchzeri* meadows or among rocks on stabilized slopes, on earth (associated with *Solenostoma subellipticum*, *Cephalozia pleniceps*, *Riccardia palmata*, *Tritomaria quinquedentata*, *Gymnomitrion concinnatum*, *Odontoschisma macounii*, and *Anthelia juratzkana;* less often with *Aneura pinguis*, *Gymnomitrion corallioides*, *Marsupella arctica*, *M. revoluta*, *Prasanthus suecicus*, and *Lophozia grandiretis*). Most often the species occurs on soil or peaty humus, frequently sheltered between rocks so that it is in diffuse light, in late snow areas, often on the borders of rills, very commonly with *Anthelia juratzkana*, *Odontoschisma macounii*, and *Cephalozia pleniceps*. Appearing somewhat indifferent to pH conditions, but apparently lacking from very acid as well as extremely calcareous sites.

In the New England mountains, where relatively frequent, occurring variously in moory sites, particularly on gravelly-peaty stream margins and on sunny, peaty, sphagnous hummocks at the edges of mountain streams (with *Pellia epiphylla*, *Lophozia alpestris*, *Cladopodiella francisci*, *Scapania irrigua*, *Odontoschisma elongatum*, *Cephalozia bicuspidata*, and such angiosperms as *Drosera rotundifolia*). Müller (1938a) also reports it from boggy areas. Perhaps even more frequent on damp, peaty or mineral soil, among boulders in rocky fell fields above treeline, where associated with *Cephalozia ambigua*, *Anthelia juratzkana*, and *Lophozia ventricosa*, less often *Chandonanthus setiformis* and *Lophozia atlantica*. The plant often occurs between large boulders, in places where snow and ice persist until very late into the season (also a typical habitat of the frequently associated *Anthelia*!); it is found here as an emarginate-leaved, green, lax, shade form. Müller (1938a) also reports the species as occurring in the European mountains in places where the snow remains the longest, forming blackish mats. In more exposed sites sometimes with *Gymnomitrion concinnatum*, *Marsupella ustulata*, and *Lophozia bicrenata*.

Although reputedly with a wide range south of such boreal and alpine situations, I am unfamiliar with the species under other conditions. At least in my experience, it is strictly an alpine or arctic plant, absent or at least very rare below 1800–2800 feet in New England.

Differentiation

The rather extensive synonymy is evidence of the involved history of the species in Europe and of its considerable variability. At least in eastern North America, it is local and shows only a limited variability, from small phases, 5–8 mm long, with brownish pigmentation and entire leaves (mod. *colorata-parvifolia*), to more robust phases, 8–10 mm long, which are soft-textured and green (mod. *viridis*) and have laxer, more spreading leaves that are often emarginate. Lindberg (1879) distinguished a var. *suberecta*, based on the strongly ascending aspect of the plants. This, reported from Nova Scotia, hardly represents more than an environmental modification. Associated with crowding, such suberect plants develop perigynia which lie almost or quite in the axis of the stem and are hardly bulbous. However, more "normal" phases, growing prostrate, develop a bulbous perigynium which lies at nearly a right angle to the stem at maturity.

Nardia geoscyphus shows a considerable range of variation in leaf form; on sterile shoots, leaves are commonly broadly orbicular and essentially entire, although occasional ones are retuse. However, on fertile plants, upper leaves and bracts show "reversion" to the ancestral, bilobed condition, ♂ and ♀ bracts always being perceptibly emarginate, or at least retuse-emarginate, occasionally bilobed to ⅙–⅕ their length. The regular emargination of upper leaves and bracts of fertile shoots separates all forms of *N. geoscyphus* from the often superficially similar *N. scalaris*, as well as suggests a bewildering similarity to *N. insecta* and *N. lescurii*.

In point of fact, *N. geoscyphus* and the latter two species are closely allied and appear to have a nearly mutually complementary range in eastern North America, almost suggesting they are subspecies of a single polymorphous species.

The Southern Appalachian *N. lescurii* can be eliminated from further consideration by three consistently diagnostic features; (*a*) the occurrence of as many as 4–5, rarely 6, oil-bodies in occasional cells; (*b*) a dioecious infloresence; (*c*) the uniformly rounded to blunt lobes of the consistently bilobed vegetative leaves.

The relationship of *N. geoscyphus* to *N. insecta* is, however, closer; the two taxa agree in the general size, the type of inflorescence, the type, size, and number of oil-bodies and are nearly similar in facies. In fact, until recently *N. insecta* was treated almost constantly as a form (Müller, 1905–16) or variety (Macvicar, 1926; Frye and Clark, 1937–47) of *N. geoscyphus*. The uniformly bilobed leaves and the difference in chromosome number, as well as the disparate geographical ranges, suggest, however, that two distinct species are at hand. These factors are discussed under *N. insecta*.

Sterile plants can be separated from our other species of *Nardia* by the (*a*) rather small cell size; (*b*) vestigial underleaves of sterile shoots; (*c*) opaque, granular-appearing, larger oil-bodies, never occurring in excess of 2–3 per cell. The last feature is of great importance in separating the species from the frequently similar *N. scalaris*, in which we find smaller, smooth, homogeneous, glistening, and pellucid oil-bodies. The presence, at least locally, of underleaves serves to distinguish *N. geoscyphus* from *Solenostoma* spp.

Variation

Nardia geoscyphus may occur as a mod. *viridis-leptoderma,* which is identical in size and often approaches in aspect *Solenostoma subellipticum* and *S. sphaerocarpum* (two taxa that may actually occur mixed with it in west Greenland). In actual practice, the fleshy, rather clavate gynoecial "heads" of *N. geoscyphus* can usually be found—the species is almost constantly fertile—and on these, at least, bracteoles and underleaves are distinct, thus separating the species from any *Solenostoma*. Such plants might be confused with *N. scalaris* (which, however has more distinct underleaves).

More often plants are red-brown, and (under the microscope) the leaf cells may be distinctly purplish-pigmented; the dull purplish pigmentation may be most distinct on the leaf margins, which (in arctic phenotypes) may become decolorate with age. In the high Arctic small, compact, and almost clavate plants occur (e.g., *Holmen 6337,* from 81°48′ N. in Greenland) that are rather laterally compressed, dense-leaved, and purplish. Such dwarf, prostrate plants, freely fertile, may show underleaves almost reduced to stalked slime papillae.

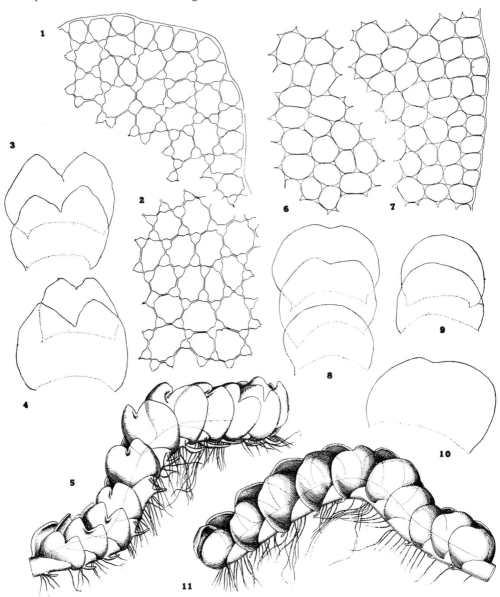

Fɪɢ. 263. *Nardia insecta* (1–5) and *N. geoscyphus* (6–11). (1) Cells of lobe apex (×230). (2) Median leaf cells (×230). (3–4) Leaves (×32). (5) Sterile shoot, lateral aspect (×22). (6) Median cells (×230). (7) Cells of leaf apex (×230). (8) Leaves (×32). (9) Leaves from small sterile shoot (×32). (10) ♀ Bract (×32). (11) Sterile shoot, lateral aspect (×22). [All drawn from intermingled plants, *RMS 15988*, NW. Basin, Mt. Katahdin, Me.; see also Fig. 262:1–3, 5–7.]

In addition to the several variations discussed, the following seems to be more marked.

NARDIA GEOSCYPHUS var. BIFIDA Schust., var. n.

Varietas var. *geoscypho* similis, omnia folia, autem, etiam in surculis sterilibus, o.1–0.2 bifida, lobis rotundatis ad obtusos; folia plerumque dorsaliter longissime decurrentia.

Similar to var. *geoscyphus*, but *all leaves*, even on sterile shoots, emarginate to bifid for o.1–0.2 their length by a blunt to sharp notch, the leaf lobes all rounded to broadly obtuse. Leaves generally exceedingly long-decurrent on the fleshy stem (which is ca. 14–15 cells high).

Type. Northeast Greenland: Mt. Zackenberg, Wollaston Foreland, 74°30′ N., 20°38′ W.; on moist soil in shady caves of rocks (*K. Holmen 5537*!; Copenhagen).

This difficult plant was initially a real enigma since, on account of the aspect and the general sterility, it could easily be confused with a *Lophozia s. lat.* The rather soft and fleshy leaves tend to be 2- or even 3-layered at the base, and are always bilobed or broadly retuse; I could find no unlobed leaves, even of sterile plants. Young gynoecia are present, and in two cases, after careful search, remnants of single antheridia could be found in bracts below the perichaetial bracts. Although the plant perhaps approaches *N. insecta* in the bilobed leaves, the marginal and subapical cells average only 24–26 μ in diam., eliminating this species.

NARDIA INSECTA Lindb.
[Figs. 258:1–4; 263:1–5; 264]

Nardia geoscyphus auct. (*p.p.*), including Evans, Rhodora 9:57, 1907.
Nardia insecta Lindb., Musci Scand., p. 8, 1879; K. Müller, Rabenh. Krypt.-Fl. ed. 3, 6:854, fig. 297, 1956; S. Arnell, Illus. Moss Fl. Fennosc. 1:132, 1956.
Alicularia geoscypha fo. *insecta* K. Müll., Rabenh. Krypt.-Fl. 6(1):519, 1909.
Alicularia geoscyphus var. *insecta* Macvicar, Studs. Hdb. Brit. Hep. ed. 2:129, 1926.
Nardia crassula Lorb., Jahresb. Wiss. Bot. 80:569, 1934 (*nomen nudum*).
Nardia geoscyphus subsp. *insecta* Joerg., Bergens Mus. Skrifter no. 16:96, 1934.
Nardia geoscyphus var. *insecta* Clark & Frye, The Bryologist 40:15, 1937; Frye & Clark, Univ. Wash. Publ. Biol. 6(2):317, (1943) 1944.

Light green, *rather translucent*, often with rhizoids and ventral stem face and leaf bases carmine, very similar to *N. geoscyphus*: with fleshy stem, suberect to erect-spreading, occasionally somewhat antically secund and somewhat concave leaves; leaves broadly rotund-quadrate to transversely quadrate, wider than long [ca. 850–960 μ wide × 600–800 μ long, up to 1000–1100(1430) μ wide × 750–800(950) μ long], rather concave, *uniformly bilobed on sterile as well as fertile shoots, the sinus descending* ¼–⅓, *rarely to* ½, *acute and sharp, often somewhat gibbous or reflexed*; lobes triangular, broad, ± sinuous, *subacute to blunt or, more rarely, narrowly rounded.* Cells distinctly *larger than in N. geoscyphus, more transparent,* thin-walled but *often with coarser trigones,* ca. 25–30 μ on margins of lobes, ca. 31–36(40) × 33–40 μ in free lobe apices, ca. (30)32–38 × 34–40(47–52) μ

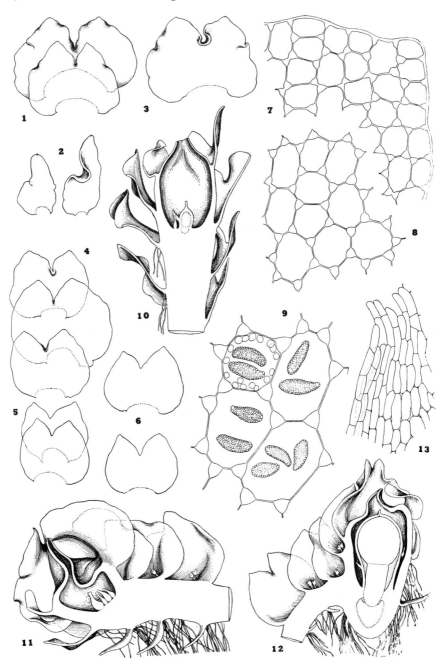

FIG. 264. *Nardia insecta*. (1) ♀ Bracts (×21). (2) ♀ Bracteoles (×21). (3) Large ♂ bract (×21). (4) Smaller ♂ bracts (×21). (5–6) Leaves of sterile shoot sectors (×21). (7) Cells of lobe apex (×240). (8) Median leaf cells (×240). (9) Median cells with oil-bodies and, upper left, chloroplasts (×480). (10) Apex of erect shoot, with erect perigynium (×27). (11) Prostrate shoot, with low perigynium at time of fertilization (×27).

in leaf middle, becoming 35–46 × 45–58 μ near leaf base; *oil-bodies 2–3(4–5) per cell, coarsely granular-segmented, opaque and grayish,* the component spherules somewhat protuberant (mostly less than 1 μ in diam.), ellipsoidal to narrowly ovoid in shape and from 6–7 × 14 μ to 6–7.5 × 18–20 μ; chloroplasts much smaller, only 3.2–4.5 μ in diam. Underleaves conspicuous, spreading away from stem, with dense rhizoid fascicles at their bases (and only sparing development of rhizoids elsewhere), often carmine-red, *present throughout on sterile shoots.*

Paroecious. Involucre varying from at a distinct angle with axis (and then with perigynium bulbous beneath, to 1300 μ high), to suberect or erect (mod. or fo. *erecta*), rhizoidous below, the axis often carmine-red postically; upper (♂) leaves subtransverse, subimbricate to imbricate, often patent, ± undulate and only weakly concave, to 1300–1400(1450) μ wide × 900–950 μ long, bifid for 0.35–0.5 their length, broadly transverse, ± undulate, the sinus sharp, *often strongly reflexed,* the leaves (and ♂ bracts) thus appearing somewhat crispate; lobes triangular, acute to subacute, occasionally blunt. ♀ Bracts broadly, transversely quadrate-reniform to reniform, *deeply bifid for 0.45–0.7 their length,* strongly crispate, with sinus sharply reflexed, with the closely juxtaposed ♂ bracts forming a *crispate, somewhat clavate, conspicuous head;* ♀ bracts from 1150 μ wide × 725 μ long up to 1550 μ wide × 700 μ long, considerably larger than vegetative leaves, but grading gradually into them. Bracteole lanceolate to lanceolate-triangular or lingulate, somewhat crispate, rather large, to 620–1000 μ long × 450–500 μ wide. Perianth conoidally, strongly contracted to the small orifice but hardly beaked, very short (ca. 400–500 μ high, as contrasted to the perigynium, which is up to 1000–1400 μ high), formed of pellucid, elongated cells, except near base, the cells distally (17)18–24 μ wide × (60)70–85 μ long, thin-walled, except at the angles; apical cells projecting slightly at their free tips, ca. 17–20 × 45–65(72) μ, the perianth mouth thus crenulate with fingerlike cells. Capsule almost spherical, the wall 2-stratose. Epidermal cells short-rectangulate or quadrate, from 35–40 to 30 × 38–40 μ, to long-rectangulate, 18–26 × 52–62 μ, the longitudinal walls with (1)2–3(4) strong, ± stalked-appearing nodular thickenings, the transverse walls with 0–1 (on longer walls 1–2) similar nodular thickenings. Inner layer of much less regular cells, often imperfectly oblong to weakly sigmoid, where rectangulate ca. (15)17–20(25) × 70–85(95) μ long, with numerous semiannular bands, almost uniformly complete on the inner (free) tangential walls. Spores ca. (18)20–22 μ (perhaps postmature), brownish, finely granulate. Elaters ca. 9–10 μ × 150–175 μ, the tapered ends not thickened, bispiral, the spiral ca. 3 μ wide . (Sporophyte from *RMS 43062.*) Chromosome no. *n* = 36.

Type. Finland (*Nyland*).

Distribution

Evidently a lowland plant, occupying in eastern North America a zone between the largely high subarctic and arctic-alpine *N. geoscyphus* (which rarely, if ever, occurs in typical form below 1800 feet in New

---

(12) Prostrate shoot with almost mature sporophyte, with erect perigynium, bulbous at base, and shoot calyptra (×24). (13) Perianth mouth cells (×112). [All from *RMS 24234* and *RMS 24244*, Forest of Dean Lake, N.Y.]

England) and the Southern Appalachian *N. lescurii*, exhibiting close systematic affinities to these two species. The range poorly known; in Europe cited as "widely distributed on the continent" and given (Müller, 1905–16) from Germany, Czechoslovakia (Bohemia), France, and Italy, northward to Norway, Sweden, and southwest Finland (Aland Isls.).

In North America previously reported only from Wyoming and Washington (*teste* Frye and Clark, 1937–47, p. 318), but also occasionally found in the East from Maine to New York (and possibly New Jersey and Pennsylvania), although the range much confused in American literature with that of *N. geoscyphus*.[310] The Newfoundland report of *N. lescurii* almost certainly represents confusion with *N. insecta* (p. 880).

NOVA SCOTIA. Wet, acid, sandy soil in ditch, 2 mi S. of Lower Argyle, Yarmouth Co. (*RMS 43061, 43062*); Cape Breton Ntl. Park: Benjy's L., French Mt., 1000 ft (*RMS 68-825*); Cheticamp R., at Salmon Pools (*RMS 68-901*). MAINE. Davis Pond, 2850 ft, Northwest Basin of Mt. Katahdin (*RMS 15985a, 15982a, 15988*, p.p.); Chimney Pond, 2800 ft, Mt. Katahdin (*RMS 32999a*). NEW HAMPSHIRE. Lake of the Clouds, White Mts., 5200 ft. (*RMS 68-2102d, 68-2118b*). NEW YORK. Forest of Dean L., near Ft. Montgomery, Rockland Co. (*RMS 24234, 24244*); near L. Montauk harbor, Long I. (*RMS 22071d, p.p.*).

Ecology

Occurring variously on moist humus or loamy banks, in shade under *Tsuga canadensis*, near the margin of a lake, associated with *Lophozia capitata, Calypogeia fissa* cf. *muelleriana, C. neesiana* s. str., *Scapania nemorosa, Cephalozia bicuspidata, Leucobryum, Ditrichum* sp., *Atrichum angustatum, Pohlia nutans*, in the Hemlock-Hardwoods Forest Region (*RMS 24234, 24244*). Also found in open "heath" on springy loam banks at Montauk, N.Y., with *Pellia epiphylla, Calypogeia sullivantii*, etc. Northward, in the Spruce-Fir Zone, found at low elevations on Mt. Katahdin (2850 feet) over peaty, exposed, strongly insolated soil along a cold stream, associated with *Nardia geoscyphus, Pellia epiphylla, Cephalozia bicuspidata, Scapania irrigua, Lophozia alpestris, L. incisa* var. *inermis, Sphagnum, Drosera rotundifolia*. Occurring under very similar conditions over peaty and stony ground at the edge of Chimney Pond, Mt. Katahdin (2800 feet), associated with *Pellia epiphylla, Cephalozia* cf. *ambigua, Scapania nemorosa, Odontoschisma elongatum, Gymnocolea inflata*, and *Lophozia alpestris*. The Nova Scotia occurrence is analogous: on sandy-loamy, moist soil in a roadside ditch, associated with some of the preceding taxa (e.g., *Pellia epiphylla, Cephalozia bicuspidata*, as well as *Cladopodiella francisci, Calypogeia sullivantii*, and *C. neesiana*.

Evidently a mesophyte, of acid, peaty, or loamy moisture-retaining

---

[310] Grolle (1964a, p. 299) still cites *N. insecta* as exclusively European.

substrates, able to tolerate much intense insolation (and then with ventral face of stem and rhizoid bases carmine-red, and with leaf lobes tending to become brown or reddish brown).

Differentiation

The combination of uniformly and deeply bilobed leaves, which average wider than long (always clearly wider than long near apices of fertile shoots), discrete underleaves, bulging trigones, and thin cell walls serves to separate the species from all but *N. lescurii*, and occasional, atypical forms of *N. geoscyphus*. Separation from these two taxa is often difficult.

The tendency toward brownish pigmentation, the relatively few, opaque oil-bodies (formed of numerous, somewhat protuberant spherules, thus coarsely papillose in appearance), and the paroecious inflorescence are shared with *N. geoscyphus*, and separate the species from *N. lescurii*. The latter has a tendency to extensively develop rose-colored pigmentation and never has subacute leaf lobes or a rectangular leaf sinus. Sterile plants (and plants with fertilized archegonia or young sporangia) are difficult to separate, since all three species have antheridia that are ephemeral and disappear (often leaving virtually no trace) soon after fertilization;[311] furthermore, ♂ bracts of all three species are so similar to vegetative leaves, lacking a distinctly ventricose base, that androecia are usually unrecognizable unless antheridial remains can be found. The few oil-bodies (usually 2, occasionally 3) of both *insecta* and *geoscyphus* afford a nearly constant separation from *N. lescurii*, which most often has 3–5, occasionally 6 or 2, oil-bodies per cell. Some cells (at least in the Nova Scotian material of *N. insecta*) may possess 4–5 oil-bodies, although 2–3 are the rule.

Separation of the species from *N. geoscyphus* is usually easily possible because all leaves—vegetative as well as those of the reproductive regions—are bifid in *N. insecta*, with the sinus usually descending ⅓, often rectangular to subacute, with the lobes often rectangular to subacute. In *N. geoscyphus* most vegetative shoots typically have round, entire leaves (occasionally retuse to emarginate with rounded sinus and lobes), while leaves of fertile regions are bilobed ⅛–⅕, with obtuse to rounded sinus and normally rounded to obtuse lobes. Both *N. insecta* and *geoscypha*, as well as *N. lescurii*, normally grow relatively depressed; they then produce a perigynium that develops at nearly right angles to the axis, forming a bulbous foot at the stem apex. However, the plants often grow crowded

---

[311] In critical cases, a series of putative ♂ bracts must be dissected free and studied under high power. In *N. insecta* the relatively persistent antheridial stalks are usually then demonstrable.

and nearly or quite erect. In such cases, the perigynium lies nearly or quite in the axis of the stem, does not develop a bulbous "foot," and closely resembles that found in *N. scalaris*. Such differences at one time were regarded as warranting separation into "species" or varieties, but appear to be due to extrinsic causes and therefore are here regarded as taxonomically irrelevant.

Nardia insecta has been regarded as of doubtful specific distinctness. Lindberg (1879) separated *N. insecta* from the allied *N. geoscyphus*, with which it agrees in the paroecious inflorescence, on the basis of the (*a*) uniformly bilobed leaves, divided for 0.25 to nearly 0.5 their length, the sinus generally sharp and acute; and (*b*) presence of underleaves throughout. These distinctions are, however, so subject to intergradation that *N. insecta* has been generally regarded as a mere variety of *N. geoscyphus* (as, i.a., by Schiffner, 1901, p. 213; Macvicar, 1926, p. 129) or even as only a forma (Müller, 1909, *in* 1905–16, p. 519), while Evans (1907b, p. 59) does not accord formal recognition to the plant with bilobed leaves. As Evans (*loc. cit.*) points out, the "lowland specimens bear leaves which are more or less bilobed and therefore represent the *Nardia insecta* of Lindberg." The implication is that the high-altitude forms are "typical" *N. geoscyphus*.

Macvicar (1926) interestingly enough, states that *insecta* does not occur in Great Britain, only "typical" *geoscyphus* being represented, suggesting that the two show differences in range. An overlapping, although largely complementary, range appears to characterize the two taxa in North America as well.

In the interpretation of *N. insecta*, data derived from two sources are pertinent; (*a*) cytology: Lorbeer showed that *N. geoscyphus* has 18 chromsomes, while *N. insecta* (= *N. crassula* Lorbeer) has 36; (*b*) the ability of the two plants to occur together. I have collected material (Davis Pond, Northwest Basin of Mt. Katahdin, Me., *RMS 15988*) in which two kinds of *Nardia* grow intermingled. One has entire-leaved sterile shoots and hardly emarginate, subinvolucral leaves; these plants have rather small cells (ca. 22–28 μ in leaf middle) and are rather deeply reddish-brown-pigmented, opaque, and freely fertile; they are also unique in having small to vestigial underleaves on sterile shoots and cells with only moderately large trigones. These are typical *N. geoscyphus*. Associated is a second plant, evidently sterile, with sharply bifid leaves, very distinct underleaves, largely green color (but rhizoids and leaf bases as well as postical stem faces more or less carmine-red), and large cells (ca. 32–36 × 35–38 μ medially) that have coarse, subconfluent trigones. These plants represent *N. insecta*, although they could easily be mistaken for the Appalachian *N. lescurii*.

On the bases outlined above, it seems probable that *N. insecta* should be given the rank of a full species. Thus delimited, *N. insecta* is difficult to separate from *N. lescurii*, which has similarly large cells and bilobed leaves. The differences essentially resolve themselves to the oil-body number [2–3 per cell in *insecta*; mostly (2)3–5 per cell in *lescurii*]; the inflorescences (paroecious in *insecta*; dioecious in *lescurii*); the form of the leaf lobes (often acute or subacute in *insecta*, although not infrequently blunt; usually blunt to broadly rounded in *lescurii*).[312]

Part of the confusion in the literature between *N. insecta* and *N. geoscyphus* lies in the variability of the latter species. *Nardia geoscyphus*, as is detailed under this species, not infrequently produces phases with emarginate leaves, even on sterile shoots. However, such plants, superficially transitional to *N. insecta*, always have rounded to blunt leaf lobes and a sinus descending usually ⅙–¼, only exceptionally to ⅖, the leaf length. It is possible to place such plants on the basis of cell size: the median cells, which are rather opaque, ranging from 21 to 25 μ in diam., rarely to 28 μ, and being nearly hexagonal and isodiametric. The marked differences in leaf shape and size and in cell size are readily evident from Fig. 263, in which the respective features are contrasted from plants collected growing intimately intermingled. In my experience, the various phases of *N. geoscyphus* are generally less robust, more opaque (because of the smaller cells), and more often strongly brownish-pigmented. The difference in robustness and cell size appears to be correlated with a discrepancy in chromosome number, *N. insecta* presumably being a polyploid derivative.

NARDIA LESCURII (Aust.) Underw.
[Figs. 265–266]

Alicularia lescurii Aust., Hep. Bor.-Amer. no. 5, 1873, and Bull. Torrey Bot. Club 6:18, 1875.
Nardia lescurii Underw., Bull. Ill. State Lab. Nat. Hist. 2:115, pl. 8E, 1884; Evans, The Bryologist 17:88–89, 1914; M. S. Taylor, *ibid.* 42:86, figs. 1–12, 1939.

In low mats or patches, bright but rather pale and pellucid green *to purplish or rose red*. Shoots prostrate, except for the ascending tips, 800–1800 μ wide × 15–25(40) mm long, sparingly branched, the branches largely intercalary, rarely terminal. Stems 250–360 μ in diam., *quite fleshy, soft*, the nearly leptodermous cortical cells ca. 22–28(35) μ wide × 32–38 μ long, hardly differing (in diam.) from the leptodermous medullary cells; medullary cells ca. 25–36 μ in diam.; mycorrhizal infection apparently lacking in both medullary and cortical cells. Rhizoids long, rather numerous, from the postical bases of leaves and the bases of underleaves, and scattered ones elsewhere from the ventral merophytes, colorless to slightly reddish. Leaves *rather soft in texture*, obliquely inserted and oriented, approximate to weakly imbricate, not decurrent antically or postically, *spreading and flat or slightly concave*, broadly rotundate, from 350–650(900) μ long × 400–1000 μ wide, *uniformly and shallowly bilobed at apex*, the sinus descending 0.1–0.25 the leaf length, usually obtuse to *often*

---

[312] The lobes of the leaves in most populations of *N. insecta* are acute or at most blunt; in the copious material from Nova Scotia they are almost constantly rounded, often broadly so. Even here, however, the sinus remains angular and most often narrowly V-shaped.

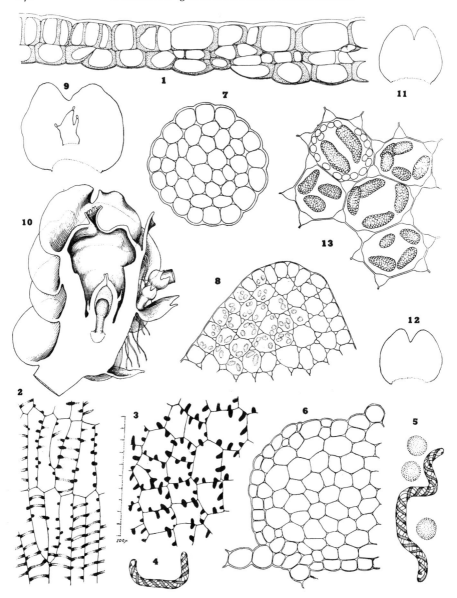

FIG. 265. *Nardia lescurii*. (1) Capsule wall cross section (×500). (2) Inner cells, capsule wall, to scale at right (×310). (3) Epidermal cells, capsule wall, to scale at left (×310). (4–5) Elaters and spores (×310). (6) Stem cross section (×178). (7) Seta cross section (×100). (8) Leaf lobe, oil-bodies drawn in in some cells (×144). (9) Leaf and underleaf (×33). (10) Apex of erect gynoecial shoot with young sporophyte in longisection (×24). (11–12). Leaves (×30). (13) Median leaf cells with oil-bodies and, upper left, chloroplasts (×565). [1, 7, *RMS 34506a*, Cypress Chapel, Va.; 2–6, 8–10, *RMS 36980*, The Cascades, Stokes Co., N.C.; 11–13, *RMS 34301*, Georgia.]

acute, often slightly gibbous; *lobes very broad, rounded or blunt at apex*, subequal or somewhat unequal with the ventral the larger. Cells averaging 20–24(30) μ on margins of leaf lobes, in leaf middle from 23–28(44) × 25–35(56) μ to 30–40 × 40–50 μ, rounded to subhexagonal, thin-walled and *with large, bulging to subconfluent trigones;* cuticle smooth or delicately verruculose; oil-bodies in part 2–3, in *part 4–5(6), per cell*, oval to ellipsoidal, rarely subspherical, large, 7–10 × 10–16(18–20) μ, *grayish and finely but distinctly granulate-papillose*, formed of numerous small spherules. *Underleaves distinct throughout*, narrowly lanceolate to triangular, acute to acuminate, entire or occasionally with a small tooth on one or both sides, free at base or sometimes narrowly connate (on one side) with the leaves.

Dioecious. Androecia becoming intercalary, *hardly differentiated* from vegetative shoot sectors; bracts in 6–15 pairs, almost identical in size and form to leaves, *little concave*, the 2–3 antheridia readily exposed to view; bracteoles like under-leaves; antheridia with stalk 2-seriate. Gynoecia usually with subfloral innovation; bracts slightly to hardly larger than leaves, otherwise similar to them, but ± undulate; bracteole large, free, lanceolate to ovate, acute to acuminate, often with 1–2 small teeth or lobes. Perianth short, 700–1000 μ long × 750–950 μ wide, *hidden between the bracts*, inflated but slightly dorso-ventrally compressed, narrowed to an obscure beak; mouth entire or shallowly 2–4-lobed at summit, crenulate; cells 2–3× as long as broad, with distinct trigones (except for apical cells). Perigynium very fleshy, either lying in axis of stem (when plants crowded and ascending) or assuming at maturity a distinct angle with axis, 700–1200 μ high (equal in height to perianth or higher), without rhizoids. Seta with ca. 16–18 epidermal cell rows, in cross section 7–8 cells in diam., ca. 320 μ in diam. Capsule 650–800 μ long, broadly ovoid. Capsule wall 2-, *locally 3-, stratose;* where 2-stratose 25–28 μ thick, where 3-stratose, 34–35 μ thick; epidermal cells averaging ca. 30 μ wide × 37 μ long and 15–16, locally 18–19, μ thick, with nodular thickenings of both transverse and longitudinal walls; inner layer with cells averaging ca. 18 μ wide × 60 μ long × 10–11 μ thick, each cell with 5–7 transverse, annular to semiannular bands. Elaters 8–10 μ in diam., bispiral; spores 16–18 μ, finely and closely granular-papillate, brown.

Type. Tallulah Falls, Rabun Co., Ga. (*Lesquereux, 1850!*).

Distribution

An Appalachian endemic occurring at median and low elevations in the mountains [in the Oak-Hickory-(Chestnut) and Mixed Mesophytic Forest Regions and also in the Hemlock-Hardwoods Forest], usually at 1500–3500 feet, southward to 4700 (rarely to 5000) feet. With a disjunct occurrence in the coastal Short Leaf-Loblolly Pine Region of southeast Virginia, and in isolated montane "islands" lying in the Piedmont (as at the Cascades, near Hanging Rock, N.C.). Also recently reported from Newfoundland, but this needs further verification.

NEWFOUNDLAND. Kitty's Brook, SE. of Avalon Pen. (Buch & Tuomikoski, 1955), associated with *Calypogeia muelleriana* and *Pellia neesiana*. This report probably represents *N. insecta*.

VIRGINIA. Mountain L., Giles Co. (Patterson, 1949, as *N. geoscyphus*!; the plants clearly dioecious and representing *N. lescurii*); Tinker Mt., Botetourt Co. (Patterson, 1950; as *N. geoscyphus* var. *insecta*, but clearly dioecious!); Giles and Spottsylvania Cos. (Patterson 1950); below the high falls, below Cascades, on Little Stony Cr., near Mountain L., Giles Co. (*RMS, s.n.*); 3–4 mi NE. of Cypress Chapel, Nansemond Co., at western edge of Dismal Swamp (*RMS & Patterson 34506, 34505;* the only report from the outermost Coastal Plain). WEST VIRGINIA. Short Mt., Hanging Rock, Hampshire Co. (*Ammons 548*); near Welch, McDowell Co. (Taylor, 1939). OHIO. Crane and Exstein Hollows, Little Rocky Fork, Queer Cr., and Red Rock Ravines, Hocking Co. (M. S. Taylor, 1939); White's Gulch, Jackson Co. (Taylor, 1939); Redrock, W. Jefferson (*M. S. Taylor 8044!*). KENTUCKY. Natural Bridge, Powell Co. (*M. S. Taylor, 1925*); Bell and McCreary Cos. (Taylor, 1939).

NORTH CAROLINA. Cascades, near Hanging Rock State Park, Stokes Co. (*RMS 28242, 28258, 28251; RMS 36980, c. caps. [Apr. 8, 1955]*); Pilot Mt., Surry Co. (*Schallert, 1925*; listed by Blomquist, 1936, as *N. geoscyphus*); Cascades, Stokes Co. (*Blomquist 10409*, cited as *N. geoscyphus* by Blomquist, 1936!); Glen Falls, E. fork of Overflow Cr., 3 mi SW. of Highlands, Macon Co. (*RMS; Anderson 10686!*). Reported by Taylor (1939) from Macon, Transylvania, Henderson, Haywood, Buncombe, Avery, and Jackson Cos., 2000–5000 ft. SOUTH CAROLINA. Oconee and Pickens Cos. (M. S. Taylor, 1939). GEORGIA. Tallulah Falls (*Lesquereux, 1850, type!; Underwood, 1891!; Schuster, 1951*, from an area 2–3 mi above Tallulah Falls proper); near summit of Brasstown Bald, Towns Co., 4600 ft (*RMS 34301, 24302, 34300*); along Chattooga R., Rabun Co. (Taylor, 1939); below High Falls of Big Cr., and along Big Cr., ca. 0.5–0.6 mi above High Falls, 4–5 mi SSE. of Highlands, N.C., Rabun Co. (*RMS, s.n.*).

The species has been widely confused with *N. geoscyphus*, even in some recent literature, in spite of the excellent study of the species by M. S. Taylor (1939). All reports of *N. geoscyphus* from the Southern Appalachians have proved referable to *N. lescurii* (at least insofar as they were correctly determined to genus).

Ecology

Usually on damp, loamy soil or sandy loam, rarely on peaty humus or on damp rocks, and most often in such sites associated with small cascades or brooks, or near dripping wet rocks, most often in deep shade. The plant is most often a "gorge species," found near, but not in the spray of, waterfalls. In such sites it occurs often with *Diplophyllum apiculatum*, *Calypogeia sullivantii*, *Pellia epiphylla*, *Pallavicinia lyellii*, and occasionally *Microlepidozia sylvatica*, *Cephalozia bicuspidata*, *Odontoschisma prostratum*. The plant also often occurs in boggy, springy sites on mountain slopes, under *Tsuga canadensis*, *Rhododendron maximum*, or *R. catawbiense*, associated with most of the preceding species. Shade forms are perhaps most frequent, although sun forms (which are often a deep rose-red) are not uncommon.

The species has a sporadic distribution: it is absent from most of the sites where one would expect to find it—yet is present in great abundance at similar places elsewhere, often forming very extended, pure mats.

The occurrence in the outer Coastal Plain, at the edge of the Dismal Swamp, is quite anomalous. The plant here grows on the sides of moist, somewhat sandy ditches, at the edge of loblolly and short-leaf pine woods, with *Scapania nemorosa*, *Telaranea nematodes*, *Calypogeia sullivantii*, *Cephalozia macrostachya*, *C. connivens*, *Odontoschisma prostratum*, *Microlepidozia sylvatica*, and *Sphagnum*.

Differentiation

A stenotypic and highly characteristic species, easily recognized in the field by the uniformly bilobed, flat, spreading leaves which have conspicuously rounded to blunt lobes, and by the fleshy, soft, thick stem. The uniformly bilobed leaves separate *N. lescurii* from other species of *Nardia*, except *N. insecta* (which is exceedingly closely allied to *N. geoscyphus*). *Nardia lescurii* differs from the latter in two respects, as regards vegetative structures: the leaves have consistently blunt or rounded, shorter lobes (acute or subacute in *N. insecta*; usually entire on sterile shoots in *N. geoscyphus*); the cells have, in part, as many as 4–5 or rarely even 6 oil-bodies apiece (consistently 2–3 per cell in *N. insecta* and *N. geoscyphus*). Although dioecious, *N. lescurii* almost always bears sex organs, and plants of the two sexes frequently occur admixed or in contiguous patches; therefore capsules are often produced (as early as April 15 on the coastal plain of Virginia; April 8 in montane parts of North Carolina). *Nardia insecta* and *N. geoscyphus* differ at once in the paroecious inflorescences. These are often difficult to demonstrate, but in no case can intercalary, loosely spicate androecia, such as are characteristic of *N. lescurii*, be found.

In the quadrate-rotundate leaves with a sharp sinus and rounded lobes, *N. lescurii* is superficially slightly similar to such species as *Lophozia obtusa*. The strongly collenchymatous cells, the few and large oil-bodies, and especially the very stout, soft, and fleshy stems prohibit confusion. Also, the lanceolate and distinct, spreading underleaves are very different from the underleaves of any Lophoziaceae.

The capsule wall in the Jungermanniaceae is supposedly generally 2-stratose, except in the deviant *Jamesoniella*. This is the basis, in part, of a distinction between the Jungermanniaceae and the Lophoziaceae. In *N. lescurii*, however, I have found that the capsule wall is locally quite often 3-stratose (Fig. 265), indicating that this distinction is much less rigid than had been assumed. Amakawa (1960a) also reports a 3- and even 3–4-stratose capsule wall in several Japanese species of *Solenostoma* (subg. *Plectocolea*) but describes the capsule wall of *Nardia* as "2-stratose."

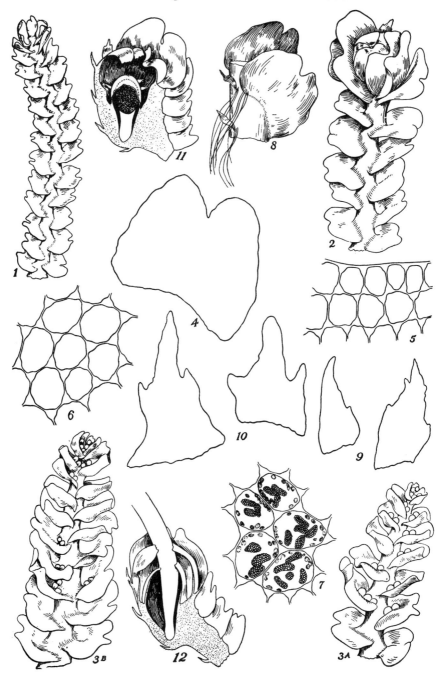

Fig. 266. *Nardia lescurii*. (1) Sterile shoot, antical view (×18). (2) Apex of ♀ shoot (×18). (3A, 3B) Apices of ♂ shoots (×18). (4) Leaf (×57). (5) Marginal leaf cells (×273). (6) Median leaf cells (×273). (7) Median leaf cells with oil-bodies and

Sectio *BREIDLERION* Grolle

Nardia sect. *Breidlerion* Grolle, Bot. Mag. Tokyo 77:299, 1964.

Minute plants, usually under 5 mm high and 0.5 mm wide with leaves; leaves consistently bilobed, the lobes subacute or blunt, sinus semilunate or angular usually; cells very small, 13–18 μ wide subapically and medially, with trigones feebly developed or absent, the walls thin or $\pm$ evenly thickened. Elaters 1- or 3–4-spiral (where known).

Type. Nardia breidleri (Limpr.) Lindb.

Including only three species: the type, which is imperfectly holarctic; *N. unispiralis* Amak. (1960a) of Japan, with unispiral elaters; *N. arnelliana* Grolle (1964a, p. 297, fig. 1) of high elevations of tropical Africa.

NARDIA BREIDLERI (*Limpr.*) *Lindb.*
[Fig. 267]

Alicularia breidleri Limpr., Jahresb. Schles. Gesell. Vaterl. Kult. 57:311, 1880; K. Müller, Rabenh. Krypt.-Fl. ed. 2, 6(1):521, fig. 269, 1909; Macvicar, Studs. Hdb. Brit. Hep., ed. 2:129, 1926.
Nardia breidleri Lindb., Meddel. Soc. F. et Fl. Fennica 6:252, 1880; Amakawa, Jour. Hatt. Bot. Lab. no. 21:285, fig. 10,1–5, 1960.
Mesophylla breidleri Boulay, Musc. France 2:141, 1904.

Minute, in reddish to purplish brown depressed patches; shoots 1–4 mm high, freely branched by ventral- or ventrolateral-intercalary branches, *often small-leaved, whitish, and stoloniform* (at least initially); stems ca. 100–155 μ in diam. Rhizoids long, colorless, to near shoot apex. Leaves of sterile shoots generally remote to approximate, small, erect-spreading, broadly ovate, concave, *only 1–1.5(2)× the diam. of the rather fleshy stem*, often only 165–200(350–400) μ wide × 160–200(300–360) μ long, larger toward androecia and gynoecia (then 210–255 μ wide × 190–200 μ long or larger), *usually 0.2–0.25 bilobed*, a few retuse-truncate or entire on sterile shoots; sinus rounded to obtuse; *lobes obtuse* usually, sometimes rather connivent. Cells *small, leptodermous* (in and toward leaf middle and base) *or rather equally thick-walled* (distally), quadrate or sub-quadrate in margins and there 12–14 μ; median cells quadrate to polygonal 13–16(18) × 15–19(24) μ; cuticle smooth; *underleaves minute, subulate*, distinct above only, ephemeral usually.

Dioecious. Androecia compact, stoutly julaceous, with (3)4–7 pairs of imbri-cate, concave bracts, each ca. 250–270 μ broad × 240–270 μ long, rounded-quadrate, ca. 0.2 bilobed, the lobes broad and triangular; antical margin often with a tooth. Antheridia 1–2, large, almost spherical, ca. 80–85 μ in diam., on a biseriate stalk 24 μ in diam. × 70 μ long. Gynoecia on a $\pm$ abbreviated,

---

chloroplasts (×273). (8) Shoot sector, lateral aspect (×28). (9) Underleaves (×57). (10) Perichaetial bracteoles (×57). (11–12) Apices of gynoecial shoots, with mature and with exserted sporophytes (×18). [All after M. S. Taylor, 1939; from plants from Red Rock Ravine, Hocking Co., Ohio.]

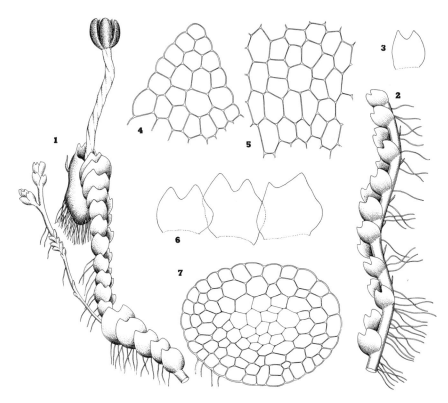

Fig. 267. *Nardia breidleri.* (1) Fertile plant (×ca. 22.5). (2) Sterile shoot, lateral aspect (×ca. 22.5). (3) Leaf (×ca. 50). (4–5) Cells of leaf apex and leaf middle, respectively (×330). (6) Three leaves (×90). (7) Stem cross section (×270). [1–3, from K. Müller, 1905–16; 4–7 from the Greenland plant.]

clavate, dense-leaved shoot, the *gynoecial complex fleshy, stout,* typically *bulbous and rhizoidous beneath,* a short marsupium at an angle with the shoot developed. Bracts larger than leaves, broader than long, sinus lunate or rounded; bracteole oblong-ovate. Perianth hidden within bracts, conical, crenulate at mouth, ca. 300 μ long × 400 μ in max. diam.; perigynium ca. 800 μ high. Spores granulate, 8–10(11) μ in diam.; elaters 3–4-spiral, 7–10 μ in diam. × 65–90 μ long.

Type. European: Austria (*Breidler*).

Distribution

Imperfectly and disjunctly holarctic. Widespread in the central European Alps (Austria, Switzerland, southern Germany, Mt. Blanc region of France, northern Italy), Great Britain (Perth, Forfar, Aberdeen,

Inverness), northward to Norway and Sweden. Eastward to Siberia (Yenisey R., *W. H. Arnell*), and recurring in Japan (Honshu: Nagano, Mt. Ontaka; see Amakawa, 1960a, p. 285).

In North America reported from the west coast and Rocky Mts., with a single previously unreported collection from our area:

W. GREENLAND. Ø ud for Sermilik Fjord, S. from Godthaab, 63°28′ N. (*Rosenvinge, May 26, 1888!*).

The Greenland collection is mixed with *Nardia scalaris* (distinct in its brownish coloration, much larger size, unlobed leaves), *Cephalozia* cf. *ambigua*, and a little *Prasanthus suecicus*. The plants are unquestionably *N. breidleri*. They are few, but ♂ inflorescences are common (♀ could not be found). The plants show the intense purplish pigmentation diagnostic for the species; they are small, only 3–4 mm high, and branch freely by means of ventral- or ventrolateral-intercalary branches; no terminal branches were seen. Branches are often leafless and colorless, stoloniferous initially, or microphyllous, and become leafy and purplish-tinged only above. Leaves of sterile shoots may be remote, are somewhat concave and quite small, averaging 1–1.5 × as wide as the rather fleshy and relatively stout stems; they range from 165–200 μ wide × 160–200 μ long, but near the androecia may become larger (210–225 μ wide × 190–200 μ long). Almost all leaves are 0.2–0.25 bilobed, but a very few entire or retuse leaves could be found.

The small cells, either leptodermous (particularly in basal and median leaf sectors) or moderately equally thick-walled, are notable. Marginal cells are quadrate or subquadrate and ca. 12–14 μ; median cells are quadrate to polygonal and only 13–15(15.5) × 15–19 μ; trigones are lacking and the cuticle is smooth. (In some cells of the shoot tip what seemed to be single, subspherical oil-bodies could still be distinguished. Whether these are oil-bodies is questionable.)

Androecia are rather compactly and stoutly julaceous, with (3)4–7 pairs of imbricate concave bracts, which are 250–270 μ wide × 240–270 μ long and, like the leaves, are purplish, rounded-quadrate, and ca. 0.2 bilobed, with broad, triangular lobes. The 1–2 antheridia, large for the bract size, are ca. 80–85 μ and almost spherical, on a 2-seriate stalk 24 μ wide and to 70 μ long. Under-leaves of both sterile sectors and androecia are linear-lanceolate and small, so that they could be overlooked (particularly because they are often appressed to the stem).

The fleshy stem is ca. 145–155 μ in diam. (9–11 cells), terete, with the cortical cells on the whole slightly larger than the medullary, ca. 16–20 μ in diam. and isodiametric in section to tangentially flattened and to 20 × 28 μ; the thin, free walls, weakly thickened internally, are notable. Medullary cells, dorsally at least, are rather thick-walled (walls often yellowish) and smaller on an average [ca. (12)14–18(19) μ in diam.], although much individual variation occurs. The cortical cells, in surface view, are short-oblong, clearly purplish dorsally and laterally.

This small species is restricted to alpine and arctic sites, usually to high elevations, often where irrigated by snow melt. It often occurs with *Anthelia juratzkana* and, as Müller states, may resemble a minute *Marsupella* or *Cephalozia*. Confusion with any other regional *Nardia* is unlikely. *Nardia breidleri* is allied only to the tropical African *N. arnelliana* (Grolle, 1964a, p. 297), from which it differs in the dioecious inflorescences and the distinct marsupium, and to the Japanese *N. unispiralis* (Amakawa, 1960a, p. 287), which is distinct in the larger spores (15 μ in diam.) and the closely unispiral elaters.

CRYPTOCOLEA Schust.
[Figs. 268–270]

Cryptocolea Schust., Amer. Midl. Nat. 49(2):414, fig. 7:7, pl. 34, 1953.

Closely prostrate in growth, erect with crowding, clear light green with ± yellow-brown pigmentation, *often stoutly julaceous or subjulaceous*, with rather thick, fleshy stems (6)8–15 mm long, subsimple or simple, but always *innovating freely beneath the ♀ inflorescence*; branches lateral-intercalary and *Frullania* type, furcate; axial differentiation absent: all stems with normal leaves, none forming stolons or flagella. Stem cross section showing no dorsiventral differentiation of the medulla, of thin-walled and only slightly collenchymatous cells, the cortical cells scarcely more thick-walled than the medullary; mycorrhizae absent in mature stem. Leaves *usually quite imbricate*, ± *antically secund*, obliquely inserted, succubous, uniformly entire, never even retuse, *imbricate and usually antically contiguous, diagnostically valvate at the shoot tips*, broad at base and attached by a broad line of insertion, concave, orbicular on sterile shoots, on ♀ shoots some-times becoming slightly longer than wide and oval to ovate, near the inflorescences. Cells moderately collenchymatous, thin-walled, with concave to slightly bulging trigones; cell walls able to develop *light golden-brown* secondary pigmentation. Oil-bodies 3–5 up to 6–12 per cell, to 5.5 × 8 μ to 6 × 9 μ, rarely a few up to 8 × 15 μ or 9 × 18 μ, formed of fine to coarse, ± strongly protruding individual globules or ± coarse segments of heterogeneous size; chloroplasts 2.5–3.5 μ. Ventral stem sectors obsolete, on most stems 1–2 cells wide; rhizoids colorless to pale brown, rarely purplish, *numerous to stem apex, rather long*, indiscriminately scattered, occurring on both the restricted ventral merophytes and ventral portions of lateral merophytes; *sterile and fertile shoots unable to develop underleaves*, but apices of ♀ shoots rarely with a small "bracteole" visible. Vegetative specialized types of reproduction (gemmae, "Bruchblätter," etc.) absent.

Dioecious. ♀ Plants larger and more robust than ♂, with inflorescences terminal on main shoots; archegonia 8 or more. Perianth reduced to a short, tubular structure, irregularly plicate or ± laterally compressed, ± sinuate-lobulate at the otherwise *truncate and crenulate mouth*, bearing marginal slime papillae, *completely hidden between the terminal perichaetial bracts*. Both perianth and terminal pair of bracts situated at apex of a low, ringlike to conspicuously tubular perigynium lying strictly in the axis of the stem, the perigynium several cell layers thick, fleshy, variably developed, often low, at times over half as high as the perianth above. ♀ Bracts in 2–3 pairs, *mutually closely involute*, forming an ovoid, compact, closely organized terminal "head," *often pinched together or pointed at the apex*, closely imbricate and erect, at most the terminal portions and lateral margins spreading to squarrose-reflexed. Terminal pairs of bracts appearing fused dorsally and/or ventrally, occasionally to the apical fourth, forming a conoidally pointed to bilabiate structure functioning as the "perianth," usually distinctly laterally compressed; dorsally and ventrally with the line of union or juxtaposition of the bracts sometimes marked by a sharp keel or wing, which gradually disappears below. Bracts below the terminal pair showing distinct development of slime papillae at bract apices. Sporophyte "boring" into stem tissue, the foot situated below the base of the perigynium. A distinct shoot calyptra sometimes present, the sterile archegonia found to within the distal 0.2 of calyptra, at other times absent. Capsule red-brown, shortly ovoid, the elliptical valves 2-, locally 2–3-, stratose; wall 42–48(52) μ thick. Epidermal cells with 1-phase development, quadrate to short-oblong, ca. 24–30(36) μ wide × 30–52 μ long, 20–25 μ high, all faces *with very many, juxtaposed strong nodular thickenings* that are not conspicuously tangentially extended; inner cells much more irregular, mostly ca. 20–26 μ wide × 52–80 μ long, 18–24 μ high, with many, often oblique and irregular, usually complete semiannular bands extended fully across the free tangential wall. Spores granulate-vermiculate, reddish brown, 14–16(17–19) μ in diam. (postmature). Elaters slender, tortuous, bispiral to their tips, attenuated, 7–10 μ in diam. ♂ Plants slender, smaller than ♀, with relatively smaller leaves. Perigonal bracts in 4–5 or more pairs, closely imbricated, forming a distinct androecial spike, at first terminal, later becoming intercalary. Antheridia 1–2 per axil; stalks biseriate.

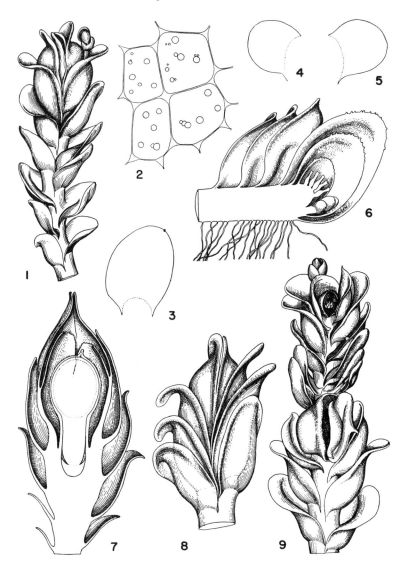

Fig. 268. *Cryptocolea imbricata*. (1) Gynoecial shoot apex at time of fertilization, with small innovation already distinct (×16). (2) Median cells with disintegrated oil-bodies, reduced to oil-droplets (×ca. 580). (3) Leaf below ♀ bract (×18). (4–5) Two leaves (×18). (6) Longisection of plant in fig. 8, showing archegonia, embryonic innovation, slime papillae on bract margins (×23). (7) Shoot apex with almost mature sporophyte, longisection (×23). (8) Apex of ♀ plant at about time of maturation of archegonia (×23). (9) ♀ Plant with old, bilabiate, unfertilized gynoecium and young gynoecium at apex of innovation (with bracts artificially parted) (×18.6). [All from the type plants, Porcupine I., Minn., *RMS*.]

♂ Bracts similar to leaves but moderately ventricose, with all but margins concave, the margins spreading to squarrose.

Type. Cryptocolea imbricata Schust. The genus is monotypic.

Cryptocolea is an isolated taxon of uncertain phylogenetic relationships. It is characterized chiefly by the involute to bilabiate gynoecial system, with the perianth reduced to a short, irregularly pleated to laterally compressed tube, totally enveloped by a pair of prominent, marginally closely contiguous or involute-appressed, concave, mussel-shaped bracts, the whole situated at the apex of a discrete perigynium. Affinities are probably most immediate to *Nardia* (which differs, i.a., in the distinct underleaves) and to two monotypic Japanese genera of uncertain affinity and status, *Cryptocoleopsis* Amak. (Amakawa, 1960a) and *Diplocolea* Amak. (Amakawa, 1962).

In the often bilabiate terminal portion of the ♀ shoot, there is a superficial approach to *Plagiochila*, *Mylia*, and *Pedinophyllum;* these genera, however, retain wider ventral merophytes and preserve discrete underleaves; they show no tendency toward the formation of a perigynium. Furthermore, they do not show reduction of the perianth to the point where it is shorter than the subtending pair of bracts, and where the bracts take over part of the function of the perianth. It is, therefore, the perianth proper, not the immediately subtending bracts, that gives the shoot apex of the Plagiochilaceae its laterally compressed appearance.[313]

The absence of underleaves, as well as development of a perigynium, allies *Cryptocolea* with the higher genera of Jungermanniaceae (such as *Nardia* and *Solenostoma*). These, however, show a well-developed perianth, which generally exceeds the bracts, is externally evident, and shows no trace of a bilabiate or imbricate, fused shell formed by the bracts.

Arnellia, of the Arnelliaceae, also shows some homoplastic similarities to *Cryptocolea;* both have entire, more or less orbicular leaves, the apical valvate like clam shells, and those near the ♀ inflorescence becoming oval to ovate and distinctly longer than wide; in both there is development of a perigynium, combined with retention of a reduced perianth whose height is inferior to that

---

[313] Furthermore, the bilabiate form of the gynoecium, although always very clearly marked in the type material and splendidly distinct in the Magdlak collections, is hardly evident in the copious Greenland material from Umîarfik Fjord. In this material, the mature gynoecia, bearing exserted capsules, are ovoid, with the apex bluntly conoidal. The gynoecial bracts may be fused only ventrally (and just for a short distance) and may be imbricate dorsally. Thus the system becomes, at least in part, imbricate rather than valvate. The perianth, then, although basically with a rather wide and shallowly plurilobulate mouth, is narrowed distally by becoming irregularly pleated. It is apparent, therefore, that the bilabiate and valvate condition of the type material—which led to the genus being originally assigned to the Plagiochilaceae—is not constant, and the perianth, in particular, is not always compressed. Hence, any similarity to the Plagiochilaceae is certainly casual at best.

of the subtending bracts. *Arnellia*, however, has opposite and dull leaves, has underleaves, and develops a vertical, strongly rhizoidous perigynium at right angles to the stem. In *Cryptocolea* the feebly nitid leaves are clearly alternate, underleaves absent; the perigynium is continuous with the stem and is non-rhizoidous. Similarities between the two genera seem essentially owing to parallel reduction of the perianth associated with parallel development of a perigynium; they do not appear to indicate any close relationship.

CRYPTOCOLEA IMBRICATA Schust.

[Figs. 268–270]

Cryptocolea imbricata Schust., Amer. Midl. Nat. 49(2):417, fig. 7:7, pl. 34, 1953.

Plants 5–9 mm long, closely adnate and creeping, usually isolated stems, rarely gregarious, quite fleshy, little or not branched, bright pellucid green, the older leaves becoming a light yellowish brown. Stems simple, with occasional axillary branches, rarely furcately (terminally) branched, and innovating almost without exception beneath the ♀ inflorescence (the inflorescence thus super-ficially appears to become dorsal); stolons absent; stem fleshy, soft, rather thick (240–300 μ), on fertile shoots to 500 μ, the cortical cells somewhat thick-walled, 20–24 μ wide × 30–40 μ long, becoming brownish-walled with age. Rhizoids scattered, partly appearing to occur in bundles, limited to postical stem surface, long, rather dense, colorless to brownish (with age). Leaves moderately to strongly imbricate, strongly concave, often spoonlike, the plants thus somewhat similar in facies to *Odontoschisma macounii*, but more *often so antically contiguous-connivent as to form, mutually, a distinct antical keel* (the plants then appearing rather *laterally compressed*), then rather resembling *Prasanthus*. Leaves entire (never retuse, even in inflorescences), generally oval or ovate, mostly slightly to distinctly longer than wide, on small sterile shoots occasionally suborbicular and as wide as long (or even slightly wider), the apices usually rather narrowly rounded; leaf insertion oblique, ventrally not at all, but dorsally quite distinctly, decurrent; leaf cells in middle 20–25 to 25–35 × (20)27–36(45) μ, near leaf base becoming elongate and 20–25 × 27–45 μ; cells very thin-walled, rectangular to polygonal, the angles with very small but dis-tinct trigones with concave or straight walls; cuticle totally smooth on most leaves, but occasional leaves locally with very low, inconspicuous, obscure papillae. *Underleaves totally lacking* even on fertile shoots (beneath perianth often a trace of a bracteole). Gemmae absent.

Unisexual (dioecious). ♂ Plants in separate patches from ♀, or occasionally in same patch. Androecia terminal, later becoming intercalary, often several in succession on one shoot, separated by several sterile leaves; ♂ bracts in 5 or more pairs, densely imbricate and forming an elongate, compact spike; an-theridia 1–2 per axil, about 175-μ long × 144 μ wide, the stalk biseriate, about 96 μ long. ♀ Inflorescence terminal on main shoot. Young inflorescences (with archegonia at time of fertilization) with leaves erect near shoot apex, closely imbricate, their basal portions strongly ventricose, the marginal portions sometimes characteristically flaring, the archegonia totally hidden between the distal several pairs of leaves and bracts, which are erect and imbricate, lie

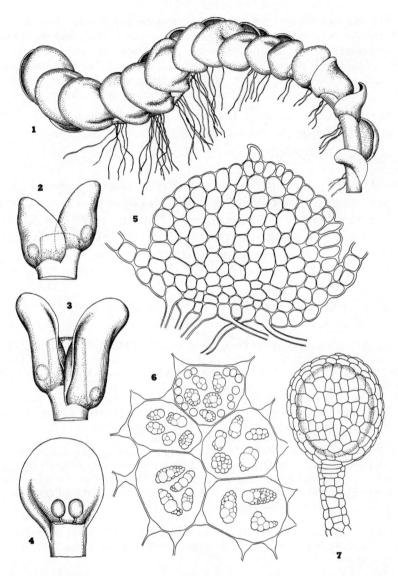

FIG. 269. *Cryptocolea imbricata.* (1) ♂ Shoot in lateral aspect (×ca. 20). (2–3) Sectors of androecia, dorsal aspect (×30). (4) ♂ Bract, adaxial aspect, slightly flattened (×30). (5) Stem cross section (×188). (6) Median cells with normal oil-bodies (×ca. 700). (7) Antheridium (×230). [All from plants from type locality, collected Sept. 8, 1951.]

parallel to the stem (except for the sometimes spreading, flaring distal margins, which are somewhat reflexed), and are inclined to each other, forming a somewhat bilabiate, closed terminal structure, in which the inflorescence lies. ♀ Bracts halfway to base with slime papillae. Perianth *totally hidden within bracts, with mouth wide* (but orifice sometimes reduced by extensive plication), shallowly plurilobulate, lobules blunt, with a few slime papillae interspersed.

Type. Porcupine I., Susie Isls., near Grand Portage, Minn. (*RMS 18592, 1950*). *Isotypes.* Same locality (*RMS 12250b, 19416, 19415*).

Distribution

Predominantly arctic in range, although first described from the Lake Superior region (where confined to damp, weakly basic ledges a few feet from the lake shore, in the microclimatic "Tundra strip"; see Schuster, 1957). In addition to the type material from the Lake Superior region, recently found as far north as northeastern Ellesmere I., northern and western Greenland, thus probably of localized distribution elsewhere in the Arctic.

Unknown outside of eastern North America, although to be expected in northern Siberia, Scandinavia, and Alaska.

ELLESMERE I. NW. slope of the Dean, about 1 mi E. of Mt. Pullen, 5 mi S. of Alert, NE. Ellesmere, ca. 82°25′ N., 62°10′ W. (*RMS 35156, June 24, 1955*); S. end of Hilgard Bay, 82°26′ N., 63°25′ W., NE. Ellesmere (*RMS 35250a, 35225, July 5, 1955*); northernmost stations for species. N. GREENLAND. Wet *Carex aquatilis* meadow, Heilprin Land, Brønlund Fjord, 82°10′ N., 31° W. (*Holmen 6749, p.p.,* among *Odontoschisma macounii, Lophozia heterocolpa* var. *harpanthoides*).[314] NW. GREENLAND. Kangerdlugssuak, Inglefield Bay, 77°23′ N. (*RMS 45831c*); Kekertat, Heilprin Gletscher, 77°31′ N. (*RMS 46046*). W. GREENLAND. Eqe, near Eqip Sermia (glacier), 69°46′ N., 50°10′ W. (*RMS & KD 66-249e,* trace; *66-289,* trace); mouth of Simiutap Kûa Valley, Umîarfik Fjord, Swartenhuk Pen., 71°59′ N., 54°35-40′ W. (*RMS & KD 66-917, 66-918*); Qalagtoq, Upernivik Ø, Inukavsait Fjord, 71°14′ N., 52°35′ W. (*RMS & KD 66-1138*); head of Marmoralik Fjord, 71°05′ N., 51°12′ W. (*RMS & KD 66-1375*); Kangerdlugssuakavsak, head of Kangerdluarssuk Fjord, 71°21′ N., 51°40′ W. (with postmature caps.; *RMS & KD 66-1288*).

MICHIGAN. Amygdaloid I., Isle Royale, Keweenaw Co. (*RMS 13221*). MINNESOTA. Porcupine I., Susie Isls., Cook Co. (*RMS 12250b, 19416, 19416a, 19409f, 19415, 19415a, 18592;* type).

Ecology

Restricted to damp humus lying over more or less basic rocks, with environmental restrictions similar to those of *Odontoschisma macounii* and

---

[314] Arnell (1960, p. 7) reported this species from Greenland for the first time, from a wet *Vaccinium uliginosum* heath in Peary Land (*Holmen 7041*). However, the plants are a small but characteristic form of *Plagiochila arctica*, as the subreniform leaves, the free intercalary branching, the sporadic occurrence of microphyllous stolons, and the laterally compressed shoots with nearly erect, subappressed leaves demonstrate. The cells are typical for *P. arctica*: marginal ca. 34–36 μ, median ca. 33–37 μ. The occurrence in a noncalcareous habitat, alone, is enough to exclude *Cryptocolea* from consideration.

Lophozia heterocolpa (with which it is found at most of the known stations). Decidedly mesophytic, but able to tolerate considerable direct sunlight when in sufficiently moist sites.

In addition to the two preceding associated species, found (in northeast Ellesmere I.) with *Tritomaria heterophylla*, *T. quinquedentata*, *Plagiochila arctica*, *Blepharostoma trichophyllum brevirete*, *Solenostoma pumilum polaris*, *Scapania polaris*, *Lophozia alpestris* subsp. *polaris*, *L. collaris*, *L. heterocolpa* var. *harpanthoides*, *L. quadriloba*, *Gymnomitrion concinnatum*, *Anthelia juratzkana*, *Cephaloziella arctica*, and *Arnellia fennica*. The plant occurs here on humus over weathered sedimentary rocks, at the foot of slopes, below persistent snow banks which "feed" into the thick moss-herb mat below. Like the type material, the plants from Ellesmere I. usually occur very scattered, although widely dispersed and not uncommon. They never appear to form pure mats or patches.

In all of the Greenland stations found over peaty but nonacid soil, usually associated with *Odontoschisma macounii*, *Lophozia heterocolpa* (sometimes var. *harpanthoides*), *Tritomaria quinquedentata*, *Blepharostoma trichophyllum* subsp. *brevirete*, and often (*RMS 45831c*) also *Aneura pinguis* and *Lophozia grandiretis*, or (*RMS & KD 66–289, 66–249e*) with *Prasanthus suecicus*, *Gymnomitrion concinnatum*, *Anthelia juratzkana*, *Cephaloziella arctica*, *Anastrophyllum minutum*, *Tritomaria heterophylla*, and *Aneura pinguis*. The species occasionally occurs in large quantities (as at the head of the Simiatup Kûa valley, Umîarfik Fjord) over moss humus, at the edges of peaty mounds, at the margins of shallow calcareous pools, associated with *Lophozia gillmani*, *L. rutheana*, *Arnellia fennica*, etc.

The type material, from humus-covered, basic, diorite ledges, occurs with *Tritomaria scitula*, *T. quinquedentata*, *Plagiochila asplenioides*, *Blepharostoma trichophyllum*, *Scapania gymnostomophila*, *Lophozia gillmani*, over shaded, northwest-facing ledges.

Differentiation

Cryptocolea imbricata is so distinctive a species that it can hardly be misidentified. Sterile plants of the rather rare mod. *laxifolia* give the superficial impression of a small, gemma-free form of *Mylia anomala*, bearing unusually strongly erect leaves. In the erect, somewhat to distinctly concave leaves, the green color (but with the marginal and submarginal portions of the leaves often a warm, light castaneous brown), and the relatively robust size (to ca. 1–2 mm wide), *C. imbricata* may approach *Arnellia* in facies. From this it differs in the feebly shiny texture (caused by the almost smooth cells), scattered rhizoids, lack of underleaves, and clearly alternate leaves. Portions of the plant which are devoid of brownish pigment are also a clear, light green in *Cryptocolea*, whereas they are a whitish, opaque gray- or bluish-green in *Arnellia*.

Confusion of compact phases (mod. *densifolia-parvifolia*) is likely to occur with *Prasanthus* and *Gymnomitrion*, because of the terete, julaceous form of phases of exposed sites. Such plants tend to be rather strongly laterally compressed (hence elliptical in cross section), with the antically connivent, concave, imbricate-contiguous leaves jointly forming an almost sharp antical "keel." The lateral compression of the julaceous sterile axes suggests *Prasanthus* (and separates such plants from *Gymnomitrion*). However, *Prasanthus* lacks the warm brown coloration of *Cryptocolea*, is smoother and shinier, and usually is denser-leaved; it also always has ventral stolons, lacking in *Cryptocolea*. Small, compact phases of *Cryptocolea* often have distal parts of the leaves decolorate, followed by a brownish zone, with only the leaf bases green.

Laxer-leaved plants are, on the basis of the alternate, orbicular, and entire leaves and the scattered rhizoids, similar to various species of *Solenostoma* and *Nardia*. The absence of underleaves eliminates confusion with the latter. The peculiar, polymorphous oil-bodies, which are generally formed of segments of uneven size, separates this plant from *Solenostoma*.[315] Fortunately plants are almost always fertile, and then distinct in the bilabiate gynoecia, with closely, marginally connivent ♀ bracts, forming a mussel-shaped structure within which the archegonia are situated. The bilabiate gynoecial system is usually ascertainable, since old, unfertilized gynoecia are mostly to be found. These invariably bear 1 (rarely 2) subfloral innovations which are soon again floriferous. ♂ Plants, and sometimes also sterile shoots from exposed sites, are quite distinctive in the imbricate, concave leaves, whose margins are slightly flaring or reflexed. The leaves are usually so perfectly imbricate that the adaxial leaf surface is quite hidden, except at the patent or flaring margins. As a consequence, xeromorphic shoots (and particularly xeromorphic ♂ plants) have a terete, or tubular, wormlike appearance which is quite unmistakable.

The leaves in *Cryptocolea imbricata* are distinctly decurrent antically, although usually relatively briefly so (as in Fig. 268:1). However, on lax, sterile stems the antical leaf bases are sometimes markedly but narrowly decurrent for much of the merophyte length. The decurrent strip is then gradually narrowed below, and throughout most of its length is only 2–3 cells broad. Branching in *Cryptocolea* occurs only sporadically and appears to be mostly furcate and terminal. The plants almost always are simple, proliferating vegetatively beyond the androecia if ♂, or invariably bearing subfloral innovations if ♀.

---

[315] Oil-bodies are often distinctly and unevenly segmented. The segmentation, however, is sometimes only slightly indicated and appears to be derived, secondarily in some cases, from initially nearly homogeneous oil-bodies. In *RMS 35235*, e.g., there are usually 4–7 ovoid to ellipsoidal oil-bodies per cell (ranging from 5×8–9μ to 7–8×9–12μ, rarely $8 \times 15 \mu$; thus essentially equal in size and number to those of the type). Oil-bodies in these plants range from virtually homogeneous and smooth to slightly (and obscurely) segmented. In some cases the major portion remains as an unsegmented, homogeneous corpus, within the rather distinct plasma membrane surrounding the oil-body, while the lesser portion undergoes dissolution into smaller oil spherules or segments. These variations in oil-body form occur in living plants and are not due to changes subsequent to death. In *RMS & KD 66-249e*, by contrast, the oil-bodies are mostly unequally fine-botryoidal.

Development of the gynoecia in this species is rather complex and has been described as follows (Schuster, 1953, pp. 416–417):

In the plants with youngest inflorescences, at about time of maturation of archegonia, there is already a distinct development of the pouchlike, bilateral terminal structure, formed by 2–3–4 pairs of subinvolucral bracts that are evident on external examination. In longitudinal section, it is seen that another pair of bracts, immature as yet, is hidden within the structure, raising the total to a maximum number of six pairs of subinvolucral leaves. The upper pair of these leaves may eventually form the perianth, but, as far as I have been able to see, there is at this point no discrete perianth as yet. The reasons why the upper, smallest pair of leaves probably does not represent the perianth will be discussed shortly in connection with differences apparent between the margins of the bracts and that of the perianth mouth.

It should also be clear . . . [Fig. 268:6] that at the time of the fertilization of the archegonia there is also no indication, as yet, of the perigynium. Apparently the perigynium and perianth are both formed subsequent to fertilization, and, indeed, fertilization appears to be a necessary prerequisite for their development. This is evident . . . [Fig. 268:9] where the old (unfertilized) archegonial involucre of the preceding season clearly shows that no perianth had been formed. . . . . Female plants, at the time apical growth is inhibited by formation of archegonia, already show innovation of shoots from below the archegonial region. This is always true (in every plant examined) of plants in which the archegonia were not fertilized; in plants with fertilized archegonia, subfloral innovations are quite unusual. . . . In the development of archegonia, the immature archegonia are protected from desiccation by relatively large quantities of slime secreted by marginal slime papillae, situated in some profusion at the apices of 2–3 pairs of bracts directly below the archegonia. When very small, these bracts have their distal margins situated only slightly above the archegonial region, and the slime they secrete appears to be of some significance in preventing drying out.

As mentioned above, no trace of any structure that can be certainly identified as the perianth is evident at the time of fertilization. The innermost . . . bracts, which are not at all connate at base, do not appear to represent the juvenile perianth, for the following reasons. Firstly, the perianth mouth does not bear marginal slime papillae; the apices of the youngest bracts seen at the time of archegonial maturation bear discrete marginal slime papillae. Secondly, the perianth is a shortly tubular structure, not showing obvious origin from a pair of leaves . . . It is, therefore, quite hidden between the latter, and is evident only upon careful dissection. Its mouth is virtually entire, and bordered by several rows of cells that have less evident trigones than the cells of the perianth middle.

The degree of elaboration of the shoot-perigynium tends to be rather variable. In the Umîarfik Fjord material, with quite young capsules, only a low, ringlike, fleshy structure had been developed; by contrast, in the Magdlak plants, with rather more developed but green sporophytes, a high and massive shoot-perigynium was present (Fig. 270), quite comparable to that found in *Marsupella emarginata*. Furthermore, there had been clear proliferation of the axial tissue below the perigynium, in and below the region of the conoidally pointed foot, and the axis was elongated between the bracts and the first pair of subfloral

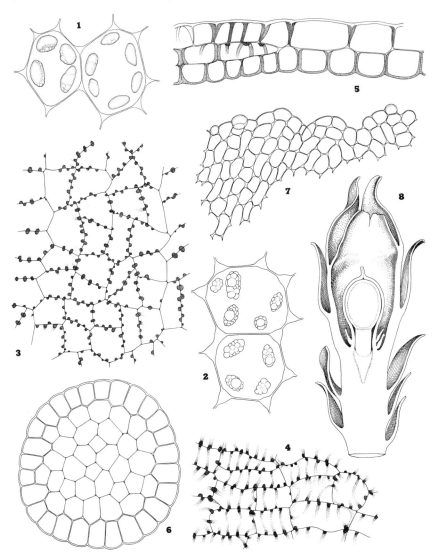

Fig. 270. *Cryptocolea imbricata.* (1) Cells with oil-bodies (×580). (2) Cells with oil-bodies (×775). (3) Epidermal cells of capsule wall (×265); (4) Inner cells, capsule wall (×265); (5) Capsule wall cross section (×308). (6) Seta cross section (×150). (7) Perianth mouth cells, with (stippled) slime papillae (×270). (8) Shoot apex longisection with immature sporophyte (×18). [1, *RMS 35235,* Hilgard Bay, near Alert, Ellesmere I.; 2, *RMS 45831c,* Kangerdlugssuak, NW. Greenland; 3–5, 7, *RMS & KD 66-1289,* Kangerdluarssuk Fjord, W. Greenland; 6, 8, *RMS & KD 66-1206,* Marmoralik, W. Greenland; figs. 3–4, one scale.]

bracts, which together formed a juxtaposed complex—the capitate and swollen gynoecial head. The second pair of subfloral bracts thus appeared somewhat stalked on mature and well-developed plants.

SOLENOSTOMA Mitt.

Gymnoscyphus Corda, *in* Sturm, Deutschlands Fl. 2:158, 1832 (*nomen rejiciendum;* proposed by Grolle, Rev. Bryol. et Lichén. 26:83, 1957).
Solenostoma Mitt., Jour. Linn. Soc. Bot. 8:51, 1865; Schuster, Amer. Midl. Nat. 49(2):389, 396, 1953; K. Müller, Rabenh. Krypt.-Fl. ed. 3, 6(2):812, 1956.
Jungermannia, in part, of numerous authors, incl. Frye & Clark, Univ. Wash. Publ. Biol. 6(2):275, 1944 (excl. *J. lanceolata*). Not *Jungermannia*, as to type.
Jungermannia sect. *Aplozia* Dumort., Syll. Jungerm. Eur., p. 47, 1831.
Aplozia Dumort., Hep. Eur., p. 55, 1874 (in part).
Haplozia K. Müll., Rabenh. Krypt.-Fl. 6(1):535, 1909.
Solenostoma subg. *Plectocolea* Mitt., Jour. Linn. Soc. Bot. 8:156, 1865; Schuster, Amer. Midl. Nat. 49(2):401, 1953.
Nardia Lindb., Acta Soc. Sci. Fennica 10:115, 1871, in part.
Nardia sect. *Eucalyx* Lindb., Bot. Notiser, p. 167, 1872.
Nardia subg. *Chascostoma* Schiffn., *in* Engler & Prantl. Nat. Pflanzenfam. 1(3):78, 1893.
Eucalyx Breidl., Mitt. Naturw. Ver. Steiermark, 30:291, 1894; K. Müller, Rabenh. Krypt.-Fl. 6(1):524, 1909; Macvicar, Studs. Hdb. Brit. Hep. ed. 2:134, 1926.
Mesophylla Corbière, Rev. Bryol. 31:13, 1904.
Plectocolea Mitt., *in* Seemann, Fl. Vitiensis, p. 405, 1871; Evans, Ann. Bryol. 10 (1937):42, 1938; Frye & Clark, Univ. Wash. Publ. Biol. 6(2):318, 1944; K. Müller, Rabenh. Krypt.-Fl. ed. 3, 6(2):839, 1956.
Solenostoma subg. *Luridaplozia* Schust., Amer. Midl. Nat. 49 (2):404, 1953.
Jungermannia subgenera *Solenostoma, Plectocolea,* and *Luridae* Amak., Jour. Hattori Bot. Lab. no. 22:2, 43, 52, 1960.

Small to medium-sized [usually 550–1900(3400) μ wide], prostrate in growth with autonomously acroscopically curved shoot apices, occasionally suberect in growth with crowding. Shoots usually sparsely laterally branched, the *branches either lateral-terminal and furcate,* subtended by a somewhat narrower than normal leaf, or *intercalary* (then normally *from the lower halves of the leaf axils*); intercalary branches rarely stolonlike and microphyllous. Stems usually 10–15 cells high; cortical cells usually 2–4 × as long as wide, slightly or hardly thick-walled, not greatly differing from medullary in form or diam.; medullary cells subequal to cortical in diam., more elongate, with age the peripheral 2–3 cell layers (or more) usually mycorrhizal, the cells infected by mycorrhizae not or hardly differing from others in diam.; usually *without postical flagella. Rhizoids frequent, scattered* over postical stem faces, but often crowded near leaf bases, sometimes a few (rarely many) from abaxial faces of lower portions of leaves, or from lower portions of perianth, colorless, pale brownish, or purplish. Leaves various in size and form, *alternate, rotundate to elliptical or ovate, unlobed* (rarely some retuse at apex), *their contours usually rounded, the margins entire, succubously obliquely inserted* (usually at an angle of ca. 45° with axis), short-decurrent to considerably decurrent antically and postically, somewhat adaxially concave, the postical margin never deflexed. *Underleaves lacking,*[316] the ventral merophytes very narrow. Cells almost always slightly to distinctly collenchymatous, medium-sized (17–40 μ wide in leaf middle), smooth or weakly verruculose, *bearing usually 2–9 (less often 1) large, finely granular, almost homogeneous-appearing oil-bodies. Asexual reproduction absent* (in our species)[317] or by means of endogenously produced gemmae (*S. caespiticium*).

Dioecious or paroecious (?never autoecious). Androecia intercalary with age. ♂ Bracts usually in several to many pairs, similar to leaves but more concave, at least at base, *unlobed,* with usually (1)2–3 antheridia. Antheridia subspherical to short-ovoid; stalks biseriate. ♀ Bracts entire, usually similar to leaves in shape or only moderately differing, their margins not bordered by slime papillae, *larger than leaves; bracteole absent. Perianth distinct,* normally *slightly to longly emergent from beyond the bracts* (bracts rarely equalling perianth or barely exceeding it; then, however, the perianth not hidden between bracts, the latter not distally connivent), very variable, but *never smoothly tubular, distal portions always variously plicate.* Perigynium absent or developed to a varying degree (cf. Figs. 271:5, 10–11 and 272:5, 8, 12–14), if present always a simple tubular extension of the stem, *lying in the stem axis.*

Sporophyte with elongate *seta of 16–20 or more epidermal cell rows and an indefinitely large number of interior cell rows,* less often with only 8–16 or 12–20 interior cell rows in 2 concentric rings (Fig. 272:6). Capsule ovoid, 4-valved to base, usually 2-stratose, rarely 2–3 or 3–4-stratose. Epidermal layer of usually relatively short-rectangular cells, wider than those of inner cell layer (as well as thicker), bearing nodular (radial) thickenings. Inner cell layer typically of narrowly rectangulate cells with semiannular bands. Spores usually 10–24 μ, finely granulate or papillate, ca. twice the diam. of the elaters. Elaters free, tapering on both ends, *usually bispiral.*

Type. Jungermannia tersa Nees (= *Jungermannia sphaerocarpa* Hook.).

I would define *Solenostoma* essentially as originally understood by its author (Mitten, 1865), i.e., to include *Plectocolea* as a subgenus, thus following the original concepts of Mitten and also my treatment (Schuster,

---

[316] *Jungermannia allenii* L. Clark, described from western North America, has underleaves. This species, known only from sterile material, is admitted by its author to be of uncertain generic position.

[317] The Japanese *Solenostoma rubripunctatum* (Hatt.), comb. n. (= *Plectocolea rubripunctata* Hatt., Jour. Hattori Bot. Lab. no. 3:41, fig. 36, 1948) bears round, unicellular, purplish gemmae. These are evidently formed singly from leaf surfaces.

No species of *Solenostoma* known to me produces fasciculate exogenous gemmae; their absence is one of the marked generic characters of *Solenostoma*.

1953).[318] Such a treatment differs in several respects from the views of Evans (1938c) and Müller (1942). Evans does not accept *Solenostoma* at all,[319] considering it a synonym of *Jungermannia* L., whereas *Solenostoma* subg. *Plectocolea* is elevated by Evans (following Mitten, 1871, *in* 1865–73) to generic status, essentially to include species having a perigynium but differing from *Nardia* in lacking underleaves. According to Evans, *Jungermannia lanceolata* is thus congeneric with a series of divergent species, including, e.g., *J. sphaerocarpa* and *J. pumila*.

Müller (1942, 1951–58), Meylan (1924), Joergensen (1934), and Schuster (1953), on the other hand, follow Nees (1844, *in* G. L. & N., 1844–47), and consider the *Jungermannia lanceolata* element as forming a discrete genus. Species which Evans places in *Jungermannia*, exclusive of *J. lanceolata*, Müller places in the genus *Solenostoma* Mitt. However, Müller agrees with Evans in segregating subg. *Plectocolea* as a distinct genus, thus accepting three genera, the monotypic *Jungermannia* and the polytypic *Solenostoma* and *Plectocolea*.

A review of the entire problem of generic lines in the underleaf-free "genera" of Jungermanniaceae led me to adopt (Schuster, 1953) a conservative view of genera in this group, which has also been adopted by Amakawa (1960a). However, I would follow such relatively conservative workers as Meylan (1924) and Müller (1942, 1951–58) in considering *J. lanceolata* as representative of a discrete genus, equivalent to *Liochlaena* of Nees (*loc. cit.*). Müller retains for this the generic name *Jungermannia* (or *Jungermania*), following Howe (1899), Joergensen (1934), et al.; I also adopt this now standard usage. Meylan, however, uses the younger name *Liochlaena* Nees for this taxon. Reasons for using the name *Jungermannia* are given on p. 835. Apart from the problematical application of the name "*J. lanceolata*," there is little doubt but that it represents a well-founded genus, even though Amakawa merges it with *Solenostoma*. Although they do not discuss the bases for their adoption of a monotypic genus for *J. lanceolata*, both Müller and Meylan stress the unique perianth form as a generic character (Fig. 257:5). The terete, eplicate, distally depressed-rostellate perianth of *J. lanceolata* does not reappear in other species assigned to *Jungermannia* by various recent authors (including Evans), although this is not the sole basis for maintaining a genus *Jungermannia*. In my opinion, the fasciculate exogenous gemmae produced from the margins of reduced leaves represent an

---

[318] Casares-Gil (1919) also proposed a conservative delimitation of *Solenostoma*, although he used for it the later invalid name of *Haplozia* Dumort. emend. K. Müller. Gil included under this genus four subgenera: *Eucalyx*, equivalent to our *Plectocolea*, i.e., the *S. hyalinum* complex; *Solenostoma*, for the *S. crenulatum-sphaerocarpum* complex, exactly equivalent to our subg. *Solenostoma*; *Plecocolea* [sic!], invalidly used for *Luridae*, for the *S. cordifolium-atrovirens* complex; and *Liochlaena* Nees, used for *Jungermannia lanceolata*.

An exactly identical subdivision has been published by Amakawa (1960a), using as the generic name *Jungermannia*, with the four subgenera designated *Plectocolea*, *Solenostoma*, *Luridae*, and *Jungermannia*. Ironically, Amakawa appears to have overlooked the earlier and taxonomically (if not nomenclaturally) identical system of Casares-Gil.

[319] Evans (1938c) gives an excellent historical account of the involved history of *Solenostoma* and *Plectocolea*.

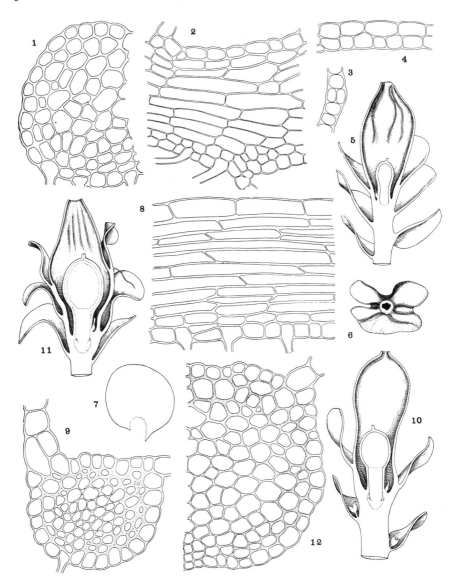

FIG. 271. *Solenostoma:* subgeneric and species criteria (compare also Fig. 272; 1–7, subg. *Solenostoma;* 8–12, subg. *Plectocolea*). (1–7) *S. pyriflorum:* 1, stem cross section (×180); 2, longisection of stem (×180); 3–4, cross sections, respectively, of distal one-third and middle of perianth (×155); 5, longisection of gynoecium, showing vestigial perigynium and no shoot calyptra (×18); 6, perianth apex cross section (×24); 7, leaf, showing typical orbicular *Solenostoma* outline (×20). (8–9) *S. crenuliformis:* 8, stem longisection, robust axis (×180); 9, cross section of weaker stem (×180). (10) *S. levieri*, longisection of paroecious shoot (×ca. 18). (11–12) *S. hyalinum* (nearctic race): 11, longisection of shoot apex with young capsule (×18); 12, stem cross section (×180). [1–7, from plants from Neddie Creek, N.C., *RMS;* 8–9, *RMS 36049*, N.C.; 10, Sand I., Wisc., *RMS;* 11–12, *RMS 36502*, Dry Falls, N.C.]

equally important generic character (Fig. 257:8–9). Gemmae of this type are almost universal in occurrence in Lophoziaceae and Scapaniaceae and in *Mylia* of the Jungermanniaceae (subf. Mylioideae), but appear to be almost or quite lacking in the Jungermanniaceae *s. str.* (subf. Jungermannioideae). The combination of perianth form, mode of asexual reproduction, and leaf shape adequately segregates *Jungermannia lanceolata* generically. In this, then, I agree with Joergensen, Jones, Meylan, and Müller, and disagree with Evans (1938c) and Frye and Clark (1937–47).[320]

If the generic name *Jungermannia* is employed for *J. lanceolata*, the oldest generic names available for the other entire-leaved species under discussion are *Gymnoscyphus* of Corda (1835) and *Solenostoma* Mitt. (1865). *Plectocolea* was established by Mitten, at the same time, as a mere subgenus of *Solenostoma*, although later (1871, *in* 1865–73) elevated to generic status by the same author. The widely used generic names *Aplozia* (or *Haplozia*, as Müller emended it) Dumort. (1874) and *Eucalyx* Breidl. (1894) are clearly of later date. Grolle (1957) and Bonner and Schuster (1964) have proposed that *Solenostoma*, which has been widely employed, be conserved over *Gymnoscyphus* [type: *Gymnoscyphus repens* = *Jungermannia pumila* With.], which has never been adopted.

A remaining problem, after exclusion of the *J. lanceolata* element, is to show whether or not the species for which these generic names have been employed are susceptible to division into well-defined genera *and, if so*, which names are applicable. As already mentioned, both Müller (1942) and Evans (1938c) split this complex of species into two genera, the first using the generic names *Solenostoma* and *Plectocolea*, the latter *Jungermannia* and *Plectocolea*, although the basis for the generic cleavage is somewhat different.[321]

Attempts, such as that of Evans (1938c), to divide this complex into two genera on the basis of presence or absence of a perigynium seem

artificial .... Both genera include forms with circular leaves, deeply 4–5-plicate perianths with a short terminal beak, i.e., species which belong to the

[320] The alternative course—inclusion of *Solenostoma* in a "portmanteau" genus *Jungermannia*—adopted by Amakawa (1960a), has been followed by Grolle. I feel that, in a taxonomically "moot" situation, it is better to avoid doing anything that could create nomenclatural chaos. I have already called attention (p. 836) to the numerous *nomina nova* that Grolle proposed subsequent to his adoption of a distended genus *Jungermannia* (e.g., *J. leiantha*, *J. karl-muelleri*, *J. jenseniana*—all for well-known European taxa). Amakawa has done the same, e.g., *J. hattorii* (for *Solenostoma* or *Pletocolea marginata* Hatt.), *J. kyushuensis* (for *Plectocolea rigidula* Hatt.), *J. cephalozielloides* (for *Plectocolea biloba* Hatt.), *J. japonica* (for *P. emarginata* Amak.).

[321] I would agree that a subdivision, if feasible, would be desirable on practical grounds, since *Solenostoma s. lat.* is a large and unwieldy genus, with ca. 45 species in Japan alone. However, a survey of all criteria that might prove significant has, to date, provided no sharp basis for subdivision. The form of the elaters initially seemed to make possible a division of the *Solenostoma-Plectocolea* complex. For example, some species of *Solenostoma s. str.* (subg. *Solenostoma*) have elaters that may be clearly 2-spiral to their relatively noncontorted and nonpointed ends [see, e.g., Fig. 280:4, *S. sphaerocarpum*, on p. 948; see also Amakawa's (1960a) fig. 38L of *J.* (*Solenostoma*) *hingaensis*]. By contrast all species of subg. *Plectocolea* studied have contorted elaters, with narrowed and thick-walled apices in which the spirals disappear (see, e.g., Fig. 293:2–3, *S. obscurum*, on p. 1018; see also Amakawa's figs. 14U, 16K, 21V, 27X, of the following *Solenostomas*: *unispiris*, *torticalyx*, *harana*, and *cephalozioides*). However, in *S.* (*Solenostoma*) *pyriflorum* and often in *S. sphaerocarpum*, the elaters are contorted, have thickened tips, and are of the "*Plectocolea*" type. Similarly, *S.* (*Solenostoma*) *gracillimum* has identical, "*Plectocolea*"-type elaters. Hence, this distinction also breaks down.

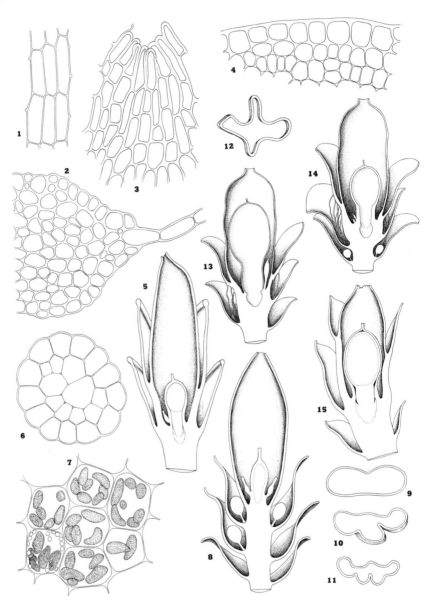

Fig. 272. *Solenostoma.* (1–5) *S. gracillimum:* 1, lateral cortical cells (×200); 2, stem cross section (×172); 3, cells of perianth mouth (×185); 4, margin of ♀ bract (×145); 5, gynoecial shoot tip (×ca. 26). (6) *S. pumilum* subsp. *polaris,* seta cross section (×200). (7) *S. crenuliformis,* median cells with oil-bodies (×450). (8–11) *S. pumilum* subsp. *pumilum:* 8, longisection fertile shoot apex (×38); 9–11, perianth cross sections, respectively 1/4, 1/6, and 1/8 from apex, showing dorsiventral compression (×50). (12–14) *S. sphaerocarpum:* 12, cross section of perianth, 1/4 from apex (×30); 13–14, longisections, fertile shoot apices, phase approaching var. *nana* (×18). (15) *S.*

J. sphaerocarpa complex. For instance, *J. crenulata* is included in *Plectocolea* because of the supposed presence of a perigynium. *J. pusilla* and *sphaerocarpa*, on the other hand, are placed in *Jungermannia*, supposedly restricted to species with the perianth not subtended by a perigynium. However, *J. crenulata* normally has a low perigynium . . . while *J. pusilla* has the bracts and perianth distinctly "united" for ¼ the length of the bracts (see K. Müller, 1916, *in* 1905–16, p. 746 and fig. 201e). Müller has stressed that *J. pusilla* exhibits distinct affinities with both "*Jungermannia sphaerocarpa* var. *nana*" (which may also, in mountainous sites, show partial formation of a perigynium) and with "*Plectocolea*" *crenulata*; compare Fig. 272:5 and 12–13. In fact, *J. crenulata*, *J. pusilla*, *J. sphaerocarpa* (and *caespiticia*, etc.) have in common . . . many characters (Schuster, 1953).

The clearly and longly emergent perianths, each with a short beak, the orbicular leaves, and the ability to develop brownish or reddish secondary pigmentation—all suggest a close affinity. "There is no difference that one can consider significant in the trivial differences between them as regards degree of development of the rudimentary perigynium: all three of these species may show a 'free' perianth (*Jungermannia*) or slight perigynium ('*Plectocolea*')" (Schuster, 1953). Therefore, I believe that a separation of *Solenostoma* into two genera, on the basis of the presence or absence of a perigynium, lacks validity. A parallel case is the traditional attempt to separate *Gymnomitrion* and *Marsupella* on a somewhat similar basis.[322]

The other basis for a generic cleavage in the complex is given by Müller (1942). He maintains *Solenostoma* for species which have well-developed, longly emergent perianths, formed of cells which are not strongly elongated (except for those fringing the mouth); a perigynium is rudimentary or absent (Fig. 272:5,8,13–14). Those species in which there is a distinct perigynium but a reduced, hardly emergent perianth formed of wholly elongated, narrow cells are assigned to *Plectocolea* (Figs. 271:10–11; 272:15). Perianths in *Plectocolea* are not beaked and do not possess 4–5 sharp, deep plicae. On this basis "*J.*" *crenulata* again proves troublesome. In some phases of this species perianths are quite distinctly beaked (Fig. 285:7); in other phases the perianth is more slender and gradually narrowed distally so that a beak is almost imperceptible

---

[322] Amakawa (1960a) has fully substantiated this viewpoint. Indeed, the type of *Plectocolea*, *J. radicellosa*, according to Amakawa, has the "perigynium hardly developed," and Amakawa's fig. 11:a,k shows a perianth that is longly emergent from between the bracts, with a hardly discernible perigynium. Since the type of *Plectocolea* approaches *Solenostoma* in regard to the perigynium (although it lacks the perianth beak of that subgenus), it affords excellent evidence of the transitional nature of the perigynium character.

As Amakawa (p. 2) tersely states, "It is not natural to separate *Plectocolea* and *Solenostoma* on the basis of the presence or absence of a perigynium, because the perigynium is present in some species of *Solenostoma* and is absent in some species of *Plectocolea* (see Schuster, 1953, pp. 387–388)."

---

subellipticum, longisection fertile shoot apex (×23.5). [1–2, *RMS 30158*, South Carolina; 3–5, *Cain 4281*, Ontario; 6, *RMS 66-283a*, Greenland; 7, Letchworth State Park, N.Y., *RMS 1949*; 8–11, *RMS 36004*, Wake Co., N.C.; 12, *RMS 66-1287b*, Greenland; 13–14, Sand I., Wisc., *RMS, Sept. 15, 1949*; 15, *RMS 34551*, Kânâk, Greenland.]

(Figs. 272:5; 285:1). Supernumerary plicae may also be developed. Similarly, *S. fossombronioides*, with the irregularly lacerate and wide-open perianth mouth can scarcely be assigned to either group on that basis. Differences in perianth form are thus neither impressive enough nor constant enough to warrant segregation into two genera. For this reason the older view of Mitten (1865) is adhered to, and *Solenostoma* is considered generically identical with *Plectocolea*.

Solenostoma, as thus broadly defined, is world-wide in distribution, ranging from the high Arctic, as far north as land occurs (ca. 82°32′N.), southward into the tropics and to the subantarctic regions. Approximately 80–90 species are known, of which about 24 are from America north of Mexico. Our species fall reasonably naturally into three groups, here considered to be subgenera:

Key to Subgenera of Solenostoma

a. Perigynium usually absent or a mere low, fleshy ring (rarely between 0.1–0.25 the length of "free" perianth); perianth at maturity 0.3–0.7 emergent from beyond the bracts, usually not closely sheathed by them; cells of perianth not strongly elongated [except those at mouth which are 2–3(4)× as long as wide], those below mouth and in perianth middle usually 1–2× as long as wide, hardly differentiated from the cells of the bracts and leaves; perianth 2-stratose below middle, firm; leaves various in shape: ovate to elliptical to cordate to rotund; our species with rhizoids and with leaf bases rarely vinaceous; oil-bodies nearly homogeneous; capsule wall 2-stratose (in our taxa) . . . . *b*.
   b. Perianth not beaked, gradually narrowed or rounded into apex, often slightly dorsiventrally compressed, with 1 or 2 deep antical sulci, but without a marked longitudinal median plica, entirely lacking any subtending perigynium; leaves elliptical to ovate to cordate, their median width usually less than their length; plants olive-green or blackish green, secondary pigments (if any) fuscous brown; cells usually small: median cells 15–22 (rarely 25) μ wide, on an average, with trigones usually minute or absent; rhizoids from abaxial faces of leaves and bracts lacking.
      . . . . subg. *Luridae* (= *Luridaplozia*) (p. 905)
   b. Perianth abruptly narrowed distally into a short, small beak, slightly laterally compressed or terete, deeply 4–5-plicate (1 fold dorsal and median), often with a low perigynial ring at base; leaves reniform to orbicular, as wide as or somewhat broader than long; plants bright green, in sun developing brownish or reddish secondary pigments; cells larger, median cells averaging 24 μ wide or more, often with larger trigones; frequently with rhizoids from abaxial faces of leaves and bracts . .
      . . . . subg. *Solenostoma* (= *Euaplozia*) (p. 942)
a. Perigynium (or a rudiment of it) present, usually 0.4–1.2× the length of perianth above it; perianth generally included or emergent for only 0.1–0.3 its length beyond the bracts, which commonly closely invest or sheath it

(particularly when young); perianth not suddenly beaked at mouth, usually ± narrowly conoidal (except *S. fossombronioides*), unistratose throughout, ± deeply plicate, formed of narrow, thin- or often thick-walled (but not collenchymatous), sometimes partially sigmoid cells 2.5–5 × as long as wide; leaves ovoid or orbicular, widest at or near middle, at most slightly longer than broad; our species with rhizoids often vinaceous or purplish red; plants always able to develop vinaceous or reddish secondary pigments; oil-bodies distinctly granular; capsule wall 2–3(4)-stratose . . . .
. . . . subg. *Plectocolea* (= *Eucalyx*) (p. 980)

SOLENOSTOMA subg. *LURIDAE* (*Spr.*) *K. Müll.*

Jungermannia sect. *Luridae* Spr., Trans. Proc. Bot. Soc. Edinb. 15:509, 1885.
Gymnoscyphus Corda *in* Sturm, Deutschlands Fl. 2:516, 1832 (*nomen rejiciendum*).
Aplozia subg. *Euaplozia* sect. *Luridae* Schiffn., *in* Engler & Prantl, Nat. Pflanzenfam. 1(3):82, 1893.
Haplozia subg. *Luridae* K. Müll., Rabenh. Krypt.-Fl. 6(1):554, 1909.
Aplozia subg. *Luridaplozia* Joerg., Bergens Mus. Skrifter 16:108, 1934.
Solenostoma subg. *Luridae* K. Müll., Hedwigia 81:117, 1942.
Solenostoma sect. *Luridae* K. Müll., Rabenh. Krypt.-Fl. ed. 3, 6:814, 1956.
Solenostoma subg. *Luridaplozia* Schust., Amer. Midl. Nat. 49(2):404, 1953.
Jungermannia subg. *Luridae* Arn., Illus. Moss Fl. Fennosc. 1:103, 1956; Amakawa, Jour. Hattori Bot. Lab. no. 22:43, 1960.

Deep to dull green, often fuscous to rarely blackish, never with reddish to vinaceous pigmentation, very small to medium-sized (0.5–3 mm wide × 5–30 mm long), often flaccid. Stem not specialized: cortical 1–2 strata of cells sometimes slightly more thick-walled than medullary, but without marked differentiation in diam., without development of a hyaloderm. Rhizoids abundant, *colorless or brownish*, very rarely dull purplish, *confined to stem*. Leaves typically ovate to *lanceolate-ovate to ovate-lingulate to broadly ovate-triangular, typically longer than broad and broadest near base*, rarely rotundate to subrotundate, their insertion oblique, obliquely to widely spreading; their marginal cells rarely forming a border, virtually *never with rhizoids attached to their bases*. Cells usually with small trigones, *usually delicate*, subquadrate to hexagonal above, *becoming elongate-hexagonal to oblong medially and near base, small* (usually 16–25 µ wide medially, or less); oil-bodies finely granular, rather large, ovoid to spherical or ellipsoidal-fusiform, 2–6(7–9) per cell, rather large or medium-sized (always larger than chloroplasts). Without asexual reproduction.

Dioecious or paroecious. Perianth usually basally sheathed by the often erect or suberect bracts, *0.3–0.6(0.7) emergent, fusiform to slenderly obovoid* or even obclavate, *without trace of perigynium at base, frontally slightly to conspicuously compressed*, the distal portion pluriplicate and ± conoidally narrowed, the *plicae not sharp; mouth never rostrate*, the cells near mouth 2–3 × as long as wide, hardly sharply differentiated, those below oblong to subquadrate, only 1.5–3 × as long as wide, *never linearly elongated and/or sigmoid, similar to median cells of leaves;* perianth 1-stratose above, 1–3-stratose below.

Type.—*Solenostoma pumilum* may be considered the type species; there had been no formal type designation before Amakawa (1960a).

The subg. *Luridae* is admitted by Amakawa (1960a, p. 44) to be a "well-defined, natural group." However, with *Solenostoma pumilum* subsp. *polaris* (p. 918) and species such as the Japanese *S. grossitexta* (Steph.) Schust., comb. n., and *S. kuwaharae* (Amak.) Schust., comb. n.,[323] we find a slight transition to subg. *Plectocolea:* these taxa all tend to develop orbicular leaves. Normally, the ovate to lanceolate leaf shape, as well as the lax texture, is diagnostic of *Luridae*.

Spruce (1884–85, p. 509), in his initial description of *Luridae*, carefully delineated the basic criteria and stated the group was notable for "the olive-green colour of the ovate-oblong, or somewhat heart-shaped leaves; for the dichotomous or lateral branching (with the addition of postical flagella in *J. riparia*); but especially for the perianth being compressed frontally instead of laterally, and having a furrow in place of a raised keel along the middle of the upper face."

The center of distribution of subg. *Luridae* appears to be eastern Asia: in Japan eight species are reported; in all of Europe, only six.

Subgenus *Luridae* is poorly represented in eastern North America; only *S. pumilum* subsp. *pumilum* occurs south of New England, and only *S. cordifolium* and *S. pumilum s. lat.* are found in the eastern U.S. The group has been badly misunderstood; e.g., all or almost all reports of *S. atrovirens* from North America, at least from the East, are based on errors. That *S. polaris* has been regarded as a synonym of *S. atrovirens*, whereas it is based on a paroecious plant at best subspecifically distinct from *S. pumilum*, has added to the confusion. I have seen no material of *S. atrovirens* from our region; hence I treat the species with brevity and in an admittedly derivative fashion, as I have *S. triste*. *Solenostoma oblongifolium* is somewhat of a mystery, and, although reluctantly accepted, I am skeptical of it as a distinct species. Finally, in my treatment, *S. schiffneri* disappears completely as a synonym of *S. pumilum* subsp. *polaris*. In effect, almost all American reports of *S. triste* I have been able to trace back to actual specimens belong to *S. pumilum* or *S. pumilum polaris*. From the foregoing (and from a perusal of the detailed treatment of *S. pumilum s. lat.*, pp. 907–929), it is clear that published reports for this entire subgenus are by and large untrustworthy.

All species, except for *S. pumilum s. lat.* in the North, are confined to rocks adjoining or even in flowing water, and even *S. pumilum* is, southward, confined to such sites.

---

[323] Basionyms: *Jungermannia grossitexta* Steph., Spec. Hep. 6:87, 1917; *Jungermannia kuwaharae* Amak., Jour. Hattori Bot. Lab. no. 22:53, 1962.

Key to Species of Solenostoma subg. Luridae

1. Leaves elliptical to oblong-lanceolate, rarely rotundate; plants minute to small, 0.2–3 cm long; paroecious or dioecious . . . . 2.
　2. Paroecious; cuticle at least weakly striolate . . . . (*S. pumilum*)　3.
　　3. Leaves of sterile stems elliptical to lanceolate-elliptical; perianth ± fusiform, gradually narrowed to a tapering apex (except in aquatic phases); capsule with epidermal cells 18–22 μ wide, typically with 2 nodular thickenings on *all* longitudinal walls; cells with (2)3–6 oil-bodies (occasionally 7–10 in isolated marginal cells); cells at apices of leaves 16–18 μ usually, those of leaf middle usually 23–25 × 25–30(35–40) μ . . . . *S. pumilum* subsp. *pumilum* (p. 912)
　　3. Leaves of sterile stems ovate to rotundate, rarely a few subelliptical; perianth obclavate to obovoid, abruptly narrowed near mouth; capsule with epidermal cells mostly 15–18 μ wide, both faces of *alternating* longitudinal walls with (1)2–4 thickenings, the intervening walls often without thickenings; cells with (1)2–4(5) oil-bodies; marginal cells at leaf apices averaging 13–17 μ; median cells averaging (12–13)14–18 μ wide × 16–22(24–26) μ long . . . .
　　　. . . . . *S. pumilum* subsp. *polaris* (p. 918)
　2. Dioecious: cuticle usually smooth (except *S. oblongifolium*); perianth ob-pyriform to oblong-ovate, rather broad at apex . . . . 4.
　　4. Plants minute, 2–10 mm long; leaves 0.3–0.5 mm long; cells of leaf margins only 12–18 μ; median cells 18–22 × 25–30 μ; oil-bodies mostly 2 per cell . . . . *S. atrovirens* (p. 929)
　　4. Plants larger, 10–15 mm long or more; leaves 0.8–1.5 mm long × 0.5–1.2 mm broad; cells of leaf margins 18–20 μ; median cells 18–25 × 30–35(40) μ . . . . 5.
　　　5. Plants delicate, 0.5–1 mm wide, 1–1.5 cm long, black-green; cuticle smooth or striolate only at base; perianth slender, sub-cylindrical to fusiform . . . . *S. oblongifolium* (p. 936)
　　　5. Plants more robust, 1.5–2 mm wide × 1–3 cm long; cuticle smooth; perianth cylindrical to obovate; oil-bodies 2–4 per cell.
　　　. . . . . *S. triste* (p. 932)
1. Leaves bluntly triangular to cordate, fully as broad at base as long, semi-amplexicaul, blackish green; perianth fusiform; large, 2–12 cm long; leaves lax, delicate; marginal cells 22 μ; median cells 20–25 × 35–40 μ. Dioecious . . . . *S. cordifolium* (p. 939)

SOLENOSTOMA (LURIDAE) PUMILUM (With.)
K. Müll.[324]
[Figs. 272:6, 8–11; 273–276]

Jungermannia pumila With., Nat. Arr. Brit. Pl., p. 866, 1776.
Aplozia rivularis Schiffn., Krit. Bemerk., *in* Lotos 59:21, 1911.
Jungermannia pumila var. *polaris* Berggr., Kgl. Sv. Vetensk.-Akad. Handl. 13(7):98, 1875.
Solenostoma pumilum K. Müll., Hedwigia 81:117, 1942.

[324] For fuller bibliographic citations, see pp. 912 and 918, under subsp. *pumilum* and subsp. *polaris*, respectively.

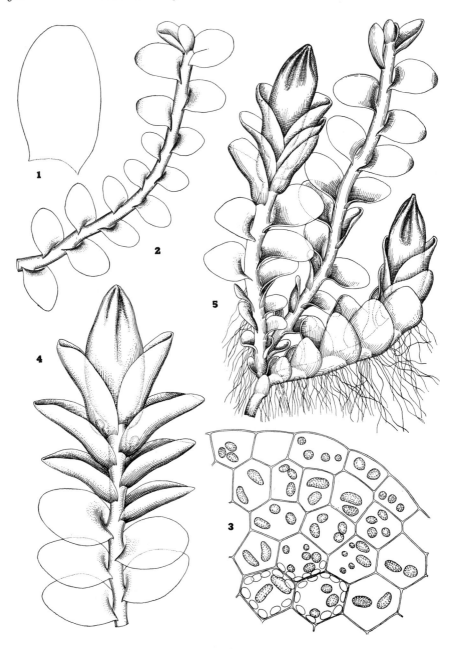

Fig. 273. *Solenostoma* (*Luridae*) *pumilum* subsp. *pumilum*. (1) Small leaf of sterile shoot (×65). (2) Sterile shoot, mod. *laxifolia* (×25). (3) Cells near leaf apex with oil-bodies and, below, chloroplasts (×610). (4) Vigorous plant, growing amidst those of the "*rivularis*" phenotype in Fig. 274, but with "*pumilum*" perianth (×23). (5) Typical

After much deliberation and study of many arctic collections, I have been forced to the conclusion that the small plants of the American Arctic which have been variously called *Jungermannia polaris* (by Berggren), *J. schiffneri* (by Steere, Schuster, and others), *J. pumilum*, *Solenostoma polaris* var. *cavifolius*, and *J. atrovirens* all being to a single, enormously complex and variable species, which must be called *S. pumilum*; for documentation, see pp. 910, 923. My reasoning is based in part on analogies obvious from other species in other families. For example, in *Tritomaria polita* the arctic plants (subsp. *polymorpha*) tend to go from the "norm" with longer than broad leaves and ♀ bracts to very broad, much wider than long leaves and ♀ bracts. Similarly, in *Tritomaria exsectiformis*, the "normal plant" has leaves that are always clearly elongated, while in subsp. *arctica* orbicular to broadly orbicular leaves are produced. In general, many Jungermanniales show a tendency under arctic conditions for phenotypic or genotypic production of shorter- and broader-leaved phenotypes. In *S. pumilum*, temperate-zone plants from North Carolina to southern New England always have spreading and elliptical leaves, usually 1.2–1.5 × *as long as wide*. By the time the arctic extremes are reached (fo. *cavifolius* Schust.) the leaves may be 1.2–1.4× *as wide as long*. The perianth response is similar: the elongated, spindle-shaped perianths of "normal" *pumilum*, temperate-zone plants from North Carolina to southern New cylindrical, shorter perianths, whose abbreviated form is achieved by contraction of the longly tapered distal part of the "*pumilum*" perianth. Consequently, maximal width of the perianth is retained to within ¼ of the perianth apex—a "*schiffneri*" characteristic, and, with the abrupt contraction thus achieved, perianths may even become obscurely beaked at the tip.

Associated with these phenomena is a general tendency toward overall reduction, which is expressed in the smaller leaf cells of the "*schiffneri*" extreme, the smaller—and especially narrower—epidermal cells of the capsule, and the usually smaller spores. Since a complex genetic system is probably involved, *S. pumilum* does not demonstrate exactly uniform and parallel variation as one goes north; furthermore, "*pumilum*"-type plants, as I have shown (Schuster et al., 1959, p. 53), may occur as far north as 82°25′ N. However, "*schiffneri*"-type plants fail to occur very far to the south, and the "*cavifolius*" extreme is purely arctic. Hence the presumption is that there has been, to some extent, genetic segregation into partly isolated populations.

---

fertile plant (×ca. 20). [1–3, *RMS 28211*, Orange Co., N.C., 4, *RMS 29440*, Neddie Creek, N.C.; 5, Grand Marais, Minn., *RMS*.]

Two collections that have been critically studied[325] may serve to demonstrate the problems involved.

The plants of *66–041a* are relatively large, with the concave and broad leaves of fo. *cavifolius* (leaves to 650–660 μ wide × 475 μ long on sterile axes). Marginal cells 16–18 μ; median cells 15–19 × 20–25 μ; 2–4(5–6) oil-bodies per cell—if 2 oil-bodies present, usually very large. Paroecious inflorescences are clearly established. The perianth is of the typical "*schiffneri*" type—widest at or above the middle, abruptly narrowed in the distal 0.15–0.25. To this point, all signs clearly are indicative of *S. pumilum* subsp. *polaris* fo. *cavifolius*. However, the capsule wall has the epidermal cells all with 1–2 strong thickenings of longitudinal walls (and many transverse walls), cited by Müller and Schiffner as diagnostic of *S. pumilum s. str.*. Furthermore, epidermal cells average 18–22, locally 20–24, μ wide × 24–36 μ long, much wider than is commonly accepted for *S.* "*schiffneri*."[326] Spores also are too large: 19–21 μ in diam. The cylindrical to, occasionally, feebly clavate perianths, without the gradually conical apex of "*S. pumilum*," effectively preclude placing these plants there. This suggests that the capsule wall and spore characters adhered to in Müller may not be valid.

In order to test this hypothesis, abundantly capsule-bearing plants (*RMS & KD 66–283*) were studied; 5 capsules of varying sizes, selected at random, illustrate the approximate variation in valve dimensions (which are a good criterion of overall capsule size). As Table 7 shows, there is a direct correlation between capsule size and epidermal-cell size. Also, the largest capsules tended to have thickenings of *all* longitudinal walls, while intermediate-sized ones usually had 3 longitudinal walls with thickenings, followed by 1 lacking thickenings, followed again by 3 with thickenings, etc. (the 0–3–0–3 formula expresses this condition). The smallest valves tended to show no thickenings of the middle 1 of the 3 "thickened" walls (of the above type) and then a 0–1–0–1–0–1–0–1 alternation prevailed but with an occasional thickening of every second "thickening-free" wall (walls *0*), leading to transitions to the more usual 0–3–0–3–0 condition. All sorts of transitions could be found, as well as many irregularities and deviations that fail to be of significance or interest. In my opinion, no consistently valid differences between *polaris* ("*schiffneri*") and *pumilum* in either capsule-wall

TABLE 7

Valve Width × Length	Epidermal Cell Width × Length	Thickenings
305 × 560 μ	18–19(24) × 25–36 μ	On nearly all long. walls
300 × 680 μ	16–18 × 24–35 μ	0–3–0–3 usually
225 × 550 μ	14–16 × 24–45 μ	0–3–0–3 or 0–2 to 3–0–2 to 3
220 × 510 μ	12–16 × 20–32 μ	± 0–3–0–3
190–200 × 460 μ	12–15 × 20–32 μ	0–1–0–1–0 to 0–3–0–3–0

[325] W. GREENLAND. Head of Sondrestrom Fjord, 66°50′ N., 50°30′W. (*RMS & KD 66–041a*; exposed sunny, dry, steep slope, in recess below rock, with *Mannia pilosa*); Blaesedalen, side of Lyngmarksfjeld, near Godhavn, Disko I. (*RMS & KD 66–283, p.p.*, with *Scapania curta*).
[326] See, e.g., Müller (1905–16).

morphology or spore size occur. For example, plants of *RMS & KD 66–283*, although mostly with the epidermal cell width of *S. polaris*, had spores 16.5–21.5 μ in diam., which is typical of *S. pumilum.*

In the final evaluation, therefore, only the perianth form is left as a criterion: this would more nearly place the cited plants in *S. polaris.* However, repeated study of a large mass of living plants has now convinced me that the entire *S. pumilum-schiffneri-polaris* complex must be treated as a single variable species, with, perhaps, recognition of several extremes as formas.[327]

All the broad- and short-leaved arctic forms assigned to subsp. *polaris* tend to develop polyoecious or heteroecious inflorescences. In most cases some gynoecial axes clearly lack subtending ♂ bracts; in other cases a leading axis is androecial, with the apex remaining sterile, but with lateral-intercalary branches that are gynoecial or paroecious (sometimes both ♀ and ♂ branches may occur on a single plant). Occasionally long androecia occur on axes to which no connection to a gynoecial axis can be determined, although such may have once occurred. Such hetero- and possibly polyoecious plants may have given rise to much of the confusion of subsp. *polaris* with *S. atrovirens.* All of these conditions are readily visible in *RMS & KD 66–283.*

The presumed partial genetic isolation—a presumption subject to further experimental proof—perhaps suffices to allow recognition of three extremes of the species:

Key to Variations of S. pumilum

1. Plants with slenderly fusiform perianths, widest typically in the median ¼, tapered gradually to a slenderly conoidal apex in the distal ⅓ or more; leaves always narrow, 1.2–1.5× as long as wide, usually shallowly spoon-shaped or almost flat, laterally patent, rarely as broad as or broader than long and concave (in arctic extremes). ♀ Bracts elongated, oval to oblong-oval or oblong-ovate. Always paroecious . . . . subsp. *pumilum*
1. Plants with perianths ovoid to ellipsoidal, slightly dorsiventrally flattened, parallel-sided for much of the median 0.35–0.6 or somewhat broadened upward to within 0.25 of the apex, rather abruptly narrowed in the distal 0.2–0.25, often into a vestigial beak; leaves ovate, and widest at base, ranging to broadly orbicular. ♀ Bracts broadly ovate to ovate-cordate or broadly ovate-orbicular, usually nearly as wide as to wider than long. Heteroecious, sporadically with long androecial branches.
   . . . . subsp. *polaris* 2.
     2. Leaves little or moderately concave, usually nearly as long as or slightly longer than wide, mostly ovate; bracts ovate, not wider than long.
     . . . . fo. *polaris*
     2. Leaves strongly concave to hemispherical, very broadly attached to stem, 1.2–1.4× as wide as long on sterile stems, reniform-orbicular to broadly

---

[327] Arnell (1956) admits all three as species and states "*J. polaris* . . . is said to be dioecious." See, however, the footnote on p. 51 in Schuster et al. (1959), where the type of *J. polaris* Lindb. (Spitsbergen, *Malmgren 1867, fide* Arnell, who sent me a single plant of the fragmentary type) is shown to be probably paroecious, as is the type of *J. pumila* var. *polaris* Berggr.

orbicular; bracts broadly ovate-orbicular to ovate-cordate, 1.0–1.3 ×
as wide as long . . . . fo. *cavifolius*

SOLENOSTOMA (LURIDAE) PUMILUM (With.)
K. Müll. subsp. PUMILUM

[Figs. 272: 8–11; 273–274]

Jungermannia pumila With., Nat. Arr. Brit. Pl., p. 866, 1776.
Gymnoscyphus repens Corda, *in* Sturm, Deutschl. Fl. 2:158, 1832.
Jungermannia zeyheri Hüben., Hep. Germ., p. 89, 1834.
Jungermannia rostellata Hüben., *ibid.*, p. 95, 1834.
Aplozia pumila var. *rivularis* Schiffn., Lotos 1900(7):326, 1900; Macvicar, Studs. Hdb. Brit. Hep.
    ed. 2:151, 1926.
Aplozia atrovirens var. *arnellii* Schiffn., Lotos 58:329, 1910.
Aplozia rivularis Schiffn., Krit. Bemerk., *in* Lotos 59:21, 1911.
Aplozia pumila Dumort., Hep. Eur., p. 59, 1874; Macvicar, Studs. Hdb. Brit. Hep. ed. 2:150,
    1926.
Solenostoma pumilum K. Müll., Hedwigia 81:117, 1942; K. Müller, Rabenh. Krypt.-Fl. ed. 3,
    6:821, fig. 280, 1956; Schuster, Amer. Midl. Nat. 49(2):404, pl. 31, fig. 4, 1953.

Small; in prostrate, flat, *thin and closely adnate, olive-green to blackish* patches,
more rarely creeping among other bryophytes. *Shoots 1.0–2.0 mm wide* (rarely
to 2.5 mm wide near apices of fertile plants) × 3.5–8 (rarely 10–15) mm long,
simple or sparsely branched, usually *without stolons;* stems brownish with age,
120–180, rarely to 220–240, μ wide, slightly flattened, to 10–12 cells high;
cortical cells scarcely differentiated from medullary, slightly and nearly equally
thick-walled, dorsally 18–21(23) μ in width; medullary cells slightly or scarcely
smaller, 14–18(21) μ, slightly to moderately, equally thick-walled, with scattered minute pits. Rhizoids scattered, colorless to brownish, numerous, elongate,
11–12 μ in diam. Leaves of sterile shoots distant, widely (75–85°) and almost
horizontally spreading, *narrowly oval to broadly elliptical*, ca. 325–375 μ wide ×
500–600 μ long, nearly flat; leaves of robust and fertile shoots larger, *oval or
ovate-oval to broadly elliptical*, to 860–1050 μ long × (600)670–680 μ wide (*length
1.25–1.55× width*),[328] becoming weakly imbricate, flat except at the slightly
concave base. Cells opaque, deep green, densely chlorophyllose, leptodermous,
and with small or vestigial trigones, or slightly and equally thick-walled;
marginal, ca. 20 μ; subapical, 21–23 μ; *median*, ca. *23–25 × 25–30(35)* μ,
subisodiametric, polygonal; cuticle delicately striolate-verruculose; *oil-bodies
2–6(7–8) per cell*, near marginal cells often small and largely spherical (then
2.5–4 μ), near leaf middle becoming larger and ovoid to ellipsoid, from 4 × 5 μ
to 4 × 8–10, occasionally 4.5 × 10 μ, formed of delicate spherules, appearing
grayish and papillose. Chloroplasts distinctly smaller than oil-bodies, ca.
5 μ long. Asexual reproduction lacking.

Paroecious and abundantly fertile. ♂ (Subinvolucral) bracts in 2–3 pairs, somewhat larger than leaves (to 1200–1400 μ long), *ovate* (width approaching length),
subtransverse to transverse, erect or erect-spreading with distal portions frequently
obliquely to widely spreading, *concave-canaliculate, saccate at base;* usually monandrous. ♀ *Bracts identical to ♂ bracts in size and form*, often bearing antheridia,

---

[328] Müller (1951–58, p. 821) states that they are 1.5–1.7 × as long as wide; I do not think that
the leaves are ever, on the average, so elongated.

to 1040 μ wide × 1050–1100 μ long, rarely to 1200–1300 × 1400 μ long, broadly ovate or orbicular ovate, *erect, usually closely sheathing the lower third of the perianth. Perianth typically fusiform,* moderately but distinctly dorsiventrally compressed, 780–850(920) μ wide × 450–500 μ thick × 1700–1900 (occasionally to 1950–2000) μ long; basal half or more of perianth smooth, virtually cylindrical, *distally distinctly pluriplicate to the acute apex;* mouth crenulate with elongate, fingerlike cells free only at the rounded apex, 10–12 μ wide × 24–28(32) μ long; subapical cells becoming irregularly polygonal, 12–14(15) μ wide × 15–23 μ long, ± thick-walled, with often distinct trigones. Capsule oval, dark brown; spores (16)17–20(22) μ, yellowish brown, finely verruculose. Elaters 6–7 μ in diam.

Type. England.

Distribution

Solenostoma pumilum subsp. *pumilum* is a widely distributed, regionally frequent, holarctic plant, occurring largely in the lowlands and in cool to temperate regions: in our area widely ranging from the Tundra and the Taiga into the Deciduous Forest Regions.

In Europe widespread from central Europe [from north Italy, Austria, Switzerland, and Germany to Czechoslovakia, Poland (Tatra Mts.), France] northward to Belgium, Great Britain, the Faroes, Denmark, Norway, Sweden, Finland, southward to Spain (Sierra Nevada), eastward to the USSR and to Siberia (Behring I.; Yenisey R.). Also in Spitsbergen (Arnell and Mårtensson, 1959) and Iceland. In the Azores (Allorge and Allorge, 1950). Apparently rare in Japan, confined to Hokkaido (Amakawa, 1960a). In North America widespread in the West (Alberta and British Columbia southward to California and Colorado) but apparently even commoner in the East:

E. GREENLAND. Kobberpynt in Scoresby Sund (Jensen, 1897). NW. GREENLAND. Nucluet, Booth Sd., Steensby Land (*RMS 46201, 46201b*; with *Lophozia quadriloba, Solenostoma pumilum trans. ad* subsp. *polaris*). W. GREENLAND. Disko I., below Skarvefjeld, E. of Godhavn (*RMS & KD 66-307, 66-308, 66-231*), and below Lyngmarksfjeld, Godhavn (*RMS & KD 66-276, 66-153b, 66-183, 66-150c*); Magdlak, Alfred Wegeners Halvö (*RMS & KD 66-1356c*); Tupilak I., W. of Egedesminde (*RMS & KD 66-026a*); Kangerdlugssuakavsak, head of Kangerdluarssuk Fjord (*RMS & KD 66-1284, 66-1290, 66-1283a*). ELLESMERE I. Mt. Pullen and hills SW. of Mt. Pullen, S. of Alert, 82°25′N. (*RMS 35783, c. caps., 35731*; see Schuster et al., 1959, p. 53); northernmost report!; Fram Harbor (Bryhn, 1906). QUEBEC. Wolstenholme (Polunin, 1947); Rupert R. above L. Némiskau; Meech L. near Chelsea; Bic (Lepage 1944–45; Macoun, 1902); N. shore of Gaspé Bay (*Williams 10819!*); N. slopes of Mt. Albert, Gaspé (*Williams 10646, p.p.*). ONTARIO. "On old logs," Seymour West in Northumberland Co. (Macoun, 1902). NOVA SCOTIA. St. Croix (Brown, 1936a); margin of shallow lake, summit of French Mt., ca. 1200 ft, Cabot Trail, Cape Breton I. (*RMS 43030a*). NEWFOUNDLAND. Squires Memorial Park, Humber R. (*RMS 68-1658*).

MAINE. Seal Harbor (*Lorenz, 1922*); Schoodic L., Piscataquis Co. (*Evans*); Mt. Desert I. (Lorenz, 1924). NEW HAMPSHIRE. Waterville (*Lorenz, 1911*); The Flume, Franconia Mts. (*Lorenz, 1908*). VERMONT. Jericho. MASSACHUSETTS. Bear R., Conway (*RMS*); Roaring Brook, Shutesbury (*RMS*); Worcester (*Greenwood, 1914*); Oxford; Leicester. RHODE ISLAND. Evans (1913). CONNECTICUT. Hamden, North Branford, New Haven Co. (Evans & Nichols, 1908).

NEW YORK. Onondaga Co. (Goodrich, 1912; see Schuster, 1949a); Pumpkin Hollow, Five Mile Rd., Cattaraugus Co. (Schuster, 1949a). NEW JERSEY. Closter (Austin, Hep. Bor.-Amer. Exsic. no. 33, 1873); Bergen Co. (Britton, 1889). PENNSYLVANIA. Fair Grounds, Bedford Co. (*Lanfear, 1931*); Milroy Gap, Center Co. (*Lanfear*). OHIO. Hocking Co. (*Taylor, 1922*); 4 mi E. of S. Bloomingville on Rte. 56 (*RMS 18370*). MARYLAND. Frye & Clark (1937–47). DISTRICT OF COLUMBIA. Banks of Potomac, opp. Georgetown (Holzinger, 1907). WEST VIRGINIA. Greenbrier, Monongalia, Pocahontas Cos. (Ammons, 1940). KENTUCKY. Powell, Lewis, Letcher Cos. (Fulford, 1934, 1936).

VIRGINIA. Mountain L., Salt Pond Mt., Giles Co. (Patterson, 1949); Rockbridge Co. (Patterson, 1950); The Cascades, Salt Pond Mt., Little Stony Cr., Giles Co. (*RMS 40294, 40276a, 40285*). NORTH CAROLINA. Jackson Co. (Blomquist, 1936); Dark Ridge Cr., Swannoa Mts., Buncombe Co. (Andrews, 1921); Crow Cr., near Cullasaja R., between Highlands and Franklin (*RMS 40860a*) and Wayah Bald, 8 mi W. of Franklin, Macon Co. (*RMS 39280*); The Caves, near Durham, New Hope Cr., Orange Co. (*RMS 28528, 28496, 28500*); Crabtree Cr. State Park, 10 mi NW. of Raleigh (*RMS 36004*), and Hemlock Bluff near Carey, 10 mi SW. of Raleigh, Wake Co. (*RMS 30002*); Neddie Cr., near jct. with E. fork of Tuckasegee R., Jackson Co. (*RMS 29378a, 29448, 29444*; also *29451, trans. ad* fo. *rivularis*). TENNESSEE. Ammons (1940); not reported from there by Sharp (1939).

Westward as follows: ILLINOIS. Rocky Ford, 5 mi NE. of Litchfield, Montgomery Co. (*Biebel 114!*). MICHIGAN. Tahaquemon Falls, Luce Co. (*Nichols, 1933*); Grand Ledge, Eaton Co. (Steere, 1940); E. of Laurium, Houghton Co.; mouth of Montreal R., Gogebic Co.; Marquette Co. (Evans & Nichols, 1935, etc.). WISCONSIN. Black R., Douglas Co. (*Conklin, 1922*); Adams, Juneau, Ashland, Douglas, Bayfield, Iron, Trempealeau Cos. (Conard, 1929); Siskiwit R., Bayfield Co., and Montreal R., Iron Co. (Conklin, 1929; as var. *rivularis*). MINNESOTA. Grand Marais, Temperance R. ravine, Lutsen and ravine near Lutsen, Rosebush Falls, Cascade R., all in Cook Co.; Manitou R., Lake Co.; Chester Cr. and Fairmount Park in Duluth, and French R., in St. Louis Co. (Schuster, 1953). IOWA. Mahaska Co. (Conard, 1945).

Ecology

Southwards always a pioneer on damp, often shaded rocks (more rarely on gravelly soil near streams), frequently on soft-textured, friable rocks, such as sandstones. The species often forms obscure, dull, olive to blackish green mats, serving as a matrix for the ecesis of larger species.

In the Spruce-Fir Region and southern edges of the Tundra, often a pioneer on basic or subbasic rocks (such as diorite and other basaltic rocks), either on damp, shaded rocks or on rocks along the borders of streams (where subject to periodic inundation).[329] The species may here

---

[329] Müller (1951–58, p. 813) separates the *S. pumilum-polaris* complex from the *S. triste-atrovirens* complex on the basis of its occurrence on "kalkfreier Unterlage." Although, toward the south, *S. pumilum* is found typically on noncalcareous rocks, toward the north the entire complex becomes strongly Ca-tolerant, as the associates cited show.

be associated with *Scapania cuspiduligera*, *S. gymnostomophila*, *Lophozia gillmani*, *Preissia quadrata*, *Pellia epiphylla* and *fabbroniana* (Schuster, 1953); in similar sites, on damp sandstone rocks, *Solenostoma crenuliformis*, *S. levieri*, *S. sphaerocarpum*, *Scapania curta*, *Tritomaria scitula*, and other species may be consociated.

From southern New England southward the species is abundant along streams (in the lower and median elevations of the Appalachian Plateau, up to 5000 feet), extending even into the Piedmont and (in North Carolina) locally into the warm Coastal Plain, wherever rock outcrops occur. Associated here are *Pellia epiphylla* and *neesiana*, *Solenostoma crenuliformis*, *S. fossombronioides*, *S. hyalinum*, and in the mountains also *S. obscurum* and *Scapania undulata*. The plant occurs there under circumneutral to medium acid conditions, usually in humid sites where periodic submersion is the rule, or where kept permanently damp by spray. Very frequently the plants form thin, deep green patches over isolated rocks at the edge of the stream bed, occasionally partially obscured by silt. Associated then (e.g., in western Massachusetts) may be *Pellia epiphylla*, *P. neesiana*, and *Conocephalum conicum*.

The occurrences in the Arctic are under conditions identical to those characterizing subsp. *polaris* (which see).

Variation

Solenostoma pumilum subsp. *pumilum* is an extremely variable plant, especially in size and in perianth form. "Typical" *S. pumilum* is a plant of rock walls above the level where regular inundation occurs. Such plants are usually somewhat xeromorphic and small (rarely over 1 mm wide on fertile shoots), often blackish in color. They have erect and sheathing bracts. Perianths are usually characteristically fusiform (Fig. 273:4, 5), but in extreme cases they may be slender and gradually drawn out to a very thin, almost beaklike apex (as figured in Macvicar, loc. cit., p. 150, fig. 1); such extremes are rare, at least in our region.

The subspecies more often occurs under more permanently moist conditions, in dull or deep green patches, then becoming much larger (1.5–2 mm wide on sterile shoots; 2.0–2.2, even 2.5, mm wide on fertile plants and to 2 cm long); the transverse or subtransverse ♂ and ♀ bracts often spread rather widely; stems frequently develop a few stolons. Leaves may have slightly larger cells, although this is scarcely a constant feature. More distinctive is the change in form of the perianth which becomes larger, to 2 mm long × 900 μ wide, cylindrical-clavate (widest in distal third), and often very long-exserted beyond the bracts—at times, indeed, quite longly stipitate. The perianth, in such cases, is more obtusely plicate near the mouth, normally with an antical, longitudinal,

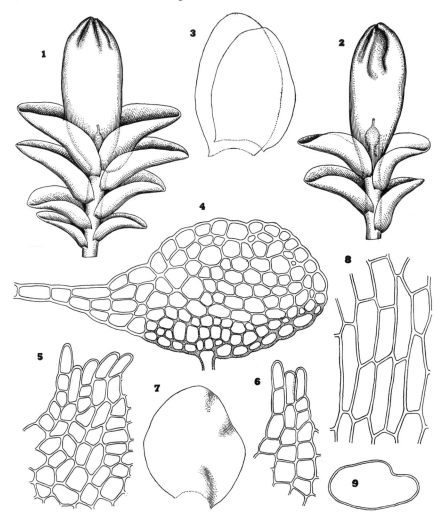

Fɪɢ. 274. *Solenostoma* (*Luridae*) *pumilum* subsp. *pumilum* ("*rivularis*" phenotype). (1–2) Fertile shoot tips, antheridia disintegrated at this stage (×18). (3) Two leaves (×27). (4) Mature stem in cross section (×212). (5–6) Sectors of perianth mouth (×300). (7) ♀ Bract (×27). (8) Dorsal cortical cells of stem (×285). (9) Cross section of upper one-third of perianth (×30). [All from *RMS 29440*, Neddie Cr., N.C.]

broad median sulcus, and is distinctly rounded to the apex (much as in *S. "schiffneri"*). Such hygric forms have been distinguished as "*Aplozia*" *pumila* var. *rivularis* Schiffn., or even separated specifically, as "*Aplozia*" *rivularis* Schiffn. The plants appear to be at best an inconstant form or perhaps an ecotype of more permanently moist sites. The imperceptible intergradation in perianth form is distinct from Figs. 273:4, 5 and 274:1, 2, in which perianth-bearing plants are illustrated. The more fusiform perianths are smaller (1750 × 780 μ) than the obovate-clavate ones (1950–2000 × 820–900 μ) and bear less-spreading, more-sheathing bracts. Macvicar (1926) has shown that the sporophytic characters used by Schiffner (1911, *in* 1901–43) are no more constant than the gametophytic features discussed above.

Differentiation

Solenostoma pumilum subsp. *pumilum* is separated from most entire-leaved taxa by the narrow, elliptical to oval leaves of sterile (and of lower portions of fertile) shoots; it is almost constantly fertile, with the paroecious inflorescences evident even after antheridia disappear (because 2–4 pairs of concave-canaliculate leaves below the perianth are subtransversely inserted and oriented, in contrast to the spreading and nearly flat, strongly oblique leaves below them; Fig. 273:4). It may be necessary, however, to dissect young fertile shoot apices to actually find archegonia and antheridia.

When paroecious inflorescences are not obvious (as on postmature plants), hygrophytic phases of *S. pumilum* ("var. *rivularis*") may be mistaken for the very similar *S. triste-atrovirens* complex. Indeed, except for the differences in sexuality, there seem to be no reasonably reliable features separating *S. pumilum* s. lat. from *S. atrovirens*.

Solenostoma pumilum subsp. *pumilum* is even more directly allied to *S. pumilum* subsp. *polaris*, differing from it essentially only in the following features: (*a*) the more fusiform perianths; (*b*) the narrower, elliptical-oval vegetative leaves, even of fertile shoots (Fig. 273:1–2, 4–5); (*c*) the somewhat larger leaf cells; (*d*) the usually larger outer cells of the capsule; but see p. 910. These differences are discussed more fully under *S. pumilum polaris* (pp. 923–924), where it is shown that none of them appears to allow a constant separation. Indeed, in much of the Arctic, under weakly to distinctly calcareous conditions, paro- to heteroecious plants exhibit a variable ensemble of characters that, in one direction, suggest *S. atrovirens*; in another, *S. pumilum* subsp. *pumilum*—and, in a third, *S. pumilum* subsp. *polaris*.[330]

---

[330] Quite typical of the broad spectrum of "intermediates" are paroecious plants from west Greenland (Tupilak I.; *RMS & KD 66-026*). These plants are an arctic phenotype with concave leaves, often spoon-shaped and broadly orbicular, little or no wider than the stem, broadly and obliquely inserted on it. Sterile plants correspond closely to *S. pumilum polaris* fo. *cavifolius* in aspect. However, perianths are long, pointed, and gradually narrowed in the distal 0.3–0.4. Cells contain usually 2–3, exceptionally 1 or 4, oil-bodies each, as in *polaris*.

Capsule characters are ambiguous: epidermal cells average (series of cells) 16.4, 17.4, 18.3, and 20.4 μ wide. Müller (*loc. cit.*, p. 184) states in his key that epidermal cells are 20 μ wide in *S. pumilum*, 12–14 μ in *S. schiffneri* (= *polaris*) (in the diagnosis, the widths are given as 18–22 μ for *S. pumilum*, 15–18 μ for *S. "schiffneri"*). However, all longitudinal walls and most or many transverse walls bear 1–2 or 3 strong thickenings, corresponding to Müller's diagnosis and fig. 280e for *S. pumilum*, but not for fig. 279g for *S. "schiffneri."* Spores are 19–21.5 μ, thus agreeing with the 17–22 μ in Müller for *S. pumilum*, but not with the 15–18 μ for *S. "schiffneri."* The bulk of characters (thickenings of capsule wall, spore size, perianth form) thus agree with *S. pumilum*, in which the plants are placed, but with clear recognition of the fact that any such placement is, in part, arbitrary.

SOLENOSTOMA (LURIDAE) PUMILUM subsp.
POLARIS (Berggr.) Schust., comb. n.

[Figs. 272: 6; 275–276]

Jungermannia pumila var. *polaris* Berggr., Kgl. Sv. Vetensk.-Akad. Handl. 13(7):98, 1875.
Jungermannia polaris Lindb., Öfversigt Kgl. Vetensk.-Akad. Förhandl. 23:560, 1876.
Aplozia schiffneri Loitlesberger, Verh. Zool.-Bot. Gesell. Wien 55:482, 1905; Schiffner,
  Hedwigia 48:184, figs. 1–13, 1909.
Aplozia polaris Bryhn, Rept. 2nd Norwegian Arctic Exped. in the "Fram" 1898–1902, 2(11):29,
  1906.
Haplozia schiffneri K. Müll., Rabenh. Krypt.-Fl. 6(1):570, fig. 285, 1909; Macvicar, Studs. Hdb.
  Brit. Hep. ed. 2:151, figs. 1–5, 1926.
Jungermannia schiffneri Evans, The Bryologist 20:21, 1917.
Solenostoma schiffneri K. Müll., Hedwigia 81:117, 1942; K. Müller, Rabenh. Krypt.-Fl., ed. 3,
  6:819, fig. 279, 1956.
Solenostoma (subg. *Luridaplozia*) *schiffneri* Schust., Amer. Midl. Nat. 49(2):405, pl. 31, figs. 1–3,
  1953.
Solenostoma polaris Schust. *in* Schust. et al., Natl. Mus. Canada Bull. 164:48, 1959.

In depressed, *dark green to blackish green or brownish black* patches, or scattered among other bryophytes (then, with crowding, sometimes suberect). Shoots very small, ca. (500–550)750–950(1100) μ wide × 4–10 mm long, usually prostrate with ascending apices, simple or sparingly furcately, terminally branched (the branches subtended by a narrowly ovate-lanceolate leaf, narrower than normal leaves), with age often with axillary intercalary branches. Stems somewhat flattened, elliptical in cross section, thick, rather fleshy, soft, (125–170)200–250 μ in width, the cortical layers deep brownish with age (medullary cells then yellowish), slightly or barely thickened, (15)17–21(23) μ wide × 23–45 μ long, rectangulate; medullary cells not or barely thick-walled, scarcely larger (22–25 μ) in diam.; some medullary cells with age brownish and mycorrhizal. Rhizoids long, hyaline to pale brownish, frequent to shoot apex. Leaves *remote on sterile stems*, to (near shoot apices) somewhat imbricate, *erect-spreading and slightly to strongly concave*, somewhat antically secund, on weak stems *often suborbicular*, on mature stems *broadly ovate to broadly ovate-cordate*, narrowly rounded at apex, from 450 μ long × 400 μ wide up to 480–520 μ long × 540–600 μ wide on mature shoots (*width-length ratio from 0.88: to 1.16:1*), but often broadly rotundate, rounded or retuse-truncate at apex on weak shoots. Cells of apex and margins *usually (12–14)15–17 μ*, rarely to 21 μ locally, *median cells (12–14)15–18 μ wide × (16)17–24(26) μ long*, hexagonal to oblong, *thin-walled* or somewhat thick-walled, *with minute to small trigones;* cuticle smooth to faintly striolate-verruculose; *oil-bodies (1)2–4(5) per cell*, subspherical to ovoid and ellipsoidal, large, 3.5 × 5–6 μ to 4.5 × 6 to 6 × 8–9(12) μ, almost homogeneous (appearing faintly granular under oil immersion), larger than chloroplasts (which are 3.6–4 μ in diam.).

Paroecious, sporadically autoecious. ♂ Bracts gradually larger toward shoot apex, in 2–4 pairs below ♀ bracts, 380–440 μ long × 420–450 μ wide (and sometimes larger), *ovate to broadly ovate*, subtransverse, strongly concave and somewhat erect, imbricate, antheridia 1–2 per axil, stalk biseriate; antheridial body ca. 100–110 μ in diam. × 120–125 μ long. ♀ Bracts similar in shape, somewhat larger, erect or suberect and sheathing perianth, their apices sometimes spreading. Perianth *ovoid to oblong-obovoid*, occasionally obovoid-clavate,

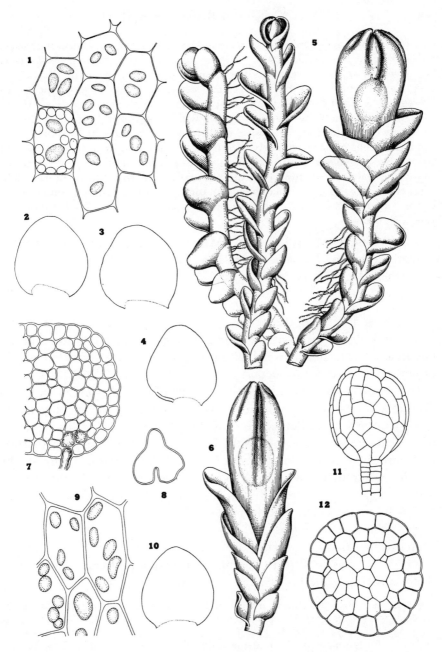

FIG. 275. *Solenostoma* (*Luridae*) *pumilum* subsp. *polaris*. (1) Median cells with oil-bodies and, lower left, chloroplasts (×670). (2–3) Mature sterile leaves (×35). (4) ♂ Bract (×44). (5) One gynoecial and two sterile plants, mod. *parvifolia* (×30). (6) Paroecious shoot apex (×25). (7) Stem cross section (×200). (8) Cross section, distal half of perianth (×23). (9) Median cells with oil-bodies (×625). (10) Upper, subinvolucral leaf (×35). (11) Antheridium (×ca. 208). (12) Seta cross section (×142). [1–8, *RMS 35051*, Alert, Ellesmere I.; 9–11, *RMS 35063*, same loc.]

scarcely dorsiventrally compressed, rather abruptly narrowed to apex, widest at or above middle, ca. 600–700 μ wide $\times$ 1200–1350 μ long, 0.5–0.65 emergent, *with a deep, long, prominent antical furrow* extending often to near base, plicate in distal 0.25–0.35; mouth narrowed but *not pointed*, crenulate with finger-shaped cells averaging ca. 12–13 $\times$ 25 μ; middle of perianth with cells somewhat thick-walled, rectangulate, averaging 15–18 μ wide $\times$ 25–32 μ long. Sporophyte with seta ca. 5 mm long. Capsule dark brown, subspherical to short-ovoid. Epidermal cells rather irregularly rectangulate, 12–14(15) μ wide, both faces of alternating *longitudinal walls generally with 2–4 nodular thickenings,* 0–2 on the transverse walls, intervening longitudinal walls often with few or none. Cells of inner layer ca. 10 μ wide, narrowly rectangular, with usually 4–6 semiannular bands. Elaters 7–8 μ in diam., bispiral; spores 15–18 μ, finely papillate, yellowish brown.

Type. Sommebay, Spitsbergen (Musci Spitsberg. Exsic. no. 176, *leg.* Berggren), type of *J. pumila* var. *polaris.* The type of *J. schiffneri,* Gorizia, in northern Italy (*Loitlesberger, 1903*).

Distribution

A local and, south of the Arctic, generally rather rare taxon, of largely arctic and alpine distribution, found throughout much of the Tundra and in at least the northern half of the Taiga. Evidently restricted largely or entirely to calcareous regions.

In Europe of sporadic distribution, ranging from Spitsbergen and from Scandinavia (Norway, Finland) to Great Britain (rare, in only a few localities, north from Perthshire to Killin, Scotland), recurring in alpine and subalpine situations southward to France, Switzerland, North Italy (the earlier "Austrian coastal region"), and the Pyrenees. Also in the USSR (Waigatsch) and Jan Mayen I. Also probably in Siberia; at least Arnell (1913) admits that plants he refers to *Aplozia tristis* are "kritisch . . . vielleicht richtiger zum Teil zu der nahestehenden *A. polaris* (Lindb.) zu bringen sein."

In North America evidently rare or of sporadic distribution, known from alpine sites in British Columbia (Brinkman; Evans, 1917e) and doubtfully from Alberta (Brinkman, 1934).

ELLESMERE I. Ravine cutting into Dumbbell Bay, Alert, NE. coast, at 82°30′ N. (*RMS 35051, 35063, 35066a*); point just E. of Colan Bay, 2.5–3 mi W. of Cape Belknap, ca. 82°33′ N., 62°35′ W. (*RMS 35281b, 35295, 35296, 35201, p.p.,* among *Athalamia hyalina*); also common elsewhere in the region of NE. Ellesmere, as at head of Parr Inlet, foot of Mt. Pullen, foot of The Dean, mouth of Parr Ravine (*RMS 35072, 35141, 35150, 35167, 35252,* etc.); S. end of Hilgard Bay (*RMS 35229, 35225b, 35235b 35246,* etc.); "Lake Anxiety," SE. side of Hilgard Bay (*RMS 35936*); see Schuster et al., 1959. Beitstad Fjord, Twin Glacier Valley, Cape Rutherford, and Fram Harbor, all in Hayes Sound region; Harbour, South Cape, and Goose Fjords, all on S. coast (Bryhn, 1906; Polunin, 1947, as *Haplozia* or *Jungermannia polaris*).

N. GREENLAND. Lemming Fjord, Cap Emory, 82°53′ N. (*Th. Wulff, June 17, 1917!*; publ. by Hesselbo, 1923a, as *S. atrovirens*); Low Pt., 83°06′ N. (*Th. Wulff, June 10–13, 1917!*; publ. by Hesselbo, 1923a, as *J. subelliptica*); Cape Benet, 83°02′ N. (*Th. Wulff, June 5, 1917!*; publ. by Hesselbo, 1923a, as *S. atrovirens*); E. coast of Peary Land, Herslufsholm Strand, 82°40′ N., ca. 21° W. (*Holmen 7581, p.p.*); Heilprin Land, Brønlund Fjord, 82°10′ N., 31° W. (*Holmen 6749e!*). NW. GREENLAND. Kangerdlugssuak, Inglefield Bay (*RMS 45592*); Kânâk, Red Cliffs Pen., Inglefield Bay (*RMS 45601*); Igdlorssuit, Bowdoin Bugt, Inglefield Bay (*RMS 45779b*). W. GREENLAND. Sondrestrom Fjord, sunny S.-facing slopes near air strip, at head of Fjord (*RMS & KD 66-041a*); Tupilak I., W. of Egedesminde (*RMS & KD 66-026*, questionable); Magdlak, Alfred Wegeners Halvö (*RMS & KD 66-1357a*); Eqe, near Eqip Sermia (glacier) ca. 1 mi from Inland Ice, 350–450 m (*RMS 66-290b*; see under fo. *cavifolius* Schust.); mouth of Blaesedalen ved Lyngmarksfjeld, Godhavn, in and near ravine running into Rodeelv (*RMS & KD 66-280 ± trans. ad polaris*; *66-283a, 66-286*); Kangerdlugssuakavsak, head of Kangerdluarssuk Fjord (*RMS & KD 66-1288*); E. side of Hollaenderbugt, NW. Nugssuaq Pen., slopes below Qarssua (*RMS & D 66-1507b*); Godhavn, Claushavn, Lerbugten, Tasiusak, Sarpiusat (Lange & Jensen, 1887; based on Berggren, as *J. pumila* var. *polaris*). E. GREENLAND. Røde-Ø′s Vestkraent, Scoresby Sund (Jensen, 1897). NEWFOUNDLAND. Squires Mem. Pk., Humber R. (*RMS 68-1646*).

QUEBEC. Pointe de la Traverse, Lac-St.-Jean (Lepage, 1944–45); The Grotto, Mt. Ste-Anne, Percé, Gaspé (*RMS 44010*); Mt. Ste-Anne, Percé (*Williams 10836!*; typical). WISCONSIN. Bark Pt., Bayfield Co. (*Conklin 1852*); Monahan Falls, (*Conklin 1590*), Black R., Amnicon Falls, and Brule R., all in Douglas Co.; Siskiwit R. and Squaw Pt., Bayfield Co.; Montreal R., Iron Co. (Conklin, 1929). MINNESOTA. Grand Marais, Cook Co. (*Conklin 3010*);[331] Lower Falls of Manitou R. (*RMS 18091*); French R. and Fairmount Park, near Duluth, St. Louis Co. (*Conklin 1777, 2052*). MICHIGAN. Copper Harbor, Keweenaw Co. (Steere, 1937).

Doubtfully reported from Pennsylvania (McKean Co.) by Lanfear (1933). This report, I think, is based on the *"rivularis"* phase of subsp. *pumilum*.

The North American distribution of this taxon is in some respects strikingly similar to that of *Cryptocolea imbricata*. Both taxa are calciphytes that recur around the Lake Superior region as rarities and then become characteristic and abundant components of the hepatic flora in the Arctic, in Greenland, and in northern Ellesmere I. Steere (1937) erroneously considers this taxon to show a "cordilleran" distribution. The ubiquitous distribution on the highly calcareous sediments of northern Ellesmere I. (and presumably equally common range elsewhere in calcareous sites in the Arctic) clearly shows that the taxon is arctic-alpine, with disjunct stations in the suitable "microtundra" zones around Lake Superior.

Ecology

Solenostoma pumilum subsp. *polaris* is the commonest and most variable member of the genus in the Arctic, with corresponding wide ability to undergo ecesis in the most diverse sites. The plant is almost "weedy" on sites where a dense luxuriant vegetation cannot (or has not yet) succeeded, often forming small, diffuse colonies on difficult or temporary sites.

---

[331] Cited, erroneously, as from "Grant Moraine" in Frye & Clark (*loc. cit.*, p. 290).

Frequently as isolated, almost indeterminable, straggly sterile stems (which have been traditionally assigned to *"Jungermannia atrovirens"*)—but whenever such plants occur with more mature ones these are clearly monoecious.

The following four general sites merely suggest the perimeters of toleration of this taxon. In general, most frequent in basic or circumneutral sites, becoming distinctly rare or lacking in areas with only acid, igneous rocks. Typical occurrences are as follows. (1) In the Tundra of Ellesmere I., common over snow-fed, damp slopes, on clayey talus slopes (basic in nature), together with *Arnellia fennica*, *Cephaloziella arctica*, *Scapania gymnostomophila*, *Blepharostoma trichophyllum brevirete*, *Athalamia hyalina*, *Lophozia collaris*, *L. gillmani*, *L. heterocolpa*, *L. quadriloba*, *L. alpestris polaris*, *L. pellucida*, *Sauteria alpina*, *Aneura pinguis*, *Cryptocolea imbricata*, and *Preissia quadrata*. Also common in the crevices of low-lying soil-polygon fields and on clay-shale solifluction slopes, associated with many of the previous species, and also *L. badensis*, *L. hyperarctica*, and *Scapania praetervisa polaris*. Usually occurring in small quantity, forming olive-green to greenish brown or blackish, loose patches, or straggling among mosses. It is almost always fertile. Associated tracheophytes under such high arctic conditions range from *Saxifraga oppositifolia* and *caespitosa* to *Salix arctica*, *Eriophorum* spp., *Carex* spp., and sometimes *Equisetum variegatum* (the last-named at its northernmost stations!). (2) In northwest Greenland sometimes on peaty-sandy moist ground near the shore of Inglefield Bay, with the annual angiosperm *Koenigia islandica*, and with *Cochlearia officinalis*, associated with *Polytrichum* and *Pohlia*, as well as *Cephaloziella arctica*, *Solenostoma pusillum*, *Nardia geoscyphus*, and *Lophozia excisa*. (3) In the same area also in irrigated, peaty grass-*Eriophorum* meadows, with *Drepanocladus*, *Distichium capillaceum*, *Fissidens*, *Scapania gymnostomophila*, *Lophozia quadriloba*, *Odontoschisma macounii*, etc. (4) In the Gaspé and Great Lakes area chiefly over basic intrusive cliffs or on damp talus slopes below them, often in the spray zone of waterfalls (as below Manitou Falls), then associated with such calciphiles as *Scapania gymnostomophila*, *Lophozia badensis*, *Preissia quadrata*, as well as *Pellia epiphylla*.

Variation and Differentiation

This rather minute plant is apt to be mistaken on one hand for *S. pumilum* subsp. *pumilum*, on the other for *S. atrovirens*. It resembles the former in the paroecious inflorescence, the latter in leaf shape and form of perianth and bracts. All three taxa are equally small, generally 1 mm wide or less, with a dark green to brownish or blackish green color, and with obliquely inserted, often quite concave leaves.

Usually S. pumilum polaris can be separated from *S. pumilum pumilum* by the shape of leaves and perianth. On sterile shoots the leaves vary from subrotundate (or broadly heart-shaped, on some compact arctic phases) to broadly ovate, with the greatest width below the middle; they range from ca. 0.8–1.2 × as long as wide; they are often nearly transversely oriented and quite concave. By contrast, on sterile stems of *S. pumilum pumilum* the leaves average 1.25–1.4 × (or even 1.5 ×) as long as wide, are nearly elliptical in shape, and are widest in the middle; they usually spread horizontally and are slightly concave. Only on the upper leaves of fertile stems does *S. pumilum pumilum* produce ovate leaves. In *S. pumilum polaris* mature leaves, and those from fertile shoots, are often slightly heart-shaped at the base; this is never the case in subsp. *pumilum*. The perianth in subsp. *polaris* is ovoid or even slightly obovoid-clavate, widest at or somewhat above the middle, and rather suddenly contracted to the mouth (its width ⅕ from the apex is almost or fully equal to the median width); it is always longly and deeply sulcate dorsally. In subsp. *pumilum* perianths are typically spindle-shaped (although ovoid to ovoid-clavate in the large, subaquatic phases, "var. *rivularis*"), and gradually narrowed to the somewhat pointed, strongly plicate mouth. Other differences between the two taxa lie supposedly in the leaf cells and epidermal cells of the capsule wall. In subsp. *polaris* median leaf cells average 15–18 μ wide × 17–24 μ long; in subsp. *pumilum* they average 23 × 30 μ or larger. Epidermal capsule cells in subsp. *polaris*, according to both Müller and Schiffner, are only 12–15 μ wide, but bear 2–4 nodular thickenings per alternating longitudinal wall and 0–1, occasionally 2, on transverse walls; in subsp. *pumilum*, epidermal cells are supposedly ca. 20 μ wide and bear (1)2 nodular thickenings on almost all longitudinal walls and only 0–1 thickening on transverse wall; see, however, p. 910. Also subsp. *polaris* (with spores 15–18 μ) is somewhat smaller-spored than subsp. *pumilum* (spores 16–20 μ). In spite of this set of differences, the two taxa occasionally become impossible to separate.

A distinction between the two becomes almost impossible to maintain in plants from the high Arctic (where the *S. polaris-pumilum* complex is, in my experience, notoriously variable!). For *RMS 35731* (southwest of Mt. Pullen, Ellesmere I.) it is almost impossible to state to which of the two the material should be relegated. In these plants the perianths are narrowed into a distinctly conical apex and leaves of robust plants may have marginal cells as high as 21–23 μ and median cells 21–24 × 23–28 μ, thus showing a close approach to "*S. pumilum.*" Yet, in the broadly ovate to subrotund leaf shape, in the leaf cells of the least robust specimens (there 17–18 μ on the margins; 14–18 × 16–23 μ medially), and in the occasional perianths which remain subequally broad to within the distal ¼ and which are obscurely conically narrowed, many of the plants clearly approach "*S. polaris*," and in extreme instances the following fo. *cavifolius*. The sexuality of these plants, like those of other high arctic phases of the complex, is variable. Almost all plants are paroecious, but occasionally perianth-bearing shoots have a single ♀ bract with one antheridium, all subtending leaves being sterile, and in a few cases ♀ shoots totally lack antheridia below. In isolated instances of this type, long, distinctly spicate ♂ branches could be demonstrated, in one case with up to 12–14 pairs of saccate bracts. Although plants of this collection (and of several similar collections) are referred to subsp. *pumilum*, this is done partly because the few capsules show clearly the characters of this taxon.

In the Arctic, *S. pumilum polaris* and *S. atrovirens* are sometimes impossible to distinguish if the inflorescences (paroecious and dioecious, respectively) cannot be ascertained. The highly modified arctic phases of these two taxa often are scarcely typical. Usually *S. atrovirens* is separable on the basis of the larger leaf cells. In *S. pumilum polaris* much of the material seen (i.e., Ellesmere I., *RMS 35051*) had slender sterile shoots, only 450–600 μ wide, with distant, suberect, concave, reduced, and broad leaves. Leaves, in such cases, were broadly rotundate, inserted by a very broad line, and totally failed to show the "normal" ovate or ovate-cordate shape usually assigned to "*S. schiffneri*." Only markedly robust and fertile plants possess leaves of the usually broadly ovate form.

In the case of such arctic extremes, clear-cut differentiation probably will have to depend upon a definite determination of the sexuality of the plant. Even this "ultimate" criterion tends to break down. I have seen individual plants in a number of Greenland and Ellesmere I. collections in which the gynoecia were not subtended by ♂ bracts, and the plants gave every evidence of being dioecious, yet occurred mixed with paroecious plants. In other cases, isolated plants were seen with gynoecia on rather short, lateral branches, without antheridia in the axils of the leaves below them, but the same plant bore an androecial branch, compactly spicate, formed of 4–6 or more pairs of bracts. Such autoecious plants are rarely produced. This high level of variability in development and location of ♂ bracts suggests that the determination of a paroecious inflorescence in this species at times entails dissection of several young inflorescences.

Variability in sexuality, although considerable, never ranges to the predominantly dioecious. The sporadically autoecious (and rarely perhaps dioecious) plants found could easily cause confusion with the otherwise exceedingly similar *S. atrovirens*.

In this conjunction "*Haplozia polaris*" [*Jungermannia pumila* var. *polaris* Berggr., Kgl. Sv. Vetensk.-Akad. Handl. 13(7):98, 1875 (= probably *J. polaris* Lindb., Öfversigt Kgl. Vetensk.-Akad. Förhandl. 23:560)] must be considered.

This "species," known from Greenland and Spitsbergen and more recently reported from numerous stations in Ellesmere I. (Bryhn, 1906, p. 29), is considered a synonym of *Jungermannia* (or *Solenostoma*) *atrovirens* by both Stephani (1902, *in* 1898–1924, p. 519) and Müller (1909, *in* 1905–16, p. 567). The interpretations of Stephani and Müller, however, are wrong; type plants I have seen were paroecious. It must be noted that Berggren (1875) considered his plant to be a mere variety of the paroecious *S. pumilum*, the implication being that "*S. polaris*" thus was also paroecious. Part of the confusion may have resulted from the fact that shoots with mature perianths usually fail to show antheridia in the subfloral leaves, these having disintegrated. However, plants with immature perianths (at and just beyond fertilization), which usually show antheridia, or remnants of them, in the subfloral leaves, are almost always to be found. *If* the antheridia are overlooked, *S. pumilum* subsp. *polaris* is quite likely to be considered dioecious and mistaken for *S. atrovirens*.[332]

Some high arctic phases not only differ in the variability of the inflorescences but also show differences in leaf shape, probably warranting separation as a distinct form, as follows:

SOLENOSTOMA PUMILUM subsp. *POLARIS* fo. *CAVIFOLIUS* (*Schust.*) Schust., comb. n.

[Figs. 275–276]

Solenostoma polaris fo. *cavifolius* Schust. *in* Schust. et al., Natl. Mus. Canada Bull. 164:51, 1959.

Green to fuscous brown, small (shoots from 450–650 to 900 μ wide × 5–8(12) mm long). Stems flexuous, 150–190(210) μ in diam., ca. 10–12 cells high, green to (with age) brownish; cortical cells thin-walled or weakly thick-walled, rectangulate, 16–21(23–25) μ wide × (32)45–72(80) μ long dorsally and laterally, 14–16 μ wide ventrally, scarcely differing in diam. from medullary

---

[332] I have studied only a single specimen of authentic "*Jungermannia pumila* var.? *polaris* Berggren," from Sommebay, Spitsbergen (*Berggren 176*, 1868; Musci Spitsberg., Esxic. no. 176!). This material deserves to be considered the basis for var. *polaris*, although the species *polaris* Lindb. is based on Greenland material (Berggren, 1870). The Spitsbergen material is wholly identical in every respect and habitually indistinguishable from the Greenland and Ellesmere I. plants that I refer to subsp. *polaris*. Furthermore, the capsules of *Berggren 176* show the following features considered "typical" for *S.* "*schiffneri*." Epidermal cells, in the middle of the capsule, bear 2–4 strong nodular thickenings on each face of alternating longitudinal walls. Walls alternating with those bearing strong thickenings either lack them or bear (on isolated walls) 1 or rarely 2 thickenings. The cells average 13–15 μ wide. Away from the central regions of the valves this condition is much less marked, and cells become irregular in size. As a consequence, I must consider the two taxa, *S. schiffneri* and *S. polaris*, identical. It was easily possible to demonstrate a paroecious inflorescence in the Spitsbergen plants, and (as in the Ellesmere plants) perianths are occasionally somewhat conoidally narrowed in their distal thirds. Furthermore, exactly as in the brownish to brownish black sun forms collected in Ellesmere I., the perianth mouth is commonly decolorate. There is, therefore, no doubt about the identity of the Spitsbergen plants, which may be considered "typical" *S. polaris*, and *S. schiffneri*. Certainly, leaf and bract shape, cell size, and form of epidermal cells of the capsule all suggest that *S.* "*polaris*" is much closer to current concepts of *S. schiffneri* than to *S. pumilum* subsp. *pumilum* or *S. atrovirens* (to both of which it has been variously assigned).

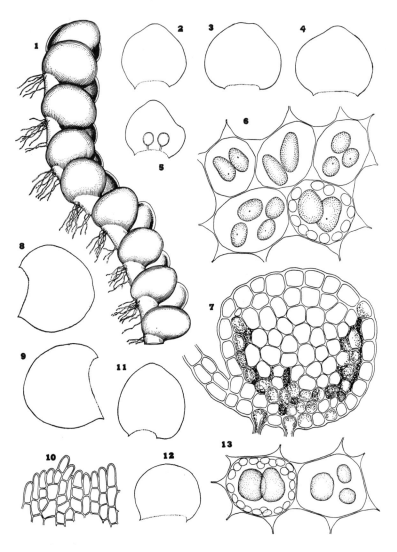

Fig. 276. *Solenostoma* (*Luridae*) *pumilum* subsp. *polaris*, fo. *cavifolius*. (1) Sterile shoot in lateral profile (×29). (2) ♂ Bract (×29). (3–4) ♀ Bracts (×29). (5) Diandrous ♂ bract (×29). (6) Median cells with oil-bodies and, lower right, chloroplasts (×ca. 1085). (7) Mature stem, in cross section, with maximal mycorrhizal development (×270). (8–9) Leaves of concave-leaved extreme (×50). (10) Mouth of immature perianth (×190). (11) ♀ Bract of ovate extreme (×25). (12) Leaf from same plant as in fig. 11 (×25). (13) Two median cells, with oil-bodies and, at left, chloroplasts (×925). [1–10, from between Parr Inlet and Mt. Pullen, Alert, Ellesmere I.; 11–12, *RMS 35079b*; 13, *RMS 35408*, all Ellesmere I.]

cells. Leaves contiguous to closely *imbricate, suberect to erect and slightly antically secund, moderately to strongly and sometimes almost hemispherically concave, broadly rotundate* and averaging slightly to greatly broader than long (ca. 350–375 μ long × 440–460 μ wide up to 525–560 μ long × 600–725 μ wide), entire-margined; insertion at ca. a 45° angle, the antical and postical leaf bases scarcely or short decurrent. Rhizoids in a brownish fascicle from a swollen area at base of postical leaf insertion, *few or virtually absent elsewhere,* 9–10 μ in diam. Cells averaging isodiametric marginally and distally, slightly elongated below leaf middle, with smooth cuticle and *small to large but concave to straight-sided trigones* (rarely faintly bulging); *marginal cells (12)13–17(18) μ; median cells (12)14–16(17–18) μ wide × 16–19(21) μ long,* the basal cells slightly larger; *oil-bodies 2–3(4) per cell,* short ovoid or ellipsoidal, very large, from 4 × 6 to 4.5 × 9 μ up to 6 × 7.5–10 μ, a few spherical and 4–5 μ, *appearing almost homogeneous* (under oil immersion formed of minute, scarcely perceptible spherules); chloroplasts smaller, ca. 3.5 μ long.

Paroecious, but occasionally dioecious or autoecious. ♂ Bracts in 1–2(3) pairs below ♀ bracts, strongly concave and more or less imbricate, subtransversely oriented, broadly ovate or ovate-orbicular, not or weakly cordate at base, widest below the middle, 580–600 μ wide × 480–500 μ long, little larger than subtending leaves, similar to them in shape but some with a rounded, obtuse, ill-defined antical lobe or angulation; usually 2-androus. Antheridia subspherical, ca. 95–100 μ in diam. when mature; stalk biseriate. ♀ Bracts broadly ovate-orbicular to ovate-cordate, usually much wider than long, ca. 700 μ wide × 540–560 μ long or slightly larger, widest in basal third, relatively narrowly rounded distally, occasionally as long as wide and 725 μ long × wide, then ovate-cordate. Perianth 4–5-plicate, at mouth narrowed; mouth crenulate with short, fingerlike cells free only at their ends and somewhat evenly thick-walled, 11–13.5 μ wide × 22–32(35) μ long, averaging 2–3× as long as wide, somewhat longer than the cells below them.

Type. Moist humus, steep, snow-fed west slope of The Dean, Alert, Ellesmere I. (*RMS 35408; c. caps.*).

Distribution

NE. ELLESMERE I. Damp, calcareous tundra slopes, between Dumbbell Bay and The Dean (*RMS 35050d*); above Parr Inlet, 250–300 ft, S. of Alert (*RMS 35070b*); foot of Mt. Pullen (*RMS 35133*); calcareous solifluction slope, head of Parr Inlet, S. of Alert (*RMS 35144*); valley between Mt. Olga and Mt. Erica, ca. 850–950 ft, ca. 5 mi SW. of end of Hilgard Bay (*RMS 35548b*); weakly calcareous weathered slopes, E. edge of U.S. Range, 9–10 mi due W. of Mt. Olga (*RMS 35999b, c. per.*); The Dean (*RMS 35175*); N. of Mt. Pullen (*RMS 35368a, c. caps.*); snow-fed slope of The Dean (*RMS 35408*). W. GREENLAND. Eqe, near Eqip Sermia (glacier), ca. 1 mi from ice cap, 350–450 m (*RMS & KD 66-290b*).

Described as a distinct form of *S. pumilum* subsp. *polaris* largely because it appears so deviant in aspect as to be more probably mistaken for other species in subg. *Solenostoma* (such as *S. sphaerocarpum*) than for *S. pumilum* subsp. *polaris*. Indeed, I was at first convinced that a distinct species was at hand. When "typical" shoots of fo. *cavifolius* are at hand, confusion is hardly possible. The leaves of such plants are very broad-based and not at all cordate; they are strongly transverse and as much as 725 μ wide but only 525 μ long (width 1.3–1.4 × the length!). Yet, in northeastern Ellesmere I., these plants can be found intergrading with more nearly typical phases of subsp. *polaris*. The extreme variation in leaf shape and the equally marked variation in shape of the ♂ and ♀ bracts, together with the often disturbing production of seemingly unisexual shoots, make recognition of the *S. pumilum-polaris* complex, in the American Arctic, difficult. It is possible that earlier reports of *S. sphaerocarpum* var. *nana* and *S. atrovirens* var. *gracilis* from Ellesmere I. may prove referable to this form.

Müller (1951–58) still refers "*Haplozia polaris*" to *Solenostoma atrovirens*, stating that the "var. *polaris*" represents an arctic phase, with more circular leaves, and that "die für *Haplozia polaris* als charakteristisch angegebene Reihe quadratischer Blattrandzellen kommt auch beim Typus (i.e., *Solenostoma atrovirens*!) vor." It is thus possible that three extremes should be recognized within *S. pumilum* subsp. *polaris*, the extreme arctic phase here assigned to fo. *cavifolius;* typical *S. pumilum* subsp. *polaris;* and a nonarctic phase with the rounded-ovate to ovate leaves of "*S. schiffneri.*" In any event, the extreme phase with concave, broader-than-long leaves, which is described at length above, can hardly be referred to typical *S. pumilum* subsp. *polaris*.

The fo. *cavifolius* occurs under much the same conditions as typical subsp. *polaris*. It was found fruiting from July 1 to 27. Plants occurred under a variety of conditions, and one cotype (*RMS 35050*) was found exposed as early as June 3, on a rocky-clayey, gentle slope at the edge of an intermittent stream, on the first day of the season with any perceptible snow melt. Associated were *Scapania gymnostomophila*, *Arnellia*, *Blepharostoma*, and a series of mosses, among them *Hypnum revolutum*, *Tortula ruralis*, *Ditrichum flexicaule*, *Distichium capillaceum*, *Barbula icmadophila*, *Timmia austriaca*, *Mnium hymenophyllum*, *Anoectangium tenuinerve*, and *Philonotis fontana*.

Differentiation

The rotundate, broad-based leaves, the sometimes distinctly collenchymatous cells, the absence of a "border" of differentiated leaf cells, and the large and almost homogeneous oil-bodies may suggest species such as *Solenostoma* (*Solenostoma*) *sphaerocarpum*, *S. levieri*, and *S.* (*Plectocolea*) *hyalinum*. The much smaller leaf cells serve to differentiate fo. *cavifolius* from all three of these species; the brownish rhizoids, occurring mostly in fascicles from the stem, at the juncture with the postical leaf base, the smaller leaf cells, and the paroecious inflorescences separate it from *S. hyalinum*. This leaves only *S. sphaerocarpum* and *levieri* to consider, with which fo. *cavifolius* agrees in the paroecious inflorescences. The much larger leaf cells at once eliminate *S. levieri* from consideration. However, *S. sphaerocarpum* and such arctic and alpine extremes as "var. *nana*" bear considerable resemblance to our plant, although rarely possessing as strongly concave leaves. From *S. sphaerocarpum* the present taxon differs in smaller cells, with the marginal 12–17 μ (18–22 μ in *S. sphaerocarpum*) and the median only 12–16 × 16–21 μ (ca. 25 × 35 μ in *S. sphaerocarpum*). Also, *S. pumilum polaris* fo. *cavifolius* possesses only 2–3, less often 4, oil-bodies per cell, whereas *S. sphaerocarpum* has usually 6–9.

SOLENOSTOMA (LURIDAE) ATROVIRENS (Schleich.) K. Müll.

[Fig. 277]

Jungermannia atrovirens Schleich. ex Dumort., Syll. Jungerm. Eur., p. 51, 1831; Frye & Clark, Univ. Wash. Publ. Biol. 6(2):290, 1944; Amakawa, Jour. Hattori Bot. Lab. no. 22:50, figs. 2e, 4j, 30:f-1, 1960.
Jungermannia sphaerocarpoidea De Not., Mem. Accad. Torino 2(18):493, 1859.
Aplozia atrovirens Dumort., Hep. Eur., p. 63, 1874.
Jungermannia pumila var. *sphaerocarpoidea* Massal., Aħn. Ist. Roma 2(2):1886 (ref. not seen; *fide* Müller, 1951–58).
Haplozia atrovirens K. Müll., Rabenh. Krypt.-Fl. 6(1):563, 1909.
Jungermannia claviflora Steph., Spec. Hep. 6:84, 1917.
Solenostoma atrovirens K. Müll., Hedwigia 81:117, 1942, and Rabenh. Krypt.-Fl. ed. 3, 6:817, fig. 278, 1956.

In small patches, creeping to ascending (with crowding suberect), green to olive-green or *becoming blackish green*,[333] *minute* (*shoots mostly 750–950 μ wide × 2–8 mm long*), distantly branched; branches either terminal (subtending leaf then narrowly ovate) or intercalary from the leaf axils. Stems ca. 100 μ in diam., creeping, ascending at apices. Rhizoids frequent to numerous, to stem apex. Leaves oblique, on sterile shoots, almost transverse on upper parts of fertile shoots, approximate to imbricate, mostly erect-spreading to (on upper parts of shoots) erect, somewhat dorsally secund, ovate or ovate-oval, widest just above the base (250–350)400–650 μ wide × (350–425)500–750 μ long, rounded at apex, *distinctly concave*, the upper strongly so. *Cells small;* marginal, (10)12–15 μ; median, ca. (14)17–22 × 20–30(35) μ and oblong to hexagonal-rectangulate, trigones small, walls thin to slightly thickened, *usually strongly pigmented; cuticle striate-verruculose.* Asexual reproduction absent.

Dioecious.[334] ♂ Plants slenderer than ♀ plants, often in separate patches;

[333] The species name refers to the blackish green color of the plant. It seems therefore singularly inappropriate to separate *S. atrovirens* on the basis of the "green to olive-green" color vs. the "blackish" color of *S. oblongifolium*, as Frye and Clark have done (1937–47; key, p. 277). No constant color difference occurs between these two taxa.

[334] Frye and Clark (1937–47, p. 291) describe the plant correctly as unisexual. Yet their illustration (fig. 4), supposedly of this species, in which the plant is paroecious, is labeled "Shoot with perianth and male bracts." The illustration, together with several others, is attributed to Pearson; it clearly represents *S. pumilum*! The two figures attributed to Hooker (1816) in Frye & Clark (*loc. cit.*, figs. 7, 10) also appear to pertain to *S. pumilum*. At least fig. 7 (p. 291), of a calyptra and young sporophyte, reappears on p. 286 (fig. 2) under *S. pumilum*, this time incorrectly tabeled "Perianth and young sporophyte."

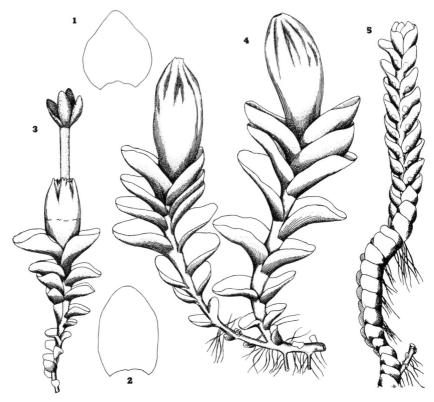

FIG. 277. *Solenostoma (Luridae) atrovirens.* (1–2) Leaves (×20). (3) plant with sporophyte (×15). (4) Fertile plant of the larger phenotype ("var. *sphaerocarpoidea*") (×20). (5) ♂ Plant of "var. *sphaerocarpoidea*" (×ca. 18). [All from K. Müller, 1905–16.]

bracts in at least 6–10 pairs, eventually becoming intercalary, contiguous to imbricate, erect but with the rounded apex ± spreading, strongly concave almost to apex; 1–2-androus. Gynoecia with bracts ovate to broadly ovate-oval, larger than leaves, semiamplexicaul, erect-spreading to spreading, except for the somewhat concave base. Perianth to 0.65 emergent, almost cylindrical but slightly dorsiventrally flattened, slenderly oblong-ovoid to clavate, smooth in basal half, the apical portion deeply 5-plicate, strongly narrowed but not or hardly tubular at mouth; mouth lobulate, the shallow lobes denticulate; cells at mouth ca. 15 × 20 μ. Capsule purplish brown, nearly spherical with seta 2–5 mm long. Epidermal cells with nodular thickenings.[335] Elaters ca. 8 μ in diam.; closely 2- (rarely 1-) spiral; spores (10)12–16(17) μ, finely papillate, reddish brown.

Type. Above Lausanne, Switzerland (*Schleicher*).

[335] Frye and Clark state incorrectly that the inner cell layer of the capsule wall is formed of cells larger than the epidermal cells, and that these cells bear spiral thickenings.

Distribution

Solenostoma atrovirens has a somewhat puzzling distribution. It is largely arctic-alpine and subalpine and appears to be holarctic in range. In Europe reported from Spitsbergen and northern Scandinavia (Sweden; Norway; Faroes) southward into alpine situations in Great Britain and in central Europe (southern Germany, Austria, Switzerland), southward occurring into the montane northern provinces of Spain (Casares-Gil, 1919). Also reported from Asia (Siberia) and from Hokkaido and Honshu, Japan (Amakawa, 1960a).

The North American distribution confused but possibly transcontinental at high latitudes, reported in the West from Alaska to British Columbia (Golden and Hector, *Brinkman;* Mt. Queest and Shawnigan Lake, *Macoun;* Shushwap Lake, *Conklin*); Washington (Aberdeen, *Foster*); Oregon (Silver Creek, Marion Co., *Foster*); eastward to Alberta, Idaho, Montana, and Colorado.[336]

In eastern North America reported from only a few sites, as follows, but possibly of wider range, at least in the Tundra.

E. GREENLAND. Danmark's Ø, Scoresby Sund (Jensen, 1897).[337]   ELLESMERE I. South Cape and Goose Fiords (*Simmons*, ex Bryhn, 1906; supposedly the typical plant); near Ft. Juliane and Beitstad Fiord, Hayes Sd. region; Goose Fiord, S. coast; Lands End, W. coast (all "var. *gracilis*"; *Simmons* ex Bryhn, 1906).

Ecology

Usually on damp basic rocks or in moist tundra, mostly over basic rocks. The species occurs at the edges of rivulets and streams and on wet, usually basic or circumneutral rocks. Generally alpine, only occasionally arctic.

Differentiation

This species is very likely to be mistaken for *S. pumilum s. lat.*, which it resembles in the minute or small size, the often blackish green color, and general leaf shape. The more elliptical leaves of typical *S. pumilum* and the distally somewhat tubularly narrowed, spindle-shaped perianths of that species in general are sufficient to separate it from all phases of *S. atrovirens*.

[336] Reported from a number of other western stations by Frye and Clark, on the basis of collections by Frye and by Svihla. These are here omitted since Frye and Clark evidently confused *S. pumilum* with *S. atrovirens*, thus making their reports valueless.

[337] There are many other reports of *S. atrovirens* from Greenland (e.g., Hesselbo, 1923a, from Cape Benet and Lemming Fjord in north Greenland; these represent *S. pumilum* subsp. *polaris* and both plants are definitely paroecious!). No Greenland specimen that I have seen is safely referable to *S. atrovirens*. Hesselbo (1948, p. 6) also reports *S. atrovirens* from Clavering Ø and (1948a, p. 4) from Rypefjeld, both in east Greenland. In view of the earlier misdeterminations, these reports are probably wrong as well.

The separation of *S. atrovirens* and *S. pumilum* subsp. *polaris* is much more difficult, if not impossible, particularly when the exact nature of the inflorescence cannot be ascertained. In *S. atrovirens* the perianth is ovoid or oblong-ovoid as in *S. pumilum* subsp. *polaris*; leaves are mostly somewhat similarly ovate in shape. A certain differentiation of the two must involve finding immature inflorescences, in which archegonia can be demonstrated distally, and ♂ bracts below them. In the absence of proof of the type of inflorescence, vegetative characters which, by their nature, are untrustworthy must be relied on.

In our region *S. atrovirens* is much rarer than published reports would suggest. In three summers of intensive field work in the Arctic, in areas extending from 68° to 82°33′ N., I was unable to collect the plant even once but saw scores of collections of *S. pumilum s. lat.* Before any collection can safely be attributed to *S. atrovirens*, definite proof that the material is unisexual must be at hand.[338]

Although *S. atrovirens* is also closely allied to the considerably larger *S. triste*, it differs from this not only in size but also in the more slender, clavate to clavate-obpyriform, and less plicate perianth. Unfortunately, large forms of *S. atrovirens* exist ["var. *sphaerocarpoidea* (De Not.) Massal."] with axes 1–2 cm long, which approach *S. triste* in vigor.

SOLENOSTOMA (LURIDAE) TRISTE (*Nees*) *K. Müll.*
[Fig. 278:6–10]

Jungermannia tristis Nees, Naturg. Eur. Leberm. 2:461, 1836; Frye & Clark, Univ. Wash. Publ. Biol. 6(2):292, figs. 1–9, 1944; Amakawa, Jour. Hattori Bot. Lab. no. 22:46, figs. 2d$_{1-3}$; 28:i–o, 1960.
Jungermannia cordifolia var. *nudifolia* Nees, Naturg. Eur. Leberm. 3:536, 1838.
Jungermannia riparia Tayl., Ann. Mag. Nat. Hist. 12:88, 1843.
Jungermannia potamophila Müller-Arg., *in* Mougeot, Nestler, & Schimper, Stirp. Crypt. Vosgeso-Rhenanae, Exsic. no. 1418, 1860, *teste* K. Müller (1909, *in* 1905–16).
Aplozia tristis and *A. riparia* Dumort., Hep. Eur., p. 63, 1874.
Jungermannia riparia var. *potamophila* Bernet, Cat. Hep. Suisse, p. 58, 1888.
Aplozia riparia var. *rivularis* Bernet, *ibid.*, p. 59, 1888.
Haplozia riparia and *H. riparia* var. *rivularis* K. Müll., Rabenh. Krypt.-Fl. 6(1):559, 561, figs. 281–282, 1909.
Aplozia tristis var. *rivularis* Joerg., Bergens Mus. Skrifter 16:110, 1934.
Solenostoma triste K. Müll., Hedwigia 81:117, 1942, and Rabenh. Krypt.-Fl. ed. 3, 6:815, fig. 277, 1956.

Dull, in olive-green to yellowish green patches. Shoots prostrate to suberect, with ascending tips, *1.6–2.2 mm wide* × 1–3 cm high, simple or sparingly branched; often with 1–2 subfloral innovations; frequently stoloniferous. Stems creeping,

---

[338] In view of the fact that *S. pumilum s. lat.* may be heteroecious, with occasional plants having long androecial branches (see, e.g., p. 924 and also Schuster et al., 1959, pp. 49–51), the very status of *S. atrovirens* needs careful reinvestigation. With regard to eastern North America, I am personally unable to recognize any such species and am skeptical that one exists.

free and ascending at least distally, ca. 150 μ in diam. Rhizoids frequent, even above, colorless to pale brownish. Leaves clearly succubously inserted and oriented below, *becoming subtransverse above*, approximate to weakly imbricate, moderately to horizontally patent, the upper semiamplexicaul and concave on fertile shoots, *ovate to oblong-ovate* (the lower sometimes elliptical), averaging 700–1200 μ wide × 800–1500 μ long on sterile shoots (1.1–1.3× as long as wide), *usually broadest in basal third, narrowly rounded at apex*, weakly concave below, the upper rather strongly concave. *Cells ca.* (16)18–21 μ on margins distally, medially ca. (15)18–22 × (22)25–40 μ to 15–22 × 30–40(70) μ at base, polygonal, thin-walled, walls usually colorless, *trigones minute or absent; cuticle smooth;* oil-bodies 2–4 per cell, spherical and 4–5 μ to ellipsoidal and 4–7 × 7–15 μ, indistinctly botryoidal. Without asexual reproduction.

Dioecious, the two sexes often in separate patches. Androecia becoming intercalary, of *(4)8–12 pairs of almost transversely inserted, contiguous to loosely imbricate bracts*; bracts suberect to erect-spreading, their apices usually more widely patent, quite concave at base, usually 1-androus. ♀ Bracts similar to upper leaves in size and form, erect- or widely spreading, rounded distally. Perianths *slender*, to 1.7 mm long × 600 μ in diam., 0.5–0.7 emergent, somewhat clavate to ellipsoidal-fusiform or pyriform, narrowed to the base, weakly dorsiventrally compressed (at least when young), sharply 5–6-plicate in distal third, strongly narrowed to mouth; mouth with shallow, finely denticulate lobes. Capsule wall 2-stratose; epidermal cells with dark, prominent nodular thickenings; inner cell layer with semiannular bands. Elaters ca. 8 μ in diam., loosely 2-spiral. Spores 14–17(18) μ in diam., finely papillate, reddish brown. Chromosome no. = 9.

Type. "Radstaedter" Tauern, western Germany *(Funk)*.

Distribution

Rather widely distributed in the cool and temperate to warmer regions of Europe; never arctic; ranging from northern Spain, Italy, and Sardinia northward to northern France, Belgium, Germany, Austria, Switzerland, Great Britain, and eastward to Bucovina, Dalmatia, then northward into Norway (to 70° N.), Sweden, and Finland; also on the Faroes. Also on the Azores and on Madeira. Widespread in Japan, from Hokkaido south to Kyushu (Amakawa, 1960a).

In North America largely "cordilleran" in range, apparently widespread and relatively common in the area from British Columbia to Washington, Oregon, and California, eastward (and southward in the Rocky Mts.) to Alberta and Montana. Very rare in our area and known with certainty only from the St. Lawrence Valley region:[339]

---

[339] Eastward reported by Frye & Clark (1937–47, p. 293) from Winston-Salem, N.C. There is virtually no possibility that this identification is correct; the report probably represents material of *S. pumilum*.

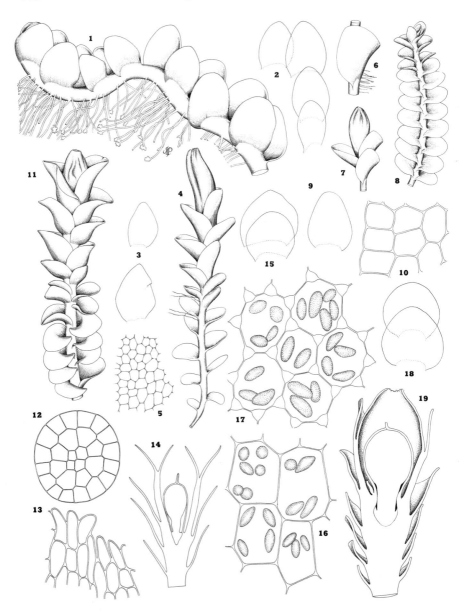

FIG. 278. *Solenostoma* (*Luridae*) *oblongifolium* (1–5), *S.* (*L.*) *triste* (6–10), and *S.* (*S.*) *subellipticum* (11–19). (1) Shoot, lateral aspect (×25). (2) Leaves from sterile shoot (×18). (3) Leaf, above, ♀ bract, below (×ca. 20). (4) Perianth-bearing shoot (×14). (5) Median cells (×110). (6) Leaf, *in situ* (×13). (7) Perianth-bearing shoot tip (×ca. 5). (8) Sterile shoot (×ca. 8). (9) Leaf (×13). (10) Apical cells (×ca. 265). (11) Fertile plant (×13.5). (12) Seta cross section (×190). (13) Perianth apex cells (×190). (14) Longisection through gynoecial shoot tip (×22). (15) Two leaves of mod.

QUEBEC. On moist rocks in Romaine R., near Havre St. Pierre, Saguenay Co. (Kucyniak, 1949). NOVA SCOTIA. Small, subcalcareous, precipitous ravine near N. end of Presque Isle, S. of Cap Rouge, Cape Breton I. (*RMS 43017a*); perhaps representing a small, narrow-leaved phase of *S. cordifolium*, under which see.

Ecology

Most common on periodically submerged stones in cold streams, often forming dingy, yellowish patches on wet rocks. The plant "habitually assumes ... [a] dismal shape ... during the course of its preparation for the herbarium" (Kucyniak, 1949), thus probably suggesting the specific epithet! Amakawa (1960a, p. 46) reports that in Japan the species "prefers calcareous rocks but is not limited to them, occurring also [on] serpentine, chert, graywacke, conglomerate, etc." He states it is "often submerged."

Differentiation

Likely to be mistaken, because of its moderately large size, for *S. cordifolium*, from which it differs in the ovate leaves, somewhat longer than wide, and in the smooth cuticle. Large forms of *S. atrovirens* may also be confused with it (p. 932), and, in my opinion, it is likely that these two plants are extremes of a single species, with *S. triste* the reduced, less lax, more terrestrial northern phase.

My experience with both taxa has been severely limited, since the *S. atrovirens-triste* complex occurs in our region so very rarely—with perhaps only two or three reports that are even probably correct. Even the report by Kucyniak (1949) and mine from Nova Scotia may represent slender and relatively narrow-leaved extremes of *S. cordifolium*.

Solenostoma triste is characterized in part by the dull, often "muddy," olive-green color; the species name, meaning "dismal," is aptly descriptive, especially of dry herbarium specimens, which are "unappetizing" objects for study. The plants are generally rather lax and shrink and turn blackish green in drying. When sterile, at times very difficult to separate from the larger, subaquatic forms of *S. pumilum* ("var. *rivularis*"), with which *S. triste* shares a nearly identical cell size, similar leaves, a like deep and dull green color. The abundant fertility of *S. pumilum* usually facilitates separation of the two species, *S. pumilum* being constantly

---

leptoderma-viridis vel colorata (×30). (16) Median cells, mod. *leptoderma-viridis vel colorata* (×705). (17) Median cells, sun form of open ground, mod. *pachyderma-colorata* (×580). (18) Two leaves, sterile shoot (×25). (19) Longisection through ♀ shoot apex (×32). [1–2, *Marr 646*, Cairn I., Quebec; 3–5, from Müller, 1905–16; 6–10, from Macvicar, 1912; 11–14, from Evans, 1919; 15–16, *RMS 45551*, Kânâk, NW. Greenland; 17–19, *RMS 45601*, same loc.]

paroecious. However, the ♂ bracts are so slightly differentiated, particularly in the larger, subaquatic phases, that the plants can easily be taken to be dioecious, especially with maturation of the perianth (and associated disintegration of antheridia). Furthermore, at times *S. pumilum* produces only 1–2 antheridia in either one (or both) of the perichaetial bracts, and 1 or 2 leaves below them; the plants may then appear to be dioecious. Therefore, care must be exercised in using the sexuality of *S. pumilum* in order to effect a separation from *S. triste*.

Solenostoma triste can also be confused with the dioecious *S. oblongifolium, S. atrovirens,* and (rarely) small forms of *S. cordifolium.* The broadly cordate leaves and much larger cells of *S. cordifolium* usually separate it easily from *S. triste.* More difficult is the separation from *S. oblongifolium* and *S. atrovirens.* Generally, both of these plants are much smaller.

Variation

In montane, acidic regions where the species is abundant, it undergoes considerable variation. In Europe a larger, laxer, aquatic, or subaquatic phase is sometimes recognized ("var. *rivularis*" of Bernet, 1888). This, like the analogous phase of *J. pumilum,* appears to be of slight or no systematic validity. In our area it is so rare that nothing can be added to the discussions of Müller (1909, *in* 1905–16) and Macvicar (1926) as regards variability.

SOLENOSTOMA (LURIDAE) OBLONGIFOLIUM
(K. Müll.) K. Müll.
[Fig. 278:1–5]

Haplozia oblongifolia K. Müll., Rabenh. Krypt.-Fl. 6(1):558, 1909, and 6(2):751, fig. 202, 1916.
Aplozia cordifolia var. *sibirica* Arn. & Jens., Ark. Bot. 13(2):19 (of reprint), 1913.
Aplozia oblongifolia Joerg., Bergens Mus. Skrifter 16:109, 1934.
Jungermannia oblongifolia Buch, Evans, & Verdoorn, Ann. Bryol. 10(1937):4, 1938; Frye & Clark, Univ. Wash. Publ. Biol. 6(2):283, figs. 1–4, 1944; Schuster, Natl. Mus. Canada Bull. 122:24 (1950)1951 [nec *J. oblongifolia* Hook. f. & Tayl., Lond. Jour. Bot. 3:563, 1844 = *Chiloscyphus oblongifolius* (Hook. f. & Tayl.) G., L., & N.].
Solenostoma oblongifolium K. Müll., Hedwigia 81:117, 1942; Paton, Trans. Brit. Bryol. Soc. 5(3):435, fig. 1, 1968.
Jungermannia karl-muelleri Grolle, Österr. Bot. Zeitschr. 111:190, 1964.

Superficially similar to *S. pumilum s. lat.,* forming small patches or creeping as isolated stems, *very small* (shoots 400–650, on fertile plants to 1000, μ wide × 3–6, occasionally 8–15, mm long), *delicate, greenish black to brownish black.* Shoots filiform, stems prostrate, simple or nearly, but often innovating beneath perianth. Rhizoids colorless or pale brownish, to stem apex. Leaves *distant to contiguous,* rarely barely imbricate, spreading or erect-spreading, sometimes somewhat antically secund, *distinctly concave, oval,* almost perfectly egg-shaped and *not cordate at base, widest somewhat above base* but well below leaf middle, *tapering to a narrowly rounded apex, ca. 1.2–1.5 ×* *as long as wide,* to 800–1000 μ long × 670–800 μ wide. Cells thin-walled, walls brownish, with *trigones minute or absent; marginal 15–19 to 20* μ, the *median 18–25* μ × *27–38(40)* μ, toward base somewhat larger and to 18–21 × 40–48 μ; *cuticle smooth* or striolate; oil-bodies 2–4(5) per cell, 4–8 × 4–12 μ, rough. Without asexual reproduction.

Dioecious. Androecia intercalary with age, of 2 or more pairs of almost transverse bracts, similar to leaves in shape but slightly more dilated antically, somewhat more strongly concave. ♀ Bracts identical to leaves in shape, little larger, broadelliptical, bluntly pointed. Perianth *fusiform to oblong-fusiform,* becoming somewhat clavate-oblong with maturity, gradually narrowed distally, with a sharp antical sulcus but otherwise hardly plicate; mouth not contracted into a beak, crenulate with fingerlike, projecting cells; perianth cells ca. 18 × 24 μ above. *Spores large,* 21–30 μ. (For sporophyte, see Paton, 1968.)

Type. Greenland (*J. Vahl, 1829*).

Distribution

An arctic-alpine species, or perhaps race of *S. cordifolium.* Rare and little known. Known from Spitsbergen (Arnell and Mårtensson, 1959), northern Europe (Norway; Arnell, 1956), the Alps (Germany; the Tyrol on the north Italian border; Müller 1905–16, p. 751), Scotland (Paton, 1968), Siberia (Arnell, 1913), and North American localities:

GREENLAND. "East coast," without other data (*Vahl, 1829; type*). QUEBEC. Hills on mainland S. of Cairn I., Richmond Gulf, E. coast of Hudson Bay (*Marr 646;* see Schuster, 1951). HUDSON BAY ISLS. Tukarak I., Belcher Isls., Northwest Territory (*Marr 665, p.p.;* "a little doubtful," Schuster, 1951).

Ecology

Plants from the two stations I have studied are from basic sites, occurring respectively at the juncture between sedimentary rocks and diabase trap (with *Anthelia, Tritomaria polita, Preissa quadrata,* and *Blepharostoma trichophyllum*) and at the borders of a pond in a diabase trap area (with *Aneura pinguis, Blepharostoma trichophyllum brevirete,* and *Lophozia collaris*).

Differentiation

Solenostoma oblongifolium is very closely allied to *S. cordifolium,* of which it is considered to be a mere variety by Arnell and Jensen. Joergensen (1934) also believes this may be its true status. However, the species is more diminutive than "normal" *S. cordifolium,* has noncordate leaves which are much narrower in shape, and has somewhat smaller leaf cells. It is also similar to *S. atrovirens,* from which it differs in the larger marginal leaf cells and the more nearly keeled, concave leaves.

Confusion is also easily possible with *S. pumilum* subsp. *polaris,* arctic phases of which may be very similar (and which occasionally produce

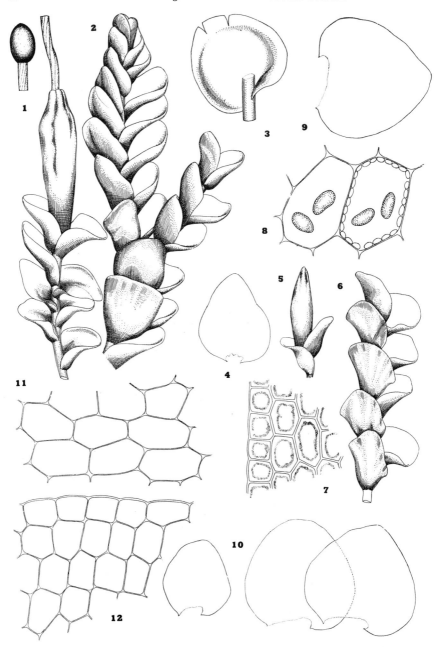

FIG. 279. *Solenostoma* (*Luridae*) *cordifolium*. (1) Sporophyte-bearing plant (×7.4). (2) Sterile plant (×7.4). (3) Leaf, *in situ* (×10). (4) Flattened leaf (×ca. 13.5). (5) Perianth with bracts (×ca. 5.5). (6) Sterile shoot sector (×ca. 8). (7) Marginal and submarginal cells (×ca. 300). (8) Two median cells with oil-bodies and, on right,

purely androecial branches). However, *S. pumilum polaris* has smaller median leaf cells, averaging no more than 18 μ wide, and tends to develop much broader leaves, averaging as wide or wider than long on mature shoots; it also has broader, often almost orbicular-ovate to ovate-reniform perichaetial bracts. When inflorescences are present, the paroecious inflorescence of *S. pumilum* at once precludes further confusion.

SOLENOSTOMA (LURIDAE) CORDIFOLIUM (Hook.)
Steph.
[Fig. 279]

Jungermannia cordifolia Hook., Brit. Jungerm., pl. 32, 1816.
Aplozia cordifolia Dumort., Hep. Eur., p. 59, 1874.
Haplozia cordifolia K. Müll., Rabenh. Krypt.-Fl 6(1):554, fig. 280, 1909.
Solenostoma cordifolium Steph., Spec. Hep. 2:61, 1901; K. Müller, Rabenh. Krypt.-Fl. ed. 3, 6:824, fig. 282, 1956.
Jungermannia exsertifolia Steph., Spec. Hep. 6:86, 1917.
Jungermannia senjoensis Amak., in Hattori, Hep. Japon. Exsic. 6:268, 1951.
Solenostoma senjoense Amak., Jour. Hattori Bot. Lab. no. 12:88, 1954.
Solenostoma exsertifolium Amak., Jour. Jap. Bot. 32:41, 1957.
Jungermannia cordifolia subsp. *exsertifolium* Amak., Jour. Hattori Bot. Lab. no. 22:44, 1960.

In *spongy*, very lax, dull or *brownish green to purplish brown* (or *purplish black*) *tufts*, aromatic when fresh. Robust, shoots 2–3.5(4) mm wide × (2)3–5(6–12) cm long, erect or suberect, simple or fasciculately branched below (branches lateral-intercalary); often stoloniferous at base. Stems 280–320 μ in diam.; cortical cells striolate, leptodermous, oblong, ca. (22)23–26 × (48)60–105 μ. Rhizoids few, absent above or almost so, colorless to pale brownish. Leaves *flaccid*, remote below, subimbricate above, obliquely inserted (exc. for acroscopically arched ventral end), semiamplexicaul, when flattened *cordate to rounded-triangular, concave and* ± *loosely sheathing stem at base*, slightly decurrent antically, *widest in basal 0.2* and there as broad or broader than long, from (1.5–1.75)2.1 mm long × (1.5)2.0–2.6 mm broad to 2.45 mm long × 2.6 mm broad, occasionally larger. Cells *lax, leptodermous*, almost lacking trigones, *rather large* [marginal subquadrate and ca. (17)20–25 μ; *median 20–29 × 30–42 μ*, rarely slightly larger; basal to 25–33 × 45–75 μ]; cuticle ± smooth distally, *striolate at least toward base*; oil-bodies 2–3, rarely in isolated cells 4–8, mostly ellipsoidal to elliptical-fusiform, 4–5 × 9–10 μ to 5–7 × 9–14 μ, hyaline, delicately granular-botryoidal; chloroplasts 3–4 μ long. No asexual reproduction.

Dioecious. ♂ Plants smaller, more slender; ♂ bracts ventricose, in 6–12 pairs, imbricate, erect or almost so, concave; monandrous; antheridia ovoid-globose, short-stalked. ♀ Bracts hardly larger than vegetative leaves, erect-spreading. Perianths fusiform to clavate, slender, longly exserted, ca. 1125–1150 μ in

chloroplasts (×565). (9) Leaf (×12). (10) Leaves, at left a juvenile leaf (×12). (11) Median cells (×336). (12) Apical cells (×336). [1–3, after K. Müller, 1905–16; 4–7, after Macvicar, 1912; 8–12, *RMS 43018*, Nova Scotia.]

diam. $\times$ 3.5–4 mm long, 3–5 stratose below, shallowly 5–6-lobed at the slightly plicate mouth and crenulate with projecting cells; otherwise *smooth*, distally $\pm$ dorsiventrally flattened. Capsule ellipsoidal; epidermal layer of large, quadrate cells with nodular thickenings; inner layer of smaller cells with semiannular bands. Elaters 8 μ in diam. $\times$ 100–140 μ long, bispiral, attenuate at one end. Spores 18–20 μ to 19–24 μ, finely granulate. Chromosome no. = 9.

Type. Scotland.

Distribution

Holarctic, ranging from the southern portions of the Tundra throughout the Taiga (*s. lat.*, including the entire Spruce-Fir Biome), with a rather general restriction to oceanic and suboceanic portions of the continents.

In Europe throughout the northwestern part (northernmost Sweden, Norway, and Finland), south to Scotland, the Faroes, England, and Ireland; absent from southern Sweden and Denmark. Reappearing, in disjunct fashion, in the central European mountains (Black Forest of Germany; Mt. Dore and Cantal, France; Switzerland; northern Italy; Tyrol; the Tatra Mts.); eastward to the Carpathians in Kom. Máramoros. Also in Belgium (Ardennes), Spain (Guipúzcoa, Sierra de Gredos, Sierra Nevada), the French Pyrenees, and eastward in the Caucasus. In Iceland.

Scattered in Japan (alpine region; Hattori, 1948, 1952)[340] and Korea (Hattori, 1948, p. 107) and reported from China and India.

In North America westward from the Aleutian Isls. (Clark and Frye, 1942, 1948) and continental Alaska to British Columbia, Alberta, Idaho, Montana, Wyoming, Washington, Oregon, and California; also in Colorado (Evans and Nichols, 1908; Evans 1915b); evidently common westward. In the East local and rare in most cases:

GREENLAND. *Fide* Macoun (1902).   LABRADOR. *Fide* Macoun (1902) and Müller (1951–58, p. 827).   ANTICOSTI I. R. Jupiter, and Baie St-Claire and Baie du Renard (Lepage, 1944–45).   QUEBEC. Ste-Anne-des-Monts R., Gaspé (Macoun, Can. Hep. no. 48!); Pont Rouge; Chute Montmorency; Mt. Albert (Lepage, 1944–45); Mt. Shefford (Fabius, 1950); N. slopes of Mt. Albert, Parc de la Gaspésie (*Williams 10663!*); 3 mi W. of Ruisseau-à-Rebous, Hwy. 6, N. shore of Gaspé Pen. (*Williams 10755!*); N. shore of Gaspé Bay (*Williams 10807, c. per.!*).   ONTARIO. Owen Sd.,

[340] Amakawa (1960a, pp. 44–46) attempts to maintain the Japanese plant as distinct at the subspecies level, as *J. cordifolia* subsp. *exsertifolia* (Steph.) Amak., stating that subsp. *cordifolia* occurs "in bogs and other acid substrata" while subsp. *exsertifolia* "grows on wet or submerged sandstones, graywacke, clay-slate . . . rarely even limestones." The latter is supposedly distinct from subsp. *cordifolia* in smaller size (2–4 cm long $\times$ 2–3 mm wide; leaves 1.5–2 mm long $\times$ 1.5–2.1 mm wide), in olive-brown color, in "imbricate leaves whose dorsal margin is not so strongly incurved and covering stem," and in "not occurring on acidic substrata." If leaf size is compared (see preceding diagnosis, based on plants from a basic ravine, associated with *Lophozia gillmani*, in Nova Scotia; *RMS4 3018*), little difference is apparent. I doubt that the criteria Amakawa exploited serve to effectively differentiate the Japanese plant.

on rocks in Sydenham R., Gray Co. (Macoun, 1902).   NOVA SCOTIA.   Cape Breton
I.: Indian Brook (*Nichols!*); Big Intervale, Margaree (Macoun, 1902); near N. end of
Presque Isle (*RMS 43018, 43014, 43017a?*); N. of Cap Rouge campground (*RMS*);
Beulach Ban Falls, near S. Aspy R. (*RMS*); Indian Brook, Cape Breton (*RMS 68-1247*);
Cheticamp R., Cape Breton (*RMS 68-960*).   NEWFOUNDLAND.   Humber R., Squires
Mem. Pk. (*RMS 68-1643*).

MAINE.   Oxford (*Greenwood 269, fide* Frye & Clark, 1944); Round Mountain L.,
Franklin Co. (*Lorenz*).   NEW HAMSPHIRE.   Waterville (*Lorenz, 1907!*).   VERMONT.
Hartland (*Lorenz, 1921!*).   MASSACHUSETTS.   Topsfield (*E. Young, 1934; fide* Frye &
Clark, 1943, *in* 1937–47; very doubtful); Green R. Gorge, N. of Greenfield (*RMS*).
CONNECTICUT.   Hartford (*Conklin, 1924; fide* Frye & Clark, *loc. cit.*); Windsor (*Lorenz;
fide* Evans & Nichols, 1908); Rainbow (Evans, 1904, p. 170).   NEW YORK.   Catskill
Mts. (Nichols, 1933, p. 70).

Reappearing as a disjunct in the L. Superior region: MICHIGAN.   Au Train Pt., Alger
Co. (*Steere, 1934!*); Pictured Rocks, Alger Co. (Nichols, 1933).   WISCONSIN.   White
R., Ashland Co. (Conklin, 1929).

Solenostoma cordifolium has a distribution nearly identical with that of *Scapania
umbrosa*; both taxa are somewhat oceanic and occur mostly in areas with acid
rocks; both are absent from the Southern Appalachian outliers of the Spruce-
Fir Zone, extending southward merely to the Catskill Mts. However, *S.
cordifolium* has a much wider range in the American West and is widespread in
continental regions. A closely allied species, *S. roridum* (Herz.), occurs in the
Andes of Columbia and Peru.[341]

Ecology

A species of insolated, wet, and sloping rocks at the edges of clear streams
and small rivers. It occurs almost exclusively in silt-free streams and is
characteristic of clear mountain brooks and streams, where often found
in swelling tufts attached to rocks and ledges. Here it is associated with
various Scapaniae (*S. undulata, S. subalpina*), *Hygrobiella laxifolia* (as on Cape
Breton I.), *Pellia epiphylla*, and Marsupellae (*M. emarginata, M. sparsifolia*).
At times the species occurs permanently or nearly permanently submerged
on rocks in rapidly flowing water (as at Beulach Ban Falls near South
Aspy R., Cape Breton); it may produce capsules under such conditions
(June 22, on Cape Breton). Much more rarely *S. cordifolium* occurs on
moist rocks above the normal flood level, over weakly basic ledges (as near
Presque Isle, Cape Breton, *RMS 43017a*), associated with *Lophozia
(Leiocolea) gillmani*. Such occurrences, associated with marked calciphytes,
are exceptional; the plant is then atypical, approaching *S. triste*.

Differentiation

A characteristic and rather stenotypic species, usually recognizable at
a glance in the field; equally distinguished by the sharply defined ecological requirements.

[341] Müller (1951–58) assumes that Spruce's (1884–85) report of *S. cordifolium* from the cordillera
of South America represents this species.

The plants usually occur in large, swelling, fuscous to brownish green tufts attached to rocks subject to inundation, in clear, rapidly flowing water. The robust size and fuscous coloration are distinctive, as are the semiamplexicaul, ovate-triangular to ovate-deltoid, somewhat cordate leaves which usually average well over 1 mm long and can attain a length of 1.8–2.5 mm; normal plants are 2–4 mm wide. The rather hollow, somewhat loosely clasping form of the leaves is quite diagnostic. Occasionally plants (especially ♂ ones) from sites above flood level are much smaller and firmer, with leaves averaging only 0.75 mm long, and may have distinct trigones and a smooth cuticle. Such plants can easily be confused with *S. triste* but differ in their ovate-deltoid leaves with a broad base but narrow insertion.

When fertile, as is frequently the case, the very slender fusiform perianth of *S. cordifolium* is distinctive, separating it from all other species of the subgenus.

Subgenus *SOLENOSTOMA* Mitt.

Solenostoma Mitt., Jour. Linn. Soc. Bot. 8:51, 1865.
Jungermannia, in part, of authors [incl. Evans, Ann. Bryol. 10:36–42, 1938; Frye & Clark, Univ. Wash. Publ. Biol. 6(2):275, 1944; Schuster, Amer. Midl. Nat. 42(3):579, 1949].
Aplozia Dumort., Hep. Eur., p. 55, 1874 (in part).
Haplozia subg. *Eu-Haplozia* K. Müll., Rabenh. Krypt.-Fl. 6(1):538, 1909.
Solenostoma subg. *Rostellatae* K. Müll., Hedwigia 81:117, 1942.
Solenostoma subg. *Eusolenostoma* Schust., Amer. Midl. Nat. 49(2):389, 397, 1953.
Jungermannia subg. *Solenostoma* Amak., Jour. Hattori Bot. Lab. no. 22:53, 1960.

Medium-sized [usually 650–1500(2000) μ wide], erect or suberect; *relatively firm*, varying from *bright green to brownish* or (in *S. gracillimum*) to reddish, *mostly without vinaceous pigmentation*. Stem lacking specialized features: the 1–2 cortical strata of cells somewhat shorter, rectangulate, hardly or somewhat thick-walled, surrounding a central cylinder of more leptodermous medullary cells which are longer and mostly somewhat larger in diam.; no hyalodermis. *Rhizoids colorless or pale brown*, rarely purplish. *Leaves essentially orbicular to reniform, varying from 1.0 to 1.25(1.41–1.75) × as wide as long*, rather strongly narrowed at base to their insertion and often subcordate, often subtransversely oriented but obliquely inserted, *frequently with rhizoids attached either to their basal* (occasionally even median) *portions or issuing in a fascicle from near their postical bases*. Cells nearly or quite isodiametric throughout leaf, virtually always quadrate to hexagonal, with thin walls and *distinct to weakly bulging trigones* (except for the marginal cells in *S. gracillimum*, and sometimes also the interior cells); oil-bodies finely granular in appearance or almost homogeneous, then glistening, 1–9 per cell, large (larger than chloroplasts). Asexual reproduction normally absent; when present by means of endogenously produced gemmae (*S. caespiticium*).

Perianth 0.5–0.7 emergent beyond the ± spreading bracts, terete or subterete below, never at all dorsiventrally flattened, distally 0.5–0.8 *deeply 4–5-plicate* (and with 4–5 sharp longitudinal keels), *one of the plicae being dorsal*, near apex *rather suddenly contracted into a short beak*, at base *without a perigynium or with a vestigial one only 0.1–0.25 the height of the perianth. Cells of perianth not or hardly elongated*, those bordering the beak 1–2 or 3–5×, those below mouth up to 1–2×, as long as wide; perianth in basal half at least locally 2-stratose. Elaters 2-spiral, the spirals ± distinct to the sharp or ± blunt ends of the elaters.

Type. *Solenostoma sphaerocarpum* (Hook.) Steph.

Subgenus *Solenostoma* occupies, in some respects, a median position between subgenera *Plectocolea* and *Luridae*. Such taxa as *S.* (*Solenostoma*) *gracillimum* and *levieri* may develop low or rather conspicuous stem perigynia, thus approaching *Plectocolea*. Occasionally *S. gracillimum* also has the perianth drawn out into a nearly conical apex, rather than into a sudden beak (Fig. 280:1), although a beak is usually distinct; it may have the distal cells of the perianth mouth unusually elongated (Fig. 280:4). Thus no sharp distinction can be drawn between subgenera *Solenostoma* and *Plectocolea*, the disagreement as to the position of *S. gracillimum* (see Müller, 1942) being a vivid demonstration of this fact. Typically, subg. *Solenostoma* differs from *Plectocolea* in (*a*) colorless rhizoids and general lack of claret-colored to vinaceous pigmentation;[342] (*b*) longly emergent, virtually "free" perianths sharply and suddenly beaked at the apex; (*c*) 4–5 pronounced longitudinal carinae and plicae of the perianth, one of the keels being antical; (*d*) leaf cells similar to those of perianth, excepting sometimes those bordering the perianth beak.

Subgenus *Solenostoma* is sufficiently closely allied to *Luridae* that no recent worker has attempted to elevate these two groups to independent generic status. This close affinity is best demonstrated by such species as *S.* (*Luridae*) *pumilum polaris*, in which certain phases (fo. *cavifolius*) have the rotundate leaves of subg. *Solenostoma* and also possess oblong-ovate perianths, at times rather abruptly contracted distally into a vestigial beak. Typically, the sharp, carinate plicae of the perianths and the obvious beak of the perianth apex, as well as the rotundate leaves and larger leaf cells, serve to separate subg. *Solenostoma* from subg. *Luridae*. Perhaps equally important in separating subg. *Solenostoma* from both subgenera *Luridae* and *Plectocolea* is the marked tendency for numerous rhizoids to develop from the proximal (and intramarginal) sectors of the leaves and bracts; at times rhizoids even develop from the perianth surface (Fig. 283:6). In the two other subgenera rhizoids are confined, almost without exception, to the stem. Also distinctive is the tendency for rhizoids to become fasciculate, running down the stem in ± distinct "bundles." This is especially

---

[342] Often vinaceous in *S. pusillum*, however.

well marked in sectio *Desmorhiza* Amak., including *S. clavellata* Mitt. ex Steph., of Japan to the Himalaya. However, such a fascicle is rather distinct even in some forms of *S. pusillum*.

The stem anatomy of species of subg. *Solenostoma* that have been investigated shows few of the specialized features frequently found in subg. *Plectocolea*. The cortical 1–2 strata of cells are not as highly differentiated, varying from the medullary cells in being somewhat shorter and occasionally somewhat more thick-walled.[343] There is never development of a large-celled epidermal layer (hyalodermis) surrounding 1–2 strata of small, thicker-walled cells. In species of *Solenostoma* that I have studied the elaters are also often relatively unspecialized and bispiral and possess relatively blunt ends with the spirals distinct to the elater apices (as, i.a., in *S. sphaerocarpum*, Fig. 283:4). In subg. *Plectocolea*, however, most species are specialized in that the spirals are confluent at the ends of the short, contorted elaters, forming thick-walled, attenuated appendices (as, i.a., in *S. obscurum*, Fig. 292:2–3).

Thus, in most characters investigated, subg. *Solenostoma* is clearly less derivative than *Plectocolea* and is here treated before the latter. The group is, as Amakawa (1960a) states, a "well-defined natural group." He places the many Japanese species into three sections: *Protosolenostoma, Desmorhiza, Solenostoma*.[344] The first section has a 3–4-plicate perianth that is hardly rostellate and has obliquely inserted leaves; the second is distinct in the tightly fasciculate rhizoids that are decurrent in a bundle on the stem. Our species, except for *S. gracillimum*, belong to the typical sectio *Solenostoma*, in which rhizoids are not conspicuously fasciculate.

The sections can be separated by the following key:

1. Leaves orbicular to ovate, quite obliquely inserted and oriented, not or weakly concave; perianth fusiform, 3–4-plicate, feebly rostellate at mouth; cells not or feebly collenchymatous, marginal often larger . . . . . . 2.
    2. Perigynium absent or vestigial; leaves mostly broadly ovate . . . . .
    . . . . sect. *Protosolenostoma* Amak.[345]
    2. Perigynium distinct, if low; leaves orbicular . . . .
    . . . . sect. *Gracillimae* Schust., sect. n. (p. 972)
1. Leaves orbicular to reniform, rarely ovate or obovate, insertion approaching subtransverse, orientation almost transverse, usually concave (concavity turned toward shoot apex); perianth distinctly and abruptly beaked; leaf cells usually with thin walls and coarse trigones, marginal ones never enlarged . . . . 3.
    3. Rhizoids not forming a dense decurrent bundle on ventral side of stem.
    . . . . 4.
        4. Leaves short-decurrent or moderately so (usually hardly so ventrally);

---

[343] Very thick-walled only in sectio *Faurianae*; see the following key.

[344] In the following key, I divide the first and the last section each into two sections.

[345] With *S. koreanum* Steph. and *S. fusiforme* (Steph.) Schust., comb. n. (basionym: *Nardia fusiformis* Steph., Bull. Herb. Boissier 5:99, 1897); both east Asiatic. See Amakawa (1960a), *sub Jungermannia* (pp. 55–57.).

stem without a rigid cortex, cortical cells in at most 1 stratum weakly
thick-walled . . . . sect. *Solenostoma* (p. 947)

4. Leaves long-decurrent both dorsally and ventrally, inserted on an
inverted U-shaped line; stem with cortex rigid, (1)2(3) layers of
cortical cells conspicuously thick-walled . . . .
. . . . sect. *Faurianae* Schust., sect. n.[346]

3. Rhizoids numerous, aggregated in a strandlike, compact, decurrent,
ventral bundle appressed to stem . . . . . . sect. *Desmorhiza* Amak.[347]

Six taxa belonging to subg. *Solenostoma* occur in our region. Except for
the widespread *S. gracillimum*, the other species are either rare and very
local (*S. pyriflorum, S. caespiticium*) or else essentially arctic-alpine in distribution. Species of this subgenus share a common basic facies, which is
due to the rotund leaves, always as broad or broader than long, and the
deeply plicate perianth, quite suddenly contracted at the apex. Only
S. gracillimum stands sharply isolated, yet this species appears to belong
more nearly in *Solenostoma* than in subg. *Plectocolea* (Müller, 1942; Schuster, 1953). Yet Evans (1938c) placed it with species now assigned to
Plectocolea (whether this group is accepted as a subgenus or genus).

Species of *Solenostoma* occur most commonly on compact, loamy or
clayey soils or soil over rocks, often closely mixed with mosses. In my
experience, *S. sphaerocarpum, S. pyriflorum*, and *S. levieri* occasionally occur
directly on rocks, often near silt-bearing streams. In contrast, the species
of subg. *Luridae* are often pioneers directly over damp to wet rocks, more
rarely occurring on soil.

Key to Species of Subgenus Solenostoma

1. Leaves not bordered with conspicuously enlarged, quadrate, thick-walled
marginal cells, the marginal cells with thin walls and trigones like the intramarginal cells and identical in size; cells collenchymatous, with thin walls
and usually distinct to bulging trigones; perigynium absent or usually
vestigial; plants in sun becoming brownish or fuscous or (*S. pyriflorum, S.
pusillum*) locally red or purplish, in shade bright green and somewhat opaque
. . . . sect. *Solenostoma*   2.

2. Paroecious. [Arctic-alpine to subarctic-subalpine species, lacking in
Southern Appalachians. Gemmae absent; cells usually with distinct
trigones, each usually with (2)3–9 oil-bodies] . . . . 3.

3. Cells of leaf margins (15)18–35 μ, of leaf middle 24–42 × 35–50 μ;

[346] Type: *Jungermannia fauriana* Beauv. *in* Stephani, Spec. Hep. 6:571, 1924 [= *Solenostoma
fauriana* (Beauv.) Schust., comb. n.]. Also here: *Solenostoma cyclops* (Hatt.) Schust., comb. n.
(basionym: *Jungermannia cyclops* Hatt., Jour. Hattori Bot. Lab. no. 3:5, 1950). Both taxa are
Japanese, the former also Korean.
Folia orbicularia, et dorsaliter et ventraliter longe decurrentia, in lineam U-formen inversam
inserta; folia numquam marginem habentia; cortex caulis rigidus, plerumque 2-stratosus, e
cellulis quae membranas crassas habent constans.

[347] With *S. clavellatum* Mitt. ex Steph.; see Amakawa (1960a, p. 69, fig. 40).

oil-bodies 3–9 per cell usually; larger plants with bracts rotundate to broadly reniform. . . . . 4.

4. Cells of leaf margins (15)18–24(25) μ, of leaf middle (24)25–30(32) × 30–35(40) μ; spores 16–20(22) μ; capsule with epidermal cells 15–20 × 30–35 μ to 22–25(30) × 22–30 μ; inner cells (8)10–15(16–17) μ wide × 45–50 μ long . . . . 5.

    5. Shoots not dimorphic: without small-leaved, flagelliform shoots; plants deep green to ± brownish; perigynium absent or vestigial; plants usually larger, to 1–3 cm long; leaves usually nearly rotundate on sterile shoot sectors, 1.0–1.2(1.3) × as wide as long . . . . S. sphaerocarpum (p. 947) *a.*

        a. Plants larger: leaves usually spreading, subrotundate and 1.0–1.2 × as broad as long; perianth bluntly 4-angulate and plicate to near middle, usually without trace of subtending perigynium . . S. sphaerocarpum var. sphaerocarpum (p. 947)

        a. Plants small, to 1 mm wide and 8–10 mm high: leaves erect to erect-appressed, broadly rotundate to reniform, usually 1.1–1.3 × as wide as long; perianth sharply 4-angulate and plicate to near base; often with slight perigynium . . . . S. sphaerocarpum var. nana (p. 952)

    5. Shoots often dimorphic: commonly with some small-leaved flagella; plants light green, the shoot apices (at least) often reddish to claret-red; perigynium usually distinct, up to 0.2–0.3 height of perianth above it; plants small, 2–5(6–8) mm high; leaves broadly reniform, to 1.4–1.75 as wide as long; on moist clayey-loamy soils. (Plants often with facies of S. crenulatum) . . . . S. pusillum (p. 953)

4. Cells of leaf margins (24)28–32(35) μ, of leaf middle mostly (28–32)30–35 × 30–40 to 34–39(42) × 35–50 μ; spores 19–24 μ; capsule with epidermal cells averaging (24)26–30 × 36–48 to 35–40 × 40–48 μ; inner capsule wall cells (13)15–16(18) × 80–90 μ usually; leaves, even of sterile shoot sectors, broadly reniform and (1.1)1.25–1.5 × as wide as long. (Often with a vestigial perigynium). . . . . S. levieri (p. 959)

3. Cells of leaf margins 13–17 μ, of leaf middle 14–18 × 16–20(26) μ; cells of perianth mouth 22–23(35) μ long × 11–13.5 μ wide; oil-bodies 2–3(4) per cell; leaves very broadly attached to stem, commonly concave; bracts broadly ovate to reniform-rotundate . . . . [See S. (Luridae) pumilum subsp. polaris (p. 918)]

2. Dioecious. [Plants either with gemmae and large-celled, then with 1(2) very large oil-bodies per cell; or tending to become deeply purplish-pigmented, at least on perianths] . . . . 6.

6. Cells with small to distinct trigones, 15–22 μ on margins, 20–26 × 23–25(28–30) μ medially, each cell with usually (3)5–8(9) oil-bodies; gemmae absent; plants in sun developing a deep purplish pigmentation of perianth apices. Southern Appalachians and Lake Superior. . . . . S. pyriflorum (p. 963)

6. Cells delicate, without trigones, 30–32 μ on margins, 32–40 × 40–60 μ medially, each cell with a single, extremely large oil-body; with endogenous 1-celled gemmae within swollen shoot apices; clear green even in sun; largely arctic-alpine and subarctic; very rare . . . . *S. caespiticium* (p. 968)

1. Leaves (at least near ♀ bracts) bordered with quadrate and subquadrate, thick-walled cells averaging 2–3× intramarginal cells in size; interior cells usually slightly thick- or thin-walled, rarely with distinct trigones; with a distinct perigynium, often 0.2–0.3× the height of the free perianth; plants pale green, in sun reddish to carmine. (Widespread, but rare in the Taiga; dioecious) . . . . [sect. *Gracillimae*] *S. gracillimum* (p. 972)

Sectio *SOLENOSTOMA*
SOLENOSTOMA (SOLENOSTOMA) SPHAEROCARPUM
(Hook.) Steph.
[Figs. 272:12–14; 280; 282:13–14]

Jungermannia sphaerocarpa Hook., Brit. Jungerm., pl. 74, 1816; Frye & Clark, Univ. Wash. Publ. Biol. 6(2):294, figs. 1–9, (1943) 1944.
Jungermannia amplexicaulis Dumort., Syll. Jungerm., p. 5, 1831.
Jungermannia confertissima, J. nana, J. tersa of Nees, Naturg. Eur. Leberm. 1:291, 317, 329, 1833.
Jungermannia scalariformis Nees, ibid., 2:463, 1836 (*fide* K. Müller, Hedwigia 81:119–122, 1942).
Aplozia sphaerocarpa Dumort., Hep. Eur., p. 61, 1874; Macvicar, Studs. Hdb. Brit. Hep. ed. 2:144, figs. 1–4, 1926.
Jungermannia goulardi Husnot, Hep. Gallica, p. 29, 1881.
Solenostoma sphaerocarpa Steph., Spec. Hep. 2:61, 1901.
Solenostoma amplexicaule Steph., ibid. 2:48, 1901.
Aplozia goulardi Boulay, Musc. France 2:129, 1904.
Haplozia sphaerocarpa K. Müll., Rabenh. Krypt.-Fl. 6(1):546, fig. 277, 1909.
Solenostoma (subg. *Eusolenostoma*) *sphaerocarpum* Schust., Amer. Midl. Nat. 49(2):397, pl. 30, figs. 7–10, 1953.
Solenostoma (subg. *Solenostoma*) *sphaerocarpum* K. Müll., Rabenh. Krypt.-Fl. ed. 3,6:827, fig. 283, 1956.
Jungermannia (subg. *Solenostoma*) *sphaerocarpa* Amak., Jour. Hattori Bot. Lab. no. 22:57, fig. 34i–o, 1960.

Usually in small patches, with crowding caespitose and tufted, *usually pure green*, more rarely olive-green or brownish-tinged (usually on distal portions of leaves and perianth), *never purplish*; medium-sized, shoots 1500–1800(2200) μ wide × 1–2(3) cm long, erect or ascending from a ± prostrate base, simple or sparingly branched. Stems ca. 200 μ in diam. Branching frequent to sparing, lateral-intercalary; no flagella or stolons. Rhizoids very long and numerous, *hyaline or pale brownish*,[348] not only from postical face of stem but occasionally *also from bases of leaves, bracts, and perianth*. Leaves weakly succubously inserted, above often only slightly so, *semiamplexicaul*, widely spreading and weakly concave on sterile shoots, strongly concave and suberect on upper leaves (or those of fertile shoots), distant to contiguous, *orbicular to weakly reniform-orbicular*, ca. 760–1200 μ wide × 660–800 μ long (averaging slightly to distinctly broader

---

[348] Amakawa (1960a, p. 59) reports them as "colorless or purple" in Japanese plants; I have seen purple rhizoids very rarely in North American material.

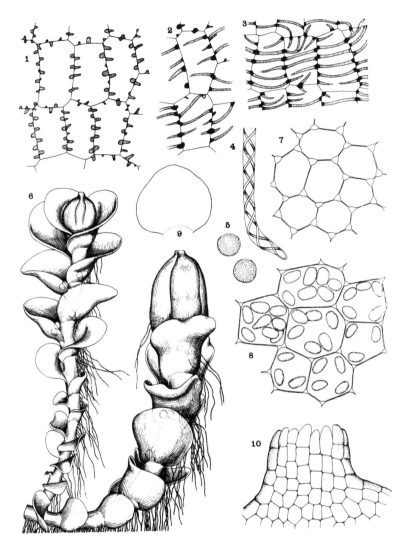

F<small>IG</small>. 280. *Solenostoma* (*Solenostoma*) *sphaerocarpum*. (1) Epidermal cells of
capsule wall (×260). (2–3) Inner cells of capsule wall, in fig. 2 with
semiannular bands largely incomplete (×260). (4–5) Elater, upper half,
and spores (×340). (6) Fertile plant, with shoot with immature perianth
at left, mature perianth at right (×14). (7) Median cells, pachydermous
phase (×260). (8) Median cells, with oil-bodies (×444). (9) Leaf of sterile
shoot sector (×17). (10) Perianth mouth (×158). [All from Belle Rose I.,
Cook Co., Minn., *RMS*.]

than long, but *rarely over 1.25× as wide as long*), narrowed basally and *rounded but hardly cordate at base, rounded at apex* or occasionally retuse; *antical base short-decurrent*. Cells distinctly collenchymatous, but the *trigones small and hardly to weakly bulging; marginal cells not forming a border*, (15)18–24(27) μ; *median cells 24–28(32) × 32–36 μ*;[349] basal cells to 25–30(36) × 36–48 μ; cuticle smooth or almost so; oil-bodies ovoid or ellipsoidal to bacilliform, (3–4)6–9(12) per cell, ca. 5 × 6–7 μ to 4.2–6(7) × 10–13(15) μ, *smooth and appearing virtually homogeneous to faintly granulate. Asexual reproduction lacking.*

Paroecious. Almost always fertile. Androecia of usually 2(3–4) pairs of bracts, immediately below perichaetial bracts; bracts like leaves but somewhat larger and more concave, usually contiguous to imbricate; antheridia 2–3, with biseriate stalk. ♀ Bracts like leaves but broader, reniform-rotundate, erect and sheathing at least at bases, the distal portions often spreading. Perianth obovate to broadly clavate, ca. 2 mm long × 800 μ in diam., to 0.5–0.65 emergent, *without a subtending perigynium* (or with a mere short trace of one), mostly obtusely 4-plicate above (sometimes with 1–2 small accessory plicae), *suddenly narrowed to a shortly beaked mouth*; mouth crenulate with rather short cells (length 1.5–2 × the width), the cells below the immediate apex almost isodiametric, gradually becoming oblong in middle and below. Sporophyte with capsule *brown*, almost spherical, the wall 2-stratose. Epidermal cells ca. 15–20 × 30–35 to 22–30 × 22–30 μ, up to 25–30 μ broad, with nodular thickenings; inner cells ca. (8)10–15(16–17) μ broad × 45–50 μ long, with narrow semiannular bands, the bands complete or *often incomplete*. Elaters 7–10 μ in diam., bispiral; spirals 4–5 μ wide. Spores 16–18 up to 18–22 μ, brown, finely verruculose. Chromosome no. $n = 18$.

Type. Dublin, Ireland (*Lyell*).

Distribution

Presumably holarctic, although its distribution in northern Asia very poorly known. Found from the northern portions of the arctic-alpine zone into the northern half of the Taiga; rarely in the southern edge of the Spruce-Fir Forest Region.

In Europe widespread from northern Scandinavia (Sweden, Norway, Finland), eastward to Russia, southward through Scotland and England and Ireland, recurring in the alpine and subalpine areas of southern Germany, Austria, Switzerland, into northern Spain (Casares-Gil, 1919) and Portugal (Stephani, 1898–1924, vol. 2:61, 1906), and to France, northern Italy, and Sardinia. Also in Iceland and Jan Mayen I. (Jensen, 1900). Extending eastward into Asia (Siberia; Honshu, Japan, *fide* Amakawa, 1960a); also in the Sikkim-Himalaya.

In North America transcontinental at high latitudes, occurring in the West from Alaska (Persson and Weber, 1958; Persson, 1952; Clark and

---

[349] Amakawa (1960a, p. 59) reports median cells only "22–30 × 15–22 μ." I have never seen them averaging so small.

Frye, 1948) to British Columbia, southward to Washington and (?)California, eastward to Alberta, and south in the Rocky Mts. to Utah, Montana, Wyoming, and Colorado (where it may ascend to 12,000 feet; north of Kiowa Peak, Boulder Co., *Weber 8591*!). Apparently relatively common in the West, whereas rather rare in the East, except for Greenland.

w. GREENLAND. SE. of Jacobshavn, Disko Bay (*RMS & KD 66-230*); Disko I., Godhavn, below Lyngmarksfjeld (*RMS & KD 66-146*); Qalagtoq, Upernivik Ø, Inukavsait Fjord (*RMS & KD 66-1101*); Kangerdlugssuakavsak, head of Kangerdluarssuk Fjord, 71°21′ N., 51°40′ W. (*RMS & KD 66-1281, 66-1289, 66-1290, 66-1287a*); Magdlak, Alfred Wegeners Halvö (*RMS & KD 66-1356c, 66-1359a*).   E. GREENLAND; Kobberpynt, Gaaseland, Gaasefjord, Jamesons Land, Danmarks Ø, Rødo-Ø's Vestkraent, all in Scoresby Sund (Jensen, 1897; var. *nana*); Hold-with-Hope (Jensen, 1897; var. *nana*); Cap Ravn (Harmsen, 1933; var. *nana*); Hold-with-Hope, 73°30′ N.; Nordost Bugt, 71°17′ N.; Hurry Inlet, 70°51′ N.; Gaaseland, Scoresby Sund (Jensen, 1897; var. *sphaerocarpa*); Hurry Inlet and Nordostbugt (Jensen, 1906a; var. *nana*). ELLESMERE I. Goose Fjord (Polunin, 1947; Bryhn, 1906; as var. *nana*).   BAFFIN I. Pond Inlet (Polunin, 1947).   MELVILLE PEN. "Cape Elizabeth," Vansittart I. (Polunin, 1947).   LABRADOR. Underwood (1892) and Macoun (1902).   QUEBEC. Montreal (Frye & Clark, 1937–47); Mingan, Côte Nord (Lepage, 1944–45); N. shore of Gaspé Bay, Gaspé Pen. (*Williams 10797*!; with spores 19–24 μ!); also in 86 ft of water in unnamed lake at 58°49′ N., 64°58′ W. (*R. B. Davis, 1968*).   NEWFOUNDLAND. Humber R., Squires Mem. Park (*RMS 68-1631a*).

MAINE. Rock wall above Davis Pond, northwest Basin, Mt. Katahdin, 2900 ft (*RMS 15984a*); Clearwater L., Megantic Preserve, Franklin Co. (*Lorenz;* see Evans, 1919a, p. 168).   VERMONT. Smuggler's Notch near Mt. Mansfield (*RMS 66-1853;* with *Scapania cuspiduligera, Preissia quadrata, Lophozia heterocolpa,* etc.).   NEW HAMPSHIRE. Waterville (Lorenz, 1908d); Huntington Ravine, Mt. Washington (*RMS*).   NEW YORK. Indian Falls, Mt. Marcy, Adirondack Mts. (*Wilson 25*!).

Westward occurring as follows: MICHIGAN. Keweenaw Co. (Steere, 1937); Luce Co.; Tahaquemon Falls (*Nichols;* Frye & Clark, 1937–47); Pictured Rocks, just E. of Miners Castle, Alger Co. (*RMS 39303a;* see p. 962).   WISCONSIN. Orienta Falls, Bayfield Co., and Apostle Isls., Ashland Co. (Conklin, 1929); Sand I., Apostle Isls. (*RMS*).   MINNESOTA. Cascade R. ravine, Belle Rose I., Rosebush Falls near Grand Marais, Lutsen, and Little Caribou R., all in Cook Co.; Gooseberry Falls and Manitou R. ravine, Lake Co.; French R., St. Louis Co. (Schuster, 1953); Hat Pt., Grand Portage, Cook Co. (*Shaw 88f*).

Also reported erroneously from Bergen Co., N.J. (Austin, Hep. Bor.-Amer. Exsic. no. 29b), and from Iowa (Frye & Clark, 1937–47, p. 295).

Ecology

Principally over damp to moist rocks, occasionally directly over bare rock but more often on thinly soil-covered rocks or in soil-filled crevices, and rarely directly on damp soil. Although generally considered an oxylophyte, it occurs at least in the Lake Superior region under subbasic conditions, being found with such calciphytes as *Scapania cuspiduligera, Lophozia heterocolpa,* and *Pellia fabbroniana* (directly on damp rocks and in their crevices) or, when established on soil-filled crevices, with *Preissia quadrata* and *Lophozia badensis*.[350] Occasionally on rather dry rock walls, in deep shade (then sometimes with *Scapania mucronata*).

In west Greenland *S. sphaerocarpum* also occurs on soil, often rather peaty soil, lying over basic or subbasic rocks, often along small rills; here it may be associated with *Lophozia heterocolpa*, *L. excisa elegans*, *Tritomaria quinquedentata*, *T. heterophylla*, *Solenostoma subellipticum*, *Scapania curta*, *S. praetervisa*, and *Anthelia juratzkana*.

Solenostoma sphaerocarpum often occurs over silt-covered rocks along small montane streams, associated with *Atrichum undulatum*, *Pellia epiphylla*, *Marsupella emarginata*, and *Conocephalum*. Under such conditions it may form large yellowgreen or pure green patches. In the Southern Appalachians *S. pyriflorum* occurs under strikingly similar conditions. Both taxa tolerate silting—in contradistinction to *S. pumilum*, which is usually found over damp rocks along rivers where silt deposits are virtually nonexistent.

Differentiation

Solenostoma sphaerocarpum is distinguished principally by the green color, in insolated sites becoming brownish or fuscous (rarely reddish); the colorless, abundant, very long rhizoids; the orbicular to broadly orbicular leaves whose marginal cells are not differentiated; the paroecious inflorescence; and the distinctly beaked perianth. The plants are generally somewhat larger and more erect than those of the *S. pumilum-polaris* complex and bear broader leaves (although those of *S. pumilum* subsp. *polaris* may become very similar in shape; in this taxon, however, the leaf cells are very distinctly smaller); they also never acquire the blackish green color of the last species. Also, *S. sphaerocarpum* often occurs some distance from running water, while *S. pumilum* and (southward) *S. pumilum* subsp. *polaris* are often plants of ravines and rocky stream walls.

Plants of this species are usually freely fertile, allowing ready determination of the paroecious inflorescence. This at once separates *S. sphaerocarpum* from such potentially confusing species as *S. hyalinum*, *S. crenuliformis*, and *S. pyriflorum*. The colorless rhizoids (rarely purplish in my experience) usually at once separate it from *S. hyalinum* and *S. crenuliformis*. Branching may be quite frequent; it is always lateral-intercalary. Confusion is most probable with *S. levieri*.[351]

However, sterile plants can be told from those of *S. levieri* by the smaller cells (18–22 μ up to 22–27 μ on the margins of the leaves, ca. 24–28 $\times$ 32–36 μ in

[350] The species also occurs under feebly basic conditions at Smuggler's Notch, Vt., associated with *Lophozia heterocolpa*, *L. gillmani*, *Scapania cuspiduligera*, and *Preissia quadrata*, as well as the northern, basophilous *Saxifraga aizoön*.

[351] The distinctions between *S. levieri*, *S. sphaerocarpum*, and *S. pyriflorum* are discussed in detail under *S. levieri;* see especially Table 8 on p. 962.

the leaf middle), by the orbicular leaves that (on sterile shoots) are rarely over 1.2 × as wide as long, and by a tendency to have the trigones less salient. Müller (1942) emphasized this [latter feature] as one of the differentiating characters, but in my experience, the degree of development of the trigones is too apt to be influenced by environmental conditions . . . . (Schuster, 1953).

Surprisingly, *S. sphaerocarpum* is reported to be diploid ($n = 18$), while the larger-celled *S. levieri* is haploid ($n = 9$). In "typical" *S. sphaerocarpum* there is no or virtually no perigynium; in *S. levieri* a distinct, if low, perigynium is present. However, some variants of the polymorphous *S. sphaerocarpum* (var. *nana*) are normally provided with distinct perigynia, reducing this difference to relatively negligible importance.

Variation

In addition to the "typical" species described above, the var. *nana*, a small, high arctic and alpine form or variety, occurs in our region. From the west of our area the "var. *amplexicaulis*" [*Aplozia amplexicaulis* Dumort., Hep. Eur., p. 60, 1874 (= *Solenostoma amplexicaulis* Steph., Spec. Hep. 2:58, 1901; = *Haplozia sphaerocarpa* var. *amplexicaulis* K. Müll., Rabenh. Krypt.-Fl. 6(1): 549, fig. 279, 1909)] has been reported. This latter variant appears to be merely a large phase of the species, occurring on wet rocks or wet, rocky ground; it has small or virtually no trigones and slightly larger cells than the type, as well as more clasping and distinctly cordate leaves.

SOLENOSTOMA SPHAEROCARPUM var. NANA
(Nees), comb. n.

[Fig. 272:13–14]

Jungermannia nana Nees, Naturg. Eur. Leberm. 1:317, 1833.
Aplozia nana Breidl., Mitt. Naturw. Ver. Steiermark 30:304, 1893.
Jungermannia sphaerocarpa var. *lurida* Pears., Hep. Brit. Isles, p. 298, 1902.
Aplozia sphaerocarpa var. *lurida* Bryhn, Rept. 2nd Norwegian Arctic Exped. in the "Fram" 1898–1902, 2(11):29, 1906.
Haplozia sphaerocarpa var. *nana* K. Müll., Rabenh. Krypt.-Fl. 6(1):548, fig. 278, 1909.
Aplozia sphaerocarpa var. *nana* Macvicar, Studs. Hdb. Brit. Hep. ed. 1:140, 1912; *ibid.*, ed. 2:145, 1926.
Jungermannia sphaerocarpa var. *nana* Frye & Clark, Univ. Wash. Publ. Biol. 6:295, (1943) 1944.

Smaller, more compact than the type, often forming dense patches or sods, brownish or olive to brownish black in color. Shoots only 8–16(20) mm long, ascending to erect. Leaves *erect to erect-appressed* on fertile stems, erect-spreading to erect on sterile shoots, *orbicular-reniform, broader than in type.* ♀ Bracts erect-appressed, sheathing. *Perianth sharply 4-angulate and plicate to near base, distinctly and tubularly beaked; often with a low, vestigial perigynium.*

Type. Riesengebirge (*Limpricht*); fide Müller, 1905–16, p. 550; this typification must be an error, since Nees (1833, *in* 1833–38, pp. 317–18) mentions no locality from the Riesengebirge. I suggest Nees's collection from near the Schlingelbande be considered the lectotype.

Distribution

Evidently an arctic and alpine form of exposed sites; perhaps not warranting nomenclatural distinction, since it seems to be a mod. *parvi-folia-densifolia*.

GREENLAND. See distribution cited under var. *sphaerocarpum*.   ELLESMERE I. Goose Fjord on S. coast (Bryhn, 1906, *sub A. sphaerocarpa* var. *lurida*).

Otherwise reported from northern Europe (Sweden and Norway), Wales and Scotland, central Europe (Germany, Switzerland, Austria, France) and Siberia (Arnell, 1928–30).

SOLENOSTOMA (SOLENOSTOMA) PUSILLUM

(Jens.) Steph.

[Figs. 281; 282:7–12]

Aplozia pusilla Jens., Rev. Bryol. 39:92, 1912; K. Müller, Rabenh. Krypt.-Fl. 6(2):746, fig. 201, 1916 (as *Haplozia p.*).
Solenostoma pusillum Steph., Spec. Hep. 6:83, 1917; K. Müller, Rabenh. Krypt.-Fl. ed. 3, 6:832, fig. 286, 1956.
Jungermannia pusilla Buch, Suomen Maksasammalet, p. 71, 1936; Arnell, Illus. Moss Fl. Fennosc. 1:102, fig. 44:1, 1956 [nec *J. pusilla* L., Spec. Pl., p. 1136, 1753 = *Fossombronia pusilla* (L.) Dumort.].
Jungermannia (subg. *Solenostoma*) *pusilla* Amak., Jour. Hattori Bot. Lab. no. 22:59, figs. 34p, 35a–h, 1960.
Jungermannia jenseniana Grolle, Österr. Bot. Zeitschr. 111:190, 1964.

Small, usually only 2–6 mm high, shoots chiefly 800–1000 to 1650 μ wide, *light green to* (at least near apices of plants) *somewhat reddish or claret colored.* Often with free development of *slender, ± small-leaved innovations.* Stem procumbent to (with crowding) erect, 250–290 μ in diam.; dorsal cortical cells short-oblong, mostly 16–20(25) × 25–40(48) μ, relatively thin-walled; *rhizoids long*, colorless to ± *reddish.* Leaves rather dense, semiamplexicaul, rather obliquely patent to erect and almost erect-appressed, obliquely inserted by a strongly arcuate line (both dorsal and postical bases of leaves of sterile stems long-decurrent, often conspicuously so), often somewhat concave, on weak sterile shoots broadly orbicular, on vigorous, sterile shoots *broader than long to* (on subinvolucral leaves) *conspicuously broader than long,* rotund-reniform to *reniform,* on sterile shoot sectors ca. 800 μ wide × 485 μ long (measured along midline), on potentially fertile shoots becoming broader and 900 × 550 μ to 980–990 × 575–595 μ up to 1000–1050 × 635–650 μ,[352] the *width up to 1.4–1.65× the length*; larger leaves of potentially fertile shoots strongly narrowed basally, not conspicuously decurrent. Cells thin-walled: the marginal (where ± isodiametric) 18–21 or 18–23 μ, rarely 20–25 μ; the median with conspicuous to rather nodular trigones,

---

[352] Dimensions from *RMS 46201*, which is typical; in the larger phenotype (*Williams 10685*) to 1550 × 880 μ (ca. 1.75× as wide as long).

FIG. 281. *Solenostoma* (*Solenostoma*) *pusillum*. (1) Three leaves, note characteristic abaxial rhizoids from lamina (×15.5). (2) ♀ Bract (×15.5). (3) Perianth mouth apex (×105). (4) Stem cross section (×135). (5) Paroecious shoot with yet immature perianth (×22). (6–7) Median and apical cells, respectively (×210). (8) Median cells with oil-bodies (×545). (9) Two subinvolucral leaves (×22). (10) Two leaves from sterile shoot sector (×22). [1–5, var. *vinaceum*, H. Williams *10685*, Quebec; 6–10, RMS *46201b*, Nucluet, Booth Sd., Greenland].

quadrate to polygonal, 20–26 μ wide × 22–28 μ long; cuticle smooth (*fide* Müller) or distinctly, if finely, verruculose to striolate; oil-bodies varying from (2)3–6(7–8) per median cell; if 2–3, elliptical and usually very large (6–7.5 × 10–12 μ), if 4–6(7–8) per cell smaller, subspherical and 5.5–6 × 5.5–7 μ to ovoid and 5.5–6 × 8–10 μ, finely granular and ± smooth externally.

Paroecious. ♂ Bracts below the perichaetial, usually in 1–2(3) pairs, rather concave at base, reniform, otherwise similar to vegetative leaves, diandrous. ♀ Bracts nearly twice as wide as long, reniform. Perianth *with a short but distinct tubular stem perigynium at its base*, the distal half emergent from between the ± erect or erect-appressed bracts; perianth somewhat ovoid in general outline, bistratose below, in the distal 0.4–0.65 with *usually 4 rounded but salient plicae*, contracted distally into a small beak crenulate with projecting fingerlike cells. Spores red-brown, 15–18(20) μ in diam.; elaters 8–9 μ in diam.

Type. Sweden.

Distribution

Imperfectly known, since in the past this taxon has been frequently confused with *S. sphaerocarpum*, especially var. *nana*. Widespread in Scandinavia (Sweden, Norway, Finland, Denmark; Arnell, 1956; Mårtensson, 1955) and extending northward to 79° N in Spitsbergen (Arnell and Mårtensson, 1959); southward in the Transylvanian Alps (at Buksoi), near Freiburg, Germany (Müller, 1951–58, p. 834), southward to the Pyrenees (Ax-les-Thermes); according to Arnell (1956), also in Siberia. Reported (Amakawa, 1960a, p. 59) from Japan (Honshu, Hokkaido).

In North America, found only four times, as follows:

NW. GREENLAND. Nucluet, Booth Sd., Steensby Land, 76°57′ N., 70°50′ W. (*RMS 46201, p. min. p.*). QUEBEC. Roggan R., Ungava, ca. 54°15′ N., 78–79°44′ W., and below Anataiatch, in same area (Lepage, 1953, p. 108; based on determinations by S. Arnell). MICHIGAN. Pictured Rocks, Alger Co. (*Hattori 389; teste* Amakawa, 1960a).

Ecology

Supposedly restricted to loamy or clayey to clayey-sandy soil, "especially on the sides of roads and ditches in the forest region," although "most abundant in the lower mountain region . . . and ascends the mountains to somewhat above the upper limits of the birch" (Arnell, 1956). Mårtensson (1955) states that the species occurs from the conifer region "up to the high alpine belt . . . at altitudes up to 1400 m." According to the latter author, chiefly "collected from 'bare,' poor, sandy soil, e.g., frost-heaving eaith, wind erosion patches and non-calcareous, late snow areas. The associated species are generally more or less common, such as *Cephalozia ambigua, Lophozia alpestris, Nardia geoscyphus . . . Pleuroclada albescens . . .* a.o." The Greenland plants were found over clayey, irrigated soil on gentle, west-facing slopes near sea level, associated with *Dryas*, *Cassiope tetragona*, grasses, and the hepatics *Lophozia quadriloba*, *Blepharostoma*, *Tritomaria quinquedentata*, and *Solenostoma pumilum polaris*; the species occurs here at the northernmost station in North America, under high arctic conditions.

The Quebec specimens were collected on "beaten tracks" and "moist banks," associated with *Scapania mucronata*, *Blepharostoma*, *Cephalozia pleniceps*, *Lophozia bicrenata*, *Cephalozia ambigua*, *Nardia geoscyphus*, *Dicranella subulata* (Lepage, 1953).

Differentiation

A small, usually compact plant, often producing reduced-leaved shoots (much as in *S. gracillimum*, from which it differs at once in the paroecious inflorescences and in the nonbordered leaves). Sun forms develop a reddish coloration; in the Greenland material that I have studied the color grades toward claret-red or even almost vinaceous; this coloration is found chiefly on the more robust and potentially fertile shoots and on the exposed shoot apices; the remainder of the plant is relatively light green in color. Rhizoids are often slightly purplish-tinged. The coloration of the leaves, when present, excludes all forms of the very closely allied *S. sphaerocarpum*, where secondary pigments always appear to be brown. With this exception the separation from *S. sphaerocarpum* (and especially the small var. *nana*, which also has 4-plicate perianths) appears to be quite subjectively conceived. Arnell is quoted (Lepage, 1953) to be of the opinion that *S. pusillum* is not specifically distinct from *S. sphaerocarpum*, although Arnell (1956) and Müller (1956, *in* 1951–58) both maintain it as distinct. According to Müller, the "lichtgrüne bis rötliche Farbe, die breiteren als langen Blätter und die noch deutlichere Marsupienbildung," together with the presence of numerous innovations, separate it specifically from *S. sphaerocarpum* var. *nana*.

When the shade form (mod. *viridis*), which may grow to a height of 12–15 mm, is at hand (as in *RMS 46201b*), it is almost impossible—and perhaps futile—to attempt a separation from *S. sphaerocarpum*. Small forms of the latter (var. *nana*), with quadriplicate perianths and a distinct, if low, shoot perigynium, can hardly be separated from *S. pusillum*. However, plants of *RMS 46201*, from a closely juxtaposed site, are "typical"; except where crowded, shoots are only 4–5(6) mm tall, with occasional small-leaved innovations; shoot apices, at least, are reddish to vinaceous, as occasionally are some lower leaves and some rhizoids; where no reddish color is developed, the shoot sectors are light green; leaves are broadly reniform and much broader than long. The coloration appears quite different from any form of "normal" *S. sphaerocarpum*. Dimensions of leaves, leaf cells, and oil-bodies in the species diagnosis are drawn from *RMS 46201*.

Referable here—yet deviant in some ways—are plants from the Gaspé (*Williams 10685*); the few individuals seen lack small-leaved innovations, are taller (5–12 mm), and have the paroecious inflorescence poorly demarcated (demonstrable on a single individual; several other plants with young perianths lacked demonstrable antheridia in the leaves below the perianth). Also, the ♀ bracts are much less broadly orbicular-reniform. Yet the beaked perianths, with 4 rounded but salient plicae; the reniform to rotundate-reniform leaves; the leaf-cell size; and, above all, the strongly vinaceous, local pigmentation of the otherwise rather pellucid, light green plants are striking. Intense vinaceous to claret-red pigmentation occurs on the postical leaf bases, spreading from there to the rhizoids (which commonly arise from the postical sector above the leaf base)[353] and even median sectors of leaves and ♂ bracts. The intense red pigmentation, coupled with the otherwise light green color, suggests *S. gracillimum* and *S. hyalinum* rather than *S. sphaerocarpum*. Sterile plants, indeed, might well be sought under *S. hyalinum*.

These plants depart so far from the "normal" concept of *S. pusillum* that they are better segregated as a separate variety:

SOLENOSTOMA (SOLENOSTOMA) PUSILLUM var. VINACEUM Schust., var. n.
[Figs. 281:1–5; 282:11–12]

Varietas maior quam var. *pusillum*, usque ad 12–15 mm alt. × 1.4–1.65 mm lat., rhizoideis basibusque foliorum intense vinaceis; ♀ bracteae minus reniformes, foliis infra non conspicue maioribus quam folia vegetativa; perigynium nullum.

Larger, 5–12 mm tall or *occasionally 15 mm tall* × 1.4–1.65 mm broad, light and clear green, except an extensive area along ventral leaf bases, extending to the rhizoids (particularly those arising from the leaf lamina), *intensely vinaceous;* erect or suberect, apparently lacking small-leaved innovations. Stems 355–410 μ in diam., 13–15 cells high; cortical cells larger and longer, ca. 25–28 × 70–110 μ. Leaves obliquely to rather widely patent, broadly rotund-reniform, to 1550 μ wide × 880 μ long (1.75× as wide as long), decurrent at both bases, with reddish to pale rhizoids arising from basal half of nonmarginal parts of lamina. ♀ Bracts less reniform, little larger than leaves (*leaf size not conspicuously increasing upward on fertile plants*), 1425–1550 μ wide × 980–1125 μ long. Leaf cells smooth, with conspicuous but not coarse, somewhat bulging trigones, relatively large: marginal cells, except where strongly elongated, 24–27 μ distally; median cells ca. 26–34 × 26–38 μ. ♂ Bracts hardly ventricose, the shallowly concave *basal sectors vinaceous;* perianth deeply 4-plicate, beaked, *without subtending perigynium.*

[353] They may arise from the nonmarginal leaf and bract bases to a point above the leaf or bract middle.

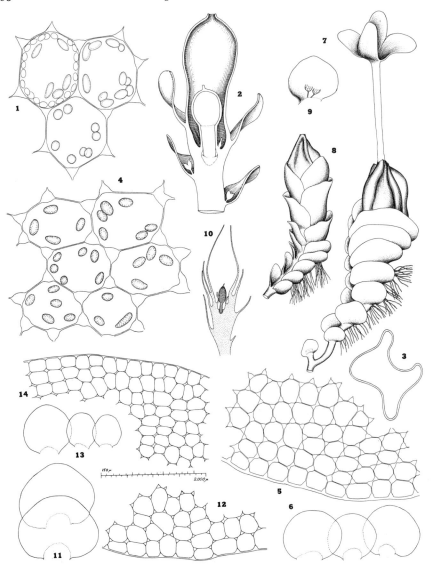

Fig. 282. *Solenostoma* (*subg. Solenostoma*) (1–6, *S. levieri* = *scalariformis;* 7–12, *S. pusillum;* 13–14, *S. sphaerocarpum*). (1) Median cells with oil-bodies and, upper left, chloroplasts (×490). (2) Shoot apex, in longisection, through immature sporophyte (×ca. 19). (3) Perianth cross section, through distal one-third (×ca. 26). (4) Median cells with oil-bodies (×610). (5) Cells along apex of leaf margin; drawn to top scale (×165). (6) Leaves, away from gynoecial regions; drawn to lower scale (×14). (7) Sporophyte-bearing shoot (×15). (8) Perianth-bearing shoot (×15). (9) ♂ Bract (×15). (10) Longisection through gynoecial shoot (×15). (11) Leaves; drawn to lower scale (×14). (12) Cells along leaf apex; drawn to upper scale (×165). (13) Leaves, the one on left from ♀ shoot, the two at right from sterile shoot; drawn to lower scale (×14). (14)

Type. Quebec: north slopes of Mt. Albert, Parc de la Gaspésie (*H. Williams 10685*). The type was mixed with *Tritomaria polita.*

SOLENOSTOMA (SOLENOSTOMA) LEVIERI (Steph.)
Steph.

[Figs. 271:10; 282:1–6]

Nardia levieri Steph., Bot. Centralbl. 50:70, 1892.
Haplozia lurida Breidl., Leberm. Steiermarks, p. 306, 1893 (not of Dumortier, 1874).
Solenostoma levieri Steph., Spec. Hep. 2:50, 1901; K. Müller, Hedwigia 81:120, fig. 7d-f, 1942;
    Schuster, Amer. Midl. Nat. 49 (2):398, 1953; K. Müller, Rabenh. Krypt.-Fl. ed. 3,
    6(1):830, fig. 285e-i, 1956.
Haplozia levieri and *H. breidleri* K. Müll., *ibid.* ed. 2, 6(1):552 and 553, 1909.
Haplozia scalariformis Schiffn., Österr. Bot. Zeitschr. 1910, no. 12:1 [of reprint] (not *Junger-
    mannia scalariformis* Nees, 1836).
Jungermannia levieri, Buch, Evans & Verdoorn, Ann. Bryol. 10(1937):4, 1938; Grolle, Österr.
    Bot. Zeitschr. 111:190, 1964.

Superficially identical to *S. sphaerocarpum,* with size and aspect the same; differing chiefly as follows (*in italics*): *leaves reniform to transversely elliptical, frequently retuse or emarginate,* bright green to somewhat olive-tinged, upper portions occasionally locally dull-purplish-tinged, varying from 970 μ wide × 650 μ long to 1050–1120 μ wide × 700–850 μ long to 1200 μ wide × 750 μ long (*width 1.3–1.6× the length*), base (on at least one side) usually *broadly rounded,* appearing conspicuously dilated. *Cells larger than in S. sphaerocarpum:* marginal, typically *28–32* μ (with extremes of 24 and 35 μ, rarely with many elongated at right angles to margin and then only 20–23 μ, tangentially measured); median, usually 34–39(42) μ wide × 35–50 μ long [rarely (28–32)30–35 × 30–40 μ; then grading in size into those of *S. sphaerocarpum*]; basal, (35)40–45(48) × (50)55–72 μ; *trigones conspicuous, usually somewhat bulging;* cuticle faintly verruculose-striolate, at least in basal portions of leaf; oil-bodies varying from ellipsoidal to bacilliform, 3.5–4.5 × 7–7.5(12) μ to 5–6 × 7–9(11) μ, in some cells spherical and 5–6 μ, from (3–4)5–8(11) per cell, appearing smooth and virtually homogeneous (under oil immersion very faintly granular), *their length averaging well below ¼ the cell length.*

Paroecious. Usually with a low, distinct perigynium. Sporangium spherical, almost blackish, in transmitted light the *wall brown.* Epidermal layer with cells varying from rectangulate and (24)26–30 × 36–48 μ to subquadrate and 35–40 × 40–48 μ, the longitudinal walls with 3–4 nodular thickenings extended as tangential spurs, the transverse walls with 0–2 similar thickenings. Inner layer with cells narrowly rectangulate, often unequally wide at the two ends, varying considerably in width, averaging (13)15–16(18) μ wide × 80–90 μ long, with numerous complete semiannular bands. *Spores 19–24* μ, pale brown, very minutely verruculose. Elaters 8–9(10) μ in diam., closely 2(3)-spiral, the

---

Cells of leaf apex; drawn to upper scale (×165). [1, *RMS 39290b*, Pictured Rocks, Mich.; 2–6, *RMS 17694*, Sand I., Bayfield Co., Wisc.; 7–10, from K. Müller, 1905–16; 11–12, *Williams 10685*, Quebec, var. *vinaceum*; 13–14, leg. H. Williams, Gaspé Bay, Quebec.]

spirals 2.5–3 μ wide, the pointed ends of the elaters unispiral with the spiral fused with the elater wall, the narrowed tips appearing thickened. Chromosome no. $n = 9$.

Type. Caucasus Mts. (west Swanetia, in the mountain pass Utbiri, *Levier*).

Distribution

Much rarer than *S. sphaerocarpum*, occurring to higher altitudes (2400 m in the Caucasus; 2667 m in Switzerland). Found in the Caucasus Mts. at 2400 m (type), in the Carpathians (Bistriz), Switzerland (in the Jura at Chasseral; Cantons Bern, Graubünden, Wallis), Germany (Black Forest in Baden; Bavaria), France (Mt. Dore, the Pyrenees), Austria (Tyrol at Geschnitztal; Steiermark); also in northwest Himalaya (Patarotal, at 4700 m).

In the United States known so far only from the Lake Superior shore, as follows:

WISCONSIN. Sand I., Apostle Isls., Bayfield Co. (*RMS 17694, 17695*). MICHIGAN. Pictured Rocks near Munising, Alger Co. (*RMS 39290b, 39294; material* at least partly transitional to *S. sphaerocarpum*!).

Ecology

Found in sites similar to those inhabited by *S. sphaerocarpum*: damp to moist, rarely rather intermittently moist, rock walls and ledges, usually in slightly acidic sites (then with *Mnium, Scapania curta, Jungermannia lanceolata, Plagiochila asplenioides, Blepharostoma trichophyllum*, as on Sand I.; or with *Cephaloziella arctica, Scapania saxicola, Solenostoma hyalinum, Anthelia juratzkana*, at Pictured Rocks). Occasionally on weakly calcareous sites, then associated with *Lophozia gillmani* and *L. heterocolpa* (Sand I.).

Differentiation

Both Müller (1942, 1951–58) and Schuster (1953) deal with the separation of this "weak" species from *S. sphaerocarpum*. It is superficially identical with the latter but has 9 chromosomes rather than 18, although the larger cells and spores suggest that the inverse relationship should occur! *Solenostoma levieri* has larger leaf cells, the marginal cells varying usually from 28–32 μ, median cells from 34–39(42) μ wide × 35–50 μ long. However, there is much variation in cell size within the same plant: the lower leaves usually have conspicuously smaller cells than the upper, larger leaves and ♂ bracts. As a consequence, differences in cell size must be used with caution. Minimal cell sizes (found on some leaves in the

Michigan plants; see figures within parentheses in the diagnosis) are often essentially those of *S. sphaerocarpum*. Oil-bodies in *S. levieri* are no larger than in *S. sphaerocarpum;* hence, on mature leaves oil-bodies average well below ¼ the length of median cells (see Schuster, 1953). Perhaps more significant is the difference in leaf shape; in *S. levieri* leaves are, on an average, 1.3–1.6 × as wide as long (length measured along midline) and are often reniform rather than broadly orbicular; associated with this, the leaves are often somewhat retuse, rarely emarginate, at the apex. Within limits, the larger cell size is also characteristic of the sporophytic cells. Müller gives the epidermal cells of the capsule wall as 40–45 μ wide, the spores as 18–24 μ,[354] and the elaters as 2–3-spiral and 10 μ in diam. The validity of these differences was tested in plants from Pictured Rocks, just east of Miners Castle, Mich. Here plants of *S. sphaerocarpum* and plants assignable to *S. levieri* (sporophyte diagnosis, above, from these plants), as well as plants of the closely allied *S. pyriflorum*, occurred under similar conditions, within 150–200 feet. Table 8 is derived from these materials.

Table 8 shows that *S. pyriflorum* is separable from the other two taxa on the basis of sexuality, violet capsule wall and contents, smaller leaf cells, ill-defined thickenings of the inner layer of the capsule wall, etc. A series of differences also emerges between plants assigned to *S. levieri* and those placed in *S. sphaerocarpum*. Unfortunately these distinctions are subject to some intergradation—plants with the spore size and other characters of *S. sphaerocarpum* (see Table 8, *RMS 39303a*) may have leaves broadly rotundate (700 μ wide × 490–600 μ long) to reniform (800 μ wide × 525 μ long to 980 μ wide × 650 μ long, particularly below the gynoecia). Such a leaf shape is supposedly characteristic of *S. levieri*.

The dubious nature of these distinctions is clear from study of Michigan plants (*RMS 39294*) which have the low perigynium of *S. levieri*, reniform leaves averaging 1.3–1.6 × as broad as long, and usually 5–11 oil-bodies per cell that average well below ⅕ the length of mature cells. Spores range from 19.2 to 24 μ, but the elaters are in part only 8–9 μ in diam., with spirals 3 μ wide, while other elaters are 10–11.5 μ wide, with spirals 4 μ wide; in all cases the elaters are bispiral. Epidermal cells of the capsule wall are slightly smaller than indicated by Müller; their dimensions are given in the diagnosis. These plants also show wide variation in cell size, many leaves having cells hardly or no larger than those of *S. sphaerocarpum*. Minimal cell sizes (of the leaves) in Table 8 are derived from these plants. Hence, on the basis of cell size the plants

---

[354] Müller, (1951–58, p. 814) states that spores are "15–18 μ" in *S. sphaerocarpum* vs. 18–24 μ for *S. levieri*. However, on p. 829 he gives the spore size of *S. sphaerocarpum* as 18–22 μ, with elaters 7–8 μ in diam. The spore-size distinction, hence, acquires a dubious status. In Gaspé plants (*Williams 10797*) spores are 19–24 μ and purplish brown, even though the plants are, otherwise, quite typical *S. sphaerocarpum*. The spores measured are from old capsules and probably had undergone pregermination swelling. I think *only* spore sizes derived from unopened capsules can have much relevance.

TABLE 8

Criterion	S. levieri (RMS 39294)	S. sphaerocarpum (RMS 39303a)	S. pyriflorum (RMS 39302)
Capsule-wall, elaters & spores	Brown	Brown	Violet to purple
Epidermal cells of capsule	(24)26–30 × 36–48 to 35–40 × 40–48 μ	15–20 × 30–35 to 22–25(30) × 22–30 μ	20–26 × (23)27–30(32) μ
Thickenings of epidermal cells	3–4 on long, 0–2 on shorter, walls	(1)2–3 on long, 0–1(2) on shorter, walls	1–2(3) on long, 0(1) on shorter, walls
Inner cells	(13)15–16(18) × 80–90 μ	(8)10–15(16–17) × 45–50 μ	(12)13–15(17) × 50–65 μ
Thickenings of inner cells	Sharply defined, semiannular, usually complete	Sharply defined, complete and *incomplete* semi-annular bands	4–6 weakly defined, complete to obsolete semiannular bands
Spores	19–24 μ, brownish	16–18 μ, brownish	15–18 μ, violet
Elater diam.	8–9(10) μ	7–9 μ	7–8 μ
Marginal leaf cells	(24)28–32(35) μ	20–24(25) μ	15–19(20–22) to 19–22 μ
Median cells	(28–32)30–35 × 30–40 μ up to 34–39(42) × 35–50 μ	25–32(33) μ wide and long	20–24(26) × (20)23–25(28) μ
Basal cells	(35)40–45(48) × (50)55–72 μ	28–36 × 40–48(50) μ	20–26 × 30–37 μ
Perianth apex	Green to brown	Green to brown	Green to violet-purple
Sexuality	Paroecious	Paroecious	Dioecious

would be placed as *S. sphaerocarpum*, yet the spores and capsule wall are typical of *S. levieri*.

The reniform leaf shape of *S. levieri* is of dubious value as a species criterion. Müller, e.g., illustrates *S. sphaerocarpum* with a distinctly reniform leaf (1951–58, fig. 283d), and admits (*loc. cit.*, p. 831) that the reniform or elliptical leaf shape also occurs, "albeit gelegentlich" in *S. sphaerocarpum*. Therefore it is not always possible to distinguish between these two species, if they are truly "good" species; their separation must involve study of a constellation of features, since no one criterion affords a constant distinction.

Both *S. sphaerocarpum* and *S. levieri* exhibit a marked and diagnostic tendency for the perianth surface and the lower portions of the leaf surfaces to give rise to occasional rhizoids. The perianths are often brownish-tinged but never acquire the intensely purplish—almost blackish purple—coloration of *S. pyriflorum*.

The *S. sphaerocarpum-levieri* complex is easily confused with *Nardia scalaris*, since these taxa share the orbicular-reniform leaves, a generally bright green to yellow-green color, and a similar leaf insertion and size. However, the constant and uniform lack of underleaves will separate all species of *Solenostoma* from *Nardia scalaris*. Furthermore, *N. scalaris* and its immediate relatives are softer plants, with thicker, fleshier stems. It is sometimes difficult to establish the sexuality of material of this species, and of *S. sphaerocarpum*, when plants are collected with mature or old perianths. In the latter case it is usually possible to find weathered perianths with subfloral innovations on which antheridial bracts are already distinct (and sometimes the apical cluster of archegonia is already initiated).

SOLENOSTOMA (SOLENOSTOMA) PYRIFLORUM
Steph.

[Figs. 271:1–7; 283]

Solenostoma pyriflorum Steph., Spec. Hep. 6:83, 1917.
Jungermannia pyriflora Steph., *ibid.* 6:90, 1917.[355]
Jungermannia monticola Hatt., Bull. Tokyo Sci. Mus. 11:33, figs. 19–22, 1944.
Plectocolea jishibana Hatt., *ibid.* 11:38, 1944 (*nomen nudum*).
Plectocolea pyriflora Hatt., *ibid.* 11:38, 1944.
Jungermannia (subg. *Solenostoma*) *pyriflora* Amak., Jour. Hattori Bot. Lab. no. 22:59, figs. 1(partim), 2 (partim), 4 (partim), 35i–r, 1960.[356]

Forming rather compact, *yellow-green* sods or small turfs, more rarely scattered among other Hepaticae, somewhat caespitose and *erect or suberect*, usually somewhat crowded, small, shoots 5–12 mm high × *1.2–1.5* mm *wide*, subsimple, rare branches intercalary. Stems to 230–270 μ in diam. × ca. 220–245 μ high, very slightly flattened and broadly elliptical in cross section, *ca. 10–12 cells high*, somewhat fleshy; cortical cells dorsally ca. (20)22–27 μ wide × 40–50(60) μ long, somewhat thick-walled, much shorter in 2 cortical strata than medullary cells *but hardly differing from these in diam.*; medullary cells somewhat thick-walled, 23–27(35) μ in diam., the *stem in cross section very uniform in structure;* medulla hardly dorsiventrally differentiated, the ventral 2–3 strata of slightly smaller cells. Rhizoids abundant, to shoot apex, ca. 12–14 μ in diam., long, colorless or pale brownish, *sometimes with light, dull purplish pigmentation.* Leaves rather soft, distant to contiguous, the lowermost clearly obliquely inserted, those above *transversely or subtransversely oriented* (although the insertion remains perceptibly oblique), *semiamplexicaul, the antical margin long-decurrent*, the postical margin less so, *suborbicular to rotundate-reniform* (ca. 725 μ long along midline × 975 μ wide to 780–820 μ long × 1025–1050 μ wide, occasionally to 900 μ long × 1200 μ wide), rounded or rounded-truncate but *never emarginate* at apex, rather strongly concave, *erect-spreading to spreading.* Cells thin-walled, with distinct, *usually somewhat bulging trigones, the marginal not differentiated;* marginal and *submarginal cells 15–19 μ, median cells 20–24(26) × 23–25(28) μ;* cuticle smooth; oil-bodies (2)*3–8(9) per median cell, finely granulose to nearly smooth* in appearance, with a

---

[355] Stephani described this taxon twice—once in each of two genera, from collections made by different persons; the first description and type cited should be regarded as the valid ones.

[356] Unaware of its similarity to the Japanese *S. pyriflorum*, I identified my initial collections of the American populations as a new species; material of these early collections has been distributed as "*Solenostoma appalachianum*," a name which has never been validated. Amakawa agrees that the American populations are identical with the Japanese *S. pyriflorum*.

small, highly refractive central globule, spherical and 6–8 μ to ellipsoidal and 5 × 8–10 to 7 × 13 μ; chloroplasts ca. 4 μ. Asexual reproduction lacking.

Dioecious, but ♂ and ♀ plants often in the same patch. ♂ Plants only slightly less robust than ♀, ca. 1200 μ wide (the androecium 900–1100 μ wide), the androecia *compact*, terminal (at least initially), *usually of 4–6 or more pairs of closely imbricate bracts;* bracts similar to leaves but strongly concave, except for the occasionally slightly flaring or reflexed margins. Gynoecia usually without subfloral innovation; bracts and subinvolucral leaves (which gradually decrease in size below) larger than vegetative leaves, subtransversely inserted and oriented, usually *quite concave* and erect-spreading to somewhat patent, the bracts not sheathing perianth; bracts usually *broadly rotundate to subreniform*, to 1400–1500 μ wide × 825–850 μ long. Perianth oblong-obovate, ca. 750–900 μ in diam. × 1.3–1.6 mm long, ⅓–⅔ emergent, usually *deeply 4(5)-plicate* in distal half, 2 of the plicae latero-antical (and with a deep sulcus lying between them), the apex rather suddenly contracted into a *short (and often almost obscure) beak;* perianth firm and 2-stratose in basal 0.4–0.6, 1-stratose distally; cells at mouth from 12 × 18 to 13 × 30 μ or 10 × 30–40 μ, fingerlike, their rounded ends free (the orifice of the beak thus crenulate), cells below the fingerlike terminal row subisodiametric, collenchymatous, ca. 18–20 × 20–30 μ; median cells of perianth *strongly collenchymatous*, ca. 22–25 × 23–30 μ, *subisodiametric; perianth apex often dull-purplish-pigmented. Perigynium usually present, but a mere low ring* (ca. 0.05–0.15 the height of the free perianth). Sporophyte with capsule, spores, and elaters ± violet-pigmented. Capsule wall 2-stratose; epidermal cells in large part subquadrate, except near base, ca. 20–26 μ broad × (23)27–30(32) μ long, occasional ones with longer axis transversely oriented, the shorter (usually transverse) walls with 0(1) thickening, the longer walls with 1–2 (rarely 3) thickenings, the thickenings nodular, with faint, tangential, spurlike extensions. Inner layer of irregular cells, often irregularly oriented, in large part narrowly rectangulate, occasionally faintly sigmoid, most often (12)13–15(17) μ wide × 50–65 μ long, each cell with usually 4–6 weakly defined, complete to *subcomplete to almost obsolete violet semiannular bands*, the bands often gradually confluent with tangential cell wall. Spores violet, delicately verruculose, and superficially appearing smooth, 15–18 μ in diam.; elaters 7–8 μ in diam., bispiral, *contorted*, violet or violet-brown, the spiral 2–2.5 μ wide; *tapered ends of elaters thickened and pointed.* (Sporophyte description based on Michigan material.)

Type. Mt. Natsuzawa, Nagano Pref., Honshu, Japan (*Ihsiba*); G.

Distribution

Widely distributed in subalpine and alpine sites in Japan, rarely descending into the lowlands; from 600 to 2800 m.

FIG. 283. *Solenostoma* (*Solenostoma*) *pyriflorum*. (1) Perianth-bearing shoot (×25). (2) ♂ Plant (×25). (3) Perianth-bearing shoot apex (×25). (4) Mature sterile shoot (×25). (5) Cells with oil-bodies and, lower left, chloroplasts (×650). (6) Cells of leaf apex (×300). (7) Two leaves (×22.5). (8–9) ♀ Bract and leaf, respectively, from one plant (×22.5). (10) Perianth mouth sector (×260). (11) Median cells of perianth (×260). [All from Neddie Creek, N.C., *RMS*; see also Fig. 271:1–7.]

Not previously reported from North America, where it is widespread from Virginia southward barely into South Carolina and Georgia, in the Escarpment gorges; mostly at 2100–4500 feet in the Southern Appalachians, thus not subalpine (absent from the Spruce-Fir Forest Zone or barely penetrating it).

VIRGINIA. Near the Cascades, Little Stony Cr., near Mountain L., Giles Co. (*RMS 40286*). NORTH CAROLINA. Neddie Cr., E. fork of Tuckaseegee R. (*RMS 29456, 29379, 29378*) and Chattooga R., 1 mi below Narrows, ledges on E. side, Jackson Co. (*RMS 39468, 39450*); same loc., Macon Co. (*RMS 39450a*); S. of Sunburst, along W. branch from Fork Ridge of W. fork of Pigeon R., near jct. with Flat Laurel Fork, 4500–4600 ft, Haywood Co. (*RMS 39379*); E. side of Whitewater R. gorge, ca. 0.5 mi below High Falls, Transylvania Co. (*RMS 40575*); E. fork, Pigeon R., S. of Sunburst (*Anderson 11135!*); Bubbling Spring, N. of Beech Gap, N. of Blue Ridge Pkwy., 5100–5200 ft (*RMS 39371, 39367, 39375a*); Middle Cr., Black Mts., 4000 ft, Yancey Co. (*Schofield 9551*); Whitewater R. gorge, between ¼ and ⅘ mi below the High Falls, both on Jackson Co. and Transylvania Co. sides of stream (*RMS 39356*). SOUTH CAROLINA. Oconee Co.: rocky moist bank along road cut, 2–3 mi E. of Jocassee (*RMS 30009, 30156*); Whitewater R. gorge, 0.3–0.4 mi below Lower Falls, 3 mi N. of Jocassee (*RMS 40955*); Thompson R. ravine, N. of Jocassee (*RMS 41009b*). GEORGIA. Below High Falls of Big Cr., N. Rabun Co., 4–5 mi SSE. of Highlands in North Carolina, 1900–2000 ft (*RMS 40031, 40714*).

Recurring as a disjunct in MICHIGAN. Moist sandstone bluff just E. of Miners Castle, Pictured Rocks, Alger Co., 15–40 ft above level of L. Superior (*RMS 39298*). The disjunct L. Superior distribution recalls the analogous range of *Diplophyllum obtusifolium* subsp. *obtusatum*, as well as that of *Marsupella paroica*.

Ecology

Solenostoma pyriflorum has ecological characteristics somewhat similar to those of other species of *Solenostoma*; it is occasionally found intermingled with *S. crenuliformis* and with *S. pumilum* and *obscurum* growing close by.[357] Typical habitats are moist rock crevices, more rarely bare rock faces, which are populated with the bright yellow-green, dense patches of the *Solenostoma*. Although extremely localized in distribution, the species is occasionally abundant below waterfalls, occurring on both partially insolated and shaded rocks. At least some of the sites where the species is found appear subject to occasional inundation.

The entirely disjunct station at Pictured Rocks, Mich., on the shore of Lake Superior, diverges somewhat ecologically. The plants occur here under much more exposed conditions, on insolated, acidic, friable sandstone bluffs and steep sandstone slopes, particularly at points where seepage occurs. Although the shaded individuals show the bright, somewhat opaque, yellow-green color of the species, ♂ plants tend to have at least the margins of bracts dull purplish,

---

[357] When associated (as in South Carolina) with *S. hyalinum* and *gracillimum*, markedly distinct from both, even to the naked eye. The South Carolina plants agree very closely with those from the other stations in both color and size (averaging less robust than associated *S. gracillimum;* much smaller than mature plants of *S. hyalinum*), as well as in the form of the oil-bodies.

while ♀ plants have the entire distal half or more of the perianth deep purplish-pigmented—superficially appearing almost black. Except for the color difference, the Michigan plants agree almost perfectly with the Appalachian material, in which secondary pigments appear confined to the very apices, often only the beaks, of the perianth. The material at Pictured Rocks occurred with the Great Lakes phase of *Solenostoma hyalinum*, with *Cephalozia bicuspidata* and *Gymnocolea inflata*, i.e., as a member of the *Gymnocolea-Cephalozia bicuspidata* Associule. A small quantity of material was also found on a damp sandstone bluff where *Anthelia juratzkana* occurred, associated with *S. hyalinum* and *Scapania saxicola*.

Differentiation

The size, the color, the lack of a differentiated border of the leaves, and the dioecious inflorescence, as well as the perianth form and the virtual lack of perigynium, suggest the rare *S. caespiticium*. However, the numerous oil-bodies (1 per cell in *S. caespiticium*), absence of gemmae, and much smaller cells with obvious trigones (cells pellucid, leptodermous, 35–40 × 40–50 μ in *S. caespiticium*) make confusion impossible.

Solenostoma pyriflorum is related closely to the northern *S. sphaerocarpum*. It agrees with the latter in (*a*) the usually 4-plicate, beaked perianth, at its base with only (*b*) a vestigial perigynium; (*c*) the broadly orbicular to orbicular-reniform leaves, inserted by a narrow line, much less than half the leaf width; (*d*) the mostly 3–6 large and conspicuous oil-bodies, formed of fine spherules lying in a matrix of virtually the same refractive index—hence the spherules are barely perceptible; (*e*) the yellow-green color; (*f*) the distinct, sometimes slightly bulging trigones, of relatively small cells; (*g*) the lack of differentiated marginal cells. However, a number of characters at once serve to separate *S. pyriflorum* from *S. sphaerocarpum*, chief among them: (*a*) a clear-cut and consistent dioecious inflorescence;[358] (*b*) a tendency for the apices of the perianths, and some of the rhizoids, to develop purplish pigmentation; (*c*) the less homogeneous appearing oil-bodies, each with a hyaline, obvious granule located in it; (*d*) the somewhat smaller leaf cells, averaging only 20–24 μ wide × 23–25 μ long.

The dioecious inflorescence, color, leaf shape, and low or vestigial perigynium, as well as perianth form, also suggest *S. gracillimum*. However, the smaller cells and lack of a border of differentiated cells, as well as the larger and more numerous oil-bodies, at once separate the two species. Furthermore, *S. gracillimum* never develops any trace of purplish pigmentation of the rhizoids.

[358] Miller (1963, p. 513, fig. 111) has briefly described a Hawaiian species, *S. hawaiicum*, which is also dioecious, as in *S. pyriflorum*, and appears to be allied in other respects as well. Indeed, from his figures and ambiguous diagnosis I can find no characters that would clearly separate the two, aside, perhaps, from the somewhat less broad, yet suborbicular leaves of *S. hawaiicum*.

Confusion of sterile plants is conceivably possible with *S. hyalinum* and *fossombronioides*. However, these are more pellucid and less yellow-green than *S. pyriflorum* and have a very different facies in the field. Furthermore, they have broadly ovate to orbicular leaves inserted by a broad line (they are not conspicuously narrowed basally); the insertion is also much more oblique. When fertile, the perianth form of both species at once isolates them from *S. pyriflorum*.

The species appears to be quite stenotypic regionally. In Japan, where evidently widespread, small forms with ovate leaves (var. *minutissima* Amak.) and large forms with more obliquely inserted leaves [var. *major* (Hatt.) Amak.] are known.

The Michigan material, originating from a relatively strongly insolated site, is deviant in the frequent blackish purple pigmentation of the distal half or more of the perianth; a similar pigmentation of the distal portions of the ♀ bracts, in some cases; and tendencies for similar, if less intense, pigmentation of some of the leaves. Such pigmentation is quite lacking in the associated plants of *S. levieri* that occur juxtaposed at Pictured Rocks, Mich. It is noteworthy that this intense violet-purple pigmentation of the gametophyte is restricted to leaves and perianths, the rhizoids remaining colorless (as contrasted to the associated *Solenostoma hyalinum*, in which rhizoids are always reddish purple). Also noteworthy is the fact that the intense purplish pigmentation also characterizes the capsule wall and its contents. By contrast, associated *S. sphaerocarpum* and *S. levieri* show a "normal" brown capsule, with brownish elaters and spores. Comparative study of the material of these three species (*S. levieri*, *sphaerocarpum*, and *pyriflorum*) brings out other differences, summarized in Table 8, p. 962.

SOLENOSTOMA (SOLENOSTOMA) CAESPITICIUM
(*Lindenb.*) *Steph.*
[Fig. 284]

Jungermannia caespititia Lindenb., Nova Acta Acad. Caes. Leop.-Carol. Nat. Cur. Suppl. 14:67, pl. 1, figs. 1–8, 1829;[359] Schuster, Amer. Midl. Nat. 42(3):580, 1949.
Aplozia caespiticia Dumort., Bull. Soc. Roy. Bot. Belg. 13:61, 1874; Macvicar, Studs. Hdb. Brit. Hep. ed. 2:143, figs. 1–4, 1926.
Jungermannia punctata G., *in* G., L., & N., Syn. Hep., p. 92, 1844.
Solenostoma caespiticia Steph., Spec. Hep. 2:57, 1901.
Haplozia caespiticia K. Müll., Rabenh. Krypt.-Fl. 6(1):544, fig. 276, 1909.

Loosely gregarious or in small scattered patches or at times forming small sods, a *pellucid pale green to pale yellowish green*, closely prostrate with ascending shoot apices, simple or rarely (then usually terminally) branched (except for subfloral innovations), soft, *small* or medium-sized, shoots 900–1250 μ wide × 3–6 mm long, sterile plants sometimes smaller, the fertile larger toward apex (there to 1500–1650 μ wide). Stems to 250 μ in diam., somewhat fleshy, tender, the dorsal and lateral *cortical cells delicate*, irregularly rectangulate, *pellucid, large*, (24)30–36 μ wide × (70)75–110 (125) μ long. Rhizoids numerous, *colorless*, dense. Leaves quite obliquely inserted, contiguous to moderately imbricate,

---

[359] The original spelling was *caespititia*. Müller (1909, *in* 1905–16, p. 535) pointed out that this was erroneous, emending it to *caespiticia*, on the assumption that the original spelling represented a mere typographic error or lapsus. He has been uniformly followed in this emendation, even though certain authors (i.e., Frye and Clark) accept it with reluctance.

patent to somewhat patent to (on fertile plants) erect or suberect and rather antically connivent, flat to somewhat concave, *rotundate to rotund-reniform*, slightly decurrent antically, 450–600 μ long (rarely to 800–850 μ on fertile plants) × 700 (on fertile shoots often to 1200) μ wide, usually *1.0–1.25× as wide as long*, rounded to occasionally retuse at apex. *Cells very large, strongly pellucid, quite leptodermous*, the *trigones absent or vestigial;* marginal cells quadrate, but not differentiated as a distinct border, ca. 32–36(40) μ; median cells ca. (32)35–40(45) × 40–60(70–80) μ; basal cells 34–48(50) × 45–65(75) μ; *cuticle smooth; oil-bodies very large*, ellipsoidal to fusiform, up to 18–20 × 10–12 μ, *singly per cell*, finely granulose and almost smooth. *Asexual reproduction commonly present, by means of endogenously formed 1-celled gemmae*, arising in large masses amidst a glutinous matrix, *within the somewhat swollen, olive to brownish apices of fertile or sterile shoots* (the swollen apex usually invested by a pair of concave leaves); gemmae extremely leptodermous, roundish quadrate or somewhat ovoid, ca. 9–12 × 12–18 μ, each with a single, glistening oil-body.

Dioecious. ♂ Plants slenderer than ♀; bracts subtransverse, in 3–4 or more pairs, becoming intercalary, commonly slightly smaller than leaves, concave at base, and with antical margin ampliate and sometimes elaborated as an obscure basal tooth, otherwise like leaves; 1–2-androus. ♀ Shoots with leaves gradually larger above, the bracts broadly rotundate to reniform, to 1400–1450 μ wide × 1000–1100 μ long, sheathing perianth at least at base, their margins rounded and entire, sometimes somewhat sinuous. Bracteole absent, or occasionally present as a slender vestige. Perianth from ovoid to oblong-obovoid, ca. 0.5–0.65 emergent, to 1700 μ long × 1000 μ in diam., strongly if obtusely 4–5-plicate in at least the distal half, rather sharply, but roundly, narrowed into a small tubular beak. Cells of mouth fingerlike, free only at their ends, leptodermous, 27–33 × 60–85 μ, much longer than cells below (which average ca. 30 × 38–48 μ). Capsule almost spherical; epidermal cells ca. 25–30 μ wide, with well-developed nodular (radial) thickenings; inner cell layer with cells ca. 20 μ wide, more narrowly rectangulate, with numerous semiannular bands, usually extending completely across interior tangential faces. Elaters ca. 7 μ in diam.; spores mostly (10)12–15 μ, delicately granulate.

Type. Near Hamburg, Germany (*Lindenberg*).

Distribution

Occurring in northern and, more rarely, central Europe, ranging from Scandinavia (Finland, Sweden, Norway) south to Belgium, France Germany, Switzerland, Czechoslovakia, eastward to Silesia (Polish occupied; formerly Germany), England, from Yorkshire southward, and also in Denmark (Jensen, 1915). Müller (1905–16) states it to be one of the rarities of central Europe, becoming more frequent in north Germany and Scandinavia.

In North America the plant is extremely rare and has a very sporadic distribution. There are only three reliable reports, those of Evans (1901) and Persson (1952) from near Nome, Alaska, and the following from our area:

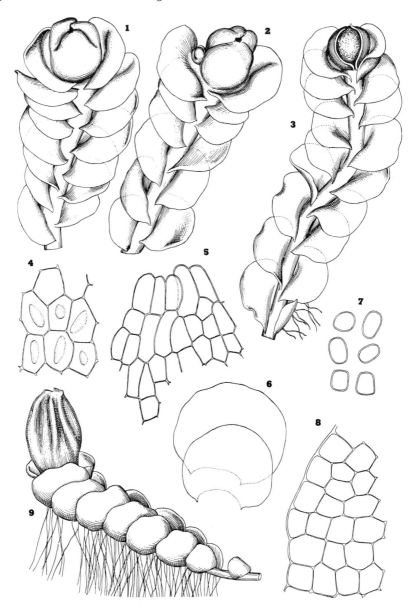

FIG. 284. *Solenostoma (Solenostoma) caespiticium.* (1–2) Plants with unfertilized perianths, interior tissue proliferated as endogenous gemmae (×21). (3) Gemmiparous shoot apex (×21). (4) Median cells with oil-bodies (×160). (5) Cells of perianth mouth (×160). (6) ♀ Bract, above, and leaf, below (×21). (7) Gemmae (×360). (8) Marginal cells of leaf apex (×160). (9) Shoot with mature perianth, lateral aspect (×20). [1–8, Briggs Gully, Honeoye Lake, N.Y., *RMS;* 9, from K. Müller, 1905–16.]

NEW YORK. Briggs Gully, SE. end of Honeoye L., Richmond Twp., Ontario Co. (*RMS H-3120*, 1945; see Schuster, 1949a).[360]

The species appears to be scarce, but hardly rare, in north and central Europe; its rarity regionally is difficult to understand, as is the extreme disjunction in range. Evidently, in North America, with a range from at least the southern portion of the Tundra to the Hemlock-Hardwoods Forest of the Deciduous Forest Region.

Ecology

Typically a species of fine-grained inorganic soils, such as moist clay and loam; rarely on soils with a high humus content. Clearly a decided mesophyte, evidently not occurring where periodic desiccation is a normal occurrence.

At the single station in our area, over moist, highly decayed shale talus, the talus detritus in a matrix of clay, at the foot of the steep north-facing side of a ravine fringed by Hemlock-Hardwoods Forest. The plants grew in small, somewhat yellowish green rosettes or clumps and, though gregarious, did not form sods (in spite of the species name), usually directly over wet clay or shale talus fragments. Associated were *Blasia pusilla, Cephalozia bicuspidata* cf. *lammersiana, Scapania nemorosa, Diplophyllum apiculatum,* and *Calypogeia muelleriana.*

Differentiation

Almost constantly with yellowish green to olive-brown, swollen shoot apices, bearing the uniquely produced, endogenous gemmae. When these are developed, confusion with any other species is hardly possible. Buch (1911) has gone into detail as to the mode of formation of gemmae. The first impression gemmiparous plants make is either of diseased plants, the shoot apices having the facies of nematode galls, or of ♀ plants with unfertilized and partially decayed or abnormal gynoecia. The latter impression is perhaps due to the fact that the gemmiparous "heads" are often closely invested by a pair of leaves.

When without gemmae, *S. caespiticium* is still easily identifiable, even if sterile. It bears a superficial resemblance to shade phases of *S. gracillimum.* The yellowish green color, soft texture, colorless rhizoids, and extremely large and leptodermous leaf cells of the unbordered leaves are distinctive. The cells are as devoid of trigones, and as large as or larger (32—40 μ wide

---

[360] There is an old and certainly incorrect report from Illinois (Wolf & Hall, 1878), unfortunately accepted in Frye & Clark (1937–47). Evans (1901) clearly emphasizes that his report is the first of the species from North America. Persson (1952, p. 10), evidently overlooking my report (1949a) of the species from New York, refers to the collection that he reports from Nome as the second from North America. There is no doubt about the determination of the New York plants, since they bore the absolutely diagnostic, swollen, brownish apical "heads," invested by a pair of leaves, within which gemmae occur.

in leaf middle) than those of *S. fossombronioides*, a species differing at once in the more ovate leaves and purplish rhizoids. No other regional species of *Solenostoma*, except *S. levieri*, has leaf cells nearly as large, and in none do the cells have a single, very large oil-body. Therefore, if living materials are available, *S. caespiticium* can be readily recognized solely on the basis of the large cells and solitary, large oil-bodies.

On the basis of color, dioecious inflorescence, and perianth form, *S. caespiticium* appears to be allied most immediately with the *S. sphaero-carpum-levieri* complex and with *S. pyriflorum*. All three of these species differ from *S. caespiticium* in the (*a*) absence of gemmae, (*b*) distinctly collenchymatous cells, (*c*) presence of ca. 3–9 oil-bodies per cell. Both *S. pyriflorum* and *S. sphaerocarpum* possess markedly smaller leaf cells, although those of *S. levieri* are nearly similar in size. *Solenostoma caespiticium* has also been considered to be closely allied to *S. gracillimum* (i.a., by Evans, 1901 and Müller, 1905–16). However, it differs from the latter species in having no large-celled border of the leaves and bracts, markedly larger cells, smooth cuticle, solitary oil-bodies of much superior size, softer texture, total lack of secondary pigmentation, and wholly "free" perianth, without trace of subtending perigynium.

In materials collected in New York the shoot tips surrounded by unfertilized perianths are regularly freely gemmiparous. As a consequence, each of the slightly reduced and abnormal perianths superficially appears to contain a globular, olive-brown or olive-green, immature capsule. However, on dissection, it is evident that the shoot tip itself has proliferated a mass of small, hyaline gemmae. Two such gemmiparous and perianth-bearing shoots are illustrated in Fig. 281:1–2, and the gemmae taken from Fig. 281:2 are shown in Fig. 281:4.

Sectio *GRACILLIMAE* Schust., sect. n.

Folia orbicularia, obliquissime inserta non aut vix concava; perianthium 3–4 plicatum, ad os vix rostellatum; cellulae non aut vix collenchymatosae, cellularibus marginalibus multo maioribus et quadratis; perigynium humile sed distinctum. Typus: *S. gracillimum* (Smith) Schust.

SOLENOSTOMA (SOLENOSTOMA) GRACILLIMUM
(*Smith*) *Schust., comb. n.*[361]
[Figs. 272:1–5; 285]

Jungermannia crenulata Smith, *in* Sowerby, Eng. Bot., pl. 1463, 1805; Frye & Clark, Univ. Wash. Publ. Biol. 6(2):326, figs. 1–15, 1944 (non *J. crenulata* Schmidel/Paver, Dissertatio de Jungermanniae charactere, p. 20, 1760 [Erlangen; *nomen dubium*]).
Jungermannia gracillima Smith, *in* Sowerby, Eng. Bot., pl. 2238, 1805.
Jungermannia crenulata var. *gracillima* Hook., Brit. Jungerm., pl. 37, 1816.

[361] Grolle (1966, p. 142) has shown that *J. crenulata* Smith is a later homonym; for that reason *J. gracillima* Smith has to be adopted.

Jungermannia genthiana Hüben., Hep. Germ., p. 107, 1834.
Solenostoma crenulatum Mitt., Jour. Linn. Soc. Bot. 8:51, 1865; K. Müller, Hedwigia 81:117, 1942.
Nardia crenulata Lindb., Bot. Notiser, p. 167, 1872.
Aplozia cristulata and *A. gracillima* Dumort., Hep. Eur., p. 57, 1874.
Aplozia crenulata Dumort., *ibid.*; Macvicar, Studs. Hdb. Brit. Hep. ed. 1:136, figs. 1–3, 1912; ed. 2:141, figs. 1–3, 1926.
Nardia gracillima Lindb., Acta Soc. Sci. Fennica 10:530, 1875.
Nardia crenulata var. *gracillima* Lindb., Musci Scand., p. 8, 1879.
Southbya crenulata Bernet, Cat. Hep. Suisse, p. 55, 1888.
Mesophylla crenulata Corbière, Rev. Bryol. 31:39, 1904.
Eucalyx crenulatus Loeske, Hedwigia 49:8, 1909.
Haplozia crenulata fo. *gracillima* K. Müll., Rabenh. Krypt.-Fl. 6(1):542, fig. 274a, 1909.
Haplozia crenulata K. Müll., *ibid.* 6(1):539, fig. 274, 1909.
Plectocolea crenulata Buch, Evans & Verdoorn, Ann. Bryol. 10(1937):4, 1938.
Plectocolea crenulata var. *gracillima* Frye & Clark, Univ. Wash. Publ. Biol. 6(2):329, figs. 1–6, (1943) 1944.
Solenostoma (subg. *Eusolenostoma*) *crenulatum* Schust., Amer. Midl. Nat. 49(2):399, pls. 31, figs. 8–10, 33, figs. 1:2, 1953.

Usually forming *prostrate, closely adnate, pale, somewhat pellucid or whitish green patches or sods*, in sun at least the perianths and upper portions of the leaves *dull carmine-red*. Shoots *laterally compressed* (at least when crowded), small to medium-sized, 0.9–1.3 mm wide (but *often with numerous, much more slender innovations*), ca. 8–15 mm long, irregularly but *freely branching*. Stems prostrate, except when crowded, with ascending apices, 160–250 μ in diam.; dorsal and lateral cortical cells slightly thick-walled, rectangulate, ca. (17)18–24 × 50–85(100) μ long, slightly smaller in diam. than the similarly thick-walled medullary cells (which are mostly 19–26 μ in diam.). Rhizoids abundant to shoot apices, long, *colorless or pale brownish with age*. Leaves generally somewhat antically secund, at times erect-appressed, the lower only weakly antically secund and sometimes erect-spreading or spreading, obliquely inserted, remote or contiguous below, becoming much larger and rather closely imbricate above on robust and fertile shoots, *orbicular, with a somewhat narrowed base and line of insertion, rounded and never retuse at apex*. Leaves bordered by a *conspicuous, complete border of large, swollen cells* (vestigial or absent on juvenile leaves or on the small leaves of slender innovations), the marginal cells ca. 30–38(44) μ, generally subquadrate, *thick-walled, averaging fully twice the area of the cells immediately within* [which average (18)20–24 μ in diam.]; median cells ca. 23–25 × 25–35 μ, at or near base up to 30–32 × 45–50 μ; walls of the intramarginal cells *thin to thickened and with minute*, occasionally distinct or weakly bulging, trigones; *cuticle distinctly papillose*. Oil-bodies absent in all but occasional marginal cells, and in many cells adjoining the marginal cells (when present very small, ca. 4–4.5 μ, spherical to shortly ovoid); in intramarginal and median cells *mostly 2, occasionally 1 or 3(4), per cell*, occasionally lacking in some cells, ovoid to ellipsoid, ca. 5.5 to 5 × 7 μ, a few to 4.5 × 9 or 5.5 × 10 μ, finely granular to papillose, larger than chloroplasts (which are 2–3.2 μ). Asexual reproduction absent.

Dioecious. ♂ Plants relatively slender; bracts in usually 4–8 pairs, forming a rather compactly spicate androecium, becoming intercalary, imbricate, quite saccate at base and erect, but with usually spreading apices or flaring margins; 1–2-androus. ♀ Bracts somewhat larger than leaves, erect, loosely sheathing

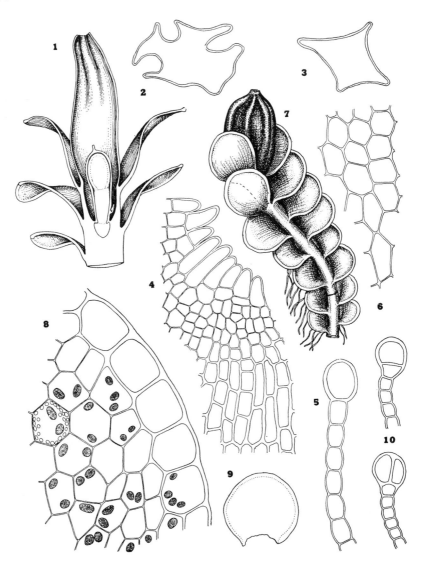

FIG. 285. *Solenostoma (Solenostoma) gracillimum.* (1) Sector through apex of gynoecial shoot (×26.5). (2–3) Cross sections through, respectively, upper half and lower half of perianth (×32). (4) Cells of perianth mouth, of phase without a beak developed (×290). (5) Section through leaf apex (×395). (6) Cells from perianth middle (+290). (7) Shoot with perianth (×16). (8) Cells from leaf apex, oil-bodies drawn in in cells where present (×340). (9) Leaf (×29). (10) Cross sections of leaf margin (×170). [1–5, *RMS 19718,* Tishomingo State Park, Miss.; 6, *Cain 4281,* Norway Lake, Ontario; 7, Ithaca, N.Y., *RMS;* 8, Letchworth State Park, N.Y., *RMS;* 9–10, after K. Müller, 1905–16.]

perianth, ovate to orbicular, margined like leaves (marginal cells ca. 35–40 μ; intramarginal ca. 25 μ). Perianth variable, usually only 0.3–0.5 emergent, ca. 1.2–1.5 mm high, ovoid to somewhat oblong, varying from rather swiftly contracted near apex (then perceptibly beaked) to relatively gradually and conoidally narrowed (then with beak hardly differentiated as such), slightly to strongly *laterally compressed, deeply 4–5-plicate* at least in distal 0.5–0.6; plicae varying from smooth to crenulate, or tuberculate (var. *cristulata*); mouth shallowly lobed, the lobes crenulate with elongated thick-walled cells [ranging from 15–18 μ wide × (37)45–65(75) μ long] whose apices are rounded and free; infra-apical cells often thin-walled and much smaller and subquadrate (14–16 × 14–18 μ), varying to thick-walled and hardly to moderately smaller than apical; median cells $\pm$ thick-walled, only 1.0–2.5× as long as wide [ca. 20–28 × (24)40–60(72) μ]. Perianth subtended by a low to distinct perigynium, varying from obsolete on one side of axis to 0.25–0.35 the height of the perianth. Sporophyte with seta 6 cells in diam., the outer ring of 16–18 rows of cells, the middle of 8–12 rows, the inner of 4 rows. Capsule short ovoid; epidermal cells with nodular, inner cells with semiannular, thickenings. Elaters contorted, short, 65–80 μ long × 6–7 μ in diam., the stiff and slender apices thick-walled, the elaters otherwise bispiral; spores 12–15 μ, finely granulate, red-brown.

Type. England.

Distribution

A widespread, abundant species, nearly holarctic in range, occurring from the Tundra margin (in which it is rare) southward into the Lower Austral regions. In Europe widespread, ranging from Scandinavia (Norway, Sweden, Finland, chiefly in the southern portions, to Great Britain and Ireland, the Faroes) southward to southern Europe (Spain, *fide* Casares-Gil, 1919; Italy, *fide* Zodda, 1934; southeast France, *fide* Allorge, 1955) and eastward to Austria, Moravia, Poland, Russia, southeastward to the Caucasus, southward into Africa (Algeria and Tunis) and to the Azores (Allorge and Allorge, 1948) and Madeira. Also in Iceland (Hesselbo, 1918). Reputedly ranging eastward into Siberia and eastern Asia (Japan, *fide* Stephani, *in* 1898–1924[362]).

In North America restricted to the eastern half of the continent; replaced in the Rocky Mts. and Far West by the closely allied *Solenostoma rubrum* (G. ex Underw.), comb. n. [= *Jungermannia rubra* Underw., Bot. Gaz. 13:113, 1888; = *Nardia rubra* Evans, The Bryologist 22:62, 1919; = *Plectocolea rubra* Evans, Ann. Bryol. 10(1937):4, 1938]. The latter, at times considered a synonym of *S.* "*crenulatum,*" has been shown by Evans (1919b) to be a distinct species, leaving *S. gracillimum* (= *crenulatum*) with the following range:

---

[362] Amakawa (1960a, p. 73) states that he has never seen material of the species from Japan!

W. GREENLAND. Frederikshaab and Tasermiut Fjord (Lange & Jensen, 1887); not reported since, and very doubtfully to be credited to Greenland. (I have seen a single specimen, clearly misdetermined.)   E. GREENLAND. Ikerasarsuk (*Vahl; fide* Lange & Jensen, 1887, p. 95); report probably based on error.   NEWFOUNDLAND. Benoit Brook, W. Nfld. (Buch & Tuomikoski, 1955); Squires Mem. Pk., C. Nfld. (*RMS*).

QUEBEC. Montreal (*RMS*); Mt. Shefford (Fabius, 1950); Spider R.; Longueuil; Rupert R., above L. Némiskau, 76°50′ W. (Lepage, 1944–45); Rimouski R., ca. 40 mi from mouth; near St. Jovite, Labelle Co. (Crum & Williams, 1960).   ONTARIO. Norway L., Algonquin Park (*Cain*); Nipissing, High Park, and York (Williams & Cain, 1959); Ottawa (Macoun, 1902); W. of Algonquin Park, at Oxbow L., Nipissing Distr., Carleton Co. (Cain & Fulford, 1948); 74 mi N. of Sault Ste Marie, Algoma Distr. (*Cain & Williams 5294*).   NOVA SCOTIA. Truro; Baddeck and Aspy Bay, Cape Breton I. (Macoun, 1902); Halifax (*Brown, 1925!*); Valley of Barrasois, Cape Breton (*Nichols*); 3–4 mi E. of Barrington, Shelburne Co. (*RMS 43066a*).   NEW BRUNSWICK. Grand Manan, North Head; Brunfoot Hill, New Galloway (*Andrew!*).   PRINCE EDWARD I. Near Brackley Pt. (Macoun, 1902).

MAINE. Prospect Harbor (*Northrup!*); Mt. Desert (*Lorenz*); Matinicus I.; near Bangor (*Pringle!*).   NEW HAMPSHIRE. Randolph, Coos Co. (Evans, 1917e).   VERMONT. Newfane (*Grout*); Brandon (*Dutton!*).   MASSACHUSETTS. Amesbury (*Huntington 5*); West Newbury (*Haynes 653*); Worcester; Oxford; Holden; Mt. Greylock; Cheshire; Hadley, Hampshire Co. (*RMS 66-1790*); Bear R., Conway (*RMS 66-1812*).   RHODE ISLAND. *Fide* Evans (1903b).   CONNECTICUT. Litchfield, Tolland, Fairfield, New Haven, and Middlesex Cos. (Evans & Nichols, 1908).

NEW YORK. Napeaque Beach (*RMS 22089a*), 4–5 mi W. of Montauk Pt. (*RMS 22071a*) and near L. Montauk Harbor (*RMS 22096*), all on Long I.; Bear Mt. (*RMS*); Cattaraugus, Tompkins, Onondaga, Wyoming Cos. (Schuster, 1949a); Fishers I. (Evans, 1926); Cold Spring Harbor, Long I.; New Dorp, Staten I. (*Grout!*); Todt Hill, Staten I. (*Britton & Hollick!*); near summit of Slide Mt., 4200 ft, Catskill Mts. (*RMS 17610*). NEW JERSEY. Highlands, Monmouth Co. (*Haynes!*); "common in northern and southern parts" (Britton, 1889); Water Witch (*Howe!*); Closter, Bergen Co. (*Rau! p.p.*, among *S. fossombronioides!*).   PENNSYLVANIA. Meadville (*Clarke 633*); Center, Alleghany, Bedford, Butler, Crawford, Elk, Fayette, Forest, Lawrence, Potter, Venango Cos. (Lanfear, 1933a).   MARYLAND. Vic. of Baltimore (Plitt, 1908).   DISTRICT OF COLUMBIA. Washington (*Waite 521; Holzinger!*); Rock Cr. (Holzinger, 1907).

VIRGINIA. Pinnacles of Dan (*RMS s.n.*; with *Diplophyllum apiculatum*); Giles, Greene, Rappahannock, Rockingham Cos. (Patterson, 1949, etc.); Cascades, Giles Co. (*Blomquist!*); Great Falls (*Svihla!*).   WEST VIRGINIA. Brooke, Marion, Marshall, Monongalia, Pocahontas, Preston, Randolph, Wyoming Cos. (Ammons, 1940).   NORTH CAROLINA. Crow Cr., Cullasaja Gorge, and Big Dog Mt., above Dry Falls, Cullasaja R. (*RMS 29484, p.p.*, with *Diplophyllum andrewsii*) and Wayah Bald, W. of Franklin (*RMS 39291a*), all in Macon Co.; Craggy Gardens, Blue Ridge Pkwy., ca. 5900 ft, Buncombe Co. (*RMS 29669, 29630*; with *Diplophyllum andrewsii*); Durham, Watauga, Haywood, Swain, Burke, Transylvania, Buncombe, and Ashe Cos.! (38 numbers); Linville Gorge (*RMS 28875*), Linville Caverns (*RMS 28804a*), 6 mi N. of Marion (*RMS 28845*), all in McDowell Co.; Neddie Cr., near E. fork of Tuckaseegee R., Jackson Co. (*RMS 29455*); Bluff Mt., 4500 ft, near W. Jefferson, Ashe Co. (*RMS 30091*).   TENNESSEE. Blount, Carter, Johnson, Knox, Sevier Cos. (Sharp, 1939); Montgomery Co. (Clebsch, 1954).

SOUTH CAROLINA. Summerville, Berkeley Co. (*Haynes 1353*). Oconee Co.: Whitewater R. gorge, ca. 1 mi below falls (*RMS 33566, 34315*, with *D. andrewsii*); 2–3 mi SE. of Jocassee (*RMS 30009, 30156*, with *S. hyalinum* and *S. pyriflorum*); Thompson R. ravine, N. of Jocassee (*RMS 41009*).   GEORGIA. Thomas Co.; Hall Co.; Thomasville (*Brown!*); Gainesville (*Howe!*); N. slope of Rabun Bald, 4000 ft, Rabun Co. (*RMS*

40689, 40678).   ALABAMA. Auburn, Lee Co. (*Underwood*, Amer. Hep. no. 61!, a phase with carmine-red calyptral wall and slender innovations, "fo. *gracillima*"). FLORIDA. Lost Cr., between Crawfordville and Arran, Wakulla Co. (Kurz & Little, 1933).   MISSISSIPPI. Tishomingo State Park, Tishomingo Co. (*RMS; plate*) ; W. of Tupelo, Lee Co. (*RMS*) ; Ocean Springs, Jackson Co. (*Pennebaker!, c.fr., Feb. 28, 1938*).

Northward and westward occurring as follows: KENTUCKY. Boone, Letcher, Lewis, Powell, McCreary Cos. (Fulford, 1934, 1936).   OHIO. Liberty Twp., Jackson Co. (*Bartley & Pontius!*).   INDIANA. Fern, Putnam Co. (*Underwood!*).   ILLINOIS. Clark Co. (Arzeni, 1947).   WISCONSIN. Arena, Iowa Co. (Conklin, 1929).   MINNE-SOTA. Lamoille, Winona Co. (*Holzinger!*; see Schuster, 1953).   IOWA. Coggon, Linn Co. (*Conard!*; reported by Conard, 1942, as *Plectocolea crenulata* var. *gracillima*, but listed by Conard, 1945, incorrectly as *Odontoschisma prostratum*).   OHIO. Athens, Cuyahoga, Hocking Cos. (Miller, 1964).

The preceding citations are merely representative; in much of temperate eastern North America the species is a common, almost weedy plant, rare only along the western border of its range (Wisconsin and Minnesota southward to Iowa) and in the northern edge of the Taiga. Evidently unknown from any area contiguous to the Tundra in continental North America, although (unconvincingly) reputed to occur in southern Greenland. Southward ranging into the warm, southeastern Coastal Plain, but hardly penetrating southward to the Sabalian region of broad-leaved evergreen forests.

Ecology

Almost uniformly restricted to sandy, loamy or clayey, inorganic com-pact soils, rarely also on porous calcareous to acidic sandstones and decay-ing schistose rocks. Generally a mesophyte, but occasionally occurring in somewhat xeric and insolated sites on relatively exposed banks, associated, e.g., with *Polygala sanguinea*. Often considered a weakly "calciphilous" species, but this is hardly consistently so, since it may occur on exposed, broken clay or loam or even moist sand, in acid, leached situations (asso-ciated with oak and pine and various ericads, such as *Gaylussacia* and *Vaccinium*, or *Rhododendron*—indicators of acid, leached, sterile soils).

The species is frequent on damp to rather dry sandy banks or on open sand (as in gravel pits), associated with *Ditrichum*, *Cephaloziella rubella* or *hampeana*, *Lophozia bicrenata*, and (in moister sites) *L. capitata*.

In weakly calcareous sites, as on calcareous loam, the species may be associated with such calciphiles as *Lophocolea minor*. On more acid, leached soils (loam or clay banks), *Lophozia bicrenata* and sometimes *L. excisa* and *Cephaloziella rubella* are associated (in relatively exposed and xeric sites) or *Diphyscium sessile* and *Diplophyllum apiculatum* (under generally more mesic and humicolous conditions). On moister, but acid, sandy soils (such as the edges of fields or in ditches at the edges of swamps), there may be associated *Lophozia capitata*, *Calypogeia fissa* (or *C. muelleriana*), and in ex-treme cases members of the *Blasia-Anthoceros* community. On shaded loam banks *S. gracillimum* is particularly frequently associated, especially northward, with members of this last community, particularly *Blasia pusilla*, *Anthoceros laevis*, and *Anthoceros punctatus*, occasionally *Notothylas orbicularis* or *Pellia epiphylla*. Associated with these is often *Pogonatum pensilvanicum* (and, in more exposed areas with lesser slope, such as the edges of fields, often various pigmy mosses, such as *Nanomitrium* and *Ephemerum*).

Southward occasionally an accessory species on loamy banks, soil-covered talus slopes, or even soil-covered cliffs, associated with *Physcomitrium* spp. and with *Diplophyllum andrewsii*. The association of the latter with *S. gracillimum* within the restricted range of the *Diplophyllum* is nearly constant. More rarely, on soil-covered cliffs, with *Ditrichum*, *Solenostoma hyalinum*, *S. pyriflorum*, and *Pellia epiphylla* (near Jocassee, S.C.).

Fundamentally a tolerant, "weedy" species, often associated with man and with disturbed conditions (such as roadside banks or eroded paths), but showing an almost consistent restriction to inorganic substrates. The rapid ecesis of this species is often remarkable. For example, within 5 years after disuse, a gravel bank was colonized by the species, associated with *Lophozia bicrenata*, *L. capitata*, *Cephaloziella rubella*, together with the angiosperms *Comptonia peregrina* and *Betula populifolia* (Hadley, Mass.).

Differentiation

Normally easily recognized; bearing a distinctive facies because of the strongly bordered, orbicular, usually erect or suberect and connivent leaves. The "border" of thick-walled, swollen cells is more sharply developed than in *S. crenuliformis*, although distinct on larger leaves only, and is readily visible with the hand lens. In addition to the unique border, the plants are distinct because of the light green color, in exposed sites commonly a very pale, somewhat bleached-appearing whitish green, except for secondary carmine-red pigmentation which often suffuses the perianths and upper leaves. The restriction to loam or clay banks, often in greatly disturbed sites (roadside banks or ditches, sides of erosion gullies, etc.), also serves to identify it.

On the basis of the swollen marginal leaf cells *S. gracillimum* may be confused with *Arnellia* (which differs at once in the opposite leaves and the occurrence of rhizoids in sharply distinct fascicles at the bases of distinct underleaves) and with *S. crenuliformis*. However, *S. crenuliformis* has less obviously inflated marginal cells, averaging under twice the area of the intramarginal cells, and almost always has distinctly purplish rhizoids. It also generally is greener, not whitish green, in color and possesses a tendency to develop purplish pigmentation of the leaf bases, never a carmine red of the distal portions of the perianth and leaves. Also, *S. crenuliformis* possesses larger and more conspicuous oil-bodies, present in marginal cells as well, and develops a high, tubular perigynium, beyond which the perianth is hardly emergent. The rather beaked, sharply 4(5–6)-carinate perianth of *S. gracillimum* is wholly different from that of *S. crenuliformis*.

Only *S. gracillimum* fo. *gracillimum* (which see below) is likely to cause difficulties in identification. Sterile plants of this forma have marginal cells hardly differentiated, but on fertile plants at least the upper leaves and ♀ bracts are normally bordered.

Perhaps *S. gracillimum* is rather closely allied to the Korean-Japanese *S. koreana* Steph. (see Amakawa, 1960a, p. 55), which also has reddish secondary coloration, bordered leaves, and a slight beak of the perianths. This species lacks a perigynium and has rather ovate leaves.

Variation

Both Schiffner (1904a, pp. 410–18; 1910, p. 271) and Evans (1919b) have carefully studied this species and its pattern of variation. Evans (*loc. cit.*) has also examined its relationships to *S. rubrum* of the West Coast. Both taxa show shallow perigynia (whose height is usually merely ca. 0.2–0.35 that of the free perianth above); both have extremely variable perianths, almost always somewhat laterally compressed. Often four sharp carinae are developed, although supplementary carinae may occur. Although the perianth mouth is fringed with fingerlike, elongated cells, *S. gracillimum* normally has the cells of the perianth body only slightly elongated (quite unlike *S. crenuliformis* and the other species of *Plectocolea*). In some cases (Fig. 285:4) the cells in several tiers immediately below the perianth mouth are leptodermous and very small, usually subquadrate. In other cases, there is no sharp differentiation of the cells below the distal ring of cells (Fig. 272:3). These extremes seem strongly marked, but connected by intermediate conditions. Equally variable is the form of the perianth mouth. Most often the apex of the perianth is rather abruptly contracted and a distinct beak is differentiated, as in other species of the subgenus. However, sometimes the perianth is somewhat conoidal and gradually narrowed to the apex, with hardly a differentiated beak. In the latter instances a close resemblance to the perianth of other taxa of subg. *Solenostoma* exists. This resemblance is further enhanced by the fact that the perianth appears to be 1-stratose to its base (the species of *Plectocolea* normally having a perianth that is firm and at least 2-stratose below).

Not only is there a wide range of variability in regard to the perianth, but also the vegetative features of *S. gracillimum* are equally polymorphous. For example, some phases (fo. *gracillimum*) fail to show the swollen, highly differentiated marginal cells of the leaves. In addition to variation in degree of differentiation of the leaf border, the leaf cells are highly polymorphous. Normally marginal cells are thick-walled, usually somewhat verruculose (even if intramarginal cells are virtually smooth), and intramarginal cells vary from thin-walled and without marked trigones (mod. *leptoderma*) to almost equally and noticeably thick-walled (mod. *pachyderma*). The equally thick-walled cells are regarded as a species characteristic in many manuals. However, as both Hesselbo (1918) and Evans (1919a) point out, distinctly collenchymatous phases occur in exposed places (mod. *pachyderma-colorata*); Evans also indicated that the less xeromorphic phases show a strong contrast between marginal cells (ca. 40 μ) and intramarginal cells (ca. 20 μ), while xeromorphic phases have a less strongly contrasting cell size (marginal cells ca. 32 μ; intramarginal, ca. 18 μ).

Solenostoma gracillimum occurs in two growth forms: the "common" one, with few or no slender, small-leaved innovations (= "*S. crenulatum*"),[363] and the plant variously distinguished as a distinct species (*Jungermannia gracillima* Smith; *Aplozia gracillima* Dumort.) or variety (*Aplozia crenulata* var. *gracillima* Heeg; *Plectocolea crenulata* var. *gracillima* Frye & Clark) or mere form (*Haplozia crenulata* fo. *gracillima* K. Müll.). This plant, the typical form, or fo. *gracillimum*, bears numerous slender innovations, mostly with small, remote to contiguous leaves devoid of the distinct border of swollen cells. Only perichaetial bracts, and usually the relatively large leaves subtending these, show a distinct border. In my opinion the plant is merely a minor environmental or genotypic phase. It has been reported from New Jersey (Austin, 1876) and Georgia (Brown, 1924) but actually occurs sporadically through the range of the species.

Perhaps more validly separated as a discrete variety is the following:

SOLENOSTOMA GRACILLIMUM var. CRISTULATUM
(*Dumort.*) Schust., comb. n.

Aplozia cristulata Dumort., Hep. Eur., p. 57, 1874.
Aplozia crenulata var. *cristulata* Massal., Ann. Istit. Bot. Roma 2(2):20, 1886.
Haplozia crenulata var. *cristulata* K. Müll., Rabenh. Krypt.-Fl. 6(1):542, fig. 275, 1909.
Plectocolea crenulata var. *cristulata* Frye & Clark, Univ. Wash. Publ. Biol. 6(2):328, (1943) 1944.

Similar in its entirety to the species, but the ridges of the perianth conspicuously tuberculate with small, irregular, projecting teeth.

Type. Germany.

Distribution

Found in a number of European localities and also incorrectly reported from Greenland and North America, supposedly by Müller (1909, *in* 1905–16, p. 543), *teste* Frye and Clark (1937–47). However, it is clear from Müller that the Greenland and North American reports he cites are for typical "*S. crenulatum*," var. *cristulata* being cited (pp. 543–544) *only* from Germany, Bohemia, Austria, and Italy.

Subgenus PLECTOCOLEA Mitt.

Nardia of authors, in part (includ. Evans, 1919).
Solenostoma subg. *Plectocolea* Mitt., Jour. Linn. Soc. Bot. 8:156, 1865; Schuster, Amer. Midl. Nat. 49(2):401, 1953.

[363] *Solenostoma gracillimum* fo. *crenulatum* (Smith) Schust., status n., for those who wish to formally recognize this extreme.

Plectocolea (as genus) Mitt., *in* Seemann, Fl. Vitiensis, p. 405, 1871; Evans, Ann. Bryol. 10 (1937): 42, 1938; Frye & Clark, Univ. Wash. Publ. Biol. 6(2):318, 1944; Hattori, Jour. Jap. Bot. 22(1–2):17–24, figs. 1–4, 1948.

Nardia subg. *?Chascostoma* Schiffn. *in* Engler & Prantl, Nat. Pflanzenfam. 1(3):79, 1893.

Eucalyx Breidl., Mitt. Naturw. Ver. Steiermark, 30:291, 1894; K. Müller, Rabenh. Krypt.-Fl. 6(1):524, 1909.

Jungermannia Steph., Spec. Hep. 2:500, 1901 (p.p.).

Mesophylla Corbière, Rev. Bryol. 31:13, 1904 (p.p.).

Jungermannia subg. *Plectocolea* Amak., Jour. Hattori Bot. Lab. no. 21:270, 1960.

Medium sized (usually 750–2800 μ wide), *rather delicate, bright or pale green, usually purplish- or vinaceous-tinged* on underside of stem and with *rhizoids similarly tinged*, at least at base; *the purplish pigmentation often extending to the postical leaf bases.* Stem nearly terete, oni, slightly elliptical in cross section, *usually with the cortical 1–2(3–4) cell layers sharply differentiated from the medullary;* often with intercalary flagella. Often with rhizoids inserted at or near the postical leaf bases, less often from scattered initials on leaf. Leaves rotundate to ovate, varying from 0.8 to 1.0(1.1) as wide as long, rather broad at base, never subcordate, rounded to rarely retuse or bilobed at apex. Cells nearly iso-diametric throughout the leaf, $\pm$ *thin-walled*, with minute to (often) somewhat *bulging trigones;* cuticle usually verruculose to striolate; oil-bodies (1)2–7(8–15) per cell, usually *very large*, granular-papillate. Asexual reproduction absent.

Antheridia 1–5 per bract. Perianth $\pm$ reduced, usually *included or emergent* for only 0.1–0.35 its length, *usually closely sheathed by the bracts* (which are *erect or suberect*, except sometimes for their spreading apices); perianth ovoid, *narrowed in a conoidal fashion distally* and contracted to a *slender, sometimes tubulose mouth* (exc. *S. fossombronioides*), *but not beaked*, usually 4-several plicate distally and shallowly 4-lobed at mouth; *perianth* usually *unistratose*, formed nearly throughout of *narrow*, often *sigmoid*, thin- or often thick-walled *cells generally 2.5–5$\times$ as long as wide*. Perigynium *distinct, shortly tubular, usually well-developed*, from 0.2 to 0.65(1.2) the height of the perianth above it, bearing commonly 2(3) pairs of bracts. Elaters *tortuous*, (1)2-spiral, the *spirals confluent with wall at slender elater apices.*

Type. Solenostoma radicellosum Mitt. of Japan (type of subg. *Plectocolea*). See Evans (1938c, p. 38). This species has been treated in detail and admirably illustrated by Hattori (Jour. Jap. Bot. 22:18–19, fig. 1, 1948) and by Amakawa (1960a, fig. 11:a-i).

Subgenus *Plectocolea* is restricted to species in which there is development of a distinct, usually conspicuous, and tubular perigynium, correlated with which the perianth is usually reduced and not or hardly emergent from between the erect or suberect, often sheathing bracts. Although the distinct perigynium suggests an affinity to *Nardia*, in which the species were placed (as sectio *Eucalyx*) by Lindberg (1872a) and others, the species of *Plectocolea* differ in several respects from *Nardia*, among them (a) the total absence of underleaves and (b) the perigynium always continuous with the axis, never becoming ventrally bulbous, fleshy, and rhizoidous. In a number of species the perigynium is feebly developed and the perianth longly emergent from between the bract apices.

Species of subg. *Plectocolea* are habitually distinct in the violet or reddish rhizoids (rarely colorless in extreme shade forms; frequently colorless in *S. subellipticum*) and in the often similar pigmentation of the postical leaf bases and perianths, a pigmentation which may extend in extreme cases over almost all the rest of the plant. The absence of such pigmentation of the rhizoids and leaf bases and the common lack of a distinct perigynium seem to warrant placing "*Jungermannia*" *gracillima* in subg. *Solenostoma* rather than in the present subgenus. Since the presence of a low perigynium might lead to its being sought in *Plectocolea*, it is also included in the following key. *Plectocolea* is also characterized by the wholly unistratose perianth, normally constricted and strongly narrowed into a slender, but not beaked, mouth; the perianth is deeply plicate and distally often somewhat twisted. The perianth cells are typically strongly elongated throughout, often averaging 3–4 ×, or even 3–5 ×, as long as wide. In the subgenera *Solenostoma* and *Luridae* and in the genus *Nardia*, as well as in *Jungermannia s. str.*, only the terminal row of cells of the perianth may be strongly elongated (and sometimes not even these), while the cells below the beak or mouth are short-oblong to virtually isodiametric.[364]

Subgenus *Plectocolea* includes at least 35–40 species, of which 26 are reported for Japan (Amakawa, 1960a); 7 occur in our area, 4 of them (*S. ontariensis*, *S. obscurum*, *S. fossombronioides*, and *S. crenuliformis*) being endemic to eastern North America.

Variation

Subgenus *Plectocolea* shows a wealth of variation in, i.a., degree of development of perigynium (already commented on); number of elater spirals; form, number, and disposition of oil-bodies; number, color, and orientation of rhizoids; stem anatomy; growth habit; and leaf form. These and other features have led Amakawa (1960a) to divide the numerous Japanese species into 3 sections, which in turn fall into 5 "groups." The much less numerous American species can be oriented only in part in the classification of Amakawa. Some of the variability is so rampant that the following brief discussion is necessary.

Oil-bodies occur 2 or more per cell in all of our species. As developed under *S. hyalinum*, there may be major differences, perhaps in part environmentally induced, in oil-body size, form, and number. However, they are always ± granular to somewhat botryoidal in form. The Japanese *S. tetragonum* is unique in having single, very large oil-bodies in scattered leaf cells—the other cells lacking oil-bodies (in this it approaches the genus *Treubia*). Other species lack such ocelli or oil-cells.

---

[364] This is decidedly the case in *Solenostoma gracillimum* (Fig. 280:4), suggesting again that this species should not go into the subg. *Plectocolea*.

The rhizoids in our species do not cohere in a decurrent conspicuous strand along the posterior face of the stem; in a variety of species, i.a., the type, *S. radicellosum*, such a specialized orientation of the rhizoids occurs. The occurrence of decurrent rhizoid bundles is largely correlated with erect growth. Although a violet or red rhizoid color is almost constantly present, some forms of *S. hyalinum* and *S. crenuliformis* may have colorless rhizoids. In some species, notably *S. tetragonum*, rhizoids may originate from scattered leaf cells.

Although leaf form, in our species, varies to a limited extent, ranging chiefly from rotundate to ovate-elliptical, some exotic species are strongly divergent. For example, in the Japanese *S. bilobum*, which should probably be segregated into a separate subgenus, the leaves are bilobed with acutely pointed leaf lobes (and the elaters are unispiral!). Our species have a generally oblique orientation of leaves; in some of the exotic species which are suberect in growth, the leaves are subtransversely oriented and laterally distichous.

In all examined species of *Plectocolea*, the elaters are strongly contorted, with each end markedly tapered and sharply pointed. These pointed ends are uniformly thickened, and, insofar as the spirals can be distinguished there, are unispiral on the ends. Such elaters are not found in *Jamesoniella* and *Nardia*[365] or in the related Lophoziaceae. The elaters are regularly bispiral in all but their distal portions, with very few exceptions, such as the Japanese *S. unispiris* Amak. (see Amakawa, 1954, 1960a).

The capsule wall in most species of *Plectocolea* is 2-stratose (see, e.g., Amakawa, 1960a, fig. 4d, m), but in certain taxa, including *S. (Plectocolea) radicellosum sensu* Amakawa, the generitype, the capsule wall is 3–4-stratose. This is associated, in the last species, with other nonderivative features (e.g., a virtually nonexistent perigynium and a massive seta, of the *Jamesoniella* type). Apparently with *S. radicellosum* we deal with a primitive taxon showing some contact points with *Jamesoniella*, as Amakawa (1960a, p. 267) suggests.

A marked feature of *Plectocolea* is the diversity of modes in which the stem has become specialized. This divergent specialization is, of course, correlated with the other specialized features (elongated perianth cells, strongly developed perigynium, etc.) noted above. Hattori (Jour. Jap. Bot. 22:17–24, figs. 1–4, 1948; *ibid*. 27:53, fig. 62, 1952, etc.) shows that in some species, such as *Solenostoma (Plectocolea) rigidulus* (Hatt.), comb. n.,[366] cortical cells may be

---

[365] I have not seen similarly specialized elaters in *Nardia*. However, Amakawa (1960a, fig. 7:z) shows a bispiral elater with confluent spirals in the tips for *Nardia sieboldii*. *Solenostoma (Plectocolea) sphaerocarpum* and other species of *Plectocolea* may have the elater apices thickened and sharp-pointed. However, *Solenostoma* subg. *Luridae* does not have specialized elater apices, at least in the taxa in which I have seen sporophytes.

[366] Basionym: *Plectocolea rigidula* Hatt., Jour. Jap. Bot. 27:53, 1952. Amakawa (1960a, p. 23), who does not recognize *Solenostoma* as a genus, proposed the *nomen novum Jungermannia kyushuensis*.

It may also be relevant to note that, associated with the similar stem anatomy, *S. rigidulus* and *S. crenuliformis* share an identical tendency for the marginal leaf cells to be thick-walled and swollen (compare fig. 62:E,F, I, in Hattori, 1952b, p. 54, with the illustrations here given of *S. crenuliformis*). Evidently also to be associated with these two species is the Japanese *Solenostoma (Plectocolea) marginatus* (Hatt.) comb. n. (= *Plectocolea marginata* Hatt., Jour. Hattori Bot. Lab. 3:40, fig. 35, 1948). This species approaches *S. crenuliformis* in leaf shape, in enlarged and swollen marginal cells, and in inflorescence; it also has (judging from Hattori's fig. 35S) enlarged hyaline cortical cells surrounding 1–2 strata of small-bored, thick-walled intracortical cells. It differs at once from *S. crenuliformis* in the lower perigynium and in the strongly elongated, longly emergent perianth.

relatively thin-walled (at least externally) and larger in diam. than 1–2 rows of thick-walled intracortical cells. In such cases, a weak indication of a hyalo-dermis is present. The North American *S. (Plectocolea) crenuliformis* is similar (Fig. 271:9), although the stem is only 10–12 cells high vs. ca. 18–21 cells in *S. rigidulus*. Furthermore, the hyalodermis appears to be even more strongly differentiated, although there is only a slight level of differentiation between the 1–2 strata of small-diameter, thick-walled intracortical cells and the central medullary cells. Hattori (1948a, fig. 1M and N) illustrates the stem anatomy of the type, *S. radicellosum* Mitt. This has the outermost cell layer of the stem less differentiated and not recognizable as a hyalodermis (and its cells are not consistently larger than in the 1–2 strata lying within).[367] A third type of stem anatomy is exemplified by *S. (Plectocolea) ovicalyx* and *S. (P.) virgatum*. In both, the unistratose cortex is formed of relatively small, slightly to somewhat thick-walled cells, surrounding a large-celled medulla (see Hattori, 1948a, figs. 2N, O, and fig. 3I, J, K, L). In this instance, there is a tendency for the medulla to be dorsiventrally differentiated, the cells of the upper strata being larger in diam. than those of the ventral strata. In *S. virgata* the relatively simple stem anatomy closely approaches that of species such as *S. (Luridae) pumilum* which are also unspecialized in lacking a perigynium. *Solenostoma virgatum*, judging from fig. 3A in Hattori (1948a), also has only a vestigial perigynium.[368]

Whatever the level of differentiation of the cells of the stem, there appears to be a general difference between the 1–2 cortical cell layers (which are ± abbreviated) and the medullary cells (which are always strongly elongated), as is clear from Fig. 271:8. This led Hattori (1948a) to speculate that "Generally speaking, bract and leaf cells of *Plectocolea* are for the most part derived from (the relatively short cells of the) 'epidermis' of the stem . . . . The perianth cells of *Plectocolea* are based upon this tissue within (thus explaining why the perianth cells of *Plectocolea* are strongly elongated and narrow, as are the medullary cells . . .)." By contrast, in the subgenera *Luridae* and *Solenostoma* (i.e., in "*Jungermannia*" of some recent authors), in which there is no perigynium, the perianth cells are not strongly elongated. Hattori presumes that in these cases the perianth is derived from elements "based mainly upon 'epidermis' " (i.e., cortex).

Key to Species of Subgenus Plectocolea

1. Leaves not bordered, the marginal cells not equally thick-walled, neither swollen nor displaced, no larger than those of the intramarginal row; paroecious or dioecious; leaves elliptical, ovate or subovate (exc. in *S. hyalinum*) . . . . 2.

---

[367] *Solenostoma obovatum* (Fig. 290:8) and *S. hyalinum* (Fig. 271:12) possess an essentially identical stem anatomy.

[368] The paroecious Japanese *Solenostoma (Plectocolea) otianus* (Hatt.), comb. n. [= *Plectocolea otiana* Hatt., Jour. Jap. Bot. 28(6):183, fig. 66, 1953] has a simply organized stem, with cortical cells thin-walled or hardly thickened, and scarcely different (in diam.) from adjacent medullary cells; there is also some, if not always marked, dorsiventral differentiation of the medulla. This species also may show a low perigynium (see fig. 66D in Hattori, 1953c). Species of this type, with a low perigynium and slight (or no) cortical differentiation, are also "low" in having non-margined leaves.

2. Dioecious; cells usually with obvious, often distinctly bulging trigones
. . . . 3.
   3. Rhizoids colorless, with age pale brownish; plants without trace of
reddish or purplish pigmentation, with facies of *S. sphaerocarpum* and
S. pyriflorum; perianth mouth rather wide open and truncate; median
cells 23–30 μ, subisodiametric, on margins 20–25 μ; upper leaves
subtransversely oriented, erect-spreading . *S. (P.) ontariensis* (p. 1004)
   3. Rhizoids normally purplish; leaf bases often purplish; plants not
with facies of *S. sphaerocarpum*; perianth mouth slender, conoidally
narrowed; upper leaves oblique, spreading . . . . 4.
      4. Perianth usually ± emergent beyond the bracts; perigynium only
moderate in height (0.2–0.5× that of the free perianth above it);
leaves often slightly undulate on ♀ plants, usually rotundate and
as broad as long, not or rarely retuse on sterile shoots; cells of
perianth in part or largely of short cells; ♂ plants with bracts
numerous, forming a slender spike; median cells 24–34 × 30–41 μ,
on margins (20)23–34 μ; plants pale green with reddish secondary
pigments; usually on soil or soil-covered rocks in the lowland . .
. . . . *S. (P.) hyalinum* (p. 993)
      4. Perianth low, included within the erect bracts; perigynium high,
tubular, its height (0.65)1.0–2.0× the height of the free but short
perianth; leaves not undulate, broadly ovate, often retuse at apex
on sterile shoots; cells of perianth strongly elongated; ♂ plants
with bracts in 2–6 pairs, forming a short, compact spike; median
cells 20–28 × 25–35 μ, on margins 15–20 μ; plants usually on
bare rock, deep, pure green with intense purplish secondary
pigmentation (at times the plants blackish purple in exposed sites);
saxicolous and montane . . . . *S. (P.) obscurum* (p. 1013)
2. Paroecious; cells usually with small, concave-sided trigones (rarely
slightly bulging) . . . . 5.
   5. Perianths immersed to barely emergent, at mouth conically narrowed,
not laciniate and lobed; perigynium high and conspicuous, equal in
height to the perianth or even higher; median cells from 20 to 30 μ
wide, the marginal averaging 18–24(25) μ; arctic and montane
species, ranging southward into New England . . . . 6.
      6. Median cells ca. 20–24 × (25)30–45 μ; plants small, 5–10(18) mm
long × 800–1800 μ wide; leaves longer than broad, oval to
elliptical, those of sterile shoots usually almost transversely oriented;
rhizoids usually colorless or brownish; perianth included or
virtually so, the bracts ± erect and sheathing, except for ± flaring
tips . . . . *S. (P.) subellipticum* (p. 1021)
      6. Median cells ca. 25–30 × 27–45(49) μ; plants usually more
robust, 2–5 cm long × 1800–2500 μ wide; leaves broadly ovate
to ovate-rotundate, ± equally broad as long, obliquely oriented;
rhizoids usually purplish; perianth often slightly emergent, the
bracts ± spreading . . . . *S. (P.) obovatum* (p. 1007)
   5. Perianths at maturity longly emergent, subcampanulate, usually wide

and subtruncate at the laciniate and lobate mouth; perigynium a low, inconspicuous ring; median cells 30–40 μ wide, the marginal ca. 22–28 μ; Appalachian, lowland species (500–2500 ft) from S. New England to Minn. and southward . . *S.(P.) fossombronioides* (p. 1027)

1. Leaves distinctly bordered, the marginal cells thick-walled, swollen, and ± turgid in appearance, on mature leaves slightly to greatly larger than the immediately adjacent interior cells; dioecious; leaves rotundate or almost so . . . . 7.

7. Marginal cells usually 1.1–1.5(2.0)× the size of the intramarginal; rhizoids (and often postical leaf bases) usually deeply reddish tinged or purplish, rest of plant dull yellowish green to green; perigynium high, tubular; perianths included (to barely emergent) within the concave, sheathing bracts; cells with distinct trigones, the median 23–28 μ . . . . *S. (P.) crenuliformis* (p. 986)

7. Marginal cells (of mature shoots) 1.5–3× the size of the interior, usually thick-walled; rhizoids and postical leaf bases not reddish tinged (although the rest of the plant may be carmine-red); perigynium low, inconspicuous; perianths distinctly emergent, the bracts not closely sheathing; cells usually without distinct trigones, the median 23–35 μ or larger . . . . [See *S. (S.) gracillimum* (p. 972)]

SOLENOSTOMA (PLECTOCOLEA) CRENULIFORMIS
(*Aust.*) *Steph.*
[Figs. 271:8–9; 272:7; 286]

Jungermannia crenuliformis Aust., Bull. Torrey Bot. Club 3:10, 1872.
Nardia crenuliformis Lindb., Acta Soc. Sci. Fennica 10:529, 1875; Evans, Rhodora 10:186, 1908; Lorenz, The Bryologist, 19:24, pl. 2, 1916.
Plectocolea crenuliformis Mitt., Trans. Linn. Soc., Ser. 2, 3:198, 1891; Frye & Clark, Univ. Wash. Publ. Biol. 6(2):331, figs. 1–10, 1944; Schuster, Amer. Midl. Nat. 42(3):706, 1949.
Solenostoma crenuliformis Steph., Spec. Hep. 2:56, 1901.
Solenostoma (subg. *Plectocolea*) *crenuliformis* Schust., Amer. Midl. Nat. 49(2):402, pl. 31, figs. 5–7, 1953.

In compact *prostrate* patches or small turfs, occasionally scattered, *rather pellucid, pale to yellowish green* (*but with rhizoids, and almost always the postical leaf bases* and often the perianth and distal halves of leaves *conspicuously reddish to purplish*). Shoots 1400–2000(2200) μ wide × 10–25 mm long, prostrate with ascending tips, simple or sparingly branched, with occasional to frequent flagelliform intercalary branches. *Often with leading shoots becoming attenuated and whiplike.* Stems 175–250 μ in diam., slender; *cortical cells only slightly thick-walled,* their external walls thin, somewhat pellucid, narrowly rectangulate [ca. (13)18–22(25) μ wide × 60–95, occasionally 95–120, μ long], *forming a ± distinct hyalodermis,* the adjacent medullary cells ± *thick-walled, often yellowish,* only 12–16(23) μ in diam.; *stem 10–12 cells high.* Leaves contiguous to weakly or moderately imbricate, obliquely inserted and oriented, distinctly decurrent antically, except on weak shoots ± *concave* (and more conspicuously so at base), especially the uppermost leaves, commonly erect-spreading to loosely spreading, often *somewhat antically secund, obliquely subrotundate, broadly rounded to occasionally retuse or emarginate at apex*, ca. 760–900 μ wide × 650–700 μ long (1.1–1.4× as wide as long) to 1350 μ long × 1350 μ wide, leaf margins entire to obscurely sinuous, *appearing conspicuously swollen* (the marginal row of cells averaging somewhat *larger than the cells immediately within and appearing slightly adaxially displaced, forming a slightly elevated border*). Marginal cells somewhat swollen, 25–27 μ *up to 30–38(40)* μ × 35–36 μ thick, quadrate, *rather distinctly thick-walled, averaging 1.1–1.5× the area of the interior adjacent cells;* median cells (23)25–27 × 25–28 μ, in hygrophytic forms rarely to 35 × 40 μ, polygonal, thin- to somewhat thick-walled and with distinct, often bulging trigones; *cuticle distinctly verruculose to striolate* (at least in basal half of leaf); oil-bodies large, usually 2–4 per cell (then 7–9 × 15–22 μ) or sometimes 4–10 smaller ones (then 6.5–9 × 6–10 μ), ellipsoidal to spherical, finely granular appearing; chloroplasts only 4–4.5 μ.

Dioecious, somewhat heteromorphic, ♂ plants more slender. Androecia becoming intercalary, the bracts in up to 5–8 pairs, forming a loosely spicate androecium; bracts subtransverse, hardly smaller than leaves but more strongly concave at base, at best the distal portions obliquely spreading, asymmetrically ovate-rotundate to broadly ovate, averaging wider than long, with the antical base somewhat dilated, otherwise similar to leaves. ♀ Bracts (and subinvolucral leaves) somewhat larger than vegetative leaves, subtransverse, broadly ovate-rotundate to rotundate, often retuse at apex, *strongly bordered; bracts closely sheathing perianth below, erect or suberect* to rarely obliquely spreading distally. Perianth *immersed but at maturity often barely emergent, often strongly reddish*, narrowly subovoid, somewhat conically narrowed to apex, 4–7-plicate above, the constricted mouth hardly beaked but somewhat tubulose, crenulate; apical row of cells narrow, 2.5–4.5× as long as wide; subapical and median cells of perianth usually similarly narrow, rectangulate. *Perigynium high but occasionally inconspicuously developed*, its height averaging 0.5–0.8 that of the free perianth arising from it. Calyptra often purplish.

Type. Near Closter, N. J. (*Coe F. Austin*).

Distribution

Broadly Appalachian in range; endemic to eastern North America.[369] Although its range, northward, impinges on the Spruce-Fir Forest Region, the species is essentially one of the temperate Deciduous Forest. It is equally frequent in the Piedmont and at low elevations in the Appalachian Mts., and occasionally occurs associated with local rock outcrops in the inner edges of the southeastern Coastal Plain, where it becomes quite rare. General on the east and south slopes of the Appalachian system, at elevations of ca. 500–3800 feet, very rarely higher.

QUEBEC. "Wet bank, below Lake Némiskau, Rupert R., 76°50′ W." (det. L. Clark; Lepage, 1944–45); near St. Jovite, Labelle Co. (Crum & Williams, 1960); reports

---

[369] Reported by Underwood from British Columbia; this report is based on a specimen collected by Macoun (Canad. Hep. no. 297, Selkirk Mts., B.C., 1890) and wholly different from *S. crenuliformis*.

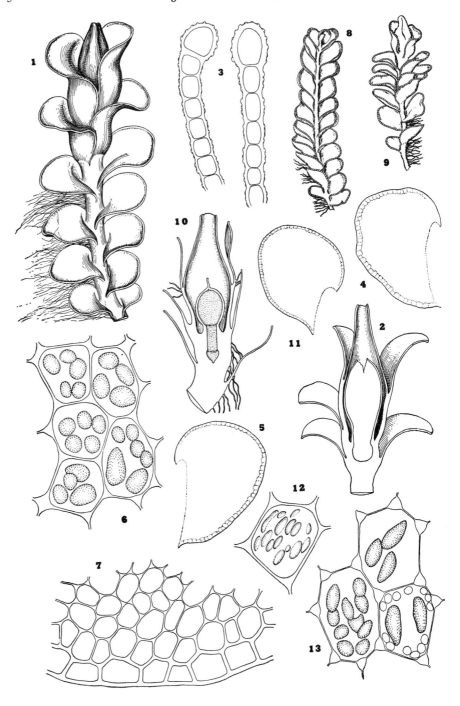

needing confirmation, as this austral species otherwise unknown N. of southern Ontario and Massachusetts. ONTARIO. Byng Inlet, Magnetawan R., Parry Sd. (*Cain 4715*). MASSACHUSETTS. Granville (*Lorenz, 1911*); Bear R. Valley, Conway, Franklin Co. (*RMS 68-203*). CONNECTICUT. Beacon Falls (Evans, 1908c; see Lorenz, 1916); S. Glastonbury (Lorenz, 1916); Granby (*Lorenz, 1924*).

NEW YORK. Gorge of Genesee R. in Letchworth State Park, Wyoming Co. (Schuster, 1949a); Thorn Mt., Palisades Interstate Park (*RMS, 1943*); Orange Co. (*Austin*); Forest of Dean L., near Ft. Montgomery (*RMS 24231*). NEW JERSEY. Closter, Bergen Co. (*Austin;* type). PENNSYLVANIA. Ohiopyle, Fayette Co. (*Lanfear*). VIRGINIA. The Cascades, on Little Stony Cr., Mountain L., Giles Co. (*RMS 40286a*); Stafford Co. (*Patterson*); Albemarle Co. (Patterson, 1955). WEST VIRGINIA. Tibb's Run, Monongalia Co. (Ammons, 1940).

NORTH CAROLINA. Below Windy Falls, Horsepasture R.; along Toxaway Cr., near Toxaway Gorge; Toxaway Gorge below L. Toxaway, all in Transylvania Co. (*Anderson 8444!, 9015!, 8785!, 11077!*); Sols Cr., near E. fork, Tuckaseegee R., Jackson Co. (*Anderson 10120!*); Melrose Falls, 4 mi NW. of Tryon, Polk Co. (*RMS 36049*); Tuckasegee Falls, Jackson Co. (*Blomquist, 1932*); The Caves, W. of Durham, Orange Co. (*RMS 28517*); Roan Mt., Mitchell Co. (*Blomquist, 1932!*); Crow Cr., Cullasaja Gorge, NW. of Highlands, Macon Co. (*RMS 29491a, 29499*); Raven Rock, Cape Fear R., near Mamers, Harnett Co. (*RMS 28683a, 28681*); Gorge of Whitewater R., below High Falls, Jackson Co. (*RMS 25009, 40585, 40572*); Hemlock Bluff near Cary, Wake Co. (*RMS 28746, 28633, 28730*); Dry Falls, NW. of Highlands, Macon Co. (*RMS 25160*); Buncombe, Mitchell Cos. (Blomquist, 1936); below The Narrows, Chattooga R., Macon Co., 1 mi N. of Georgia border (*RMS 39475*); same loc., E. side of Chattooga R., Jackson Co. (*RMS 39453, 39454*); Thompson R., ca. 1600 ft, just N. of South Carolina border, Transylvania Co. (*RMS 45172*); Neddie Cr., near jct. of E. fork of Tuckasegee R., Jackson Co. (*RMS 29477, 29430a, 29456*). SOUTH CAROLINA. S. end of Whitewater R. gorge, below Lower Falls, near Jocassee (*RMS 40938a, 40891, 40955, 40954*) and Thompson R. gorge, ca. 2 mi above jct. with Whitewater R. (*RMS 41006*), all in Oconee Co.; gorge of Estatoe R., 2–3 mi N. of Rte. 11, Pickens Co. (*RMS 44629b*); along Rte. 178, N. of Rocky Bottom, Pickens Co. (*RMS 40496*).

GEORGIA. Stone Mt. (Clebsch, 1954); W. side of Chattooga R., NE. corner of Rabun Co. (*RMS 39863, 39865*). TENNESSEE. "Moist rocks, Sevier Co." (Sharp, 1939); branch of W. fork of Little Pigeon R., below Indian Gap, ca. 4000 ft, Sevier Co. (*RMS 36529a, p.p.,* with *S. obscurum*). KENTUCKY. Greenbrier Church, Muhlenberg Co. (*Quenby, 1936*); Natural Bridge (*Taylor, 1924*); McCreary, Carter, Letcher, Lewis, Morgan, Powell Cos. (Fulford, 1934, 1936).

ARKANSAS. Camp Bard, S. of Mena, Ouachita Mts. (*Anderson 11422, 11423, 11426!*); Howard's Bluff, E. of Springdale, Washington Co. (*Anderson 12217!*). OHIO. Coshocton Co. (Evans, 1908c); 4 mi E. of S. Bloomingville (*RMS 18370, 18350, 18353a, 18367c*); Darwin, Meigs Co. (*Taylor, 1925*); Hocking Co. (*Taylor, 1923*); Whites Gulch, Jackson Co. (*Taylor, 1925*). INDIANA. Putnam Co.; Crawford, Fountain,

---

FIG. 286. *Solenostoma* (*Plectocolea*) *crenuliformis.* (1) Perianth-bearing shoot (×15). (2) Apex of ♀ shoot, longisection (×18). (3) Longisections through upper parts of leaves (×275). (4–5) Leaves (×32). (6) Median cells with oil-bodies (×560). (7) Cells of leaf apex (×245). (8–9) Sterile and fertile shoots (×ca. 6). (10) Longisection through apex of shoot with immature sporophyte (×18). (11) Leaf (×16). (12) Cell, with cuticular papillae (×440). (13) Median cells with oil-bodies and, lower right, chloroplasts (×440). [1–2, Letchworth State Park, N.Y., *RMS;* 3–7, *RMS 18350,* S. Bloomingville, Ohio; 8–9, after Lorenz; 10, *RMS 1540b,* Sand I., Wisc.; 11–13 from, Forest of Dean Lake, N.Y., *RMS;* extreme with pale rhizoids; see also Fig. 271:8–9.]

Perry, Dubois Cos. (Welch, 1937, 1941).   ILLINOIS. Clark Co. (Vaughn, 1947).
KANSAS. Woodson Co. (McGregor, 1955).

MICHIGAN.   Au Train Pt., Alger Co. (*Steere 460*); report needing verification, compare
with *S. hyalinum*.[370]   WISCONSIN.   Bark Pt., Bayfield Co. (*Conklin 1980*); Superior Bay
Pt., Superior Co. (*Conklin 1969*); Sand I., Apostle Isls., Bayfield Co. (*RMS 15480b,
15481*); material doubtful, compare with *S. hyalinum*.

Evidently quite rare in the Lake Superior extension of its range, which is
reminiscent of that of *Tsuga canadensis* and *Diplophyllum apiculatum*. In the
Southeast clearly less able to penetrate into the Coastal Plain than the frequently associated *Solenostoma pumilum* and *S. fossombronioides*.

Ecology

Usually associated with noncalcareous rocks, often schistose rocks or
sandstone, usually along shaded rocky streams, less often on seepage-
moist sandstone ledges in relatively insolated sites. The species typically
forms more closely prostrate mats than *S. hyalinum*, occurs either as a
pioneer directly on rocks, or is found on rocks after deposition of only a
very thin soil layer. Associated usually are *Scapania nemorosa, Blasia
pusilla, Anthoceros laevis, Conocephalum conicum, Pellia epiphylla*, as well as
Solenostoma pumilum, S. fossombronioides (and, with deposition of considerable humus, *S. hyalinum*), more rarely *Geocalyx*. Less often with *Fossombronia cristula, F. brasiliensis*, etc. At the northern edge of its range plants
intermediate to *S. hyalinum* occur, sometimes with *S. levieri* and *pumilum,
Scapania curta* and *saxicola* (Sand I., Wisc.), under weakly calcareous conditions also with *Preissia quadrata*; on acid rocks with *Lophozia alpestris,
Gymnocolea inflata, Scapania nemorosa, S. curta, Cephalozia* cf. *lammersiana*
(Genesee R., N.Y.).

The plant is a clear green in shaded sites, but usually somewhat dull or
even slightly olive green, except for the deep violet-red rhizoids and ventral
cortical cells. The violet-red pigmenttaion occasionally spreads to the
postical leaf bases, but sun forms are more often characterized by a bright
reddish color. Frequently perianths are strongly reddish-tinged while the
vegetative leaves are virtually green; in extreme cases both perianths
and leaves are bright reddish. Usually such sun forms are compact and
unusually small, with suberect and antically secund, concave leaves;
they form very thin layers over exposed, only intermittently damp sandstone.

The near-pioneer nature and general restriction to rocks contrast strongly
with the typical occurrence of *S. gracillimum* and *S. hyalinum*, species which are

---

[370] The Michigan and Wisconsin material needs revision; as is shown on p. 1001, Lake
Superior populations of *S. hyalinum* approach *S. crenuliformis* in ecology and morphology.

usually restricted to loose inorganic substrates, such as loamy or clay banks. However, *S. crenuliformis* occasionally is found on moist, peaty soil at the edges of lakes, in shade, with *Calypogeia fissa*, *Scapania nemorosa*, and *Sphagnum* (under a canopy of oak and hemlock; as at the edge of Forest of Dean Lake, N.Y., *RMS 24231*). In such sites the plants are quite atypical, lax and green, and may have colorless rhizoids.[371]

Variation and Differentiation

Usually readily recognizable because of the characteristically concave, usually somewhat antically secund, broadly asymmetrically ovate-orbic-ular leaves, whose margins appear tumid and narrowly incurved—the border of the leaf appearing somewhat raised above the surface of the rest of the leaf (Fig. 286:1, 8).[372] Associated with this, marginal cells are thick-walled (Fig. 286:7) and very often, although hardly universally, somewhat larger than the interior cells, which become gradually more distinctly collenchymatous and thin-walled in the intramarginal regions of the leaf (Fig. 286:6, 13). Commonly, thin-walled polygonal median cells, with somewhat bulging trigones (Fig. 286:13), are associated with quadrate to rectangulate, thick-walled marginal cells. When this feature is well developed (as is the case particularly on the larger, upper leaves and on ♀ bracts), the species is at once distinct from *S. hyalinum*, which it otherwise approaches in most features, among them (*a*) large and con-spicuous oil-bodies; (*b*) generally reddish to purplish rhizoids; (*c*) rather pellucid and elongated cortical cells that average only 18–25 μ wide but often 5 × as long as wide, or longer; (*d*) a conoidally narrowed perianth, formed of slender, elongated, hyaline cells surmounting a distinct peri-gynium; (*e*) dioecious inflorescences. Small, xeromorphic, pigmented, sun phases of *S. hyalinum* (mod. *colorata-parvifolia*) often have somewhat highly colored, concave leaves and may closely approach *S. crenuliformis* in facies, as well as in the only slightly emergent perianths. The marginal

---

[371] When sterile, such plants are quite likely to be mistaken for *S. gracillimum*, although the marginal leaf cells are consistently less differentiated, and the leaves possess the peculiar concave form and orientation of those of *S. crenuliformis*. Plants growing in such atypical sites also may have larger than normal cells (ca. 25–30 × 35 μ, locally to 35–38 × 36–45 μ) and approach the polymorphic *S. hyalinum*. With this, the present phenotype also agrees in the 2–10 large (to 7.5–8 × 17–20 μ) oil bodies per cell. However, the somewhat concave leaves, with a tumid, narrowly incurved margin (and strongly thick-walled marginal cells), serve to separate such plants from *S. hyalinum*. Admittedly, when material is so thoroughly atypical, separation of these three species becomes very difficult. In such lax extremes of *S. crenuliformis*, the differentiation of the hyalodermis of the stem is also hardly discernible, and the stem anatomy closely approaches that of *S. hyalinum* (although the cortical cells are only ca. 75–110 μ long × 22–25 μ wide).

[372] The elevated border of the leaves is, paradoxically, often much more readily evident under the low-power binocular, when the leaves are in *in situ*, than when free-dissected leaves are studied under higher power.

cells in these are, however, collenchymatous and thin-walled like the rest
of the leaf cells.[373]

Solenostoma crenuliformis generally, but not universally, exhibits other
peculiarities that allow ready recognition of typical forms. (1) Sterile
shoots often run out into slender flagelliform, small-leaved, attenuated
apices; this peculiarity is not always present, but appears universally
absent in *S. hyalinum*. (2) ♀ Plants have broadly ovate bracts, never
reniform, and show a very conspicuous, swollen "border"; they are never
undulate, unlike in *S. hyalinum*. (3) Median cells average less than 28 μ
wide usually. (4) ♀ Bracts are concave and almost always more closely
sheathing, with often only their very apices erect-spreading to somewhat
patent. (5) The generally reddish violet perianth, even at maturity, is
hardly emergent—indeed, usually nearly or quite immersed. (6) Cortical
cells of the stem are pellucid, forming a discernible hyalodermis, surrounding a central cylinder of slightly smaller, more thick-walled cells.
The last two features deserve special comment.

The closely sheathed perianth, with the narrow apex hardly emergent, often
suffices to separate *S. crenuliformis* from *S. hyalinum*. However, the latter species
occasionally has almost immersed perianths (Fig. 287:1–2), usually associated
with undulate bracts. A series of specimens of *S. crenuliformis* has been studied
with regard to stem anatomy. In all cases, the cortical cells have thin, free
(external) walls, with the inner and radial walls more or less thick. In normal
cases the stem shows a high level of internal differentiation, with the 2–3 strata
of intracortical cells smaller in diam. (ca. 12–16, rarely to 23, μ) and much more
thick-walled, with walls often yellowish. Only the central medullary cells
gradually become somewhat larger and less thick-walled, although still perceptibly smaller in diam. than the cortical cells. In such cases, there is a discrete
hyalodermis. In *S. hyalinum* there is never such a high level of axial differentiation (compare Fig. 271:9 and 12), cortical cells not forming a perceptible
hyalodermis, and intracortical cells not being distinctly smaller in diam.,
although sometimes slightly thick-walled in 1–3 strata. When stems of the two
species are typically developed, therefore, cross sections reveal considerable
differences. Unfortunately, *S. crenuliformis* is apparently quite variable in the
degree to which the intracortical cells are differentiated from the cortical. For
example, in otherwise totally "typical" plants (*Anderson 11077*, Transylvania
Co., N.C.) the hyalodermis is less discrete, and 2–3 intracortical cell strata,
though thick-walled, may be of cells 18–23 μ in diam. and often hardly smaller
than those of the cortex. In extreme cases (as in lax, hygromorphic extremes
from Forest of Dean Lake, N.Y., *RMS 24231*), in which the rhizoids are colorless, the "hyalodermis" is undifferentiated, intracortical (and other medullary)

---

[373] This is occasionally also the case in small sun forms that approach *S. crenuliformis* but appear
to belong to *S. hyalinum*, such as plants from Sand I., Wisc. (*RMS 15484*), in which the marginal
cells are somewhat tumid, but collenchymatous and hardly larger than the intramarginal;
the cells here are coarsely collenchymatous. Such forms could be assigned to *S. crenuliformis*—
although with difficulty—on the basis of the smaller cells, averaging only 23–26 × 25–28 μ
medially (see p. 1001).

cells being thin-walled, like the cortical, and equal to them in diam. Such hygric shade forms, fortunately, are rare, although they admittedly give only an incomplete idea of the highly differentiated nature of both cortical stem and marginal leaf cells of this species. However, one feature of the cortical cells appears at all times to be distinctive: they are only moderately elongated, varying mostly from 18–24 μ wide × 60–100 (rarely in isolated instances to 120) μ long, their length averaging ca. 4–6× their width. In *S. hyalinum* the cortical cells, although hardly wider, average 120–240 μ long (ca. 6–10× their width in most cases).

It is thus clear that extreme forms of *S. crenuliformis* and *S. hyalinum* are to be separated with care (see also p. 1001). A similarly close relationship occurs between some forms of *S. gracillimum* and *S. crenuliformis*. Indeed, Austin (1872) recognized the immediate affinity of the two, and Lindberg (1875) was reluctant to accord to *S. crenuliformis* the status of a distinct species. Evans (1908c, p. 186) stated that the two species "possess in common . . . the border of enlarged and thick-walled cells, which usually form a conspicuous feature of the leaves and bracts" and believed "this border is even more constant in *N. crenuliformis*" than in *S. gracillimum*, although the former often has it developed to only a slight extent. The border in the two species differs in subtle, but constant, ways, that of *S. gracillimum* being formed of cells mostly 2–3 × the intramarginal in size, with the swollen marginal cells *lying strictly in the plane of the leaf*. In *S. crenuliformis*, as has been emphasized, the marginal cells lie somewhat above the others, forming an elevated ridge which is usually quite conspicuous, although the contrast in size between marginal and intramarginal cells is often much less marked. *Solenostoma crenuliformis* differs also from *S. gracillimum* in the smaller, often quite verruculose leaf cells, the much larger and more conspicuous oil-bodies, the usually purplish rhizoids, the frequent occurrence of flagella, and (when fertile) the nearly or quite immersed perianths.

The oil-body number of *S. crenuliformis* is subject to considerable variation. On mature fertile plants there may be 2–10 oil-bodies per cell (Fig. 286:13). However, in plants from a deeply shaded ravine (Whitewater R., *RMS 40891*) there are only (1)2–3(4) oil-bodies per cell on ♀ plants, and only 1–2(3) per cell on slender, sterile plants.

SOLENOSTOMA (PLECTOCOLEA) HYALINUM
(Lyell in Hook.)   Mitt.
[Figs. 271: 11-12; 287-288]

Jungermannia hyalina Lyell, *in* Hooker, Brit. Jungerm., pl. 63, [1814] 1816.
Jungermannia schmideliana Hüben., Hep. Germ., p. 99, 1834.
Jungermannia biformis Aust., Proc. Acad. Sci. Phila. 21 (1869):220, 1870.
Solenostoma hyalinum Mitt., *in* Godman, Nat. Hist. Azores, p. 319, 1870.
Southbya biformis Aust., Hep. Bor.-Amer. no. 26, 1873.
Aplozia hyalina Dumort., Hep. Eur., p. 58, 1874.

Southbya hyalina Husnot, Hep. Gallica, p. 16, 1875.
Nardia biformis Lindb., Acta Soc. Sci. Fennica 10:530, 1875.
Nardia hyalina Carr., Brit. Hep., p. 35, pl. 11, fig. 36, 1875; Evans, Rhodora 21:149, figs. 1-9, 1919.
Plectocolea hyalina Mitt., Trans. Linn. Soc. Bot., Ser. 2(3):198, 1891; Frye & Clark, Univ. Wash. Publ. Biol. 6(2):333, figs. 1-9, (1943) 1944.
Eucalyx hyalinus Breidl., Mitt. Naturw. Ver. Steiermark 30:292, 1894; K. Müller, Rabenh. Krypt.-Fl. 6(1):531, fig. 272, 1909; Macvicar, Studs. Hdb. Brit. Hep. ed. 2:138, figs. 1-5, 1926.
Nardia muelleriana Schiffn., Österr. Bot. Zeitschr. 1904(4):1, 1904; K. Müller, Rabenh. Krypt.-Fl. 6(1):533, 1909.
Mesophylla hyalina Corbière, Rev. Bryol. 31:13, 1904.
Solenostoma (subg. *Plectocolea*) *hyalinum* Schust., Amer. Midl. Nat. 49(2):401, pl. 32, figs. 1-7, 1953.

In prostrate to ascending tufts, occasionally creeping, pale to yellowish to pure green, often somewhat glistening, *frequently carmine or reddish-pigmented on the leaf bases and postical sides* of the stem (pigmentation sometimes extending to the distal portions of leaves and perianths). Shoots mostly 1.5-2.2(3.0) mm wide × 10-15 mm long, rather soft-textured, sparingly branched; branches all intercalary; often with subfloral innovations, particularly when gynoecia are unfertilized. Stems rather soft, 200-400 μ wide × 180-300 μ high, 9-10 *up to 14-15 cells high;* cortical cells from slightly smaller to subequal to medullary in diam., ca. 20-25 μ in diam., *usually 120-240 μ long (6-10:1),* their outer walls very thin, their radial and inner walls sometimes slightly thickened, *not forming a large-celled hyalodermis;* intracortical 1-2-cell strata with cells equal in diam. to those of cortex, often ± *thick-walled* (sometimes ± yellowish-pigmented); medullary cells slightly to hardly thick-walled, 23-30(35) μ in diam.; rarely with stolons. Rhizoids *long, dense, numerous, usually reddish,* exceptionally colorless, often partially parallel to stem, a few frequently from postical leaf bases and from perigynium. Leaves slightly to moderately imbricate, typically *inserted by a long and oblique,* acroscopically arched line which bends abruptly down at the postical end, *long-decurrent antically, nearly horizontally oriented* and widely patent on the lower leaves and vegetative shoots, only upper leaves erect-spreading, *usually little concave* except at the very base, the larger leaves usually *somewhat undulate,* suborbicular to broadly reniform-orbicular (on weak and ♂ shoots often somewhat ovate), *widest just above base,* from a minimum of 750 μ wide and long to, more usually, (900-1000)1200-1375 μ long × 1500-1900 μ wide (*averaging from 1.2-1.55×* as wide as long on mature, robust shoots), broad and rounded to occasionally retuse at apex, entire-margined. Cells collenchymatous, thin-walled; trigones generally small but *slightly bulging* (rarely strongly so), the *marginal cells not larger, not swollen, or thick-walled,* leaves thus not bordered; marginal cells averaging 23-34 μ; median cells ca. (24)26-34(36) μ wide × 32-41 μ long; cuticle smooth to occasionally faintly verruculose; *oil-bodies typically large* and few: usually 2-7(8-15) *per cell,* ellipsoidal to fusiform and from 4-7 × 9 μ to 5-7 × 12-14(15-22) μ, finely granulose in appearance; chloroplasts much smaller, only 2.2-3.6 μ long. Underleaves lacking.[374] Asexual reproduction lacking.

[374] Although underleaves, *s. str.,* are lacking, clusters of ephemeral slime papillae are present. As Leitgeb (Untersuch. über Leberm. 2:8, pl. 9, fig. 10, 1875) has shown, there may be 4-5 of such papillae at one point, and they may occur on the margins of a rudimentary few-celled, vestigial underleaf.

Dioecious. ♂ Plants slightly smaller than gynoecial, *bracts loosely imbricated,* usually in 3–6 or more pairs, becoming intercalary; bracts somewhat shorter than leaves, strongly dilated and inflexed near the antical base (which often has the appearance of a rounded lobe) and only slightly decurrent, strongly concave except for the spreading to flaring apical ¼; (1)2–3-androus. ♀ Plants with upper leaves grading into the bracts in size and shape, gradually broadening and becoming subreniform, *usually strongly undulate;* bracts typically undulate, slightly spreading or erect-spreading from a sheathing base, often larger than vegetative leaves, *somewhat reniform* or rarely subrotundate, from 1400–1500 μ wide × 650–900 μ long up to 2125 μ wide × 1150 μ long, sometimes broadly ovate and ca. 1200 μ wide × 900–1100 μ long, then narrowly rounded at apex. Perianth variable in degree of development (if without fertilization often remaining short and immersed), usually *slightly to distinctly emergent upon fertilization,* often up to 0.3–0.5 its length, delicate, pellucid, unistratose, narrowly ovoid, strongly and evenly conoidally narrowed distally, deeply and irregularly 4–6-plicate distally; mouth fringed by fingerlike elongated cells whose rounded ends are free, the cells ca. 16–18 × (45)60–80 μ long, averaging 3.5–5× as long as wide; perianth cells strongly elongated throughout,[375] typically 18–22 μ wide × 50–100 μ long, often slightly sigmoid-rectangulate, thin-walled, without obvious trigones. *Perigynium usually only moderately high,* from 600 to 800 (rarely to 1000) μ: on one side from 0.3 to 0.6, on the other side from 0.15 to 0.4, the height of the perianth, usually bearing only one pair of bracts inserted on it; the free perianth beyond 1000–1200 (rarely to 1500–2000) μ high.[376] Sporophyte with seta fundamentally of 3 concentric cell layers: the outer of ca. 20, the middle of ca. 12, the innermost of ca. 4, cell rows, but often with considerable deviation in the number of cell rows. Capsule ovoid; epidermal cells averaging 25 × 35 μ, with nodular thickenings of both longitudinal and transverse walls; inner cells more narrowly rectangulate, ca. 13 × 30–50 μ, each with 4–10 semiannular bands. *Spores 14–16(17) μ,* granular-papillate, brownish; elaters 9 μ thick, bispiral.

Type. England.

Distribution

A common, widespread, polymorphous species, occurring almost throughout Europe (from Scandinavia, the Faroes, Scotland, and England, southward to Spain and Italy), south to North Africa (Algeria, Tunis) peripheral to the Mediterranean; also on the Azores, Madeira, the Canary Isls. Eastward extending to the south shore of the Black Sea, to the Baltic region, northern Russia, the Caucasus. Absent from westernmost

[375] Evans (1919a) states, "Especially if fertilization has taken place, the basal part (of the perianth) is composed of short, almost isodiametric cells, and these occasionally extend almost to the mouth."

[376] According to Evans (1919a), the upper extreme is a height of ⅓ and ⅙ of the perianth, on each of the two sides; this figure is often exceeded, as the illustrations given demonstrate. The lower extreme is represented by var. *heteromorpha* (G.) Schiffn., in which the perigynium is ca. ⅙ the height of the perianth on one side and obsolete on the other.

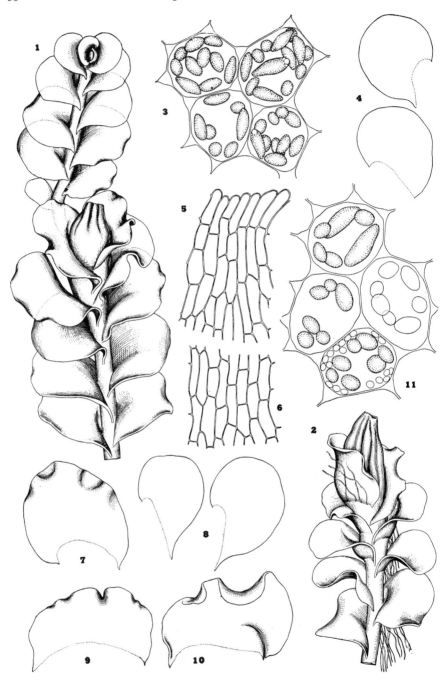

FIG. 287. *Solenostoma* (*Plectocolea*) *hyalinum*. (1) Perianth-bearing shoot with subfloral innovation (×17). (2) Apex of perianth-bearing stem, with maximal emergence of perianth (×17). (3) Cells with oil-bodies (×470). (4) Leaves (×13). (5) Apex of

Asia, except for the sole report from "the southern shores of the Black Sea," which must refer to Turkey. Evidently absent from most of eastern Asia, although very closely allied species occur there; Amakawa (1960a, p. 40) reports it from Japan only from Hokkaido.

In the New World first reported from Orizaba, Mexico (Gottsche, 1863), and somewhat doubtfully from Bogota, Columbia (Gottsche, Ann. Sci. Nat. Bot. V, 1:119, 1864), and later from the Andes of Ecuador (Spruce, 1884–85). Evans (1919a) comments that Spruce's specimens differ considerably from *S. hyalinum*, although "certainly closely related," and that "similar specimens from Mexico and the West Indies have also been examined, but whether these should be regarded as a well-marked variety of *N. hyalina* or as a distinct species is not yet clear." Evans concludes that none of the specimens he has seen represent "unquestioned" *S. hyalinum*. The species, however, is widespread in the more temperate portions of North America, where it is transcontinental, ranging in the West from British Columbia and Alberta, south to Washington, Oregon, California, Arizona, Colorado (doubtfully), and New Mexico. In our area as follows:

QUEBEC. Rupert R., above L. Némiskau; R. à Martre, aobve Lac aux Sables; Montreal; Como; Rimouski R., ca. 40 mi above mouth (Lepage, 1944–45); George R., near 55°05′ N. (Kucyniak, 1949); near St. Jovite, Terrebonne Co. (Crum & Williams, 1960). Not reported from NEWFOUNDLAND (in the careful study of Buch & Tuomikoski, 1955). ONTARIO. Pottageville and High Park, York (Williams & Cain, 1959). MAINE. St.. John R., Ft. Kent; St. John Pond, upper reaches St. John R. (Evans, 1919a); Jerusalem (Collins, 1896; *teste* Frye and Clark, 1937–47). NEW HAMPSHIRE. Randolph (Evans, 1919a); Mt. Madison, White Mts. (*Evans 1917*, *teste* Frye & Clark). VERMONT. Newfane; Quechee Gulf, Hartford; Brandon; Rochester (Evans, 1919a). MASSACHUSETTS. West Newbury; Mt. Greylock; Granville (Evans, 1919a); Oxford; Chesterfield Gorge, near W. Chesterfield, Hampshire Co. (*RMS 41302*); Bear R. above jct. with Deerfield R., Conway, Franklin Co. (*RMS*). CONNECTICUT. Ansonia; Middletown; Naugatuck, Canterbury; Watertown (Evans, 1919a); Higgins Gorge (Lorenz, 1923; *fide* Frye & Clark).

NEW YORK. Lime Kiln Falls, Herkimer Co. (Evans, 1919a); Allegany, Cattaraugus Co. (*Boehner*; see Schuster, 1949a). NEW JERSEY. Closter, Bergen Co. (*Austin; fide* Evans, 1919a; distributed in Austin's Hep. Bor.-Amer. Exsic. no. 28, in part); Deal Beach (*Haynes, 1905*; *fide* Frye & Clark); Delaware Water Gap (*Austin;* distr. in Hep. Bor.-Amer. Exsic. no. 26, as *Southbya biformis!;* type station of that entity). PENNSYL-VANIA. Milford (Evans, 1919a); Clarion, Lawrence, Venango, Warren Cos. (*Lanfear*).

DISTRICT OF COLUMBIA. "Banks of Potomac, opposite Georgetown" (Holzinger, 1907; perhaps collected in Virginia). VIRGINIA. Giles Co. (Patterson, 1949). WEST VIRGINIA. Bretz (Sheldon, 1910); Hancock, Monongalia, Wetzel Cos. (Ammons, 1940). KENTUCKY. Elliot, Lewis, Morgan, Powell Cos. (Fulford, 1934, 1936).

NORTH CAROLINA. Bladen, Forsyth, Jackson, Polk, Randolph, Stokes, Swain, Warren Cos. (Blomquist, 1936); also Eno R., 5–6 mi NW. of Durham, Orange Co. (*RMS 37454*); Glenn Falls, SW. of Highlands, 3600 ft, Macon Co. (*RMS 40605a*); Bubbling Spring, W. side of road N. from Beech Gap, N. of Blue Ridge Pkwy., Haywood Co.

---

perianth mouth (×150). (6) Cells of middle of perianth (×150). (7) ♀ Bract, unusually narrow (×23). (8) Leaves (×13). (9–10) Normal ♀ bracts (×23). (11) Median cells with oil-bodies and, lower cell, chloroplasts (×650). [1–10, *RMS 36502*, Dry Falls, N.C.; 11, *RMS 14612*, Taylors Falls, Minn.]

(*RMS 39371*); Linville Gorge, Burke Co. (*RMS 29077, 28974a, 28959, 28921*; atypical! form with numerous oil-bodies); Hemlock Bluff near Cary, Wake Co. (*RMS 28708a, 28631, 36890*); Eno R., Durham Co. (*RMS & Blomquist 28416*); Dry Falls near Highlands, Macon Co. (*RMS 30153, 36502*; atypical, form with numerous oil-bodies); W. side of Chattooga R. gorge above jct. with Ammons Branch, SE. of Highlands, Macon Co. (*RMS 40831a*).   SOUTH CAROLINA. 2–3 mi E. of Jocassee, Oconee Co. (*RMS 30154*; form with numerous oil-bodies).   GEORGIA. Summit of Rabun Bald, 4700 ft, Rabun Co. (*RMS 40682*).   ALABAMA. Auburn (Evans, 1919a).

Westward occurring as follows: OHIO: Urbana; Thompson Ledge, Geauga Co.; Adams, Athens, Cuyahoga, Hocking, Jackson, Scioto Cos. (Miller, 1964; Evans, 1919a).   INDIANA. Putnam Co.; Dubois, Jasper, Montgomery, Perry, Ripley Cos. (*Welch; Wagner*).   ILLINOIS. Clark Co. (Arzeni, 1947).   MICHIGAN. Pictured Rocks, Munising Co. (*RMS 39290c*); Grand Ledge; Au Train Pt.; Montreal R., Gogebic Co.; E. of Laurium, Houghton Co.; Luce Co.; Ontonagon Co. (*Evans & Nichols; Steere*).

WISCONSIN. Lafayette, Sauk, Wood, Adams, Juneau, Iron, Bayfield, Trempealeau, Ashland, Buffalo, Grant Cos. (Conklin, 1929).   MINNESOTA. Curtain Falls, Taylor Falls, Chisago Co. (*RMS*); Grand Marais, Cook Co. (*Conklin*); Big Falls, Koochiching Co. (*Nelson*); Two Island R. and Gooseberry R., Lake Co. (*Conklin*); St. Paul, Ramsey Co. (*Lapham*); Lairds Spring, Winona Co. (*Holzinger*); see Schuster (1953).   IOWA. Allamakee, Webster, Linn, Jasper, Marion, Mahaska, Muscatine, Lee Cos. (Conard, 1945); Cliffland, 3 mi E. of Ottumwa, Wapello Co. (*Conard!*).   MISSOURI. St. Francis R. (*Savage & Stull 139*); Pacific (Evans, 1919a); Franklin Co. (Gier, 1955). KANSAS. Chatauqua, Douglas, Leavenworth, Montgomery, Woodson Cos. (McGregor, 1955); 5 mi W. of Baldwin, Douglas Co. (*McGregor 4782!*).

In general, a lowland species in eastern North America; in the Southern Appalachians rare or absent at above 2500 feet (but represented at from 2500 to 4000 feet and sometimes at lower elevations by a form with numerous oil-bodies, which is treated subsequently).

Ecology

Solenostoma hyalinum is widely dispersed at median and low elevations in eastern North America, ranging from the Taiga (where rare) throughout much of the Deciduous Forest Region. In the Southeast almost restricted to the Piedmont and the lower elevations of the mountains, rarely above 2500 feet. It is particularly frequent on shaded bluffs (under stands of hemlock, *Kalmia latifolia*, and *Rhododendron catawbiense*) in the lower mountains and Piedmont of the Southeast, along streams that cut through rock. *Solenostoma hyalinum* forms extensive patches on soil-covered rocks as well as more often on elevated loamy or clayey river banks. The correlation in distribution with moving water is striking, and I have yet to see the species at any considerable distance from streams.

At such sites *S. hyalinum* occurs most commonly with *Plagiochila asplenioides* (more rarely *P. columbiana*), *Chiloscyphus pallescens*, *Scapania nemorosa*, *Pellia* spp. (*epiphylla, neesiana*), occasionally *Calypogeia* (*fissa, muelleriana*), and *Geocalyx graveolans*, *Jungermannia lanceolata*, and *Trichocolea tomentella*.

Although most often on shaded banks, where the plants are a pure, some-what pale green (except for the violet rhizoids), the species occasionally is found under intermittently sunny conditions; then the reddish pigmentation of the rhizoids becomes more general, extending over at least the leaf bases and bases of the perianths. Highly pigmented plants (which are also developed in the closely related *P. crenuliformis*) are among the most hand-some of all regional leafy Hepaticae. Although the species almost invariably has purplish red rhizoids, it may, under exceptional conditions, produce phenotypes with colorless rhizoids. This does not appear to be correlated with light intensity, since I have seen forms with colorless rhizoids from areas with relatively high insolation, and forms with purplish rhizoids from sites with deep, constant shade. Apparently genetic differences are involved.

Solenostoma hyalinum is rarely found directly on rocks (a habitat much more characteristic of the related *S. crenuliformis*, *S. fossombronioides*, and, in particular, *S. obscurum*) but occurs almost always over fine-grained soils. Rarely on "loose, friable, often calcareous sandstone . . . with *Conocephalum conicum* and *Anthoceros laevis* . . . with a pH of about 6.0–6.5 or even up to 7.3" (Schuster, 1953). Similarly, on damp, shaded cliffs (as in Burke Co., N.C.), associated with *Diplophyllum apiculatum*, a form of *Calypogeia muelleriana*, and *Pellia epiphylla*. All such epipetric populations deviate to some extent (as was noted by Schuster, 1953; footnote, p. 401) and often give the impression of being intermediate between *S. hyalinum* and either *S. crenuliformis* or *S. obscurum*, particularly in the suberect and somewhat concave leaves, in the broadly ovate, rather than reniform, ♀ bracts, and in the generally nonundulate leaves and bracts.

Differentiation and Variation

This widespread species is easily subject to misidentification. "In the case of sterile material, the difficulties of determination are still further increased, and species of other genera with undivided leaves, such as *Jungermannia*, *Jamesoniella* and *Odontoschisma*, are likely to be mistaken" (Evans, 1919a) for the present species. Fortunately, the normal presence of reddish rhizoids separates it from all of these taxa and shows its clear affinity with other species of *Solenostoma* subg. *Plectocolea*. The rather pellucid, thin-walled, extremely narrow, and elongated dorsal cortical cells (often averaging 130–160 μ long) are also diagnostic.

Within *Plectocolea*, *S. hyalinum* is rather closely related to *S. obscurum*, *S. fossombronioides*, and *S. crenuliformis* (among regional species). It has been held to be a close ally of the Japanese *S. radicellosum* Mitt. (= *Plectocolea radicellosa* Mitt.), upon which Mitten (1865a) based *Plectocolea*. The resemblance to this east Asiatic species is remote, but there is a distinct affinity to the Japanese *S. tsukushiensis* (Amak.) Schust., comb. n. (basionym

Plectocolea tsukushiensis Amak., Jour. Jap. Bot. 34: 113, 1959). As Amakawa (1960a) points out, this is the plant which for long was erroneously considered identical with *S. radicellosum.*

Solenostoma hyalinum differs from most regional species with which it might be confused by the obliquely orbicular-ovate, broad, decurrent leaves; large, distinctly collenchymatous cells with large and conspicuous oil-bodies; generally reddish or purplish rhizoids, which are long, dense, and prominent (exceptionally colorless!); absence of underleaves; and conically narrowed perianth formed of narrow, elongate cells situated at the apex of a distinct perigynium. These features are ± shared by *S. obscurum* and *S. crenuliformis* (to a smaller extent also by *S. fossombronioides* and *S. obovatum*).

The lower perigynium (varying from ¼ the total height of the perianth to absent on one side, but as high as ⅓–½ on the other side), together with the fact that, after fertilization, the perianth typically extends at least slightly beyond the bracts at maturity, separates *S. hyalinum* from *S. obscurum.* In *S. hyalinum*, also, we usually find more nearly orbicular leaves, although, as is indicated below, this character must be used with caution. The flat leaf margins, with the marginal row of cells neither larger nor conspicuously swollen, separates *S. hyalinum* from *crenuliformis.*[377] Other differences occur which, though generally evident, are not constant, among them: (*a*) the usually somewhat undulate ♀ bracts; (*b*) the usually more longly exserted perianth (to ¼–½ beyond the bracts); (*c*) the fewer oil-bodies per cell, of *S. hyalinum.* Most important, perhaps, is stem anatomy (compare Fig. 271:9 and 12).

In much nearctic material of the widespread *S. hyalinum* we find that some of these differences nearly or quite disappear. In plants from Wake Co., N.C. (*RMS 28581, 28708, 28761*) the ♀ plants have more or less ovate leaves below the perichaetial bracts, and the bracts themselves may be ovate and narrowly rounded at the apex (although the lower leaves of such plants are quadrate-orbicular); these bracts and subtending leaves are not at all undulate. By contrast, note the very broad, reniform, broadly rounded bracts of plants from Macon Co., N.C. (*RMS 36502*) and of European material; these bracts and subtending leaves are abundantly undulate, as is evident from the figures.[378] Even though Macvicar (1926, p. 138) and Müller (1939), as well as Schuster (1953, pl. 32:2), indicate that *S. hyalinum* has 2–5 oil-bodies in most leaf cells, which are very large in size, there is much variation in this respect also.

---

[377] There is also a general difference in facies. *Solenostoma hyalinum* has the lower leaves, at least, nearly flat and spreading laterally, and very obliquely inserted, while the upper leaves are more erect-spreading, less horizontally inserted, often somewhat undulate. In *S. crenuliformis*, leaves are (in all cases studied) ± uniformly concave and more erect-spreading, often somewhat dorsally secund. The border of swollen cells in *S. crenuliformis* is distinct at lower magnification; it may be poorly or not developed on juvenile shoots. The often difficult distinction between these two taxa is discussed under *S. crenuliformis* at greater length, p. 992.

[378] The variation in shape of the ♀ bracts is probably meaningless taxonomically, since it is extensive within a single population (compare Fig. 287:7, 9–10, drawn from plants from Dry Falls, N.C., growing under identical conditions at the same site).

Indeed, there appears to be variation (within the same population) between ♂ and ♀ plants. In *RMS 28581* (Wake Co., N.C.), e.g., ♀ plants may have up to 7–15 ellipsoidal oil-bodies per cell, almost obscuring the lumen, but more often only 5–8 per cell; associated ♂ plants have generally 2–5, occasionally 5–7, per cell. In contrast, Minnesota material (Fig. 288:2) shows ♀ plants with 2–4 much larger oil-bodies in each cell. When *S. hyalinum* occurs as a form with few oil-bodies, it can be differentiated on the basis of the vegetative plant from *S. crenuliformis*, where there often appear to be 4–8 or more oil-bodies per cell. Further study of much living material, however, strengthens my earlier opinion (1953, p. 404): "The shape, size and number of oil-bodies per cell are so variable in these two species that this character must be used with caution." Although *S. hyalinum* is ordinarily described as having a smooth cuticle (as in Müller, loc. cit., Macvicar, 1926, and Schuster, 1953—however, I qualified this by "usually"), it is often fully and closely verruculose-striolate (as in *RMS 28581*). This, judging from the descriptions of European material and from study of European specimens, does not appear to be often the case with Old World material. Evans (1919a) also notes that American plants tend more often to possess a verruculose cuticle.

The variation in (*a*) form of the bracts, (*b*) presence or absence of undulation of them, (*c*) color of the rhizoids, (*d*) number of oil-bodies per cell, and (*e*) presence or absence of cuticular papillae or striae—all suggest that we deal with an old, polymorphic species. In the concave, ovate, distally narrowed, and narrowly rounded upper leaves and ♀ bracts (of such specimens as *RMS 28581*), the nearctic material at times approaches *S. obscurum, crenuliformis,* and *fossombronioides,* all of which are endemics of eastern North America. At the same time, such plants differ sufficiently from "typical" *S. hyalinum,* which appears confined to Europe, that it may prove necessary to treat the nearctic material as a different subspecies.

The preceding account of variability deals almost exclusively with the eastern population of *S. hyalinum.* With study of the Lake Superior populations, new problems arise, chiefly in the separation of *S. hyalinum* from *S. crenuliformis.*[379] Lake Superior populations differ ecologically: they occur as pioneers or near pioneers on damp sandstone, rather than on soil, approaching *S. crenuliformis* in this respect. Associated with this occurrence, these populations tend to have somewhat concave leaves whose margins are slightly swollen and appear tumid under low power; perianths are often hardly emergent from the bracts, although there is a good deal of variability in this respect. Perhaps most significantly, the leaf cells are intermediate in size: the marginal average (20)22–26(30) μ, tangentially measured; the median average (23)26–32 × (25)30–36 μ. The cuticle appears, in all cases, to be distinctly verruculose, even on marginal cells. This Lake Superior phenotype, possibly forming a discrete subspecies of *S. hyalinum,* is separable only with great difficulty from

[379] The Lake Superior reports of *S. crenuliformis,* which I have attempted to check, all appear to represent *S. hyalinum.* I have pointed out (Schuster, 1953) that the Minnesota material "seen of this species is not identical with 'typical' European material in the form of the leaves and female bracts .... Some of our plants give the impression of being nearly intermediate between typical *S. hyalinum* and the endemic *S. crenuliformis.* The relationships between these taxa deserve further study."

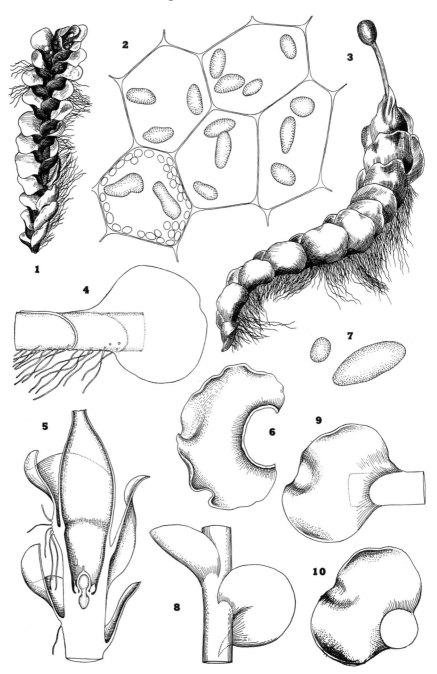

FIG. 288. *Solenostoma (Plectocolea) hyalinum.* (1) ♂ Shoot (×12). (2) Median cells with oil-bodies and, lower left, chloroplasts (×685). (3) Sporophyte-bearing shoot (×12). (4) Shoot sector with leaf and rhizoids, showing leaf insertion (×ca. 20). (5) Shoot apex,

S. crenuliformis, from which it differs primarily in the (*a*) stem anatomy typical of *S. hyalinum*, with the cortical cells averaging no larger, usually slightly smaller, than the 2–3 strata of intracortical cells; (*b*) marginal cells, although slightly tumid, never distinctly larger than intramarginal. The Lake Superior phase of *S. hyalinum* is usually highly pigmented, with leaf bases deeply reddish purple and ♂ bracts often pigmented; otherwise the plants are quite pellucid, unlike the Southern Appalachian phase of the species, which is quite opaque.

The three phases of *S. hyalinum* found in North America, together with a fourth form so far reported only from Europe, may be separated by the subjoined key. Until the necessary culture experiments are carried through, it appears wisest to refrain from taxonomic recognition of our endemic phases.

Key to Variations of S. hyalinum

1. Perigynium distinct, averaging at least 0.25 the height of the free perianth; perianth varying from included to slightly or moderately emergent . .   2.
   - 2. Leaf cells relatively small: the marginal usually (20)22–26(30) μ, the median (23)26–32 × (25)30–36 μ; marginal cells somewhat swollen; plants with concave, often somewhat erect leaves and with bracts closely sheathing the (usually weakly exserted) perianth; with the facies of *S. crenuliformis* (but without a distinct hyalodermis); pioneer on damp to dry sandstone rocks; Lake Superior region . . *S. hyalinum* [saxicolous phase]
   - 2. Leaf cells larger: marginal averaging 23–34 μ, usually 25–30 μ, median (24)26–34(36) × 32–41 μ; marginal cells not tumid; leaves usually nearly flat and laterally patent, with bracts spreading from perianth, at least distally; almost exclusively on soil or soil-covered rocks . . . 3.
      - 3. Light green to green, often with a "peculiar glistening appearance"; cells rather translucent, not whitish and opaque by reflected light, with usually only 2–5(5–10) oil-bodies per cell; with perigynium commonly less than 0.5 the height of the perianth above it; perianth commonly ¼–½ emergent . . . . *S. hyalinum* [typical]
      - 3. Somewhat opaque, slightly bluish green, usually dull; perigynium commonly 0.6–0.9 the height of the perianth above it, the perianths varying from included to 0.2–0.4 emergent; cells opaque and whitish by reflected light, with usually 5–18 oil-bodies per cell . . . . *S. hyalinum* [southeastern phase]
1. Perigynium virtually absent; perianth long-emergent . . . . *S. hyalinum* var. *heteromorpha*[380]

Several collections of *S. hyalinum* that have been studied represent the extreme nearctic form of the species. These are from Linville Falls (*RMS 28921, 28974a, 29077*) and from Dry Falls, N.C. (*RMS 30153, 36502*), and from east of Jocasee, Oconee Co., S.C. (*RMS 30154*). In these collections, most plants

---

[380] Known, to date, only from Europe. Perhaps this is merely a small and poorly developed form; see Müller (1905–16).

---

through perigynium and perianth (×18). (6) ♀ Bract (×18). (7) Two oil-bodies (×1195). (8) Sterile shoot sector, ventral aspect (×18). (9–10). Leaves of ♀ plants, subgynoecial, showing insertion (×18). [1, 3, after K. Müller, 1905–16; 2, 7, from Taylors Falls, Minn., *RMS;* remainder from English material.]

have an opaque appearance—the cells often look somewhat whitish by reflected light because of the abundance of the subopaque oil-bodies. The plants, furthermore, are somewhat bluish green in color (but have the typical reddish rhizoids); they are often considerably smaller than typical English *S. hyalinum*, ranging up to 2.2 mm wide, and differ from the latter not only in facies but also in the 5–18 oil-bodies per cell, the much less exserted perianth, and the usually striolate to verruculose cuticle.

However, in *RMS 30153* and *36502* large plants closely approach robust European *S. hyalinum* in the strongly undulate bracts, the exserted perianths, and reniform bracts, wider than long, usually retuse or emarginate. Such plants also usually have a triangular or lanceolate bracteole (as do English specimens studied). From a study of these plants, it must be concluded that the majority of nearctic plants, though not as robust as is often the case with the species, is referable to *S. hyalinum*. The plants, however, tend to have a large number of oil-bodies (often 7–15 per cell), which when fresh, or even when disintegrated, form an opaque mass, resulting in a peculiar, opaque appearance.

SOLENOSTOMA (PLECTOCOLEA) ONTARIENSIS
Schust., sp. n.

[Fig. 289]

Planta caespitosa, erecta suberectave, viridis aut flavo-viridis; rhizoidea incolorata; folia in ♀ plantis subtransversa, late rotundata, paululum asymmetrica; cellulae marginis non maiores quam cellulae folii interiores; guttae olei 3–5 vel plures omni in cellula; planta dioecia; perigynium bene evolutum; os perianthii latius apertum, invaginatum.

Caespitose, erect to ascending in growth, *pure to somewhat yellowish green*, without secondary pigments (or with slight brownish pigment), *quite opaque, with facies of S. sphaerocarpum*. Shoots strongly dimorphous, the ♂ plants slender (to 1000–1200 μ wide; stems ca. 150–160 μ in diam.), the ♀ plants more robust (to 1200–1500 μ wide × 8–15 mm high; stems ca. 200 μ in diam.), very sparingly and irregularly branched with intercalary axillary branches. Stems pale, somewhat translucent, the cortical cells nearly leptodermous, elongated, ca. 20–23 μ wide × 65–95 μ long. *Leaves somewhat antically secund* to weakly laterally patent, quite obliquely inserted but the upper several pairs on ♀ plants *appearing subtransversely oriented*, not or hardly undulate, somewhat concave, when mature *broadly and somewhat asymmetrically rotundate* (commonly 975–1000 μ wide × 725–800 μ long up to 1200 μ wide × 900 μ long, measured along midline), averaging clearly broader than long except on weak (and on ♂) shoots. *Rhizoids numerous, colorless or (with age) pale brownish*. Cells thin-walled, polygonal to (near margins) quadrate, with distinct to slightly bulging trigones; marginal cells ca. 20–26 μ, *not differentiated* from interior cells; median cells ca. 23–28 × 28–35 μ; cuticle nearly smooth to delicately verruculose (at least in basal portions of leaves); oil-bodies finely granular-papillose, 3–5 or more per cell, medium-sized. Asexual reproduction absent.

Dioecious, dimorphic (♂ plants much more slender). ♂ Plants with androecia with 2–4 pairs of bracts, laxly or hardly imbricate to contiguous, similar to leaves except that the basal portions are strongly concave (only the distal portions being somewhat spreading); androecia becoming intercalary; 1–2-androus. Gynoecia usually without subfloral innovation; bracts (and subtending 1–2 pairs of leaves) *erect to suberect and somewhat sheathing*, similar to leaves but somewhat larger and relatively broader (at times nearly reniform), entire-margined, *not or hardly undulate*. Perianth relatively short, the *apex hardly projecting beyond the bracts* (or for less than 0.2 its length), *closely sheathed by them* and often hardly evident, 600–750 μ in diam. and 650–1000 μ high, *situated at the apex of a well-developed perigynium*. Perianth *truncate at mouth*, somewhat but not strongly conoidally narrowed to the rather *broad and open apex, the apical cells inflexed into the orifice;* perianth delicate, hardly to weakly plicate, uni-stratose throughout; *apical cells often somewhat clavate, appearing commonly inflated,* curved inward and distinctly inflexed, ca. 20–22 μ in diam. $\times$ (42–45)54–64 μ long; cells below apical cells often slightly smaller, the median cells (and those below middle) again becoming slightly larger [ca. 18–21 $\times$ (35)48–60 μ], *distinctly elongated*, rectangulate. Perigynium commonly 450–550 μ high (averaging *0.4–0.6 the height of the free perianth above it*). Sporophyte unknown.

Type. Coopers Falls, Ontario Co., Ontario (*Cain 4262, p. p., Sept. 16, 1952*). The type material grew on moist soil, together with a slender phase of *Calypogeia muelleriana* and with *Scapania irrigua, Mnium*, and other mosses. Known only from the type collection, which is, unfortunately, rather sparse.

Differentiation

A species with distressingly close similarities to the *Solenostoma (Solenostoma) sphaerocarpum-pyriflorum* complex, agreeing with it in (*a*) the rather bright green, chlorophyllose coloration; (*b*) several oil-bodies per cell; (*c*) the nearly rotundate leaves, quite oblique and decurrent on the lower portions of the stem but becoming subtransversely oriented, somewhat concave, and suberect on the upper portions of ♀ shoots—in the last respect the plants bear an overwhelming similarity to the *S. sphaerocarpum-pyriflorum* complex; (*d*) cells ca. 24–32 μ in the leaf middle, with rather sharply defined and feebly bulging trigones; (*e*) lack of pronounced secondary pigments, together with brownish (never purplish) color of the older rhizoids, the younger being colorless. The rather short internodes of upper portions of ♀ shoots, together with the suberect, subtransversely oriented, and somewhat concave upper leaves suggest *S. sphaerocarpum*, to which an effort was made to refer the plants. However, several much smaller, more laxly leafed, ♂ plants were found, but no trace whatsoever of old antheridia in the leaves below the gynoecia. On that basis, the plants are clearly dioecious, suggesting the Appalachian, dioecious *S. pyriflorum.* In spite of the obvious superficial similarities to this species,

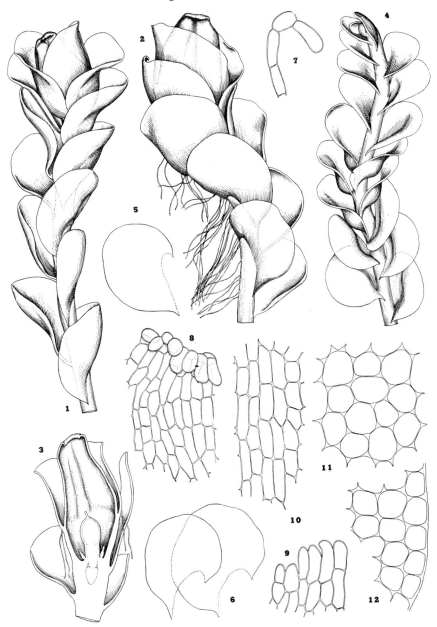

FIG. 289. *Solenostoma* (*Plectocolea*) *ontariensis*. (1) Anterolateral aspect of perianth-bearing shoot (×15.5). (2) Lateral aspect, perianth-bearing shoot (×21.5). (3) Section through apex of gynoecial plant (×21.5). (4) ♂ Shoot, dorsal aspect (×21.5). (5–6) Leaves (×22.5). (7) Longisection of invaginated perianth mouth apex (×175). (8) Ental aspect of perianth mouth apex (×170). (9) Perianth mouth sector, flattened (×170). (10) Cells of perianth middle (×170). (11) Median cells (×250). (12) Apical cells (×250). [All from type, *Cain 4262*, Ontario.]

S. ontariensis differs from it in three fundamental respects. (1) A well-developed perigynium is present at or just beyond the period of fertilization; it attains a height equal to at least 0.5–0.6 that of the free perianth above it. (2) The free perianth is of delicate, strongly elongated cells, unistratose to its base. (3) The perianth mouth is not contracted into a beak.

In these features, *S. ontariensis* is clearly distinct from the *Solenostoma sphaerocarpum-pyriflorum* complex (and, indeed, from all species of subg. *Solenostoma*). These characters place it within subg. *Plectocolea*, in which it appears quite isolated. Indeed, the series of features emphasized as giving the species a likeness to *S. sphaerocarpum* result in a facies distinct from that of our species of *Plectocolea*. On the basis of the dioecious inflorescences and the unmargined leaves, it might be confused with *S. hyalinum* and *S. obscurum*, both of which usually bear pigmented rhizoids, have perianths with the mouth conoidally contracted, rather than open, and bear ± horizontally patent leaves—not even the leaves below the gynoecia being transversely oriented. The marked habitual characters of the species thus prohibit its confusion with the *S. hyalinum-obscurum* complex. No other alternative, therefore, to treatment of these plants as a discrete species is possible.

Solenostoma ontariensis appears to be allied also to the western *S. rubrum* (G.), comb. n.,[381] a close relative of *S. gracillimum*. It agrees with this in the colorless rhizoids, dioecious inflorescence, rather low perigynium, cell size, distinct trigones, lack of differentiated marginal cells, subtransversely oriented upper leaves, and other characters. However, *S. rubrum* has longly emergent perianths often slightly beaked at the apex (or, at any rate, with a narrowly contracted mouth), which are relatively sharply plicate. A close affinity between the two species fails to exist.

The plants bear perianths with the apical 1–2 cell rows invaginated. Furthermore, the terminal row of cells usually consists of elongated, often somewhat clavate and swollen cells (Fig. 289:8). In no other species of the genus *Solenostoma* known to me does a similar perianth apex occur. Should this feature be constant (and the type plants are uniform in showing it), it will prove to be an obvious and important differentiating character.

SOLENOSTOMA (PLECTOCOLEA) OBOVATUM
(Nees) Schust., comb. n.

[Figs. 290, 291]

Jungermannia obovata Nees, Naturg. Eur. Leberm. 1:332, 1833.
Jungermannia flaccida Hüben., Hep. Germ., p. 87, 1834.
Southbya obovata Lindb., *in* Hartmann's Skand. Fl. ed. 10, 2:130, 1871.
Nardia obovata Lindb., Bot. Notiser 1872:167; Carrington, Brit. Hep., p. 32, 1874–75; Evans, Rhodora 21:162, figs. 10–14, 1919.
Eucalyx obovatus Breidl., Mitt. Naturw. Ver. Steiermark 30:291, 1893; K. Müller, Rabenh. Krypt.-Fl. 6(1):525, fig. 270, 1909; Meylan, Beitr. Krypt.-Fl. Schweiz 6(1):141, fig. 76, 1924; Macvicar, Studs. Hdb. Brit. Hep. ed. 2:135, figs. 1–3, 1926.

[381] *Jungermannia rubra* G. in Bolander, California Med. Gaz. 1870: 184 (*nomen nudum*); Underwood, Bot. Gaz. 13:113, pl. 4, 1888.

Aplozia obovata Loeske, Moosfl. Harzes, p. 59, 1903.
Mesophylla obovata Corbière, Rev. Bryol. 31:13, 1904.
Plectocolea obovata Mitt., *in* Seemann, Fl. Vitiensis, p. 405, 1871; Evans, Ann. Bryol. 10(1937):42,
    1938; Frye & Clark, Univ. Wash. Publ. Biol. 6(2):319, figs. 1–9, (1943) 1944.

Medium-sized, strongly fragrant, in loose tufts, patches, or mats, *deep* green to reddish brown or even dark purplish brown, almost blackish when dry. *Shoots 1.8–2.5 mm wide* (on ♀ shoot apices to 2.8 mm wide) × 2–5 cm long, varying from suberect to prostrate, simple or nearly so, *not or rarely with subinvolucral innovations*, frequently with stolons or flagelliform branches; branches intercalary from ventral half of leaf axil. Rhizoids copious, *purplish* or less often colorless, some arising from near the postical leaf base. Stem flexuous, 250–300(325) μ in diam.; dorsal cortical cells rectangulate, quite leptodermous to very slightly thick-walled, striolate, ca. (21)24–30 μ wide × (65)75–110(120) μ long (width:length 1:3–5); one intracortical layer of distinctly, evenly thick-walled (often slightly yellowish) cells similar in diam. to cortical cells, grading into the quite leptodermous, somewhat larger (diam. 28–35 μ, rarely 38 μ), pellucid medullary cells. Leaves varying from strongly obliquely oriented on vegetative regions to subtransverse below the gynoecia, *antical half subtransversely inserted*, varying from subremote to contiguous or barely imbricate, shortly to rather longly decurrent antically, widely spreading, broadly *ovate or less often rotundate* in part, ca. 500–700 μ wide × 600–800 μ long, up to 1080–1275 μ wide × 800–1100 μ long, concave at leaf base but plane and abruptly spreading above the base; *apex rounded* or rarely retuse; *leaves varying from 1.0–1.1× to (robust extremes) 0.95–0.75× as long as wide. Cells large*, thin-walled, with small but distinct, rarely bulging, trigones, the marginal only 18–21(25) μ; median cells ca. 30 × 45 μ or 24–28 × 30–45 μ; basal cells to 35–38 × 55–96 μ; *cuticle distinctly verruculose to striate;* oil-bodies 3–5 per cell, spherical to ovoid, almost smooth in appearance (formed of minute, hardly perceptible granules), ranging from 6 × 6–8 μ to 7–8 × 9 μ, a few up to 9 × 12 μ in basal half of leaf.

Paroecious. ♂ Bracts in 2–3(4–5) pairs immediately below perichaetial bracts, strongly concave at base but with *apices ± strongly reflexed to spreading*, broadly ovate, as large as or larger than leaves, similar to them in shape, generally 2-androus. ♀ Bracts sheathing, strongly concave at base, erect except for the *spreading to squarrose* apices. *Perianth included or the apex barely emergent, slender, conically* narrowed, 4–6-plicate for its whole length, the lobulate mouth crenulate to ciliolate with projecting cells; perianth cells elongated. *Perigynium well developed, subequal in height* to the free perianth. Capsule with epidermal cells rectangulate to subquadrate, ca. 20 × 55 μ, 2–3× the area of the inner cells, with nodular thickenings; inner cells narrow, rectangulate, ca. 11 × 40–70 μ, with semiannular bands. Seta of ca. 16–18 epidermal cell rows, 8–12 rows immediately within, and 4–7 central rows. Elaters 2-spiral. Spores 16–21 μ, yellowish brown, granulate.

Type. The Riesengebirge of Silesia and Bohemia.

Distribution

Regionally infrequent or even rare. Essentially montane, both alpine and subalpine, with a restricted range in the Arctic; lacking in the

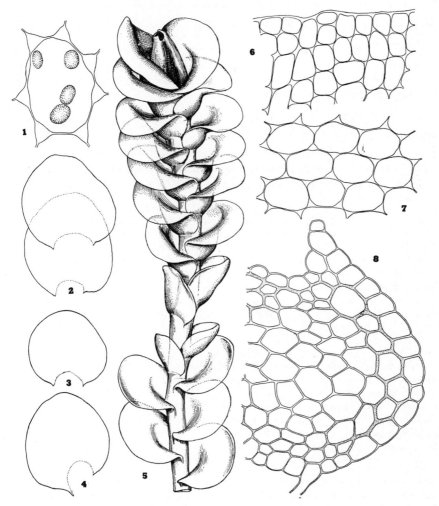

Fig. 290. *Solenostoma (Plectocolea) obovatum.* (1) Cells with oil-bodies (×560). (2–4) Leaves (×20). (5) Mature, vigorous shoot with perianth and androecium (×15.2). (6) Marginal cells near leaf apex (×255). (7) Median cells (×255). (8) Stem cross section (×245; stippled, 1–2 intracortical strata brownish or yellowish, otherwise pellucid). [All from plants from Scotland, leg. E. W. Jones, June 31, 1953.]

lowlands but occasionally descending along the borders of alpine streams; ascending to 2400 m.

Rather widespread in the mountains of western, northern, and central Europe from Scandinavia (Norway, Sweden, Finland; the Faroes), southward to England and Scotland, Ireland. Abundant throughout the central European igneous mountain ranges (Czechoslovakia, Germany, Austria, Switzerland), south to the Pyrenees of Spain (Casares-Gil, 1919) and to northern Italy (Zodda, 1934). Also in west Spitsbergen (Arnell & Mårtensseon, 1959).

Also reported from South America (Colombia) by Gottsche (Frye & Clark, 1937–47, but this is certainly incorrect, as suggested by Evans, 1919a, p. 163, the plant probably representing a discrete species).

In North America reported from the Far West and the Rocky Mt. region, from Attu I. (Clark & Frye, 1946) to Alaska, British Columbia, Washington, and Oregon south possibly to California[382] and in the Rocky Mts. supposedly to Alberta and Montana (latter report doubtful).

In our area restricted to the mountains of New England and northward:

E. GREENLAND. Between 69°30′ and 65°37′ N., on E. coast (Harmsen, 1933); Scoresby Sund (Danmarks Ø and Hekla Havn); Jeran Ø; Elvbakker (Jensen, 1897, 1906a). W. GREENLAND. Igdlorssuit, Ameralik Fjord (Lange & Jensen, 1887); Tunugdliarfik Fjord (Lange & Jensen, 1887; as *J. schraderi*).

NEWFOUNDLAND. Squires Mem. Pk. (*RMS 68-1649*), and Steady Brook, W. Nfld. (*RMS 68-1393*); Mt. Sykes (*Northrup 16*); Table Mt. of Cape Ray, South Branch, Steady Brook, and Benoit Brook, all in W. Nfld.; Doctor Hill, near St. John, N. Pen. (Buch & Tuomikoski, 1955). QUEBEC. Roggan R., Ungava, 54°15′ N. and 78–79°44′ W. (Lepage, 1953); N. slopes of Mt. Albert, Gaspé (*Williams 10655!*, with *Hygrobiella*); Fort Prével, S. shore, Gaspé Bay (*Williams 10833!*); N. shore, Gaspé Bay (*Williams 10793*); Cascapédia R., Marcil Twp., Bonaventure Co. (*Williams 10954!*). NOVA SCOTIA. Barrasois R. valley, Cape Breton (*Nichols*; Evans, 1919a); Aspy Bay (Brown, 1936a); Beulach Ban Falls, SW. of Big Intervale, Cape Breton Natl. Park (*RMS 42909*); Indian Brook, Cheticamp R., Corney Brook, all on Cape Breton (*RMS 68-1246 68-971, 68-902*, etc.).

MAINE. Chimney Pond and E. slope of the "Saddle," Mt. Katahdin (*Lorenz, 1916; Collins; fide* Evans, 1919a). NEW HAMPSHIRE. Thompson's Falls, White Mts. (*Underwood, 1889*); Tuckerman's Ravine, Oakes Gulf, and elsewhere on Mt. Washington (those from Oakes Gulf collected by Evans, distr. as *Jungermannia cordifolia*, *in* Underwood & Cook, Hep. Amer. no. 113). VERMONT. Smugglers Notch.

The range of this species is still imperfectly known in North America, partly because of earlier confusion with related taxa (such as *S. obscurum*); a report from Indiana (Turkey Run State Park, *teste* Frye & Clark, 1937–47, p. 322) cannot be credited. Similarly, reports from Maryland (Plitt, 1908; Frye & Clark, *loc. cit.*) are also certainly unfounded. Both the erroneous Maryland and Indiana reports are perpetuated in Buch & Tuomikoski (1955).

Ecology

An epipetric, pioneer species, found on both acidic and basic rocks, usually on spray-wet boulders along rocky, shaded, cascading streams and rivulets on mountain slopes, or in mats over wet rocks where subject to inundation; associated with various Scapaniae (particularly often *S. subalpina, S. undulata*), *Pellia neesiana, Hygrobiella laxifolia*, etc. Sometimes growing submerged, forming erect, swelling green tufts ("var. *rivularis*

[382] Evans (1919a) cast serious doubts about the correctness of the reports from California.

Schiffn.," Lotos, 1905, p. 1). The species is characteristic of acidic, wet
rocks in the mountains but also occurs commonly in areas with weakly
calcareous rocks. Near the southern edge of its range confined to high
elevations, often well developed above timber line.

Differentiation

Sterile material is difficult to identify with any certainty, but the paroecious inflorescences of fertile plants, together with the extremely restricted
habitat of the species serve to identify it easily from most allied taxa. A
close affinity exists with *S. subellipticum* (which see), *S. hyalinum*, and *S.
obscurum;* it has been confused with all three of these taxa.

Solenostoma obovatum shares the virtually included short perianth and
high perigynium with *S. obscurum*, a species which is, however, dioecious.
Evans (1919a) emphasized that the

general habitat is not unlike that of *N. obscura;* the older stems cling to the
substratum, while the younger stems are free or nearly so. There is therefore a
tendency for the rhizoids to be restricted to the older stems and to the stolons,
which are usually produced in abundance. The color of the plants is originally
a deep dull green, but brownish or reddish pigmentation is often present, and
in extreme cases a distinct crimson hue becomes apparent in transmitted light.
The plants never show the deep purple color which is often associated with *N.
obscura.*

However, *S. obovatum* may have the postical leaf bases somewhat purplish
tinged, and the rhizoids are almost invariably deep purplish. The leaf
bases never achieve the intensity of pigmentation characteristic of *S.
obscurum*, and the coloring is never as beautifully vinaceous as is characteristic of the latter species.

There is an immediate affinity to *S. subellipticum*, which is also paroecious.
However, *S. subellipticum* usually has colorless to brownish rhizoids,
generally slightly smaller median cells of the leaves, and a smooth or only
delicately verruculose cuticle. It is considerably smaller in size and has
erect to erect-spreading ♀ and ♂ bracts, the latter rarely showing distinctly
spreading or reflexed apices. Typically, *S. obovatum* also has leaves which
are considerably broader. They range from slightly longer than wide to
distinctly broader than long and may actually be rotundate, although more
often very broadly ovate. In *S. subellipticum* leaves average $1.15-1.25\times$ as
long as wide and are narrowly ovate to subelliptical in outline. In spite
of these differences, the closest affinity occurs between these two species
(as is discussed in detail under *S. subellipticum*).

Although sometimes confused with *S. hyalinum*, *S. obovatum* differs from it in
the (*a*) much deeper color; (*b*) paroecious inflorescence; (*c*) much deeper

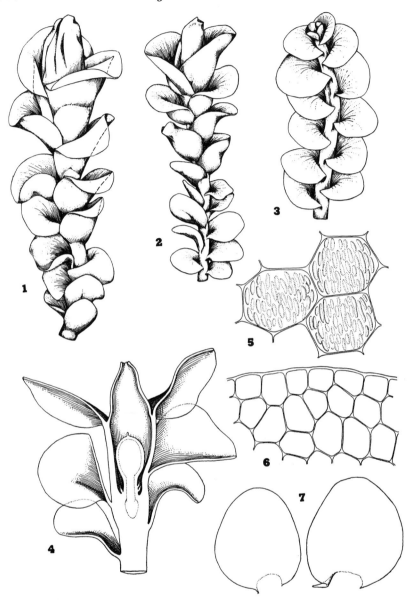

Fig. 291. *Solenostoma (Plectocolea) obovatum.* (1) Perianth-bearing shoot (×14). (2) Perianth-bearing shoot, androecial region obvious (×14). (3) Sterile shoot apex (×14). (4) Longisection through shoot apex (×18). (5) Cells from above leaf middle, showing cuticular papillae (×415). (6) Marginal apical cells (×240). (7) Leaves (×15.3). [1–3, after Evans (1, 3 from Oakes Gulf, Mt. Washington, N.H., *Evans;* 2, Thompson's Falls, White Mts. N.H., *Underwood*); 4–7, from Swedish plants, leg. Arnell & Persson, 1950.]

perigynium and more reduced perianth, the latter not or hardly exceeding the bract apices; (*d*) strongly verrucose-striolate cuticle; (*e*) leaves with a peculiar mode of attachment. The ventral half of the leaf is inserted quite obliquely, the antical half subtransversely, except for the distinctly decurrent and often reflexed inner extreme, resulting in a leaf which is quite concave at the base antically (with almost the suggestion of a shallow basal cup), although the distal portion is nearly plane and widely spreading. Because of this mode of insertion, mature leaves often appear narrowed, hence obovate, when *in situ*, leading to the species name. When flattened, of course, the leaves are clearly widest in their basal half. A similarly subtransverse insertion of the dorsal half of the vegetative leaves characterizes *S. subellipticum* and to a lesser extent *S. obscurum*. Evans (1919a, p. 166) states that the trigones of the leaf cells are usually much smaller than in *S. hyalinum*, but this is not invariably the case.

Solenostoma obovatum is polymorphous, showing considerable differences in size (small forms often being distressingly close to *S. subellipticum*), pigmentation, and leaf shape. Jensen (1901, p. 135) has described these variations well:

Near the coast, the small forms—fruiting abundantly during spring—are frequent in crevices and on the ground among rocks. In the mountains the plants gradually become stouter in size and habit, but fruit less abundantly, until they form, above 300–400 m., large, deep, dark purplish-brown, barren tufts in the rills and around springs.

The most marked variety of the species is var. *bipartita* K. Müll. [Rabenh. Krypt.-Fl. 6(1):528, fig. 270e, 1909], in which the leaves are sharply emarginate. Such extremes have not been found in North America, although occasional plants show isolated retuse leaf apices.

SOLENOSTOMA OBSCURUM (*Evans*) *Schust., comb. n.*
[Figs. 292–294]

Nardia obscura Evans, Rhodora 21:159, pl. 126, 1919.
Plectocolea obscura Evans, *in* Buch, Evans, & Verdoorn, Ann. Bryol. 10(1937):4, 1938; Frye & Clark, Univ. Wash. Publ. Biol. 6(2):325, figs. 1–5, (1943) 1944.

Loosely prostrate to ascending, in patches or mats, or often creeping and scarcely gregarious, rather deep but pellucid green, to (in sun) purplish black, the *postical leaf bases* (usually) *and rhizoids* (always) *an intense, vinaceous or purplish red*. Sterile shoots 1.2–1.75(2.4) mm wide × 1–2 cm long, the fertile shoots to 3.25 mm wide, sparingly branched, *usually freely stoloniferous;* both branches and stolons originating from lower portions of leaf axils, intercalary. Stems ca. 160–225(300) μ in diam., the cortical cells narrowly rectangulate, their free faces *thin-walled, forming a distinct hyalodermis* [cells (18–21)23–26(32) μ in diam. × 80–100(125) μ long], surrounding 2–3 strata of *thick-walled, often yellowish cells* only 12–20(23) μ in diam., which grade into the central medullary cells [which are 20–25(30) μ in diam. and more nearly leptodermous, hyaline, and colorless]. Rhizoids long, usually abundant on prostrate shoot sectors and

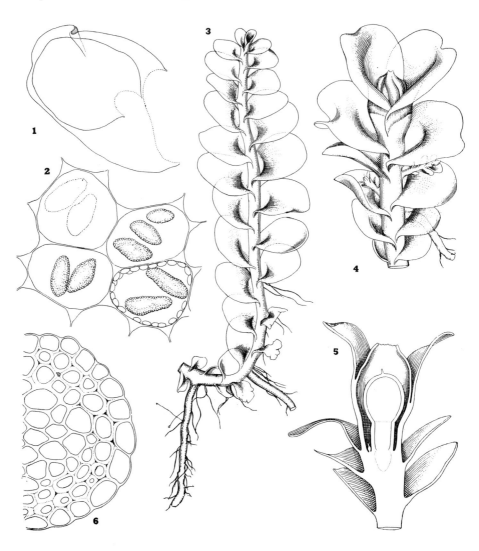

FIG. 292. *Solenostoma* (*Plectocolea*) *obscurum*. (1) Leaf from ♀ shoot, drawn within a ♀ bract (×21). (2) Median cells with oil-bodies and lower right, chloroplasts (×810). (3) Sterile shoot, with stolons (×12). (4) Apex of perianth-bearing shoot, with young lateral-intercalary branches and a postical stolon (×12). (5) Longisection through apex of ♀ plant (×12). (6) Stem cross section (×268). [1, 3–4, 6, *RMS 36110*, Soco Falls, N.C.; 2, *RMS 45119*, Standing Indian Mt., N.C.; 5, *RMS 35625*, Tennessee.]

stolons, 13–14 μ in diam. Leaves somewhat distant to laxly imbricate, strongly succubous except immediately below gynoecia, somewhat *abruptly decurrent antically from a subtransverse antical base*, horizontally spreading to erect-spreading, *ovate-oblong to broadly ovate, averaging 1.05–1.2×* as long as broad, on weak sterile shoots occasionally even narrower, usually ca. 900–1500 μ long × 750–1300, (1500) μ wide, rounded or *sometimes retuse* at apex, not margined, little to moderately concave at base, nearly plane elsewhere, *not undulate*. Cells thin-walled, with small to barely bulging trigones and a smooth to weakly striolate to verruculose cuticle; *marginal cells 15–20 μ*; median cells 20–28 μ wide × 25–35 μ long; oil-bodies 2–10, usually 2–5, but sometimes largely 5–8, per cell, usually ovoid to ellipsoid and 5–7 × 6–9 to 7–9 × 11–13(14) μ, some spherical and 5–6 μ, finely granular in appearance, pale grayish, opaque; chloroplasts 2–3 μ. Without gemmae.

Dioecious. Androecia becoming intercalary, loosely spicate, usually in (2)4–6 pairs. ♂ Bracts ± imbricate, deeply concave but with abruptly spreading apices; 1–2-androus. ♀ *Bracts obliquely ovate, erect*, frequently somewhat undulate, *often retuse* at apex, usually greatly exceeding perianth, closely *sheathing the usually fully included perianth*, (1450)1650–2000 μ wide × 1500–1750 μ long (measured along midline; maximal length 2100–2500 μ). Perianth and subtending perigynium 1.5–2 mm high, the *perigynium from (1)1.5 to 2×* as high as the free *perianth*, firm, 700–1000 μ wide; free perianth conical, short, plicate above, its cells elongated and rectangulate; mouth finely crenulate to crenulate-ciliolate with cells whose distal portions are free and rounded. Seta ca. 280–320 μ in diam., of ca. 24–27 epidermal cell rows surrounding an infinite number of inner cell rows, the epidermal and inner cells subequal in diam., leptodermous. Capsule ovoid, yellowish brown; wall 2-stratose. Epidermal cells subquadrate to short-rectangulate, ranging from 15–16 × 24–27 μ up to 19–22 × 27–32(35) μ, each long wall with (0)1–2 nodular thickenings, each transverse wall with 0–1 thickening. Inner cells much more irregular, averaging narrowly rectangulate but often sigmoid, 12–15(16–19) μ wide × (35)50–75 μ long, with slender, occasionally subcomplete, yellowish brown, semiannular bands. Elaters 8–9 μ in diam., commonly 100–135 μ long, bispiral, the yellowish brown spirals 3–4 μ wide, the *ends of the elaters sharply tapering, often tortuous, evenly or irregularly thick-walled, not* distinctly spirally thickened. Spores yellowish brown, very finely verruculose, 14–15 μ in diam.[383]

Type. Waterville, N. H., at the "V" (*Lorenz 203!*).

Distribution

Until recently known only from New England and northern New York, with the exception of a single report from Oregon (Sanborn, 1929) which remains questionable because of lack of subsequent substantiation

---

[383] The description of the previously unknown sporophyte is based on plants from Lyon Mt., N.Y. (*RMS A-266*). The sporophytes in this collection are barely mature. Therefore the spore diameter may prove to extend up to 16 μ, and the wall cells to average slightly larger on fully mature capsules.

from other western stations. More recently shown to have an essentially Appalachian distribution, southward into North Carolina and Tennessee.

MAINE. Mt. Desert I. (*Lorenz*); Carrabasset R., Jerusalem (Evans, 1919a).　　NEW HAMPSHIRE. Tylers Spring near Waterville (*Lorenz, 1907*); The "V," Waterville (*Lorenz 203*, type!); Ellis R., Jackson (Evans, 1919a!; distr. in Underwood & Cook, Hep. Amer. no. 83 as *Nardia crenuliformis*); Profile Brook and the Flume, Franconia Mts.; Coosauk Falls and Ice Gulch, Randolph; Huntington Ravine, Mt. Washington (Evans, 1919a; some of these reports earlier listed by Evans, 1908c, and Lorenz, 1908d, as *Nardia hyalina*).　　VERMONT. Downer's Glen, Manchester (*Grout*!; listed by Evans, 1905, as *Nardia obovata*).　　MASSACHUSETTS. Mt. Greylock (*Andrews*; Evans, 1919a); Roaring Brook, Shutesbury (*RMS*); Bear R., Conway (*RMS*); Tannery Falls (*RMS*). CONNECTICUT. Beacon Falls (Evans, 1905); Salisbury (*Lorenz, 1920*).

NEW YORK. Rainbow Falls, Adirondack Mts. (Evans, 1919a); foot of Slide Mt., above L. Winnisink, Ulster Co., Catskill Mts., 2800 ft (*RMS 17648, 17649*); Lyon Mt., Adirondack Mts., 2500–3000 ft (*RMS A-266, c. caps.*!).　　VIRGINIA. White Top Mt., 5400 ft, Smyth Co. (*RMS & Patterson 38073, 38098, 38055a, 38090*); The Cascades, Little Stony Cr., near Mountain L., Giles Co. (*RMS 20276*).

NORTH CAROLINA. Jackson Co.: Soco Falls, near Maggie (*RMS 36109, 24209, 24806*); below Bald Rock Mt., 3.5 mi E. of Cashiers (*Anderson 10299*!); Neddie Cr., E. fork of Tuckasegee R. (*RMS 29430, 29444c*); Upper Falls, Whitewater R. (*Anderson 8772*!). Macon Co.: Big Dog Mt., above Dry Falls, Cullasaja Gorge, NW. of Highlands (*RMS 29520*); Wayah Bald, 8 mi W. of Franklin (*RMS 39282*). Linville Gorge, McDowell Co. (*RMS 28921*); W. branch of W. fork of Pigeon R., above jct. with Flat Laurel Cr., 4500 ft, S. of Sunburst, Haywood Co. (*RMS 39377, 39383, 39385*); Upper Bunch Cr., Swain Co. (*Blomquist, 1932*!; det. Fulford as *Jungermannia lanceolata*, and basis for report of that species in Blomquist, 1936); recess under summit of Standing Indian Mt., 5480 ft, Nantahala Mts., Clay Co. (*RMS 45119b*).　　TENNESSEE. Sevier Co.: below "The Chimneys," on trail to Indian Gap, Smoky Mt. Natl. Park (*RMS 35629, 35629a, 35625*); Dry Sluice Gap, 5350 ft, Appalachian Trail, Smoky Mt. Natl. Park (*Schofield*!).　　GEORGIA. NE. corner of Rabun Co., on W. side of Chattooga R. (*RMS 39865*); below High Falls of Big Cr., ca. 2200 ft, Rabun Co. (*RMS 40032*).

Ecology

A pioneer species usually found on wet rocks, often subject to inundation, lying in or at the edges of rivulets and mountain streams; essentially montane or submontane, confined to areas with igneous or metamorphic, acidic rocks. Usually in deep shade (and then a deep to rather olive green, although somewhat pellucid by transmitted light, except for the intensely and characteristically vinaceous or violet-purple rhizoids and, usually, leaf bases), but occasionally growing in partial sun (and then forming scorched-appearing, blackish purple patches).

Occurring often with *Marsupella* (*M. emarginata*, rarely *M. sphacelata;* southward also *M. paroica*), *Riccardia multifida*, *Pellia epiphylla*, *Plagiochila asplenioides* (southward sometimes *P. sullivantii* and *austini*), *Cephalozia bicuspidata*, *Scapania nemorosa* (and sometimes *S. undulata*), *Jubula pennsylvanica*, *Solenostoma pumilum*, and (southward) *Metzgeria hamata*.

The species is not infrequent in the Southern Appalachians at high elevations (usually 3800–5000 feet or even higher), forming thin, dark patches over strongly shaded boulders and rocks lying in intermittent rivulets and small, steep brooks. It sometimes grows submerged in 1–2 inches of water.

Solenostoma obscurum often occurs in deep shade under *Rhododendron maximum* or hemlock. In its ecology, it is quite distinct from *S. hyalinum*.

Differentiation

Often approaching in size *S. hyalinum*, with which it agrees in the dioecious inflorescence, the general presence of pigmentation of the leaf bases and rhizoids, the somewhat pellucid green color, the frequently retuse leaves, the similar oil-bodies, etc. Differing, however, in the restriction to bare rocks (*S. hyalinum* usually found on soil or soil-covered rocks); the intense vinaceous or violet-red pigmentation of the rhizoids and, usually, leaf bases (reddish in *S. hyalinum*); the often somewhat verruculose cuticle; the smaller cells; the included perianths with the subtending perigynium at maturity of sporophyte always higher than the perianth proper (in *S. hyalinum* the perianth usually distinctly emergent and the perigynium relatively low); and the narrower oval leaves (of sterile shoots), some of which are often retuse. In most of these features, particularly in leaf shape and insertion, *S. obscurum* closely approaches the northern *S. obovatum*, with which it was once confused. The more intense pigmentation, less strongly marked cuticle of the leaves, smaller cells, and dioecious inflorescences serve adequately to separate *S. obscurum* from all phases of *S. obovatum*.

Although fertile plants are easily determined, sterile *S. obscurum* is sometimes difficult to separate from the two preceding species and from *S. fossombronioides*. There is a distinct tendency for sterile plants to be bright green, however (rather than pale green, as in *S. hyalinum*), and to show an intense claret or vinaceous pigmentation of the postical leaf bases, which in exposed situations spreads from the bases of the leaves and extends over all or most of the plant.[384]

When fertile, the exceptionally deep perigynium is distinctive. It is higher than the perianth beyond it, at least on one side. In some material (i.e., *Anderson 10299*), the height on one side varies from 0.6 to 0.7 the total length of perigynium + free perianth; on the other side the perigynium is often lower and 0.3–0.4 (at times only 0.1) the total height. This is true of material with

---

[384] Evans (1919a) also calls attention to the slightly different leaf insertion of this species. He states: "Although the leaves are attached in both [*S. hyalinum* and *obscurum*] by long oblique lines, the basal part of the leaf in . . . *hyalinum* is usually plane or convex, somewhat as it is in *Plagiochila*; in *N. obscura* [= *S. obscurum*], on the other hand, it shows a tendency to be appressed to the stem for a short distance and then sharply revolute, giving the effect of a more transverse attachment." In our material the antical base of the leaf shows this peculiarity very well.

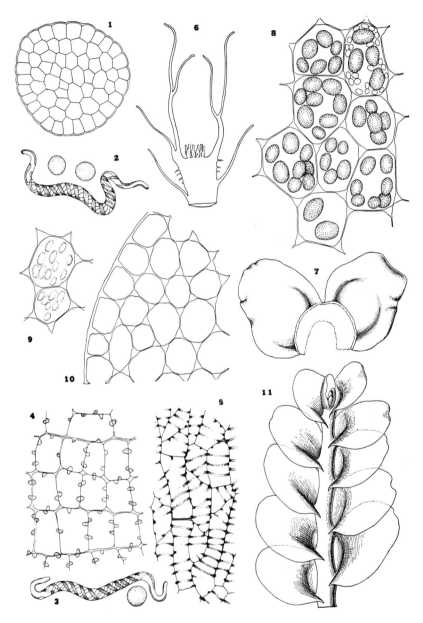

FIG. 293. *Solenostoma* (*Plectocolea*) *obscurum*. (1) Seta cross section (×100). (2–3) Spores and elaters (×325). (4) Epidermal cells of capsule wall (×325). (5) Inner cells, capsule wall (×325). (6) Longisection through shoot apex at time of fertilization; note the already massive perigynium development (×15). (7) ♀ Bracts (×11). (8) Median cells with oil-bodies and, upper right, chloroplasts (×500). (9) Two cells, showing cuticular papillae (×435). (10) Cells of leaf apex (×312). (11) Sterile shoot,

mature archegonia; with sporophyte development there is further increase in height of the perigynium. For instance, in plants from Big Dog Mt., N.C. (*RMS 29520*) the perigynium is 880 μ on both sides, with the perianth above it only 650 μ high on one side, 850 μ high on the other side. The perianth, even at maturity of the sporophyte, does not extend beyond the bracts. The cells of the perianth (and interior cell layer of the perigynium) are strongly elongate, averaging 2.5–4× as long as wide, and, like the rest of the plant, often show the characteristic violet or violet-purple pigmentation.

In the related *S. hyalinum*, the perigynium is much shorter (Fig. 271), at most 0.25–0.45 the total height of the perigynium + perianth—i.e., the perianth is always higher than the subtending perigynium and normally extends for some distance beyond the apices of the bracts. Also the lower portion of the perianth, at least on fertilized plants with mature perianths, is of cells that are less elongate. ♀ Bracts of *S. hyalinum* are nearly invariably broader than long, usually somewhat reniform in shape. In contrast, those of *S. obscurum* typically are rather obliquely, asymmetrically ovate, and either virtually as long as wide or longer than wide. The ovate form of the bracts is correlated with an essentially oblong-ovate form of the leaves in *S. obscurum*, in which we generally find the greatest width attained below the leaf middle. *Solenostoma hyalinum*, however, has essentially orbicular leaves, often considerably wider than long. These differences, although subject to admitted variation, serve to give so distinct a facies to the two species that a separation, even on the basis of sterile material, is usually feasible.

Solenostoma obscurum is also related to *S. crenuliformis*. The distinction between the two is most clear when they grow intermingled (as in *RMS 29430*). *Solenostoma obscurum* has ovate-elliptical leaves of sterile shoots (virtually orbicular in *S. crenuliformis*), and, particularly on sterile shoots, pronounced, beautiful vinaceous pigmentation of the basal portions of the leaves. Such pigmentation does not occur in either *S. crenuliformis* or *hyalinum* (even though the rhizoids and postical stem surface are usually reddish in color). The distinctly ovate to ovate-elliptical leaf form, with the leaves clearly widest below the middle, contrasts strongly with the nearly orbicular leaves of *S. crenuliformis* (and *hyalinum*), which, if slightly longer than wide, are not narrowed distally. The lack of a border of swollen marginal leaf cells of *S. obscurum* and the more nearly plane and laterally spreading leaves (rather than somewhat dorsally connivent and often suberect leaves of *S. crenuliformis*) serve to give these two species rather different facies.

Solenostoma obscurum and *S. crenuliformis* share a similar stem anatomy. In both there is a pellucid hyalodermis, consisting of relatively large cells (averaging to 20–26 μ in diam.), surrounding a 2–3-stratose layer of rather yellowish, thick-walled chlorophyllose cells (averaging only 12–20 μ). In this respect, both species are distinct from *S. hyalinum*, in which the cortical cell layer is not differentiated as a similarly discrete hyalodermis.

---

showing the diagnostically partly retuse leaves ($\times$19). [1–5, *RMS A-266*, Lyon Mt., N.Y.; 6–7, *Anderson 10299*, N.C.; 8, Slide Mt., Catskill Mts., N.Y., *RMS;* 9–11, from Underwood & Cook, Hep. Amer. no. 83, "*Nardia crenuliformis*".]

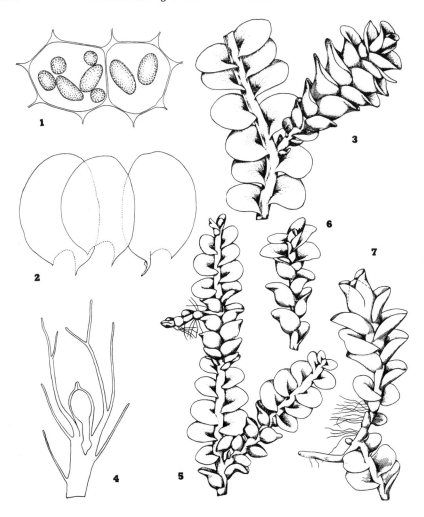

FIG. 294. *Solenostoma (Plectocolea) obscurum.* (1) Median cells with oil-bodies (×710). (2) Leaves from sterile shoot (×23). (3) Plant with lateral-intercalary androecial branch (×12). (4) Longisection through apex of gynoecial shoot (×20). (5–6) Parts of ♂ shoots; in fig. 5 with a flagelliform branch (×12). (7) Gynoecial shoot, with perianth and stolons (×12). [1–2, *RMS 36110*, Soco Falls, N.C.; 3–7, from Evans, 1919 (3, 6, from Beacon Falls, Conn., *Evans;* 5, 7, Waterville, N.H., *Evans*).]

The spores and elaters of *S. obscurum* agree closely with those of *S. hyalinum.* In both species the spores average 14–15 μ, and the elaters are ca. 9 μ in diam. Furthermore, the elaters are relatively short and suddenly attenuate at their ends (or at least at one end), which are somewhat thickened and show only vestigial spirals. In the smaller spores, *S. obscurum* clearly differs from *S. obovatum* (in which the spores are commonly 17–21 μ in diam.).

Variation

The preceding account is based on what may be considered typical plants. On White Top Mt., Smyth Co., Va., a deviant phase (*RMS & Patterson 38073, 38051, 38098, 38090*) occurs which is difficult to separate from *S. hyalinum.* It agrees with the latter in that it grows on thin soil layers, over rocks, however, associated with *Scapania nemorosa* and *Calypogeia fissa neogaea.* It is also similar to *S. hyalinum* in the broadly rotundate leaves (on sterile shoot sectors oval-ovate only on small-leaved juvenile sectors) and in the restriction of bright pigmentation to the rhizoids. However, an unusually luxuriant phase of *S. obscurum* is certainly at hand because of the (*a*) somewhat transverse and abruptly decurrent antical bases of the leaves, which are rather squarrose, standing away from the stem; (*b*) much smaller cells, averaging 15–20 μ or less on the margins, and 21–24 × 25–30 μ medially; (*c*) more chlorophyllose, opaque, bright green leaves; (*d*) exceedingly intensely, vividly vinaceous rhizoids; (*e*) mature perianths far exceeded by the ovate and emarginate bracts. When sterile, such plants are exceptionally difficult to recognize if the extraordinary pigmentation of the rhizoids (not an ordinary reddish color, as in *S. hyalinum*), the peculiar orientation of the antical leaf bases, and the smaller cells are overlooked.

SOLENOSTOMA (PLECTOCOLEA) SUBELLIPTICUM

(*Lindb.*) *Schust., comb. n.*

[Figs. 272:15; 278:11–19; 295]

Nardia subelliptica Lindb., Meddel. Soc. F. et Fl. Fennica 4:182, 1883; Evans, The Bryologist 22:69, figs. 8–15, 1919.
Eucalyx subellipticus Breidl., Mitt. Naturw. Ver. Steiermark 30:291, 1894; K. Müller, Rabenh. Krypt.-Fl. 6(1):529, fig. 271, 1909; Macvicar, Studs. Hdb. Brit. Hep. ed. 2:136, figs. 1–4, 1926.
Southbya subelliptica Lett, List Spec. Hep. Brit. Isles, p. 140, 1902.
Jungermannia subelliptica Levier, Bull. Soc. Bot. Ital. 1905:211, 1905; Amakawa, Jour. Hattori Bot. Lab. no. 22:21, 1960.
Haplozia subelliptica Casares-Gil, Fl. Iberica. Hep., p. 424, 1919.
Plectocolea subelliptica Evans, in Buch, Evans, & Verdoorn, Ann. Bryol. 10 (1937):4, 1938; Frye & Clark, Univ. Wash. Publ. Biol. 6(2):324, figs. 1–9, 1944; Schuster, Natl. Mus. Canada Bull. 122:23, (1950) 1951.

Very small, green to sometimes brownish, rarely locally purplish-tinged. Shoots from 650–800 to 800–1200 up to 1500–1800 μ wide × 5–12(12–18) mm long, prostrate or ascending, with crowding suberect, sparingly branched; occasionally *with stolons;* often with subfloral innovations. Stems (170)180–250 μ in diam., flexuous, rather stout, cortical cells 19–24 μ wide × 75–110 μ long (1:4–6), medullary cells 20–27 μ in diam., thin-walled. Rhizoids frequent, *usually colorless to pale brownish* or pale reddish, *never deep violet.* Leaves only weakly oblique in insertion and orientation, *upper leaves and those of sterile shoots subtransversely oriented*, the lower remote to contiguous, spreading, the upper subimbricate and erect-spreading, elliptical to broadly *oval to ovate*, shortly or not decurrent, *widest a little above the base*, 250–350 μ wide × 300–400 μ long up to 650–740 μ wide × 770–860 μ long (1.15–1.25× as long as wide),

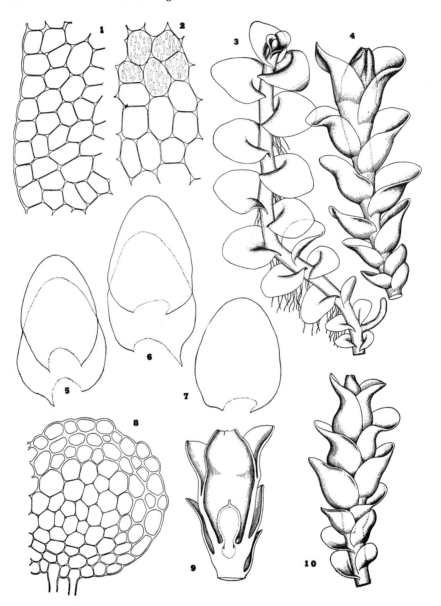

FIG. 295. *Solenostoma* (*Plectocolea*) *subellipticum*. (1) Cells of leaf apex (×260). (2) Median cells, at top with cuticular papillae drawn in (×260). (3–4) Dorsal aspects of sterile and fertile shoots (×16). (5–7) Leaves (×32). (8) Stem cross section (×227). (9) Longisection through apex of gynoecium with young sporophyte (×21). (10) Paroecious shoot apex (×19). [1–8, *Marr 656b*, Quebec; see Schuster, 1951; 9–10, from Schiffner's Hep. Eur. Exsic. no. 1372, Jämtland, Sweden.]

strongly concave at base, flat to squarrose distally, rounded at apex.[385] Cells thin-walled, with *small to minute*, rarely weakly to (arctic phenotypes) strongly bulging trigones, rounded-hexagonal except near margins, where usually subquadrate; *marginal cells 18–20(23) μ; median somewhat elongated, ca. 20–24(27) × 30 or up to 38–46 μ; cuticle smooth or delicately verruculose.* Oil-bodies (2)3–5(6) per cell, granulate, 4–5(5.5) × 6–10 μ. Without gemmae.

Paroecious, or exceptionally and sporadically autoecious. ♂ Bracts usually in 3–4 pairs, *almost transversely inserted and oriented*, usually slightly to distinctly imbricate, similar in shape to leaves but usually somewhat larger, quite evidently *concave*, except for the spreading or recurved apices; 2-androus. ♀ Bracts transversely oriented, erect or suberect, loosely sheathing the perianth and nearly *equaling or exceeding it in length*, roundish-ovate to broadly oval, somewhat undulate, the apices often spreading. *Perianth very short, 525–880 μ high × 480–525 μ in diam., essentially immersed*, broadly conical to ovoid, plicate, almost beaked at mouth when juvenile, crenulate at mouth; cells of mouth and of several tiers below them averaging 2.5–4× as long as wide, at mouth ca. 13–17×40–52 μ, fingerlike. *Perigynium very high, firm, its height 1.0–1.5(2.0)* that of the free perianth (and bracts) inserted at its summit. Sporophyte with seta ca. 150 μ in diam., of ca. 16 rows of epidermal cells, surrounding ca. 12 rows of cells within, which in turn surround 4 rows of central cells. Capsule nearly globose to short-ovoid; epidermal cells 3–4× as broad as the narrow inner cells, subquadrate, with nodular thickenings; inner cells narrowly rectangulate, with semiannular bands. Elaters ca. 8 μ in diam., laxly 2-spiral. Spores 12–14 μ, delicately papillate, red-brown.

Type. Near Dovre, Norway (*Lindberg, 1882*).

Distribution

A rather rare arctic-alpine and subarctic species, found on both sides of the Atlantic, in Europe from Spitsbergen (Arnell & Mårtensson, 1959) and Scandinavia (Sweden!, Norway, Faroe Isls.) southward to Britain (Scotland) and recurring in central Europe at high elevations (Switzerland, *fide* Meylan, 1924; Bavaria; the Harz Mts. of Germany; Austria, in the Steiermark, Tyrol, and Lower Austria); also in the Auvergne Distr. of France and in northern Italy (Zodda, 1934). Also in Iceland.[386]

In North America evidently very rare and local, confined apparently to the coldest portion of the Atlantic coastline and to the Hudson Bay region.

NW. GREENLAND.[387] Kangerdlugssuak, Inglefield Bay (*RMS 45983a*); Kangerdluarssuk, Inglefield Bay (*RMS 45777*). W. GREENLAND. Near shore of L. Ferguson, head of Sondrestrom Fjord (*RMS & KD 66-060, 66-059, 66-088a*); Qalagtoq, Upernivik Ø,

---

[385] In arctic phenotypes (e.g., *RMS 45601*, Kânâk, Greenland) the leaves may become shorter and rotundate, and then measure 550–710 μ wide × 440–520 μ long.

[386] Amakawa (1960a, p. 21) reports *S. subellipticum* (as "var. *nana* Amakawa") from Hokkaido, Japan. This plant, described as dioecious, is not identical with *S. subellipticum*.

[387] A single prior report of the species from Greenland (Hesselbo, 1923a) represents an error; the plants (Copenhagen) are clearly *S. pumilum* subsp. *polaris*.

Inukvsait Fjord (*RMS & KD 66-1101b*); Kangerdlugssuakavsak, head of Kangerdluarssuk Fjord, 71°21′ N., 51°40′ W. (*RMS & KD 66-1287, p.p.*, with *S. sphaerocarpum*, etc.; *66-1281, 66-1289, 66-1287c*); Eqe, near Eqip Sermia (glacier), 350–450 m, 69°46′ N., 50°10′ W. (*RMS & KD 66-326, 66-340, 66-340a, 66-322a*); Magdlak Alfred Wegeners Halvö (*RMS & KD 66-1356c*).

QUEBEC. Mouth of Great Whale R., E. coast of Hudson Bay (Schuster, 1951); N. shore of Gaspé Bay, Gaspé Pen. (*Williams 10819!*). NOVA SCOTIA. Valley of the Barrasois, and mountains W. of Ingonish, Cape Breton I. (*Nichols 1269, 1478, 1725;, teste* Evans, 1919b; these reports cited by Nichols, 1916, as *N. obovata*).

Also at Kânâk, NW. Greenland (*RMS 45601*) as a deviant mod. *colorata-densifolia*.

Ecology

Usually a pioneer on damp, often sandy-gravelly, soil, with *Scapania curta*, but also on moist rocks, or in crevices of moist granite (with other Hepaticae, such as *Scapania subalpina, S. irrigua*, and *S. paludicola*).

The northwest Greenland collections, from a high latitude, are in part critical. The plants occur variously mixed with *Cephalozia pleniceps, C. bicuspidata, Anthelia juratzkana, Nardia geoscyphus, Scapania perssonii, S. praetervisa, Tritomaria heterophylla, Cephaloziella arctica, C. grimsulana, Anastrophyllum minutum grandis, Prasanthus suecicus, Gymnomitrion corallioides, G. concinnatum, Lophozia excisa, Odontoschisma macounii*. The species may occur near sea level, on sandy soil above the beach (with *Koenigia islandica, Cochlearia officinalis*), on peaty soil in a grassy *Eriophorum* meadow, or on damp soil on rocky slopes. Less often found on soil between rocks, adjoining rills draining permanent snow and ice fields, associated with *Anastrophyllum assimile, Marsupella revoluta, M. arctica, Scapania crassiretis, S. serrulata*.

In west Greenland (where more frequent than the few reports suggest), often over moist soil between or over rocks, associated with *Scapania praetervisa, Tritomaria scitula, T. heterophylla, T. quinquedentata, Solenostoma sphaerocarpum, Anthelia juratzkana*, less often *Lophozia excisa, L. quadriloba, Cephalozia pleniceps*, and *Odontoschisma macounii*. Occasionally, when on soil over basalt, *Peltolepis grandis* may be associated.

The associates are in most cases Ca-tolerant when not actually calciphilous. Most often, *S. subellipticum* occurs as erect, scattered, light green plants admixed with several of the foregoing taxa, so that even small pure patches are rare. Admixture, in West Greenland, with *S. sphaerocarpum* is quite frequent; with *Nardia geoscyphus*, less common. The separation of the three taxa must be carried through with some care.

Differentiation

The major differentiating characters of this species are the small size (no more than 1.5–1.8 mm wide × 5–12, rarely 18, mm long); paroecious inflorescence; high and obvious perigynium, conically narrowed and nearly included perianths; usually brownish (in Greenland plants exceptionally purplish!) rhizoids; and rather large cells (18–20 μ marginally, 20–24 × ca. 30–46 μ medially).

Of approximately the same size as *S. pumilum polaris* (= *S. schiffneri*) and with the leaf shape often somewhat similar, as well as a similar variation from a paroecious to (occasionally) autoecious inflorescence.[388] Differing, however, in the larger cell size and the presence of a short and conically narrowed perianth, situated at the summit of a high and obvious perigynium. The similarities to *S. pumilum s. lat.* are thus essentially superficial in nature.

Solenostoma subellipticum, however, is closely allied to *S. obscurum* and *S. obovatum*, as has been pointed out by Evans (1919b), from whose detailed study much of the present account of the species is drawn. All three species share a strongly developed perigynium, in most cases fully equaling or exceeding the free perianth in height, and possess a very short perianth which either is included within the bracts or projects to a negligible degree beyond their apices. In two significant respects *S. subellipticum* differs at once from *S. obscurum*: (*a*) the usually paroecious, never dioecious inflorescences; (*b*) the generally paler, often yellowish green color and the usual absence of vinaceous pigmentation of the leaf bases and rhizoids. Also, *S. subellipticum* is generally a considerably smaller species. In other respects (cell size, size of trigones, free development of stolons, leaf shape) the two species are often quite similar.

Although *S. subellipticum* agrees with *S. obovatum* in the paroecious inflorescence, it differs in the much inferior size, the smaller cells of the leaf middle, the generally colorless to faintly brownish (rarely reddish) rhizoids, and, usually, the smooth or delicately verruculose cuticle. None of these differences is totally reliable or absolute. However, *S. subellipticum* has sterile shoots with characteristically subtransversely inserted and oriented leaves, the basal portions of which are concave and suberect while the distal portions are strongly spreading or even subsquarrose. Furthermore, *S. subellipticum* normally bears 3–4 pairs of perigonal bracts (*S. obovatum*, 2–3 pairs) and may have erect or suberect perichaetial bracts (in *S. obovatum* usually somewhat spreading). These latter distinctions are not invariably present, but in *S. subellipticum* there are very often subfloral innovations (while in the allied *S. obovatum* such innovations are normally absent).

To do justice to Stephani's viewpoint, one must admit that plants which are "intermediate" between these two species are quite frequent, and that, as

---

[388] Exactly as in that species, the high arctic phenotypes may develop broad, orbicular leaves (see, i.a., Schuster, Steere, & Thomson, 1959).

Müller (1909, *in* 1905–16, p. 530) says, "the differences [between them] are very small." The plants from Quebec, referred with some hesitation to *S. subellipticum*, are a case in point. They are somewhat large for this species (robust shoots 1.5–1.8 mm wide × 8–12 mm long, isolated shoots to 18 mm; stems to 250 μ wide) and bear leaves which are narrowly ovate to elliptical, ranging from 650 μ wide × 770 μ long (along midline) up to 740 μ wide × 860 μ long. Small phases of *S. obovatum* are no larger, and on the basis of leaf size the Quebec plants might well be referred to a small form of *S. obovatum*, differing from the latter, however, in the elliptical to rather narrowly ovate form of the leaves (Fig. 295:5–7). Leaf cells vary from 18–20 up to 23 μ on the margins and sub-apically and from 20–24 × 38–46 to 23–26 × 30–35 μ medially. As in *S. obovatum*, the cuticle is striolate in the basal halves of the leaves, verruculose distally. Evans (1919a), however, has shown that this may also be the case in *S. subellipticum*. The rhizoids are pale brownish, locally slightly violet or purplish tinged, but are never deep purplish. ♀ Bracts are erect or suberect, with only the apices flaring outward. In the latter characters, as well as in leaf form, the plants clearly agree with *S. subellipticum*, to which, by and large, they appear to belong as a lax, hygromorphic extreme.

The opposite extreme of *S. subellipticum* is illustrated by Swedish plants (Schiffner no. 1372, Jämtland, Sweden, *H. W. Arnell*). The plants are a compact soil form, only 5–8 mm long and from 700–1250 μ wide, with strongly imbricate, concave-erect ♀ and ♂ bracts, whose tips only are spreading. The plants are evidently a sun form, with the distal portions of the leaves and bracts yellowish brown; rhizoids are distinctly purplish!

Greenland plants may show additional deviations: the leaves may become broadly rotundate; cells may develop sharply defined, nodulose trigones; rhizoids may become purplish. These plants represent extreme phenotypes adapted to conditions of intense illumination (Fig. 278:17–19).

In my Greenland collections two contrasted phenotypes, separable by the following key, are found.

a. Plants thick, firm, fleshy, compact; cells strongly collenchymatous, walls pigmented (brownish to purplish brown to vinaceous, often admixed on a single plant); leaves of sterile axes orbicular, often wider than long; cells mostly with 2–4(5), rarely more, oil-bodies, large. Perianth mouth with cells irregular, mostly 24–27 × 26–28 μ to 18–21 × 35–42 μ . . . . mod. *colorata-densifolia*[389]

a. Plants, especially sterile ones, relatively slender, green to weakly brownish, rhizoids usually colorless (sometimes purplish basally, but with no other vinaceous pigmentation); leaves of sterile axes spreading, elliptical-ovate, longer than wide; cells mostly with 5–6 oil-bodies each, never with 2, rarely 3–4. Perianth mouth with fingerlike cells 13–16(17) × 30–40–52 μ . . . . mod. *viridis-laxifolia*

Amakawa (1960a) describes a dioecious "*J. subelliptica* var. *nana* Amakawa." Grolle (1964j, p. 237) believes that this plant belongs, not

[389] See i.a., *RMS 45601*, Kânâk, northwest Greenland.

to *S. subellipticum,* but rather to the east Asiatic *Nardia subtilissima* Schiffn., Ann. k.k. Naturhist. Hofmus. Wien 23:137, 1909.[390]

SOLENOSTOMA (PLECTOCOLEA) FOSSOMBRONIOIDES (*Aust.*) *Schust., comb. n.*

[Figs. 296–297]

Jungermannia fossombronioides Aust., Proc. Acad. Nat. Sci. Phila. 21(1869):220, 1870.
Nardia (sect. *Chascostoma) fossombronioides* Lindb., Acta Soc. Sci. Fennica 10:529, 1875.
Plectocolea fossombronioides Mitt., Trans. Linn. Soc. Bot., Ser. 2, 3:198, 1891; Evans, Ann. Bryol.
    10(1937):42, 1938.
Nardia fossombronioides Evans, The Bryologist 22:59, figs. 1–7, 1919.
Solenostoma (Plectocolea) paroicum Schust., Amer. Midl. Nat. 49(2):403, pl. 32, figs. 8–9, 1953
    (not *Nardia paroica* Schiffn., 1910).

In prostrate patches or merely loosely gregarious, *light green to pure or slightly yellowish green, rather soft and tender, pellucid, the rhizoids and rarely postical leaf bases purplish.* Shoots (1200)1500–2000 μ wide × 25(30) mm long, to 3100 μ on upper part of ♀ shoots, prostrate with ascending apices, very sparingly branched, the branches exclusively *intercalary,* often in part or largely reduced-leaved and *flagelliform* or *stolon-like.* Stem rather soft and fleshy, 200–325 μ in diam., light green, *not* (or rarely) *pigmented ventrally,* 8–9 cells high, of delicate, soft, thin-walled cells throughout, the dorsal cortical ca. (20–25)30–35 μ wide × (60–80)85–125(150) μ long. Rhizoids long and frequent, *purplish, abundant to the shoot apex.* Leaves contiguous to loosely imbricate, obliquely inserted and oriented (but the uppermost often almost subtransverse), semiamplexicaul, their distal portions spreading but hardly squarrose, at base weakly concave, obliquely *ovate to oblong-ovate,* slightly narrowed to apex and narrowly rounded distally, to 1000 μ wide × 1100–1225 μ long up to 1100–1150 μ wide × 1125–1275 μ long (length 1.05–1.2× width), on lax sterile shoots smaller and narrower, entire-margined and unlobed, or rarely isolated leaves retuse, apex usually rounded. *Cells large and pellucid, delicate,* marginal cells ca. 22–28 μ, median 30–35 × 30–40 μ, up to 34–40 × 36–46 μ, occasionally 36–43 × 45–55 μ, *thin-walled, trigones virtually absent or very small; marginal cells not swollen; cuticle smooth;* oil-bodies large, ovoid to fusiform and ellipsoidal, appearing very finely granular but almost smooth, 2–5(6–8) per cell, to ca. 7–8 × 19–22 μ (when only 2–3 per cell) or somewhat smaller when more numerous, then varying from 7 × 8–13 to 6–7 × 15–18 μ; chloroplasts 4–5 μ.

Paroecious, almost always freely fertile; rarely, in isolated instances, with androecium not followed by a gynoecium, the plant then autoecious. ♂ Bracts in 3–5(6) pairs, contiguous to weakly imbricate, subtransversely inserted and oriented, similar to leaves in size and form or smaller, but with base strongly ventricose, only the distal portions spreading (or slightly reflexed); ♀ bracts often also with antheridia; antheridia 2–3 usually. Gynoecia with *perigynium low,* a low ring 0.15–0.35(0.55) the height of the perianth above it; bracts suberect to loosely spreading, their apices often reflexed, broadly ovate to ovate-oblong, wide-based, as wide as to wider than long, somewhat narrowed to the rounded

---

[390] *Jungermannia subtilissima* Grolle, Jour. Jap. Bot. 39:237, 1964 = *Solenostoma subtilissimum* Schust., comb. n.

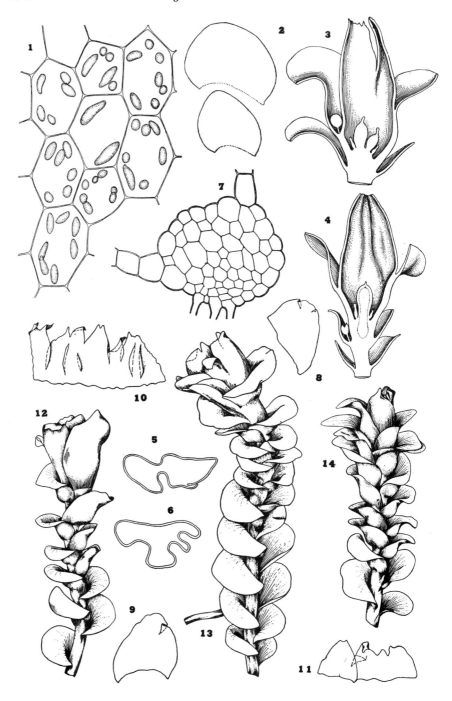

apex. Perianth pellucid, pale green, firm (2-stratose from middle down), usually *appearing abnormally developed*, often superficially *similar to that of Fossombronia, then subcampanulate*, 0.3–0.6 emergent at maturity, in the distal 0.5–0.8 deeply and irregularly plicate, often *wide and irregularly lobulate and laciniate at the mouth*, less often narrowed to the mouth and hardly laciniate, the laciniae irregular and polymorphous, variously erect, involute or revolute, triangularly lobelike to toothlike, entire or crenulate or denticulate; mouth sometimes contracted and crenulate; cells of perianth usually pellucid and strongly elongated, the marginal and subapical cells 18–22 × (50)60–75(90–100) μ, those of middle of perianth ca. 25–30(32) × 90–125 μ, averaging 3.5–5× as long as wide (only isolated cells nearly isodiametric). Calyptra pale green to vinaceous. Seta with ca. 24 rows of epidermal cells, surrounding ca. 18 rows of cells in the second tier, in turn surrounding ca. 10–12 cells in an inner group.

Type. Near Closter, N.J. (*Austin*); issued in Hep. Bor.-Amer. Exsic. no. 32 as *Jungermannia fossombronioides.*

Distribution

Broadly Appalachian in range; an endemic species, occurring through much of the eastern Deciduous Forest Region, but totally lacking in the Spruce-Fir Region. Absent or rare in most of the Coastal Plain (although found in isolated stations at its inner periphery), but common in the Piedmont and at low elevations in the mountains. Rare or absent above 2500 feet, even in the Southern Appalachians, thus not a mountain species, *s. str.*

CONNECTICUT. Washington (Evans, 1921). NEW JERSEY. Near Closter (*Austin;* type; *Rau!p.p.*, among *S. gracillimum!*). DISTRICT OF COLUMBIA. Washington (Evans, 1905). VIRGINIA. Bedford, Chesterfield, and Craig Cos. (Patterson, 1949, etc.). WEST VIRGINIA. Preston and Ritchie Cos. (Ammons, 1940).

NORTH CAROLINA. Raven Rock, near Mamers, Harnett Co. (*RMS 28689, 28670a, 36000*); "The Caves," SW. of Durham, Orange Co. (*RMS 28490, 28528, 28564, 28554, 28561, 28557, p.p.,* among *S. pumilum*); 10 mi NW. of Raleigh (*RMS 36004a, 30002, 28490*) and Hemlock Bluff near Cary, Wake Co. (*RMS 30002, p.p.,* among *S. pumilum*); on Broad R. 8 mi S. of Blue Ridge, Buncombe Co. (*Anderson 9871!*); Crow Cr., Cullasaja Gorge NW. of Highlands, Macon Co., 2500 ft (*RMS 29489*); Durham, Durham Co. (*Blomquist*); Winston-Salem; Forsyth, Burke, Franklin, and Randolph Cos. (Blomquist, 1936); Melrose Falls, 4 mi NW. of Tryon, Polk Co. (*RMS 36049, 36050*); Crabtree Cr.

FIG. 296. *Solenostoma (Plectocolea) fossombronioides.* (1) Median cells (×410). (2) ♀ (upper) and ♂ (lower) bracts (×17.2). (3–4) Longisections of gynoecia, in fig. 3 with two fertilized archegonia (×19.5). (5–6) Transverse sections of perianths, antical face below in each case (×ca. 24). (7) Stem cross section (×140). (8–9) Innermost ♀ bracts (×12). (10) Perianth apex, spread out (×12). (11) Malformed perianth apex, spread out (×12). (12–14) Paroecious shoots, with variously malformed perianths (×12). [1, Taylors Falls, Minn. (*RMS*); 2–7, *RMS 36050*, Melrose Falls, N.C.; 8–14, after Evans, 1919 (taken from plants from Closter, N.J., from Austin's Hep. Bor.-Amer. no. 32).]

State Park (*RMS*).   SOUTH CAROLINA.   Lower Falls, Whitewater R., near Jocassee, Oconee Co. (*Anderson 8491*)!.

GEORGIA.   Macon (Underwood & Cook, Hep. Amer. no. 39; sterile, doubtful); NE. corner of Rabun Co., on W. side of Chattooga R. (*RMS 39863*).   MISSISSIPPI.   Tishomingo State Park, Tishomingo Co. (*RMS*; southernmost report!).

Extending westward as follows.

OHIO.   Little Rock Ford, Hocking Co. (*Taylor, 1924*).   ILLINOIS.   Canton (Evans, 1919b); Clark Co. (Arzeni, 1947).   INDIANA.   Fern, Putnam Co. (*Underwood, 1893!*; *p.p.*, among *Scapania nemorosa, Geocalyx, Plagiochila asplenioides*).   MINNESOTA.   Curtain Falls, Taylors Falls, Chisago Co., pH 7.3 (*RMS 14605, 14608;* reported by Schuster, 1953, as *S. paroicum*).   KANSAS.   Chatauqua, Douglas, Elk, Leavenworth, and Woodson Cos. (McGregor, 1955).

The plants from Minnesota are atypical in some respects, grading into the European *S. paroicum* in that the perianth mouth in many cases is contracted.

Ecology

Most frequently over thinly soil-covered rocks along streams, but occasionally a pioneer on sedimentary or slightly metamorphosed rocks (such as sandstones or schists); absent, or at least rare, in calcareous districts. The plant almost habitually is found near stream banks, most often in a zone just below to just above the point of annual flooding. It occurs here most often with *Pellia epiphylla* (less frequently also *P. neesiana*), *Conocephalum, Solenostoma hyalinum* and/or *pumilum, S. crenuliformis, Riccardia multifida, Chiloscyphus pallescens,* less often *Anthoceros laevis, Asterella tenella, Dumortiera hirsuta, Jubula pennsylvanica,* and other mesophytic or meso-hygrophytic species, among them *Fossombronia cristula*.

Most often under constantly or intermittently shaded conditions, and then a soft, pure to slightly yellowish, pellucid green (but with the rhizoids retaining their vinaceous coloration!). Less frequently in somewhat insolated sites, and then with the leaf bases slightly tinged like the rhizoids, at least postically. The restriction to rocks on shaded banks of rivulets (rather than soil) is extremely distinctive.

Differentiation

When occurring with the "typical," teratological-appearing perianths, this species is hardly to be confused, since the perianths are without a parallel in other members of the genus *Solenostoma*. Indeed, Lindberg (1875) was so impressed with the distinctive perianth form of this species that he placed it in a discrete section, *Chascostoma* Lindb. Schiffner (1893–95, p. 79) was similarly impressed by the anomalous perianth and placed the species in subg. *Chascostoma*, retained questionably in the genus *Nardia* (*s. lat.*; then including *Solenostoma*). Evans (1919b) carefully studied this anomalous species with particular reference to the exceptionally polymorphous and often abnormal-appearing perianths. The latter are typically broadened near the wide and often distinctly flaring, ir-regularly truncate mouth, and are thus somewhat campanulate, much as in *Fossombronia* (as is implied by the appropriate species name). As in this genus, perianths may be incised along one side for a variable distance and then appear to be incompletely formed. In almost all cases the mouth is extremely irregularly lobed, lobes and teeth being often variously inflexed or reflexed, the mouth then appearing crispate and undulate.

In spite of the extraordinary form of such perianths, Evans (1919b, 1938c) was unwilling to grant generic or subgeneric status to this species, disagreeing with Schiffner and Lindberg. Study of a series of specimens (Raven Rock, N.C., *RMS 36000*) confirms the widsom of such a conservative position. In-deed, the material shows full gradation from the "typical" *S. fossombronioides* perianth, with the truncate to flaring, irregularly incised mouth, to an ovoid to ovoid-oblong perianth, strongly narrowed distally, more regularly and deeply plicate, and with a relatively narrow and merely crenulate mouth (Fig. 297:1), thus approaching the condition in *S. hyalinum*. Such plants occur admixed with others having irregularly developed perianths. (Figure 297:3 is taken from the same patch of plants as Fig. 297:1.) Plants with normal, *Solenostoma*-type perianths may have them deeply and strongly plicate, almost exactly as in *S. hyalinum* and the allied *S. paroicum*, and conically narrowed to the apex (Fig. 297:1). In such cases they may be incised on one side (Fig. 297:2; the underlying free edge indicated by a dotted line) or may be merely crenulate and shallowly lobulate at the mouth (Fig. 297:7); in cross section through the distal third of the perianth, these respective conditions become quite clear. The form of the perianth as a specific character is thus very much diminished. On the basis of the large leaf cells, the shallow perigynium and plicate, ovoid, distally contracted perianths, and the paroecious inflorescence, such individual plants closely approach the rare European *Solenostoma paroicum* (Schiffn.) Schust.[391]

Almost equally illuminating is the study of plants from the northern edge of the range of *S. fossombronioides* (Indiana to Minnesota). For instance, in sparse material from Fern, Putnam Co., Ind. (*Underwood, Sept. 1893;* traces mixed with *Scapania nemorosa*, etc.) the following deviant features are evident: (*a*) perianth rather sharply 4-plicate, as is characteristic of other members of subg. *Plectocolea* and of the related *Solenostoma*; Austin had described the perianths as "multi-plicate;" (*b*) apex of perianth quite distinctly narrowed, 0.25–0.5 the width of the perianth at the middle, slightly lobed, the apical cells all elongate, protruding and free at their ends (but without contraction into a distinct beak);

[391] When I erroneously reported this species from Minnesota (Schuster, 1953), my determination was based on the paroecious inflorescences, the large cell size, and the plicate, distally conically narrowed perianths of the Minnesota material. With study of the critical series of specimens from Raven Rock, N.C. (*RMS 36000, Oct. 1955*), a distinction between the Minnesota and North Carolina plants becomes impossible to maintain. In the Minnesota plants no campanulate, laciniate-tipped perianths were found; in the North Carolina plants variation occurred from the type uniformly developed in the Minnesota material (e.g., a *S. hyalinum-paroicum* type of perianth) to the "normal" *S. fossombronioides* perianth.

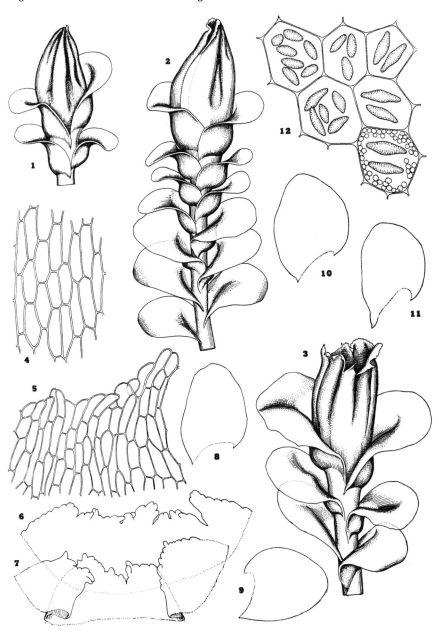

FIG. 297. *Solenostoma (Plectocolea) fossombronioides*. (1–3) Apices of paroecious shoots, with variation in perianth form from the normal (fig. 1) to the moderately modified, antically incised (fig. 2), to the wholly aberrant (fig. 3) (×18). (4) Cells of perianth middle (×120). (5) Cells of perianth apex, from sector of that in fig. 6 (×120). (6) Apex from perianth in fig. 2, the perianth once-incised (×60). (7) Apex of perianth

(c) ♀ bracts sometimes each with a single antheridium at base. The Minnesota plants [reported (Schuster, 1953) as *S. paroicum*] are even more extreme, with distally contracted, almost beaked perianths. Thus ± normal, *Solenostoma*-type perianths can clearly be demonstrated in *S. fossombronioides*, bringing this species into immediate contact with the *S. hyalinum-paroicum* complex. As in *S. hyalinum* and *S. paroicum*, a perigynium is present. However, *S. fossombronioides* sometimes has a much less distinct perigynium than is commonly found in the other two species, the perigynium being largely confined to one side of the axis in some cases. "Even when fertilization has taken place [the perigynium] is exceedingly shallow" (Evans, 1919b). However, in cases where the perianth is similar to that of such typical species of *Solenostoma* as *S. hyalinum*, the perigynium is commonly higher and may be from 0.35 to 0.7 the height of the free perianth above it (Fig. 296:4).

Thus, with examination of a series of specimens of *S. fossombronioides* distinctions between it and the *S. hyalinum-paroicum* complex become much less marked. The larger leaf cells, more pellucid appearance, and paroecious inflorescence adequately separate the species from *S. hyalinum*. In exactly these features, however, *S. fossombronioides* agrees with the rare *S. paroicum*. The latter, known only from a few European localities, is a more strongly pigmented plant, with more rotundate leaves, and usually more collenchymatous cells (the trigones commonly being slightly nodulose). In spite of these differences, an extremely close affinity between *S. fossombronioides* and *S. paroicum* appears to exist. *If*, as I suspect is the case, the *Fossombronia*-like perianths of *S. fossombronioides* are partly a consequence of the habitat (rocks subject to inundation), these distinctions may fail to be adequate to separate the two taxa.

Evans (1919b) also called attention to similarities to the northern *S. obovatum* in regard to vegetative organs. Both taxa are similar in size, bear purplish rhizoids and long stolons, appear to possess exclusively intercalary branching, and appear to lack subfloral innovations. Furthermore, both species are paroecious. However, *S. fossombronioides* has, on an average, larger cells with much less discrete trigones; a smooth cuticle; denser rhizoids, forming a ventral mat almost to the shoot apex; and usually a larger number of ♂ bracts (ordinarily 3–6 pairs). When fertile, the peculiar, strongly emergent perianths and the low perigynium separate *S. fossombronioides* from all phases of the variable *S. obovatum*.

Subfamily *MYLIOIDEAE* Grolle

Jungermanniaceae subf. Mylioideae Grolle, Nova Acta Leopoldina, N. F. 25:15, 1963.

Prostrate or (with crowding) erect, vigorous, with sparse, ± furcate, terminal branching; bearing long, exserted *perianths that are sharply*

---

in fig. 1, the perianth nonincised (×60). (8–11) Leaves (×18). (12) Median cells with oil-bodies and, lower right, chloroplasts (×382). [1–11, *RMS 36050*, Melrose Falls near Tryon, N.C.; 12, New Hope Creek, N.C., *RMS*.]

laterally compressed and wide at the vertical, truncate, bilabiate mouth; no subtending perigynium. Small, subulate underleaves distinct. *Asexual reproduction by 2-celled gemmae* from apices of leaves that are not prominently reduced in size, and that bear enlarged cells. Antheridia long-stalked, the *stalk 1-seriate,* the spherical body with jacket cells *few and very large.* Capsule wall 3–5-stratose, the epidermal cell undergoing a ± distinct "two-phase" development: primary longitudinal and normally almost all transverse walls lacking secondary thickenings; secondary (usually longitudinal) walls that typically bisect the primary cells, with few, coarse "nodular" (columnar) thickenings that are dilated on the free tangential walls and are ± coalescent there with juxtaposed thickenings.

The Mylioideae, with only the genus *Mylia,* are a remote element in the Jungermanniaceae. Previously they have usually been placed in the Plagiochilaceae (Müller, 1951–58; following him, Inoue, 1958; Schuster, 1958c, 1959–60; Amakawa, 1960a, and others) but are a foreign element there. Antheridial structure and the epidermal cells of the capsule serve to isolate the group from both Plagiochilaceae and Jungermanniaceae *s. str.* Perhaps the best solution is to place them, as did Grolle (1963), in a separate subfamily of the Jungermanniaceae.

Although the laterally compressed, bilabiate perianths suggest some affinity to *Leptoscyphus* (Geocalycaceae) and *Pedinophyllum* (Plagiochilaceae), similarities to these genera are superficial only; the nature of the antheridia, of the epidermal cells of the capsule, and of the asexual reproduction clearly isolates *Mylia* as a genus (see, e.g., Schuster, 1959–60, and Grolle, 1963). The presence of gemmae is shared with the otherwise very different *Gottschelia* Grolle (1968), which has dorsally interlocking merophytes.

MYLIA S. F. Gray emend. Carr.

Mylia S. F. Gray, Nat. Arr. Brit. Pl. 1:693, 1821 (as *Mylius*); Carrington, Trans. Proc. Bot. Soc. Edinb. 10:305, 1870.

Subsimple to sparingly branched, with lateral, terminal and axillary branches, *robust,* 2–10 cm long × *1.5–3 mm wide,* decumbent to moderately ascending in growth (when crowded). Stems uniform; *without differentiation into leafless stoloniferous stems and leafy stems,* bearing clearly alternate leaves. Stems rather fleshy, almost *uniform in structure,* the cortical cells hardly differentiated from the medullary, collenchymatous like the medullary cells and not tangentially flattened, not equally thick-walled, *strongly elongated* (ventral cortical cells ca. 23–27 μ wide × 140–260 μ long, leptodermous and convex; dorsal cortical cells similar, 23–35(40) μ

wide × (100)150–400 μ long). *Rhizoids in dense masses*, occurring largely in a fascicle at bases of leaves and of underleaves, *a few indiscriminately over the postical stem surface*. Leaves alternate, ± *orbicular*, entire (with development of gemmae becoming somewhat to strongly elongate), with *antical margin not reflexed*, a cnemis thus absent, inserted by an oblique and scarcely arched line, the dorsal and postical bases both essentially nondecurrent. *Cells very large*, usually 45–55 μ in the leaf middle, becoming larger in gemmiparous regions, *strongly collenchymatous, with bulging trigones;* oil-bodies several (5–12) per cell, each formed of many small or few large, protruding globules, papillose or segmented in appearance. *Underleaves distinct, rather large, entire, lanceolate, quite free on both sides from the lateral leaves*, often ± hidden in the feltlike mat of rhizoids. *Asexual reproduction by 2-celled gemmae* produced in fascicles from the upper leaves.

Dioecious. ♀ Plants with terminal inflorescences; with suborbicular bracts; perianth terminal, entire to ciliate at mouth, strongly laterally compressed but the keels not winged, the wide mouth essentially truncate; usually with subfloral innovations. ♂ Plants with androecia becoming intercalary, on main shoots; antheridia 1–2 per bract, stalk 1-seriate; bracts somewhat similar to leaves, little smaller in size, but saccate at base. Capsule short-ovoid, the wall 3–5-stratose; epidermal layer usually with nodular thickenings (e.g., radial bands), which may extend onto the tangential walls (then with incomplete semiannular thickenings); inner layer with semiannular thickenings. Spores rather large (15–20 μ), their diam. ca. 1.5 × that of the bispiral elaters.

Type. *Mylia taylori* (Hook.) Gray (= *Jungermannia taylori* Hook.).

Schuster (1959–60) has shown that *Mylia s. lat.* (including *Anomylia* and the extraterritorial *Leptoscyphus*) is a heterogeneous assemblage, differing widely in some critical features. *Mylia s. str.*, e.g., produces true gemmae (*Anomylia* bears caducous leaves; *Leptoscyphus* usually lacks asexual reproductive modes but may possess fragmenting leaves); *Mylia* has clearly alternate leaves never united at the base with the underleaves, and unlobed underleaves (in *Leptoscyphus* the typically quadrilobed underleaves are decurrent laterally and usually united, at least narrowly, with the often subopposite lateral leaves). *Mylia* also has scattered rhizoids, whereas *Anomylia* has them restricted to the underleaf bases and *Leptoscyphus* usually has them ± restricted to the underleaf bases and peripheral portions of the ventral merophytes. Finally, the exceedingly large, strongly collenchymatous cells of *Mylia* are very distinctive, *vis-à-vis* both *Anomylia* and *Leptoscyphus*. Other differential features are mentioned under the discussion of *Leptoscyphus* (see Vol. III). The basic distribution

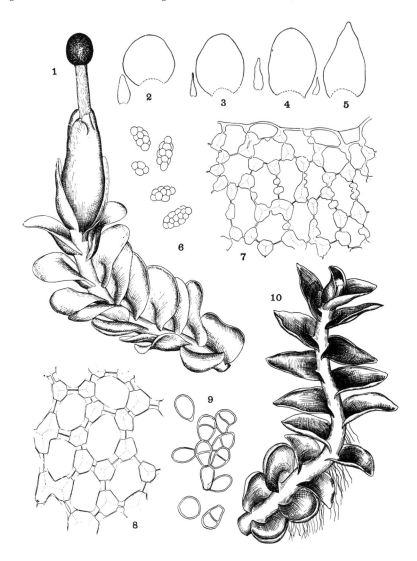

Fig. 298. *Mylia anomala*. (1) Shoot with sporophyte (×8.7). (2–5) Leaves, with adjacent underleaves, in a series from shoot base to gemmiparous shoot apex (×10). (6) Oil-bodies (×580). (7) Cells of perianth mouth (×175). (8) Median leaf cells of mod. *pachyderma* (×ca. 325). (9) Gemmae, partly *in situ* on the fascicle (×155). (10) Gemmiparous shoot (×8.7). [1, 7–8, from near Orr, Minn., *RMS;* 2–5, 9–10, Plymouth Bog, Chenango Co., N.Y., *RMS;* 6, Great Palisade, Lake Co., Minn., *RMS;* all from Schuster, 1959–60.]

patterns of the three groups also are indicative of a long history of separate evolution.

The genus *Mylia*, as thus restricted, is essentially subarctic and low arctic in distribution, with two species (*M. anomala*, *M. taylori*) in North America and Europe, the latter extending eastward to Japan; a third species, *M. verrucosa*, occurs in Siberia to Japan. The few species appear to fall into two sharply discrete sections (Schuster, 1959–60) as follows:

SECTIO 1. ANOMALAE: Oil-bodies coarsely segmented, relatively small; cuticle smooth; perianth mouth subentire; gemmiparous leaves elongate, lanceolate, pointed (*M. anomala*).

SECTIO 2. VERRUCOSAE: Oil-bodies ellipsoidal, formed of numerous minute, not protuberant globules, opaque in appearance; cuticle coarsely, obviously sculptured; perianth mouth shortly to longly ciliate; gemmiparous leaves short-ellipsoidal, never pointed (*M. taylori*, *M. verrucosa*; also *M. nuda*, an endemic of Taiwan [see Inoue & Yang, 1966]).

Key to Nearctic Species of Mylia

1. Cuticle smooth; oil-bodies coarsely segmented, irregular in shape, the individual segments protuberant; gemmiparous leaves becoming lanceolate-pointed; pigmentation fulvous to brownish, with little or no red; perianth mouth subentire; over *Sphagnum* . . . . *M. anomala* (p. 1037)
1. Cuticle sculptured into polygonal, coarse plates, appearing fissured; oil-bodies nearly smooth, opaque, nearly filling lumen, formed of numerous small, nonprotuberant globules; gemmiparous leaves becoming slightly ellipsoidal to rounded-rectangulate, never pointed; perianth mouth more or less ciliate; over shaded rocks and decaying logs . . *M. taylori* (p. 1042)

MYLIA ANOMALA (*Hook.*) S. F. Gray
[Figs. 298-299; 300:8–10]

Jungermannia anomala Hook., Brit. Jungerm., pl. 34, 1816.
Mylia anomala (Hook.) S. F. Gray, Nat. Arr. Brit. Pl. 1:693, 1821; Schuster, Amer. Midl. Nat. 49(2):410, fig. 12:1, 13:8, pl. 35:4–8, 1953; Schuster, *ibid.* 62(1):36, figs. 2–3, 1959.
Jungermannia taylori var. *anomala* Nees, Naturg. Eur. Leberm. 2:455, 1836.
Leptoscyphus anomalus Mitt., Lond. Jour. Bot. 3:358, 1851; Lindberg, Acta Soc. Sci. Fennica 10:40, 1875; K. Müller, Rabenh. Krypt.-Fl. 6(1):788, 1911.
Mylia taylori var. *anomala* Carr., Brit. Hep., p. 68, 1875.
Coleochila anomala Dumort., Hep. Eur., p. 106, 1874.
Aplozia anomala Warnst., Krypt.-Fl. Mark Brandenburg 1:144, 1902.
Leioscyphus anomalus Steph., Spec. Hep. 3:16, 1905; Bull. Herb. Boissier, Ser. 2, 5:1144, 1905.

Plants ± prostrate, creeping over *Sphagnum* or forming turflike mats over peat, often in dense, luxurious patches, pure green to (in sun) *yellowish brown to fulvous* to somewhat reddish brown. Shoots usually simple, usually innovating below perianths, 2.4–3 mm wide × 2–3 cm long. Stem stout, 440–600 μ wide, rather fleshy, with poorly developed cortex, the cortical cells only little smaller than the medullary in cross section (28–36 μ in diam. × 140–350(400) μ long),

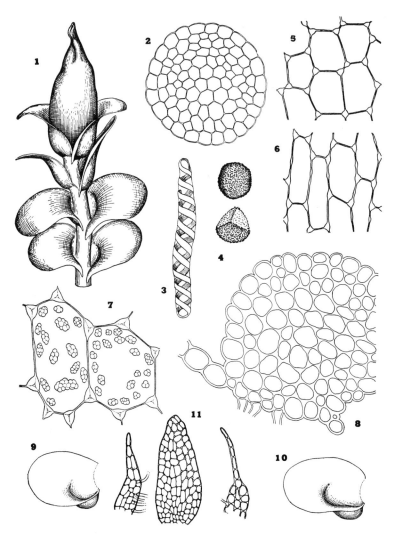

Fig. 299. *Mylia anomala*. (1) Perianth-bearing shoot apex, dorsal aspect (×ca. 9). (2) Seta cross section (×53). (3–4) Elater and spores (×540). (5) Median leaf cells, mod. *mesoderma* (×ca. 135). (6) Cells near apex of gemmiparous leaf (×ca. 135). (7) Median cells with oil-bodies (×345). (8) Cross section of medium-sized stem (×112). (9–10) ♂ Bracts (×ca. 11). (11) Underleaves, showing the origin, in part, of rhizoids from near underleaf bases (×40). [1–4, 9–10, bog near Orr, Minn., *RMS*; 5–6, 11, Plymouth Bog, Chenango Co., N.Y., *RMS*; 7, Great Palisade, Lake Co., Minn., *RMS*; 8, Malloryville Bog, Tompkins Co., N.Y., *RMS*; all from Schuster, 1959–60.]

relatively thin-walled; medullary cells strongly elongate, thin-walled but with distinct trigones, 32–48 μ in diam. Rhizoids abundant, *in dense felty masses*, long, *often tufted*, arising largely near postical leaf bases and from underleaf bases, some indiscriminately from postical stem surface, occasional ones from surface or margins of the basal $\frac{1}{5}$–$\frac{1}{2}$ of underleaves. Leaves approximate to moderately imbricate, somewhat to strongly narrowed basally, not decurrent postically, short-decurrent antically, the lower leaves usually spreading or erect-spreading, the upper spreading to nearly erect. Lower, nongemmiparous leaves almost perfectly orbicular, ca. 1450–1500 μ wide × long; *upper leaves (of gemmiparous shoots) progressively somewhat more elongate*, at first ovate and to ca. 1300 μ wide × 1440 μ long, *becoming distinctly ovate-lanceolate to lanceolate distally* (and ca. 1100–1125 μ wide × 1500–1750 μ long); nongemmiparous leaves varying from nearly flat to somewhat adaxially concave and saucerlike but gemmiparous leaves often canaliculate, with both postical and antical margins adaxially curved. Cells strongly collenchymatous, the trigones often bulging, sometimes so strongly that the cell lumens are connected merely by pits; cells of leaf apex and leaf middle from 45–50 × 50–60 μ, polygonal, of the base 50–55 × 60–70 μ or even larger, of the margins mostly 44–50 μ; *cuticle smooth;* oil-bodies 5–8, occasionally in the larger cells 12–20, per cell, irregularly spherical to ovate, *hyaline, distinctly formed of relatively few, coarse, protuberant globules or segments* (each ca. 1.5–2 μ), varying from 5–7 × 7–8 μ to a maximum of 7–8 × 10–12, rarely 14–18, μ. *Underleaves relatively large*, from 0.25 to 0.5 the leaf length, the larger 500–680 μ long × 250–300 μ wide, mostly *lanceolate to narrowly lanceolate*, some tipped by a cilium 2–5 cells long, undivided, usually nearly obscured by the rhizoids. Asexual reproduction usually abundantly present by means of *fasciculate gemmae, mostly 2-celled, from apices of upper, lanceolate leaves* (the cells of these apices large and elongate), to 32–40 × 50 μ, greenish, spherical to broad-ellipsoidal.

Dioecious. ♂ Inflorescences becoming intercalary, on main stems, of 4–7 pairs of bracts; bracts similar to leaves but somewhat ovate, the basal half strongly concave and suberect, the distal portion spreading; 2 antheridia, subglobular to short-ovoid, the stalk uniseriate. ♀ Inflorescences terminal on main shoots; bracts like leaves, ovate to roundish-ovate, erect and somewhat sheathing perianth at base, erect-spreading distally, with the tips often somewhat re-curved; bracteole distinct, large, lanceolate. Perianth at least $\frac{1}{2}$ emergent at maturity, weakly compressed below but strongly laterally compressed and flattened above, somewhat (but little) narrowed to the bilabiate, relatively wide mouth; *mouth subentire to crenulate.* Capsule short-ovoid; valves 3–4-stratose, outer layer with radial (nodular) thickenings, inner layer with incomplete to complete semiannular bands. Elaters 12–14 μ thick; *spores granulose*, 15–20 μ.

Type. Holt, England (presumably in the Hooker Herbarium).

Distribution

Almost ubiquitous in bogs in the Coniferous Zone of northern North America and Europe, eastward into Siberia, becoming rarer in

"Hochmoore" and on the peaty ridges of rocky, mountain Tundra.[392] Relatively rare in the isolated bogs of the Transition Zone and in the Arctic Tundra. According to Buch & Tuomikoski (1955), circumboreal, although in east Asia known only from Kamchatka (Arnell, 1927).

Westward ranging from Alaska to the Yukon, southward to British Columbia, Alberta, and Washington. Eastward as follows:[393]

W. GREENLAND. Ikertoq Fjord, 66°45′ N. (*J. Vahl 192, p.p.*, with *Calypogeia sphagnicola, Lophozia binsteadii*); Tupilak I., Egedesminde, 68°42′ N., 52°55′ W. (*Holmen, June 29, 1956!, p.m.p.*, among *Lophozia binsteadii*);[394] Lyngmarksbugt, Godhavn, Disko I. (*RMS & KD 66-561*); tarn, summit of narrow neck at E. end of Agpat I., 70°53′ N., 51°40′ W. (*RMS & KD 66-1422*); boggy tarn edges, Egedesminde (*RMS & KD 66-172, 66-170*). SW. GREENLAND. Kangerdluarssuakasik, 61°55′ N., 49°18′ W. (*Damsholt, Aug. 13, 1965; with Lophozia ventricosa*). S. GREENLAND. Frederickshaab, 62°N., 49°40′ W. (*Damsholt, July 25, 1963, with Cephalozia leucantha*); Narssalik, 61°40′ N., 49°19′ W. (*Damsholt, July 26, 1963;* with *Cephalozia leucantha, Riccardia sinuata, Calypogeia sphagnicola s. lat., Nardia scalaris, Odontoschisma elongatum, Diplophyllum albicans*, etc.); Agdluitsoq Fjord, Sletten, 60°33′ N., 45°30′ W. (*Joergensen, Aug. 9, 1958!*; trace on *Sphagnum*, among *Cephalozia leucantha, Lophozia cf. opacifolia*).

QUEBEC. Cairn I.; Richmond Gulf; Great Whale R. on E. coast of Hudson Bay; R. à Martre, above Lac aux Sables; La Tuque at Lac Bourgeois; Montmorency R., St.-Arsène, near R.-du-Loup; Mt. Albert; Lac Salé, Anticosti I.; Rupert R. near Sand L., 76°20′ W.; L. Hubbard; Shigawake, Gaspé Co. (*Kucyniak & Ray 56-060*); Granby (*Fabius 3662*). NEWFOUNDLAND. Hoggan's Pond, Avalon Distr.; Push-through and Port aux Basques, S. coast; Port au Choix, Cook Harbor, and St. Anthony, N. Pen.; Kitty's Brook, C. and NE. Nfld. NOVA SCOTIA. Between Campbell and Craigmore, Inverness Co. (*RMS 43073, 43080, 43075a*); 2 mi S. of Lower Argyle, Yarmouth Co. (*RMS 43061c*); Barrasois, Barrasois Barrens, and Ingonish, Cape Breton I.; Halifax Co. ONTARIO. L. of Two Rivers, Algonquin Park (*Cain 4319!*); Chaffey Lock, Leeds Co. (*Cain 4736!*); mouth of Montreal R., Algoma Distr. (*Cain & Williams 5243!*); Cochrane; Sudbury; L. Timagami; Thunder Bay.

MAINE. Matinicus I.; Round Mountain L., Franklin Co.; Mt. Desert I.; Mt. Katahdin (29 collections). NEW HAMPSHIRE. Mt. Chocorua; Franconia Mts.; Lonesome L., Mt. Lafayette; edge of "Hochmoor," at ca. 3200 ft, Mt. Monadnock. VERMONT. Near summit, Mt. Mansfield (*RMS 43820*); Jericho; Willoughby; Brandon. MASSACHUSETTS. Hawley Bog, Franklin Co. (*RMS*); Holden; Swamp S. of Sesachacha Pond, Nantucket I. CONNECTICUT. Woodbury, Litchfield Co.; Bethany and New Haven, New Haven Co.

NEW YORK. McLean and Malloryville bogs, both in Tompkins Co.; swamp near Crystal L., Cattaraugus Co.; Bergen Swamp, Genesee Co.; Junius Peat Bogs, Seneca Co.; Jamesville Rd., Onondaga Co.; Plymouth Bog and Reservoir, near S. Plymouth, Chenango Co.; L. George; southern West Ft. Ann; sphagnum marsh N. of Glen L., Warren Co. PENNSYLVANIA. Near Sulphur Spring, Warren Co. NEW JERSEY. near Closter, Bergen Co., in peat bog. WEST VIRGINIA. Pocahontas Co.

[392] As on Garfield Ridge, Mt. Katahdin, Me. (there with *Calypogeia sphagnicola* and at Caribou Springs, Mt. Katahadin (there with *Lopholia wenzelii, L. kunzeana, Cephalozia*).

[393] Distribution data, recently accumulated citations excepted, for *M. anomala* and *M. taylori* condensed from Schuster (1959–60).

[394] The Holmen specimen from Tupilak I. is very sparing but of utmost interest in that, unlike other phases of this species seen, yellow-brown pigments normally present in *M. anomala* are supplemented or replaced on the underleaves and postical leaf bases by a vinaceous-red pigment.

Westward occurring as follows. MICHIGAN. Burt L., Cheboygan Co.; Eagle Harbor, Isle Royale, Keweenaw Co.; Scotts, Kalamazoo Co.; Luce and Chippewa Cos.; Pictured Rocks, Alger Co. WISCONSIN. Black R., Superior, Stone's Bridge on Brule R., in Douglas Co.; E.-facing edge of Sand I., Apostle Isls., Bayfield Co.; Superior. MINNESOTA. Pigeon R., L. Superior; Beaver Dam on Hungry Jack Trail; Sailboat I., Lucille I., Belle Rose I., Long I., Big Susie I., Porcupine I., all in Susie Isls., Grand Portage marl bog, Big Bay at Hoveland, all in Cook Co.; bog near Jaynes; 5 mi W. of Togo, both in Itasca Co.; bog 4 mi SE. of Ericsburg; bog at Black Bay, near Island View, Koochiching Co.; bog at Kerrick, Pine Co.; 8 mi SW. of Gheen; bog 2 mi N. of Orr; bog 1–2 mi E. of Celina, all in Pine Co.

Ecology

The ecology of this species is treated in detail by Schuster (1953, 1959–60). It occurs under analogous conditions in two very different habitats. (1) In peat bogs, almost invariably over *Sphagnum*, which it often kills by its luxuriant growth, then forming thick, peaty, pure mats; here it is a member of the *Mylia-Cladopodiella* Associule. Under wet conditions, *Cladopodiella fluitans*, *Cephaloziella elachista*, and various *Cephalozia* species (*compacta*, *connivens*, *pleniceps*, etc.) are consociated, occasionally with *Scapania irrigua* and *paludicola*. In more xeric sites, such as the upper slopes of large *Sphagnum* hummocks, in sun, *Mylia* grows with *Microlepidozia setacea*, *Calypogeia sphagnicola*, *Cephalozia loitlesbergeri*, occasionally *Cephaloziella subdentata*. (2) Over *Sphagnum* and peat on moist ledges, especially in very damp environments (as on the north- or northwest-facing sides of cold lakes). Here it is a member of the *Mylia-Odontoschisma* facies of the *Mylia-Cladopodiella* Associule. Consociated are *Odontoschisma denudatum*, *Cephaloziella subdentata*, *Lophozia incisa* and *ventricosa silvicola* (occasionally the pigmented, *porphyroleuca*-simulating forms), occasionally *L. wenzelii* (often the "*confertifolia*" type), and rarely *L. kunzeana; Cephalozia loitlesbergeri* and occasionally *C. leucantha* or *media* are also consociated.

At the northern edge of its range, at the juncture with the Tundra in northern Quebec, it is found over peat on margins of pools, in raised sand beaches, with *Cephaloziella elachista*, *Cephalozia leucantha*, *Lophozia wenzelii*, *L. grandiretis*, *L. kunzeana*, *Cephalozia loitlesbergeri*, *C. pleniceps*, *Scapania irrigua*, and *Geocalyx graveolans*, and "above a temporary snowbank," among *Sphagnum*, *Lophozia atlantica*, and *Calypogeia* (Schuster, 1951).

In south and west Greenland, at the northern periphery of its range, the species occurs among *Sphagnum* at the edges of tarns, usually with *Calypogeia sphagnicola*, *Lophozia binsteadii*, *Cephalozia pleniceps*, *Riccardia sinuata*, less often *Cephalozia leucantha*, *Nardia scalaris*, *Odontoschisma elongatum*.

Differentiation

A strongly isolated species that need not be confused with any other. At one time it was considered merely a variety of *M. taylori* but is not closely related to that species, from which it differs abundantly in ecology, as well as in the smooth leaves, the development of acute apices to the leaves with gemma formation, the brownish (never reddish purple) pigmentation, the segmented and hyaline oil-bodies, and the subentire perianth mouth.

Variation

Although widespread and locally abundant, the ecological tolerances are sufficiently narrow so that the species does not exhibit a disturbing amplitude of variation. Shade forms are greenish and somewhat lax-leaved, with relatively long internodes. Sun forms develop shorter internodes, denser leaves, and a brownish pigmentation; furthermore, they become very strongly pachydermous. Müller (*loc. cit.*, p. 788) states that, as a rule, the trigones are smaller in *M. anomala* than in *M. taylori*. However, I find in the form of sunny bogs (mod. *colorata-densifolia-pachyderma*) the trigones may be extremely large, with the lumen consequently reduced to pits along the intervening walls. Schiffner has grouped the various "forms" of *M. anomala* into two series (gemma-producing forms that are habitually sterile, and round-leaved forms, devoid of gemmae, reproducing sexually), which are subdivided into forms on the basis of adaptation to varying combinations of moisture and light. There is no basis for the primary division; I have often found perianth and capsule-bearing specimens growing intermingled with gemmiparous material (see Schuster, 1959–60). Second, the various "forms," as Müller (*loc. cit.*, p. 791) indicates, are modifications resulting from differing environmental conditions. The series of forms and varieties proposed by Schiffner, therefore, loses taxonomic significance.

MYLIA TAYLORI (*Hook.*) *S. F. Gray*
[Figs. 300:1–7; 301]

Jungermannia taylori Hook., Brit. Jungerm., pl. 57, 1816.
Mylia taylori (Hook.) S. F. Gray, Nat. Arr. Brit. Pl. 1:693, 1821 (as *Mylius*).
Mylia taylori Lindb., Hepaticae *in* Hibern. Lect., p. 525, 1874.
Jungermannia sect. *Aplozia taylori* Dumort., Rec. d'Obs., p. 16, 1835.
Leptoscyphus taylori Mitt., London Journ. Bot. 3:358, 1851.
Leioscyphus taylori Mitt., *in* Hooker, Fl. Antarct. 2(2):134, 1855.
Coleochila taylori Dumort., Hep. Eur., p. 106, 1874.
Jungermannia reticulato-papillata Steph., Mem. Soc. Nat. Sci. Cherbourg 29:215, 1892.

Robust, usually in thick (3–10 cm) tufts or sods, almost invariably *carmine-red to purplish brown pigmented* (mod. *colorata*) at least on distal parts of upper leaves, even in very diffuse light. Plants recumbent to strongly ascending (when crowded), up to 3–4(4–5) mm wide × 3–6(8–12) cm long, the ♂ plants slightly smaller (2.7–3.5 mm wide); stems simple or sparingly branched, flexuous, rather stout (480–550 μ in diam.); branches in part terminal; cortex of rather pellucid, ± thin-walled, rectangulate cells, (28)30–40 μ wide × (80)100–180(200) μ long, whose external walls are convex, very thin; internal walls and those of adjacent medullary cells slightly thicker; medulla ca. 10 cells high, cells subequal in diam. to the cortical (32–40, a few to 48 μ); no mycorrhizae. *Rhizoids forming a dense, felted mat*, usually brownish, *long*, scattered, densest at postical bases of leaves and below underleaves, often several inserted on under-leaves. Leaves at maturity nearly horizontally spreading, *the upper characteristic-ally erect-appressed*, subimbricate to moderately imbricate, inserted by a moder-ately oblique line, very short-decurrent antically, from broadly ovate (ca. 1500–1600 μ wide × 1600–1800 μ long) to orbicular (to 2400 × 2400 μ), slightly concave, at least at median base, to somewhat convex above the base (because of the somewhat reflexed antical margin and *often deflexed apices*). Cells very large, 34–38 to 40–45 μ on margins, 40–45 × 45–60(75) μ in leaf middle, somewhat larger usually near base, strongly collenchymatous with bulging trigones, the *trigones usually nodulose*, sharply defined, with intervening walls very thin; *cuticle fissured, divided into irregular, coarsely polyhedral plates;* oil-bodies opaque, grayish or *brownish gray*, often obscuring most of lumen, 7–12 per cell, ellipsoidal or, less often, subspherical, *appearing papillose* because of the numerous minute (less than 1.3–1.5 μ) included globules that slightly protrude externally, very large, 9–10 to 7–10 × 18–20 μ. *Underleaves slenderly lanceolate to subulate*, ca. 425–500 μ long × 90–100 μ wide, usually nearly hidden in the rhizoid mat. Asexual reproduction frequent but not copious, by means of fasciculate gemmae arising from the margins of more or less ovate (*never lanceolate) leaves with rounded apices;* gemmae 1–2-celled, usually thin-walled, oblong to oval, 30 to 25–30 × 40–45 μ, green to somewhat reddish brown (in mod. *colorata*).

♂ Plants with androecia becoming intercalary, of 4–8 pairs of imbricate bracts, similar to those of *M. anomala;* saccate base of each bract with 2 (more rarely 1) antheridia, the distal half of bracts strongly spreading. ♀ Plants with perianth terminal; often with 1 or 2 subfloral innovations; bracts somewhat larger than leaves, broadly ovate to subrotundate, ca. 2250 μ wide × 2150 μ long or larger, usually sheathing perianth at base and concave and erect, spreading distally and sometimes reflexed apically; bracteole large, lanceolate. Perianth as in *M. anomala*, but the *mouth ciliate, with cilia (2–3)4–6 cells long;* cells of cilia nearly or quite smooth, the lower 20–25 × 22–35 μ, the terminal cells commonly somewhat clavate and larger, ca. 20–28 × 40–45(48) μ; perianth surface and cilia with cuticle like that of leaves. Capsule as in *M. anomala*, purplish brown, the *wall 3–5-stratose*, epidermal cells with sometimes partial tangential extensions of the radial (nodular) bands; inner cell layer with narrow, complete, *often 2–3-furcate tangential bands.* Elaters 10–12(14) μ in diam., the two purplish spirals tightly wound; spores (16–17)18–20(21) μ, *finely aerolate on outer faces.* Seta of numerous cell rows, the epidermal in 20–25 rows, often slightly larger than interior cell rows.

Type. Toulagee Mt., Wicklow Co., Ireland (*Dr. Thomas Taylor*), presumably in the Hooker Herbarium.

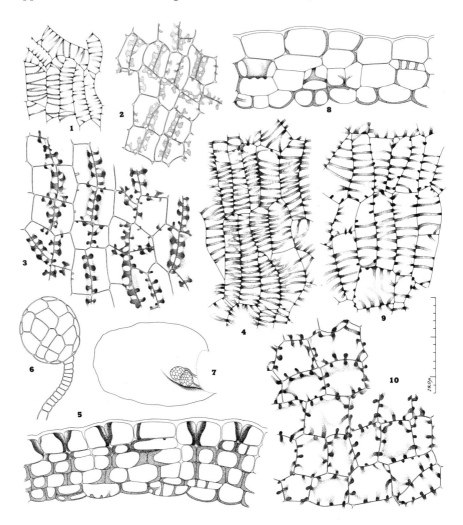

FIG. 300. *Mylia:* capsule walls and antheridia. (1–7) *M. taylori;* (8–10) *M. anomala.* (1) Inner cells, capsule wall (×125). (2) Same, epidermal cells (×125). (3) Epidermal cells, capsule wall (×215). (4) Inner cells, capsule wall (×215). (5) Cross section, capsule wall (×300). (6) Antheridium (×92). (7) ♂ Bract (×20). (8) Capsule wall cross section (×245). (9) Inner cells, capsule wall (×215). (10) Epidermal cells, capsule wall (×215). [1–2, 6–7, after Grolle, 1963; 3–5, *RMS 17591*, Wittenberg Mt., N.Y.; 8–10, *RMS 5724b*, N. of Orr, Minn.; 3–4, 9–10, all to scale at right.]

Distribution

Widespread, circumboreal, found from northern Europe to the central European Alps, southward to Dalmatia, but largely western in range; very common in montane situations on the Atlantic coast and on the Faroes, but not appearing to extend southward to the Pyrenees (Allorge, 1955); recurring in the Azores (Allorge and Allorge, 1950). Also in Japan (alpine region, *fide* Hattori, 1952), in China, and in Sikkim and East Nepal (Grolle, 1965, p. 215). The species extends northward to the edge of the Tundra but is nearly absent in the main body of the Tundra (it is frequent, however, in alpine Tundra); it is suboceanic in range and evidently absent from the interior portions of the continents of the Northern Hemisphere, as well as from Iceland and all but the southern tip of Greenland.[395]

In North America restricted largely to the Appalachians and the Atlantic Coast region, southward as a relict in the southern Appalachians, rarely northward to the Hudson Bay region and then recurring in the Pacific Coast region. In the West found from Alaska and British Columbia and reported from Alberta[396] and from Washington (Röll, 1893; this report needs verification).

In eastern North America not reported to the west of the Appalachian system, except for stations north of Lake Superior and on the east coast of Hudson Bay (Williams and Cain, 1959; Schuster, 1951); the report from Illinois in Frye and Clark (*loc. cit.*), based on the old list of Wolf and Hall (1878), is without merit.

S. GREENLAND. Kangerdluarssukasik, 61°55′ N., 49°18′ W. (*Damsholt 651321*). The first credible report for Greenland; the plants are typical and show the diagnostically fissured cuticle; associated were *Lophozia atlantica* and *Blepharostoma t.* subsp. *trichophyllum*. QUEBEC. Manitounuck Sd. and mouth of Great Whale R., E. coast of Hudson Bay; Rupert R., above L. Némiskau; Ste-Anne de la Pocatière; Mt. Albert; R. Ste-Anne des Monts; Mt. La Table; Lac Ste-Anne, Laurentide Park (*Cain 5631*). LABRADOR. Without locality. NEWFOUNDLAND. Channel; Avalon Distr., S. coast, W. Nfld., N. Pen., and NE. Nfld. MIQUELON I. NOVA SCOTIA. Pirate's Cove, Canso; Margaree, Halfway-House, and Louisburg, Cape Breton I.; Kearney's Rd., Halifax. MAINE. Mt. Katahdin: Northwest Basin, ca. 2800–3200 ft; Hamline Ridge, on North Basin trail; Cleft Rock pool, above Chimney Pond, ca. 3000–3200 ft; Cathedral Trail, ca. 4500–4800 ft; N. slope of Baxter Peak, ca. 5000 ft; Saddle Slide, ca. 5000 ft

---

[395] *Mylia taylori* is listed from Greenland in Müller (1911, *in* 1905–16), Frye & Clark (1937–47), and Buch & Tuomikoski (1955), evidently on the basis of a report by Jensen (1897). However, Jensen (1906a, p. 303) states that it does not occur in east Greenland, his own report of it in "Mosser fra Øst-Grønland" referring to *Arnellia fennica*. Macoun (1902) also cites an old ambiguous collection from Greenland by Vahl.

[396] "In a bog at House Mountain, Lesser Slave Lake" (Macoun, 1902); this almost surely is an error for *M. anomala*, judging both from the continental position of the station and from the habitat (Schuster, 1959–60).

(*RMS*).   NEW HAMPSHIRE.   Franconia Mts.; Mt. Lafayette; very widespread in the Presidential Range, White Mts., from Mt. Clinton and Mt. Monroe to Mt. Adams; Crystal Cascade, Mt. Washington.   VERMONT.   Mt. Mansfield (*RMS 43844a*) and *fide* Evans (1913).

NEW YORK.   Catskill Mts.: near summit, Wittenberg Mt.; near summit, Cornell Mt.; slope of Slide Mt., leading down to Cornell Mt., ca. 3700–3900 ft (*RMS*).   Arnold L., 0.5 mi above Avalanche Gap, Mt. Marcy; L. Placid, Adirondack Mts.; Opalescent R., near Buckley's Clearing, Essex Co., Adirondack Mts.; Sams Point, SE. of Ellenville, Ulster Co. (N. G. Miller, 1966).

Occurring as a disjunct as follows.   NORTH CAROLINA.   Grandfather Mt., ca. 5800 ft, Avery Co. (*Andrews!*).   TENNESSEE.   Mt. Kephart, at The Jumpoff, 5950–6100 ft, and below Myrtle Pt., Mt. LeConte (*RMS 45385a, 45243, 45242*).

Westward and northward as a disjunct in the L. Superior area: ONTARIO.   Agawa Pt., W. of Hwy. 17 (*Cain & Williams 5276!*); S. of MacGregor Cove (*Cain & Williams 5171!*); near Mile 104, Algoma Central Railway (*Cain 5136!*), all in Algoma Distr.

The report of this species from Bergen Co., N.J. ("among peat mosses in bog near Closter" *Austin, in* Britton, 1889) refers to *M. anomala* (Schuster, 1959–60). A report from Rhode Island (cited in Frye and Clark, 1937–47) must be in error, since Evans (1913, 1923g) refused to accept it.

The species is an excellent indicator of the Taiga Biome, dropping out almost completely before the lower edge of the Spruce-Fir Forest is reached. In the Catskill Mts., where it is abundant locally, strictly limited to the Spruce-Fir Zone. In Ulster Co., New York, restricted to ice caves (N. G. Miller, 1966).

Ecology

Largely restricted to montane areas and to acidic rocks, although Müller (1905–16) states that it also occurs over limestone cliffs, where he admits it is not found directly on rock, but is confined to a humus layer. In my experience, it is a suboceanic species which is lacking from calcareous sites.

Northward *M. taylori* is reported from moist granite, often on vertical siliceous walls or in their crevices, or near margins of pools in granite, associated with *Lophozia atlantica, L. kunzeana, L. wenzelii, L. ventricosa, Ptilidium ciliare, Scapania parvifolia, Cephalozia ambigua, C. bicuspidata,* and *C. leucantha, Pleuroclada albescens* var. *islandica, Chandonanthus setiformis,* and *Gymnocolea inflata* (Schuster, 1951). In more oceanic sites, as in Newfoundland, it is commonly with *Cephalozia leucantha, C. media, Microlepidozia setacea, Lophozia attenuata, Bazzania trilobata, Lepidozia reptans,* and *Diplophyllum albicans,* and variously on moist or wet siliceous rocks, where it often forms thick cushions on wet soil and ocasionally on decaying logs (Buch and Tuomikoski, 1955). Very abundant in the New England mountains, as on Mt. Katahdin, where found most often on moist rock walls or large damp boulders and crags, with *Anastrophyllum michauxii, Scapania umbrosa, Cephalozia bicuspidata* (incl. *ambigua*), *Lophozia ventricosa silvicola*, *L. atlantica*, *Blepharostoma*, *Chandonanthus setiformis*, *Tritomaria quinquedentata*, and *Gymnomitrion concinnatum* (in drier than normal areas). Also over peat on banks, as at the bases of boulders, or on exposed, seepage-moistened, *Sphagnum*-covered ledges, or on sunny, peat-covered boulders, there with *Lophozia attenuata*, *L. kunzeana*, *L. atlantica*, *Ptilidium ciliare*, *Polytrichum*, etc.

Remaining frequent southward into the Catskill Mts., where forming swelling, spongy tufts on intermittently moist cliff walls and crests; there often with gemmae, occasionally perianths, rarely capsules. There associated with *Anastrophyllum michauxii* and *A. minutum*, *Lophozia ventricosa silvicola*, *L. attenuata*, *Bazzania trilobata*, *B. tricrenata*, *B. denudata*, *Diplophyllum taxifolium*, *Scapania nemorosa*, rarely *Calypogeia integristipula*, *Cephalozia media*, and *Lepidozia reptans*.

Although typically epilithic, the species is often a transient element in the rapidly succeeding facies of the *Nowellia-Jamesoniella* Associule. It apparently depends upon a reservoir population on rocks, since it is quite lacking on decaying logs away from the "normal" montane habitats of the species on ledges. On logs a wide variety of associated species occurs, i.a., *Cephalozia media*, *Bazzania trilobata*, *Anastrophyllum michauxii*, *Nowellia*, *Jamesoniella*, *Calypogeia suecica*, and *Geocalyx graveolans*.

Differentiation

Easily recognized in the field. The robust size, strong tendency for a purplish to purplish brown pigmentation, at least of the upper halves of the leaves (never the warm brown of the related *M. anomala*!), and orbicular leaves eliminate most species occurring in similar habitats. *Jamesoniella autumnalis*, which may be associated, is never more than half the size and therefore should not be confused with it, even in the field.

Under the microscope, the most diagnostic feature which at once characterizes this species is the very curious manner in which the cuticle of the leaves is adorned. The cuticle is coarsely roughened by what are almost irregular, jagged plates, giving the appearance of a smooth cuticle that has become irregularly fissured. Sterile plants also have distinct lanceolate underleaves (usually largely hidden among the very characteristically abundant, dense rhizoids), large leaf cells which always bear bulging trigones, and a strong tendency toward purplish red to reddish brown pigmentation. The plants often bear gemmae, if not abundantly so, on the margins of the upper leaves (which may become slightly longer with gemmae production, but never attain the characteristically ovatelanceolate form of the related *M. anomala*). Fertile plants are frequent, especially on rock walls, and the strongly laterally compressed perianths with a truncate mouth at once eliminate most superficially similar species.

Mylia taylori is so different from *M. anomala* in color, in the absence of

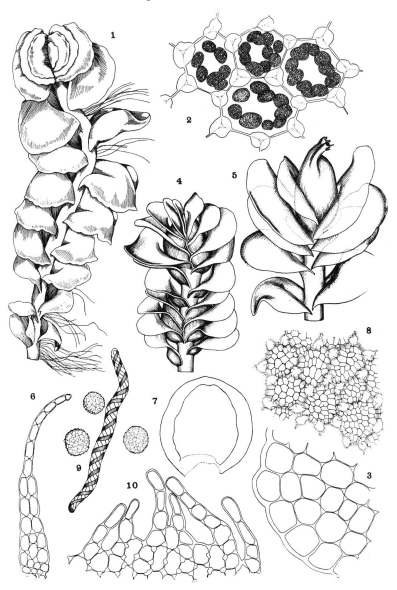

FIG. 301. *Mylia taylori*. (1) Sterile plant (×ca. 10). (2) Median cells with oil-bodies (×325). (3) Part of seta cross section (×150). (4–5) Apices of ♂ and ♀ plants, respectively (×10.7). (6) Underleaf (×104). (7) ♀ Bract and, drawn within, sterile leaf (×14). (8) Median cells in surface view, to show cuticular ornamentation (×215). (9) Spores and elater (×340). (10) Part of perianth mouth (×150). [1, after K. Müller, 1905–16; 2–3, *Marr 656c*, Great Whale R., Quebec; 4–10, *RMS 17030*, Mt. Katahdin, Me.; all from Schuster, 1959–60.]

ovate-lanceolate gemmiparous leaves, in the rough cuticle, and in the oil-bodies that confusion between the two hardly seems possible. The restriction of *M. taylori* to rock walls and to logs near montane rock outcrops also separates it from the helophytic *M. anomala*. The oval to ellipsoidal oil-bodies formed of numerous minute, little-protruding spherules, are also very different from the few-segmented oil-bodies of *M. anomala*.

The capsule wall in *M. taylori* varies locally from 3–5-stratose; the total thickness is usually 45–78 μ. The epidermal cells, when not irregular, average 23–28 μ wide × 50–75 μ long; they are adorned with thickenings that are extremely polymorphous but nevertheless quite diagnostic, reminding one strongly of those of *Aneura pinguis* in some respects. The nodular thickenings, which are largely confined to both longitudinal faces of alternating walls, appear somewhat stalked in profile. The highly pigmented, distal, expanded portion of these thickenings is sometimes extended as a vague transverse band (then becoming incompletely semiannular), but more often the expanded portions of the thickenings are connected to each other by longitudinal, pigmented thickenings lying on the tangential walls parallel with but some distance out from the radial walls. The innermost cell layer is equally distinctive in that the numerous, complete to subcomplete, tangential bands show an unusually high level of anastomosis with each other, occasionally merging, at other times being furcate into two or three slightly divergent bands that often do not attain the opposite side of the cell. The inner cells average, where regular, ca. 23–28 μ wide × 75–120 μ long; in some cases, through subdivision, occasional cells are only 18–20 μ wide. The seta is ca. 540 μ in diam. and has the epidermal cell rows slightly larger (1–1.5×) in diam. than the inner cell rows; the epidermal average 55–60 μ in diam. This description contrasts considerably with the statements and figures in Müller (1911, *in* 1905–16, fig. 343c).

Branching in *M. taylori* appears to be nearly exclusively terminal, except below the gynoecia. Terminal branching may occur at the apex of a shoot which immediately produces, on both main axis and branch, perichaetial bracts and then a perianth, resulting in closely approximated, geminate gynoecia, each of which produces a single subfloral innovation. Subfloral innovations, which appear to be almost universally produced, occur either singly or in pairs; they are apparently the only intercalary branches that are regularly produced.

Variation

A characteristic and stenotypic species, therefore not offering any problems in its recognition. The plants are almost always of the mod. *colorata-pachyderma-densifolia*.

Schiffner has described a "var. *uliginosa*"—a modification with long internodes and often no secondary pigmentation, growing in water. Such exceptional forms still retain the characteristic ornamentation of the leaf cells by which the species can be distinguished from other regional Hepaticae.

INDEX

Valid scientific names are set in roman type; synonyms are set in italics. Abbreviations used are: diff. = differentiation; fo. = forma or formae; relat. = relationships; sect. = sectio or sectios; spp. = species (plural). Under each genus are listed first, alphabetically, subgenera (if any). Then sections are listed (if any) and only finally species. Subspecies, varieties, and forms, if any, are listed only under the species.